Angiogenesis

An Integrative Approach From Science to Medicine

Contents

Preface. v

Contributors xv

Chapter 1   History of Angiogenesis 1
Judah Folkman

Section I. Physiological & Pathological Angiogenesis: Biology of the Angiogenic Process

Chapter 2   Angiogenesis and Vascular Remodeling in Inflammation and Cancer:
Biology and Architecture of the Vasculature 17
Donald M. McDonald

Chapter 3   Endothelial Cell Activation. 35
M. Luisa Iruela-Arispe

Chapter 4   Pericytes, the Mural Cells of the Microvascular System 45
Gabriele Bergers

Chapter 5   Matrix Metalloproteinases and Their Endogenous Inhibitors 55
Liliana Guedez and William G. Stetler-Stevenson

Chapter 6   Integrins in Angiogenesis 63
Alireza S. Alavi and David A. Cheresh

Section II. Angiogenesis and Regulatory Proteins

Chapter 7   Fibroblast Growth Factor-2 in Angiogenesis 77
*Marco Presta, Stefania Mitola, Patrizia Dell'Era, Daria Leali, Stefania Nicoli,
Emanuela Moroni, and Marco Rusnati*

Chapter 8   Vascular Permeability/Vascular Endothelial Growth Factor. 89
Masabumi Shibuya

Chapter 9   Platelet-Derived Growth Factor 99
Andrius Kazlauskas

Chapter 10   Angiopoietins and Tie Receptors 113
Pipsa Saharinen, Lauri Eklund, and Kari Alitalo

Chapter 11   Basement Membrane Derived Inhibitors of Angiogenesis. 121
Michael B. Duncan and Raghu Kalluri

Chapter 12   Angiostatin and Endostatin: Angiogenesis Inhibitors in Blood and Stroma 129
Judah Folkman

Chapter 13   Thrombospondins: Endogenous Inhibitors of Angiogenesis   147
                Paul Bornstein

Section III. Molecular & Cellular Mechanisms of the Angiogenic Process

Chapter 14   Overview of Angiogenesis During Tumor Growth   161
                Domenico Ribatti and Angelo Vacca

Chapter 15   Hypoxic Regulation of Angiogenesis by HIF-1   169
                Philip J.S. Charlesworth and Adrian L. Harris

Chapter 16   Regulation of Angiogenesis by von Hippel Lindau Protein and HIF2   181
                Donald P. Bottaro, Nelly Tan, and W. Marston Linehan

Chapter 17   Nitric Oxide in Tumor Angiogenesis   193
                L. Morbidelli, S. Donnini, and M. Ziche

Chapter 18   VEGF Signal Tranduction in Angiogenesis   205
                Harukiyo Kawamura, Xiujuan Li, Michael Welsh, and Lena Claesson-Welsh

Chapter 19   Delta-like Ligand 4/Notch Pathway in Tumor Angiogenesis   217
                Gavin Thurston, Irene Noguera-Troise, Ivan B. Lobov, Christopher Daly,
                John S. Rudge, Nicholas W. Gale, Stanley J. Wiegand, and George D. Yancopoulos

Chapter 20   Immune Cells and Inflammatory Mediators as Regulators of Tumor Angiogenesis   225
                Michele De Palma and Lisa M. Coussens

Chapter 21   Contribution of Endothelial Progenitor Cells to the Angiogenic Process   239
                Marco Seandel, Andrea T. Hooper, and Shahin Rafii

Chapter 22   Tumor Angiogenesis and the Cancer Stem Cell Model   249
                Chris Folkins and Robert S. Kerbel

Chapter 23   Targeting the Tumor Microenvironment (Stroma) for Treatment of Metastasis   259
                Isaiah J. Fidler, Cheryl Hunt Baker, Kenji Yokoi, Toshio Kuwai, Toru Nakamura,
                Monique Nilsson, J. Erik Busby, Robert R. Langley, and Sun-Jin Kim

Section IV. Functional Assessments of Angiogenesis

Chapter 24   Normalization of Tumor Vasculature and Microenvironment   273
                Rakesh K. Jain, Dan G. Duda, Tracy T. Batchelor, A. Gregory Sorensen,
                and Christopher G. Willett

Chapter 25   Targeted Drug Delivery to the Tumor Neovasculature   283
                Grietje Molema

Chapter 26   Models for Angiogenesis   299
                Robert Auerbach

Chapter 27   Surrogates for Clinical Development   313
                Sylvia S. W. Ng and Kim N. Chi

Chapter 28   Imaging of Angiogenesis   321
                Tristan Barrett and Peter L. Choyke

Chapter 29   Tumor Endothelial Markers   333
                Janine Stevens and Brad St.Croix

Section V. Clinical Translation of Angiogenesis Inhibitors

Chapter 30   Overview and Clinical Applications of VEGF-A   345
                Napoleone Ferrara

Chapter 31   Protein Tyrosine Kinase Inhibitors as Antiangiogenic Agents   353
                Alexander Levitzki

Chapter 32   Therapeutic Strategies that Target the HIF System 359
              Kristina M. Cook and Christopher J. Schofield

Chapter 33   The Clinical Utility of Bevacizumab 375
              Jeanny B. Aragon-Ching, Ravi A. Madan, and James L. Gulley

Chapter 34   Development of Thalidomide and Its IMiD Derivatives 387
              Cindy H. Chau, William Dahut, and William D. Figg

Chapter 35   TNP-470: The Resurrection of the First Synthetic Angiogenesis Inhibitor....................... 395
              Hagit Mann-Steinberg and Ronit Satchi-Fainaro

Chapter 36   Clinical Development of VEGF Trap.. 415
              John S. Rudge, Ella Ioffe, Jingtai Cao, Nick Papadopoulos,
              Gavin Thurston, Stanley J. Wiegand, and George D. Yancopoulos

Chapter 37   Recent Advances in Angiogenesis Drug Development 421
              Cindy H. Chau and William D. Figg

Chapter 38   Combination of Antiangiogenic Therapy with Other Anticancer Therapies........................ 431
              Beverly A. Teicher

Chapter 39   Immunotherapy of Angiogenesis with DNA Vaccines..................................... 451
              Chien-Fu Hung, Archana Monie, and T.-C. Wu

Chapter 40   Challenges of Antiangiogenic Therapy of Tumors.. 461
              Roberta Sarmiento, Raffaele Longo, and Giampietro Gasparini

Chapter 41   Pharmacogenetics of Antiangiogenic Therapy... 477
              Guido Bocci, Giuseppe Pasqualetti, Antonello Di Paolo,
              Mario Del Tacca, and Romano Danesi

Section VI. Angiogenesis in Health & Disease

Chapter 42   Angiogenesis in the Central Nervous System 489
              Carmen Ruiz de Almodovar, Serena Zacchigna, and Peter Carmeliet

Chapter 43   Lymphatic Vascular System and Lymphangiogenesis 505
              Leah N. Cueni and Michael Detmar

Chapter 44   Ocular Neovascularization 517
              Peter A. Campochiaro

Chapter 45   Angiogenesis and Pathology in the Oral Cavity... 533
              Luisa A. DiPietro

Chapter 46   Revascularization of Wounds: The Oxygen-Hypoxia Paradox 541
              Thomas K. Hunt, Michael Gimbel, and Chandan K. Sen

Chapter 47   Journeys in Coronary Angiogenesis.. 561
              Julie M.D. Paye, Chohreh Partovian, and Michael Simons

Chapter 48   Perspectives on the Future of Angiogenesis Research.................................... 575
              Douglas Hanahan

Index 585

Dr. Judah Folkman
1933 – 2008

Tribute

"I spent a memorable year with Judah Folkman in his Boston laboratory 1973/1974. Unforgettable to this day were his enthusiasm and his eagerness to learn from others, as well as to communicate his own ideas on angiogenesis as an integral role in the progression and metastasis of tumors; his unusual concern for his patients; and his generosity towards students and co-workers. His sudden unexpected death was a blow to me and many others, and a grievous loss to the fields of cancer treatment and research to which he had contributed so much."

ROBERT AUERBACH
UNIVERSITY OF WISCONSIN, MADISON, WI

"Dr. Folkman was the kindest person that I knew. He would listen to everyone who wanted to speak with him from a medical student to a Nobel Prize winner. He would make the person feel comfortable and engage them in deep communication. Without a doubt after speaking with him one would leave with a clearer vision of the topic of the day. That is a skill that is innate and cannot be learned nor taught."

KEVIN CAMPHAUSEN
NATIONAL CANCER INSTITUTE, BETHESDA, MD

"Dr. Folkman was a unique scientific mentor for me. At the beginning of the 90s, I joined Dr. Folkman's lab. Working in Dr. Folkman's lab, I had not only learned angiogenesis research, but learned the way of novel thinking and the way to approaching unusual medical problems. This has affected my entire research career. He usually taught me by saying that doing research is like playing chess. There are many ways to go. One has to find the smartest way to do research. For me, Dr. Folkman is still alive and his scientific spirit will stay with us for generations."

YIHAI CAO
KAROLINSKA INSTITUTET, MTC, STOCKHOLM, SWEDEN

"Judah Folkman was an individual like none other. He was too busy to eat and seldom took a vacation but always had time to stop and chat with an excited student or patient who needed hope. In addition to his innate brilliance, he had three characteristics that set him apart: his genuine and unbridled curiosity, his perseverance and his ability to synthesize information. I joined Judah's Laboratory for Surgical Research right out of my postdoctoral fellowship and have remained associated with him all of my professional life. I consider myself truly blessed to have known him and been mentored by him."

PAT D'AMORE
HARVARD MEDICAL SCHOOL AND SCHEPENS EYE RESEARCH INSTITUTE, BOSTON, MA

"Judah Folkman was one of the most creative scientists that I have ever known, an exceptionally wise and caring physician, and an outstanding teacher. As a result of his scientific creativity, he often stirred up controversy, because it often takes decades to determine whether highly original ideas are correct. He has literally thousands of scientific children. He was also a persuasive speaker and for many years the leading spokesman for the fields of angiogenesis and anti-angiogenesis. Judah was also an exceedingly wise and compassionate doctor who was 110% devoted to his work, yet generous with his time. In sum, Judah Folkman was a giant as a scientist, physician and teacher. He is sorely missed."

HAROLD DVORAK
BETH ISRAEL DEACONESS MEDICAL CENTER, BOSTON, MA

"Even though I did not have the opportunity and the honor of working directly with Dr. Folkman, I learned a great deal from him through our interactions at various meetings over the last twenty years. The first was that Dr. Folkman remained humble despite all of his wonderful contributions and the world-wide respect he commanded. A second lesson I learned from Dr. Folkman was the fact that he took the time to respond to each and every patient that contacted him. Lastly, Dr. Folkman taught me the value of translating simple observations into clinical practice. Dr. Folkman's vision, persistence, elegance and mentorship will be missed by colleagues, patients, and trainees. Current and future investigators are encouraged to pursue science and clinically relevant studies with the same vigor and enthusiasm of Dr. Folkman himself."

LEE M. ELLIS
UNIVERSITY OF TEXAS M.D. ANDERSON CANCER CENTER,
HOUSTON, TX

"I first met Judah Folkman in 1986 at a cancer symposium in Oakland, CA, where I spoke about my then new genetically engineered mouse model of cancer (RIP-Tag2) and Judah spoke about tumor angiogenesis. Our intersection there seeded a dialog and then collaboration and in turn an enduring friendship that has flourished over the years. The first milestone was our discovery (reported in Nature in 1989) that angiogenesis was activated as a discrete pre-malignant event in that model of pancreatic islet cell carcinogenesis; a second was our synthesis of the "angiogenic switch" concept in a perspective published in Cell (1996). I admired and even envied his boundless curiosity and his biological intuition, and all his knowledge of human biology and disease, and yet I prided myself in providing a balance of critical counterpoint, anchored in the possibility of testing mechanistic hypotheses in increasingly convincing ways using the tools of molecular genetics. And thus my career was deeply touched and inspired, both by our collaborative successes as well as the lessons of those that never reached fruition, and evermore so by the special experience of knowing this remarkable man."

DOUGLAS HANAHAN
HELEN DILLER FAMILY COMPREHENSIVE CANCER CENTER,
UCSF, SAN FRANCISCO, CA

"Dr. Folkman's brilliance was in paying attention to the novel and not so obvious. My first memory of him is that of a gentle, kind professor descending from a podium amidst applause of world renowned scientists and stopping to speak to a medical student. He looked at science with fresh eyes; he looked for new explanations of old phenomena and unorthodox connections between previously unrelated fields. He was a crusader for new ideas, and persevered until they were published and established."

GIANNOULA KLEMENT
CHILDREN'S HOSPITAL BOSTON AND DANA-FARBER CANCER
INSTITUTE, BOSTON, MA

"Judah Folkman was simply an amazing man - brilliant, thoughtful, generous. I remember hearing him speak at the Whitehead Symposium in 1996 and it changed my life forever. I will always cherish having worked with him, and for introducing me to this wonderful field of angiogenesis."

CALVIN KUO
STANFORD UNIVERSITY SCHOOL OF MEDICINE, STANFORD, CA

"I applied for lots of jobs and was lucky enough to receive 20 job offers, but I only thought one might fulfill that dream. That was with Dr. Folkman. I remember writing Dr. Folkman and him calling me, and my visit in 1974 to Children's Hospital. He asked if I'd help isolate the first angiogenesis inhibitor. That job changed my life. Not only did I help isolate that angiogenesis inhibitor, but also we were able to develop principles that would lead to new controlled release medications that would affect millions of people. Dr. Folkman was the greatest role model a young scientist could have. He was a fantastic mentor, a superb role model, one of nicest human being I've ever met, and a truly great man."

ROBERT LANGER
MASSACHUSETTS INSTITUTE OF TECHNOLOGY, CAMBRIDGE,
MA

"Dr. Folkman was an inspiration for me early on in my training. As a surgical resident, after reading his papers on tumor angiogenesis, I knew that I wanted to pursue a career in science and clinical medicine focused on cancer and the role of the tumor vasculature. As my career progressed, I had the privilege of meeting Dr. Folkman and then interacting with him on several projects over the years. I considered him a great mentor and colleague. He was unselfish with his time and his ideas and a true champion for the development of novel treatments for patients suffering from cancer. I will miss his energy, enthusiasm and intellect. The field of angiogenesis research has lost its most passionate leader."

STEVEN K. LIBUTTI
NATIONAL CANCER INSTITUTE, BETHESDA, MD

"He lived his life exactly as he wanted. He did what he loved most and he died in the best way a man can. It was just too early for those of us left behind. My colleagues in the vascular biology program in Boston, cancer researchers around the world, and scientists everywhere are bereft of a great spirit and a giant in his field."

RONIT SATCHI-FAINARO
SACKLER SCHOOL OF MEDICINE, TEL AVIV UNIVERSITY, TEL
AVIV, ISRAEL

"I had the honor and privilege to work with Dr. Folkman for the past 27 years. The impact of his life on me is beyond what I can express. He was a great scientific pioneer and a true leader. He was a very kind person."

YUEN SHING
HARVARD MEDICAL SCHOOL AND CHILDREN'S HOSPITAL,
BOSTON, MA

"I came in contact with Dr. Folkman's ideas when I was setting my up my first lab at the Beth Israel Hospital in Boston. What made the most impact was the obvious passion he had for science and the everyday desire to ask how can we translate new science advances into the clinical practice. In that he was, and still is, far ahead of his time."

MICHAEL SIMONS
DARTMOUTH MEDICAL SCHOOL, ANGIOGENESIS RESEARCH
CENTER, HANOVER, NH

"Dr. Folkman was a great spokesperson for cancer research. He very effectively communicated the promise and hurtles of cancer drug discovery to the public. He was ever optimistic that we would be successful in impacting therapeutically on malignant diseases. Personally, I will miss his 'voice' which gently and effectively motivated all of us to continue our efforts in drug discovery even though the journey is long and challenging."

BEVERLY TEICHER
ONCOLOGY RESEARCH, GENZYME CORPORATION,
FRAMINGHAM, MA

"Dr. Folkman leaves behind a legacy of years of well documented scientific discoveries and achievements. However, his greatest attribute was his ability to mentor and inspire those outside of his own laboratory. This was clearly the case for me. I met Dr. Folkman in 1992 and he has had a tremendous influence on my laboratory's research direction ever since. His ability to stimulate basic science hypotheses from clinical observations was exceptional."

WILLIAM DOUGLAS FIGG, SR.
MEDICAL ONCOLOGY BRANCH, CENTER FOR CANCER
RESEARCH, NATIONAL CANCER INSTITUTE, BETHESDA, MD

Contributors

Alireza S. Alavi, Ph.D.
Moores Cancer Center, University of California, San Diego, La Jolla, CA, USA

Kari Alitalo, M.D., Ph.D.
Molecular/Cancer Biology Laboratory and Ludwig Institute for Cancer Research, Biomedicum Helsinki, University of Helsinki, Haartmaninkatu 8, P.O. Box 63, 00014 Helsinki, Finland

Jeanny B. Aragon-Ching, M.D.
Medical Oncology Branch, National Cancer Institute, National Institutes of Health, Bethesda, MD, USA

Robert Auerbach, Ph.D.
University of Wisconsin, Madison, WI 53706, USA

Cheryl Hunt Baker, Ph.D.
Department of Cancer Biology, The University of Texas M.D. Anderson Cancer Center, Houston, TX, USA

Tristan Barrett, M.B.B.S.
National Cancer Institute, Building10, Bethesda, MD 20892-1088, USA

Tracy T. Batchelor, M.D., M.P.H.
Department of Neuro-Oncology, Massachusetts General Hospital, 55 Fruit Street, Yawkey-9E, Boston, MA, USA

Gabriele Bergers, Ph.D.
Department of Neurological Surgery, Brain Tumor Research Center and UCSF Comprehensive Cancer Center, University of California, 513 Parnassus Avenue, San Francisco, CA 94143-0520, USA

Guido Bocci, M.D., Ph.D.
Division of Pharmacology and Chemotherapy, Department of Internal Medicine, University of Pisa, Via Roma, 55, 56126 Pisa, Italy

Paul Bornstein, M.D.
Departments of Biochemistry and Medicine, University of Washington, Seattle, WA, USA

Donald P. Bottaro, Ph.D.
Urologic Oncology Branch, Center for Cancer Research, National Cancer Institute, National Institutes of Health, Bethesda, MD 20892-1107, USA

J. Erik Busby, M.D.
Department of Cancer Biology, The University of Texas M.D. Anderson Cancer Center, Houston, TX, USA

Peter A. Campochiaro, M.D.
The Departments of Ophthalmology and Neuroscience, The Johns Hopkins University School of Medicine, Maumenee 719, 600 N. Wolfe Street, Baltimore, MD 21287-9277, USA

Jingtai Cao, Ph.D.
Regeneron Pharmaceuticals, 777 Old Saw Mill River Road Tarrytown, NY 10591, USA

Peter Carmeliet, M.D., Ph.D.
Department for Transgene Technology and Gene Therapy, VIB, 912 3000 Leuven, Belgium
The Center for Transgene Technology and Gene Therapy (CTG), K.U. Leuven, 912 3000 Leuven, Belgium

Philip J.S. Charlesworth, B.M., M.R.C.S
Laboratory of Molecular Oncology, Weatherall Institute of Molecular Medicine, University of Oxford, John Radcliffe Hospital, Oxford, UK

Cindy H. Chau, Pharm.D., Ph.D.
Molecular Pharmacology Section, Medical Oncology Branch, Center for Cancer Research, National Cancer Institute, Bethesda, MD, USA

David A. Cheresh, Ph.D.
Moores Cancer Center, University of California, San Diego, La Jolla, CA, USA

Kim N. Chi, M.D.
Departments of Advanced Therapeutics and Medical Oncology, Faculty of Medicine, University of British Columbia, Vancouver, BC, Canada

Peter L. Choyke, M.D.
Molecular Imaging Program, National Cancer Institute,
Building10, Bethesda, MD 20892-1088, USA

Lena Claesson-Welsh, Ph.D.
Department of Genetics and Pathology, Uppsala University,
Dag Hammarskjöldsv. 20, 751 85 Uppsala, Sweden

Kristina M. Cook, B.S.
The Chemistry Research Laboratory, University of Oxford,
Mansfield Road, Oxford, OX1 3TA, UK

Lisa M. Coussens, Ph.D.
Department of Pathology and Comprehensive Cancer Center,
University of California, 2340 Sutter St, San Francisco,
CA 94143, USA

Leah N. Cueni, M.Sc.
Institute of Pharmaceutical Sciences, Swiss Federal Institute
of Technology, ETH Zurich, Zurich, Switzerland

William L. Dahut, M.D.
Medical Oncology Branch, Center for Cancer Research,
National Cancer Institute, Bethesda, MD, USA

Christopher Daly, Ph.D.
Regeneron Pharmaceuticals, 777 Old Saw Mill River Road,
Tarrytown, NY 10591, USA

Romano Danesi, M.D., Ph.D.
Division of Pharmacology and Chemotherapy,
Department of Internal Medicine, University of Pisa,
Via Roma, 55, 56126 Pisa, Italy

Patrizia Dell'Era, Ph.D.
Unit of General Pathology and Immunology,
Department of Biomedical Sciences and Biotechnology,
University of Brescia, 25123 Brescia, Italy

Mario Del Tacca, M.D.
Division of Pharmacology and Chemotherapy,
Department of Internal Medicine, University of Pisa, Via Roma,
55, 56126 Pisa, Italy

Michele De Palma, Ph.D.
Angiogenesis and Tumour Targeting Research Unit
and Telethon Institute for Gene Therapy,
San Raffaele Scientific Institute, via Olgettina,
58, 20132 Milan, Italy

Michael Detmar, M.D.
Institute of Pharmaceutical Sciences, Swiss Federal Institute
of Technology, ETH Zurich, Zurich, Switzerland

Antonello Di Paolo, M.D., Ph.D.
Division of Pharmacology and Chemotherapy, Department
of Internal Medicine, University of Pisa, Via Roma,
55, 56126 Pisa, Italy

Luisa A. DiPietro, D.D.S., Ph.D.
Center for Wound Healing and Tissue Regeneration,
College of Dentistry, University of Illinois at Chicago,
USA

S. Donnini, Ph.D.
Section of Pharmacology, Department of Molecular Biology
and C.R.I.S.M.A., Pharmacy School, University of Siena,
Via A. Moro 2, 53100 Siena, Italy

Dan G. Duda, D.M.D., Ph.D.
Steele Laboratory, Department of Radiation Oncology,
Massachusetts General Hospital, 100 Blossom Street,
Cox-734, Boston, MA 02114, USA

Michael B. Duncan, Ph.D.
Division of Matrix Biology, Department of Medicine,
Beth Israel Deaconess Medical Center and Harvard
Medical School, Boston, MA 02215, USA

Lauri Eklund, Ph.D.
Collagen Research Unit, Biocenter Oulu and
Department of Medical Biochemistry and Molecular Biology,
University of Oulu, P.O. Box 5000, 90014 Oulu,
Finland

Napoleone Ferrara, M.D.
Genentech, Inc., 1 DNA Way, South San Francisco,
CA 94080, USA

Isaiah J. Fidler, D.V.M., Ph.D.
Department of Cancer Biology, The University of Texas
M.D. Anderson Cancer Center, Houston, TX, USA

William D. Figg, Pharm.D., M.B.A.
Molecular Pharmacology Section, Medical Oncology
Branch, Center for Cancer Research, National Cancer
Institute, Bethesda, MD, USA

Chris Folkins, BSc.
Department of Molecular and Cellular Biology Research,
Sunnybrook Health Sciences Centre, Toronto, ON,
Canada Department of Medical Biophysics,
University of Toronto, Toronto, ON, Canada

Judah Folkman, M.D.
Department of Surgery, Harvard Medical School and
Vascular Biology Program, Children's Hospital Boston,
Boston, MA, USA

Nicholas W. Gale, Ph.D.
Regeneron Pharmaceuticals, 777 Old Saw Mill River Road,
Tarrytown, NY 10591, USA

Giampietro Gasparini, M.D.
Division of Medical Oncology, Azienda Ospedaliera
"San Filippo Neri", Via Martinotti, 20, 00135
Rome, Italy

Michael Gimbel, M.D.
Department of Plastic Surgery, University of Pittsburgh,
Pittsburgh, PA, USA

Liliana Guedez, Ph.D.
Cell and Cancer Biology Branch, Extracellular
Matrix Pathology Section, Center for Cancer
Research, National Cancer Institute, Bethesda,
MD 20892-4605, USA

James L. Gulley, M.D., Ph.D., F.A.C.P.
Laboratory of Tumor Immunology and Biology, National
Cancer Institute, National Institutes of Health, Bethesda,
MD, USA

Douglas Hanahan, Ph.D.
Department of Biochemistry and Biophysics, Helen Diller
Family Comprehensive Cancer Center, and Diabetes Center,
UCSF, San Francisco, CA 94143, USA

Adrian L. Harris, B.Sc. Hons, M.A., D.Phil., M.B.Ch.B.,
F.R.C.P.
Laboratory of Molecular Oncology, Weatherall Institute of
Molecular Medicine, University of Oxford, John Radcliffe
Hospital, Oxford, UK

Andrea T. Hooper, B.S.
Department of Genetic Medicine, Howard Hughes Medical
Institute, Weill Medical College of Cornell University,
New York, NY 10021, USA
Department of Physiology, Biophysics and Systems Biology,
Weill Cornell Graduate School of Medical Sciences,
New York, NY 10021, USA

Chien-Fu Hung, Ph.D.
Departments of Pathology, Johns Hopkins School of
Medicine, Baltimore, MD, USA

Thomas K. Hunt, M.D.
Department of Surgery, University of California,
San Francisco, CA, USA

Ella Ioffe, Ph.D.
Regeneron Pharmaceuticals, 777 Old Saw Mill River Road,
Tarrytown, NY 10591, USA

M. Luisa Iruela-Arispe, Ph.D.
Department of Molecular, Cell and Developmental Biology,
Molecular Biology Institute and Jonsson Comprehensive
Cancer Center, University of California, LA 90095, USA
UCLA, 615 Charles Young Drive South, Los Angeles,
CA 90095, USA

Rakesh K. Jain, Ph.D.
Edwin L. Steele Laboratory for Tumor Biology, Department
of Radiation Oncology, Massachusetts General Hospital,
100 Blossom Street, Cox-734, Boston,
MA, USA

Raghu Kalluri, Ph.D.
Division of Matrix Biology, Department of Medicine,
Beth Israel Deaconess Medical Center and Harvard Medical
School, Boston, MA 02215, USA
Department of Biological Chemistry and Molecular
Pharmacology, Harvard Medical School, Boston,
MA 02215, USA
Harvard-MIT Division of Health Sciences and Technology,
Boston, MA 02215, USA

Harukiyo Kawamura, M.D., Ph.D.
Department of Genetics and Pathology, Uppsala University,
Dag Hammarskjöldsv. 20, 751 85 Uppsala, Sweden

Andrius Kazlauskas, Ph.D.
Schepens Eye Research Institute, Harvard Medical School,
20 Staniford Street, Boston, MA 02114, USA

Robert S. Kerbel, Ph.D.
Department of Molecular and Cellular Biology Research,
Sunnybrook Health Sciences Centre, Toronto, ON, Canada
Department of Medical Biophysics, University of Toronto,
Toronto, ON, Canada

Sun-Jin Kim, M.D., Ph.D.
Department of Cancer Biology, The University of Texas
M.D. Anderson Cancer Center, Houston, TX, USA

Toshio Kuwai, Ph.D.
Department of Cancer Biology, The University of Texas
M.D. Anderson Cancer Center, Houston, TX, USA

Robert R. Langley, Ph.D.
Department of Cancer Biology, The University of Texas
M.D. Anderson Cancer Center, Houston, TX, USA

Daria Leali, Ph.D.
Unit of General Pathology and Immunology, Department of
Biomedical Sciences and Biotechnology, University
of Brescia, 25123 Brescia, Italy

Alexander Levitzki, Ph.D.
Unit of Cellular Signaling, Department of Biological
Chemistry, The Hebrew University of Jerusalem,
Jerusalem, Israel

Xiujuan Li, Ph.D.
Department of Genetics and Pathology, Uppsala University,
Dag Hammarskjöldsv. 20, 751 85 Uppsala, Sweden

W. Marston Linehan, M.D.
Urologic Oncology Branch, Center for Cancer Research,
National Cancer Institute, National Institutes of Health,
Bethesda, MD 20892, USA

Ivan B. Lobov, Ph.D.
Regeneron Pharmaceuticals, 777 Old Saw Mill River Road,
Tarrytown, NY 10591, USA

Raffaele Longo, M.D.
Division of Medical Oncology, Azienda Ospedaliera
"San Filippo Neri", Rome, Italy

Ravi A. Madan, M.D.
Laboratory of Tumor Immunology and Biology, National
Cancer Institute, National Institutes of Health, Bethesda,
MD, USA

Hagit Mann-Steinberg, Ph.D.
Department of Physiology and Pharmacology, Sackler
School of Medicine, Tel Aviv University, Ramat Aviv,
Tel Aviv 69978, Israel

Donald M. McDonald, M.D., Ph.D.
Comprehensive Cancer Center, Cardiovascular Research
Institute, and Department of Anatomy, University
of California, San Francisco, CA, USA

Stefania Mitola, Ph.D.
Unit of General Pathology and Immunology, Department
of Biomedical Sciences and Biotechnology, University
of Brescia, 25123 Brescia, Italy

Grietje Molema, Ph.D.
Department of Pathology and Medical Biology, Medical
Biology section, Laboratory for Endothelial Biomedicine
and Vascular Drug Targeting Research, University Medical
Center Groningen (UMCG), University of Groningen,
Groningen, The Netherlands

Archana Monie, M.Sc.
Departments of Pathology, Johns Hopkins School
of Medicine, Baltimore, MD, USA

L. Morbidelli, Ph.D.
Section of Pharmacology, Department of Molecular Biology
and C.R.I.S.M.A., Pharmacy School, University of Siena,
Via A. Moro 2, 53100 Siena, Italy

Emanuela Moroni, Ph.D.
Unit of General Pathology and Immunology, Department
of Biomedical Sciences and Biotechnology, University
of Brescia, 25123 Brescia, Italy

Toru Nakamura, Ph.D.
Department of Cancer Biology, The University of Texas
M.D. Anderson Cancer Center Houston, TX, USA

Sylvia S.W. Ng, Ph.D.
Departments of Advanced Therapeutics, British Columbia
Cancer Agency, Faculty of Pharmaceutical Sciences,
University of British Columbia, Vancouver, BC, Canada

Stefania Nicoli, Ph.D.
Unit of General Pathology and Immunology, Department
of Biomedical Sciences and Biotechnology, University
of Brescia, 25123 Brescia, Italy

Monique Nilsson, Ph.D.
Department of Cancer Biology, The University of Texas
M.D. Anderson Cancer Center, Houston, TX, USA

Irene Noguera-Troise, Ph.D.
Regeneron Pharmaceuticals, 777 Old Saw Mill River Road,
Tarrytown, NY 10591, USA

Nick Papadopoulos, Ph.D.
Regeneron Pharmaceuticals, 777 Old Saw Mill River Road
Tarrytown, NY 10591, USA

Chohreh Partovian, M.D., Ph.D.
Angiogenesis Research Center, Section of Cardiology,
Departments of Medicine and Pharmacology
and Toxicology, Dartmouth Medical School, Lebanon,
NH, USA

Giuseppe Pasqualetti, M.D.
Division of Pharmacology and Chemotherapy, Department
of Internal Medicine, University of Pisa, Via Roma, 55,
56126 Pisa, Italy

Julie M.D. Paye, Ph.D.
Angiogenesis Research Center, Section of Cardiology,
Departments of Medicine and Pharmacology and
Toxicology, Dartmouth Medical School, Lebanon,
NH, USA

Marco Presta, Ph.D.
Unit of General Pathology and Immunology, Department
of Biomedical Sciences and Biotechnology, University
of Brescia, 25123 Brescia, Italy

Shahin Rafii, M.D.
Department of Genetic Medicine, Howard Hughes Medical
Institute, Weill Medical College of Cornell University,
New York, NY 10021, USA
Department of Physiology, Biophysics and Systems Biology,
Weill Cornell Graduate School of Medical Sciences,
New York, NY 10021, USA
Weill Cornell Medical College, HHMI, 1300 York Ave,
NY 10021, USA

Domenico Ribatti, M.D.
Department of Human Anatomy and Histology,
University of Bari Medical School,
Bari, Italy

John S. Rudge, Ph.D.
Regeneron Pharmaceuticals, 777 Old Saw Mill River Road
Tarrytown, NY 10591, USA

Carmen Ruiz de Almodovar, Ph.D.
Department for Transgene Technology and Gene Therapy,
VIB, 3000 Leuven, Belgium
The Center for Transgene Technology and
Gene Therapy (CTG), K.U. Leuven, 3000 Leuven,
Belgium

Marco Rusnati, Ph.D.
Unit of General Pathology and Immunology,
Department of Biomedical Sciences and Biotechnology,
University of Brescia, 25123 Brescia,
Italy

Pipsa Saharinen, Ph.D.
Molecular/Cancer Biology Laboratory and Ludwig Institute
for Cancer Research, Biomedicum Helsinki, University of
Helsinki, Haartmaninkatu 8, P.O. Box 63, 00014 Helsinki,
Finland

Roberta Sarmiento, M.D.
Division of Medical Oncology, Azienda Ospedaliera
"San Filippo Neri", Rome, Italy

Ronit Satchi-Fainaro, Ph.D.
Department of Physiology and Pharmacology,
Sackler School of Medicine, Tel Aviv University, Ramat
Aviv, Tel Aviv 69978, Israel

Christopher J. Schofield, B.Sc., M.A., D.Phil.
The Chemistry Research Laboratory,
University of Oxford, Mansfield Road, Oxford,
OX1 3TA, UK

Marco Seandel, M.D., Ph.D.
Department of Genetic Medicine, Howard Hughes
Medical Institute, Weill Medical College
of Cornell University, New York, NY 10021, USA
Division of Medical Oncology, Department of Medicine,
Memorial Sloan-Kettering Cancer Center, New York,
NY 10021, USA

Chandan K. Sen, Ph.D., F.A.C.N., F.A.C.S.M.
Comprehensive Wound Center, The Ohio State University
Medical Center, Columbus, OH, USA

Masabumi Shibuya, M.D., Ph.D.
Department of Molecular Oncology, Tokyo Medical
and Dental University, 1-5-45 Yushima, Bunkyo-ku,
Tokyo 113-8519, Japan

Michael Simons, M.D.
Angiogenesis Research Center, Section of Cardiology,
Departments of Medicine and Pharmacology and
Toxicology, Dartmouth Medical School, Lebanon,
NH, USA

A. Gregory Sorensen, M.D.
Department of Radiology, A.A. Martinos Imaging
Center, Massachusetts General Hospital and Harvard
Medical School, Charletown, MA 02129, USA

Brad St. Croix, Ph.D.
Tumor Angiogenesis Section, Mouse Cancer Genetics
Program, National Cancer Institute at Frederick, Frederick,
MD 21702, USA

William G. Stetler-Stevenson, M.D., Ph.D.
Cell and Cancer Biology Branch, Extracellular Matrix
Pathology Section, Center for Cancer Research,
National Cancer Institute, Bethesda, MD
20892-4605, USA

Janine Stevens, Ph.D.
Tumor Angiogenesis Section, Mouse Cancer Genetics
Program, National Cancer Institute at Frederick,
Frederick, MD 21702, USA

Nelly Tan, M.D.
Urologic Oncology Branch, Center for Cancer Research,
National Cancer Institute, National Institutes of Health,
Bethesda, MD 20892, USA

Beverly A. Teicher, Ph.D.
Genzyme Corporation, 1 Mountain Road, Framingham,
MA 01701, USA

Gavin Thurston, Ph.D.
Regeneron Pharmaceuticals, 777 Old Saw Mill River Road,
Tarrytown, NY 10591, USA

Angelo Vacca, M.D.
Department of Biomedical Sciences and Human Oncology,
University of Bari Medical School, Bari, Italy

Michael Welsh, Ph.D.
Department of Genetics and Pathology, Uppsala University,
Biomedical Center, Box 571, 751 23 Uppsala, Sweden

Stanley J. Wiegand, Ph.D.
Regeneron Pharmaceuticals, 777 Old Saw Mill River Road,
Tarrytown, NY 10591, USA

Christopher G. Willett, M.D.
Department of Radiation Oncology, Box 3085, Duke University Medical Center, Durham, NC 27710, USA

T.-C. Wu, M.D., Ph.D.
Departments of Pathology, Oncology, Molecular Microbiol-
ogy and Immunology, and Obstetrics and Gynecology, Johns
Hopkins School of Medicine, Baltimore, MD, USA

George D. Yancopoulos, M.D., Ph.D.
Regeneron Pharmaceuticals, 777 Old Saw Mill River Road,
Tarrytown, NY 10591, USA

Kenji Yokoi, M.D., Ph.D.
Department of Cancer Biology, The University of Texas
M.D. Anderson Cancer Center, Houston,
TX, USA

Serena Zacchigna, M.D., Ph.D.
Department for Transgene Technology and Gene Therapy,
VIB, B-3000 Leuven, Belgium
The Center for Transgene Technology and Gene Therapy
(CTG), K.U. Leuven, B-3000 Leuven, Belgium

M. Ziche, M.D.
Section of Pharmacology, Department of
Molecular Biology and C.R.I.S.M.A., Pharmacy School,
University of Siena, Via A. Moro 2, 53100 Siena,
Italy

Chapter 1
History of Angiogenesis

Judah Folkman

Keywords: angiogenesis, tumor, blood vessel formation, angiogenesis inhibitors, VEGF, bFGF, polymer delivery

Introduction

The first use of the term angiogenesis was in 1787 by John Hunter, a British surgeon [1]. However, there were very few reports of tumor angiogenesis until almost 100 years later, and these were mainly anatomical studies. For example, the vascular morphology of tumors was studied in considerable detail beginning in the 1860s [2, 3]. By 1907, the vascular network in human and animal tumor specimens was visualized by intra-arterial injections of bismuth in oil [4]. The vascular morphology was studied in both human and animal tumors in the first half of the 20th century, mainly to determine if vascular patterns could distinguish benign from malignant tumors, to understand the shedding of tumor emboli into the circulation, or to interpret the delivery of active agents into specific tumors [5–8].

By the late 1930s and early 1940s, several investigators began to study the events of neovascularization in experimental tumors, instead of simply observing anatomical specimens. For the first time, these reports described ongoing neovascularization of tumors implanted subcutaneously, or in transparent chambers, or in the hamster cheek pouch [9–19]. The experimental study of new blood vessel formation, i.e., angiogenesis, began in the late 1930s and early 1940s by Ide et al., and Algire et al. [9, 20, 21]. Experimental tumors were separated from host tissue by a micropore filter to demonstrate that an unknown diffusible substance was released from the tumor that could stimulate new blood vessel growth [22]. In other reports, the onset of tumor neovascularization was elucidated [17, 23–29].

The cause of tumor neovascularization in these studies was not clear. It was attributed variously to inflammation, vasodilation, increased tumor metabolism, overproduction of specific metabolites such as lactic acid, or to hypoxia from "tumors outgrowing their blood supply [30]." Prior to 1970, a prevailing belief was that tumor angiogenesis was a side-effect of dying tumor cells.

Tumor Growth in Isolated Perfused Organs

In the early 1960s, Frederick Becker and I perfused hemoglobin solutions into the carotid artery of rabbit and canine thyroid glands isolated in glass chambers while studying hemoglobin as a possible blood substitute. When mouse melanomas were implanted into the glands, tiny tumors grew up to ~1 mm³ [31–35]. All tumors stopped expanding at the same size. When these microscopic-sized tumors were transplanted to syngeneic mice, the tumors grew more than 1,000 times their original volume in the perfused thyroid gland. Large tumors in mice were highly neovascularized, in contrast to tumors in the isolated organs, which were viable, but not vascularized. This difference suggested that in the absence of neovascularization, tumors would stop growing at approximately ~1 mm³.

The hemoglobin solution was acellular. It did not contain red cells, leukocytes, or platelets. My student, Michael Gimbrone, and I returned to this experiment 6 years later and perfused isolated thyroid glands with platelet-rich plasma. Endothelial vascular integrity was preserved in the isolated organs perfused with platelet-rich plasma. However, in organs perfused with platelet-poor plasma, endothelial cells were disrupted within 5 h, and basement membrane was exposed [35] (Fig. 1.1). We did not implant tumors in this system because of the technical difficulties of prolonged isolated perfusion with platelets. Nevertheless, this experiment implied that absence of platelets was a possible mechanism for lack of neovessels in the earlier experiments of thyroid glands perfused only with hemoglobin solution.

Hypothesis that Tumor Growth is Angiogenesis-dependent

In 1969, a clinical clue revealed additional evidence to me that tumor growth was angiogenesis-dependent. I saw a child with a retinoblastoma in the eye. A large tumor (> 1 cm³) protruded from the retina into the vitreous (Fig. 1.2A) [41]. It was

Department of Surgery, Harvard Medical School and Vascular Biology Program, Children's Hospital, Boston, MA, USA

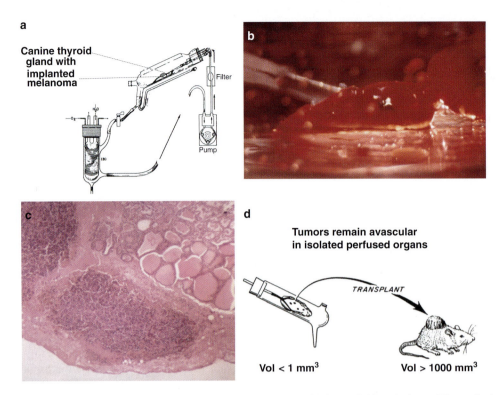

FIG. 1.1A–D. Isolated canine thyroid gland perfused through the carotid artery with hemoglobin solution. **a** The perfusion circuit includes a silicone rubber oxygenator, and a roller pump with silicone rubber tubing. **b** Transilluminated canine thyroid gland in the perfusion chamber, containing a transplanted murine melanoma that grew to ~1 mm³ and stopped expanding. **c** Histologic section of thyroid gland showing viable tumor embedded among viable thyroid follicles. **d** When the tiny, non-expanding tumor was transplanted to a syngeneic mouse, it grew more than 1,000 times its volume in the perfused thyroid gland. The large tumor in the mouse was highly neovascularized, in contrast to its precursor tumor which was not vascularized. The hemoglobin solution was acellular, i.e., it did not contain red cells, leukocytes, or platelets. Reprinted from (Folkman 2007 J Pediatr Surg, with permission of the publisher). Also, see [30–32]. Reprinted from [145] with permission from the publisher.

highly neovascularized. Dozens of tiny metastases had grown from tumor cells shed into the vitreous and into the aqueous humour of the anterior chamber. They were all approximately the *same* size ~ 1 mm³. They were all avascular and pale white (Fig. 1.2B). Histologic sections showed that they averaged about 1.25 mm diameter, that the tumor cells were viable in a rim of 200–250 µm thickness, and that the center was necrotic (Fig. 1.2C) [41]. The metastases could not become neovascularized, in part because they were too far removed from the nearest vascular bed. Fifteen years later, the first antiangiogenic protein of vitreous was discovered [35a].

From the organ perfusion experiments, taken together with clinical observations that metastases of retinoblastoma to the vitreous and aqueous humour of the human eye remained avascular and less than 1 mm diameter, I developed the concept that tumors could not grow beyond approximately 1–2 millimeters without recruiting new blood vessels. This hypothesis that "tumor growth is angiogenesis-dependent" was published in 1971 [35]. This paper also: (1) predicted that virtually all tumors would be restricted to a microscopic size in the absence of angiogenesis; (2) suggested that tumors would be found to secrete *diffusible angiogenic* molecules; (3) described a model of tumor *dormancy* due to blocked angiogenesis; (4) proposed the term *antiangiogenesis* to mean the prevention of new capillary *sprouts* from being recruited into an early tumor implant; (5) envisaged the future discovery of *angiogenesis inhibitors*; and (6) presented the idea that an *antibody* to a tumor angiogenic factor (TAF), could be an anti-cancer drug. The hypothesis was further supported by subsequent experiments in my laboratory reported a year later [33]. These experiments demonstrated that tumor dormancy at a microscopic size was due to blocked angiogenesis of tumors in the aqueous humour of the anterior chamber of the rabbit eye [33]. We also demonstrated DNA synthesis induced in endothelial cells of a tumor bed in vivo, by autoradiography [36]. The hypothesis that tumor growth is angiogenesis-dependent was extended and supported in subsequent invited reviews [37–44].

Nevertheless, for at least a decade after the hypothesis was published, very few scientists believed that tumors needed new blood vessels, or that specific angiogenic molecules could be expressed by tumors [45]. The conventional wisdom was that tumor vascularity was caused by non-specific inflammation.

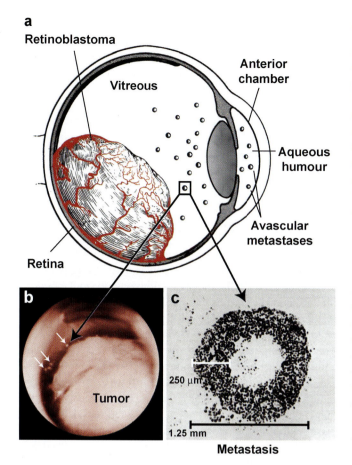

a

Retinoblastoma

Vitreous

Anterior chamber

Aqueous humour

Avascular metastases

Retina

b

Tumor

c

250 µm

1.25 mm

Metastasis

Fig.1.2a–c. A human retinoblastoma that is highly neovascularized. Its metastases in the vitreous and the aqueous humour are not neovascularized because they are floating at too great a distance from the nearest vascular bed (**a–c**). Histology of the metastatic tumors (**c**) shows viable tumor cells in a rim of approximately 250 µm thickness. This is the oxygen diffusion limit. When cryotherapy was used to regress retinoblastomas, the tiny metastases fell on the vascular bed vacated by the primary tumor, and became neovascularized themselves. The entire concept that tumors are angiogenesis-dependent and that tumor dormancy can be due to blocked angiogenesis, can be visualized in this single clinical specimen. With kind permission of Springer Science and Business Media. See reference [41].

Many skeptics challenged the hypothesis that tumor growth depended on angiogenesis. The lack of bioassays for angiogenesis, the inability to culture endothelial cells in vitro in the early 1970s, and the absence of any angiogenesis regulatory molecules did not help matters.

Development of Bioassays for Angiogenesis Research

Before angiogenic or antiangiogenic molecules could be found, it was necessary to develop bioassays for angiogenesis.

Corneal Neovascularization

In the early 1970s, a challenging problem was how to maintain an in vivo tumor separate from its vascular bed in order to prove that tumors secreted diffusible "angiogenic" molecules. Michael Gimbrone, a post-doctoral fellow, and I implanted tumors (of approximately 0.5 mm³) into the stromal layers of the rabbit cornea at distances of up to 2 mm from the limbal edge (Fig. 1.3). New capillary blood vessels grew from the limbus, invaded the stroma of the avascular corneas, and reached the edge of the tumor over a period of approximately 8–10 days [46]. The neovascularization was not due to inflammation and the cornea did not become opaque or edematous, as was the case with inflammatory agents (i.e., silver nitrate).

Vascularized tumors grew exponentially in three dimensions, and became exophytic and protruded from the cornea within 2–3 weeks. Non-vascularized tumors in the center of the cornea expanded slowly in two dimensions, as thin, flat, translucent, intracorneal lesions until one edge extended to within ~2 mm of the limbus and recruited new blood vessels [44]. This method demonstrated that a diffusible "angiogenic factor" existed, and secondly, that such a putative angiogenic molecule could possibly be isolated from tumors. But, when tumor extracts were implanted into the cornea to mimic a tumor implant, the extracts rapidly diffused away into the cornea. A focal steady-state concentration gradient of angiogenic activity, similar to a tumor implant, could not be established. Silicone rubber capsules that we had previously found to steadily release small molecules (< 500 Daltons) by diffusion through the polymer itself [47] could not release proteins.

Robert Langer, a post-doctoral fellow, solved the problem by dissolving the polymer polyhydroxy ethylmethacrylate (PolyHEMA) in alcohol and adding the lyophilized protein to be tested [48]. After evaporation of the solvent, the protein remained trapped in a rubbery polymeric pellet. When the pellet was implanted into the cornea, water diffused into the pellet. This caused the formation of microchannels around the protein. Protein diffused out from these channels at zero-order kinetics for weeks to months [49]. Another polymer, ethylene vinyl acetate copolymer (ELVAX), dissolved in ethylene chloride was also used. These polymers did not irritate the cornea. Robert Auerbach had come to my laboratory on a sabbatical. He subsequently showed that tumors, or the polymer pellets could also be implanted in the mouse cornea [50]. This advance permitted genetic experiments, and the model is now routinely employed.

The corneal neovascularization bioassay and the technology of sustained-release corneal implants have also been pivotal for the study of the molecular regulation of angiogenesis, and for the development of new drugs that inhibit angiogenesis.

The corneal neovascularization model also permitted the demonstration that removal of an angiogenic stimulus was followed by a sequential series of steps beginning with occlusion of the new capillary vessels by platelets, and then desquamation of endothelium, followed by complete *regression of neovasculature*

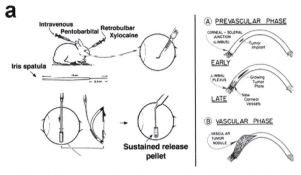

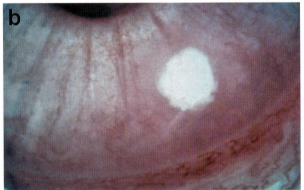

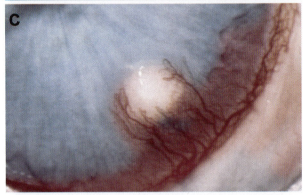

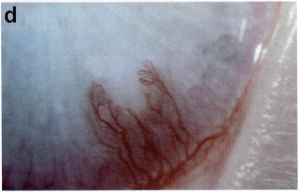

FIG. 1.3. **a** Diagram of implantation of a sustained release polymer pellet into the corneal of the rabbit eye. **b** The pellet causes no irritation, inflammation or cornea edema. **c** When the pellet contains an angiogenic protein, new vessels grow from the limbal edge of the cornea between the lamellar layers of the cornea. The new vessels have extended approximately 2.0 mm to reach the edge of the pellet. **d** After a small incision is made over the pellet it is easily removed. Blood vessels then undergo complete regression by approximately 10 weeks.

within a few weeks [51]. This finding revealed that newly induced neovasculature does not become 'established.' The results also indicated that, if angiogenesis inhibitors could be developed, they could possibly cause newly induced tumor vessels to regress. These results provided a compelling rationale for future attempts to discover and develop angiogenesis inhibitors. Of interest is that, almost three decades later, the corneal model was first employed to dissociate lymphangiogenesis from angiogenesis, by simply implanting a significantly lower concentration of bFGF to induce lymphatic growth [52].

Vascular Endothelial Cells In Vitro

The history of the field of angiogenesis research coincides with the development of long-term in vitro cultures of vascular endothelial cells. In 1973, Gimbrone, in my laboratory, [46, 53] and Eric Jaffe's laboratory at Cornell [54] were independently the first to successfully grow and passage vascular endothelial cells in vitro (from human *umbilical veins*). The first long-term passage of cloned *capillary* endothelial cells came later and was reported in 1979 [55].

It is impossible to recount the ridicule from the scientific community that greeted these early papers. It was widely believed that endothelial cells could never be cultured outside the body, because they "live in blood." Therefore, it was believed that our reported endothelial cells must be some unknown "contaminants." Another problem was that we had suggested that these endothelial cells could be useful to guide purification of novel endothelial mitogens or growth inhibitors. But cultured vascular endothelial cells become *refractory* to virtually any mitogen once the cells had reached *confluence,* … in contrast to confluent fibroblasts which still responded to mitogens. We soon found that vascular endothelial cells were among *the most stringently regulated cells at high cell density*. Until the mid-1970s, it was conventional practice to guide the purification of growth factors with fibroblasts in cell culture. Fibroblasts (3T3 cells) were grown to confluence. When a putative growth factor was added, one or two additional rounds of DNA synthesis ensued. In contrast, when *endothelial* cells were used to guide purification of endothelial mitogens, confluent endothelial cells did *not* undergo additional DNA synthesis, and investigators assumed that their tumor extracts were inactive. We were eventually able to report in 1976 [56] that in order to test a putative endothelial mitogen, it was necessary to incubate endothelial cells with the mitogen when the cells were *sparse, not when they were confluent*. This was a critical detail, just the opposite of employing 3T3 fibroblasts to purify a mitogen for fibroblasts, but essential for testing endothelial mitogens or inhibitors in vitro. If the writer of a history of the field of angiogenesis research is obligated to emphasize certain crucial findings without which development of the whole field would have been slowed or stopped, I submit that the 1976 paper [56] is one of them.

Furthermore, the difficulty in the 1970s of persuading the scientific community to accept the validity and usefulness of endo-

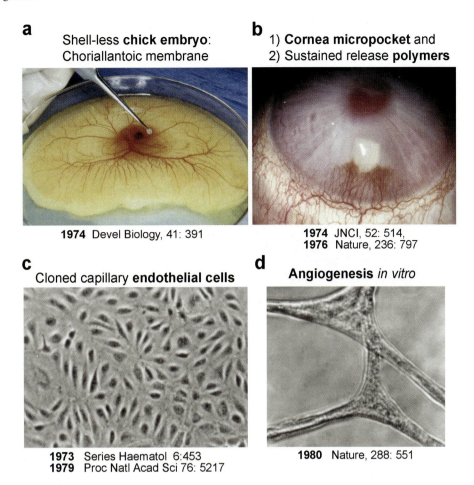

a Shell-less **chick embryo**: Choriallantoic membrane

1974 Devel Biology, 41: 391

b 1) **Cornea micropocket** and 2) Sustained release **polymers**

1974 JNCI, 52: 514, **1976** Nature, 236: 797

c Cloned capillary **endothelial cells**

1973 Series Haematol 6:453 **1979** Proc Natl Acad Sci 76: 5217

d **Angiogenesis** *in vitro*

1980 Nature, 288: 551

Fig. 1.4. Bioassays for angiogenesis developed during the 1970s. **a** The chick embryo chorioallantoic membrane. **b** Corneal neovascularization stimulated by an implanted polymer releasing an angiogenic protein. **c** Capillary endothelial cells in vitro. **d** Angiogenesis in vitro. Reprinted from [146] with permission of the publisher. With kind permission of Springer Science and Business Media.

thelial cells in vitro deserves mention here, because there is no written record of the misperceptions of the time. This chapter on the 'history of angiogenesis' may be helpful today because those who are unaware of this history risk being misled by their experiments, just as researchers were 30 years ago. By the 1980s, endothelial cells were employed in laboratories world-wide, and they have served to guide isolation and purification of angiogenesis regulatory molecules, to elucidate endothelial metabolism, to identify endothelial receptors, to discover that endothelial cells respond to certain mitogens or inhibitors of proliferation with a biphasic, *U*-shaped response, and to uncover the mechanism by which confluent endothelial cells are refractory to mitogens.

Cell Shape and Cell Growth Control

It was unclear why *confluent* endothelial cells in vitro should be refractory to mitogens, until we subsequently found that changes in cell shape of endothelial cells during confluence in vitro, *revealed* a central mechanism of suppression of DNA synthesis in endothelial cells [57]. In fact, shape control of DNA synthesis [57], especially by the crowding of cells in confluence appears to have eluded discovery until the advent of successful

in vitro growth of vascular endothelial cells. This mechanism was further elucidated by Donald Ingber who showed how changes in cell shape can signal through integrins to regulate gene expression and DNA synthesis. He went on to develop an entirely new field of investigation of cell biology based on the mechanisms by which mechanical forces modify DNA synthesis and gene expression [58–62]. The experiments of Mina Bissell on cell shape and differentiation of function also informed the role of cell shape in cell growth [63].

Angiogenesis In Vitro

When angiogenesis in vitro was demonstrated [64] (Fig. 1.4d), it became possible to elucidate the morphologic and molecular events of lumen formation in microvessels. Lumens developed from vesicles that formed within individual endothelial cells. Vesicles became connected to form tubes and branches. Branches formed from the vesicles. A similar mechanism has been found in the development of vasculature in zebra fish [65, 66]. Tube formation arising from endothelial cell monolayers in vitro, may more accurately represent vasculogenesis *in vivo* than angiogenesis (in which sprouts arise from pre-existing vasculature).

Chick Embryo Chorioallantoic Membrane

By culturing fertilized chick embryos in petri dishes beginning at day 3, tumors and fractions from protein purification were implanted on the chorioallantoic membrane (Fig. 1.4a). However, the optimum time for testing of proangiogenic or antiangiogenic molecules on this vascular substratum was day 6 to approximately day 8 [67–69]. Normal and neoplastic tissues revealed different mechanisms of vascularization after being grafted to the chorioallantoic membrane [70]. The chick chorioallantoic membrane became a quantitative angiogenesis bioassay when test proteins were implanted in a white opaque gel sandwiched between two squares of nylon mesh [71]. New microvessel sprouts grew *vertically* through the mesh. When two sprouts anastomosed to form a capillary loop enveloping a nylon thread, this was compelling proof that the microvessels were new (and not just dilated vessels). New microvessels could be accurately quantified by the ratio of squares of mesh containing a microvessel vs empty squares. Norrby et al. also developed a quantitative angiogenesis bioassay using spreads of intact rat mesentery windows [71a].

Ribatti et al. further improved the chick chorioallantoic membrane bioassay [72], and also studied the modification of angiogenesis on the chorioallantoic membrane, by heparin [73], bFGF [74], and interferon α [75]. For recent reviews of in vivo models of angiogenesis see [76, 77].

Discovery of Angiogenic Molecules

The first proangiogenic molecules to be isolated from tumors were isolated and purified to homogeneity beginning in the early 1980s by Yuen Shing and Michael Klagsbrun in my laboratory [78]. Purification was guided by the bioassays described above using heparin-sepharose chromatography. This technique was subsequently employed by many other investigators to purify angiogenic regulatory molecules. Over the ensuing years, heparin affinity turned out to be a fundamental property of the majority of angiogenesis regulatory molecules. When Esch et al. [79] subsequently purified an identical protein from bovine pituitary, the amino acid sequence revealed it to be basic fibroblast growth factor (bFGF). Gospodarowicz was the first to isolate fibroblast growth (from pituitary) and had shown it to be a mitogen for 3T3 fibroblasts and vascular endothelial cells [80, 81], but it was not purified to homogeneity.

The discovery of a second angiogenic protein began in 1983, with a report by Senger and Dvorak of the purification of a vascular permeability factor (VPF) from tumor cells that promoted accumulation of ascites [82]. By 1989, Rosenthal in my laboratory had isolated and purified to homogeneity a second angiogenic protein from a tumor that did not express bFGF. We had not yet sequenced this new protein when we received a call from Napoleone Ferrara of Genentech, who had purified a novel angiogenic protein from pituitary cells. He had heard about the new angiogenic protein in our labora-tory, and suggested that the two laboratories compare their proteins, because Ferrara had already sequenced his protein. The two proteins were identical and were named vascular endothelial growth factor by Ferrara (VEGF). Ferrara's report was published in mid-1989 [83], and our paper reporting the first VEGF from a tumor was published in 1990, with Ferrara as a co-author [84]. By 1990, it was clear that VPF was also the same as VEGF. Thus, VEGF had a propitious start, having been purified from three different sources, but first sequenced in Ferrara's lab.

Many other proangiogenic molecules have since been discovered and are discussed by other authors in this book. Recently, Klagsbrun discovered that neuropilin-1 is another receptor for VEGF and stimulates angiogenesis [85].

Heparin Affinity Mediates Sequestration of an Angiogenic Protein in Extracellular Matrix

In 1987, Vlodavsky et al. reported that bFGF was stored in extracellular matrix, where it was bound to heparan sulfate proteoglycans [86, 87]. The storage of angiogenesis regulatory proteins in extracellular matrix has become a general property for this class of molecules [88]. In fact, the majority of angiogenesis regulatory molecules, both proangiogenic and antiangiogenic, have been reported to have high affinity for heparin. This property of angiogenesis regulatory molecules also provides a mechanism: (1) to protect these proteins from degradation; (2) to protect vascular endothelial cells from exposure to these biologically proteins; and (3) a mechanism of rapid release of endothelial mitogens localized to a wound site where extracellular matrix is disrupted [88].

Discovery of Angiogenesis Inhibitors

From 1971 to 1980, there was no method of inhibiting tumor angiogenesis except by mechanical separation of a tumor from its vascular bed, for example by implanting a tumor in the rabbit cornea [46], or by floating it in the aqueous humour of the anterior chamber of the eye [33], or in the vitreous [89]. Therefore, a search was begun for molecules that could inhibit angiogenesis, because stronger evidence was needed to support the hypothesis that tumors are angiogenesis-dependent.

Low concentrations of interferon α in vitro were found to specifically suppress migration of endothelial cells in vitro [90]. Subsequently, Dvorak [91], and also Auerbach [50] reported that interferon α inhibited angiogenesis in experimental animals. By 1988, daily low dose interferon α was used to inhibit angiogenesis in a teenager dying of progressive pulmonary hemangiomatosis of both lungs with hemoptysis. He had failed all conventional therapy. The patient's physician, Carl White, a pulmonary specialist at Denver Jewish Hospital,

TABLE 1.1. Drugs with antiangiogenic activity which have received FDA approval for the treatment of cancer or age-related macular degeneration [45].

32,598 papers published on angiogenesis from 1971 -September 2007

Date approved	Drug Disease Place
May 2003	**Velcade** (Bortezomib) U.S. (FDA) **Multiple myeloma**
December 2003	**Thalidomide** Australia **Multiple myeloma**
February 2004	**Avastin** (Bevacizumab) U.S. (FDA) **Colorectal cancer**
February 2004	**Erbitux** U.S. (FDA) **Colorectal cancer**
November 2004	**Tarceva** (Erlotinib) **Lung cancer** U.S. (FDA)
December 2004	**Avastin** Switzerland **Colorectal cancer**
December 2004	**Macugen** (Pegaptanib) U.S. (FDA) **Macular degeneration**
January 2005	**Avastin** E. U. (27 countries) **Colorectal cancer**
September 2005	**Endostatin** (Endostar) China (SFDA) **Lung cancer**
December 2005	**Nexavar** (Sorafenib) U.S. (FDA) **Kidney cancer**
December 2005	**Revlimid** (Lenalidomide) U.S. (FDA) **Myelodysplastic syn.**
January 2006	**Sutent** (Sunitinib) U.S. (FDA) GIST
June 2006	**Lucentis** (Ranibizumab) U.S. (FDA) **Macular degeneration**
June 2006	**Revlimid** U.S. (FDA) **Multiple myeloma**
August 2006	**Lucentis** Switzerland **Macular degeneration**
September 2006	**Lucentis** India **Macular degeneration**
October 2006	**Avastin** U.S. (FDA) **Lung cancer**
January 2007	**Lucentis** E. U. (27 countries) **Macular degeneration**
March 2007	**Avastin** E. U. Iceland, Norway **Metastatic breast**
April 2007	**Avastin** Japan **Colorectal cancer**
May 2007	**Torisel** (CCI-779) U.S. (FDA) **Kidney cancer**

TABLE. 1.2. Publications with "angiogenesis" in the title from 1971 through September 2007.

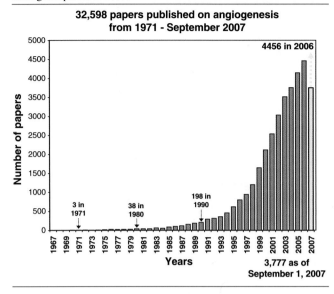

32,598 papers published on angiogenesis from 1971 - September 2007

called me. I suggested a trial of daily or every other day *low-dose* interferon α, based on our experimental elucidation of its anti-endothelial properties, its antiangiogenic activity in animals [91], and because it was an FDA-approved drug. The hemangiomatosis regressed completely by 7 months and the patient went home. He was treated for an additional 5 years (by subcutaneous self-injection), while he completed his education and obtained a job [92, 93]. He has a normal chest film and is in good health today, 18 years later. This is the first recorded case of antiangiogenic therapy. It was subsequently found in other patients that these low doses are antiangiogenic, but are not cytotoxic, nor immunosuppressive. Fidler and colleagues reported that low-dose interferon α down-regulates expression of basic fibroblast growth factor (bFGF) in human tumor cells [94]. In 1982, we found that protamine and platelet factor 4 inhibited angiogenesis in the chick embryo [95]. In 1985, we reported a new class of corticosteroids, called 'angiostatic' steroids [96]. For example, tetrahydrocortisol is a potent angiogenesis inhibitor that has neither glucocorticoid nor mineralocorticoid activity. It is in clinical trials for the treatment of ocular neovascularization. The antiangiogenic activity of angiostatic steroids is potentiated by an arylsulfatase inhibitor (synthesized by Professor E.J. Corey), that inhibits desulfation of endogenous heparin [97]. Fotsis et al. [98] reported that the steroid 2-methoxyestradiol inhibited angiogenesis, and Robert

D'Amato in my laboratory reported that this angiostatic steroid inhibited tubulin polymerization by interacting at the colchicine site [99].

From 1980 to 2005, eleven angiogenesis inhibitors were identified or discovered in the Folkman laboratory (Table 1.1). The majority of these (eight), were found to be in the blood or tissues. Some of them were previously unknown molecules, such as angiostatin and endostatin (see Chapter 12), while for others, angiogenesis inhibition was a new function, such as interferon α [100] and platelet factor 4 [95]. Other laboratories joined this research effort, and at this writing there are 28 known endogenous angiogenesis inhibitors [101, 102]. As of September 2007, 32,598 papers on angiogenesis have been published since 1971 (Table 1.2).

Angiogenesis Inhibitors in the Clinic

The accumulating literature on angiogenesis inhibitors reported during the 1980s and early 1990s led to the development by pharmaceutical and biotechnology companies of angiogenesis inhibitors for clinical use. At the time of writing, ten drugs with antiangiogenic activity have received FDA approval in the U.S. and in more than 30 other countries by their regulatory agencies (Table 1.1). They are approved mainly for cancer, but also for treatment of age-related macular degeneration. A total of 1,257,600 patients received prescriptions for FDA-approved angiogenesis inhibitors during 2006.[1]

Antiangiogenic therapies include either relatively pure angiogenesis inhibitors such as Avastin (Bevacizumab), or

---

[1] From a professional search of world-wide databases by "info2go", using Thomson Pharma, Chemical Market Reporter, & Business Communications Co.

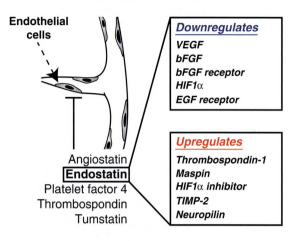

"Direct" angiogenesis inhibitors target activated endothelium directly, and inhibit _multiple_ angiogenic proteins.

FIG. 1.5. Direct angiogenesis inhibitors block activated endothelium directly and inhibit multiple angiogenic proteins.

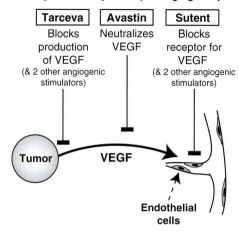

"Indirect" angiogenesis inhibitors target tumor cell products and usually inhibit only 1 or 2 pro-angiogenic proteins.

FIG. 1.6. Indirect angiogenesis inhibitors mainly target tumor cell products, and usually inhibit a more narrow spectrum of angiogenic proteins than "direct" angiogenesis inhibitors.

drugs that are antiangiogenic in addition to other activities such as inhibition of tumor cell proliferation [i.e., Tarceva (erlotinib)]. Drugs with antiangiogenic activity can also be classified as either "direct" or "indirect" angiogenesis inhibitors. Direct angiogenesis inhibitors inhibit activated cells directly. Endothelial cells are usually prevented from responding to a wide spectrum of proangiogenic proteins (Fig. 1.5). Many of the endogenous angiogenesis inhibitors such as thrombospondin-1, angiostatin, endostatin, and tumstatin, are "direct" angiogenesis inhibitors. For example, endostatin not only down-regulates endothelial receptors for a range of proangiogenic ligands, but simultaneously up-regulates endothelial expression of other endogenous angiogenesis inhibitors such as thrombospondin-1 and maspin [45, 103]. Endostatin, approved in China as Endostar, is a "direct" angiogenesis inhibitor. In contrast, "indirect" angiogenesis inhibitors block mainly tumor cell products. Examples are: Tarceva, which blocks tumor cell production of VEGF and other proangiogenic proteins; Avastin, which neutralizes VEGF; and Sutent (Sunitinib) which blocks endothelial receptors for VEGF and other proangiogenic proteins (Fig. 1.6).

The angiogenesis inhibitors that are currently FDA-approved are "indirect." They also have narrow targets. Avastin blocks only VEGF. Others block 2 or 3 proangiogenic proteins. However, most human tumors can express up to 6 or more angiogenic proteins [104]. Furthermore, long-term suppression of a single proangiogenic pathway (i.e., VEGF) in experimental animals can result in emergence of increased expression of other proangiogenic proteins, such as bFGF or PLGF. This phenomenon has been named tumor "evasion" of antiangiogenic therapy to signify its difference from acquired drug resistance to cytotoxic chemotherapy [105]. It is not clear whether "evasion" of mono-antiangiogenic therapy is due to overexpression of additional proangiogenic proteins from expansion of new clones of tumor cells, or from the same tumor cells in which the original proangiogenic protein was suppressed [106–110]. Nevertheless, it is becoming clear that "evasion" of antiangiogenic therapy against tumor angiogenesis can be prevented by _combinations_ of up to three different angiogenesis inhibitors in experimental systems [111]. It remains to be seen whether "evasion" may become less of a problem when broad-spectrum angiogenesis inhibitors (such as endogenous antiangiogenic proteins) are introduced into clinical application, or when combinations of angiogenesis inhibitors are employed clinically.

Angiogenesis as an Organizing Principle

Certain general principles of the angiogenic process have been identified that underlie many phenomena in the field of angiogenesis research and its clinical application. An understanding of these may help to guide future research and future development of angiogenesis based therapies. In this sense, angiogenesis itself may be considered as an organizing principle in biomedicine [45].

Normalization of Angiogenesis

Jain et al. have shown that various antiangiogenic therapies can decrease vessel permeability and interstitial fluid pressure in neovascularized tumors [112–115]. Leaky tumor blood vessels diffuse albumin and other proteins across their vascular wall and cannot sustain oncotic pressure gradients. In addition, lymphatics are either absent in some tumors or dysfunctional in others. Angiogenesis inhibitors, especially those that block VEGF activity, "normalize" tumor blood vessels [116] and reduce intratumoral fluid pressure. This has been shown in animal and in human tumors. The cellular and molecular mechanisms of "normalization" are currently being worked out. However, certain antiangiogenic therapies can temporarily increase blood flow into a tumor because of decreased intratumoral tissue pressure. This permits increased delivery of intravenous therapeutic agents, and leads to increased oxygenation and increased radiosensitivity. Even as capillary drop-out occurs from the antiangiogenic therapy, microvessels may remain normalized. Furthermore, antiangiogenic therapy that interferes with VEGF activity can decrease ascites from ovarian cancer, alleviate brain tumor edema, and may reduce metastatic shedding of tumor cells into lymphatics, or into the general circulation.

Biphasic Efficacy of Antiangiogenic Therapy

The efficacy of many angiogenesis inhibitors can often be quantified as a biphasic, U-shaped curve. In its simplest terms, higher doses are often less effective than lower doses. This phenomenon was first demonstrated when mice bearing human bladder cancer were treated with low-dose interferon [117]. Frequent lower doses of interferon-α decreased tumor weight, bFGF serum levels, and tumor microvessel density, more than 2-fold more effectively than higher doses. Other angiogenesis inhibitors that show biphasic efficacy are rosiglitazone [118] endostatin protein therapy [119], and endostatin gene therapy [120].

Gene expression in microvascular endothelial cells treated with endostatin in vitro at increasing doses and times is also biphasic and U-shaped [103]. Also, cytotoxic chemotherapy, administered at lower doses and more frequently, is more effective against experimental tumors than maximum tolerated dose chemotherapy, even if the tumors are drug-resistant [121–123]. This is called antiangiogenic chemotherapy [121], which describes the target of low dose chemotherapy, or metronomic chemotherapy [123], which describes how it is administered. Of interest are recent reports that administration of low-dose cyclophosphamide at frequent intervals increases circulating levels of thrombospondin [123a, 124a].

The Switch to the Angiogenic Phenotype

Newly-arising human tumors usually stop expanding at a microscopic size (millimeters) and remain as non-angiogenic, dormant, in situ tumors. Only a small percentage switch to the angiogenic phenotype [45]. The angiogenic switch has been studied in animal models. For example, spontaneous tumors, under the control of a specific oncogene and arising in transgenic mice, are not angiogenic when they first appear [124]. The majority of tumors are non-angiogenic and remain in that state at a microscopic size of 1–2 mm³. After a predictable time [i.e., ~6–7 weeks for islet cell carcinomas driven by the large T antigen hybridized to the rat insulin promoter (RIP-

Tag)], approximately 10% of tumors switch to the angiogenic phenotype, expand rapidly and are lethal to the mice. The angiogenic switch is accompanied by a shift in the balance of positive and negative regulators of angiogenesis [125].

In another animal model, a similar angiogenic switch is observed in human tumors grown in SCID immunodeficient mice. Single non-angiogenic tumor cells can be cloned from human tumors discarded in the operating room or obtained from human tumor cell lines. A subpopulation (~5%) of human tumor cells is non-angiogenic. The non-angiogenic tumor cells are expanded in vitro and infected with a reporter gene (luciferase). When non-angiogenic tumors cells are inoculated into SCID mice, the tumors remain dormant at a microscopic size of ~ 1 mm³. The microscopic tumors have a high proliferation rate and a high apoptotic rate. They are harmless until they switch to the angiogenic phenotype, which occurs at a predictable time from approximately 3 months to >1 year, depending on the tumor type. The percentage of non-angiogenic tumors that undergo the angiogenic switch is dependent on tumor type and is highly predictable. For example, about 95% of human liposarcomas switch at approximately 133 ± 25 days, while only 5% of human osteosarcomas switch to the angiogenic phenotype after more than a year. Tumor cells that become angiogenic proliferate at a similar rate to non-angiogenic tumor cells, but have a significantly lower rate of apoptosis [126–130]. Under some conditions, for example when the gene PPAR-α is deleted, tumors remain non-angiogenic and harmless throughout the life of the host [131]. These animal models demonstrate a novel mechanism of prolonged dormancy of human tumors. They also provide models for antiangiogenic therapy that could in the future prevent the switch to the angiogenic phenotype.

Angiogenesis Regulatory Proteins in Platelets

Increasing platelet counts in cancer patients have been reported to correlate with a worse prognosis [132]. A recent finding is that platelets carry a set of angiogenesis regulatory proteins in their α granules [133–137]. Furthermore, positive and negative regulators of angiogenesis are segregated into different α granules. This new biology may have implications for release of proangiogenic proteins at the initiation of wound healing, followed by release of antiangiogenic proteins at a later stage of wound healing. Platelets accumulate in some tumors and release of proangiogenic proteins could further stimulate tumor growth.

Angiogenesis-dependent Disease

Certain non-neoplastic diseases appear to be angiogenesis-dependent. The two most common are diabetic retinopathy and age-related macular degeneration. Currently, three FDA approved angiogenesis inhibitors that block VEGF are being used as intravitreal injections by ophthalmologists to treat age-related macular degeneration. Macugen (pegaptanib) and Lucentis (ranibizumab) are approved for macular degeneration and Avastin (bevacizumab) is being used off-label. Results with Lucentis in large populations of patients reveal that with progressive disease, visual acuity is significantly improved in ~95% of patients. For blind patients, 40% have their sight restored [138].

However, it was not always clear that VEGF could be a drug target in age-related macular degeneration. In a series of reports starting in 1993, Folkman and collaborators, Anthony Adamis, Patricia D'Amore, and Joan Miller, demonstrated that VEGF was a major mediator of ocular neovascularization in the non-human primate. Correlative evidence was also reported for humans. They showed in 1993 that VEGF was expressed and secreted by human retinal cells in vitro [139]. They subsequently showed that iris neovascularization correlated spatially and temporally with intraocular VEGF levels [140], and that expression of VEGF in the retina was upregulated by ischemia. Adamis et al. reported that the vitreous of human eyes with proliferative diabetic retinopathy harbored significantly increased levels of VEGF [141]. Aiello et al. also reported high VEGF levels in diabetic vitreous [142]. In 1995, the Folkman laboratory, in collaboration with Napoleone Ferrara of Genentech, reported that VEGF was the major endothelial mitogen produced by hypoxic retinal cells [143]. A subsequent collaboration between Adamis, D'Amore, Ferrara, Folkman, and Miller demonstrated that intravitreal injection of a neutralizing antibody to VEGF, produced by Ferrara at Genentech, inhibited retinal-ischemia-associated neovascularization in a non-human primate [144]. Eight and 10 years later, Macugen and Lucentis, respectively, were approved by the FDA for the treatment of age-related macular degeneration.

Conclusion

This is a brief history of the development of the field of angiogenesis research as it originated from studies of tumor angiogenesis. I have attempted to put in perspective certain obstacles that had to be overcome, and specific technical problems that had to be solved, before further advances were possible. Where I have moved fast forward to citations of current work, or to clinical applications, the purpose was to illustrate the successful result of figuring out a solution to some seemingly intractable biological problem. The field of angiogenesis continues to grow and to be very exciting. New and interesting obstacles have emerged because of the successes of the past. For example, there is an urgent need for a quantitative biomarker of angiogenesis activity in the body that could be detected in the blood or urine. Such detection methods may also be invaluable to rapidly determine efficacy of angiogenesis inhibitors in patients. Once this problem is solved, it may then become possible in the future to guide antiangiogenic anti-cancer therapy with a blood or urine biomarker that detects angiogenesis in recurrent tumors years before symptoms or before anatomical location is possible by conventional methods. Such an advance would also facilitate the combination of antiangiogenic therapy with other modalities, such as various forms of immunotherapy, telomerase inhibitor therapy, and others.

References

1. Hall AP. The role of angiogenesis in cancer. Comp Clin Path 2005; 13:95–99.
2. Virchow R. Die Krankhaften Geschwuste, August Hirschwald, Berlin, 1863; cited by Rogers W. et al., Surg Clin N Am 1967; 47:1473.
3. Thiersch C. Der Epithelialkrebs namentlich der Haut mit Atlas, Lepzig, 1865, as cited in Rogers W, Edlich RF, Lewis DV, Aust JB. Surg Clin North Am 1967; 47:1473.
4. Goldmann E. Growth of malignant disease in man and the lower animals with special reference to vascular system. Proc R Soc Med 1907; 1:1.
5. Thiessen NW. The vascularity of benign and malignant lesions of the stomach. Surg Gynecol Obstet 1936; 2:149.
6. Wright RD. The blood supply of abnormal tissues in the lung. J Pathol Bacteriol 1938; 47:489.
7. Lindgren AGH. The vascular supply of tumours with special references to the capillary angioarchitecture. Acta Pathol Microbiol Scand 1945; 22:493.
8. Peterson HI (ed). Tumor Blood Circulation: Angiogenesis, Vascular Morphology and Blood Flow of Experimental and Human Tumors. CRC Press, Boca Raton, Florida 1979.
9. Ide AG, Baker NH, Warren SL. Vascularization of the Brown-Pearce rabbit epithelioma transplant as seen in the transparent ear chamber. Am J Roentgenol 1939; 42:891.
10. Cowdry DR, Sheldon WF. The significance of hyperaemia around tumor transplants. Am J Pathol 1946; 22:821.
11. Lutz BR, Patt DI, Handler AH, Stevens DF. Serial sarcoma transplantation in the hamster cheek pouch and the effects of advanced neoplasia on the small blood vessels. Anat Rec 1950; 108:545.
12. Toolan HW. Proliferation and vascularization of adult human epithelium in subcutaneous tissues of x-irradiated heterologous hosts. Proc Soc Exp Biol Med 1951; 78:540.
13. Green HSN. The significance of the heterologous transplantability of human cancer. Cancer (Philadelphia) 1952; 5:24.
14. Wood S Jr. Pathogenesis of metastasis formation observed in vivo in the rabbit ear chamber. AMA Arch Pathol 1958; 66: 550–568.
15. Day ED, Planinsek JA, Pressman D. Localization of radio iodinated rat fibrogen in transplanted rat tumors. J Natl Cancer Inst 1959; 23:799.
16. Bierman HR, Kelly KH, Dod KS, Byron RL Jr. Studies on the blood supply of tumor in man. I. Fluorescence of cutaneous lesions. J Natl Cancer Inst 1951; 11:877.
17. Gullino PM, Grantham FH, Clark SH. The collagen content of transplanted tumors. Cancer Res 1962; 22:1031.
18. Urbach F, Graham JH. Anatomy of human skin tumour capillaries. Nature 1962; 194:652.
19. Rubin R, Casarett G. Microcirculation of tumors. I. Anatomy, function and necrosis. Clin Radiol 1966; 17:220.
20. Algire GH. Microscopic studies of the early growth of a transplantable melanoma of the mouse, using the transparent chamber technique. J Natl Cancer Inst 1943; 4:1.

21. Algire GH, Legallais FY, Park HD. Vascular reactions of normal and malignant tissues in vivo. II. The vascular reaction of normal and neoplastic tissues of mice to a bacterial polysaccharide from *Serratia Marcescens (Bacillus prodigius)* culture filtrates. J Natl Cancer Inst 1947; 8:53.

22. Ehrmann RL, Knoth M. Choriocarcinoma: Transfilter stimulation of vasoproliferation in the hamster cheek pouch – Studied by light and electron microscopy. J Natl Cancer Inst 1968; 41:1329–1341.

23. Goldacre RJ, Sylven B. On the access of blood borne dyes to various tumor regions. Br J Cancer 1962; 16:306.

24. Delarue J, Mignot J, Caulet T. Modifications vasculaires de la poche jugale du hamster doré au dours du développement de greffes d'une tumeur mélanique. C. R. Seances Soc Biol Patis 1963; 157 :69.

25. Day ED. Vascular relationships of tumor and host. Prog Exp Tumor Res 1964; 4:57.

26. Goodall CM, Sanders AG, Shubik P. Studies of vascular patterns in living tumors with a transparent chamber inserted in hamster cheek pouch. J Natl Cancer Inst 1965; 34:497.

27. Warren BA, Shubik P. The growth of the blood supply to melanoma transplants in the hamster cheek pouch chamber. Lab Invest 1966; 15:464.

28. Tannock IF. Population kinetics of carcinoma cells, capillary endothelial cells, and fibroblasts in a transplanted mouse mammary tumor. Cancer Res 1970; 30:2470.

29. Folkman J. Tumor angiogenesis: From bench to bedside. In: Marme D, Fusenig N, eds. Tumor Angiogenesis: Basic Mechanisms and Cancer Therapy. Springer-Verlag, Heidelberg, Germany, 2008:3–28.

30. Folkman MJ, Long DM, Becker FF. Tumor growth in organ culture. Surg Forum 1962; 13:164.

31. Folkman J, Long DM, Becker FF. Growth and metastasis of tumor in organ culture. Cancer 1963; 16.

32. Folkman J, Cole P, Zimmerman S. Tumor behavior in isolated perfused organs: *in vitro* growth and metastases of biopsy material in rabbit thyroid and canine intestinal segment. Ann Surg 1966; 164:491.

33. Gimbrone MA Jr, Leapman SB, Cotran RS, Folkman J. Tumor dormancy *in vivo* by prevention of neovascularization. J Exp Med 1972; 136(2):261.

34. Gimbrone MA Jr, Aster RH, Cotran RS, Corkery J, Jandl JH, Folkman J. Preservation of vascular integrity in organs perfused in vitro with a platelet-rich medium. Nature 1969; 222:33.

35. Folkman J. Tumor angiogenesis: therapeutic implications. N Engl J Med 1971; 285:1182.

35a. Lutty GA, Thompson DC, Gallup JY, Mello RJ, Patz A, Fenselau A. Vitreous: an inhibitor of retinal extract-induced neovascularization. Invest Ophthalmol Vis Sci 1983;24:52.

36. Cavallo T, Sade R, Folkman J, Cotran RS. Tumor angiogenesis: rapid induction of endothelial mitoses demonstrated by autoradiography. J Cell Biol 1972; 54:408.

37. Folkman J. Tumor angiogenesis. In Klein G, Weinhouse S, eds. Advances in Cancer Research. New York: Academic Press, 1974; Volume 19, 331.

38. Folkman J. Tumor angiogenesis: role in regulation of tumor growth. In: Hay, ED, King TJ, Papaconstantinou J, eds. Macromolecules Regulating Growth and Development. New York: Academic Press, 1974; 43.

39. Brem H, Folkman J. Inhibition of tumor angiogenesis mediated by cartilage. J Exp Med 1975; 141(2):427.

40. Folkman J, Gimbrone MA Jr. Perfusion of the thyroid gland. In: Hardman JG, O'Malley BW, eds. Methods in Enzymology, Hormone Action, Part D, Isolated Cells, Tissues and Organ Systems. New York: Academic Press 1975; 39:359.

41. Folkman J. Tumor angiogenesis. In: Becker FF, ed. Cancer: Comprehensive Treatise. New York: Plenum Press 1975; 3:355.

42. Folkman J, Klagsbrun M. Tumor angiogenesis: effect on tumor growth and immunity. In: Gottlieb AA, Plescia OJ, Bishop DHL, eds. Fundamental Aspects of Neoplasia. New York: Springer-Verlag 1975; 401.

43. Folkman J, Cotran RS. Relation of vascular proliferation to tumor growth. In: Richter GW, ed. International Review of Experimental Pathology. New York: Academic Press 1976; 16:207.

44. Folkman J. Tumor angiogenesis and tumor immunity. In: Castro JE, ed. Immunological Aspects of Cancer. Lancaster: MTP Press, Limited 1978; 267.

45. Folkman J. Angiogenesis: an organizing principle for drug discovery? Nat Rev Drug Discov. 2007; 6:273.

46. Gimbrone MA Jr, Cotran RS, Leapman SB, Folkman J. Tumor growth and neovascularization: an experimental model using rabbit cornea. J Natl Canc Inst 1974; 52(2):413.

47. Folkman J, Long DM. The use of silicone rubber as a carrier for prolonged drug therapy. J Surg Res 1964; 4:139.

48. Langer R, Folkman J. Polymers for the sustained release of proteins and other macromolecules. Nature 1976; 263(5580):797.

49. Brown LR, Wei CL, Langer R. In vivo and in vitro release of macromolecules from drug delivery systems. J Pharm Sci 1983; 72:1181.

50. Muthukkaruppan V, Auerbach R. Angiogenesis in the mouse cornea. Science 1979; 205:1416.

51. Ausprunk DH, Falterman K, Folkman J. The sequence of events in the regression of corneal capillaries. Lab Invest 1978; 38:284.

52. Chang LK, Garcia-Cardena G, Farnebo F, Fannon M, Chen EJ, Butterfield C,Moses MA, Mulligan RC, Folkman J, Kaipainen A. Dose-dependent response of FGF-2 for lymphangiogenesis. Proc Natl Acad Sci U S A 2004; 101:11658.

53. Gimbrone MA Jr, Cotran RS, Folkman J. Endothelial regeneration: Studies with human endothelial cell cultures. Series Haematol 1973; 6:453.

54. Jaffe EA, Nachman RL, Becker, CG & Minick CR. Culture of human endothelial cells derived from umbilical veins. Ientification by morphologic and immunologic criteria. J Clin Invest 1973; 52:2745.

55. Folkman J, Haudenschild CC, Zetter BR. Long-term culture of capillary endothelial cells. Proc Natl Acad Sci USA 1979; 76:5217.

56. Haudenschild CC, Zahniser D, Folkman J, Klagsbrun M. Human vascular endothelial cells in culture. Lack of response to serum growth factors. Exp Cell Res 1976; 98:175.

57. Folkman J, Moscona A. Role of cell shape in growth control. Nature 1978; 273:345.

58. Ingber DE, Madri JA, Folkman J. Endothelial growth factors and extracellular matrix regulate DNA synthesis through modulation of cell and nuclear expansion. In Vitro Cell Dev Biol 1987; 23(5):387.

59. Ingber DE, Folkman J. Mechanochemical switching between growth and differentiation during fibroblast growth factor-stimulated angiogenesis *in vitro*: role of extracellular matrix. J Cell Biol 1989; 109:317.

60. Ingber DE, Folkman J. How does extracellular matrix control capillary morphogenesis? Cell 1989; 58:803.

61. Ingber DE, Folkman J. Tension and compression as basic determinants of cell form and function: utilization of a cellular tensegrity mechanism. In: Stein W, Bronner F, eds. Cell Shape: Determinants, Regulation, and Regulatory Role. New York: Academic Press, 1989; 3.

62. Huang S, Ingber DE. Cell tension, matrix mechanics, and cancer development. Cancer Cell 2005; 8:175.

63. Bissell MJ, Farson D, Tung AS. Cell shape and hexose transport in normal and virus-transformed cells in culture. J Supramolecular Structure 1977; 6:1.

64. Folkman J, Haudenschild C. Angiogenesis *in vitro*. Nature 1980; 288:551.

65. Kamei M, Saunders WB, Bayless KJ, Dye L, Davis GE, Weinstein BM. Endothelial tubes assemble from intracellular vacuoles in vivo. Nature 2006; 442:453–456.

66. Bayliss PE, Bellavance KL, Whithead GG, Abrams JM, Aegerter S, Robbins HS, Cowan DB, Keating MT, O'Reilly T, Wood JM, Roberts TM, Chan J. Chemical modulation of receptor signaling inhibits regenerative angiogenesis in adult zebrafish. Nat Chem Biol 2006; 2(5):265–273.

67. Auerbach R, Kubai L, Knighton D, Folkman J. A simple procedure for the long-term cultivation of chicken embryo. Dev Biol 1974; 41:391.

68. Ausprunk DH, Knighton DR, Folkman J. Differentiation of vascular endothelium in the chick chorioallantois: a structural and autoradiographic study. Dev Biol 1974; 38:237.

69. Folkman J. Angiogenesis and its inhibitors. In: DeVita VT Jr, Hellman S, Rosenberg SA, eds. Important Advances in Oncology. Philadelphia: J.B. Lippincott, 1985; 42–62.

70. Ausprunk DH, Knighton DR, Folkman J. Vascularization of normal and neoplastic tissues grafted to the chick chorioallantois. Am J Pathol 1975 79(3):597.

71. Nguyen M, Shing Y, Folkman J. Quantitation of angiogenesis and antiangiogenesis in the chick embryo chorioallantoic membrane. Microvasc Res 1994; 47:31–40.

71a. Norrby K, Jakobsson A, Sorbo J. Quantitative angiogenesis in spreads of intact rat mesenteric windows. Microvasc Res 1990; 39(3): 341–348.

72. Ribatti D, Gualandris A, Bastaki M, Vacca A, Lurlaro M, Roncali L, Presta M. New model for the study of angiogenesis and antiangiogenesis in the chick embryo chorioallantoic membrane: the gelatin ponge/chorioallantoic membrane assay. J Vasc Res 1997; 34:455.

73. Ribatti D, Locci P, Marinucci L, Lilli C, Roncali L, Becchetti E. Exogenous heparin induces an increase in glycosaminoglycans of the chick embryo chorioallantoic membrane : its possible role in the regulation of angiogenic processes. Int J Microcirculation Clin Exp 1995; 15:181.

74. Ribatti D, Nico B, Bertossi M, Roncali L, Presta M. Basic fibroblast growth factor-induced angiogenesis in the chick embryo chorioallantoic membrane: an electron microscopy study. Microvascular Research. 1997;53:187.

75. Ribatti D, Crivellato E, Candussio L, et al. Angiogenic activity of rat mast cells in the chick embryo chorioallantoic membrane is down-regulated by treatment with recombinant human a2a interferon and partly mediated by fibroblast growth factor-2. Haematologica. 2002;87:465.

76. Murray JC. Angiogenesis Protocols, Humana Press, Totowa, New Jersey 2001.

77. Norrby K. In vivo models of angiogenesis. J Cell Mol Med 2006; 10:588.

78. Shing Y, Folkman J, Sullivan R, Butterfield C, Murray J, Klagsbrun M. Heparin affinity: purification of a tumor-derived capillary endothelial cell growth factor. Science 1984; 223:1296.

79. Esch F, Esch F, Baird A, Ling N, Ueno N, Hill F, Denoroy L, Klepper R, Gospodarowicz D, Bohlen P, Guillemin R. Primary structure of bovine pituitary basic fibroblast growth facto (FGF) and comparison with the amino-terminal sequence of bovine brain acidic FGF. Proc Natl Acad Sci USA 1985; 82:6507.

80. Gospodarowicz D. Localization of a fibroblast growth factor and its effect with hydrocortisone on 3T3 cell growth. Nature 1974; 249:123.

81. Gospodarowicz D, Moran J, Braun D, Birdwell C. Clonal growth of bovine endothelial cells: fibroblast growth factor as a survival agent. Proc Natl Acad Sci USA 1976; 73:4120.

82. Senger DR, Galli SJ, Dvorak AM, perruzzi CA, Harvey VS, Dvorak HF. Tumor cells secrete a vascular permeability factor that promotes accumulation of ascites fluid. Science 1983; 219:983.

83. Ferrara N, Henzel WJ. Pituitary follicular cells secrete a novel heparin-binding growth factor specific for vascular endothelial cells. Biochem Biophys Res Commun 1989; 161:851.

84. Rosenthal RA, Megyesi JF, Henzel WJ, Ferrara N, Folkman J. Conditioned medium from mouse sarcoma 180 cells contains vascular endothelial growth factor. Growth Factors 1990; 4:53.

85. Miao HQ, Lee P, Lin H, Soker S, Klagsbrun M. Neuropilin-1 expression by tumor cells promotes tumor angiogenesis and progression. FASEB J. 2000; 14:2532.

86. Vlodavsky I, Folkman J, Sullivan R, Fridman R, Ishai-Michaeli R, Sasse J, Klagsbrun M. Endothelial cell-derived basic fibroblast growth factor: Synthesis and deposition into subendothelial extracellular matrix. Proc Natl Acad Sci USA 1987; 84:2292.

87. Folkman J, Klagsbrun M, Sasse J, Wadzinski M, Ingber D, Vlodavsky I. A heparin-binding angiogenic protein - basic fibroblast growth factor - is stored within basement membrane. Am J Pathol 1988; 130(2):393.

88. Bashkin P, Doctrow S, Klagsbrun M, Svahn CM, Folkman J, Vlodavsky I. Basic fibroblast growth factor binds to subendothelial extracellular matrix and is released by heparitinase and heparin-like molecules. Biochem 1989; 28:1737.

89. Brem S, Brem H, Folkman J, Finkelstein D, Patz A. Prolonged tumor dormancy by prevention of neovascularization in the vitreous. Canc Res 1976; 36:2807.

90. Brouty-Boye D, Zetter B. Inhibition of cell motility by interferon. Science 1980; 208: 516.

91. Dvorak HF, Gresser I. Microvascular injury in pathogenesis of interferon-induced necrosis of subcutaneous tumors in mice. J Natl Cancer Inst 1989; 81:497.

92. White CW, Sondheimer HM, Crouch EC, Wilson H, Fan LL. Treatment of pulmonary hemangiomatosis with recombinant interferon alfa-2a. N Engl J Med 1989; 320:1197.

93. Folkman J. Successful treatment of an angiogenic disease. N Engl J Med 1989; 320:1211.

94. Singh RK, Gutman M, Bucana CD, Sanchez R, Llansa N, Fidler IJ. Interferons alpha and beta down-regulate the expres-

sion of basic fibroblast growth factor in human carcinomas. Proc Natl Acad Sci U S A 1995; 92: 4562.

95. Taylor S, Folkman J. Protamine is an inhibitor of angiogenesis. Nature 1982; 297: 307.

96. Crum R, Szabo S, Folkman J. A new class of steroids inhibits angiogenesis in the presence of heparin or a heparin fragment. Science 1985; 230:1375.

97. Chen NT, Corey EJ, Folkman J. Potentiation of angiostatic steroids by a synthetic inhibitor of arylsulfatase. Lab Invest 1988; 59:453.

98. Fotsis T, Zhang Y, Pepper MS, Adlercreutz H, Montesano R, Nawroth PP, Schweigerer L.The endogenous oestrogen metabolite 2-methoxyestradiol inhibits angiogenesis and suppresses tumor growth. Nature 1994; 368:237.

99. D'Amato RJ, Lin CM, Flynn E, Folkman J, Hamel E. 2-Methoxyestradiol, an endogenous mammalian metabolite, inhibits tubulin polymerization by interacting at the colchicine site. Proc Nat Acad Sci USA 1994; 91(9):3964.

100. Folkman J, Mulliken JB, Ezekowitz RAB. Antiangiogenic therapy of haemangiomas with interferon A. In: Stuart-Harris R, Penny R, eds. The Clinical Applications of the Interferons. Chapman & Hall Medical, London 1997; 255.

101. Folkman J. Endogenous angiogenesis inhibitors. Acta Pathologica, Microbiologica, et Immunologica Scandinavica 2004; 112:496.

102. Nyberg P, Xie L, Kalluri R. Endogenous inhibitors of angiogenesis. Cancer Res 2005; 65:3967.

103. Abdollahi A, Hahnfeldt P, Maercker C, Gröne, HJ, Debus J, Ansorge W, Folkman J, Hlatky L, Huber PE. Endostatin's antiangiogenic signaling network. Molecular Cell 2004; 13:649.

104. Relf M, LeJeune S, Scott PA, Fox S, Smith K, Leek R, Moghaddam A, Whitehouse R, Bicknell R, Harris AL. Expression of the angiogenic factors vascular endothelial cell growth factor, acidic and basic fibroblast growth factor, tumor growth factor beta-1, platelet-derived endothelial cell growth factor, placenta growth factor, and pleiotrophin in human primary breast cancer and its relation to angiogenesis. Cancer Res 1997; 57:963.

105. Casanovas O, Hicklin DJ, Bergers G, Hanahan D. Drug resistance by evasion of antiangiogenic targeting of VEGF signaling in late-stage pancreatic islet tumors. Cancer Cell 2005; 8:299.

106. Viloria-Petit AM, Kerbel RS. Acquired resistance to EGFR inhibitors: mechanisms and prevention strategies. Int J Radiat Oncol Biol Phys 2004; 58:914.

107. Bianco C, Strizzi L, Ebert A, Chang C, Rehman A, Normanno N, Guedez L, Salloum R, Ginsburg E, Sun Y, Khan N, Hirota M, Wallace-Jones B, Wechselberger C, Vonderhaar BK, Tosato G, Stetler-Stevenson WG, Sanicola M, Salomon DS. Role of human cripto-1 in tumor angiogenesis. J Natl Cancer Inst 2005; 97:132–141.

108. Dor Y, Djonov V, Abramovitch R, Itin A, Fishman GI, Carmeliet P, Goelman G, Keshet E. Conditional switching of VEGF provides new insights into adult neovascularization and pro-angiogenic therapy. Embo J 2002; 21:1939.

109. Reynolds LE, Wyder L, Lively JC, Taverna D, Robinson SD, Huang X, Sheppard D, Hynes RO, Hodivala-Dilke KM. Enhanced pathological angiogenesis in mice lacking beta3 integrin or beta3 and beta5 integrins. Nat Med 2002; 8:27.

110. Mizukami Y, Jo WS, Duerr EM, Gala M, Li J, Zhang X, Zimmer MA, Iliopoulos O, Zukerberg LR, Kohgo Y, Lynch MP, Rueda

BR, Chung DC. Induction of interleukin-8 preserves the angiogenic response in HIF-1alpha-deficient colon cancer cells. Nat Med 2005; 11:992.

111. Dorrell MI, Aguilar E, Scheppke L, Barnett FH, Friedlander M. Combination angiostatic therapy completely inhibits ocular and tumor angiogenesis. Proc Natl Acad Sci U S A 2007; 104:967.

112. Jain RK, Tong RT, Munn LL. Effect of vascular normalization by antiangiogenic therapy on interstitial hypertension, peritumor edema, and lymphatic metastasis: insights from a mathematical model. Cancer Res 2007; 67:2729.

113. Fukumura D, Jain RK. Tumor microvasculature and microenvironment: Targets for anti-angiogenesis and normalization. Microvasc Res 2007; 74(2–3):72–84.

114. Jain RK. Barriers to drug delivery in solid tumors. Sci Am. 1994;271:58–65.

115. Boucher Y, Jain RK. Microvascular pressure is the principal driving force for interstitial hypertension in solid tumors: implications for vascular collapse. Cancer Res 1992;52:5110.

116. Jain RK. Normalizing tumor vasculature with anti-angiogenic therapy: a new paradigm for combination therapy. Nat Med 2001; 7:987.

117. Slaton JW, Perrotte P, Inoue K, Dinney CP, Fidler IJ. Interferon-alpha-mediated down-regulation of angiogenesis-related genes and therapy of bladder cancer are dependent on optimization of biological dose and schedule. Clin Cancer Res 1999; 5:2726.

118. Panigrahy D, Singer S, Shen LQ, Butterfield CE, Freedman DA, Chen EJ, Moses MA, Kilroy S, Duensing S, Fletcher C, Fletcher JA, Hlatky L, Hahnfeldt P, Folkman J, Kaipainen A. PPARgamma ligands inhibit primary tumor growth and metastasis by inhibiting angiogenesis. J Clin Invest 2002; 110:923.

119. Celik I, Surucu O, Dietz C, Heymach JV, Force J, Hoschele I, Becker CM, Folkman J, Kisker O. Therapeutic efficacy of endostatin exhibits a biphasic dose-response curve. Cancer Res 2005; 65:11044.

120. Tjin Tham Sjin RM, Naspinski J, Birsner AE, Li C, Chan R, Lo KM, Gillies S, Zurakowski D, Folkman J, Samulski J, Javaherian K. Endostatin therapy reveals a U-shaped curve for antitumor activity. Cancer Gene Ther. 2006; 13(6):619–627.

121. Browder T, Butterfield CE, Kräling BM, Shi B, Marshall B, O'Reilly MS, Folkman J. Antiangiogenic scheduling of chemotherapy improves efficacy against experimental drug-resistant cancer. Cancer Res 2000; 60:1878.

122. Klement G, Baruchel S, Rak J, Man S, Clark K, Hicklin DJ, Bohlen P, Kerbel RS. Continuous low-dose therapy with vinblastine and VEGF receptor-2 antibody induces sustained tumor regression without overt toxicity. J Clin Invest 2000; 105:R15.

123. Hanahan D, Bergers G, Bergsland E. Less is more, regularly: metronomic dosing of cytotoxic drugs can target tumor angiogenesis in mice. J Clin Invest 2000; 105:1045.

123a. Bocci G, Francia G, Man S, Lawler J, Kerbel RS. Thrombospondin 1, a mediator of the antiangiogenic effects of low-dose metronomic chemotherapy. Proc Natl Acad Sci USA 2003; 100(22):12917–12922.

124. Hanahan D, Folkman J. Patterns and emerging mechanisms of the angiogenic switch during tumorigenesis. Cell 1996; 86:353.

124a. Damber JE, Valbo C, Albertsson P, Lennernas B, Norrby K. The anti-tumor effect of low-dose continuous chemotherapy may partly be mediated by thrombospondin. Cancer Chemother Pharmacol 2006; 58(3):354–60.

125. Bouck N. Tumor angiogenesis: the role of oncogenes and tumor suppressor genes. Cancer Cells. 1990; 2:179.

126. Achilles E-G, Fernandez A, Allred EN, Kisker O, Udagawa T, Beecken W-D, Flynn E, Folkman J. Heterogeneity of angiogenic activity in a human liposarcoma: A proposed mechanism for 'no take' of human tumors in mice. J Natl Cancer Inst 2001; 93:1075.

127. Udagawa T, Fernandez A, Achilles EG, Folkman J, D'Amato RJ. Persistence of microscopic human cancers in mice: alterations in the angiogenic balance accompanies loss of tumor dormancy. FASEB J 2002; 16:1361.

128. Almog N, Henke V, Flores L, Hlatky L, Kung AL, Wright RD, Berger R, Hutchinson L, Naumov GN, Bender E, Akslen LA, Achilles EG, Folkman J. Prolonged dormancy of human liposarcoma is associated with impaired tumor angiogenesis. FASEB J 2006; 20: 947.

129. Naumov GN, Bender E, Zurakowski D, Kang SY, Sampson D, Flynn E, Watnick RS, Straume O, Akslen LA, Folkman J, Almog N. A model of human tumor dormancy: an angiogenic switch from the nonangiogenic phenotype. J Natl Cancer Inst 2006; 98: 316–325.

130. Naumov GN, Folkman J. Strategies to prolong the nonangiogenic dormant state of human cancer. In: Davis DW, Herbst RS, Abbruzzese JL, eds. Antiangiogenic Cancer Therapy, CRC Press, Boca Raton, FL, 2007; 3.

131. Kaipainen A, Kieran MW, Huang S, Butterfield C, Bielenberg D, Mostoslavsky G, Mulligan R, Folkman J, Panigrahy D. PPARα deficiency in inflammatory cells suppresses tumor growth. PLoS ONE 2007; 2:e260.

132. Verheul HM, Pinedo HM. Tumor Growth: A putative role for platelets? Oncologist 1998; 3(2):II.

133. Folkman J, Browder T, Palmblad J. Angiogenesis research: Guidelines for translation to clinical application. Thromb Haemost 2001; 86:23–33.

134. Klement G, Kikuchi L, Kieran M, Almog N, Yip TT, Folkman J. Early tumor detection using platelets uptake of angiogenesis regulators. Proc 47th American Society of Hematology. Blood; December 2004; 104:239a, abstract 839.

135. Klement G, Cervi D, Yip T, Folkman J, Italiano J. Platelet PF-4 is an early marker of tumor angiogenesis. Blood 2006; 108:426a, abstract 1476.

136. Cervi D, Yip T-T, Bhattacharya N, Podust VN, Peterson J, Abou-slaybi A, Naumov GN, Bender E, Almog N, Italiano JE Jr., Folkman J, Klement GL. Platelet-associated PF-4 as a biomarker of early tumor detection. Blood; 2007; in press.

137. Italiano J, Richardson JL, Folkman J, Klement G. Blood platelets organize pro- and anti-angiogenic factors into separate, distinct alpha granules: implications for the regulation of angiogenesis. Blood 2006; 108:120a, abstract 393.

138. Stone EM. A very effective treatment for neovascular macular degeneration. N Engl J Med 2006; 355: 1493–5.

139. Adamis, AP, Shima DT, Yeo K-T, Yeo T-K, Brown LF, Berse B, D'Amore PA, Folkman J. Synthesis and secretion of vascular permeability factor/vascular endothelial growth factor by human retinal pigment epithelial cells. Biochem Biophys Res Commun, 1993; 193:631–638.

140. Miller JW, Adamis AP, Shima DT, D'Amore PA, Moulton RS, O'Reilly MS, Folkman J, Dvorak HF, Brown LF, Berse B, Yeo T-K, Yeo K-T. (1994). Vascular endothelial growth factor/vascular permeability factor is temporally and spatially correlated with ocular angiogenesis in a primate model. Am J Pathol; 145:574–584.

141. Adamis AP, Miller JW, Bernal M-T, D'Amico DJ, Folkman J, Yeo T-K, Yeo K-T. Increased vascular endothelial growth factor levels in the vitreous of eyes with proliferative diabetic retinopathy. Am J Ophthal; 1994; 118(4):445–450.

142. Aiello LP, Avery RL, Arrigg PG, et al. Vascular endothelial growth factor in ocular fluid of patients with diabetic retinopathy and other retinal disorders. N Engl J Med. 1994; 331:1480–1487.

143. Shima DT, Adamis AP, Ferrara N, Yeo K-T, Yeo T-K, Allende R, Folkman J, D'Amore PA. Hypoxic induction of endothelial cell growth factors in retinal cells: Identification and characterization of vascular endothelial growth factor (VEGF) as the mitogen. Mol Med; 1995; 1:182–193.

144. Adamis AP, Shima DT, Tolentino MJ, Gragoudas ES, Ferrara N, Folkman J, D'Amore PA, Miller JW. Inhibition of vascular endothelial growth factor prevents retinal ischemia-associated iris neovascularization in a nonhuman primate. Arch Ophthalmol; 1996; 114:66–71.

145. Folkman J. Is angiogenesis an organizing principle in biology and medicine? J Pediatr Surg. 2007; 42:1–11.

146. Folkman J. Historical Overview. In: Marme D, Fusenig N, eds. Tumor Angiogenesis: From Bench to Bedside. Springer-Verlag, Germany 2008:1–28.

Section I
Physiological & Pathological Angiogenesis:
Biology of the Angiogenic Process

Chapter 2
Angiogenesis and Vascular Remodeling in Inflammation and Cancer: Biology and Architecture of the Vasculature

Donald M. McDonald

Keywords: angiogenesis, angiogenesis inhibitors, basement membrane, endothelial cells, pericytes, tumors, plasma leakage, vascular permeability, vascular remodeling, VEGF

Abstract: Blood vessels proliferate by sprouting from existing vessels (angiogenesis) and undergo changes in phenotype (vascular remodeling) in inflammatory diseases, tumors, and many other chronic conditions. Changes in newly formed and remodeled blood vessels are disease-specific, as they reflect vascular adaptations to environmental cues unique to each condition. In inflamed tissues, vascular remodeling expands the vasculature and increases blood flow, plasma leakage, and inflammatory cell influx, which contribute to the pathophysiology and clinical manifestations of the disease. Remodeling of endothelial cells into a venular phenotype, typical of sustained inflammation, is accompanied by expression of molecules that promote endothelial gap formation and leukocyte rolling, attachment, and migration. Blood vessels in tumors differ from those in inflammation. Endothelial cells in tumors undergo disorganized sprouting, proliferation and regression, and become dependent on vascular endothelial growth factor (VEGF) or other factors for survival. The growing vasculature enables tumor enlargement, but structural defects impair endothelial barrier function and increase interstitial pressure and luminal resistance, diminish blood flow, and alter immune cell traffic. Oxygen delivery may be inadequate for tumor cell viability despite the rich vascularity. Inhibition of VEGF signaling in tumors stops sprouting angiogenesis and triggers regression of some tumor vessels while normalizing others. Some capillaries in normal thyroid, pancreatic islets, and intestine may also regress after VEGF blockade, but most remodeled vessels at sites of inflammation do not. Pericytes and empty sleeves of vascular basement membrane persist after endothelial cells regress and provide a scaffold for blood vessel regrowth, which can occur within days after the inhibition ends. The clinical efficacy of VEGF signaling inhibitors in cancer and age-related macular degeneration provides proof of concept and stimulates the search for even more effective agents. Further advances in vascular biology will lead to more powerful strategies for controlling blood vessel growth and regression in health and disease.

Introduction

During normal body growth and in many disease processes, blood vessels proliferate by angiogenesis, where new vessels sprout from existing ones. The expanding vasculature provides nutrients to enlarging tissues and routes for cells to leave or enter the circulation. Blood vessels can also undergo remodeling, whereby they acquire a new phenotype manifested by changes in structural and functional properties.

Normal blood vessels are lined by a monolayer of thin, smooth, tightly joined endothelial cells that form the barrier that controls the transendothelial flux of water, solutes, and cells. The endothelial cells rarely sprout or divide. Vascular stability results in part from the intimate association of mural cells (pericytes or smooth muscle cells) with the abluminal surface of endothelial cells. Endothelial cells and mural cells, which together generate a tight envelopment of basement membrane, form a stable, functional unit.

Blood vessels of the normal microcirculation are organized into a hierarchy of arterioles, capillaries, and venules, each having distinctive structural and functional characteristics [1]. This hierarchical organization is superimposed on organ-specific specializations of the blood vessels. As a result, blood vessels of different organs have some features in common and some that are unique to each organ.

Newly formed or remodeled blood vessels in most pathological conditions differ from normal vessels at multiple levels.

Comprehensive Cancer Center, Cardiovascular Research Institute, and Department of Anatomy, University of California, San Francisco, CA, USA

Address for Correspondence:
Department of Anatomy, University of California, Room S-1363, 513 Parnassus Avenue, San Francisco, CA, USA
E-mail: donald.mcdonald@ucsf.edu

Abnormalities range from altered activity of receptors, adhesion molecules, or signaling pathways, to loss of hierarchical organization, and to widespread changes in three-dimensional architecture [2]. In inflammation, the types of blood vessels that are sources of plasma leakage and leukocyte efflux expand by remodeling and angiogenesis [3–5]. Vascular specialization progresses as inflammation evolves and becomes an integral part of the inflammatory response. Of the four cardinal signs of inflammation—rubor, tumor, calor, and dolor—the first three reflect changes in the vasculature.

Abnormalities in tumor blood vessels differ from the changes in inflammation. The endothelial cells in tumors undergo disorganized sprouting, proliferation and regression, and may be dependent on growth factors for survival [1, 2]. Despite extensive angiogenesis in tumors, structural defects lead to leakiness, high luminal resistance, poor blood flow, altered trafficking of immune cells, and elevated interstitial pressure. Blood flow may be inadequate to support tumor cell viability despite dense vascularity.

This chapter reviews the evolution of approaches we have developed or adapted to obtain a better understanding of changes in three-dimensional vascular architecture, vascular phenotype and hierarchy, and the cell biology of endothelial cells and pericytes in vivo. The strategy was first to understand the properties of a relatively simple vascular network under baseline conditions and then to determine how the vasculature changed under pathological conditions. Models in mice or rats were used to compare angiogenesis and vascular remodeling in chronic inflammation and cancer. As expected, the learning process has been evolutionary, with each stage building on previous ones. Progress has enabled, step by step, the development of more informative methods, analysis of more complex systems, and investigation of more complex issues in the in vivo setting.

Approaches for Detecting Changes in Vascular Architecture

With the goal of developing more effective ways of detecting alterations in the three-dimensional architecture and cellular changes in blood vessels under conditions that lead to angiogenesis and vascular remodeling, we sought better approaches for visualizing the microvasculature [6, 7]. As a starting point, we adapted the historical method of using silver nitrate to stain the borders of endothelial cells of blood vessels in situ (Fig. 2.1A, B) and exploited the simple segmented vascular architecture of the murine trachea as a test system (Fig. 2.1B inset) [6]. Once stained, the preparations could be examined as three-dimensional whole mounts, where all segments of the microvasculature—arterioles, capillaries, and venules—could be identified. This system had the additional attribute of making it possible to pinpoint sites of leakage at the cellular level. Leaky sites were found to coincide with gaps between endothelial cells of postcapillary venules (Fig. 2.1C, D) [6].

As a technically easier and more flexible approach, we explored the binding properties of plant lectins for labeling the vascular endothelium in situ, again using the vasculature of the rat trachea as a test system [7]. *Lycopersicon esculentum* (LEA) lectin, which has a primary specificity for *N*-acetyl-D-glucosamine oligomers, was found to bind strongly and uniformly to the luminal surface of endothelial cells when administered by intravenous (iv) injection or vascular perfusion. Of the 20 lectins tested in this system, *Solanum tuberosum* (STL) and *Datura stramonium* (DSL), which also bind *N*-acetyl-D-glucosamine oligomers, had properties similar to LEA. Later experiments revealed that *Griffonia (Bandeiraea) simplicifolia* I (GSL-I) isolectin B4 effectively labels the vasculature of mice after iv injection with less toxicity than LEA. Biotinylated lectins were visualized by avidin-peroxidase histochemistry (Fig. 2.2A, B) and fluorescent lectins by fluorescence microscopy or confocal microscopy. Because of uniform binding, LEA lectin revealed the overall architecture of the vasculature (Fig. 2A, B) and amazingly detailed features of the luminal surface [3, 7]. *Triticum vulgaris* lectin (Wheat germ agglutinin, WGA) did not bind uniformly but had the distinctive property of binding strongly to endothelial cells of capillaries and arterioles but not postcapillary venules (Fig. 2A inset) [7]. By comparison, *Ricinus communis* agglutinin I lectin bound only weakly to the vasculature but clearly marked leaky sites by binding intensely to the extracellular matrix in regions of extravasation [4, 7].

Vascular Remodeling and Angiogenesis in Inflammation

Staining the luminal surface of the endothelium with biotinylated or fluorescent LEA lectin proved useful in characterizing vascular changes in chronic inflammation in the respiratory tract of rats (Fig. 2.2A, B) and mice (Fig. 2.2C, D). This approach provides an overview of the vascular architecture and the number and location of adherent intravascular leukocytes [5, 8, 9]. In this way, angiogenesis and vascular remodeling were found to be prominent features of the inflammatory response of the airway mucosa after infection by the respiratory pathogen *Mycoplasma pulmonis* [5, 8, 9]. This infection is known to lead to chronic airway inflammation in rats and mice [10–12]. Unlike normal vasculature, blood vessels that undergo remodeling after *M. pulmonis* infection are leaky, and the leakiness is greatly exaggerated by the inflammatory mediator substance P (Fig. 2.2B inset), because of up-regulation of neurokinin-1 receptors on endothelial cells [13, 14]. Remodeled capillaries acquire a venular phenotype, with increased expression of molecules typical of inflamed venules, including P-selectin, E-selectin, EphB4, and ICAM-1. These changes lead to selective expansion of a specialized population of blood vessels (venules) that support plasma leakage and leukocyte influx (Fig. 2.2D). The model of experimental *M. pulmonis* infection makes it possible to examine the mechanism, consequences, and reversibility of angiogenesis and vascular remodeling in sustained inflammation.

The overall architecture of the tracheal vasculature changes dramatically after *M. pulmonis* infection. The magnitude of

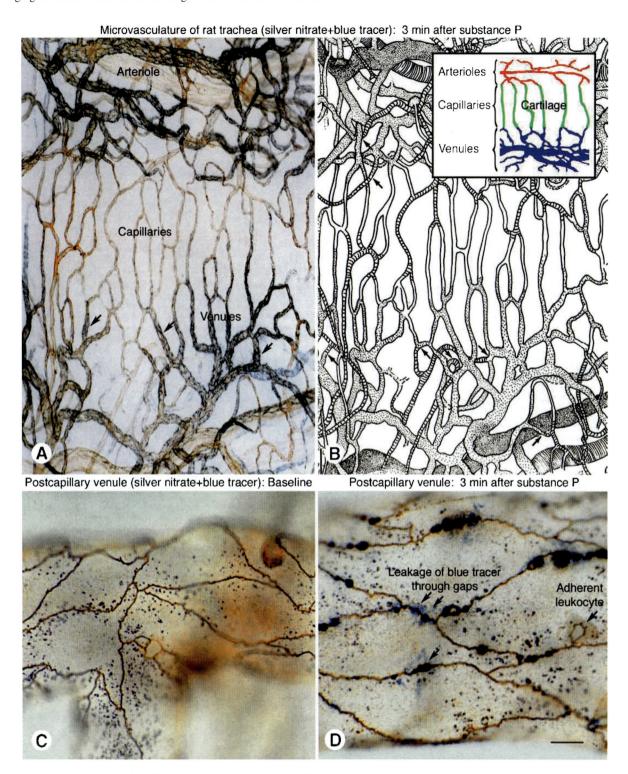

FIG. 2.1A–D. Light micrograph (**A**) and corresponding drawing (**B**) of vasculature of rat tracheal mucosa after injection of particulate blue tracer, Monastral blue, and leak-inducing substance P (5μg/kg iv) followed 3min later by removal of intravascular tracer by vascular perfusion of fixative and staining of endothelial cells in situ with silver nitrate (from [6]). Postcapillary venules in (**A**) are labeled by extravasated Monastral blue, reflecting sites of substance P-induced leakage. Capillaries are not labeled. The drawing identifies six types of blood vessels distinguished by endothelial cell morphology: segmental arterioles *(long contour lines)*, terminal arterioles *(short contour lines)*, arteriovenous anastomoses *(arrows)*, capillaries *(unmarked)*, postcapillary venules *(light stipple)*, and collecting venules *(heavy stipple)*. **B** *(inset)* Schematic of vascular architecture, with most capillaries over cartilage rings and other vessels between the rings. **C, D** Light micrographs showing borders of endothelial cells of postcapillary venule of rat trachea after silver nitrate staining under baseline conditions (**C**) and after substance P (**D**) (5μg/kg iv, 3min) (from [3]). Dot-like silver deposits *(arrows)* at endothelial cell borders mark intercellular gaps (**D**). Patches of extravasated Monastral blue *(arrows)* mark sites of leakage (**D**). Adherent leukocyte (**D** *arrowhead*). *Scale bar*: 100μm in (**A,B**); 10μm in (**C, D**).

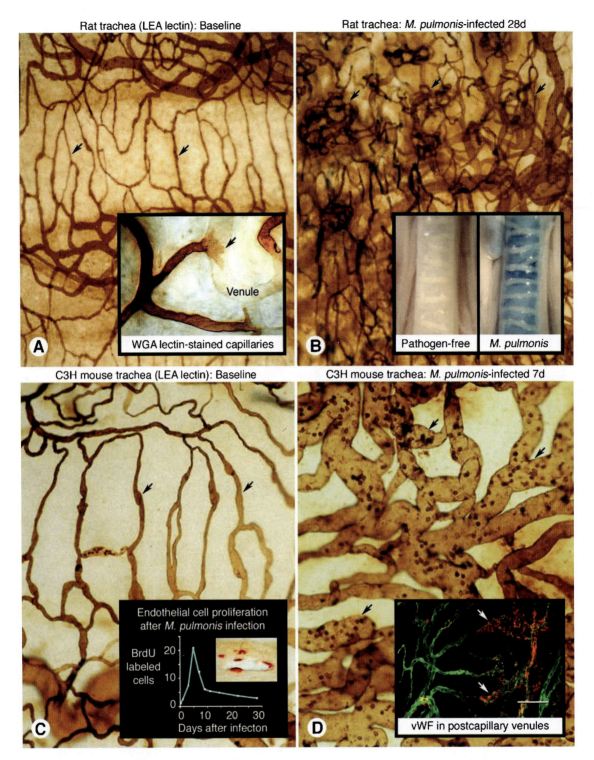

Fig. 2.2A–D. Comparison of vasculature in tracheal mucosa under baseline conditions (pathogen-free) and after *M. pulmonis* infection in rats (**A,B**) and mice (**C,D**). Blood vessels stained by perfusion of LEA lectin. Simple pattern of tracheal vasculature in pathogen-free rat, with relatively straight capillaries (**A** *arrows*), compared to tortuous networks of new vessels (**B** *arrows*) in trachea of rat infected with *M. pulmonis* for 4 weeks. **A** (*inset*) Strong binding of WGA lectin to endothelial cells of capillary (*arrow*) next to weak binding to endothelial cells of venule in rat trachea (from [7]). **B** (*inset*) Exaggerated substance P-induced plasma leakage in rat trachea after *M. pulmonis* infection. Pale trachea reflects little Monastral blue leakage in pathogen-free trachea (*left*) compared to extensive leakage after *M. pulmonis* infection for 4 weeks (*right*). Both tracheas prepared 2 min after iv injection of capsaicin (75 μg/kg) to evoke release of endogenous substance P (from [12]). Simple network of tracheal capillaries (**C** *arrows*) in pathogen-free mouse compared to capillaries enlarged and remodeled into venules with abundant adherent leukocytes (**D** *arrows*) in C3H mouse infected with *M. pulmonis* for 8 weeks (from [5]). **C** (*inset*) Time course of endothelial cell proliferation assessed by BrdU labeling of tracheal mucosa of *M. pulmonis*-infected C3H/HeN mice (from [18]). Micrograph shows BrdU-labeled cells localized by immunohistochemistry (*red*) in C3H mouse infected for 1 week (from [9]). **D** (*inset*) Confocal micrograph showing von Willebrand factor (vWf)-immunoreactivity (*red*) in endothelial cells of tracheal blood vessels in C3H mouse infected with *M. pulmonis* for 4 weeks (from [8]). Vasculature stained by perfusion of green fluorescent LEA lectin. vWf immunoreactivity is strongest in venules (*arrows*). *Scale bar*: 150 μm in (**A,B**); 75 μm in (**C,D**); 50 μm in (**A** *inset*); 5 mm in (**B** *inset*); 50 μm in (**C** *inset*); 100 μm in (**D** *inset*).

the changes corresponds to the severity of the infection. Some strains of mice develop more severe disease than others [15, 16]. Relative amounts of angiogenesis and vascular remodeling also vary among strains [8]. Vascular remodeling, in which mucosal blood vessels can double in size without much increase in number, predominates in C3H mice (Fig. 2.2C, D), but angiogenesis and vascular remodeling are both conspicuous in C57BL/6 mice and in rats after *M. pulmonis* infection [8, 14, 17].

Vascular remodeling after *M. pulmonis* infection is manifested by enlargement of capillaries and venules, and less so arterioles. The enlargement results from endothelial cell proliferation, not vasodilatation. Endothelial cell size is preserved. The rate of endothelial cell proliferation, determined by BrdU uptake, begins to increase a few days after infection, peaks at 5 days with a value 18 times baseline (pathogen-free), declines somewhat through day 9, and remains at 3 times the pathogen-free value for at least 28 days (Fig. 2.2C inset) [18]. Through selective expression of adhesion molecules, remodeled vessels become sites of leukocyte adherence (Fig. 2.2D). This remodeling of the microvasculature occurs at an early stage of inflammation, considerably before widespread tissue remodeling.

Staining of blood vessels with a lectin is made more informative by concurrent immunohistochemical staining of endothelial cells, pericytes, and adjacent cells in three-dimensional tracheal whole mounts [4, 19–23]. This approach has the attribute of showing the amount and location of angiogenesis and vascular remodeling in the context of inflammatory cells, lymphatic vessels, and specific enzymes, growth factors, receptors, adhesion molecules, or matrix elements [21–23]. For example, preferential expression of von Willebrand factor (vWf) in venules is apparent when vWf immunoreactivity is combined with lectin staining (Fig. 2.2D inset).

Staining for platelet endothelial cell adhesion molecule-1 (PECAM-1, CD31) and lymphatic vessel endothelial hyaluronan receptor (LYVE-1) reveals the three-dimensional architecture of blood vessels and lymphatic vessels and the prominent changes both undergo after *M. pulmonis* infection (Fig. 2.3A–C) [23]. In comparison to blood vessel remodeling, which is greatest during the first week after infection, lymphatics begin to grow during the second week (Fig. 2.3B) [23], and by 28 days are more abundant than blood vessels in the inflamed airway mucosa (Fig. 2.3C) [23]. Growth of lymphatic vessels can be blocked by inhibition of vascular endothelial growth factor (VEGF) receptor-3 signaling. However, remodeling (enlargement) of blood vessels is not prevented by inhibition of VEGFR1, VEGFR-2, and/or VEGFR-3 [23].

Reversibility of Angiogenesis and Vascular Remodeling in Inflammation

Vascular remodeling after *M. pulmonis* infection is not blocked by inhibition of VEGF signaling [23]. However, it can be reversed by reducing the inflammatory and immunologic response with dexamethasone or by reducing the number of organisms with oxytetracycline [17, 19]. After infection with *M. pulmonis* for 6 weeks,

the simple vasculature of the rat tracheal mucosa (Fig. 2.3D) is unrecognizably changed by chaotic angiogenesis (Fig. 2.3E). But daily treatment with dexamethasone or oxytetracycline for 4 weeks leads to almost complete resolution of the angiogenesis and vascular remodeling (Fig. 2.3F) [17].

Similarly, tracheal capillaries of C3H mice infected with *M. pulmonis* change from their conventional narrow caliber and absence of adherent leukocytes (Fig. 2.3G) to conspicuously enlarged vessels that have a venular phenotype, high P-selectin expression, and abundant adherent leukocytes (Fig. 2.3H) [19]. Treatment with dexamethasone for a week reverses the vascular enlargement, venular transformation, and leukocyte adherence (Fig. 2.3I) [19].

The driving force for vascular remodeling after *M. pulmonis* infection has not been identified, but existing evidence indicates that VEGF receptor activation is not essential [23]. Perhaps relevant to the alternatives, airway capillaries become conspicuously enlarged in mice treated with angiopoietin-1, its mimic, COMP-Ang1, or angiopoietin-2 [24–26]. These ligands signal through Tie2 receptors. The transformed vessels have a venular phenotype with similarities to the remodeled vasculature after *M. pulmonis* infection [8, 9]. This phenotype has also been reported in venous malformations resulting from point mutations accompanied by constitutive Tie2 activation [27, 28]. How the role, if any, of Tie2 activation after *M. pulmonis* infection fits with the actions of tumor necrosis factor, interleukins, and other cytokines [29, 30] must still be determined.

Abnormalities of Blood Vessels in Tumors

Identification of tumor vessels

Blood vessels in tumors have bizarre defects unlike those in inflammation. Endothelial cells, pericytes, and the vascular basement membrane are all abnormal [1, 2, 31, 32]. Tumor vessels are so unusual that even their identification as blood vessels can be challenging. Do tumor vessels include strands of endothelial cells regardless of whether they have a lumen and are routes for blood flow? Are all routes of blood flow in tumors lined by endothelial cells? Do collections of erythrocytes without accompanying endothelial cells mark sites of blood flow or sites of hemorrhage? These questions have stimulated innovative hypotheses, novel experiments, and much debate [33–38].

From the perspective of mammalian vascular biology, blood vessels in tumors are the channels for blood circulation. Channels continuously connected to the circulation do not include extravascular pathways for movement of extravasated fluid or collections of extravasated erythrocytes. With this definition, how can capillary-sized blood vessels be identified unambiguously in the abnormal setting of tumors? The most straightforward method is intravital imaging of flowing blood through vessels in tumors [39]. Intravital imaging is nicely complemented by confocal microscopic imaging of tissues preserved by perfusion fixation and stained by immunohistochemistry, which

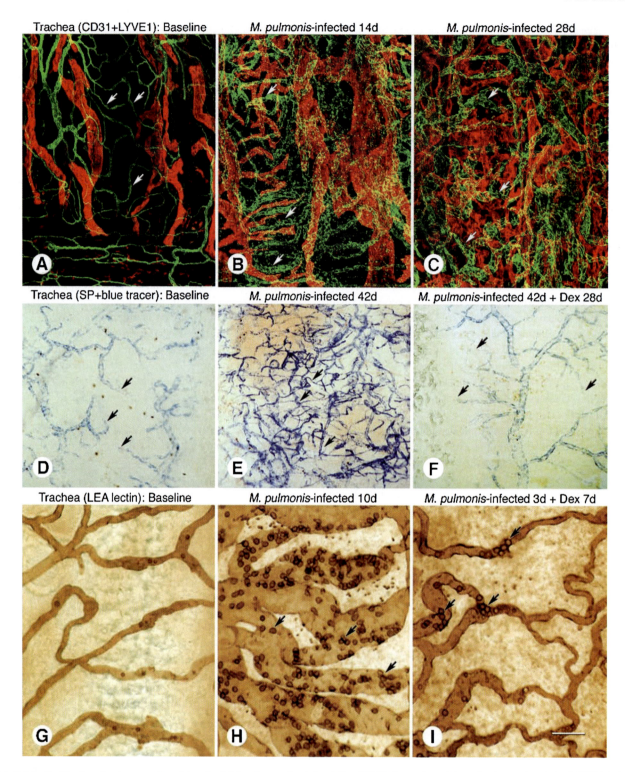

FIG. 2.3A–C. Confocal micrographs of tracheal blood vessels stained for CD31 (*green*) and lymphatics stained for LYVE-1 (*red*) showing angiogenesis and lymphangiogenesis after *M. pulmonis* infection in C3H mice: pathogen-free (**A**), infected for 14 days (**B**), or infected for 28 days (**C**) (from [23]). **D–F** Light micrographs comparing number of leaky tracheal blood vessels, marked by extravasated Monastral blue after substance P (5 µg/kg iv), in three groups of rats: pathogen-free (**D**), infected with *M. pulmonis* for 6 weeks and then saline-treated for 4 weeks (**E**), or infected with *M. pulmonis* for 6 weeks and then treated with dexamethasone (0.5 mg/ kg/day ip) for 4 weeks (**F**) (from [17]). In pathogen-free rat, blue labeling begins abruptly (*arrows*) at the junction of capillaries and postcapillary venules (**D**). After infection, the number of blue vessels is strikingly increased (**E**), but blue vessels return to baseline after treatment with dexamethasone (**F**) (from [17]). **G–I.** Tracheal vasculature perfused with biotinylated LEA lectin to show vascular enlargement after *M. pulmonis* infection and reversal by dexamethasone in C3H mice: pathogen-free (**G**), infected with *M. pulmonis* for 10 days (**H**), or infected for 10 days and treated with dexamethasone (0.2 mg/day, ip) for final 7 days (**I**) (from [19]). *Arrows* mark adherent intravascular leukocytes (**H, I**). *Scale bar*: 100 µm in (**A–F**); 50 µm in (**G–I**).

make it easy to identify changes at the cellular level and to localize specific molecules [31, 38].

As another approach, fluorescent lectins that selectively bind to the luminal surface of blood vessels after iv injection can unambiguously mark the vasculature of tumors (Fig. 2.4A) [7, 40]. Fluorescent 50–100 nm cationic liposomes also label endothelial cells of tumor vessels [41]. Functional blood vessels can be distinguished from lumenless endothelial sprouts and extravascular blood lakes by using a sequence of lectin labeling, vascular perfusion of fixative, and ex vivo immunohistochemistry (Fig. 2.4A, B). The lectin labels functional vessels, vascular perfusion washes blood from the circulation, fixing extravasated erythrocytes in place, and CD31 staining marks all endothelial cells regardless of whether they are organized into blood vessels and have a lumen [40].

The distribution of endothelial cells in tumors, determined by CD31 immunoreactivity, is more complex than the vessel pattern shown by lectin binding (Fig. 2.4A, B). Although the overall arrangement of vessels is similar, more structures have CD31 immunoreactivity than are stained by lectin (Fig. 2.4A, B). Thin sprouts, not detectable by lectin staining (Fig. 2.4A), radiate from the wall of some vessels (Fig. 2.4B). Similar sprouts are evident by scanning electron microscopy (EM) in perivascular sleeves of viable tumor (Fig. 2.4C).

Vascular architecture, caliber, and density differ markedly in different types of tumors, but a consistent feature is defective or absent arteriole–capillary–venule hierarchy typical of normal organs. RIP-Tag2 tumors have unusually abundant, densely packed, anastomotic, capillary-size blood vessels (mean diameter, 8 μm) [42]. In MCa-IV tumors, capillary-size vessels are mixed with extremely large vessels; vessel diameters range from 8 μm to 294 μm (mean diameter, 45 μm) [42]. Lewis lung carcinomas have blood vessels intermediate in size between the other tumors (mean diameter, 31 μm) [42].

Endothelial Cells of Tumor Vessels

Blood vessels of tumors are lined by endothelial cells that have diverse abnormalities in gene expression, structure, and function [2, 32, 40]. The endothelial cells vary in size and thickness, and some have an irregular shape, ruffled margins, or cytoplasmic processes (Fig. 2.4D). Long projections of some endothelial cells overlap other endothelial cells, span the luminal surface of neighboring cells, or bridge the lumen (Fig. 2.4D).

Defects in the endothelial monolayer make tumor vessels leaky. Some endothelial cells are partially detached, do not form a uniformly intact barrier, or form multiple incomplete layers. Extravasated erythrocytes are a prominent manifestation of the defect in endothelial barrier function. In RIP-Tag2 tumors, erythrocyte extravasation is extensive, and blood lakes form [40]. Blood lakes resemble large, sack-like blood vessels or sinusoids, but are not continuously connected to the circulation, as shown by lack of labeling after LEA lectin and cationic liposomes were injected as tracers [40]. Round or oval openings, as large as an erythrocyte (Fig. 2.4E), and narrow

slit-like spaces, are present between endothelial cells of many tumor vessels [40]. Transcellular holes 200–900 nm in diameter (Fig. 2.4E) and diaphragm-covered fenestrae 50–80 nm in diameter are also present.

Endothelial sprouts, rarely seen in quiescent blood vessels, are abundant in tumor vessels, where sprouts as long as 70 μm project from the endothelium. Sprouts are broadest at their base, taper toward a blind ending, and have filopodia at the tip (Fig. 4F-H). LEA lectin injected into the bloodstream stains only the proximal portion of sprouts (Fig. 2.4A). Filopodia are stained by GSL-I isolectin B4 [43].

Pericytes of Tumor Vessels

All normal blood vessels have mural cells that are tightly associated with endothelial cells within a common sleeve of basement membrane. Pericytes are the mural cells of capillaries and venules; smooth muscle cells are the mural cells of other blood vessels. Smooth muscle cells are readily identified by the presence of α-SMA and other muscle-related proteins, but pericytes are heterogeneous and do not consistently express a single protein that can be used for identification. Most pericytes have immunoreactivity for one or more of four markers: α-smooth muscle actin (α-SMA), platelet derived growth factor receptor-β (PDGFR-β), desmin, or chondroitin sulfate proteoglycan NG2 (NG2, CSPG4, HMW-MAA) [42, 44]. Pericytes on normal capillaries typically cover only a small proportion of the endothelial surface, have multiple long, branched cytoplasmic processes, are oriented along the vessel's longitudinal axis, and have desmin and/or NG2 but not α-SMA immunoreactivity (Fig. 2.5A).

Pericytes on tumor vessels have an abnormally loose association with endothelial cells (Fig. 2.5B) [42]. The difference in pericyte-endothelial cell association is particularly conspicuous in scanning EM views of the cell surface (Fig. 2.5C,D). Pericytes in tumors have an irregular shape (Fig. 2.5D) and may have cytoplasmic processes that accompany endothelial sprouts or project away from the vessel wall into the tumor parenchyma [42].

Pericytes in many tumors have abnormal expression of α-SMA. Pericytes on capillaries of normal pancreatic islets do not express α-SMA but do express desmin. However, pericytes in RIP-Tag2 tumors, which develop from islets, express α-SMA as well as desmin [42]. Some tumors contain cells that are morphologically similar to pericytes and are immunoreactive for α-SMA, desmin, or PDGFR-β but have no apparent association with blood vessels. Pericytes that lack a vessel association but are accompanied by basement membrane may be remnants of degenerated tumor vessels.

Basement Membrane of Tumor Vessels

Vascular basement membrane, marked by immunohistochemical staining for type IV collagen, laminin, fibronectin, or nidogen (entactin), is present on all normal blood vessels

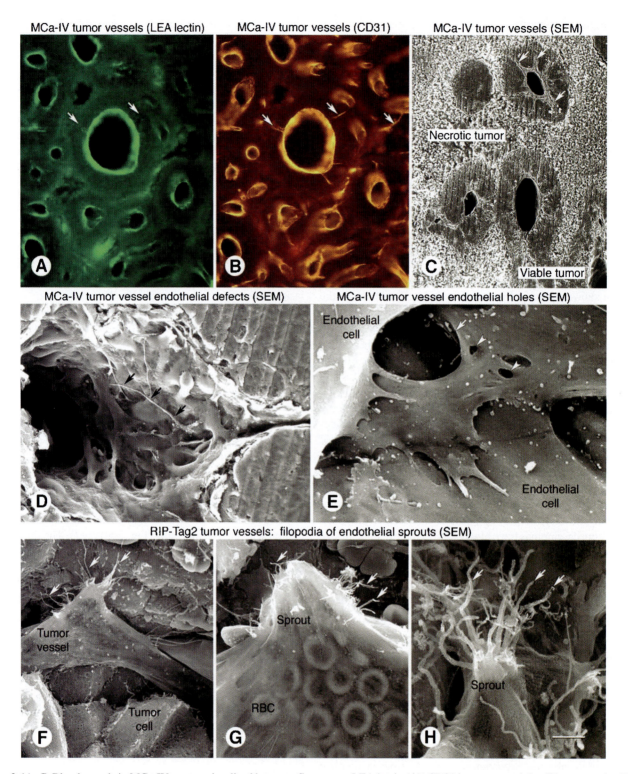

FIG. 2.4A–C. Blood vessels in MCa-IV tumors visualized by green fluorescent LEA lectin (**A**), CD31 immunoreactivity (**B**), or scanning EM (**C**) (from [40]). Both lectin and CD31 immunoreactivity mark most of the tumor vessels, but, unlike LEA lectin, CD31 also marks lumenless endothelial sprouts (*arrows*) radiating away from the vessel wall. Scanning EM shows similar sprouts (*arrows*) radiating into perivascular sleeve of tumor tissue. Necrotic tumor surrounds the perivascular sleeves of viable tumor. Scanning EM of luminal surface of blood vessel in MCa-IV tumor showing multiple abnormalities (**D**), including disorganized endothelial cells with bridges, tunnels, and 50-μm long cellular projection (*arrows*) (from [40]). Pathways for extravasation (**E**) through intercellular openings (*arrow*) and transcellular holes (*arrowheads*) in endothelium of MCa-IV tumor vessel (from [40]). Scanning EM views of sprouts with filopodia (**F–H** *arrows*) projecting from the abluminal surface of endothelial cells of blood vessels in RIP-Tag2 tumors (from [32]). *Scale bar* 150 μm in (**A,B**); 100 μm in (**C**); 15 μm in (**D**); 2.5 μm in (**E**); 10 μm in (**F**); 5 μm in (**G**); 2 μm in (**H**).

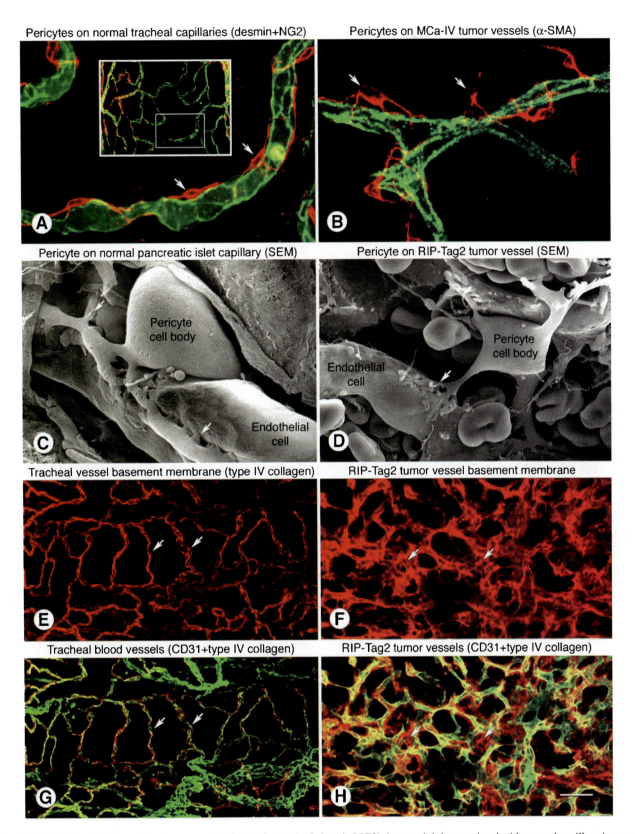

Pericytes on normal tracheal capillaries (desmin+NG2)

Pericytes on MCa-IV tumor vessels (α-SMA)

Pericyte on normal pancreatic islet capillary (SEM)

Pericyte cell body

Endothelial cell

Pericyte on RIP-Tag2 tumor vessel (SEM)

Pericyte cell body

Endothelial cell

Tracheal vessel basement membrane (type IV collagen)

RIP-Tag2 tumor vessel basement membrane

Tracheal blood vessels (CD31+type IV collagen)

RIP-Tag2 tumor vessels (CD31+type IV collagen)

Fig. 2.5A,B. Confocal microscopic images contrasting pericytes (*red*, desmin/NG2) that are tightly associated with normal capillary in mouse trachea (**A**) (from [20]), and pericytes (*red*, α-SMA) that are loosely associated with tumor vessels in Lewis lung carcinoma (**B**) (from [49]). Endothelial cells (*green*, CD31). **A** (*inset*). Overview of tracheal vasculature showing region (*box*) enlarged in (**A**). **C,D** Scanning EM images of a pericyte closely associated with the endothelium of a normal capillary in mouse pancreatic islet (**C**) and a pericyte loosely associated with a blood vessel in RIP-Tag2 tumor (**D**). **E–H.** Confocal micrographs comparing basement membrane (*red*) of capillaries of normal adult mouse trachea (**E,G** *arrows*) (from [49]) and tortuous, anastomotic vasculature of RIP-Tag2 tumor (**F,H** *arrows*) after staining for type IV collagen immunoreactivity alone (**E,F** *red*) or with CD31 immunoreactivity (**G,H** *green*). Some type IV collagen in the tumor does not colo-calize with CD31 (*arrows*). *Scale bar*: 20 μm in (**A,B**); 5 μm in (**C,D**); 50 μm in (**E–H**).

(Fig. 2.5E) and on most blood vessels in tumors (Fig. 2.5F) [45]. Type IV collagen immunoreactivity tends to be the most selective for marking vascular basement membrane in tumors [45]. Because of the close association of basement membrane with the abluminal surface of endothelial cells, sleeves of type IV collagen faithfully match the pattern of CD31 in normal organs and thereby reflect the vascular architecture (Fig. 2.5G). A uniform sleeve of type IV collagen tightly envelopes endothelial cells and pericytes, forming a sandwich with little or no space between the layers.

The basement membrane of tumor vessels has multiple abnormalities. The layer of type IV collagen is variable in thickness and has broad extensions and other irregularities not found in normal vessels. Still, tumor vessels are almost completely covered by basement membrane [45]. Interruptions in type IV collagen represent less than 2% of the endothelial surface. The largely complete basement membrane visible by immunofluorescence contrasts with reports of incomplete basement membrane on tumor vessels from EM studies. The difference may be based on methods of preservation, staining, or sampling.

Some tumor vessels have spikes or extra layers of basement membrane not in close contact with endothelial cells, which give a fuzzy outline to tumor vasculature after type IV collagen staining (Fig. 2.5F, H) [45, 46]. Spaces between layers of type IV collagen, CD31, and α-SMA reflect the loose association of endothelial cells and pericytes (Fig. 2.5B, D). Projections of basement membrane away from tumor vessels may accompany endothelial sprouts or pericyte processes [45, 46]. Multiple layers of basement membrane on tumor vessels probably result from repeated cycles of disorganized vascular growth and regression.

Cellular Actions of VEGF Inhibitors on Tumor Blood Vessels

With the goal of obtaining a better understanding of the cellular actions of angiogenesis inhibitors on tumor vessels, we took advantage of the imaging methods described earlier in the chapter and the known abnormalities of endothelial cells, pericytes, and vascular basement membrane [40, 42, 45]. Agents that block VEGF signaling were used to exploit the availability of well characterized inhibitors and detailed knowledge of VEGF ligands, VEGF receptors, and downstream signaling pathways [47, 48]. Two agents that inhibit VEGF signaling through different mechanisms had comparable effects: AG-013736 is a small molecule inhibitor of VEGF receptors and related tyrosine kinase receptors; VEGF-Trap is a decoy construct of the extracellular domain of VEGFR1 and VEGFR-2 that inhibits VEGF signaling by selectively binding the ligands [49]. Changes in tumor vessels during the first week of treatment were sought to distinguish direct effects of the agents from secondary changes.

Tumor Vascularity

Treatment with AG-013736 or VEGF-Trap causes rapid and robust changes in endothelial cells of blood vessels of RIP-Tag2 tumors (Fig. 2.6A) and in Lewis lung carcinomas. Vascular sprouting is suppressed, endothelial fenestrations disappear, patency is lost and blood flow ceases in some vessels, and tumor vascularity decreases. A slight decrease in vascularity is evident at 1 day (Fig. 6B). The reduction is larger at 2 days (Fig. 2.6C), much greater at 7 days (Fig. 2.6D), and tends to plateau thereafter. The amount of vascular regression at 7 days is greater in RIP-Tag2 tumors (70% reduction of tumor vessels) than in Lewis lung carcinomas (50% reduction). Prolongation of treatment to 21 days does not further reduce vascularity but can reduce tumor volume of RIP-Tag2 tumors [49].

Vessel Patency

How can endothelial cells of tumor vessels die without causing hemorrhage? One explanation is that vessels close and lose patency before endothelial cells regress. Under baseline conditions, most blood vessels in RIP-Tag2 tumors are patent and functional as shown by LEA lectin labeling (Fig. 2.6E) [49]. After 1 day of treatment with AG-013736 or VEGF-Trap, about 30% of tumor vessels lack lectin staining despite the continued presence of endothelial cells (Fig. 2.6F) [49]. The number of non-functional tumor vessels is maximal at 2 days and falls nearly to zero by 7 days [49]. At 7 days, only about 30% of tumor vessels remain, and the amounts of lectin staining and CD31 immunoreactivity again match. These findings fit a model whereby tumor vessels lose patency before they degenerate, and most vessels still present after a week are functional.

Endothelial Sprouts

Endothelial sprouts tipped by filopodia are abundant on blood vessels in untreated tumors (Fig. 4F-H), but are rare on tumor vessels after treatment with AG-013736. Tumor vessels that survive the treatment are less tortuous, more uniform in caliber, and have fewer sprouts and branches.

Endothelial Fenestrations

Blood vessels of many tumors including those in RIP-Tag2 mice have abundant endothelial fenestrations [40, 49, 50]. These diaphragm-covered pores are also common in normal endocrine glands, intestinal mucosa, choriocapillaris, and certain other organs [51]. The number of fenestrations in tumor vessels rapidly decreases after inhibition of VEGF signaling. Treatment with AG-013736 or VEGF-Trap reduces the number of endothelial fenestrations by about 90% at 1 day and 98% at 7 days [49]. The greater effect of inhibitors of VEGF signaling on heavily fenestrated blood vessels in RIP-Tag2 tumors, compared to non-fenestrated blood vessels in Lewis lung carcinoma, raises the possibility that

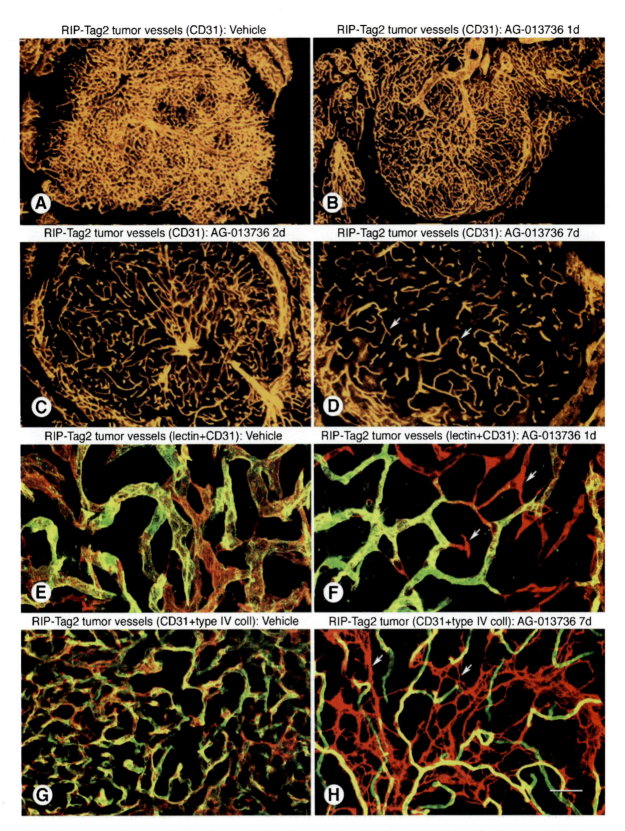

FIG. 2.6A–D. Time course of decrease in vascularity of RIP-Tag2 tumors after inhibition of VEGF signaling by AG-013736. Fluorescence microscopic images of CD31 immunoreactivity showing dense tumor vasculature under baseline conditions (**A,** vehicle treatment) and decreasing vascularity after AG-013736 for 1 day (**B**), 2 days (**C**), or 7 days (**D**) (from [49]). **E,F** Confocal microscopic images of blood vessels in RIP-Tag2 tumors after vehicle (**E**) or AG-013736 (**F**) for 1 day. LEA lectin injected iv before fixation (from [49]). Lectin staining (*green*) of blood vessels in vehicle-treated tumor largely coincides with CD31 immunoreactivity (*red*). However, after treatment with AG-013736, some vessels (*arrows*) have CD31 immunoreactivity but lack lectin staining, and thus are *red* (**F**), indicating loss of vessel patency (from [49]). **G,H** Persistence of vascular basement membrane after endothelial cells degenerate. Confocal micrographs show colocalization of CD31 (*green*) and type IV collagen (*red*) in vasculature of RIP-Tag2 tumor at baseline (**G**). After AG-013736 for 7 days, sleeves of type IV collagen devoid of CD31 (*arrows*) are abundant between surviving, normalized blood vessels (*yellow*) in RIP-Tag2 tumor (**H** *arrows*). *Scale bar*: 200 μm in (**A–D**); 25 μm in (**E,F**); 50 μm in (**G,H**).

presence of endothelial fenestrations is predictive of response to inhibition of VEGF.

VEGFR Immunoreactivity

Another consistent effect of inhibitors of VEGF signaling is reduction in expression of VEGFR-2 and VEGFR-3, reflected by reduced immunofluorescence [20, 46, 49, 51]. This change does not reflect generalized effects on membrane proteins because immunoreactivities for CD31, CD105, and PDGFR-β do not change significantly [49]. The decrease in VEGFR-2 fluorescence appears to reflect a change in epitope density on individual vessels rather than reduced vessel caliber [49]. The reduction of VEGFR-2 in endothelial cells of Lewis lung carcinoma is smaller than RIP-Tag2 tumors. To the extent that brightness of fluorescence reflects amount of expression, vessels with the highest initial VEGFR-2 expression are most likely to be destroyed or normalized into vessels with lower VEGFR-2 expression [49].

Pericytes

Inhibition of VEGF signaling results in a much larger reduction in tumor vessels than in pericytes [49]. After treatment of RIP-Tag2 tumors and Lewis lung carcinomas with AG-013736 or VEGF-Trap, twice as many pericytes remain as tumor vessels. Surviving pericytes have two fates in the short-term. Some become more tightly associated with endothelial cells, even oriented circumferentially around vessels in a fashion resembling smooth muscle cells on arterioles [49]. Others, identified as cells expressing pericyte markers surrounded by basement membrane, have no apparent association with tumor vessels [49]. The long-term fate of the latter cells is unknown.

Basement Membrane

Blood vessels of RIP-Tag2 tumors, like those in most other tumors, are covered by basement membrane (Fig. 6G) [45, 49]. After treatment with AG-013736 or VEGF-Trap for 7 days, distinctive strands of type IV collagen unaccompanied by endothelial cells are scattered throughout the tumor (Fig. 6H). When tumor vascularity decreases more than 75%, basement membrane decreases at most 30%. Empty strands of basement membrane become more abundant as tumor vessels regress during treatment and remain for at least three weeks [49]. Empty sleeves of basement membrane are not stained by lectin or endothelial cell markers and are not perfused with blood [49].

Rapid Regrowth of Tumor Vessels After Cessation of VEGF Inhibitor

Tumor vessels regrow rapidly after cessation of VEGF inhibition. In RIP-Tag2 tumors, overall vascularity decreases 60–75% during treatment with AG-013736 for 7 days (Fig. 2.7A, B) [46,

49]. After the inhibitor is stopped, tumor vascularity nearly doubles during the first 4 days and returns to the baseline condition in 7 days (Fig. 2.7C, D) [46]. Tumor vascularity does not exceed the baseline even after longer periods of regrowth. Vascular regrowth is at least as rapid in Lewis lung carcinomas, where the 50% reduction in vascularity is completely reversed within 4 days [46].

Regrowing tumor vessels become functional almost as rapidly as they regrow [46]. LEA lectin injected iv to label patent blood vessels marks nearly all vessels in untreated RIP-Tag2 tumors (Fig. 2.7E). After AG-013736 for 7 days, surviving tumor vessels are sparse but all have lectin labeling (Fig. 2.7F). One day after the treatment ends, sprouts are present (Fig. 2.7G, arrows). By 7 days, lectin stains almost all of the regrown vasculature (Fig. 2.7H). Lectin labeling and vascularity increase in parallel as tumor vascularity returns to baseline [46].

Tumor vessels sensitive to VEGF inhibitors have unusually high expression of VEGFR-2. Inhibition of VEGF signaling by AG-013736 leads to a reduction in VEGFR-2 immunofluorescence [46, 49]. After the inhibition is withdrawn, VEGFR-2 immunofluorescence returns to baseline in regrown tumor vessels during the first week.

Regrown tumor vessels reacquire VEGF dependence within a week after treatment ends [46]. Treatment with a second round of AG-013736 beginning 7 days after the first round reduces tumor vascularity as much as the first round.

Treatment with AG-013736 reduces by about one-third the population of α-SMA-positive pericytes in RIP-Tag2 tumors [46, 49]. This reduction reverses rapidly and is fully back to baseline within 4 days after the treatment ends. PDGFR-β immunoreactivity does not undergo a corresponding reduction during treatment, suggesting that AG-013736 induces a reversible change in pericyte phenotype, characterized by reduction in α-SMA expression, rather than a decrease in number of pericytes. AG-028262, a potent small molecule inhibitor of VEGFR-2 phosphorylation with little action on PDGFR-β, has similar effects on α-SMA expression [46]. These results fit with the change in pericyte phenotype being a downstream consequence of inhibition of VEGFR signaling, rather than a direct effect on PDGF signaling.

Regression of endothelial cells after inhibition of VEGF signaling leaves pericytes in otherwise empty sleeves of basement membrane [46]. In untreated tumors, endothelial cells and vascular basement membrane have similar distributions (Fig. 7I). After 7 days of AG-013736, the overall pattern of basement membrane changes little, despite the marked decrease in tumor vascularity (Fig. 2.7J). By 7 days after the treatment ends, tumor vasculature returns to baseline and empty sleeves of basement membrane disappear (Fig. 7K) [46].

The abundance of empty sleeves of basement membrane left behind by regressing tumor vessels and the apparent lack of duplication of vascular basement membrane during regrowth point to the possibility that the sleeves serve as a scaffold for regrowing tumor vessels. Consistent with this process, the rate of vessel regrowth is similar to the rate of disappearance of the

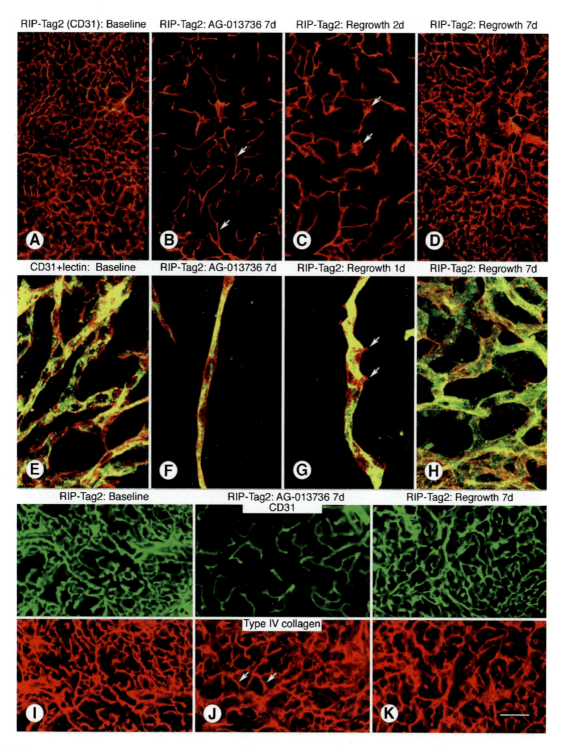

FIG. 2.7A–D. Regrowth of tumor vessels after cessation of VEGF inhibition. Confocal micrographs of RIP-Tag2 tumors stained for CD31 immunoreactivity (*red*) comparing vascularity of untreated tumor (**A**), tumor after AG-013736 for 7 days (**B**), and tumor after AG-013736 was stopped for 2 days (**C**) or 7 days (**D**) (from [46]). By 7 days, tumor vessels are as abundant as at baseline. **E–H** Confocal micrographs of RIP-Tag2 tumors comparing vessels, labeled by iv green fluorescent LEA lectin and then stained for CD31 immunoreactivity, at baseline (**E**), AG-013736 for 7 days (**F**), or after AG-013736 was stopped for 1 day (**G**) or 7 days (**H**) (from [46]). In the untreated tumor (**E**), almost all blood vessels are labeled with lectin. After AG-013736 for 7 days (**F**), surviving tumor vessels have lectin labeling. At 1 day after treatment ended nearly all vessels are stained with lectin, but lectin-negative sprouts (*red*) are present (**G** *arrows*). At 7 days after treatment ended (**H**), lectin and CD31 staining resembles the baseline (**E**). **I–K.** Fluorescence microscopic images of tumors showing CD31-positive endothelial cells (*green, upper*) and type IV collagen-positive basement membrane (*red, lower*) at baseline (**I**), after AG-013736 for 7 days (**J**), and 7 days after end of 7-day treatment (**K**) (from [46]). The two markers have similar distributions in untreated RIP-Tag2 tumors (**I**), but after AG-013736, CD31-positive vessels are sharply reduced but basement membrane is not (**J**). By 7 days after treatment ended, tumor vascularity is restored, and patterns of CD31 and type IV collagen again match (**K**). *Scale bar*: 150 μm in (**A–D**); 25 μm in (**E–H**); 120 μm in (**I–K**).

empty sleeves [46]. Sleeves of basement membrane, which are sites of bound VEGF, may facilitate the rapid revascularization of tumors [46].

Effects of VEGF Inhibitors on Normal Blood Vessels

The essential role of VEGF during embryonic development has been assumed not to persist into adult life. Yet VEGF has actions in normal organs of the adult, including effects on the structure, function, and survival of some blood vessels, on blood pressure regulation, and on renal, neurological, and hepatic function [51, 52]. Functions of VEGF in normal organs provide insight into mechanisms of side effects in cancer patients treated with VEGF inhibitors [52].

Inhibition of VEGF signaling is accompanied by multiple changes in normal capillaries of certain organs. These include reduction in endothelial fenestrations, reduction in endothelial cell expression of VEGFR-2 and VEGFR-3, and regression of capillaries [51]. Endothelial fenestrations are a feature of capillaries in gastrointestinal tract, many endocrine organs, kidney, liver, choroid plexus, and choriocapillaris [51]. Fenestrations begin to go away within 24h of inhibition of VEGF signaling [49]. After 2 or 3 weeks, the number of fenestrations is conspicuously reduced in capillaries of pancreatic islets (Fig. 2.8A, B), renal glomerulus (Fig. 2.8C, D), and thyroid, where fenestrations decrease by as much as 88% [51].

Inhibition of VEGF signaling leads to regression of capillaries in the trachea (Fig. 2.8E), villi of small intestine (Fig. 2.8F, G), pancreatic islets, thyroid (Fig. 2.8H, I), adrenal cortex, pituitary, choroid plexus, and adipose tissue [51, 53]. The amount of regression is dose-dependent and varies from organ to organ, with a maximum of 68% in thyroid [51]. In these experiments, little or no capillary regression was found in brain, retina, skeletal muscle, cardiac muscle, or lung [49, 51].

After inhibition of VEGF signaling for only 1 day, fibrin accumulates and patency is lost in some capillaries [20, 49, 53]. By 2 days, endothelial cells undergo apoptosis and regression. The magnitude of capillary loss decreases with age, ranging from 39% at 4 weeks of age, 28% at 8 weeks, to 14% at 16 weeks [20]. Empty sleeves of basement membrane persist for weeks after endothelial cells regress (Fig. 2.8E) [46].

Most capillaries in the thyroid grow back within 1 or 2 weeks (Fig. 2.8H-J) [46]. Tracheal capillaries also regrow rapidly [53]. As in tumors, regrowth appears to be facilitated by if not dependent on empty sleeves of basement membrane that provide a scaffold for revascularization [46].

Together, these findings show the dependency on VEGF signaling of endothelial fenestrations and survival of normal fenestrated capillaries [49, 51]. They also show the potential plasticity of the microvasculature [51]. Importantly, these observations raise the possibility that blood vessels having abundant endothelial fenestrations and high VEGFR-2 expression are especially sensitive to inhibitors of VEGF signaling. Reduction in endothelial fenestrations and VEGFR-2 are

responses to VEGF inhibition that may serve as surrogates for predicting therapeutic efficacy [52].

Conclusions

The expanding diversity of preclinical models makes it possible to examine angiogenesis and vascular remodeling under many defined experimental conditions. Imaging techniques are being improved to complement the experimental models and make it easier to elucidate changes in endothelial cells and pericytes, obtain detailed cellular information on blood vessels of the microcirculation, and characterize alterations in the architecture of three-dimensional vascular networks. These approaches are rapidly advancing the understanding of the normal microvasculature and the diverse changes blood vessels undergo in disease.

With the recognition that angiogenesis and vascular remodeling occur under many conditions, it seems reasonable to assume that they are generic processes with similar properties regardless of disease pathophysiology. While this probably is true at some level, blood vessels change in strikingly different ways in different diseases. New and remodeled blood vessels at sites of inflammation acquire features specialized to support increased blood flow and regulated influx of inflammatory cells. Capillaries in inflamed tissues develop a phenotype resembling that of venules, with up-regulation of endothelial cell receptors for inflammatory mediators that initiate vessel leakiness and leukocyte influx. Leakage from these vessels occurs though focal gaps that form reversibly between endothelial cells. Adhesion molecules expressed on the remodeled endothelial cells participate in the orchestrated process of leukocyte rolling, attachment, and migration.

Unlike the changes in inflammation, endothelial cells of tumor vessels undergo disorganized sprouting and proliferation, overexpress or underexpress membrane receptors, and may become dependent on VEGF or other growth factors for survival. Structural defects result in impaired endothelial barrier function, poor blood flow, high interstitial pressure, and altered immune cell entry. Functional abnormalities of tumor vessels can restrict blood flow to growing tumor cells despite rich vascularity. Poor blood flow can promote invasion and result in hypoxia and cell death. Robust angiogenesis and expansion of tumor cells can paradoxically occur adjacent to regions of necrosis.

Abnormalities of blood vessels can be exploited in diagnosis and treatment of inflammation and tumors. Reversal of environmental factors that induce the abnormal vascular phenotype can lead to clinical improvement. When factors that promote VEGF signaling contribute to blood vessel abnormalities in tumors, inhibition of VEGF signaling can stop angiogenesis and promote vascular regression or normalization. Empty sleeves of basement membrane left behind after tumor vessels regress provide a scaffold for revascularization. As understanding of the vascular biology of inflammation and tumors advances, so will strategies for effectively blocking

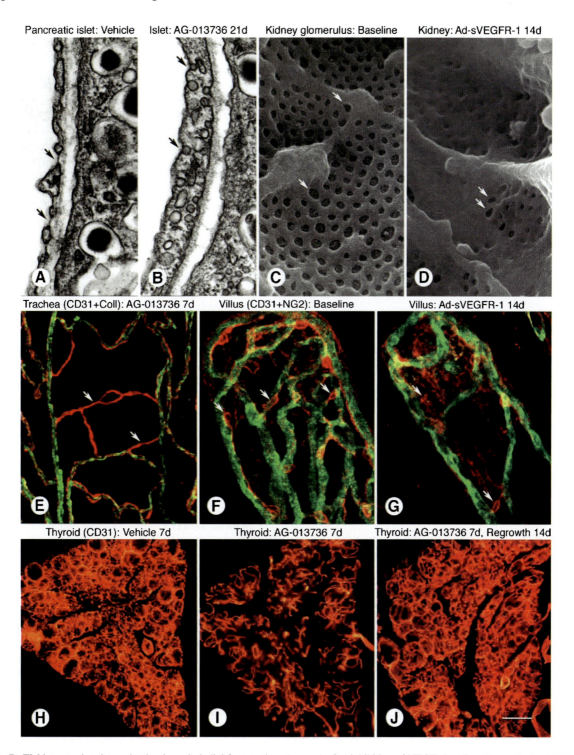

FIG. 2.8A–D. EM images showing reduction in endothelial fenestrations (arrows) after inhibition of VEGF signaling. Transmission EM images of pancreatic islet capillaries showing thin endothelium and abundant fenestrations with diaphragms under baseline conditions (**A**) compared to thick endothelium, few fenestrations, and abundant caveolae after AG-013736 for 21 days (**B**) (from [51]). **C,D** Scanning EM images of luminal surface of renal glomerular capillaries showing abundant endothelial fenestrations under baseline conditions (**C**) and few fenestrations after adenovirally delivered soluble VEGFR1 (Ad-sVEGFR-1) for 14 days (**D**) (from [51]). **E** Confocal micrograph showing capillary regression in mouse trachea after inhibition of VEGF signaling by AG-013736 for 7 days (from [49]). Type IV collagen (*red*); CD31 (*green*). Empty sleeves of basement membrane (*red, arrows*) are left behind where capillaries regressed (**E**). **F,G** Confocal microscopic images of capillaries in intestinal villi of normal adult mice under baseline conditions (**F**) and after inhibition of VEGF signaling by Ad-sVEGFR-1 for 14 days (**G**) (from [51]). Pericytes (*red*, NG2, *arrows*) are located on normal capillaries (**F**) and at sites where capillaries have regressed (**G**). **H–J**. Regression of thyroid capillaries after inhibition of VEGF signaling followed by rapid regrowth after end of treatment (from [51]). Fluorescence micrographs of CD31 immunoreactivity show dense vascularity of thyroid follicles under baseline conditions (**H**), loss of half of the capillaries after AG-013736 for 7 days (**I**), and complete vascular regrowth over 14 days after end of treatment (**J**). *Scale bar*: 0.3 μm in (**A,B**); 0.5 μm in (**C,D**); 50 μm in (**E**); 25 μm in (**F,G**); 160 μm in (**H–J**).

angiogenesis and destroying pathological vessels or reversing the abnormal phenotype without impacting the normal vasculature.

Acknowledgments. The author thanks Hiroya Hashizume for preparing the scanning electron micrographs, Amy Haskell for the transmission electron micrographs, and Fabienne Baffert, Peter Baluk, Jeffrey Bowden, Tetsuichiro Inai, Tomomi Kamba, Michael Mancuso, Shunichi Morikawa, Scott Norberg, and Gavin Thurston for fluorescence and confocal microscopic images. The research described in this review was supported in part by National Institutes of Health grants HL24136 and HL59157 from the National Heart, Lung, and Blood Institute and CA82923 from the National Cancer Institute, and by funding from AngelWorks Foundation.

References

1. Baluk P, Falcón BL, Hashizume H, et al. Cellular actions of angiogenesis inhibitors on blood vessels. In: Marmé D, Fusenig N, eds. Tumor Angiogenesis: Basic Mechanisms and Cancer Therapy. New York: Springer; 2007:557–76.

2. Baluk P, Hashizume H, McDonald DM. Cellular abnormalities of blood vessels as targets in cancer. Curr Opin Genet Dev 2005;15(1):102–11.

3. McDonald DM, Thurston G, Baluk P. Endothelial gaps as sites for plasma leakage in inflammation. Microcirculation 1999;6(1):7–22.

4. Thurston G, Baluk P, McDonald DM. Determinants of endothelial cell phenotype in venules. Microcirculation 2000;7(1):67–80.

5. McDonald DM. Angiogenesis and remodeling of airway vasculature in chronic inflammation. Am J Respir Crit Care Med 2001;164(10 Pt 2):S39–45.

6. McDonald DM. Endothelial gaps and permeability of venules in rat tracheas exposed to inflammatory stimuli. Am J Physiol 1994;266(1 Pt 1):L61–83.

7. Thurston G, Baluk P, Hirata A, et al. Permeability-related changes revealed at endothelial cell borders in inflamed venules by lectin binding. Am J Physiol 1996;271(6 Pt 2):H2547–62.

8. Thurston G, Murphy TJ, Baluk P, et al. Angiogenesis in mice with chronic airway inflammation: strain-dependent differences. Am J Pathol 1998;153(4):1099–112.

9. Murphy TJ, Thurston G, Ezaki T, et al. Endothelial cell heterogeneity in venules of mouse airways induced by polarized inflammatory stimulus. Am J Pathol 1999;155(1):93–103.

10. Lindsey JR, Baker HJ, Overcash RG, et al. Murine chronic respiratory disease. Significance as a research complication and experimental production with Mycoplasma pulmonis. Am J Pathol 1971;64(3):675–708.

11. Lindsey JR, Cassell H. Experimental Mycoplasma pulmonis infection in pathogen-free mice. Models for studying mycoplasmosis of the respiratory tract. Am J Pathol 1973;72(1):63–90.

12. McDonald DM, Schoeb TR, Lindsey JR. Mycoplasma pulmonis infections cause long-lasting potentiation of neurogenic inflammation in the respiratory tract of the rat. J Clin Invest 1991;87(3):787–99.

13. Baluk P, Bowden JJ, Lefevre PM, et al. Upregulation of substance P receptors in angiogenesis associated with chronic airway inflammation in rats. Am J Physiol 1997;273(3 Pt 1):L565–71.

14. Kwan ML, Gomez AD, Baluk P, et al. Airway vasculature after mycoplasma infection: chronic leakiness and selective hypersensitivity to substance P. Am J Physiol Lung Cell Mol Physiol 2001;280(2):L286–97.

15. Davidson MK, Lindsey JR, Parker RF, et al. Differences in virulence for mice among strains of Mycoplasma pulmonis. Infection and immunity 1988;56(8):2156–62.

16. Cartner SC, Simecka JW, Lindsey JR, et al. Chronic respiratory mycoplasmosis in C3H/HeN and C57BL/6N mice: lesion severity and antibody response. Infection and immunity 1995;63(10):4138–42.

17. Bowden JJ, Schoeb TR, Lindsey JR, et al. Dexamethasone and oxytetracycline reverse the potentiation of neurogenic inflammation in airways of rats with Mycoplasma pulmonis infection. Am J Respir Crit Care Med 1994;150(5 Pt 1):1391–401.

18. Ezaki T, Baluk P, Thurston G, et al. Time course of endothelial cell proliferation and microvascular remodeling in chronic inflammation. Am J Pathol 2001;158(6):2043–55.

19. Thurston G, Maas K, Labarbara A, et al. Microvascular remodelling in chronic airway inflammation in mice. Clin Exp Pharmacol Physiol 2000;27(10):836–41.

20. Baffert F, Thurston G, Rochon-Duck M, et al. Age-related changes in vascular endothelial growth factor dependency and angiopoietin-1-induced plasticity of adult blood vessels. Circ Res 2004;94(7):984–92.

21. Baluk P, Lee CG, Link H, et al. Regulated angiogenesis and vascular regression in mice overexpressing vascular endothelial growth factor in airways. Am J Pathol 2004;165(4):1071–85.

22. Baluk P, Raymond WW, Ator E, et al. Matrix metalloproteinase-2 and -9 expression increases in Mycoplasma-infected airways but is not required for microvascular remodeling. Am J Physiol Lung Cell Mol Physiol 2004;287(2):L307–17.

23. Baluk P, Tammela T, Ator E, et al. Pathogenesis of persistent lymphatic vessel hyperplasia in chronic airway inflammation. J Clin Invest 2005;115(2):247–57.

24. Thurston G, Wang Q, Baffert F, et al. Angiopoietin 1 causes vessel enlargement, without angiogenic sprouting, during a critical developmental period. Development 2005;132(14):3317–26.

25. Cho CH, Kim KE, Byun J, et al. Long-term and sustained COMP-Ang1 induces long-lasting vascular enlargement and enhanced blood flow. Circ Res 2005;97(1):86–94.

26. Baffert F, Le T, Thurston G, et al. Angiopoietin-1 decreases plasma leakage by reducing number and size of endothelial gaps in venules. Am J Physiol Heart Circ Physiol 2006;290(1):H107–18.

27. Calvert JT, Riney TJ, Kontos CD, et al. Allelic and locus heterogeneity in inherited venous malformations. Hum Mol Genet 1999;8(7):1279–89.

28. Vikkula M, Boon LM, Carraway KL, 3rd, et al. Vascular dysmorphogenesis caused by an activating mutation in the receptor tyrosine kinase TIE2. Cell 1996;87(7):1181–90.

29. Nishimoto M, Akashi A, Kuwano K, et al. Gene expression of tumor necrosis factor alpha and interferon gamma in the lungs of Mycoplasma pulmonis-infected mice. Microbiol Immunol 1994;38(5):345–52.

30. Faulkner CB, Simecka JW, Davidson MK, et al. Gene expression and production of tumor necrosis factor alpha, interleukin 1, interleukin 6, and gamma interferon in C3H/HeN and C57BL/6N mice in acute Mycoplasma pulmonis disease. InfectImmun 1995;63(10):4084–90.

31. McDonald DM, Choyke PL. Imaging of angiogenesis: from microscope to clinic. Nat Med 2003;9(6):713–25.

32. Ocak I, Baluk P, Barrett T, et al. The biologic basis of in vivo angiogenesis imaging. Front Biosci 2007;12:3601–16.

33. Maniotis AJ, Folberg R, Hess A, et al. Vascular channel formation by human melanoma cells in vivo and in vitro: vasculogenic mimicry. Am J Pathol 1999;155(3):739–52.

34. Chang YS, di Tomaso E, McDonald DM, et al. Mosaic blood vessels in tumors: frequency of cancer cells in contact with flowing blood. Proc Natl Acad Sci U S A 2000;97(26):14608–13.

35. McDonald DM, Munn L, Jain RK. Vasculogenic mimicry: how convincing, how novel, and how significant? Am J Pathol 2000;156(2):383–8.

36. Folberg R, Maniotis AJ. Vasculogenic mimicry. Apmis 2004;112(7–8):508–25.

37. Hendrix MJ, Seftor EA, Hess AR, et al. Vasculogenic mimicry and tumour-cell plasticity: lessons from melanoma. Nature Rev 2003;3(6):411–21.

38. di Tomaso E, Capen D, Haskell A, et al. Mosaic tumor vessels: cellular basis and ultrastructure of focal regions lacking endothelial cell markers. Cancer Res 2005;65(13):5740–9.

39. Jain RK, Munn LL, Fukumura D. Dissecting tumour pathophysiology using intravital microscopy. Nature Rev 2002;2(4):266–76.

40. Hashizume H, Baluk P, Morikawa S, et al. Openings between defective endothelial cells explain tumor vessel leakiness. Am J Pathol 2000;156(4):1363–80.

41. Thurston G, McLean JW, Rizen M, et al. Cationic liposomes target angiogenic endothelial cells in tumors and chronic inflammation in mice. J Clin Invest 1998;101(7):1401–13.

42. Morikawa S, Baluk P, Kaidoh T, et al. Abnormalities in pericytes on blood vessels and endothelial sprouts in tumors. Am J Pathol 2002;160(3):985–1000.

43. Gerhardt H, Golding M, Fruttiger M, et al. VEGF guides angiogenic sprouting utilizing endothelial tip cell filopodia. J Cell Biol 2003;161(6):1163–77.

44. Sennino B, Falcon BL, McCauley D, et al. Sequential loss of tumor vessel pericytes and endothelial cells after inhibition of platelet-derived growth factor-beta by selective aptamer AX102. Cancer Res 2007;67(15):7358–67.

45. Baluk P, Morikawa S, Haskell A, et al. Abnormalities of basement membrane on blood vessels and endothelial sprouts in tumors. Am J Pathol 2003;163(5):1801–15.

46. Mancuso MR, Davis R, Norberg SM, et al. Rapid vascular regrowth in tumors after reversal of VEGF inhibition. J Clin Invest 2006;116(10):2610–21.

47. Ferrara N. Vascular endothelial growth factor: basic science and clinical progress. Endocr Rev 2004;25(4):581–611.

48. Ferrara N, Mass RD, Campa C, et al. Targeting VEGF-A to treat cancer and age-related macular degeneration. Ann Rev Med 2007;58:491–504.

49. Inai T, Mancuso M, Hashizume H, et al. Inhibition of vascular endothelial growth factor (VEGF) signaling in cancer causes loss of endothelial fenestrations, regression of tumor vessels, and appearance of basement membrane ghosts. Am J Pathol 2004;165(1):35–52.

50. Pasqualini R, Arap W, McDonald DM. Probing the structural and molecular diversity of tumor vasculature. Trends Mol Med 2002;8(12):563–71.

51. Kamba T, Tam BY, Hashizume H, et al. VEGF-dependent plasticity of fenestrated capillaries in the normal adult microvasculature. Am J Physiol Heart Circ Physiol 2006;290(2):H560–76.

52. Kamba T, McDonald DM. Mechanisms of adverse effects of anti-VEGF therapy for cancer. British journal of cancer 2007;96(12):1788–95.

53. Baffert F, Le T, Sennino B, et al. Cellular changes in normal blood capillaries undergoing regression after inhibition of VEGF signaling. Am J Physiol Heart Circ Physiol 2006;290(2):H547–59.

Chapter 3
Endothelial Cell Activation

M. Luisa Iruela-Arispe

Keywords: endothelial cell activation, developmental angiogenesis, endothelial tip cells, endothelial stalk cells

Abstract: The initiation of the angiogenic cascade from a pre-existent vascular network requires the selective departure of individual endothelial cells from differentiated capillaries. The process entails the activation of specific signaling pathways that enable endothelial cells to exit their vessel of origin, invade the underlying stroma and initiate a new vascular sprout. Two major signaling pathways: VEGF and Notch, coordinate this process to select a subset of leading endothelial cells, referred to as tip cells. These cells display long filopodia and are highly migratory, but remain linked to their followers, the stalk cells. The stalk cells constitute the body of the sprout and proliferate in response to VEGF increasing the length of the incipient capillary. It is the coordination of Notch and VEGF signaling that regulates the extent to which cells become leaders (tip cells) and which become followers (stalk cells). Activation of Notch represses the tip cell in favor of the stalk cell phenotype, in part, by regulating the levels of VEGFR2. The resolution of the endothelial activation phase requires synthesis and organization of the basement membrane and the recruitment of pericytes and smooth muscle cells. This chapter focuses on the molecular regulation of these signaling pathways, and it contrasts our current understanding of endothelial cell activation in development and in disease.

Department of Molecular, Cell and Developmental Biology, Molecular Biology Institute and Jonsson Comprehensive Cancer Center, University of California, Los Angeles, CA 90095, USA

Address for Correspondence: UCLA, 615 Charles Young Drive South, Los Angeles, CA 90095, USA

Introduction

The term endothelial cell activation makes reference to the series of events by which a fully differentiated, non-motile and non-proliferative cell acquires an angiogenic phenotype. The process entails the development of invasive, migratory, and proliferative capacities by the endothelial cell. This same term has also been used to describe the phenotypic alterations of the endothelium in response to inflammatory mediators and that result in the retention and recruitment of inflammatory cells from the blood stream into the stroma. In this chapter, we will focus on angiogenic endothelial cell activation.

Endothelial cells are the basic and constant component of the vascular system. These are also the cells that initiate the angiogenic response and are responsible for establishing the pattern of the future capillary plexus. Once specified as endothelial, these cells enclose the genetic information that pre-determines their contribution to either veins or arteries, as well as their association with presumptive smooth muscle cells. Thus, early decisions in vascular morphogenesis carry important consequences for the overall formation of the vascular tree. During the last decade, the implementation of targeted gene inactivation in whole animals has provided an explosion of information regarding the genetic circuitry that mediates endothelial cell activation. The vascular system is one of the first fully functional organs to be established in vertebrate embryos and it is essential for viability and survival. This dependency has been extremely advantageous to vascular biologists through the recent rush of genetic knockout models. The unsuspected contribution of several regulatory molecules has been revealed by phenotypes that include hemorrhage and embryonic lethality. Indeed, genetic inactivation in whole animals has provided major breakthroughs in our understanding of vascular development. Based on information from loss- and gain-of-function studies, today we know that the key signaling pathways in endothelial cell activation include vascular endothelial growth factor (VEGF) and Notch. Subsequently, Slit, Ephrins, Cadherins, Wnts and Angiopoetins, Transforming

Growth factor β and integrins participate in stages post-activation, to guide, remodel, stabilize and differentiate the newly formed vessels [1,2]. This chapter will focus on the cellular events that regulate endothelial cell activation during development and in pathological conditions.

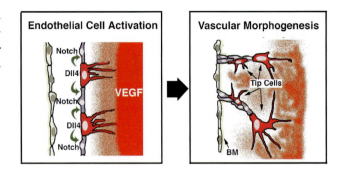

FIG. 3.1. *Endothelial Cell Activation* (*left*) results in the departure of a subset of endothelial cells from parental vessels. The process requires the acquisition of a tip cell phenotype whereby some endothelial cells are specified to become "leaders" in the sprout and their immediate neighbors are the "followers" or simply remain in the original vessel. Several loss- and gain-of-function studies in mouse and zebrafish have determined that the Notch signaling pathway is critical for this process. Thus, cells expressing the ligand Delta 4 (*Dll4*) activate the *Notch* 1 receptor in adjacent cells. Once activated, Notch mediates down-regulation of VEGFR2 and increased levels of VEGFR1. As consequence, the Dll4-expressing cells are more susceptive than their neighbors in sensing gradients of VEGF. The outcome is the formation of a vascular sprout (*Vascular Morphogenesis*; *right*) with highly migratory *tip cells* and proliferating stalk cells.

Activation of Endothelial Cells During Developmental Angiogenesis

The elucidation of the molecular underpins that regulate endothelial cell activation are critical to a concrete understanding of how blood vessels are formed. After the formation of the primary vascular plexus, which arises from the direct differentiation of mesenchymal cells into endothelial cells (vasculogenesis); the subsequent expansion of the vascular system occurs through angiogenic growth. That is, endothelial cells depart from their vascular beds and sprout into the avascular stroma.

Initiation of the Angiogenic Response

What initiates the vascular sprout? Current experimental evidence indicates that VEGF, through activation of its tyrosine kinase receptors, VEGFR1 and VEGFR2, is likely to be the initiating factor [3–8]. Activation of VEGFR2 results in significant cytoskeletal changes with extension of filopodia and acquisition of a migratory phenotype [9–16] (Fig. 3.1). As the vascular sprout continues to grow it is followed by a solid cord of cells which differ from the initial sprouting cells, as they display less filopodia and are not as migratory. Instead, these "followers" appear to proliferate more frequently than the sprouting cell [17,18]. Thus, the just beginning vessel is formed by two morphologically and functionally distinct cell types: the *tip cell* that provides directional migration and the *stalk cells* that compose the body of the rapidly expanding capillary (Fig. 3.1). Although both cell types have been shown to respond to VEGF, the tip cell appears to migrate and not proliferate, while the stalk cells mostly proliferate in response to this growth factor [17]. What mediates such alternative responses? While the answer to this question is unclear, it is likely that the selection of a particular outcome lies on either the phosphorylation of distinct tyrosine residues in the VEGFR2 with consequent recruitment of alternative second messengers; or the contribution of additional signaling pathways or a combination thereof.

Studies conducted in the retina indicate that the presentation of VEGF is likely aided by astrocytes [18–21]. These cells provide spatial guidance into pre-determined tracks. In other organs, this function might be provided by functionally analogous cell types or by the nature and composition of the matrix. Nonetheless, it has been considered that the presentation of VEGF, either bound to the matrix (or to the surface of adjacent cells) or in a soluble form, alters the responses of endothelial cells to the growth factor [22–24]. Thus, the

ability of VEGF to be immobilized allows for the formation of a gradient that is sensed by filopodia on the tip cell and provides directional migratory cues. In contrast, when VEGF is soluble there is no gradient formed and endothelial cells tend to lose directionality and be less migratory [23,24].

The ability of VEGF to interact with the extracellular matrix is regulated by two mechanisms: (1) splicing, and (2) extracellular processing. Encoded by a single gene, VEGF-A can originate multiple transcripts as the result of alternative splicing. Seven isoforms have been identified: the most frequently detected forms are VEGF 121, 165, and 189 (the names represent number of amino acids) [22]. Interestingly, the differences amongst the isoforms reside in a region coded by exons 6a, 6b and 7 and that is targeted by the splicing machinery. These exons code for domains that interact with heparin and other matrix proteins. There is a direct correlation between the ability of VEGF to bind to the matrix and the extent of the carboxy-terminal tail coded by the exons mentioned above. Thus, VEGF 189 binds more avidly to matrix proteins than VEGF 165. In contrast to these, VEGF 121 is considered to be the soluble VEGF form.

The biological significance of each VEGF isoform has only been revealed recently and has relevance to our understanding of endothelial cell activation. Using a knockin strategy, a group of investigators decided to integrate the cDNAs of VEGF 121, 165 and 189 into the VEGF locus, i.e., under the regulatory control of its promoter [23]. The resulting mice were only able to generate 121, 165 or 189. The approach not only restricted the production of one particular transcript, it also resulted in an overexpressor for such isoforms, as the full activity of the promoter was confined to only one isoform. The findings were remarkable as they clarified the relevance of matrix-bound VEGF: the longest VEGF isoform (able to bind tightly to matrix) was essential for directional filopodial growth. Mice exhibited increased vascular density and thinner vessels than wild-type mice. In contrast, expression of 121 resulted in lower capillary density, enlarged vessels, and poor directionality of the vascular sprouts [23]. VEGF 121 mice exhibited patterning anomalies in larger vessels, including Tetralogy of Fallot [25].

The second mechanism for alteration of matrix binding is proteolytic processing. Several enzymes including plasmin and a cohort of matrix metalloproteinases (MMPs) are able to cleave VEGF in the extracellular space [24]. Processing of the growth factor occurs at aa113 and severs the molecule to separate the receptor binding domain from the extracellular binding region. This intramolecular processing event is extremely effective at dissociating VEGF from its matrix anchorage, and capable of releasing a soluble form fully able to activate VEGF tyrosine kinase receptors. Thus, depending upon the availability of enzymes, the extracellular environment can interfere with the well-orchestrated control provided by alternative splicing [24]. The contribution of extracellular enzymes to VEGF processing is likely to be a process associated with inflammation and other pathological events, such as cancer and less likely to be an active participant of developmental angiogenesis.

In addition to matrix-bound VEGF, the direction of migration is aided and subsequently regulated by plexins, slits, and semaphorins [26]. The contribution of these molecules to the process of endothelial cell activation will be discussed later.

Acquisition of Tip *Versus* Stalk Endothelial Cell Identities

As mentioned previously, VEGF is essential for endothelial cell activation, as it regulates both migratory and proliferative activities. However, signaling via VEGF alone is not sufficient to organize a well-orchestrated vasculature. A critical step in endothelial cell activation is to establish leadership: Who will be the leader cell that initiates the vascular sprout and who will follow? Recent studies have demonstrated that the Notch signaling pathway is critical for specifying stalk versus tip and for generating the required functional hierarchy that allows a vascular cord to emerge from a field of equivalent endothelial cells (Fig. 3.1) [27–29].

Prior to angiogenic growth, the local (in situ) differentiation of mesenchymal cells into endothelial cells results in the formation of a homogenous capillary plexus (vasculogenesis). This *"vascular rete"* expands quickly and remodels into a hierarchic vascular tree consisting of arteries, veins and interconnecting capillaries. Thus, angiogenesis, and therefore endothelial cell activation, are the first steps towards achieving vascular remodeling. However, to be functional, only a subgroup of endothelial cells must lead (i.e., be activated). The obvious question is how can hierarchic leadership be established in the context of a primary plexus where all endothelial cells are equal? Furthermore, how can this be accomplished if all these cells are exposed to the same VEGF gradient? A recent "boon" in the literature has shown that the Notch pathway enables endothelial cells to differentially "read" the same VEGF gradient by altering the levels of VEGF receptors [30–36]. Activation of the Notch receptor represses VEGFR2 and increases VEGFR1. The outcome are cells with a lower ability to "sense" VEGF [33]. In this manner, Notch provides suppressive signals that enable only a few cells to respond more avidly to the VEGF gradient (those in which Notch was not activated) and initiate the vascular sprout.

The first piece of evidence implicating the Notch pathway in the suppression of sprouts came from expression studies. Delta-like 4 (Dll4), one of the five mammalian Notch ligands, is specifically and conspicuously expressed by tip cells [37,38]. While the majority of endothelial cells within the vascular plexus display some degree of Notch receptor at their cell surface, expression of the ligands is not detected prior to tip cell specification [37,38]. Presence of Dll4 in the incipient tip cell rapidly results in the activation of Notch in the immediately adjacent neighbors (Fig. 3.1). The process leads to a reduction in their ability to detect VEGF signals and the suppression of the tip cell phenotype. This interpretation is consistent with findings from genetic deletion of Dll4 in both mouse and zebra fish [30,31,35]. Lack of Dll4 results in excessive sprouting and capillary hyperfusion during active angiogenesis, indicating that activation of a Notch receptor via Dll4 is necessary for inhibition of excessive sprouting events. The large number of sprouts in these mutants is not compatible with the organization of interconnected patented vessels. The outcome is the formation of a non-functional vascular bed that precipitates in embryonic lethality, despite the excessive number of activated endothelial cells.

Additional genetic loss- and gain-of-function has identified Notch1 as the primary receptor of Dll4 during these events [32]. Although, in contrast to Dll4, no haploinsufficiency was observed in Notch1, inactivation of this gene or pharmacological inhibition of the pathway also leads to excessive sprouting events. In the case of targeted inactivation, mice die at E9.5 with absence of vascular remodeling [39]. Interestingly, excess of Notch also results in embryonic lethality at a similar point in time. In this case, however, mice showed enlarged vessels [40], a phenotype that is consistent with the absence of tip cells and with the increase in stalk cells that proliferate but are unable to coordinate the organization of vascular sprouts.

Considering the critical requirement for Notch signaling for stalk / tip cell specification, it is not surprising that genetic ablation of genes involved in the regulation of this pathway, as well as major downstream targets, all lead to embryonic lethality. Interestingly, inactivation of all these molecules (a total of 14 KOs) die between E9.5 and E11.5 with no or extremely poor vascular remodeling [29,41].

While it is likely that Notch also regulates later aspects in vascular morphogenesis and homeostasis, its ability to suppress the tip cell phenotype is essential for productive vascular growth and it is critical for endothelial cell activation.

Guiding Cues for Activated Endothelial Cells

The highly ordered pattern of a fully developed vascular tree has implied the existence of a well-orchestrated molecular machinery able to provide guidance cues at the onset of vascular remodeling. In fact, the activated endothelial tip cell follows a VEGF gradient, but it is aided by attractive and repellent factors that fine-tuned its directionality. Originally identified as axon guidance molecules, semaphorins, plexins and slits are currently known to be more widely expressed and to play significant roles in vascular patterning [26,42–44].

Semaphorins comprise a family of membrane bound or secreted proteins that provide signals to facility navigational control during neuronal growth and, more recently, also acknowledged to provide vascular directionality [42–45]. They signal through plexins and neuropilins. In general, membrane-bound semaphorins bind to plexins, whereas secreted semaphorins bind to neuropilins [45,46]. A large cohort of genetic studies in Drosophila indicate that semaphorin signaling acts as a repulsive cue in axon guidance, in addition to suppressing neuronal migration. Nonetheless, other studies showed that these same molecules might also provide stimulatory signals depending upon the levels of intracellular cGMP [47–50]. These two functions, attraction and repulsion, are a theme of the so-called "guidance molecules" and offer a Yin and Yang balance essential for the fine-tune trajectory of endothelial navigations.

Semaphorin4A (Sema4A) suppresses VEGF-mediated endothelial cell migration and angiogenesis in vivo. Genetic targeting of Sema4A in mice results in enhanced angiogenesis in response to VEGF or inflammatory stimuli [51]. The effects of Sema4A on endothelial cells are mediated by Plexin D1 that blocks VEGF-mediated Rac activation and integrin-dependent cell adhesion. Combined, the findings indicated that Sema4A-Plexin-D1 signaling negatively regulates angiogenesis [51]. In addition to Sema4A, Sema3A has been shown not to compete with VEGF165 for binding to neuropilin1, functioning as an antagonist for the VEGF-VEGFRs proangiogenic signals [52–57].

The second group of ligands and receptor molecules involved in vascular patterning are the Netrins and UNC5 / DCC receptor families [58]. In neurons, Netrins have been shown to attract and repel neurons depending upon the nature of the receptor that is receiving the signal. Thus, attraction is generally mediated by DCC, while repulsion is conveyed by UNC5 [59]. Consistent with this notion, genetic deletion of UNC5B leads to excessive vascular branching and increased filopodia, particularly in the tip cells, suggesting a role for this receptor in vascular retraction [60]. Furthermore, exposure of growing sprouts to Netrin 1 results in retraction of filopodia in

wild type mice, but not in UNC5B knockout mice [60]. These findings are also supported by studies in zebrafish. Morpholino knockdown of UNC5B results in excessive capillary branching and aberrant vessel patterning. Intersomitic vessels migrate laterally, invading somites, instead of migrating dorsally [60–62]. This is perhaps the most clear demonstration that the activated endothelial cell requires Netrin-UNC5B for directionality.

Slits and roundabouts (Robo) are the last family of ligand/receptor that contributes to neuronal and vascular patterning [63,64]. Signaling through slit has been shown to act as a repulsive factor, preventing axons that have crossed the midline from re-crossing [65]. Four Robo receptors (named 1–4) have been identified in mammals, and from these, Robo-4 appears to be endothelial-specific. The contribution of Slit-Robo to the guidance of the tip cell is controversial in vitro with reports demonstrating promigratory and others inhibitory activity [66–68]. Morpholino knockdown of Robo4 in zebrafish, however, results in spatio-temporal disruption of intersomitic vessels. The outcome includes vessels sprouting from the aorta in the wrong direction and premature interruption of their trajectory [68]. Together, the data indicate that Robo4 functions to direct vessel growth to the correct path.

Formation of the Vascular Lumen

The resolution of endothelial cell activation requires the differentiation of endothelial cells and acquisition of a lumen. This is perhaps the step in the angiogenic cascade that is least understood. As of now, genetic analysis using targeting inactivation has been unable to identify molecules responsible for lumen formation. However, as previously discussed, Notch contributes to lumen diameter, by regulating the ratio of stalk to tip cells. Thus, more tip cells (less Notch) reduces vascular lumens [39], while excess of stalk cells (more Notch) leads to vascular hyperplasia and distended lumens [40]. In addition to Notch, it has been shown that soluble VEGF favors enlarged vessels, in contrast to bound VEGF (both during development and in the adult) [23,24].

More recently, elegant morphological descriptions of lumen formation have been reported in zebrafish. These combined with in vitro analysis indicate that formation of vacuoles precedes lumen development within a vessel and that flow is not required for the event, but it facilitates the process [69].

Activation of Endothelial Cells in Pathological Conditions

Angiogenesis induced during pathological events results in vessels that are structurally and functionally altered, when compared to capillary beds from normal organs and tissues [70–73]. In contrast to developmental angiogenesis, the growth of capillaries during pathology is disorganized, excessive, and dysfunctional. Endothelial cells under pathology become activated by an irregular set of stimuli, too much VEGF, altered levels of Notch and Notch ligands, and variations in the levels of guiding molecules (plexins, slits and semaphorins) [73].

Excessive tissue growth, such as tumors, results in decreased oxygen tension that increases production of a number of genes, including VEGF [74,75]. A strong mitogen in vivo, VEGFA induces proliferation and permeability. Extravasation of plasma provides both matrix components, such as vitronectin, fibrinogen and fibronectin, and an additional cohort of growth factors. These include TGF-β, FGF and PDGF, all of which contribute to vascular growth and to further up-regulate VEGF expression [76]. The final outcome is an irregular and dysfunctional vascular plexus. Specifically, tumor blood vessels differ from their normal counterparts by altered morphology and blood flow, enhanced leakiness, abnormal pericytes and basement membrane [71]. Many of these phenotypes have been associated with excess of VEGF. In particular, vascular tortuosity, dilation and permeability mimic situations when VEGF has been locally delivered to an otherwise "normal" tissue [24,77,78].

In contrast to developmental vascular growth, activation of the endothelial cell in pathological conditions requires the digestion of a well-organized and cross-linked basement membrane. Thus, MMPs and their inhibitors are essential [79–81]. However, as could be expected, the system is extremely redundant. Multiple MMPs are able to perform the job, i.e., digest the basement membrane. In fact, genetic inactivation of most MMPs has been innocuous to post-natal angiogenesis [79–81]. MMPs also modulate exposure of cryptic extracellular matrix domains and regulate growth factor function. We have shown, for example, that MMP-mediated proteolytic processing of VEGF alters its association with the matrix and it induces distinct modes of vascular expansion [24]. Specifically, excess of MMPs result in VEGF cleavage, increasing the levels of its soluble form. This leads to the formation of highly tortuous and hyperplastic vessels that are unable to perfuse tissues with the same effectiveness as thin vessels [24].

The information gathered from development studies has significantly helped in generating therapeutic strategies for suppression of vascular growth that target, in particular, the activated endothelial cell. For example, suppression of VEGF through the monoclonal specific antibody bevacizumab has resulted in increased survival and reduction of tumor growth [82]. Excess of Semaphorin3F, as a means to modulate the function of the activated endothelial cell, has also been employed for vascular suppression [83,84].

The Notch pathway provides another example of harnessing signaling molecules towards therapeutic exploitation. Pharmacological suppression of Dll4 signaling in tumors results in a dramatic enhancement of tip cells unable to interconnect and organize functional vascular networks. The end result is poor blood perfusion and tumor mass

reduction [33,34,85]. It is interesting to consider that an enhancement in the number of tip cells (activated endothelial cells) could lead to such an outcome; i.e., vessel suppression in the context of excessive endothelial cells [86]. These results bring to light the exquisite balance between tip and stalk cells and their relevance to the organization of a functional vasculature.

An Alternative Mode for Endothelial Cell Activation: Mechanical Forces and Angiogenesis / Arteriogenesis

There is vast experimental support for the concept that endothelial cells can sense changes in blood flow and pressure [87–90]. More importantly, these physical forces appear to dynamically transmit this information to the cytoskeleton and surrounding extracellular matrix [91]. The relevance of this statement stems from the fact that the level of flow and shear stress can result in either an angiogenic or an arteriogenic event that is triggered at the time of endothelial cell activation [89,92,93]. The distinction lies in whether the resulting sprout will recruit smooth muscle cells (arteriogenic event) or remain as a single capillary, with or without pericytes (angiogenic event). Several studies have now demonstrated that multiple physical forces participate to maintain homeostatic balance in the vascular endothelium. They also serve to maintain endothelial responsiveness while preserving the integrity of the endothelial monolayer and barrier properties. Shear stress triggers arteriogenesis events, including remodeling of arterioarteriolar anastomoses and enlargement of vascular wall [94,96]. However, at which point do physical forces contribute to the angiogenic event? The answer to this question is not clear, and while not necessarily initiators, hemodynamic forces play a role in remodeling events during postnatal angiogenesis, although the molecular details remain unclear.

In addition to physical forces, it appears that the contribution of monocytes is required for arteriogenesis [97,98]. These cells release specific chemokines, growth factors and proteases that work to mediate vascular growth and contribute to the formation of new arterioles. The process occurs at sites of pre-existing arterio-arteriolar anastomoses [89]. The initial trigger appears to be altered shear stress within the collateral arteriole after an increase in blood flow. Subsequently, large pressure differences in pre-existing arterioles connecting up- and downstream leads to induction of cell proliferation, migration and vascular remodeling [99]. The increased diameter of collateral arterioles to arteries proceeds as an active growth rather than a passive dilatation [100,101].

Fluid forces also contribute to the primary triggering events associated with endothelial cell activation. Integrins, ion channels and tyrosine kinase receptors are the initial sensors for changes in physical forces [87,91,102]. The combination of initiating signaling events and transmission of information via the cytoskeleton to the nucleus culminates in the activation of a subset of shear stress responsive genes [103,104]. The cellular responses to shear stress include endothelial swelling and [102] and changes in the profile of cell surface/chemokine production that eventually result in recruitment of monocytes, as well as production of MMPs that initiate the digestion of the basement membrane [79,80].

Termination of the Angiogenic Endothelial Activation

Timely termination of the angiogenic response is as important as its initiation. A persistent or exaggerated angiogenic growth may lead to detrimental effects. Therefore, and in accordance with the complex and highly coordinated activation phase, negative regulatory processes have evolved and function at multiple levels to imposse termination of vascular sprouting.

Several mechanisms are operative in endothelial cells to shut down the activity of proangiogenic signaling pathways and transcription factors. Combined, they provide the stage for termination that is already set early in the activation phase of the angiogenic response. Unfortunately, little is known about these events, yet their further molecular elucidation might provide novel strategies for therapeutic intervention and suppression of vascular growth.

A key step during the termination phase is the formation of the basement membrane and the incorporation of pericytes and / or smooth muscle cells into the recently formed endothelial tubes (Fig. 3.2) [105,106]. Pericyte-induced stabilization appears to involve inhibitors of matrix metalloproteinases. In particular, endothelial cell-derived tissue inhibitor of metalloproteinase-2 (TIMP-2) and pericyte-derived TIMP-3 are shown to co-regulate human capillary tube stabilization following endothelial-pericyte interactions, through a combined ability to block tube morphogenesis and regression in three-dimensional collagen matrices [107]. TIMP-3 expression by pericytes is only induced upon association with endothelial cells. Blockade of TIMP3 leads to capillary tube regression, but it also requires MMP-1, MMP-10-, and ADAM-15 (a disintegrin and metalloproteinase-15). It has been demonstrated the proteinase inhibitory function of TIMP3 is essential for its capillary-stabilizing activity. These findings indicate that vascular networks are predisposed to undergo regression unless they acquire a vascular coat able to produce TIMP3 [107]. A large number of pharmacological studies concur with a key role of pericytes in vascular stability [105]. However, this vast cohort of data does not explain why certain capillaries remain highly stable in the absence of pericytes. Are there other cells responsible for this function or is the endothelium induced

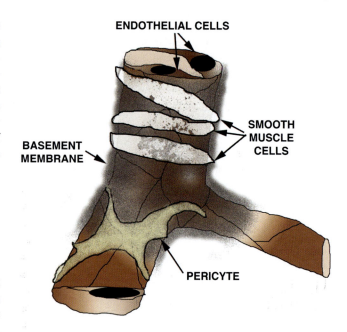

FIG. 3.2. Termination of the *endothelial cell* activation state requires the acquisition of a differentiated phenotype. In addition, the deposition of a highly organized *basement membrane* made by the contribution of both endothelial and mural cells (*smooth muscle* and *pericytes*) is an indication of vascular stability and prevents regression of newly formed vessels. The association of a coat of mural cells (common in most vessels) also prevents vascular regression and marks the end of the angiogenic cascade.

to secrete TIMP3 in the absence of pericytes? Clearly much remains to be understood within this particular step.

Concluding Thoughts

The consequences of vascular occlusion are devastating for organ function. Today, this pathology remains the most significant cause for morbidity and death in the industrialized world. Current therapies associated with myocardial infarction, stroke, and peripheral artery disease are limited to angioplasty, palliative interventions, and / or bypass. Therefore, a comprehensive understanding of how to modulate vascular growth in a manner that is appropriate for the resolution of a particular pathology can bring unequivocal value to a large number of diseases. Furthermore, the implementation of tissue engineering for wound healing and organ regeneration requires a sophisticated understanding of vessel growth and stabilization.

The last decade has marked the initiation of the molecular era in vascular biology. The advent of target genetic manipulation combined with the interdependency of the vascular system for embryonic survival have led to a remarkable expansion in our mechanistic understanding of how blood vessels are formed. Key genes and central signaling pathways have been identified, and the evolution of this field has allowed for the implementation of therapeutic strategies that aim at suppressing or enhancing the vasculature. Yet, much remains to be learned. While we can therapeutically induce vessels, they tend to be unstable or lack the hierarchic structure essential for function. On the other side of the coin, while we have been able to suppress neovascularization, the strategy is not as effective as predicted and it frequently leads to the development of resistance.

As we improve our knowledge of how endothelial cells are activated and guided to navigate in different tissues; we must strive to think therapeutically and translate the information into meaningful tools that would enable the modulation of vascular growth during pathological conditions.

Acknowledgements. Dr. Iruela-Arispe is supported by grants from the National Institutes of Health R01 HL074455, HL 085618 and CA126935.

Bibliography

1. Rossant J, Howard L. Signaling pathways in vascular development. Annu Rev Cell Dev Biol 2002; 18:541–573.
2. Iruela-Arispe. Vascular Development and Angiogenesis. In: Encyclopedia of Molecular Cell Biology. Ed: Robert A. Meyers 2005; 15:201–232.
3. Achen MG, Stacker SA. The vascular endothelial growth factor family; proteins which guide the development of the vasculature. Int J Exp Pathol, 1998; 79(5):255–265.
4. Matsumoto T, Claesson-Welsh L. VEGF Receptor Signal Transduction. Sci. STKE 2001; 112: 1223–1245.
5. Ferrara N, Carver-Moore K, Chen H et al. Heterozygous embryonic lethality induced by targeted inactivation of the VEGF gene. Nature. 1996; 380:439–442.
6. Carmeliet P, Ferreira V, Breier G et al. A.bnormal blood vessel development and lethality in embryos lacking a single VEGF allele. Nature. 1996 380:435–439.
7. Fong GH, Rossant J, Gertsenstein M et al. Role of the Flt-1 receptor tyrosine kinase in regulating the assembly of vascular endothelium. Nature 1995; 376:66–70.
8. Shalaby F, Rossant J, Yamaguchi TP et al. Failure of blood-island formation and vasculogenesis in Flk-1-deficient mice. Nature 1995; 376:62–66.
9. Lamalice L, Houle F, Huot J. Phosphorylation of Tyr[1214] within VEGFR-2 triggers the recruitment of Nck and activation of Fyn leading to SAPK2/p38 activation and endothelial cell migration in response to VEGF. J Biol Chem. 2006; 281(45): 34009–34020.
10. Lamalice L, Houle F, Jourdan G et al. Phosphorylation of tyrosine 1214 on VEGFR2 is required for VEGF-induced activation of Cdc42 upstream of SAPK2/p38. Oncogene 2004; 23(2): 434–445.
11. Rousseau S, Houle F, Huot J. Integrating the VEGF Signals Leading to Actin-Based Motility in Vascular Endothelial Cells. Trends Cardiovasc Med. 2000; 10(8):321–327.
12. Rousseau S, Houle F, Landry J et al. p38 MAP kinase activation by vascular endothelial growth factor mediates actin reorganization and cell migration in human endothelial cells. Oncogene 1997; 15(18):2169–2177.
13. Sakurai Y, Ohgimoto K, Kataoka Y et al. Essential role of Flk-1 (VEGF receptor 2) tyrosine residue 1173 in vasculogenesis in mice. Proc Natl Acad Sci USA 2005; 102(4): 1076–1081.
14. McMullen M, Keller R, Sussman M et al. Vascular endothelial growth factor-mediated activation of p38 is dependent upon Src and RAFTK/Pyk2. Oncogene 2004; 23(6): 1275–1282.
15. Meyer RD, Dayanir V, Majnoun F et al. The Presence of a Single Tyrosine Residue at the Carboxyl Domain of Vascular Endothelial Growth Factor Receptor-2/FLK-1 Regulates Its Autophosphorylation and Activation of Signaling Molecules. J Biol Chem 2002; 277(30): 27081–27087.
16. Holmqvist K, Cross MJ, Rolny C et al. The Adaptor Protein Shb Binds to Tyrosine 1175 in Vascular Endothelial Growth Factor (VEGF) Receptor-2 and Regulates VEGF-dependent Cellular Migration. J Biol Chem 2004; 279(21): 22267–22275.
17. Gerhardt H, Betsholtz C. How do endothelial cells orientate? EXS. 2005; 94:3–15.
18. Gerhardt H, Golding M, Fruttiger M et al. VEGF guides angiogenic sprouting utilizing endothelial tip cell filopodia. J Cell Biol 2003; 161(6):1163–1177.
19. Fruttiger M. Development of the retinal vasculature. Angiogenesis. 2007;10:77–88
20. West H, Richardson WD, Fruttiger M. Stabilization of the retinal vascular network by reciprocal feedback between blood vessels and astrocytes. Development. 2005; 132:1855–1862.
21. Chow J, Ogunshola O, Fan SY, et al. Astrocyte-derived VEGF mediates survival and tube stabilization of hypoxic brain microvascular endothelial cells in vitro. Brain Res Dev Brain Res. 2001 Sep 23;130(1):123–132.
22. Tischer E, Mitchell R, Hartman T et al. The human gene for vascular endothelial growth factor. Multiple protein forms are encoded through alternative exon splicing. J Biol Chem. 1991; 266(18):11947–11954.
23. Ruhrberg C, Gerhardt H, Golding M, et al. Spatially restricted patterning cues provided by heparin-binding VEGF-A control blood vessel branching morphogenesis. Genes Dev. 2002; 16:2684–2698.
24. Lee S, Jilani SM, Nikolova GV, et al. Processing of VEGF-A by matrix metalloproteinases regulates bioavailability and vascular patterning in tumors. J Cell Biol. 2005; 169:681–691.
25. Van der Akker NM, Molin DG, Peters PP, et al. Tetralogy of fallot and alterations in vascular endothelial growth factor-A signaling and notch signaling in mouse embryos solely expressing the VEGF120 isoform. Circ Res. 2007; 100:842–849.
26. Suchting S, Bicknell R and Eichmann A. Neuronal clues to vascular guidance. Exp Cell Res. 2006; 312: 668–675.
27. Alva JA, Iruela-Arispe ML. Notch signaling in vascular morphogenesis. Curr Opin Hematol. 2004; 11(4):278–283.
28. Gridley T. Notch signaling in vascular development and physiology. Development. 2007; 134:2709–2718.
29. Hofmann JJ, Iruela-Arispe ML. Notch signaling in blood vessels: who is talking to whom about what? Circ Res. 2007; 100: 1556–1568.
30. Gale NW, Dominguez MG, Noguera I et al. Haploinsufficiency of delta-like 4 ligand results in embryonic lethality due to major defects in arterial and vascular development. PNAS 2004; 101:15949–954.
31. Duarte A, Hirashima M, Benedito R et al. Dosage-sensitive requirement for mouse Dll4 in artery development. Genes Dev. 2004; 18:2474–2478.

32. Hellström M, Phng L-K, Hofmann JJ et al. Dll4 signalling through Notch1 regulates formation of tip cells during angiogenesis. Nature 2007; 15:776–780.

33. Suchting S, Freitas C, le Noble F, et al. The Notch ligand Delta-like 4 negatively regulates endothelial tip cell formation and vessel branching. Proc Natl Acad Sci USA. 2007; 104: 3225–3230

33. Noguera-Troise I, Daly C, Papadopoulos NJ et al. Blockade of Dll4 inhibits tumour growth by promoting non-productive angiogenesis. Nature 2006; 444:1032–1037.

34. Ridgway J, Zhang G, Wu Y et al. Inhibition of Dll4 signalling inhibits tumour growth by deregulating angiogenesis. Nature. 2006; 444:1083–1087.

35. Siekmann AF, Lawson ND. Notch signalling limits angiogenic cell behaviour in developing zebrafish arteries. Nature. 2007; 445:781–784.

36. Lobov IB, Renard AA, Papadopoulos N, et al. Delta-like ligand 4 (Dll4) is induced by VEGF as a negative regulator of angiogenic sprouting. Proc Natl Acad Sci USA. 2007; 104: 3219–3224.

37. Claxton S, Fruttiger M. Periodic Delta-like 4 expression in developing retinal arteries. Gene Expr Patterns. 2004; 5:123–127.

38. Hofmann JJ, Iruela-Arispe ML. Notch expression patterns in the retina: An eye on receptor–ligand distribution during angiogenesis. Gene Expr Patterns 2007; 7:461–467.

39. Conlon RA, Reaume AG, Rossant J. Notch1 is required for the coordinate segmentation of somites. Development 1995; 121:1533–1545.

40. Uyttendaele H, Ho J, Rossant J, Kitajewski J. Vascular patterning defects associated with expression of activated Notch4 in embryonic endothelium. PNAS 2001; 98:5643–5648.

41. Tsunematsu R, Nakayama K, Oike Y et al. Mouse Fbw7/Sel-10/Cdc4 is required for notch degradation during vascular development. J Biol Chem. 2004; 279:9417–9423.

42. Eichmann A, Makinen T, Alitalo K. Neural guidance molecules regulate vascular remodeling and vessel navigation. Genes Dev. 2005; 19:1013–1021.

43. Klagsbrun M, Eichmann A. A role for axon guidance receptors and ligands in blood vessel development and tumor angiogenesis. Cytokine Growth Factor Rev. 2005; 16:535–548.

44. Weinstein BM. Vessels and nerves: marching to the same tune. Cell. 2005; 120:299–302.

45. Bagri A, Tessier-Lavigne M. Neuropilins as Semaphorin receptors: in vivo functions in neuronal cell migration and axon guidance. Adv Exp Med Biol. 2002; 515:13–31.

46. Gu C, Yoshida Y, Livet J, et al. Semaphorin 3E and plexin-D1 control vascular pattern independently of neuropilins. Science 2005; 307:265–268.

47. Song H, Ming G, He Z, et al. Conversion of neuronal growth cone responses from repulsion to attraction by cyclic nucleotides. Science 1998; 281:1515–1518.

48. de Castro F, Hu L, Drabkin H, et al. Chedotal, Chemoattraction and chemorepulsion of olfactory bulb axons by different secreted semaphorins, J Neurosci. 1999; 19: 4428–4436.

49. Wong JT, Wong ST, O'Connor TP. Ectopic semaphorin-1a functions as an attractive guidance cue for developing peripheral neurons. Nat Neurosci. 1999; 2:798–803.

50. Bagnard D, Lohrum M, Uziel D, et al. Semaphorins act as attractive and repulsive guidance signals during the development of cortical projections. Development 1998; 125: 5043–5053.

51. Toyofuku T, Kikutani H. Semaphorin signaling during cardiac development. Adv Exp Med Biol. 2007; 600:109–117.

52. Kawasaki Y, Kitsukawa T, Bekku Y, et al. A requirement for neuropilin-1 in embryonic vessel formation, Development 1999; 126:4895–4902.

53. Soker S, Takashima S, Miao HQ, et al. Neuropilin-1 is expressed by endothelial and tumor cells as an isoform-specific receptor for vascular endothelial growth factor, Cell 1998; 92:735–745.

54. Gu C, Rodriguez E.R., Reimert DV, et al. Neuropilin-1 conveys semaphorin and VEGF signaling during neural and cardiovascular development, Dev Cell 2003; 5:45–57.

55. Miao HQ, Soker S, Feiner L, et al. Neuropilin-1 mediates collapsin-1/semaphorin III inhibition of endothelial cell motility: functional competition of collapsin-1 and vascular endothelial growth factor-165, J Cell Biol. 1999; 146:233–242.

56. Gagnon ML, Bielenberg DR, GechtmanZ, et al. Identification of a natural soluble neuropilin-1 that binds vascular endothelial growth factor: in vivo expression and antitumor activity, Proc Natl Acad Sci USA. 2000: 97:2573–2578.

57. Torres-Vazquez J, Gitler AD, Fraser SD, et al. Semaphorin–plexin signaling guides patterning of the developing vasculature, Dev Cell 2004; 7:117–123.

58. Chisholm A, Tessier-Lavigne M. Conservation and divergence of axon guidance mechanisms, Curr Opin Neurobiol. 1999; 9: 603–615.

59. Kennedy TE. Cellular mechanisms of netrin function: long-range and short-range actions, Biochem. Cell Biol. 2000; 78: 569–575.

60. Lu X, Le Noble F, Yuan L, et al. The netrin receptor UNC5B mediates guidance events controlling morphogenesis of the vascular system. Nature 2004; 432:179–186.

61. Serafini T, Colamarino SA, Leonardo ED, et al. Netrin-1 is required for commissural axon guidance in the developing vertebrate nervous system, Cell 1996; 87:1001–1014.

62. Park KW, Crouse D, Lee M, et al. The axonal attractant Netrin-1 is an angiogenic factor, Proc Natl Acad Sci USA. 2004; 101:16210–16215.

63. Brose K, Bland KS, Wang KH, et al. Slit proteins bind Robo receptors and have an evolutionarily conserved role in repulsive axon guidance. Cell 1999; 96:795–806.

64. Stein E, Tessier-Lavigne M. Hierarchical organization of guidance receptors: silencing of netrin attraction by slit through a Robo/DCC receptor complex, Science 2001; 291:1928–1938.

65. Huminiecki L, Gorn M, Suchting S, et al. Magic roundabout is a new member of the roundabout receptor family that is endothelial specific and expressed at sites of active angiogenesis, Genomics 2002; 79:547–552.

66. Park KW, Morrison M, Sorensen LK, et al. Robo4 is a vascular-specific receptor that inhibits endothelial migration, Dev Biol. 2003; 261:251–267.

67. Suchting, S, Heal P, Tahtis K, et al. Soluble Robo4 receptor inhibits in vivo angiogenesis and endothelial cell migration, FASEB J. 2005; 19:121–123.

68. Bedell VM,. Yeo SY, Park KW, et al. Roundabout4 is essential for angiogenesis in vivo, Proc Natl Acad Sci.USA. 2005; 102:6373–6378.

69. Kamei M, Saunders WB, Bayless KJ, et al. Endothelial tubes assemble from intracellular vacuoles in vivo. Nature 2006;442:453–446.

70. Ellis LM. Angiogenesis and its role in colorectal tumor and metastasis formation. Semin Oncol. 2004; 31:3–9.

71. Baluk P, Hashizume H, McDonald DM. Cellular abnormalities of blood vessels as targets in cancer. Curr Opin Genet Dev. 2005; 15:102–111.

72. Bussolati B, Deambrosis I, Russo S et al. Altered angiogenesis and survival in human tumor-derived endothelial cells. FASEB J, 2003; 17:1159–1161.

73. Neufeld G, Kessler O. Pro-angiogenic cytokines and their role in tumor angiogenesis. Cancer Metastasis Rev. 2006; 25: 373–385.

74. Gruber M, Simon MC. Hypoxia-inducible factors, hypoxia, and tumor angiogenesis. Curr Opin Hematol. 2006; 13:169–174

75. Pradeep CR, Sunila ES, Kuttan G. Expression of vascular endothelial growth factor (VEGF) and VEGF receptors in tumor angiogenesis and malignancies. Integr Cancer Ther. 2005; 4:315–321.

76. Sasano H, Suzuki T. Pathological evaluation of angiogenesis in human tumor. Biomed Pharmacother. 2005; 59 Suppl 2: S334–336.

77. Nagy JA, Vasile E, Feng D, et al. Vascular permeability factor/vascular endothelial growth factor induces lymphangiogenesis as well as angiogenesis. J Exp Med. 2002; 196:1497–1506.

78. Nagy JA, Vasile E, Feng D, et al. VEGF-A induces angiogenesis, arteriogenesis, lymphangiogenesis, and vascular malformations. Cold Spring Harb Symp Quant Biol. 2002; 67:227–237.

79. Genis L, Galvez BG, Gonzalo P, Arroyo AG. MT1-MMP: universal or particular player in angiogenesis? Cancer Metastasis Rev. 2006; 25:77–86.

80. van Hinsbergh VW, Engelse MA, Quax PH. Pericellular proteases in angiogenesis and vasculogenesis. Arterioscler Thromb Vasc Biol. 2006; 26:716–728.

81. Roy R, Zhang B, Moses MA. Making the cut: protease-mediated regulation of angiogenesis. Exp Cell Res. 2006; 312: 608–622.

82. Ferrara N, Hillan KJ, Gerber HP et al. Discovery and development of bevacizumab, an anti-VEGF antibody for treating cancer. Nat Rev Drug Discov. 2004; 3:391–400.

83. Kaipainen K, Kreuter M, Kim CC, et al. Semaphorin 3 F, a chemorepulsant for endothelial cells, induces a poorly vascularized, encapsulated, nonmetastatic tumor phenotype, J. Clin Invest. 2004; 114:1260–1271.

84. Kessler O, Shraga-Heled N, Lange T, et al. Semaphorin-3 F is an inhibitor of tumor angiogenesis, Cancer Res. 2004; 64: 1008–1015.

85. Scehnet JS, Jiang W, Kumar SR, et al. Inhibition of Dll4-mediated signaling induces proliferation of immature vessels and results in poor tissue perfusion. Blood. 2004; 109: 4753–4760.

86. Thurston G, Noguera-Troise I, Yancopoulos GD. The Delta paradox: DLL4 blockade leads to more tumour vessels but less tumour growth. Nat Rev Cancer. 2007; 7:327–331.

87. Gimbrone MA Jr, Nagel T, Topper JN. Perspectives Series: Cell Adhesion in Vascular Biology- Biomechanical Activation: an Emerging Paradigm in Endothelial Adhesion Biology. J Clin Invest. 1997; 99(8):1809–1813.

88. Heil M, Schaper W. Insights into pathways of arteriogenesis. Curr Pharm Biotechnol. 2007; 8(1): 35–42.

89. Heil M, Schaper, W. Cellular mechanisms of arteriogenesis, in Clauss M, Breier G. (eds): Mechanisms of Angiogenesis. Basel: Birkhauser; 2005: 181–191.

90. le Noble F, Fleury V, Pries A et al. Control of arterial branching morphogenesis in embryogenesis: go with the flow. Cardiovasc Res. 2005; 65:619–628.

91. Davies PF, Barbee KA, Volin MV et al. Spatial Relationships in Early Signaling Events of Flow-Mediated Endothelial Mechanotransduction. Annu Rev Physiol. 1997; 59:527–549.

92. Peirce SM, Skalak TC. Microvascular Remodeling: A Complex Continuum Spanning Angiogenesis to Arteriogenesis. Microcirculation 2003; 10:99–111.

93. Resnick N, Yahav H, Shay-Salit A et al. Fluid shear stress and the vascular endothelium: for better and for worse. Prog Biophys Mol Biol. 2003; 81(3):177–199.

94. Ito WD, Arras M, Scholz D et al. Angiogenesis but not collateral growth is associated with ischemia after femoral artery occlusion. Am J Physiol Heart Circ Physiol 1997; 273: H1255–1265.

95. Skalak TC, Price RJ. The role of mechanical stresses in microvascular remodeling. Microcirculation. 1996; 3:143–165.

96. Wei-Jun C, Elisabeth K, Xiaoqiong W et al. Remodeling of the vascular tunica media is essential for development of collateral vessels in the canine heart. Mol Cell Biochem 2004; 264:201–210.

97. Arras M, Ito WD, Scholz D et al. Monocyte Activation in Angiogenesis and Collateral Growth in the Rabbit Hindlimb. J Clin Invest. 1998; 1:40–50.

98. Heil M, Ziegelhoeffer T, Pipp F et al. Blood monocyte concentration is critical for enhancement of collateral artery growth. Am J Physiol Heart Circ Physiol *2002;* 283:H2411–2419.

99. Ito WD, Arras M, Winkler B et al. Monocyte Chemotactic Protein-1 Increases Collateral and Peripheral Conductance After Femoral Artery Occlusion. Circ Res. 1997; 80:829–37.

100. Kusch A, Tkachuk S, Lutter S et al. Monocyte-expressed urokinase regulates human vascular smooth muscle cell migration in a coculture model. Biol Chem. 2002; 383(1):217–221.

101. Schaper J, König R, Franz D et al. The endothelial surface of growing coronary collateral arteries. Intimal margination and diapedesis of monocytes. Virchows Archiv. 1976; 370(3):193–205.

102. Barakat AI. Responsiveness of vascular endothelium to shear stress: Potential role of ion channels and cellular cytoskeleton. Intl J of Mol Med 1999; 4:323–332.

103. Ingber DE. Cellular Basis of Mechanotransduction. Biol Bull. 1998; 194:323–327.

104. Topper JN, Gimbrone MA Jr. Blood flow and vascular gene expression: fluid shear stress as a modulator of endothelial phenotype. Mol Med Today. 1999; 5:40–46.

105. Bergers G, Song S. The role of pericytes in blood-vessel formation and maintenance. Neuro Oncol. 2005; 7:452–464.

106. Armulik A, Abramsson A, Betsholtz C. Endothelial/pericyte interactions. Circ Res. 2005; 97:512–523.

107. Davis GE, Saunders WB. Molecular balance of capillary tube formation versus regression in wound repair: role of matrix metalloproteinases and their inhibitors. J Investig Dermatol Symp Proc. 2006; 11: 44–56.

Chapter 4
Pericytes, the Mural Cells of the Microvascular System

Gabriele Bergers

Keywords: Pericytes, mural cells, vascular smooth muscle cells, vascular stability and maturation, microvascular homeostasis, angiogenesis, vasculogenesis, diabetic retinopathy, tumorigenesis, PDGFRβ, TGFβ, angiopoietins

Abstract: Endothelial cells and pericytes regulate blood vessel formation, maturation and specification, all of which requires the orchestration of tightly regulated molecules. Communication between these two distinct vascular cell types occurs by direct cell contact and by paracrine signaling pathways. Pericytes and endotheslial cells are interdependent and defects in either can affect the vascular system. Loss of pericytes can lead to hyperdilated and hemorrhagic blood vessels, which lead to conditions such as edema, diabetic retinopathy, and even embryonic lethality. In tumors, although pericytes are less abundant and more loosely attached, pericyte dysfunction can result in increased endothelial cell apoptosis and metastatic spread, providing evidence that tumor pericytes are implicated in vessel maintenance, endothelial cell survival and potentially tumor dormancy. Based on their functional importance, pericytes present a complimentary target to endothelial cells in tumors. Therefore, combinatorial targeting of both cell types might have the potential to more efficiently diminish tumor vessels and halt subsequent tumor growth.

Characteristics of Pericytes

Blood vessels consist of endothelial cells that form the inner lining of the vessel wall and of perivascular cells that wrap around blood vessels. Charles Rouget described perivascular cells more than 100 years ago in amphibia and named them pericytes, i.e., cells that envelop blood capillaries (*peri*, around; *cyte*, cell). Pericytes have also been referred to as Rouget cells, mural cells, and, because of their contractile fibers, vascular smooth muscle cells (vSMCs) [1]. A hallmark of pericytes is their location within the basement membrane of capillaries, postcapillary venules, and collecting venules. Pericytes possess a cell body with a prominent nucleus and envelop the abluminal endothelial wall with several of their long cytoplasmic processes (Fig. 4.1) [2]. Thereby, pericytes make focal contacts with numerous endothelial cells through specialized junctions to integrate signals along the length of the vessel, but can also extend to more than one capillary in the vasculature (Fig. 4.1). Gap junctions provide direct connections between the cytoplasm of pericytes and endothelial cells, and they enable the exchange of ions and small molecules. Adhesion plaques anchor pericytes to endothelial cells, while peg-and-socket contacts enable the cells to penetrate through discontinuities in the vessel basement membrane and touch each other [3, 4]. These junction complexes support transmission of mechanical contractile forces from the pericytes to the endothelium and contain N-cadherin, cell-adhesion molecules, ®-catenin-based adherent junctions, and extracellular matrix (ECM) molecules such as fibronectin [5]. Interestingly, cell–cell contact appears necessary for the activation of the latent growth factor TGF-β1, which induces pericyte differentiation in vitro [6], supporting the notion that direct cell contact is a crucial communication tool for vessel maintenance and formation.

Identification of Pericytes

Besides identifying pericytes by their distinct localization within the vascular basement membrane, pericytes can also be visualized with various dynamic molecular markers that, although not exclusively, detect pericytes in a tissue-specific manner or based on the developmental or angiogenic stage of the organ [7]. Desmin and α-smooth-muscle actin

Department of Neurological Surgery, Brain Tumor Research Center and UCSF Comprehensive Cancer Center, University of California San Francisco, 513 Parnassus Avenue, San Francisco, CA 94143-0520, USA

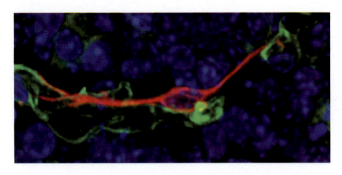

FIG. 4.1. Pericytes possess a cell body with a prominent nucleus and long cytoplasmic processes that envelop the surface of the vascular tube and extend to neighboring capillaries. Blood vessels were visualized with FITC-labeled tomato lectin (green), and pericytes were stained with red-labeled antibodies for NG2.

(α-SMA) are contractile filaments. While desmin is also expressed in mature skeletal, cardiac and smooth-muscle cells [8, 9], α-SMA is more restricted to the smooth-muscle cell lineage and myofibroblasts [10]. NG2, a chondroitin sulfate proteoglycan, and PDGFRβ, a tyrosine-kinase receptor, are cell-surface proteins. The name NG2 stems from the identification of this molecule on immature neural cells that can differentiate into either neurons or glial cell (the human equivalent is named high-molecular-weight-melanoma associated antigen (HMWAAA). Congruently, NG-2 is expressed in glial precursor O-2A cells that give rise to oligodendrocytes or type II astrocytes which turn down NG2 expression [11]. Interestingly, NG-2 displays angiogenic properties by binding angiogenic regulators such as basic-fibroblast growth factor (b-FGF), PDGF-AA and angiostatin [12]. In agreement, NG-2 knockout mice exhibit signs of reduced capabilities to promote neovascularization when tissues are substantiated to ischemic and hypoxic conditions [12]. The receptor tyrosine kinase PDGFRβ (platelet-derived growth factor receptor β) is expressed in a variety of stromal cells, including developing pericytes and pericyte progenitors, but is down-regulated in mature pericytes [13, 14]. The functional importance of PDGFRβ signaling in pericytes is underscored by the fact that mice deficient in PDGFRβ or its ligand PDGF-B, have severely reduced numbers of pericytes and subsequent hyperdilation of blood vessels, which causes edema formation and embryonic lethality [15, 16]. Taking advantage of the fact that developing blood vessels in the central nervous system are almost completely devoid of pericytes in PDGF-B deficient embryos, additional pericyte markers were recently identified by gene expression profiling from isolated brain microvascular fragments of PDGF-BB deficient embryos and the wild-type counterparts [17, 18]. Among the differentially expressed factors that appeared to be expressed in pericytes are RGS-5, a GTPase-activating protein, Kir6.1 and SUR2, which form a hetero-octameric ATP-sensitive potassium channel, and DLK1/Pref-1, which is one of several membrane–bound ligands for notch receptors. Notably, RGS-5 expression is induced during pathological and physiological angiogenesis in the adult [19], and like Kir6.1, which is expressed specifically in brain pericytes [17], is also more restricted to certain tissues such as brain and pancreas [7].

Pericytes in Microvascular Homeostasis

Historically, pericytes have been associated with stabilization of microvessels, although the molecular mechanisms by which pericytes enable vessel stability are not well understood. Because capillaries lack smooth-muscle cells, it has been believed that blood flow is regulated by precapillary arterioles. However, recent data have demonstrated that pericytes can control capillary diameter in whole retina and cerebellar slices, and that blood flow control is indeed initiated in capillaries [20]. In support of these findings, pericytes are known to express contractile proteins such as α-SMA, tropomyosin and myosin. Moreover, several regulators have been identified that control pericyte-dependent vasoconstriction and vasodilation of capillaries [4]. Pericytes, for instance, express cholinergic and adrenergic (α-2 and β-2) receptors. While the β-adrenergic response leads to relaxation in pericytes, α-2 responses initiate contraction. Another molecule with vasoactive properties is endothelin-1 which, like the vasodilator nitric oxide, is induced in endothelial cells [21–23]. Thus, these molecules regulate the pericyte contractile tone by paracrine signaling circuits and therefore support the notion that endothelial cells and pericytes interact in the regulation of blood flow [4]. Notably, oxygen levels are also capable of regulating pericyte contraction because hyperoxia increases contraction whereas elevated levels of carbon oxide, indicating low oxygen tension, initiate relaxation of pericytes. Thus, the rate of blood flow appears to be directly coupled to the metabolic state of the tissue [1, 24].

Tissue-specific Functions of Pericytes

Pericytes are not randomly distributed around microvessels but are positioned in a regional and tissue-specific manner. They are more abundant on small venules and arterioles than on capillaries and appear to preferentially conceal endothelial cell junctions [25]. Pericytes have also acquired specialized characteristics in brain, liver, and kidney based on the specific function of these organs. It is believed that pericytes in the brain are essential for vessel integrity, and form a blood-brain-barrier with endothelial cells and astrocytes to protect brain cells from potentially toxic blood-derived factors [26, 27]. Therefore, it is not surprising that the highest pericyte density in the body has been observed in neural tissues such as brain and the retinas. Interestingly, it has been postulated that pericytes are progenitors of brain macrophages because, like macrophages, pericytes have pinocytic and phagocytic activities [28]. Further support stems from the observations that pericytes possess numerous macrophage markers, such as CD4 and CR3 complement receptors and Fc receptors, which are essential for antibody-antigen complex recognition to trigger antibody-dependent phagocytosis [29]. In the kidney, pericytes of the glomerular capillaries are referred to as mesengial cells. They are essential in the proper formation of glomeruli because they enable intussusceptive branching of a single vascular loop into several glomerular capillaries to generate a large capillary surface area for blood ultrafiltration [30]. Pericytes in the liver are named hepatic stellate cells (HSC) or Ito cells (after their discoverer, Toshio Ito) [31]. They are found between the parenchymal cell plates and sinusoidal endothelial cells. Because liver endothelial cells line the hepatic sinusoids to mediate processing and exchange of metabolites between the portal blood, Kupffer cells, and hepatocytes, a dense basement structure between hepatic epithelial cells and endothelial cells does not exist, but pericytes/HSC are still in direct contact with endothelial cells through incomplete basement-membrane components and interstitial collagen fibers [32]. Interestingly, pericytes/HSC are implicated in Vitamin A metabolism and contain more than 80% of the total vitamin A in the body [32].

Pericytes in Vascular Development and Angiogenesis

Pericytes are not only involved in microvascular homeostasis, but are also instrumental in embryonic and adult blood vessel formation. Blood vessels develop early in the embryo from mesodermal (angioblasts) or hematopoietic precursors (hemangioblasts), and first assemble into a primary capillary plexus in a process referred to as vasculogenesis [33, 27]. In contrast to endothelial cells, pericytes have a complex ontogeny, because they can develop from various cells as a function of their location in the embryo. They can develop from the neurocrest in the forebrain and cardiac tracts [34, 35] or more commonly originate from mesenchymal stem cells [36]. TGF-β1 appears to be instrumental for the de novo induction of vSMC/pericytes by regulating differentiation of pericyte progenitors. TGF-β1 initiates differentiation of PDGFRβ^+ pericyte progenitor cells that are then chemotactically attracted by PDGF-B-secreting endothelial cells in the capillary plexus [37], and it facilitates differentiation of neurocrest- or mesenchyme-derived progenitors into smooth-muscle-like cells [38, 39]. Genetic disruption of TGFβ1 (Dickson et al., 1995) or genes encoding its receptors such as TGF-β receptor II [41] and endoglin (type III TGFβ-receptor) [42] result in cardio-vascular failure with vSMC/pericyte differentiation defects and subsequent embryonic lethality. The primary defects, however, seem to occur in the endothelium. This is supported by the observation that disturbed TGFβ signaling in endothelial cells in the embryo affects TGFβ signaling in the adjacent mesenchymal progenitor cells by blocking their differentiation into vSMC/pericytes [43]. There have also been findings demonstrating that endothelial cells and pericytes can arise from a common vascular endothelial growth factor receptor 2$^+$ (VEGFR2$^+$) vascular-progenitor. When these cells were derived from embryonic stem cells, they were able to differentiate into endothelial cells in the presence of VEGF or into vascular smooth-muscle cells when PDGF-B is added [44, 45]. In addition, pericytes have been found to transdifferentiate from endothelial cells in a TGF-β3 dependent manner in the dorsal aorta [46] and cardiac valves [47]. It is important to note that pericytes themselves appear to have the potential to differentiate into other cell types, such as osteoblasts, chondrocytes and adipocytes, and as described above into macrophages [48–50, 28]. More recently, pericytes from human skeletal muscle blood vessels were shown to give rise to muscle fibers in vitro [51].

When the primitive capillary plexus has been assembled in the embryo, it is refined into a functional network by angiogenesis, undergoing extensive remodeling in which endothelial sprouting, intussusception, and pruning ensures appropriate vascularization of growing organs. In vessel sprouting, endothelial cells invade the surrounding extracellular matrix and form a column consisting of proliferating stack cells and, at the very tip, a migrating tip endothelial cell which guides the column toward a VEGF gradient [52]. Studies of the corpus luteum have suggested that pericytes are also capable of guiding sprouting processes by migrating ahead of endothelial cells and expressing VEGF [53–55]. As soon as new vessels are formed, endothelial cells secrete factors, partly to recruit pericytes that promote vessel maturation. Thereby, pericytes induce endothelial cell differentiation and growth arrest [5, 56]. The most prominent and crucial recruitment factor is platelet-derived growth factor PDGF-B, which signals through its receptor PDGFR-β expressed by pericytes, resulting in proliferation and recruitment of pericytes to the newly formed vascular tube. Based on the paracrine signaling circuit, genetic ablations of PDGF-B or PDGFRβ in mice produce identical phenotypes, and reveal the significant role of PDGFRβ signaling in pericyte proliferation and recruitment to blood vessels. PDGF-B- or PDGFRβ-mutants die during late gestation from cardiovascular complications, including widespread microvascular leakage and edema, arterial smooth-muscle cell hypoplasia, and abnormal kidney glomeruli [13, 15, 16, 37, 57]. Blood-vessel dilation and microaneurysms in mutant embryos correlate with severe reduction or even total loss of pericytes on the affected vessels, most prominently in the brain and heart [37]. Notably, PDGF-B expression in the endothelium is essential for proper pericyte coverage on blood vessels, because pericyte deficiency is still observed when PDGF-B is deleted in the endothelial cells of mice [58]. The lack of pericytes due to disruption in PDGF-B/PDGFRβ signaling appears not to be caused by impaired development of pericytes, but the cells are unable to expand and spread along the newly formed vessels because of their reduced proliferative capability and, likely, reduced migratory capability [59]. Specific mutations in the PDGFR-β gene in mice, leading to disruption of distinct and different signaling pathways, revealed that PDGFRβ-dependent signaling circuits are rather additive than specific for pericyte recruitment [60].

Another reciprocal communication pathway that is instrumental in vessel stabilization includes the angiopoietins and their Tie2 receptor. In this signaling circuit, perivascular cells and pericytes secrete the ligand angiopoietin-1 (Ang1), while the receptor tyrosine kinase tie2 is preferentially expressed in endothelial cells [61, 62]. Similar to the observations in PDGF-B- and PDGFRβ-deficient mice, Ang1 and Tie2 knockout mice are embryonic lethal and die from cardiovascular failure, displaying instable blood vessels with a poorly organized vascular membrane and severely reduced pericyte coverage [63, 64, 62]. It is not well understood, however, whether loss of pericyte coverage is a direct or indirect defect, due to the inability of the endothelium to produce pericyte-recruiting factors, because Tie2 is also found on cultured aorta-derived vSMC cells that respond to Ang1 [3, 65]. Overexpression of Ang1 in mice leads to increased vascularization, but blood vessels are mature, stabilized and leakage-resistant [66, 67]. Further rationale for Ang1 as a microvessel stabilizing factor stems from the observation that recombinant Ang1 is sufficient to induce vessel maturation in pericyte-deficient blood vessels in the retina [68]. Angiopoietin-2 (Ang2), on the other hand, has been proposed as an antagonistic ligand for Tie2 because it drives destabilization of blood vessels and loosens the direct contacts between pericytes and endothelial cells. Overexpression of Ang2 in mice recapitulates the defects observed in Ang1 and Tie2 knockout mice, whereas Ang2 deficiency has no impact on vascular development in the embryo and is more required in angiogenesis and intestinal lymphangiogenesis in the adult [69, 70]. Up-regulation of Ang2 in tumors and administration of recombinant Ang2 in the retina leads to reduction in pericyte coverage [71, 72]. Based on these findings, it has been postulated that Ang2 promotes blood-vessel growth and sprouting in the presence of VEGF, whereas when VEGF is absent, Ang2 leads to endothelial cell death and vessel regression [70, 73]. In contrast to Ang1, Ang2 is expressed in endothelial cells indicative of an autocrine signaling pathway, and therefore its effects on pericyte coverage are very likely to be a secondary event [3]. Activating mutations in the human Tie2 gene have been identified that cause vascular dysmorphogenesis and abnormal vSMC coverage [74].

Vessel maturation and stability is also affected by $S1P_1$, a G-protein-coupled receptor of sphingosine-1-phosphate (S1P) that affects proliferation, survival, and migration of cells [75]. $S1P_1$ deficiency in mice causes embryonic lethality due to aberrant recruitment of vSMCs in the aorta. This defect, however, occurs only when endothelial cells, but not vSMCs, are depleted of $S1P_1$, indicating that vSMC defect is secondary to endothelial dysfunction [76].

In summary, the studies described above support the notion that pericytes and endothelial cells intimately communicate by paracrine signaling circuits and direct cell contacts to facilitate vascular growth and homeostasis.

Historically, the formation of new vessels has been solely associated with activation of existing endothelial cells within the injured tissue or tumor. However, several reports describe that angiogenesis in the adult is also supported by the mobilization and functional incorporation of bone marrow-derived circulating endothelial progenitor cells (CEP) and by the recruitment of pericyte progenitors and other accessory bone marrow-derived cells [77–79]. One of the first studies that proposed the existence of bone marrow-derived pericyte progenitors found CD11b+ and CD45+ hematopoietic cells, expressing the pericyte marker NG2, in close proximity to blood vessels in a subcutaneous Bl6-F1 melanoma model [80]. Further support of bone marrow-derived pericyte progenitor cells has been substantiated by the identification of bone marrow-derived PDGFR-β+/Sca-1+pericyte progenitors (PPPs) in an endogenous mouse model of pancreatic islet cell tumorigenesis [14]. PPPs in these tumors were able to differentiate into mature pericytes expressing the markers NG2, a-SMA, and desmin, as well as regulate vessel stability and endothelial cell survival of tumors [14]. Bone marrow-derived pericyte progenitors were also identified in the brain after middle cerebral artery occlusion in an experimental mouse model of stroke. PPPs were observed around growing blood vessels in ischemic areas and developed into desmin+ pericytes that express TGF-β and VEGF [81]. Furthermore, bone marrow-derived cells were observed in angiogenic vessels of the cornea after b-FGF- (basic-fibroblast growth factor) induced neovascularization, and differentiated into NG2+/ PDGFR-β+ perivascular cells [82]. Taken together, there is emerging evidence that a subset of bone marrow-derived cells are recruited to active sites of vascular remodeling to differentiate into pericytes and thereby facilitate vessel maturation of growing vessels.

Pericytes in Vascular Disease

The vasculature is usually quiescent in the adult, but becomes activated during wound healing and in the female reproductive system during the menstrual cycle and pregnancy. In addition, angiogenesis is also initiated under pathological conditions like diabetic retinopathy and tumor growth. In contrast to physiological neovascularization, in which newly formed vessels rapidly mature, become stable, and cease proliferation, blood vessel growth under pathological conditions loses the appropriate balance between proangiogenic and antiangiogenic factors [83–86]. As a consequence, blood vessels during pathological angiogenesis do not stop growing and are under constant reconstruction, leading to an aberrant vascular system.

Pericytes in Diabetic Retinopathy

Diabetic retinopathy is a leading cause of blindness, and an early hallmark of this condition is the loss of pericytes in the retina [87], likely to be caused by chronic hyperglycemia. Consequently, capillaries become hyperdilated and form microaneurysms [87, 88]. Ultrastructural analyses of retinal microaneurysms reveal a consistent absence of pericytes, suggesting that the loss of vessel integrity due to the absence of pericytes may indeed render vessels vulnerable to aneurysms. When progressive vascular occlusions in the human diabetic eye lead to blindness, the retina responds with either a progressive increase of vascular permeability, leading to macula edema, or the formation of new proliferating immature vessels [87, 89, 90]. Congruently, increased expression of VEGF-A and its receptors has been demonstrated in diabetic retinas [91]. VEGF-A is also very likely responsible for the induced vascular leakage, and antagonists of VEGF and its receptors have been shown to reduce retinopathy in animal models [92–94]. Although the underlying mechanisms of pericyte loss in the retina are still unknown, PDGFB$^{ret/ret}$ mice that show impaired PDGF-B signaling when they are adults also develop retinopathy concomitant with severe pericyte loss [30]. PDGF-B$^{ret/ret}$ mice lack the C-terminal retention motif in PDGF-B that mediates PDGF-B binding to proteoglycans at the cell surface and in the ECM. These mice are viable but contain fewer pericytes, as they lack proper recruitment and integration of pericytes within the vessel wall, particularly in the retina and kidney [95]. Pharmacological inhibition of PDGF signaling with Gleevec (imatinib) promoted pericyte apoptosis and exacerbated angiogenesis by further inducing VEGF and VEGFR2 in the rodent ROP model [88]. The correlation between lack of pericytes and onset of neovascularization in retinopathy, as well as the pericyte association with developing capillaries and cessation of vessel growth, supports the concept that pericytes have suppressive influence on capillary growth.

Tumor Pericytes

Angiogenesis in tumors leads to tumor vessels with multiple functional and structural abnormalities. Tumors consist of a chaotic, poorly organized vasculature with tortuous, irregularly shaped, and leaky vessels that are often unable to support efficient blood flow [96, 97]. The vasculature appears to be in a constant state of remodeling that involves simultaneous formation and regression of vascular tubes. It is believed that the imbalance of pro- and antiangiogenic factors and of endothelial-pericyte signaling circuits is the main cause for endothelial cells and pericytes to move into a continuously

activated state and exhibit atypical behavior [83, 98]. As a result, tumor pericytes have abnormal shapes, are less abundant, and are rather loosely attached to endothelial cells, even extending cytoplasmic processes away from the vessel wall and into the tumor stroma [97, 99]. Several genetic and pharmacological studies convey that loss or even impaired attachment of pericytes has functional consequences, including instability of vessels, higher risk of hemorrhage formation, and even increased metastatic potential. Implanted tumors in PDGF-B$^{ret/ret}$ mice, that lack proper recruitment and integration of pericytes within the vessel wall, contain hyperdilated and hemorrhagic blood vessels with very little pericyte coverage [98, 100]. Ectopic expression of PDGF-B in tumor cells has been shown to facilitate an increase in pericyte density, but is unable to cause pericytes to attach more firmly to blood vessels. This further supports the notion that PDGF-B produced by endothelial cells, and retained by heparan sulfate, is essential for proper pericyte adhesion to the vessel wall [100–102]. Alternatively, ablation of pericytes with a neutralizing antibody against PDGFRβ in pancreatic islet tumors of the Rip1Tag2 transgenic mice also causes hyperdilation of tumor vessels and increased endothelial cell apoptosis [14]. Similarly, the receptor tyrosine kinase (RTK) inhibitor SU6668, which also affects PDGFRβ signaling, detaches and diminishes pericytes in Rip1Tag2 and xenotransplant tumors and thereby restricts tumor growth [103, 104]. It is important to note that pericytes are a major source of the endothelial survival factor VEGF [14, 105]. Therefore, depletion of pericytes might not only directly affect vessel stability, but also survival. Indeed, blood vessels are more dependent on VEGF when they are depleted of pericytes because they are selectively eliminated when VEGF is withdrawn from the tumors [105]. In addition to affecting primary tumor growth, recent studies of pancreatic β-islet cell tumorigenesis in PDGFRb$^{ret/ret}$ mice reveal that pericyte dysfunction can also lead to metastases in distant organs and local lymph nodes [106]. In agreement with these findings, absence of α-SMA-positive pericyte coverage of tumor vessels correlates with metastastic spread and poor prognosis in patients with colorectal cancer [107]. Further evidence for the significance of pericytes in metastatic spread was obtained by gene expression profiling of various primary tumors that revealed a 17-gene molecular signature of metastasis, in which four of nine downregulated genes were markers of smooth-muscle cells (actinγ2, myosin light chain kinase, myosin heavy chain 11, and calponin h1) [108].

Taken together, these results suggest that tumors use the same signal mechanisms that are used in developmental angiogenesis. They further underscore the functional significance of tumor pericytes which, although less abundant and more loosely attached than normal pericytes, are still instrumental in regulating vessel integrity, maintenance, and function. Importantly, given that tumor vessels devoid of pericytes appear to be more vulnerable, it is likely that they may be more responsive to anti-endothelial drugs. In support of this hypothesis, combinations of RTK inhibitors that target endothelial cells and pericytes by blocking VEGF and PDGF signaling, respectively, more efficiently diminished tumor blood vessels and tumors than any of the inhibitors individually [83]. Similarly, PDGF inhibitors in combination with an antiangiogenic chemotherapy regimen that targeted endothelial cells provide similar benefits [109]. Targeting PDGFR signaling disrupts pericyte–endothelial cell interaction, while the antiangiogenic chemotherapy targets the sensitized endothelial cells, collectively destabilizing and regressing the pre-existing tumor vasculature. This information provides evidence that pericytes are potentially important and functional vascular cell components in tumors that elicit survival mechanisms to establish and maintain tumor vessels. Therefore, combinations of anti-pericyte and anti-endothelial drugs might act synergistically in antiangiogenic therapy and improve outcome.

Acknowledgements. This work was supported by a grant from the National Institutes of Health (RO1 CA109390). I thank Ilona Garner for help with the manuscript preparation.

References

1. Hirschi, K. K., and D'Amore, P. A. (1996). Pericytes in the microvasculature. Cardiovasc Res *32*, 687.

2. Mandarino, L. J., Sundarraj, N., Finlayson, J., and Hassell, H. R. (1993). Regulation of fibronectin and laminin synthesis by retinal capillary endothelial cells and pericytes in vitro. Exp Eye Res *57*, 609–621.

3. Armulik, A., Abramsson, A., and Betsholtz, C. (2005). Endothelial/pericyte interactions. Circ Res *97*, 512–523.

4. Rucker, H. K., Wynder, H. J., and Thomas, W. E. (2000). Cellular mechanisms of CNS pericytes. Brain Res Bull *51*, 363–369.

5. Gerhardt, H., and Betsholtz, C. (2003). Endothelial-pericyte interactions in angiogenesis. Cell Tissue Res.

6. Orlidge, A., and D'Amore, P. A. (1987). Inhibition of capillary endothelial cell growth by pericytes and smooth muscle cells. J Cell Biol *105*, 1455–1462.

7. Bergers, G., and Song, S. (2005). The role of pericytes in blood-vessel formation and maintenance. Neuro-oncol *7*, 452–464.

8. Li, Z., Colucci-Guyon, E., Pincon-Raymond, M., Mericskay, M., Pournin, S., Paulin, D., and Babinet, C. (1996). Cardiovascular lesions and skeletal myopathy in mice lacking desmin. Dev Biol *175*, 362–366.

9. Milner, D. J., Weitzer, G., Tran, D., Bradley, A., and Capetanaki, Y. (1996). Disruption of muscle architecture and myocardial degeneration in mice lacking desmin. J Cell Biol *134*, 1255–1270.

10. Ronnov-Jessen, L., and Petersen, O. W. (1996). A function for filamentous alpha-smooth muscle actin: retardation of motility in fibroblasts. J Cell Biol *134*, 67–80.

11. Stallcup, W. B. (2002). The NG2 proteoglycan: past insights and future prospects. J Neurocytol *31*, 423–435.

12. Ozerdem, U., and Stallcup, W. B. (2004). Pathological angiogenesis is reduced by targeting pericytes via the NG2 proteoglycan. Angiogenesis *7*, 269–276.

13. Lindahl, P., Johansson, B., Leveen, P., and Betsholtz, C. (1997). Pericyte loss and microaneurysm formation inPDGF-B-deficient mice. Science *126*, 3047–3055.

14. Song, S., Ewald, A. J., Stallcup, W., Werb, Z., and Bergers, G. (2005). PDGFRbeta+ perivascular progenitor cells in tumours regulate pericyte differentiation and vascular survival. Nat Cell Biol *7*, 870–879.

15. Leveen, P., Pekny, M., Gebre-Medhin, S., Swolin, B., Larsson, E., and Betsholtz, C. (1994). Mice deficient for PDGF B show renal, cardiovascular, and hematological abnormalities. Genes Dev *8*, 1875–1887.

16. Soriano, P. (1994). Abnormal kidney development and hematological disorders in PDGF beta-receptor mutant mice. Genes Dev *8*, 1888–1896.

17. Bondjers, C., He, L., Takemoto, M., Norlin, J., Asker, N., Hellstrom, M., Lindahl, P., and Betsholtz, C. (2006). Microarray analysis of blood microvessels from PDGF-B and PDGF-Rbeta mutant mice identifies novel markers for brain pericytes. Faseb J *20*, 1703–1705.

18. Bondjers, C., Kalen, M., Hellstrom, M., Scheidl, S. J., Abramsson, A., Renner, O., Lindahl, P., Cho, H., Kehrl, J., and Betsholtz, C. (2003). Transcription profiling of platelet-derived growth factor-B-deficient mouse embryos identifies RGS5 as a novel marker for pericytes and vascular smooth muscle cells. Am J Pathol *162*, 721–729.

19. Berger, M., Bergers, G., Arnold, B., Hammerling, G. J., and Ganss, R. (2005). Regulator of G-protein signaling-5 induction in pericytes coincides with active vessel remodeling during neovascularization. Blood *105*, 1094–1101.

20. Peppiatt, C. M., Howarth, C., Mobbs, P., and Attwell, D. (2006). Bidirectional control of CNS capillary diameter by pericytes. Nature *443*, 700–704.

21. Endemann, D. H., and Schiffrin, E. L. (2004a). Endothelial dysfunction. J Am Soc Nephrol *15*, 1983–1992.

22. Endemann, D. H., and Schiffrin, E. L. (2004b). Nitric oxide, oxidative excess, and vascular complications of diabetes mellitus. Curr Hypertens Rep *6*, 85–89.

23. Vanhoutte, P. M. (2003). Endothelial control of vasomotor function: from health to coronary disease. Circ J *67*, 572–575.

24. Tilton, R. G., Kilo, C., and Williamson, J. R. (1979). Pericyte-endothelial relationships in cardiac and skeletal muscle capillaries. Microvasc Res *18*, 325–335.

25. Sims, D. E. (2000). Diversity within pericytes. Clin Exp Pharmacol Physiol *27*, 842–846.

26. Ballabh, P., Braun, A., and Nedergaard, M. (2004). The blood-brain barrier: an overview: structure, regulation, and clinical implications. Neurobiol Dis *16*, 1–13.

27. Cleaver, O., and Melton, D. A. (2003). Endothelial signaling during development. Nat Med *9*, 661–668.

28. Thomas, W. E. (1999). Brain macrophages: on the role of pericytes and perivascular cells. Brain Res Brain Res Rev *31*, 42–57.

29. Balabanov, R., Washington, R., Wagnerova, J., and Dore-Duffy, P. (1996). CNS microvascular pericytes express macrophage-like function, cell surface integrin alpha M, and macrophage marker ED-2. Microvasc Res *52*, 127–142.

30. Betsholtz, C. (2004). Insight into the physiological functions of PDGF through genetic studies in mice. Cytokine Growth Factor Rev *15*, 215–228.

31. Suematsu, M., and Aiso, S. (2001). Professor Toshio Ito: a clairvoyant in pericyte biology. Keio J Med *50*, 66–71.

32. Sato, M., Suzuki, S., and Senoo, H. (2003). Hepatic stellate cells: unique characteristics in cell biology and phenotype. Cell Struct Funct 28, 105–112.

33. Carmeliet, P. (2003). Angiogenesis in health and disease. Nat Med 9, 653–660.

34. Bergwerff, M., Verberne, M. E., DeRuiter, M. C., Poelmann, R. E., and Gittenberger-de Groot, A. C. (1998). Neural crest cell contribution to the developing circulatory system: implications for vascular morphology? Circ Res 82, 221–231.

35. Etchevers, H. C., Couly, G., and Le Douarin, N. M. (2002). Morphogenesis of the branchial vascular sector. Trends Cardiovasc Med 12, 299–304.

36. Creazzo, T. L., Godt, R. E., Leatherbury, L., Conway, S. J., and Kirby, M. L. (1998). Role of cardiac neural crest cells in cardiovascular development. Annu Rev Physiol 60, 267–286.

37. Hellstrom, M., Kalen, M., Lindahl, P., Abramsson, A., and Betsholtz, C. (1999). Role of PDGF-B and PDGFR-beta in recruitment of vascular smooth muscle cells and pericytes during embryonic blood vessel formation in the mouse. Development 126, 3047–3055.

38. Chen, S., and Lechleider, R. J. (2004). Transforming growth factor-beta-induced differentiation of smooth muscle from a neural crest stem cell line. Circ Res 94, 1195–1202.

39. Darland, D. C., and D'Amore, P. A. (2001). TGF beta is required for the formation of capillary-like structures in three-dimensional cocultures of 10T1/2 and endothelial cells. Angiogenesis 4, 11–20.

40. Dickson, M. C., Martin, J. S., Cousins, F. M., Kulkarni, A. B., Karlsson, S., and Akhurst, R. J. (1995). Defective haematopoiesis and vasculogenesis in transforming growth factor-beta 1 knock out mice. Development 121, 1845–1854.

41. Oshima, M., Oshima, H., and Taketo, M. M. (1996). TGF-beta receptor type II deficiency results in defects of yolk sac hematopoiesis and vasculogenesis. Dev Biol 179, 297–302.

42. Li, D. Y., Sorensen, L. K., Brooke, B. S., Urness, L. D., Davis, E. C., Taylor, D. G., Boak, B. B., and Wendel, D. P. (1999). Defective angiogenesis in mice lacking endoglin. Science 284, 1534–1537.

43. Carvalho, R. L., Jonker, L., Goumans, M. J., Larsson, J., Bouwman, P., Karlsson, S., Dijke, P. T., Arthur, H. M., and Mummery, C. L. (2004). Defective paracrine signalling by TGFbeta in yolk sac vasculature of endoglin mutant mice: a paradigm for hereditary haemorrhagic telangiectasia. Development 131, 6237–6247.

44. Carmeliet, P. (2004). Manipulating angiogenesis in medicine. J Intern Med 255, 538–561.

45. Yamashita, J., Itoh, H., Hirashima, M., Ogawa, M., Nishikawa, S., Yurugi, T., Naito, M., and Nakao, K. (2000). Flk1-positive cells derived from embryonic stem cells serve as vascular progenitors. Nature 408, 92–96.

46. Gittenberger-de Groot, A. C., DeRuiter, M. C., Bergwerff, M., and Poelmann, R. E. (1999). Smooth muscle cell origin and its relation to heterogeneity in development and disease. Arterioscler Thromb Vasc Biol 19, 1589–1594.

47. Nakajima, Y., Mironov, V., Yamagishi, T., Nakamura, H., and Markwald, R. R. (1997). Expression of smooth muscle alpha-actin in mesenchymal cells during formation of avian endocardial cushion tissue: a role for transforming growth factor beta3. Dev Dyn 209, 296–309.

48. Dayoub, S., Devlin, H., and Sloan, P. (2003). Evidence for the formation of metaplastic bone from pericytes in calcifying fibroblastic granuloma. J Oral Pathol Med 32, 232–236.

49. Doherty, M. J., Ashton, B. A., Walsh, S., Beresford, J. N., Grant, M. E., and Canfield, A. E. (1998). Vascular pericytes express osteogenic potential in vitro and in vivo. J Bone Miner Res 13, 828–838.

50. Farrington-Rock, C., Crofts, N. J., Doherty, M. J., Ashton, B. A., Griffin-Jones, C., and Canfield, A. E. (2004). Chondrogenic and adipogenic potential of microvascular pericytes. Circulation 110, 2226–2232.

51. Dellavalle, A., Sampaolesi, M., Tonlorenzi, R., Tagliafico, E., Sacchetti, B., Perani, L., Innocenzi, A., Galvez, B. G., Messina, G., Morosetti, R., et al. (2007). Pericytes of human skeletal muscle are myogenic precursors distinct from satellite cells. Nat Cell Biol 9, 255–267.

52. Gerhardt, H., Golding, M., Fruttiger, M., Ruhrberg, C., Lundkvist, A., Abramsson, A., Jeltsch, M., Mitchell, C., Alitalo, K., Shima, D., and Betsholtz, C. (2003). VEGF guides angiogenic sprouting utilizing endothelial tip cell filopodia. J Cell Biol 161, 1163–1177.

53. Ozerdem, U., Grako, K. A., Dahlin-Huppe, K., Monosov, E., and Stallcup, W. B. (2001). NG2 proteoglycan is expressed exclusively by mural cells during vascular morphogenesis. Dev Dyn 222, 218–227.

54. Ozerdem, U., and Stallcup, W. B. (2003). Early contribution of pericytes to angiogenic sprouting and tube formation. Angiogenesis 6, 241–249.

55. Reynolds, L. P., Grazul-Bilska, A. T., and Redmer, D. A. (2000). Angiogenesis in the corpus luteum. Endocrine 12, 1–9.

56. Hirschi, K. K., Rohovsky, S. A., and D'Amore, P. A. (1998). PDGF, TGF-beta, and heterotypic cell-cell interactions mediate endothelial cell-induced recruitment of 10T1/2 cells and their differentiation to a smooth muscle fate. J Cell Biol 141, 805–814.

57. Lindahl, P., Hellstrom, M., Kalen, M., and Betsholtz, C. (1998). Endothelial-perivascular cell signaling in vascular development: lessons from knockout mice. Curr Opin Lipidol 9, 407–411.

58. Enge, M., Bjarnegard, M., Gerhardt, H., Gustafsson, E., Kalen, M., Asker, N., Hammes, H. P., Shani, M., Fassler, R., and Betsholtz, C. (2002). Endothelium-specific platelet-derived growth factor-B ablation mimics diabetic retinopathy. Embo J 21, 4307–4316.

59. Betsholtz, C., Karlsson, L., and Lindahl, P. (2001). Developmental roles of platelet-derived growth factors. Bioessays 23, 494–507.

60. Tallquist, M. D., French, W. J., and Soriano, P. (2003). Additive effects of PDGF receptor beta signaling pathways in vascular smooth muscle cell development. PLoS Biol 1, E52.

61. Sundberg, C., Kowanetz, M., Brown, L. F., Detmar, M., and Dvorak, H. F. (2002). Stable expression of angiopoietin-1 and other markers by cultured pericytes: phenotypic similarities to a subpopulation of cells in maturing vessels during later stages of angiogenesis in vivo. Lab Invest 82, 387–401.

62. Suri, C., Jones, P. F., Patan, S., Bartunkova, S., Maisonpierre, P. C., Davis, S., Sato, T. N., and Yancopoulos, G. D. (1996). Requisite role of angiopoietin-1, a ligand for the TIE2 receptor, during embryonic angiogenesis. Cell 87, 1171–1180.

63. Dumont, D. J., Gradwohl, G., Fong, G. H., Puri, M. C., Gertsenstein, M., Auerbach, A., and Breitman, M. L. (1994). Dominant-negative and targeted null mutations in the endothelial receptor tyrosine kinase, tek, reveal a critical role in vasculogenesis of the embryo. Genes Dev 8, 1897–1909.

64. Sato, T. N., Tozawa, Y., Deutsch, U., Wolburg-Buchholz, K., Fujiwara, Y., Gendron-Maguire, M., Gridley, T., Wolburg, H., Risau, W., and

Qin, Y. (1995). Distinct roles of the receptor tyrosine kinases Tie-1 and Tie-2 in blood vessel formation. Nature *376*, 70–74.

65. Iurlaro, M., Scatena, M., Zhu, W. H., Fogel, E., Wieting, S. L., and Nicosia, R. F. (2003). Rat aorta-derived mural precursor cells express the Tie2 receptor and respond directly to stimulation by angiopoietins. J Cell Sci *116*, 3635–3643.

66. Suri, C., McClain, J., Thurston, G., McDonald, D. M., Zhou, H., Oldmixon, E. H., Sato, T. N., and Yancopoulos, G. D. (1998). Increased vascularization in mice overexpressing angiopoietin-1. Science *282*, 468–471.

67. Thurston, G., Suri, C., Smith, K., McClain, J., Sato, T. N., Yancopoulos, G. D., and McDonald, D. M. (1999). Leakage-resistant blood vessels in mice transgenically overexpressing angiopoietin-1. Science *286*, 2511–2514.

68. Uemura, A., Ogawa, M., Hirashima, M., Fujiwara, T., Koyama, S., Takagi, H., Honda, Y., Wiegand, S. J., Yancopoulos, G. D., and Nishikawa, S. (2002). Recombinant angiopoietin-1 restores higher-order architecture of growing blood vessels in mice in the absence of mural cells. J Clin Invest *110*, 1619–1628.

69. Gale, N. W., Thurston, G., Hackett, S. F., Renard, R., Wang, Q., McClain, J., Martin, C., Witte, C., Witte, M. H., Jackson, D., *et al.* (2002). Angiopoietin-2 is required for postnatal angiogenesis and lymphatic patterning, and only the latter role is rescued by Angiopoietin-1. Dev Cell *3*, 411–423.

70. Maisonpierre, P. C., Suri, C., Jones, P. F., Bartunkova, S., Wiegand, S. J., Radziejewski, C., Compton, D., McClain, J., Aldrich, T. H., Papadopoulos, N., *et al.* (1997). Angiopoietin-2, a natural antagonist for Tie2 that disrupts in vivo angiogenesis. Science *277*, 55–60.

71. Hammes, H. P., Lin, J., Wagner, P., Feng, Y., Vom Hagen, F., Krzizok, T., Renner, O., Breier, G., Brownlee, M., and Deutsch, U. (2004). Angiopoietin-2 causes pericyte dropout in the normal retina: evidence for involvement in diabetic retinopathy. Diabetes *53*, 1104–1110.

72. Zhang, L., Yang, N., Park, J. W., Katsaros, D., Fracchioli, S., Cao, G., O'Brien-Jenkins, A., Randall, T. C., Rubin, S. C., and Coukos, G. (2003). Tumor-derived vascular endothelial growth factor upregulates angiopoietin-2 in host endothelium and destabilizes host vasculature, supporting angiogenesis in ovarian cancer. Cancer Res *63*, 3403–3412.

73. Hanahan, D. (1997). Signaling vascular morphogenesis and maintenance. Science *277*, 48–50.

74. Vikkula, M., Boon, L. M., Carraway, K. L., 3rd, Calvert, J. T., Diamonti, A. J., Goumnerov, B., Pasyk, K. A., Marchuk, D. A., Warman, M. L., Cantley, L. C., *et al.* (1996). Vascular dysmorphogenesis caused by an activating mutation in the receptor tyrosine kinase TIE2. Cell *87*, 1181–1190.

75. Hla, T. (2001). Sphingosine 1-phosphate receptors. Prostaglandins *64*, 135–142.

76. Allende, M. L., Yamashita, T., and Proia, R. L. (2003). G-protein-coupled receptor S1P1 acts within endothelial cells to regulate vascular maturation. Blood *102*, 3665–3667.

77. De Palma, M., Venneri, M. A., Galli, R., Sergi, L. S., Politi, L. S., Sampaolesi, M., and Naldini, L. (2005). Tie2 identifies a hematopoietic lineage of proangiogenic monocytes required for tumor vessel formation and a mesenchymal population of pericyte progenitors. Cancer Cell *8*, 211–226.

78. Grunewald, M., Avraham, I., Dor, Y., Bachar-Lustig, E., Itin, A., Yung, S., Chimenti, S., Landsman, L., Abramovitch, R., and Keshet, E. (2006). VEGF-induced adult neovasculariza-

tion: recruitment, retention, and role of accessory cells. Cell *124*, 175–189.

79. Jin, D. K., Shido, K., Kopp, H. G., Petit, I., Shmelkov, S. V., Young, L. M., Hooper, A. T., Amano, H., Avecilla, S. T., Heissig, B., *et al.* (2006). Cytokine-mediated deployment of SDF-1 induces revascularization through recruitment of CXCR4(+) hemangiocytes. Nat Med.

80. Rajantie, I., Ilmonen, M., Alminaite, A., Ozerdem, U., Alitalo, K., and Salven, P. (2004). Adult bone marrow-derived cells recruited during angiogenesis comprise precursors for periendothelial vascular mural cells. Blood *104*, 2084–2086.

81. Kokovay, E., Li, L., and Cunningham, L. A. (2006). Angiogenic recruitment of pericytes from bone marrow after stroke. J Cereb Blood Flow Metab *26*, 545–555.

82. Ozerdem, U., Alitalo, K., Salven, P., and Li, A. (2005). Contribution of bone marrow-derived pericyte precursor cells to corneal vasculogenesis. Invest Ophthalmol Vis Sci *46*, 3502–3506.

83. Bergers, G., and Benjamin, L. E. (2003). Tumorigenesis and the angiogenic switch. Nat Rev Cancer *3*, 401–410.

84. Bergers, G., Song, S., Meyer-Morse, N., Bergsland, E., and Hanahan, D. (2003). Benefits of targeting both pericytes and endothelial cells in the tumor vasculature with kinase inhibitors. J Clin Invest *111*, 1287–1295.

86. Hanahan, D., and Folkman, J. (1996). Patterns and emerging mechanisms of the angiogenic switch during tumorigenesis. Cell *86*, 353.

85. Folkman, J. (2000). Tumor angiogenesis, In Cancer Medicine, H. et al., ed. (Hamilton, Ontario: B C Decker).

87. Cai, J., and Boulton, M. (2002). The pathogenesis of diabetic retinopathy: old concepts and new questions. Eye *16*, 242–260.

87. Hammes, H. P., Lin, J., Renner, O., Shani, M., Lundqvist, A., Bets holtz, C., Brownlee, M., and Deutsch, U. (2002). Pericytes and the pathogenesis of diabetic retinopathy. Diabetes *51*, 3107–3112.

88. Wilkinson-Berka, J. L., Babic, S., De Gooyer, T., Stitt, A. W., Jaworski, K., Ong, L. G., Kelly, D. J., and Gilbert, R. E. (2004). Inhibition of platelet-derived growth factor promotes pericyte loss and angiogenesis in ischemic retinopathy. Am J Pathol *164*, 1263–1273.

89. Campochiaro, P. A. (2004). Ocular neovascularisation and excessive vascular permeability. Expert Opin Biol Ther *4*, 1395–1402.

90. Miller, J. W., Adamis, A. P., and Aiello, L. P. (1997). Vascular endothelial growth factor in ocular neovascularization and proliferative diabetic retinopathy. Diabetes Metab Rev *13*, 37–50.

91. Benjamin, L. E. (2001). Glucose, VEGF-A, and diabetic complications. Am J Pathol *158*, 1181–1184.

92. Aiello, L. P., Pierce, E. A., Foley, E. D., Takagi, H., Chen, H., Riddle, L., Ferrara, N., King, G. L., and Smith, L. E. (1995). Suppression of retinal neovascularization in vivo by inhibition of vascular endothelial growth factor (VEGF) using soluble VEGF-receptor chimeric proteins. Proc Natl Acad Sci USA *92*, 10457–10461.

93. McLeod, D. S., Taomoto, M., Cao, J., Zhu, Z., Witte, L., and Lutty, G. A. (2002). Localization of VEGF receptor-2 (KDR/Flk-1) and effects of blocking it in oxygen-induced retinopathy. Invest Ophthalmol Vis Sci *43*, 474–482.

94. Robbins, S. G., Rajaratnam, V. S., and Penn, J. S. (1998). Evidence for upregulation and redistribution of vascular endothelial growth factor (VEGF) receptors flt-1 and flk-1 in the oxygen-injured rat retina. Growth Factors *16*, 1–9.

95. Lindblom, P., Gerhardt, H., Liebner, S., Abramsson, A., Enge, M., Hellstrom, M., Backstrom, G., Fredriksson, S., Landegren, U., Nystrom, H. C., *et al.* (2003). Endothelial PDGF-B retention is required for proper investment of pericytes in the microvessel wall. Genes Dev *17*, 1835–1840.

96. Jain, R. K. (2003). Molecular regulation of vessel maturation. Nat Med *9*, 685–693.

97. Morikawa, S., Baluk, P., Kaidoh, T., Haskell, A., Jain, R. K., and McDonald, D. M. (2002). Abnormalities in pericytes on blood vessels and endothelial sprouts in tumors. Am J Pathol *160*, 985–1000.

98. Abramsson, A., Berlin, O., Papayan, H., Paulin, D., Shani, M., and Betsholtz, C. (2002). Analysis of mural cell recruitment to tumor vessels. Circulation *105*, 112–117.

99. Baluk, P., Hashizume, H., and McDonald, D. M. (2005). Cellular abnormalities of blood vessels as targets in cancer. Curr Opin Genet Dev *15*, 102–111.

100. Abramsson, A., Lindblom, P., and Betsholtz, C. (2003). Endothelial and nonendothelial sources of PDGF-B regulate pericyte recruitment and influence vascular pattern formation in tumors. J Clin Invest *112*, 1142–1151.

101. Abramsson, A., Kurup, S., Busse, M., Yamada, S., Lindblom, P., Schallmeiner, E., Stenzel, D., Sauvaget, D., Ledin, J., Ringvall, M., *et al.* (2007). Defective N-sulfation of heparan sulfate proteoglycans limits PDGF-BB binding and pericyte recruitment in vascular development. Genes Dev *21*, 316–331.

102. Kurup, S., Abramsson, A., Li, J. P., Lindahl, U., Kjellen, L., Betsholtz, C., Gerhardt, H., and Spillmann, D. (2006). Heparan sulphate requirement in platelet-derived growth factor B-mediated pericyte recruitment. Biochem Soc Trans *34*, 454–455.

103. Reinmuth, N., Liu, W., Jung, Y. D., Ahmad, S. A., Shaheen, R. M., Fan, F., Bucana, C. D., McMahon, G., Gallick, G. E., and Ellis, L. M. (2001). Induction of VEGF in perivascular cells defines a potential paracrine mechanism for endothelial cell survival. Faseb J *15*, 1239–1241.

104. Shaheen, R. M., Tseng, W. W., Davis, D. W., Liu, W., Reinmuth, N., Vellagas, R., Wieczorek, A. A., Ogura, Y., McConkey, D. J., Drazan, K. E., *et al.* (2001). Tyrosine kinase inhibition of multiple angiogenic growth factor receptors improves survival in mice bearing colon cancer liver metastases by inhibition of endothelial cell survival mechanisms. Cancer Res *61*, 1464–1468.

105. Benjamin, L. E., Golijanin, D., Itin, A., Pode, D., and Keshet, E. (1999). Selective ablation of immature blood vessels in established human tumors follows vascular endothelial growth factor withdrawal [see comments]. J Clin Invest *103*, 159–165.

106. Xian, X., Hakansson, J., Stahlberg, A., Lindblom, P., Betsholtz, C., Gerhardt, H., and Semb, H. (2006). Pericytes limit tumor cell metastasis. J Clin Invest *116*, 642–651.

107. Yonenaga, Y., Mori, A., Onodera, H., Yasuda, S., Oe, H., Fujimoto, A., Tachibana, T., and Imamura, M. (2005). Absence of smooth muscle actin-positive pericyte coverage of tumor vessels correlates with hematogenous metastasis and prognosis of colorectal cancer patients. Oncology *69*, 159–166.

108. Ramaswamy, S., Ross, K. N., Lander, E. S., and Golub, T. R. (2003). A molecular signature of metastasis in primary solid tumors. Nat Genet *33*, 49–54.

109. Pietras, K., and Hanahan, D. (2004). A Multitargeted, Metronomic, and Maximum-Tolerated Dose "Chemo-Switch" Regimen is Antiangiogenic, Producing Objective Responses and Survival Benefit in a Mouse Model of Cancer. J Clin Oncol.

Chapter 5
Matrix Metalloproteinases and Their Endogenous Inhibitors

Liliana Guedez and William G. Stetler-Stevenson

Keywords: matrix metalloproteinase, tissue inhibitor of metalloproteinase, extracellular matrix, protease

Abstract: The extracellular matrix (ECM) plays a central role in maintaining the homeostatic balance, structure and function of normal tissues. The ECM consists of a meshwork of proteins that self assemble into different organized structures, such as the interstitial matrix or basement membranes. Components of the ECM include members of the collagen and laminin families, thrombospondins, fibronectins, and a variety of proteoglycans and other glycoproteins. The loss of structural integrity in normal tissues is heralded by the disruption of the ECM scaffolding that supports the cellular components. This is a critical step in the tissue remodeling that accompanies many physiological as well as pathological processes [1–4]. In tumor growth, the role of the ECM is complex and involves cell–cell and cell–ECM interactions, generation of biologically active peptides from ECM components and release of sequestered growth factors [1, 5]. For instance, the turnover of the ECM during malignant progression alters the behavior of both tumor and stromal cells, and also results in the recruitment of a variety of host cells that combine with the altered ECM composition and structure to constitute the tumor microenvironment [1–4]. One essential component of this tumor microenvironment is the angiogenic response, composed primarily of endothelial cells and pericytes.

Matrix Metalloproteinases

Tumor-associated angiogenesis is mediated principally through stimulation and migration of quiescent endothelial cells from pre-existing mature blood vessels towards the tumor mass,

Cell and Cancer Biology Branch, Extracellular Matrix Pathology Section, Center for Cancer Research, National Cancer Institute, Bethesda, MD 20892-4605, USA

although in some tumors bone marrow-derived or circulating endothelial precursors may also contribute to tumor vessel development [6, 7]. The process of endothelial cell invasion demonstrates many mechanistic similarities to the invasion of metastatic tumor cells [8]. These similarities are significant alterations in cell adhesion as well as invasion of both basement membrane and stromal ECMs.

One class of proteases that have long been associated with alterations in the ECM during both tumor progression and angiogenesis are the matrix metalloproteinases, or MMPs. The MMPs are members of the metzincin superfamily of metalloenzymes that also include ADAM (a metalloproteinase with a disintegrin and metalloproteinase domain), ADAM-TS (ADAM with a thrombospondin-like-domain), and astacin proteases. The MMPs constitute a multi-gene family that collectively can degrade all components of the ECM, as well as cell surface-associated receptors and adhesion proteins [9, 10]. The MMP family is defined by a characteristic domain structure that involves: (1) the primary catalytic site defined by the conserved sequence HexGHxxGxxHS/T in which the histidine residues coordinate the catalytic Zn atom; (2) a prodomain that contains another conserved "activation locus" sequence surrounding a critical cysteine residue that occupies the fourth coordination site of the Zn atom in the latent or proenzyme form; and (3) a signal peptide domain which directs the eventual secretion of the proenzymes into the pericellular milieu. These three domains constitute the minimal MMP domain structure. However, most members of the MMP family have additional domains that influence substrate specificity, such as a carboxyl-terminal hemopexin domain or fibronectin-like repeats associated with gelatin binding activity of MMP-2 and MMP-9. Finally, a subgroup of the MMP family contains several members which are cell surface-associated either by introduction of a true transmembrane domain or through a phosphotidyl inositol linkage site. These members are referred to as membrane type- or MT-MMPs [9, 10].

Most members of the MMP family, with the principal exception of the MT-MMPs, are synthesized and secreted in an inactive pro-enzyme form. These secreted pro-MMPs are activated extracellularly by serine- or metalloproteinase-mediated cleavage of their pro-domains. Proteolytic activation of MMP family members disrupts the critical interaction of the highly conserved amino-terminal cysteine of the activation locus with the catalytic zinc atom. Activation of soluble MMPs is associated with a decrease in molecular mass corresponding to the loss of the pro-fragment (~10–12 kilodaltons) [9, 10]. Recently, it has been demonstrated that the small integrin binding ligand N-linked glycoproteins, also known as the SIBLING family, can bind and activate members of the MMP family. Furthermore, this activation can occur reversibly and without proteolytic processing or removal of the MMP pro-fragment [11]. These SIBLING family members can also reverse the inhibition of MMPs by enhancing the dissociation of endogenous MMP inhibitors [11]. The literature on MMPs and angiogenesis is extensive, and therefore this chapter will highlight a few basic concepts. For more details concerning specific members of the MMP family, the reader is referred to any of a number of recent comprehensive reviews [9, 10, 12].

Endogenous Inhibitors of Metalloproteinases

The proteolytic activities of the MMP family members in the tumor microenvironment are regulated by at least two types of endogenous inhibitors. The best characterized of these is the tissue inhibitor of metalloproteinase family or TIMPs [13–16]. The TIMP family consists of four members with homologues in *Xenopus* and *Drosophila*. The TIMP family members are relatively small proteins with molecular weight of the core proteins in the range of 21 kilodaltons. However, they contain 12 cysteine residues that must correctly pair into six disulfide bonds to give the correct secondary/tertiary structure required for MMP inhibitory activity. The disulfide-bonding pattern is conserved among all members of the family and results in the formation of six disulfide loops. Loops 1–3 form the amino-terminal domain that is the most highly conserved domain and is sufficient for inhibition of MMP activity. Loops 4–6 form a second domain that can mediate interaction of TIMP family members with select members of the MMP family. In some cases, this second carboxyl-terminal domain of TIMPs can promote cell-surface binding of soluble pro-MMPs as well as cell surface activation of select members of the MMP family.

The TIMPs are somewhat unusual proteins in that the N-terminal amino acid of all members is a cystine residue. Interestingly, the mechanism of MMP inhibition, as demonstrated by X-ray crystal analysis of TIMP interactions with the MMP active site, involves the amino group of this critical amino-terminal cysteine with the catalytic zinc [17]. The selectivity of TIMPs for inhibiting MMP activity is greatly influenced by amino acid residues that surround the N-terminal cysteine, as revealed by site-directed mutagenesis studies. TIMP family members are distinguished by differential expression in various

tissues. TIMP-2 is unique in that expression of this inhibitor is constitutive in most tissues, whereas TIMP-1, TIMP-3 and TIMP-4 protein expression is inducible. Recently, it has been reported that TIMP-4 is unique among members of the TIMP family in that it does not inhibit angiogenesis in vivo [18]. However, the ability of the other three members of the TIMP family to inhibit angiogenesis demonstrates the importance of this protein family, as well as the MMPs, in regulating tumor angiogenesis. For more details on the TIMP family members, the reader is referred to several excellent reviews [9, 10, 16].

The second class of MMP inhibitors that are known to modulate the angiogenic response consist of a novel membrane associated MMP inhibitor that has been recently identified. This inhibitor is known as RECK, for reversion-inducing, cysteine-rich protein with Kazal motifs [19]. RECK has been shown to inhibit several members of the MMP family including MMP-2, MMP-9 and MT1-MMP. RECK-deficient mice are embryonic lethal at E10.5 due to vascular defects [20]. Interestingly, this phenotype is partially reversed by knock down of MMP-2 expression.

Proangiogenic Activities of Matrix Metalloproteinases

It has long been recognized that angiogenic factor stimulation of endothelial cells results in enhanced production of MMPs by these cells, in particular MMP-2 and MT1-MMP [8, 12, 21, 22]. However, genetic studies have clearly demonstrated a role for select MMPs in tumor angiogenesis that may not be produced directly by the endothelial cells. For example, genetic manipulation of MMP-9 expression has demonstrated that this MMP is critical for induction of the angiogenic switch. This mechanism specifically involves the ability of MMP-9 from stromal components to mobilize matrix-bound vascular endothelial growth factor-A (VEGF-A) into the tumor microenvironment [23]. It has also been reported that MMP-9 again produced by the stroma is important in the recruitment of pericytes to the tumor microenvironment [24]. In MMP-2 deficient mice, tumor-induced angiogenesis was markedly reduced in a dorsal sac assay using B16-B6 tumor cells and in a quantitative in vivo angiogenesis assay that demonstrated a 30% reduction in angiogenesis in response to VEGF when compared with wild type control mice [25, 26]. However, the mechanism and/or substrate targets involved in MMP-2 contribution to tumor angiogenesis are not well understood, other than generalized cleavage of ECM components.

During the first stages of tumor-associated angiogenesis, there is an increase in vascular permeability leading to deposition of a provisional matrix composed principally of thrombin, as well as disruption of the sub-endothelial basement membrane [27–29]. Thus, endothelial cells in the tumor microenvironment encounter a variety of types of ECM during tumor angiogenesis [30, 31]. Studies in vivo and in vitro demonstrate that this fibrinolysis is not carried out by the fibronolytic/

plasminogen activator system, but by endothelial MMPs [32]. Thus, invading endothelial cells mobilize proteolytic enzymes whose activities are limited to pericellular compartments [29, 33]. The most potent pericellular fibrinolysin is endothelial MT1-MMP that confers activated endothelial cells with the ability to invade fibrin and promotes tube formation [34–37]. Moreover, the collagenase activity of MT1-MMP is also central in angiogenesis, by providing endothelial cells with the ability to invade collagen-rich matrices [38, 39]. MT-1-MMP can directly process collagen type I, or indirectly other collagens and gelatins in the tumor microenvironment by activating MMP-2 in the presence of TIMP-2 [10, 40]. MT1-MMP together with the proteolytic activities of MMP-2 and MMP-9 play a key role in driving the early events during tumor neovascularization [31, 41–43]. Thus, numerous studies report on the requirement of MMPs for tumor angiogenesis, with MMP-9 mobilizing VEGF from the stroma and MT1-MMP being the most closely associated with early stages of endothelial cell invasion. However, it is not clear where, when and how the activity of other MMPs may modulate the different stages of matrix remodeling and modulation of the microenvironment during pathological angiogenesis.

In addition to their proteolytic modification of structural matrix proteins, MMPs are also involved in the regulation of cytokines and growth factors as well as their concomitant receptors needed for endothelial cell functions [44, 45]. Cell surface localized MMP-9 can mediate IGF-I-triggered-cell migration [23, 46, 47]. MMP-9 can also release activated latent transforming growth factor-β (TGF-β) through degradation of its latency associated peptide and release ECM-bound VEGF to promote angiogenesis [23, 48]. The fibroblast growth factor (FGF) receptor, FGFR1, is cleaved by MMP-2 [49]. The soluble receptor can still bind FGF and modulate mitogenic and angiogenic activities. MMPs are also associated with the cleavage of TGF-α and interleukin-6 (IL-6) receptors [50]. The MMP cleavage sites for many ECM components have been identified, but no strong pattern has emerged for the primary peptide sequences cleaved by MMPs in these cytokines during tumor-associated angiogenesis. The amino acid sequence ELR (glutamate, leucine, arginine), identified by phage display technology, has recently emerged as the peptide substrate for MMP-7 and as a conserved motif in the CXC chemokines [51]. Further investigation is necessary to determine the specificity of MMPs in modulation of cytokines and receptors during neovessel formation [52].

During physiologic angiogenesis, the balance between MMPs and inhibitors must be restored during the maturation stage of neovessels to favor basement membrane assembly and the formation of a functional mature blood vessels [12]. However, this balance between proteolysis and inhibition is perturbed in tumor angiogenesis as tumor blood vessels are leaky, dilated and have a decrease in associated pericytes [27, 29, 30]. A considerable body of data clearly supports the role of MMPs in endothelial cell invasion and basement membrane degradation during the sprouting phase of tumor

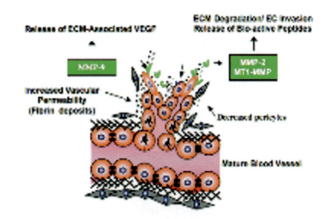

FIG. 5.1. Proangiogenic activities of MMPS. During early phases of tumor-associated angiogenesis, endothelial cells require membrane-bound MT1-MMP for fibrinolysis of the provisional matrix and invasion of the collagen I rich stromal compartment. Secreted MMPs at the perivascular compartment reinforced the initial angiogenic responses. For example, MMP-9 mediates the release of ECM-bound VEGF, and MMP-2 induces mobilization of bioactive factors from basement membrane and stromal matricies.

angiogenesis [53]. Based on these studies, it was hypothesized that co-administration of synthetic MMP inhibitor, in addition to conventional chemotherapy, would have a clinical impact by preventing sprouting and inducing stabilized blood vessels [8, 54–56]. However, this has not yet been successfully achieved in the clinical setting. These proangiogenic activities are briefly outlines in Fig. 5.1.

Antiangiogenic Activities of Matrix Metalloproteinases

Activation of MMPs can also be antiangiogenic as these enzymes are capable of the metabolizing a variety of ECM proteins to generate potent endogenous angiogenesis inhibitors such as Angiostatin and Endostatin [57–60]. These are generated from their precursors plasminogen and collagen XVIII, respectively. For example, MMP-9 is capable of generating Tumstatin, a potent antiangiogenic peptide derived from the NC1 domain of the α3 chain of type IV collagen (α3NC1). Tumstatin, not only inhibits angiogenesis but can also inhibit the growth of melanoma and other epithelial tumor cells [61]. One mechanism mediating these effects involves binding of Tumstatin to the $\alpha v\beta 3$ and $\alpha v\beta 5$ integrin receptors [5, 62]. However, using a novel flow cytometric assay, investigators have demonstrated that Tumstatin/α3NC1 can also directly bind to $\alpha 3\beta 1$, and suggested that this binding might play a role

Anti-angiogenic Activities of MMPs

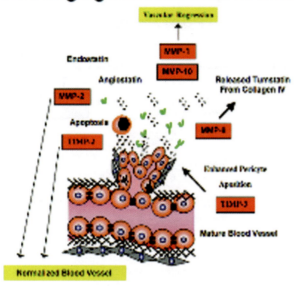

FIG. 5.2. Antiangiogenic activities of MMPs. Activation of secreted MMPs have also been shown to have antiangiogenic effects and/or promote vascular regression. Soluble MMPs, such as MMP-2 and MMP-9 are capable of the degrading the type IV collagen of the basement membrane, as well as other ECM proteins, to generate potent endogenous inhibitors. For example, MMP-9 releases Tumstatin from the α 3 chain of collagen IV, while a variety of MMPs have been demonstrated to generatet Angiostatin and Endostatin. Activation of MMP-1 and MMP-10 directly can induce vascular regression. TIMP-2 and TIMP-3 inhibit proliferation of endothelial cells, induce apoptosis and/or promote close association of pericytes with endothelial cells, respectively. Thus, TIMPs may promote normalization of tumor blood vessels.

in the tumor suppressor effects of Tumstatin/α3NC1 by either directly affecting tumor cell proliferation or by changing the affinity of the αv integrin receptors [62]. Tumstatin/α3NC1 inhibits both tumor cell proliferation and angiogenesis, and these effects appear to be mediated by binding to α3β1. It should also be noted that thrombospondin, the first naturally occurring angiogenesis inhibitor, also contains an α3β1 binding site [63], and that α3β1 mediates tumor cell arrest and metastasis formation in the lung [64]. Collectively, these data suggest that α3β1 should be an important target in developing new cancer therapies.

More recently, MMP-1 and MMP-10 activities have been implicated in the matrix remodeling associated with vascular regression. Expression and activation of MMP-10 is required for regression. Its expression is repressed by histone deacetylase 7 in endothelial cells preventing vascular regression and endothelial cell death [65]. Moreover, inhibition of these MMPs with TIMP-1, for instance, does not alter tube formation, but selectively prevents vascular regression. Thus, activation of MMP-10 is an important condition for vascular regression. Mechanisms that regulate MMP-10 expression or

its activation are the subject of intense research and are pivotal targets for antiangiogenic therapies. A brief overview of these antiangiogenic activities for select MMPs are presented in Fig. 5.2.

Antiangiogenic Activity of the Tissue Inhibitors of Matrix Metalloproteinases

Based on the extensive data demonstrating the proangiogenic roles for many MMPs, one might assume that all TIMPs would be antiangiogenic through their MMP-inhibitory activity. Numerous studies have demonstrated that with the exception of TIMP-4, all three TIMPs (TIMP-1, -2, -3) inhibit angiogenesis in vivo [18]. Although TIMP-4 inhibits endothelial cell migration, it failed to inhibit FGF-induced angiogenesis in the chick chorioallantoic membrane assay. However, the mechanisms utilized by each of the TIMPs with in vivo antiangiogenic activity may have unique aspects. For example, TIMP-3 has been shown to bind directly to the VEGFR-2 receptor and possibly antagonize VEGF binding [66], but it is not known what contribution this mechanism makes during the antiangiogenic activity of TIMP-3. Endothelial cell interaction with pericytes has been shown to strongly induce TIMP-3 expression by the pericyte, whereas TIMP-2 expression remains restricted to the endothelial cell [67]. These findings suggest that endothelial-pericyte interactions are important in controlling TIMP-2/3 expression which in turn may function to stabilize newly formed vessels. TIMP-3 also inhibits endothelial cell sprouting morphogenesis. The ability of TIMP-3 to bind heparan sulfate allows this inhibitor to directly interact with the basement membrane proteoglycans such as perlacan, that are highly expressed by pericytes, again promoting endothelial-pericyte interactions [68]. Vascular stabilization also involves the integrity of the junctions between endothelial cells. Proteases associated with disassembly of endothelial cell junctions can also be inhibited by TIMPs [69]. TIMP-3 is unique in its effects on vascular stabilization since it can inhibit not only MMPs, but also ADAMs (or A Disintegrin And Metalloproteinase), in particular ADAM-15, which has been shown to disassemble cell-cell junctions [12, 70]. Thus, TIMP-3 expression could be essential for vascular "normalizaton" through its ability to inhibit MMPs such as MT-MMPs, MMP-3, MMP-1 and MMP-7 that are involved in the disintegration of basement membrane, as well as ADAM-15.

TIMP-2 is the only member of this family of metalloproteinase inhibitors that can efficiently inhibit endothelial cell proliferation [71, 72]. Although TIMP-2 is the only member of the TIMP family with this biological activity, it took many years to demonstrate that inhibition of endothelial cell mitogenesis was independent of the MMP inhibitory activity [73, 74]. Some details of the mechanism of this effect have been worked out. It has been shown that TIMP-2, like Tumstatin, can bind to the integrin α3β1 [74, 75], resulting in inactivation of the cognate tyrosine kinase receptor of the angiogenic factor used to stimulate endothelial cell proliferation in vitro or angiogenesis in vivo [74]. This was the first demonstration of heterologous receptor inactivation initiated by a member of the integrin family. Further study has shown that receptor tyrosine kinase inactivation is mediated by dephosphorylation of critical tyrosine residues in the cytoplasmic domain of the receptors by the phosphotyrosine phosphatase known as Shp-1. In Shp-1 deficient mice (known as moth-eaten viable mice), the antiangiogenic activity of TIMP-2 in vivo is completely abrogated [76]. This study also demonstrated that TIMP-2 induces endothelial cell arrest in G1/G0, as evidenced by hypophosphorylation of pRb and involves de novo synthesis of the cyclin-dependent kinase inhibitor p27^{Kip1}.

The role of TIMP-1 in angiogenesis is complex. Recent in vitro studies suggest that TIMP-1 inhibits endothelial cell migration by both MMP-dependent and MMP-independent pathways [77]. Interestingly, the MMP-independent pathway involves phosphatase activity, similar to the TIMP-2 pathway described previously, except TIMP-1 inhibition of endothelial cell migration involves the phosphatase known as PTEN. However, in many tumor xenograft models, TIMP-1 inhibits angiogenesis [78–81]. Although one recent report suggests TIMP-1 may also promote angiogenesis, this was not in a tumor-related model [82].

These findings suggest that, like the MMPs, the role of TIMPs in regulating tumor angiogenesis is not completely well understood. The observation that TIMP-4 did not inhibit angiogenesis in vivo [18] could be an indication that the selective MMP inhibitor profile of the particular TIMP may dictate its antiangiogenic activity. Alternatively, this observation concerning TIMP-4, and those reported for TIMP-2 [74], might also suggest that the metalloproteinase inhibitory activity of the TIMPs are a minor component of their reported antiangiogenic activity.

Concluding Remarks

There is substantial evidence supporting the concept that MMPs play an essential role in promoting tumor-associated angiogenesis. We have attempted to highlight the roles of MMPs, such as MMP-1, MMP-7, MMP-9, MMP-10 and MT1-MMP, which are better defined than those of other members of the MMP family. MT-MMPs in general are associated with the early morphogenic and invasive events in the angiogenic response with concomitant ECM degradation and endothelial cell invasion. Soluble MMPs released from stromal elements clearly re-enforce the initial angiogenic stimulus, as in MMP-9 mediated release of VEGF from the tumor microenvironment, whereas activation of MMP-1 and MMP-10 appear critical for stabilization of the perivascular compartment. However, the roles of other MMP-s during tumor-associated angiogenesis are not fully understood. The difficulty of dissecting the roles of various MMPs in promoting tumor-associated angiogenesis is further compounded by the complexity of the tumor micro-environment in which a variety of cell types, such as macrophages, activated fibroblasts and leukocytes, and possible endothelial progenitor cells, may all make contributions to the proteolytic environment. These contributions certainly include not only introduction of MMPs but also other classes of proteases, in particular the ADAMS and ADAM-TS type enzymes. The potential interaction of these various classes of matrix-degrading proteases may be coordinated, and by focusing on one class we may miss the important inter-relationship between proteases of different classes.

The complexity of the role of MMPs in tumor-associated angiogenesis is further complicated by the dual functions MMPs are known to play. For instance, MMP activities can also negatively affect the development of the tumor angiogenic response, as clearly evidenced by their ability to generate angiogenesis inhibitors, such as Endostatin and Angiostatin.

With respect to the role of TIMPs in controlling tumor-associated angiogenesis, the emerging picture is clearly more complicated than their role simply as MMP inhibitors. Although some TIMPs, such as TIMP-2, are clearly antiangiogenic and anti-tumorigenic, other members of the family have more complicated mechanisms. This is best evidenced by the contradictory roles that TIMP-1 plays in tumor progression, functioning to inhibit tumor angiogenesis but also promoting tumor cell growth [78].

Finally, although we are beginning to understand the roles of specific MMPs and TIMPs during physiologic angiogenesis, we must also realize that, unlike physiologic angiogenesis, tumor progression results in a continuously changing tumor microenvironment, and that it may be possible for the same MMP or TIMP to promote angiogenesis during one phase of tumor development and then play an alternative or even antiangiogenic role at a later stage of tumor-associated angiogenesis. Further progress in understanding the role of MMPs and TIMPs in tumor-associated angiogenesis will require development of new models that account for the contributions not only of tumor cells, endothelial cells and pericytes, but also cells of the associated immune response. Finally, an understanding of the temporal and spatial interaction of MMPs and TIMPs with other protease systems will be critical to the development of new and effective antiangiogenic therapies that are substantive and useful with significant impact in the clinical setting.

References

1. Bhowmick NA, Neilson EG, Moses HL: Stromal fibroblasts in cancer initiation and progression. Nature 2004;432:332–337.
2. Bissell MJ, Kenny PA, Radisky DC: Microenvironmental regulators of tissue structure and function also regulate tumor induction and progression: the role of extracellular matrix and its degrading enzymes. Cold Spring Harbor Symposia on Quantitative Biology 2005;70:343–356.
3. Gupta GP, Massague J: Cancer metastasis: building a framework. Cell 2006;127:679–695.
4. Gupta GP, Minn AJ, Kang Y, Siegel PM, Serganova I, Cordon-Cardo C, Olshen AB, Gerald WL, Massague J: Identifying site-

specific metastasis genes and functions. Cold Spring Harbor Symposia on Quantitative Biology 2005;70:149–158.

5. Kalluri R: Basement membranes: structure, assembly and role in tumour angiogenesis. Nat Rev Cancer 2003;3:422–433.

6. Carmeliet P: Mechanisms of angiogenesis and arteriogenesis. Nat Med 2000;6:389–395.

7. Grunewald M, Avraham I, Dor Y, Bachar-Lustig E, Itin A, Jung S, Chimenti S, Landsman L, Abramovitch R, Keshet E: VEGF-induced adult neovascularization: recruitment, retention, and role of accessory cells.[see comment][erratum appears in Cell. 2006 Aug 25;126(4):811 Note: Yung, Steffen [corrected to Jung, Steffen]]. Cell 2006;124:175–189.

8. Stetler-Stevenson WG: Matrix metalloproteinases in angiogenesis: a moving target for therapeutic intervention. J Clin Invest 1999;103:1237–1241.

9. Nagase H, Visse R, Murphy G: Structure and function of matrix metalloproteinases and TIMPs. Cardiovascular Research 2006;69:562–573.

10. Visse R, Nagase H: Matrix metalloproteinases and tissue inhibitors of metalloproteinases: structure, function, and biochemistry. Circ Res 2003;92:827–839.

11. Fedarko NS, Jain A, Karadag A, Fisher LW: Three small integrin binding ligand N-linked glycoproteins (SIBLINGs) bind and activate specific matrix metalloproteinases. FASEB Journal 2004;18:734–736.

12. Roy R, Zhang B, Moses MA: Making the cut: protease-mediated regulation of angiogenesis. Exp Cell Res 2006;312:608–622.

13. Brew K, Dinakarpandian D, Nagase H: Tissue inhibitors of metalloproteinases: evolution, structure and function. Biochim Biophys Acta 2000;1477:267–283.

14. Chirco R, Liu XW, Jung KK, Kim HR: Novel functions of TIMPs in cell signaling. Cancer Metastasis Rev 2006;25:99–113.

15. Jiang Y, Goldberg ID, Shi YE: Complex roles of tissue inhibitors of metalloproteinases in cancer. Oncogene 2002;21:2245–2252.

16. Lambert E, Dasse E, Haye B, Petitfrere E: TIMPs as multifacial proteins. Crit Rev Oncol Hematol 2004;49:187–198.

17. Gomis-Rüth FX, Maskos K, Betz M, Bergner A, Huber R, Suzuki K, Yoshida N, Nagase H, Brew K, Bourenkov G, Bartunik H, Bode W: Mechanism of inhibition of the human matrix metalloproteinase stromelysin-1 by TIMP-1. Nature 1997;Vol 389.

18. Fernandez CA, Moses MA: Modulation of angiogenesis by tissue inhibitor of metalloproteinase-4. Biochem Biophys Res Commun 2006;345:523–529.

19. Noda M, Oh J, Takahashi R, Kondo S, Kitayama H, Takahashi C: RECK: a novel suppressor of malignancy linking oncogenic signaling to extracellular matrix remodeling. Cancer Metastasis Rev 2003;22:167–175.

20. Oh J, Takahashi R, Kondo S, Mizoguchi A, Adachi E, Sasahara RM, Nishimura S, Imamura Y, Kitayama H, Alexander DB, Ide C, Horan TP, Arakawa T, Yoshida H, Nishikawa S, Itoh Y, Seiki M, Itohara S, Takahashi C, Noda M: The membrane-anchored MMP inhibitor RECK is a key regulator of extracellular matrix integrity and angiogenesis. Cell 2001;107:789–800.

21. Cornelius LA, Nehring LC, Roby JD, Parks WC, Welgus HG: Human dermal microvascular endothelial cells produce matrix metalloproteinases in response to angiogenic factors and migration. J Invest Dermatol 1995;105:170–176.

22. Haas TL: Endothelial cell regulation of matrix metalloproteinases. Can J Physiol Pharmacol 2005;83:1–7.

23. Bergers G, Brekken R, McMahon G, Vu TH, Itoh T, Tamaki K, Tanzawa K, Thorpe P, Itohara S, Werb Z, Hanahan D: Matrix metalloproteinase-9 triggers the angiogenic switch during carcinogenesis. Nat Cell Biol 2000;2:737–744.

24. Chantrain CF, Shimada H, Jodele S, Groshen S, Ye W, Shalinsky DR, Werb Z, Coussens LM, DeClerck YA: Stromal matrix metallo proteinase-9 regulates the vascular architecture in neuroblastoma by promoting pericyte recruitment. Cancer Res 2004;64:1675–1686.

25. Itoh T, Tanioka M, Yoshida H, Yoshioka T, Nishimoto H, Itohara S: Reduced angiogenesis and tumor progression in gelatinase A-deficient mice. Cancer Res 1998;58:1048–1051.

26. Guedez L, Rivera AM, Salloum R, Miller ML, Diegmueller JJ, Bungay PM, Stetler-Stevenson WG: Quantitative assessment of angiogenic responses by the directed in vivo angiogenesis assay. Am J Pathol 2003;162:1431–1439.

27. Davis GE, Senger DR: Endothelial extracellular matrix: biosynthesis, remodeling, and functions during vascular morphogenesis and neovessel stabilization. Circ Res 2005;97:1093–1107.

28. Hornebeck W, Emonard H, Monboisse JC, Bellon G: Matrix-directed regulation of pericellular proteolysis and tumor progression. Semin Cancer Biol 2002;12:231–241.

29. van Hinsbergh VW, Engelse MA, Quax PH: Pericellular proteases in angiogenesis and vasculogenesis. Arterioscler Thromb Vasc Biol 2006;26:716–728.

30. Chang C, Werb Z: The many faces of metalloproteases: cell growth, invasion, angiogenesis and metastasis. Trends Cell Biol 2001;11:S37–43.

31. Masson V, de la Ballina LR, Munaut C, Wielockx B, Jost M, Maillard C, Blacher S, Bajou K, Itoh T, Itohara S, Werb Z, Libert C, Foidart JM, Noel A: Contribution of host MMP-2 and MMP-9 to promote tumor vascularization and invasion of malignant keratinocytes. Faseb J 2005;19:234–236.

32. Zucker S, Mirza H, Conner CE, Lorenz AF, Drews MH, Bahou WF, Jesty J: Vascular endothelial growth factor induces tissue factor and matrix metalloproteinase production in endothelial cells: conversion of prothrombin to thrombin results in progelatinase A activation and cell proliferation. Int J Cancer 1998;75:780–786.

33. Sounni NE, Noel A: Membrane type-matrix metalloproteinases and tumor progression. Biochimie 2005;87:329–342.

34. Chun TH, Sabeh F, Ota I, Murphy H, McDonagh KT, Holmbeck K, Birkedal-Hansen H, Allen ED, Weiss SJ: MT1-MMP-dependent neovessel formation within the confines of the three-dimensional extracellular matrix. J Cell Biol 2004;167:757–767.

35. Itoh Y, Seiki M: MT1-MMP: a potent modifier of pericellular microenvironment. J Cell Physiol 2006;206:1–8.

36. Holmbeck K, Bianco P, Birkedal-Hansen H: MT1-mmp: a collagenase essential for tumor cell invasive growth. Cancer Cell 2003;4:83–84.

37. Zhang W, Matrisian LM, Holmbeck K, Vick CC, Rosenthal EL: Fibroblast-derived MT1-MMP promotes tumor progression in vitro and in vivo. BMC Cancer 2006;6:52.

38. Collen A, Hanemaaijer R, Lupu F, Quax PH, van Lent N, Grimbergen J, Peters E, Koolwijk P, van Hinsbergh VW: Membrane-type matrix metalloproteinase-mediated angiogenesis in a fibrin-collagen matrix. Blood 2003;101:1810–1817.

39. Koike T, Vernon RB, Hamner MA, Sadoun E, Reed MJ: MT1-MMP, but not secreted MMPs, influences the migration of human microvascular endothelial cells in 3-dimensional collagen gels. J Cell Biochem 2002;86:748–758.

40. Sounni NE, Janssen M, Foidart JM, Noel A: Membrane type-1 matrix metalloproteinase and TIMP-2 in tumor angiogenesis. Matrix Biol 2003;22:55–61.

41. Lambert V, Wielockx B, Munaut C, Galopin C, Jost M, Itoh T, Werb Z, Baker A, Libert C, Krell HW, Foidart JM, Noel A, Rakic JM: MMP-2 and MMP-9 synergize in promoting choroidal neovascularization. Faseb J 2003;17:2290–2292.

42. Langlois S, Di Tomasso G, Boivin D, Roghi C, Murphy G, Gingras D, Beliveau R: Membrane type 1-matrix metalloproteinase induces endothelial cell morphogenic differentiation by a caspase-dependent mechanism. Exp Cell Res 2005;307:452–464.

43. Nabha SM, Bonfil RD, Yamamoto HA, Belizi A, Wiesner C, Dong Z, Cher ML: Host matrix metalloproteinase-9 contributes to tumor vascularization without affecting tumor growth in a model of prostate cancer bone metastasis. Clin Exp Metastasis 2006.

44. Li A, Dubey S, Varney ML, Dave BJ, Singh RK: IL-8 directly enhanced endothelial cell survival, proliferation, and matrix metalloproteinases production and regulated angiogenesis. J Immunol 2003;170:3369–3376.

45. Park HY, Kwon HM, Lim HJ, Hong BK, Lee JY, Park BE, Jang Y, Cho SY, Kim HS: Potential role of leptin in angiogenesis: leptin induces endothelial cell proliferation and expression of matrix metalloproteinases in vivo and in vitro. Exp Mol Med 2001;33:95–102.

46. Yu Q, Stamenkovic I: Cell surface-localized matrix metalloproteinase-9 proteolytically activates TGF-beta and promotes tumor invasion and angiogenesis. Genes Dev 2000;14:163–176.

47. Hiratsuka S, Nakamura K, Iwai S, Murakami M, Itoh T, Kijima H, Shipley JM, Senior RM, Shibuya M: MMP9 induction by vascular endothelial growth factor receptor-1 is involved in lung-specific metastasis. Cancer Cell 2002;2:289–300.

48. Annes JP, Munger JS, Rifkin DB: Making sense of latent TGF-beta activation. J Cell Sci 2003;116.

49. Levi E, Fridman R, Miao HQ, Ma YS, Yayon A, Vlodavsky I: Matrix metalloproteinase 2 releases active soluble ectodomain of fibroblast growth factor receptor 1. Proc Natl Acad Sci USA 1996;93:7069–7074.

50. Paduch R, Walter-Croneck A, Zdzisinska B, Szuster-Ciesielska A, Kandefer-Szerszen M: Role of reactive oxygen species (ROS), metalloproteinase-2 (MMP-2) and interleukin-6 (IL-6) in direct interactions between tumour cell spheroids and endothelial cell monolayer. Cell Biol Int 2005;29:497–505.

51. Smith MM, Shi L, Navre M: Rapid identification of highly active and selective substrates for stromelysin and matrilysin using bacteriophage peptide display libraries. J Biol Chem 1995;270:6440–6449.

52. Haque NS, Fallon JT, Pan JJ, Taubman MB, Harpel PC: Chemokine receptor-8 (CCR8) mediates human vascular smooth muscle cell chemotaxis and metalloproteinase-2 secretion. Blood 2004;103:1296–1304.

53. Rundhaug JE: Matrix metalloproteinases and angiogenesis. J Cell Mol Med 2005;9:267–285.

54. Caudroy S, Polette M, Nawrocki-Raby B, Cao J, Toole BP, Zucker S, Birembaut P: EMMPRIN-mediated MMP regulation in tumor and endothelial cells. Clin Exp Metastasis 2002;19:697–702.

55. Burbridge MF, Coge F, Galizzi JP, Boutin JA, West DC, Tucker GC: The role of the matrix metalloproteinases during in vitro vessel formation. Angiogenesis 2002;5:215–226.

56. Kondo M, Asai T, Katanasaka Y, Sadzuka Y, Tsukada H, Ogino K, Taki T, Baba K, Oku N: Anti-neovascular therapy by liposomal drug targeted to membrane type-1 matrix metalloproteinase. Int J Cancer 2004;108:301–306.

57. O'Reilly MS, Boehm T, Shing Y, Fukai N, Vasios G, Lane WS, Flynn E, Birkhead JR, Olsen BR, Folkman J: Endostatin: an endogenous inhibitor of angiogenesis and tumor growth. Cell 1997;88:277–285.

58. O'Reilly MS, Holmgren L, Shing Y, Chen C, Rosenthal RA, Moses M, Lane WS, Cao Y, Sage EH, Folkman J: Angiostatin: a novel angiogenesis inhibitor that mediates the suppression of metastases by a Lewis lung carcinoma. Cell 1994;79:315–328.

59. O'Reilly MS, Wiederschain D, Stetler-Stevenson WG, Folkman J, Moses MA: Regulation of angiostatin production by matrix metalloproteinase-2 in a model of concomitant resistance. J Biol Chem 1999;274:29568–29571.

60. Patterson BC, Sang QA: Angiostatin-converting enzyme activities of human matrilysin (MMP-7) and gelatinase B/type IV collagenase (MMP-9). J Biol Chem 1997;272:28823–28825.

61. Kalluri R: Discovery of type IV collagen non-collagenous domains as novel integrin ligands and endogenous inhibitors of angiogenesis. Cold Spring Harbor Symposia on Quantitative Biology 2002;67:255–266.

62. Borza CM, Pozzi A, Borza DB, Pedchenko V, Hellmark T, Hudson BG, Zent R: Integrin alpha3beta1, a novel receptor for alpha3(IV) noncollagenous domain and a trans-dominant Inhibitor for integrin alphavbeta3. J Biol Chem 2006;281:20932–20939.

63. Rodrigues RG, Guo N, Zhou L, Sipes JM, Williams SB, Templeton NS, Gralnick HR, Roberts DD: Conformational regulation of the fibronectin binding and alpha 3beta 1 integrin-mediated adhesive activities of thrombospondin-1. J Biol Chem 2001;276:27913–27922.

64. Wang H, Fu W, Im JH, Zhou Z, Santoro SA, Iyer V, DiPersio CM, Yu QC, Quaranta V, Al-Mehdi A, Muschel RJ: Tumor cell alpha3beta1 integrin and vascular laminin-5 mediate pulmonary arrest and metastasis. J Cell Biol 2004;164:935–941.

65. Chang S, Young BD, Li S, Qi X, Richardson JA, Olson EN: Histone deacetylase 7 maintains vascular integrity by repressing matrix metalloproteinase 10. Cell 2006;126:321–334.

66. Qi JH, Ebrahem Q, Moore N, Murphy G, Claesson-Welsh L, Bond M, Baker A, Anand-Apte B: A novel function for tissue inhibitor of metalloproteinases-3 (TIMP3): inhibition of angiogenesis by blockage of VEGF binding to VEGF receptor-2. Nature Med 2003;9:407–415.

67. Saunders WB, Bohnsack BL, Faske JB, Anthis NJ, Bayless KJ, Hirschi KK, Davis GE: Coregulation of vascular tube stabilization by endothelial cell TIMP-2 and pericyte TIMP-3. J Cell Biol 2006;175:179–191.

68. Cruz-Munoz W, Kim I, Khokha R: TIMP-3 deficiency in the host, but not in the tumor, enhances tumor growth and angiogenesis. Oncogene 2006;25:650–655.

69. Nisato RE, Hosseini G, Sirrenberg C, Butler GS, Crabbe T, Docherty AJ, Wiesner M, Murphy G, Overall CM, Goodman SL, Pepper MS: Dissecting the role of matrix metalloproteinases (MMP) and integrin alpha(v)beta3 in angiogenesis in vitro: absence of hemopexin C domain bioactivity, but membrane-Type 1-MMP and alpha(v)beta3 are critical. Cancer Res 2005;65:9377–9387.

70. Jones GC, Riley GP: ADAMTS proteinases: a multi-domain, multi-functional family with roles in extracellular matrix turnover and arthritis. Arthritis Res Ther 2005;7:160–169.

71. Moses MA, Sudhalter J, Langer R: Identification of an inhibitor of neovascularization from cartilage. Science 1990;248:1408–1410.

72. Murphy AN, Unsworth EJ, Stetler-Stevenson WG: Tissue inhibitor of metalloproteinases-2 inhibits bFGF-induced human microvascular endothelial cell proliferation. J Cell Physiol 1993;157:351 358.

73. Fernandez CA, Butterfield C, Jackson G, Moses MA: Structural and functional uncoupling of the enzymatic and angiogenic inhibitory activities of tissue inhibitor of metalloproteinase-2 (TIMP-2): loop 6 is a novel angiogenesis inhibitor. J Biol Chem 2003;278:40989–40995.

74. Seo DW, Li H, Guedez L, Wingfield PT, Diaz T, Salloum R, Wei BY, Stetler-Stevenson WG: TIMP-2 mediated inhibition of angiogenesis: an MMP-independent mechanism. Cell 2003;114:171–180.

75. Perez-Martinez L, Jaworski DM: Tissue inhibitor of metalloproteinase-2 promotes neuronal differentiation by acting as an anti-mitogenic signal. J Neurosci 2005;25:4917–4929.

76. Seo DW, Li H, Qu CK, Oh J, Kim YS, Diaz T, Wei B, Han JW, Stetler-Stevenson WG: Shp-1 mediates the antiproliferative activity of tissue inhibitor of metalloproteinase-2 in human microvascular endothelial cells. J Biol Chem 2006;281: 3711–3721.

77. Akahane T, Akahane M, Shah A, Connor CM, Thorgeirsson UP: TIMP-1 inhibits microvascular endothelial cell migration by MMP-dependent and MMP-independent mechanisms. Exp Cell Res 2004;301:158–167.

78. Guedez L, McMarlin AJ, Kingma DW, Bennett TA, Stetler-Stevenson M, Stetler-Stevenson WG: Tissue inhibitor of metalloproteinase-1 alters the tumorigenicity of Burkitt's lymphoma via divergent effects on tumor growth and angiogenesis.[see comment]. Am J Pathol 2001;158:1207–1215.

79. Ikenaka Y, Yoshiji H, Kuriyama S, Yoshii J, Noguchi R, Tsujinoue H, Yanase K, Namisaki T, Imazu H, Masaki T, Fukui H: Tissue inhibitor of metalloproteinases-1 (TIMP-1) inhibits tumor growth and angiogenesis in the TIMP-1 transgenic mouse model. Int J Cancer 2003;105:340–346.

80. Johnson MD, Choi H-R, Chesler L, Tsao-Wu G, Bouck N, Polverini PJ: Inhibition of Angiogenesis by Tissue Inhibitor of Metalloproteinase. J Cell Physiol 1994;160:194–202.

81. Thorgeirsson UP, Yoshiji H, Sinha CC, Gomez DE: Breast cancer; Tumor neovasculature and the effect of tissue inhibitor of metalloproteinases-1 (TIMP-1) on angiogenesis. In Vivo 1996;10:137–144.

82. Yamada E, Tobe T, Yamada H, Okamoto N, Zack DJ, Werb Z, Soloway PD, Campochiaro PA: TIMP-1 promotes VEGF-induced neovascularization in the retina. Histol Histopathol 2001;16:87–97.

Chaper 6
Integrins in Angiogenesis

Alireza S. Alavi and David A. Cheresh

Keywords: Integrin, Angiogenesis, Cell Invasion, Apoptosis, Extracellular Matrix

Abstract: Angiogenesis, the formation of new blood vessels from pre-existing vessels, plays a prominent role in a variety of pathological conditions, including tumor growth, ocular disease and arthritis. Endothelial cell adhesion to the extracellular matrix (ECM) permits transduction of signals initiated by angiogenic growth factors leading to activation of distinct signaling pathways that promote endothelial cell proliferation, migration/invasion and survival. Integrins play a pivotal role in promoting the transfer of chemical and mechanical signals from the extracellular environment into the intracellular compartment. Here, we review recent advances in our understanding of the role of distinct integrins in mediating endothelial cell invasion, migration, and importantly, their role in regulating the survival of invasive cells as they encounter distinct new microenvironments. Progress in the use of integrin antagonists as antiangiogenic agents in the clinic will be summarized.

Introduction

Cell adhesion mechanisms facilitating migration/invasion through extracellular matrix (ECM) are critical for the growth of new blood vessels. In fact, during the process of neovascularization, endothelial and stromal cells depend on adhesion receptors for survival, growth and differentiation [1–3]. Integrins, the primary cell surface receptors for the ECM, are a family of heterodimeric transmembrane glycoproteins composed of one α and one β subunit [4]. Noncovalent asso-

ciations between 18 different α and 8 different β subunits form at least 24 distinct cell surface receptors each capable of binding to a subset of cell surface or ECM ligands (Table 6.1). A single ECM ligand can bind to several different integrins (for example, binding of laminin to α1β1, α2β1, α3β1, α6β1, α7β1), while a single integrin can anchor the cell to a variety of different ECM proteins (for example, binding of αvβ3 to vitronectin, fibrinogen, vonWillebrand's factor, proteolyzed collagen and laminin, thrombospondin among others). In addition to ECM proteins, integrins facilitate binding to matrix metalloproteinases (MMPs) and cell surface immunoglobulin-type receptors such as ICAMs (intercellular cell adhesion molecule) and VCAMs (vascular cell adhesion molecule). Integrin-mediated migration and invasion of endothelial cells has emerged as a critical regulator of developmental and pathological angiogenesis.

Cellular adhesion to the interstitial and basement membrane ECM is required for transduction of growth factor-initiated signals that lead to a variety of cellular responses including proliferation, migration and invasion, gene expression, cell survival and ultimately angiogenesis or lymphangiogenesis. While integrins have no intrinsic catalytic activity, they readily sense and integrate signals from the cellular microenvironment and promote cellular responses. In this case, engagement and clustering of integrin receptors promotes the formation of cell-ECM contact sites called focal adhesions. The focal adhesion sites serve to attract a number of important signaling molecules, adaptor proteins and cytoskeletal elements. For example, signaling complexes composed of growth factor receptors, intracellular nonreceptor protein kinases such as focal adhesion kinase (FAK), integrin-linked kinase (ILK), Src and phosphoinositide 3-kinase (PI3K), actin-associated cytoskeletal, adapter and docking proteins, are found in integrin-mediated focal contacts [5]. In fact, integrin-mediated activation of the Rho family of GTPases triggers immediate cellular remodeling events, such as cell spreading, formation of lamellipodia at the leading edge of the cell and release of ECM contacts at the trailing edge to facilitate cell migration [6]. Activation of the Ras GTPase family transduces growth factor-generated

Moores Cancer Center, University of California San Diego, La Jolla, CA, USA

TABLE 6.1. Cell surface receptors formed by noncovalent associations between 18 different α and 8 different β subunits each capable of binding to a subset of cell surface or ECM ligands.

Integrin	Ligands
α1β1	CN, LN
α2β1	CN, LN, FN, TSP
α3β1	CN, LN, FN, TSP
α4β1	FN, VCAM-1, TSP, OST
α4β7	FN, MAdCAM-1, VCAM-1, OST
α5β1	CN, FN, OST
α6β1	LN
α6β4	LN
α7β1	LN
α8β1	FN, VN, TNS, OST
α9β1	TNS, OST
α10β1	LN, CN
α11β1	CN
αIIbβ3	VN, FN, vWf, FG, TSP, CN
αvβ1	FN, VN, OST
αvβ3	VN, FN, vWf, FG, TSP, CN (proteolyzed), MMP-2, TNS, OST, MFG-E8, Del-1, Fibrillin, PECAM-1
αvβ5	VN, OST, MFG-E8, Del-1
αvβ6	FN, TNS, OST
αvβ8	VN
αEβ7	E-Cadherin
αLβ2	ICAM-1, ICAM-2, ICAM-3
αMβ2	ICAM-1, FG, Factor X, iC3b
αXβ2	ICAM-1, FG, LPS, iC3b, CN
αdβ2	ICAM-3

Abbreviations: Collagen (*CN*), Laminin (*LN*), Fibronectin (*FN*), Thrombospondin (*TSP*), Vitronectin (*VN*), von Willebrand's factor (*vWf*), Fibrinogen (*FG*), Osteopontin (*OST*), Tenascin (*TNS*), developmental endothelial locus-1 (*Del-1*), intercellular cell adhesion molecule (*ICAM*), inactivated complement component C3b (*iC3b*), mucosal addressin cell adhesion molecule (*MAdCAM*), milk fat globule EGF factor 8 (*MFG-E8*), platelet endothelial cell adhesion molecule (*PECAM*), vascular cell adhesion molecule (*VCAM*), lipopolysaccharide (*LPS*)

signals to the PI3K/Akt and Ras/mitogen-activated protein (MAP) kinase cascades that activate transcription factors, such as NF-κB, Hox D3 [7], Id1 and Id3 [8] that regulate cell cycle progression, ECM production and cell survival events that are critical for angiogenesis to occur. Therefore, outside-in signaling by integrins regulates cytoskeletal organization, cell migration and survival. Conversely, in circulating cells, such as leukocytes and platelets, agonists induce an inside-out signaling mechanism, leading to changes in the conformation of the cytoplasmic domain of integrins which propagate to the extracellular domain and modulate the ligand-binding affinity of integrins [9].

Integrins in Endothelial Cell Invasion and Angiogenesis

Only a subset of all integrins that match the ligands present within the local tissue microenvironment are expressed on the cell surface. For example, quiescent endothelial cells adhere to ECM proteins collagen or laminin using a specific repertoire

of integrins which includes α1β1, α2β1, α3β1, α5β1, α6β1, α6β4 and αvβ5. This repertoire can be dramatically altered to match the anticipated requirements of a cell exiting its initial niche to invade sites of angiogenesis. Exposure of endothelial cells to angiogenic growth factors within tumors, wounds or inflammatory tissues leads to a general up-regulation of integrins on the cell surface, in particular integrins that bind to provisional ECM components fibronectin, fibrinogen, vitronectin and osteopontin, including α1β1, α2β1 [10], α5β1 [11] and αvβ3 [12, 13].

The angiogenic process has been shown to be impacted by genetic ablations of at least 9 different integrins including α1β1, α2β1 [10], α4β1 [14] and α6β4 [15]. The α9β1 integrin which mediates adhesion to tenascin C [16], osteopontin [17] and vascular cell adhesion molecule-1 (VCAM-1) [18] has been shown to play a critical role in lymphangiogenesis [19]. Targeted disruption of the α5 integrin leads to early embryonic lethality resulting from defects in patterning of the yolk sac and embryonic vasculature [20, 21]. Consistently, inhibition of α5β1 or fibronectin by antibodies or by blocking peptides has been shown to block growth factor-induced angiogenesis [11]. Targeted disruption of the αv class of integrins (αvβ1, αvβ3, αvβ5, αvβ6, αvβ8) has also been shown to impact angiogenesis, but the role of these integrins in endothelial cell invasion and vascular development is more complex. Targeted disruption αv [22] and β8 [23] does not impact embryonic vasculogenesis or angiogenesis since the majority of these embryos survive to term. Mice with targeted disruption of β3, β5, β6 and combinations of β3/β5, β3/β6 and β5/β6 are viable and exhibit no overt angiogenesis defects [24–26]. However, loss of β3 expression and αIIbβ3 function leads to impaired platelet aggregation and bleeding defects and was originally described as a model of Glanzmann thrombaesthenia in humans [24]. Interestingly, mice lacking either β3 or β3 and β5 integrins not only develop normally, but exhibit enhanced tumor growth and growth factor-induced angiogenesis as well as hypersensitivity of endothelial cells to proliferation and permeability responses downstream of vascular endothelial growth factor (VEGF) [27]. This phenotype has been attributed to enhanced expression of VEGF receptor Flk-1/KDR and wound-healing responses and decreased macrophage infiltration [27–29]. In this respect, we have recently shown that the enhanced VEGF/Flk-1 signaling in β3-null mice prevents maturation of coronary capillaries in adult males leading to a thickened endothelium, luminal filopodia or expanded vacuoles [30]. These and earlier results suggest that αvβ3 expression can act to repress proangiogenic signaling. This may be mediated via a direct interaction with growth factor receptors [31, 32] or via regulation of cell survival.

In addition to integrin-mediated signaling, the composition of the ECM plays an important role in tumor angiogenesis. Endothelial and immune cells, as well as surrounding stromal and tumor cells, produce new ECM proteins, as well as matrix metalloproteinases such as MMP-9 and MMP-2, that remodel and clear the existing ECM for cellular invasion during angiogenesis [33]. Interestingly, MMP-2 has been shown to interact with the $\alpha v\beta 3$ integrin, a process believed to promote localization and activation of MMP-2 metalloproteinase activity to discrete regions of the cell surface to enhance directed cellular invasion [34]. Negative regulators of MMPs, such as the autoproteolytic hemopexin fragment of MMP-2 (PEX), accumulate during angiogenesis to prevent excessive ECM degradation which may reduce cellular traction and inhibit migration [35]. Cleavage and degradation of the tumor ECM components by MMPs not only modifies the structural configuration of the ECM to release sequestered angiogenic growth factors, such as VEGF and basic fibroblast growth factor (bFGF), but also reveals new integrin adhesion sites that support cellular migration and survival during angiogenesis [36]. For example, cleavage of fibrillar collagens, fibronectin and laminin during tissue remodeling reveals cryptic sites recognized by $\alpha v\beta 3$ and $\alpha 3\beta 1$ integrins. This has been shown to be critical for endothelial cell growth, differentiation and tube formation during angiogenesis.

The $\alpha v\beta 3$ Integrin

Integrin $\alpha v\beta 3$ is one of the most intensively studied and best characterized integrins and was the first adhesion receptor to be characterized as a marker of angiogenic blood vessels [13]. Quiescent endothelial cells express little or no $\alpha v\beta 3$ on the cell surface. However, stimulation of endothelial cells by growth factors promotes the expression of high levels of $\alpha v\beta 3$ on newly-formed blood vessels. Integrin $\alpha v\beta 3$ is the most promiscuous receptor in the integrin family. This enables endothelial cells expressing this integrin to attach to and migrate on a wide range of provisional matrix components, such as fibrin, vitronectin, fibronectin, osteopontin, proteolyzed collagen, von Willebrand factor and other ECM components found in high concentrations within tissues undergoing remodeling, such as wounds, inflammatory sites and tumors [37]. Dvorak and colleagues have referred to tumors as "wounds that do not heal" [38–40] since proangiogenic provisional matrix components are highly enriched within tumors. These deposits provide a supportive scaffold that facilitates endothelial cell adhesion, migration and invasion leading to enhanced angiogenesis [38, 41, 42]. In contrast, cellular invasion of the basement membrane and interstitial space is dependent on MMP-mediated proteolysis of collagen and laminin to expose cryptic RGD motifs bound by $\alpha v\beta 3$. In this case, quiescent cells adhering to native collagen in an $\alpha 2\beta 1$-dependent manner switch to $\alpha v\beta 3$-dependent adhesion to RGD sites present in proteolyzed collagen to promote cellular invasion [43, 44].

Extensive evidence suggests that antagonism of $\alpha v\beta 3$ blocks angiogenesis. Ligation of $\alpha v\beta 3$ enhances cellular invasion through the ECM and promotes MMP-2 localization and activation at the leading edge of the invading cell [34]. In addition, ligated $\alpha v\beta 3$ induces cellular proliferation by promoting sustained MAPK signaling [45] and enhances cell survival via suppression of caspase 8 [46], activation of the NF-κB cascade [47], and inhibition of p53 activity [48]. In contrast, inhibition of $\alpha v\beta 3$ function using function-blocking antibodies or RGD-containing peptide mimetics suppresses corneal vascularization [49], hypoxia-induced retinal neovascularization [50] and tumor angiogenesis in the mouse [13]. Several antagonists of $\alpha 5\beta 1$ and αv integrins are currently in clinical trials as inhibitors of angiogenesis and tumor cell metastasis (see below). Nevertheless, integrin $\alpha v\beta 3$-deficient mice survive with most of their blood vessels intact [24, 30] suggesting that mice lacking $\alpha v\beta 3$ may compensate for the loss of this integrin. Alternatively, $\alpha v\beta 3$ may act as a negative regulator of blood vessel growth [51].

Integrin-mediated Regulation of Cell Survival

During angiogenesis, activated endothelial cells exit their local microenvironment to invade new tissues and are subject to apoptotic controls designed to maintain tissue homeostasis. Cell survival is influenced by cues from matrix proteins and growth factors within the extracellular matrix which can provide protection from distinct pro-apoptotic mechanisms. Therefore, survival is an intrinsic and critical component of cell invasion and angiogenesis [52–54] (Fig. 6.1). Integrins, which have been intimately associated with cell adhesion and migration, also play a role in regulating cell survival by regulating survival signaling events such as activation of Akt, FAK, Src, PI3K, ERK and mTOR (mammalian target of rapamycin) to regulate CAP-dependent protein synthesis [46, 54–56].

Several studies indicate that integrins $\alpha v\beta 3$ and $\alpha 5\beta 1$ not only appear to contribute to vascular cell invasion and survival but under some circumstances can actually promote cell death [46, 57–59]. In this respect, when in the unligated state, these integrins can directly promote apoptosis through caspase-dependent [60] and independent [58] mechanisms. Furthermore, RGD-based peptidomimetic antagonists and blocking antibodies of αv integrins have been shown to block growth factor- and tumor-induced angiogenesis in the retina, in the chick chorioallantoic membrane (CAM), and in tumors by inducing apoptosis of invasive endothelial cells without affecting the quiescent vasculature [13, 50, 61, 62]. This inhibition of angiogenesis depends on p53 function since the same αv integrin antagonists have minimal effects on angiogenesis in p53-deficient mice [63]. Consistent with these observations, unligated $\alpha v\beta 3$ in adherent cells [46] or $\alpha v\beta 3$ antagonized by soluble ligands [51, 55, 64–67] promotes programmed cell death. At the molecular level, unligated $\alpha v\beta 3$ on adherent cells was shown to recruit and cluster initiator caspase 8 at the membrane leading to its cleavage and activation, triggering apoptosis [46]. This process, termed integrin-mediated death (IMD), is not dependent on death receptor ligation [46] and is in contrast

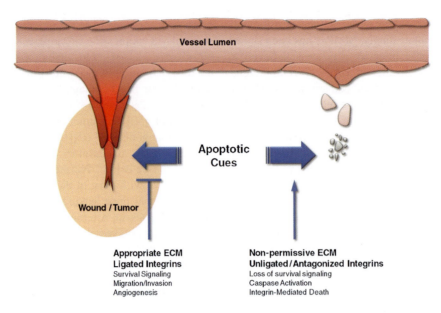

FIG. 6.1. Influence of integrin ligation state on cell survival. Integrin ligation promotes signaling cascades that promote survival as the cell invades foreign tissue microenvironments. Lack of integrin ligation disrupts cell survival signaling and can promote caspase activation leading to programmed cell death.

to 'anoikis' or cell death due to loss of adhesion [60]. During IMD, integrins behave as dependence receptors that promote survival in the presence of ligands while inducing apoptosis in the absence of the trophic survival factor (ECM ligation) [68]. In this case, loss of the dependence receptor (integrin) leads to enhanced survival, while overexpression in the absence of ligation promotes cell death. Therefore, lack of integrin-mediated cell-ECM interaction can promote cell death through loss of survival signals and/or active initiation of the apoptotic cascade. This may explain why mice lacking αv integrins show increased angiogenesis during pathological angiogenesis [27].

Integrins αvβ3 and αvβ5 are highly homologous integrins that share a number of their ligand binding properties. These integrins were shown to regulate distinct pathways of cell survival and angiogenesis. Specifically, antagonists of αvβ3 preferentially suppress bFGF-mediated angiogenesis while antagonists of αvβ5 disrupt VEGF-dependent blood vessel growth [49]. The VEGF/αvβ5 pathway appeared to depend on Src kinase while the bFGF/αvβ3 did not [69]. Importantly, both pathways differentially regulate Ras-Raf-Mek-Erk signaling contributing to distinct mechanisms of endothelial cell survival [70, 71] (Fig. 6.2). bFGF together with αvβ3 activates Raf promoting its localization to the mitochondria and protection of cells from stress-mediated death. Importantly, this does not involve Raf activation of its effectors Mek or Erk. In contrast, VEGF/αvβ5 signaling activates Raf in a Src-dependent manner promoting endothelial cell protection from apoptosis due to death ligands such as tumor necrosis factor-α (TNF-α) and Fas ligand (FasL). These distinct angiogenic pathways may enable endothelial cells to survive and thereby undergo angiogenesis in distinct proapoptotic environments that might otherwise cause cells to undergo apoptosis. For example, new

blood vessel growth at sites of inflammation may depend on resisting the apoptotic effects of inflammatory mediators such as TNF-α, whereas in ischemic tissue endothelial cells must overcome the hypoxic stress response. However, in tumors it is likely that both of these pathways are activated enabling endothelial cells to survive in the face of both stress (hypoxia, regions of nutrient deprivation, chemotherapeutic intervention) and the presence of death ligands.

Proteolysis of the ECM Generates Integrin-Binding Antiangiogenic Peptide Fragments

Proteolytic remodeling of the extracellular matrix has been associated with angiogenesis. Both serine proteases and matrix metalloproteinases (MMPs) are known to facilitate this process [72]. However, MMPs appear to have a dual role in regulating angiogenesis. Not only do they remodel the collagenous extracellular matrix/basement membrane thereby promoting cell invasion but they also degrade collagen to reveal cryptic integrin binding sites. These cryptic sites reveal adhesive sequences that bind integrins and likely contribute to the invasive properties of angiogenic endothelial cells [36]. In fact, a number of antibodies directed to the integrin binding domains of these cyptic collagen fragments are themselves antiangiogenic [13, 61]. Once these proteolyzed protein fragments are released into the circulation or local tissue microenvironment they have potent antiangiogenic activity based on the ability of these peptides to bind to and antagonize endothelial cell integrins. Thus, proteases not only promote angiogenesis by facilitating endothelial cell invasion but also facilitate

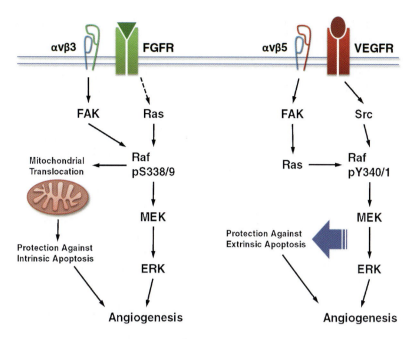

FIG. 6.2. Specific integrin/growth factor signaling complexes promote protection of cells against distinct pathways of apoptosis. Raf-1 is differentially phosphorylated downstream of αvβ3/bFGF or αvβ5/VEGF signaling cascades leading to differential protection from intrinsic or extrinsic apoptotic pathways, respectively.

the release of naturally occurring inhibitors of angiogenesis. In the last several years, a number of such naturally occurring proteolytic antiangiogenic protein fragments have been identified. These inhibitory fragments include: endostatin, tumstatin, arrestin, canstatin and the plasminogen-derived angiostatin [73–81] as listed below.

Tumstatin is derived from MMP-9-mediated cleavage of the C-terminal non-collagenous (NC1) collagen α3 (IV) subunit and has been shown to bind to endothelial cells via αvβ3 and α6β1 and promote apoptosis via negative regulation of mTOR signaling leading to inhibition protein synthesis [55, 67, 82, 83]. These effects are mediated through binding of tumstatin to αvβ3 although the antiangiogenic segment of tumstatin lacks an RGD sequence and thus does not appear to directly compete with vitronectin. Mice lacking Collagen IV α3 show enhanced angiogenesis leading to accelerated tumor growth, a process which can be corrected by exogenous supplementation of mice with recombinant tumstatin [64]. Consistently, mice lacking αvβ3 are resistant to the antiangiogenic effects of tumstatin [27, 64, 84].

Endostatin, a C-terminal fragment of the α1-chain of type XVIII collagen, has been shown to bind to cell surface integrins αvβ3, αvβ5 and α5β1 in an RGD-independent manner and blocks endothelial cell proliferation and promote apoptosis [59, 75, 85–87]. Binding of endostatin to α5β1 inhibits the MAPK pathway but has no effect on the PI3K/Akt pathway [66]. Endostatin has also been shown to block the function of MMP-2, -9, -13 and MT1-MMP [74, 81, 82, 88, 89].

Arresten is a collagen α1 (IV) NC1 proteolytic fragment that can bind to α1β1 and block endothelial cell tube forma-

tion, proliferation and migration in vitro and tumor growth in vivo [73].

Canstatin is a collagen α2 (IV) NC1 domain fragment which, similar to Arresten, is capable of inhibiting endothelial cell proliferation, migration and tube formation in vitro and growth factor-induced angiogenesis and tumor growth in vivo [77]. Canstatin has been shown to bind to αvβ3, αvβ5 and α3β1 integrins on endothelial cells [78].

PEX, the noncatalytic C-terminal hemopexin C domain of MMP-2 mediates a direct RGD-independent interaction with αvβ3 [34]. Soluble autolytic fragment PEX is a naturally occurring peptide that blocks cells surface activation of MMP-2 by competing with MMP-2 for binding to integrin αvβ3. Treatment of growth factor-stimulated vasculature with PEX inhibits angiogenesis and tumor growth [35]. Virally-expressed PEX inhibits bFGF-induced MMP-2 activation leading to inhibition of tumor-induced angiogenesis and tumor growth [90]. Activation of MMP-2 by αvβ3 may not be direct and may depend on functional cooperation between MT1-MMP and αvβ3 [91].

Integrin Interactions During Vessel Maturation

During angiogenesis, the interaction between pericytes and vascular smooth muscle cells (PC/VSMC) with endothelial cells plays an important role in remodeling and stabilization of the primitive vascular plexus, composed of nascent endothelial tubes, into a complex network of arterial and venous blood vessels and small microcapillaries [92, 93]. PC/VSMC are localized on the abluminal endothelial surface and regulate vessel diameter and endothelial cell proliferation and migration via physical and juxtacrine interactions [94, 95]. PC/VSMCs produce and organize a fibronectin-rich basement membrane which is shared with the underlying endothelial cells [96, 97]. The interaction between PC/VSMC and endothelial cells is critical for the formation and stabilization of functional vasculature during development and tumorigenesis.

The fibronectin-binding integrin $\alpha 4\beta 1$ (VLA-4) is expressed as early as E9.5 in endothelial cells and pericytes and plays a critical role in cell migration [98–103] and embryonic vascular development [14, 104]. Disruption of $\alpha 4\beta 1$ perturbs the distribution of PC/VSMC along angiogenic vessels leading to defects in blood vessel diameter in vivo and abnormal migratory capacity on fibronectin in vitro [98]. Interestingly, disruption of the $\beta 4$ integrin does not impact the process of vasculogenesis but is limited to angiogenic vessels [98]. Thus, the $\alpha 4\beta 1$ integrin appears to play an important role in angiogenesis, by regulating PC/VSMC spreading and distribution along vessels, and in vascular contraction and/or intussusceptive branching.

Another major ligand for $\alpha 4\beta 1$ is vascular cell adhesion molecule-1 (VCAM-1) [105]. Expression of VCAM-1 on the activated endothelium in inflamed tissues promotes adhesion and extravasation of $\alpha 4\beta 1$-expressing circulating lymphocytes [106, 107]. It has recently been shown that $\alpha 4\beta 1$ is expressed on growth factor-stimulated activated endothelium but not on quiescent normal vessels [14]. Furthermore, $\alpha 4\beta 1$ expression is upregulated in the vasculature of breast and colon carcinoma and in melanoma tumors. In addition to fibronectin, the expression of VCAM-1 is also strongly upregueated in VSMCs of the neovasculature within tumors but not in those of normal vessels. Function-blocking antibodies to $\alpha 4\beta 1$ or VCAM-1 disrupt pericyte/endothelial interactions in vivo and promote apoptosis of both cell types leading to inhibition of angiogenesis and tumor growth. Thus, $\alpha 4\beta 1$/VCAM-1-mediated association of PC/VSMCs appears to promote cell survival during angiogenesis [14].

Circulating bone marrow-derived myeloid progenitors have been shown to promote tumor angiogenesis and inflammatory disease by either differentiating into endothelial cells and directly participating in vessel sprouting [108, 109] or by expressing angiogenic growth factors [110, 111]. Integrins have been shown to play an important role in trafficking of these cells into and out of the tumor microenvironment. Expression of $\alpha 4\beta 1$ is required for macrophage/monocyte trafficking to the tumor vasculature [112]. Furthermore, bone marrow-derived CD34$^+$ endothelial progenitor cell recruitment to the tumor is mediated by $\alpha 4\beta 1$ binding to VCAM-1 and cellular fibronectin, which are both highly expressed in the activated and proliferating endothelium within tumors [113]. This is significant in light of the observation that circulating bone marrow-derived myeloid progenitors can give rise to up to 15% of the tumor neovasculature during angiogensis [114, 115].

Integrin Antagonists in Clinical Trials

Selective up-regulation of specific integrins on the surface of activated angiogenic endothelium make these receptors useful biomarkers for targeting and early diagnostic imaging of malignant tumors. Antagonists of $\alpha v\beta 3$ include function-blocking humanized antibodies, RGD mimetic peptides [116, 117] and small molecule inhibitors [118, 119]. A humanized form of the function-blocking monoclonal anti-$\alpha v\beta 3$ antibody LM609 (Vitaxin®) and Abegrin®, an affinity-matured form of Vitaxin, have been used in phase I and II clinical trials [120] and have been associated with disease stabilization in a number of patients [121]. In phase II trials of stage IV melanoma patients, Abegrin when administered as a single agent significantly prolonged patient survival (by 53%). Based on these studies Abegrin is now being tested in Phase III trials.

Volociximab (M200) is a humanized form of the function-blocking mouse monoclonal IIA1 anti-$\alpha v\beta 1$ antibody which promotes apoptosis of actively proliferating endothelial cells and inhibits in vitro endothelial cell tube formation [122]. Volociximab is currently in phase II clinical trials for patients with metastatic pancreatic cancer, renal cell carcinoma and late stage melanoma. CNTO 95 is a fully humanized antibody that binds to and inhibits the αv family of integrins, including $\alpha v\beta 3$ and $\alpha v\beta 5$, leading to inhibition of integrin-mediated tumor growth and angiogenesis in mice [123]. CNTO 95 is currently in phase I clinical trials for patients with advanced melanoma [124]. A cyclic RGD peptide inhibitor of $\alpha v\beta 3$ and $\alpha v\beta 5$, EMD 121974 or Cilengitide™, is in Phase II clinical trials as an angiogenesis inhibitor for glioblastoma and other cancers [125, 126]. In addition, a number of other integrin-blocking peptides with antiangiogenic and anti-tumor effects are currently in preclinical development [72, 127, 128].

Expression of $\alpha v\beta 3$ on the angiogenic vasculature within tumors, inflammatory disease and atherosclerotic lesions has also been used for noninvasive molecular imaging of angiogenesis. The anti-$\alpha v\beta 3$ antibody LM609 conjugated to gadolinium containing paramagnetic liposomes has been used for imaging of the angiogenic vasculature by magnetic resonance imaging (MRI) [129]. In a different study, a paramagnetic nanoparticle was directed to the $\alpha v\beta 3$ integrin epitopes of atherosclerotic plaque-associated neovasculature using a peptidomimetic vitronectin antagonist [130]. In addition, a number of radiolabeled monomeric and dimeric cyclic RGD mimetic peptides have been evaluated by positron emission tomography (PET) [131–135], single-photon emission computed tomography (SPECT) [136, 137] and near-infrared (700–900 nm) fluorescence (NIRF) imaging [138] as noninvasive αv-targeted radiotracers with highly selective accumulation and image contrast in several preclinical xenograft models. In clinical studies, radiolabeled RGD peptides may be used to document $\alpha v\beta 3$ expression of the tumors before the administration of therapeutic doses of $\alpha v\beta 3$ antagonists, thus allowing appropriate selection of patients for clinical trials. Furthermore, radiolabeled RGD peptides may be used to assess the inhibition of the αvβ3 integrin by specific antagonists and to define in vivo pharmacokinetics and dose optimization. Accordingly, a reduction in the uptake of radiolabeled RGD peptides may be used as a marker for successful inhibition of angiogenesis by other therapeutic agents.

The use of αvβ3-targeted agents have important clinical relevance as effective vehicles for drug delivery to the angiogenic vasculature as well as invasive tumors with upregulated αvβ3 expression. RGD-containing peptides and liposomes carrying the genotoxic agent doxorubicin have been used to target the angiogenic vasculature [139, 140]. Adenoviral particles modified to encode the RGD integrin-binding motif on the surface significantly enhanced gene transfer to αvβ3 expressing cells [141]. Recently, target tissue selectivity of avβ3 ligand-conjugated nanoparticles has been used in combination with gene target selectivity of plasmid DNA encoding a dominant negative form of Raf-1 [71] or siRNA sequences for VEGF-R2 knockdown [142] leading to inhibition of angiogenesis and tumor growth.

Conclusions

Integrins play a pivotal role in endothelial cell invasion events that promote neovascularization by regulating growth factor-induced cellular responses such as proliferation, migration and survival. The presence of naturally occurring endogenous integrin antagonists that negatively regulate angiogenesis underscores the significance of integrin-mediated endothelial cell-ECM interactions in tumor angiogenesis and justifies the use of integrin antagonists as anti-cancer therapeutics. Integrin antagonists, conventional chemotherapeutics and other inhibitors of angiogenesis in combination with αvβ3-targeted delivery to the angiogenic vasculature promise to significantly enhance the efficacy of anti-cancer strategies. In addition, targeting of integrins and thus inhibition of growth factor-induced survival signals can be used to enhance the chemosensitivity of the angiogenic vasculature as well as tumor cells.

References

1. Folkman J, D'Amore PA. Blood vessel formation: what is its molecular basis? Cell 1996;87(7):1153–5.
2. Ingber DE, Folkman J. Mechanochemical switching between growth and differentiation during fibroblast growth factor-stimulated angiogenesis in vitro: role of extracellular matrix. The Journal of cell biology 1989;109(1):317–30.
3. Kalebic T, Garbisa S, Glaser B, Liotta LA. Basement membrane collagen: degradation by migrating endothelial cells. Science 1983;221(4607):281–3.
4. Hynes RO. Integrins: versatility, modulation, and signaling in cell adhesion. Cell 1992;69(1):11–25.
5. Sastry SK, Burridge K. Focal adhesions: a nexus for intracellular signaling and cytoskeletal dynamics. Experimental cell research 2000;261(1):25–36.
6. DeMali KA, Wennerberg K, Burridge K. Integrin signaling to the actin cytoskeleton. Current opinion in cell biology 2003;15(5):572–82.
7. Boudreau N, Andrews C, Srebrow A, Ravanpay A, Cheresh DA. Induction of the angiogenic phenotype by Hox D3. The Journal of cell biology 1997;139(1):257–64.
8. Lyden D, Young AZ, Zagzag D, et al. Id1 and Id3 are required for neurogenesis, angiogenesis and vascularization of tumour xenografts. Nature 1999;401(6754):670–7.
9. Ginsberg MH, Du X, Plow EF. Inside-out integrin signalling. Current opinion in cell biology 1992;4(5):766–71.
10. Senger DR, Claffey KP, Benes JE, Perruzzi CA, Sergiou AP, Detmar M. Angiogenesis promoted by vascular endothelial growth factor: regulation through alpha1beta1 and alpha2beta1 integrins. Proceedings of the National Academy of Sciences of the United States of America 1997;94(25):13612–7.
11. Kim S, Bell K, Mousa SA, Varner JA. Regulation of angiogenesis in vivo by ligation of integrin alpha5beta1 with the central cell-binding domain of fibronectin. The American journal of pathology 2000;156(4):1345–62.
12. Okada Y, Copeland BR, Hamann GF, Koziol JA, Cheresh DA, del Zoppo GJ. Integrin alphavbeta3 is expressed in selected microvessels after focal cerebral ischemia. The American journal of pathology 1996;149(1):37–44.
13. Brooks PC, Clark RA, Cheresh DA. Requirement of vascular integrin alpha v beta 3 for angiogenesis. Science 1994;264(5158):569–71.
14. Garmy-Susini B, Jin H, Zhu Y, Sung RJ, Hwang R, Varner J. Integrin alpha4beta1-VCAM-1-mediated adhesion between endothelial and mural cells is required for blood vessel maturation. The Journal of clinical investigation 2005;115(6):1542–51.
15. Hiran TS, Mazurkiewicz JE, Kreienberg P, Rice FL, LaFlamme SE. Endothelial expression of the alpha6beta4 integrin is negatively regulated during angiogenesis. Journal of cell science 2003;116(Pt 18):3771–81.
16. Yokosaki Y, Palmer EL, Prieto AL, et al. The integrin alpha 9 beta 1 mediates cell attachment to a non-RGD site in the third fibronectin type III repeat of tenascin. The Journal of biological chemistry 1994;269(43):26691–6.
17. Smith LL, Cheung HK, Ling LE, et al. Osteopontin N-terminal domain contains a cryptic adhesive sequence recognized by alpha9beta1 integrin. The Journal of biological chemistry 1996;271(45):28485–91.
18. Taooka Y, Chen J, Yednock T, Sheppard D. The integrin alpha-9beta1 mediates adhesion to activated endothelial cells and transendothelial neutrophil migration through interaction with vascular cell adhesion molecule-1. The Journal of cell biology 1999;145(2):413–20.
19. Huang XZ, Wu JF, Ferrando R, et al. Fatal bilateral chylothorax in mice lacking the integrin alpha9beta1. Molecular and cellular biology 2000;20(14):5208–15.
20. Francis SE, Goh KL, Hodivala-Dilke K, et al. Central roles of alpha5beta1 integrin and fibronectin in vascular development in mouse embryos and embryoid bodies. Arteriosclerosis, thrombosis, and vascular biology 2002;22(6):927–33.
21. Yang JT, Rayburn H, Hynes RO. Embryonic mesodermal defects in alpha 5 integrin-deficient mice. Development (Cambridge, England) 1993;119(4):1093–105.
22. Bader BL, Rayburn H, Crowley D, Hynes RO. Extensive vasculogenesis, angiogenesis, and organogenesis precede lethality in mice lacking all alpha v integrins. Cell 1998;95(4):507–19.

23. Zhu J, Motejlek K, Wang D, Zang K, Schmidt A, Reichardt LF. beta8 integrins are required for vascular morphogenesis in mouse embryos. Development (Cambridge, England) 2002;129(12):2891–903.

24. Hodivala-Dilke KM, McHugh KP, Tsakiris DA, et al. Beta3-integrin-deficient mice are a model for Glanzmann thrombasthenia showing placental defects and reduced survival. The Journal of clinical investigation 1999;103(2):229–38.

25. Huang X, Griffiths M, Wu J, Farese RV, Jr., Sheppard D. Normal development, wound healing, and adenovirus susceptibility in beta5-deficient mice. Molecular and cellular biology 2000;20(3):755–9.

26. Huang XZ, Wu JF, Cass D, et al. Inactivation of the integrin beta 6 subunit gene reveals a role of epithelial integrins in regulating inflammation in the lung and skin. The Journal of cell biology 1996;133(4):921–8.

27. Reynolds LE, Wyder L, Lively JC, et al. Enhanced pathological angiogenesis in mice lacking beta3 integrin or beta3 and beta5 integrins. Nature medicine 2002;8(1):27–34.

28. Reynolds LE, Conti FJ, Lucas M, et al. Accelerated re-epithelialization in beta3-integrin-deficient- mice is associated with enhanced TGF-beta1 signaling. Nature medicine 2005;11(2):167–74.

29. Taverna D, Moher H, Crowley D, Borsig L, Varki A, Hynes RO. Increased primary tumor growth in mice null for beta3- or beta3/beta5-integrins or selectins. Proceedings of the National Academy of Sciences of the United States of America 2004;101(3):763–8.

30. Weis SM, Lindquist JN, Barnes LA, et al. Cooperation between VEGF and {beta}3 integrin during cardiac vascular development. Blood 2006.

31. Schneller M, Vuori K, Ruoslahti E. Alphavbeta3 integrin associates with activated insulin and PDGFbeta receptors and potentiates the biological activity of PDGF. The EMBO journal 1997;16(18):5600–7.

32. Borges E, Jan Y, Ruoslahti E. Platelet-derived growth factor receptor beta and vascular endothelial growth factor receptor 2 bind to the beta 3 integrin through its extracellular domain. The Journal of biological chemistry 2000;275(51):39867–73.

33. Bergers G, Brekken R, McMahon G, et al. Matrix metalloproteinase-9 triggers the angiogenic switch during carcinogenesis. Nature cell biology 2000;2(10):737–44.

34. Brooks PC, Stromblad S, Sanders LC, et al. Localization of matrix metalloproteinase MMP-2 to the surface of invasive cells by interaction with integrin alpha v beta 3. Cell 1996;85(5):683–93.

35. Brooks PC, Silletti S, von Schalscha TL, Friedlander M, Cheresh DA. Disruption of angiogenesis by PEX, a noncatalytic metalloproteinase fragment with integrin binding activity. Cell 1998;92(3):391–400.

36. Xu J, Rodriguez D, Petitclerc E, et al. Proteolytic exposure of a cryptic site within collagen type IV is required for angiogenesis and tumor growth in vivo. The Journal of cell biology 2001;154(5):1069–79.

37. Clark RA, Tonnesen MG, Gailit J, Cheresh DA. Transient functional expression of alphaVbeta 3 on vascular cells during wound repair. The American journal of pathology 1996;148(5):1407–21.

38. Nagy JA, Brown LF, Senger DR, et al. Pathogenesis of tumor stroma generation: a critical role for leaky blood vessels and fibrin deposition. Biochimica et biophysica acta 1989;948(3):305–26.

39. Dvorak HF, Harvey VS, Estrella P, Brown LF, McDonagh J, Dvorak AM. Fibrin containing gels induce angiogenesis. Implications for tumor stroma generation and wound healing. Laboratory investigation; a journal of technical methods and pathology 1987;57(6):673–86.

40. Dvorak HF. Tumors: wounds that do not heal. Similarities between tumor stroma generation and wound healing. The New England journal of medicine 1986;315(26):1650–9.

41. Bayless KJ, Salazar R, Davis GE. RGD-dependent vacuolation and lumen formation observed during endothelial cell morphogenesis in three-dimensional fibrin matrices involves the alpha(v)beta(3) and alpha(5)beta(1) integrins. The American journal of pathology 2000;156(5):1673–83.

42. van Hinsbergh VW, Collen A, Koolwijk P. Role of fibrin matrix in angiogenesis. Annals of the New York Academy of Sciences 2001;936:426–37.

43. Montgomery AMP, Reisfeld RA, Cheresh DA. Integrin {alpha}v{beta}3 Rescues Melanoma Cells from Apoptosis in Three-Dimensional Dermal Collagen. In; 1994:8856–60.

44. Senger DR, Perruzzi CA, Streit M, Koteliansky VE, de Fougerolles AR, Detmar M. The alpha(1)beta(1) and alpha(2)beta(1) integrins provide critical support for vascular endothelial growth factor signaling, endothelial cell migration, and tumor angiogenesis. The American journal of pathology 2002;160(1):195–204.

45. Eliceiri BP, Klemke R, Stromblad S, Cheresh DA. Integrin alphavbeta3 requirement for sustained mitogen-activated protein kinase activity during angiogenesis. The Journal of cell biology 1998;140(5):1255–63.

46. Stupack DG, Puente XS, Boutsaboualoy S, Storgard CM, Cheresh DA. Apoptosis of adherent cells by recruitment of caspase-8 to unligated integrins. The Journal of cell biology 2001;155(3):459–70.

47. Scatena M, Almeida M, Chaisson ML, Fausto N, Nicosia RF, Giachelli CM. NF-kappaB mediates alphavbeta3 integrin-induced endothelial cell survival. The Journal of cell biology 1998;141(4):1083–93.

48. Stromblad S, Becker JC, Yebra M, Brooks PC, Cheresh DA. Suppression of p53 activity and p21WAF1/CIP1 expression by vascular cell integrin alphaVbeta3 during angiogenesis. The Journal of clinical investigation 1996;98(2):426–33.

49. Friedlander M, Brooks PC, Shaffer RW, Kincaid CM, Varner JA, Cheresh DA. Definition of two angiogenic pathways by distinct alpha v integrins. Science 1995;270(5241):1500–2.

50. Hammes HP, Brownlee M, Jonczyk A, Sutter A, Preissner KT. Subcutaneous injection of a cyclic peptide antagonist of vitronectin receptor-type integrins inhibits retinal neovascularization. Nature medicine 1996;2(5):529–33.

51. Maeshima Y, Yerramalla UL, Dhanabal M, et al. Extracellular matrix-derived peptide binds to alpha(v)beta(3) integrin and inhibits angiogenesis. The Journal of biological chemistry 2001;276(34):31959–68.

52. Stupack DG, Cheresh DA. Get a ligand, get a life: integrins, signaling and cell survival. Journal of cell science 2002;115(Pt 19):3729–38.

53. Mehlen P, Puisieux A. Metastasis: a question of life or death. Nature reviews 2006;6(6):449–58.

54. Stupack DG, Teitz T, Potter MD, et al. Potentiation of neuroblastoma metastasis by loss of caspase-8. Nature 2006;439(7072):95–9.

55. Maeshima Y, Colorado PC, Kalluri R. Two RGD-independent alpha vbeta 3 integrin binding sites on tumstatin regulate dis-

tinct anti-tumor properties. The Journal of biological chemistry 2000;275(31):23745–50.

56. Giancotti FG, Ruoslahti E. Integrin signaling. Science 1999;285(5430):1028–32.

57. Storgard CM, Stupack DG, Jonczyk A, Goodman SL, Fox RI, Cheresh DA. Decreased angiogenesis and arthritic disease in rabbits treated with an alphavbeta3 antagonist. The Journal of clinical investigation 1999;103(1):47–54.

58. Jan Y, Matter M, Pai JT, et al. A mitochondrial protein, Bit1, mediates apoptosis regulated by integrins and Groucho/TLE corepressors. Cell 2004;116(5):751–62.

59. Dhanabal M, Ramchandran R, Waterman MJ, et al. Endostatin induces endothelial cell apoptosis. The Journal of biological chemistry 1999;274(17):11721–6.

60. Frisch SM, Screaton RA. Anoikis mechanisms. Current opinion in cell biology 2001;13(5):555–62.

61. Brooks PC, Stromblad S, Klemke R, Visscher D, Sarkar FH, Cheresh DA. Antiintegrin alpha v beta 3 blocks human breast cancer growth and angiogenesis in human skin. The Journal of clinical investigation 1995;96(4):1815–22.

62. Friedlander M, Theesfeld CL, Sugita M, et al. Involvement of integrins alpha v beta 3 and alpha v beta 5 in ocular neovascular diseases. Proceedings of the National Academy of Sciences of the United States of America 1996;93(18):9764–9.

63. Stromblad S, Fotedar A, Brickner H, et al. Loss of p53 compensates for alpha v-integrin function in retinal neovascularization. The Journal of biological chemistry 2002;277(16):13371–4.

64. Hamano Y, Zeisberg M, Sugimoto H, et al. Physiological levels of tumstatin, a fragment of collagen IV alpha3 chain, are generated by MMP-9 proteolysis and suppress angiogenesis via alphaV beta3 integrin. Cancer cell 2003;3(6):589–601.

65. Maeshima Y, Manfredi M, Reimer C, et al. Identification of the anti-angiogenic site within vascular basement membrane-derived tumstatin. The Journal of biological chemistry 2001;276(18):15240–8.

66. Sudhakar A, Sugimoto H, Yang C, Lively J, Zeisberg M, Kalluri R. Human tumstatin and human endostatin exhibit distinct anti-angiogenic activities mediated by alpha v beta 3 and alpha 5 beta 1 integrins. Proceedings of the National Academy of Sciences of the United States of America 2003;100(8):4766–71.

67. Maeshima Y, Sudhakar A, Lively JC, et al. Tumstatin, an endothelial cell-specific inhibitor of protein synthesis. Science 2002;295(5552):140–3.

68. Stupack DG. Integrins as a distinct subtype of dependence receptors. Cell death and differentiation 2005;12(8):1021–30.

69. Eliceiri BP, Paul R, Schwartzberg PL, Hood JD, Leng J, Cheresh DA. Selective Requirement for Src Kinases during VEGF-Induced Angiogenesis and Vascular Permeability. Molecular Cell 1999;4(6):915–24.

70. Alavi A, Hood JD, Frausto R, Stupack DG, Cheresh DA. Role of Raf in vascular protection from distinct apoptotic stimuli. Science 2003;301(5629):94–6.

71. Hood JD, Bednarski M, Frausto R, et al. Tumor regression by targeted gene delivery to the neovasculature. Science 2002;296(5577):2404–7.

72. Kalluri R. Basement membranes: structure, assembly and role in tumour angiogenesis. Nature reviews 2003;3(6):422–33.

73. Colorado PC, Torre A, Kamphaus G, et al. Anti-angiogenic cues from vascular basement membrane collagen. Cancer research 2000;60(9):2520–6.

74. Kim YM, Jang JW, Lee OH, et al. Endostatin inhibits endothelial and tumor cellular invasion by blocking the activation and catalytic activity of matrix metalloproteinase. Cancer research 2000;60(19):5410–3.

75. O'Reilly MS, Boehm T, Shing Y, et al. Endostatin: an endogenous inhibitor of angiogenesis and tumor growth. Cell 1997;88(2):277–85.

76. O'Reilly MS, Holmgren L, Shing Y, et al. Angiostatin: a novel angiogenesis inhibitor that mediates the suppression of metastases by a Lewis lung carcinoma. Cell 1994;79(2):315–28.

77. Kamphaus GD, Colorado PC, Panka DJ, et al. Canstatin, a novel matrix-derived inhibitor of angiogenesis and tumor growth. The Journal of biological chemistry 2000;275(2):1209–15.

78. Petitclerc E, Boutaud A, Prestayko A, et al. New functions for non-collagenous domains of human collagen type IV. Novel integrin ligands inhibiting angiogenesis and tumor growth in vivo. The Journal of biological chemistry 2000;275(11):8051–61.

79. Nyberg P, Xie L, Kalluri R. Endogenous inhibitors of angiogenesis. Cancer research 2005;65(10):3967–79.

80. Kim YM, Hwang S, Kim YM, et al. Endostatin blocks vascular endothelial growth factor-mediated signaling via direct interaction with KDR/Flk-1. The Journal of biological chemistry 2002;277(31):27872–9.

81. Rehn M, Veikkola T, Kukk-Valdre E, et al. Interaction of endostatin with integrins implicated in angiogenesis. Proceedings of the National Academy of Sciences of the United States of America 2001;98(3):1024–9.

82. Ferreras M, Felbor U, Lenhard T, Olsen BR, Delaisse J. Generation and degradation of human endostatin proteins by various proteinases. FEBS letters 2000;486(3):247–51.

83. Maeshima Y, Colorado PC, Torre A, et al. Distinct antitumor properties of a type IV collagen domain derived from basement membrane. The Journal of biological chemistry 2000;275(28):21340–8.

84. Hynes RO. A reevaluation of integrins as regulators of angiogenesis. Nature medicine 2002;8(9):918–21.

85. Dhanabal M, Volk R, Ramchandran R, Simons M, Sukhatme VP. Cloning, expression, and in vitro activity of human endostatin. Biochemical and biophysical research communications 1999;258(2):345–52.

86. Yamaguchi N, Anand-Apte B, Lee M, et al. Endostatin inhibits VEGF-induced endothelial cell migration and tumor growth independently of zinc binding. The EMBO journal 1999;18(16):4414–23.

87. Sund M, Hamano Y, Sugimoto H, et al. Function of endogenous inhibitors of angiogenesis as endothelium-specific tumor suppressors. Proceedings of the National Academy of Sciences of the United States of America 2005;102(8):2934–9.

88. Lee SJ, Jang JW, Kim YM, et al. Endostatin binds to the catalytic domain of matrix metalloproteinase-2. FEBS letters 2002;519(1–3):147–52.

89. Nyberg P, Heikkila P, Sorsa T, et al. Endostatin inhibits human tongue carcinoma cell invasion and intravasation and blocks the activation of matrix metalloprotease-2, -9, and -13. The Journal of biological chemistry 2003;278(25):22404–11.

90. Pfeifer A, Kessler T, Silletti S, Cheresh DA, Verma IM. Suppression of angiogenesis by lentiviral delivery of PEX, a non-catalytic fragment of matrix metalloproteinase 2. Proceedings of the National Academy of Sciences of the United States of America 2000;97(22):12227–32.

91. Deryugina EI, Bourdon MA, Jungwirth K, Smith JW, Strongin AY. Functional activation of integrin alpha V beta 3 in tumor cells expressing membrane-type 1 matrix metalloproteinase. International journal of cancer 2000;86(1):15–23.

92. Gerhardt H, Betsholtz C. Endothelial-pericyte interactions in angiogenesis. Cell and tissue research 2003;314(1):15–23.

93. Jain RK, Booth MF. What brings pericytes to tumor vessels? The Journal of clinical investigation 2003;112(8):1134–6.

94. Hellstrom M, Kalen M, Lindahl P, Abramsson A, Betsholtz C. Role of PDGF-B and PDGFR-beta in recruitment of vascular smooth muscle cells and pericytes during embryonic blood vessel formation in the mouse. Development (Cambridge, England) 1999;126(14):3047–55.

95. Hellstrom M, Gerhardt H, Kalen M, et al. Lack of pericytes leads to endothelial hyperplasia and abnormal vascular morphogenesis. The Journal of cell biology 2001;153(3):543–53.

96. Mecham RP, Stenmark KR, Parks WC. Connective tissue production by vascular smooth muscle in development and disease. Chest 1991;99(3 Suppl):43S-7S.

97. Tan EM, Glassberg E, Olsen DR, et al. Extracellular matrix gene expression by human endothelial and smooth muscle cells. Matrix (Stuttgart, Germany) 1991;11(6):380–7.

98. Grazioli A, Alves CS, Konstantopoulos K, Yang JT. Defective blood vessel development and pericyte/pvSMC distribution in alpha 4 integrin-deficient mouse embryos. Developmental biology 2006;293(1):165–77.

99. Goldfinger LE, Han J, Kiosses WB, Howe AK, Ginsberg MH. Spatial restriction of alpha4 integrin phosphorylation regulates lamellipodial stability and alpha4beta1-dependent cell migration. The Journal of cell biology 2003;162(4):731–41.

100. Sengbusch JK, He W, Pinco KA, Yang JT. Dual functions of [alpha]4[beta]1 integrin in epicardial development: initial migration and long-term attachment. The Journal of cell biology 2002;157(5):873–82.

101. Kassner PD, Hemler ME. Interchangeable alpha chain cytoplasmic domains play a positive role in control of cell adhesion mediated by VLA-4, a beta 1 integrin. The Journal of experimental medicine 1993;178(2):649–60.

102. Pinco KA, He W, Yang JT. alpha4beta1 integrin regulates lamellipodia protrusion via a focal complex/focal adhesion-independent mechanism. Molecular biology of the cell 2002;13(9):3203–17.

103. Nishiya N, Kiosses WB, Han J, Ginsberg MH. An alpha4 integrin-paxillin-Arf-GAP complex restricts Rac activation to the leading edge of migrating cells. Nature cell biology 2005;7(4):343–52.

104. Sheppard AM, Onken MD, Rosen GD, Noakes PG, Dean DC. Expanding roles for alpha 4 integrin and its ligands in development. Cell adhesion and communication 1994;2(1):27–43.

105. Osborn L, Hession C, Tizard R, et al. Direct expression cloning of vascular cell adhesion molecule 1, a cytokine-induced endothelial protein that binds to lymphocytes. Cell 1989;59(6):1203–11.

106. Elices MJ, Osborn L, Takada Y, et al. VCAM-1 on activated endothelium interacts with the leukocyte integrin VLA-4 at a site distinct from the VLA-4/fibronectin binding site. Cell 1990;60(4):577–84.

107. Guan JL, Hynes RO. Lymphoid cells recognize an alternatively spliced segment of fibronectin via the integrin receptor alpha 4 beta 1. Cell 1990;60(1):53–61.

108. Jain RK, Duda DG. Role of bone marrow-derived cells in tumor angiogenesis and treatment. Cancer cell 2003;3(6):515–6.

109. Taguchi A, Soma T, Tanaka H, et al. Administration of CD34+ cells after stroke enhances neurogenesis via angiogenesis in a mouse model. The Journal of clinical investigation 2004;114(3):330 8.

110. Cursiefen C, Chen L, Borges LP, et al. VEGF-A stimulates lymphangiogenesis and hemangiogenesis in inflammatory neovascularization via macrophage recruitment. The Journal of clinical investigation 2004;113(7):1040–50.

111. Scapini P, Morini M, Tecchio C, et al. CXCL1/macrophage inflammatory protein-2-induced angiogenesis in vivo is mediated by neutrophil-derived vascular endothelial growth factor-A. J Immunol 2004;172(8):5034–40.

112. Jin H, Su J, Garmy-Susini B, Kleeman J, Varner J. Integrin alpha4beta1 promotes monocyte trafficking and angiogenesis in tumors. Cancer research 2006;66(4):2146–52.

113. Jin H, Aiyer A, Su J, et al. A homing mechanism for bone marrow-derived progenitor cell recruitment to the neovasculature. The Journal of clinical investigation 2006;116(3):652–62.

114. Ruzinova MB, Schoer RA, Gerald W, et al. Effect of angiogenesis inhibition by Id loss and the contribution of bone-marrow-derived endothelial cells in spontaneous murine tumors. Cancer cell 2003;4(4):277–89.

115. Dietrich J, Lacagnina M, Gass D, et al. EIF2B5 mutations compromise GFAP+ astrocyte generation in vanishing white matter leukodystrophy. Nature medicine 2005;11(3):277–83.

116. Haubner R, Schmitt W, Holzemann G, Goodman SL, Jonczyk A, Kessler H. Cyclic RGD Peptides Containing β-Turn Mimetics. In; 1996:7881–91.

117. Mitjans F, Meyer T, Fittschen C, et al. In vivo therapy of malignant melanoma by means of antagonists of alphav integrins. International journal of cancer 2000;87(5):716–23.

118. Miller WH, Keenan RM, Willette RN, Lark MW. Identification and in vivo efficacy of small-molecule antagonists of integrin alphavbeta3 (the vitronectin receptor). Drug Discov Today 2000;5(9):397–408.

119. Goodman SL, Holzemann G, Sulyok GA, Kessler H. Nanomolar small molecule inhibitors for alphav(beta)6, alphav(beta)5, and alphav(beta)3 integrins. Journal of medicinal chemistry 2002;45(5):1045–51.

120. Gutheil JC, Campbell TN, Pierce PR, et al. Targeted antiangiogenic therapy for cancer using Vitaxin: a humanized monoclonal antibody to the integrin alphavbeta3. Clin Cancer Res 2000;6(8):3056–61.

121. McNeel DG, Eickhoff J, Lee FT, et al. Phase I trial of a monoclonal antibody specific for alphavbeta3 integrin (MEDI-522) in patients with advanced malignancies, including an assessment of effect on tumor perfusion. Clin Cancer Res 2005;11(21): 7851–60.

122. Ramakrishnan V, Bhaskar V, Law DA, et al. Preclinical evaluation of an anti-alpha5beta1 integrin antibody as a novel antiangiogenic agent. Journal of experimental therapeutics & oncology 2006;5(4):273–86.

123. Trikha M, Zhou Z, Nemeth JA, et al. CNTO 95, a fully human monoclonal antibody that inhibits alphav integrins, has antitumor and antiangiogenic activity in vivo. International journal of cancer 2004;110(3):326–35.

124. Jayson GC, Mullamitha S, Ton C, et al. Phase I study of CNTO 95, a fully human monoclonal antibody (mAb) to {alpha}v integrins, in patients with solid tumors. In; 2005:3113-.

125. Smith JW. Cilengitide Merck. Curr Opin Investig Drugs 2003;4(6):741–5.

126. Eskens FA, Dumez H, Hoekstra R, et al. Phase I and pharma-cokinetic study of continuous twice weekly intravenous admin-istration of Cilengitide (EMD 121974), a novel inhibitor of the integrins alphavbeta3 and alphavbeta5 in patients with advanced solid tumours. Eur J Cancer 2003;39(7):917–26.

127. Carron CP, Meyer DM, Pegg JA, et al. A peptidomimetic antag-onist of the integrin alpha(v)beta3 inhibits Leydig cell tumor growth and the development of hypercalcemia of malignancy. Cancer research 1998;58(9):1930–5.

128. Reinmuth N, Liu W, Ahmad SA, et al. Alphavbeta3 integ-rin antagonist S247 decreases colon cancer metastasis and angiogenesis and improves survival in mice. Cancer research 2003;63(9):2079–87.

129. Sipkins DA, Cheresh DA, Kazemi MR, Nevin LM, Bednarski MD, Li KC. Detection of tumor angiogenesis in vivo by alphaV-beta3-targeted magnetic resonance imaging. Nature medicine 1998;4(5):623–6.

130. Winter PM, Morawski AM, Caruthers SD, et al. Molecular Imaging of Angiogenesis in Early-Stage Atherosclerosis With {alpha}v{beta}3-Integrin-Targeted Nanoparticles. In; 2003:2270–4.

131. Chen X, Park R, Shahinian AH, et al. 18F-labeled RGD peptide: initial evaluation for imaging brain tumor angiogenesis. Nuclear medicine and biology 2004;31(2):179–89.

132. Chen X, Park R, Tohme M, Shahinian AH, Bading JR, Conti PS. MicroPET and autoradiographic imaging of breast cancer alpha v-integrin expression using 18F- and 64Cu-labeled RGD peptide. Bioconjugate chemistry 2004;15(1):41–9.

133. Haubner R, Kuhnast B, Mang C, et al. [18F]Galacto-RGD: synthesis, radiolabeling, metabolic stability, and radiation dose estimates. Bioconjugate chemistry 2004;15(1):61–9.

134. Haubner R, Wester HJ, Weber WA, et al. Noninvasive imaging of alpha(v)beta3 integrin expression using 18F-labeled RGD-containing glycopeptide and positron emission tomography. Cancer research 2001;61(5):1781–5.

135. Cai W, Wu Y, Chen K, Cao Q, Tice DA, Chen X. In vitro and In vivo Characterization of 64Cu-Labeled AbegrinTM, a Human-ized Monoclonal Antibody against Integrin {alpha}v{beta}3. Cancer research 2006;66(19):9673–81.

136. Janssen M, Oyen WJ, Massuger LF, et al. Comparison of a monomeric and dimeric radiolabeled RGD-peptide for tumor targeting. Cancer biotherapy & radiopharmaceuticals 2002;17(6):641–6.

137. Janssen ML, Oyen WJ, Dijkgraaf I, et al. Tumor targeting with radiolabeled alpha(v)beta(3) integrin binding peptides in a nude mouse model. Cancer research 2002;62(21):6146–51.

138. Chen X, Conti PS, Moats RA. In vivo near-infrared fluorescence imaging of integrin alphavbeta3 in brain tumor xenografts. Cancer research 2004;64(21):8009–14.

139. Arap W, Pasqualini R, Ruoslahti E. Cancer treatment by tar-geted drug delivery to tumor vasculature in a mouse model. Science 1998;279(5349):377–80.

140. Schiffelers RM, Koning GA, ten Hagen TL, et al. Anti-tumor efficacy of tumor vasculature-targeted liposomal doxorubicin. J Control Release 2003;91(1–2):115–22.

141. Kasono K, Blackwell JL, Douglas JT, et al. Selective gene delivery to head and neck cancer cells via an integrin targeted adenoviral vector. Clin Cancer Res 1999;5(9):2571–9.

142. Schiffelers RM, Ansari A, Xu J, et al. Cancer siRNA therapy by tumor selective delivery with ligand-targeted sterically stabilized nanoparticle. Nucleic acids research 2004;32(19): e149.

Section II
Angiogenesis and Regulatory Proteins

Chapter 7
Fibroblast Growth Factor-2 in Angiogenesis

Marco Presta*, Stefania Mitola, Patrizia Dell'Era, Daria Leali, Stefania Nicoli, Emanuela Moroni, and Marco Rusnati

Keywords: angiogenesis, cancer, endothelium, extracellular matrix, inflammation, integrins, intracellular signaling, receptors, vasculogenesis

Abstract: Fibroblast growth factor-2 (FGF2) is a heparin-binding growth factor endowed with a potent angiogenic activity in vitro and in vivo. Due to their role in the neovascularization process, FGF2 and its receptors are viewed as targets for the development of antiangiogenic strategies to be exploited in cancer treatment.

FGF2-dependent neovascularization is the outcome of a complex network of interactions among FGF2 and a variety of free and extracellular matrix-associated molecules that modulate the bioavailability and biological activity of the growth factor. Also, to exert its angiogenic potential, FGF2 interacts with multiple endothelial cell surface receptors, including tyrosine kinase receptors, integrins, gangliosides, and heparan-sulfate proteoglycans. This complex network of extracellular interactions is mirrored intracellularly by the activation of various signal transduction pathways. Further complexity is added by the observation that FGF2 may act in synergy with other angiogenic growth factors and cytokines.

This chapter will focus on the mechanism of action of FGF2 in endothelial cells and its role in vasculogenesis and angiogenesis that occur under physiological and pathological conditions, including inflammation and tumor growth.

Unit of General Pathology and Immunology, Department of Biomedical Sciences and Biotechnology, University of Brescia, 25123 Brescia, Italy

*Corresponding Author:

Marco Presta, General Pathology and Immunology, Department of Biomedical Sciences and Biotechnology, Viale Europa 11, 25123 Brescia, Italy
E-mail: presta@med.unibs.it

Introduction

In 1984, J. Folkman and coworkers discovered a rodent tumor-derived factor able to bind with high affinity to heparin so that it could be purified 200,000-fold by a single passage over a heparin-affinity column. This protein, identified as basic fibroblast growth factor (FGF2), had a molecular mass of 18.8 kilodaltons (kD) and stimulated capillary endothelial cells (ECs) proliferation in vitro and angiogenesis in vivo [1, 2]. FGF2 was then purified from human normal cells/tissues [3] and from their transformed counterparts [4]. Purified FGF2 stimulated DNA synthesis, motility, and protease production [urokinase-type plasminogen activator (uPA) and metalloproteases (MMPs)] in ECs and induced angiogenesis in vivo [4]. Since then, FGF2 has been shown to exert neovascularization in a variety of in vivo animal models, including the chick embryo chorioallantoic membrane (CAM) assay, the rodent cornea assay, and the s.c. Matrigel plug assay in mice [5].

The single copy human FGF2 gene encodes four co-expressed isoforms (24, 22.5, 22, and 18 kD). The latter isoform is trans-lationally initiated at a classical AUG codon while the high molecular weight FGF2 isoforms are colinear NH_2-terminal extensions initiated at novel CUG codons [6]. FGF2 lacks a classic signal peptide for secretion [7]. However, cell damage and alternative mechanisms of exocytosis have been proposed to cause its release in the extracellular environment [7]. Even though experimental evidences point to significant differences in intracellular and extracellular fate and mechanisms of action of low and high molecular weight FGF2 isoforms [8], both 24 kD and 18 kD isoforms induce neovascularization in vitro and in vivo [9].

In cultured ECs, FGF2 triggers a complex "proangiogenic phenotype" that recapitulates the neovascularization process. It consists of extracellular matrix (ECM) degradation following uPA and MMPs up-regulation, EC proliferation and chemotactic migration [10]. Accordingly, FGF2 stimulates ECs to invade and to organize themselves into capillary-like structures with a hollow lumen in permissive three-dimensional matrices[5]. Also, FGF2 regulates the expression of cadherins, integrins, and various ECM components that contribute to the maturation of the new blood vessels by regulating lateral cell-cell and substrate adhesions of ECs [5].

Mechanisms of Action of FGF2

Extracellular Interactions

The angiogenic process is commonly assumed to represent the outcome of the apparently straightforward interaction between free angiogenic growth factors present in body fluids and their cognate tyrosine kinase (TK) receptors expressed on the luminal aspect of the EC surface. Actually, FGF2 binds different EC receptors and a variety of free or immobilized proteins, polysaccharides, and complex lipids present in the extracellular milieu. This network of interactions affects FGF2 integrity, stability, bioavailability, and diffusion, thus exerting a deep impact on the angiogenic potential of the growth factor.

Four TK FGF receptors (TK-FGFRs) have been identified, whose structural variability is increased by alternative splicing: FGFR1 (*flg*), FGFR2 (*bek*), FGFR3, and FGFR4. They belong to the subclass IV of membrane-spanning receptors [11] and are encoded by distinct genes. All TK-FGFRs bind FGF2, with preferential activation for the alternative spliced IIIc form in FGFRs 1–3 [12]. FGFR1 [13], and less frequently FGFR2 [14] are expressed by ECs, whereas the expression of FGFR3 or FGFR4 has never been reported in the endothelium. FGF2/TK-FGFRs interaction causes receptor dimerization and autophosphorylation of specific tyrosine residues located in the intra-cytoplasmic tail of the receptor. This in turn leads to complex signal transduction pathways (see below). Actually, a productive FGF2/TK-FGFR interaction requires the binding of the growth factor also to low affinity heparan-sulfate proteoglycans (HSPGs) [15]. Typical FGF2-binding HSPGs expressed at the EC surface are syndecans and glypicans. They consist of a core protein and of glycosaminoglycan (GAG) chains formed by repeating disaccharides units of sulfated uronic acid and hexosamine residues [16]. Interaction of FGF2 with specific oligosaccharide sequences of the GAG chain leads to the formation of HSPG:FGF2:TK-FGFR ternary complexes on the EC surface, leading to receptor activation and intracellular signaling [15]. Beside TK-FGFRs and HSPGs, other cell surface receptors have direct or indirect effects on EC responses to FGF2. Indeed, neuropilin-1 [17], nucleolin [18], ganglioside GM_1 [19], and $\alpha_v\beta_3$ integrin [20] have been shown to act as functional FGF2 co-receptors on the EC surface.

In body fluids, heparin, α_2-macroblobulin, platelet factor 4, platelet-derived growth factor (PDGF)-BB, CXCL13 chemokine, pentraxin 3 (PTX3), and a soluble form of the extracellular portion of FGFR1 bind FGF2, thus preventing its interaction with cellular receptors and inhibiting its biological activity [21]. Relevant to this point, the concentration of FGF2 in blood ranges between 0.6 and 6.0 pM [22], whereas the concentration of its binding partners can be up to 1,000,000-fold higher [5]. This suggests that FGF2 may exist in body fluids mainly as circulating complexes rather than as a free molecule.

FGF2 also interacts with various extracellular matrix (ECM) components that differently modulate its biological activity [21]. Indeed, thrombospondin-1, fibronectin (FN) fragments, and vitronectin (VN) bind FGF2 and inhibit FGF2-dependent EC proliferation, migration, and tubulogenesis in vitro and angiogenesis and tumor growth in vivo. At variance, fibrin(ogen) binds FGF2 without affecting FGF2/FGFR1 interaction. Consequently, FGF2 bound to immobilized fibrin(ogen) supports EC proliferation and protease production. Similarly, FGF2 bound to collagen type I is protected from proteolysis and is spontaneously released in a bioactive form [21]. HSPGs of the ECM, including perlecan [23], bind FGF2 and act as a reservoir for the growth factor to sustain a localized, long-term stimulation of ECs [24].

Taken together, these observations indicate that the alternative binding of FGF2 to free, ECM-associated, or cell-surface associated binders (Fig. 7.1) may result in a fine control of its bioavailability and cell interactions [15].

ECM-bound FGF2: an Immobilized Angiogenic Stimulus?

As described above, FGF2 interacts with a variety of ECM components, leading to the formation of immobilized ECM-bound complexes. Indeed, FGF2 is found mainly associated

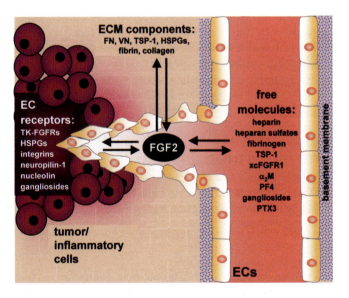

Fig. 7.1. The network of interactions among FGF2, cell surface EC receptors, ECM components, and free molecules. FGF2 is at the center of a complex network of interactions with different molecules that regulate its bioavailability, cell interaction, and biological activity. α_2M, α_2-macroglobulin; xcFGFR1, soluble FGFR1; PF4, platelet factor-4; TSP-1, thrombospondin-1.

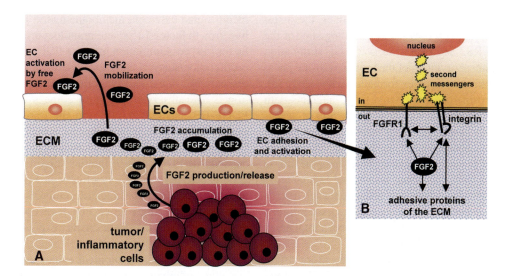

FIG. 7.2. Biological activity of ECM-immobilized FGF2. **A** Once released by tumor/inflammatory cells, FGF2 accumulates in the ECM from where it can be mobilized by various proteases and glycosidases. Alternatively, ECM-immobilized FGF2 promotes cell adhesion and activates adherent ECs. **B** EC adhesion to substrate-immobilized FGF2 induces the recruitment of TK-FGFR1 and $\alpha_v\beta_3$ integrin at focal adhesion contacts, leading to the activation of intracellular signals.

with ECM in vitro [24] and located in the blood vessel basement membranes in vivo [25]. Although various proteases and glycosidases can mobilize ECM-associated FGF2 in a biologically active form [15], experimental evidences indicate that FGF2 induces ECs activation also in an ECM-immobilized form (Fig. 7.2). Similar to classical adhesive proteins (e.g., VN and FN), substrate-immobilized FGF2 binds integrin $\alpha_v\beta_3$, thus promoting EC adhesion and spreading [20]. Moreover, immobilized FGF2/$\alpha_v\beta_3$ interaction leads to the activation of focal adhesion kinase pp125^{FAK}, pp60src, and of the mitogen-activated protein kinase (MAPK)/extracellular signal-regulated kinase$_{1/2}$ (ERK$_{1/2}$) pathway [26]. Consequently, ECs adherent to the immobilized growth factor are characterized by enhanced cell proliferation and motility [20], protease up-regulation [27], and by the capacity to form capillary-like structures [26].

Adhesion to immobilized FGF2 induces the co-recruitment of FGFR1 and $\alpha_v\beta_3$ integrin in cell-substratum contacts on the basal aspect of ECs [26]. Relevant to this point, a direct, specific association between the two receptors has been demonstrated under defined experimental conditions [28]. Thus, multiple interactions may occur at the basal aspect of ECs (Fig. 7.2) where the binding of FGF2 to ECM components may represent the appropriate scaffold to present the immobilized growth factor to FGFR1 and $\alpha_v\beta_3$ receptors, thus allowing their organization at the focal adhesion contacts. This is reflected at the intracellular level by the cytoskeleton organization and second messenger activation that occur following engagement of both receptors.

In conclusion, FGF2 produced and released by tumor and inflammatory cells is anticipated to bind various components of the subendothelial matrix where it will represent both an adhesion substrate and an angiogenic stimulus for ECs. Most of the in vitro studies aimed at characterizing the biological activity of FGF2 on ECs are performed by adding the growth factor to EC culture medium, implying its interaction with the luminal aspect of the endothelium. The capacity of immobilized FGF2 to activate ECs calls for further experiments addressing the mechanism of action of ECM-bound FGF2 and the characterization of its interaction(s) with the basal aspect of the endothelium.

FGF2-mediated Intracellular Signaling in Endothelial Cells

The complexity of the extracellular interactions engaged by FGF2 is mirrored by the complexity of the network of the intracellular signal transduction pathways activated by the growth factor in ECs [29]. Following ligand binding and receptor dimerization, a number of autophosphorylation sites on specific tyrosine residues have been identified in TK-FGFRs with possible different biological significances [30, 31]. In FGFR1, tyrosine residues Y653 and Y654 located in the intracellular kinase domain are essential for the enzymatic activity of the receptor, whereas phosphorylated Y463 and Y766 serve as docking sites for intracellular transducers. Phosphorylated Y463 forms a stable complex with the SH2 domain of the small adaptor protein Crk that is involved in ERK$_2$ and Jun kinase activation and mitogenic signaling [32]. Phosphorylated Y766 serves as a docking site for phospholipase C-γ (PLC-γ) [33] and for the adaptor protein Shb [34]. Whereas PLC-γ activity is not required for FGFR1-mediated cell proliferation [35], Shb regulates FGF-mediated mitogenicity via FRS2 phosphorylation and subsequent activation of the Ras/MEK/MAPK pathway [34].

In addition to tyrosine phosphorylated residues, the juxta-membrane domain of FGFR1 serves as a binding site for the docking proteins FRS2 and FRS3 [36, 37]. FGF stimulation leads to tyrosine phosphorylation of the protein tyrosine phosphatase Shp2, resulting in complex formation with Grb2 [38]. Grb2 bound to tyrosine-phosphorylated FRS2 interacts with Gab1 and with the nucleotide exchange factor Son of sevenless (Sos). Assembly of the FRS2:Grb2:Gab1 complex results in the activation of phosphoinositide 3-kinase (PI3-K) and downstream effector proteins such as the serine/threonine kinase Akt, whose cellular localization and activity are regulated by PI3-K products [39]. Meanwhile, Sos activation leads to activation of the Ras/Raf-1/ERK$_{1/2}$ signaling pathway. Accordingly, PI3-K [40] and ERK$_{1/2}$ [41, 42] activation have been implicated in the angiogenic activity of FGF2.

Beside TK-FGFRs, other cell surface receptors affect FGF2-mediated intracellular signaling in ECs. FGF2/HSPG interaction modulates angiogenesis by direct activation of phosphatidylinositol 4,5-bisphosphate and protein kinase C (PKC)-α [43] that eventually leads to MAPK activation [44]. As stated above, GM$_1$ ganglioside acts as a functional FGF2 co-receptor in ECs [19] and ganglioside-rich lipid rafts have been implicated in the modulation of signal transduction and biological activity of different growth factors [45]. Accordingly, the specific GM$_1$ ligand cholera toxin B subunit acts as a FGF2 antagonist in ECs [19]. Finally, numerous experimental evidences demonstrate the cross-talk between growth factor receptors and integrins to support cell proliferation, migration, and invasion [46]. FGF2/α$_v$β$_3$ integrin interaction leads to pp125^{FAK}, pp60src and ERK$_{1/2}$ activation in focal adhesion contacts [26]. Also, integrins and TK-FGFRs share PLC-γ [47, 48] and PKC [49, 50] activation, inositol lipid turnover [47, 48], and calcium influx [51, 52] as common intracellular signaling events. Accordingly, integrin inhibitors block FGF2-induced angiogenesis in different experimental models [53, 54].

FGF2 Cross-talk with other Angiogenic Growth Factors

Physiological and pathological modulation of the angiogenic process is dependent on the presence of different angiogenic growth factors and cytokines and on their synergism within the vascular microenvironment. FGF2 sets up a variety of synergies/antagonisms with other angiogenic growth factors/cytokines. For instance, various inflammatory mediators upregulate FGF2 expression/release (see below). Similarly, hepatocyte growth factor (HGF) increases FGF2 levels in cholesteatoma cell cultures [55]. The existence of a HGF/FGF2 cross-talk is supported by the observation that FGF2 induces HGF up-regulation via a hypoxia-independent mechanism [56], and that neutralizing anti-HGF antibodies diminish the therapeutic effect of FGF2 gene transfer in ischemic limbs [57]. Placental growth factor positively modulates FGF2-induced capillary-like tube formation and secondary sprouting on a reconstituted basement membrane matrix [58]. Also, insulin-like growth factor-1 [58] and granulocyte colony-stimulating factor [59] co-stimulation increase FGF2-induced retinal EC proliferation/survival and mouse ischemic limb angiogenesis, respectively. A marked synergistic effect on neovascularization is also observed in mouse corneas co-implanted with PDGF-BB and FGF2 [60], and the transfer of the PDGF gene to ischemic flaps promotes the expression of FGF2 [61]. On the other hand, experimental evidences demonstrate the capacity of PDGF-BB to interact directly with FGF2 [62], leading to potent inhibitory effects on FGF2-dependent EC migration, proliferation, and differentiation [63]. The hypothesis of an opposing role of PDGF receptor-α (PDGFR-α) and PDGFR-β in angiogenesis may explain these apparently contradictory data [64].

Genetic and pharmacological evidences demonstrate that vascular endothelial growth factor (VEGF) plays a major role in the neovascularization process in a variety of physiological and pathological processes [65]. An intimate cross-talk exists between FGF2 and VEGF during angiogenesis and vasculogenesis. For instance, EC tube formation stimulated by VEGF in murine embryonic explants depends on endogenous FGF2 [66]. On the other hand, the study of the transcriptional changes occurring in cultured ECs reveals that, together with a cluster of angiogenesis-related genes that are similarly modulated by FGF2 and VEGF, the two growth factors affect the expression of distinct subsets of transcripts [67, 68]. Indeed, FGF2 and VEGF may exert a synergistic effect in different angiogenesis models [58, 69]. Finally, several experimental evidences point to the possibility that FGF2 induces neovascularization indirectly by activating the VEGF/VEGF receptor (VEGFR) system: (1) VEGFR-2 antagonists inhibit not only VEGF- but also FGF2-induced angiogenesis [70]; (2) expression of dominant-negative FGFR1 or FGFR2 in glioma cells results in a decrease in tumor vascularization paralleled by VEGF down-regulation [71]; (3) endogenous and exogenous FGF2 modulate VEGF expression in ECs [72]; (4) in the mouse cornea, the quiescent endothelium of vessels of the limbus express VEGF only after FGF2 treatment and systemic administration of anti-VEGF neutralizing antibodies dramatically reduces FGF2-induced vascularization [72]; (5) VEGFR1-blocking antibodies or the expression of a dominant-negative VEGFR1 result in a reduction of FGF2-induced capillary morphogenesis [73]; and (6) FGF2 upregulates the expression of both FGFRs and VEGFRs in ECs [74].

Interestingly, a FGF2/VEGF cross-talk appears to occur also during lymphangiogenesis, the lymphatic system playing a critical role in the metastatic tumor spread [75].

FGF2 in Vasculogenesis

The development of the vascular system from embryonic mesoderm involves the differentiation and assembly of EC precursors (angioblasts) into the initial vascular pattern [76].

The formation of embryonic vasculature occurs primarily within the mesoderm and is positively regulated by the endoderm that plays a critical role in angioblast specification [77]. FGF2 is expressed in the endoderm and plays an important role in mesoderm induction [78]. In vitro, FGF2 is able to rescue the expression of the angioblast expression marker QH-1 in isolated quail embryo mesoderm, thus mimicking the inductive signals from the endoderm [79].

The quail/chick embryo chimera has represented a useful tool to address the role of FGF2 in vasculogenesis. When a donor quail somite, a mesodermal tissue that can be isolated free of angioblasts, is transplanted into the head mesoderm of the chicken host, the number of angioblasts originating from the donor somite is far greater than from a donor somite transplanted into the trunk region of the chick embryo. This reflects the higher concentration of FGF2 in the head as compared to the trunk [80]. Accordingly, neutralizing anti-FGF2 antibodies decrease the number of angioblasts from the donor somite when co-injected into the head [81]. Moreover, implantation of FGF2-coated beads into somatic mesoderm induces the formation of ectopic vessels that often contain red blood cells and are not connected to the existing vasculature [81].

The CAM assay is a well-established assay for studying the effects of growth factors on blood vessel growth during chick embryo development [82]. Neutralizing anti-FGF2 antibodies prevent CAM neovascularization, supporting the role of endogenous FGF2 in the development of the vascular system in avian embryos [83]. Accordingly, TK-FGFRs are expressed in the CAM until E10, when the angiogenic process is switched off [84].

At variance with the avian systems, the role of endogenous FGF2 in mammalian embryo angiogenesis remains uncertain. Indeed, *fgf2* knockout mice are morphologically normal [85] and do not show differences in neovascularization following injury [86] or hypoxia [87]. Conversely, transgenic overexpression of FGF2 does not result in spontaneous or inherent vascular defects, even though an amplified angiogenic response can be observed after wounding or s.c. implantation of a Matrigel plug in adult animals [88]. The apparently normal vascularization in *fgf2*$^{-/-}$ mice as well as in double *fgf2*$^{-/-}$/*fgf1*$^{-/-}$ mice may reflect the wide redundancy in the FGF family [89] and the contribution to the angiogenesis process of several other angiogenic growth factors, including VEGF. On the other hand, adenovirus-mediated gene transfer of dominant-negative TK-FGFR [90] or of FGF2 antisense cDNA [91] causes abnormal embryonic and extraembryonic vascular development in murine embryos.

Hematopoietic and endothelial systems have a close association during ontogeny, hematopoietic stem cells and angioblasts originating from a common progenitor, the hemangioblast [92]. FGF2-mediated signaling stimulates the growth of early hematopoietic progenitors [93] and is essential for hemangioblast proliferation [94]. In the adult, the CD34$^+$ cell population contains both hematopoietic stem cells and endothelial progenitor cells (EPCs). FGF2 has both synergistic and direct effect on progenitor cell proliferation of primitive hematopoietic cell lines and stimulates the growth of CD34$^+$/FGFR1$^+$ EPCs (see [95] and references therein). Also, recent observations have shown that 17β-estradiol-induced EPC mobilization is lost when FGF2 expression is specifically abolished in bone marrow-derived cells [96].

Taken together, these observations indicate that FGF2 and its receptors are relevant in hemangioblast, early hematopoietic cell, and EPC biology.

Inflammation and FGF2-dependent Angiogenesis

Inflammation may promote FGF2-dependent angiogenesis in different pathological conditions, including cancer. Inflammatory cells express FGF2, including mononuclear phagocytes, mast cells, CD4$^+$ and CD8$^+$ T lymphocytes [5]. Moreover, osmotic shock and shear stress induce the release of FGF2 from ECs [97, 98]. FGF2 production and release from ECs are also triggered by interferon (IFN)-α *plus* IL-2, IL-1β, and nitric oxide (NO) [5]. Accordingly, the proangiogenic effects exerted by NO and NO-inducing molecules are due, at least in part, to NO-mediated FGF2 up-regulation in ECs [99]. Thus, inflammatory mediators can activate the endothelium to synthesize and release FGF2 that, in turn, will stimulate angiogenesis by an autocrine mechanism of action. In agreement with a possible role of inflammatory cells in FGF2-mediated neovascularization, a significant inhibition of the angiogenic response to FGF2 is observed in neutropenic mice [100], and the monocyte chemoattractant protein-1 (MCP-1)/CC chemokine receptor (CCR2) system plays a critical role in FGF2-mediated therapeutic neovascularization [101]. On the other hand, the soluble pattern recognition receptor long-pentraxin PTX3, synthesized locally by ECs in response to IL-1β and TNF-α, binds FGF2 and acts as a natural angiogenesis inhibitor [102], thus allowing a fine tuning of FGF2 proangiogenic activity in inflammation. The inflammatory response may also cause hypoxia and cell damage. Hypoxia upregulates the production of FGF2 in ECs and vascular pericytes [103–105] and increases EC responsiveness to FGF2 by promoting HSPG synthesis [106], whereas EC damage results in increased FGF2 production and release [107].

Short-term exposure to FGF2 may amplify the inflammatory and angiogenic response by inducing vasoactive effects and the recruitment of an inflammatory infiltrate whereas long-term stimulation by FGF2 may have anti-inflammatory effects (reviewed in [5]). These observations suggest that the pro- or anti-inflammatory activity of FGFs may be contextual and may explain, at least in part, the reduced leukocyte adhesion and transendothelial migration observed in experimental tumors [108] that, nevertheless, are characterized by the presence of proangiogenic tumor-associated macrophages [109].

FGF2 in Tumor Angiogenesis

Experimental Tumors

Several experimental evidences suggest that FGF2 production and release may occur in vivo and influence the growth and neovascularization of tumor xenografts. Various tumor cell lines express FGF2 [4, 110], and the appearance of an angiogenic phenotype correlates with the export of FGF2 during the development of fibrosarcoma in a transgenic mouse model [111]. Antisense cDNAs for FGF2 and FGFR1 inhibit neovascularization and growth of human melanomas in nude mice [112]. Also, neutralizing anti-FGF2 antibodies or soluble TK-FGFR affect tumor growth under defined experimental conditions [113–116]. Accordingly, down-regulation of the FGF-binding protein (FGF-BP) inhibits the growth and vascularization of xenografted tumors in mice [117] despite the high levels of VEGF produced by these cells [118]. Interestingly, FGF-BP may exert its biological function via a paracrine stimulation on both tumor and ECs [119]. Indeed, given the pleiotropic activity of FGF2, it is not always possible to dissociate its effects on tumor angiogenesis from those exerted directly on tumor cells. For instance, inhibition of the FGF/TK-FGFR system in glioma cells by dominant negative TK-FGFR transfection [71] or in prostate cancer cells by *fgf2* gene knockout [120] results in inhibition of tumor growth by both angiogenesis-dependent and -independent mechanisms.

Constitutive [121, 122] or tetracycline-regulated [123] FGF2 overexpression causes a significant increase of the angiogenic activity and tumorigenic capacity of VEGF-producing human cancer cells without affecting their proliferation in vitro [123]. FGF2 and VEGF exert a synergistic effect on tumor blood vessel density even though they differently affect blood vessel maturation and functionality (see also [124]). Also, conditional switching of FGF2 expression indicates that late down-regulation of the growth factor does not affect the growth of large tumors although it results in a marked inhibition of small tumors [123]. Similar results have been obtained after conditional switching of VEGF or of FGF-BP [125, 126]. Accordingly, early initiation of angiostatic therapies can be more efficacious than late initiation in reducing tumor growth [127–129].

In keeping with the role for FGFs in experimental tumor angiogenesis, adenoviral expression of a soluble form of FGFR2 impairs the maintenance of tumor angiogenesis in spontaneous β-cell pancreatic tumors in *Rip1Tag2* mice whereas soluble VEGFR1 affects the initial stages of tumor angiogenesis. The combination of the two soluble receptors exerted a synergistic inhibitory effect [116]. Interestingly, the hypoxia-mediated induction of the FGF2/TK-FGFR system may cause reactivation of tumor angiogenesis and drug resistance to VEGF blockade in this model [130]. In addition, expression of a dominant-negative FGFR1 in the retina of *Tryp1*-Tag mice that develop early vascularized tumors of the retinal pigment epithelium results in a significant decrease in tumor burden and vascularity [131]. Also, inactivation of even one FGF2 allele leads to increased survival, a significant decrease in metastasis,

and inhibition of tumor progression in transgenic primary adenocarcinomas of the mouse prostate [120].

Human Tumors

The possibility that FGF2 may play a role in human tumor vascularization represents an important issue in tumor biology and for the development of antiangiogenic therapies. FGF2 is produced by various tumor and normal cell types, including ECs and inflammatory cells. Accordingly, FGF2 has been detected not only in neoplastic cells, but also in infiltrating cells within human tumors of different origin [132–136]. As stated above, FGF2 may target not only stromal cells but also the parenchyma of the neoplastic tissue [71, 120]. Indeed, FGF2 induces proliferation/survival [137], migration/invasiveness [138, 139], and protease production [139] in tumor cells and promotes their shift to a more metastatic potential [140]. This indicates that FGF2 may act on both ECs and tumor cells via autocrine and paracrine mechanisms (Fig. 7.3). FGF2 is overexpressed in melanoma, renal, gastric, pancreatic, bladder, hepatocellular, prostate, pituitary, thyroid, and hematological tumors (reviewed in [5, 141]). However, the relationship between dysregulation of FGF2 expression and tumor progression remains unclear [142, 143]. This may be due to differences in the source of FGF2 within the tumor (ECs, tumor cells, inflammatory cells, etc.) and in its intracellular or extracellular fate. Evaluation of intratumoral microvessel density (MVD) may have prognostic significance and numerous studies have attempted to establish a correlation between intratumoral levels of FGF2 transcript or protein and MVD in cancer patients (summarized in [5]). With a few exceptions (e.g., melanomas) FGF2 levels do not correlate consistently with MVD. It is

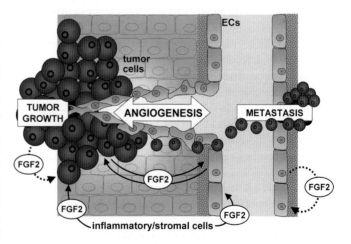

Fig. 7.3. Autocrine and paracrine mechanisms of action of FGF2. FGF2 is produced by various cell types within the tumor (including ECs, tumor cells, inflammatory/stroma cells) and exerts both autocrine (*dotted arrows*) and paracrine (*straight arrows*) activities on the different cell types, thus affecting tumor growth, angiogenesis, and the metastatic process.

interesting to note that in some tumor types (e.g., breast and hepatocellular carcinomas) intratumoral levels of FGF2 correlate with the clinical outcome but not with MVD. Relevant to this point, clonal differences exist within a tumor for the capacity to release FGF2. Indeed, FGF2 overexpression confers an increased angiogenic and tumorigenic potential in tumor cells only when paralleled by an increased export of the growth factor [121]. Accordingly, the discrete switch to the angiogenic phenotype which occurs in the multistep development of fibrosarcoma correlates with the export of FGF2 [111]. Thus, FGF2 overexpression in tumors does not necessarily result in an increase in the release of the growth factor that will occur only in those cell clones that have acquired a heretofore unrecognized capacity to export FGF2 during tumor progression. These considerations call for extreme caution in the extrapolation of the autocrine and paracrine properties of a neoplasm from the data obtained in situ on the production and localization of FGF2 in tumor biopsies. Also, as stated above, the pleiotropic activity of FGF2 may affect both tumor vasculature and tumor parenchyma. Thus, FGF2 may contribute to cancer progression not only by inducing neovascularization, but also by acting directly on tumor cells.

Serum concentrations of angiogenic growth factors increase with tumor progression [144] and decrease in response to treatment and long-term disease control [145]. Thus, apart from providing prognostic information in early detection of primary tumors or to follow tumor progression, measurement of these circulating factors may be used to monitor tumor regression during therapy and for the selection of patients at high risk of recurrences after treatment [146]. However, the prognostic significance of FGF2 levels in biological fluids of cancer patients is controversial. Early studies showed that elevated levels of FGF2 in urine samples collected from cancer patients were significantly correlated with the status and the extent of disease [147]. However, no association between increased serum levels of FGF2 and tumor type was observed in later studies on a large spectrum of metastatic carcinomas even though two-thirds of the patients that showed progressive disease had increasing serum levels of FGF2 compared with less than one-tenth of the patients showing response to therapy [148]. The clinical significance of circulating FGF2 in individual types of cancer has been reviewed [141]. Briefly, the levels of circulating FGF2 may have prognostic significance in head and neck cancer, lymphoma, leukemia, prostate carcinoma, and soft tissue sarcoma, but they do not correlate with breast cancer progression and their significance in colorectal carcinoma remains unclear. Moreover, serum FGF2 may not entirely derive from the neoplastic tissue in cancer patients [149].

In conclusion, clinical reports have not yet established a clear relationship among FGF2, tumor angiogenesis, and tumor progression/prognosis. Further studies assessing the possibility to utilize FGF2 levels at the tumor site and/or in body fluids as a prognostic indicator and/or surrogate marker of angiogenesis in cancer patients are eagerly awaited.

FGF2 as a Target for Antiangiogenic/Anti-Cancer Regimens

FGF2 may represent a target for the development of novel therapeutics for cancer treatment. The various anti-cancer strategies based on the inhibition of FGF2 have been reviewed extensively elsewhere (see [150] and references therein). Briefly, FGF2 can be neutralized at different levels by: (1) inhibition of FGF2 production/release; (2) inhibition of the expression of the various FGF2 receptors in ECs (including TK-FGFRs, HSPGs, integrins, gangliosides); (3) engagement of the same FGF2 receptors by selected antagonists; (4) sequestration of FGF2 in the extracellular environment; (5) inhibition of the signal transduction pathways triggered by FGF2 in ECs; and (6) neutralization of the FGF2-induced effectors/biological responses in ECs (Fig. 7.4).

Using a genetically-modified experimental tumor model [151], we have observed that the selective inhibition of FGF2 or of VEGF expression results in a similar decrease in blood vessel density. However, the two growth factors differently affect tumor blood vessel maturation and functionality, with different consequences on tumor oxygenation and viability. Inhibition of the FGF2/FGFR system results in a significant reduction of tumor size and vascularization that is slightly less efficacious than that caused by the inhibition of the VEGF/VEGFR system. Inhibition of both systems caused a further, albeit limited decrease in the rate of tumor growth and angiogenesis. However, none of the treatments was able to fully suppress tumor growth in this model. Nevertheless, the high percentage of necrotic tumor parenchyma following VEGF blockade in respect to the limited tumor necrosis observed

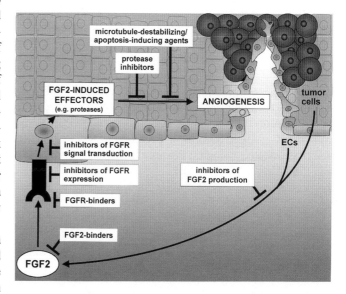

FIG. 7.4. Anti-FGF2 strategies for the development of antiangiogenic therapies. See text for further details.

after FGF2 blockade suggests that inhibition of the VEGF/ VEGFR system may represent a more efficacious antineoplastic approach. On the other hand, drug resistance to VEGF blockade may occur following reactivation of tumor angiogenesis triggered by compensatory up-regulation of FGF2/ FGFR [130].

It must be pointed out that various anti-FGF2 strategies may affect tumor growth by acting on ECs, thus inhibiting angiogenesis, and by acting on tumor cells, thus inhibiting directly tumor cell proliferation. Also, the impairment of receptors characterized by a broad spectrum of ligands, like HSPGs or integrins, may result in the simultaneous inhibition of various angiogenic growth factors [21].

Concluding Remarks

FGF2 is at the center of a complex network of interactions that deeply affect tumor biology. The bulk of experimental data indicate that FGF2 and/or FGF2 receptors may represent a target for antiangiogenic strategies in cancer. At present, cancer clinical trials are in progress to assess the safety and efficacy of various compounds with a potential capacity to affect the FGF2/FGFR system at different levels [152, 153]. In several cases, however, the main rationale for testing these compounds was independent of their putative FGF2 antagonist activity. For instance, heparin derivatives have been tested in cancer patients because of their anti-thrombotic effect rather than for their capacity to bind FGF2. Similarly, the humanized monoclonal anti-$\alpha_v\beta_3$ antibody vitaxin [154, 155] has been investigated for its ability to affect the cell-adhesive function of this integrin receptor rather than for its potential role in angiogenesis and FGF2 activity. Also, numerous cytotoxic drugs can affect the FGF2/FGF receptor system and angiogenesis [150]. Novel strategies aimed at inhibiting multiple targets, including FGF2, may represent an efficacious approach for the treatment of angiogenesis-dependent diseases, including cancer.

Acknowledgements. Limitations of space preclude extensive citation of the literature; we apologize to those whose work is not mentioned herein. This work was supported by grants from AIRC, MIUR (Centro di Eccellenza "IDET", Cofin), and ISS (Oncotechnological Program) to M.P. and from AIRC and ISS (AIDS Project) to M.R.

References

1. Shing Y, Folkman J, Sullivan R, et al. Heparin affinity: purification of a tumor-derived capillary endothelial cell growth factor. Science 1984; 223:1296–9.
2. Shing Y, Folkman J, Haudenschild C, et al. Angiogenesis is stimulated by a tumor-derived endothelial cell growth factor. J.Cell Biochem. 1985; 29:275–287.
3. Bohlen P, Baird A, Esch F, et al. Isolation and partial molecular characterization of pituitary fibroblast growth factor. Proc Natl Acad Sci USA 1984; 81:5364–8.
4. Presta M, Moscatelli D, Joseph-Silverstein J, et al. Purification from a human hepatoma cell line of a basic fibroblast growth factor-like molecule that stimulates capillary endothelial cell plasminogen activator production, DNA synthesis, and migration. Mol Cell Biol 1986; 6:4060–6.
5. Presta M, Dell'Era P, Mitola S, et al. Fibroblast growth factor/ fibroblast growth factor receptor system in angiogenesis. Cytokine Growth Factor Rev 2005; 16:159–78.
6. Florkiewicz RZ, Sommer A. Human basic fibroblast growth factor gene encodes four polypeptides: three initiate translation from non-AUG codons. Proc Natl Acad Sci U S A 1989; 86:3978–81.
7. Mignatti P, Morimoto T, Rifkin DB. Basic fibroblast growth factor, a protein devoid of secretory signal sequence, is released by cells via a pathway independent of the endoplasmic reticulum-Golgi complex. J Cell Physiol 1992; 151:81–93.
8. Bikfalvi A, Savona C, Perollet C, et al. New insights in the biology of fibroblast growth factor-2. Angiogenesis 1998; 1:155–73.
9. Gualandris A, Urbinati C, Rusnati M, et al. Interaction of high-molecular-weight basic fibroblast growth factor with endothelium: biological activity and intracellular fate of human recombinant M(r) 24,000 bFGF. J Cell Physiol 1994; 161:149–59.
10. Taraboletti G, D'Ascenzo S, Borsotti P, et al. Shedding of the matrix metalloproteinases MMP-2, MMP-9, and MT1-MMP as membrane vesicle-associated components by endothelial cells. Am J Pathol 2002; 160:673–80.
11. Fantl WJ, Escobedo JA, Martin GA, et al. Distinct phosphotyrosines on a growth factor receptor bind to specific molecules that mediate different signaling pathways. Cell 1992; 69:413–423.
12. Ornitz DM, Xu J, Colvin JS, et al. Receptor specificity of the fibroblast growth factor family. J Biol Chem 1996; 271:15292–7.
13. Javerzat S, Auguste P, Bikfalvi A. The role of fibroblast growth factors in vascular development. Trends Mol Med 2002; 8: 483–9.
14. Dell'Era P, Belleri M, Stabile H, et al. Paracrine and autocrine effects of fibroblast growth factor-4 in endothelial cells. Oncogene 2001; 20:2655–63.
15. Rusnati M, Presta M. Interaction of angiogenic basic fibroblast growth factor with endothelial cell heparan sulfate proteoglycans. Biological implications in neovascularization. Int J Clin Lab Res 1996; 26:15–23.
16. Lindahl U, Lidholt K, Spillmann D, et al. More to "heparin" than anticoagulation. Thromb Res 1994; 75:1–32.
17. West DC, Rees CG, Duchesne L, et al. Interactions of multiple heparin binding growth factors with neuropilin-1 and potentiation of the activity of fibroblast growth factor-2. J Biol Chem 2005; 280:13457–64.
18. Bonnet H, Filhol O, Truchet I, et al. Fibroblast growth factor-2 binds to the regulatory beta subunit of CK2 and directly stimulates CK2 activity toward nucleolin. J Biol Chem 1996; 271:24781–7.
19. Rusnati M, Urbinati C, Tanghetti E, et al. Cell membrane GM1 ganglioside is a functional coreceptor for fibroblast growth factor 2. Proc Natl Acad Sci USA 2002; 99:4367–72.

20. Rusnati M, Tanghetti E, Dell'Era P, et al. alphavbeta3 integrin mediates the cell-adhesive capacity and biological activity of basic fibroblast growth factor (FGF-2) in cultured endothelial cells. Mol Biol Cell 1997; 8:2449–61.

21. Rusnati M, Presta M. Extracellular angiogenic growth factor interactions: an angiogenesis interactome survey. Endothelium 2006; 13:93–111.

22. Sahni A, Altland OD, Francis CW. FGF-2 but not FGF-1 binds fibrin and supports prolonged endothelial cell growth. J Thromb Haemost 2003; 1:1304–10.

23. Lopez-Casillas F, Cheifetz S, Doody J, et al. Structure and expression of the membrane proteoglycan betaglycan, a component of the TGF-beta receptor system. Cell 1991; 67:785–95.

24. Presta M, Maier JA, Rusnati M, et al. Basic fibroblast growth factor is released from endothelial extracellular matrix in a biologically active form. J Cell Physiol 1989; 140:68–74.

25. Folkman J, Klagsbrun M, Sasse J, et al. A heparin-binding angiogenic protein–basic fibroblast growth factor–is stored within basement membrane. Am J Pathol 1988; 130:393–400.

26. Tanghetti E, Ria R, Dell'Era P, et al. Biological activity of substrate-bound basic fibroblast growth factor (FGF2): recruitment of FGF receptor-1 in endothelial cell adhesion contacts. Oncogene 2002; 21:3889–97.

27. Presta M, Rusnati M, Urbinati C, et al. Basic fibroblast growth factor bound to cell substrate promotes cell adhesion, proliferation, and protease production in cultured endothelial cells. Exs 1992; 61:205–9.

28. Sahni A, Francis CW. Stimulation of endothelial cell proliferation by FGF-2 in the presence of fibrinogen requires alphavbeta3. Blood 2004; 104:3635–41.

29. Eswarakumar VP, Lax I, Schlessinger J. Cellular signaling by fibroblast growth factor receptors. Cytokine Growth Factor Rev 2005; 16:139–49.

30. Mohammadi M, Dikic I, Sorokin A, et al. Identification of six novel autophosphorylation sites on fibroblast growth factor receptor 1 and elucidation of their importance in receptor activation and signal transduction. Mol Cell Biol 1996; 16: 977–89.

31. Dell'Era P, Mohammadi M, Presta M. Different tyrosine autophosphorylation requirements in fibroblast growth factor receptor-1 mediate urokinase-type plasminogen activator induction and mitogenesis. Mol Biol Cell 1999; 10:23–33.

32. Larsson H, Klint P, Landgren E, et al. Fibroblast growth factor receptor-1-mediated endothelial cell proliferation is dependent on the Src homology (SH) 2/SH3 domain- containing adaptor protein Crk. J Biol Chem 1999; 274:25726–25734.

33. Mohammadi M, Honegger AM, Rotin D, et al. A tyrosine-phosphorylated carboxy-terminal peptide of the fibroblast growth factor receptor (Flg) is a binding site for the SH2 domain of phospholipase C-gamma 1. Mol Cell Biol 1991; 11:5068–78.

34. Cross MJ, Lu L, Magnusson P, et al. The Shb adaptor protein binds to tyrosine 766 in the FGFR-1 and regulates the Ras/ MEK/MAPK pathway via FRS2 phosphorylation in endothelial cells. Mol Biol Cell 2002; 13:2881–93.

35. Mohammadi M, Dionne CA, Li W, et al. Point mutation in FGF receptor eliminates phosphatidylinositol hydrolysis without affecting mitogenesis. Nature 1992; 358:681–4.

36. Ong SH, Guy GR, Hadari YR, et al. FRS2 proteins recruit intracellular signaling pathways by binding to diverse targets on fibroblast growth factor and nerve growth factor receptors. Mol Cell Biol 2000; 20:979–89.

37. Dhalluin C, Yan KS, Plotnikova O, et al. Structural basis of SNT PTB domain interactions with distinct neurotrophic receptors. Mol Cell 2000; 6:921–9.

38. Hadari YR, Kouhara H, Lax I, et al. Binding of Shp2 tyrosine phosphatase to FRS2 is essential for fibroblast growth factor-induced PC12 cell differentiation. Mol Cell Biol 1998; 18:3966–73.

39. Ong SH, Hadari YR, Gotoh N, et al. Stimulation of phosphatidylinositol 3-kinase by fibroblast growth factor receptors is mediated by coordinated recruitment of multiple docking proteins. Proc Natl Acad Sci USA 2001; 98:6074–9.

40. Rieck PW, Cholidis S, Hartmann C. Intracellular signaling pathway of FGF-2-modulated corneal endothelial cell migration during wound healing in vitro. Exp Eye Res 2001; 73: 639–50.

41. Giuliani R, Bastaki M, Coltrini D, et al. Role of endothelial cell extracellular signal-regulated kinase1/2 in urokinase-type plasminogen activator upregulation and in vitro angiogenesis by fibroblast growth factor-2. J Cell Sci 1999; 112:2597–606.

42. Eliceiri BP, Klemke R, Stromblad S, et al. Integrin alphavbeta3 requirement for sustained mitogen-activated protein kinase activity during angiogenesis. J Cell Biol 1998; 140:1255–63.

43. Horowitz A, Tkachenko E, Simons M. Fibroblast growth factor-specific modulation of cellular response by syndecan-4. J Cell Biol 2002; 157:715–25.

44. Chua CC, Rahimi N, Forsten-Williams K, et al. Heparan sulfate proteoglycans function as receptors for fibroblast growth factor-2 activation of extracellular signal-regulated kinases 1 and 2. Circ Res 2004; 94:316–23.

45. Miljan EA, Bremer EG. Regulation of growth factor receptors by gangliosides. Sci STKE 2002; 2002:RE15.

46. Eliceiri BP. Integrin and growth factor receptor crosstalk. Circ Res 2001; 89:1104–10.

47. Banga HS, Simons ER, Brass LF, et al. Activation of phospholipases A and C in human platelets exposed to epinephrine: role of glycoproteins IIb/IIIa and dual role of epinephrine. Proc Natl Acad Sci USA 1986; 83:9197–201.

48. Peters KG, Marie J, Wilson E, et al. Point mutation of an FGF receptor abolishes phosphatidylinositol turnover and Ca2+ flux but not mitogenesis. Nature 1992; 358:678–81.

49. Vuori K, Ruoslahti E. Activation of protein kinase C precedes alpha 5 beta 1 integrin-mediated cell spreading on fibronectin. J Biol Chem 1993; 268:21459–62.

50. Presta M, Maier JA, Ragnotti G. The mitogenic signaling pathway but not the plasminogen activator-inducing pathway of basic fibroblast growth factor is mediated through protein kinase C in fetal bovine aortic endothelial cells. J Cell Biol 1989; 109: 1877–84.

51. Pelletier AJ, Bodary SC, Levinson AD. Signal transduction by the platelet integrin alpha IIb beta 3: induction of calcium oscillations required for protein-tyrosine phosphorylation and ligand-induced spreading of stably transfected cells. Mol Biol Cell 1992; 3:989–98.

52. Schwartz MA. Spreading of human endothelial cells on fibronectin or vitronectin triggers elevation of intracellular free calcium. J Cell Biol 1993; 120:1003–10.

53. Brooks PC, Clark RA, Cheresh DA. Requirement of vascular integrin alpha v beta 3 for angiogenesis. Science 1994; 264: 569–71.

54. Kumar CC, Malkowski M, Yin Z, et al. Inhibition of angiogenesis and tumor growth by SCH221153, a dual alpha(v)beta3 and alpha(v)beta5 integrin receptor antagonist. Cancer Res 2001; 61:2232–8.

55. Naim R, Chang RC, Sadick H, et al. Influence of hepatocyte growth factor/scatter factor (HGF/SF) on fibroblast growth factor-2 (FGF-2) levels in external auditory canal cholesteatoma (EACC) cell culture. In Vivo 2005; 19:599–603.

56. Onimaru M, Yonemitsu Y, Tanii M, et al. Fibroblast growth factor-2 gene transfer can stimulate hepatocyte growth factor expression irrespective of hypoxia-mediated downregulation in ischemic limbs. Circ Res 2002; 91:923–30.

57. Masaki I, Yonemitsu Y, Yamashita A, et al. Angiogenic gene therapy for experimental critical limb ischemia: acceleration of limb loss by overexpression of vascular endothelial growth factor 165 but not of fibroblast growth factor-2. Circ Res 2002; 90: 966–73.

58. Castellon R, Hamdi HK, Sacerio I, et al. Effects of angiogenic growth factor combinations on retinal endothelial cells. Exp Eye Res 2002; 74:523–35.

59. Jeon O, Hwang KC, Yoo KJ, et al. Combined sustained delivery of basic fibroblast growth factor and administration of granulocyte colony-stimulating factor: synergistic effect on angiogenesis in mouse ischemic limbs. J Endovasc Ther 2006; 13:175–81.

60. Cao R, Brakenhielm E, Pawliuk R, et al. Angiogenic synergism, vascular stability and improvement of hind-limb ischemia by a combination of PDGF-BB and FGF-2. Nat Med 2003; 9: 604–13.

61. Wang XT, Liu PY, Tang JB. PDGF gene therapy enhances expression of VEGF and bFGF genes and activates the NF-kappaB gene in signal pathways in ischemic flaps. Plast Reconstr Surg 2006; 117:129–37.

62. Russo K, Ragone R, Facchiano AM, et al. Platelet-derived growth factor-BB and basic fibroblast growth factor directly interact in vitro with high affinity. J Biol Chem 2002; 277:1284–91.

63. De Marchis F, Ribatti D, Giampietri C, et al. Platelet-derived growth factor inhibits basic fibroblast growth factor angiogenic properties in vitro and in vivo through its alpha receptor. Blood 2002; 99:2045–53.

64. Yu J, Deuel TF, Kim HR. Platelet-derived growth factor (PDGF) receptor-alpha activates c-Jun NH2-terminal kinase-1 and antagonizes PDGF receptor-beta -induced phenotypic transformation. J Biol Chem 2000; 275:19076–82.

65. Ferrara N, Gerber HP, LeCouter J. The biology of VEGF and its receptors. Nat Med 2003; 9:669–76.

66. Tomanek RJ, Sandra A, Zheng W, et al. Vascular endothelial growth factor and basic fibroblast growth factor differentially modulate early postnatal coronary angiogenesis. Circ Res 2001; 88:1135–41.

67. Jih YJ, Lien WH, Tsai WC, et al. Distinct regulation of genes by bFGF and VEGF-A in endothelial cells. Angiogenesis 2001; 4:313–21.

68. Ho M, Yang E, Matcuk G, et al. Identification of endothelial cell genes by combined database mining and microarray analysis. Physiol Genomics 2003; 13:249–62.

69. Pepper MS, Mandriota SJ. Regulation of vascular endothelial growth factor receptor-2 (Flk-1) expression in vascular endothelial cells. Exp Cell Res 1998; 241:414–25.

70. Tille JC, Wood J, Mandriota SJ, et al. Vascular endothelial growth factor (VEGF) receptor-2 antagonists inhibit VEGF- and basic fibroblast growth factor-induced angiogenesis in vivo and in vitro. J Pharmacol Exp Ther 2001; 299:1073–85.

71. Auguste P, Gursel DB, Lemiere S, et al. Inhibition of fibroblast growth factor/fibroblast growth factor receptor activity in glioma cells impedes tumor growth by both angiogenesis-dependent and -independent mechanisms. Cancer Res 2001; 61:1717–26.

72. Seghezzi G, Patel S, Ren CJ, et al. Fibroblast growth factor-2 (FGF-2) induces vascular endothelial growth factor (VEGF) expression in the endothelial cells of forming capillaries: an autocrine mechanism contributing to angiogenesis. J Cell Biol 1998; 141:1659–73.

73. Kanda S, Miyata Y, Kanetake H. Fibroblast growth factor-2-mediated capillary morphogenesis of endothelial cells requires signals via Flt-1/vascular endothelial growth factor receptor-1: possible involvement of c-Akt. J Biol Chem 2004; 279:4007–16.

74. Gabler C, Plath-Gabler A, Killian GJ, et al. Expression pattern of fibroblast growth factor (FGF) and vascular endothelial growth factor (VEGF) system members in bovine corpus luteum endothelial cells during treatment with FGF-2, VEGF or oestradiol. Reprod Domest Anim 2004; 39:321–7.

75. Chang L, Kaipainen A, Folkman J. Lymphangiogenesis new mechanisms. Ann NY Acad Sci 2002; 979:111–9.

76. Ferguson JE, 3rd, Kelley RW, Patterson C. Mechanisms of endothelial differentiation in embryonic vasculogenesis. Arterioscler Thromb Vasc Biol 2005; 25:2246–54.

77. Risau W, Flamme I. Vasculogenesis. Annu Rev Cell Dev Biol 1995; 11:73–91.

78. Bikfalvi A, Klein S, Pintucci G, et al. Biological roles of fibroblast growth factor-2. Endocr Rev 1997; 18:26–45.

79. Poole TJ, Finkelstein EB, Cox CM. The role of FGF and VEGF in angioblast induction and migration during vascular development. Dev Dyn 2001; 220:1–17.

80. Kubota Y, Ito K. Chemotactic migration of mesencephalic neural crest cells in the mouse. Dev Dyn 2000; 217:170–9.

81. Cox CM, Poole TJ. Angioblast differentiation is influenced by the local environment: FGF-2 induces angioblasts and patterns vessel formation in the quail embryo. Dev Dyn 2000; 218:371–82.

82. Ribatti D, Vacca A, Roncali L, et al. The chick embryo chorioallantoic membrane as a model for in vivo research on anti-angiogenesis. Curr Pharm Biotechnol 2000; 1:73–82.

83. Ribatti D, Presta M. The role of fibroblast growth factor-2 in the vascularization of the chick embryo chorioallantoic membrane. J Cell Mol Med 2002; 6:439–46.

84. Parsons-Wingerter P, Elliott KE, Clark JI, et al. Fibroblast growth factor-2 selectively stimulates angiogenesis of small vessels in arterial tree. Arterioscler Thromb Vasc Biol 2000; 20:1250–6.

85. Zhou M, Sutliff RL, Paul RJ, et al. Fibroblast growth factor 2 control of vascular tone. Nat Med 1998; 4:201–7.

86. Tobe T, Ortega S, Luna JD, et al. Targeted disruption of the FGF2 gene does not prevent choroidal neovascularization in a murine model. Am J Pathol 1998; 153:1641–6.

87. Ozaki H, Okamoto N, Ortega S, et al. Basic fibroblast growth factor is neither necessary nor sufficient for the development of retinal neovascularization. Am J Pathol 1998; 153:757–65.

88. Fulgham DL, Widhalm SR, Martin S, et al. FGF-2 dependent angiogenesis is a latent phenotype in basic fibroblast growth factor transgenic mice. Endothelium 1999; 6:185–95.

89. Miller DL, Ortega S, Bashayan O, et al. Compensation by fibroblast growth factor 1 (FGF1) does not account for the mild phenotypic defects observed in FGF2 null mice. Mol Cell Biol 2000; 20:2260–8.

90. Lee SH, Schloss DJ, Swain JL. Maintenance of vascular integrity in the embryo requires signaling through the fibroblast growth factor receptor. J Biol Chem 2000; 275:33679–87.

91. Leconte I, Fox JC, Baldwin HS, et al. Adenoviral-mediated expression of antisense RNA to fibroblast growth factors disrupts murine vascular development. Dev Dyn 1998; 213:421–30.

92. Hamaguchi I, Huang XL, Takakura N, et al. In vitro hematopoietic and endothelial cell development from cells expressing TEK receptor in murine aorta-gonad-mesonephros region. Blood 1999; 93:1549–56.

93. Moroni E, Dell'Era P, Rusnati M, et al. Fibroblast growth factors and their receptors in hematopoiesis and hematological tumors. J Hematother Stem Cell Res 2002; 11:19–32.

94. Faloon P, Arentson E, Kazarov A, et al. Basic fibroblast growth factor positively regulates hematopoietic development. Development 2000; 127:1931–41.

95. Burger PE, Coetzee S, McKeehan WL, et al. Fibroblast growth factor receptor-1 is expressed by endothelial progenitor cells. Blood 2002; 100:3527–35.

96. Fontaine V, Filipe C, Werner N, et al. Essential role of bone marrow fibroblast growth factor-2 in the effect of estradiol on reendothelialization and endothelial progenitor cell mobilization. Am J Pathol 2006; 169:1855–62.

97. Gloe T, Sohn HY, Meininger GA, et al. Shear stress-induced release of basic fibroblast growth factor from endothelial cells is mediated by matrix interaction via integrin alpha(v)beta3. J Biol Chem 2002; 277:23453–8.

98. Hartnett ME, Garcia CM, D'Amore PA. Release of bFGF, an endothelial cell survival factor, by osmotic shock. Invest Ophthalmol Vis Sci 1999; 40:2945–51.

99. Ziche M, Parenti A, Ledda F, et al. Nitric oxide promotes proliferation and plasminogen activator production by coronary venular endothelium through endogenous bFGF. Circ Res 1997; 80:845–52.

100. Shaw JP, Chuang N, Yee H, et al. Polymorphonuclear neutrophils promote rFGF-2-induced angiogenesis in vivo. J Surg Res 2003; 109:37–42.

101. Fujii T, Yonemitsu Y, Onimaru M, et al. Nonendothelial mesenchymal cell-derived MCP-1 is required for FGF-2-mediated therapeutic neovascularization: critical role of the inflammatory/arteriogenic pathway. Arterioscler Thromb Vasc Biol 2006; 26:2483–9.

102. Rusnati M, Camozzi M, Moroni E, et al. Selective recognition of fibroblast growth factor-2 by the long pentraxin PTX3 inhibits angiogenesis. Blood 2004.

103. Kuwabara K, Ogawa S, Matsumoto M, et al. Hypoxia-mediated induction of acidic/basic fibroblast growth factor and platelet-derived growth factor in mononuclear phagocytes stimulates growth of hypoxic endothelial cells. Proc Natl Acad Sci USA 1995; 92:4606–10.

104. Wang L, Xiong M, Che D, et al. The effect of hypoxia on expression of basic fibroblast growth factor in pulmonary vascular pericytes. J Tongji Med Univ 2000; 20:265–7.

105. Calvani M, Rapisarda A, Uranchimeg B, et al. Hypoxic induction of an HIF-1alpha-dependent bFGF autocrine loop drives angiogenesis in human endothelial cells. Blood 2006; 107:2705–12.

106. Li J, Shworak NW, Simons M. Increased responsiveness of hypoxic endothelial cells to FGF2 is mediated by HIF-1alpha-dependent regulation of enzymes involved in synthesis of heparan sulfate FGF2-binding sites. J Cell Sci 2002; 115:1951–9.

107. Gajdusek CM, Carbon S. Injury-induced release of basic fibroblast growth factor from bovine aortic endothelium. J Cell Physiol 1989; 139:570–9.

108. Griffioen AW, Damen CA, Blijham GH, et al. Tumor angiogenesis is accompanied by a decreased inflammatory response of tumor-associated endothelium. Blood 1996; 88:667–73.

109. Mantovani A, Allavena P, Sica A. Tumour-associated macrophages as a prototypic type II polarised phagocyte population: role in tumour progression. Eur J Cancer 2004; 40:1660–7.

110. Moscatelli D, Presta M, Joseph-Silverstein J, et al. Both normal and tumor cells produce basic fibroblast growth factor. J Cell Physiol 1986; 129:273–6.

111. Kandel J, Bossy-Wetzel E, Radvanyi F, et al. Neovascularization is associated with a switch to the export of bFGF in the multistep development of fibrosarcoma. Cell 1991; 66:1095–104.

112. Wang Y, Becker D. Antisense targeting of basic fibroblast growth factor and fibroblast growth factor receptor-1 in human melanomas blocks intratumoral angiogenesis and tumor growth. Nat Med 1997; 3:887–93.

113. Baird A, Mormede P, Bohlen P. Immunoreactive fibroblast growth factor (FGF) in a transplantable chondrosarcoma: inhibition of tumor growth by antibodies to FGF. J Cell Biochem 1986; 30:79–85.

114. Gross JL, Herblin WF, Dusak BA, et al. Effects of modulation of basic fibroblast growth factor on tumor growth in vivo. J Natl Cancer Inst 1993; 85:121–31.

115. Hori A, Sasada R, Matsutani E, et al. Suppression of solid tumor growth by immunoneutralizing monoclonal antibody against human basic fibroblast growth factor. Cancer Res 1991; 51:6180–4.

116. Compagni A, Wilgenbus P, Impagnatiello MA, et al. Fibroblast growth factors are required for efficient tumor angiogenesis. Cancer Res 2000; 60:7163–9.

117. Czubayko F, Liaudet-Coopman ED, Aigner A, et al. A secreted FGF-binding protein can serve as the angiogenic switch in human cancer. Nat Med 1997; 3:1137–40.

118. Rak J, Kerbel RS. bFGF and tumor angiogenesis–back in the limelight? Nat Med 1997; 3:1083–4.

119. Aigner A, Butscheid M, Kunkel P, et al. An FGF-binding protein (FGF-BP) exerts its biological function by parallel paracrine stimulation of tumor cell and endothelial cell proliferation through FGF-2 release. Int J Cancer 2001; 92:510–7.

120. Polnaszek N, Kwabi-Addo B, Peterson LE, et al. Fibroblast growth factor 2 promotes tumor progression in an autochthonous mouse model of prostate cancer. Cancer Res 2003; 63:5754–60.

121. Coltrini D, Gualandris A, Nelli EE, et al. Growth advantage and vascularization induced by basic fibroblast growth factor overexpression in endometrial HEC-1-B cells: an export-dependent mechanism of action. Cancer Res 1995; 55:4729–38.

122. Konerding MA, Fait E, Dimitropoulou C, et al. Impact of fibroblast growth factor-2 on tumor microvascular architecture. A tridimensional morphometric study. Am J Pathol 1998; 152:1607–16.

123. Giavazzi R, Giuliani R, Coltrini D, et al. Modulation of tumor angiogenesis by conditional expression of fibroblast growth factor-2 affects early but not established tumors. Cancer Res 2001; 61:309–17.

124. Cao R, Eriksson A, Kubo H, et al. Comparative evaluation of FGF-2-, VEGF-A-, and VEGF-C-induced angiogenesis, lymphangiogenesis, vascular fenestrations, and permeability. Circ Res 2004; 94:664–70.

125. Liaudet-Coopman ED, Schulte AM, Cardillo M, et al. A tetracycline-responsive promoter system reveals the role of a secreted binding protein for FGFs during the early phase of tumor growth. Biochem Biophys Res Commun 1996; 229:930–7.

126. Yoshiji H, Harris SR, Thorgeirsson UP. Vascular endothelial growth factor is essential for initial but not continued in vivo growth of human breast carcinoma cells. Cancer Res. 1997; 57:3924–3928.

127. Teicher BA. A systems approach to cancer therapy. (Antioncogenics + standard cytotoxics→mechanism(s) of interaction). Cancer Metastasis Rev 1996; 15:247–72.

128. Chirivi RG, Garofalo A, Crimmin MJ, et al. Inhibition of the metastatic spread and growth of B16-BL6 murine melanoma by a synthetic matrix metalloproteinase inhibitor. Int J Cancer 1994; 58:460–4.

129. Bergers G, Javaherian K, Lo KM, et al. Effects of angiogenesis inhibitors on multistage carcinogenesis in mice. Science 1999; 284:808–12.

130. Casanovas O, Hicklin DJ, Bergers G, et al. Drug resistance by evasion of antiangiogenic targeting of VEGF signaling in late-stage pancreatic islet tumors. Cancer Cell 2005; 8:299–309.

131. Rousseau B, Larrieu-Lahargue F, Javerzat S, et al. The tyrp1-Tag/tyrp1-FGFR1-DN bigenic mouse: a model for selective inhibition of tumor development, angiogenesis, and invasion into the neural tissue by blockade of fibroblast growth factor receptor activity. Cancer Res 2004; 64:2490–5.

132. Zagzag D, Miller DC, Sato Y, et al. Immunohistochemical localization of basic fibroblast growth factor in astrocytomas. Cancer Res 1990; 50:7393–8.

133. Nakamoto T, Chang CS, Li AK, et al. Basic fibroblast growth factor in human prostate cancer cells. Cancer Res 1992; 52:571–7.

134. Schulze-Osthoff K, Risau W, Vollmer E, et al. In situ detection of basic fibroblast growth factor by highly specific antibodies. Am J Pathol 1990; 137:85–92.

135. Ohtani H, Nakamura S, Watanabe Y, et al. Immunocytochemical localization of basic fibroblast growth factor in carcinomas and inflammatory lesions of the human digestive tract. Lab Invest 1993; 68:520–7.

136. Takahashi JA, Mori H, Fukumoto M, et al. Gene expression of fibroblast growth factors in human gliomas and meningiomas: demonstration of cellular source of basic fibroblast growth factor mRNA and peptide in tumor tissues. Proc Natl Acad Sci USA 1990; 87:5710–4.

137. Wesley UV, McGroarty M, Homoyouni A. Dipeptidyl peptidase inhibits malignant phenotype of prostate cancer cells by blocking basic fibroblast growth factor signaling pathway. Cancer Res 2005; 65:1325–34.

138. Fontijn D, Duyndam MC, van Berkel MP, et al. CD13/Aminopeptidase N overexpression by basic fibroblast growth factor mediates enhanced invasiveness of 1F6 human melanoma cells. Br J Cancer 2006; 94:1627–36.

139. Yang C, Zeisberg M, Lively JC, et al. Integrin alpha1beta1 and alpha2beta1 are the key regulators of hcpatocarcinoma cell invasion across the fibrotic matrix microenvironment. Cancer Res 2003; 63:8312–7.

140. Suyama K, Shapiro I, Guttman M, et al. A signaling pathway leading to metastasis is controlled by N-cadherin and the FGF receptor. Cancer Cell 2002; 2:301–14.

141. Poon RT, Fan ST, Wong J. Clinical implications of circulating angiogenic factors in cancer patients. J Clin Oncol 2001; 19:1207–25.

142. Landriscina M, Cassano A, Ratto C, et al. Quantitative analysis of basic fibroblast growth factor and vascular endothelial growth factor in human colorectal cancer. Br J Cancer 1998; 78:765–70.

143. Burian M, Quint C, Neuchrist C. Angiogenic factors in laryngeal carcinomas: do they have prognostic relevance? Acta Otolaryngol 1999; 119:289–92.

144. Folkman J. Angiogenesis-dependent diseases. Semin Oncol 2001; 28:536–42.

145. Ferrara N, Alitalo K. Clinical applications of angiogenic growth factors and their inhibitors. Nat Med 1999; 5:1359–64.

146. Ria R, Portaluri M, Russo F, et al. Serum levels of angiogenic cytokines decrease after antineoplastic radiotherapy. Cancer Lett 2004; 216:103–7.

147. Nguyen M, Watanabe H, Budson AE, et al. Elevated levels of an angiogenic peptide, basic fibroblast growth factor, in the urine of patients with a wide spectrum of cancers. J Natl Cancer Inst 1994; 86:356–61.

148. Dirix LY, Vermeulen PB, Pawinski A, et al. Elevated levels of the angiogenic cytokines basic fibroblast growth factor and vascular endothelial growth factor in sera of cancer patients. Br J Cancer 1997; 76:238–43.

149. Salgado R, Benoy I, Vermeulen P, et al. Circulating basic fibroblast growth factor is partly derived from the tumour in patients with colon, cervical and ovarian cancer. Angiogenesis 2004; 7:29–32.

150. Rusnati M, Presta M. Fibroblast growth factors/fibroblast growth factor receptors as targets for the development of anti-angiogenesis strategies. Current Pharma Design 2007; in press.

151. Giavazzi R, Sennino B, Coltrini D, et al. Distinct role of fibroblast growth factor-2 and vascular endothelial growth factor on tumor growth and angiogenesis. Am J Pathol 2003; 162:1913–26.

152. Hagedorn M, Bikfalvi A. Target molecules for anti-angiogenic therapy: from basic research to clinical trials. Crit Rev Oncol Hematol 2000; 34:89–110.

153. Ziche M, Donnini S, Morbidelli L. Development of new drugs in angiogenesis. Curr Drug Targets 2004; 5:485–93.

154. Patel SR, Jenkins J, Papadopolous N, et al. Pilot study of vitaxin–an angiogenesis inhibitor-in patients with advanced leiomyosarcomas. Cancer 2001; 92:1347–8.

155. Posey JA, Khazaeli MB, DelGrosso A, et al. A pilot trial of Vitaxin, a humanized anti-vitronectin receptor (anti alpha v beta 3) antibody in patients with metastatic cancer. Cancer Biother Radiopharm 2001; 16:125–32.

Chapter 8
Vascular Permeability/Vascular Endothelial Growth Factor

Masabumi Shibuya

Keywords: endothelial cells, vascular permeability, tumor angiogenesis, metastasis, ascites, tyrosine kinase receptor

Abstract: The vascular permeability factor (VPF)/vascular endothelial growth factor (VEGF) family has more than seven members including VEGF-A, VEGF-B, VEGF-C, VEGF-D, VEGF-E, PlGF, and *Trimeresurus flavoviridis* (*T. f.*) svVEGFs. Except for VEGF-E and *T.f.* svVEGFs, all members are encoded in the mammalian genome and involved in angiogenesis and/or lymphangiogenesis. Among these five gene products, VEGF-A (also known as VEGF and VPF) binds two receptor-type tyrosine kinases, VEGFR1 and VEGFR2, and transduces major signals for angiogenesis and vascular permeability. VEGF-A expression is efficiently induced by hypoxia, and regulates not only physiological but also most of the pathological angiogenesis, such as tumor angiogenesis. Since VEGF-A utilizes VEGFR2 as a direct stimulator for angiogenesis, this VEGF-VEGFR2 system represents an ideal pharmaceutical target for suppressing various diseases. Interestingly, VEGFR1 has also been shown to be deeply involved in various pathological processes in cancer as well as inflammatory diseases via a mechanism different from VEGFR2, suggesting that VEGF-VEGFR1 is another attractive target for suppressing human diseases. VEGF-C/D and their receptor VEGFR3 play a central role in lymphangiogenesis, and the blocking of this system significantly decreases lymph node metastasis in animal models of cancer. VEGF-E, a VEGFR2-specific ligand, induces angiogenesis with fewer side effects such as edema and inflammatory responses which are commonly observed on treatment with VEGF-A. Thus, VEGF-E is a useful candidate for proangiogenic therapy.

Department of Molecular Oncology, Tokyo Medical and Dental University, 1-5-45, Yushima, Bunkyo-ku, Tokyo, 113-8519, Japan
E-mail: shibuya@ims.u-tokyo.ac.jp

Introduction

The blood vessel system is essential for the development and maintenance of the tissues in the body, supplying oxygen and nutrition in vertebrates. This system is also involved in a variety of diseases including cancer (Fig. 8.1) [1–3]. In 1983, Senger et al. isolated a protein with strong vascular permeability activity, and designated it vascular permeability factor (VPF) [4]. Some years later, Ferrara and Henzel purified a protein with growth-promoting activity for vascular endothelial cells (ECs) named VEGF [5]. Surprisingly, molecular cloning revealed that the two proteins are identical and encoded by a single gene which is now known as *VEGF* (or *VEGF-A*) [6,7].

Extensive studies on the VEGF family have to date revealed more than seven members, with VEGF-A essential not only for vasculogenesis, the formation of new blood vessels from endothelial progenitor cells in embryogenesis, but also for angiogenesis, the formation of new blood vessels from the pre-existing vasculature [8–11]. Furthermore, VEGF-A was demonstrated to be a key player for tumor angiogenesis [12–14], and anti-human VEGF-A neutralizing antibody in combination with chemotherapy has recently been approved by the U.S. Food and Drug Administration (FDA) for the treatment of late-stage colorectal cancer [15] and non-squamous lung cancer.

VEGF-C mostly binds VEGFR3, and this system is the first to be shown to directly regulate lymphangiogenesis [16–18]. Strong suppression of VEGF-C/VEGFR3 signaling induces dysfunction and loss of the lymphatic system, resulting in lymphedema and a poor lipid-absorbance [19]. Tumor cells that express *VEGF-C or D* have extensive potential to metastasize to the lymph nodes, strongly suggesting that VEGF-C/D-VEGFR3 signaling is an important target for decreasing lymph node metastasis in cancer patients [17,20].

VEGF and its receptor are considered fundamental regulators of angiogenesis/lymphangiogenesis in vertebrates, and also closely linked to vascular permeability (Fig. 8.2) [14,21]. Because of these biological activities, VEGF-related molecules have developed in various organisms including viruses, and

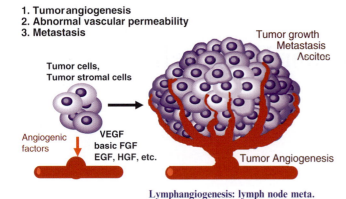

1. Tumor angiogenesis
2. Abnormal vascular permeability
3. Metastasis

Tumor cells,
Tumor stromal cells

Tumor growth
Metastasis
Ascites

Angiogenic
factors

VEGF
basic FGF
EGF, HGF, etc.

Tumor Angiogenesis

Lymphangiogenesis: lymph node meta.

FIG. 8.1. Involvement of the vascular system in tumor progression. Tumor cells and tumor stromal cells such as macrophages, smooth muscle cells and fibroblasts secrete various angiogenic factors, and stimulate tumor angiogenesis as well as vascular permeability. Blood vessels in the tumor enhance tumor-growth and metastasis. Lymphangiogenesis significantly increases lymph-node-oriented metastasis.

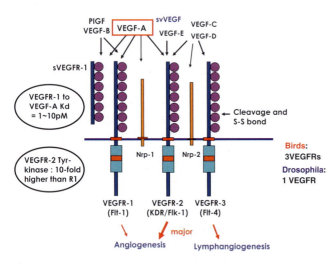

FIG. 8.2. VEGF and its receptor system. Major signals of angiogenesis are generated from VEGFR2. Although VEGFR1 has a weak tyrosine kinase activity, it also stimulates angiogenesis via recruitment of bone marrow-derived mononuclear cells. VEGFR1 plays an important role in inflammation and atherosclerosis. Soluble VEGFR1 is involved in placental regulation and avascularity in the cornea.

are utilized for specialized purposes. The VEGF-E family encoded by the Orf viral genome [22] has a similar structure to VEGF-A, but activates only VEGFR2, and efficiently induces angiogenesis in virally infected tissue in the skin [23–25]. This angiogenic response may facilitate viral replication and production in the host by supplying nutrition and oxygen. On the other hand, snake venom VEGF-like proteins such as the *T.f.* svVEGF (snake venom-derived VEGF in *T.f.* snake in southern Japan) are a family secreted from the venom tissue, and significantly increase vascular permeability [26–30]. This activity may be used to efficiently distribute the snake venom-derived toxins into the body of target animals.

During the past two decades, many angiogenic factors such as VEGF, fibroblast growth factor (FGF), Angiopoietin, hepatocyte growth factor (HGF), and epidermal growth factor (EGF) have been described [3,31,32]. To understand more deeply the molecular basis underlying the formation and regression of blood vessels in our body, further extensive studies on the characteristics of each factor as well as the interrelationship among these factors, particularly between VEGF and the others, appear to be important.

VEGF-A

Function of VEGF-A

VEGF-A promotes the differentiation of endothelial progenitor cells in the early embryo, and stimulates vascular endothelial cell growth, survival, tubular formation, and migration [8,13,14]. VEGF-A is a strong vascular permeability factor with a high specific activity [21]. VEGF-A stimulates the secretion of growth/survival factors from ECs towards surrounding cells such as smooth muscle cells and hepatocytes [33]. Furthermore, VEGF-A induces the expression of factors related to coagulation such as tPA and PAI-1, suggesting it to be an important regulator of blood coagulation [34]. VEGFR2 is also expressed to some extent on lymphatic endothelial cells; thus, VEGF-A could be one of the regulators in lymphangiogenesis [17,35]. Recent reports suggest that VEGFR is expressed on neuronal cells or oligodendrocytes, raising the possibility that under certain conditions, the VEGF-VEGFR system is a direct regulator of cell growth/survival in the nervous system [36–38].

Embryonic Lethality of *VEGF-A* (+/−) Heterozygotic Mice

VEGF-A shows endothelial cell-specific growth-promoting activity, implying a crucial role in blood vessel formation during embryogenesis. Carmeliet et al. [39] and Ferrara et al. [40] demonstrated that the knockout of *VEGF-A* in mice is embryonic lethal even among heterozygotes. They confirmed that these mice are not functionally null for the *VEGF-A* gene via a gene-silencing of the wild-type locus, and that these mice had multiple defects in angiogenesis, such as a disconnection of the heart with the aorta and a poor development of the dorsal aorta. These results strongly suggest that the concentration of VEGF-A in tissues is crucial for normal development of the closed circulatory system in embryos, and that half the normal level of VEGF-A is insufficient to complete morphogenesis in the vascular system. Heterozygotic lethality is extremely rare among mammals.

Isoform of *VEGF-A* Gene Products.

The human *VEGF-A* gene encodes at least 9 different products (isoforms) consisting of 121 to 206 amino acids due to alternative splicing [14, 41]. Three major isoforms, 121, 165 and 189-amino-acids long exist, which are well conserved from mammals to other vertebrates [29]. The most abundant isoform in vivo is the 165-amino-acid type, which is expressed in a variety of cells in the body.

VEGF-A belongs to the VEGF/PDGF (platelet-derived growth factor) super-gene family [6–8, 42], whose major characteristics are (1) growth factor with a homodimeric structure, (2) eight conserved cysteines in a monomer at the same positions, and (3) three intramolecular S-S bonds to form three loop structure within the monomeric peptide.

A major difference between the 165- and 189-amino acid isoforms from the 121-isoform is the presence of a basic stretch of residues that bind heparin and heparan-sulfate-containing acidic molecules. The affinity of heparin for the basic stretch in the 165-amino-acid isoform is weaker than that in the 189-isoform [14]. In addition, VEGF-A$_{165}$ binds neuropilin-1 (NRP-1), a co-receptor for VEGF-A through this same basic stretch, which is expressed on the cell surface [43,44]. Isoform-specific mutant mice bearing only VEGF-A$_{120}$ (VEGF-A120/120) or only VEGF-A$_{188}$ (VEGF-A188/188) die in the embryonic stage due to multiple defects in angiogenesis, whereas VEGF-A164/164 mice are healthy, indicating that the 164-isoform (165-isoform in humans) of VEGF-A is essential and sufficient for the basic development and morphogenesis of the closed circulatory system [45,46]. Two major reasons for the importance of VEGF-A$_{165}$ could be as follows: (1) because of the mild affinity of the basic stretch for heparin, VEGF-A$_{165}$ has an appropriate balance between free and bound forms, resulting in a proper gradient in the concentration of this angiogenic factor surrounding the VEGF-A$_{165}$–secreting cells [47]; and (2) in association with NRP-1 through the basic stretch, VEGF-A$_{165}$ binds with higher affinity than VEGF-A$_{121}$ to the receptor, and efficiently activates the tyrosine kinase of VEGFR2 to transduce angiogenic signals. VEGF-A$_{189}$ binds heparin in the extracellular matrix and NRP-1 on the cell surface. However, due to an extremely high affinity for these molecules, the VEGF-A$_{189}$ isoform does not diffuse efficiently, making a narrow-range gradient of VEGF-A$_{189}$ [47].

The isoform VEGF-A165b, which carries a carboxy terminal sequence different from VEGF-A$_{165}$, has less affinity for VEGFR; thus, it might be a negative regulator of angiogenesis under certain conditions.

Regulation of *VEGF-A* Gene Expression.

The *VEGF-A* gene is regulated at both the transcriptional and post-transcriptional levels. Growth factors operating via the transcription factors Fos/Jun complex and nuclear factor κB (NF-κB), and hormones such as estrogen, appear to be the major stimulators of *VEGF-A* gene expression under normoxic condition [48–50]. In addition, hypoxic stress blocks the function of von Hippel-Lindau (VHL), a component of the ubiquitin-ligase system, and stabilizes the transcrip-

tion factor hypoxia-inducible factor complex (HIFα/HIFβ) important for *VEGF-A* gene induction [51–57]. The HIF complex binds at a hypoxia-responsive element (HRE) site in the *VEGF-A* gene and upregulates transcription of *VEGF-A*. Furthermore, hypoxic conditions increase the stability of *VEGF-A* mRNA post-transcriptionally, resulting in the production of more VEGF-A protein.

Involvement of VEGF-A in Pathological Angiogenesis and Vascular Permeability.

VEGF-A levels are increased in a variety of diseases such as cancer, rheumatoid arthritis, diabetic retinopathy, age-related macular degeneration, and atherosclerosis [58–61]. In tumor tissues, VEGF-A is secreted not only from the tumor cells themselves, but also from infiltrating macrophage-lineage cells and mesenchymal cells [62,63]. Blocking of the VEGF-A/VEGFR system with VEGF-neutralizing antibody, soluble VEGFR1 including 'VEGF-Trap', and a low-molecular weight chemical tyrosine kinase inhibitor in a tumor-implanted mouse system significantly decreased tumor growth and metastasis [12, 64–68]. Furthermore, clinical trials for the treatment of colorectal [15], breast, renal, and non-small cell lung cancer with anti-VEGF-A/VEGFR therapy showed a statistically significant increase in disease-free survival with minimal side effects [68a]. Based on these results, the FDA has recently granted approvals to angiogenic inhibitors with anti-VEGF-VEGFR activity (VEGF-A neutralizing antibody and VEGFR tyrosine kinase inhibitor) to treat colorectal, renal, and a part of lung cancer patients (Nonsquamous non-small cell lung cancer [68b]).

VEGF-A has potent vascular permeability activity. Senger et al. [4] and Luo et al. [69] demonstrated that VEGF-A is highly accumulated in ascites fluid of the ascites-tumor model, and that the blocking of VEGF-A strongly suppressed the volume of ascites, number of tumor cells, and hemorrhagic tendency [70]. Thus, the VEGF-VEGFR system is a good target for decreasing symptoms in patients bearing tumor-induced ascites.

Rheumatoid arthritis (RA) models in mice using a variety of inducers revealed that suppression of the VEGF-A/VEGFR system significantly decreased the clinical as well as pathological scores in arthritis, suggesting that anti-VEGF-A/VEGFR therapy could be beneficial for RA-patients [71–73].

Age-related macular degeneration is also related to an increase in VEGF levels in the eye. An aptamer, which is a short RNA molecule specifically blocking the VEGF-A$_{165}$ isoform, and an anti-VEGF-A neutralizing antibody were shown to be effective at suppressing macular degeneration.

PlGF

The placenta growth factor (PlGF) is a member of the VEGF family, highly expressed in the placenta, and binds and activates only VEGFR1 [11,74]. PlGF has a few isoforms with or without the basic stretch, and a longer form with the basic stretch that binds NRP-1 similar to VEGF-A$_{165}$ and VEGF-A$_{189}$. Because of the weak tyrosine kinase activity of VEGFR1, PlGF has a limited effect on angiogenesis in vitro such as the proliferation of ECs. Carmeliet et al. [75] showed that *PlGF*-gene knockout mice are basically healthy and fertile, but under certain ischemic conditions, the *PlGF* (-/-) mice are impaired in angiogenesis, wound healing, and cancer. These results imply that PlGF might be another target for suppressing tumor growth.

VEGF-B

VEGF-B also binds and activates only VEGFR1 similar to PlGF. VEGF-B is expressed in a wide variety of tissues, but particularly in heart and skeletal muscle. *VEGF-B* gene knockout mice have no abnormalities in the embryonic stages, but after birth, have smaller hearts, demonstrating an insufficient recovery from an experimentally induced myocardial ischemia [76].

VEGF-C and VEGF-D

VEGF-C and -D are related in structure with approximately 30 and 100 amino-acid sequences in the amino- and carboxy-terminal regions, respectively [16–19]. These extra sequences are proteolytically removed, and the shortest form thus generated has the greatest ability to bind and activate VEGFR3, turning on the signaling cascade for lymphangiogenesis. This short form of VEGF-C also binds and activates VEGFR2 to some extent, suggesting angiogenic activity under certain conditions.

VEGF-D (-/-) mice exhibit no phenotype, but *VEGF-C* (-/-) mice die late in embryogenesis due to a defect of lymph vessel formation [19]. Furthermore, mice heterozygous at the *VEGF-C* gene locus often die during the perinatal stage. These mutant embryos show severe lymphedema and chylous ascites in the abdominal cavity, clearly indicating that a proper concentration of VEGF-C is essential for the development and normal function of lymph vessels. The *VEGF-C* gene is at first expressed in the mesenchymal cells near the budding of lymph vessels from the vein in embryos; thus, the VEGF-C and VEGFR3 system is used in a paracrine manner for the development of lymph vessels [19].

A high level of VEGF-C and –D in tumor cells is a prognostic factor for cancer patients [17,20]. Also, tumor cells exogenously expressing VEGF-C or VEGFD show a high degree of potential for lymph node metastases in mice. These results strongly suggest that blocking of the VEGF-C/D system, using inhibitors such as soluble VEGFR3-Fc, is important for suppressing lymph node metastasis. VEGF-C

and -D are potential candidates in the treatment of lymph edema caused by a VEGFR3-inactivation mutation [77] or by postnatal lymph vessel deficiency.

VEGF-E (Orf-VEGF)

Orf virus, a parapox virus, infects sheep, goats, and sometimes humans, and induces a local and transient angiogenesis in skin. Lyttle et al. [22] identified a sequence in the viral genome which could encode a VEGF-related molecule. Ogawa et al. found that VEGF-E$_{NZ7}$ protein encoded in the Orf-viral strain NZ7 binds and activates only VEGFR2, and strongly induces proliferation of vascular endothelial cells [23]. Essentially the same results were obtained in other VEGF-Es encoded in the strains NZ2 and D1701 [24,25]. Furthermore, Kiba et al. and Zheng et al. clearly showed that VEGF-E$_{NZ7}$ and its chimeric forms, together with the human PlGF sequence, strongly induced angiogenesis in subcutaneous tissues in K14-promoter transgenic mice with less edema or inflammatory responses [78,79]. On the other hand, K14- and related promoter-driven *VEGF-A* transgenic mice have a variety of side effects such as severe edema, hemorrhage, and inflammation [80,81]. These results suggest that VEGF-E may be an attractive molecule to use for proangiogenic therapy in the clinic.

T.f. svVEGF

The venom of a snake named "Habu", *Trimeresurus flavoviridis* (*T. f.*), targets blood vessels as well as muscle tissues in animals. Takahashi et al. [26] isolated a VEGF-like protein from this snake venom, referred to as *T.f.* svVEGF, which has weak endothelial cell proliferating activity but strong vascular permeability activity. Surprisingly, *T.f.* svVEGF significantly activates VEGFR1 but only weakly activates VEGFR2, and this coordinated activation of two VEGFRs appears to induce permeability-oriented signaling within the vascular ECs. Takahashi et al. also isolated the snake *VEGF-A* gene that encodes a protein highly homologous to human VEGF-A (amino acid identity: 71%). Snake *VEGF-A* mRNA is expressed essentially in all the tissues of this animal with three representative isoforms (121, 165, and 189 amino acids), whereas *T.f. svVEGF* mRNA is expressed specifically in the venom tissue. Similar VEGF-like proteins were reported in other snake venoms although the affinity of these proteins for VEGFRs is not well characterized yet [26,28]. Snake venom VEGF is unique among the VEGF family in terms of being an exocrine-type protein and of its permeability-oriented activity. In phylogenetical development, it is most likely that snakes possess the gene for a permeability-dominant, VEGF-like protein to enhance the efficacy of toxins by increasing the permeability of blood vessels in the targeted animals.

VEGF-receptor

Receptor for VEGF and Signaling Within Endothelial Cells

The VEGF family has three high-affinity receptors, VEGFR1 (Flt-1: Fms-like tyrosine kinase-1), VEGFR2 (KDR: Kinase-insert Domain-containing Receptor in humans; Flk-1: Fetal-liver kinase-1 in mice), and VEGFR3 (Flt-4) [16,82–87]. These receptors are structurally highly related to each other, and conserve the seven-Ig (Immuno-globulin) like domain-containing extracellular domain, the tyrosine kinase domain with about a 70-amino acid-long kinase-insert sequence, and the carboxy terminal tail. VEGFRs are distantly related to 5-Ig domain-containing tyrosine kinase receptors such as the PDGF receptor (PDGFR) [42]. Thus, the VEGFR and PDGFR families belong to a super-family of tyrosine kinase genes. Mammals and birds conserve three VEGFR systems, and reptiles and amphibians are also suggested to keep this set. However, the zebrafish has four genes for VEGFR, indicating a redundancy via gene duplication [88]. The ligand-receptor relationship among VEGFs-VEGFRs is shown in Fig. 8.2.

VEGFR1

VEGF-A, a key player for angiogenesis in vivo, binds VEGFR1 and VEGFR2. VEGFR1 has a higher affinity for VEGF-A with a Kd of 2~10 pM, which is about 10-fold that of VEGFR2, whereas the kinase activity of VEGFR1 is one order of magnitude weaker than that of VEGFR2 [89–91]. In addition, the *VEGFR1* gene encodes not only a full-length receptor but also a soluble form that carries the first 6-Ig domains without a transmembrane domain or tyrosine kinase domain [82,92]. This characteristic, that *VEGFR1* encodes both forms, is conserved not only in mammals but also in birds and frogs, indicating that it has been established at an early stage in the phylogenetic development of vertebrates [93].

During embryogenesis in mammals, the *VEGFR1* gene is essential for the normal development of blood vessels, and mutant mice without *VEGFR1* die due to the overgrowth and disorganization of blood vessels [94]. This negative role of VEGFR1 in vascular development is exerted by tight-binding at the extracellular domain of VEGFR1 since tyrosine kinase-domain-deficient (VEGFR1 TK-/-) mice are healthy and develop an almost normal circulatory system [95].

Interestingly, VEGFR1 is expressed in adulthood not only in vascular ECs but also in monocyte / macrophage lineage cells [96–98], playing an important role in the progression of various diseases such as cancer, rheumatoid arthritis, and atherosclerosis (Fig. 8.3) [71,99–104]. Furthermore, soluble VEGFR1 was found to be abnormally expressed in high amounts in preeclampsia patients, such that the levels of soluble VEGFR1 in the maternal serum were well correlated with the degree of preeclampsia, including hypertension and proteinurea [105–107]. The intravenous injection of soluble

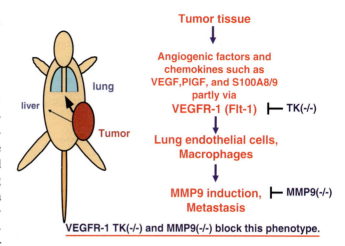

Fig. 8.3. VEGFR1 stimulates tumor metastasis via pre-metastatic induction of MMP9 and other factors in tissues such as lung. Primary tumors stimulate secretion of a variety of factors such as MMP9, SDF-1, S100A8 and S100A9 from lung, spleen and other tissues before metastasis, and enhance tumor metastasis (ref. 101, 102, 125). Lung is the highest tissue for metastasis. Blocking of VEGFR1 signaling by anti-VEGFR1 neutralizing antibody or VEGFR1 TK-deficiency in mice significantly suppresses the process and hence metastasis.

VEGFR1 into normal pregnant rats induced symptoms similar to preeclampsia in humans, strongly suggesting that an excess amount of soluble VEGFR1 abnormally traps the physiologically required VEGF-A in the body, particularly in the kidney, resulting in the dysfunction and apoptosis of vascular endothelial cells in the glomeruli.

Soluble VEGFR1 is also highly expressed in the cornea of mammals. Ambati et al. [108] showed that the avascularity of the cornea is maintained by soluble VEGFR1, and aniridia patients with vascularized cornea lost expression of soluble VEGFR1 in corneal epithelial cells.

VEGFR2

In the embryonic stage, *VEGFR2/flk-1*-gene minus mutant mice die due to a lack of vasculogenesis [109]. This indicates that VEGF-A-VEGFR2 is essential for the differentiation of hemangioblasts into ECs as well as for the proliferation and morphogenesis of ECs. Activation of VEGFR2 tyrosine kinase results in the autophosphorylation of several tyrosine (Y) residues in its intracellular domain, and Y951, Y1054, Y1059, Y1175, and Y1214 were highly phosphorylated [110,111]. Among them, Y1175 is important for triggering the downstream signaling from the receptor. Phorphorylation of Y1175 recruits PLCγ (phospholipaseCγ) and activates the PLCγ-PKC-Raf-MEK-MAP-kinase pathway resulting in DNA synthesis and angiogenesis (Fig. 8.4) [110,112]. Surprisingly, unlike other representative tyrosine kinase receptors such as EGFR, the activation of Ras (Ras-GTP formation) is

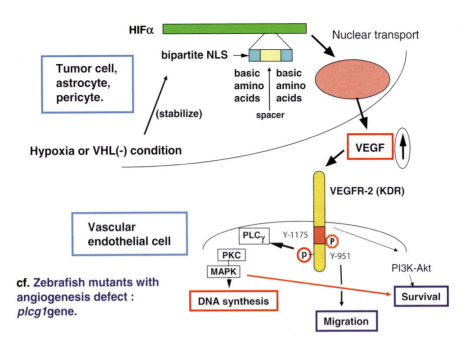

F ig. 8.4. Expression of *VEGF-A* gene and signal transduction from VEGFR2. Hypoxia is a crucial inducer for the *VEGF-A* gene expression in tumor tissue. VEGF-A activates VEGFR2 tyrosine kinase, and stimulates the PLCγ-PKC-Raf-MAP-kinase pathway toward angiogenesis. The TSAd pathway is a regulator for cell migration.

very weak downstream from VEGFR2. Although PDGFR is structurally related to VEGFR, it acquired a kinase-insert (KI) sequence different from that in VEGFR, and PDGFR-KI contains phosphoinositide 3-kinase (PI3K)-Akt activation motifs which connect to the Ras pathway. Therefore, the signaling system of VEGFR appears very unique, much milder than that of regular tyrosine kinase signaling, most likely for keeping the vascular structure stable during angiogenesis. For the migration signal from VEGFR2, the adaptor protein TSAd is reported to be involved in this process by its binding to the phosphorylated Y951 residue of VEGFR2 [113].

Phosphorylation of Y1175 is essential for the function of VEGFR2 in vasculogenesis, since a single tyrosine to phenylalanine mutation at Y1173 in murine VEGFR2 (corresponding to Y1175 in humans) results in embryonic death due to a lack of blood vessel formation [114]. Other crucial signaling molecules, such as VE-cadherin, integrins, and c-Src, were also reported to bind VEGFR2, regulating endothelial cell-cell interaction, cell-matrix adhesion, and vascular permeability [115–117]. Nitric oxide (NO) synthesizing system such as eNOS and iNOS appears to be partly related to the downstream signaling from VEGFR2 [118].

VEGFR2 is directly involved in many forms of pathological angiogenesis such as tumor angiogenesis and the formation of ascites. Thus, in addition to VEGF-A, VEGFR2 is also an important target in the pharmacological development of anti-cancer drugs. A VEGFR-tyrosine kinase inhibitor was recently approved by the FDA as a therapeutic agent to treat renal cell carcinoma patients [119], and many other tyrosine

kinase inhibitors such as ZD4190 and PTK787/ZK 222584 are currently being evaluated in clinical trials [67,68,120].

VEGFR3

As described in detail in Chapter 43, the VEGF-C/D and VEGFR3 system is the major regulator for lymphangiogenesis in vertebrates. A deficiency of this system induces a severe defect of angiogenesis and lymphangiogenesis, and embryos die in the middle stages of pregnancy, E10.5 in mice [121]. Other signaling systems such as Angiopoietin-Tie2 cooperate with VEGF-C/D-VEGFR3 for the physiological development of lymph vessels (refer to Chapter 10 for a discussion on Angiopoietins) [122]. Lymphangiogenesis is highly related to lymph node metastasis in cancer, indicating that blocking of VEGF-C/D and/or suppression of VEGFR3 signaling is an effective way to decrease metastasis and malignancy in cancer.

Co-Receptor: Neuropilin-1 and Neuropilin–2.

Vascular and lymphatic endothelial cells express the membrane proteins neuropilin-1 and −2 (NRP-1, NRP-2), respectively. They function as a co-receptor for the VEGF family. Particularly, VEGF-A$_{165}$ binds NRP-1 via the basic stretch, and this association increases significantly the affinity of VEGF-A for VEGFR2, stimulating its signaling of angiogenesis [44]. The association of VEGF-A$_{165}$ with NRP-1 is essential for embryogenesis, since a lack of the VEGF-A$_{165}$ isoform or a lack of NRP-1 results in similar embryonic lethality due to poor development of the dorsal aorta, insufficient aorticopulmonary truncus, and a lack of remodeling in angiogenesis in the yolk sac [123]. NRP-2 also associates with VEGF-C/D, and is suggested to increase in the signaling of lymphangiogenesis [124].

Prospective

Recently, the VEGF-VEGFR system was reported to be involved in the neuronal system and some neuronal disorders in mice (see Chapter 42). VEGF and its receptor are basically used in tissues in a paracrine manner, where the cells adjacent to vascular endothelial cells such as smooth muscle cells and astrocytes secrete VEGF, and activate VEGFR on endothelial cells. On the other hand, endothelial cells secrete cytokine(s) other than VEGF to communicate to adjacent cells. However, such a paracrine mechanism might be disrupted in pathological situations, and an autocrine type activation or a reverse activation may occur. Thus, poor signaling from VEGFRs might directly induce severe cellular damage not only in vascular endothelial cells but also in other cell types such as neurons. Further studies are necessary to fully understand how deeply VEGF/VEGFR is involved in a variety of systems in the body, and how it is linked to various diseases.

Acknowledgements. This work was supported by Grants-in-aid for Special Project Research on Cancer-Bioscience (12215024, 17014020) from the Ministry of Education, Science, Sports and Culture of Japan and for the program 'Research for the Future' of the Japan Society for the Promotion of Science, and the program 'Promotion of Fundamental Research in Health Sciences' from the Organization for Pharmaceutical Safety and Research (OPSR).

References

1. Risau W. Mechanism of angiogenesis. Nature 1997; 386:671–4.
2. Folkman J. Tumor angiogenesis: therapeutic implications. N Engl J Med 1971; 285:1182–6.
3. Folkman J. Angiogenesis in cancer, vascular, rheumatoid and other disease. Nat Med 1995; 1:27–31.
4. Senger DR, Galli SJ, Dvorak AM, et al. Tumor cells secrete a vascular permeability factor that promotes accumulation of ascites fluid. Science 1983; 219:983–5.
5. Ferrara N, Henzel WJ. Pituitary follicular cells secrete a novel heparin -binding growth factor specific for vascular endothelial cells. Biochem Bophys Res Commun1989; 161:851–8.
6. Keck PJ, Hauser SD, Krivi G, et al. Vascular permeability factor, an endothelial cell mitogen related to PDGF. Science 1989; 246:1309–12.
7. Leung DW, Cachianes G, Kuang W-J, et al. Vascular endothelial growth factor is a secreted angiogenic mitogen. Science 1898; 246:1306–9.
8. Ferrara N, Davis-Smyth T. The biology of vascular endothelial growth factor. Endocrine Rev 1997; 18:4–25.
9. Shibuya M. Role of VEGF-Flt receptor system in normal and tumor angiogenesis. Adv Cancer Res 1995; 67:281–316.
10. Matsumoto T, Claesson-Welsh L. VEGF receptor signal transduction. Sci STKE 2001; RE21.
11. Shibuya M, Claesson-Welsh L. Signal transduction by VEGF receptors in regulation of angiogenesis and lymphangiogenesis. Exp Cell Res 2006; 312:549–60.
12. Kim KJ, Li B, Winer J, et al. Inhibition of vascular endothelial growth factor-induced angiogenesis suppresses tumour growth in vivo. Nature 1993; 362:841–4.
13. Hanahan D, Folkman J. Patterns and emerging mechanisms of the angiogenic switch during tumorigenesis. Cell 1996; 86:353–64.
14. Ferrara N. VEGF and the quest for tumour angiogenesis factors. Nat Rev Cancer 2002; 2:795–803.
15. Hurwitz H, Fehrenbacher L, Novotny W, et al. Bevacizumab plus irinotecan, fluorouracil, and leucovorin for metastatic colorectal cancer. N Engl J Med 2004; 350:2335–42.
16. Joukov V, Pajusola K, Kaipainen A, et al. A novel vascular endothelial growth factor, VEGF-C, is a ligand for the Flt4 (VEGFR-3) and KDR (VEGFR-2) receptor tyrosine kinases. EMBO J 1996; 15:290–8.
17. Alitalo K, Carmeliet P. Molecular mechanisms of lymphangiogenesis in health and disease. Cancer Cell 2002; 1:219–27.
18. Veikkola T, Jussila L, Makinen T, et al. Signalling via vascular endothelial growth factor receptor-3 is sufficient for lymphangiogenesis in transgenic mice. EMBO J 2001; 20:1223–31.
19. Karkkainen MJ, Haiko P, Sainio K, et al. Vascular endothelial growth factor C is required for sprouting of the first lymphatic vessels from embryonic veins. Nat Immunol 2004; 5:74–80.
20. Stacker SA, Caesar C, Baldwin ME, et al. VEGF-D promotes the metastatic spread of tumor cells via the lymphatics. Nature Med 2001; 7:186–91.
21. Dvorak HF. Vascular permeability factor/vascular endothelial growth factor: a critical cytokine in tumor angiogenesis and a potential target for diagnosis and therapy. J Clin Oncol 2002; 20:4368–80.
22. Lyttle DJ, Fraser KM, Fleming SB, et al. Homologs of vascular endothelial growth factor are encoded by the poxvirus orf virus. J Virol 1994; 68:84–92.
23. Ogawa S, Oku A, Sawano A, et al. A novel type of Vascular Endothelial Growth Factor: VEGF-E (NZ-7 VEGF) preferentially utilizes KDR/Flk-1 receptor and carries a potent mitotic activity without heparin-binding domain. J Biol Chem 1998; 273:31273–82.
24. Meyer M, Clauss M, Lepple-Wienhues A, et al. A novel vascular endothelial growth factor encoded by Orf virus, VEGF-E, mediates angiogenesis via signalling through VEGFR-2 (KDR) but not VEGFR-1 (Flt-1) receptor tyrosine kinases. EMBO J 1999; 18:363–74.
25. Wise LM, Veikkola T, Mercer AA, et al. Vascular endothelial growth factor (VEGF)-like protein from orf virus NZ2 binds to VEGFR2 and neuropilin-1. Proc Natl Acad Sci USA 1999; 96:3071–6.
26. Junqueira de Azevedo IL, Farsky SH, Oliveira ML, et al. Molecular cloning and expression of a functional snake venom vascular endothelium growth factor (VEGF) from the Bothrops insularis pit viper. A new member of the VEGF family of proteins. J Biol Chem 2001; 276:39836–42.
27. Komori Y, Nikai T, Taniguchi K, et al. Vascular endothelial growth factor VEGF-like heparin-binding protein from the

venom of Vipera aspis aspis (Aspic viper). Biochemistry 1999; 38:11796–803.

28. Gasmi A, Bourcier C, Aloui Z, et al. Complete structure of an increasing capillary permeability protein (ICPP) purified from Vipera lebetina venom. ICPP is angiogenic via vascular endothelial growth factor receptor signalling. J Biol Chem 2002; 277:29992–8.

29. Takahashi H, Hattori S, Iwamatsu A, et al. A novel snake venom vascular endothelial growth factor (VEGF) predominantly induces vascular permeability through preferential xignaling via VEGF receptor-1. J Biol Chem 2004; 279:46304–14.

30. Yamazaki Y, Takani K, Atoda H, et al. Snake venom vascular endothelial growth factors (VEGFs) exhibit potent activity through their specific recognition of KDR (VEGF receptor 2). J Biol Chem 2003; 278:51985–8.

31. Yancopoulos GD, Davis S, Gale NW, et al. Vascular-specific growth factors and blood vessel formation. Nature 2000; 407:242–8.

32. Funakoshi H, Nakamura T. Hepatocyte growth factor: from diagnosis to clinical applications. Clin Chim Acta 2003; 327:1–23.

33. LeCouter J, Moritz DR, Li B, et al. Angiogenesis-independent endothelial protection of liver: role of VEGFR-1. Science 2003; 299:890–3.

34. Pepper MS, Wasi S, Ferrara N, et al. In vitro angiogenic and proteolytic properties of bovine lymphatic endothelial cells. Exp Cell Res 1994; 210:298–305.

35. Hirakawa S, Kodama S, Kunstfeld R, et al. VEGF-A induces tumor and sentinel lymph node lymphangiogenesis and promotes lymphatic metastasis. J Exp Med 2005; 201:1089–99.

36. Storkebaum E, Lambrechts D, Dewerchin M, et al. Treatment of motoneuron degeneration by intracerebroventricular delivery of VEGF in a rat model of ALS. Natute Neurosci 2005; 8:85–92.

37. Azzouz M, Ralph GS, Storkebaum E, et al. VEGF delivery with retrogradely transported lentivector prolongs survival in a mouse ALS model. Nature 2004; 429:413–7.

38. Le Bras B, Barallobre MJ, Homman-Ludiye J, et al. VEGF-C is a trophic factor for neural progenitors in the vertebrate embryonic brain. Nature Neurosci 2006; 9:340–8.

39. Carmeliet P, Ferreira V, Breier G, et al. Abnormal blood vessel development and lethality in embryos lacking a single VEGF allele. Nature 1996; 380:435–9.

40. Ferrara N, Carver-Moore K, Chen H, et al. Heterozygous embryonic lethality induced by targeted inactivation of the VEGF gene. Nature 1996; 380:439–42.

41. Takahashi H, Shibuya M. The vascular endothelial growth factor (VEGF)/VEGF receptor system and its role under physiological and pathological conditions. Clin Sci (Lond) 2005; 109:227–41.

42. Heldin CH, Westermark B. Mechanism of action and in vivo role of platelet-derived growth factor. Physiol Rev 1999; 79:1283–316.

43. Soker S, Fidder H, Neufeld G, et al. Characterization of novel vascular endothelial growth factor (VEGF) receptors on tumor cells that bind VEGF165 via its exon 7-encoded domain. J Biol Chem 1996; 271:5761–7.

44. Soker S, Takashima S, Miao HQ, et al. Neuropilin-1 is expressed by endothelial and tumor cells as an isoform-specific receptor for vascular endothelial growth factor. Cell 1998; 92:735–45.

45. Maes C, Carmeliet P, Moermans K, et al. Impaired angiogenesis and endochondral bone formation in mice lacking the vascular endothelial growth factor isoforms VEGF164 and VEGF188. Mech Dev 2002; 111:61–73.

46. Carmeliet P, Ng YS, Nuyens D, et al. Impaired myocardial angiogenesis and ischemic cardiomyopathy in mice lacking the vascular endothelial growth factor isoforms VEGF164 and VEGF188. Nature Med 1999; 5:495–502.

47. Ruhrberg C, Gerhardt H, Golding M, et al. Spatially restricted patterning cues provided by heparin-binding VEGF-A control blood vessel branching morphogenesis. Genes Dev 2002; 16:2684–98.

48. Pertovaara L, Kaipainen A, Mustonen T, et al. Vascular endothelial growth factor is induced in response to transforming growth factor-β in fibroblastic and epithelial cells. J Biol Chem 1994; 269:6271–4.

49. Warren RS, Yuan H, Matli MR, et al. Induction of vascular endothelial growth factor by insulin-like growth factor 1 in colorectal carcinoma. J Biol Chem 1996; 271:29483–8.

50. Ferrara N, Chen H, Davis-Smyth T, et al. Vascular endothelial growth factor is essential for corpus luteum angiogenesis. Nature Med 1998; 4:336–40.

51. Safran M, Kaelin WG Jr. HIF hydroxylation and the mammalian oxygen-sensing pathway. J Clin Invest 2003; 111:779–83.

52. Dor Y, Porat R, Keshet E. Vascular endothelial growth factor and vascular adjustments to perturbations in oxygen homeostasis. Am J Physiol Cell Physiol 2001; 280:C1367–74

53. Liu Y, Cox SR, Morita T, et al. Hypoxia regulates vascular endothelial growth factor gene expression in endothelial cells. Identification of a 5′ enhancer. Circ Res 1995; 77:638–43.

54. Ema M, Taya S, Yokotani N, et al. A novel bHLH-PAS factor with close sequence similarity to hypoxia-inducible factor 1alpha regulates the VEGF expression and is potentially involved in lung and vascular development. Proc Natl Acad Sci USA 1997; 94:4273–8.

55. Maxwell PH, Wiesener MS, Chang GW, et al. The tumour suppressor protein VHL targets hypoxia-inducible factors for oxygen-dependent proteolysis. Nature 1999; 399:271–5.

56. Ivan M, Kondo K, Yang H, et al. HIFalpha targeted for VHL-mediated destruction by proline hydroxylation: implications for O2 sensing. Science 2001; 292:464–8.

57. Pugh CW, Ratcliffe PJ. Regulation of angiogenesis by hypoxia: role of the HIF system. Nature Med 2003; 9:677–84.

58. Ferrara N, Houck K, Jakeman L, et al. Molecular and biological properties of the vascular endothelial growth factor family of proteins. Endocr Rev 1992; 13(1):18–32.

59. Volm M, Koomagi R, Mattern J. Prognostic value of vascular endothelial growth factor and its receptor Flt-1 in squamous cell lung cancer. Int J Cancer 1997; 74:64–8.

60. Yoshiji H, Gomez DE, Shibuya M, et al. Expression of vascular endothelial growth factor, its receptor, and other angiogenic factors in human breast cancer. Cancer Res 1996; 56:2013–6.

61. Fava RA, Olsen NJ, Spencer-Green G, et al. Vascular permeability factor/endothelial growth factor (VPF/VEGF): accumulation and expression in human synovial fluids and rheumatoid synovial tissue. J Exp Med 1994; Jul 1 180(1):341–6.

62. Dong J, Grunstein J, Tejada M, et al. VEGF-null cells require PDGFR alpha signaling-mediated stromal fibroblast recruitment for tumorigenesis. EMBO J 2004; 23:2800–10.

63. Liang WC, Wu X, Peale FV, et al. Cross-species vascular endothelial growth factor (VEGF)-blocking antibodies completely inhibit the growth of human tumor xenografts and measure the contribution of stromal VEGF. J Biol Chem 2006; 281:951–61.

64. Asano M, Yukita A, Matsumoto T, et al. Inhibition of tumor growth and metastasis by an immunoneutralizing monoclonal antibody to human vascular endothelial growth factor/vascular permeability factor121. Cancer Res 1995; 55:5296–301.

65. Kong HL, Hecht D, Song W, et al. Regional suppression of tumor growth by in vivo transfer of a cDNA encoding a secreted form of the extracellular domain of the flt-1 vascular endothelial growth factor receptor. Hum Gene Ther 1998; 9:823–33.

66. Holash J, Davis S, Papadopoulos N, et al. VEGF-Trap: a VEGF blocker with potent antitumor effects. Proc Natl Acad Sci USA 2002; 99:11393–8.

67. Wedge SR, Ogilvie DJ, Dukes M, et al. ZD4190: an orally active inhibitor of vascular endothelial growth factor signaling with broad-spectrum antitumor efficacy. Cancer Res 2000; 60:970–5.

68. Wood JM, Bold G, Buchdunger E, et al. PTK787/ZK 222584, a novel and potent inhibitor of vascular endothelial growth factor receptor tyrosine kinases, impairs vascular endothelial growth factor-induced responses and tumor growth after oral administration. Cancer Res 2000; 60:2178–89.

68a. Sandler A, Gray R, Perry MC, et al. Paclitaxel-carboplatin alone or with bevacizumab for non-small-cell lung cancer. N Engl J Med 2006; 355:2542–50.

68b. Cohen MH, Gootenberg J, Keegan P, et al. FDA drug approval summary: bevacizumab (Avastin) plus Carboplatin and Paclitaxel as first-line treatment of advanced/metastatic recurrent nonsquamous non-small cell lung cancer. Oncologist 2007;12:713–8.

69. Luo JC, Yamaguchi S, Shinkai A, et al. Significant expression of vascular endothelial growth factor/vascular permeability factor in mouse ascites yumors. Cancer Res 1998; 58:2652–60.

70. Luo J-C, Toyoda M, Shibuya M. Differential inhibition of fluid accumulation and tumor growth in two mouse ascites tumors by an anti-vascular endothelial growth factor/permeability factor neutralizing antibody. Cancer Res 1998; 58:2594–2600.

71. Luttun A, Tjwa M, Moons L, et al. Revascularization of ischemic tissues by PlGF treatment, and inhibition of tumor angiogenesis, arthritis and atherosclerosis by anti-Flt1. Nature Med 2002; 8:831–40.

72. De Bandt M, Ben Mahdi MH, Ollivier V, et al. Blockade of vascular endothelial growth factor receptor I (VEGF-RI), but not VEGF-RII, suppresses joint destruction in the K/BxN model of rheumatoid arthritis. J Immunol 2003; 171:4853–9.

73. Murakami M, Iwai S, Hiratsuka S, et al. Signaling of vascular endothelial growth factor receptor-1 tyrosine kinase promotes rheumatoid arthritis through activation of monocyte/macrophages. Blood 2006; 108:1849–56.

74. Maglione D, Guerriero V, Viglietto G, et al. Isolation of a human placenta cDNA coding for a protein related to the vascular permeability factor. Proc Natl Acad Sci USA 1991; 88:9267–71.

75. Carmeliet P, Moons L, Luttun A, et al. Synergism between vascular endothelial growth factor and placental growth factor contributes to angiogenesis and plasma extravasation in pathological conditions. Nature Med 2001; 7:575–83.

76. Aase K, von Euler G, Li X, et al. Vascular endothelial growth factor-B-deficient mice display an atrial conduction defect. Circulation 2001; 104:358–64.

77. Karkkainen MJ, Ferrell RE, Lawrence EC, et al. Missense mutations interfere with VEGFR-3 signalling in primary lymphoedema. Nature Genet 2000; 25:153–9.

78. Kiba A, Sagara H, Hara T, et al. VEGFR-2-specific ligand VEGF-E induces non-edematous hyper-vascularization in mice. Biochem Biophys Res Commun 2003; 301:371–7.

79. Zheng Y, Murakami M, Takahashi H, et al. Chimeric VEGF-E_{NZ7}/PlGF promotes angiogenesis via VEGFR-2 without significant enhancement of vascular permeability and inflammation. Arterioscler Thromb Vasc Biol 2006; 26:2019–26.

80. Larcher F, Murillas R, Bolontrade M, et al. VEGF/VPF overexpression in skin of transgenic mice induces angiogenesis, vascular hyperpermeability and accelerated tumor development. Oncogene 1998; 17:303–11.

81. Detmar M, Brown LF, Schon MP, et al. Increased microvascular density and enhanced leukocyte rolling and adhesion in the skin of VEGF transgenic mice. J Invest Dermatol 1998; 111:1–6.

82. Shibuya M, Yamaguchi S, Yamane A, et al. Nucleotide sequence and expression of a novel human receptor-type tyrosine kinase gene (flt) closely related to the fms family. Oncogene 1990; 5:519–24.

83. De Vries C, Escobedo JA, Ueno H, et al. The fms-like tyrosine kinase, a receptor for vascular endothelial growth factor. Science 1992; 255:989–91.

84. Terman BI, Carrion ME, Kovacs E, et al. Identification of a new endothelial cell growth factor receptor tyrosine kinase. Oncogene 199; 6:1677–83.

85. Matthews W, Jordan CT, Gavin M, et al. A receptor tyrosine kinase cDNA isolated from a population of enriched primitive hematopoietic cells and exhibiting close genetic linkage to c-kit. Proc Natl Acad Sci USA 1991; 88:9026–30.

86. Millauer B, Wizigmann-Voos S, Schnurch H, et al. High affinity VEGF binding and developmental expression suggest flk-1 as a major regulator of vasculogenesis and angiogenesis. Cell 1993; 72:835–46.

87. Galland F, Karamysheva A, Pebusque MJ, et al. The FLT4 gene encodes a transmembrane tyrosine kinase related to the vascular endothelial growth factor receptor. Oncogene 1993; 8:1233–40.

88. Covassin LD, Villefranc JA, Kacergis MC, et al. Distinct genetic interactions between multiple Vegf receptors are required for development of different blood vessel types in zebrafish. Proc Natl Acad Sci USA 2006; 103:6554–9.

89. Waltenberger J, Claesson-Welsh L, Siegbahn A, et al. Different signal transduction properties of KDR and Flt1, two receptors for Vascular Endothelial Growth Factor. J Biol Chem 1994; 269:26988–95.

90. Seetharam L, Gotoh N, Maru Y, et al. A unique signal transduction from FLT tyrosine kinase, a receptor for vascular endothelial growth factor VEGF. Oncogene 1995; 10:135–47.

91. Sawano A, Takahashi T, Yamaguchi S, et al. Flt-1 but not KDR/Flk-1 tyrosine kinase is a receptor for Placenta Growth Factor (PlGF), which is related to vascular endothelial growth factor (VEGF). Cell Growth Differ 1996; 7:213–21.

92. Kendall RL, Thomas KA. Inhibition of vascular endothelial cell growth factor activity by an endogenously encoded soluble receptor. Proc Natl Acad Sci USA 1993; 90:10705–9.

93. Yamaguchi S, Iwata K, Shibuya M. Soluble Flt-1 (soluble VEGFR-1), a potent natural anti-angiogenic molecule in mammals,

is phylogenetically conserved in avians. Biochem Biophys Res Commun 2002; 291:554–9.

94. Fong GH, Rossant J, Gertsentein M, et al. Role of the Flt-1 receptor tyrosine kinase in regulating the assembly of vascular endothelium. Nature 1995; 376:66–70.

95. Hiratsuka S, Minowa O, Kuno J, et al. Flt-1 lacking the tyrosine kinase domain is sufficient for normal development and angiogenesis in mice. Proc Natl Acad Sci USA 1998; 95:9349–54.

96. Clauss M, Weich H, Breier G, et al. The vascular endothelial growth factor receptor Flt-1 madiates biological activities. J Biol Chem 1996; 271:17629–34.

97. Barleon B, Sozzani S, Zhou D, et al. Migration of human monocytes in response to vascular endothelilal growth factor (VEGF) is mediated via the VEGF receptor flt-1. Blood 1996; 87:3336–43.

98. Sawano A, Iwai S, Sakurai Y, et al. Vascular endothelial growth factor receptor-1 (Flt-1) is a novel cell surface marker for the lineage of monocyte-macrophages in humans. Blood 2001; 97:785–91.

99. Hiratsuka S, Maru Y, Okada A, et al. Involvement of Flt-1 tyrosine kinase (vascular endothelial growth factor receptor-1) in pathological angiogenesis. Cancer Res 2001; 61:1207–13.

100. Lyden D, Hattori K, Dias S, et al. Impaired recruitment of bone-marrow-derived endothelial and hematopoietic precursor cells blocks tumor angiogenesis and growth. Nature Med 2001; 7:1194–201.

101. Hiratsuka S, Nakamura K, Iwai S, et al. MMP9 induction by vascular endothelial growth factor receptor-1 is involved in lung specific metastasis. Cancer Cell 2002; 2:289–300.

102. Kaplan RN, Riba RD, Zacharoulis S, et al. VEGFR1-positive haematopoietic bone marrow progenitors initiate the pre-metastatic niche. Nature 2005; 438:820–7.

103. Zhao Q, Egashira K, Hiasa KI, et al. Essential role of vascular endothelial growth factor and Flt-1 signals in neointimal formation after periadventitial injury. Arterioscler Thromb Vasc Biol 2004; 24:2284–9.

104. Ohtani K, Egashira K, Hiasa KI, et al. Blockade of vascular endothelial growth factor suppresses experimental restenosis after intraluminal injury by inhibiting recruitment of monocyte lineage cells. Circulation 2004; 110:2444–52.

105. Koga K, Osuga Y, Yoshino O, et al. Elevated serum soluble vascular endothelial growth factor receptor 1 (sVEGFR-1) levels in women with preeclampsia. J Clin Endocrinol Metab 2003; 88:2348–51.

106. Maynard SE, Min JY, Merchan J, et al. Excess placental soluble fms-like tyrosine kinase 1 (sFlt1) may contribute to endothelial dysfunction, hypertension, and proteinuria in preeclampsia. J Clin Invest 2003; 111:649–58.

107. Levine RJ, Maynard SE, Qian C, et al. Circulating angiogenic factors and the risk of preeclampsia. N Engl J Med 2004; 350:672–83.

108. Ambati BK, Nozaki M, Singh N, et al. Corneal avascularity is due to soluble VEGF receptor-1. Nature 2006; 443:993–7.

109. Shalaby F, Rossant J, Yamaguchi TP, et al. Failure of blood-island formation and vasculogenesis in Flk-1-deficient mice. Nature 1995; 376:62–6.

110. Takahashi T, Yamaguchi S, Chida K, et al. A Single autophosphorylation site on KDR/Flk-1 is essential for VEGF-A-dependent activation of PLC-γ and DNA synthesis in vascular endothelial cells. EMBO J 2001; 20:2768–78.

111. Zeng H, Sanyal S, Mukhopadhyay D. Tyrosine residues 951 and 1059 of vascular endothelial growth factor receptor-2 (KDR) are essential for vascular permeability factor/vascular endothelial growth factor-induced endothelium migration and proliferation, respectively. J Biol Chem 2001; 276:32714–9.

112. Takahashi T, Ueno H, Shibuya M. VEGF activates Protein kinase C-dependent, but Ras-independent Raf-MEK-MAP kinase pathway for DNA synthesis in primary endothelial cells. Oncogene 1999; 18:2221–30.

113. Matsumoto T, Bohman S, Dixelius J, et al. VEGF receptor-2 Y951 signaling and a role for the adapter molecule TSAd in tumor angiogenesis. EMBO J 2005; 24:2342–53.

114. Sakurai Y, Ohgimoto K, Kataoka Y, et al. Essential role of Flk-1 (VEGF receptor 2) tyrosine residue 1173 in vasculogenesis in mice. Proc Natl Acad Sci USA 2005; 102:1076–81.

115. Carmeliet P, Lampugnani MG, Moons L, et al. Targeted deficiency or cytosolic truncation of the VE-cadherin gene in mice impairs VEGF-mediated endothelial survival and angiogenesis. Cell 1999; 98:147–157.

116. Shay-Salit A, Shushy M, Wolfovitz E, et al. VEGF receptor 2 and the adherens junction as a mechanical transducer in vascular endothelial cells. Proc Natl Acad Sci USA 2002; 99: 9462–7.

117. Stupack, DG, Cheresh DA. Integrins and angiogenesis. Curr Top Dev Biol 2004; 64:207–38.

118. Fukumura D, Gohongi T, Kadambi A, et al. Predominant role of endothelial nitric oxide synthase in vascular endothelial growth factor-induced angiogenesis and vascular permeability. Proc Natl Acad Sci USA 2001; 98:2604–9.

119. van Spronsen DJ, De Mulder PH. Targeted approaches for treating advanced clear cell renal carcinoma. Onkologie 2006; 29:394–402.

120. Nakamura K, Taguchi E, Miura T, et al. KRN951, a highly potent inhibitor of vascular endothelial growth factor receptor tyrosine kinases, has antitumor activities and affects functional vascular properties. Cancer Res 2006; 66:9134–42.

121. Dumont DJ, Jussila L, Taipale J, et al. Cardiovascular failure in mouse embryos deficient in VEGF receptor-3. Science 1998; 282:946–9.

122. Morisada T, Oike Y, Yamada Y, et al. Angiopoietin-1 promotes LYVE-1-positive lymphatic vessel formation. Blood 2005; 105:4649–56.

123. Kawasaki T, Kitsukawa T, Bekku Y, et al. A requirement for neuropilin-1 in embryonic vessel formation. Development 1999; 126:4895–902.

124. Yuan L, Moyon D, Pardanaud L, et al. Abnormal lymphatic vessel development in neuropilin 2 mutant mice. Development 2002; 129:4797–806.

125. Hiratsuka S, Watanabe A, Aburatani H, eta al. Tumour-mediated upregulation of chemoattractants and recruitment of myeloid cells predetermines lung metastasis. Nature Cell Biol 2006; 8:1369–75.

Chapter 9
Platelet-Derived Growth Factor

Andrius Kazlauskas

Keywords: platelet-derived growth factor (PDGF), PDGF receptor, pericytes, vascular progenitor cells

Abstract: In the approximately 40 years since the initial report of the existence of PDGF we have learned a great deal regarding this family of five growth factors. We know at least something about how they initiate signaling events leading to cellular responses, and how these events contribute to biological processes essential for development, physiology and pathology. The discoveries to date indicate that we have not learned all there is to know. Additional reagents (to selectively block the action of PDGFs) and approaches (to detect PDGF-specific contributions) will profoundly accelerate future experimentation. This additional information is likely to provide new opportunities to use PDGF-based approaches to correct a wide variety of pathological afflictions including angiogenesis-dependent (solid tumors, age-related macular degeneration, diabetic complications) and proliferation-driven (atherosclerosis, solid tumors) diseases.

The PDGF Family Has 5 Members

Platelet-derived growth factor (PDGF) was purified to homogeneity in the late 1970s as a factor from platelets that promoted the proliferation of mesenchymal cells [1–4]. Over the next 10–15 years, additional studies revealed that there are two PDGF genes (A and B), which homo- or hetero-dimerize into 3 biologically active forms of PDGF: PDGF-A (AA homodimer), PDGF-B (BB homodimer) and PDGF-AB (AB heterodimer) [5]. In 2000, three groups reported the discovery of an additional PDGF family member (PDGF-C; CC homodimer) [6–8]. The following year, one more member, PDGF-D (DD homodimer),

was discovered [9–11]. The absence of any additional reports in the 7 years since the discovery of PDGF-D suggests that the PDGF family consists of 5 members [12, 13] (Fig. 9.1).

The X-ray crystal structure for PDGF-B has been solved [14] and is found to be very similar to the structure of vascular endothelial growth factor (VEGF) [15, 16]. In both cases, 8 cysteine residues form intra or inter-disulfide bonds that organize the dimer into a cysteine knot motif that is common to many growth factors, even those that share little homology at the level of the primary amino acid sequence [17]. While the structure of the other PDGF family members has not been solved, a model for PDGF-C has been proposed based on the PDGF-B and VEGF structures [18]. Surprisingly, the predicted structure of PDGF-C is closer to VEGF than to PDGF-B [18]. Thus, the conformation of all PDGF family members is likely to be a cysteine knot. A biochemical curiosity of PDGFs is their thermo stability; PDGF-AB retains its biological activity even after being heated to 100°C [19]. While this is likely to be due to the inherent stability of the cysteine knot motif, the physiological relevance of PDGFs extraordinary stability remains a mystery.

The two newest members of the PDGF family, PDGF-C and PDGF-D, differ from the others in that they are secreted in a latent form. Proteolytic cleavage of the amino-terminal CUB domain permits the growth factor domain to bind and activate PDGF receptors (PDGFRs) [12, 13] (Fig. 9.2). "CUB" is an acronym for Clr/Cls, urchin endothelial growth factor-like protein and bone morphogenic protein 1 [20]. These domains are found in a wide variety of proteins, and are believed to mediate protein–protein and protein–carbohydrate interactions. In the context of PDGF-C and D, the CUB domain appears to prevent activation of the PDGFR [6, 9, 11]. In the context of the VEGF family, CUB domains promote interaction of ligands and receptors. Neuropilins 1 and 2 are CUB-domain containing transmembrane proteins that facilitate that binding of VEGF165 with their signaling receptors [21–26]. While there are some reports that the liberated CUB domains are biologically active [27], the prevailing thought is that they function to localize the secreted PDGF-C and –D and repress their activity.

Schepens Eye Research Institute, Harvard Medical School, 20 Staniford Street, Boston, MA 02114, USA
E-mail: ak@eri.harvard.edu

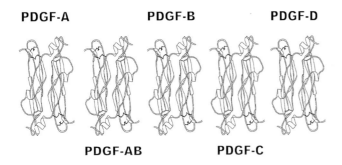

FIG. 9.1. The 5 members of the PDGF family. The PDGF family consists of 5 family members, PDGF-A, -B, -AB, -C, and -D. With the exception of PDGF-AB, all are homodimers. It is not known if PDGF-AC, PDGF-BC, PDGF-AD, PDGF-BD or PDGF-CD exists.

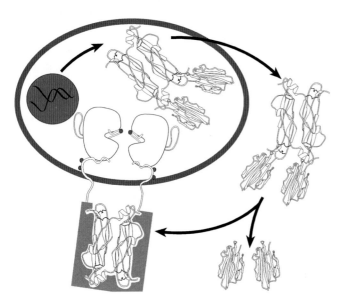

FIG. 9.2. PDGF-C and -D are secreted in a latent state. Proteolytic cleavage of the CUB domain permits the growth factor domain to bind and activate PDGFRs. The diagram shows an autocrine relationship between the cell that produces and responds to PDGF, as this appears to be the case for mesenchymal cells [28, 29]. PDGF-C and -D are also likely to activate cells in a paracrine manner.

TABLE 9.1. Proteases capable of processing PDGF-C or -D.

Protease	PDGF-C	PDGF-D	References
Plasmin	+ (RKSR234)	+ (RKSK257)	[6, 11, 28]
tPA	+ (RKSR234)	–	[29, 30]
uPA	–	+ (RGRS250)	[31]
Unidentified	+ (?)	ND	[32]

The stretch of amino acids between the CUB and growth factor domain (called the 'hinge') contains proteolytic cleavage sites, which are predicted to be exposed to proteases [12, 13]. Fetal bovine serum (present in the culture medium of PDGF-overexpressing cells) was noted to have the ability to liberate the growth factor domain from the CUB domain [9, 28]. Plasmin was subsequently identified as a protease capable of process-

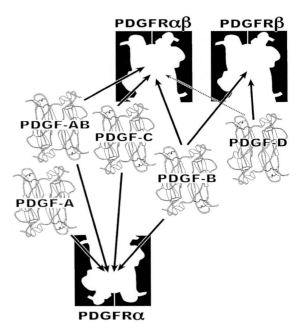

FIG. 9.3. The preferences of PDGF family members for PDGFRs. There are two PDGFR subunits that are homo- or heterodimerized by PDGF to form a functional PDGFR. PDGF-B is the universal ligand, i.e. it activates all 3 PDGFRs. PDGF-AB and PDGF-C assemble and activate PDGFRα and PDGFRαβ. PDGF-DD and PDGF-AA are the most selective family members as the activate only PDGFRα and PDGFRβ, respectively. PDGFRα is activated by the greatest number of PDGF family members (4/5), next is PDGFRαβ, which is activated by 3/5. PDGFRβ is the most selective receptor as only 2/5 PDGF family members are capable of activating it.

ing both PDGF-C and-D [6, 11] (Table 9.1). Subsequent studies demonstrated that tissue plasminogen activator (tPA) and uro-kinase plasminogen activator (uPA) are specific for PDGF-C and -D, respectively [29–31]. The liberated CUB domain was capable of suppressing tPA activity, which reveals an additional physiological function for the CUB domain [30]. More recent studies suggest that there are other proteases capable of processing PDGF-C [32]. Importantly, the proteases that liberate active PDGF-C and –D are regulated by pathological conditions, suggesting a mechanism by which these PDGF family members contribute to disease progression [143].

There Are 3 Types of PDGFRs

Two genes encode the PDGF receptor (PDGFR) subunits, which hetero or homodimerize into 3 different PDGFRs: PDGFRα (αα homodimers), PDGFRβ, (ββ homodimers) and PDGFRαβ, (αβ heterodimers) [33, 34] (Fig. 9.3). Each half of the dimeric PDGF ligand recruits one receptor subunit to assemble the PDGFR dimer. The intrinsic affinity of the PDGF family members for the two receptor subunits determines the composition of the assembled PDGFR [12, 13]. PDGF-B has a high affinity for both subunits and hence it is the universal ligand; it assembles and activates all 3 types of PDGFRs. Most, but not all [9], groups report that PDGF-DD bind only the β subunit with high affinity, hence it is a PDGFRβ-selective ligand. Similarly, PDGF-A is a PDGFRα-specific ligand. PDGF-C and PDGF-AB have the same specificity; they assemble and activate PDGFRα and PDGFRαβ forms of the receptor.

Activation of PDGFRs

While the activation mechanism for the PDGFRs has not been reported, it is thought to proceed by the same general mechanism by which the kinase activity of other receptor tyrosine kinases is regulated [35] (Fig. 9.4). The kinase activity of the receptor subunits that are not assembled into dimers is suppressed by at least two mechanisms: (1) the conformation of the kinase is unfavorable for catalysis, and (2) the catalytic cleft is sterically hindered. PDGF-induced dimerization correctly orients two receptor subunits such that they can cross phosphorylate on a key tyrosine residue in the activation segment (Y857 in the PDGFRβ; the homologous residue is Y849 for PDGFRα) [36, 37]. Phosphorylating this residue promotes reorganization of the kinase into a catalytically favorable conformation. In addition, it relieves steric hindrance by the activation segment and thereby clears the catalytic cleft for entry of substrates. At least for some receptors (including PDG-FRs), dimerization also reverses a juxtamembrane-based catalytically unfavorable conformation [35, 38]. Thus, activation of the PDGFRs is thought to be initiated by ligand-induced dimerization of subunits, which leads to a phosphorylation-enhanced conformational change in the kinase domain that enables efficient catalysis, i.e. phosphorylation of proteins on tyrosine residues.

While PDGF promotes activation of the PDGFR's intrinsic kinase activity, there are a variety of mechanisms suppressing this event. Most notable are the phosphotyrosine phosphatase (PTPs) which dephosphorylate the tyrosine phosphorylated receptor. PDGF overcomes the suppressive influence of PTPs by generating hydrogen peroxide, which inactivates the PTPs [39]. More recent studies provide additional insight into this hydrogen peroxide-based mechanism. Upon activation, the PDGFRs recruit peroxiredoxin, which reduces the level of hydrogen peroxide and thereby attenuates the extent of PTP inhibition and suppress the activation of PDGFRs [40]. The existence of several mechanisms in regulating the level of hydrogen peroxide (and hence PTP activity) provides a potential explanation for why not all investigators report a requirement to reduce the level of hydrogen peroxide in order to fully activate the PDGFR [41].

Signal Transduction Pathways Triggered by PDGF

Once the PDGFR is activated (i.e. its intrinsic kinase activity is elevated), then it phosphorylates proteins on tyrosine residues. One of the most important substrates is the receptor itself, which becomes a docking site for numerous signaling enzymes. The interaction of signaling enzymes with the activated receptor is dependent on tyrosine phosphorylation of the receptor. This interaction is specific, and the specificity is determined by both the receptor and the signaling enzyme [42, 43]. The amino acid context surrounding the tyrosine phosphorylation site is the receptor's contribution to specificity. The phosphotyrosine binding (PTB) or Src homology 2 (SH2) domain of the signaling enzymes has an intrinsic preference for binding partners, and hence makes a contribution to the specificity of the interaction. There is a long list of proteins that associate with the PDGFRs [44], and these same signaling enzymes can also associate with many other activated receptor tyrosine kinases. The change in subcellular location that results from association with an activated receptor tyrosine kinase has important consequences with respect to signaling. The new location of the signaling enzymes typically alters their access to upstream regulators, substrates, and downstream effectors.

In addition to experiencing a change in their subcellular location, stable association with the PDGFR can result in tyrosine phosphorylation of signaling enzymes. While some of these proteins are direct substrates of the receptor, they may also be phosphorylated by kinases that are activated by the PDGFR. For instance, Src family kinases (SFKs) are activated in PDGF-stimulated cells [45–49], and they phosphorylate many proteins. Using a variety of approaches some (but not all [50]) investigators found that PDGF-dependent

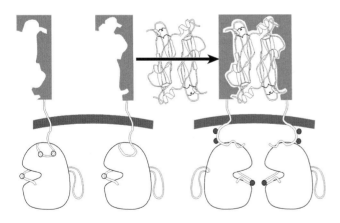

FIG. 9.4. Activation of the PDGFR. In the absence of PDGF, the PDGFRs are disassembled into monomeric subunits, which have low kinase activity because the conformation of the kinase domain is unfavorable for catalysis. PDGF induces dimerization, which promotes phosphorylation of the receptor (represented by filling of the circles in the juxtamembrane and activation segment) that is predicted to stabilize a catalytically favorable conformation of the kinase domain. Phosphorylation is also believed to reorient the activation segment outside of the catalytic cleft in order to clear the mouth of the kinase and thereby facilitate entry of substrates.

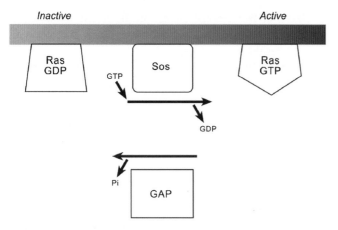

FIG. 9.5. Activation of Ras. The nucleotide exchange factor Sos catalyzes the exchange of GDP with GTP and thereby activates Ras. Like many small G proteins, Ras has intrinsic GTPase activity, which can inactivate Ras, i.e. hydrolyze the GTP to GDP. This reaction can be greatly accelerated by RasGAP, a GTPase activating protein that is specific for Ras.

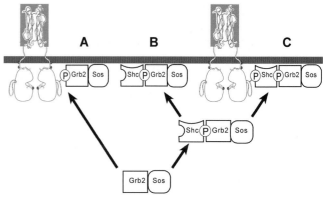

FIG. 9.6. Multiple mechanism to recruit Sos. The constitutive Grb2/Sos complex is recruited to the membrane in one of three ways. (A) The SH2 domain of Grb2 binds directly to the tyrosine phosphorylated PDGF. (B/C) PDGF promotes tyrosine phosphorylation of Shc and hence its association with Grb2 via the SH2 domain of Grb2. The resulting trimer is recruited to the membrane in one of two ways. The SH2 domain of Shc mediates binding to the tyrosine phosphorylated PDGFR (C). Alternatively, the PTB/PH domain of Shc interacts with lipids in the membrane and hence translocates the Shc/Grb2/Sos complex to the membrane. These membrane lipids include products of PI3K, which is activated in response to PDGF. The three recruitment scenarios are not mutually exclusive, and may be occurring simultaneously in PDGF-stimulated cells. As noted in Fig. 9.5, the functional consequence of recruiting Sos to the membrane is activation of Ras.

phosphorylation of certain substrates (Shc, c-Cbl, protein kinase C delta) was dependent on SFKs [49, 51, 52]. Hence tyrosine phosphorylation of signaling enzymes may proceed via more than one PDGF-activated kinase.

One of the signaling proteins activated in response to PDGF, as well as many other growth factors is Ras [53]. Activated Ras functions as a cofactor for a variety of signaling enzymes [54]. The nucleotide exchange factor, Sos promotes the exchange of GDP for GTP on Ras, which converts Ras to its active state (Fig. 9.5). Inactivation of Ras proceeds by hydrolysis of the GTP to GDP, and is promoted by GTP-ase activating proteins called GAPs.

The mechanism by which receptor tyrosine kinases activate Ras was intensely investigated, and a combination of genetic and biochemical approaches revealed the answer [53]. Sos is a constitutively active cytoplasmic enzyme, whereas Ras is anchored to the membrane. Consequently, activation of Ras requires translocation of Sos from the cytoplasm to the membrane, which is mediated by adapter proteins. There are several ways by which Ras can be activated in growth factor-stimulated cells, and they all appear to involve a change in the subcellular localization of the constitutive Grb2/Sos complex. Grb2 is an SH3-SH2-SH3 adapter protein that mediates binding of Sos to activated receptors. Tyrosine phosphorylation of the receptor enables the SH2 domain of Grb2 to stably associate with receptor tyrosine kinase receptors. This relocalizes Sos to the membrane, the cellular compartment in which its substrate, Ras resides (Fig. 9.6). Alternative scenarios include association of the Grb2/Sos complex with other adapter proteins such as Shc. This trimeric complex can relocate to the membrane via Shc's SH2 or PTB/PH domains. The functional consequence of these changes is the same as when Grb2/Sos associates with a tyrosine phosphorylated growth factor receptor: Sos gains access to Ras and activates it.

The mechanism by which phosphoinositide 3 kinase (PI3K) is activated in response to acute exposure to PDGF has been extensively investigated and is well understood. The PDGFR is among the first model systems in which the mechanism of PI3K activation is studied and hence is a source of early insights. Stimulation of serum-arrested cells with growth factors such as PDGF results in a rapid increase in the amount of PI3K lipid products [55–57]. Ligand-induced tyrosine phosphorylation of two tyrosine residues in the kinase insert of the PDGF receptor (tyr 740 and 751 for PDGFRβ; tyr 731 and 742 for PDGFRα) enables stable association of Class IA PI3K with the PDGFR [42, 58, 59]. Class IA PI3K consists of a p110 catalytic subunit and an SH2 domain-containing regulatory subunit [60, 61]. It is the SH2 domains within the regulatory subunit that mediates the interaction with the tyrosine phosphorylated PDGFR. This interaction accomplishes two steps in the activation process: (1) it translocates the catalytic domain of PI3K to the plasma membrane, which is where its lipid substrate(s) resides, and (2) it relieves inhibition of the catalytic subunit by the regulatory subunit [62]. The third step needed for full activation of PI3K is accumulation of active Ras, which binds to the catalytic domain of PI3K [63, 64]. Thus, activation of PI3K in acutely stimulated cells involves three steps: engaging the SH2 domains in the regulatory subunit, translocating to the plasma membrane, and binding of active Ras to the catalytic subunit (Fig. 9.7).

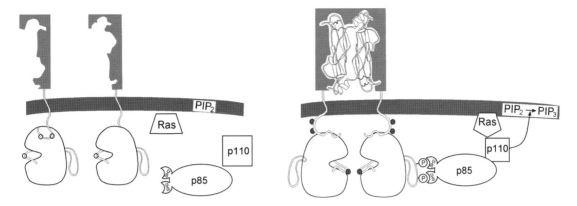

FIG. 9.7. PDGF-dependent activation of PI3K. PDGF induces dimerization of the PDGFR and subsequent tyrosine phosphorylation at numerous sites within the catalytic domain; only the two phosphorylation sites within the kinase insert that are required for association of PI3K are shown. The SH2 domains of the regulatory subunit of PI3K (p85) stably associates with the tyrosine phosphorylated PDGFR. This interaction relieves inhibition of the regulatory subunit on the catalytic subunit, and localizes PI3K close to its substrates in the plasma membrane. Binding activated Ras (shown as a *pentagon*) to the catalytic subunit (p110) fully activates PI3K.

Cellular Responses Initiated by PDGF

PDGF-dependent signaling is readily detectable within seconds to minutes after exposure to PDGF. Investigators have focused on these early time points because at later times it is necessary to distinguish the primary PDGF-dependent events from the accumulating secondary events. For instance, PDGF induces expression of immediate early genes, whose protein products include those that influence the primary signaling events. MKP-1 is the product of an immediate early gene that is a phosphatase, which dephosphorylates MAP kinase [65]. Even the signaling molecules such as PI3K and protein kinase C family members that are activated early are also activated at later time points [66], and this further complicates the elucidation of signaling pathways in cells that have been stimulated for hours instead of seconds or minutes.

Cultured cells such as fibroblasts proliferate by moving through 4 phases of the cell cycle: G1, S, G2 and M (Fig. 9.8). Each of these phases is regulated by the coordinated action of kinases and proteases [67, 68]. Progression through these stages of the cell cycle requires either serum or purified growth factors. If the mitogen is removed, then cells accumulate in the G0 state [69, 70].

While cell cycle progression requires mitogens, they must be present for only a portion of the cycle. The growth factor-dependent segments include exit from G0 and the majority of G1. While the entire cell cycle is approximately 24 h in fibroblasts, exposure to growth factors is necessary for only the first 8–10 h. This time point in the later portion of G1 is termed "R" [69, 70]. A similar concept of a G1 restriction point emerged from elegant studies using time-lapse cinematography of cells that were deprived of serum for a 1-h interval at selected periods throughout the cell cycle [71]. The "R" point of the cell cycle was molecularly defined following the discovery of Rb and its role in regulating exit out of G1. We now understand that "R" represents inactivation of the retinoblastoma protein Rb [72].

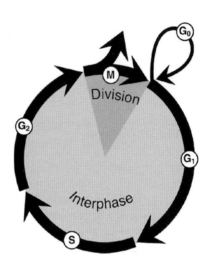

FIG. 9.8. Phases of the cell cycle. In the presence of serum, mammalian cells progress through the 4 stage of the cell cycle indicated by the heavy arrows. Cells withdraw from the cell cycle into G0 when starved of serum. The entire cell cycle takes approximately 24 h for fibroblasts. (Reprinted with permission from *Growth Factors* 2005; 23(3): 203–210).

Competence and Progression

As purified growth factors became available, investigators noted that cell cycle progression required sequential exposure to more than one type of growth factor. Using Balb/c 3T3 cells Pledger, Stiles, Antoniades and Sher [73–75] reported that PDGF made cells competent (i.e. endowed them with the capability of exiting G0), although it did not appreciably advance them through the G1 phase of the cell cycle. A different growth factor, which they termed a progression factor (Fig. 9.9), was needed to propel the competent cell past the R point. Once these events had occurred, subsequent cell cycle progression was independent of extracellular growth factors.

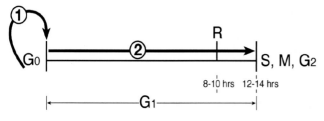

① Competence Factors enable the cell to exit G0

② Progression Factors drive the cell out of G0 and
   through most of G1, including R

FIG. 9.9. The competence and progression phenomenon. Sequential
exposure to different classes of growth factors is required to advance
Balb/c 3T3 cells through the cell cycle. The cells must first be stimulated with a competence factor (such as PDGF or fibroblast growth
factor [FGF]), which prepares them to exit G0 and/or enter G1. Subsequent addition of a progression factor (plasma-derived factors such
as insulin-like growth factor-1 [IGF-1]) [75] advances cells out of G0
and through G1. After 8–10 h of continuous stimulation, cells pass
the R point and commit to one full round of the cell cycle [69, 70],
i.e. they undergo one round of replication even if all growth factors
are removed. (Reprinted with permission from *Growth Factors* 2005;
23(3):203–210).

Priming and Completion

Recent studies have revealed that most cells do not follow
the principles of competence and progression [76]. A single
growth factor (such as PDGF) is typically capable of inducing cell cycle progression in a large variety of cultured cells.
Since cultured cells can produce growth factors [77], it is possible,
even likely, that cell cycle progression results from the combined action of exogenous and endogenous mitogens. Yet
this possibility does not counter the main point that cell cycle
progression does not require the addition of multiple growth
factors as it does for Balb/c 3T3 cells used in the competence/
progression studies.

A second important point that has come to light recently is
that continual stimulation with mitogens during most of G1 is
not necessary to drive quiescent cells into S phase [76]. Two
appropriately timed pulses of growth factor were sufficient to
mediate the G0 to S transition. This finding suggested that
there were two distinct growth factor-dependent phases of the
G0 to R interval of the cell cycle: "priming", which prepared
cells to engage the cell cycle program, and "completion" in
which the cell cycle program was active and advanced cells
past R. This updated version of how growth factors regulate cell cycle progression is shown in Fig. 9.10. Our current
understanding for how PDGF (and other growth factors) drive
cell cycle progression is that the G1 phase of the cell cycle
consists of two interlinked segments in which priming sets the
stage for the cell cycle program, which is executed during the
completion phase [66].

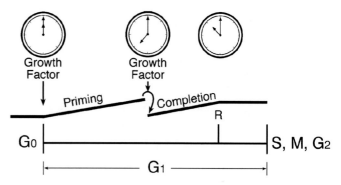

FIG. 9.10. Updated version of how growth factors regulate the cell
cycle. Exposing a quiescent cell to PDGF primes it, i.e. makes it
capable of engaging the cell cycle program (increasing cyclin D1,
decreasing Cdk inhibitors, phosphorylating Rb, etc.). Although
exposure to PDGF can be brief (as little as 30 min), priming takes
5–8 h, (depending on the cell type). Once the cells are primed, then
they initiated the cell cycle program in response to a second exposure
to PDGF. If primed cells do not encounter growth factor, they return
to G0 [76]. (Reprinted with permission from *Growth Factors* 2005;
23(3):203–210).

PDGFRα Versus PDGFRβ

The discovery of a second PDGFR gene [78] initiated an
ongoing quest for PDGF-mediated events that are unique to
each of the PDGFRs. With respect to acute signaling events,
activation of either PDGFRα or PDGFRβ triggers a similar
cascade of events, including recruitment and/or activation
of PI3K, phospholipase C γ (PLCγ), SFKs, the phosphotyrosine phosphatase SHP-2, Shc, Shb and the MAPK pathway
[44]. A notable exception is the GTPase activation protein of
Ras (RasGAP), which is recruited to PDGFRβ and tyrosine
phosphorylated; RasGAP is not tyrosine phosphorylated in
response to activation of PDGFRα, and it is not recruited to
the activated PDGFRα [59]. Activation of either of the PDG-
FRs induces comparable cellular responses (migration and
proliferation) [59], which further frustrated efforts of those
looking to explain the existence of multiple PDGFRs.

The most convincing evidence that activation of different
PDGFRs results in unique sets of directions came from the
characterization of mice that lacked either of the two *Pdgfr*
genes. While both genes were required for embryogenesis,
there were distinct phenotypes when each *Pdgfr* gene was
ablated. Knocking out *Pdgfrb* reduced the number of contractile vascular smooth muscle cells (vSMC) and capillary pericytes [79–81], which resulted in hemorrhaging and
microaneurisms in late gestation. Furthermore, there was
reduced recruitment of mesangial cells to the kidney [82].
Mice lacking *Pdgfra* displayed defects in skeletal development and patterning that led to spina bifida and clefting of
the craniofacial skeleton [83]. In addition, the amount of various mesenchymal cell populations was reduced, including
chondrocytes, placental spongiotrophoblasts, oligodendrocytes, non-neuronal neural crest and lung alveolar smooth

muscle cells and intestinal, kidney and dermal mesenchyme [83–89]. Thus, while vSMC and pericytes were the only two cell types dependent on the PDGFRβ, there were many (albeit primarily mesenchymal) types of cells that required the PDGFRα during development [90, 91].

In addition to the possibility that the PDGFRs are capable of sending unique instructions, other explanations for the distinct phenotypes of the mice lacking either of the PDGFR genes include the timing, location or amplitude of expression of the PDGFRs or their ligands. To address these issues, mice were generated in which the cytoplasmic domains of the PDGFRs were interchanged. This experimental setting preserved all the above-mentioned variables and changed only the signaling output of the receptor. When PDGFRα signaled using a PDGFRβ cytoplasmic domain, there was no detectable phenotype, which strongly suggested that PDGFRβ was capable of sending all PDGFRα signals [92]. In contrast, the cytoplasmic domain of PDGFRα was not a functional substitute for the PDGFRβ cytoplasmic domains; mice expressing the β/α chimeric PDGFR displayed multiple vSMC abnormalities [92]. Thus at least during development, PDGFRβ communicates with vSMCs in a way that cannot be reproduced by PDGFRα [91].

The fact that the cytoplasmic domain (and presumably the signaling output) of closely related receptor tyrosine kinases were not functionally interchangeable speaks to the intriguing question of how receptor tyrosine kinases evoke specific responses. Using a gene profiling approach, we found that activation of different receptor tyrosine kinases (PDGF and fibroblast growth factor) induced expression of the same set of immediate early genes (IEGs) [93], which suggests that the signaling specificity is not at the level of which IEGs are expressed. More recently IEGs have been implicated as a key determinant of signaling specificity downstream of PDGF [94]. Mice that lacked a single PDGF-inducible IEG showed a phenotype that partially phenocopied the corresponding PDGFR knockout [94]. The most likely resolution of these reports is that specificity can originate at the level of the IEGs. However, there may be additional mechanisms to regulate the nature of the response that results from exposure to PDGF.

PDGF-B-Mediated Recruitment of Pericytes

The observation that endothelial cells produce PDGF or PDGF-like factors [95–97] was the first hint that PDGF participates in angiogenesis. Since most endothelial cells do not express PDGFRs [98, 99] (see the "PDGF-dependent recruitment of vascular progenitor cells" section below), the role of PDGF would probably involve an additional cell type, most likely one that expressed PDGFRs. Indeed, D'Amore and colleagues reported that in an in vitro setting, endothelial-derived PDGF recruited 10T1/2 cells, a presumptive mural/pericyte precursor. The subsequent interaction of these two cell types activated a pool of latent transforming growth factor β (TGFβ), which inhibited

proliferation of both cell types and induced expression of smooth muscle cell markers in the 10T1/2 cells [100–102]. These studies revealed that PDGF contributes to vascular development by recruiting pericytes to immature vessel and promoting their stabilization/maturation.

The findings from these in vitro studies resonate well with a large body of in vivo studies. For instance, in the developing embryo, the relevant players are expressed in the appropriate cell types: PDGF-B is expressed in endothelial cells and PDGFRβ in the vSMCs and pericytes [80]. Similarly, mice that lacked the players implicated in the in vitro setting (PDGF-B, PDGFRβ) had vessels with reduced coverage of pericytes and leaky vessels [79–81, 103]. Subsequent studies demonstrated that the functional interaction between endothelial cells and pericytes required that the endothelial cells produce the PDGF-B and that the PDGF-B remain cell-associated [104, 105].

The fact that the simple in vitro models can mimic at least a portion of the complex in vivo setting predicts that the in vitro systems will be an invaluable approach to define the molecular mechanism by which pericytes stabilizes nascent endothelial tubes. It seems plausible that the pericytes stabilize vessels by changing the microenvironment (contact with the pericyte and extracellular matrix), which reprograms how the endothelial cells respond to this microenvironmental input. For instance, VEGF promotes angiogenesis of endothelial cells in immature vessels, and anti-VEGF/VEGFR strategies typically attenuate angiogenic responses. Thus, there is a strong positive correlation between activation of VEGFRs and angiogenesis. However, recent studies indicate that VEGF-dependent activation of VEGFR does not always lead to an angiogenic response. The aorta of an adult mouse expresses both VEGF and VEGFRs, and the VEGFRs are activated [106], yet the adult aorta of a healthy mouse is not a site of active angiogenesis. One explanation for these observations is that mature vessels, invested with pericytes and embedded in the appropriate extracellular matrix, have been reprogrammed to respond to VEGF differently from the immature vessel. Molecularly defined what "reprogramming" details will require the use of both in vitro and in vivo approaches.

While the focus of this section has been on the role of PDGF-B/PDGFRβ in the pericyte/endothelial relationship, there are a number of excellent review articles focused on additional factors that also contribute to this relationship [107–109]. They include sphingosine 1-phosphate (S1P), angiopoetin 1 (Ang1), TGFβ and proteases such as matrix metalloproteinases (MMPs). The current understanding of how the various players mediate the pericyte/endothelial relationship is as follows. Both PDGF-B and S1P promote recruitment of mural cells capable of differentiating into pericytes, PDGF-B also induces proliferation of these cells; Ang1 facilitates the physical interaction between endothelial cells and the mural cells, whereas TGFβ drives the production of ECM and differentiation of the recruited cells into pericytes. Finally, MMPs remodel the ECM and release matrix-associated growth factors.

The Contribution of PDGF to Tumor Angiogenesis

There is convincing evidence that PDGF is required for pericyte investment in tumor vessels. In mice, expression of PDGFRβ in pericyte was required for proper pericyte coverage of the tumor vessels [110]. Within a tumor, the endothelial cells produced PDGF, and in some instances the tumor cells also secreted PDGF. While either source of PDGF functioned to recruit pericytes, it was the endothelial-derived PDGF-B (which remained in close contact with the endothelial cells) that was required for proper association of the pericyte with the vessels [110]. In addition to these genetic studies, pharmacological approaches have also concluded that PDGFR activity was essential for pericyte coverage of vessels in tumors. The pericyte number of tumor vessels and vascularity of the tumors declined when mice were treated with inhibitors that blocked PDGFR [111, 112]. Importantly, the tumor burden was more effectively reduced in mice treated with a combination of drugs that included a PDGFR inhibitor [111, 113, 114–116]. Thus elucidating the endothelial/pericyte relationship is providing guidance for developing anti-tumor therapies.

The pericytes impact tumor vessel in a variety of ways. First, they suppress proliferation of the endothelial cells, which appears to be mediated by TGFβ and promotes vessel stability [117, 118]. In addition, pericytes released endothelial cells from their dependence on proangiogenic factors such as VEGF [119]. Furthermore, pericyte-dependent maturation of vessels within a tumor boosted the vascularity of tumor [120, 121]. Finally, pericytes reduced the diameter of vessels and thereby contributed to blood flow [81]. Thus, pericytes promote the stability and functionality of vessels in a tumor in much the same way as they do in normal tissue. However, it is important to note that the tumor vasculature is typically abnormal, and this includes endothelial/pericyte coverage and adherence [122, 123]. In light of the findings that proper investment of pericytes required that the PDGF be physically associated with the endothelial cells [105], high levels of PDGF, perhaps generated by the tumor cells, may be at least part of the reason of the apparent pericyte dysfunction in tumor vessels.

PDGF in Proliferative Diabetic Retinopathy

An additional pathology associated with the loss of pericytes is the proliferative phase of diabetic retinopathy. The ratio of pericytes to endothelial cells is higher in the retina than in any other organ [124], and loss of pericytes is the earliest morphological indicator of retinopathy [125, 126]. Additional features of this background or non-proliferative phase of diabetic retinopathy are microaneurisms, which indicate localized capillary hemorrhaging and/or leakage. Rodent models of diabetes never progress to the neovascular (proliferative) form of diabetic retinopathy (PDR), and only 20% of patients

with diabetes develop PDR [127]. Why 100% of rodents and 80% of humans afflicted with diabetes do not progress to PDR is a long-standing unanswered question.

A potential answer has recently emerged, and it involves PDGF. Mice bearing an endothelial cell-specific knockout in PDGF-B developed retinopathy (as well as other abnormalities), and the severity correlated with the magnitude of pericyte deficiency [104]. In mice with retinal vessels having 50% or more of the normal amount of pericytes, retinopathy was non-existent or encompassed only aneurisms [104]. All animals with less than 50% the normal level of pericytes displayed proliferative retinopathy, which included the growth of new vessels [104]. The concept that pericytes are dependent on PDGF and suppress retinal angiogenesis emerged from a pharmacological study in which inhibiting PDGFR reduced pericyte coverage and intensified ischemia-induced retinal neovascularization [128]. While these findings are with non-diabetic animals, a recent study reported that diabetes resulted in a decline in PDGF-B, whereas many growth factors increased in response to diabetes [129]. In summary, recent studies indicate that diabetes results in a fall in the level of PDGF-B, and this reduction may contribute to the loss of pericyte coverage of retinal vessels. Furthermore, the loss of pericytes may result in an angiogenic switch that sets the stage for the neovascular response of PDR.

PDGF-Dependent Recruitment of Vascular Progenitor Cells

Recent studies indicate that PDGF plays a role in recruitment of bone marrow-derived cells that contribute to the angiogenesis. Administration of PDGF-C promoted post-ischemic revascularization of the heart and limb [130]. The mechanism appeared to involve the recruitment and maturation of both endothelial and pericyte precursors. In contrast, PDGF-B was unable to mobilize the precursors [130], and this is consistent with observation that administration of PDGF-B alone was insufficient to induce revascularization in the ischemic hind limb setting [131].

In at least certain types of experimental tumors, PDGFRβ (and hence PDGF-B and/or PDGF-D) makes an essential contribution to recruitment of bone marrow progenitors that populate the tumor vasculature [132]. Bone marrow-derived cells that expressed PDGFRβ are recruited to the tumor vasculature and differentiate into pericytes. Taken together, these studies indicate that PDGF plays a role in the recruitment of bone marrow-derived cells that incorporate into newly forming vessels. PDGF-C acts on both endothelial and pericyte progenitors, whereas, surprisingly, PDGF-B is restricted to the pericyte progenitors.

These observations raise a number of intriguing questions. First, do endothelial cells or their precursors express PDGFRs? While pretty much every mesenchymal cell responds to PDGF, it is difficult to find PDGF-responsive endothelial cells. So entrenched is this concept that many investigators have used endothelial cells as PDGFR-negative cell type in which to introduce and characterize PDGFR mutants [58]. Yet there are excellent papers reporting both biochemical and functional evidence for PDGFRs on endothelial cells. These studies typically report that endothelial cells from small vessels express PDGFR [133–138], although this is not always the case [139–141]. Furthermore, some groups found that PDGFRs are present in the normal setting [135, 138], whereas others report that expression is detected only following a pathological insult [140, 142]. Taken together, these publications demonstrate that while endothelial cells typically do not express PDGFRs, endothelial cells do have the capacity to express PDGFRs and respond to PDGF.

A second question that emerges is why PDGF-B fails to mimic the action of PDGF-C. PDGF-B activates both PDGFRα and PDGFRαβ (the receptors activated by PDGF-C, see Fig. 9.3), and hence one would anticipate that the universal ligand (PDGF-B) would induce the same responses as any of the other family members (including PDGF-C). Yet PDGF-B also activates PDGFRβ, which may suppress or simply dominate the response. The inability to explain these observations underscores that fact that the molecular understanding of how the various PDGFRs signal remains incomplete. Likely explanations might include the nature and/or timing of PDGF-dependent signaling. Only the acute signaling output of the PDGFRs has been compared; later signaling events are required for responses such as proliferation [66], and may proceed differently for the various types of PDGFRs.

In summary, endothelial cells secrete PDGF, and it functions to promote the recruitment of pericytes, which express receptors for PDGF (Fig. 9.11). Pericytes drive stabilization and quiescence of the vasculature. PDGF is also involved with recruitment of bone marrow-derived endothelial and pericyte precursors (Fig. 9.11). The signaling pathways responsible for these angiogenic events remain largely unaddressed and exiting areas of future investigation.

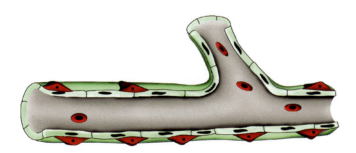

FIG. 9.11. Cells that participate in angiogenesis and either secrete or respond to PDGF. Shown in *red* are the cells that express PDGFRs and hence respond to PDGF; they include the circulating progenitor cells (both endothelial and pericyte) and the pericytes. Endothelial cells produce PDGF, and they are shown in *green*. The vessel is undergoing angiogenesis as indicated by the newly-formed branch that is devoid of pericytes.

References

1. Antoniades HN, Scher CD, Stiles CD. Purification of human platelet-derived growth factor. Proc Natl Acad Sci USA 1979;76(4):1809–13.
2. Heldin CH, Westermark B, Wasteson A. Platelet-derived growth factor: purification and partial characterization. Proc Natl Acad Sci USA 1979;76(8):3722–6.
3. Raines EW, Ross R. Platelet-derived growth factor I. High yield purification and evidence for multiple forms. J Biol Chem 1982;257:5154–60.
4. Deuel TF, Huang JS, Proffitt RT, Baenziger JU, Chang D, Kennedy BB. Human platelet-derived growth factor. Purification and resolution into two active protein fractions. J Biol Chem 1981;256(17):8896–9.
5. Heldin CH, Wasteson A, Westermark B. Platelet-derived growth factor. Molecular and Cellular Endocrinology 1985;39:169–87.
6. Li X, Ponten A, Aase K, et al. PDGF-C is a new protease-activated ligand for the PDGF alpha-receptor [see comments]. Nat Cell Biol 2000;2(5):302–9.
7. Hamada T, Ui-Tei K, Miyata Y. A novel gene derived from developing spinal cords, SCDGF, is a unique member of the PDGF/VEGF family. FEBS Lett 2000;475(2):97–102.
8. Tsai YJ, Lee RK, Lin SP, Chen YH. Identification of a novel platelet-derived growth factor-like gene, fallotein, in the human reproductive tract. Biochim Biophys Acta 2000;1492(1):196–202.
9. LaRochelle WJ, Jeffers M, McDonald WF, et al. PDGF-D, a new protease-activated growth factor. Nat Cell Biol 2001;3(5):517–21.
10. Hamada T, Ui-Tei K, Imaki J, Miyata Y. Molecular cloning of SCDGF-B, a novel growth factor homologous to SCDGF/PDGF-C/fallotein. Biochem Biophys Res Commun 2001;280(3):733–7.
11. Bergsten E, Uutela M, Li X, et al. PDGF-D is a specific, protease-activated ligand for the PDGF beta-receptor. Nat Cell Biol 2001;3(5):512–6.
12. Fredriksson L, Li H, Eriksson U. The PDGF family: four gene products form five dimeric isoforms. Cytokine Growth Factor Rev 2004;15(4):197–204.
13. Reigstad LJ, Varhaug JE, Lillehaug JR. Structural and functional specificities of PDGF-C and PDGF-D, the novel members of the platelet-derived growth factors family. Febs J 2005;272(22):5723–41.
14. Oefner C, D'Arcy A, Winkler FK, Eggimann B, Hosang M. Crystal structure of human platelet-derived growth factor BB. The EMBO Journal 1992;11:3921–6.
15. Muller YA, Christinger HW, Keyt BA, de Vos AM. The crystal structure of vascular endothelial growth factor (VEGF) refined to 1.93 A resolution: multiple copy flexibility and receptor binding. Structure 1997;5(10):1325–38.
16. Muller YA, Li B, Christinger HW, Wells JA, Cunningham BC, de Vos AM. Vascular endothelial growth factor: crystal structure and functional mapping of the kinase domain receptor binding site. Proc Natl Acad Sci USA 1997;94(14):7192–7.
17. Murray-Rust J, McDonald NQ, Blundell TL, et al. Topological similarities in TGF-beta 2, PDGF-BB and NGF define a superfamily of polypeptide growth factors. Structure 1993;1(2):153–9.
18. Reigstad LJ, Sande HM, Fluge O, et al. Platelet-derived growth factor (PDGF)-C, a PDGF family member with a vas-

cular endothelial growth factor-like structure. J Biol Chem 2003;278(19):17114–20.

19. Raines EW, Ross R. Purification of human platelet-derived growth factor. Methods Enzymol 1985;109:749–73.

20. Bork P, Beckmann G. The CUB domain. A widespread module in developmentally regulated proteins. J Mol Biol 1993;231(2):539–45.

21. Soker S, Fidder H, Neufeld G, Klagsbrun M. Characterization of novel vascular endothelial growth factor (VEGF) receptors on tumor cells that bind VEGF165 via its exon 7-encoded domain. J Biol Chem 1996;271(10):5761–7.

22. Soker S, Takashima S, Miao HQ, Neufeld G, Klagsbrun M. Neuropilin-1 is expressed by endothelial and tumor cells as an isoform-specific receptor for vascular endothelial growth factor. Cell 1998;92:735–45.

23. Makinen T, Olofsson B, Karpanen T, et al. Differential binding of vascular endothelial growth factor B splice and proteolytic isoforms to neuropilin-1. J Biol Chem 1999;274(30):21217–22.

24. Gluzman-Poltorak Z, Cohen T, Herzog Y, Neufeld G. Neuropilin-2 is a receptor for the vascular endothelial growth factor (VEGF) forms VEGF-145 and VEGF-165. J Biol Chem 2000;275(38):29922.

25. Gluzman-Poltorak Z, Cohen T, Herzog Y, Neufeld G. Neuropilin-2 is a receptor for the vascular endothelial growth factor (VEGF) forms VEGF-145 and VEGF-165 [corrected]. J Biol Chem 2000;275(24):18040–5.

26. Mamluk R, Gechtman Z, Kutcher ME, Gasiunas N, Gallagher J, Klagsbrun M. Neuropilin-1 binds vascular endothelial growth factor 165, placenta growth factor-2, and heparin via its b1b2 domain. J Biol Chem 2002;277(27):24818–25.

27. Dijkmans J, Xu J, Masure S, et al. Characterization of platelet-derived growth factor-C (PDGF-C): expression in normal and tumor cells, biological activity and chromosomal localization. Int J Biochem Cell Biol 2002;34(4):414–26.

28. Gilbertson DG, Duff ME, West JW, et al. Platelet-derived growth factor C (PDGF-C), a novel growth factor that binds to PDGF alpha and beta receptor. J Biol Chem 2001;276(29):27406–14.

29. Fredriksson L, Li H, Fieber C, Li X, Eriksson U. Tissue plasminogen activator is a potent activator of PDGF-CC. Embo J 2004;23(19):3793–802.

30. Fredriksson L, Ehnman M, Fieber C, Eriksson U. Structural requirements for activation of latent platelet-derived growth factor CC by tissue plasminogen activator. J Biol Chem 2005;280(29):26856–62.

31. Ustach CV, Kim HR. Platelet-derived growth factor D is activated by urokinase plasminogen activator in prostate carcinoma cells. Mol Cell Biol 2005;25(14):6279–88.

32. Lei H, Hovland P, Velez G, et al. A potential role for PDGF-C in experimental and clinical proliferative vitreoretinopathy. IOVS 2007;48(5):2335–42.

33. Heldin C-H, Westermark B. Platelet-derived growth factor: mechanism of action and possible in vivo function. Cell Regulation 1990;1:555–66.

34. Raines EW, Bowen-Pope DF, Ross R. Peptide growth factors and their receptors; 1990.

35. Hubbard SR. Juxtamembrane autoinhibition in receptor tyrosine kinases. Nat Rev Mol Cell Biol 2004;5(6):464–71.

36. Kazlauskas A, Durden DL, Cooper JA. Functions of the major tyrosine phosphorylation site of the PDGF receptor β subunit. Cell Regulation 1991;2:413–25.

37. Fantl WJ, Escobedo JA, Williams LT. Mutations of the platelet-derived growth factor receptor that cause a loss of ligand-induced conformational change, subtle changes in kinase activity, and impaired ability to stimulate DNA synthesis. Mol Cell Biol 1989;9:4473–8.

38. Cools J, Stover EH, Gilliland DG. Detection of the FIP1L1-PDGFRA fusion in idiopathic hypereosinophilic syndrome and chronic eosinophilic leukemia. Methods Mol Med 2006;125:177–87.

39. Sundaresan M, Yu ZX, Ferrans VJ, Irani K, Finkel T. Requirement for generation of H2O2 for platelet-derived growth factor signal transduction. Science 1995;270(5234):296–9.

40. Choi MH, Lee IK, Kim GW, et al. Regulation of PDGF signalling and vascular remodelling by peroxiredoxin II. Nature 2005;435(7040):347–53.

41. Bae YS, Sung JY, Kim OS, et al. Platelet-derived growth factor-induced H(2)O(2) production requires the activation of phosphatidylinositol 3-kinase. J Biol Chem 2000;275(14):10527–31.

42. Kazlauskas A. Receptor tyrosine kinases and their targets. Curr Opin Gen Dev 1994;4:5–14.

43. Heldin CH, Ostman A, Ronnstrand L. Signal transduction via platelet-derived growth factor receptors. Biochim Biophys Acta 1998;1378(1):F79–113.

44. Kovalenko MV, Kazlauskas A. Signaling by the platelet-derived growth factor receptor family. In: Bradshaw RA, Dennis EA, eds. Handbook of Cell Signaling: Elsevier; 2003:397–404.

45. Gould K, Hunter T. Platelet-derived growth factor induces multisite phosphorylation of pp60c-src and increases its protein tyrosine kinase activity. Mol Cell Biol 1988;8:3345–56.

46. Kypta RM, Goldberg Y, Ulug ET, Courtneidge SA. Association between the PDGF receptor and members of the src family of tyrosine kinases. Cell 1990;62:481–92.

47. Mori S, Rönnstrand L, Yokote K, et al. Identification of two juxtamembrane autophosphorylation sites in the PDGF β-receptor: invovlement in the interaction with src family tyrosine kinases. EMBO J 1993;12:2257–64.

48. Hooshmand-Rad R, Yokote K, Heldin CH, Claesson-Welsh L. PDGF alpha-receptor mediated cellular responses are not dependent on Src family kinases in endothelial cells. J Cell Sci 1998;111(Pt 5):607–14.

49. Gelderloos JA, Rosenkranz S, Bazenet C, Kazlauskas A. A role for Src in signal relay by the platelet-derived growth factor alpha receptor. J Biol Chem 1998;273(10):5908–15.

50. Klinghoffer RA, Sachsenmaier C, Cooper JA, Soriano P. Src family kinases are required for integrin but not PDGFR signal transduction. Embo J 1999;18(9):2459–71.

51. DeMali KA, Kazlauskas A. Activation of Src family members is not required for the platelet-derived growth factor beta receptor to initiate mitogenesis. Mol Cell Biol 1998;18(4):2014–22.

52. Blake RA, Broome MA, Liu X, et al. SU6656, a selective src family kinase inhibitor, used To probe growth factor signaling. Mol Cell Biol 2000;20(23):9018–27.

53. Schlessinger J. How tyrosine kinases activate ras. Trends Biol Sci 1993;18:273–6.

54. Marshall CJ. Ras effectors. Curr Opin Cell Biol 1996;8(2):197–204.

55. Auger KR, Serunian SA, Soltoff SP, Libby P, Cantley LC. PDGF-dependent tyrosine phosphorylation stimulates production of novel polyphosphoinositides in intact cells. Cell 1989;57:167–75.

56. Whiteford CC, Best C, Kazlauskas A, Ulug ET. D-3 phosphoinositide metabolism in PDGF-treated cell. Biochem J 1996;319:851–60.

57. Jones SM, Klinghoffer R, Prestwich GD, Toker A, Kazlauskas A. PDGF induces an early and late wave of PI3-kinase activity, and only the late wave is required for progression through G1. Curr Biol 1999;9:512–21.

58. Claesson-Welsh L. Platelet-derived growth factor receptor signals. J Biol Chem 1994;269:32023–6.

59. Rosenkranz S, Kazlauskas A. Evidence for distinct signaling properties and biological responses induced by the PDGF receptor alpha and beta subtypes. Growth Factors 1999;16(3):201–16.

60. Katso R, Okkenhaug K, Ahmadi K, White S, Timms J, Waterfield MD. Cellular function of phosphoinositide 3-kinases: implications for development, homeostasis, and cancer. Annu Rev Cell Dev Biol 2001;17:615–75.

61. Engelman JA, Luo J, Cantley LC. The evolution of phosphatidylinositol 3-kinases as regulators of growth and metabolism. Nat Rev Genet 2006;7(8):606–19.

62. Yu J, Zhang Y, McIlroy J, Rordorf-Nikolic T, Orr GA, Backer JM. Regulation of the p85/p110 phosphatidylinositol 3′-kinase: stabilization and inhibition of the p110alpha catalytic subunit by the p85 regulatory subunit. Mol Cell Biol 1998;18(3):1379–87.

63. Rodriguez-Viciana P, Warne PH, Vanhaesebroeck B, Waterfield MD, Downward J. Activation of phosphoinositide 3-kinase by interaction with Ras and by point mutation. Embo J 1996;15(10):2442–51.

64. Klinghoffer RA, Duckworth B, Valius M, Cantley L, Kazlauskas A. Platelet-derived growth factor-dependent activation of phosphatidylinositol 3-kinase is regulated by receptor binding of SH2-domain-containing proteins which influence Ras activity. Mol Cell Biol 1996;16:5905–14.

65. Sun H, Charles CH, Lau LF, Tonks NK. MKP-1 (3CH134), an immediate early gene product, is a dual specificity phosphatase that dephosphorylates MAP kinase in vivo. Cell 1993;75:487–93.

66. Kazlauskas A. The priming/completion paradigm to explain growth factor-dependent cell cycle progression. Growth Factors 2005;23(3):203–10.

67. Kirschner MW. The biochemical nature of the cell cycle. Important Adv Oncol 1992:3–16.

68. King RW, Deshaies RJ, Peters JM, Kirschner MW. How proteolysis drives the cell cycle. Science 1996;274(5293):1652–9.

69. Pardee AB. A restriction point for control of normal animal cell proliferation. Proc Natl Acad Sci USA 1974;71(4):1286–90.

70. Pardee AB. G$_1$ Events and Regualtion of Cell Proliferation. Science 1989;240(November):603–8.

71. Zetterberg A, Larsson O. Kinetic analysis of regulatory events in G1 leading to proliferation or quiescence of swiss 3T3 cells. Proc Natl Acad Sci USA 1985;82:5365–9.

72. Planas-Silva MD, Weinberg RA. The restriction point and control of cell proliferation. Curr Opin Cell Biol 1997;9(6):768–72.

73. Pledger WJ, Stiles CD, Antoniades HN, Scher CD. Induction of DNA synthesis in BALB/c 3T3 cells by serum components: reevaluation of the commitment process. Proc Natl Acad Sci USA 1977;74(10):4481–5.

74. Pledger WJ, Stiles CD, Antoniades HN, Scher CD. An ordered sequence of events is required before BALB/c-3T3 cells become committed to DNA synthesis. Proc Natl Acad Sci USA 1978;75(6):2839–43.

75. Stiles CD, Capone GT, Scher CD, Antoniades HN, Van Wyk JJ, Pledger WJ. Dual control of cell growth by somatomedins and platelet-derived growth factor. Proc Natl Acad Sci USA 1979;76:1279–83.

76. Jones SM, Kazlauskas A. Growth-factor-dependent mitogenesis requires two distinct phases of signalling. Nat Cell Biol 2001;3(2):165–72.

77. Janes KA, Albeck JG, Gaudet S, Sorger PK, Lauffenburger DA, Yaffe MB. A systems model of signaling identifies a molecular basis set for cytokine-induced apoptosis. Science 2005;310(5754):1646–53.

78. Matsui T, Heidaran M, Miki T, et al. Isolation of a novel receptor cDNA establishes the existence of two PDGF receptor genes. Science 1989;243:800–4.

79. Soriano P. Abnormal kidney development and hematological disorders in PDGF β-receptor mutant mice. Genes Dev 1994;8:1888–96.

80. Lindahl P, Johansson BR, Leveen P, Betsholtz C. Pericyte loss and microaneurysm formation in PDGF-B-deficient mice. Science 1997;277(5323):242–5.

81. Hellstrom M, Gerhardt H, Kalen M, et al. Lack of pericytes leads to endothelial hyperplasia and abnormal vascular morphogenesis. J Cell Biol 2001;153(3):543–53.

82. Lindahl P, Hellstrom M, Kalen M, et al. Paracrine PDGF-B/PDGF-Rbeta signaling controls mesangial cell development in kidney glomeruli. Development 1998;125(17):3313–22.

83. Soriano P. The PDGF alpha receptor is required for neural crest cell development and for normal patterning of the somites. Development 1997;124:2691–700.

84. Hamilton TG, Klinghoffer RA, Corrin PD, Soriano P. Evolutionary divergence of platelet-derived growth factor alpha receptor signaling mechanisms. Mol Cell Biol 2003;23(11):4013–25.

85. Tallquist MD, Weismann KE, Hellstrom M, Soriano P. Early myotome specification regulates PDGFA expression and axial skeleton development. Development 2000;127(23):5059–70.

86. Fruttiger M, Karlsson L, Hall AC, et al. Defective oligodendrocyte development and severe hypomyelination in PDGF-A knockout mice. Development 1999;126(3):457–67.

87. Morrison-Graham K, Schatteman GC, Bork T, Bowen-Pope DF, Weston JA. A PDGF receptor mutation in the mouse (Patch) perturbs the development of a non-neuronal subset of neural crest-derived cells. Development 1992;115:133–43.

88. Boström H, Willetts K, Pekny M, et al. PDGF-A signaling is a critical event in lung alveolar myofibroblast development and alveogenesis. Cell 1996;85:863–73.

89. Karlsson L, Bondjers C, Betsholtz C. Roles for PDGF-A and sonic hedgehog in development of mesenchymal components of the hair follicle. Development 1999;126(12):2611–21.

90. Lindahl P, Betsholtz C. Not all myofibroblasts are alike: revisiting the role of PDGF-A and PDGF-B using PDGF-targeted mice. Curr Opin Nephrol Hypertens 1998;7(1):21–6.

91. Tallquist M, Kazlauskas A. PDGF signaling in cells and mice. Cytokine Growth Factor Rev 2004;15(4):205–13.

92. Klinghoffer RA, Mueting-Nelsen PF, Faerman A, Shani M, Soriano P. The two PDGF receptors maintain conserved signaling in vivo despite divergent embryological functions. Mol Cell 2001;7(2):343–54.

93. Fambrough D, McClure K, Kazlauskas A, Lander ES. Diverse signaling pathways activated by growth factor receptors induce broadly overlapping, rather than distinct, sets of genes. Cell 1999;97:727–41.

94. Schmahl J, Raymond CS, Soriano P. PDGF signaling specificity is mediated through multiple immediate early genes. Nat Genet 2006.

95. DiCorleto PE, Bowen-Pope DF. Cultured endothelial cells produce a platelet-derived growth factor-like protein. Proc Natl Acad Sci USA 1983;80(7):1919–23.

96. Collins T, Ginsburg D, Boss JM, Orkin SH, Pober JS. Cultured human endothelial cells express platelet-derived growth factor B chain: cDNA cloning and structural analysis. Nature 1985;316:748–50.

97. Collins T, Pober JS, Gimbrone MA, Jr., et al. Cultured human endothelial cells express platelet-derived growth factor A chain. Am J Pathol 1987;126(1):7–12.

98. Kazlauskas A, DiCorleto PE. Cultured endothelial cells do not respond to a platelet-derived growth-factor-like protein in an autocrine manner. Biochem Biophys Acta 1985;846:405–12.

99. Heldin C-H, Westermark B, Wasteson A. Specific receptors for platelet-derived growth factor on cells derived from connective tissue and glia. Proc Natl Acad Sci USA 1981;78:3664–8.

100. Antonelli-Orlidge A, Saunders KB, Smith SR, D'Amore PA. An activated form of transforming growth factor beta is produced by cocultures of endothelial cells and pericytes. Proc Natl Acad Sci USA 1989;86(12):4544–8.

101. Hirschi KK, Rohovsky SA, D'Amore PA. PDGF, TGF-beta, and heterotypic cell-cell interactions mediate endothelial cell-induced recruitment of 10T1/2 cells and their differentiation to a smooth muscle fate. J Cell Biol 1998;141(3):805–14.

102. Hirschi KK, Rohovsky SA, Beck LH, Smith SR, D'Amore PA. Endothelial cells modulate the proliferation of mural cell precursors via platelet-derived growth factor-BB and heterotypic cell contact. Circ Res 1999;84(3):298–305.

103. Leveen P, Pekny M, Gebre-Medhin S, Swolin B, Larsson E, Betsholtz C. Mice deficient for PDGF B show renal, cardiovascular, and hematological abnormalities. Genes Dev 1994;8:1875–87.

104. Enge M, Bjarnegard M, Gerhardt H, et al. Endothelium-specific platelet-derived growth factor-B ablation mimics diabetic retinopathy. Embo J 2002;21(16):4307–16.

105. Lindblom P, Gerhardt H, Liebner S, et al. Endothelial PDGF-B retention is required for proper investment of pericytes in the microvessel wall. Genes Dev 2003;17(15):1835–40.

106. Maharaj AS, Saint-Geniez M, Maldonado AE, D'Amore PA. Vascular endothelial growth factor localization in the adult. Am J Pathol 2006;168(2):639–48.

107. Jain RK. Molecular regulation of vessel maturation. Nat Med 2003;9(6):685–93.

108. von Tell D, Armulik A, Betsholtz C. Pericytes and vascular stability. Exp Cell Res 2006;312(5):623–9.

109. Chantrain CF, Henriet P, Jodele S, et al. Mechanisms of pericyte recruitment in tumour angiogenesis: a new role for metalloproteinases. Eur J Cancer 2006;42(3):310–8.

110. Abramsson A, Lindblom P, Betsholtz C. Endothelial and nonendothelial sources of PDGF-B regulate pericyte recruitment and influence vascular pattern formation in tumors. J Clin Invest 2003;112(8):1142–51.

111. Bergers G, Song S, Meyer-Morse N, Bergsland E, Hanahan D. Benefits of targeting both pericytes and endothelial cells in the tumor vasculature with kinase inhibitors. J Clin Invest 2003;111(9):1287–95.

112. Shaheen RM, Tseng WW, Davis DW, et al. Tyrosine kinase inhibition of multiple angiogenic growth factor receptors improves survival in mice bearing colon cancer liver metastases by inhibition of endothelial cell survival mechanisms. Cancer Res 2001;61(4):1464–8.

113. Pietras K, Rubin K, Sjoblom T, et al. Inhibition of PDGF receptor signaling in tumor stroma enhances antitumor effect of chemotherapy. Cancer Res 2002;62(19):5476–84.

114. Erber R, Thurnher A, Katsen AD, et al. Combined inhibition of VEGF and PDGF signaling enforces tumor vessel regression by interfering with pericyte-mediated endothelial cell survival mechanisms. Faseb J 2004;18(2):338–40.

115. Uehara H, Kim SJ, Karashima T, et al. Effects of blocking platelet-derived growth factor-receptor signaling in a mouse model of experimental prostate cancer bone metastases. J Natl Cancer Inst 2003;95(6):458–70.

116. Kim SJ, Uehara H, Yazici S, et al. Simultaneous blockade of platelet-derived growth factor-receptor and epidermal growth factor-receptor signaling and systemic administration of paclitaxel as therapy for human prostate cancer metastasis in bone of nude mice. Cancer Res 2004;64(12):4201–8.

117. Gee MS, Procopio WN, Makonnen S, Feldman MD, Yeilding NM, Lee WM. Tumor vessel development and maturation impose limits on the effectiveness of anti-vascular therapy. Am J Pathol 2003;162(1):183–93.

118. Darland DC, D'Amore PA. Blood vessel maturation: vascular development comes of age. J Clin Invest 1999;103(2):157–8.

119. Benjamin LE, Golijanin D, Itin A, Pode D, Keshet E. Selective ablation of immature blood vessels in established human tumors follows vascular endothelial growth factor withdrawal. J Clin Invest 1999;103(2):159–65.

120. Chantrain CF, Shimada H, Jodele S, et al. Stromal matrix metalloproteinase-9 regulates the vascular architecture in neuroblastoma by promoting pericyte recruitment. Cancer Res 2004;64(5):1675–86.

121. Spurbeck WW, Ng CY, Strom TS, Vanin EF, Davidoff AM. Enforced expression of tissue inhibitor of matrix metalloproteinase-3 affects functional capillary morphogenesis and inhibits tumor growth in a murine tumor model. Blood 2002;100(9):3361–8.

122. Benjamin LE, Hemo I, Keshet E. A plasticity window for blood vessel remodelling is defined by pericyte coverage of the preformed endothelial network and is regulated by PDGF-B and VEGF. Development 1998;125(9):1591–8.

123. Morikawa S, Baluk P, Kaidoh T, Haskell A, Jain RK, McDonald DM. Abnormalities in pericytes on blood vessels and endothelial sprouts in tumors. Am J Pathol 2002;160(3):985–1000.

124. Sims DE. The pericyte–a review. Tissue Cell 1986;18(2):153–74.

125. Cogan DG, Toussaint D, Kuwabara T. Retinal vascular patterns. IV. Diabetic retinopathy. Arch Ophthalmol 1961;66:366–78.

126. Speiser P, Gittelsohn AM, Patz A. Studies on diabetic retinopathy. 3. Influence of diabetes on intramural pericytes. Arch Ophthalmol 1968;80(3):332–7.

127. Arfken CL, Reno PL, Santiago JV, Klein R. Development of proliferative diabetic retinopathy in African-Americans and whites with type 1 diabetes. Diabetes Care 1998;21(5):792–5.

128. Wilkinson-Berka JL, Babic S, De Gooyer T, et al. Inhibition of platelet-derived growth factor promotes pericyte loss and angio-

genesis in ischemic retinopathy. Am J Pathol 2004;164(4):1263–73.

129. Tanii M, Yonemitsu Y, Fujii T, et al. Diabetic microangiopathy in ischemic limb is a disease of disturbance of the platelet-derived growth factor-BB/protein kinase C axis but not of impaired expression of angiogenic factors. Circ Res 2006;98(1):55–62.

130. Li X, Tjwa M, Moons L, et al. Revascularization of ischemic tissues by PDGF-CC via effects on endothelial cells and their progenitors. J Clin Invest 2005;115(1):118–27.

131. Cao R, Brakenhielm E, Pawliuk R, et al. Angiogenic synergism, vascular stability and improvement of hind-limb ischemia by a combination of PDGF-BB and FGF-2. Nat Med 2003;9(5):604–13.

132. Song S, Ewald AJ, Stallcup W, Werb Z, Bergers G. PDGFRbeta+ perivascular progenitor cells in tumours regulate pericyte differentiation and vascular survival. Nat Cell Biol 2005;7(9):870–9.

133. Beitz JG, Kim I-S, Calabresi P, Frackelton ARJ. Human microvascular endothelial cells express receptors for platelet-derived growth factor. Proc Natl Acad Sci USA 1991;88:2021–5.

134. Bar RS, Boes M, Booth BA, Dake BL, Henley S, Hart MN. The effects of platelet-derived growth factor in cultured microvessel endothelial cells. Endocrinology 1989;124(4):1841–8.

135. Edelberg JM, Aird WC, Wu W, et al. PDGF mediates cardiac microvascular communication. J Clin Invest 1998;102(4):837–43.

136. Koyama N, Watanabe S, Tezuka M, Morisaki N, Saito Y, Yoshida S. Migratory and proliferative effect of platelet-derived growth factor in rabbit retinal endothelial cells: evidence of an autocrine pathway of platelet-derived growth factor. J Cell Physiol 1994;158:1–6.

137. Marx M, Perlmutter RA, Madri JA. Modulation of platelet-derived growth factor receptor expression in microvascular endothelial cells during in vitro angiogenesis. J Clin Invest 1994;93(1):131–9.

138. Smits A, Hermansson M, Nister M, et al. Rat brain capillary endothelial cells express functional PDGF B-type receptors. Growth Factors 1989;2(1):1–8.

139. Thommen R, Humar R, Misevic G, et al. PDGF-BB increases endothelial migration on cord movements during angiogenesis in vitro. J Cell Biochem 1997;64(3):403–13.

140. Lindner V, Reidy MA. Platelet-derived growth factor ligand and receptor expression by large vessel endothelium in vivo. Am J Pathol 1995;146(6):1488–97.

141. Battegay EJ, Rupp J, Iruela-Arispe L, Sage EH, Pech M. PDGF-BB modulates endothelial proliferation and angiogenesis in vitro via PDGF beta-receptors. J Cell Biol 1994;125(4):917–28.

142. Plate KH, Brier G, Farrell CL, Risau W. Platelet-derived growth factor receptor-β is induced during tumor development and upregulated during tumor progression in endothelial cells in human gliomas. Lab Invest 1992;67:529–34.

143. Lei H, Velez G, Horland P, Hirose T, Kazlauskas A. Plasmin is the major protease responsible for processing PDGF-C in the vitreous of patients with proliferative vitreoretinopathy. IOVS 2008;49(1):42–48.

Chapter 10
Angiopoietins and Tie Receptors

Pipsa Saharinen[1], Lauri Eklund[2], and Kari Alitalo[1, *]

Keywords: angiopoietins, tie receptors, venous malformations, tie receptor signaling, tyrosine phosphastase

Abstract: The Tie1 and Tie2 receptor tyrosine kinases and the angiopoietin growth factor ligands, Ang1-4, are essential for vascular maturation. Targeted deletion of any of the *Tie1*, *Tie2* or *Ang1* genes in mice results in embryonic lethality during embryonic days 9.5–13.5. The receptors are expressed mainly in endothelial cells, while Ang1 is produced by perivascular cells and thought to stabilize quiescent endothelium. In contrast, Ang2 is secreted by endothelial cells in angiogenic vasculature, such as in tumors, leading to destabilization of the endothelium. Ang1 multimers stimulate the phosphorylation of Tie1 and Tie2, while Ang2 functions as a context-dependent agonist/antagonist for Tie2. Ang1 has promising vascular protective effects as an anti-permeability, anti-inflammatory and cell survival factor, but it can also induce vessel remodelling. The angiopoietin-Tie signalling pathway may be a therapeutically useful target in the treatment of a number of diseases, including oedema, endotoxaemia, transplant arteriosclerosis and cancer.

Introduction

A search for protein tyrosine kinases expressed in endothelial cells as well as during murine cardiogenesis and hematopoiesis resulted in the isolation of the Tie1 and Tie2 receptor tyrosine kinases (RTKs) in the beginning of the 1990s [1–4]. Tie1 and Tie2 were found to possess a unique extracellular structure consisting of epidermal growth factor, immunoglobulin and fibronectin type III domains, and they constitute a distinct family among the 20 RTK subfamilies formed by 57 genes [5]. Angiopoietin1 (Ang1, ANGPT1) was isolated as a ligand for Tie2 using secretion-trap cloning approach [6], and subsequently, low stringency DNA homology cloning was used to identify other members of the angiopoietin family, namely Ang2 (ANGPT2), Ang3 and Ang4 (ANGPT4) [7–10].

Gene targeting studies of *Tie1* and *Tie2* in mice showed that both RTKs are essential for vascular maturation [11–14]. The similarities between the phenotypes of Ang1- and Tie2-deficient as well as Ang2-overexpressing mice, supported a widely publicized model where Ang1 was the activating ligand for Tie2, while Ang2 was able to block the stimulatory effect of Ang1, thereby acting as a natural antagonist for Ang1 [7, 11, 13, 15–17]. However, since then, results were published that did not fit into this concept, and recent findings indeed confirm that Ang2 may instead function as a weak agonist of Tie2 stimulation [18–21]. The functions of Ang3 and Ang4 are far less characterized, but they can bind and activate Tie2 in context-dependent manner [9, 10].

Angiopoietins are unique among other RTK ligands as only multimeric forms induce Tie signalling [6, 22, 23]. However, none of the angiopoietins binds to Tie1, and its role has remained elusive despite its evident role in angiogenesis during development [12, 13]. Very recent data now indicate that angiopoietins also activate Tie1, most likely via its interaction with Tie2 [24]. The angiopoietin–Tie system has been found to mediate endothelial cell survival, sprouting, migration, periendothelial cell recruitment, anti-permeability and anti-inflammatory effects, but the exact signalling mechanisms involved in these processes have remained largely unknown [25, 26]. This may be due to the complexity of the interactions between various growth factors and of the specialized interactions of endothelial cells with perivascular cells, which are likely to regulate the angiogenic processes *in vivo*, but which are poorly amenable for studies *in vitro*.

[1] Molecular/Cancer Biology Laboratory and Ludwig Institute for Cancer Research, Biomedicum Helsinki, University of Helsinki, Haartmaninkatu 8, P.O.B. 63, 00014 Helsinki, Finland

[2] Collagen Research Unit Biocenter Oulu and Department of Medical Biochemistry and Molecular Biology, University of Oulu, P.O.B. 5000, 90014 Oulu, Finland

*Corresponding Author:
E-mail: Kari.Alitalo@Helsinki.fi

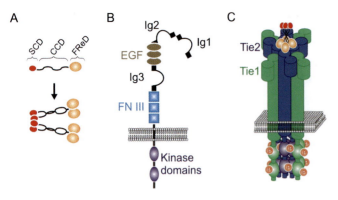

Fig. 10.1. Domain organization and complex formation of the angio-poietins and Tie receptors. **A** Angiopoietins contain N-terminal super clustering domain (*SCD*, residues 20–158), central coiled coil domain (*CCD*, residues 159–255), and C-terminal fibrinogen-related domain (*FReD*, residues 284–498). Ang1 forms dimers, trimers and tetramers through its *CCD* domain, and further assemble into multi-mers via *SCD* domain. **B** Tie receptors are composed of intracellular kinase domains, and extracellular immunoglobulin homology (*Ig1-Ig3*), epidermal growth factor homology (*EGF*), and fibronectin type III -domains (*FN III*). **C** Angiopoietin multimers bind to the second Ig domain in Tie2 via their FRed domains and induces clustering of the preformed Tie1–Tie2 heteromers, resulting in autophosphorylation of tyrosine residues (P) in the C-terminal receptor tails and initiation of downstream signaling cascades.

Angiopoietins

Structure

Ang1, 2 and 4 constitute the angiopoietin family of growth factors [6–9], which bind to Tie2 and activate both Tie1 and Tie2. The mouse ortholog of Ang4 was termed Ang3, an unfortu-nately somewhat confusing nomenclature. Ang2 and Ang3/4 share approximately 65% and 50% amino acid sequence iden-tity with Ang1, respectively [7, 9].

Ang1 is a secreted glycoprotein; the molecular weight of the monomer subunit is approximately 70 kilodaltons (kD). All angio-poietins contain a unique amino (N)-terminal domain, a coiled coil domain, a linker region and a carboxyl (C)-terminal fibrino-gen like domain (Fig. 10.1A). Ang1 forms dimers, trimers and tetramers through its coiled coil domains. They further assemble into higher order multimers via the N-terminus, also called as superclustering domain, as determined by rotary shadowing transmission electron microscopy [22, 23, 27] .

The multimeric nature of the angiopoietins results in their aggregation and poor solubility, thus making production of recombinant angiopoietins complicated. In addition, due to their oligomerization potential, preparations of recombinant angiopoietins are heterogenic in regard to the degree of mul-timerization, reflecting the natural properties of these recom-binant proteins. Thus, modified, more soluble, angiopoietin variants have been developed, which retain many of the bio-logical activities of angiopoietins, or are even more potent than their native counterparts. Ang1* is an Ang1/Ang2 chi-meric protein [28], while in COMP-Ang1 and BOW-Ang1

the oligomerization domains of Ang1 have been replaced by oligomerization domains from heterologous proteins [27, 29, 30]. These angiopoietin recombinant proteins have been widely used to study the effects of the angiopoietins.

Expression

Ang1 is expressed during early development in the mouse myocardium and likely signals to the heart endocardium in a paracrine manner [6]. Later in development, Ang1 becomes more widely expressed, being highest in the mesenchymal tissue surrounding endothelial cells of developing vessels [6]. Ang1 is also expressed in periendothelial cells (pericytes and smooth muscle cells) within the adult vasculature, and it is believed that constitutive Ang1 expression maintains vessel quiescence in the mature vasculature [31].

Contrary to the paracrine expression pattern of Ang1, Ang2 is expressed in an autocrine manner by endothelial cells. While Ang2 is expressed at low levels in most normal adult tissues, it is strongly upregulated during vessel remodelling, for example during ovarian follicle maturation and tumor angio-genesis [7, 16]. A broad range of factors are known to upregulate Ang2 mRNA in endothelial cells, including hypoxia, vascular endothelial growth factor (VEGF), basic fibroblast growth factor (bFGF), angiotensin II, leptin and tumor necrosis factor-α (TNF-α) [32]. Hypoxia also upregulates Ang2 expression *in vivo*, for example in rat dorsal skin flaps that are ischemic, or in the brains of rats during hypoxic conditions [33]. The Ang2 synthesized by endothelial cells is stored in Weibel-Palade bodies, from where it can be rapidly released upon stimulation [34]. This may be especially important during inflammatory responses, where Ang2 has been found to promote signals for leukocyte adhesion to endothelial cells [35]. Alternatively spliced mRNAs exist for Ang1 and Ang2, but the functional significance of these isoforms is not known [36, 37].

The expression pattern of Ang4 is not well known, although the lungs appear to express relatively high levels [9]. Ang4 is upregulated by hypoxia and VEGF in endothelial cells [38].

Ang1, but not Ang2, is incorporated into the extracellular matrix via its linker domain [39], while Ang3 is bound to the cell surface via heparan sulfate proteoglycans [40]. The differ-ent cellular localization may further contribute to the activity and bioavailability of the various angiopoietins.

The Tie Receptor Tyrosine Kinase Family

Expression

Tie1 and Tie2 are type 1 transmembrane receptor protein tyrosine kinases that are almost exclusively expressed in endothelial cells. The expression of the *Tie1* gene is restricted to endothelial cells and to some hematopoietic cell lineages, and it is upregulated during wound healing, ovarian follicle maturation and tumor angiogenesis [1, 41-46]. Interestingly, Tie1 is induced by dis-turbed flow in atherogenic vascular niches such as in vascular bifurcations and branching points along the arterial tree [47, 48].

Tie2 is prominent in the endocardium, leptomeninges and endothelial cells from the earliest stages of their development [4, 44, 46]. Notably, Tie2 is expressed prior to von Willebrand factor and may mark the embryonic progenitors of mature endothelial cells [2]. In addition, Tie2 is expressed in a subpopulation of hematopoietic stem cells (HSCs), and it helps to maintain these cells in a quiescent state in the bone marrow niche [49].

Structure

The amino acid sequence identity between Tie1 and Tie2 intracellular domains is 76%, and 31%, respectively, between the extracellular domains [25]. The intracellular tyrosine kinase domains of Tie1 and Tie2 are interrupted by a short kinase insert and the carboxyl (C)-terminal tail has several autophosphorylation sites. The three-dimensional structure of the Tie2 tyrosine kinase domain shows a typical kinase fold with an amino-terminal lobe containing mostly β-sheets and a mainly α-helical C-terminal lobe [50]. However, several features rarely seen in other kinases make the Tie2 structure unique. The activation loop is found in an active conformation even in the unphosphorylated Tie2 kinase domain, while the nucleotide binding loop occupies an inhibitory conformation preventing ATP binding [50]. The C-terminal tail is positioned close to the substrate binding pocket and the phosphorylation sites of tyrosine (Y) residues, Y1101 and Y1112, are buried. This suggests that the C-terminal tail must undergo a conformational change upon activation of the protein and expose both the substrate binding site and Y1101 and Y1112 for phosphorylation and subsequent signalling [50].

The Tie2 extracellular domain is composed of two immunoglobulin (Ig) homology domains followed by three epidermal growth factor (EGF)-homology domains, a third Ig domain and three fibronectin type-III domains (Fig. 10.1B) [51]. The crystal structure of Tie2 shows that the Ig and EGF homology domains fold together into a compact, arrowhead-shaped structure [51]. Ang2 binds to Tie2 at the tip of the arrowhead formed by the second Ig domain [51]. The binding involves a lock-and-key mode of ligand recognition, which is unique among RTKs and induces little conformational change in either molecule [51]. The binding site for Ang1 has also been mapped to the Tie2 N-terminus, encompassing the two Ig motifs and EGF-homology regions [52]. As both Ang1 and Ang2 bind to Tie2 via their C-terminal fibrinogen-like domains, the data collectively suggests that Ang1 and Ang2, and possibly Ang3/4, interact with Tie2 in a structurally similar manner [51].

Angiopoietin-Receptor Complex

The angiopoietin fibrinogen-like domain has been found to bind Tie2 in a 1:1 stoichiometry [22, 53]. However, in their native forms Ang1 and Ang2 exist as oligomers with 2 or more receptor binding sites, and at least Ang1 is further clustered into multimers [22, 23, 27]. Thus, the expected stoichiometry of the angiopoietin-receptor complex *in vivo* would be 2:2 or higher. Interestingly, Ang1 dimers fail to activate Tie2 in

endothelial cells, while they can activate ectopically expressed Tie2 in fibroblasts [22]. In contrast, at least a tetrameric Ang1 ligand is required for Tie2 activation in endothelial cells [22]. Ang2 oligomers activate Tie2 only weakly. Using chimeric molecules, it has been shown that the weaker Tie2 activation by Ang2 compared to Ang1 in part depends on differences in the fibrinogen-like domains between Ang1 and Ang2 [22].

Despite the close homology between Tie1 and Tie2, Tie1 does not significantly bind angiopoietins. However, stimulation of endothelial cells with angiopoietins results in induction of Tie1 tyrosine phosphorylation [24]. This occurs with the same kinetics and doses of angiopoietins as Tie2 phosphorylation, but is clearly weaker in intensity [24]. Tie1 activation is most likely mediated via Tie2 as coexpression of Tie2 can enhance the Tie1 response to angiopoietins and Tie1 and Tie2 were found to co-immunoprecipitate from endothelial cells (Fig. 10.1C) [24].

Angiopoietin and *Tie* Gene-Modified Mice

Tie2⁻/⁻, *Tie1⁻/⁻* and Double Deficient Mice

Gene targeting studies in mice have revealed that Tie1, Tie2, Ang1 and Ang2 are required for angiogenic remodelling and vessel stabilization after the initial assembly of vasculature, which requires VEGF. *Tie2* gene-targeted embryos die between embryonic day 9.5 (E9.5) and E12.5 as a consequence of impaired cardiac function, vascular rupture and hemorrhage [11, 13]. The heart is able to form, but myocardial trabecular projections are absent from Tie2-deficient embryos and the endocardial cells remain poorly attached to the myocardium [11, 13]. Vascular network formation is impaired; the blood vessels throughout the embryo apparently do not remodel or form normal hierarchical networks, after the initial phases of angiogenesis [13]. In addition, the number of endothelial cells is reduced when compared to wild-type littermates, and peri-endothelial support cells are either lacking or rare in regions of deficient vessel branching [11, 54]. Collectively, these data support a role for Tie2 in vessel stability and endothelial cell survival. However, the exact function of Tie2 during later stages of blood vessel development has not been investigated due to the early embryonic lethality of Tie2-deficient mice.

Targeted disruption of the mouse *Tie1* gene results in lethality after E13.5 because of severe edema, hemorrhages and defective microvessel integrity [12, 13]. *Tie1⁻/⁻* mice have small hearts with endocardial defects, but the major blood vessels appear normal [12, 13]. *Tie1⁻/⁻* endothelial cells in chimeric mice are not incorporated into new vessels of tissues, such as brain and kidney, which undergo angiogenesis during late embryogenesis [55]. This indicates that Tie1 is required autonomously for endothelial cell survival and extension of the vascular network, particularly in regions undergoing angiogenic growth of capillaries [55].

The phenotype of embryos lacking both Tie1 and Tie2 is similar to that of the Tie2-deficient embryos, but more

severe [14]. Mosaic analysis of the Tie1/Tie2 double deficient embryos have revealed an absolute requirement for
Tie2 in the endocardium at E10.5, whereas both receptors
were dispensable for the initial assembly of the rest of the
vasculature [14]. In contrast, both receptors were required in
the microvasculature during late organogenesis and in essentially all blood vessels of the adult [14].

Mosaic analysis of Tie1/Tie2 double knockout mice also
showed that the Tie receptors are not required for differentiation
and proliferation of definitive hematopoietic cell lineages in
the embryo or fetus, but they are specifically required during postnatal bone marrow hematopoiesis [56]. In the double
deficient cells, this effect is likely due to the lack of Tie2 since
Tie1-deficient cells expressing normal levels of Tie2 can contribute to hematopoiesis [55].

Venous Malformations Induced by Tie2 Mutations

Missense mutations resulting in single amino acid substitutions (R849W and Y897S) in the Tie2 kinase domain have
been identified in some families with inherited mucocutane-
ous venous malformations [57, 58]. These mutations increase
the basal activity of Tie2 [57, 58], resulting in abnormally
large, thinned-walled, venous-like vascular channels. The
affected vessels also have disorganized smooth muscle cell
coverage when compared to normal veins [57]. The R849W
substitution may result in abnormal Tie2 signalling [59–61],
however, the exact mechanism of how the hyperactive Tie2
variant causes vascular lesions is not known.

Ang1-Deficient and -Overexpressing Mice

Mice deficient of Ang1 have a similar phenotype as Tie2-
deficient mice, and die by E12.5 [15]. They have dilated vessels and decreased complexity of the vascular network [15].
In addition, the number of endothelial cells is reduced and
they are poorly associated with perivascular support cells and
with the basement membrane [15]. It is possible that other
members of the angiopoietin family compensate for the loss
of Ang1, which would explain why Ang1-deficient embryos
survive longer than Tie2-deficient embryos.

Overexpression of Ang1 in the skin under the K14 keratin promoter resulted in enlargement of vessel diameter,
mainly capillaries, but only slightly increased vessel number
(Fig. 10.2) [62, 63]. The vessels contained increased number
of endothelial cells, and were covered by pericytes similarly
to venules despite their location in the capillaries [62, 63].
The phenotype suggested that Ang1 was required for endothelial cell survival or proliferation and for endothelial-pericyte
interactions. The K14-Ang1/VEGF double transgenic mice
showed both significantly increased number of small blood
vessels and enlarged vessels. This indicated that VEGF and
Ang1 most likely regulated blood vessel growth via distinct
pathways, with VEGF inducing sprouting and vessel growth
while Ang1 mediating vessel remodeling [63]. However,
while VEGF-induced vessels were leaky and the vascularised

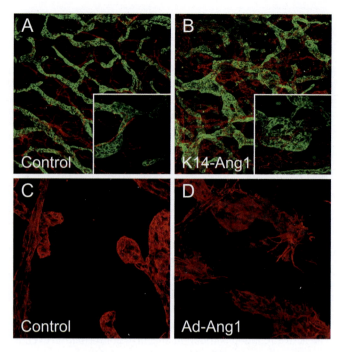

FIG. 10.2. Angiopoietin induces blood and lymphatic vessel enlargement
as well as lymphatic sprouting. *A,B* Confocal microscopic images of
lymphatic and blood vessels in the skin of wild type (*Control*) and
K14-Ang1 transgenic mice stained with fluorochrome conjugated
antibodies specific to blood endothelial (PECAM-1, red) and lymphatic endothelial (LYVE-1, *green*) cells. The square shows magnifi-
cation of a different area. *C,D* Confocal microscopic images of dermal
lymphatic (LYVE-1, *red*) vessels of mice treated locally with Ang1
(*Ad-Ang1*) or LacZ (*Control*) expressing adenovirus. At the 1-week
time point lymphatic vessels in Ang1 adenovirus treated mice show
significantly increased sprouting, when compared to control mice.

skin showed signs of inflammation, the combined expression
of VEGF with Ang1 induced non-leaky, and even leakage-
resistant blood vessels with little inflammation [63] .

Ang2-Deficient and -Overexpressing Mice

Ang2 gene-targeted mice are born normally, but have defects
in the lymphatic vasculature, chylous ascites and edema, and
remodeling of the retinal vessels is impaired [19]. Many of
them die by postnatal day 14 (P14) [19].

The hierarchy of large and small lymphatic vessels is retained
in the *Ang2⁻/⁻* mice, but the organization of the lymphatic vasculature is altered [19]. In addition, lymphatic central lacteals
in the intestinal villi are missing, and the association of smooth
muscle cells with the lymphatic vessels is disrupted [19]. The
Ang2 knock-in mice, which express Ang1 in place of Ang2,
have fairly normal lymphatic vasculature, indicating that Ang1
is able to rescue the *Ang2⁻/⁻* phenotype [19]. However, the retinal
defects found in *Ang2⁻/⁻* mice are not compensated by expression
of Ang1 [19]. This suggests that in the lymphatic endothelium
Ang2 mediates an agonistic signal, which can be mimicked by
Ang1, while only Ang2 can contribute to the proper remodeling
of the blood vasculature in the eye.

In addition to lymphatic and vascular defects, $Ang2^{-/-}$ mice show impaired inflammatory responses [35]. Specifically, TNF-α failed to induce leukocyte adhesion to activated endothelium, although the rolling of leucocytes along the endothelium that precedes adhesion occurred normally [35]. Ang2 sensitized endothelial cells in culture to TNF-α and increased the expression of several adhesion molecules, suggesting a mechanism for the Ang2 effects *in vivo* [35]. Thus, endogenous Ang2 secreted by endothelial cells, may act in an autocrine manner to regulate endothelial cell responses to proinflammatory cytokines [35].

Overexpression of Ang2 in endothelial cells results in mouse embryonic lethality at E9.5-10.5 [7]. The phenotype of these mice is very similar to that of $Tie2^{-/-}$ and $Ang1^{-/-}$ mice, but more severe [7]. Similarly, overexpression of Ang2 in the dermis under the K14 promoter resulted in embryonic lethality (T. Veikkola and K.Alitalo, unpublished data), most likely due to systemic effects of Ang2, by blocking the required stimulatory effect of Ang1. The embryos were pale and exhibit problems with hematopoiesis.

Potential Clinical Applications of the Angiopoietin-Tie System

Disruption of the microvasculature, plasma leakage, edema, vascular inflammation and neovascularisation are involved in the initiation and/or progression of many common human diseases, including diabetic retinopathy, cancer, stroke, and transplant arteriosclerosis. In many experimental settings, the angiopoietins have shown biological effects that may have significant clinical value in the treatment of vascular diseases.

Ang1 treatment may be useful to prevent vascular inflammation and leakage in various conditions [29, 63–68], and to increase survival of endothelial cells after stress [30, 69, 70]. Since Ang1 also stimulates lymphangiogenesis, it may be useful in attempts to treat tissue edema [71–73]. Although the exact roles of Ang1 and Ang2 in the tumor vasculature are not clear, the change in the balance between angiopoietins, especially high concentrations of Ang2 relative to Ang1 may promote the growth and vascularisation of tumors [16, 74–79]. Treatments targeted against Ang2 [77, 80] and competitive inhibition using soluble ligand binding domains of Tie2 may be useful to inhibit undesirable angiogenesis [81–83]. However, Ang1 treatments may also have potential harmful side effects. For example, since Ang1 induces vessel remodelling [84–87], it may promote pulmonary hypertension with prolonged treatment [88, 89].

Angiopoietin Signal Transduction

The Tie receptors contain tyrosine residues in their cytoplasmic domains, which are autophosphorylated following ligand stimulation. The phosphorylated tyrosine residues serve as docking sites for a number of effector proteins that couple the activated receptors to downstream signaling pathways. In Tie2, tyrosine residues in the juxtamembrane and kinase domains and three in the C-terminal tail are phosphorylated in response to Ang1. Several downstream effectors have been identified that associate with the activated Tie2 receptor and those have been linked to downstream signaling cascades in vitro [25, 26, 90]. The results suggest that signaling through phosphatidylinositol 3' kinase (PI3K) is a major pathway that may mediate Ang1 induced endothelial cell survival, motility, tube formation, and sprouting [31, 91–97]. Activated Tie2 also associates with the Dok-R/Nck/PAK pathway [98–100], adaptor proteins SchA, Grp7, Grb14, and Shp2 [101–103], and regulates transcription factor NF-κB [104] and signal transducers and activators of transcription factors (STATs) [59]. The physiological significance of these signaling cascades has remained mostly unclear. Using gene-targeted mice, however, it has been demonstrated that signaling through tyrosine residue 1100, which serves as a docking site for the p85 subunit of PI3K and for adaptor proteins SchA and Grp7, is required for heart development and early hematopoiesis but not for perivascular cell recruitment [105]. This suggests that multiple tyrosine residues in Tie2 may mediate biological effects of angiopoietins [105].

Little is known about the regulation of Tie signaling. There is some evidence that ligand binding may induce internalization and degradation of Tie2 [20] and that vascular endothelial protein tyrosine phosphatase (VE-PTP) dephosphorylates Tie2 [106]. It is also possible that certain integrins, including $\alpha5\beta1$ and $\alpha_v\beta5$, may modulate the actions of angiopoietins [107–110].

Because of the lack of a ligand to directly bind and activate Tie1, its function has remained enigmatic. It has been proposed that a possible function of Tie1 is to modulate signaling through the formation of heterodimeric complexes with Tie2 [24, 47, 111–113]. In such heteromeric complexes Tie1 can be activated by the angiopoietins [24]. In addition, Tie1 may have a ligand-independent function, involving shedding of the receptor [47, 111, 113, 114].

Conclusions

The endothelial expressed Tie receptors and their ligands, the angiopoietin growth factors, are essential for remodeling of blood and lymphatic vessels. However, their exact roles in these processes are not well understood.

Changes in expression of Tie receptors and angiopoietins occur in many pathological conditions including psoriasis [115], pulmonary hypertension [89, 116], infantile hemangiomas [80, 117], different tumors [78] and mutations in the Tie2 gene cause vascular anomalies [57, 58, 118]. Recent findings in experimental animal models suggest that modulation of Tie/angiopoietin signaling has many potential therapeutic applications. To maximize their clinical benefits yet avoid the harmful side effects their mechanisms of action need to be better characterized.

Several important questions remain unanswered. The role of Tie1 during development is not understood. What is the specificity of the active Tie1–Tie2 signaling complex, and if additional binding partners, such as integrins and phosphatases, are recruited to the signaling complex to generate all the effects of angiopoetins? What are the

angiopoietin-mediatedcellularmechanismsthatregulateendothelial–
pericyte interactions or those behind the differential responses
of lymphatic and blood vessels to angiopoietins? What is the
mechanism by which Tie2 may contribute to disease states,
e.g. in venous malformations or tumorigenesis? The high
interest in the Tie receptors and angiopoietins will no doubt
guarantee that some of the answers to these questions will be
obtained in the near future.

References

1. Partanen, J. *et al*. A novel endothelial cell surface receptor tyrosine kinase with extracellular epidermal growth factor homology domains. *Mol Cell Biol* 12, 1698–1707 (1992).
2. Dumont, D.J., Yamaguchi, T.P., Conlon, R.A., Rossant, J. & Breitman, M.L. tek, a novel tyrosine kinase gene located on mouse chromosome 4, is expressed in endothelial cells and their presumptive precursors. *Oncogene* 7, 1471–1480 (1992).
3. Dumont, D.J., Gradwohl, G.J., Fong, G.H., Auerbach, R. & Breitman, M.L. The endothelial-specific receptor tyrosine kinase, tek, is a member of a new subfamily of receptors. *Oncogene* 8, 1293–1301 (1993).
4. Iwama, A. et al. Molecular cloning and characterization of mouse TIE and TEK receptor tyrosine kinase genes and their expression in hematopoietic stem cells. *Biochem Biophys Res* Commun 195, 301–309 (1993).
5. Manning, G., Whyte, D.B., Martinez, R., Hunter, T. & Sudarsanam, S. The protein kinase complement of the human genome. *Science* 298, 1912–1934 (2002).
6. Davis, S. et al. Isolation of angiopoietin-1, a ligand for the TIE2 receptor, by secretion-trap expression cloning. *Cell* 87, 1161–1169 (1996).
7. Maisonpierre, P.C. et al. Angiopoietin-2, a natural antagonist for Tie2 that disrupts in vivo angiogenesis. *Science* 277, 55–60 (1997).
8. Kim, I. et al. Molecular cloning and characterization of a novel angiopoietin family protein, angiopoietin-3. *FEBS Lett* 443, 353–356 (1999).
9. Valenzuela, D.M. et al. Angiopoietins 3 and 4: diverging gene counterparts in mice and humans. *Proc Natl Acad Sci USA* 96, 1904–1909 (1999).
10. Lee, H.J. et al. Biological characterization of angiopoietin-3 and angiopoietin-4. *Faseb J* 18, 1200–1208 (2004).
11. Dumont, D.J. et al. Dominant-negative and targeted null mutations in the endothelial receptor tyrosine kinase, tek, reveal a critical role in vasculogenesis of the embryo. *Genes Dev* 8, 1897–1909 (1994).
12. Puri, M.C., Rossant, J., Alitalo, K., Bernstein, A. & Partanen, J. The receptor tyrosine kinase TIE is required for integrity and survival of vascular endothelial cells. *Embo J* 14, 5884–5891 (1995).
13. Sato, T.N. et al. Distinct roles of the receptor tyrosine kinases Tie-1 and Tie-2 in blood vessel formation. *Nature* 367, 70–74 (1995).
14. Puri, M.C., Partanen, J., Rossant, J. & Bernstein, A. Interaction of the TEK and TIE receptor tyrosine kinases during cardiovascular development. *Development* 126, 4569–4580 (1999).
15. Suri, C. et al. Requisite role of angiopoietin-1, a ligand for the TIE2 receptor, during embryonic angiogenesis. *Cell* 87, 1171–1180 (1996).
16. Holash, J. et al. Vessel cooption, regression, and growth in tumors mediated by angiopoietins and VEGF. *Science* 284, 1994–1998 (1999).
17. Yancopoulos, G.D. et al. Vascular-specific growth factors and blood vessel formation. *Nature* 407, 242–248 (2000).
18. Teichert-Kuliszewska, K. et al. Biological action of angiopoietin-2 in a fibrin matrix model of angiogenesis is associated with activation of Tie2. *Cardiovasc Res* 49, 659–670 (2001).
19. Gale, N.W. et al. Angiopoietin-2 is required for postnatal angiogenesis and lymphatic patterning, and only the latter role is rescued by Angiopoietin-1. *Dev Cell* 3, 411–423 (2002).
20. Bogdanovic, E., Nguyen, V.P. & Dumont, D.J. Activation of Tie2 by angiopoietin-1 and angiopoietin-2 results in their release and receptor internalization. *J Cell Sci* 119, 3551-3560 (2006).
21. Daly, C. et al. Angiopoietin-2 functions as an autocrine protective factor in stressed endothelial cells. *Proc Natl Acad Sci USA* 103, 15491–15496 (2006).
22. Davis, S. et al. Angiopoietins have distinct modular domains essential for receptor binding, dimerization and superclustering. *Nat Struct Biol* 10, 38–44 (2003).
23. Kim, K.T. et al. Oligomerization and multimerization is critical for angiopoietin-1 to bind and phosphorylate tie2. *J Biol Chem* 280, 20126–20131 (2005).
24. Saharinen, P. et al. Multiple angiopoietin recombinant proteins activate the Tie1 receptor tyrosine kinase and promote its interaction with Tie2. *J Cell Biol* 169, 239–243 (2005).
25. Brindle, N.P., Saharinen, P. & Alitalo, K. Signaling and functions of angiopoietin-1 in vascular protection. *Circ Res* 98, 1014–1023 (2006).
26. Eklund, L. & Olsen, B.R. Tie receptors and their angiopoietin ligands are context-dependent regulators of vascular remodeling. *Exp Cell Res* 312, 630–641 (2006).
27. Cho, C.H. et al. COMP-Ang1: a designed angiopoietin-1 variant with nonleaky angiogenic activity. *Proc Natl Acad Sci USA 101*, 5547–5552 (2004).
28. Koblizek, T.I., Weiss, C., Yancopoulos, G.D., Deutsch, U. & Risau, W. Angiopoietin-1 induces sprouting angiogenesis in vitro. *Curr Biol* 8, 529–532 (1998).
29. Zhang, Z.G., Zhang, L., Croll, S.D. & Chopp, M. Angiopoietin-1 reduces cerebral blood vessel leakage and ischemic lesion volume after focal cerebral embolic ischemia in mice. *Neuroscience* 113, 683–687 (2002).
30. Cho, C.H. et al. Designed angiopoietin-1 variant, COMP-Ang1, protects against radiation-induced endothelial cell apoptosis. *Proc Natl Acad Sci USA* 101, 5553–5558 (2004).
31. Kim, I. et al. Angiopoietin-1 regulates endothelial cell survival through the phosphatidylinositol 3'-Kinase/Akt signal transduction pathway. *Circ Res* 86, 24–29 (2000).
32. Jones, P.F. Not just angiogenesis--wider roles for the angiopoietins. *J Pathol* 201, 515–527 (2003).
33. Mandriota, S.J. et al. Hypoxia-inducible angiopoietin-2 expression is mimicked by iodonium compounds and occurs in the rat brain and skin in response to systemic hypoxia and tissue ischemia. *Am J Pathol* 156, 2077–2089 (2000).
34. Fiedler, U. et al. The Tie-2 ligand angiopoietin-2 is stored in and rapidly released upon stimulation from endothelial cell Weibel-Palade bodies. *Blood* 103, 4150–4156 (2004).
35. Fiedler, U. et al. Angiopoietin-2 sensitizes endothelial cells to TNF-alpha and has a crucial role in the induction of inflammation. *Nat Med* 12, 235–239 (2006).
36. Huang, Y.Q., Li, J.J. & Karpatkin, S. Identification of a family of alternatively spliced mRNA species of angiopoietin-1. *Blood* 95, 1993–1999 (2000).
37. Kim, I. et al. Characterization and expression of a novel alternatively spliced human angiopoietin-2. *J Biol Chem* 275, 18550–18556 (2000).

38. Yamakawa, M. et al. Hypoxia-inducible factor-1 mediates activation of cultured vascular endothelial cells by inducing multiple angiogenic factors. *Circ Res* 93, 664–673 (2003).

39. Xu, Y. & Yu, Q. Angiopoietin-1, unlike angiopoietin-2, is incorporated into the extracellular matrix via its linker peptide region. *J Biol Chem* 276, 34990–34998 (2001).

40. Xu, Y., Liu, Y.J. & Yu, Q. Angiopoietin-3 is tethered on the cell surface via heparan sulfate proteoglycans. *J Biol Chem* 279, 41179–41188 (2004).

41. Korhonen, J. et al. Enhanced expression of the tie receptor tyrosine kinase in endothelial cells during neovascularization. *Blood* 80, 2548–2555 (1992).

42. Kaipainen, A. et al. Enhanced expression of the tie receptor tyrosine kinase mesenger RNA in the vascular endothelium of metastatic melanomas. *Cancer Res* 54, 6571–6577 (1994).

43. Korhonen, J., Polvi, A., Partanen, J. & Alitalo, K. The mouse tie receptor tyrosine kinase gene: expression during embryonic angiogenesis. *Oncogene* 9, 395–403 (1994).

44. Dumont, D.J. et al. Vascularization of the mouse embryo: a study of flk-1, tek, tie, and vascular endothelial growth factor expression during development. *Dev Dyn* 203, 80–92 (1995).

45. Hashiyama, M. et al. Predominant expression of a receptor tyrosine kinase, TIE, in hematopoietic stem cells and B cells. *Blood* 87, 93–101 (1996).

46. Yano, M. et al. Expression and function of murine receptor tyrosine kinases, TIE and TEK, in hematopoietic stem cells. *Blood* 89, 4317–4326 (1997).

47. Chen-Konak, L. et al. Transcriptional and post-translation regulation of the Tie1 receptor by fluid shear stress changes in vascular endothelial cells. *Faseb J* 17, 2121–2123 (2003).

48. Porat, R.M. et al. Specific induction of tie1 promoter by disturbed flow in atherosclerosis-prone vascular niches and flow-obstructing pathologies. *Circ Res* 94, 394–401 (2004).

49. Arai, F. et al. Tie2/angiopoietin-1 signaling regulates hematopoietic stem cell quiescence in the bone marrow niche. *Cell* 118, 149–161 (2004).

50. Shewchuk, L.M. et al. Structure of the Tie2 RTK domain: self-inhibition by the nucleotide binding loop, activation loop, and C-terminal tail. *Structure* 8, 1105–1113 (2000).

51. Barton, W.A. et al. Crystal structures of the Tie2 receptor ectodomain and the angiopoietin-2-Tie2 complex. *Nat Struct Mol Biol* 13, 524–532 (2006).

52. Fiedler, U. et al. Angiopoietin-1 and angiopoietin-2 share the same binding domains in the Tie-2 receptor involving the first Ig-like loop and the epidermal growth factor-like repeats. *J Biol Chem* 278, 1721–1727 (2003).

53. Barton, W.A., Tzvetkova, D. & Nikolov, D.B. Structure of the angiopoietin-2 receptor binding domain and identification of surfaces involved in Tie2 recognition. *Structure* 13, 825–832 (2005).

54. Patan, S. TIE1 and TIE2 receptor tyrosine kinases inversely regulate embryonic angiogenesis by the mechanism of intussusceptive microvascular growth. Microvasc Res 56, 1–21 (1998).

55. Partanen, J. et al. Cell autonomous functions of the receptor tyrosine kinase TIE in a late phase of angiogenic capillary growth and endothelial cell survival during murine development. *Development* 122, 3013–3021 (1996).

56. Puri, M.C. & Bernstein, A. Requirement for the TIE family of receptor tyrosine kinases in adult but not fetal hematopoiesis. *Proc Natl Acad Sci USA* 100, 12753–12758 (2003).

57. Vikkula, M. et al. Vascular dysmorphogenesis caused by an activating mutation in the receptor tyrosine kinase TIE2. *Cell* 87, 1181–1190 (1996).

58. Calvert, J.T. et al. Allelic and locus heterogeneity in inherited venous malformations. *Hum Mol Genet* 8, 1279–1289 (1999).

59. Korpelainen, E.I., Karkkainen, M., Gunji, Y., Vikkula, M. & Alitalo, K. Endothelial receptor tyrosine kinases activate the STAT signaling pathway: mutant Tie-2 causing venous malformations signals a distinct STAT activation response. *Oncogene* 18, 1–8 (1999).

60. Morris, P.N. et al. Functional analysis of a mutant form of the receptor tyrosine kinase Tie2 causing venous malformations. *J Mol Med* 83, 58–63 (2005).

61. Morris, P.N., Dunmore, B.J. & Brindle, N.P. Mutant Tie2 causing venous malformation signals through Shc. *Biochem Biophys Res Commun* 346, 335–338 (2006).

62. Suri, C. et al. Increased vascularization in mice overexpressing angiopoietin-1. *Science* 282, 468–471 (1998).

63. Thurston, G. et al. Leakage-resistant blood vessels in mice transgenically overexpressing angiopoietin-1. *Science* 286, 2511–2514 (1999).

64. Thurston, G. et al. Angiopoietin-1 protects the adult vasculature against plasma leakage. *Nat Med* 6, 460–463 (2000).

65. Nambu, H. et al. Angiopoietin 1 inhibits ocular neovascularization and breakdown of the blood-retinal barrier. *Gene Ther* 11, 865–873 (2004).

66. Joussen, A.M. et al. Suppression of diabetic retinopathy with angiopoietin-1. Am J Pathol 160, 1683–1693 (2002).

67. Nykanen, A.I. et al. Angiopoietin-1 protects against the development of cardiac allograft arteriosclerosis. *Circulation 107*, 1308–1314 (2003).

68. Witzenbichler, B., Westermann, D., Knueppel, S., Schultheiss, H.P. & Tschope, C. Protective role of angiopoietin-1 in endotoxic shock. *Circulation* 111, 97–105 (2005).

69. Zhao, Y.D., Campbell, A.I., Robb, M., Ng, D. & Stewart, D.J. Protective role of angiopoietin-1 in experimental pulmonary hypertension. *Circ Res* 92, 984–991 (2003).

70. Jung, H., Gurunluoglu, R., Scharpf, J. & Siemionow, M. Adenovirus-mediated angiopoietin-1 gene therapy enhances skin flap survival. *Microsurgery* 23, 374–380 (2003).

71. Morisada, T. et al. Angiopoietin-1 promotes LYVE-1-positive lymphatic vessel formation. Blood 105, 4649–4656 (2005).

72. Tammela, T. et al. Angiopoietin-1 promotes lymphatic sprouting and hyperplasia. *Blood* 105, 4642–4648 (2005).

73. Shimoda, H. et al. Abnormal recruitment of periendothelial cells to lymphatic capillaries in digestive organs of angiopoietin-2-deficient mice. *Cell Tissue Res* 328, 329–337 (2007).

74. Holash, J., Wiegand, S.J. & Yancopoulos, G.D. New model of tumor angiogenesis: dynamic balance between vessel regression and growth mediated by angiopoietins and VEGF. *Oncogene* 18, 5356–5362 (1999).

75. Zagzag, D. et al. In situ expression of angiopoietins in astrocytomas identifies angiopoietin-2 as an early marker of tumor angiogenesis. *Exp Neurol* 159, 391–400 (1999).

76. Ahmad, S.A. et al. The effects of angiopoietin-1 and -2 on tumor growth and angiogenesis in human colon cancer. *Cancer Res* 61, 1255–1259 (2001).

77. Oliner, J. et al. Suppression of angiogenesis and tumor growth by selective inhibition of angiopoietin-2. Cancer Cell 6, 507–516 (2004).

78. Tait, C.R. & Jones, P.F. Angiopoietins in tumours: the angiogenic switch. *J Pathol* 204, 1–10 (2004).

79. Tanaka, S., Wands, J.R. & Arii, S. Induction of angiopoietin-2 gene expression by COX-2: a novel role for COX-2 inhibitors during hepatocarcinogenesis. *J Hepatol* 44, 233–235 (2006).

80. Perry, B.N. et al. Pharmacologic blockade of angiopoietin-2 is efficacious against model hemangiomas in mice. *J Invest Dermatol* 126, 2316–2322 (2006).

81. Lin, P. et al. Antiangiogenic gene therapy targeting the endothelium-specific receptor tyrosine kinase Tie2. *Proc Natl Acad Sci USA* 95, 8829–8834 (1998).

82. Hangai, M. et al. Systemically expressed soluble Tie2 inhibits intraocular neovascularization. *Hum Gene Ther* 12, 1311–1321 (2001).

83. Das, A. et al. Angiopoietin/Tek interactions regulate mmp-9 expression and retinal neovascularization. *Lab Invest* 83, 1637–1645 (2003).

84. Baffert, F. et al. Age-related changes in vascular endothelial growth factor dependency and angiopoietin-1-induced plasticity of adult blood vessels. *Circ Res* 94, 984–992 (2004).

85. Ward, N.L., Van Slyke, P. & Dumont, D.J. Functional inhibition of secreted angiopoietin: a novel role for angiopoietin 1 in coronary vessel patterning. *Biochem Biophys Res Commun* 323, 937–946 (2004).

86. Cho, C.H. et al. Long-term and sustained COMP-Ang1 induces long-lasting vascular enlargement and enhanced blood flow. *Circ Res* 97, 86–94 (2005).

87. Thurston, G. et al. Angiopoietin 1 causes vessel enlargement, without angiogenic sprouting, during a critical developmental period. *Development* 132, 3317–3326 (2005).

88. Thistlethwaite, P.A. et al. Human angiopoietin gene expression is a marker for severity of pulmonary hypertension in patients undergoing pulmonary thromboendarterectomy. *J Thorac Cardiovasc Surg* 122, 65–73 (2001).

89. Sullivan, C.C. et al. Induction of pulmonary hypertension by an angiopoietin 1/TIE2/serotonin pathway. *Proc Natl Acad Sci USA* 100, 12331–12336 (2003).

90. Peters, K.G. et al. Functional significance of Tie2 signaling in the adult vasculature. *Recent Prog Horm Res* 59, 51–71 (2004).

91. Kontos, C.D. et al. Tyrosine 1101 of Tie2 is the major site of association of p85 and is required for activation of phosphatidylinositol 3-kinase and Akt. *Mol Cell Biol* 18, 4131–4140 (1998).

92. Fujikawa, K. et al. Role of PI 3-kinase in angiopoietin-1-mediated migration and attachment-dependent survival of endothelial cells. *Exp Cell Res* 253, 663–672 (1999).

93. Papapetropoulos, A. et al. Angiopoietin-1 inhibits endothelial cell apoptosis via the Akt/survivin pathway. *J Biol Chem* 275, 9102–9105 (2000).

94. Babaei, S. et al. Angiogenic actions of angiopoietin-1 require endothelium-derived nitric oxide. *Am J Pathol* 162, 1927–1936 (2003).

95. Harfouche, R. et al. Angiopoietin-1 activates both anti- and pro-apoptotic mitogen-activated protein kinases. *Faseb J* 17, 1523–1525 (2003).

96. Saito, M., Hamasaki, M. & Shibuya, M. Induction of tube formation by angiopoietin-1 in endothelial cell/fibroblast co-culture is dependent on endogenous VEGF. *Cancer Sci* 94, 782–790 (2003).

97. DeBusk, L.M., Hallahan, D.E. & Lin, P.C. Akt is a major angiogenic mediator downstream of the Ang1/Tie2 signaling pathway. *Exp Cell Res* 298, 167–177 (2004).

98. Jones, N. & Dumont, D.J. The Tek/Tie2 receptor signals through a novel Dok-related docking protein, Dok-R. *Oncogene* 17, 1097–1108 (1998).

99. Master, Z. et al. Dok-R plays a pivotal role in angiopoietin-1-dependent cell migration through recruitment and activation of Pak. *Embo J* 20, 5919–5928 (2001).

100. Jones, N. et al. A unique autophosphorylation site on Tie2/Tek mediates Dok-R phosphotyrosine binding domain binding and function. *Mol Cell Biol* 23, 2658–2668 (2003).

101. Huang, L., Turck, C.W., Rao, P. & Peters, K.G. GRB2 and SH-PTP2: potentially important endothelial signaling molecules downstream of the TEK/TIE2 receptor tyrosine kinase. *Oncogene* 11, 2097–2103 (1995).

102. Jones, N. et al. Identification of Tek/Tie2 binding partners. Binding to a multifunctional docking site mediates cell survival and migration. *J Biol Chem* 274, 30896–30905 (1999).

103. Audero, E. et al. Adaptor ShcA protein binds tyrosine kinase Tie2 receptor and regulates migration and sprouting but not survival of endothelial cells. *J Biol Chem* 279, 13224–13233 (2004).

104. Hughes, D.P., Marron, M.B. & Brindle, N.P. The antiinflammatory endothelial tyrosine kinase Tie2 interacts with a novel nuclear factor-kappaB inhibitor ABIN-2. *Circ Res* 92, 630-636 (2003).

105. Tachibana, K., Jones, N., Dumont, D.J., Puri, M.C. & Bernstein, A. Selective role of a distinct tyrosine residue on Tie2 in heart development and early hematopoiesis. *Mol Cell Biol* 25, 4693–4702 (2005).

106. Fachinger, G., Deutsch, U. & Risau, W. Functional interaction of vascular endothelial-protein-tyrosine phosphatase with the angiopoietin receptor Tie-2. *Oncogene* 18, 5948–5953 (1999).

107. Carlson, T.R., Feng, Y., Maisonpierre, P.C., Mrksich, M. & Morla, A.O. Direct cell adhesion to the angiopoietins mediated by integrins. *J Biol Chem* 276, 26516–26525 (2001).

108. Cascone, I., Napione, L., Maniero, F., Serini, G. & Bussolino, F. Stable interaction between {alpha}5{beta}1 integrin and tie2 tyrosine kinase receptor regulates endothelial cell response to ang-1. *J Cell Biol* 170, 993–1004 (2005).

109. Dallabrida, S.M., Ismail, N., Oberle, J.R., Himes, B.E. & Rupnick, M.A. Angiopoietin-1 promotes cardiac and skeletal myocyte survival through integrins. *Circ Res* 96, e8–24 (2005).

110. Weber, C.C. et al. Effects of protein and gene transfer of the angiopoietin-1 fibrinogen-like receptor-binding domain on endothelial and vessel organization. *J Biol Chem* 280, 22445–22453 (2005).

111. McCarthy, M.J. et al. Potential roles of metalloprotease mediated ectodomain cleavage in signaling by the endothelial receptor tyrosine kinase Tie-1. *Lab Invest* 79, 889–895 (1999).

112. Marron, M., Hughes, D.P., Edge, M.D., Forder, C.L. & Brindle, N. Evidence for heterotypic interaction between the receptor tyrosine kinases TIE-1 and TIE-2. *J Biol Chem* 275, 39741–39746 (2000).

113. Tsiamis, A.C., Morris, P.N., Marron, M.B. & Brindle, N.P. Vascular endothelial growth factor modulates the Tie-2:Tie-1 receptor complex. *Microvasc Res* 63, 149–158 (2002).

114. Yabkowitz, R. et al. Regulation of tie receptor expression on human endothelial cells by protein kinase C-mediated release of soluble tie. *Blood* 90, 706–715 (1997).

115. Voskas, D. et al. A cyclosporine-sensitive psoriasis-like disease produced in Tie2 transgenic mice. *Am J Pathol* 166, 843-855 (2005).

116. Du, L. et al. Signaling molecules in nonfamilial pulmonary hypertension. *N Engl J Med* 348, 500–509 (2003).

117. Yu, Y., Varughese, J., Brown, L.F., Mulliken, J.B. & Bischoff, J. Increased Tie2 expression, enhanced response to angiopoietin-1, and dysregulated angiopoietin-2 expression in hemangioma-derived endothelial cells. *Am J Pathol* 159, 2271–2280 (2001).

118. Wang, H., Zhang, Y., Toratani, S. & Okamoto, T. Transformation of vascular endothelial cells by a point mutation in the Tie2 gene from human intramuscular haemangioma. *Oncogene* 23, 8700-8704 (2004).

Chapter 11
Basement Membrane Derived Inhibitors of Angiogenesis

Michael B. Duncan[1] and Raghu Kalluri[1,2,3]

Keywords: vascular basement membrane, endostatin, tumstatin
Vascular basement membranes

Abstract: Blood vessel growth during development and disease is likely governed by the balance between pro- and antiangiogenic factors. Numerous reports have focused on the important role of various growth factors during angiogenesis. Peptide fragments derived from basement membranes constitute a relatively new and expanding class of antiangiogenic factors with a potential for clinical relevance. These factors have been studied in a variety of disease models, and genetic evidence for their role in controlling angiogenesis is beginning to be realized. This chapter highlights several of these factors and their mechanism of action as we understand them to date. Gaining additional insight into the full compliment of these antiangiogenic fragments from basement membranes, how they are derived, and their full mechanism of action represents an important challenge in vascular biology today.

The vascular basement membrane (VBM) is a complex structure composed of a variety of functionally diverse glycoproteins and proteoglycans (Fig. 11.1) [1]. The primary molecules detected in the VBM are the heparan sulfate proteoglycans (such as perlecan and type XVIII collagen), laminin, niodogen, entactin, fibulin, as well as type IV collagen. While providing structural support to the vessel, these molecules also serve important functional roles in endothelial cell (EC) signaling and adhesion. Therefore, these basement membrane molecules play a critical role in EC proliferation, migration, morphogenesis, survival, and vessel stability [2]. Much of the interactions and subsequent biological activity of VBM macromolecules are due to their ability to engage endothelial cell surface receptors, namely the integrins [3]. In some instances, evidence suggests that the antiangiogenic functionality of basement membrane molecules is elicited only when cryptic fragments of these molecules are revealed through proteolytic processing [4]. These findings highlight an essential role for (1) understanding the structural organization of the VBM in normal and disease states, (2) identifying the full complement of antiangiogenic entities derived from the VBM and other basement membranes, and (3) defining the mechanisms by which changes in the vascular microenvironment, and subsequent changes in the balance of angiogenic mediators, occur. This chapter highlights the origin and biological activity of some recently discovered basement membrane derived antiangiogenic molecules.

Arresten

Arresten is a 26 kDa molecule derived from the COOH-terminal noncollagenous (NC) 1 domain of the α 1 chain of type IV collagen. An early report demonstrated that α 1 (as well as α 2 type IV) collagen derived from the Engelbreth-Holm-Swarm sarcoma tumor possess inhibitory effects towards capillary endothelial cell proliferation [5]. Subsequent studies revealed a dose dependent inhibition of FGF2-dependent endothelial cell proliferation as well as inhibition of endothelial tube formation and VEGF stimulated migration in vitro and inhibition of angiogenesis and tumor growth in vivo [6]. These inhibitory effects are mediated through the FAK/c-RAF/MEK1/2/p38/ERK1 MAPK pathway via engagement of the α1β1 integrin receptor, while no effect on the PI3K/Akt pathway is observed [7]. The cysteine protease cathepsin S, which is highly expressed in malignant tissues, and its inhibitor cystatin C appear to have the capacity to regulate the levels of arresten and the antiangiogenic peptide canstatin (described below) [8].

[1] Division of Matrix Biology, Department of Medicine, Beth Israel Deaconess Medical Center and Harvard Medical School, Boston, MA 02215, USA

[2] Department of Biological Chemistry and Molecular Pharmacology, Harvard Medical School, Boston, MA 02215, USA

[3] Harvard-MIT Division of Health Sciences and Technology, Boston, MA 02215, USA

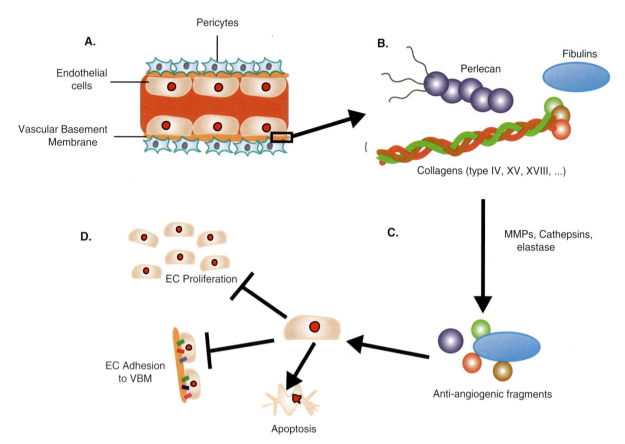

FIG. 11.1. **A** The vasculature is composed of endothelial cells supported by perictyes. **B** These two cell populations synthesize and share the vascular basement membrane, which is composed of matrix molecules including the collagens, fibulins, and proteoglycans. **C** These matrix molecules are degraded by matrix metalloproteinases, cathepsins, and elastase into a variety of antiangiogenic fragments. **D** These fragments affect endothelial cell viability by inhibiting adhesion, proliferation, and inducing apoptosis.

Canstatin

Canstatin is a 24 kDa fragment derived from the COOH-terminal NC1 domain of the α 2 chain type IV collagen [9]. In vitro, canstatin inhibits endothelial cell tube formation on Matrigel as well as VEGF-meditated cell migration. Endothelial cells adhere to canstatin via the αvβ3, αvβ5, and α3β1 integrin receptors. Several studies have shown that canstatin promotes apoptosis in endothelial cells [9–11]. Canstatin mediates apoptosis through the PI3K/Akt pathway and Fas/FasL signaling. Exposure of HUVECs to canstatin inhibits phosphorylation of Akt, FAK, mTOR, eIF4E binding protein1, and ribosomal S6 kinase. Fas/FasL mediated apoptosis results in the activation of procaspases 8 and 9, a decrease in mitochondrial membrane potential, and ultimately leads to increased endothelial cell death [11]. Studies have also shown that canstatin mediated apoptosis may in part be regulated by FLIP [9]. Canstatin inhibits tumor growth in several murine models [9, 12–14]. Histological analyses of these tumors reveal substantially reduced angiogenesis. In a mouse melanoma tumor model, intraperitoneally injected recombinant canstatin was not only shown to reduce angio-

genesis, but it also induced cancer cell apoptosis and cellular senescence [15].

In addition to its ligand dependent effects on endothelial cells, studies suggest that canstatin can directly act on some cancer cells. While no effects were observed when the human prostate cancer cell line PC3 or renal cell carcinoma line 786-0 were exposed to high doses of canstatin [9], integrin specific inhibition of adhesion has been observed against the M21 and SKOV-3 cancer cell lines [15], as well as MDA-MB-231 cells [10].

Endorepellin

Endorepellin is an 85 kDa laminin-like domain derived from the COOH terminus (domain V) of the proteoglycan perlecan. Perlecan is a large (~470 kDa) modular protein that is composed of five structural domains and plays critical roles in vascular growth and the structural integrity of the VBM [16–18]. Endorepellin has been shown to be a potent inhibitor of several key aspects of angiogenesis including endothelial cell migration, tube morphogenesis, and vessel growth in several *in vivo* models [19]. Structurally, endorepellin is composed of three laminin-type globular domains that are separated by four EGF-like domains. Interestingly, the LG3 domain of endorepellin retains antiangiogenic activity and the LG2 domain binds to and limits the antiangiogenic activity of endostatin derived from type XVIII collagen.

At the molecular level endorepellin disrupts actin stress fibers and focal adhesions in a heparan sulfate-independent, but calcium-dependent fashion [20]. These effects are mediated through the collagen integrin receptor $\alpha2\beta1$. When endorepellin engages this receptor, several effects are observed including a rapid rise in intracellular cAMP levels, activation of cAMP-dependent protein kinase A, FAK, p38 MAP kinase and heat shock protein 27.

In two mouse cancer models, an orthotopic squamous carcinoma and the syngeneic Lewis lung carcinoma, systemic administration of endorepellin targeted the tumor vasculature resulting in an inhibition of angiogenesis, increased tumor hypoxia, and a decrease in tumor metabolism [21]. The LG3 domain of endorepellin, which has been detected in urine of patients with end-stage renal disease and in the amniotic fluid of pregnant women with premature rupture of fetal membranes, is proteolytically processed by metalloproteinases of the bone morphogenic protein-1/Tolloid family [22–24].

Endostatin

Endostatin is a 20 kDa fragment derived from the $\alpha1$ chain of collagen XVIII. This fragment was originally purified from the conditioned media of the murine hemangioendothelioma (EOMA) cell line [25]. Recombinant endostatin blocks angiogenesis and suppresses both primary tumor growth and metastasis in various experimental animal models [26–28].

The molecular mechanisms underlying the antiangiogenic activity of endostatin are manifold. Endostatin has been shown to interfere with FGF2-mediated signal transduction resulting in the blockade of endothelial cell motility [29]. Endostatin induces apoptosis [30], G1 arrest of endothelial cells through inhibition of cyclin D1 [31], blocks VEGF-mediated signaling [32], and inhibits tumor necrosis factor induced activation of c-Jun NH_2-terminal kinase and c-Jun NH_2-terminal kinase-dependent proangiogenic gene expression [33]. Endostatin downregulates many genes in endothelial cells, including immediate early response genes, cell cycle related genes, genes regulating apoptosis inhibitors, MAPKs, FAKs, G protein coupled receptors mediating endothelial cell growth, mitogenic factors, adhesion molecules, and cell structure components [34]. The signaling pathway(s) whereby endostatin modulates gene expression is not yet clear. However, endostatin has been shown to increase intracellular Ca^{2+} levels via the influx of extracellular Ca^{2+} and activation of PKA [34].

Endostatin binds to $\alpha5$ and αv integrins on the surface of human endothelial cells (EC) and this interaction is involved in cell migration and survival [35]. Sudhakar et al. established that endostatin binds to the $\alpha5\beta1$ integrin and inhibits migration of endothelial cells by blocking signaling pathways via Ras and Raf and further downstream via ERK1 or p38 [36]. In ECs, endostatin induces rapid clustering of $\alpha5\beta1$ integrin associated with actin stress fibers and induces co-localization with the membrane anchor protein caveolin-1, which couples integrins to cytoplasmic signaling cascades [37]. In these experiments, endostatin was bound to both $\alpha5\beta1$ and caveolin-1 and induced phosphatase dependent activation of caveolin associated Src family kinases. The disassembly of actin stress fibers/focal adhesions by endostatin affected cell-matrix interaction and cell motility via activation of Src and in a tyrosyl phosphatase dependent manner. A contrasting finding by Eriksson et al. concludes that endostatin inhibits chemotaxis without affecting the common intracellular pathways known to regulate endothelial cell migration, proliferation, and survival [38]. Instead, they propose that the antiangiogenic activity of endostatin is derived from its ability to engage cell surface integrin receptors and disrupt cell-matrix interactions.

Endostatin binds to heparin and cell surface heparan sulfate proteoglycans, including the glypicans [39]. The heparin-binding capacity of endostatin depends on its interactions with discontinuous sulfated domains in the polysaccharide and arginine clusters at the endostatin surface [40]. A study suggests that endostatin can compete with bFGF for heparin binding, which may in part explain its effects on bFGF-mediated EC motility [41]. An arginine rich motif of endostatin has been shown to interact with EC $\beta1$ integrin and heparan proteoglycans and is sufficient for inhibiting endothelial cell migration and tube formation [42]. This RGD independent sequence may represent the minimum sequence requirement for the antiangiogenic activity of endostatin.

Endostatin inhibits the activation and activity of certain matrix metalloproteinases and it binds directly to MMP-2 and MMP-9 [43–45]. Endostatin has also been shown to interfere with the actions of other proteases, including the plasminogen activator system [46]. MMPs can also generate endostatin containing peptides differing in molecular size (20–30 kDa) from human type XVIII collagen [47].

The physiologic levels of circulating endostatin in the serum are in the range of 40 to 100 ng/ml compared with the low mg/ml concentrations of endostatin that have been utilized for the inhibition of tumor growth in many experiments. Recently, Sund et al. reported that a less than two-fold increase in endostatin expression could suppress tumor growth [48]. A potential clinical correlation with this data is that cancer occurrence tends to be lower in individuals with Down's syndrome. This is due to trisomy 21 providing an extra copy of the type XVIII collagen gene. Individuals with Down's syndrome have circulating endostatin levels of 38.6 ± 20 ng/ml compared with usual levels of 20.3 ± 11.5 ng/ml [49].

Fibulins

Proteolytic digestion of basement membrane preparations by elastases and cathepsins liberates antiangiogenic fragments [50]. The full-length version of fibulin 5 has been shown to antagonize VEGF signaling and inhibit endothelial angiogenic sprouting in an RGD/integrin binding independent fashion [51]. Fibulin 5 also binds to extracellular superoxide dismutase and upregulates the expression of the matricellular antiangiogenic protein thrombospondin-1 [51, 52]. In contrast, another report recently shows that retroviral delivery of fibulin 5 promotes wound healing in vivo [53]. Since wound healing is an angiogenesis-dependent process, these contrasting findings suggest that the angiogenic (either pro- or anti-) activity of fibulin 5 may be context dependent.

NC1 Domain of α 6 Chain Type IV Collagen

The NC1 domain of the α 6 chain of type IV collagen has demonstrated antiangiogenic activity and can regulate endothelial cell adhesion and migration (Mundel and Kalluri, unpublished data) [12].

NC1 Domain of Type XIX Collagen

Type XIX collagen is found in a variety of basement membranes and has been shown to co-localize with other collagens including type IV, XV, and XVIII [54, 55]. Recently, the COOH-terminal the α 1 NC1 domain of type XIX collagen was shown to have antiangiogenic effects in a murine melanoma model [56]. In an in vitro model, this NC1 domain inhibits EC migration, Matrigel invasion, and tube formation. Ramont et al. also report that this NC1 domain can reduce the expression of MT1-MMP and VEGF. The molecular mechanisms for the activity of this fragment are as yet undefined.

Endostatin Like Domain of Type XV Collagen (Restin)

Restin is a 22-kDa, antiangiogenic peptide fragment derived from the NC1 domain of collagen XV. This peptide fragment shares high homology with endostatin, derived from collagen XVIII [57]. Restin inhibits the migration of endothelial cells but has no effect on proliferation. Systemic administration of this fragment suppresses the growth of tumors in a xenograft renal carcinoma model. While sharing a high degree of sequence homology and similar antiangiogenic effects, restin and endostatin do display some functional differences [58]. For example, the relative binding affinities and profile of extracellular matrix molecules that engage restin and endostatin

are unique. Additionally, restin unlike endostatin cannot bind heparin or heparan sulfate proteoglycans. This suggests that there is a difference in the mechanism of action between these two molecules, since the heparin-binding capacity is important for the antiangiogenic activity of endostatin.

Tumstatin

Tumstatin is a 28 kDa peptide fragment derived from the α 3 chain of type IV collagen. This molecule has been shown to have anti-proliferative effects on melanoma cells as well as other carcinomas and antiangiogenic effects on a variety of endothelial cell types [12, 59–61]. The anti-tumor activity of tumstatin is localized to amino acids 185 to 203 [62], and the antiangiogenic activity to amino acids 54 to 132 [63]. Regarding its antiangiogenic effect, tumstatin inhibits EC proliferation, causes G1 cell cycle arrest and induces apoptosis through the up-regulation of caspase 3 [64]. Tumstatin engages the αvβ3 integrin receptor and induces the inhibition of protein synthesis [63, 65, 66]. The molecular mechanism for reduced protein synthesis involves the inhibition of FAK/PI3-kinase and Akt signaling pathways [66]. In addition, tumstatin suppresses mTOR kinase activity and prevents the dissociation of eukaryotic initiation factor 4E protein from 4E-BPI leading to the inhibition of cap-dependent protein synthesis. Tumstatin also has the capacity to inhibit VEGF-mediated endothelial cell adhesion to the extracelllular matrix via integrin receptors (not VEGF receptor mediated) [67]. Interruption of EC adhesion to immobilized VEGF by tumstatin promotes apoptosis.

The role of αvβ3 integrin as a receptor for tumstatin has been validated using a β3 integrin knockout mouse [68]. Angiogenesis was significantly reduced in this knockout mouse in a Matrigel plug assay. Additionally, when treated with tumstatin, the proliferative index of mouse lung endothelial cells derived from the β3 knockout is significantly reduced. Recently, Kawaguchi et al. demonstrated that the inability of the antiangiogenic portion of tumstatin to directly suppress the growth of tumor cells expressing the αvβ3 is a result of the constitutive activation of the Akt/mTOR signaling pathway [69]. These results suggest that receptor binding of the tumstatin fragment is insufficient for its antiangiogenic activity, and that cellular response to this molecule is dependent on the capacity of cell signaling machinery.

Circulating levels of tumstatin can be generated by a variety of matrix metalloproteinases including MMP-2, -3, -9, and -13 [70]. MMP-9 knockout mice have reduced levels of circulating tumstatin and tumors grow at a more rapid rate compared to wild type [68]. As determined by ELISA, the circulating level of tumstatin is approximately 300 ng/mL. The pharmacological doses of tumstatin administered in tumor studies are well above this concentration. This raises the question of whether the endogenous levels of tumstatin can impact tumor growth. Evidence from a knockout mouse deficient in the α 3 chain of type IV collagen suggests that circulating tumstatin levels can impede tumor (Lewis lung carcinoma) growth [68]. Tumors on the knockout mice grew at a more rapid rate when compared to wild type. Subsequent administration of the physiologic amount of tumstatin to the knockout mouse was able to slow tumor growth to a rate comparable with wild type.

Concluding Remarks

In addition to their role as a structural support for cells and tissues, the glycoproteins and proteoglycans of the basement membrane provide molecular signals that impact normal and pathological angiogenesis. These signals can be regulated at the genetic level, as evidenced by the regulation of endostatin and tumstatin (and potentially other collagen derived antiangiogenic molecules) levels by the tumor suppressor gene p53, as well as through the protease dependent turnover of the BM [71, 72]. Mouse models and clinical data point to an important role for BM derived antiangiogenic molecules in affecting pathological angiogenesis. Many studies have focused on the role of BM derived antiangiogenic molecules in the cancer setting, nevertheless emerging data suggest that these molecules can affect other angiogenic dependent diseases, including diabetic nephropathy [73, 74], age-related macular degeneration [75], and various forms of arthritis [33, 76]. While the application of these BM-derived molecules to treat cancer has met some setbacks [77], a variety of therapeutic strategies including gene therapy and combination therapy are being explored [78–81]. These efforts combined with a more precise understanding of how BM-derived antiangiogenic molecules function, should further expand their relevance in medicine.

Acknowledgements. The research work in the laboratory of the authors is supported by the NIH (grants DK 55001, DK 62987, DK 61688, and AA 13913) and program funds provided to the Division of Matrix Biology by the Beth Israel Deaconess Medical Center. Michael B. Duncan is supported by the NIH (DK055001-07S1) and the UNCF-Merck Postdoctoral Science Research Fellowship program.

References

1. Kalluri R. Basement membranes: structure, assembly and role in tumour angiogenesis. Nat Rev Cancer 2003;3(6):422–33.
2. Davis GE, Senger DR. Endothelial extracellular matrix: biosynthesis, remodeling, and functions during vascular morphogenesis and neovessel stabilization. Circ Res 2005;97(11):1093–107.
3. Giancotti FG, Ruoslahti E. Integrin signaling. Science 1999;285 (5430):1028–32.
4. Xu J, Rodriguez D, Petitclerc E, et al. Proteolytic exposure of a cryptic site within collagen type IV is required for angiogenesis and tumor growth in vivo. J Cell Biol 2001;154(5):1069–79.
5. Madri JA. Extracellular matrix modulation of vascular cell behaviour. Transpl Immunol 1997;5(3):179–83.
6. Colorado PC, Torre A, Kamphaus G, et al. Anti-angiogenic cues from vascular basement membrane collagen. Cancer Res 2000;60(9):2520–6.
7. Sudhakar A, Nyberg P, Keshamouni VG, et al. Human alpha1 type IV collagen NC1 domain exhibits distinct antiangiogenic activity mediated by alpha1beta1 integrin. J Clin Invest 2005;115(10):2801–10.
8. Wang B, Sun J, Kitamoto S, et al. Cathepsin S controls angiogenesis and tumor growth via matrix-derived angiogenic factors. J Biol Chem 2006;281(9):6020–9.
9. Kamphaus GD, Colorado PC, Panka DJ, et al. Canstatin, a novel matrix-derived inhibitor of angiogenesis and tumor growth. J Biol Chem 2000;275(2):1209–15.
10. Magnon C, Galaup A, Mullan B, et al. Canstatin acts on endothelial and tumor cells via mitochondrial damage initiated through interaction with alphavbeta3 and alphavbeta5 integrins. Cancer Res 2005;65(10):4353–61.
11. Panka DJ, Mier JW. Canstatin inhibits Akt activation and induces Fas-dependent apoptosis in endothelial cells. J Biol Chem 2003;278(39):37632–6.
12. Petitclerc E, Boutaud A, Prestayko A, et al. New functions for non-collagenous domains of human collagen type IV. Novel integrin ligands inhibiting angiogenesis and tumor growth in vivo. J Biol Chem 2000;275(11):8051–61.
13. He GA, Luo JX, Zhang TY, Hu ZS, Wang FY. The C-terminal domain of canstatin suppresses in vivo tumor growth associated with proliferation of endothelial cells. Biochem Biophys Res Commun 2004;318(2):354–60.
14. He GA, Luo JX, Zhang TY, Wang FY, Li RF. Canstatin-N fragment inhibits in vitro endothelial cell proliferation and suppresses in vivo tumor growth. Biochem Biophys Res Commun 2003;312(3):801–5.
15. Roth JM, Akalu A, Zelmanovich A, et al. Recombinant alpha2(IV)NC1 domain inhibits tumor cell-extracellular matrix interactions, induces cellular senescence, and inhibits tumor growth in vivo. Am J Pathol 2005;166(3):901–11.
16. Iozzo RV. Matrix proteoglycans: from molecular design to cellular function. Annu Rev Biochem 1998;67:609–52.
17. Iozzo RV, San Antonio JD. Heparan sulfate proteoglycans: heavy hitters in the angiogenesis arena. J Clin Invest 2001; 108(3):349–55.
18. Zhou Z, Wang J, Cao R, et al. Impaired angiogenesis, delayed wound healing and retarded tumor growth in perlecan heparan sulfate-deficient mice. Cancer Res 2004;64(14):4699–702.
19. Mongiat M, Sweeney SM, San Antonio JD, Fu J, Iozzo RV. Endorepellin, a novel inhibitor of angiogenesis derived from the C terminus of perlecan. J Biol Chem 2003;278(6):4238–49.
20. Bix G, Fu J, Gonzalez EM, et al. Endorepellin causes endothelial cell disassembly of actin cytoskeleton and focal adhesions through alpha2beta1 integrin. J Cell Biol 2004;166(1):97–109.
21. Bix G, Castello R, Burrows M, et al. Endorepellin in vivo: targeting the tumor vasculature and retarding cancer growth and metabolism. J Natl Cancer Inst 2006;98(22):1634–46.
22. Oda O, Shinzato T, Ohbayashi K, et al. Purification and characterization of perlecan fragment in urine of end-stage renal failure patients. Clin Chim Acta 1996;255(2):119–32.
23. Vuadens F, Benay C, Crettaz D, et al. Identification of biologic markers of the premature rupture of fetal membranes: proteomic approach. Proteomics 2003;3(8):1521–5.

24. Gonzalez EM, Reed CC, Bix G, et al. BMP-1/Tolloid-like metalloproteases process endorepellin, the angiostatic C-terminal fragment of perlecan. J Biol Chem 2005;280(8):7080–7.

25. O'Reilly MS, Boehm T, Shing Y, et al. Endostatin: an endogenous inhibitor of angiogenesis and tumor growth. Cell 1997;88(2):277–85.

26. Sorensen DR, Read TA, Porwol T, et al. Endostatin reduces vascularization, blood flow, and growth in a rat gliosarcoma. Neuro-oncol 2002;4(1):1–8.

27. Jia YH, Dong XS, Wang XS. Effects of endostatin on expression of vascular endothelial growth factor and its receptors and neovascularization in colonic carcinoma implanted in nude mice. World J Gastroenterol 2004;10(22):3361–4.

28. Li XP, Li CY, Li X, et al. Inhibition of human nasopharyngeal carcinoma growth and metastasis in mice by adenovirus-associated virus-mediated expression of human endostatin. Mol Cancer Ther 2006;5(5):1290–8.

29. Dixelius J, Cross M, Matsumoto T, Sasaki T, Timpl R, Claesson-Welsh L. Endostatin regulates endothelial cell adhesion and cytoskeletal organization. Cancer Res 2002;62(7):1944–7.

30. Dhanabal M, Ramchandran R, Waterman MJ, et al. Endostatin induces endothelial cell apoptosis. J Biol Chem 1999; 274(17):11721–6.

31. Hanai J, Dhanabal M, Karumanchi SA, et al. Endostatin causes G1 arrest of endothelial cells through inhibition of cyclin D1. J Biol Chem 2002;277(19):16464–9.

32. Kim YM, Hwang S, Kim YM, et al. Endostatin blocks vascular endothelial growth factor-mediated signaling via direct interaction with KDR/Flk-1. J Biol Chem 2002;277(31):27872–9.

33. Yin G, Liu W, An P, et al. Endostatin gene transfer inhibits joint angiogenesis and pannus formation in inflammatory arthritis. Mol Ther 2002;5(5 Pt 1):547–54.

34. Shichiri M, Hirata Y. Antiangiogenesis signals by endostatin. Faseb J 2001;15(6):1044–53.

35. Rehn M, Veikkola T, Kukk-Valdre E, et al. Interaction of endostatin with integrins implicated in angiogenesis. Proc Natl Acad Sci USA 2001;98(3):1024–9.

36. Sudhakar A, Sugimoto H, Yang C, Lively J, Zeisberg M, Kalluri R. Human tumstatin and human endostatin exhibit distinct anti-angiogenic activities mediated by alpha v beta 3 and alpha 5 beta 1 integrins. Proc Natl Acad Sci USA 2003;100(8):4766–71.

37. Wickstrom SA, Alitalo K, Keski-Oja J. Endostatin associates with integrin alpha5beta1 and caveolin-1, and activates Src via a tyrosyl phosphatase-dependent pathway in human endothelial cells. Cancer Res 2002;62(19):5580–9.

38. Eriksson K, Magnusson P, Dixelius J, Claesson-Welsh L, Cross MJ. Angiostatin and endostatin inhibit endothelial cell migration in response to FGF and VEGF without interfering with specific intracellular signal transduction pathways. FEBS Lett 2003;536(1–3):19–24.

39. Karumanchi SA, Jha V, Ramchandran R, et al. Cell surface glypicans are low-affinity endostatin receptors. Mol Cell 2001;7(4):811–22.

40. Kreuger J, Matsumoto T, Vanwildemeersch M, et al. Role of heparan sulfate domain organization in endostatin inhibition of endothelial cell function. Embo J 2002;21(23):6303–11.

41. Reis RC, Schuppan D, Barreto AC, et al. Endostatin competes with bFGF for binding to heparin-like glycosaminoglycans. Biochem Biophys Res Commun 2005;333(3):976–83.

42. Wickstrom SA, Alitalo K, Keski-Oja J. An endostatin-derived peptide interacts with integrins and regulates actin cytoskeleton and migration of endothelial cells. J Biol Chem 2004;279(19):20178–85.

43. Kim YM, Jang JW, Lee OH, et al. Endostatin inhibits endothelial and tumor cellular invasion by blocking the activation and catalytic activity of matrix metalloproteinase. Cancer Res 2000;60(19):5410–3.

44. Lee SJ, Jang JW, Kim YM, et al. Endostatin binds to the catalytic domain of matrix metalloproteinase-2. FEBS Lett 2002;519(1–3): 147–52.

45. Nyberg P, Heikkila P, Sorsa T, et al. Endostatin inhibits human tongue carcinoma cell invasion and intravasation and blocks the activation of matrix metalloprotease-2, -9, and -13. J Biol Chem 2003;278(25):22404–11.

46. Wickstrom SA, Veikkola T, Rehn M, Pihlajaniemi T, Alitalo K, Keski-Oja J. Endostatin-induced modulation of plasminogen activation with concomitant loss of focal adhesions and actin stress fibers in cultured human endothelial cells. Cancer Res 2001;61(17):6511–6.

47. Ferreras M, Felbor U, Lenhard T, Olsen BR, Delaisse J. Generation and degradation of human endostatin proteins by various proteinases. FEBS Lett 2000;486(3):247–51.

48. Sund M, Hamano Y, Sugimoto H, et al. Function of endogenous inhibitors of angiogenesis as endothelium-specific tumor suppressors. Proc Natl Acad Sci USA 2005;102(8):2934–9.

49. Zorick TS, Mustacchi Z, Bando SY, et al. High serum endostatin levels in Down syndrome: implications for improved treatment and prevention of solid tumours. Eur J Hum Genet 2001;9(11):811–4.

50. Kalluri R. Discovery of type IV collagen non-collagenous domains as novel integrin ligands and endogenous inhibitors of angiogenesis. Cold Spring Harb Symp Quant Biol 2002;67: 255–66.

51. Albig AR, Schiemann WP. Fibulin-5 antagonizes vascular endothelial growth factor (VEGF) signaling and angiogenic sprouting by endothelial cells. DNA Cell Biol 2004;23(6):367–79.

52. Nguyen AD, Itoh S, Jeney V, et al. Fibulin-5 is a novel binding protein for extracellular superoxide dismutase. Circ Res 2004;95(11):1067–74.

53. Lee CH, Wu CL, Shiau AL. Endostatin gene therapy delivered by Salmonella choleraesuis in murine tumor models. J Gene Med 2004;6(12):1382–93.

54. Myers JC, Li D, Bageris A, Abraham V, Dion AS, Amenta PS. Biochemical and immunohistochemical characterization of human type XIX defines a novel class of basement membrane zone collagens. Am J Pathol 1997;151(6):1729–40.

55. Sumiyoshi H, Inoguchi K, Khaleduzzaman M, Ninomiya Y, Yoshioka H. Ubiquitous expression of the alpha1(XIX) collagen gene (Col19a1) during mouse embryogenesis becomes restricted to a few tissues in the adult organism. J Biol Chem 1997;272(27):17104–11.

56. Ramont L, Brassart-Pasco S, Thevenard J, et al. The NC1 domain of type XIX collagen inhibits in vivo melanoma growth. Mol Cancer Ther 2007;6(2):506–14.

57. Ramchandran R, Dhanabal M, Volk R, et al. Antiangiogenic activity of restin, NC10 domain of human collagen XV: comparison to endostatin. Biochem Biophys Res Commun 1999;255(3):735–9.

58. Sasaki T, Larsson H, Tisi D, Claesson-Welsh L, Hohenester E, Timpl R. Endostatins derived from collagens XV and XVIII differ in structural and binding properties, tissue distribution and anti-angiogenic activity. J Mol Biol 2000;301(5):1179–90.

59. Monboisse JC, Garnotel R, Bellon G, et al. The alpha 3 chain of type IV collagen prevents activation of human polymorphonuclear leukocytes. J Biol Chem 1994;269(41):25475–82.

60. Han J, Ohno N, Pasco S, Monboisse JC, Borel JP, Kefalides NA. A cell binding domain from the alpha3 chain of type IV collagen inhibits proliferation of melanoma cells. J Biol Chem 1997;272(33):20395–401.

61. Maeshima Y, Colorado PC, Torre A, et al. Distinct antitumor properties of a type IV collagen domain derived from basement membrane. J Biol Chem 2000;275(28):21340–8.

62. Floquet N, Pasco S, Ramont L, et al. The antitumor properties of the alpha3(IV)-(185–203) peptide from the NC1 domain of type IV collagen (tumstatin) are conformation-dependent. J Biol Chem 2004;279(3):2091–100.

63. Shahan TA, Ziaie Z, Pasco S, et al. Identification of CD47/ integrin-associated protein and alpha(v)beta3 as two receptors for the alpha3(IV) chain of type IV collagen on tumor cells. Cancer Res 1999;59(18):4584–90.

64. Maeshima Y, Yerramalla UL, Dhanabal M, et al. Extracellular matrix-derived peptide binds to alpha(v)beta(3) integrin and inhibits angiogenesis. J Biol Chem 2001;276(34):31959–68.

65. Maeshima Y, Colorado PC, Kalluri R. Two RGD-independent alpha vbeta 3 integrin binding sites on tumstatin regulate distinct anti-tumor properties. J Biol Chem 2000;275(31):23745–50.

66. Maeshima Y, Sudhakar A, Lively JC, et al. Tumstatin, an endothelial cell-specific inhibitor of protein synthesis. Science 2002;295(5552):140–3.

67. Hutchings H, Ortega N, Plouet J. Extracellular matrix-bound vascular endothelial growth factor promotes endothelial cell adhesion, migration, and survival through integrin ligation. Faseb J 2003;17(11):1520–2.

68. Hamano Y, Zeisberg M, Sugimoto H, et al. Physiological levels of tumstatin, a fragment of collagen IV alpha3 chain, are generated by MMP-9 proteolysis and suppress angiogenesis via alphaV beta3 integrin. Cancer Cell 2003;3(6):589–601.

69. Kawaguchi T, Yamashita Y, Kanamori M, et al. The PTEN/ Akt Pathway Dictates the Direct {alpha}V{beta}3-Dependent Growth-Inhibitory Action of an Active Fragment of Tumstatin in Glioma Cells In vitro and In vivo. Cancer Res 2006;66 (23):11331–40.

70. McCawley LJ, Matrisian LM. Matrix metalloproteinases: they're not just for matrix anymore! Curr Opin Cell Biol 2001;13(5):534–40.

71. Rundhaug JE. Matrix metalloproteinases and angiogenesis. J Cell Mol Med 2005;9(2):267–85.

72. Teodoro JG, Parker AE, Zhu X, Green MR. p53-mediated inhibition of angiogenesis through up-regulation of a collagen prolyl hydroxylase. Science 2006;313(5789):968–71.

73. Yamamoto Y, Maeshima Y, Kitayama H, et al. Tumstatin peptide, an inhibitor of angiogenesis, prevents glomerular hypertrophy in the early stage of diabetic nephropathy. Diabetes 2004;53(7):1831–40.

74. Ichinose K, Maeshima Y, Yamamoto Y, et al. Antiangiogenic endostatin peptide ameliorates renal alterations in the early stage of a type 1 diabetic nephropathy model. Diabetes 2005;54(10):2891–903.

75. Mori K, Ando A, Gehlbach P, et al. Inhibition of choroidal neovascularization by intravenous injection of adenoviral vectors expressing secretable endostatin. Am J Pathol 2001;159(1):313–20.

76. Matsuno H, Yudoh K, Uzuki M, et al. Treatment with the angiogenesis inhibitor endostatin: a novel therapy in rheumatoid arthritis. J Rheumatol 2002;29(5):890–5.

77. Kulke MH, Bergsland EK, Ryan DP, et al. Phase II Study of Recombinant Human Endostatin in Patients With Advanced Neuroendocrine Tumors. J Clin Oncol 2006;24(22):3555–61.

78. Li X, Fu GF, Fan YR, et al. Potent inhibition of angiogenesis and liver tumor growth by administration of an aerosol containing a transferrin-liposome-endostatin complex. World J Gastroenterol 2003;9(2):262–6.

79. Subramanian IV, Bui Nguyen TM, Truskinovsky AM, Tolar J, Blazar BR, Ramakrishnan S. Adeno-associated virus-mediated delivery of a mutant endostatin in combination with carboplatin treatment inhibits orthotopic growth of ovarian cancer and improves long-term survival. Cancer Res 2006;66(8):4319–28.

80. Itasaka S, Komaki R, Herbst RS, et al. Endostatin improves radioresponse and blocks tumor revascularization after radiation therapy for A431 xenografts in mice. Int J Radiat Oncol Biol Phys 2007;67(3):870–8.

81. Xu YF, Zhu LP, Hu B, et al. A new expression plasmid in Bifidobacterium longum as a delivery system of endostatin for cancer gene therapy. Cancer Gene Ther 2007;14(2):151–7.

Chapter 12
Angiostatin and Endostatin: Angiogenesis Inhibitors in Blood and Stroma

Judah Folkman

Keywords: endothelium, neoplasia inflammation, ocular neovascularization, gene therapy

Introduction

A clinical clue led to the discovery of angiostatin [1], a 38-kD internal peptide of plasminogen, and endostatin [2], an 18-kD internal peptide of collagen XVIII.

Clinical Observations that Tumor Mass can Suppress Tumor Growth

Surgeons know that the removal of certain types of primary tumors may be followed by rapid growth of distant metastases [3–6]. In patients with melanoma, partial spontaneous regression of the primary tumor may be followed by rapid growth of metastases. Cancer biologists report similar phenomena in tumor-bearing animals. Some primary tumors may inhibit the growth, but not the number of remote metastases [7–12]. Partial removal of a tumor can increase the growth rate in the residual tumor [13]. Metastatic growth can suppress the growth of a primary tumor [14]. Many primary tumors can suppress the growth of a second tumor inoculum [15–17]. This "resistance" to a second tumor challenge is inversely proportional to the size of the tumor inoculum, and directly proportional to the size of the first tumor. A threshold tumor size is necessary for the inhibitory effect to occur. Furthermore, some tumors can inhibit a secondary tumor of a different type [15]. For reviews see [18] and [19].

Traditional Explanations for Inhibition of Tumor Growth by Tumor Mass

Three hypotheses have been proposed to explain how certain tumors can inhibit their metastases, or how tumor mass in general can inhibit tumor growth: (1) "concomitant immunity," a primary tumor is thought to induce an immunological response against a metastasis, or against a secondary tumor; (2) a primary tumor depletes available nutrients; and (3) a primary tumor produces antimitotic factors that inhibit proliferation of tumor cells within a metastasis. However, none of these ideas had provided a molecular mechanism to explain how tumor growth is suppressed by tumor mass [18]. (For review, see [1, 20]).

Angiostatin

An Alternative Hypothesis: Certain Primary Tumors Release Angiogenesis Inhibitors into the Circulation

In 1994, we proposed an alternative hypothesis, namely that some primary tumors upon reaching a threshold size, could release angiogenesis inhibitors into the circulation [1]. Because this idea was counterintuitive, a brief background of its origin may be helpful. In 1990, Harold Brem, a post-doctoral fellow in my laboratory noticed a puzzling phenomenon. He was testing a novel angiogenesis inhibitor AGM-1470 (subsequently called TNP-470, a synthetic analogue of fumagillin [21]) in tumor-bearing mice [22]. When mice were treated with the angiogenesis inhibitor, the primary tumor underwent partial regression, but lung metastases grew more rapidly. In cancer patients, clinicians call this outcome a "mixed result." I hypothesized that perhaps the primary tumor was secreting antiangiogenic activity into the circulation, which, when taken together with the administered exogenous angiogenesis inhibitor, was sufficient to suppress growth of remote metastases. However, as the primary tumor was regressed below some threshold, I speculated that its production of "antiangiogenic activity" would decrease, so that at the dose being used the drug could not sustain suppression of metastatic growth. I did not think that 'antimitotic factors' from the primary tumor could explain growth suppression of metastasis, because tumor cells in the stable metastases, in the presence of the primary tumor or in growing metastases during regression of the primary

Department of Surgery, Harvard Medical School and Vascular Biology Program, Children's Hospital Boston, MA, USA

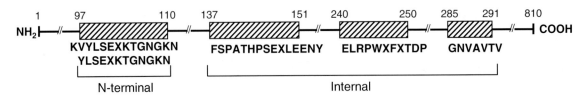

FIG. 12.1. Amino acid sequence of the angiostatin from human plasminogen digested with elastase. The N-terminal sequence and sequences of tryptic fragments of angiostatin purified from human plasminogen are noted. Unidentified amino acid residues are denoted by Xs. (From [1])

tumor or after its removal, all had a similar proliferation rate to each other and to the primary tumor. Furthermore, there was no evidence of immune cell infiltration in any tumors. Also, the animals were gaining weight, making 'nutritional depletion' an unlikely explanation.

When Brem returned to his surgical training, Michael O'Reilly took Brem's place in the laboratory. I suggested to O'Reilly that we try to isolate a putative angiogenesis inhibitor from tumor-bearing animals in which growth of metastasis was suppressed. Importantly, both Brem and O'Reilly were young surgeons who had seen patients in whom occult metastases grew explosively after a primary tumor was removed. While most colleagues in the lab thought we were on a wild-goose chase, or called the project "career knockout," once O'Reilly had isolated a growth suppressor of metastasis from serum and urine of tumor-bearing mice, biochemists in the lab helped him purify the protein to homogeneity (Yuen Shing, Rosalind Rosenthal, Marsha Moses, and Helene Sage [who was on sabbatical in the lab]). Yihai Cao carried out antibody depletion studies and William Lane of Harvard did the microsequencing [1].

Discovery of Angiostatin

An animal model was developed in which Lewis lung carcinoma was grown subcutaneously. After it had reached 1 to 2 cm³, it was surgically removed. The number of lung metastases increased 10-fold by day 13, compared to controls in which the primary tumor was not removed [23]. The lung metastases were highly neovascularized. Purification from blood and urine by heparin affinity chromatography yielded a 38 kD internal peptide of plasminogen, containing the kringle structures 1 through 4 (Fig. 12.1), which we named angiostatin [1].

The purified protein specifically inhibited endothelial cell proliferation in vitro, tumor angiogenesis in vivo, and tumor metastases. It was a 38-kD internal fragment of plasminogen, corresponding to kringles 1–4, which we named angiostatin [1] (Fig. 12.1). Endothelial cells treated with angiostatin in the absence of growth factors, increased their apoptotic index, without changing their proliferation index [24]. However, angiostatin inhibited proliferation of bFGF-stimulated endothelial cells, but did not inhibit proliferation of fibroblasts, epithelial cells, smooth muscle cells or tumor cells [1]. Endothelial migration and tube formation are also inhibited by angiostatin. Systemic administration of angiostatin purified from mouse urine significantly inhibited angiogenesis in lung metastases and restricted their growth to a microscopic size [1].

An antibody against the 1–2 kringle region of plasminogen depleted urine of angiostatin activity [1]. Recombinant angiostatin was orginally produced in *E. coli* [25], and subsequently in yeast [26]. Recombinant angiostatin potently inhibited growth of the primary Lewis lung carcinoma and other tumor types [27]. Gene transfer of cDNA for murine angiostatin into murine fibrosarcoma cells (T241) suppressed primary tumors by up to 78%. Even after removal of the primary tumors, ~70% of lung metastases inhibited their own angiogenesis and remained in a microscopic dormant state. In contrast, when tumor cells lacked the transfected angiostatin, or were transfected with vector or with mock plasmids, surgical removal of the primary tumor was followed by rapid growth of all lung metastases [28]. This experimental model was subsequently confirmed in other tumor-bearing animals [29] and in man [30].

Eradication of a primary Lewis lung carcinoma by curative local radiotherapy, known to generate angiostatin, was followed by rapid growth of lung metastases which killed all animals within 18 days [31]. The lungs of treated animals averaged >50 surface lung metastases, compared to 5 in the untreated animals, and lung weight was 300% greater in the treated than in untreated animals (Fig. 12.2). Administration of angiostatin prevented the growth of metastases after radiotherapy. Urinary levels of matrix metalloproteinase-2, one of the enzymes responsible for angiostatin processing in this tumor model, correlated with the viability and size of the primary tumor. This enzyme was significantly decreased in irradiated animals in which tumor was eradicated. This experiment suggests that combining angiogenesis inhibitors with radiation therapy may control distant metastases. The experiment also implies that it may someday be possible to employ urine biomarkers to determine which angiogenesis inhibitor (if any) should be administered to obtain optimal prevention of metastatic growth after radiotherapy.

Different forms of Angiostatin

Tumors generate angiostatin by a stepwise enzymatic cleavage of circulating plasminogen, beginning with urokinase from tumor cells that liberates plasmin from plasminogen, and ending with metalloproteinases 2 and 9 [32–36] (Fig. 11.3). Moreover, several different forms of angiostatin have been produced, depending on which cleavage sites in plasminogen were enzymatically targeted to produce

A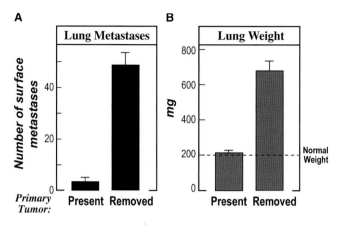

B

FIG. 12.2. Inhibition of the growth of metastases by the presence of a primary tumor. Thirty mice were implanted with Lewis lung carcinoma. When tumors were ~1,500 mm³ they were removed from half of the mice. Within 15 days of tumor removal, the number of surface lung metastases (**A**) and lung weight (**B**) had significantly increased, as compared with mice bearing an intact primary tumor. In mice with an intact primary tumor, lung weight (which correlates with tumor burden), was not significantly different from that of normal lung. (From [1])

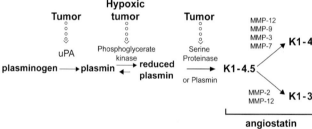

FIG. 12.3. Model of angiostatin formation by sequential enzymatic cleavage of plasminogen and plasmin. (From [36])

angiostatin. Plasminogen contains five kringle domains, each of which contains three characteristic disulfide bonds [37]. In the presence of plasminogen activators, plasminogen is cleaved to form plasmin [32]. Plasmin is reduced in the presence of phosphoglycerate kinase [36]. It is then cleaved to form angiostatin kringles 1–4.5 by plasmin or other serine proteases, and then differentially cleaved to either angiostatin kringles 1–3 or kringles 1–4 by the action of various metalloproteinases [36, 37]. These differences may explain the occasional different conclusions reached by different investigators who use different angiostatins. Furthermore, the kringle 4 domain contains two different glycosylation sites [37]. As a result, angiostatin fragments prepared from purified plasminogen glycoforms revealed that only the 2episilon glycoform inhibited proliferation of endothelial cells in vitro [37, 38].

Different Receptors for Angiostatin on the Endothelial Cell Surface

Although angiostatin (K1–4) was originally discovered as an internal fragment of plasminogen, angiostatin inhibited endothelial cell proliferation in vitro. However, plasminogen in the serum of tissue culture media did not appear to compete with angiostatin [1]. In 1995, Moser and Pizzo began a systematic study of endothelial surface binding sites for angiostatin [37]. Scatchard analysis of angiostatin K1–3 binding revealed an apparent K_d of 245 nM and 38,000 sites per human umbilical vein endothelial cell, compared to plasminogen, which revealed an apparent K_d of 158 nM and 870,000 sites per cell. The kinetics of plasminogen binding were consistent with its previously identified binding site, annexin II.

ATP Synthase

In 1999, Moser and Pizzo reported that ATP synthase is a receptor for angiostatin on the endothelial cell surface [37, 39, 40]. ATP synthase is found on the endothelial cell surface, despite the fact that it is also located on the mitochondrial inner membrane. Endothelial cell proliferation is inhibited by angiostatin when it binds to ATP-synthase. Others subsequently confirmed this finding [41] and also demonstrated expression of ATP synthase on the surface of some tumor cells. Venous endothelial cells, and especially endothelial cells in veins that drain tumors, live at lower levels of oxygen and pH than most other cells in the body. Therefore, ATP synthase on the endothelial cell surface may have a survival value for endothelial cells. Also, many of the enzymes and components of the mitochondrial electron transport chain and ATP synthesis generating mechanisms, are located on the plasma membrane of the endothelial cell [41, 42]. The average tumor extracellular pH (5.6–7.6) is lower and more variable than in normal tissues (7.2–7.6), yet tumor cells have a normal average intracellular pH [37, 43]. Angiostatin induces intracellular acidosis in endothelial cells and cell death in vitro, when these cells are incubated at low pH in the range of tumor extracellular pH [44]. Endothelial cells can maintain a relatively normal intracellular pH, even though the extracellular pH has been decreased to tumor-like conditions of pH 6.5 [44]. However, the addition of angiostatin to endothelial cells incubated at tumor-like conditions of pH ~6.5, causes a significant decrease of intracellular pH and death of these cells. In contrast, angiostatin appears to exert no deleterious effect on endothelial cells cultured at normal pH. Therefore, it has been suggested that angiostatin-mediated inhibition of proton flux via F_1F_0ATP synthase, may be a major mechanism of the increased sensitivity of endothelial cells to pH stress in the presence of angiostatin [37]. A specific monoclonal antibody directed against the β-catalytic subunit of ATP synthase, bound to the F(1) domain of ATP synthase with 25-fold higher affinity than angiostatin. It also inhibited activity of the F(1) domain of ATP [45]. The antibody disrupted tube formation and decreased intracellular pH in endothelial cells exposed to low pH. Both angiostatin and the anti-ATP syn-thase antibody inhibited angiogenesis in the low-pH environ-ment of the chicken chorioallantoic membrane, but neither inhibited angiogenesis in the corneal neovascularization assay. Such an antibody may be used in the antiangiogenic therapy of cancer. Therefore, it can be speculated that combinations of anti-VEGF angiogenesis inhibitors and angiostatin may have maximum anti-tumor efficacy if angiostatin is adminis-tered *before* the anti-VEGF inhibitor. The rationale for this idea is that certain angiogenesis inhibitors such as AZ2171 (AstraZeneca) can reduce vascular leakage and intratumoral pressure, i.e., by "normalization" of tumor vessels [46] lead-ing to increased blood flow [47] and temporary restoration of intratumoral oxygen and pH. Endostatin also potently blocks vascular leakage induced by VEGF [48]. Angiostatin admin-istered in combination with endostatin had synergistic anti-tumor activity [2, 49]. However, the effect of administering angiostatin before endostatin remains to be determined. Also, piceatannol and resveratrol, found in grapes and red wine, are non-specific inhibitors of ATP synthase [37]. Therefore, it also remains to be determined whether these compounds would synergize or counteract the antiangiogenic efficacy of angiostatin.

Angiomotin

In 2001, Lars Holmgren (a former post-doctoral fellow in the Folkman lab) and his associates discovered a protein which they named angiomotin because it facilitates motility, invasion and morphogenetic changes of endothelial cells that resem-ble activated endothelium in a tumor bed [50]. Angiomotin is preferentially expressed on endothelial cells. Treatment of angiomotin-transfected cells with angiostatin, resulted in up-regulation of focal adhesion kinase (FAK) activity and inhibition of endothelial migration and tube formation in vitro. Endothelial cell migration and tube formation were also blocked by a specific antibody to recombinant angiomo-tin [40], or when endothelial cells were engineered to express a truncated form of angiomotin (deletion of three amino acids in the carboxyterminus) [51]. Angiostatin had no effect on cells that were not expressing angiomotin. Angiomotin nor-mally regulates endothelial cell migration during embryonic angiogenesis [52]. Transgenic mice expressing this same mutated form of angiomotin, died at embryonic day 9.5 with absence of migration of endothelial cells into the neuroecto-derm and intersomitic regions [37, 53]. Human angiomotin expressed in endothelial cells stabilized established tubes in vitro and enhanced invasion by these endothelial cells [53]. When angiomotin was targeted by therapeutic antibodies, angiogenesis was inhibited in vivo [54].

The Hepatocyte Growth Factor (HGF) Receptor, c-met

Wajih and Sane reported that recombinant angiostatin kringles 1–3 inhibited HGF-induced phosphorylation of c-met (the HGF receptor), and also inhibited downstream signaling mediators

Akt and EK1/2 in human endothelial cells in vitro [37, 55]. Other receptor tyrosine kinase activators such as VEGF, bFGF and IGF-1 were not blocked by angiostatin in the phorphoryla-tion of Akt or ERK1/2. These results, taken together with the fact that plasminogen had no effect, indicates that angiostatin's effect was specific to c-met. Furthermore, HGF inhibited the binding of angiostatin to human endothelial cells, and angio-statin inhibited HGF-induced proliferation of these endothelial cells. Binding of angiostatin to c-met was dose-responsive and saturable with an apparent K_d of ~5 nM and 12,044 sites per cell. Exelixus 880, is an angiogenesis inhibitor currently in Phase II clinical trials for the treatment of cancer. It inhibits c-met activ-ity. It can be speculated that angiostatin and Exelixus 880 may synergize each other's antiangiogenic activity.

$\alpha_v\beta_3$ Integrin and CD26

The antiangiogenic effect of angiostatin has also been attributed to its binding to $\alpha_v\beta_3$ integrin [56] and to CD26 [38].

NG2 Proteoglycan

NG2 proteoglycan is a chondroitin sulfate that is upregulated in the tumor vascular bed. It binds angiostatin [1–3] and plas-minogen with dose-responsive and saturable kinetics [57].

Despite the multiple binding sites reported for angiostatin, at least two of which are functional, no single target can as yet account for all of the anti-endothelial and antiangiogenic activities of angiostatin. This is not dissimilar from the history of plasminogen, its affinity for other coagulation proteins, and its multiple binding sites on endothelium.

Nucleolin

Recently, it has been reported that nucleolin, a nuclear protein, is expressed on the surface of proliferating human microvascular endothelial cells, but not on the surface of quiescent endothe-lium [58, 59]. In endothelial cells in an angiogenic vascular bed, the cell surface nucleolin binds endostatin and transports it to the nucleus, where endostatin inhibits phosphorylation of nucleolin. Phosphorylation of nucleolin induced by VEGF or bFGF has been reported to be essential for cell prolifera-tion. Furthermore, endostatin does not inhibit proliferation of many types of tumor cells per se, possibly because, while they express nucleolin on their surfaces, they do not internalize it in the presence of endostatin. The heparin binding sites on nucleolin were critical for endostatin affinity. Increasing con-centrations of exogenous heparin dissociated the binding of endostatin to nucleolin. Will the endostatin–nucleolin connec-tion lead to the uncovering of a mechanism for the biphasic, U-shaped, dose-efficacy of endostatin's bioactivity [60]? This is also a common property of other angiogenesis inhibitors, and was first described for interferon α [61]. Is the increased expression of endostatin that is mediated by p53 [62, 63] also regulated by nucleolin? High circulating levels of endostatin do not delay angiogenesis in wound healing or pregnancy, as for example in individuals with Down syndrome, or in animals receiving endostatin therapy [64] (see below).

Mechanisms of Action of Angiostatin

During embryonic development, angiostatin appears to have no effect on embryonic vasculogenesis, despite the presence of known angiostatin binding sites [65]. However, angiostatin inhibits vascular endothelial growth factor (VEGF)-induced sprouting from embryoid bodies in vitro, indicating a selective inhibitory activity on sprouting angiogenesis. Angiostatin (and endostatin) inhibited bFGF- and VEGF- induced migraton of primary human microvascular endothelial cells, but without affecting intracellular signaling pathways known to regulate endothelial cell migration and proliferation. These include phospholipase C-γ, Akt/PKB, p4/42 mitogen associated kinase (MAPK), and p38 MAPK and p21- activated kinase (PAK) [66].

Angiostatin has no effect on growth factor-induced signal transduction, but leads to an RGD-independent induction of the kinase activity of focal adhesion kinase. This suggests that the biological effects of angiostatin relate in part to subversion of adhesion plaque formation in endothelial cells [24]. Human angiostatin (hK1–3), also inhibits monocyte/macrophage migration via disruption of the actin cytoskeleton [67]. Because macrophages infiltrate tumors as well as sites of non-neoplastic angiogenesis, immobilization of macrophages by angiostatin, may contribute to its antiangiogenic activity.

Angiostatin's presence in the eye has been suggested to be a mechanism of suppression of neovascularization in the normal eye. For example, angiostatin was detected by electron immunocytochemistry in the corneal epithelium of the cat, dog, horse, pig and rat, although not in that of cow [68]. In the diabetic eyes of mice [69] and rats [70], angiostatin significantly reduced VEGF expression and reduced vascular leakage, but did not have this effect in normal eyes.

Angiostatin(4.5) induces endothelial cell apoptosis in vitro, by a process that appears to involve both the intrinsic and extrinsic apoptosis pathways [71].

Healthy human platelets contain angiostatin, which is released in active form during platelet aggregation. Platelet-derived angiostatin has the capacity to inhibit angiogenesis [72]. When this finding is taken together with the recent discovery that angiogenesis regulatory proteins, both proangiogenic and antiangiogenic are segregated into separate α granules in platelets [73, 74], it can be speculated that angiostatin may contribute to down-regulation of angiogenesis at the end stages of wound healing.

Angiostatin is increased by 62% in diabetic internal mammary arteries collected from patients undergoing coronary artery bypass surgery [75]. This increased angiostatin is associated with increased up-regulation of matrix metalloproteinases-2 and -9 by these diabetic vessels. Non-diabetic arteries did not release metalloproteinases-2 and -9. VEGF expression in these vessels was reduced by 48%. These results may explain why angiostatin is negatively associated with coronary artery

growth in diabetics [76]. However, angiostatin negatively regulated endothelial-dependent vasodilation in rat aortas and arterioles by allowing endothelal cells to generate O2 on activation [77].

Endothelial cell expression of p53 was increased by angiostatin K1–3, K1–4, and K1–4.5 [78]. Thus, angiostatin may induce an antiangiogenic cascade, because increased expression of p53 controls tumor angiogenesis by: (1) increasing expression of three endogenous angiogenesis inhibitors, endostatin, tumstatin [62], and thrombospondin-1 [79]; and (2) by downregulating expression of hypoxia-inducible factor-1 α (HIF-1 α) [80], suppressing expression of VEGF [81], and suppressing bFGF-binding protein [82]. Angiostatin also decreases activation of AKT, enhances FasL-mediated signaling pathways [78], and induces increased expression of p53 [83]. There are very few molecules that induce p53. If this latter finding is confirmed, it will be of interest because p53 also maintains normally elevated levels of endostatin and tumstatin [62, 63]. (For a review of mechansims of angiostatin see [84] and for a review of structure - activity relationships of angiostatin, see [85].)

Anti-inflammatory Activity of Angiostatin

In *acute* inflammation, the microvasculature is vasodilated, which accounts for increased blood flow, redness, pain, and swelling, at the inflammatory site. In *chronic* inflammation, there is continuous neovascularization. New blood vessels provide an expanding conduit for the delivery of inflammatory cells to the site [86]. Angiostatin has been shown to suppress chronic inflammation by inhibiting leukocyte recruitment [87]. Angiostatin (k1–4) specifically binds to leukocyte β1- and β2-integrins. This inhibits leukocyte adhesion to extracellular matrix proteins and to endothelium. Leukocyte transmigration through endothelium in vitro is also suppressed by this mechanism. In addition, through its interaction with Mac-1, angiostatin reduces activation of the proinflammatory transcription factor NFκB.

In a mouse model of cystitis induced by cyclophosphamide chemotherapy, angiostatin prevented this inflammatory side effect [88]. Both endostatin and TNP-470 (an antiangiogenic synthetic analogue of fumagillin [89] also suppressed cyclophosphamide cystitis, suggesting that certain angiogenesis inhibitors could reduce chemotherapy-related side effects, while adding an anti-tumor effect.

Angiostatin also suppresses experimental arthritis. Angiostatin gene therapy injected into the knee joints of mice with collagen-induced arthritis led to significant improvement of the clinical, radiological, and histological findings in these mice [90].

Induction of Angiostatin In Vivo

Angiostatin is one of 28 "endogenous" angiogenesis inhibitory molecules in the stroma and/or in the circulation [91, 92].

Biotechnology and pharmaceutical companies have begun to search for small molecules that could potentially elevate levels of endogenous antiangiogenic proteins [93]. A clear example is that when N-acetyl-cysteine was administered for up to 8 weeks to nude mice bearing human breast cancer (MDA-MB-435), there was significant endothelial cell apoptosis, marked reduction of microvessel density, and subsequent tumor cell necrosis. Elevated levels of angiostatin accumulated in the tumors, but not in untreated controls. Moreover, in vitro studies showed that angiostatin formed in endothelial cultures in a VEGF-dependent and N-acetyl-cysteine dependent manner, a process that requires endothelial surface plasminogen activation [94]. Cyclooxygenase-2 overexpression inhibits cathepsin-D mediated cleavage of plasminogen to angiostatin [95]. A cathepsin D inhibitor, pepstatin A, also inhibited the cleavage of plasminogen to angiostatin. Therefore, it is possible that other cyclooxygenase inhibitors such as celebrex, could increase angiostatin levels in vivo.

Angiostatin Gene Therapy

There is a large literature on angiostatin gene therapy. This therapy can induce significant and often dramatic inhibition of growth of experimental tumors. Salient examples that illustrate different therapeutic approaches are summarized here. In this section, I have also summarized examples of angiostatin therapy of *non*-neoplastic diseases, to emphasize that angiostatin often has a therapeutic effect on other diseases. For example, it does not cause hypertension, but in fact can return kidney function toward normal in diabetic animals. Thus, endogenous angiogenesis inhibitors when used for tumor therapy are likely to have less side effects than angiogenesis inhibitors that are antibodies or aptamers designed to block VEGF or other proangiogenic proteins.

When murine angiostatin cDNA was stably expressed in human hepatocellular carcinoma implanted in nude mice, angiogenesis as determined by microvessel density was reduced by 49% compared to untreated controls. As a result, tumor growth rate was reduced by 78%, and tumor mass by 34%. There were no side-effects [96].

Experimental Anti-tumor Therapy

An angiostatin–endostatin fusion protein expressed by a prostate-restricted replicative adenovirus, co-administered with a prostate-restricted adenovirus, resulted in complete durable regression of seven out of eight treated androgen-independent cancers in mice [97]. Tumor growth in vivo was inhibited by 95% in Morris hepatomas overexpressing angiostatin in experimental animals [98]. The authors also noted increased tumor cell apoptosis, but increased blood flow as measured by PET in the residual tumor and increased microvessel density, which they interpreted as increased vascularization. However, quantification of microvessesl density in a treated tumor can be misleading [99]. Microvessel density is dictated by *intercapillary* distance. If a treated tumor undergoes significant regression with associated dropout of

microvessels, in some tumors the few remaining microvessels become more closely packed, microvessel density then increases, and this can be misinterpreted as "increased vascularization." When the treatment is an angiogenesis inhibitor which decreases vascular permeability (e.g., angiostatin), intratumoral tissue pressure drops and blood flow can temporarily increase. Jain has termed this process "normalization" of residual tumor vessels [46].

Administration of an adeno-associated virus vector encoding mouse angiostatin into the portal vein caused stable transgene expression of angiostatin for up to 6 months, and significantly prolonged survival in mice bearing primary and metastatic lymphoma in the liver [100]. Growth of neovessels was inhibited significantly in the tumor beds of the treated animals.

In another report, co-administration of plasmids encoding mouse angiostatin (K1–3) together with mouse endostatin, inhibited growth of mouse melanoma by 75% compared to 46% and 52% inhibition for the angiostatin gene and endostatin genes alone [101]. The co-treatment inhibited pulmonary metastasis by 80%, compared to 68% and 71% for angiostatin and endostatin alone.

A recombinant adenoviral vector of angiostatin (k1–3) significantly inhibited growth of a murine colorectal carcinoma, and a human hepatocellular carcinoma in mice, but was ineffective against human colorectal carcinoma in mice. This angiostatin also inhibited Lewis lung carcinoma in immunocompetent mice, and was even more effective in athymic mice [102]. An adenoviral vector encoding angiostatin and endostatin gave complete protection against recurrence of mouse tumor xenografts, whereas treatment with either gene alone gave only moderate protection [103]. Adeno-associated viral gene delivery of angiostatin produced sustained serum levels in mice for up to 6 months after a single injection. Growth of murine melanoma and Lewis lung carcinoma were significantly inhibited and survival was significantly increased. Antitumor efficacy was consistently observed when angiostatin serum levels were 15–20 ng/ml, but the effect was minimal when the levels were lower or higher than this range [104].

Adenovirus-delivered angiostatin (ADK3) delivered as a single intratumoral dose in combination with a single intravenous injection of docetaxel (Taxotere), caused tumor regression in *all* human prostate tumors growing in mice and *complete* tumor regression in 40–83% of tumors, but not with either therapy alone [105].

Angiostatin Gene Therapy in Ocular Neovascularization

Angiogenesis inhibitors, developed from principles of antiangiogenic therapy in cancer research, have recently been approved by the FDA for the treatment of age-related neovascular macular degeneration [106]. Therefore, in a review such as this, it is informative to know that the first successful report of antiangiogenic therapy of murine proliferative retinopathy was with lentivirus-mediated expression of angiostatin [107].

Angiostatin levels are abnormally low in the kidneys of streptozotocin-induced diabetic rats. Adenovirus-mediated delivery of angiostatin significantly alleviated albuminuria and attenuated the glomerular hypertrophy in diabetic rats [108]. Angiostatin significantly downregulated VEGF and TGF β1, two major pathogenic factors in diabetic nephropathy of diabetic kidneys that are induced by high glucose. Angiostatin also effectively inhibited the high glucose- and TGFβ1-induced overproduction of pro-inflammatory factors and extracellular matrix proteins via blockade of the Smad signaling pathway. These results suggest that the low levels of angiostatin in diabetic kidneys may contribute to pathologic changes such as inflammation and fibrosis.

Angiostatin (and endostatin) gene therapy, administered into the subretinal space of mice with laser-induced choroidal neovascularization, significantly inhibited hyperpermeability and angiogenesis. There were no deleterious effects on neurosensory retinal cells, or on mature retinal vasculature non-lasered eyes [109, 110]. Angiostatin significantly reduced vascular leakage in diabetic macular edema in rats with streptozotocin-induced diabetes. This was mediated in part by blockade of VEGF overexpression [111].

Production of Recombinant Angiostatin

Small quantities of murine angiostatin were originally (1994) isolated and purified from the urine of tumor-bearing mice [1]. Larger scale production of recombinant human angiostatin was subsequently produced in *E. coli* and in yeast (*Pichia pastoris*). Fermentation methods have continued to be improved. A recent advance for production of angiostatin by *Pichia pastoris* has been the addition of controlled methanol and glycerol feeding, guided by oxygen sensing. [112]. Cell density reached 150 g/l and angiostatin reached 108 mg/l after an expression period of 96 h. The mean specific angiostatin productivity was improved to 0.02 mg/(g × h). The apparent cell yield on methanol and glycerol were respectively 0.69 g/g and 0.93 g/g higher than without using this feeding strategy.

Delivery of Angiostatin

Angiostatin can be administered parenterally to animals and to patients, but by a variety of different methods. For example, cells genetically engineered to produce and secrete angiostatin have been encapsulated in alginate beads which maintain viable cells for up to 6 months. Angiostatin is continuously released [113].

Anti-tumor Activity of Angiostatin Protein

A wide variety of different tumors have been treated by angiostatin in animals. Low dose (100 ml per day of 0.1 mg/ml), but not high dose (100 ml of 0.3 mg/ml) angiostatin effectively decreased the number and size of hepatic micrometastases by three different lines of murine melanoma implanted into the eye [114]. Angiostatin dose-efficacy follows a biphasic, (U-shaped) curve in which lower doses are commonly more effective than higher doses [115]. U-shaped dose-efficacy was first demonstrated for the endogenous angiogenesis inhibitor, interferon-α [61], and subsequently for other molecules that inhibit angio-

genesis, including thrombospondin-1 [116], endostatin [60, 117], 2-methoxyestradiol [118], and rosiglitazone [119]. Angiostatin inhibited bone metastases from human breast cancer in nude mice. In addition to its antiangiogenic activity, angiostatin also inhibited bone destruction through a direct anti-osteoclastic activity [120]. Angiostatin also inhbited the growth of murine squamous cell carcinoma in vivo [121].

Certain Tumors that were not Inhibited by Angiostatin

A human neuroblastoma cell line produced tumors in mice that did not respond to an adenoviral vector of angiostatin fused to albumin [121].

Adverse Effects of Angiostatin

In experimental colon surgery in mice, angiostatin appeared to decrease neovascularization and increase inflammation at anastomotic sites a few days after surgery. However, at 1 week after surgery there was no histological difference between angiostatin treated mice and untreated controls [122].

Physiological Angiogenesis

Angiostatin has generally had little or no effect on physiological angiogenesis. For example, it did not suppress angiogenesis in a human placenta vein disk implanted in a fibrin-thrombin clot [123]. However, angiostatin inhibited angiogenesis in the chick embryo chorioallantoic membrane from day 7 to 9, but not after day 11 [124].

Clinical Studies

Angiostatin was discovered because it was generated in mice bearing Lewis lung carcinoma [1]. Subsequently, certain human tumors were found to generate angiostatin by similar enzymatic cleavage of plasminogen as animal tumors. For example, some human ovarian cancers generate angiostatin, where it has been used as a prognostic indicator of survival [125]. The presence of angiostatin expression and absence of VEGF expression are favorable prognostic indicators in ovarian cancer.

Crystallography

The crystallographic structure of human angiostatin derived from plasminogen has been refined to 2.3 A resolution [126].

Endostatin

Discovery of Endostatin

Endostatin was discovered three years after the discovery of angiostatin, also by Michael O'Reilly in the Folkman laboratory, using a similar rationale and strategy as for angiostatin. A murine hemangioendothelioma implanted into mice

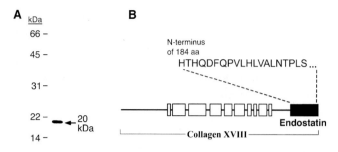

FIG. 12.4. Purification of a 20-kD inhibitor (**A**) of endothelial proliferation (endostatin) from conditioned media of murine hemangioendothelioma, (EOMA). Amino acid sequence (**B**) showing the relationship of endostatin in the C-terminal fragment of collagen XVIII. (From [2])

suppressed growth of a second tumor implant. The conditioned medium from an in vitro culture of murine hemangioendothelioma cells reversibly inhibited proliferation of capillary endothelial cells [2]. A 20-kD protein was isolated and completely purified (Fig. 12.4). It was named endostatin. Amino acid analysis revealed that the precursor of endostatin is the NC1 domain of collagen XVIII, which contains three globular endostatin domains [127].

Endostatin Affinity for Zinc

Crystallography revealed one atom of zinc per endostatin molecule [128]. Replacement of histidines 1 and 3 by alanines prevented binding of zinc and eliminated the anti-tumor activity of endostatin [129]. The first recombinant endostatin that elicited strong anti-tumor activity and tumor regression in mice was expressed in *E. coli* and administered as a poorly soluble suspension [2]. Under these conditions, the protein retained its full anti-tumor activity, indicating that a fold structure was not required. A 27 amino acid peptide corresponding to the NH2-terminal domain of endostatin retained the anti-tumor activity of endostatin, as well as its inhibitory activity against endothelial cell migration and its ability to inhibit vascular leakage [130]. The peptide contains three histidines that are responsible for zinc bonding. Mutations of the zinc binding histidines abolished its anti-tumor and anti-migration activities, but not its anti-permeability activity.

Binding of Specific Heparan Sulfates to Endostatin

Like angiostatin, endostatin was first purified by heparin affinity chromatography [2]. It was found to contain a discontinuous sequence of 11 basic arginine residues, clustered on the protein surface, which are highly conserved between species [131]. Substitutions of alanine for critical arginine residues impaired both heparin affinity and bioactivity [132,133]. When binding of endostatin to heparan sulfate from microvascular and macrovascular endothelium was analysed, two types of endostatin-binding regions were identified: one comprising sulfated domains of five or more disaccharides in length,

enriched in 6-O-sulfate groups, and the other containing long heparinase I-resistant fragments [127]. The binding of endostatin to heparan sulfate appears to depend critically on 6-O sulfates and not on 2-O-sulphates. Chemical depletion of only 56% of the 6-O sulfate groups of heparan sulphate virtually eliminates endostatin binding activity. Endostatin's heparin-binding site has also been shown by others to be essential for inhibition of angiogenesis, and enhances in situ binding to capillary-like structures in bone explants [134].

The mobilization of endostatin from collagen XVIII is mediated in part by an elastase like activity which cleaves the NC1 domain of collagen XVIII at the Ala-His site [135]. A prior step requires cathepsin [136], and a metal dependent early step [135].

Crystal Structure

The crystal structure of human endostatin was determined with a resolution of 3.0 A [128] and of murine endostatin at 1.5 A [131]. Atomic absorption spectroscopy of human endostatin indicted that a zinc site is formed by three histidines at the N terminus, residues 1, 2, and 11, and an aspartic acid at residue 76. Human and murine endostatin both contain an arginine rich surface that is a potential heparin binding site. Dimerization of endostatin may be zinc-dependent. Zinc content may not have been quantified in the murine endostatin because analysis was carried out at a low pH.

Mechanisms of the Antiangiogenic Activity of Endostatin

Biphasic U-Shaped Dose-Response Curve of Endostatin

Endostatin and other angiogenesis inhibitors have revealed a dose-response curve that is biphasic and U-shaped instead of linear. Optimum doses in vivo are lower than maximum doses [60] (Fig. 12.5).

Molecular Mechanisms

Endostatin binds to the integrin $\alpha_5\beta_3$ on endothelial cells [137–139]. Endostatin bioactivity also appears to be dependent on binding to E-selectin [140]. Endostatin also blocks activity of metalloproteinases 2,9, and 13 [141]. Shichiri and Hirata demonstrated that endostatin initiated intracellular signaling that down-regulated a set of growth-associated genes in a wide range of endothelial lineage cells [142]. Abdollahi et al., in collaboration with the Folkman lab, analyzed gene expression in fresh human microvascular endothelial cells incubated at different times and with different concentrations of endostatin employing custom microarrays covering over 90% of the genome [143, 144]. They reported that ~12% of all genes are significantly regulated in human microvascular endothelial cells exposed to endostatin. The upregulated genes as a group included angiogenesis inhibitors, such as thrombospondin-1, maspin, and sphingomyelinase. Proangio-

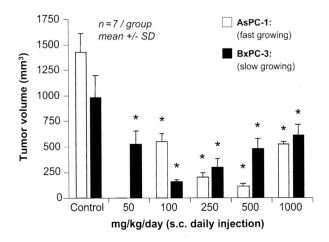

FIG. 12.5. Biphasic, U-shaped dose response curve of endostatin. The dose and efficacy of certain endogenous (interferon alpha) and synthetic (rosiglitazone) angiogenesis inhibitors follows a biphasic, U-shaped relationship. (From [60])

TABLE 12.1. Endogenous inhibitors of angiogenesis (from 139).

Matrix derived
Arresten
Canstatin
Collagen fragments
EFC-XV
Endorepellin
Endostatin
Fibronection fragments
Fibulin
Thrombospondin-1 and -2
Tumstatin
Non-matrix derived

Growth factors and cytokines
Interferons
Interleukins
PEDF
Platelet factor-4
Other
Angiogstatin
Antithrombin III (cleaved)
Chondromodulin
2-Methoxyestradiol
PEX
Plasminogen Kringle 5
Prolactin fragments
Prothrombin Kringle 2
sFlt-1
TIMPs
Troponin I
Vasostatin

genic genes were downregulated by endostatin, such as VEGF receptor, bFGF receptor, and hypoxia inducible factor-1 (HIF-1 α) (Table 12.1). On hundreds of genetic pathway fronts, endostatin suppresses one process—angiogenesis. Therefore, one would not expect the molecular mechanism of endostatin to be reduced to a single gene response, or even to a few

signaling cascades. More surprising is that there are so many routes to angiogenesis suppression, and that a single molecule can regulate these routes en masse to control the switch to the angiogenic phenotype in tumors. The majority of the more than 1,000 papers published to date on endostatin reveal that it suppresses mainly pathological angiogenesis. Endostatin appears to have little or no activity against wound healing, reproduction or development.

Endostatin and also K5 angiostatin induce autophagy in addition to apoptosis in endothelial cells [145]. Autophagy is a cellular degradation pathway for the clearance of damaged or superfluous proteins and organelles. The recycling of these intracellular constituents also serves as an alternative energy source during periods of metabolic stress to maintain homeostasis and viability [146]. A common feature of the two treatments was the up-regulation of Beclin 1 levels, leading to alterations in the Beclin 1-Bcl-2 complex. Interestingly, angiogenesis inhibitor-induced autophagy in endothelial cells was independent of nutritional or hypoxic stress and was initiated even in the presence of endothelial-specifc survival factors such as VEGF. Interfering with the autophagic response by knocking down Beclin 1 levels dramatically increased apoptosis of endothelial cells. Therefore, it can be speculated that autophagy in vascular endothelium may be a novel mechanism of evasion or escape from endostatin antiangiogenic therapy. Ramakrishnan et al. [145] have proposed that the autophagic response may be a novel target for enhancing the therapeutic efficacy of angiogenesis inhibitors.

Endostar is a new recombinant human endostatin expressed and purified in *E. coli*. Endostar contains an additional nine-amino acid sequence (MGGSHHHHH) at the N-terminus which simplifies purification and improves stability of the protein. It was approved for the treatment of lung cancer in China in 2005. Endostar suppresses the VEGF-induced tyrosine phosphorylation of KDR/Flk-1 (VEGFR-2) as well as the overall VEGFR-2 expression and activation of ERK, p38, MAPK, and AKT in human endothelial cells [147].

Endostatin binding to human ovarian cancer cells inhibits cell attachment to the peritoneum (but not cell migration) [148]. Down-regulation of the $\alpha_5\beta_1$ integrin by siRNA, abrogated the binding of human ovarian cancer cells and human endothelial cells to endostatin. This finding supports previous evidence that $\alpha_5\beta_1$ has high affinity for endostatin.

The role of nitric oxide in endostatin's antiangiogenic activity is, at this writing, confusing. For example, some investigators find that endostatin produced acute release of nitric oxide from endothelial cells in vitro [149], while others report that endostatin downregulates nitric oxide by its dephosphorylation of endothelial nitric oxide synthase [150]. However, these experiments mainly employ a single cell in vitro system or aortic rings in vitro. Therefore, it is difficult to predict the end result of endostatin's effect on nitric oxide in a given tissue, or in the circulation, because endothelial cells, myocytes, platelets and other cells all produce nitric oxide and many cell types express nitric oxide synthase.

TABLE 12.2. Tumors Inhibited by Recombinant Endostatin Protein.

Murine tumors

Human tumors in mice
Ovarian
Acute myeloenous leukemia (chloroma)
Colorectal carcinoma
Spontaneous mammary carcinoma
B-16 melanoma
Transplantable mammary carcinoma
Hepatoma
Lung Adenocarcinoma
Lewis lung carcinoma
Rat glioma in brain (continuous endostatin + PKC
Murine colorectal liver metastases
B-16 melanoma + endostatin fusion with angiostatin
Human myeloid leukemia in SCID rats
Lewis lung carcinoma
Rat carcinogen induced mammary cancer
Pancreatic insulinoma
Laryngeal squamous cell carcinoma
Glioblastoma (U87)
Prostate carcinoma (PC3)
Neuroblastoma
Testicular carcinoma
Breast carcinoma
Head and neck squamous cell carcinoma
Kaposi's sarcoma
Pancreatic carcinoma
Human non-small cell lung cancer
Human pancreatic carcinoma
Brain tumors (U87)
Non-Hodgkin lymphoma (high grade)
Renal cell cancer
Bladder cancer
Murine metastatic tumors
Lung adenocarcinoma (completely inhibited)

Experimental anti-cancer activity

Endostatin significantly inhibits the growth of >65 different tumor types (Table 12.2), and targets angiogenesis regulatory genes on more than 12% of the human genome [143]. Yet, it is the least toxic anti-cancer drug in mice (and in humans) [151].

Endostatin Protein Therapy

Reports selected from more than 100 publications, which show significant inhibition of growth of animal and human tumors and their metastases, illustrate the broad spectrum of anti-cancer activity of endostatin. All reports emphasize the lack of toxicity. Tumor inhibition ranged from 47% to 91% by endostatin doses of 10 mg/kg/day to 100 mg/kg/day [151]. For example, human ovarian cancer was inhibited by 73% by day 41 [152]. Human brain tumors in the brains of nude mice were inhibited by 74% by direct microinfusion at 2 mg/kg/day) accompanied by a significant increase in survival [153]. Human breast cancer in nude mice was inhibited by 80% when treated with a fusion protein of endostatin (at 5 mg/kg/day), compared to only 60% inhibition for endostatin alone [154]. This report illustrates how increasing the half-life of endostatin increases its efficacy.

Human pancreatic cancer that was p53−/− regressed almost completely (97% inhibition of tumor growth) when endostatin was administered *continuously* by a micro-osmotic pump in the peritoneal cavity, but was inhibited by only 66% when the same dose (20 mg/kg/day) was injected into the peritoneal cavity as a *bolus* once/day [155, 156]. These experiments illustrate the significantly increased efficacy of continuous versus bolus dosing of endostatin. Continuously elevated levels of circulating endostatin achieve optimum inhibition and regression of different tumor types (Fig. 12.6). In contrast, when human testicular cancer was treated by continuous administration of endostatin at *one-half the dose* (10 mg/kg/day) that was used for pancreatic cancer, and from a micro-osmotic pump implanted *subcutaneously* instead of in the peritoneal cavity, there was no effect on growth of the primary tumor, nor on metastases that occurred in all animals by 6 months.

Also, carboplatin alone or thrombospondin-1 alone had no effect on a primary tumor or its metastases. However, a combination of endostatin plus thrombospondin-1, or a combination of endostatin plus carboplatin, prevented all metastases, significantly inhibited primary tumors, decreased tumor vascularity,

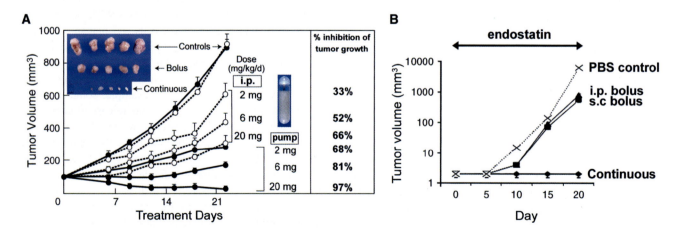

FIG. 12.6. Continuous administration of endostatin provides more effective anti-tumor activity than repeated bolus doses. For example, (**A**) a p53−/− human pancreatic cancer is inhibited by 66% when a dose of 20 mg/kg/day of endostatin is administered once a day as a bolus into the peritoneal cavity. In contrast, the same dose administered continuously over 24 h (**B**), by infusion from an implanted micro-osmotic pump, results in almost complete tumor regression and 97% inhibition of tumor growth. (From (**A**) [155]; (**B**) [156])

decreased tumor cell expression of VEGF-A and increased tumor cell apoptosis [157]. These results illustrate another important principle of antiangiogenic therapy: a tumor refractory to three drugs administered as single agents, can respond to a *combination* of two angiogenesis inhibitors, or to a combination of an angiogenesis inhibitor and a cytotoxic chemotherapeutic agent. Endostatin therapy at 20 mg/kg/day inhibited the primary tumors of a murine lung cancer growing subcutaneously in mice, and completely inhibited lung metastases.

Another general rule about antiangiogenic therapy is illustrated by endostatin therapy of a pair of human pancreatic cancers in SCID immunodeficient mice. A slowly growing pancreatic cancer (BXPC3) was inhibited by 91%. In contrast, a rapidly growing variant of this cancer was inhibited by only 69% at the same dose. Endostatin significantly inhibited microvascular density by 66% for both tumors ($P < 0.001$ for both tumors) [158]. These results indicate that slowly growing tumors are more responsive to antiangiogenic therapy than rapidly growing tumors, i.e., just the opposite of cytotoxic therapy. A similar result was observed with a pair of human bladder cancers, one of which grows 10 times faster and is more highly vascularized than the other. However, the slowly growing tumor that was less well vascularized was the most responsive to antiangiogenic therapy [159].

Endostatin Improves Radiotherapy

Several reports show that endostatin can potentiate the effect of radiotherapy of cancer. Recently Itasaka et al. showed that a mechanism by which endostatin improves radiotherapy is that it blocks revascularization of tumors after irradiation [160].

Endostatin Gene Therapy

Endostatin appears to be an ideal candidate for gene therapy. It is a highly conserved protein, being found in evolution as early as *C. elegans*. It has virtually no toxicity in animals or in humans, including 4 patients who received daily endostatin protein subcutaneously for >3.5 years. In transgenic mice overexpressing endostatin in vascular endothelium, only a small increase of circulating endostatin of approximately 1.6-fold above normal, slowed tumor growth by 300% to 400% [139], analogous to the protection against cancer in individuals with Down syndrome. In more than 60 reports since the first report of endostatin in 1997, various forms of endostatin gene therapy have significantly inhibited growth of many different primary tumors and their metastases. In animals, inhibition was up to 86–91% reduction in tumor volume and/or complete prevention of metastases. Lowest inhibition was ~40–45% (Table 12.3). Some examples of the efficacy of gene therapy are summarized here. When murine endostatin was transfected into murine renal carcinoma or human colon cancer cells so that endostatin was continuously secreted, flank tumors were inhibited by 73–91% and liver and lung metastases were inhibited or completely prevented [161].

In another study of human glioblastoma xenografts in nude mice, a combination of three angiogenesis inhibitors was administered by intratumoral injection of plasmids containing

TABLE 12.3. Tumors Significantly Inhibited by Endostatin Gene Therapy.

Murine primary tumors
Renal cell carcinoma
Brain tumors
Renal cell carcinoma
Breast cancer and brain tumor (FM3A P-15) metastasis
Breast cancer (Mid-T2-1)
Breast cancer (spontaneous)
Lewis lung carcinoma
Leukemia (L1210)
Myeloproliferative disease (resembling human chronic myelogenous leukemia)
Melanoma (K1735)
Melanoma (B16F10)
Bladder MBT-2
Colon cancer (colon 26)
Colon adenocarcinoma MC38
Hepatocarcinoma (H22)
Hepatoma (Hepa1c1c7)
Hepatocarcinoma
Melanoma (B16F10) (and metastases)
Spontaneous tongue carcinoma
Spontaneous breast cancer in C3(1)/T mice
Mammary carcinoma MCa-4
Brain tumor
Murine mammary ascites (TA3)
Neuroblastoma NXS2
Murine pulmonary metastases
Fibrosarcoma
Fibrosarcoma (NFsa Y83)
Melanoma (B16F10)
Rat tumors
Morris hepatoma
Hepatoma (orthotopic)
Gliosarcoma (9 L)
Osteosarcoma
Hamster
Pancreatic cancer (orthotopic) and liver metastases
Human tumors in mice
Colon cancer (SW620)
Colorectal cancer (HT29)
Colorectal cancer (HT29)
Colorectal
Colorectal advanced stage IV (T3N1M1)
Colorectal cancer (LoVo)
Glioblastoma
Lung cancer
Non-small-cell lung cancer(KNS 62) (& metastases)
Hepatocellular carcinoma
Hepatocellular carcinoma (BEL-7402)
Hepatocellular carcinoma Hep3B
Hepatocellular carcinoma HepG2
Hepatocellular carcinoma HepGH
Hepatocellular carcinoma HepG2
Hepatocellular carcinoma (SMMC7721)
Ovarian cancer
Ovarian carcinoma
Ovarian carcinoma (SKOV3)
Tongue squamous cell carcinoma
Bladder carcinoma (KU-7) orthotopic
Lack of inhibition of angiogenesis, tumor growth and/or metastases.
Murine primary tumors
Fibrosarcoma T241
Murine lung cancer
Lewis lung carcinoma (weak anti-tumor activity)
Human tumors
Acute lymphocytic leukemia
Breast cancer (MDA-MB-231) (minimal effect)
Neuroblastoma (SKNAS)

an angiostatin-endostatin fusion gene (statin-AE) and a soluble VEGF receptor (sFlt). Microvessel density was significantly reduced. Tumors were eradicated in up to 50% of mice and survival was prolonged by 4-fold. 50% of the mice were still living at the end of the experiment (200 days) [162]. Both intracranial and subcutaneous tumors were successfully treated.

These results illustrate the benefits of combinatorial antiangiogenic gene therapy, especially for brain tumors, where intratumoral therapy is more feasible. Furthermore, intratumoral antiangiogenic gene therapy of brain tumors may facilitate sustained levels of inhibitor at the tumor site, but by reverse diffusion toward neighboring capillaries. Other combinations of endostatin gene therapy with conventional modalities are noteworthy. Endostatin gene therapy enhanced the efficacy of ionizing radiation on Lewis lung carcinomas. Tumor volumes were reduced by as much as 50% with the combination therapy [163]. Endostatin gene therapy also enhanced the anti-tumor efficacy of gemcitabine and produced a significant decrease of vascularization and of tumor volume without added toxicity in a human lung cancer model in mice [164].

Indraccolo has written a thoughtful review of antiangiogenic gene therapy [165]. Tabruyn and Griffioen have written a scholarly review of molecular pathways of angiogenesis inhibition [166].

In six reports, endostatin gene therapy failed to inhibit tumor growth. Two of these [167, 168] were possibly due to circulating levels of endostatin that were too high, because of the biphasic, U-shaped dose-efficacy curve for endostatin (and other angiogenesis inhibitors) described above. Other possibilities are discussed in [151] (See Table 12.3).

Endostatin in Down Syndrome

Individuals with Down syndrome are arguably the most protected against cancer of all humans. The overall incidence of cancer is less than 0.1 the expected incidence, with the exception of testicular cancer and a megakaryocytic leukemia [169]. For example: in a study of 17,897 individuals with Down syndrome, 4,914 malignancies were expected, but only 344 were observed [169]. In a study of 6,724 Down syndrome individuals, 11 cases of neuroblastoma were expected, but no cases were observed [170]. In 3,581 individuals with Down syndrome, 56 tumors were expected, but only 32 were observed [171]. In a study of 860 individuals with Down syndrome, 69 breast cancer deaths were expected, but only 5 were observed [172]. Because these individuals now live to age 60 or 70, short lifespan is no longer a likely cause for this 'resistance' to cancer. Endostatin is. These individuals have an extra copy of collagen XVIII on chromosome 21 (trisomy). This contributes to a serum level of endostatin that is approximately one-third higher than normal [173].

Also, individuals with Down syndrome have a lower incidence of diabetic retinal neovascularization than other diabetics, even though diabetes in the Down syndrome individuals is as common as in the rest of the population [174].

This additional correlative evidence suggests that suppressed angiogenesis is a major contributor to the decreased cancer incidence associated with the Down syndrome. Collagen XVIII may not be the only source of antiangiogenic activity on chromosome 21. Nevertheless, the clinical clue from individuals with Down syndrome suggests that very small elevations of one or more endogenous angiogenesis inhibitors in the circulation could protect the rest of the human population against cancer. Such a preventive therapy would be especially useful in patients at high risk for cancer, such as women with the breast cancer gene.

Platelets

Endostatin is concentrated in the α granules of platelets. Endostatin is segregated together with other endogenous angiogenesis inhibitors, such as thrombospondin-1 and platelet factor-4, in one set of α granules. Another set of α granules contain proangiogenic proteins, such as bFGF and VEGF [175]. Endostatin and other angiogenesis regulators are sequestered by platelets and retained at higher concentrations than in plasma or serum [176]. A platelet angiogenesis proteome has been uncovered that has been reported to undergo changes in its content of angiogenesis regulatory molecules (such as platelet factor 4), in the presence of microscopic-sized human tumors in mice [177]. It remains to be determined whether analysis of the platelet angiogenesis proteome can be used as a sensitive biomarker for ultra-early detection of recurrent cancer, before symptoms, or before anatomical location of such a tumor is feasible by conventional methods [173].

Principles of Antiangiogenic Therapy Learned from Angiostatin and Endostatin

Endostatinuria

Angiostatin and endostatin were both originally identified in serum of tumor-bearing animals, but at much greater concentrations in urine. Both proteins were purified to homogeneity from urine. Recently, I hypothesized that this "angiostatinuria" and "endostatinuria" could be based on glomerular leakage of proteins from VEGF from the tumor. Ofer Fainaru and I have shown that the kidney can be subjected to a Miles test by intravenous injection of Evans blue which binds to albumin and is not excreted. However, after a single intravenous injection of VEGF, the urine turns blue (Fainaru, Benny and Folkman, unpublished).

At this writing, we are determining whether a continuous elevation in circulating VEGF could lead to depletion of endostatin or angiostatin. An analogy would be the loss of antithrombin III in children with proteinuria from the nephrotic syndrome [178]. We speculate that depletion of endostatin or other endogenous angiogenesis inhibitors could in part be a basis for rapid progression of cancer in a patient.

Conclusion

Angiostatin and endostatin represent the first endogenous angiogenesis inhibitors discovered to be internal peptides of a circulating protein, plasminogen, and a matrix protein, collagen XVIII, neither of which is itself antiangiogenic. Angiostatin and endostatin have opened up new directions to understand how physiologic and pathologic angiogenesis are regulated. In more than 1,700 publications since both proteins were discovered in 1994 and 1997, respectively, many other investigators including ourselves, have asked: (1) How are these proteins mobilized from their parent proteins? (2) What are their endothelial cell receptors? (3) What are their intracellular signaling pathways? (4) How are they carried in the circulation? (5) What is their function in neutrophils, platelets, and in other cells that are not endothelium? (6) Do they defend the host against the switch to the angiogenic phenotype by microscopic, non-angiogenic dormant, in situ tumors? (7) Do they cooperate with other endogenous angiogenesis inhibitors, such as thrombospondin-1, platelet factor 4, and tumstatin? (8) Are they (and possibly other endogenous angiogenesis inhibitors), depleted by urinary loss when tumors are secreting VEGF, which we have shown can cause vascular leakage in the glomerulus, and also induce proteinuria? (9) How can angiostatin and endostatin have such a broad spectrum of angiogenic targets, affecting hundreds of genes involved in angiogenesis regulation, without causing side-effects? (10) Will angiostatin and endostatin eventually be employed as long-term maintenance angiogenesis inhibitors to prevent tumor recurrence, and to protect against cancer in individuals at high risk? (11) Finally, the hypothesis that led to the discovery of angiostatin and endostatin was based on a clinical clue that removal of a primary tumor can often be followed by rapid growth of remote metastases. Is the suppression of growth of metastases by other primary tumors mediated by angiostatin or endostatin, or possibly by novel undiscovered angiogenesis inhibitors? These insights and others could not have been possible before angiostatin and endostatin were discovered.

Acknowledgements. Supported in part by a grant to J. Folkman from the Breast Cancer Research Foundation; NIH grant PO1 CA45548, and Department of Defense Innovator Award W81XWH-04-1-0316. I thank Caitlin Welsh for editorial assistance.

References

1. O'Reilly MS, Holmgren L, Shing Y, Chen C, Rosenthal RA, Moses M, Lane WS, Cao Y, Sage EH, Folkman J. Angiostatin: A novel angiogenesis inhibitor that mediates the suppression of metastases by a Lewis lung carcinoma. Cell 1994; 79:315–328.

2. O'Reilly MS, Boehm T, Shing Y, Fukai N, Vasios G, Lane WS, Flynn E, Birkhead JR, Olsen BR, Folkman J. Endostatin: An endogenous inhibitor of angiogenesis and tumor growth. Cell 1997; 88:277–285.

3. Sugarbaker EV, Thornthwaite J, Ketchan AS. Inhibitory effect of a primary tumor on metastasis. In: DAY SB, Myers WPL, Stansly P, Gerattini S, Lewis MG, eds. Progress in Cancer Research and Therapy, Raven Press, New York, 1977; pp 227–240.

4. Woodruff M. The interactions of cancer and the host. Grune and Stratton, New York, 1980.

5. Clark WH, Jr., Elder DE, Guerry DIV, Braitman LE, Trock BJ, Schultz D, Synnestevdt M, Halpern AC. Model predicting survival in stage 1 melanoma based on tumor progression. J Natl Cancer Inst 1989; 81:1893–1904.

6. Lange PH, Hekmat K, Bosl G, Kennedy BJ, Fraley EE. Accelerated growth of testicular cancer after cytoreductive surgery. Cancer 1980; 45:1498–1506.

7. Marie P, Clunet J. Fréquence des metastases viscérales chez les souris cancéreyses après ablation chirurgicale de leur tumeur. Bull Assoc Fraç l'Etude Cancer 1910; 3:19–23.

8. Tyzzer EE. Factors in the production and growth of tumor metastases. J Med Res 1913; 28:309–333.

9. Schatten WE. An experimental study of postoperative tumor metastases. Cancer 1958; 11:455–459.

10. Greene HSN, Harvey EK. The inhibitory influence of a transplanted hamster lymphoma on metastasis. Cancer Res 1960; 20:1094–1103.

11. Gorelik E, Segal S, Feldman M. On the mechanism of tumor "Concomitant immunity." Int J Cancer 1981; 27:847–856.

12. Bonfil RD, Ruggiero RA, Bustuoabad OD, Meiss RP, Pasqualini CD. Role of concomitant resistance in the development of murine lung metastases. Int J Cancer 1988; 41:415–422.

13. Fisher B, Gunduz N, Saffer EA. Influence of the interval between primary tumor removal and chemotherapy on kinetics and growth of metastases. Cancer Res 1983; 43:1488–1492.

14. Yuhas JM, Pazmiño NH. Inhibition of subcutaneously growing line 1 carcinomas due to metastatic spread. Cancer Res 1974; 34:2005–2010.

15. Gorelik E. Resistance of tumor-bearing mice to a second tumor challenge. Cancer Res 1983; 43:138.145.

16. Lausch RN, Rapp F. Concomitant immunity in hamsters bearing DMBA-induced tumor transplants. Int J Cancer 1969; 4:226–231.

17. Prehn RT. The inhibition of tumor growth by tumor mass. Cancer Res 1991; 51:2–4.

18. Prehn RT. Two competing influences that may explain concomitant tumor resistance. Cancer Res 1993; 53:3266–3269.

19. Folkman J. Angiogenesis in cancer, vascular, rheumatoid and other disease. Nature Med 1995; 1:27–31.

20. Ruggiero RA, Bustuoabad OD, Bonfil RD, Meiss RP, Pasqualini CD. "Concomitant immunity" in murine tumours of non-detectable immunogenicity. Br J Cancer. 1985;51:37–48.

21. Ingber D, Fujita T, Kishimoto S, Sudo K, Kanamaru T, Brem H, Folkman J. Synthetic analogues of fumagillin that inhibit angiogenesis and suppress tumour growth. Nature 1990; 348:555–557.

22. Brem H, Ingber D, Blood CH, Bradley D, Urioste S, Folkman J. Suppression of tumor metastasis by angiogenesis inhibition. Surg Forum 1991; XLII:439–441.

23. O'Reilly M, Rosenthal R, Sage EH, Smith S, Holmgren L, Moses M, Shing Y, Folkman J. The suppression of tumor metastases by a primary tumor. Surg Forum 1993; XLIV:474–476.

24. Claesson-Welsh L, Welsh M, Ito N, Anand-Apte B, Soker S, Zetter B, O'Reilly M, Folkman J. Angiostatin induces endothelial cell apoptosis and activation of focal adhesion kinase independently

of the integrin-binding motif RGD. Proc Natl Acad Sci USA. 1998;95:5579–5583.

25. O'Reilly MS, Holmgren L, Chen C, Folkman J. Angiostatin induces and sustains dormancy of human primary tumors in mice. Nat Med. 1996;2:689–692.

26. Wu Z, O'Reilly MS, Folkman J, Shing Y. Suppression of tumor growth with recombinant murine angiostatin. Biochem Biophys Res Commun 1997; 236:651–654.

27. Lin J, Panigraphy D, Trinh LB, Folkman J, Shiloach J. Production process for recombinant human angiostatin in *Pichia pastoris*. J Industrial Microbiol Biotech 2000; 24:31–35.

28. Cao Y, O'Reilly MS, Marshall B, Flynn E, Ji RW, Folkman J. Expression of angiostatin cDNA in a murine fibrosarcoma suppresses primary tumor growth and produces long-term dormancy of metastases. J Clin Invest. 1998;101:1055–1063.

29. Sun X, Qiao H, Jiang H, Zhi X, Liu F, Wang J, Liu M, Dong D, Kanwar JR, Xu R, Krissansen GW. Intramuscular delivery of antiangiogenic genes suppresses secondary metastases after removal of primary tumors. Cancer Gene Ther. 2005;12:35–45.

30. Peeters CF, de Geus LF, Westphal JR, de Waal RM, Ruiter DJ, Wobbes T, Oyen WJ, Ruers TJ. Decrease in circulating anti-angiogenic factors (angiostatin and endostatin) after surgical removal of primary colorectal carcinoma coincides with increased metabolic activity of liver metastases. Surgery. 2005;137:246–249.

31. Camphausen K, Moses MA, Beecken WD, Khan MK, Folkman J, O'Reilly MS. Radiation therapy to a primary tumor accelerates metastatic growth in mice. Cancer Res. 2001;61:2207–2211.

32. Gately S, Twardowski P, Stack MS, Patrick M, Boggio L, Cundiff DL, Schnaper HW, Madison L, Volpert O, Bouck N, Enghild J, Kwaan HC, Soff GA. Human prostate carcinoma cells express enzymatic activity that converts human plasminogen to the angiogenesis inhibitor, angiostatin. Cancer Res. 1996;56:4887–4890.

33. Stathakis P, Fitzgerald M, Matthias LJ, Chesterman CN, Hogg PJ. Generation of angiostatin by reduction and proteolysis of plasmin. Catalysis by a plasmin reductase secreted by cultured cells. J Biol Chem. 1997;272:20641–20645.

34. Dong Z, Kumar R, Yang X, Fidler IJ. Macrophage-derived metalloelastase is responsible for the generation of angiostatin in Lewis lung carcinoma. Cell. 1997;88:801–810.

35. O'Reilly MS, Wiederschain D, Stetler-Stevenson WG, Folkman J, Moses MA. Regulation of angiostatin production by matrix metalloproteinase-2 in a model of concomitant resistance. J Biol Chem. 1999;274:29568–29571.

36. Lay AJ, Jiang XM, Kisker O, Flynn E, Underwood A, Condron R, Hogg PJ. Phosphoglycerate kinase acts in tumour angiogenesis as a disulphide reductase. Nature. 2000;408:869–873.

37. Wahl ML, Kenan DJ, Gonzalez-Gronow M, Pizzo SV. Angiostatin's molecular mechanism: aspects of specificity and regulation elucidated. J Cell Biochem. 2005;96:242–261.

38. Gonzalez-Gronow M, Grenett HE, Gawdi G, Pizzo SV. Angiostatin directly inhibits human prostate tumor cell invasion by blocking plasminogen binding to its cellular receptor, CD26. Exp Cell Res. 2005;303:22–31.

39. Moser TL, Stack MS, Asplin I, Enghild JJ, Hojrup P, Everitt L, Hubchak S, Schnaper HW, Pizzo SV. Angiostatin binds ATP synthase on the surface of human endothelial cells. Proc Natl Acad Sci U S A. 1999;96:2811–2816.

40. Moser TL, Kenan DJ, Ashley TA, Roy JA, Goodman MD, Misra UK, Cheek DJ, Pizzo SV. Endothelial cell surface F1-F0 ATP synthase is active in ATP synthesis and is inhibited by angiostatin. Proc Natl Acad Sci U S A. 2001;98:6656–6661.

41. Arakaki N, Nagao T, Niki R, Toyofuku A, Tanaka H, Kuramoto Y, Emoto Y, Shibata H, Magota K, Higuti T. Possible role of cell surface H+ -ATP synthase in the extracellular ATP synthesis and proliferation of human umbilical vein endothelial cells. Mol Cancer Res. 2003;1:931–939.

42. Yegutkin GG, Henttinen T, Samburski SS, Spychala J, Jalkanen S. The evidence for two opposite, ATP-generating and ATP-consuming, extracellular pathways on endothelial and lymphoid cells. Biochem J. 2002;367:121–128.

43. Yamagata M, Tannock IF. The chronic administration of drugs that inhibit the regulation of intracellular pH: in vitro and antitumour effects. Br J Cancer. 1996;73:1328–1334.

44. Wahl ML, Owen CS, Grant DS. Angiostatin induces intracellular acidosis and anoikis in endothelial cells at a tumor-like low pH. Endothelium. 2002;9:205–216.

45. Chi SL, Wahl ML, Mowery YM, Shan S, Mukhopadhyay S, Hilderbrand SC, Kenan DJ, Lipes BD, Johnson CE, Marusich MF, Capaldi RA, Dewhirst MW, Pizzo SV. Angiostatin-like activity of a monoclonal antibody to the catalytic subunit of F1F0 ATP synthase. Cancer Res. 2007;67:4716–4724.

46. Jain RK. Normalization of tumor vasculature: an emerging concept in antiangiogenic therapy. Science. 2005;307:58–62.

47. Batchelor TT, Sorensen AG, di Tomaso E, Zhang WT, Duda DG, Cohen KS, Kozak KR, Cahill DP, Chen PJ, Zhu M, Ancukiewicz M, Mrugala MM, Plotkin S, Drappatz J, Louis DN, Ivy P, Scadden DT, Benner T, Loeffler JS, Wen PY, Jain RK. AZD2171, a pan-VEGF receptor tyrosine kinase inhibitor, normalizes tumor vasculature and alleviates edema in glioblastoma patients. Cancer Cell. 2007;11:83–95.

49. Boehm T, Folkman J, Browder T, O'Reilly MS. Antiangiogenic therapy of experimental cancer does not induce acquired drug resistance. Nature. 1997;390:404–407.

50. Troyanovsky B, Levchenko T, Mansson G, Matvijenko O, Holmgren L. Angiomotin: an angiostatin binding protein that regulates endothelial cell migration and tube formation. J Cell Biol. 2001;152:1247–1254.

51. Levchenko T, Aase K, Troyanovsky B, Bratt A, Holmgren L. Loss of responsiveness to chemotactic factors by deletion of the C-terminal protein interaction site of angiomotin. J Cell Sci. 2003;116:3803–3810.

52. Aase K, Ernkvist M, Ebarasi L, Jakobsson L, Majumdur A, Yi C, Birot O, Ming Y, Kvanta A, Edholm D, Aspenström P, Kissil J, Claesson-Welsh L, Shimono A, Holmgren L. Angiomotin regulates endothelial cell migration during embroni angiogenesis. Genes Development 2007, 21:2055–2068.

53. Levchenko T, Bratt A, Arbiser JL, Holmgren L. Angiomotin expression promotes hemangioendothelioma invasion. Oncogene. 2004;23:1469–1473.

54. Levchenko T, Veitonmäki N, Lundkvist A, Gerhardt H, Ming Y, Berggren K, Kvanta A, Carlsson R, Holmgren L. Therapeutic antibodies targeting angiomotin inhibit angiogenesis in vivo. FASEB J. 2007 Nov 7, (Epub ahead of print).

55. Wajih N, Sane DC. Angiostatin selectively inhibits signaling by hepatocyte growth factor in endothelial and smooth muscle cells. Blood. 2003;101:1857–863.

56. Tarui T, Miles LA, Takada Y. Specific interaction of angiostatin with integrin alpha(v)beta(3) in endothelial cells. J Biol Chem. 2001;276:39562–39568.

57. Goretzki L, Lombardo CR, Stallcup WB. Binding of the NG2 proteoglycan to kringle domains modulates the functional properties of angiostatin and plasmin(ogen). J Biol Chem. 2000;275:28625–28633.

58. Shi H, Huang Y, Zhou H, Song X, Yuan S, Fu Y, Luo Y. Nucleolin is a receptor that mediates antiangiogenic and antitumor activity of endostatin. Blood. 2007;110:2899–2906.

59. Folkman, J. Endostatin finds a new partner: nucleolin. Blood 2007; 110(8): 2786 – 2787.

60. Celik I, Surucu O, Dietz C, Heymach JV, Force J, Hoschele I, Becker CM, Folkman J, Kisker O. Therapeutic efficacy of endostatin exhibits a biphasic dose-response curve. Cancer Res. 2005;65:11044–11050.

61. Slaton JW, Perrotte P, Inoue K, Dinney CP, Fidler IJ. Interferon-alpha-mediated down-regulation of angiogenesis-related genes and therapy of bladder cancer are dependent on optimization of biological dose and schedule. Clin Cancer Res. 1999;5:2726–2734.

62. Teodoro JG, Parker AE, Zhu X, Green MR. p53-mediated inhibition of angiogenesis through up-regulation of a collagen prolyl hydroxylase. Science. 2006;313:968–971.

63. Folkman J. Tumor Suppression by p53 is Mediated in Part by the Antiangiogenic Activity of Endostatin and Tumstatin. Sci STKE. 2006;2006:pe35.

64. Becker CM, Sampson DA, Rupnick MA, Rohan RM, Efstathiou JA, Short SM, Taylor GA, Folkman J, D'Amato RJ. Endostatin inhibits the growth of endometriotic lesions but does not affect fertility. Fertil Steril. 2005;84 Suppl 2:1144–1155.

65. Prandini MH, Desroches-Castan A, Feraud O, Vittet D. No evidence for vasculogenesis regulation by angiostatin during mouse embryonic stem cell differentiation. J Cell Physiol. 2007;213(1):27–35.

66. Eriksson K, Magnusson P, Dixelius J, Claesson-Welsh L, Cross MJ. Angiostatin and endostatin inhibit endothelial cell migration in response to FGF and VEGF without interfering with specific intracellular signal transduction pathways. FEBS Lett. 2003;536:19–24.

67. Perri SR, Annabi B, Galipeau J. Angiostatin inhibits monocyte/macrophage migration via disruption of actin cytoskeleton. Faseb J. 2007;21(14):3928–3936.

68. Pearce JW, Janardhan KS, Caldwell S, Singh B. Angiostatin and integrin alphavbeta3 in the feline, bovine, canine, equine, porcine and murine retina and cornea. Vet Ophthalmol. 2007;10:313–319.

69. Shyong MP, Lee FL, Kuo PC, Wu AC, Cheng HC, Chen SL, Tung TH, Tsao YP. Reduction of experimental diabetic vascular leakage by delivery of angiostatin with a recombinant adeno-associated virus vector. Mol Vis. 2007;13:133–141.

70. Sima J, Ma J, Zhang SX, Guo J. Study of the influence of angiostatin intravitreal injection on vascular leakage in retina and iris of the experimental diabetic rats. Yan Ke Xue Bao. 2006;22:252–258.

71. Hanford HA, Wong CA, Kassan H, Cundiff DL, Chandel N, Underwood S, Mitchell CA, Soff GA. Angiostatin(4.5)-mediated apoptosis of vascular endothelial cells. Cancer Res. 2003;63:4275–4280.

72. Jurasz P, Alonso D, Castro-Blanco S, Murad F, Radomski MW. Generation and role of angiostatin in human platelets. Blood. 2003;102:3217–3223.

73. Cervi D, Yip TT, Bhattacharya N, Podust VN, Peterson J, Abou-Slaybi A, Naumov GN, Bender E, Almog N, Italiano JE, Jr, Folkman J, Klement GL. Platelet-associated PF-4 as a biomarker of early tumor growth. Blood. 2007;111:1201–1207.

74. Italiano JE, Jr, Richardson JL, Patel-Hett S, Battinelli E, Zaslavsky A, Short S, Ryeom S, Folkman J, Klement GL. Angiogenesis is regulated by a novel mechanism: Pro- and anti-angiogenic proteins are organized into separate platelet {alpha}-granules and differentialy released. Blood. 2007;111:1227–1233.

75. Chung AW, Hsiang YN, Matzke LA, McManus BM, van Breemen C, Okon EB. Reduced expression of vascular endothelial growth factor paralleled with the increased angiostatin expression resulting from the upregulated activities of matrix metalloproteinase-2 and -9 in human type 2 diabetic arterial vasculature. Circ Res. 2006;99:140–148.

76. Matsunaga T, Chilian WM, March K. Angiostatin is negatively associated with coronary collateral growth in patients with coronary artery disease. Am J Physiol Heart Circ Physiol. 2005;288: H2042–2046.

77. Koshida R, Ou J, Matsunaga T, Chilian WM, Oldham KT, Ackerman AW, Pritchard KA, Jr. Angiostatin: a negative regulator of endothelial-dependent vasodilation. Circulation. 2003;107:803–806.

78. Chen YH, Wu HL, Li C, Huang YH, Chiang CW, Wu MP, Wu LW. Anti-angiogenesis mediated by angiostatin K1–3, K1–4 and K1–4.5. Involvement of p53, FasL, AKT and mRNA deregulation. Thromb Haemost. 2006;95:668–677.

79. Dameron KM, Volpert OV, Tainsky MA, Bouck N. Control of angiogenesis in fibroblasts by p53 regulation of thrombospondin-1. Science. 1994;265:1582–1584.

80. Ravi R, Mookerjee B, Bhujwalla ZM, Sutter CH, Artemov D, Zeng Q, Dillehay LE, Madan A, Semenza GL, Bedi, A. Regulation of tumor angiogenesis by p53-induced degradation of hypoxia-inducible factor 1alpha. Genes Dev. 2000;14:34–44.

81. Jouanneau E, Alberti L, Nejjari M, Treilleux I, Vilgrain I, Duc A, Combaret V, Favrot M, Leboulch P, Bachelot R. Lack of antitumor activity of recombinant endostatin in a human neuroblastoma xenograft model. J Neuro-oncol 2001; 51:11–18.

82. Sherif ZA, Nakai S, Pirollo KF, Rait A, Chang EH. Downmodulation of bFGF-binding protein expression following restoration of p53 function. Cancer Gene Ther. 2001;8:771–782.

83. Chen YH, Wu HL, Chen CK, Huang YH, Yang BC, Wu LW. Angiostatin antagonizes the action of VEGF-A in human endothelial cells via two distinct pathways. Biochem Biophys Res Commun. 2003;310:804–810.

84. Cao Y, Xue L. Angiostatin. Semin Thromb Hemost. 2004;30:83–93.

85. Geiger JH, Cnudde SE. What the structure of angiostatin may tell us about its mechanism of action. J Thromb Haemost. 2004;2:23–34.

86. Folkman J. Angiogenesis in arthritis. In: Smolen JS, Lipsky PE, eds. Targeted Therapies in Rheumatology, Martin Dunitz, London, United Kingdom, 2003; pp 111–131.

87. Chavakis T, Athanasopoulos A, Rhee JS, Orlova V, Schmidt-Woll T, Bierhaus A, May AE, Celik I, Nawroth PP, Preissner KT. Angiostatin is a novel anti-inflammatory factor by inhibiting leukocyte recruitment. Blood. 2005;105:1036–1043.

88. Beecken WD, Engl T, Blaheta R, Bentas W, Achilles EG, Jonas D, Shing Y, Camphausen K. Angiogenesis inhibition by angiostatin, endostatin and TNP-470 prevents cyclophosphamide induced cystitis. Angiogenesis. 2004;7:69–73.

89. Satchi-Fainaro R, Puder M, Davies JW, Tran HT, Sampson DA, Greene AK, Corfas G, Folkman J. Targeting angiogenesis with a conjugate of HPMA copolymer and TNP-470. Nat Med. 2004;10:255–261.

90. Kato K, Miyake K, Igarashi T, Yoshino S, Shimada T. Human immunodeficiency virus vector-mediated intra-articular expression of angiostatin inhibits progression of collagen-induced arthritis in mice. Rheumatol Int. 2005;25:522–529.

91. Folkman J. Endogenous angiogenesis inhibitors. Acta Pathologica, Microbiologica, et Immunologica Scandinavica. 2004;112:496–507.

92. Nyberg P, Xie L, Kalluri R. Endogenous inhibitors of angiogenesis. Cancer Res. 2005;65:3967–3979.

93. Folkman J. Angiogenesis: an organizing principle for drug discovery? Nat Rev Drug Discov. 2007;6:273–286.

94. Agarwal A, Munoz-Najar U, Klueh U, Shih SC, Claffey KP. N-acetyl-cysteine promotes angiostatin production and vascular collapse in an orthotopic model of breast cancer. Am J Pathol. 2004;164:1683–1696.

95. Perchick GB, Jabbour HN. Cyclooxygenase-2 overexpression inhibits cathepsin D-mediated cleavage of plasminogen to the potent antiangiogenic factor angiostatin. Endocrinology. 2003;144:5322–5328.

96. Tao KS, Dou KF, Wu XA. Expression of angiostatin cDNA in human hepatocellular carcinoma cell line SMMC-7721 and its effect on implanted carcinoma in nude mice. World J Gastroenterol. 2004;10:1421–1424.

97. Li X, Raikwar SP, Liu YH, Lee SJ, Zhang YP, Zhang S, Cheng L, Lee SD, Juliar BE, Gardner TA, Jeng MH, Kao C. Combination therapy of androgen-independent prostate cancer using a prostate restricted replicative adenovirus and a replication-defective adenovirus encoding human endostatin-angiostatin fusion gene. Mol Cancer Ther. 2006;5:676–684.

98. Schmidt K, Hoffend J, Altmann A, Strauss LG, Dimitrakopoulou-Strauss A, Engelhardt B, Koczan D, Peter J, Dengler TJ, Mier W, Eisenhut M, Haberkorn U, Kinscherf R. Angiostatin overexpression in Morris hepatoma results in decreased tumor growth but increased perfusion and vascularization. J Nucl Med. 2006;47:543–551.

99. Hlatky L, Hahnfeldt P, Folkman J. Clinical application of antiangiogenic therapy: microvessel density, what it does and doesn't tell us. J Natl Cancer Inst. 2002;94:883–893.

100. Xu R, Zhang X, Zhang W, Fang Y, Zheng S, Yu XF. Association of human APOBEC3 cytidine deaminases with the generation of hepatitis virus B x antigen mutants and hepatocellular carcinoma. Hepatology. 2007;46:1810–1820.

101. Kim KS, Park YS. Antitumor effects of angiostatin K1–3 and endostatin genes coadministered by the hydrodynamics-based transfection method. Oncol Res. 2005;15:343–350.

102. Schmitz V, Wang L, Barajas M, Gomar C, Prieto J, Qian C. Treatment of colorectal and hepatocellular carcinomas by adenoviral mediated gene transfer of endostatin and angiostatin-like molecule in mice. Gut. 2004;53:561–567.

103. Ponnazhagan S, Mahendra G, Kumar S, Shaw DR, Stockard CR, Grizzle WE, Meleth S. Adeno-associated virus 2-mediated antiangiogenic cancer gene therapy: long-term efficacy of a vector encoding angiostatin and endostatin over vectors encoding a single factor. Cancer Res. 2004;64:1781–1787.

104. Lalani AS, Chang B, Lin J, Case SS, Luan B, Wu-Prior WW, VanRoey M, Jooss K. Anti-tumor efficacy of human angiostatin using liver-mediated adeno-associated virus gene therapy. Mol Ther. 2004;9:56–66.

105. Galaup A, Opolon P, Bouquet C, Li H, Opolon D, Bissery MC, Tursz T, Perricaudet M. Griscelli F. Combined effects of docetaxel and angiostatin gene therapy in prostate tumor model. Mol Ther. 2003;7:731–740.

106. Stone EM. A very effective treatment for neovascular macular degeneration. N Engl J Med. 2006;355:1493–1495.

107. Igarashi T, Miyake K, Kato K, Watanabe A, Ishizaki M, Ohara K, Shimada T. Lentivirus-mediated expression of angiostatin efficiently inhibits neovascularization in a murine proliferative retinopathy model. Gene Ther. 2003;10:219–226.

108. Zhang SX, Wang JJ, Lu K, Mott R, Longeras R, Ma JX. Therapeutic potential of angiostatin in diabetic nephropathy. J Am Soc Nephrol. 2006;17:475–486.

109. Balaggan KS, Binley K, Esapa M, MacLaren RE, Iqball S, Duran Y, Pearson RA, Kan O, Barker SE, Smith AJ, Bainbridge JW, Naylor S, Ali RR. EIAV vector-mediated delivery of endostatin or angiostatin inhibits angiogenesis and vascular hyperpermeability in experimental CNV. Gene Ther. 2006;13:1153–1165.

110. Drixler TA, Rinkes IH, Ritchie ED, Treffers FW, van Vroonhoven TJ, Gebbink MF, Voest EE. Angiostatin inhibits pathological but not physiological retinal angiogenesis. Invest Ophthalmol Vis Sci. 2001;42:3325–3330.

111. Sima J, Zhang SX, Shao C, Fant J, Ma JX. The effect of angiostatin on vascular leakage and VEGF expression in rat retina. FEBS Lett. 2004;564:19–23.

112. Xie JL, Zhou QW, Zhang L, Ye Q, Xin L, Du P, Gan RB. [Feeding of mixed-carbon-resource during the expression phase in cultivation of recombinant Pichia pastoris expressing angiostatin]. Sheng Wu Gong Cheng Xue Bao. 2003;19:467–470.

113. Visted T, Furmanek T, Sakariassen P, Foegler WB, Sim K, Westphal H, Bjerkvig R, Lund-Johansen M. Prospects for delivery of recombinant angiostatin by cell-encapsulation therapy. Hum Gene Ther. 2003;14:1429–1440.

114. Yang H, Akor C, Dithmar S, Grossniklaus HE. Low dose adjuvant angiostatin decreases hepatic micrometastasis in murine ocular melanoma model. Mol Vis. 2004;10:987–995.

115. Benelli R, Morini M, Brigati C, Noonan DM, Albini A. Angiostatin inhibits extracellular HIV-Tat-induced inflammatory angiogenesis. Int J Oncol. 2003;22:87–91.

116. Motegi K, Harada K, Pazouki S, Baillie R, Schor AM. Evidence of a bi-phasic effect of thrombospondin-1 on angiogenesis. Histochem J. 2002;34:411–421.

117. Davis DW, Shen Y, Mullani NA, Wen S, Herbst RS, O'Reilly M, Abbruzzese JL, McConkey DJ. Quantitative analysis of biomarkers defines an optimal biological dose for recombinant human endostatin in primary human tumors. Clin Cancer Res. 2004;10:33–42.

118. Banerjee SN, Sengupta K, Banerjee S, Saxena NK, Banerjee SK. 2-Methoxyestradiol exhibits a biphasic effect on VEGF-A in tumor cells and upregulation is mediated through ER-alpha: a possible signaling pathway associated with the impact of 2-ME2 on proliferative cells. Neoplasia. 2003;5:417–426.

119. Panigrahy D, Singer S, Shen LQ, Butterfield CE, Freedman DA, Chen EJ, Moses MA, Kilroy S, Duensing S, Fletcher C, Fletcher JA, Hlatky L, Hahnfeldt P, Folkman J, Kaipainen A. PPARgamma ligands inhibit primary tumor growth and metastasis by inhibiting angiogenesis. J Clin Invest. 2002;110:923–932.

120. Peyruchaud O, Serre CM, NicAmhlaoibh R, Fournier P, Clezardin P. Angiostatin inhibits bone metastasis formation in nude mice through a direct anti-osteoclastic activity. J Biol Chem. 2003;278:45826–45832.

121. Matsumoto G, Ohmi Y, Shindo J. Angiostatin gene therapy inhibits the growth of murine squamous cell carcinoma in vivo. Oral Oncol. 2001;37:369–378.

122. te Velde EA, Kusters B, Maass C, de Waal R, Borel Rinkes IH. Histological analysis of defective colonic healing as a result of angiostatin treatment. Exp Mol Pathol. 2003;75:119–123.

123. Jung SP, Siegrist B, Wang YZ, Wade MR, Anthony CT, Hornick C, Woltering EA. Effect of human Angiostatin protein on human angiogenesis in vitro. Angiogenesis. 2003;6:233–240.

124. Seidlitz E, Korbie D, Marien L, Richardson M, Singh G. Quantification of anti-angiogenesis using the capillaries of the chick chorioallantoic membrane demonstrates that the effect of human angiostatin is age-dependent. Microvasc Res. 2004;67:105–116.

125. Yabushita H, Noguchi M, Obayashi Y, Kishida T, Noguchi Y, Sawaguchi K, Noguchi M. Angiostatin expression in ovarian cancer. Oncol Rep. 2003;10:1225–1230.

126. Cnudde SE, Prorok M, Castellino FJ, Geiger JH. X-ray crystallographic structure of the angiogenesis inhibitor, angiostatin, bound to a peptide from the group A streptococcal surface protein PAM. Biochemistry. 2006;45:11052–11060.

127. Blackhall FH, Merry CLR, Lyon M, Jayson GC, Folkman J, Javaherian K, Gallagher JT. Binding of endostatin to endothelial heparan sulphate shows a differential requirement for specific sulphates. Biochem J 2003; 375:131–139.

128. Ding YH, Javaherian K, Lo KM, Chopra R, Boehm T, Lanciotti J, Harris BA, Li Y, Shapiro R, Hohenester E, Timpl R, Folkman J, Wiley DC. Zinc-dependent dimers observed in crystals of human endostatin. Proc Natl Acad Sci U S A. 1998;95:10443–10448.

129. Boehm T, O'Reilly M S, Keough K, Shiloach J, Shapiro R, Folkman J. Zinc-binding of endostatin is essential for its antiangiogenic activity. Biochem Biophys Res Commun. 1998;252:190–194.

130. Tjin Tham Sjin RM, Satchi-Fainaro R, Birsner AE, Ramanujam VM, Folkman J, Javaherian K. A 27-amino-acid synthetic peptide corresponding to the NH2-terminal zinc-binding domain of endostatin is responsible for its antitumor activity. Cancer Res. 2005;65:3656–3663.

131. Hohenester E, Sasaki T, Olsen BR, Timpl R. Crystal structure of the angiogenesis inhibitor endostatin at 1.5 A resolution. Embo J. 1998;17:1656–1664.

132. Sasaki T, Larsson H, Kreuger J, Salmivirta M, Claesson-Welsh L, Lindahl U, Hohenester E, Timpl R. Structural basis and potential role of heparin/heparan sulfate binding to the angiogenesis inhibitor endostatin. Embo J. 1999;18:6240–6248.

133. Dixelius J, Larsson H, Sasaki T, Holmqvist K, Lu L, Engstrom A, Timpl R, Welsh M, Claesson-Welsh L. Endostatin-induced tyrosine kinase signaling through the Shb adaptor protein regulates endothelial cell apoptosis. Blood. 2000;95:3403–3411.

134. Gaetzner S, Deckers MM, Stahl S, Lowik C, Olsen BR, Felbor U. Endostatin's heparan sulfate-binding site is essential for inhibition of angiogenesis and enhances in situ binding to capillary-like structures in bone explants. Matrix Biol. 2005;23:557–561.

135. Wen W, Moses MA, Wiederschain D, Arbiser JL, Folkman J. The generation of endostatin is mediated by elastase. Cancer Res. 1999;59:6052–6056.

136. Felbor U, Dreier L, Bryant RA, Ploegh HL, Olsen BR, Mothes W. Secreted cathepsin L generates endostatin from collagen XVIII. Embo J. 2000;19:1187–1194.

137. Wickstrom SA, Alitalo K, Keski-Oja J. Endostatin associates with integrin alpha5beta1 and caveolin-1, and activates Src via a tyrosyl phosphatase-dependent pathway in human endothelial cells. Cancer Res. 2002;62:5580–5589.

138. Sudhakar A, Sugimoto H, Yang C, Lively J, Zeisberg M, Kalluri R. Human tumstatin and human endostatin exhibit distinct antiangiogenic activities mediated by alpha v beta 3 and alpha 5 beta 1 integrins. Proc Natl Acad Sci U S A. 2003;100:4766–4771.

139. Sund M, Hamano Y, Sugimoto H, Sudhakar A, Soubasakos M, Yerramalla U, Benjamin LE, Lawler J, Kieran M, Shah A, Kalluri R. Function of endogenous inhibitors of angiogenesis as endothelium-specific tumor suppressors. Proc Natl Acad Sci U S A. 2005;102:2934–2939.

140. Yu Y, Moulton KS, Khan MK, Vineberg S, Boye E, Davis VM, O'Donnell PE, Bischoff J, Milstone DS. E-selectin is required for the antiangiogenic activity of endostatin. Proc Natl Acad Sci U S A. 2004;101:8005–8010.

141. Nyberg P, Heikkila P, Sorsa T, Luostarinen J, Heljasvaara R, Stenman UH, Pihlajaniemi T, Salo T. Endostatin inhibits human tongue carcinoma cell invasion and intravasation and blocks the activation of matrix metalloprotease-2, -9 and -13. J Biol Chem. 2003;278(25):22404–22411.

142. Shichiri M, Hirata Y. Antiangiogenesis signals by endostatin. Faseb J. 2001;15:1044–1053.

143. Abdollahi A, Hahnfeldt P, Maercker C, Grone HJ, Debus J, Ansorge W, Folkman J, Hlatky L, Huber PE. Endostatin's antiangiogenic signaling network. Mol Cell. 2004;13:649–663.

144. Abdollahi A, Schwager C, Kleeff J, Esposito I, Domhan S, Peschke P, Hauser K, Hahnfeldt P, Hlatky L, Debus J, Peters JM, Friess H, Folkman J, Huber PE. Transcriptional network governing the angiogenic switch in human pancreatic cancer. Proc Natl Acad Sci U S A. 2007;104:12890–12895.

145. Ramakrishnan S, Nguyen TM, Subramanian IV, Kelekar A. Autophagy and angiogenesis inhibition. Autophagy. 2007;3:512–515.

146. Mathew R, Karantza-Wadsworth V, White E. Role of autophagy in cancer. Nat Rev Cancer. 2007;7:961–967.

147. Ling Y, Yang Y, Lu N, You QD, Wang S, Gao Y, Chen Y, Guo QL. Endostar, a novel recombinant human endostatin, exerts antiangiogenic effect via blocking VEGF-induced tyrosine phosphorylation of KDR/Flk-1 of endothelial cells. Biochem Biophys Res Commun. 2007;361:79–84.

148. Yokoyama Y, Ramakrishnan S. Binding of endostatin to human ovarian cancer cells inhibits cell attachment. Int J Cancer. 2007;121:2402–2409.

149. Li C, Harris MB, Venema VJ, Venema RC. Endostatin induces acute endothelial nitric oxide and prostacyclin release. Biochem Biophys Res Commun. 2005;329:873–878.

150. Urbich C, Reissner A, Chavakis E, Dernbach E, Haendeler J, Fleming I, Zeiher AM, Kaszkin M, Dimmeler S. Dephosphorylation of endothelial nitric oxide synthase contributes to the antiangiogenic effects of endostatin. FASEB J. 2002;16:706–708.

151. Folkman J. Antiangiogenesis in cancer therapy–endostatin and its mechanisms of action. Exp Cell Res. 2006;312:594–607.

152. Yokoyama Y, Ramakrishnan S. Addition of an aminopeptidase N-binding sequence to human endostatin improves inhibition of ovarian carcinoma growth. Cancer. 2005;104:321–331.

153. Schmidt NO, Ziu M, Carrabba G, Giussani C, Bello L, Sun Y, Schmidt K, Albert M, Black PM, Carroll RS. Antiangiogenic therapy by local intracerebral microinfusion improves treatment efficiency and survival in an orthotopic human glioblastoma model. Clin Cancer Res. 2004;10:1255–1262.

154. Beck MT, Chen NY, Franek KJ, Chen WY. Prolactin antagonist-endostatin fusion protein as a targeted dual-functional therapeutic agent for breast cancer. Cancer Res. 2003;63:3598–3604.

155. Kisker O, Becker CM, Prox D, Fannon M, D'Amato R, Flynn E, Fogler WE, Sim BK, Allred EN, Pirie-Shepherd SR, Folkman J. Continuous administration of endostatin by intraperitoneally implanted osmotic pump improves the efficacy and potency of therapy in a mouse xenograft tumor model. Cancer Res. 2001;61:7669–7674.

156. Capillo M, Mancuso P, Gobbi A, Monestiroli S, Pruneri G, Dell'Agnola C, Martinelli G, Shultz L, Bertolini F. Continuous infusion of endostatin Inhibits differentiation, mobilization, and clonogenic potential of endothelial cell progenitors. Clin Cancer Res. 2003;9:377–382.

157. Abraham D, Abri S, Hofmann M, Holtl W, Aharinejad S. Low Dose Carboplatin Combined With Angiostatic Agents Prevents Metastasis in Human Testicular Germ Cell Tumor Xenografts. J Urol. 2003;170:1388–1393.

158. Prox D, Becker C, Pirie-Shepherd SR, Celik I, Folkman J, Kisker O. Treatment of human pancreatic cancer in mice with angiogenic inhibitors. World J Surg. 2003;27:405–411.

159. Beecken WD, Fernandez A, Joussen AM, Achilles EG, Flynn E, Lo KM, Gillies SD, Javaherian K, Folkman J, Shing Y. Effect of antiangiogenic therapy on slowly growing, poorly vascularized tumors in mice. J Natl Cancer Inst. 2001;93:382–387.

160. Itasaka S, Komaki R, Herbst RS, Shibuya K, Shintani T, Hunter NR, Onn A, Bucana CD, Milas L, Ang KK, O'Reilly MS. Endostatin improves radioresponse and blocks tumor revascularization after radiation therapy for A431 xenografts in mice. Int J Radiat Oncol Biol Phys. 2007;67:870–878.

161. Yoon SS, Eto H, Lin CM, Nakamura H, Pawlik TM, Song SU, Tanabe KK. Mouse endostatin inhibits the formation of lung and liver metastases. Cancer Res. 1999;59:6251–6256.

162. Ohlfest JR, Demorest ZL, Motooka Y, Vengco I, Oh S, Chen E, Scappaticci FA, Saplis RJ, Ekker SC, Low WC, Freese AB, Largaespada DA. Combinatorial Antiangiogenic Gene Therapy by Nonviral Gene Transfer Using the Sleeping Beauty Transposon Causes Tumor Regression and Improves Survival in Mice Bearing Intracranial Human Glioblastoma. Mol Ther. 2005;12(5):778–788.

163. Luo X, Slater JM, Gridley DS. Enhancement of radiation effects by pXLG-mEndo in a lung carcinoma model. Int J Radiat Oncol Biol Phys. 2005;63:553–564.

164. Wu Y, Yang L, Hu B, Liu JY, Su JM, Luo Y, Ding ZY, Niu T, Li Q, Xie XJ, Wen YJ, Tian L, Kan B, Mao YQ, Wei YQ. Synergistic anti-tumor effect of recombinant human endostatin adenovirus combined with gemcitabine. Anticancer Drugs. 2005;16:551–557.

165. Indraccolo S. Undermining tumor angiogenesis by gene therapy: an emerging field. Curr Gene Ther. 2004;4:297–308.

166. Tabruyn SP, Griffioen AW. Molecular pathways of angiogenesis inhibition. Biochem Biophys Res Commun. 2007;355:1–5.

167. Kuo CJ, Farnebo F, Yu EY, Christofferson R, Swearingen RA, Carter R, von Recum HA, Yuan J, Kamihara J, Flynn E, D'Amato R, Folkman J, Mulligan RC. Comparative evaluation of the antitumor activity of antiangiogenic proteins delivered by gene transfer. Proc Natl Acad Sci U S A. 2001;98:4605–4610.

168. Pawliuk R, Bachelot T, Zurkiya O, Eriksson A, Cao Y, Leboulch P. Continuous intravascular secretion of endostatin in mice from transduced hematopoietic stem cells. Mol Ther. 2002;5:345–351.

169. Yang Q, Rasmussen SA, Friedman JM. Mortality associated with Down's syndrome in the USA from 1983 to 1997: a population-based study. Lancet. 2002;359:1019–1025.

170. Satge D, Sasco AJ, Carlsen NL, Stiller CA, Rubie H, Hero B, de Bernardi B. de Kraker J, Coze C, Kogner P, Langmark F, Hakvoort-Cammel FG, Beck D, von der Weid N, Parkes S, Hartmann O, Lippens RJ, Kamps WA, Sommelet D. A lack of neuroblastoma in Down syndrome: a study from 11 European countries. Cancer Res. 1998;58:448–452.

171. Patja K, Pukkala E, Sund R, Iivanainen M, Kaski M. Cancer incidence of persons with Down syndrome in Finland: a population-based study. Int J Cancer. 2006;118:1769–1772.

172. Satge D, Sasco AJ, Pujol H, Rethore MO. [Breast cancer in women with trisomy 21]. Bull Acad Natl Med. 2001;185:1239–1252; discussion 1252–1234.

173. Zorick TS, Mustacchi Z, Bando SY, Zatz M, Moreira-Filho CA, Olsen B, Passos-Bueno MR. High serum endostatin levels in Down syndrome: implications for improved treatment and prevention of solid tumours. Eur J Hum Genet. 2001;9:811–814.

174. Fulcher T, Griffin M, Crowley S, Firth R, Acheson R, O'Meara N. Diabetic retinopathy in Down's syndrome. Br J Ophthalmol. 1998;82:407–409.

175. Italiano J, Richardson JL, Folkman J, Klement G. Blood Platelets Organize Pro- and Anti-Angiogenic Factors into Separate, Distinct Alpha Granules: Implications for the Regulation of Angiogenesis. Blood; November 2006; 108:120a, abstract 393.

176. Klement G, Kikuchi L, Kieran M, Almog N, Yip TT, Folkman J. Early tumor detection using platelets uptake of angiogenesis regulators. Proc 47th American Society of Hematology. Blood; December 2004; 104:239a, abstract 839.

177. Klement G, Cervi D, Yip T, Folkman J, Italiano J. Platelet PF-4 Is an Early Marker of Tumor Angiogenesis. Blood; November 2006; 108:426a, abstract 1476.

178. Thaler E. [Pathogenetic mechanism and clinical relevance of acquired anti-thrombin III deficiency in internal medicine (author's transl)]. Wien Klin Wochenschr. 1981;93:563–572.

Chapter 13
Thrombospondins: Endogenous Inhibitors of Angiogenesis

Paul Bornstein

Keywords: angiogenesis, matricellular, extracellular matrix, mouse knockouts, apoptosis, transcriptional regulation, endothelial cells, integrins, domain structure, matrix metalloproteinases, TGFβ, cell adhesion, nitric oxide, tumor growth and metastases

Abstract: The thrombospondin (TSP) gene family consists of five members, two of which, TSP1 and TSP2, have been shown to play important roles in the regulation of angiogenesis. While TSP1 and TSP2 are secreted into the extracellular environment, they do not play structural roles but rather function to regulate cellular behavior by interaction with numerous cell-surface receptors, proteases, cytokines, and other bioactive proteins. As a consequence, these TSPs are termed matricellular proteins. TSP1, but not TSP2, is capable of activating latent TGFβ, and can thereby stimulate the production of extracellular matrix. Matricellular TSPs inhibit angiogenesis both by causing apoptosis of endothelial cells (EC) and by inhibiting their proliferation. However, under some circumstances TSP1 has also been shown to be proangiogenic. TSPs interact with matrix metalloproteinases (MMPs) 2 and 9 and function as clearance factors by directing these MMPs to the scavenger receptor, LRP1, and thence to lysosomal degradation. As a consequence, TSPs function to regulate cell adhesion. TSP1 also inhibits nitric oxide-stimulation of EC proliferation by interaction with the CD47/integrin-associated protein receptor. Finally, by virtue of their ability to inhibit angiogenesis, TSPs have the potential to inhibit tumor growth and metastases, a property that may have clinical applications.

Introduction

Thrombospondins (TSPs) comprise a small family of extracellular proteins that are present in all vertebrates and in some invertebrates [1,2]. The first member of the family, TSP1, was shown to be released from thrombin-treated platelets and was initially

termed thrombin-sensitive protein. It was subsequently renamed thrombospondin to indicate that its release from the α granules of platelets in response to thrombin treatment did not necessarily require proteolysis of the protein. Vertebrates express five paralogous genes whose translation products are assembled as homotrimers of 145 kDa chains (TSP1 and TSP2) or homopentamers of ~110 kDa chains (TSPs 3–5; Fig. 13.1). TSP5 was first identified in cartilage and is still referred to as cartilage oligomeric matrix protein (COMP) in many publications. There is some evidence that TSP4 and TSP5 exist as both homo- and heteropentamers [3], but these findings require confirmation.

The structures of the five thrombospondin monomers are shown schematically in Fig. 13.1. Trimeric TSPs are composed of a globular, heparin-binding NH_2-terminal (N-terminal) domain followed by a procollagen homology domain, three type I thrombospondin or properdin repeats, three type II or EGF-like repeats, seven type III or Ca^{2+}-binding repeats and a COOH-terminal (C-terminal) globular domain. The three monomers are linked by interchain disulfide bonds that are placed between the N-terminal domain and the type I repeats. Pentameric TSPs differ from their trimeric paralogues in that they lack the procollagen homology domain and type I repeats and contain four instead of three type II repeats (Fig. 13.1). The numbers of amino acids in the pentameric N-terminal domains also differ substantially among the three proteins, whereas TSP1 and TSP2 have N-terminal domains that are very similar in size.

The macromolecular structure of TSP1 has been studied by rotary-shadowing electron microscopy and by X-ray crystallography. Rotary shadowing reveals a bola-like structure in which a large globule, representing the three N-terminal domains, is connected by strands to each of the three individual C-terminal globules [4]. This structure is supported by the crystal structures of the three type I repeats of TSP1 [5] and of a fragment consisting of the three type II and type III repeats and the C-terminal globule of TSP2 [6]. Several reviews provide additional information regarding the structures of the thrombospondins [7–9].

Departments of Biochemistry and Medicine, University of Washington, Seattle, WA, USA

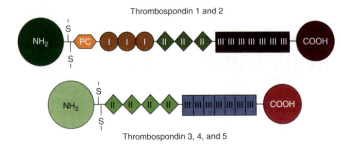

FIG. 13.1. A schematic representation of the structures of the individual chains of the trimeric thrombospondins 1 and 2, and the pentameric thrombospondins 3–5. In the trimeric proteins, the NH2-terminal heparin-binding domain is followed by an interchain disulfide knot linking all three chains, a procollagen domain (PC) that is homologous to sequences in the N-propeptides of types I-III procollagens, three type I repeats, also known as properdin or TSR domains, three type II or EGF-like repeats, seven type III or calcium-binding repeats, and a globular COOH-terminal domain. The pentameric thrombospondins differ from the trimeric proteins in that the sizes of the pentameric N-terminal domains vary considerably among the three proteins. These proteins also lack type I repeats, and contain four type three repeats. (Reproduced with permission from Bornstein, P. 'Matricellular Proteins' in G.J. Laurent and S.D. Shapiro, Eds, Encyclopedia of Respiratory Medicine, London, UK. Elsevier. 2006, pp 175–183)

Thrombospondins as Matricellular Proteins

Thrombospondins are frequently referred to as 'extracellular matrix' proteins. While this is valid in the sense that these proteins are secreted and function in the extracellular environment in close association with macromolecular structural elements such as collagen fibrils, there is no good evidence that thrombospondins actually serve in a structural capacity. TSP5 represents a possible exception in that although the phenotype of the TSP5 knockout mouse is grossly normal [10], mutations in the *THBS5* gene are responsible for the dwarfing syndromes, pseudoachondroplasia and multiple epiphyseal dysplasia [11]. Instead, TSP1 and TSP2 function as extracellular modulators of cellular function. These properties are achieved by the ability of these proteins to interact with a wide variety of cell-surface signaling receptors, as well as with growth factors, cytokines, and other bioactive molecules such as matrix metalloproteinases (MMPs).

A consequence of these properties is that the functions of TSP1 and TSP2 are highly complex and context-dependent, that is they are subject to differences that reflect the tissue and cellular environments in which these proteins are expressed. As a general rule, these proteins are not prevalent in normal adult animals, but their genes are induced during development and growth, and in response to injury. These characteristics have led to the application of the term 'matricellular proteins' to TSP1 and TSP2, as well as to SPARC, osteopontin, and to some members of the tenascin and CCN (Cyr, Connective tissue growth factor, Nov) families, which are unrelated struc-

turally but function similarly to thrombospondins [12, 13]. However, it should be noted that while matricellular proteins do not appear to play structural roles, structural proteins such as the collagens and fibronectin are clearly capable of engaging signaling receptors, and can thereby affect cell function.

Since relatively little is known concerning the functions of the pentameric thrombospondins, and particularly in regard to their role in angiogenesis if any, this chapter will focus on TSP1 and TSP2.

The Functions of Thrombospondins in Wound Healing

As matricellular proteins, thrombospondins are involved in a wide range of functions, including a role in synapse formation in the central nervous system [14, 15], but their participation in wound healing, and in the organization of the extracellular matrix (ECM), are of particular relevance to their functions as modulators of angiogenesis. Studies of wound healing in mice have presented investigators with an apparent paradox. Although the intrinsic functional properties of TSP1 and TSP2 proteins are quite similar, the phenotypes of mice that lack functional TSP1 or TSP2 genes (knockout mice) are very different [16, 17]. This paradox can be resolved by the realization that the promoter sequences upstream from the start of transcription in the two genes differ considerably. As a result, the spatial patterns of synthesis of the two encoded proteins in different cells and tissues, and the temporal program of synthesis during development, growth, and in response to injury, are also quite different. This dichotomy can best be illustrated by a study of excisional skin wound healing in TSP1-null, TSP2-null, and TSP1/TSP2 double-null mice. Mice that lack a functional TSP2 gene close skin wounds more rapidly and with less scarring. Histological examination of the wound bed as a function of time revealed prolonged vascularity and an abnormal organization of collagen fibers in its granulation tissue [18]. These findings reflect the known properties of TSP2, as indicated by the phenotype of TSP2 knockout mice [17]. By contrast, wound healing was delayed in TSP1-null mice and was accompanied by a reduction in blood vessels and inflammatory cells, relative to that in TSP2-null mice. [19].

Surprisingly, the healing response in TSP1/TSP2 double-null mice resembled that in TSP1-null animals, despite the fact that double-null mice also lack TSP2 [19]. These findings can be explained by the fact that TSP1 is normally released from platelets and secreted by inflammatory cells early in the wound healing process, and is strongly chemotactic for neutrophils and monocytes [20, 21] that are needed for the normal progression of the healing process. Thus, the presence or absence of TSP1 dictates the course of wound healing. A similar conclusion was reached by the use of antisense TSP1 oligonucleotides in wildtype (WT) mice [22].The importance of a normal spatial and temporal expression of TSP1 is further demonstrated by the finding that over-expression of TSP1 under the control of a keratin 14 promoter in transgenic mice also inhibited wound healing [23]. This finding would not have been predicted by the phenotype of the TSP1-null mouse and may reflect the extended window of expression provided by the K14 promoter in contrast to that of the endogenous TSP1 promoter [19, 24].

The Functions of TSP2 in the Organization of the ECM

The phenotype of the TSP1-null mouse shows relatively little in the way of connective tissue abnormalities, other than a mild spinal lordosis and reduced dermal matrix [16, 25]. This is true even though TSP1 is a major activator of latent TGFβ1 [25, 26], which in turn stimulates the expression of a number genes encoding matrix macromolecules, including type I procollagen and fibronectin. In contrast, abnormalities of the ECM represent a prominent feature of the phenotype of TSP2 knockout mice [17]. This difference reflects the cells that predominantly synthesize and secrete these TSPs: epithelial, endothelial, and related cells in the case of TSP1, and other mesenchymal cells, i.e., fibroblasts, smooth muscle cells, etc., in the case of TSP2.

TSP2-null mice have connective tissue defects that are obvious when the animal is examined. Thus, the skin is unusually stretchable and tendons and ligaments are loose, permitting the tail to be tied into a knot [17]. Tensile strength measurements revealed that TSP2-null skin ruptures at a lower load and shows increased ductility. At the light microscope level, collagen fibers in skin appear disorganized and lack the normal predominant parallel orientation to the epidermal surface; by electron microscopy, fibrils are larger and have irregular contours. Dermal fibroblasts also show a defect in adhesion [17].

Although the biochemical basis for the changes in collagen fibrillogenesis in TSP2-null mice is not thoroughly understood, it is likely that alterations in pericellular levels of MMP2, and possibly MMP9, are responsible. Both TSP2 and TSP1 have been shown to bind MMP2 [27, 28], and both TSPs are recognized and endocytosed by the scavenger receptor, low density lipoprotein receptor-related protein 1 (LRP1) [29–31]. It has been shown that MMP2 levels are increased in cultures of TSP2-null fibroblasts, and that this increase is due to the failure of the MMP2-TSP2 complex to be removed by LRP1 [30]. Thus, TSP2 acts as a clearance factor that modulates the levels of MMP2 in the pericellular environment; in the absence of TSP2, increased proteolysis of nascent collagen fibrils and cell-surface proteins is likely to occur. Although type I collagen molecules themselves should be resistant to the action of MMP2 at body temperature, fibril-associated molecules, such as type V collagen and decorin that are important for organization of a collagenous matrix, are susceptible, and proteolysis of these molecules could therefore contribute to the phenotype of the TSP2-null mouse. In addition, direct evidence for reduced activity of a cell-surface-associated enzyme, tissue transglutaminase (tTG), in TSP2-null mice has been reported [32]. tTG functions as a co-receptor for β1 and β3 integrins and introduces covalent isopeptide cross-links in many constituents of the ECM [33]. Isopeptide cross-links were reduced in both uninjured skin and in healing wounds of TSP2-null mice. Thus, reduced activity of tTG could account for both the reduced adhesion of TSP2-null fibroblasts and the compromised integrity of connective tissues in the TSP2-null mouse. Since fibroblasts synthesize both TSP2 and TSP1, why is there little or no connective tissue phenotype in the TSP1-null mouse? A likely answer is that fibroblasts, at least in culture, produce much more TSP2 than TSP1. Therefore, TSP2 could compensate for the lack of TSP1, but the reverse may not be effective.

Thrombospondins as Endogenous Inhibitors of Angiogenesis

Introduction

The process of angiogenesis, the growth of blood vessels from extant vessels, is complex and involves not only the proliferation of endothelial cells (EC), but also their migration and formation of tube-like structures, the association of pericytes and the production of an extracellular matrix by these cells, and the elaboration of a basal lamina by both EC and pericytes. TSP1 was the first endogenous inhibitor of angiogenesis to be described. It was identified as a 140-kDa fragment of a TSP1 chain (gp140) in the culture medium of baby hamster kidney (BHK) cells [34], and its expression was linked to that of a tumor suppressor gene in these cells [35]. Since its N-terminal sequence matched a sequence in the type I repeats of human TSP1, gp140 was presumably derived from secreted TSP1 by limited proteolysis. In the interval since the discovery of gp140 and its relation to TSP1, a number of other endogenous inhibitors of angiogenesis have been described (see related chapters for discussion of endogenous inhibitors of angiogenesis), and several reviews have been published on the role of thrombospondins in the inhibition of angiogenesis [36–43]. This chapter will focus on some of the more recent aspects of the complex physiology and molecular biology of TSP1 and TSP2 as inhibitors of angiogenesis.

Transcriptional Regulation of Thrombospondins

Knowledge of the regulation of expression of TSP1 and TSP2 is central to an understanding of their roles in modulating the complex process of angiogenesis. As might be expected from the intricate regulation of cell functions and cell-matrix interactions performed by thrombospondins, this control is also likely to be highly varied. In a survey of many cytokines and bioactive compounds, both positive and negative effects on the expression of TSP1 were observed, and in many cases different cells in culture responded differently

[44]. Current evidence indicates that regulation occurs predominantly at the transcriptional level; in the few cases in which increases in protein levels occurred in the absence of increases in mRNA levels, an increase in mRNA stability was not excluded.

Early studies of the TSP1 promoter identified transcription factor-binding sequences upstream from the transcription start site that were responsible for the regulation of expression of the rodent and human genes [7, 45,46]. Transcriptional repression of TSP1 has been substantiated in rodent cells transformed by *v-src* [47], and in keeping with its function as an inhibitor of angiogenesis, silencing by methylation of the TSP1 gene promoter has been documented in neuroblastomas [48] and gastric carcinomas [49]. Id1, a member of the helix-loop-helix family of transcription factors, also represses transcription of the *Thbs1* gene [50], and this effect is likely to be important in the regulation of angiogenesis in embryonic development and tumorigenesis [51]. Additional consideration of the roles played by transcriptional regulation of TSPs in tumor growth and metastasis will be given later in this chapter. On the other hand, the increase in TSP1 mRNA and protein in response to hypoxia has been attributed to stabilization of mRNA levels; the biological rationale for an increase in TSP1 under conditions of hypoxia is complex and may be unrelated to its function as an inhibitor of angiogenesis, because these experiments were performed in human umbilical vein and aortic endothelial cells [52].

It is of interest that TSP1 and TSP2 are not necessarily coordinately expressed in response to transcriptional activators or repressors. Rac^{V12}, a constitutively active mutant of Rac, whose function is dependent on the production of reactive oxygen species, induces TSP2 in human aortic endothelial cells [53]. However there is no effect on expression of TSP1. This finding undoubtedly reflects the fact that the promoters of the two genes differ substantially in their nucleotide sequences, and it underscores the importance of distinguishing between the conclusions drawn from the results of experiments in vivo and from those in which purified thrombospondins are added to systems in vitro.

Cellular Mechanisms in the Inhibition of Angiogenesis

TSP1 and TSP2 are capable of modulating the adhesion and migration of EC, and can inhibit tube formation and other responses to many different angiogenic stimuli in experiments both in vivo and ex vivo [37, 42, 54]. There is also ample evidence that these TSPs can function as endogenous inhibitors of angiogenesis in a variety of circumstances [36–43], but the mechanisms that are involved in inhibition in vivo have not been well- defined. TSP1 causes apoptosis in numerous cells in culture, including immune cells [55], promyelocytic leukemia cells [56], fibroblasts [57], and microvascular and aortic EC [55, 58–60]. However, a role for thrombospondins in causing apoptosis of normal EC in vivo is more difficult to ascertain and has not been well-documented. In a study of human melanoma in nude mice treated with TSP1, apoptotic EC were detected within the tumor, but not more frequently in the surrounding normal tissue than was observed in untreated mice [58]. Some indication that apoptosis of EC may not be a major cause of inhibition of tumor growth and angiogenesis in mice is also provided by studies of A431 squamous cell carcinoma cells in nude mice [61], and of chemically induced squamous cell carcinomas in normal mice and in mice that over-expressed TSP1 transgenically in skin [62]. In A431 cells that over-expressed TSP2, a marked decrease in tumor growth and blood vessel density was observed without an increase in apoptosis in tumor cells or EC [61]; in mice that over-expressed TSP1, apoptosis was observed in tumor epithelial cells, but again not in EC, despite diminished tumor angiogenesis [62].

TSP1 and TSP2 Cause Apoptosis of EC in Culture Via their Interaction with CD 36

The contributions of several laboratories have established that TSP1 is capable of causing apoptosis by its activation of the caspase proteolytic pathway, and that this activation occurs subsequent to its interaction with the scavenger receptor, CD36 [58, 60, 63, 64]. The signal transduction pathway responsible for causing apoptosis in EC consists of the activation of the *src* family member, p59^{fyn}, which leads to the phosphorylation of the stress-responsive kinases, p38MAPK and c-Jun N-terminal kinase (JNK). These kinases, in turn, stimulate the transcription of a number of genes, including the gene encoding a membrane-bound protein, Fas ligand. Fas ligand then binds to membrane-bound Fas on an adjacent cell, thereby launching the caspase death pathway, in which caspase 3 plays a prominent role [58]. The ability of natural inhibitors of angiogenesis, such as TSPs, to target pathological angiogenesis selectively, as occurs in tumors [58], results from the fact that rapidly proliferating EC markedly increase their expression of Fas. Fas then sensitizes these cells to apoptosis, which is a consequence of its interaction with TSP-stimulated Fas ligand [65].

TSP2, like TSP1, inhibits bFGF-induced angiogenesis in a corneal pocket assay in mice [66]. This inhibition depends on the presence of CD36 because the ability of TSP2 to inhibit bFGF-induced angiogenesis is lost in CD36-null mice; additional evidence for the binding of TSP2 to CD36 is provided by its reduced binding to macrophages in the presence of anti-CD36 antibody [66]. It is therefore likely that TSP2 will also cause apoptosis in cells that express CD36. The activities of both TSP1 and TSP2 are inhibited by histidine-rich glycoprotein (HRGP), which also serves as a ligand for CD36 and therefore acts as a decoy receptor for TSPs [66, 67].

It is of interest that human umbilical vein endothelial cells (HUVECs), seeded on silicone sheets and subjected to pulsatile uniaxial stretching, do not proliferate or undergo apoptosis. In contrast, in the absence of a mechanical stimulus, these cells divide, express integrin-associated protein (IAP, also known as CD47), integrin αvβ3, and TSP1, and undergo a degree of apoptosis [68]. Similar findings were reported in HUVECs subjected to irregular flow, in comparison to laminar flow. Since TSP1 binds to the CD47/ integrin αvβ3 complex [69], it is likely that signaling initiated by activation of this complex merges with that generated by activation of CD36.

Antiproliferative Functions of TSP1 and TSP2

In addition to the ability of TSPs to cause endothelial cell death, there is emerging evidence for an anti-proliferative effect, resulting from an inhibition of cell cycle progression. Such an effect is biologically reasonable since there are circumstances under which homeostasis would be more desirable than cell death, e.g., maintenance of an intact vascular endothelium in the absence of growth, or repair of an injury. The ability of TSPs to restrain EC growth can be achieved indirectly by their binding to growth factors such as VEGF, bFGF and hepatocyte growth factor/scatter factor [70, 71]. The fate of the resulting complexes is likely to be endocytosis by the scavenger receptor, LRP1 (see section on NH2-Terminal Domain, below). Alternatively, TSPs can act directly on EC by activating a signal transduction pathway [72]. In the latter case it was shown that the growth of human microvascular endothelial cells (HMVECs) was inhibited by a caspase-independent mechanism, probably by inhibition of cell cycle progression, although the cell surface receptor(s) and signaling pathway responsible for this effect have not yet been identified [72].

Angiogenic or Biphasic Functions of TSP1

Despite the overwhelming preponderance of data supporting the angio-inhibitory functions of TSPs, there are a number of reliable studies that are also consistent with a proangiogenic or biphasic function for these proteins [73–76]. In support of an angiogenic function for TSPs, capillary outgrowth in the rat aortic ring model of angiogenesis was measured directly and was found to be stimulated by TSP1. This stimulation was dependent on increased proliferation of myofibroblasts, but the putative agent responsible for the proliferation of EC was not fully characterized [73]. In a subsequent study, the capacity of bovine aortic endothelial (BAE) cells to invade and form tubes in collagen gels was increased at low concentrations of TSP1, but inhibited at higher concentrations [74]. By way of explanation, the authors propose that increased microvessel-like tubes can result from the initial increase in MMP9 concentrations, but that further increases were inhibitory due to proteolysis of the ECM. It has also been proposed that the dual behavior of TSP1 could reflect its complex domain structure, which could include both negatively and positively acting motifs [75]. Support for this suggestion was provided by the finding that a 25-kDa heparin-binding, N-terminal fragment of TSP1 promoted angiogenesis in the rabbit cornea, whereas the 140-kDa C-terminal fragment was inhibitory [75]. A biochemical basis for both positive and negative responses to TSP1 has been provided by studies

of EC-dependent activation of integrin α3β1 [76]. Whereas interaction of soluble TSP1 with CD47, CD36, or heparan sulfate proteoglycan receptors on several types of EC inhibited angiogenesis in the chicken allantoic membrane (CAM) assay, the interaction of immobilized TSP1 with integrin α3β1, which involved VE-cadherin in cell-cell interactions, stimulated angiogenesis [76]. These findings are consistent with the matricellular nature of TSPs, which enables different domains of the proteins to interact with different receptors on EC, and thereby elicit very different responses [12]. The findings also underscore the importance of the ECM in influencing the physiological responses of EC.

Specific Domains in Thrombospondins Determine their Functions in Angiogenesis

The complex domain structure of TSP1 and TSP2, illustrated in Figs. 13.1 and 13.2, forms the basis for the multiple, sometimes opposing, functions attributable to these proteins. As matricellular proteins, TSPs serve as multifunctional regulators of cell-cell and cell-matrix interactions, and in addition bind directly to bioactive molecules such as growth factors, cytokines, and proteases [8, 13, 41, 77]. This section will focus on interactions that influence the process of angiogenesis, but Table 13.1 also lists other interactions

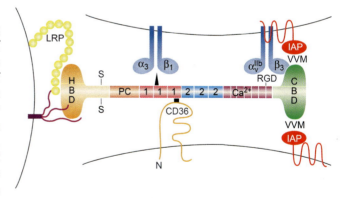

FIG. 13.2. A schematic representation of a generic TSP1 or TSP2 monomer together with the location of the binding sites for some of the major receptors with which the protein interacts. The structure of the chain is described in the legend to Fig. 1. The N-terminal heparin-binding domain (*HBD*) interacts with cell-surface heparan sulfate proteoglycans (*purple*), and the low density lipoprotein receptor-like protein, *LRP1*, (*yellow circles*); the type 1 repeats bind to α3β1 integrin and *CD36*; the *RGD* sequence in the last Ca^{2+}-binding type 3 repeat interacts with integrins αvβ3 and α2bβ3; and the COOH-terminal cell-binding domain (*CBD*) contains two valine-valine-methionine sequences that can interact with two integrin-associated protein (IAP/CD47) receptors on different cells. More information on these interactions and on interactions with other receptors is provided in the text. Modified from Brown and Frazier [69] by permission of *Trends in Cell Biology*.

TABLE 13.1. Interactions and functions of specific domains in TSP1 and 2 in the regulation of angiogenesis.

Domain	Binding receptors or molecules	Functions	References
NH₂-Terminal	HSPGs	Adhesion of EC; co-receptor for LRP1	[8, 30, 77]
	LRP1	Clearance of MMPs 2 and 9	[79,80]
	α3β1	Stimulates angiogenesis; adhesion spreading and chemotaxis of cells[a]	[82]
	α6β1	Adhesion of HUMVEC to immobilized TSPs; chemotaxis of EC to soluble TSPs	[83]
Type I Repeats	Small latent TGFβ1 complex	Activation of latent TGFβ1[a]	[85,86]
	CD36	Induces apoptosis in EC in vitro and in tumor-derived blood vessels in vivo	[58,64]
	α4β1	Supports adhesion of venous EC and mediates chemotaxis of microvascular EC	[88]
Types I and II Repeats	pan-β1 integrins	Modulate EC adhesion, inhibit EC migration	[89,102]
Type III Repeats	αvβ3, αIIbβ3	Adhesion of EC	[8,93]
COOH-Terminal	CD47/IAP	Antagonizes NO-mediated vasodilation	[100]
	αvβ3, αIIbβ3	Adhesion and spreading of EC and platelets	[69,95]

HSPGs Heparan sulfate proteoglycans; *HUMVEC* human umbilical vein endothelial cells; EC, endothelial cells
[a] Function not shared with TSP2

that are important for an understanding of the functions of thrombospondins.

NH₂-Terminal Domain

The ability to synthesize fragments of TSPs that encompass one or a few domains recombinantly in mammalian, insect, or bacterial cell systems, and to generate function-blocking monoclonal antibodies to specific domains [78], has made it possible to assign particular functions of TSPs to specific domains in the proteins, and to identify peptide sequences that mediate these functions. The N-terminal domain, also known as the heparin-binding domain, serves to bind cell-surface heparan sulfate proteoglycans, which function as co-receptors for the low density lipoprotein-related protein (LRP1) (Fig. 13.2; Table 13.1). LRP1 is a scavenger receptor that is capable of clearing complexes of TSP1 or TSP2 with MMP2, MMP9, or VEGF from the pericellular environment of fibroblasts and other mesenchymal cells [71, 79, 80], and the reduction in protease activity has an inhibitory effect on angiogenesis [81]. The N-terminal domain of TSP1 also binds the integrins α3β1, α4β1, and α6β1. In addition to stimulating angiogenesis [76], TSP1-bound α3β1 mediates the adhesion, spreading, and chemotaxis of a number of different cells [82]. This function is not shared by TSP2 because the recognition sequence for this integrin differs between the two proteins. Integrin α6β1 is involved in capillary morphogenesis and in adhesion of microvascular EC to immobilized TSPs, as well as in chemotaxis to soluble TSPs [83]. In addition to interactions that affect angiogenesis directly, the N-terminal domain of TSPs performs other functions that can influence the growth of blood vessels indirectly [84].

Types I, II, and III Repeats

The type I or thrombospondin structural homology repeats (TSR) mediate some of the more important interactions of the TSPs, including those with the small latent TGFβ1 complex [85, 86] and with CD36 [58, 64]. As described earlier in this chapter, CD36, which recognizes the sequence CSVTCG in both TSP1 and TSP2, mediates the proapoptotic function of TSPs on rapidly dividing EC such as cells in culture and in tumors in vivo. The ability of TSP1 to generate active TGFβ1 from the small latent complex enables this protein to elicit some of the panoply of effects produced by this cytokine [26, 87]. It is of interest that the activating sequence, KRFK, which is located between the first and second type I repeats in TSP1, is replaced in TSP2 by KRIR, which does not activate latent TGFβ1 [85]. This difference in sequence between the two paralogues accounts for a significant fraction of the differences in their properties.

The interaction of the type I repeats with α4β1 supports the adhesion of venous EC and mediates chemotaxis of microvascular EC. This interaction may also contribute to the pro-angiogenic functions of TSPs [88]. In addition to sites in the NH₂-terminal domains of TSPs that are relatively specific for individual β1 integrins, some pan-β1-specific sites in the type I and II repeats recognize multiple β1 integrins and serve to modulate EC adhesion [89]. Subdomains in the type I repeats that discriminate in their ability to inhibit angiogenesis induced by FGF-2 and VEGF have also been identified [90]. It is of interest that ADAMTS I, a member of a family of proteases that contain one or more TSP type I repeats [91], is capable of cleaving both TSP1 and TSP2 into fragments that inhibit proliferation of EC [92]. Finally, both TSP1 and TSP2 contain an RGD sequence in the last of the type III repeats. This 'classical' integrin-binding site is recognized by integrins αIIbβ3 and αvβ3 and serves to promote adhesion of the TSPs to EC, platelets, and a number of other cells [8, 93]. However, the extent of adhesion is dependent on the conformation of this domain, which can vary as a function of intramolecular thiol-disulfide exchange, catalyzed by protein disulfide isomerase, and the resulting differences in the patterns of intrachain disulfide bonds [8, 94].

COOH-Terminal Domain

CD47/IAP represents the major receptor that influences the process of angiogenesis in the COOH-terminal domain of TSP1 and TSP2 [69]. This cell-binding region is localized to two sequences, each of which shares the adhesion motif, VVM [95]. The first of these peptide sequences is often referred to as 4N1K. CD47 associates with integrin αvβ3 on EC to promote adhesion and spreading on vitronectin and other substrates, and with αIIbβ3 on platelets to stimulate spreading and aggregation. 4N1K inhibits FGF2-induced tube formation by a murine brain capillary EC line [96]. The mechanisms by which CD47 and associated integrins influence angiogenesis are not fully understood, but there is evidence in C32 human melanoma cells, which also express αvβ3 and CD47, that signal transduction involves a heterotrimeric G$_i$ protein, since the formation of a stable CD47/αvβ3/G$_i$ protein complex can be inhibited by pertussis toxin [97].

Very recently, evidence has been presented that implicates CD47 in the inhibition of EC and smooth muscle cell outgrowth from murine muscle biopsies that were stimulated with nitric oxide (NO) [98–100]. (Refer to the chapter on nitric oxide and angiogenesis for further discussion.) Low to moderate concentrations of NO stimulate angiogenesis by acting on EC to activate soluble guanyl cyclase (sGC), which increases cellular levels of cGMP and leads, in turn, to activation of several phosphorylation-driven pathways that culminate in increased angiogenesis [99, 100]. The mechanism by which ligation of CD47 leads to an inhibition of sGC is not known. The stimulation of EC proliferation by VEGF also involves an increase in endogenous NO, which is achieved by activation of Akt, and subsequently of endothelial nitric oxide synthase (eNOS) [100].

It is notable that the crystal structure of a fragment of human TSP1 containing the type III repeats and the C-terminal domain predicts that the two peptide VVM adhesion sequences are buried within the three-dimensional structure of the protein and are not exposed to the surface of the protein, as might be expected if these motifs functioned to interact with cell-surface receptors [101]. This apparent contradiction could be resolved if one takes into account the finding that the structure of the type III repeats/ C-terminal domain of the protein is highly sensitive to Ca^{2+} ion concentration [101], and that its structure also varies significantly with changes in the patterns of intrachain disulfide bonds [8, 94].

The Role of Thrombospondins in Tumor Growth and Metastases

The identification of TSP1 as an inhibitor of angiogenesis, and an appreciation of its regulation by an incompletely characterized tumor-suppressor gene in BHK cells, were achieved 18 years ago [34, 35]. Subsequently, it was demonstrated that the transfection of either the cDNA for intact TSP1, or for its N-terminal domain into src-NIH 3T3 cells markedly suppressed angiogenesis and tumorigenicity when the transfected cells were injected into nude mice [103]. The potential ability of TSP1 and TSP2 to inhibit the growth of tumors and the spread of metastases is clearly one of the most important clini-

cal applications of studies designed to elucidate the mechanisms by which these endogenous inhibitors of angiogenesis function. [104]. However, a combination angiostatic therapy, which targets different aspects of the angiogenic process (105), including those used by TSPs and their derivatives, may eventually prove most effective in inhibiting tumor angiogenesis.

The Regulation of Tumor Growth and Metastases by Thrombospondins

There are a number of considerations that underscore the difficulty in arriving at a definitive conclusion regarding the ability of TSPs to inhibit tumor growth and metastases. Thus, there are several different sources of TSPs in tumors, including the malignant cells, and the host-derived vasculature and stroma, and these cellular compartments are subject to different and complex regulation. Furthermore, TSPs are multi-domain proteins that can function either positively or negatively in modulating angiogenesis (Table 13.1). It is therefore not surprising that both inhibitory [106–109] and stimulatory [106, 110, 111] effects have been attributed to the role of TSP expression in influencing the extent of tumor growth and metastatic spread.

There are a number of experimental systems in which TSPs have been shown to be at least partially inhibitory for angiogenesis and carcinogenesis. When human squamous cell carcinoma cell lines, stably transfected to express human TSP1, were transplanted into nude mice, a reduction in tumor growth or abolition of tumor formation was observed [112]. These effects were indirect, possibly resulting from reduced vascularization by the host, since the intrinsic properties of the transfected cells, such as proliferation rates and anchorage-independent growth, were unchanged from those of control cells. In another study, squamous cell carcinomas were generated by a chemical skin carcinogenesis regimen in control mice and in transgenic mice in which over-expression of TSP1 had been targeted to the epidermis [62]. In this study, local over-expression of TSP1 delayed and reduced the development of premalignant lesions, but failed to inhibit their conversion to squamous cell carcinomas, or the spread of tumors to regional lymph nodes. In a related study, a continuous source of circulating TSP2 was provided by a peritoneal biodegradable implant, consisting of retrovirally transduced fibroblasts that expressed high levels of the protein in nude mice [113]. These bioimplants inhibited both angiogenesis and growth of several different tumors implanted at distant sites.

Mechanisms of Action of Thrombospondins in Regulating Tumor Growth

Since TSPs are capable of inhibiting angiogenesis in many different ways (see section on Cellular Mechanisms in the Regulation of Angiogenesis in this chapter), it stands to reason that these mechanisms should also apply to the regulation of tumor growth. In addition, there are other mechanisms by which TSPs could contribute to the inhibition of tumor growth. For example, TSP1 binds VEGF directly and the complex can then be endocytosed by the scavenger receptor, LRP1, which reduces the activity of this proangiogenic cytokine (71). This clearance function is analogous to that by which TSPs reduce pericellular levels of MMP2 [30]. In another instance, TSP1 has been shown to inhibit NO-mediated vascular smooth muscle relaxation [114]; because NO stimulates angiogenesis in EC (see section on COOH-Terminal Domain in this chapter) this activity would also be expected to inhibit angiogenesis.

The tumor suppressor gene, p53, inhibits angiogenesis by stimulating the production of TSP1, whereas loss of the functional gene induces VEGF [115, 116]. Similarly, the expression of a p53-activated gene, BAI1 (brain-specific angiogenesis inhibitor1), was absent or significantly reduced in 8 out of 9 glioblastoma cell lines, and a recombinant protein containing the five TSP type I repeats present in BAI1 inhibited the neovascularization stimulated by bFGF in the rat cornea [117]. Additional support for a correlation between the loss of expression of TSP1 and the malignancy of glioblastomas is provided by an experiment in which a WT chromosome 10, containing many of the tumor suppressor genes that are lost in glioblastomas, was reintroduced into glioblastoma cell lines. The resulting cells lost their ability to form tumors in nude mice and increased their expression of TSP1 [118]. More recently, BAI1 has been shown to be a transmembrane protein with a conserved G-protein-coupled receptor proteolytic cleavage site in its extracellular domain. Cleavage at this site released a protein fragment, termed vasculostatin, which contains the five TSP type I repeats in BAI1, and was shown to inhibit migration of EC in vitro and tumor angiogenesis and growth in mice [119].

The Potential for the Clinical use of TSPs in Inhibition of Tumor Growth

Based on the studies that have been described in this chapter, the potential for the clinical use of TSPs, or derivatives thereof, as inhibitors of tumor growth and metastatic spread is good. Evidence has been presented showing that expression of TSP1 in tumor and perivascular cells contributes to the suppression of tumor growth and EC apoptosis by low-dose cyclophosphamide in mice injected with Lewis lung carcinoma or B16F10 melanoma cells [120], and in a prostate tumor model in rats [121]. Thus, it is reasonable to propose that the delivery of exogenous bioactive fragments of TSP to tumors might constitute an effective mode of therapy.

Because the large size and protease sensitivity of the thrombospondin protein, ~435 kDa, limit its production and hamper its use as a therapeutic agent, efforts have been made to exploit its domain structure and our knowledge of the function of specific domains and sequences, in order to design fragments and peptides as therapeutic agents that are easier to deliver, less immunogenic, and less susceptible to proteolysis. Retro-

and adenoviral vectors expressing an N-terminal fragment of murine TSP2, which contained both the N-terminal globular domain and the type I repeats, were used to transduce syngeneic cell lines. Injection of these cells directly into several different tumors in mice led to a substantial reduction in the sizes and vascularity of the tumors [122]. In another study, a fragment of TSP1, containing only the three type I repeats, was successful in inhibiting the angiogenesis that accompanied the infection of human glioma cells by an oncolytic herpes simplex virus in athymic mice [123].

Finally, mimetic peptides, i.e., peptides which mimic sequences in the type I repeats of TSP1 that have been implicated in its antiangiogenic activity, are being tested for their anti-tumor activity and for their utility in complementing established anti-tumor agents. ABT-526 and ABT-510 are two chemically modified nonapeptides, chosen from several hundred that were designed based on the heptapeptide Gly-Val-Ile-Thr-Arg-Ile-Arg, which represents a sequence in the second type I repeat of TSP1 [124]. Modifications were performed to increase the stability of these peptides in vivo. ABT-526 was found to reduce bFGF-induced angiogenesis in the rat cornea by 92%. ABT-510, while less active than ABT-526 in inhibiting EC migration, was more active in a tube formation assay. In a Phase I study, ABT-510, when added to 5-fluorouracil and leucovorin in the treatment of patients with solid tumors, showed no pharmacokinetic interactions and no ill effects [125]. Two trials of the use of these peptide mimetics have recently been reported in dogs. In one study of spontaneously arising malignant tumors in 242 pet dogs, no dose-limiting toxicity in any of the dogs was observed and 42 dogs showed a greater than 50% reduction in tumor size. The two peptides were equally effective [126]. In a second prospective, randomized, placebo-controlled study of pet dogs with relapsed non-Hodgkin's lymphoma, ABT-526 was added to chemotherapy with lomusine (CeeNu). While no difference was observed in the response rate, there was a significant increase in duration of the response and in the time to progression of responding cases. No toxicity related to ABT-526 was observed [127]. These encouraging findings, together with others documented in this chapter, make it likely that the efforts of many basic scientists to understand the complex mechanisms underlying the functions of TSPs will be translated into effective therapies for the treatment of invasive tumors.

References

1. Adams JC, Monk R, Taylor, AL, et al. Characterisation of *Drosophila* thrombospondin defines an early origin of pentameric thrombospondins. J Mol Biol 2003; 328:479–94.
2. McKenzie P, Chadalavada SC, Bohrer J, et al. Phylogenomic analysis of vertebrate thrombospondins reveals fish-specific paralogues, ancestral gene relationships and a tetrapod innovation. BMC Evol Biol 2006; 6:33–48
3. Sodersten F, Ekman S, Schmitz M, et al. Thrombospondin-4 and cartilage oligomeric matrix protein form heterooligomers in equine tendon. Connect Tissue Res 2006; 47:85–91.

4. Lawler J, Derick LH, Connolly JE, et al. The structure of human platelet thrombospondin. J Biol Chem 1985; 260:3762–72.

5. Tan K, Duquette M, Liu JH, et al. Crystal structure of the TSP-1 type 1 repeats: a novel layered fold and its biological implication. J Cell Biol 2002; 159:373–82.

6. Carlson CB, Bernstein DA, Annis DS, et al. Structure of the calcium-rich signature domain of human thrombospondin-2. Nat Struct Mol Biol 2005; 12:910–14.

7. Bornstein P. Thrombospondins: structure and regulation of expression. FASEB J 1992; 6:3290–99.

8. Adams JC. Thrombospondins: Multifunctional regulators of cell interactions. Annu Rev Cell Dev Biol 2001; 17:25–51.

9. Adams JC, Lawler J. The thrombospondins. Int J Biochem 2004; 36:961–68.

10. Svensson L, Aszodi A, Heinegard D, et al. Cartilage oligomeric matrix protein-deficient mice have normal skeletal development. Mol Cell Biol 2002; 22:4366–71.

11. Briggs MD, Chapman KL. Pseudoachondroplasia and multiple epiphyseal dysplasia: mutation review, molecular interactions, and genotype to phenotype correlations. Hum Mutat 2002; 19:465–78.

12. Bornstein P. Thrombospondins as matricellular modulators of cell function. J Clin Invest 2001; 107:929–34.

13. Bornstein P, Sage EH. Matricellular proteins: extracellular modulators of cell function. Current Opin Cell Biol 2002; 14:608–16.

14. Christopherson KS, Ullian EM, Stokes CCA, et al. Thrombospondins are astrocyte-secreted proteins that promote CNS synaptogenesis. Cell 2005; 120:421–33.

15. Cáceres M, Suwyn C, Maddox M, et al. Increased cortical expression of two synaptogenic thrombospondins in human brain evolution. Cerebral Cortex 2007; 17:2312–21.

16. Lawler J, Sunday M, Thibert V, et al. Thrombospondin-1 is required for normal murine pulmonary homeostasis and its absence causes pneumonia. J Clin Invest. 1998; 101: 982–92.

17. Kyriakides TR, Zhu Y-H, Smith LT, et al. Mice that lack thrombospondin 2 display connective tissue abnormalities that are associated with disordered collagen fibrillogenesis, an increased vascular density, and a bleeding diathesis. J Cell Biol 1998; 140: 419–30.

18. Kyriakides TR, Tam JWY, Bornstein P. Accelerated wound healing in mice with a disruption of the thrombospondin 2 gene. J Invest Dermatol 1999; 113:782–87.

19. Agah A, Kyriakides TR, Lawler J, et al. The lack of thrombospondin-1 (TSP1) dictates the course of wound healing in double-TSP1/TSP2-null mice. Am J Pathol 2002; 161:831–39.

20. Mansfield PJ, Suchard SJ. Thrombospondin promotes both chemotaxis and haptotaxis in neutrophil-like HL-60 cells. J Immunol 1993; 150:1959–70.

21. Mansfield PJ, Suchard SJ. Thrombospondin promotes chemotaxis and haptotaxis of human peripheral blood monocytes. J Immunol 1994; 153:4219–29.

22. DiPietro LA, Nissen NN, Gamelli RL, et al. Thrombospondin 1 synthesis and function in wound repair. Am J Pathol 1996; 148:1851–60.

23. Streit M, Velasco P, Riccardi L, et al. Thrombospondin-1 suppresses wound healing and granulation tissue formation in the skin of transgenic mice. EMBO J 2000; 19:3272–82.

24. Kyriakides TR, Bornstein P. Matricellular proteins as modulators of wound healing and the foreign body response. Thromb Haemost 2003; 90:986–92.

25. Crawford SE, Stellmach V, Murphy-Ullrich JE, et al. Thrombospondin-1 is a major activator of TGF-β1 in vivo. Cell 1998; 93:1159–70.

26. Murphy-Ullrich JE, Poczatek M. Activation of latent TGF-beta by thromospondin-1: mechanisms and physiology. Cytokine Growth Factor Rev 2000; 11:59–69.

27. Yang Z, Kyriakides TR, Bornstein P. Matricellular proteins as modulators of cell-matrix interactions: Adhesive defect in thrombospondin 2-null fibroblasts is a consequence of increased levels of matrix metalloproteinase-2. Mol Biol Cell 2000; 11:3353–64.

28. Bein K, Simmons M. Thrombospondin type 1 repeats interact with matrix metalloproteinase 2: Regulation of metalloproteinase activity. J Biol Chem 2000; 275:32167–73.

29. Chen H, Strickland DK, Mosher DF. Metabolism of thrombospondin 2: Binding and degradation by 3T3 cells and glycosaminoglycan-variant Chinese hamster ovary cells. J Biol Chem 1996; 271:15993–99.

30. Yang Z, Strickland DK, Bornstein P. Extracellular matrix metalloproteinase 2 levels are regulated by the low density lipoprotein-related scavenger receptor and thrombospondin 2. J Biol Chem 2001; 276:8403–08.

31. Strickland DK, Kounnas MZ, Argraves WS. LDL receptor-related protein: a multiligand receptor for lipoprotein and proteinase catabolism. FASEB J 1995; 9:890–98.

32. Agah A, Kyriakides TR, Bornstein, P. Proteolysis of cell-surface tissue transglutaminase by matrix metalloproteinase-2 contributes to the adhesive defect and matrix abnormalities in thrombospondin-2-null fibroblasts and mice. Am J Pathol 2005; 167:81–88.

33. Lorand L, Graham RM. Transglutaminases: Crosslinking Enzymes with pleiotropic functions. Nat Rev Nat Cell Biol 2003; 4:140–55.

34. Good DJ, Polverini PJ, Rastinejad F, et al. A tumor suppressor-dependent inhibitor of angiogenesis is immunologically and functionally indistinguishable from a fragment of thrombospondin. Proc Natl Acad Sci USA 1990; 87: 6624–28.

35. Rastinejad F, Polverini PJ, Bouck NP. Regulation of the activity of a new inhibitor of angiogenesis by a cancer suppressor gene. Cell 1989; 56:345–55.

36. DiPietro LA. Thrombospondin as a regulator of angiogenesis. In: Goldberg LD, Rosen EM, editors. Regulation of Angiogenesis. Basel: Birkhauser, 1997:295–14.

37. Dawson DW, Bouck NP. Thrombospondin as an inhibitor of angiogenesis. In: Teicher BA, editor. Antiangiogenic Agents in Cancer Therapy. Totowa: Humana Press, 1999:185–03.

38. Sheibani N, Frazier WA. Thrombospondin-1, PECAM-1, and regulation of angiogenesis. Histol Histopathol 1999: 14:285–94.

39. Carpizo D, Iruela-Arispe ML. Endogenous regulators of angiogenesis – emphasis on proteins with thrombospondin – type I motifs. Cancer Metast Rev 2000; 19:159–65.

40. de Fraipont F, Nicholson AC, Feige J-J, et al. Thrombospondins and tumor antiogenesis. Trends Mol Med. 2001; 7: 401–07.

41. Lawler J. Thrombospondin-1 as an endogenous inhibitor of angiogenesis and tumor growth. J Cell Mol Med 2002; 6:1–12.

42. Armstrong LC, Bornstein P. Thrombospondins 1 and 2 function as inhibitors of angiogenesis. Matrix Biol 2003; 22:63–71.

43. Ren B, Yee KO, Lawler J, Khosravi-Far R. Regulation of tumor angiogenesis by thrombospondin-1. Biochimica Biophysics Acta 2005; 1765:178–88.

44. Kim S-A, Kang J-H, Cho I, et al. Cell-type specific regulation of thrombospondin-1 expression and its promoter activity by regulatory agents. Exp Mol Med 2001; 33:117–23.

45. Framson P, Bornstein P. A serum response element (SRE) and a binding site for NF-Y mediate the serum response of the human thrombospondin I gene. J Biol Chem 1993; 268:4989–96.

46. Shingu T, Bornstein P. Overlapping Egr-1 and SPI sites function in the regulation of transcription of the mouse thrombospondin I gene. J Biol Chem 1994; 269: 32551–57.

47. Slack JL, Bornstein P. Transformation by v-src causes transient induction followed by repression of mouse thrombospondin-1. Cell Growth Differ. 1994; 5:1373–80.

48. Yang Q-W, Liu S, Tian Y, et al. Methylation-associated silencing of the thrombospondin-1 gene in human neuroblastoma. Cancer Res 2003; 63:6299–10.

49. Oue N, Matsumura S, Nakayama H, et al. Reduced expression of the TSP1 gene and its association with promoter hypermethylation in gastric carcinoma. Oncology 2003; 64:423–29.

50. Volpert OV, Pili R, Sikder HA, et al. Id1 regulates angiogenesis through transcriptional repression of thrombospondin-1. Cancer Cell 2002; 2:473–83.

51. Benezra R, Rafii S, Lyden D. The Id proteins and angiogenesis. Oncogene 2001; 20:8334–41.

52. Phelan MW, Forman LW, Perrine SP, et al. Hypoxia increases thrombospondin-1 transcript and protein in cultured endothelial cells. J Lab Clin Med 1998; 132:519–29.

53. Lopes N, Gregg D, Vasudevan S, et al. Thrombospondin 2 regulates cell proliferation induced by Rac1 redox-dependent signaling. Mol Cell Biol 2003; 23: 5401–08.

54. Bornstein P. Thrombospondins as matricellular modulators of cell function. J Clin Invest 107:929–34.

55. Friedl P, Vischer P and Freyberg MA. The role of thrombospondin-1 in apoptosis. Cell Mol Life Sci. 2002; 59:1347–57.

56. Bruel A, Touhami-Carrier M, Thomaidis A, et al. Thrombospondin-1 (TSP-1) and TSP-1-derived heparin-binding peptides induce promyelocytic leukemia cell differentiation and apoptosis. Anticancer Res 2005; 25:757–64.

57. Graf R, Freyberg M, Kaiser D, et al. Mechanosensitive induction of apoptosis in fibroblasts is regulated by thrombospondin-1 and integrin associated protein (CD47). Apoptosis 2002; 7:493–98.

58. Jimenez B, Volpert OV, Crawford SE, et al. Signals leading to apoptosis-dependent inhibition of neovascularization by thrombospondin-1. Nature Med. 2000; 6:41–48.

59. Guo N-h, Krutzsch HC, Inman JK, et al. Thrombospondin I and type I repeat peptides of thrombospondin I specifically induce apoptosis of endothelial cells. Cancer Res. 1997; 57:1735–42.

60. Nör, JE, Mitra RS, Sutorik MM, et al. Thrombospondin-1 induces endothelial cell apoptosis and inhibits angiogenesis by activating the caspase death pathway. J Vasc Res 2000; 37:209–18.

61. Streit M, Riccardi L, Velasco P, et al. Thrombospondin-2: A potent endogenous inhibitor of tumor growth and angiogenesis. Proc Natl Acad Sci USA 1999; 96:14888–93.

62. Hawighorst T, Oura H, Streit M, et al. Thrombospondin-1 selectively inhibits early-stage carcinogenesis and angiogenesis but not tumor lymphangiogenesis and lymphatic metastasis in transgenic mice. Oncogene 2002; 21:7945–56.

63. Jiménez, B, Volpert OV, Reiher F, et al. c-Jun N-terminal kinase activation is required for the inhibition of neovascularization by thrombospondin-1. Oncogene 2001; 20:3443–48.

64. Dawson DW, Pearce SFA, Zhong R, et al. CD36 mediates the in vitro inhibitory effects of thrombospondin-1 on endothelial cells. J Cell Biol 1997; 138:707–17.

65. Volpert OV, Zaichuk T, Zhou W, et al. Inducer-stimulated Fas targets activated endothelium for destruction by anti-angiogenic thrombospondin-1 and pigment epithelium-derived factor. Nature Med 2002; 8:349–57.

66. Simantov R, Febbraio M, Silverstein RL. The antiangiogenic effect of thrombospondin-2 is mediated by CD36 and modulated by histidine-rich glycoprotein. Matrix Biol 2005; 24:27–34.

67. Simantov R, Febbraio M, Crombie R, et al. Histidine-rich glycoprotein inhibits the antiangiogenic effect of thrombospondin-1. J Clin Invest 2001; 107:45–52.

68. Graf R, Apenberg S, Freyberg M, et al. A common mechanism for the mechanosensitive regulation of apoptosis in different cell types and for different mechanical stimuli. Apoptosis 2003; 8:531–38.

69. Brown EJ, Frazier WA. Integrin-associated protein (CD47) and its ligands. Trends Cell Biol 2001; 11:130–35.

70. Margosio, B, Marchetti, D, Vergani, V, et al. Thrombospondin 1 as a scavenger for matrix-associated fibroblast growth factor 2. Blood; 102: 4399–4406

71. Greenaway J, Lawler J, Moorehead R, et al. Thrombospondin-1 inhibits VEGF levels in the ovary directly by binding and internalization via the low density lipoprotein receptor-related protein-1 (LRP-1). J Cell Physiol 2007; 210:807–18.

72. Armstrong, LC, Björkblom B, Hankenson KD, et al. Thrombospondin 2 inhibits microvascular endothelial cell proliferation by a caspase-independent mechanism. Mol Biol Cell 2002; 13:1893–05.

73. Nicosia RF, Tuszynski GP. Matrix-bound thrombospondin promotes angiogenesis in vitro. J Cell Biol 1994;124:183–93.

74. Qian X, Wang TN, Rothman VL, et al. Thrombospondin-1 modulates angiogenesis in vitro by up-regulation of matrix metalloproteinase-9 in endothelial cells. Exp Cell Res 1997; 235: 403–12.

75. Taraboletti G, Morbidelli L, Donnini S, et al. The heparin binding 25 kDa fragment of thrombospondin-1 promotes angiogenesis and modulates gelatinase and TIMP-2 production in endothelial cells. FASEB J. 2000; 14: 1674–76.

76. Chandrasekaran L, He C-Z, Al-Barazi H, et al. Cell contact-dependent activation of $\alpha 3 \beta 1$ integrin modulates endothelial cell responses to thrombospondin-1. Mol Biol Cell 2000; 11: 2885–00.

77. Lawler J. The functions of thrombospondin-1 and -2. Curr Opin Cell Biol 2000; 12:634–40.

78. Annis DS, Murphy-Ullrich JE, Mosher DF. Function-blocking antithrombospondin-1 monoclonal antibodies. J Thromb Haemost 2006; 4:459–68.

79. Yang Z, Strickland DK, Bornstein P. Extracellular matrix metalloproteinase 2 levels are regulated by the low density lipoprotein-related scavenger receptor and thrombospondin 2. J Biol Chem 2001; 276:8403–08.

80. Hahn-Dantona E, Ruiz JF, Bornstein P, et al. The low density lipoprotein receptor-related protein modulates levels of matrix metalloproteinase 9 (MMP-9) by mediating its cellular catabolism. J Biol Chem 2001; 276:15498–03.

81. Rodriguez-Manzaneque JC, Lane TF, Ortega MA, et al. Thrombospondin-1 suppresses spontaneous tumor growth and inhibits activation of matrix metalloproteinase-9 and mobilization of vascular endothelial growth factor. Proc Natl Acad Sci USA 2001; 98:12485–90.

82. Krutzsch HC, Choe BJ, Sipes JM, et al. Identification of an $\alpha_3 \beta_1$ integrin recognition sequence in thrombospondin-1. J Biol Chem 1999; 274:24080–86.

83. Calzada MJ, Sipes JM, Krutzsch HC, et al. Recognition of the N-terminal modules of thrombospondin-1 and thrombospondin-2 by $\alpha_6\beta_1$ integrin. J Biol Chem 2003; 278: 40679–87.

84. Elzie CA, Murphy-Ullrich JE. The N-terminus of thrombospondin: the domain stands apart. Int J Biochem Cell Biol 2004; 36:1090–01.

85. Schultz-Cherry S, Chen H, Mosher DF, et al. Regulation of transforming growth factor-β activation by discrete sequences of thrombospondin 1. J Biol Chem 1995; 270: 7304–10.

86. Ribeiro SMF, Poczatek M, Schultz-Cherry S, et al. The activation sequence of thrombospondin-1 interacts with the latency-associated peptide to regulate activation of latent transforming growth factor-b. J Biol Chem 1999; 274:13586–93.

87. Murphy-Ullrich JE, Poczatek M. Activation of latent TGF-β by thrombospondin-1: mechanisms and physiology. Cytokine Growth Factor Rev 2000; 11:59–69.

88. Calzada MJ, Zhou L, Sipes JM, et al. $\alpha_4\beta_1$ integrin mediates selective endothelial cell responses to thrombospondins 1 and 2 in vitro and modulates angiogenesis in vivo. Circ Res 2004; 94:462–70.

89. Calzada MJ, Annis DS, Zeng B, et al. Identification of novel β1 integrin binding sites in the type 1 and type 2 repeats of thrombospondin-1. J Biol Chem 2004; 279:41734–43.

90. Iruela-Arispe ML, Lombardo M, Krutzsch HC, et al. Inhibition of angiogenesis by thrombospondin-1 is mediated by independent regions within the type 1 repeats. Circulation 1999; 100:1423–31.

91. Iruela-Arispe ML, Luque A, Lee N. Thrombospondin modules and angiogenesis. Int J Biochem Cell Biol 2004; 36:1070–78.

92. Lee NV, Sato M, Annis DS, et al. ADAMTS1 mediates the release of antiangiogenic polypeptides from TSP1 and 2. EMBO J 2006; 25:5270–83.

93. Lawler J, Hynes RO. An integrin receptor on normal and thrombasthenic platelets that binds thrombospondin. Blood 1989; 74:2022–27.

94. Hotchkiss KA, Chesterman CN, Hogg PJ. Catalysis of disulfide isomerization in thrombospondin 1 by protein disulfide isomerase. Biochemistry 1996; 35:9761–67.

95. Kosfeld MD, Frazier WA. Identification of a new cell adhesion motif in two homologous peptides from the COOH-terminal cell binding domain of human thrombospondin. J Biol Chem 1993; 268:8808–14.

96. Kanda S, Shono T, Tamasini-Johansson B, et al. Role of thrombospondin-1-derived peptide, 4N1K, in FGF-2-induced angiogenesis. Exp Cell Res 1999; 252:262–72.

97. Frazier WA, Gao A-G, Dimitry J, et al. The thrombospondin receptor integrin-associated protein (CD47) functionally couples to heterotrimeric G$_i$. J Biol Chem 1999; 274:8554–60.

98. Ridnour LA, Isenberg JS, Espey MG, et al. Nitric oxide regulates angiogenesis through a functional switch involving thrombospondin-1. Proc Natl Acad Sci USA 2005; 102:13147–52.

99. Isenberg JS, Ridnour LA, Perruccio EM, et al. Thrombospondin-1 inhibits endothelial cell responses to nitric oxide in a cGMP-dependent manner. Proc Natl Acad Sci USA 2005; 102:13141–46.

100. Isenberg JS, Ridnour LA, Dimitry J, et al. CD47 is necessary for inhibition of nitric oxide-stimulated vascular cell responses by thrombospondin-1. J Biol Chem 2006; 281:26069–80.

101. Kvansakul M, Adams JC, Hohenester E. Structure of a thrombospondin C-terminal fragment reveals a novel calcium core in the type 3 repeats. EMBO J 2004; 23:1223–33.

102. Short SM, Derrien A, Narsimhan RP, et al. Inhibition of endothelial cell migration by thrombospondin-1 type 1 repeats is mediated by β$_1$ integrins. J Cell Biol 2005; 168:643–53.

103. Castle VP, Dixit VM, Polverini PJ. Thrombospondin-1 suppresses tumorigenesis and angiogenesis in serum- and anchorage-independent NIH 3T3 cells. Lab Invest 1997; 77:51–61.

104. Folkman J. Angiogenesis. Annu Rev Med 2006; 57:1–18.

105. Dorrell MI, Aguilar E, Scheppke L, Barnett FH, Friedlander M. Combination angiostatic therapy completely inhibits ocular and tumor angiogenesis. Proc Natl Acad Sci USA 2007; 104:967–72.

106. Roberts DD. Regulation of tumor growth and metastasis by thrombospondin-1. FASEB J 1996; 10:1183–91.

107. Tokunaga T, Nakamura M, Oshika Y, et al. Thrombospondin 2 expression is correlated with inhibition of angiogenesis and metastasis of colon cancer. Brit J Cancer 1999; 79:354–59.

108. Lawler J, Detmar M. Tumor progression: the effects of thrombospondin-1 and -2. Int J Biochem Cell Biol 2004; 36:1038–45.

109. de Fraipont F, Nicholson AC, Feige J-J, et al. Thrombospondins and tumor angiogenesis. Trends Mol Med 2001; 7:401–07.

110. Tuszynski GP, Nicosia RF. The role of thrombospondin-1 in tumor progression and angiogenesis. BioEssays 1996; 18:71–76.

111. Sargiannidou I, Qiu C, Tuszynski GP. Mechanisms of thrombospondin-1 – mediated metastasis and angiogenesis. Semin Thromb Hemost 2004; 30:127–36.

112. Streit M, Velasco P, Brown LF, et al. Overexpression of thrombospondin-1 decreases angiogenesis and inhibits the growth of human cutaneous squamous cell carcinomas. Am J Pathol 1999; 155:441–52.

113. Sreit M, Stephen AE, Hawighorst T, et al. Systemic inhibition of tumor growth and angiogenesis by thrombospondin-2 using cell-based antiangiogenic gene therapy. Cancer Res. 2002; 62:2004–12.

114. Isenberg JS, Hyodo F, Matsumoto K-I, et al. Thrombospondin-1 limits ischemic tissue survival by inhibiting nitric oxide-mediated vascular smooth muscle relaxation. Blood, in press

115. Dameron KM, Volpert OV, Tainsky MA, et al. Control of angiogenesis in fibroblasts by p53 regulation of thrombospondin-1. Science 1994; 265:1582–84.

116. Bouck N. P53 and angiogenesis. Biochim Biophys Acta 1996; 1287:63–66.

117. Nishimori H, Shiratsuchi T, Urano T, et al. A novel brain-specific p53-target gene, BAI1, containing thrombospondin type 1 repeats inhibits experimental angiogenesis. Oncogene 1997; 15:2145–50.

118. Hsu SC, Volpert OV, Steck PA, et al. Inhibition of angiogenesis in human glioblastomas by chromosome 10 induction of thrombospondin-1. Cancer Res 1996; 56:5684–91.

119. Kaur B, Brat DJ, Devi NS, et al. Vasculostatin, a proteolytic fragment of brain angiogenesis inhibitor 1, is an antiangiogenic and antitumorigenic factor. Oncogene 2005; 24: 3632–42.

120. Hamano Y, Sugimoto H, Soubasakos MA, et al. Thrombospondin-1 associated with tumor microenvironment contributes to low-dose cyclophosphamide- mediated endothelial cell apoptosis and tumor growth suppression. Cancer Res 2004; 64:1570–74.

121. Damber J-E, Vallbo C, Albertsson P, et al. The anti-tumour effect of low-dose continuous chemotherapy may partly be mediated by thrombospondin. Cancer Chemother Pharmacol 2006; 58:354–60.

122. Hahn W, Ho S-H, Jeong J-G, et al. Viral vector-mediated transduction of a modified thrombospondin-2 cDNA inhibits tumor growth and angiogenesis. Gene Therapy 2004; 11:739–45.

123. Aghi M, Rabkin SD, Martuza RL. Angiogenic response caused by oncolytic herpes simplex virus-induced reduced thrombospondin expression can be prevented by specific viral mutations or by administering a thrombospondin-derived peptide. Cancer Res 2007; 67:440–44.

124. Haviv F, Bradley MF, Kalvin DM, et al. Thrombospondin-1 mimetic peptide inhibitors of angiogenesis and tumor growth: Design, synthesis, and optimization of pharmacokinetics and biological activities. J Med Chem 2005; 48:2838–46.

125. Hoekstra R, de Vos FYFL, Eskens FALM, et al. Phase I study of the thrombospondin-1-mimetic angiogenesis inhibitor ABT-510 with 5-fluorouracil and leucovorin: A safe combination. Eur J Cancer 2006; 42:467–72.

126. Rusk A, McKeegan E, Haviv F, et al. Preclinical evaluation of antiangiogenic thrombospondin-1 peptide mimetics, ABT-526 and ABT-510, in companion dogs with naturally occurring cancers. Clin Cancer Res 2006; 12:7444–55.

127. Rusk A, Cozzi E, Stebbins M, et al. Cooperative activity of cytotoxic chemotherapy with antiangiogenic thrombospondin-1 peptides, ABT-526 in pet dogs with relapsed lymphoma. Clin Cancer Res 2006; 12:7456–64.

Section III
Molecular & Cellular Mechanisms of the Angiogenic Process

Chapter 14
Overview of Angiogenesis During Tumor Growth

Domenico Ribatti[1] and Angelo Vacca[2]

Keywords: angiogenesis; angiogenic switch; endothelial precursor cells; hematopoietic cells; hypoxia; inflammatory cells; pericytes; tumor progression; vascular cooption

Abstract: In tumors, the phenotypic switch to angiogenesis involves more than simple up-regulation of angiogenic activity and is thought to be the result of a net balance of positive and negative regulators. Tumor angiogenesis is regulated by several factors, including growth factors for the endothelial cells secreted by both the tumor and stromal inflammatory cells, and mobilized from extracellular matrix stores by proteases secreted by tumor cells. Regulatory factors also include the extracellular matrix components and endothelial cell integrins, hypoxia, oncogenes and tumor suppressor genes. Angiogenesis is mandatory for tumor progression, in the form of growth, invasion and metastasis; hence it has prognostic value.

Isolation of the First Angiogenic Tumor Factor

In 1971, Judah Folkman first advanced the hypothesis that tumor growth depends on the formation of new blood vessels from the preexisting vascular bed [1]. According to this hypothesis, endothelial cells (EC) may be switched from a resting state to a rapid growth phase by a diffusible chemical signal emanating from the tumor cells. Moreover, an angiogenic factor, named "tumor angiogenesis factor" (TAF), was also isolated by Folkman and co-workers in 1971 [2]. TAF has since been non-destructively extracted from several tumor cell lines, and several low molecular weight angiogenic factors have been isolated, all from the Walker 256 carcinoma. These factors induced a vasoproliferative response in vivo

[1] Department of Human Anatomy and Histology, University of Bari Medical School, Bari, Italy

[2] Department of Biomedical Sciences and Human Oncology, University of Bari Medical School, Bari, Italy

when tested on rabbit cornea or chick chorioallantoic membrane (CAM), and in vitro on cultured EC [3–5].

The Avascular and the Vascular Phases and the Concept of the "Angiogenic Switch"

Angiogenesis and the production of angiogenic factors are fundamental for tumor progression in the form of growth, invasion and metastasis (Fig. 14.1) [6]. The process of angiogenesis begins with local degradation of the basement membrane surrounding the capillaries, which is followed by invasion of the surrounding stroma by the underlying EC, in the direction of the angiogenic stimulus. EC migration is accompanied by the proliferation of EC and their organization into three dimensional structures that join with other similar structures to form a network of new blood vessels.

New vessels promote growth by conveying oxygen and nutrients and removing catabolites [7]. These requirements vary, however, among tumor types, and change over the course of tumor progression [8]. EC secrete growth factors for tumor cells and a variety of matrix-degrading proteinases that facilitate tumor invasion [9]. An expanding endothelial surface also gives tumor cells more opportunities to enter the circulation and metastasize [10].

Solid tumor growth occurs by means of an avascular phase followed by a vascular phase. Assuming that such growth is dependent on angiogenesis and that this depends on the release of angiogenic factors, the acquisition of an angiogenic ability can be seen as an expression of progression from neoplastic transformation to tumor growth and metastasis [6].

The avascular phase appears to correspond to the histopathological picture presented by a small colony of neoplastic cells that reaches a steady state before it proliferates and becomes rapidly invasive. In this scenario, metabolites and catabolites are transferred by simple diffusion through the surrounding tissue. The cells at the periphery of the tumor continue to reproduce, whereas those in the deeper portion die away. Dormant tumors have been discovered during autopsies

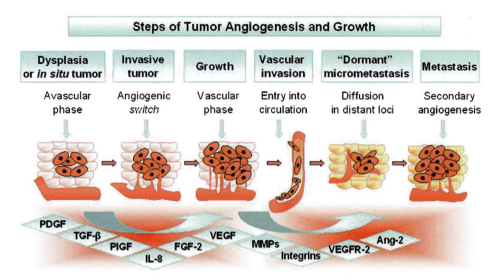

FIG. 14.1. Steps of tumor angiogenesis and growth.

of individuals who died of causes other than cancer [11]. Carcinoma in situ is found in 98% of individuals aged 50 to 70 years who died of trauma, but is diagnosed in only 0.1% during life. Malignant tumors can grow beyond the critical size of 2 mm at their site of origin by exploiting the host's pre-existing vessels. This occurs in tumors implanted in the rat brain [12] and in naturally occurring human lung carcinomas [13]. These findings support the notion that only a very small subset of dormant tumors enters the vascular phase.

Practically all solid tumors, including those of the colon, lung, breast, cervix, bladder, prostate and pancreas, progress through these two phases. The role of angiogenesis in the growth and survival of leukemias and other hematological malignancies has only become evident since 1994, thanks to a series of studies demonstrating that progression in several forms is clearly related to their degree of angiogenesis [14].

Tumor angiogenesis is linked to a switch in the balance between positive and negative regulators, and mainly depends on the release by neoplastic cells of specific growth factors for EC, that stimulate the growth of the host's blood vessels or the down-regulation of natural angiogenesis inhibitors. In normal tissues, vascular quiescence is maintained by the dominant influence of endogenous angiogenesis inhibitors over angiogenic stimuli.

The mechanism of this switch was classified by Hanahan, who developed transgenic mice in which the large "T" oncogene is hybridized to the insulin promoter [15]. In this model for β-islet cell tumorigenesis (RIP-Tag model), all islet cells in a transgenic mouse line express the large T antigen at birth. By 12 weeks, 75% of islets have progressed to small foci of proliferating cells, but only 4% are angiogenic and their number is closely correlated with the incidence of tumor formation [15].

The switch depends on increased production of one or more positive regulators of angiogenesis, such as vascular endothelial growth factor (VEGF), fibroblast growth factor-2 (FGF-2), interleukin-8 (IL-8), placental growth factor (PlGF), transforming growth factor-β (TGF-β), platelet derived growth factor (PDGF), pleiotrophins, and others [16]. These can be exported from tumor cells, mobilized from the extracellular matrix, or released from host cells recruited to the tumor. The switch clearly involves more than a simple up-regulation of angiogenic activity and has thus been regarded as the result of the net balance between positive and negative regulators.

Another variant of angiogenesis, different from sprouting, is called intussusceptive microvascular growth (IMG), or nonsprouting angiogenesis. It occurs through the splitting of the existing vasculature by transluminal pillars or transendothelial bridges [17]. It is thought that the pillars then increase in diameter and become a capillary mesh. IMG has been implicated in tumor growth and can explain a rapid remodeling of the vasculature [18, 19]. In a model of human colon cancer xenografted in mouse, it has been demonstrated that 50% of the vasculature is made by IMG [20].

Angiogenesis is not Necessarily Involved in Tumor Progression

Angiogenesis in human tumors is considerably less active than in a physiological condition such as the formation of granulation tissue and in the reproductive organs. In fact, the endothelial cell proliferation index value is 0.15% for the human prostate or breast cancer compared to 6.7% in granulation tissue and 36% in the corpus luteum. Moreover, the microvessel densities in human lung, mammary, renal cell and colon carcinomas, glioblastoma and pituitary adenomas are lower than those in their normal counterparts [21, 22]. In lung carcinoma, for example, the microvessel density (MVD) was found to be only 29% that of normal lung tissue. In glioblastoma, MVD was found to be 78% that of normal brain tissue. This apparent paradox is partially explained by the lower oxygen consumption rate of tumor cells [23], which are also known to tolerate oxygen deprivation [24]. As a result, the intercapillary distance in tumors is greater than in their normal tissue counterparts.

Vascular Cooption

Holash et al.[12] reported that tumor cells migrate toward existing host organ blood vessels in sites of metastases, or in vascularized organs such as the brain, to trigger blood vessel-dependent tumor growth as opposed to classic angiogenesis. These vessels then regress owing to apoptosis of the constituent EC, apparently mediated by angiopoietin-2 (Ang2). Finally, at the periphery of the growing tumor mass angiogenesis occurs by the cooperative interaction of VEGF and Ang-. Tumor cells often appear to have immediate access to blood vessels, such as when they metastasize to or are implanted within a vascularized tissue [12, 25]. They immediately coopt existing adjacent vessels and often grow as cuffs around them. A robust host defense mechanism is activated, in which the coopted vessels initiate an apoptotic cascade, probably by autocrine induction of Ang2, followed by regression of the coopted vessels, that carries off much of the dependent tumor and results in massive tumor death. However, successful tumors overcome this vessel regression by initiating neoangiogenesis.

Many solid tumors may fail to form a well-differentiated and stable vasculature because their newly formed tumor vessels continue to overexpress Ang2. Ang2 induction in host vessels in the periphery of experimental C6 glioma precedes VEGF up-regulation of tumor cells, and causes regression of coopted vessels [26, 27].

Vajkoczy et al. [28] have demonstrated a parallel induction of Ang2 and VEGFR-2 in quiescent host endothelial cells, suggesting that their simultaneous activity is critical for the induction of tumor angiogenesis during vascular initiation of microtumors. Consequently, the simultaneous expression of VEGFR-2 and Ang2, rather than the expression of Ang2 alone, may indicate the EC angiogenic phenotype and thus provide an early marker of activated host vasculature. The VEGF/Ang2 balance may determine whether the new tumor vessels will continue to expand when the ratio of VEGF to Ang2 is high, or regress when it is low during remodeling of the tumor microvasculature.

Cooption depends on the site of tumor development. Rat mammary carcinoma is vascularized by cooption only if cells are injected inside the brain. Lewis lung carcinoma and melanoma cells metastasized in the lung or brain, respectively, and are partially vascularized by cooption [12, 29]. Astrocytomas first acquire their blood supply by coopting existing normal brain vessels, growing alongside blood vessels, without a tumor capsule, and eliciting an invasive character. They can enlarge as much as some angiogenic tumors [28].

Phenotypic and Genotypic Characteristics of Tumor Vessels

Tumor EC may divide up to 50 times more frequently than EC of normal tissues. Considerable differences exist between normal and tumor vasculature. The immaturity of tumor vessels led H. Dvorak to define a tumor as "a wound that never heals" [30]. Although the tumor vasculature originates from the host

vessels and the mechanisms of angiogenesis are similar, the organization may differ dramatically depending on the tumor type and its location.

The blood vessels of tumors display many structural and functional abnormalities [31]. Their unusual leakiness, potential for rapid growth and remodeling, and expression of distinctive surface molecules, mediate the dissemination of tumor cells in the bloodstream and maintain the tumor microenvironment. Like normal blood vessels, they consist of EC, pericytes and their enveloping basement membrane. Common features, regardless of their origin, size and growth pattern, include the absence of a hierarchy, the formation of large-caliber sinusoidal vessels and a markedly heterogenous density. Low permeability tumors overexpress Ang1 and/or underexpress VEGF. Conversely, those with high permeability may lack Ang- or overexpress its antagononist, Ang2 [32]. Tumor EC are most markedly activated. Many of the surface markers they express are more strongly expressed by tumoral than normal vessels. For instance, they express much more VEGFR-2, Tie1, Tie2, the integrins $\alpha v\beta 3$ and $\alpha v\beta 5$ [33] or the alternative, spliced variant of fibronectin ED-B [34], than resting EC do in normal tissues. Furthermore, tumor EC express different levels of adhesion molecules for circulating leukocytes and high levels of E-selectin [35].

St Croix et al. [36] compared the gene expression patterns of EC derived from normal and malignant colorectal tissues. Differential expression was present in 79 transcripts: 46 were elevated at least 10-fold while 33 were expressed at substantially lower levels in tumor-associated EC. Most of the differentially expressed genes have also been found during luteal angiogenesis and wound healing, which suggests that in tumor angiogenesis the same signaling pathways are involved as in physiologic angiogenesis. These data clearly delineate the effects of the local microenvironment on gene expression patterns in endothelial cells and support the critical role of the microenvironment in defining the angiogenic phenotype.

After the work by St. Croix, only a limited number of studies have characterized the gene expression profile of freshly isolated tumor EC [37, 38]. More recently, van Beijnum et al. [39] compared the transcriptional profiles of angiogenic EC isolated from both malignant and non-malignant tissues with those of resting EC and identified 17 genes that show specific overexpression in tumor endothelium, but not in the angiogenic endothelium of normal tissues. Moreover, antibody targeting of four cell-surface expressed or secreted products (vimentin, CD59, HMGB1 and IGFBP7) inhibited angiogenesis in vitro and in vivo.

Hypoxic Regulation of Tumor Angiogenesis

There is a complex interrelationship between tumor hypoxia and tumor angiogenesis. Hypoxia in tumors develops in the form of chronic hypoxia, resulting from long diffusion distances between tumor vessels, and/or of acute hypoxia, resulting from a transient collapse of tumor vessels. Many tumors contain a hypoxic microenvironment, a condition that is associated with poor prognosis and resistance to treatment. The production of several angiogenic cytokines, such as FGF-2, VEGF, TGF-β, TNF-α and IL-8, is regulated by hypoxia. VEGF-mRNA expression is rapidly and reversibly induced by exposure of cultured EC to low PO$_2$ [40]. Many tumor cell lines have been reported to show hypoxia-induced expression of VEGF [7, 41–44]. In a rat glioma model, VEGF gene expression was activated in a distinct tumor cell subpopulation by two distinct hypoxia-driven mechanisms [45]. Hypoxia-inducible factor (HIF)-1 helps to restore oxygen homeostasis by inducing glycolysis, erythropoiesis and angiogenesis [46] and tumor vascularization is largely controlled by HIF-1, partly as a result of VEGF up-regulation [47].

The Role of Pericytes in Tumor Blood Vessels

Among the pathways involved in pericyte recruitment during embryonic development, the contribution of PDGF-B is confirmed in tumor angiogenesis. PDGF-B expressed by tumor cells increased pericyte recruitment in several in vivo tumor models, but failed to correct their detachment in PDGF-B retention motif deficient mice [48, 49]. Genetic abolition of the PDGF-B receptor expressed by embryonic pericytes decreased their recruitment in tumor [48]. In Lewis lung carcinoma tumors implanted in mice, inhibition by RNA interference of endothelial differentiation gene-1 (EDG-1) expression in EC strongly reduced pericyte coverage [50].

In a human glioma model developed in rat, Ang1 led to enhanced pericyte recruitment and increased tumor growth, presumably by favoring angiogenesis [51]. In contrast, in a colon cancer model, overexpression of Ang1 led to smaller tumors with

fewer blood vessels and a higher degree of pericyte coverage, resulting in a decreased vascular permeability and reduced hepatic metastasis [52, 53]. In a human neuroblastoma xenotransplanted model, pericyte coverage along tumor microvessels is decreased by half in tumors grafted to matrix metalloproteinase-9 (MMP-9) deficient mice, and transplantation with MMP-9-expressing bone marrow cells restores the formation of mature tumor vessels [54]. Overexpression of the tissue inhibitor of MMP-3 (TIMP-3) results in decreased pericyte recruitment in neuroblastoma and melanoma models [55].

The Role of Inflammatory Cells in Tumor Angiogenesis

Tumor cells are surrounded by an infiltrate of inflammatory cells, such as lymphocytes, neutrophils, macrophages and mast cells. These cells communicate by means of a complex network of intercellular signaling pathways mediated by surface adhesion molecules, cytokines and their receptors [56]. It is becoming clear that stromal cells cooperate with endothelial and cancer cells in promoting angiogenesis, secreting a varied repertoire of growth factors and proteases that enable them to enhance tumor growth (Fig. 14.2).

Tumor-associated macrophages accumulate in poorly vascularized hypoxic or necrotic areas [57], and respond to experimental hypoxia by increasing the release of VEGF and FGF2 and a broad range of other factors, such as PDGF, tumor necrosis factor-α (TNF-α), FGF2, VEGF, urokinase-type plasminogen activator and MMP [58]. Moreover, activated macrophages synthesize and release inducible nitric oxide synthase (NOS), which increases blood flow and promotes

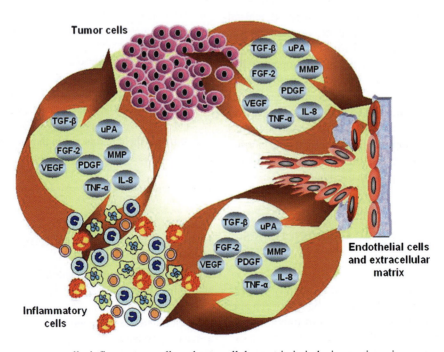

Fig. 14.2. Interplay between tumor cells, inflammatory cells and extracellular matrix in inducing angiogenic response.

angiogenesis [59]. The angiogenic factors secreted by macrophages stimulate mast cell migration [60]. Lin and Pollard [61] showed that in the PyMT model of mammary carcinogenesis, macrophages are recruited to premalignant tumors immediately before the angiogenic switch that precedes the transition to a malignant phenotype. Depletion of these macrophages resulted in a decreased vascular density in tumors.

There is overwhelming evidence that the density of mast cells is strictly correlated with the extent of tumor angiogenesis [62]. Mast cells accumulate at premalignant stages of tumor progression and at the periphery of invasive tumors. In experimentally-induced tumors, an increasing number of mast cells has been demonstrated, before the onset of angiogenesis, in the proximity of tumor cells [63]. In tumors induced in mast cell-deficient mice, a reduced angiogenesis and metastatic potential have been reported [64]. Mast cells contain several angiogenic factors, including tryptase, chymase, heparin and histamine, TGF-β, TNF-α, IL-8, FGF2 and VEGF. Mast cell accumulation has been associated with enhanced angiogenesis in both solid and hematological tumors [62].

Genetic Evidence that Tumors are Angiogenesis-Dependent

It is increasingly recognized that oncogenes, such as mutant RAS or SRC, may also contribute to tumor angiogenesis by enhancing the production of VEGF [65, 66]. Down-regulation of the RAS-oncogene in a melanoma driven by doxycycline-inducible ras led to tumor regression within 12 days [67]. Cells that expressed low levels of RAS were dormant and non-angiogenic, whereas cells that expressed high levels of RAS developed into full-grown tumors [68]. These authors demonstrated that whereas VEGF levels increased only modestly in tumors that expressed high levels of RAS, thrombospondin-1 (TSP-1) levels increased markedly in these cells.

Tumors that express bcl-2 escape mitomycin C therapy and grow ~1,000 mm^3. When bcl-2-expressing tumors are treated with an angiogenesis inhibitor (TNP-470) which selectively inhibits the proliferation of tumor cells or fibroblasts, the bcl-2 effect is annulled, and tumor growth is restricted to <10–15% of the growth observed in untreated bcl-2 expressing tumors [69].

Alternative Mechanisms of Tumor Angiogenesis

Vasculogenic Mimicry

Maniotis et al. [70] described a new model of formation of vascular channels by human melanoma cells, and called it "vasculogenic mimicry" to emphasize the de novo generation of blood vessels without the participation of EC and independent of angiogenesis. The word "vasculogenic" was selected to indicate the de novo generation of the pathway, and "mimicry" was used because the tumor uses cell pathways serving to transport fluid in tissues that were clearly not blood vessels.

Microarray gene chip analysis of highly aggressive human cutaneous melanoma cell lines compared with poorly aggressive lines revealed a significant increase in the expression of laminin 5 and MMP-1, -2, and –9 and MT1-MMP in the highly aggressive cells [71], suggesting that they interact with and alter their extracellular environment in a different way as compared with poorly aggressive cells, and that increased expression of MMP-2 and MT1-MMP, along with matrix deposition of laminin 5, are required to achieve their "vasculogenic mimicry". These data have been vigorously disputed by Mc Donald et al. [72], who consider that the evidence presented is neither persuasive nor novel. In their opinion, the data are not convincing because three key questions were not addressed: (1) if erythrocytes are used as markers, are they located inside or outside blood vessels; (2) where is the interface between EC and tumor cells in the blood vessel wall; and (3) how extensive is the presumptive contribution of tumor cells to the lining of blood vessels?

Mosaic Vessels

Another possibility is that the EC lining is replaced by tumor cells, resulting in so-called "mosaic vessels", where both endothelial and tumor cells contribute to form the vascular tube [73]. These authors used CD31 and CD105 to identify EC and endogenous green fluorescent protein (GFP) labeling of tumor cells, and showed that approximately 15% of perfused vessels of a colon carcinoma xenografted at two sites in mice were mosaic, with focal regions where no CD31/CD105 immmoreactivity was detected and tumor cells were in contact with the vessel lumen.

Participation of Hematopoietic Cells (HC) and Endothelial Precursor Cells (EPC) to Tumor Angiogenesis

High levels of VEGF produced by tumors may result in mobilization of EPC in the peripheral circulation and enhance their recruitment into the tumor vasculature [74]. Hypoxia mobilizes EPC from the bone marrow in the same way as it does hematopoietic cytokines, such as granulocyte macrophage-colony stimulating factor (GM-CSF) [75]. Malignant tumor growth results in neoplastic tissue hypoxia, and may mobilize bone-marrow-derived EPC in a paracrine fashion and thus contribute to promote the sprouting of new vessels.

By introducing a suicide gene with a lentiviral vector, De Palma et al. [76, 77] conditionally eliminated bone marrow-derived HC during the early phases of tumor growth and showed that the impaired recruitment of these cells inhibited tumor angiogenesis and progression.

Carcinoma-associated fibroblasts isolated from breast cancer secrete high levels of stromal cell derived factor-1 (SDF-1), a molecule that functions as a potent chemoattractant for endothelial cells and HC and enhances tumor angiogenesis [78].

The importance of EPC mobilization in tumor vascularization has been demonstrated in a mouse mutant for Id proteins [79]. Tumors xenografted in lethally irradiated Id1+/–Id3–/– mice and reconstituted with wild type EPC grow as well in control mice and most of the EC are also Id3+. Moreover, in wild type mice, EPC recruitment to the tumor vasculature is completely inhibited by the injection of neutralizing antibody against VEGFR-2, but not against VEGFR1.

The extent of EPC recruitment into the tumor vasculature may depend on the tumor type, varying from 90% in lymphoma to 5% in neuroblastoma implanted subcutaneously in mice [80].

The Prognostic Significance of Tumor Angiogenesis

In 1991, Weidner and co-workers used specific anti-endothelial antibodies to highlight the tumor vasculature, in order to demonstrate that MVD was a prognostic marker for human breast cancer [81]. Since then, the majority of reports have confirmed that MVD is a powerful and often independent prognostic indicator for many different types of human cancers, such as breast cancer, prostate cancer, melanoma, ovarian carcinoma, gastric carcinoma, and colon carcinoma. However, a few other reports failed to show that MVD is a prognostic indicator, because human tumors are heterogeneous and consist of subpopulations of cells having different biological properties. Moreover, MVD is determined by the intercapillary distance. During tumor regression induced by an angiogenesis inhibitor, MVD may decrease if capillary dropout exceeds tumor cell dropout, increase if tumor cell dropout exceeds capillary dropout, or remain the same if a parallel disappearance of capillaries and tumor cells parallel each other [82]. Therefore, the detection of a decrease in MVD during treatment with an angiogenesis inhibitor, suggests that the agent is active. However, the absence of a decrease in MVD does not correspondingly suggest that the agent is ineffective.

Acknowledgements. This study was supported by Associazione Italiana per la Ricerca sul Cancro (AIRC, National and Regional Funds), Milan, the Ministry for Education, the Universities and Research (FIRB 2001, PRIN 2005, Progetto Carso 72/2), Rome, and Fondazione Italiana per la Lotta al Neuroblastoma, Genoa, Italy.

References

1. Folkman J. Tumour angiogenesis: therapeutic implications. N Engl J Med 1971; 285: 1182–1186.
2. Folkman J, Merler E, Abernathy C, et al. Isolation of a tumor fraction responsible for angiogenesis. J Exp Med 1971; 133: 275–288.
3. Mc Auslan BR, Hoffman H. Endothelium stimulating factor from Walker carcinoma cells. Relation to tumor angiogenic factor. Exp Cell Res 1979; 119: 181–190.
4. Weiss JB, Brown RA, Kumar S, et al. An angiogenic factor isolated from tumours: a potent low-molecular weight compound. Brit J Cancer1979; 40: 493–496.
5. Fenselau A, Kaiser D, Wallis K. Nucleoside requirements for the *in vivo* growth of bovine aortic endothelial cells. J Cell Physiol 1981; 108: 375–384.
6. Ribatti D, Vacca A, Dammacco F. The role of the vascular phase in solid tumor growth: a historical review. Neoplasia 1999; 1: 293–302.
7. Papetti M, Herman IM. Mechanisms of normal and tumor-derived angiogenesis. Am J Physiol Cell Physiol 2002; 282: C947–C970.
8. Hlatky L, Hahnfeldt P, Folkman J. Clinical application of antiangiogenic therapy: microvessel density, what it does and doesn't tell us. J Natl Cancer Inst 2002; 94: 883–893.
9. Mignatti P, Rifkin DB. Biology and biochemistry of proteinases in tumor invasion. Physiol Rev 1993; 73: 161–195.
10. Aznavoorian S, Murphy AN, Stetler-Stevenson WG, Liotta LA. Molecular aspects of tumor cell invasion and metastasis. Cancer 1993; 71: 1638–1683.
11. Black WC, Welch HG. Advances in diagnostic imaging and overestimations of disease prevalence and the benefits of therapy. N Engl J Med 1993; 328: 1237–1243.
12. Holash J, Maisonpierre PC, Compton D, et al. Vessel cooption, regression and growth in tumors mediated by angiopoietins and VEGF. Science 1999; 284: 1994–1998.
13. Pezzella F, Pastorino U, Tagliabue E, et al. Non-small-lung carcinoma tumor growth without morphological evidence of neo-angiogenesis. Am J Pathol 1997; 151: 1417–1423.
14. Vacca A, Ribatti D. Bone marrow angiogenesis in multiple myeloma. Leukemia 2006; 20: 193–199.
15. Hanahan D. Heritable formation of pancreatic beta-cell tumors in transgenic mice expressing recombinant unsylin/simian virus 40 oncogene. Nature 1985; 315: 115–122.
16. Ribatti D, Nico B, Crivellato E, et al. The history of the angiogenic switch comcept. Leukemia 2007; 21: 44–52.
17. Burri PH, Djonov V. Intussusceptive angiogenesis, the alternative to capillary sprouting. Mol Aspects Med 2002; 23: S1–S27.
18. Patan S, Munn LL, Jain RK. Intussusceptive microvascular growth in a human colon adenocarcinoma xenograft: a novel mechanism of tumor angiogenesis. Microvasc Res 1996; 51: 260–272.
19. Djonov V Andres AC, Ziemiecki A. Vascular remodelling during the normal and malignant life cycle of the mammary gland. Microsc Res Techn 2001; 52: 182–189.
20. Patan, S, Tanda S, Roberge S, et al. Vascular morphogenesis and remodeling in a human tumor xenograft: blood vessel formation and growth after ovariectomy and tumor implantation. Circ Res 2001; 89: 732–739.
21. Eberhard A, Kahlert S, Goede V, et al. Heterogeneity of angiogenesis and blood vessel maturation in human tumors: implications for antiangiogenic tumor therapies. Cancer Res 2002; 60: 1388–1393.
22. Turner HE, Nagy Z, Gatter KC, et al. Angiogenesis in pituitary adenomas and in the normal pituitary gland. J Clin Endocrinol Metab 2000; 85: 1159–1162.
23. Sterinberg F, Rohrborn HJ, Otto T, et al. NIR reflection measurements of hemoglobin and cytochrome da3 in healthy tissue and tumors. Correlation to oxygen consuption: preclinical and clinical data. Adv Exp Med Biol 1997; 428: 69–77.

24. Graeber TG, Osmanian C, Jacks T, et al. Hypoxia-mediated selection of cells with diminished apoptotic potential in solid tumours. Nature 1996; 379: 88–91.

25. Zagzag D, Hooper A, Friedlander DR, et al. *In situ* expression of angiopoietins in astrocytomas identifies angiopoietin-2 as an early marker of tumor angiogenesis. Exp Neurol 1999; 159: 391–400.

26. Holash J, Wiegand SJ, Yancopoulos GF. New model of tumor angiogenesis: dynamic balance between vessel regression and growth mediated by angiopoietins and VEGF. Oncogene 1999; 18: 5356–5362.

27. Yancopoulos GD, Davis S, Gale NW, et al. Vascular specific growth factors and blood vessels formation. Nature 2000; 407: 242–248.

28. Vajkoczy P, Farhadi M, Gaumann A, et al.. Microtumor growth initiates angiogenic sprouting with simulatenous expression of VEGF, VEGF receptor-2, and angiopoietin-2. J Clin Invest 2002; 109: 777–785.

29. Kusters B, Leenders WP, Wesseling P, et al. Vascular endothelial growth factor-A(165) induces progression of melanoma brain metastases without induction of sprouting angiogenesis. Cancer Res 2002; 62: 341–345.

30. Dvorak H. Tumors: wounds that not heal. Similarities between tumor stroma generation and wound healing. New Engl J Med 1986; 315: 1650–1659.

31. Ribatti D, Nico B, Crivellato E, et al. The structure of the vascular networks of tumors. Cancer Letters; 2007; 248: 18–23.

32. Jain RK, Munn LL. Leaky vessels? Call Ang1! Nat Med 2000; 6: 131–132.

33. Stromblad S, Cheresh DA. Cell adhesion and angiogenesis. Trends Cell Biol 1996; 6: 462–468.

34. Neri D, Caremolla B, Nissim A, et al. Targeting by affinity-matured recombinant antibody fragments of an angiogenesis associated fibronectin isoform. Nat Biotechnol 1997; 15: 1271–1275.

35. Bischoff J. Approaches to studying cell adhesion molecules in angiogenesis. Trends Cell Biol 1995; 5: 69–74.

36. St Croix B, Rago C, Velculescu V, et al. Genes expressed in human tumor endothelium. Science 2000; 289: 1197–1202.

37. Madden SL, Cook BP, Nacht M, et al. Vascular gene expression in nonneoplastic and malignant brain. Am J Pathol 2004; 165: 601–608.

38. Parker BS, Argani P, Cook BP, et al. Alterations in vascular gene expression in invasive breast carcinoma. Cancer Res 2004; 64: 7857–7866.

39. Van Beijnum J, Dings RP, van der Linden E, et al.. Gene expression of tumor angiogenesis dissected; specific targeting of colon cancer angiogenic vasculature. Blood 2006; 108: 2339–2348.

40. Levy AP, Levy NS, Wegner S, et al. Transcriptional regulation of the rat vascular endothelial growth factor gene by hypoxia. J Biol Chem 1995; 270: 13333–13340.

41. Plate KH, Breier G, Widch HA, et al. Vascular endothelial growth factor is a potent tumour angiogenesis factor in human gliomas *in vivo*. Nature 1992; 359: 845–848.

42. Shweiki D, Itin A, Soffer D, et al. Vascular endothelial growth factor induced by hypoxia may mediate hypoxia-initiated angiogenesis. Nature 1992; 359: 843–845.

43. Potgens AJ, Lubsen NH, van Altena MC, et al. Vascular permeability factor expression influences tumor angiogenesis in human melanoma lines xenografted to nude mice. Am J Pathol 1995; 146: 197–209.

44. Claffey KP, Brown LF, Del Aguila LF, et al. Expression of vascular permeability factor/vascular endothelial growth factor by melanoma cells increases tumor growth, angiogenesis, and experimental metastasis. Cancer Res 1996; 56: 172–181.

45. Damert A, Machein M, Breier G, et al. Up-regulation of vascular endothelial growth factor expression in a rat glioma is conferred by two distinct hypoxia-driven mechanisms. Cancer Res 1997; 57: 3860–3864.

46. Semenza GL. Transcriptional regulation by hypoxia-inducible factor-1. Trends Cardiovasc Med 1996; 6: 151–157.

47. Carmeliet P, Dor Y, Herbert JM, Fukumura D, et al. Role of HIF-1 in hypoxia-mediated apoptosis, cell proliferation and tumor angiogenesis. Nature 1998; 394: 485–490.

48. Abramsson A, Lindblom P, Betshowz C. Endothelial and non-endothelial sources of PDGF-B regulate pericyte recruitment and influence vascular pattern formation in tumors. J Clin Invest 2003; 112: 1142–1151.

49. Guo P, Hu B, Gu W, Xu L, et al. Platelet-derived growth factor-B enhances glioma angiogenesis by stimulating vascular endothelial growth factor expression in tumor endothelia and by promoting pericyte recruitment. Am J Pathol 2003; 162: 1083–1093.

50. Chae SS, Paik JH, Furneaux H, et al. Requirement for sphingosine 1-phosphate receptor-1 in tumor angiogenesis demonstrated by in vivo RNA interference. J Clin Invest 2004; 114: 1082–1089.

51. Machein MR, Knedla A, Knoth R, et al. Angiopoietin-1 promotes tumor angiogenesis in a rat glioma model. Am J Pathol 2004; 165: 1557–1570.

52. Ahmad SA, Liu W, Jung YD, et al. The effects of angiopoietin-1 and -2 on tumor growth and angiogenesis in human colon cancer. Cancer Res 2001; 61: 1255–1259.

53. Stoeltzing O, Ahmad SA, Liu W, et al. Angiopoietin-1 inhibits vascular permeability, angiogenesis, and growth of hepatic colon cancer tumors. Cancer Res 2003; 63: 3370–3377.

54. Chantrain CF, Shimada H, Jodele S, et al. Stromal matrix metalloproteinase-9 regulates the vascular architecture in neuroblastoma by promoting pericyte recruitment. Cancer Res 2004; 64: 1675–1686.

55. Spurbeck WW, NG CY, Strom TS, et al.. Enforced expression of tissue inhibitor of matrix metalloproteinase-3 affects functional capillary morphogenesis and inhibits tumor growth in a murine tumor model. Blood 2002; 100: 3361–3368.

56. Park CC, Bissell MJ, Barcellos-Hoff MH. The influence of the microenvironment on the malignant phenotype. Mol Med Today 2000; 6: 324–329.

57. Leek RD, Lander RJ, Harris AL, et al. Necrosis correlates with high vascular density and focal macrophages infiltration in invasive carcinoma of the breast. Br J Cancer 1999; 79: 991–995.

58. Bingle L, Brown NJ, Lewis CE. The role of tumor associated macrophages in tumor progression; implications for new anticancer therapies. J Pathol 2002; 196: 254–265.

59. Jenkins DC, Charles IG, Thomsen LL, et al. Role of nitric oxide in tumor growth. Proc Natl Acad Sci USA 1995; 92: 4392–4396.

60. Gruber BL, Marchase MJ, Kaw R. Angiogenic factors stimulate mast cell migration. Blood, 86: 2488–2493.

61. Lin EY, Pollard JW. Role of infiltrated leucocytes in tumour growth and spread. Br J Cancer 2004; 90: 2053–2058.

62. Ribatti D, Crivellato E, Roccaro AM, et al. Mast cell contribution to angiogenesis related to tumor progression. Clin Exp Allergy 2004; 34: 1660–1664.

63. Kessler DA, Langer RS, Pless NA, et al. Mast cell and tumor angiogenesis. Int J Cancer 1977; 18: 703–709.

64. Starkey JR, Crowle PK, Taubenberger S. Mast cell-deficient W/ W mice exhibit a decreased rate of tumor angiogenesis. Int J Cancer 1998; 42: 48–52.

65. Rak J, Mitsuhashi Y, Bayko L et al. Mutant ras oncogenes upregulate VEGF/VPF expression: implications for induction and inhibition of tumor angiogenesis. Cancer Res 1995; 55: 4575–4580.

66. Ellis LM, Staley CA, Liu W et al. Downregulation of vascular endothelial growth factor in a human colon carcinoma cell line transfected with an antisense expression vector specific for c-src. J Biol Chem 1998; 273: 1052–1057.

67. Tang Y, Kim M, Carrasco D et al. In vivo assessment of RAS-dependent maintenance of tumor angiogenesis by real-time magnetic resonance imaging. Cancer Res 2005; 65: 8324–8330.

68. Watnick RS, Cheng YN, Rangarajan A, et al. Ras modulates Myc activity to repress thrombospondin-1 expression and increase tumor angiogenesis. Cancer Cell 2003; 3: 219–231.

69. Fernandez A, Udagawa T, Schwesinger C, et al.. Angiogenic potential of prostate carcinoma cells overexpressing bcl-2. J Natl Cancer Inst 2001; 93: 208–213.

70. Maniotis AJ, Folberg R, Hess A, et al. Vascular channel formation by human melanoma *cells in vivo* and *in vitro*: vasculogenic mimicry. Am J Pathol 1999; 155: 739–752.

71. Seftor RB, Seftor EA, Koshikawa N, et al Cooperative interactions of laminin 5 gamma-2 chain, matrix metalloproteinase-2, and membrane type-1-matrix/metalloproteinase are required for mimicry of embryonic vasculogenesis by aggressive melanoma. Cancer Res 2001; 61: 6322–6327.

72. Mc Donald DM, Munn L, Jain RK. Vasculogenic mimicry: how convincing, how novel, and how significant. Am J Pathol 2000; 156: 383–388.

73. Chang YS, di Tomaso E, Mc Donald DM, et al. Mosaic blood vessels in tumors: frequency of cancer cells in contact with flowing blood. Proc Natl Acad Sci USA 2000; 97: 14608–14613.

74. Asahara T, Takahashi T, Masuda H, et al. VEGF contributes to postnatal neovascularization by mobilizing bone marrow-derived endothelial progenitor cells. EMBO J 1999; 18: 3964–3972.

75. Takahashi T, Kalka C, Masuda H, et al. Ischemia- and cytokine-induced mobilization of bone marrow-derived endothelial progenitor cells for neovascularization. Nat Med 1999; 5: 434–438.

76. De Palma M, Venneri MA, Naldini L. In vivo targeting of tumor endothelial cells by systemic delivery of lentiviral vectors. Hum Gene Ther 2003; 14: 1193–1206.

77. De Palma M, Venneri MA, Roca C, et al. Targeting exogenous genes to tumor angiogenesis by transplantation of genetically modified hematopoietic stem cells. Nat Med 2003; 9: 789–795.

78. Orimo A, Gupta PB, Sgroi DC, et al. Stromal fibroblasts present in invasive human breast carcinomas promote tumor growth and angiogenesis through elevated SDF-1/CXCL12 secretion. Cell 2005; 121: 335–348.

79. Lyden D, Hattori K, Dias S, et al. Impaired recruitment of bone-marrow-derived endothelial and hematopoietic precursor cells blocks tumor angiogenesis and growth. Nat Med, 7: 1194–1201.

80. Davidoff AM, Ng CY, Brown P, et al. Bone marrow-derived cells contribute to tumor neovasculature and, when modified to express an angiogenesis inhibitor, can restrict tumor growth in mice. Clin Cancer Res 2001; 7: 2870–2879.

81. Weidner N, Sample JP, Welch WR, et al. Tumor angiogenesis correlates with metastasis in invasive breast carcinoma. N Engl J Med 1991; 324: 1–8.

82. Kerbel R, Folkman J. Clinical translation of angiogenesis inhibitors. Nat Rev Cancer 2002; 2: 727–739.

Chapter 15
Hypoxic Regulation of Angiogenesis by HIF-1

Philip J.S. Charlesworth and Adrian L. Harris

Keywords: hypoxia-inducible factor, HIF-1, von Hippel Lindau protein, tumor-associated macrophages, hypoxic regulation

Abstract: Oxygen homeostasis and protection from episodes of low oxygen tension in human tissues is important for cell survival. There are a number of physiological and pathological scenarios that place the cell in hypoxic conditions warranting adaptation to the stressful environment.

In hypoxia, oxidative phosphorylation is deceased with subsequent reduction of ATP production. A responsive increase in glycolysis compensates for this ATP reduction to some degree, but many ATP-dependent processes such as protein translation are decreased in hypoxic cells. Despite this, the cell must adapt to the hypoxic environment via increased oxygen delivery systemically and locally, as well as protect itself from secondary effects of hypoxia, such as decreased pH. Therefore, in a background of decreased total protein translation, specific upregulation of protective mechanisms safeguard the cell from hypoxic stress. This process is primarily regulated by a transcription factor known as Hypoxia Inducible Factor, HIF.

Background

Hypoxia-Inducible Factor 1, HIF-1, was initially identified as a transcriptional activator of erythropoietin (a glycoprotein regulating red cell production in the kidney), under hypoxic conditions [1, 2]. Numerous factors have subsequently been identified to be induced by HIF (discussed later in this chapter), and many of these genes are also known to be similarly regulated in cancer, suggesting an important regulatory role of HIF in tumors. HIF is highly conserved in eukaryotes from C elegans to humans, highlighting its fundamental role in normal physiological cell functions.

Structure

HIF1 is composed of a 120-kDa HIF-1α subunit complexed with a 91- to 94-kDa HIF-β subunit (Fig. 15.1). They are both basic-helix-loop helix proteins containing a PAS domain. HIF-1α located on chromosome 14q21–q24 [3] is closely related to Sim, and HIF-β is the aryl hydrocarbon receptor nuclear translocator (ARNT), which heterodimerizes with either HIF-1α or AHR. HIF-1α protein levels increase under hypoxic conditions and subsequently return to normal state under normoxia [4]. HIF-1α contains an amino-terminus bHLH domain that mediates binding to consensus DNA sequences (CTACGTGCT) in the promoter region of target genes [5, 1, 4, 6].

HIF Degradation Pathway

In normal physiological conditions, HIF-1α undergoes rapid degradation. The proline residues (Pro402 and Pro564 in HIF-1; Pro 406 and Pro 531 in HIF-2) [7, 8] within its oxygen-dependent degradation domain (ODD) are hydroxylated by one of a 3 member family of prolyl hydroxylase domain containing proteins 1–3 (PHDs 1–3). These enzymes are oxygen-, Fe^{2+}-, ascorbate- and 2 oxoglutarate-dependent. Their action allows HIF-α interaction with von Hippel Lindau protein (pVHL), (the recognition component of an E3 ubiquitin ligase) subsequently promoting larger complex formation with elongin-B, elongin-C and cullin-2, which is rapidly destroyed by 26S proteasomes (Fig. 15.2).

Molecular Oncology Laboratories, Department of Medical Oncology, Weatherall Institute of Molecular Medicine, John Radcliffe Hospital, Oxford OX3 9DS, UK

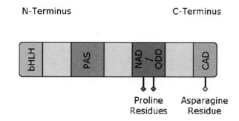

Fɪɢ. 15.1. HIF-α structure.

HIF Activation in Hypoxia

In times of insufficient oxygen, the prolyl hydroxylation of HIF-α is reduced, allowing the HIF α and β subunits to combine at nuclear HREs (hypoxic response elements) of target genes [9, 10]. Subsequent interaction with coactivators CBP (CREB binding protein) and p300 activates transcription and thus gene expression [11].

In many target genes HIF binding to HREs is sufficient for gene induction, although in others synergistic co-operation with other transcription factors such as Smad3, HNF4, ATF1/CREB1, AP1 and Ets-1 is required [12] (recently reviewed in refs. [13, 14]) (Fig. 15.3).

Alternative Pathways of HIF Activation

An alternative mechanism for HIF-α activation in hypoxia is regulated by Factor-Inhibiting HIF (FIH). HIF-α has two transactivation domains, NTAD (N-terminal transactivation domain) and CTAD (C-terminal transactivation domain). The CAD overlaps with the ODD (which binds the pVHL complex) and the NAD offers the alternative regulatory mechanism. Hydroxylation of the asparagine residue (Asn 803 – HIF1, Asn 851 – HIF2) on the CTAD blocks interaction with the transcriptional coactivators CBP and p300, as previously mentioned. This hydroxylation is enabled through the enzyme FIH that requires oxygen as a co-factor. Therefore, in hypoxic conditions, hydroxylation of this asparaginyl residue is avoided allowing HIF transcriptional activation [15] (Fig. 15.4).

HIF can also be activated by interaction with reactive oxygen species (ROS) possible via inhibition of PHD hydroxylation [15], the source of which may be from mitochondria as oxygen tension drops. Increased ROS production is also seen in response to radiation treatment, where there is subsequent increased HIF expression and VEGF production [16]. The source is probably inflammatory cells such as macrophages.

While low oxygen tension is the primary activator of the HIF pathway, growth factors (e.g., insulin, EGF, PDGF) and cytokines (e.g., TNF-α, interleukin-1β) have also been shown to initiate the pathway although the amplitude of induction is lower [17, 18]. Furthermore, HIF degradation is dependent upon sufficient levels of iron, ascorbate and 2-oxoglutarate, therefore deficiency of these co-factors can activate the HIF system. HIF activity is also enhanced by oncogenes such as Human Epidermal Growth Factor Receptor 2 (HER2), H-ras and v-Src [19–21]. Every oncogene pathway investigated has been reported to enhance HIF function [22, 23]. This is likely related to a fundamental link between proliferation, the generation of hypoxia, metabolite consumption and the need to increase blood flow and oxygen delivery locally and systemically [24].

Regulation of HIF Synthesis

Although not mediating rapid responses to hypoxia, regulation of HIF mRNA translation to protein is still necessary and is essential to maintain protein synthesis. It can also be an important contribution to the mechanisms by which oncogenes activate HIF. Thus, HER2 can activate two intracellular signalling pathways PI3K and MAPK, which in turn lead to activation of the eukaryotic translation initiation factor 4E (eIF-4E) which increases HIF mRNA translation [25]. These pathways are subject to positive feedback loops as HIF can induce expression of many growth factors and their receptors.

HIF Subfamily

HIF family comprises: HIF-1α (aka Member of PAS1 (MOP1)); HIF-2α (aka endothelial PAS domain protein 1 (EPAS1)); HIF-3α (aka MOP3); HIF-1α-like protein (MOP2 / HLF); and

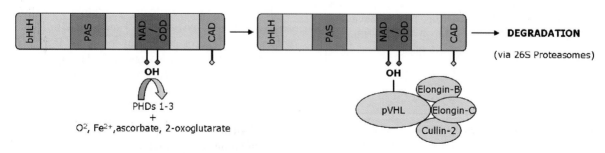

Fɪɢ. 15.2. HIF-α degradation pathway; oxygen dependent.

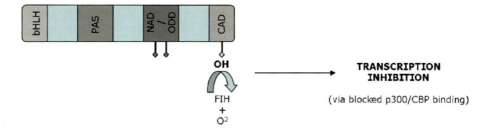

FIG. 15.3. HIF-α inhibition pathway; oxygen dependent.

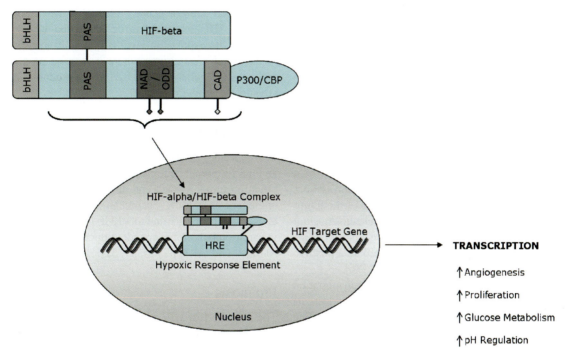

FIG. 15.4. HIF-α activation. If above pathways avoided due to hypoxia / loss of pVHL / insufficient Fe²⁺ or ascorbate or 2-oxoglutarate.

HIF-related factor (HRF). HIF-1α and HIF-2α both dimerise with HIF-β and subsequently bind to nuclear HREs. HIF-1 and HIF-2 contain 48% sequence homology, but have significant differences in their carboxyl-terminal sequences. This may explain the findings in specific knockdown studies of HIF-1 and HIF-2 in embryonic stem cells, mouse embryos and human cancer cells that there is differential function and target gene expression between the two isoforms [13, 26, 27]. This is particularly apparent in the difference between HIF-2 predominance in renal cancer which promotes tumor growth [28] and breast cancer where growth is inhibited [22].

An alternative splice variant of the HIF-3 locus, termed inhibitory PAS, negatively regulates HIF-induced gene expression, in particular the vascular endothelial growth factor (VEGF) gene. This HIF-3 α isoform may have a role in maintaining an avascular environment in certain tissues such as the cornea, and possibly forms a negative feedback loop to regulate HIF hypoxic response [29].

HIF in Cancer

HIF and VHL

The loss of function of the protein product of VHL allows HIF-α to avoid degradation and leads to its constitutive activation, resulting in increased expression of its target genes that control angiogenesis, proliferation, glucose metabolism, and pH regulation.

VHL syndrome is an autosomal dominant loss of function of the VHL gene and includes the clinical features of angiomatosis, haemangioblastomas, phaeochromocytomas and clear cell tumors of the kidney. The VHL gene is also inactivated in a large percentage of hemangioblastomas and renal cell carcinomas (RCCs) via mutation, large scale deletions, or methylation [30]. An unanswered problem is why malignant tumors form in the kidney, but with the same mutations only hemangiomas form in the central nervous system.

That a hereditary cancer syndrome results directly in HIF upregulation has provided strong evidence for the role of HIF in cancer.However, it remains a puzzle as to why there are no mutations in HIF leading directly to its stabilization. VHL has additional functions and partners that may be required, which is a subject of much investigation. Nevertheless, there is a striking correlation of genes induced by hypoxia and HIF as well as those regulated by VHL [31].

HIF and Tumor-associated Macrophages

Large numbers of circulating monocytes are attracted to hypoxic tumors along chemotaxic gradients [32]. They migrate across the vascular endothelium becoming tumour-associated macrophages (TAMs), a process regulated to a large degree by intra-tumoral hypoxia. Their presence within tumors can be anti- or pro-tumorigenic due to their secondary effects on antigen presenting abilities to cytotoxic T-cells, or their ability to secrete tumor mitogens and angiogenic promoters (including HIF) [33].

HIF activation within the tumor induces secretion of a number of factors that aid monocyte adhesion and extravasation, as well as TAM migration to the hypoxic center of the tumor [34]. The most widely studied hypoxia-induced chemoattractant is VEGF, which not only triggers the VEGF receptor, VEGF-R1, on the TAM cell surface [35], but also initiate a positive feedback mechanism that amplifies the signal [36].

The phenotype of these macrophages changes upon migration into the tumor where they then function to promote angiogenesis [36], remodel the surrounding matrix [37], and suppress the adaptive immunity [38]. This leads to a predominant proangiogenic phenotype in tumors with high a TAM component [39].

Expression

Identification of HIF-α within normal and cancerous tissues with immunohistochemistry has shown that there is much variation between different tissue types. Immunostaining is predominantly nuclear and generally strongest in perinecrotic regions (where presumably hypoxia is maximal), and there is often co-localization of HIF and VEGF [40] expression.

The majority of benign tissues do not express stabilized HIF-α. However, it is commonly overexpressed in malignant tissues. The pattern of expression within tumors appears partly dependent upon the mechanism of activation. Diffuse expression is seen in tumors with mutations within the HIF degradation pathway, such as clear cell RCC with its common mutation in the VHL gene. High and more localized expression is seen in peri-necrotic areas where HIF activation occurs via hypoxia [41, 42].

Hypoxia is the primary activator of HIF in tumors, but other mechanisms, such as mutations in tumor suppressor genes (e.g., VHL, p53, PTEN) [43–45] or activation of oncogenes (e.g., v-SRC, EGFR, and HER2) with subsequent signaling via the phosphatidyllinositol-3-kinase (PI3K) and mitogen-activated protein kinase (MAPK) pathway [20, 21], can increase or amplify HIF activation.

Interaction with the microenvironment and stroma is also important. HIF-1-deficient mouse astrocytoma cells when placed in a vessel-poor subcutaneous environment resulted in necrosis and poor growth and vessel development. However, when these same cells were placed in a vessel-rich environment (brain parenchyma) they grew quickly with extensive invasion [46].

HIF-1α protein is significantly increased in all human cancers studied and correlates with highly vascularized tumors [47, 48]. HIF-2α can also be identified in these tumors. However, its expression level is generally lower, apart from hepatocellular carcinoma where it is more pronounced [49] and in RCC where HIF-2 seems to predominate in more advanced tumors [50].

Xenograft Studies

In vivo studies determining the effect of HIF-1 in experimental xenograft tumors, demonstrate a HIF-1 dependent stimulatory effect on tumor growth and angiogenesis. The HIF effect on angiogenesis is more distinct in studies with human cancer cells and is primarily due to VEGF production, although there is variation, which is dependent upon the specific microenvironment and cell type [46, 51, 52].

Prognosis

The net effect of HIF activation on outcome appears to vary depending upon the origin of the primary tumor. Positive correlation with pathological features and poor prognosis have been reported in brain [53], breast [54], bladder [55], ovarian [56] and cervical tumors [57], while HIF-1 overexpression has been shown to be a predictor of better prognosis in kidney [58]. This probably reflects the balance between HIF1 and HIF2, which is unique for this VHL-mutated tumor. The majority of studies in lung cancer shows that high HIF-1 expression is associated with worse prognosis [59–61], but there is an exception [62]. These differences in HIF expression as it correlates to prognosis can be attributed to several factors: the difficulties and reliability of HIF immunostaining of paraffin-embedded tissues due to its inherent instability; the possible differential effects of HIF-1 and HIF-2 [27]; or the variability of activity of HIF depending on its microenvironment [46]. Overexpression of HIF in tumors has also been shown to correlate with resistance to drug, radiation and hormone treatment as well as photodynamic therapy [63–66].

HIF-1 Regulated Gene Expression

As mentioned earlier in the chapter, EPO was the first target gene associated with HIF. Since then many more genes have been discovered to be regulated by HIF-1 (Table 15.1).

Table 15.1 HIF-1 target genes.

Secreted factors	Adrenomedullin
	Angiopoietins[a]
	Angiopoietin 2[a]
	Angiopoietin like 4[a]
	Connective Tissue Growth Factor, CTGF
	Endothelin 1
	Endothelin 2
	Erythropoetin EPO
	Hepatocyte Growth Factor, HGF
	Interleukin-8, IL-8
	Inducible Nitric Oxide Sythase, iNOS[a]
	ORP150 VEGF chaperone[a]
	Osteopontin
	Platelet Derived Growth Factor β, PDGF β[a]
	Stromal Cell-Derived Factor-1, SDF1 (CXCL12)
	Spingosine 1 Phosphate, S1P
	Stanniocalcin 1
	Stanniocalcin 2
	Transforming Growth Factor beta 2, TGF beta 2
	Transforming Growth Factor beta 3, TGF beta 3
	Tie2[a]
	Vascular Endothelial Growth Factor, VEGF[a]
Tethered ligand Cell adhesion	Delta-like 4, DLL4 - Notch signaling[a]
	Integins[a]
	Tenascin (Hexabrachion)
Transmembrane receptors	VEGF Receptor 1, VEGFR 1[a]
	VEGF Receptor 2, VEGFR 2[a]
	CXCR4, Fusin
	Transferrin receptors
	Adenosine receptors
Copper pathways	Caeruloplasmin
Extracellular matrix proteases	Lysyl Oxidase, LOX
	Matrix Metalloproteinase 1,MMP 1[a]
	Matrix Metalloproteinase 2,MMP 2[a]
	Matrix Metalloproteinase 9,MMP 9[a]
	Plasminogen activator inhibitor-1, PAI1[a]
	Urokinase receptor[a]

[a]Covered in other chapters

The methods that have been used for evaluating HIF-dependent gene expression include screening for:

1. Genes with increased expression secondary to HIF-1 activated cells (e.g. VHL mutated cells, hypoxic cells).
2. Genes with decreased expression secondary to targeted HIF-1 inactivation.
3. Genes that contain the HIF binding site 5′-RCGTG-3′ at cis-acting HREs.

There may be as many as 1–5% of all genes that can be expressed in response to hypoxia in a cell-/tissue-specific manner [25]. This may reflect the fact that although the main role of HIF is the direct regulation of primary transcripts, there is also interaction with other signaling pathways that may amplify the hypoxic signal. Examples of this include HIF enhancing notch signaling which subsequently inhibits differentiation of certain cell types [67]; HIF decreasing the effect of c-myc on p21, resulting in cell cycle arrest [68]; as well as other effects on p53, c-Jun and NF-κB [69].

Angiogenic Factors Induced by HIF

The remainder of the chapter will focus on angiogenic factors induced by HIF, with a focus on those not covered in other chapters.

Vascular Endothelial Growth Factor

Vascular endothelial growth factor (VEGF) is probably the most important HIF-1-dependent target gene, playing a major role in hypoxia-induced angiogenesis. It has been shown to be expressed in many tumors, including breast, colon, kidney, bladder, and prostate, as well as many other cancer types.

VEGF is a growth factor that is vital in blood and vessel development [70]. It exists in four forms of variable amino acid length, which modulate their effects via tyrosine kinase receptors VEGF receptor 1 (also known as fms-like tyrosine kinase receptor 1, or Flt-1) and VEGF receptor 2 (or kinase insert domain receptor, a product of the KDR gene). Further VEGF signal transduction occurs via other receptors including neuropilin-1 and -2.

In hypoxia, activation of HIF directly increases VEGF and VEGFR1 transcription [71–73], as well as enhancing the stability of VEGF mRNA via HIF activation of HuR protein, which binds to VEGF 3′UTR [74]. VEGF translation is protected from the general reduction in protein synthesis secondary to hypoxia, by the presence of an internal ribosomal entry site within the 5′UTR of VEGF. This allows efficient cap-independent translation to be maintained [75].

Adrenomedullin

Adrenomedullin (ADM), a 52 amino acid peptide belonging to the calcitonin gene peptide superfamily, is secreted by many human tissues and is involved in the pathophysiology of several diseases including cancer. It has been shown to promote angiogenesis via its effects on endothelial cell growth, apoptosis and migration. Overexpression of ADM is induced by hypoxia (via HIF-1) and has been demonstrated in several tumor types, including brain, breast, colon, prostate, endometrial, ovarian, adrenal and lung. In endometrial, ovarian and breast tumors, this overexpression has been associated with increased vascularity and poor prognosis [76, 77].

Connective Tissue Growth Factor

Connective tissue growth factor (CTGF) is a member of a family of regulatory proteins termed CCN (Cysteine rich protein / connective tissue growth factor / nephroblastoma overexpressed gene). It promotes growth and angiogenesis, is overexpressed in invasive and metastatic breast cancer, and is involved in estrogen and growth factor-dependent cancer progression. It is induced by HIF-1α and promotes angiogenesis as well as increasing cell growth and migration mediated via platelet-derived growth factor (PDGF) and basic fibroblast growth factor (bFGF) [78].

Endothelin 1 and 2

Endothelins are potent vasoconstrictor peptide produced predominantly from vascular endothelial cells in hypoxic conditions due to induction by HIF-1α [79, 80]. Ligand-receptor binding with Endothelin G-protein-coupled transmembrane receptors promotes tumor cell growth, invasion, angiogenesis and metastasis. Endothelin-1 and -2 have been shown to be overexpressed in several tumor cell lines such as breast cancer, with elevated plasma concentrations found in patients with various solid tumors [81–85].

Erythropoietin

Erythropoietin (EPO) is a glycoprotein hormone produced in the kidney that regulates red blood cell production, and consequently increases systemic oxygen delivery. It is angiogenic, and its receptors are increased in tumors and tumor vessels. It was the first HIF-1 target gene to be discovered when cis-acting DNA sequences required for transcriptional activation in response to hypoxia were present in the EPO 3$'$ flanking region. The trans-acting factor that bound to the enhancer regulated EPO transcription in hypoxic conditions and was subsequently termed hypoxia-inducible factor-1 [1].

Hepatocyte Growth Factor

Hepatocyte growth factor (HGF) is a cytokine that has an important role in tumor growth, invasion, metastasis and angiogenesis via numerous pathways including the activation of CXCR4 and modulation by NF-κB [85]. Met protein, HGF's high affinity receptor, is induced by HIF-1, and triggers a signal transduction cascade involving PI3K when activated by HGF binding [86]. Hypoxia promotes HGF anti-apoptosis, invasion and metastasis via HIF-1α [87].

Interleukin-8

Interleukin-8 (IL-8 / Chemokine (C-X-C motif) ligand 8 / CXCL8) is a cytokine that was initially identified as a neutrophil chemotactic factor. It has subsequently been shown to promote angiogenesis, tumorigenesis and metastasis, and is often modulated by NF-κB [88]. Within the vascular endothelium, transcriptional activation of IL-8 is induced by hypoxia via ras, PI3K/Akt, p38, MAPK and HIF-1α pathways [89].

Notch Signaling

Signaling via Notch receptors (1–4) and their ligands Delta-like-1, -3 and -4 / Jagged1 and Jagged2 is important in a num-

ber of cellular functions including differentiation, proliferation and apoptosis. Disruption of Notch signaling is implicated in a range of tumors, and usually promotes tumor growth. Hypoxia (via HIF-1α stabilization) induces the expression of Delta-like-4 in endothelial cells which leads to increased Notch signaling and the transcriptional activation of Notch targets [90, 91].

Osteopontin

Osteopontin (also known as sialoprotein I / 44KBPP / Bone Phosphoprotein) is an extracellular phosphoglycoprotein making up the organic component of bone. It has been shown to be overexpressed in a number of different tumor types, and is associated with tumor progression and survival. It enhances invasion via induction of proteases, mediates proliferation via interaction with growth factors, promotes angiogenesis and regulates cytokine secretion by TAMs [92]. It is induced by hypoxia (HIF-1α) and its plasma levels can be used as a surrogate marker of tumor hypoxia [93, 94].

Stromal Cell-derived Factor-1 and Fusin

The cytokine stromal cell-derived factor-1 (SDF-1 / Chemokine (C-X-C motif) ligand 12 / CXCL12) and its receptor fusin (CXCR4) play an essential role in cellular chemotaxis. Their expression is complimentary and is induced by hypoxia (via HIF-1α) on monocytes, monocyte-derived macrophages, TAMs, bone marrow-derived circulating endothelial cells and metastatic cancer cells [85, 95]. These complimentary expression patterns increase the adhesion, migration and homing of circulating CXCR4-positive cells, guiding them to the hypoxic areas, thereby enhancing tumor progression, angiogenesis, and metastasis [96].

Stanniocalcin 1 and 2

Stanniocalcins are endocrine hormones that play an important role in normal development and physiology as well as numerous pathophysiological processes including cancer, although their full role remains to be fully elucidated. Stanniocalcin 1 and 2 are structurally similar proteins and are encoded on a region of chromosome 8. Their expression has been shown to be increased in many different tumors including breast, colon, head and neck, and ovarian. This expression has also be demonstrated to be partially regulated by hypoxia and is HIF-1α-dependent [97–99].

Transforming Growth Factor-β

Transforming growth factor-β (TGF-β) has three isoforms TGF-β 1, 2 and 3. They are polypeptide growth factors involved in many cancers through their effects on cellular differentiation, tissue regeneration, angiogenesis and modulation of the immune system. TGF-β signaling is mediated by SMAD proteins which act synergistically with HIF-1α at the promoter regions of VEGF and EPO toregulate their induction [100]. HIF-1α binds to the promoter region of TGF-β and directly regulates its expression [101].

Tie 2

The Tie family of endothelial cell specific tyrosine kinase receptors has two members, Tie1 and Tie2. They have important roles in developmental vessel formation, maintenance of adult vasculature and pathological angiogenesis including cancer. Tie1, important in vessel maturation, has no known ligands but may act via dimerization with Tie2 to modulate its signal. Tie2 has four ligands called Angiopoietins that have varying roles from receptor activation (proangiogenic – Ang1) to receptor inhibition (antiangiogenic – Ang2) [102]. Ang4 and Tie2 have increased expression in hypoxia as their transcription is induced by HIF-1α stabilization [103].

Tenascin-C (Hexabrachion)

Tenascin-C is a large extracellular matrix glycoprotein that is highly expressed in the stroma of most solid tumors and is generally associated with poor prognosis. It has anti-adhesive properties via its ability to bind fibronectin and block its interaction with specific syndecans. These actions and others (not yet fully elucidated) promote tumorigenesis, angiogenesis, metastasis and modulation of the immune response [104]. Tenascin-C can be induced by numerous factors including pro- and anti-inflammatory cytokines, various growth factors, ROS, NF-κB, and mechanical stress as well as hypoxia (via HIF-1α) [99].

Transferrin Receptor

Iron containing proteins are key elements in cellular energy metabolism and DNA synthesis. Cancer cells utilize more iron than normal cells and are subsequently more sensitive to iron chelation treatment. The activation of HIF-1 via PHDs is dependent on sufficient oxygen and iron levels. Therefore, HIF-1 not only regulates oxygen-dependent transcriptional gene expression but also iron-dependent gene expression. Transferrin is a plasma glycoprotein that binds two iron ions (Fe^{3+}) and transports them to cell surface transferrin receptors. The expression of the transferrin receptors mediate cellular iron uptake and are regulated by iron regulatory proteins (IRPs) (including HIF-1) which have the ability to sense iron levels and induce expression via binding to iron responsive elements (IREs) in the promoter regions of the transferrin receptor gene. Expression is also induced in hypoxia via binding of HIF-1 to adjacent HREs [105]. Transferrin receptor expression has been shown to be increased on tumor endothelium, and their blockade may be of future therapeutic use [106].

Adenosine Receptor

Adenosine receptors (A1, A2A, A2B, and A3) are a family of G-protein coupled receptors for which adenosine is their primary endogenous ligand that generally exert a cytoprotective effect. Adenosine receptors can couple with MAPK resulting in changes in cell growth, survival, death, and differentiation. A2B and A3 receptors have been shown to be overexpressed in some tumors, and have been targeted in some novel therapies, which have shown inhibition of tumor growth, induction of apoptosis, and the down-regulation of estrogen receptors in breast cancer [107]. Adenosine signaling is increased in hypoxia where A2B receptors are induced by functional binding of HIF-1 to the A2B promoter [108].

Ceruloplasmin

Copper is an essential trace element important for the function of many enzymes and catalytic cellular reactions. In tumors, it has a role in promoting angiogenesis via stimulation of endothelial cells, and its serum levels have been correlated with tumor burden and recurrence [109]. Ceruloplasmin is an enzyme containing 8 copper atoms that oxidizes iron ($Fe2+$) into a transportable and usable state ($Fe3+$) allowing cells to take up iron (essential for cellular proliferation as well as HIF degradation). It is induced in response to HIF-1 binding at its promoter [110], and its increased expression in a number of tumors has been associated with poor prognosis. Treatment of tumors with copper chelators in animal models has shown significant reduction in tumor growth and microvessel density [111].

Insulin-like Growth Factor-2

The insulin-like growth factors (IGF) are polypeptides with high sequence similarity to insulin that form part of the IGF axis. This axis is pro-tumorigenic in function acting to promote cellular proliferation and inhibit apoptosis. It includes two membrane bound receptors (IGF1R and IGF2R), two ligands (IGF-1 and IGF-2) and six IGF binding proteins (IGFBP 1–6). IGF has a reciprocal relationship with HIF-1, whereby IGF increases production and stabilization of the HIF-1α constituent which can subsequently bind to HREs within the promoter of IGF-2, IGFBP-2 and IGFBP-3, inducing their expression [112].

Lysyl Oxidase

Lysyl oxidase (LOX) is an extracellular matrix protease that covalently cross-links collagen and elastin in the extracellular matrix. It is overexpressed in hypoxic tumors and is positively correlated with increasing stage, invasion and metastasis [113]. The induction of LOX by hypoxia is via HIF-1 binding to HREs within its promoter region [114].

Therapeutic Implications

The therapeutic approaches targeting angiogenic pathways is covered in other sections of this textbook, and has been reviewed recently [25, 69]. However, it is worth mentioning briefly a few of these and how they relate to HIF-1.

Bevacizumab is a monoclonal antibody directed to the primary HIF-1 gene target, VEGF. It has been shown to increase survival from metastatic colorectal cancer and NSCLC when administered with standard chemotherapy [115–118]. Another therapy targeting VEGF is the VEGF trap, a soluble chimeric VEGF receptor [119].

Sorafenib is a tyrosine kinase inhibitor targeting angiogenic receptors (VEGF-R, PDGF-R, and EGFR) as well as the Raf/MEK/ERK signalling pathway. It has been shown to increase progression free survival in advanced RCC [120].

Direct HIF-1 inhibition can be achieved with antisense HIF-1 treatment. This has been shown to block T-cell lymphoma growth [121] and, together with VHL overexpression increased apoptosis and decreased angiogenesis in gliomas [122]. Blockade of HIF and CBP/p300 interaction is another therapeutic avenue, and has been shown to decrease tumor growth in xenografts [51].

As previously mentioned, HIF can be induced by ROS formation secondary to radiotherapy. Inhibition of HIF with a small molecule, YC-1, slowed tumor growth after radiation exposure [123–125].

It is however apparent from this chapter that there are multiple angiogenic pathways regulated by hypoxia, so it is not surprising that single agents are active in only a small proportion of patients. One of the major challenges clinically is to develop markers or profiles to determine which tumor will respond to which targeted drug or combination.

References

1. Semenza, G. L. & Wang, G. L. A nuclear factor induced by hypoxia via de novo protein synthesis binds to the human erythropoietin gene enhancer at a site required for transcriptional activation. Mol Cell Biol 12, 5447–54 (1992).
2. Wang, G. L. & Semenza, G. L. Purification and characterization of hypoxia-inducible factor 1. J Biol Chem 270, 1230–7 (1995).
3. Semenza, G. L. et al. Structural and functional analysis of hypoxia-inducible factor 1. Kidney Int 51, 553–5 (1997).
4. Wang, G. L., Jiang, B. H., Rue, E. A. & Semenza, G. L. Hypoxia-inducible factor 1 is a basic-helix-loop-helix-PAS heterodimer regulated by cellular O2 tension. Proc Natl Acad Sci USA 92, 5510–4 (1995).
5. Murre, C. et al. Interactions between heterologous helix-loop-helix proteins generate complexes that bind specifically to a common DNA sequence. Cell 58, 537–44 (1989).
6. Semenza, G. L., Koury, S. T., Nejfelt, M. K., Gearhart, J. D. & Antonarakis, S. E. Cell-type-specific and hypoxia-inducible expression of the human erythropoietin gene in transgenic mice. Proc Natl Acad Sci USA 88, 8725–9 (1991).
7. Bruick, R. K. & McKnight, S. L. A conserved family of prolyl-4-hydroxylases that modify HIF. Science 294, 1337–40 (2001).
8. Epstein, A. C. et al. C. elegans EGL-9 and mammalian homologs define a family of dioxygenases that regulate HIF by prolyl hydroxylation. Cell 107, 43–54 (2001).
9. Ivan, M. et al. HIFalpha targeted for VHL-mediated destruction by proline hydroxylation: implications for O2 sensing. Science 292, 464–8 (2001).
10. Jaakkola, P. et al. Targeting of HIF-alpha to the von Hippel-Lindau ubiquitylation complex by O2-regulated prolyl hydroxylation. Science 292, 468–72 (2001).
11. Ruas, J. L., Poellinger, L. & Pereira, T. Role of CBP in regulating HIF-1-mediated activation of transcription. J Cell Sci 118, 301–11 (2005).
12. Bracken, C. P., Whitelaw, M. L. & Peet, D. J. The hypoxia-inducible factors: key transcriptional regulators of hypoxic responses. Cell Mol Life Sci 60, 1376–93 (2003).
13. Pugh, C. W. & Ratcliffe, P. J. Regulation of angiogenesis by hypoxia: role of the HIF system. Nat Med 9, 677–84 (2003).
14. Hirota, K. & Semenza, G. L. Regulation of angiogenesis by hypoxia-inducible factor 1. Crit Rev Oncol Hematol 59, 15–26 (2006).
15. Liu, L. & Simon, M. C. Regulation of transcription and translation by hypoxia. Cancer Biol Ther 3, 492–7 (2004).
16. Moeller, B. J., Cao, Y., Li, C. Y. & Dewhirst, M. W. Radiation activates HIF-1 to regulate vascular radiosensitivity in tumors: role of reoxygenation, free radicals, and stress granules. Cancer Cell 5, 429–41 (2004).
17. Treins, C., Giorgetti-Peraldi, S., Murdaca, J., Semenza, G. L. & Van Obberghen, E. Insulin stimulates hypoxia-inducible factor 1 through a phosphatidylinositol 3-kinase/target of rapamycin-dependent signaling pathway. J Biol Chem 277, 27975–81 (2002).
18. Zhou, J., Schmid, T. & Brune, B. Tumor necrosis factor-alpha causes accumulation of a ubiquitinated form of hypoxia inducible factor-1alpha through a nuclear factor-kappaB-dependent pathway. Mol Biol Cell 14, 2216–25 (2003).
19. Chen, C., Pore, N., Behrooz, A., Ismail-Beigi, F. & Maity, A. Regulation of glut1 mRNA by hypoxia-inducible factor-1. Interaction between H-ras and hypoxia. J Biol Chem 276, 9519–25 (2001).
20. Jiang, B. H., Agani, F., Passaniti, A. & Semenza, G. L. V-SRC induces expression of hypoxia-inducible factor 1 (HIF-1) and transcription of genes encoding vascular endothelial growth factor and enolase 1: involvement of HIF-1 in tumor progression. Cancer Res 57, 5328–35 (1997).
21. Laughner, E., Taghavi, P., Chiles, K., Mahon, P. C. & Semenza, G. L. HER2 (neu) signaling increases the rate of hypoxia-inducible factor 1alpha (HIF-1alpha) synthesis: novel mechanism for HIF-1-mediated vascular endothelial growth factor expression. Mol Cell Biol 21, 3995–4004 (2001).
22. Blancher, C., Moore, J. W., Talks, K. L., Houlbrook, S. & Harris, A. L. Relationship of hypoxia-inducible factor (HIF)-1alpha and HIF-2alpha expression to vascular endothelial growth factor induction and hypoxia survival in human breast cancer cell lines. Cancer Res 60, 7106–13 (2000).
23. Vogelstein, B. & Kinzler, K. W. Cancer genes and the pathways they control. Nat Med 10, 789–99 (2004).
24. Maxwell, P. H., Pugh, C. W. & Ratcliffe, P. J. Activation of the HIF pathway in cancer. Curr Opin Genet Dev 11, 293–9 (2001).
25. Semenza, G. L. Targeting HIF-1 for cancer therapy. Nat Rev Cancer 3, 721–32 (2003).
26. Hu, C. J., Wang, L. Y., Chodosh, L. A., Keith, B. & Simon, M. C. Differential roles of hypoxia-inducible factor 1alpha (HIF-1alpha) and HIF-2alpha in hypoxic gene regulation. Mol Cell Biol 23, 9361–74 (2003).
27. Raval, R. R. et al. Contrasting properties of hypoxia-inducible factor 1 (HIF-1) and HIF-2 in von Hippel-Lindau-associated renal cell carcinoma. Mol Cell Biol 25, 5675–86 (2005).

28. Maranchie, J. K. et al. The contribution of VHL substrate binding and HIF1-alpha to the phenotype of VHL loss in renal cell carcinoma. Cancer Cell 1, 247–55 (2002).
29. Makino, Y., Kanopka, A., Wilson, W. J., Tanaka, H. & Poellinger, L. Inhibitory PAS domain protein (IPAS) is a hypoxia-inducible splicing variant of the hypoxia-inducible factor-3alpha locus. J Biol Chem 277, 32405–8 (2002).
30. Kaelin, W. G., Jr. Molecular basis of the VHL hereditary cancer syndrome. Nat Rev Cancer 2, 673–82 (2002).
31. Wykoff, C. C., Pugh, C. W., Maxwell, P. H., Harris, A. L. & Ratcliffe, P. J. Identification of novel hypoxia dependent and independent target genes of the von Hippel-Lindau (VHL) tumour suppressor by mRNA differential expression profiling. Oncogene 19, 6297–305 (2000).
32. Leek, R. D., Harris, A. L. & Lewis, C. E. Cytokine networks in solid human tumors: regulation of angiogenesis. J Leukoc Biol 56, 423–35 (1994).
33. Leek, R. D. & Harris, A. L. Tumor-associated macrophages in breast cancer. J Mammary Gland Biol Neoplasia 7, 177–89 (2002).
34. Murdoch, C., Giannoudis, A. & Lewis, C. E. Mechanisms regulating the recruitment of macrophages into hypoxic areas of tumors and other ischemic tissues. Blood 104, 2224–34 (2004).
35. Barleon, B. et al. Migration of human monocytes in response to vascular endothelial growth factor (VEGF) is mediated via the VEGF receptor flt-1. Blood 87, 3336–43 (1996).
36. Lewis, J. S., Landers, R. J., Underwood, J. C., Harris, A. L. & Lewis, C. E. Expression of vascular endothelial growth factor by macrophages is up-regulated in poorly vascularized areas of breast carcinomas. J Pathol 192, 150–8 (2000).
37. Burke, B. et al. Hypoxia-induced gene expression in human macrophages: implications for ischemic tissues and hypoxia-regulated gene therapy. Am J Pathol 163, 1233–43 (2003).
38. Sica, A., Schioppa, T., Mantovani, A. & Allavena, P. Tumour-associated macrophages are a distinct M2 polarised population promoting tumour progression: potential targets of anti-cancer therapy. Eur J Cancer 42, 717–27 (2006).
39. Leek, R. D., Lewis, C. E. & Harris, A. L. The role of macrophages in tumour angiogenesis (ed. Bicknell R, L. C.) (Oxford University Press, 1997).
40. Talks, K. L. et al. The expression and distribution of the hypoxia-inducible factors HIF-1alpha and HIF-2alpha in normal human tissues, cancers, and tumor-associated macrophages. Am J Pathol 157, 411–21 (2000).
41. Krieg, M. et al. Up-regulation of hypoxia-inducible factors HIF-1alpha and HIF-2alpha under normoxic conditions in renal carcinoma cells by von Hippel-Lindau tumor suppressor gene loss of function. Oncogene 19, 5435–43 (2000).
42. Wykoff, C. C. et al. Hypoxia-inducible expression of tumor-associated carbonic anhydrases. Cancer Res 60, 7075–83 (2000).
43. Maxwell, P. H. et al. The tumour suppressor protein VHL targets hypoxia-inducible factors for oxygen-dependent proteolysis. Nature 399, 271–5 (1999).
44. Ravi, R. et al. Regulation of tumor angiogenesis by p53-induced degradation of hypoxia-inducible factor 1alpha. Genes Dev 14, 34–44 (2000).
45. Zundel, W. et al. Loss of PTEN facilitates HIF-1-mediated gene expression. Genes Dev 14, 391–6 (2000).
46. Blouw, B. et al. The hypoxic response of tumors is dependent on their microenvironment. Cancer Cell 4, 133–46 (2003).
47. Brahimi-Horn, M. C. & Pouyssegur, J. The hypoxia-inducible factor and tumor progression along the angiogenic pathway. Int Rev Cytol 242, 157–213 (2005).
48. Semenza, G. L. HIF-1 and tumor progression: pathophysiology and therapeutics. Trends Mol Med 8, S62–7 (2002).
49. Bangoura, G., Yang, L. Y., Huang, G. W. & Wang, W. Expression of HIF-2alpha/EPAS1 in hepatocellular carcinoma. World J Gastroenterol 10, 525–30 (2004).
50. Mandriota, S. J. et al. HIF activation identifies early lesions in VHL kidneys: evidence for site-specific tumor suppressor function in the nephron. Cancer Cell 1, 459–68 (2002).
51. Kung, A. L., Wang, S., Klco, J. M., Kaelin, W. G. & Livingston, D. M. Suppression of tumor growth through disruption of hypoxia-inducible transcription. Nat Med 6, 1335–40 (2000).
52. Maxwell, P. H. et al. Hypoxia-inducible factor-1 modulates gene expression in solid tumors and influences both angiogenesis and tumor growth. Proc Natl Acad Sci USA 94, 8104–9 (1997).
53. Korkolopoulou, P. et al. Hypoxia-inducible factor 1alpha/vascular endothelial growth factor axis in astrocytomas. Associations with microvessel morphometry, proliferation and prognosis. Neuropathol Appl Neurobiol 30, 267–78 (2004).
54. Schindl, M. et al. Overexpression of hypoxia-inducible factor 1alpha is associated with an unfavorable prognosis in lymph node-positive breast cancer. Clin Cancer Res 8, 1831–7 (2002).
55. Theodoropoulos, V. E. et al. Hypoxia-inducible factor 1 alpha expression correlates with angiogenesis and unfavorable prognosis in bladder cancer. Eur Urol 46, 200–8 (2004).
56. Birner, P., Schindl, M., Obermair, A., Breitenecker, G. & Oberhuber, G. Expression of hypoxia-inducible factor 1alpha in epithelial ovarian tumors: its impact on prognosis and on response to chemotherapy. Clin Cancer Res 7, 1661–8 (2001).
57. Birner, P. et al. Overexpression of hypoxia-inducible factor 1alpha is a marker for an unfavorable prognosis in early-stage invasive cervical cancer. Cancer Res 60, 4693–6 (2000).
58. Lidgren, A. et al. The expression of hypoxia-inducible factor 1alpha is a favorable independent prognostic factor in renal cell carcinoma. Clin Cancer Res 11, 1129–35 (2005).
59. Giatromanolaki, A. et al. DEC1 (STRA13) protein expression relates to hypoxia-inducible factor 1-alpha and carbonic anhydrase-9 overexpression in non-small cell lung cancer. J Pathol 200, 222–8 (2003).
60. Giatromanolaki, A. et al. Relation of hypoxia inducible factor 1 alpha and 2 alpha in operable non-small cell lung cancer to angiogenic/molecular profile of tumours and survival. Br J Cancer 85, 881–90 (2001).
61. Kim, S. J. et al. Expression of HIF-1alpha, CA IX, VEGF, and MMP-9 in surgically resected non-small cell lung cancer. Lung Cancer 49, 325–35 (2005).
62. Volm, M. & Koomagi, R. Hypoxia-inducible factor (HIF-1) and its relationship to apoptosis and proliferation in lung cancer. Anticancer Res 20, 1527–33 (2000).
63. Koukourakis, M. I. et al. Hypoxia-inducible factor (HIF1A and HIF2A), angiogenesis, and chemoradiotherapy outcome of squamous cell head-and-neck cancer. Int J Radiat Oncol Biol Phys 53, 1192–202 (2002).
64. Generali, D. et al. Hypoxia-inducible factor-1alpha expression predicts a poor response to primary chemoendocrine therapy and disease-free survival in primary human breast cancer. Clin Cancer Res 12, 4562–8 (2006).

65. Koukourakis, M. I. et al. Endogenous markers of two separate hypoxia response pathways (hypoxia inducible factor 2 alpha and carbonic anhydrase 9) are associated with radiotherapy failure in head and neck cancer patients recruited in the CHART randomized trial. J Clin Oncol 24, 727–35 (2006).

66. Koukourakis, M. I. et al. Hypoxia inducible factor (HIF-1a and HIF-2a) expression in early esophageal cancer and response to photodynamic therapy and radiotherapy. Cancer Res 61, 1830–2 (2001).

67. Gustafsson, M. V. et al. Hypoxia requires notch signaling to maintain the undifferentiated cell state. Dev Cell 9, 617–28 (2005).

68. Koshiji, M. et al. HIF-1alpha induces cell cycle arrest by functionally counteracting Myc. Embo J 23, 1949–56 (2004).

69. Hickey, M. M. & Simon, M. C. Regulation of angiogenesis by hypoxia and hypoxia-inducible factors. Curr Top Dev Biol 76, 217–57 (2006).

70. Weinstein, B. M. What guides early embryonic blood vessel formation? Dev Dyn 215, 2–11 (1999).

71. Liu, Y., Cox, S. R., Morita, T. & Kourembanas, S. Hypoxia regulates vascular endothelial growth factor gene expression in endothelial cells. Identification of a 5′ enhancer. Circ Res 77, 638–43 (1995).

72. Forsythe, J. A. et al. Activation of vascular endothelial growth factor gene transcription by hypoxia-inducible factor 1. Mol Cell Biol 16, 4604–13 (1996).

73. Gerber, H. P., Condorelli, F., Park, J. & Ferrara, N. Differential transcriptional regulation of the two vascular endothelial growth factor receptor genes. Flt-1, but not Flk-1/KDR, is up-regulated by hypoxia. J Biol Chem 272, 23659–67 (1997).

74. Levy, N. S., Chung, S., Furneaux, H. & Levy, A. P. Hypoxic stabilization of vascular endothelial growth factor mRNA by the RNA-binding protein HuR. J Biol Chem 273, 6417–23 (1998).

75. Stein, I. et al. Translation of vascular endothelial growth factor mRNA by internal ribosome entry: implications for translation under hypoxia. Mol Cell Biol 18, 3112–9 (1998).

76. Nikitenko, L. L., Fox, S. B., Kehoe, S., Rees, M. C. & Bicknell, R. Adrenomedullin and tumour angiogenesis. Br J Cancer 94, 1–7 (2006).

77. Garayoa, M. et al. Hypoxia-inducible factor-1 (HIF-1) up-regulates adrenomedullin expression in human tumor cell lines during oxygen deprivation: a possible promotion mechanism of carcinogenesis. Mol Endocrinol 14, 848–62 (2000).

78. Higgins, D. F. et al. Hypoxic induction of Ctgf is directly mediated by Hif-1. Am J Physiol Renal Physiol 287, F1223–32 (2004).

79. Hu, J., Discher, D. J., Bishopric, N. H. & Webster, K. A. Hypoxia regulates expression of the endothelin-1 gene through a proximal hypoxia-inducible factor-1 binding site on the antisense strand. Biochem Biophys Res Commun 245, 894–9 (1998).

80. Grimshaw, M. J., Naylor, S. & Balkwill, F. R. Endothelin-2 is a hypoxia-induced autocrine survival factor for breast tumor cells. Mol Cancer Ther 1, 1273–81 (2002).

81. Grant, K., Loizidou, M. & Taylor, I. Endothelin-1: a multifunctional molecule in cancer. Br J Cancer 88, 163–6 (2003).

82. Grimshaw, M. J. et al. A role for endothelin-2 and its receptors in breast tumor cell invasion. Cancer Res 64, 2461–8 (2004).

83. Fan, S. et al. Role of NF-kappaB signaling in hepatocyte growth factor/scatter factor-mediated cell protection. Oncogene 24, 1749–66 (2005).

84. Tacchini, L., De Ponti, C., Matteucci, E., Follis, R. & Desiderio, M. A. Hepatocyte growth factor-activated NF-kappaB regulates HIF-1 activity and ODC expression, implicated in survival, differently in different carcinoma cell lines. Carcinogenesis 25, 2089–100 (2004).

85. Maroni, P., Bendinelli, P., Matteucci, E. & Desiderio, M. A. HGF induces CXCR4 and CXCL12-mediated tumor invasion through Ets1 and NF-{kappa}B. Carcinogenesis 28, 267–79 (2007).

86. Pennacchietti, S. et al. Hypoxia promotes invasive growth by transcriptional activation of the met protooncogene. Cancer Cell 3, 347–61 (2003).

87. Hara, S. et al. Hypoxia enhances c-Met/HGF receptor expression and signaling by activating HIF-1alpha in human salivary gland cancer cells. Oral Oncol 42, 593–8 (2006).

88. Karashima, T. et al. Nuclear factor-kappaB mediates angiogenesis and metastasis of human bladder cancer through the regulation of interleukin-8. Clin Cancer Res 9, 2786–97 (2003).

89. Kim, K. S., Rajagopal, V., Gonsalves, C., Johnson, C. & Kalra, V. K. A novel role of hypoxia-inducible factor in cobalt chloride- and hypoxia-mediated expression of IL-8 chemokine in human endothelial cells. J Immunol 177, 7211–24 (2006).

90. Sainson, R. C. & Harris, A. L. Hypoxia-regulated differentiation: let's step it up a Notch. Trends Mol Med 12, 141–3 (2006).

91. Shi, W. & Harris, A. L. Notch signaling in breast cancer and tumor angiogenesis: cross-talk and therapeutic potentials. J Mammary Gland Biol Neoplasia 11, 41–52 (2006).

92. Rittling, S. R. & Chambers, A. F. Role of osteopontin in tumour progression. Br J Cancer 90, 1877–81 (2004).

93. Zhu, Y. et al. Hypoxia upregulates osteopontin expression in NIH-3T3 cells via a Ras-activated enhancer. Oncogene 24, 6555–63 (2005).

94. Le, Q. T. et al. Identification of osteopontin as a prognostic plasma marker for head and neck squamous cell carcinomas. Clin Cancer Res 9, 59–67 (2003).

95. Schioppa, T. et al. Regulation of the chemokine receptor CXCR4 by hypoxia. J Exp Med 198, 1391–402 (2003).

96. Ceradini, D. J. et al. Progenitor cell trafficking is regulated by hypoxic gradients through HIF-1 induction of SDF-1. Nat Med 10, 858–64 (2004).

97. Ito, D. et al. Characterization of stanniocalcin 2, a novel target of the mammalian unfolded protein response with cytoprotective properties. Mol Cell Biol 24, 9456–69 (2004).

98. Yeung, H. Y. et al. Hypoxia-inducible factor-1-mediated activation of stanniocalcin-1 in human cancer cells. Endocrinology 146, 4951–60 (2005).

99. Lal, A. et al. Transcriptional response to hypoxia in human tumors. J Natl Cancer Inst 93, 1337–43 (2001).

100. Sanchez-Elsner, T. et al. A cross-talk between hypoxia and TGF-beta orchestrates erythropoietin gene regulation through SP1 and Smads. J Mol Biol 336, 9–24 (2004).

101. Schaffer, L. et al. Oxygen-regulated expression of TGF-beta 3, a growth factor involved in trophoblast differentiation. Placenta 24, 941–50 (2003).

102. Jones, N., Iljin, K., Dumont, D. J. & Alitalo, K. Tie receptors: new modulators of angiogenic and lymphangiogenic responses. Nat Rev Mol Cell Biol 2, 257–67 (2001).

103. Yamakawa, M. et al. Hypoxia-inducible factor-1 mediates activation of cultured vascular endothelial cells by inducing multiple angiogenic factors. Circ Res 93, 664–73 (2003).

104. Orend, G. & Chiquet-Ehrismann, R. Tenascin-C induced signaling in cancer. Cancer Lett 244, 143–63 (2006).
105. Lok, C. N. & Ponka, P. Identification of a hypoxia response element in the transferrin receptor gene. J Biol Chem 274, 24147–52 (1999).
106. Jones, D. T., Trowbridge, I. S. & Harris, A. L. Effects of transferrin receptor blockade on cancer cell proliferation and hypoxia-inducible factor function and their differential regulation by ascorbate. Cancer Res 66, 2749–56 (2006).
107. Jacobson, K. A. & Gao, Z. G. Adenosine receptors as therapeutic targets. Nat Rev Drug Discov 5, 247–64 (2006).
108. Kong, T., Westerman, K. A., Faigle, M., Eltzschig, H. K. & Colgan, S. P. HIF-dependent induction of adenosine A2B receptor in hypoxia. Faseb J 20, 2242–50 (2006).
109. Coates, R. J., Weiss, N. S., Daling, J. R., Rettmer, R. L. & Warnick, G. R. Cancer risk in relation to serum copper levels. Cancer Res 49, 4353–6 (1989).
110. Martin, F. et al. Copper-dependent activation of hypoxia-inducible factor (HIF)-1: implications for ceruloplasmin regulation. Blood 105, 4613–9 (2005).
111. Lowndes, S. A. & Harris, A. L. The role of copper in tumour angiogenesis. J Mammary Gland Biol Neoplasia 10, 299–310 (2005).
112. Feldser, D. et al. Reciprocal positive regulation of hypoxia-inducible factor 1alpha and insulin-like growth factor 2. Cancer Res 59, 3915–8 (1999).
113. Erler, J. T. & Giaccia, A. J. Lysyl oxidase mediates hypoxic control of metastasis. Cancer Res 66, 10238–41 (2006).
114. Erler, J. T. et al. Lysyl oxidase is essential for hypoxia-induced metastasis. Nature 440, 1222–6 (2006).
115. Hurwitz, H. Integrating the anti-VEGF-A humanized monoclonal antibody bevacizumab with chemotherapy in advanced colorectal cancer. Clin Colorectal Cancer 4 Suppl 2, S62–8 (2004).
116. Hurwitz, H. et al. Bevacizumab plus irinotecan, fluorouracil, and leucovorin for metastatic colorectal cancer. N Engl J Med 350, 2335–42 (2004).
117. Belani, C. P. & Ramalingam, S. Bevacizumab extends survival for patients with nonsquamous non-small-cell lung cancer. Clin Lung Cancer 6, 267–8 (2005).
118. Kerr, C. Bevacizumab and chemotherapy improves survival in NSCLC. Lancet Oncol 6, 266 (2005).
119. Lau, S. C., Rosa, D. D. & Jayson, G. Technology evaluation: VEGF Trap (cancer), Regeneron/sanofi-aventis. Curr Opin Mol Ther 7, 493–501 (2005).
120. Kane, R. C. et al. Sorafenib for the treatment of advanced renal cell carcinoma. Clin Cancer Res 12, 7271–8 (2006).
121. Sun, X. et al. Gene transfer of antisense hypoxia inducible factor-1 alpha enhances the therapeutic efficacy of cancer immunotherapy. Gene Ther 8, 638–45 (2001).
122. Sun, X. et al. Overexpression of von Hippel-Lindau tumor suppressor protein and antisense HIF-1alpha eradicates gliomas. Cancer Gene Ther 13, 428–35 (2006).
123. Moeller, B. J. et al. Pleiotropic effects of HIF-1 blockade on tumor radiosensitivity. Cancer Cell 8, 99–110 (2005).
124. Yeo, E. J. et al. YC-1: a potential anticancer drug targeting hypoxia-inducible factor 1. J Natl Cancer Inst 95, 516–25 (2003).
125. Semenza, G. L. Development of novel therapeutic strategies that target HIF-1. Expert Opin Ther Targets 10, 267–80 (2006).

Chapter 16
Regulation of Angiogenesis by von Hippel Lindau Protein and HIF2

Donald P. Bottaro, Nelly Tan, and W. Marston Linehan

Keywords: von Hippel Lindau, tumor suppressor, renal cell carcinoma, oncogenic mechanisms, tumor angiogenesis

Abstract: von Hippel-Lindau (VHL) Syndrome is a rare, autosomal dominant, hereditary neoplastic disorder characterized by the development of hemangioblastomas, retinal angiomas and solid tumors in several organs. VHL-associated renal tumors are highly vascular, malignant and very often fatal. Affected individuals inherit an altered copy of the *VHL* tumor suppressor gene, and the wild type copy is later inactivated in somatic cells. *VHL* loss of function also occurs in the majority of sporadic cases of renal cell carcinoma. Thus, understanding the molecular basis of VHL syndrome has proven to be relevant to both patient populations. The *VHL* gene encodes the substrate recognition component (pVHL) of a ubiquitin ligase complex that targets hypoxia inducible factors (HIFs) for proteosome-mediated degradation in normoxic conditions. HIFs are ubiquitous transcriptional regulators that provide an essential oxygen sensing function. Failure to degrade HIFs, such as in hypoxia or in disease, results in the increased expression of a large collection of genes that regulate cellular energy metabolism, migration, proliferation and angiogenesis. In particular, because of its historic relevance to the pathogenesis of acute, chronic and inherited kidney diseases, the study of angiogenic regulation by the pVHL/HIF axis has contributed to fundamentally important advances in basic biology and medicine.

Urologic Oncology Branch, Center for Cancer Research, National Cancer Institute, National Institutes of Health, Bethesda, MD 20892, USA

*Corresponding Author:
Donald P. Bottaro, Urologic Oncology Branch, National Cancer Institute, Building 10 - Hatfield CRC, Room 1-3961, 10 Center Drive, MSC 1107, Bethesda, MD 20892-1107, USA
E-mail: dbottaro@helix.nih.gov

von Hippel-Lindau Syndrome and Renal Cell Carcinoma: an Overview

von Hippel-Lindau (VHL) Syndrome is an autosomal dominant hereditary neoplastic disorder characterized by the development of hemangioblastomas in the cerebellum and spine, retinal angiomas, endolymphatic sac tumors, pheochromocytomas and cystic and solid tumors of the pancreas and kidneys [1, 2]. VHL associated bilateral, multifocal renal tumors are highly vascular, malignant and can metastasize; historically up to 40% of untreated patients with VHL have died of advanced clear cell renal cell carcinoma (RCC) [1, 3]. Affected individuals inherit an altered copy of the *VHL* tumor suppressor gene, located on the short arm of chromosome 3 (3p25–26), and the wild type copy is later inactivated in somatic cells, most often by loss of chromosome 3p or *VHL* gene hypermethylation [1, 2]. Alterations or deletions of 3p also occur in the majority of sporadic clear cell RCC cases. Thus, while VHL is a rare familial disorder, knowledge obtained from the study of these patients has proven to be highly relevant to the molecular etiology of its histologically and cytogenetically similar sporadic cancer counterpart [1]. For example, reconstitution of wild type *VHL* expression in RCC derived cell lines has been shown to regulate tumorigenesis in athymic nude mice, confirming a fundamental role for *VHL* in clear cell RCC oncogenesis [4].

Soon after the discovery of the *VHL* gene, investigations into the molecular basis of the angiogenic nature of several VHL symptoms revealed significantly elevated levels of vascular endothelial growth factor (VEGF; vascular permeability factor) [5]. In subsequent studies, human RCC cells that lacked wild-type *VHL* or were transfected with an inactive mutant *VHL* displayed deregulated VEGF expression that was reverted upon restoration of wild type *VHL* gene expression. Vascular endothelial cell (EC) proliferation driven by RCC cell conditioned medium was substantially blocked by selectively neutralizing VEGF activity [5, 6]. These results established that *VHL* regulated *VEGF* expression and that VEGF was a critical tumor angiogenesis factor in VHL disease [6, 7].

The *VHL* Tumor Suppressor Gene and its Product, pVHL

VHL Gene Structure

The *VHL* gene consists of three exons encoding a 4.7-kb mRNA, comprising approximately 20 kb of genomic DNA on chromosome 3p25–p26 (NCBI Reference Sequence Accession NM_000551, variant 1; NM_198156, variant 2). The largest presumed coding region, 642 nucleotides, predicts a 213 residue protein, while a second translation initiation site at codon 54 of this region predicts a protein of 160 amino acids. The observed sizes of proteins recognized by pVHL-specific antibodies suggests that both codons 1 and 54 may be used as start codons. Since no pathogenic mutations have been identified in codons 1–53, this region might not be important for tumor suppressor function. An alternatively spliced mRNA lacking exon 2 is predicted to produce an in-frame deletion of 41 amino acids [8]. The *VHL* gene sequence is highly conserved in primates and rodents [9, 10], and homologs have been identified in *C. elegans* [9] as well as *Drosophila* [11, 12]. The most highly conserved regions are involved in binding to other proteins or in maintaining pVHL structure [8]. The *VHL* promoter is GC-rich, lacking TATA or CCAAT motifs. Transcription initiates around a putative Sp1-binding site approximately 60 nucleotides upstream of codon 1 [13]. The promoter (NCBI Locus AF010238, HSU97187) contains numerous predicted binding sites for transcription factors, although little is known about the regulation of *VHL* expression.

pVHL Structure and Function

The *VHL* coding sequence is organized in 3 exons: exons 1 and 2 form β-sheets in the translated VHL protein (pVHL), and exon 3 forms α-helices. These secondary structural features are the basis of the two major structural domains (designated β and α) identified from the three-dimensional structure of its heterotrimeric complex with elongins B and C [14]. The smaller α-domain (residues 155–192) contains the elongin C binding site as well as residues that provide intra-domain stability, and is a hot spot for missense mutations (e.g. Arg167).

Missense mutations also occur frequently on the surface of the β-domain opposite the binding site for elongin C (Trp88, Asn90, Gln96, Tyr98, Tyr112); this surface may represent another site of protein interaction [14]. VHL disease is subdivided broadly on the basis of pheochromocytoma frequency into Type 1 (low) and Type 2 (high). Type 2 VHL is further classified on the basis of RCC frequency (2A, low risk for RCC and 2B, high risk for RCC). Analysis of different *VHL* mutations among VHL disease subtypes showed that most Type

2 mutations map to the elongin C binding site or the opposite surface in the β-domain, or are likely to disrupt these structures. In contrast, the few missense mutations associated with Type 1 disease map to residues in the β-domain hydrophobic core and are predicted to completely disrupt pVHL structure [14].

pVHL forms a complex with the Elongin C and B proteins [15–17], the so-called VCB complex, that targets regulatory proteins for ubiquitin-mediated proteolysis. The role of this function in the regulation of angiogenesis is discussed in further detail below. In addition to regulating angiogenesis through E3 ligase activity, pVHL binds fibronectin, an extracellular matrix component that binds to cell-surface integrins and controls tumor cell migration and metastasis as well as the recruitment and organization of vascular ECs during angiogenesis. Fibronectin matrix assembly is defective in *VHL*-negative RCC cells and partially restored upon expression of wild type pVHL [18]. *VHL*-negative mouse embryo fibroblasts also display defective fibronectin matrix assembly compared with wild-type counterparts [18]. These findings support a direct role for pVHL in fibronectin matrix assembly and may be relevant to some of the pathologic microvascular features associated with VHL disease.

VCB E3 ligase structure and function

The VCB complex containing pVHL, Elongin C and B proteins is implicated in tumor suppression because most of VHL RCC tumor-associated mutations destabilize it [15–17, 19]. Through Elongin C, the VCB complex also binds Cul2, an association that is required for the pVHL-mediated regulation of *VEGF* gene transcription [19, 20]. In addition to pVHL, the Elongin C–Elongin B complex also binds to the suppressor of cytokine signaling (SOCS) superfamily of proteins that share the 40-residue SOCS-box motif [21–26], and to the Elongin A transcriptional elongation factor [27]. The Elongins were initially characterized as part of a family of genes that are critical for transcriptional regulation, leading to early theories that pVHL might act through inhibition of transcriptional elongation [15, 16].

The VHL–Elongin C/B complex also binds to Cul-2, a member of the multigene cullin family. Together these protein interactions provided additional clues about possible functions of pVHL: the yeast Cul-2 homolog CDC53 was known to be part of the SCF multiprotein complex that targets cell cycle regulatory proteins for ubiquitin-mediated proteolysis [28]. Indeed, the VCB–CUL2 complex exhibits E3 ubiquitin ligase activity, with Ubc5 a, b and c as the ubiquitin-conjugating E2 enzyme, and with another protein (Rbx1) enhancing the ligase activity [29–32].

E3 ubiquitin ligase complexes, including their substrate recognition components, have been classified based on the included cullin subunit. Cul-1 forms an E3 ligase together with Roc1, Skp1 and one of many F-box proteins, to form what is called an SCF (Skip/Cullin/F-box) E3 ubiquitin ligase complex [33]. Cul-2 forms an E3 ligase complex with Roc1, Elongin B, Elongin C (a protein similar to Skp1) and VHL or one of many SOCS (suppressor of cytokine signaling)-box-containing subunits [33]. In the VCB ligase, VHL and SOCS proteins carry out the role of the substrate-targeting subunit performed by F-box proteins in SCF complexes.

The pVHL/HIF Regulatory Pathway

HIFs as Critical VCB E3 Ligase Targets

As the substrate recognition component of the VCB E3 ubiquitin ligase complex, pVHL targets hypoxia inducible factors (HIFs) 1α and 2α for ubiquitin-mediated degradation under normoxic conditions [34, 35]. HIF-1α, HIF-2α and a recently identified HIF-3α are the α subunit isoforms of a heterodimeric transcription factor that is essential for cellular and systemic homeostatic responses to hypoxia such as angiogenesis and erythropoeisis. They are members of the Per-ARNT-Sim (PAS) family of heterodimeric basic helix-loop-helix (bHLH) transcription factors, where the α-subunits are oxygen-sensitive and the β-subunits are constitutively expressed (reviewed in [36, 37]). The HIF β-subunit is also known as the aryl hydrocarbon receptor nuclear translocator (ARNT) [38, 39]. Transcriptional regulation occurs through the binding of HIF heterodimers to hypoxia-response elements (HREs), which are present in regulatory regions of hypoxia-sensitive genes [39]. HIF-1α and HIF-2α proteins share 48% amino acid identity, both contain a conserved 15 amino acid minimal pVHL binding domain and both bind and activate HREs [40, 41]. HIF-1α and HIF-2α are the best characterized substrates of the VHL E3 ubiquitin ligase, but others, such as β-catenin, may also contribute to RCC oncogenesis and angiogenesis [42].

HIFα Activity in Hypoxia and RCC

Under normoxic conditions, specific proline residues of HIFα within the oxygen-dependent degradation (ODD) domain are hydroxylated by a family of oxygen and iron-dependent prolyl-4-hydroxylases (PHDs) [43]. HIFα hydroxylation is required for binding to the pVHL-E3-ubiquitin ligase complex, polyubiquitination and subsequent proteasomal degradation. Three major mammalian HIF prolyl-hydroxylases have been identified, and differential effects of individual PHDs on HIF-1α and HIF-2α hydroxylation have been reported [44]. Under hypoxic conditions, prolyl-hydroxylases are inactive, unhydroxylated HIFα does not bind pVHL and thus accumulates in cells [45, 46]. HIFα accumulation and subsequent translocation to the nucleus allows dimerization with HIFβ and recruitment of transcriptional cofactors including CBP and p300 [38, 39]. Most, if not all, RCC-associated pVHL mutants result in defective ubiquitination of HIFα [34, 35].

PHD catalyzed HIFα hydroxylation is a complex, irreversible reaction involving oxygen, ferrous iron, 2-oxoglutarate and reactive oxygen species (ROS) scavengers such as ascorbate; succinate and carbon dioxide are released as by-products.

The complexity of this reaction enables several different pathways to modulate net activity; for example, ascorbate [47], transition metals [48, 49], ROS including nitric oxide (NO) [50, 51], and Krebs cycle intermediates [52, 53] all influence PHD activity. The inactivity of PHDs in hypoxia and consequent stabilization of HIFα subunits involves PHD inhibition by ROS generated by mitochondrial complex III [54–56]. ROS inhibit PHD activity most likely by changing the redox state of enzyme-bound iron that is required for catalysis from $Fe(II)$ to $Fe(III)$ [50]. Unlike heme-containing proteins, the $Fe(II)$ in 2-oxoglutarate-dependent oxygenases can be chelated or substituted by $Co(II)$, rendering the enzyme inactive [57]. Hydroxylation of an asparagine residue in the carboxyl-terminal transactivation domain of HIFα by Factor-Inhibiting-HIF (FIH) constitutes a second hypoxic switch [58]. Inhibition of asparagine hydroxylation in hypoxia facilitates CBP/p300 recruitment, enhancing target gene transcription [58].

Increased HIF transcriptional activity in the absence of hypoxia has been shown to be mediated by NO, growth factors and cytokines including tumor necrosis factor-α [59], interleukin 1 [60, 61], angiotensin II [62], epidermal growth factor (EGF), insulin, and insulin-like growth factors [61, 63–66]. Growth factor-induced Ras pathway may inhibit HIF prolyl-hydroxylation [38, 67], and, in addition, concurrent phosphoinositide 3-kinase (PI3K)/Akt/mammalian target of rapamycin pathway activation may lead to increased HIF activation through increased HIFα protein translation [68–70]. It is easy to envision a scenario in which increased HIFα activity associated with loss of pVHL function in precursor RCC lesions may be compounded by the induction of growth factor/Ras/Akt pathway activation, such as through transforming growth factor-α (TGF-α)/EGF receptor autocrine loop formation (Fig. 16.1).

HIF-1α and HIF-2α: Common and Distinct Features

HIF-1α and HIF-2α proteins are widely expressed in normal tissues, although HIF-2α expression is less ubiquitous, being expressed primarily in hepatocytes, cardiomyocytes, glial cells, type II pneumocytes and vascular ECs [71]. In normal primary renal tubular epithelial cells, HIF-1α expression predominates [72, 73]; HIF-2α is expressed in renal interstitial fibroblasts and renal ECs under conditions of toxic injury or ischemia [71, 73]. HIF-1α is overexpressed in a broad spectrum of solid tumors, suggesting that increased HIF-1α expression may be as much a consequence of solid tumor growth as its cause. Increased HIF-1α levels in these tumors may have a positive or negative effect on tumor growth depending on the individual tumor cell type [74–76].

The organ- and cell type-dependent biology of HIF function makes dissection of the downstream molecular events affected by multiple HIFs complex. Empirical observations suggest that genes encoding glycolytic enzymes appear to be predominantly regulated by HIF-1α [77]; both HIF-1α and 2α regulate the hypoxic induction of *VEGF*, *Epo*, and other

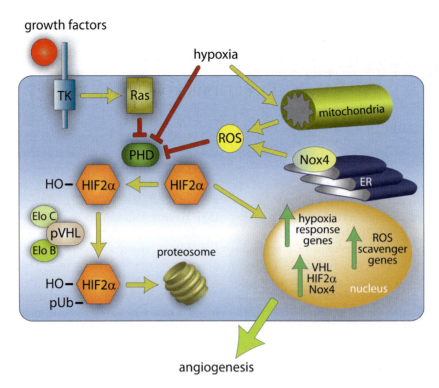

FIG. 16.1 pVHL and Elongins (*Elo*) C and B form the substrate recognition component of an E3 ubiquitin ligase complex. Under normoxic conditions, specific residues of *HIFα* are hydroxylated (*OH*) by prolyl-4-hydroxylase (*PHD*), which enables binding to the *pVHL* complex, polyubiquitination (*pUb*) and subsequent proteasomal degradation. Several different pathways modulate *PHD* catalyzed HIFα hydroxylation. Under hypoxic conditions, *PHD* is deprived of required cofactors, and *ROS* generated by mitochondrial complex III and *Nox4* in the endoplasmic reticulum inhibit *PHD* activity. Increased *HIFα* transcriptional activity in the absence of hypoxia is also mediated by *ROS*, growth factors and cytokines. Unhydroxylated *HIFα* does not bind *pVHL* and translocates to the nucleus, where it recruits transcriptional cofactors and activates a large set of genes, including many that strongly stimulate angiogenesis. *Nox4* activity also promotes the transcription of *HIF-2α* and *VHL* genes. See text for references and additional details.

downstream targets in tissues that express both isoforms, although there is evidence of an essential role for HIF-2α in this process [78–80] (Fig. 1). In RCC cell lines that are HIF-1α-deficient, regardless of *VHL* gene mutation status, HIF-2α regulates *VEGF* expression and angiogenesis [81]. To address the possible functions of HIFs systematically in the context of human renal cortical tumorigenesis, Kim et al. analyzed HIF isoforms expression in patients with well-characterized hereditary renal tumor syndromes [82]. Overexpression of HIF-1α and HIF-2α was documented in almost all nascent clear cell renal tumors from patients with VHL syndrome, suggesting that each has a role in clear cell renal tumorigenesis. The contributions made by HIF-1α, 2α and possibly other HIFs may be synergistic or antagonistic in promoting renal tumor growth. The consistent finding of HIF-2α overexpression in clear cell, chromophobe and papillary renal tumors in vivo suggests that HIF-2α may be critical to overall renal cortical tumorigenesis [82].

Global deletion of *Hif2α* results in severe phenotypes that vary in pathology and embryonic viability depending on the mouse strain used; nonetheless these studies consistently demonstrate an important role for HIF-2α in normal mouse devel-

opment [83, 84]. Embryonically viable *Hif2α* knockout mice display multiple-organ pathology, metabolic abnormalities and altered gene expression patterns [83]. Notably, these mice display reduced expression of HIF-2α target genes encoding several critical antioxidant enzymes, as well as increased ROS generation, and many aspects of the HIF-2α null phenotype are reversed by administration of a super oxide dismutase mimetic [83]. Together, these observations suggest the existence of a negative feedback loop where ROS-mediated PHD inhibition leading to increased HIF-2α activity induces the transcription of antioxidant enzymes and ultimately ROS depletion.

Adding further complexity to the balance of PHD and HIF-2α activities through ROS, an important source of ROS production in the distal renal tubules, NADP(H) oxidase-4 (Nox4), also appears to promote HIF-2α transcription [85–87]. Nox4 is also widely expressed at lower levels in vascular endothelium, heart, pancreas, ovary, testis, osteoclasts, placenta, and astrocytes [88]. Small inhibitory RNA silencing of Nox4 in *VHL* negative or positive RCC cells resulted in substantially decreased intracellular ROS, HIF-2α mRNA, HIF-2α protein and expression of the HIF target genes *VEGF, TGF-α, and glut-1* [87]. These results suggest that renal Nox4 expression is essential for full HIF-2α expression and activity in RCC cells, even in the absence of functional pVHL [87]. Interestingly, suppression of Ras transformation and associated tumor angiogenesis by the AP-1 transcription factor family member JunD also occurs through the Nox4/PHD/HIFα axis [50]. Deletion of the *junD* gene in mice resulted in increased Nox4 expression, increased intracellular ROS, decreased PDH2 activity, accumulation of HIF-1α and 2α and increased *VEGF* gene transcript [50]. *junD*$^{-/-}$ mice do not develop tumors spontaneously, suggesting that the protective effect of JunD may only be uncovered under conditions of stress [50]. Nevertheless, these findings highlight the general relevance of ROS metabolism and HIFα activity in the control of *VEGF* expression during pathological angiogenesis (Fig. 1).

HIF-2α Activated Genes that Contribute to Angiogenesis

The list of known HIF-regulated genes has grown rapidly and clearly indicates that HIFs represent an adaptive "master switch" in adult homeostasis [39]. Several expression profile studies in recent years offer comprehensive lists of hypoxia-inducible and HIF-regulated genes [89–91]. For example, HIFs are involved in the regulation of a multitude of biological processes that are essential for normal kidney function under physiological conditions, and their dysregulated function is broadly manifested in disease. Normal processes including glucose and energy metabolism, angiogenesis, erythropoiesis and iron homeostasis, cell migration, and cell–cell and cell–matrix interactions are tightly regulated through HIFs. HIF-regulated genes with direct impact on the pathogenesis of acute, chronic and inherited kidney diseases include those encoding heme oxygenase-1 (HO-1), VEGF, plasminogen activator inhibitor-1 (PAI-1), tissue inhibitor of metalloproteinase-1 (TIMP-1), connective tissue growth factor (CTGF), EPO, the Wilms' tumor suppressor protein (WT-1), and others [38, 39]. HIF target genes regulating angiogenesis have been particularly well characterized because of their historic relevance to these pathologies. The individual role of HIF-2α in activating angiogenesis-regulatory genes has come primarily from the study of pathological conditions where HIF-2α expression prevails, including clear cell RCC as already noted. Clearly, many of the HIF-2α target genes relevant to angiogenesis also function outside of these contexts, have been studied extensively, and are described in considerable detail in other chapters in this volume. For similar reasons, the regulation of angiogenesis by HIF-1α warrants its own review (see chapter 15). With this in mind, we focus here on a subset of genes regulating angiogenesis where a well-defined or unique connection to pVHL and HIF-2α has been established. These examples illustrate the breadth and intricacy of angiogenesis regulation by this specific axis of a ubiquitous oxygen sensing pathway.

Direct Regulators of Angiogenesis

VEGF and Angiopoietins

No review of this topic would be complete without acknowledging the fundamental importance of *VEGF* as a downstream target of HIFs in the regulation of angiogenesis in normal development and homeostasis, VHL syndrome and many other diseases. The properties of VEGF family members and the mechanisms by which they regulate angiogenesis are reviewed extensively elsewhere in this volume (refer to chapter 8 on VEGF). VEGF was linked to tumor angiogenesis characteristic of the VHL pathway through investigations into the molecular basis of the capillary hemangioblastomas so frequently manifested in patients with VHL disease [5]. Investigations of *VHL* tumor suppressor function also clearly demonstrated reversion of elevated *VEGF* expression and tumorigenesis upon restoration of *VHL* expression in RCC cells [6, 7, 92, 93]. Contemporaneous, independent investigations firmly established the role of hypoxia and HIF in the regulation of VEGF expression [94–97]. Recent reviews are available that provide additional details of the molecular basis of VEGF regulation downstream of the VHL/HIF pathway [98, 99] as well as the specific role of VEGF in the molecular pathology of VHL-associated eye disease [100].

Angiopoietins (Ang) are proteins with important roles in vascular development, angiogenesis and tumor neovascularization. Their properties and the mechanisms by which they regulate these processes are also reviewed extensively in this volume (for an in-depth discussion, see chapter 10). A significant increase in Ang2 expression in the tumor vascular endothelium, but not in tumor cells, has been reported in RCC; this increase may contribute to vessel stability and suggests that the angiopoietin receptor, Tie2, may be a viable target for antiangiogenic therapy in clear cell RCC [101]. Ang-4 was found to be overexpressed in VHL-negative RCC cells expressing predominantly HIF-2α [102], and expression profiling of VHL-associated pheochromocytomas also revealed up-regulation of Ang2, as well as Tie1 [103]. While these results are intriguing, more work is needed to define significance of the Ang/Tie pathway activation downstream of pVHL and HIF.

Endoglin

Endoglin (Eng) is a homodimeric transmembrane glycoprotein that is a component of the transforming growth factor-β (TGF-β) receptor complex and is highly expressed in normal EC as well as tumor blood vessels [104, 105]. Inherited mutations that result in premature termination of Eng are the basis for hereditary hemorrhagic telangiectasia (HTT), an autosomal dominant disorder characterized by multisystemic vascular dysplasia and recurrent hemorrhage [106]. The role of Eng in tumor angiogenesis is well-illustrated by several recent studies. Eng$^{+/-}$ haploinsufficient mice displayed reduced tumor capillary density, perfusion, hemoglobin content and VCAM-1 expression relative to Eng$^{+/+}$ littermates [105]. Eng has been used as a marker of the extent and impact of tumor angiogenesis: high expression level is a significant predictor of poor prognosis in disease-free or overall survival in a range of cancers including brain, lung, breast, stomach, colon, gynecologic, and liver carcinomas [107–110]. Eng transcript is up-regulated in hypoxia through cooperative signaling by the TGFβ and HIF pathways [111]. A discrete connection between HIF-2α and Eng has been found in polycystic kidney disease (PKD), where pericystic hypoxia is thought to cause increased EPO production and interstitial hypervascularity [112]. In a rodent PKD model, the distinct expression patterns of HIFα isoforms closely resembled the pattern displayed in normal kidneys under systemic hypoxia; VEGF and Glut-1 staining correlated with HIF-1α expression, whereas EPO and Eng staining correlated with regions of HIF-2α expression [112].

Endothelins

Endothelins (ETs) are potent vasoconstrictive, mitogenic and motogenic peptides that exist in 3 isoforms, ET-1 being the most abundant. Two ET receptors (ETA and ETB) have been characterized in mammals, classified according to the relative binding affinities of the three peptide ligands: ETA which binds ET-1>ET-2>ET-3 and ETB which binds all three equally well. Endothelins appear to act mainly as local paracrine or autocrine peptides, but increased circulating levels of endothelins also have biological significance in many important pathological conditions [113]. The pathway is critically involved in pulmonary arterial hypertension, acute renal failure and diabetic angiopathy. Given this disease spectrum, it is not surprising that endothelins play a fundamental role in the pathogenesis of fibrosis [114], suggestive of a homeostatic and/or developmental function in matrix remodeling. Activation of the ETA/ET-1 axis also promotes tumorigenesis in several human cancers through induction of MMP secretion, disruption of cell junctions and increased cell migration and invasion [115, 116]. ET-1 and ET-2 are also target genes in the pVHL/HIF pathway. Significant gene induction was observed in expression profiling comparisons of *VHL*-negative RCC cell lines under hypoxic relative to normoxic conditions, and reciprocal suppression in the same cell lines where pVHL function had been restored [89, 91]. Induction of ETs in the RCC cell line 786-0, which expresses little HIF-1α, indicates the capacity HIF-2α to mediate ET gene activation.

Placental Growth Factor

Placental growth factor (PlGF) is a member of the VEGF family and activates VEGFR1 [117, 118]. The role of PlGF in angiogenesis is complex. In some settings, PlGF binding to VEGFR1 does not result in EC growth or angiogenesis, while in others, PlGF/VEGFR1 signaling promotes EC viability as well as angiogenesis [119]. In placenta and PlGF-expressing tumors, increased PlGF levels inhibit EC growth [120]. PlGF appears to stimulate unique signals in EC as well as amplify VEGF-driven angiogenesis. The latter may occur through multiple mechanisms, including displacement of VEGF-A from VEGFR1 to R2, by inducing transactivation of R2 through R1 binding, and or via VEGF-A/PlGF heterodimerization and consequent activation of an R1/R2 complex [121]. In VHL syndrome and in hypoxic states, HIF stabilization results in significant up-regulation of PlGF [103, 122, 123]. Regulation of PlGF induction by HIF-2α occurs in the initiation of blood vessel formation during chondrocytic growth plate formation [124]. During the ordered maturation of a murine chondrocyte cell line, increased HIF-2α expression was paralleled by the increased expression of VEGF, PlGF and Glut-1 [124]. These observations were consistent with immunohistochemical studies of mouse bone, which showed HIF-2α staining in hypertrophic growth plate chondrocytes [124]. Together these results suggest that HIF-2α is involved in the initiation of blood vessel formation and a metabolic shift in the growth plate, processes that are crucial for endochondral ossification [124].

Genes Encoding Indirect Regulators of Angiogenesis

Erythropoietin

Erythropoietin (EPO) is a secreted, glycosylated plasma protein that regulates red blood cell production by promoting erythroid differentiation and initiating hemoglobin synthesis; recent studies also suggest that EPO is a pleiotropic cytokine that acts on non-hematopoietic cells, including tumor cells [125]. In VHL-associated renal cysts, RCC, pheochromocytoma and endolymphatic sac tumors, there is increased co-expression of EPO and its receptor (EPOR) [126–128]. EPO expression is likely to be due to *VHL* loss of function and increased HIF-1α activity, whereas EPOR expression normally occurs in the angioblast stage of embryonic development and persists in tumor precursor cells [128]. The presence of an autocrine or paracrine EPO/EPOR system in tumors and potential effects of EPO on tumor microenvironment and angiogenesis are consistent with a complex biology for EPO signaling in cancer that warrants further research.

Chuvash polycythemia, an autosomal recessive disorder in which a missense mutation in *VHL* results in a single amino acid substitution in pVHL, is also associated with defective pVHL/HIF interaction and target gene induction (reviewed in [129]). But in contrast to classical VHL syndrome, Chuvash polycythemia is associated with lower peripheral blood pressures, varicose veins, vertebral hemangiomas, lower white blood cell and platelet counts, arterial and venous thrombosis, major bleeding episodes, cerebral vascular events, and premature mortality, and not with RCC, pheochromocytomas, or any other malignancy. It is not clear whether the Chuvash mutation in pVHL elicits induction of a unique subset of the HIF targets or retains another pVHL function that, when lost, predisposes classical VHL syndrome patients to cancer. Individuals with Chuvash polycythemia do have elevated circulating levels of VEGF, PAI-1 and most notably EPO.

Very recently Gruber et al. uncovered an important connection between the regulation of EPO expression and HIF2α function in adult homeostasis [84]. To circumvent embryonic lethality and other defects associated with complete deletion of the murine *Hif2α* gene, Gruber et al. generated a conditional *Hif2α* allele and compared the impact of strictly controlled postnatal deletion with global *Hif2α* deletion. They showed that HIF2α plays a critical role in adult erythropoiesis, with acute deletion leading to anemia, and concluded that while HIF1α is a critical regulator of EPO expression during embryogenesis, HIF2α is the critical *Epo* regulator under physiologic and stress conditions in adults [84].

Plasminogen Activators

Plasminogen activators are serine proteases which catalyze the conversion of plasminogen to its activate form, plasmin. Of the two types of plasminogen activator (PA), tissue plasminogen activator (tPA) is mainly responsible for thrombolysis, while urokinase plasminogen activator (urokinase, uPA) is important for extracellular matrix degradation and implicated broadly in cancer metastasis. Intimately involved with PA function are two serpin PA inhibitors (PAI): PAI-1 and PAI-2. The urokinase receptor (uPAR) is a cell-membrane anchored protein for uPA and controls the accumulation of plasminogen activity at cell surfaces.

The uPA system is involved in cancer progression at multiple levels. For both tumor cells and adjacent microvessel EC, uPA has been implicated in remodeling of the extracellular matrix required for enhanced cell proliferation, adhesion, migration and angiogenesis (reviewed in [130, 131]). Multiple studies have shown that patients whose primary breast cancer contains high levels of uPA have a significantly worse outcome than patients with low levels. Paradoxically, high levels of PAI-1 also predict for aggressive disease [130]. Because the spatial distribution of PAI-1 on the surface of motile cells, as well as complex formation with other proteins, are critical for organized movement and selective matrix degradation, elevated PAI-1 levels cannot be taken as a simple indicator of a net reduction in uPa activity, matrix invasion or angiogenesis. This may explain, at least in part, the counterintuitive correlation between elevated PAI-1 expression and poor prognosis for many cancers [132, 133]. Elevated expression of both uPAR and PAI-1 has been reported for hypoxic and pVHL-defective RCC cells; including RCC cell lines expressing primarily HIF2α [134]. A later study using siRNA to alternately silence HIF-1α and HIF-2α in pVHL-defective RCC cells constitutively expressing both isoforms functionally demonstrated the predominant role of HIF-2α in mediating uPAR and PAI-1 overexpression [135].

Matrix Metalloproteinases

Matrix metalloproteinases (MMPs) are a group of zinc endopeptidases with multiple functions in normal morphogenesis, angiogenesis and tissue remodeling, as well in tumor growth, angiogenesis and metastasis (see chapter 5). Characterization of the aberrant invasiveness of *VHL*-negative RCC cells has revealed the overexpression of MMP-2 and -9 [136], and later expression profiling studies of RCC cells demonstrated the up-regulation of MMP-1 and -13 due to loss of pVHL function or hypoxia [90, 91]. The predominant expression of HIF-2α in many of the RCC cell lines analyzed in these studies indicates the capacity of this isoform to mediate MMP gene activation. Membrane-type matrix metalloproteinases (MT-MMPs) are a more recently characterized subgroup of MMPs that also function in neovascularization during development and in tumor progression via matrix degradation and through their ability to cleave adhesion molecules and other MMPs [137, 138]. Membrane-type 1 matrix metalloproteinase (MT1-MMP) has been found to be up-regulated in *VHL*-null RCC cells through the cooperative effects of HIF-2α and Sp1 [138]. A subsequent study to assess the specific contribution of MT1-MMP to the invasive potential of RCC cells showed that overexpression of either HIF-2α or MT1-MMP in a poorly invasive, *VHL*-positive RCC cell line promoted collagen degradation and invasiveness characteristic of *VHL*-negative cells, while RNAi silencing of MT1-MMP in the latter suppressed invasiveness [139]. These studies suggest that suppression of HIF-2α or MT1-MMP may represent a viable strategy to counter RCC tumor invasion and metastasis.

Acknowledgments. We thank Jean-Baptiste Lattouf for thoughtful discussions. This work was supported by the Intramural Research Program of the NIH, National Cancer Institute, Center for Cancer Research.

References

1. Linehan WM, Vasselli J, Srinivasan R et al. Genetic basis of cancer of the kidney: disease-specific approaches to therapy. Clin Cancer Res 2004; 10(18 Pt 2):6282S–6289S.
2. Kaelin WG Jr. Molecular basis of the VHL hereditary cancer syndrome. Nat Rev Cancer 2002; 2(9):673–682.
3. Jemal A, Clegg LX, Ward E et al. Annual report to the nation on the status of cancer, 1975–2001, with a special feature regarding survival. Cancer 2004; 101(1):3–27.
4. Iliopoulos O, Kibel A, Gray S et al. Tumour suppression by the human von Hippel-Lindau gene product. Nat Med 1995; 1(8):822–826.
5. Wizigmann-Voos S, Breier G, Risau W et al. Up-regulation of vascular endothelial growth factor and its receptors in von Hippel-Lindau disease-associated and sporadic hemangioblastomas. Cancer Res 1995; 55(6):1358–1364.
6. Siemeister G, Weindel K, Mohrs K et al. Reversion of deregulated expression of vascular endothelial growth factor in human

renal carcinoma cells by von Hippel-Lindau tumor suppressor protein. Cancer Res 1996; 56(10):2299–2301.

7. Gnarra JR, Zhou S, Merrill MJ et al. Post-transcriptional regulation of vascular endothelial growth factor mRNA by the product of the VHL tumor suppressor gene. Proc Natl Acad Sci USA 1996; 93(20):10589–10594.

8. Richards FM. Molecular pathology of von HippelLindau disease and the VHL tumour suppressor gene. Expert Rev Mol Med 2001; 2001:1–27.

9. Woodward ER, Buchberger A, Clifford SC et al. Comparative sequence analysis of the VHL tumor suppressor gene. Genomics 2000; 65(3):253–265.

10. Gao J, Naglich JG, Laidlaw J et al. Cloning and characterization of a mouse gene with homology to the human von Hippel-Lindau disease tumor suppressor gene: implications for the potential organization of the human von Hippel-Lindau disease gene. Cancer Res 1995; 55(4):743–747.

11. Adryan B, Decker HJ, Papas TS et al. Tracheal development and the von Hippel-Lindau tumor suppressor homolog in Drosophila. Oncogene 2000; 19(24):2803–2811.

12. Aso T, Yamazaki K, Aigaki T et al. Drosophila von Hippel-Lindau tumor suppressor complex possesses E3 ubiquitin ligase activity. Biochem Biophys Res Commun 2000; 276(1):355–361.

13. Kuzmin I, Duh FM, Latif F et al. Identification of the promoter of the human von Hippel-Lindau disease tumor suppressor gene. Oncogene 1995; 10(11):2185–2194.

14. Stebbins CE, Kaelin WG, Jr., Pavletich NP. Structure of the VHL-ElonginC-ElonginB complex: implications for VHL tumor suppressor function. Science 1999; 284(5413):455–461.

15. Duan DR, Pause A, Burgess WH et al. Inhibition of transcription elongation by the VHL tumor suppressor protein. Science 1995; 269(5229):1402–1406.

16. Kibel A, Iliopoulos O, DeCaprio JA et al. Binding of the von Hippel-Lindau tumor suppressor protein to Elongin B and C. Science 1995; 269(5229):1444–1446.

17. Kishida T, Stackhouse TM, Chen F et al. Cellular proteins that bind the von Hippel-Lindau disease gene product: mapping of binding domains and the effect of missense mutations. Cancer Res 1995; 55(20):4544–4548.

18. Ohh M, Yauch RL, Lonergan KM et al. The von Hippel-Lindau tumor suppressor protein is required for proper assembly of an extracellular fibronectin matrix. Mol Cell 1998; 1(7):959–968.

19. Lonergan KM, Iliopoulos O, Ohh M et al. Regulation of hypoxia-inducible mRNAs by the von Hippel-Lindau tumor suppressor protein requires binding to complexes containing elongins B/C and Cul2. Mol Cell Biol 1998; 18(2):732–741.

20. Pause A, Lee S, Worrell RA et al. The von Hippel-Lindau tumor-suppressor gene product forms a stable complex with human CUL-2, a member of the Cdc53 family of proteins. Proc Natl Acad Sci USA 1997; 94(6):2156–2161.

21. Starr R, Willson TA, Viney EM et al. A family of cytokine-inducible inhibitors of signalling. Nature 1997; 387(6636):917–921.

22. Endo TA, Masuhara M, Yokouchi M et al. A new protein containing an SH2 domain that inhibits JAK kinases. Nature 1997; 387(6636):921–924.

23. Naka T, Narazaki M, Hirata M et al. Structure and function of a new STAT-induced STAT inhibitor. Nature 1997; 387(6636):924–929.

24. Hilton DJ, Richardson RT, Alexander WS et al. Twenty proteins containing a C-terminal SOCS box form five structural classes. Proc Natl Acad Sci USA 1998; 95(1):114–119.

25. Kamura T, Sato S, Haque D et al. The Elongin BC complex interacts with the conserved SOCS-box motif present in members of the SOCS, ras, WD-40 repeat, and ankyrin repeat families. Genes Dev 1998; 12(24):3872–3881.

26. Zhang JG, Farley A, Nicholson SE et al. The conserved SOCS box motif in suppressors of cytokine signaling binds to elongins B and C and may couple bound proteins to proteasomal degradation. Proc Natl Acad Sci USA 1999; 96(5):2071–2076.

27. Aso T, Lane WS, Conaway JW et al. Elongin (SIII): a multisubunit regulator of elongation by RNA polymerase II. Science 1995; 269(5229):1439–1443.

28. Kipreos ET, Lander LE, Wing JP et al. cul-1 is required for cell cycle exit in C. elegans and identifies a novel gene family. Cell 1996; 85(6):829–839.

29. Lisztwan J, Imbert G, Wirbelauer C et al. The von Hippel-Lindau tumor suppressor protein is a component of an E3 ubiquitin-protein ligase activity. Genes Dev 1999; 13(14):1822–1833.

30. Iwai K, Yamanaka K, Kamura T et al. Identification of the von Hippel-lindau tumor-suppressor protein as part of an active E3 ubiquitin ligase complex. Proc Natl Acad Sci U S A 1999; 96(22):12436–12441.

31. Tyers M, Rottapel R. VHL: a very hip ligase. Proc Natl Acad Sci USA 1999; 96(22):12230–12232.

32. Kamura T, Sato S, Iwai K et al. Activation of HIF1alpha ubiquitination by a reconstituted von Hippel-Lindau (VHL) tumor suppressor complex. Proc Natl Acad Sci U S A 2000; 97(19):10430–10435.

33. DeSalle LM, Pagano M. Regulation of the G1 to S transition by the ubiquitin pathway. FEBS Lett 2001; 490(3):179–189.

34. Cockman ME, Masson N, Mole DR et al. Hypoxia inducible factor-alpha binding and ubiquitylation by the von Hippel-Lindau tumor suppressor protein. J Biol Chem 2000; 275(33):25733–25741.

35. Ohh M, Park CW, Ivan M et al. Ubiquitination of hypoxia-inducible factor requires direct binding to the beta-domain of the von Hippel-Lindau protein. Nat Cell Biol 2000; 2(7):423–427.

36. Haase VH. Hypoxia-inducible factors in the kidney. Am J Physiol Renal Physiol 2006; 291(2):F271–F281.

37. Haase VH. The VHL/HIF oxygen-sensing pathway and its relevance to kidney disease. Kidney Int 2006; 69(8):1302–1307.

38. Schofield CJ, Ratcliffe PJ. Oxygen sensing by HIF hydroxylases. Nat Rev Mol Cell Biol 2004; 5(5):343–354.

39. Wenger RH, Stiehl DP, Camenisch G. Integration of oxygen signaling at the consensus HRE. Sci STKE 2005; 2005(306):re12.

40. Tian H, McKnight SL, Russell DW. Endothelial PAS domain protein 1 (EPAS1), a transcription factor selectively expressed in endothelial cells. Genes Dev 1997; 11(1):72–82.

41. Flamme I, Frohlich T, von RM et al. HRF, a putative basic helix-loop-helix-PAS-domain transcription factor is closely related to hypoxia-inducible factor-1 alpha and developmentally expressed in blood vessels. Mech Dev 1997; 63(1):51–60.

42. Peruzzi B, Athauda G, Bottaro DP. The von Hippel-Lindau tumor suppressor gene product represses oncogenic beta-catenin signaling in renal carcinoma cells. Proc Natl Acad Sci USA 2006; 103(39):14531–14536.

43. Epstein AC, Gleadle JM, McNeill LA et al. C. elegans EGL-9 and mammalian homologs define a family of dioxygenases that regulate HIF by prolyl hydroxylation. Cell 2001; 107(1): 43–54.

44. Appelhoff RJ, Tian YM, Raval RR et al. Differential function of the prolyl hydroxylases PHD1, PHD2, and PHD3 in

the regulation of hypoxia-inducible factor. J Biol Chem 2004; 279(37):38458–38465.

45. Jaakkola P, Mole DR, Tian YM et al. Targeting of HIF-alpha to the von Hippel-Lindau ubiquitylation complex by O2-regulated prolyl hydroxylation. Science 2001; 292(5516):468–472.

46. Ivan M, Kondo K, Yang H et al. HIFalpha targeted for VHL-mediated destruction by proline hydroxylation: implications for O2 sensing. Science 2001; 292(5516):464–468.

47. Knowles HJ, Raval RR, Harris AL et al. Effect of ascorbate on the activity of hypoxia-inducible factor in cancer cells. Cancer Res 2003; 63(8):1764–1768.

48. Martin F, Linden T, Katschinski DM et al. Copper-dependent activation of hypoxia-inducible factor (HIF)-1: implications for ceruloplasmin regulation. Blood 2005; 105(12):4613–4619.

49. Hirsila M, Koivunen P, Xu L et al. Effect of desferrioxamine and metals on the hydroxylases in the oxygen sensing pathway. FASEB J 2005; 19(10):1308–1310.

50. Gerald D, Berra E, Frapart YM et al. JunD reduces tumor angiogenesis by protecting cells from oxidative stress. Cell 2004; 118(6):781–794.

51. Metzen E, Zhou J, Jelkmann W et al. Nitric oxide impairs normoxic degradation of HIF-1alpha by inhibition of prolyl hydroxylases. Mol Biol Cell 2003; 14(8):3470–3481.

52. Dalgard CL, Lu H, Mohyeldin A et al. Endogenous 2-oxoacids differentially regulate expression of oxygen sensors. Biochem J 2004; 380(Pt 2):419–424.

53. Selak MA, Armour SM, MacKenzie ED et al. Succinate links TCA cycle dysfunction to oncogenesis by inhibiting HIF-alpha prolyl hydroxylase. Cancer Cell 2005; 7(1):77–85.

54. Brunelle JK, Bell EL, Quesada NM et al. Oxygen sensing requires mitochondrial ROS but not oxidative phosphorylation. Cell Metab 2005; 1(6):409–414.

55. Guzy RD, Hoyos B, Robin E et al. Mitochondrial complex III is required for hypoxia-induced ROS production and cellular oxygen sensing. Cell Metab 2005; 1(6):401–408.

56. Mansfield KD, Guzy RD, Pan Y et al. Mitochondrial dysfunction resulting from loss of cytochrome c impairs cellular oxygen sensing and hypoxic HIF-alpha activation. Cell Metab 2005; 1(6):393–399.

57. Semenza GL. Hydroxylation of HIF-1: oxygen sensing at the molecular level. Physiology (Bethesda) 2004; 19:176–182.

58. Stolze IP, Tian YM, Appelhoff RJ et al. Genetic analysis of the role of the asparaginyl hydroxylase factor inhibiting hypoxia-inducible factor (HIF) in regulating HIF transcriptional target genes. J Biol Chem 2004; 279(41):42719–42725.

59. Sandau KB, Zhou J, Kietzmann T et al. Regulation of the hypoxia-inducible factor 1alpha by the inflammatory mediators nitric oxide and tumor necrosis factor-alpha in contrast to desferroxamine and phenylarsine oxide. J Biol Chem 2001; 276(43):39805–39811.

60. Hellwig-Burgel T, Rutkowski K, Metzen E et al. Interleukin-1beta and tumor necrosis factor-alpha stimulate DNA binding of hypoxia-inducible factor-1. Blood 1999; 94(5):1561–1567.

61. Stiehl DP, Jelkmann W, Wenger RH et al. Normoxic induction of the hypoxia-inducible factor 1alpha by insulin and interleukin-1beta involves the phosphatidylinositol 3-kinase pathway. FEBS Lett 2002; 512(1–3):157–162.

62. Richard DE, Berra E, Pouyssegur J. Nonhypoxic pathway mediates the induction of hypoxia-inducible factor 1alpha in vascular smooth muscle cells. J Biol Chem 2000; 275(35):26765–26771.

63. Feldser D, Agani F, Iyer NV et al. Reciprocal positive regulation of hypoxia-inducible factor 1alpha and insulin-like growth factor 2. Cancer Res 1999; 59(16):3915–3918.

64. Jiang BH, Jiang G, Zheng JZ et al. Phosphatidylinositol 3-kinase signaling controls levels of hypoxia-inducible factor 1. Cell Growth Differ 2001; 12(7):363–369.

65. Treins C, Giorgetti-Peraldi S, Murdaca J et al. Insulin stimulates hypoxia-inducible factor 1 through a phosphatidylinositol 3-kinase/target of rapamycin-dependent signaling pathway. J Biol Chem 2002; 277(31):27975–27981.

66. Zelzer E, Levy Y, Kahana C et al. Insulin induces transcription of target genes through the hypoxia-inducible factor HIF-1alpha/ARNT. EMBO J 1998; 17(17):5085–5094.

67. Kaelin WG. Proline hydroxylation and gene expression. Annu Rev Biochem 2005; 74:115–128.

68. Fukuda R, Hirota K, Fan F et al. Insulin-like growth factor 1 induces hypoxia-inducible factor 1-mediated vascular endothelial growth factor expression, which is dependent on MAP kinase and phosphatidylinositol 3-kinase signaling in colon cancer cells. J Biol Chem 2002; 277(41):38205–38211.

69. Laughner E, Taghavi P, Chiles K et al. HER2 (neu) signaling increases the rate of hypoxia-inducible factor 1alpha (HIF-1alpha) synthesis: novel mechanism for HIF-1-mediated vascular endothelial growth factor expression. Mol Cell Biol 2001; 21(12):3995–4004.

70. Zhong H, Chiles K, Feldser D et al. Modulation of hypoxia-inducible factor 1alpha expression by the epidermal growth factor/phosphatidylinositol 3-kinase/PTEN/AKT/FRAP pathway in human prostate cancer cells: implications for tumor angiogenesis and therapeutics. Cancer Res 2000; 60(6):1541–1545.

71. Wiesener MS, Jurgensen JS, Rosenberger C et al. Widespread hypoxia-inducible expression of HIF-2alpha in distinct cell populations of different organs. FASEB J 2003; 17(2):271–273.

72. Higgins DF, Biju MP, Akai Y et al. Hypoxic induction of Ctgf is directly mediated by Hif-1. Am J Physiol Renal Physiol 2004; 287(6):F1223–F1232.

73. Rosenberger C, Mandriota S, Jurgensen JS et al. Expression of hypoxia-inducible factor-1alpha and -2alpha in hypoxic and ischemic rat kidneys. J Am Soc Nephrol 2002; 13(7):1721–1732.

74. Ryan HE, Poloni M, McNulty W et al. Hypoxia-inducible factor-1alpha is a positive factor in solid tumor growth. Cancer Res 2000; 60(15):4010–4015.

75. Carmeliet P, Dor Y, Herbert JM et al. Role of HIF-1alpha in hypoxia-mediated apoptosis, cell proliferation and tumour angiogenesis. Nature 1998; 394(6692):485–490.

76. Blancher C, Moore JW, Talks KL et al. Relationship of hypoxia-inducible factor (HIF)-1alpha and HIF-2alpha expression to vascular endothelial growth factor induction and hypoxia survival in human breast cancer cell lines. Cancer Res 2000; 60(24):7106–7113.

77. Hu CJ, Wang LY, Chodosh LA et al. Differential roles of hypoxia-inducible factor 1alpha (HIF-1alpha) and HIF-2alpha in hypoxic gene regulation. Mol Cell Biol 2003; 23(24):9361–9374.

78. Morita M, Ohneda O, Yamashita T et al. HLF/HIF-2alpha is a key factor in retinopathy of prematurity in association with erythropoietin. EMBO J 2003; 22(5):1134–1146.

79. Rankin EB, Higgins DF, Walisser JA et al. Inactivation of the arylhydrocarbon receptor nuclear translocator (Arnt) suppresses von Hippel-Lindau disease-associated vascular tumors in mice. Mol Cell Biol 2005; 25(8):3163–3172.

80. Warnecke C, Zaborowska Z, Kurreck J et al. Differentiating the functional role of hypoxia-inducible factor (HIF)-1alpha and HIF-2alpha (EPAS-1) by the use of RNA interference: erythropoietin is a HIF-2alpha target gene in Hep3B and Kelly cells. FASEB J 2004; 18(12):1462–1464.

81. Shinojima T, Oya M, Takayanagi A et al. Renal cancer cells lacking hypoxia inducible factor (HIF)-1{alpha} expression maintain vascular endothelial growth factor expression through HIF-2{alpha}. Carcinogenesis 2006.

82. Kim CM, Vocke C, Torres-Cabala C et al. Expression of hypoxia inducible factor-1alpha and 2alpha in genetically distinct early renal cortical tumors. J Urol 2006; 175(5):1908–1914.

83. Scortegagna M, Ding K, Oktay Y et al. Multiple organ pathology, metabolic abnormalities and impaired homeostasis of reactive oxygen species in Epas1−/− mice. Nat Genet 2003; 35(4): 331–340.

84. Gruber M, Hu CJ, Johnson RS et al. Acute postnatal ablation of Hif-2{alpha} results in anemia. Proc Natl Acad Sci USA 2007; 104(7):2301–2306.

85. Shiose A, Kuroda J, Tsuruya K et al. A novel superoxide-producing NAD(P)H oxidase in kidney. J Biol Chem 2001; 276(2):1417–1423.

86. Geiszt M, Kopp JB, Varnai P et al. Identification of renox, an NAD(P)H oxidase in kidney. Proc Natl Acad Sci USA 2000; 97(14):8010–8014.

87. Maranchie JK, Zhan Y. Nox4 is critical for hypoxia-inducible factor 2-alpha transcriptional activity in von Hippel-Lindau-deficient renal cell carcinoma. Cancer Res 2005; 65(20):9190–9193.

88. Krause KH. Tissue distribution and putative physiological function of NOX family NADPH oxidases. Jpn J Infect Dis 2004; 57(5): S28–S29.

89. Wykoff CC, Pugh CW, Maxwell PH et al. Identification of novel hypoxia dependent and independent target genes of the von Hippel-Lindau (VHL) tumour suppressor by mRNA differential expression profiling. Oncogene 2000; 19(54):6297–6305.

90. Koong AC, Denko NC, Hudson KM et al. Candidate genes for the hypoxic tumor phenotype. Cancer Res 2000; 60(4):883–887.

91. Jiang Y, Zhang W, Kondo K et al. Gene expression profiling in a renal cell carcinoma cell line: dissecting VHL and hypoxia-dependent pathways. Mol Cancer Res 2003; 1(6):453–462.

92. Iliopoulos O, Levy AP, Jiang C et al. Negative regulation of hypoxia-inducible genes by the von Hippel-Lindau protein. Proc Natl Acad Sci USA 1996; 93(20):10595–10599.

93. Levy AP, Levy NS, Goldberg MA. Hypoxia-inducible protein binding to vascular endothelial growth factor mRNA and its modulation by the von Hippel-Lindau protein. J Biol Chem 1996; 271(41):25492–25497.

94. Minchenko A, Salceda S, Bauer T et al. Hypoxia regulatory elements of the human vascular endothelial growth factor gene. Cell Mol Biol Res 1994; 40(1):35–39.

95. Minchenko A, Bauer T, Salceda S et al. Hypoxic stimulation of vascular endothelial growth factor expression in vitro and in vivo. Lab Invest 1994; 71(3):374–379.

96. Wood SM, Gleadle JM, Pugh CW et al. The role of the aryl hydrocarbon receptor nuclear translocator (ARNT) in hypoxic induction of gene expression. Studies in ARNT-deficient cells. J Biol Chem 1996; 271(25):15117–15123.

97. Forsythe JA, Jiang BH, Iyer NV et al. Activation of vascular endothelial growth factor gene transcription by hypoxia-inducible factor 1. Mol Cell Biol 1996; 16(9):4604–4613.

98. Brugarolas J, Kaelin WG, Jr. Dysregulation of HIF and VEGF is a unifying feature of the familial hamartoma syndromes. Cancer Cell 2004; 6(1):7–10.

99. Liu L, Simon MC. Regulation of transcription and translation by hypoxia. Cancer Biol Ther 2004; 3(6):492–497.

100. Chan CC, Collins AB, Chew EY. Molecular pathology of eyes with von Hippel-Lindau (VHL) Disease: a review. Retina 2007; 27(1):1–7.

101. Currie MJ, Gunningham SP, Turner K et al. Expression of the angiopoietins and their receptor Tie2 in human renal clear cell carcinomas; regulation by the von Hippel-Lindau gene and hypoxia. J Pathol 2002; 198(4):502–510.

102. Yamakawa M, Liu LX, Belanger AJ et al. Expression of angiopoietins in renal epithelial and clear cell carcinoma cells: regulation by hypoxia and participation in angiogenesis. Am J Physiol Renal Physiol 2004; 287(4):F649–F657.

103. Eisenhofer G, Huynh TT, Pacak K et al. Distinct gene expression profiles in norepinephrine- and epinephrine-producing hereditary and sporadic pheochromocytomas: activation of hypoxia-driven angiogenic pathways in von Hippel-Lindau syndrome. Endocr Relat Cancer 2004; 11(4):897–911.

104. Cheifetz S, Bellon T, Cales C et al. Endoglin is a component of the transforming growth factor-beta receptor system in human endothelial cells. J Biol Chem 1992; 267(27):19027–19030.

105. Duwel A, Eleno N, Jerkic M et al. Reduced tumor growth and angiogenesis in endoglin-haploinsufficient mice. Tumour Biol 2007; 28(1):1–8.

106. McAllister KA, Grogg KM, Johnson DW et al. Endoglin, a TGF-beta binding protein of endothelial cells, is the gene for hereditary haemorrhagic telangiectasia type 1. Nat Genet 1994; 8(4):345–351.

107. Yao Y, Pan Y, Chen J et al. Endoglin (CD105) Expression in Angiogenesis of Primary Hepatocellular Carcinomas: Analysis using Tissue Microarrays and Comparisons with CD34 and VEGF. Ann Clin Lab Sci 2007; 37(1):39–48.

108. Taskiran C, Erdem O, Onan A et al. The prognostic value of endoglin (CD105) expression in ovarian carcinoma. Int J Gynecol Cancer 2006; 16(5):1789–1793.

109. Erdem O, Taskiran C, Onan MA et al. CD105 expression is an independent predictor of survival in patients with endometrial cancer. Gynecol Oncol 2006; 103(3):1007–1011.

110. Sugita Y, Takase Y, Mori D et al. Endoglin (CD 105) is expressed on endothelial cells in the primary central nervous system lymphomas and correlates with survival. J Neurooncol 2006.

111. Sanchez-Elsner T, Botella LM, Velasco B et al. Endoglin expression is regulated by transcriptional cooperation between the hypoxia and transforming growth factor-beta pathways. J Biol Chem 2002; 277(46):43799–43808.

112. Bernhardt WM, Wiesener MS, Weidemann A et al. Involvement of hypoxia-inducible transcription factors in polycystic kidney disease. Am J Pathol 2007; 170(3):830–842.

113. Inoue A, Yanagisawa M, Kimura S et al. The human endothelin family: three structurally and pharmacologically distinct isopeptides predicted by three separate genes. Proc Natl Acad Sci USA 1989; 86(8):2863–2867.

114. Clozel M, Salloukh H. Role of endothelin in fibrosis and antifibrotic potential of bosentan. Ann Med 2005; 37(1):2–12.

115. Rosano L, Spinella F, Di C, V et al. Integrin-linked kinase functions as a downstream mediator of endothelin-1 to promote invasive behavior in ovarian carcinoma. Mol Cancer Ther 2006; 5(4):833–842.

116. Rosano L, Spinella F, Di C, V et al. Endothelin-1 is required during epithelial to mesenchymal transition in ovarian cancer progression. Exp Biol Med (Maywood) 2006; 231(Maywood):1128–1131.

117. Maglione D, Guerriero V, Viglietto G et al. Isolation of a human placenta cDNA coding for a protein related to the vascular permeability factor. Proc Natl Acad Sci USA 1991; 88(20):9267–9271.

118. Roy H, Bhardwaj S, Yla-Herttuala S. Biology of vascular endothelial growth factors. FEBS Lett 2006; 580(12):2879–2887.

119. Roy H, Bhardwaj S, Babu M et al. Adenovirus-mediated gene transfer of placental growth factor to perivascular tissue induces angiogenesis via upregulation of the expression of endogenous vascular endothelial growth factor-A. Hum Gene Ther 2005; 16(12):1422–1428.

120. Ahmed A, Dunk C, Ahmad S et al. Regulation of placental vascular endothelial growth factor (VEGF) and placenta growth factor (PlGF) and soluble Flt-1 by oxygen–a review. Placenta 2000; 21 Suppl A:S16–S24.

121. Autiero M, Waltenberger J, Communi D et al. Role of PlGF in the intra- and intermolecular cross talk between the VEGF receptors Flt1 and Flk1. Nat Med 2003; 9(7):936–943.

122. Kasper LH, Brindle PK. Mammalian gene expression program resiliency: the roles of multiple coactivator mechanisms in hypoxia-responsive transcription. Cell Cycle 2006; 5(2):142–146.

123. Patel TH, Kimura H, Weiss CR et al. Constitutively active HIF-1alpha improves perfusion and arterial remodeling in an endovascular model of limb ischemia. Cardiovasc Res 2005; 68(1):144–154.

124. Stewart AJ, Houston B, Farquharson C. Elevated expression of hypoxia inducible factor-2alpha in terminally differentiating growth plate chondrocytes. J Cell Physiol 2006; 206(2):435–440.

125. Hardee ME, Arcasoy MO, Blackwell KL et al. Erythropoietin biology in cancer. Clin Cancer Res 2006; 12(2):332–339.

126. Vogel TW, Brouwers FM, Lubensky IA et al. Differential expression of erythropoietin and its receptor in von hippel-lindau-associated and multiple endocrine neoplasia type 2-associated pheochromocytomas. J Clin Endocrinol Metab 2005; 90(6):3747–3751.

127. Vogel TW, Vortmeyer AO, Lubensky IA et al. Coexpression of erythropoietin and its receptor in endolymphatic sac tumors. J Neurosurg 2005; 103(2):284–288.

128. Lee YS, Vortmeyer AO, Lubensky IA et al. Coexpression of erythropoietin and erythropoietin receptor in von Hippel-Lindau disease-associated renal cysts and renal cell carcinoma. Clin Cancer Res 2005; 11(3):1059–1064.

129. Gordeuk VR, Prchal JT. Vascular complications in Chuvash polycythemia. Semin Thromb Hemost 2006; 32(3):289–294.

130. Duffy MJ. Urokinase-type plasminogen activator: a potent marker of metastatic potential in human cancers. Biochem Soc Trans 2002; 30(2):207–210.

131. Duffy MJ. The urokinase plasminogen activator system: role in malignancy. Curr Pharm Des 2004; 10(1):39–49.

132. Sandberg T, Casslen B, Gustavsson B et al. Human endothelial cell migration is stimulated by urokinase plasminogen activator:plasminogen activator inhibitor 1 complex released from endometrial stromal cells stimulated with transforming growth factor beta1; possible mechanism for paracrine stimulation of endometrial angiogenesis. Biol Reprod 1998; 59(4):759–767.

133. Andreasen PA, Kjoller L, Christensen L et al. The urokinase-type plasminogen activator system in cancer metastasis: a review. Int J Cancer 1997; 72(1):1–22.

134. Los M, Zeamari S, Foekens JA et al. Regulation of the urokinase-type plasminogen activator system by the von Hippel-Lindau tumor suppressor gene. Cancer Res 1999; 59(17):4440–4445.

135. Carroll VA, Ashcroft M. Role of hypoxia-inducible factor (HIF)-1alpha versus HIF-2alpha in the regulation of HIF target genes in response to hypoxia, insulin-like growth factor-I, or loss of von Hippel-Lindau function: implications for targeting the HIF pathway. Cancer Res 2006; 66(12):6264–6270.

136. Koochekpour S, Jeffers M, Wang PH et al. The von Hippel-Lindau tumor suppressor gene inhibits hepatocyte growth factor/scatter factor-induced invasion and branching morphogenesis in renal carcinoma cells. Mol Cell Biol 1999; 19(9):5902–5912.

137. Zhou Z, Apte SS, Soininen R et al. Impaired endochondral ossification and angiogenesis in mice deficient in membrane-type matrix metalloproteinase I. Proc Natl Acad Sci USA 2000; 97(8):4052–4057.

138. Petrella BL, Lohi J, Brinckerhoff CE. Identification of membrane type-1 matrix metalloproteinase as a target of hypoxia-inducible factor-2 alpha in von Hippel-Lindau renal cell carcinoma. Oncogene 2005; 24(6):1043–1052.

139. Petrella BL, Brinckerhoff CE. Tumor cell invasion of von Hippel Lindau renal cell carcinoma cells is mediated by membrane type-1 matrix metalloproteinase. Mol Cancer 2006; 5:66.

Chapter 17
Nitric Oxide in Tumor Angiogenesis

L. Morbidelli, S. Donnini, and M. Ziche

Keywords: nitrix oxide, nitric oxide synthase (NOS), COX-2

Abstract: Nitric oxide (NO), produced from L-arginine by NO synthases (NOS), is a short-lived molecule required for many physiological functions and contributing to different pathological conditions. In the last decade, we and others contributed to demonstrate that NO stimulates angiogenesis and mediates the effect of different angiogenic molecules. In human tumors, NOS expression and activity correlate with tumor growth and aggressiveness through angiogenesis stimulation and regulation of angiogenic factor expression. Inter-relations among the NOS pathway, prostanoids and tyrosine kinase receptors have been reported in regulating tumor progression and malignancy. Drugs affecting the NOS pathway may be forseen as anti-tumor strategies able to reduce edema, inhibit angiogenesis and facilitate the delivery of chemotherapeutical agents. Recent developments include research on NOS gene polymorphisms which might become useful biomarkers for predicting cancer susceptibility as well as the role of NO in chemopreventive strategies.

Nitric Oxide: Synthesis and Roles

The discovery in 1987 that nitric oxide (NO) accounted for the bioactivity of the endothelium-derived relaxing factor (EDRF) [1,2] rapidly led to a burst of information on the physiological and pathological roles of this molecule. Although known for its role in vasorelaxation, neurotransmission, inhibition of platelet aggregation, and immune defense, NO also acts as an intracellular messenger for various cells in the body. Moreover, its up- or down-regulation is documented in different pathological conditions.

This review focuses on the role of NO in tumor angiogenesis and the potential therapeutic interventions based on targeting NO in angiogenesis.

NO is a short-lived gas (half-life 3–30 s), moderately soluble in water (up to 2 mmol/l), but highly soluble in organic solvents [1,2]. Due to its lipophilic nature, it can diffuse very easily and rapidly among cells. NO is generated from the terminal guanido nitrogen atom of L-arginine by various NADPH-dependent enzymes called NO synthases (NOS) [3]. Three main isoforms are known: neuronal (nNOS), inducible (iNOS), and endothelial (eNOS). Generally, nNOS and eNOS are expressed constitutively in neurons and endothelial cells, respectively, though they can also be expressed by other cells. Activation of these two isoforms depends on calcium ions and calmodulin, resulting in NO production in the nanomolar concentration range. Conversely, expression of iNOS typically requires induction by inflammatory cytokines or bacterial products. Activation of iNOS neither requires calcium ions nor calmodulin and leads to the sustained production of high micromolar concentrations of NO.

After the discovery of NO, several actions have been attributed to this mediator. Most of them are divergent, depending on NO concentration, the duration of its release, the cell type and the presence of scavengers or other reactive molecules present in the microenvironment which may impair or otherwise amplify NO effects.

Elucidation comes from mice specifically knocked-out for NOS isoforms. eNOS knockout mice show systemic hypertension, consistent with the role of endothelial NO in reducing vascular tone [4], showing also impairment of wound healing and angiogenesis [5]. iNOS knockout mice are prone to infections, and their macrophages exhibit poor cytotoxicity against parasites and tumor cells, consistent with the recognized roles of NO derived from neutrophils and macrophages in killing bacteria, parasites, and tumor cells [6]. Consistent with the role of neuron-derived NO in relaxing pyloric sphincter muscles, nNOS knockout mice display hypertrophic pyloric stenosis. Additionally, given the role of NO production by nNOS in neurotrasmission, aberrant behavior has been reported in male siblings [7].

Section of Pharmacology, Department of Molecular Biology and C.R.I.S.M.A., Pharmacy School, University of Siena, Via A. Moro 2, 53100 Siena, Italy

193

Role of NO in Angiogenesis

Starting from the observation that angiogenesis is accompanied by vasodilation and that many angiogenic molecules possess vasodilating properties, the existence of a close molecular/biochemical link among vasodilation, NO production and angiogenesis has been established [8]. The critical role played by the eNOS during all the steps of the angiogenesis process has also been firmly illustrated [9,10]. Several evidences suggest a stimulatory role of NO in angiogenesis. NO and cGMP increase the replication of endothelial cells [10]. Angiogenesis elicited in vivo by the vasoactive molecules substance P and prostaglandin E was blocked by systemic NOS inhibition [8]. Similarly, NOS inhibitors and genetic models of NOS knocking-out showed reduced neovascularisation and wound healing in different tissues [5, 11–13].

NO is the final mediator of angiogenesis stimulated by vascular endothelial growth factor (VEGF) [14], the major factor implicated in therapeutic angiogenesis and tumor neovascularization. Functional eNOS is required for endothelial cell migration and proliferation induced by this growth factor [14–16], as well as for its vasorelaxing properties [17]. Stimulation of eNOS by VEGF is mediated by several mechanisms. First is up-regulation of eNOS mRNA and protein [18].

Second, eNOS can be activated through increased association with heat-shock protein 90 (Hsp90) [19], activation of phosphatidylinositol-3OH-kinase (PI3K)/protein kinase B (PKB/Akt), leading to phosphorylation of eNOS [20], and activation of mitogen-activated protein kinase (MAPK)/phospholipase C-γ, leading to increased phosphatidylinositol triphosphate and intracellular calcium ions [21].

Hsp90 seems crucial also for the intracellular activity of NO, since it is complexed with both eNOS and soluble guanylate cyclase (sGC) in endothelial cells [22,23]. sGC derived cGMP is the intracellular mediator of neovessel formation [10,14,15,24] (Fig. 17.1).

NO acts as an autocrine regulator of endothelial cell function/survival. In microvascular endothelium, exogenous administered and endogenously produced NO up-regulates the expression of the endogenous angiogenic factors fibroblast growth factor (FGF-2) [25–27].

The concept of the proangiogenic role of NO, however, is not univocal. Depending on the angiogenesis model, the species, the drugs used and their concentrations, opposing results have been reported in the literature. In fact, NO behaves as an antiangiogenic mediator in the chick chorioallantoic membrane model, acting as an endogenous brake to control tissue vascularization [28].

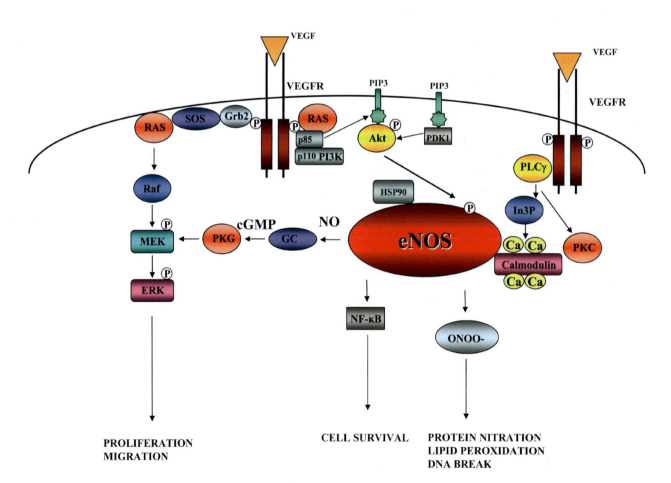

FIG. 17.1. Intracellular cross talk between eNOS signaling and VEGF receptor activation.

Role of NO in Tumor Angiogenesis

A solid tumor consists of cancer cells and host-derived cells, including tumor-infiltrating leucocytes and cells of the tumor vasculature, especially endothelial cells. One or more of these cellular constituents may express any of the active NOS isoforms, serving as a source of NO in the tumor microenvironment [29]. Functional roles of tumor-derived NO in cancer progression represent a combination of NO-mediated effects on: (1) tumor-cell proliferation, survival, migration, and invasiveness; (2) the function of immune cells infiltrating tumors; and (3) recruitment and activation of endothelial-cell progenitors able to induce angiogenesis. NO has been proposed to promote tumor growth by regulating tumor blood flow [30]. Indeed, NO has been shown to be important for maintaining the vasodilatated tone of tumors. In addition to angiogenesis stimulation, NO can promote metastasis by increasing vascular permeability and up-regulating matrix metalloproteinases (MMPs), contributing to tumor invasion and spreading [30]. Recently, NO released by metastatic tumor clones has been found to impair the immune system, facilitating the escape from immunosurveillance and spread of metastastic tumor cells [29,30].

NO differentially regulates pro- and antiangiogenic factors and chemokines in tumor cells. The expression of both VEGF and IL-8 is under the regulation of NO. In contrast, the expression of tumor-suppressive interferon-inducible protein-10 (IP-10) and monokine induced by interferon-γ (MIG) are suppressed by NO [31]. The resulting altered balance leads to stimulation of tumor angiogenesis.

A positive correlation of NO with tumor progression emerges from human and experimental tumors. The amount of tumor immunoreactive NOS protein, its activity, or both, have been positively related to the degree of malignancy for tumors of the human reproductive tract, breast and central nervous system [9,32–34]. In breast cancer, iNOS expression by macrophages, stromal and tumor cells accounted for most of the NOS activity. In gastric carcinomas, iNOS was detected in stromal cells and eNOS in the tumor vasculature [35]. iNOS expression was higher in prostatic carcinomas than in benign prostatic hyperplasia [36]. Similarly, total NOS activity was higher in lung adenocarcinomas [37] and carcinomas of the larynx, oropharynx, and oral cavity compared with healthy control tissue [38]. NOS expression has been positively associated with tumor microvessel density, and transplant of tumor samples and cell line from squamous cell carcinoma into the rabbit cornea produced angiogenesis, which regressed by treating the animals with NOS inhibitors [38].

Experimental tumor models have provided clear evidence for a direct role of NO in tumor growth and metastasis. In a rat adenocarcinoma model, in which NOS was abundantly expressed in tumor vasculature, NG-nitro-L-arginine methyl ester (L-NAME) treatment, decreased NO production and tumor growth [39]. Similarly, iNOS upregulation in a human colonic adenocarcinoma cell line led to stimulation of tumor growth and vascularity in nude mice [39], blocked by the selective iNOS inhibitor, 1400W [40]. NO-mediated stimulation of tumor growth and metastasis was also documented in a murine mammary adenocarcinoma model [41]. Treatment with exogenous NO increased VEGF and angiogenesis in ovarian cancer cell lines, whereas inhibition with L-NAME significantly reduced VEGF levels and inhibited angiogenesis [42]. Experiments studying endogenous inhibitors of NOS, such as asymmetric dimethyl-arginine (ADMA), are fully consistent with the paradigm that increased availability of NO positively correlates with tumor progression and angiogenesis. ADMA is metablized by dimethylarginine dimethylaminohydrolase (DDAH), thus its levels can be controlled regulating this enzyme expression. Overexpression of DDAH in glioma tumor models results in increased NO/cGMP levels, VEGF production and tumor vascularization and growth rate [43,44].

Wild-type P53 is a negative regulator of angiogenesis for its ability to down-regulate VEGF [45] and promote the effects of thrombospondin 1, a potent inhibitor of angiogenesis. A positive association has been made between loss of P53 function and vascularity of iNOS-expressing human tumor xenografts [46], and P53 mutation is associated with iNOS upregulation and angiogenesis. Additionally, in P53 knockout mice, the higher incidence of cancer occurrence is paralleled by iNOS up-regulation [47].

NO contributes to induced angiogenesis in tumors either in a direct or indirect mode, although the distinction appears rather artificial. When NO originates from tumor or tumor-associated cells, the molecule directly promotes vessel sprouting. Instead, when NO is derived from endothelial cells primed by VEGF that is abundantly secreted by the tumor microenvironment, the angiogenesis process is indirectely promoted by the mediator. In addition, NO stabilizes the transcription of hypoxia inducible factor-1 (HIF1-α) which, in turn, regulates VEGF production providing a further control point for the the NO-VEGF loop [48]. Other transcription factors are also involved in NO-related angiogenesis in human cancer, such as the nuclear factor-κB (NF-kB) and Sp1 [49].

In contrast to the findings described above, properties of NO potentially beneficial for the host have also been reported, which may be linked to elevated NO levels exerting cytotoxic activity on neoplastic cells. In fact, induction of apoptosis by endogenous NO has been observed in pancreatic, breast and colon cancer [50,51].

Cross-talk among NOS, COX and Growth Factor Receptors

The relationship between inflammation and tumors is a century-old research topic that has progressed from the seminal work by Virchow, illustrating the inflammatory cell infiltration in the peritumoral area, to the finely-tuned molecular mechanism regulating tumor growth. The notion that inflammation provides the fuel for malignancy has been substantiated by the recognition that many human tumors (colon, breast, head and neck, and pancreas) contain high expression levels of cyclo-oxygenase-2 (COX-2) and prostaglandin E2 (PGE2), its major metabolic product [52]. The interaction between inflammatory mediators and the NOS pathway is complex, being characterized by a number of conflicting, often controversial, results. NO has been found to stimulate COX-2 activity in the majority of inflammation/tumor models [53,54]. On the other hand, prostanoids can also modulate NOS [55].

To further document the complexity of angiogenesis mediators in neoplastic lesions, a correlation among iNOS, COX-2 and VEGF expression or microvessel density has been linked to the clinical outcome of different solid tumors [56,57]. A particular example of COX-2 and iNOS interaction has been observed in a collaborative work with our laboratory investigating the hepatocellular carcinoma (HCC) cell lines in which elevated levels of the two enzymes promote the development of the multiple drug resistance (MDR) phenotype. The acquisition of the MDR phenotype is, in turn, associated with a higher angiogenic potential of these cells [58].

PGE2, derived from COX-2, exerts its autocrine/paracrine effects by coupling to four subtypes (EP1-4) of G-protein-coupled receptors [59], with EP2 or EP4 mediating tumor progression and/or tumor-associated angiogenesis [60]. One mechanism, among those proposed, is that PGE2 stimulation of both EP2 and EP4 receptors involves the transactivation of the epidermal growth factor receptor (EGFR) signaling pathway to promote tumorigenesis [61,62]. EGFR phosphorylation leads to activation of downstream signaling molecules involved in cell proliferation and survival pathways of the ras/raf/MAPK and PI3K/Akt, respectively [63].

Recently, we provided evidence for the NOS/cGMP pathway involvement in prostaglandin-induced EGFR transactivation in squamous carcinoma cells, by showing that cell growth, invasiveness, and angiogenic output are dependent on NO availability [64] (Fig. 17.2).

Genetic Polymorphism of NOS Isoforms and Cancer Hazard

Single nucleotide polymorphisms, described for both eNOS and iNOS, correlates with increased risk for developing tumors. Most studies relate to eNOS. The eNOS exon 7 Glu-298Asp polymorphism predisposes women to ovarian and breast cancer [65,66] and men to prostate cancer [67,68]. eNOS intron 4 27-bp-tandem repeat polymorphism (and possibly exon 7 Glu298Asp) has been associated with breast cancer recurrence and death, particularly in women with estrogen receptor positive tumors [66,69]. eNOS intron 4 27-bp-repeat (but not exon 7 Glu298Asp) polymorphism influences the length of disease-free survival of patients with vulvar cancer [70]. Also, the association of different genes has been considered, such as CG (in eNOS) and TiA (in caveolin-1) haplotypes, which, together, increase colorectal cancer heritability [71].

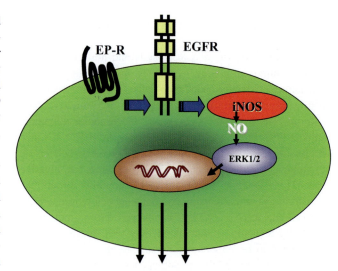

Tumor cell growth, invasiveness Angiogenic factor production

Fig. 17.2. Schematic representation of the hypothesized NOS-COX-EGFR relationship in the control of tumor cell growth and invasion, and angiogenic output. [64, and unpublished data]

In contrast, no clear data have been reported for iNOS polymorphisms. iNOS Ser608Leu allele has been associated with increased susceptibility to gastric cancer [72], whereas no significant association between iNOS polymorphism and colorectal cancer has been reported [73]. However, the field of genetic polymorphisms still has too many uncertainties, and larger studies are needed to validate the above observations.

The mechanism underlying genetic variability and cancer hazard may be linked to elevated NO levels causing cancer progression and shedding of tumor cells in the blood [67], although a direct link between NOS polymorphism and angiogenesis and/or COX-2 cannot be excluded.

Pharmacological Interventions and Perspectives

There are essentially two interventional approaches to target the NOS cascade for anti-tumor therapy: to directly affect growth and/or neovascularization in the tumor mass, or to facilitate the intra-tumoral delivery of anti-tumor agents. However, given the profound systemic effects produced by interferring with the NOS cascade, the clinical development of these strategies has been delayed.

NOS Inhibitors

The reduction of elevated NO levels can be obtained either through NOS inhibition or through the binding and scavenging of NO (Tables 17.1–17.3). Systemic NOS inhibition, through L-arginine derivatives in experimental tumor models, has been demonstrated to exert anti-tumoral activity, by reducing tumor volume and increasing tumor necrosis [74]. However, a serious problem of NOS inhibitors is their lack of selectivity for the three isoforms of the enzyme. Alhtough, numerous selective inhibitors have been identified (Table 17.1), and tested in the clinic, none have yet been approved [75,76].

Endogenous inhibitors of the NOS pathway have been identified as ADMA [43,44] and caveolin peptides. Indeed, recently, Gratton and coworkers described a cell permeable peptide derived form caveolin-1, namely cavtratin, which reduces vessel density, microvascular permeability and tumor progression in human hepatoma xenografts and murine lung cancers through inhibition of eNOS [77]. Others have shown that intravenous administration of a cationic-lipid coated plasmid encoding caveolin-1 caused selective expression in tumor vessels of recombinant caveolin-1, resulting in a reduced tumor blood flow and decreased tumor growth. This presumably occurs through inhibition of the proangiogenic and vasodilatatory effect of NO [78]. Although cavtratin does not inhibit iNOS, in human colon cancer cells ectopic expression of caveolin-1 decreases iNOS activity by direct association

with the enzyme, causing its degradation through the proteasome pathway [79]. In addition to eNOS, caveolin-1 activity extends to other G protein and tyrosine kinase receptors [80], suggesting that further studies are required to define its primary target for the clinical application in oncology.

In light of the known close relationship between NOS and COX, chemotherapeutic or chemopreventive drugs able to inhibit both COX-2 and iNOS may become important therapeutic adjuvants. Indeed, a combination of COX-2 and iNOS inhibitors has been shown to produce promising chemopreventive effect against colon carcinogenesis [81]. In addition, NOS inhibitors have been shown to retard the growth of tumors that are resistant to COX inhibitors [82].

NO Scavengers and Arginine Depletion

The selectivity of NO scavengers (Table 17.2) is based on the rate of reaction with NO, which is dependent on the concentration of NO and the scavenger. Thus, when NO concentrations are elevated, scavenging would proceed at a faster rate for a given concentration of scavenger. This is in contrast to

TABLE 17.1. NOS inhibitors (see references 75–79,122–125).

L^w-nitro-L-arginine (L-NNA) (selective for ecNOS and nNOS)	L^w-nitro-L-arginine methyl ester (L-NAME) (non selective)
L^w-nitro-monomethyl-L-arginine (L-NMMA) (selective for ecNOS and nNOS)	N-iminoethyl-L-ornithine (L-NIO) (more selective for ecNOS)
	(S,S'-1,3-phenylene-bis(1,2-ethanediyl)-bis-isothiourea) (1,3-PBITU) (more selective for iNOS)
L-N6-(1-Iminoethyl)lysine hydrocloride (L-NIL) (selective for iNOS)	N^w—propyl-L-arginine (selective for nNOS)
N-[3-(aminomethyl)benzylacetamidine (1400W) (selective for iNOS)	S-ethyl-N-[4(trifluoro-methyl)phenil]isothiuorea (selective for nNOS)
	Asymmetric dimethylarginine (ADMA) (non selective)
Guanidine derivatives: 7-nitro-indazole (selective for ecNOS and nNOS) Aminoguanidine (iNOS selective)	Tripeptides containing L-Arginine Caveolin peptides (Cavtratin) Transduction of caveolin-1 gene Intracellular domains of G-protein-coupled receptors NOS interacting protein (NOSIP)

TABLE 17.2. NO antagonists and scavengers (see references 83, 84, 122, 125).

TABLE 17.3. NOS/cGMP pathway inhibitors (see references 122–125).

TABLE 17.4. NO donors or potentiators (see references 86, 87, 122, 123, 125).

Nitro derivatives	
Sodium Nitroprusside (NaNP)	Nitroglycerin (TNG)
$Na_2[Fe(CN)_5]NO \cdot 2H_2O$	Isosorbide Dinitrate
	Amyl Nitrate
Linsidomine, SIN-1	Spermine NONOate, Diethyl-amineNONOate
Nitrosoglutathione, S-nitrosocysteine (natural carrier, unstable)	S-Nitroso-N-acetylpenicellamine (SNAP) (Higher stability)
Glyco-SNAPs (more stable, release of NO over 24–30 h)	NONOates or Diazeniumdiolates: NOC-5,-7,-9,-12,-18 NOR-1, -3 (different half.lives) (spontaneous NO release)
NO potentiators	
Antioxidants (N-Acetyl-L-cysteine, L-Ascorbic Acid, (–)-Epigallocat-echin Gallate, Glutathione, a-Lipoic Acid, Luteolin, Melatonin, Resveratrol, TEMPOL, DL-a-Tocopherol, Trolox, …)8-Br-cGMP (cGMP analog)	Inhibitors of phosphodiesterase V (Sildenafil, MY-5445)
8-Br-PET-cGMP (membrane soluble cGMP analog)	3-(5-hydroxymethyl-2-furyl)-1-benzylindazole (YC-1) (sGC activator)

NOS inhibitors, which inhibit NO synthesis equally in regions of high, normal and low NO production. A series of ruthenium compounds have been designed to act as NO scavengers with favorable in vivo properties and low toxicity. These compounds have been shown to bind NO with rapid kinetics, exhibiting activity in in vitro and in vivo models for various diseases including cancer [83]. For example, the compound termed NAMI-A and other ruthenium(III) compounds exhibited great efficacy in blocking angiogenesis elicited by VEGF and by exogenously administered NO [84].

Arginine deiminase (ADI), an enzyme which depletes plasma arginine via its hydrolysis to citrulline, represents an alternative therapeutic approach. Its anti-proliferative and anti-angiogenic effects, attributed to suppression of NO synthesis [85], have been tested in phase I and II studies, demonstrating a partial efficacy in metastatic melanoma [86].

NO Donors and Potentiators

Despite the overwhelming evidence correlating increased NOS with tumor progression, which provides the rationale for its inhibition for controlling malignancy, as stated earlier elevated NO levels might also cause tumor cell death (Table 17.4). The dual effects of NO on tumor biology might be interpreted as related to its concentration within the tumor microenvironment [87]. Low levels of NO, likely produced by tumor cells themselves, act in favor of tumor progression, while high levels of NO are tumoricidal. These observations suggest an alternative approach to anti-cancer therapy, i.e. forcing tumor cells to produce NO in excessive amounts. For example, transfection of tumor cells with the iNOS gene suppressed tumorigenicity and metastasis in murine melanoma and human renal carcinoma cell lines [88,89]. Alternatively increased NO production can be induced in lymphocytes, which alter iNOS expression [90,91].

In the same context, phosphodiesterase-5 inhibitors (such as sildenafil) have been demonstrated to increase endogenous anti-tumor immunity by reducing myeloid-derived suppressor cell function, potentially forecasting these drugs as adjunct in tumor-specific immune therapy [92].

The efficient delivery of therapeutic agents and oxygen to tumors can also be achieved by targeting the NOS cascade [93]. NO-mediated recruitment of perivascular cells to the tumor vasculature [94] is instrumental for "normalizing" the chaotic tumor vasculature. Therefore, the administration of NO donor drugs or the activation of vascular eNOS, might be considered a strategy to control tumor vessel structure, thus favoring the efficacy of cytotoxic and radiation therapies [95,96]. An example is represented by TX-1877, a bifunctional hypoxic cell radiosensitizer, which, by increasing NO production, enhances anti-cancer host response [97].

NO Releasing NSAIDs

Several epidemiological, clinical and experimental studies established nonsteroidal anti-inflammatory drugs (NSAIDs) as potential cancer chemopreventive agents. Long-term use of aspirin and other NSAIDs has been shown to reduce the risk of cancer in colon and in the gastrointestinal tract as well as in breast, prostate, lung, and skin cancer [98]. NSAIDs restore normal apoptosis reducing cell proliferation in human and experimental colorectal tumors, and in several cancer cell lines lacking genes critical for normal cell function. NSAIDs, particularly selective COX-2 inhibitors such as celecoxib, have been shown to inhibit angiogenesis in cell culture and in rodent models of angiogenesis. However, concerns on the safety of these drugs currently limit their clinical application in cancer prevention. Moreover, introducing the safe and effective use of NSAIDs into the clinic for chemoprevention is complicated by a potential rare yet serious toxicity that may offset the benefit of treating healthy individuals who are at low risk of developing the disease.

New insights are coming from combining NSAIDs with other agents. An example is the development of nitric oxide-releasing NSAIDs, which induce tumor cell apoptosis and thus reduce their inherent toxicity [99]. Nitrosulindac (NCX1102) and the NO-donating aspirin derivative NCX4040 inhibit proliferation and induce apoptosis in human prostatic and pancreatic adenocarcinoma cell lines [100]. Cytotoxic

mechanims involve overexpression and activation of Bax, release of cytochrome C and activation of caspases-9 and -3. NCX4040 is also proposed in combination with 5-fluorouracil or oxalilplatin in human colon cancer models [101]. Treatment of ovarian cancer cells with NCX4016, which releases NO in a sustained fashion for several hours, is reported to sensitize cells to cisplatin [102], suggesting that the kinetics of NO release might concur for the efficacy of these molecules. The clinical relevance of these new drugs however remains to be established.

Antiangiogenic/Anti-tumor Drugs and the NOS Pathway

The relevance of antiangiogenic drugs in cancer therapy is now well established, and the growing clinical use of these drugs is opening new scenarios in cancer treatment approaches and fostering novel considerations. Many of the molecules introduced into the clinic were designed as blockers of growth factors or as inhibitors of their corresponding receptors, which specifically decrease angiogenesis rather than affecting tumor cells. The NOS pathway, although only indirectly involved in the overall efficacy of some of these agents, is nonetheless a pivotal point of several growth factor mechanisms of action. At the forefront of this class of compounds is bevacizumab (Avastin), a humanized anti-VEGF monoclonal antibody, which contrasts the abundant output of VEGF and the consequent angiogenic drive of human and experimental tumors. Along a similar line is the clinical development of soluble VEGFR1 (sVEGFR1) regarded as a negative regulator of VEGF-A activity in angiogenesis [103]. Other agents include small-molecule receptor tyrosine kinase (RTK) inhibitors (sunitinib and sorafenib) which target VEGFRs, PDGFR, c-kit and Flt-3 [104]. A more comprehensive review of these drugs is discussed in later sections of this book.

Concern over the use of antiangiogenic drugs targeting VEGF and its NOS-centered cascade has emerged from evidence showing increased incidence of side effects possibly related to the impairment of vascular function mantainance. Conceivably, the incidence of cardiovascular events (stroke and hypertension), described in patients treated with antiangiogenic agents, may be causally related to the dearrangment of the NOS pathway. Similar concern may apply to molecules proposed for development, such as 16 kDa prolactin, found to down-regulate iNOS expression in endothelial cells [105], and TNP-470 which inhibits NF-kB activation and iNOS expression in experimental model of hepatocarcinogenesis [106].

Recent studies on conventional anti-tumor agents suggest for this class of drugs that clinical efficacy may involve NOS inhibition. The topoisomerase I inhibitor, camptothecin, exherts its anti-tumor activity in part through inhibition of NO biosynthesis [107], and irinotecan has been associated with down-regulation of COX-2, iNOS, HIF-1α and angiogenesis [108]. Also, the anti-tumor activity of selected conventional drugs, such as the tubulin binding agent combretastatin-A4 phosphate (CA4-P), melphalan, tumor necrosis factor (TNF) and mytomicin, appears to be considerably potentiated by the simultaneous administration of L-NNA, L-NAME or the iNOS selective inhibitor aminoguanidine, respectively [109,110].

An interesting facet of the NOS pathway, relevant to anti-tumurol therapy, is found in the investigation showing that NO upregulation is associated with resistance to some classical or novel cytotoxic drugs, as seen in lung carcinoma for cisplatin and colon cancer cell resistance to ZD6126 [111]. The molecular mechanisms underlying cisplatin resistance are S-nitrosylation, inhibition of Bcl-2 ubiquitination and proteasomal degradation [112].

Chemoprevention with Natural Derivatives Affecting NO Availability

A recent interesting addition to the anti-tumor armamentarium is represented by plant compounds that block the NOS pathway and reduce the tissue level of NO. These actions, together with other crucial cellular and biochemical mechanisms, are responsible for the anti-tumor and antiangiogenic activity of selected compounds or extracts from plants or beverages.

Multiple mechanisms of cancer prevention have been described for green and black tea polyphenols [113], which reduce NO production via blocking NF-kB nuclear translocation and suppression of iNOS. Polyphenols, such as epigallocatechine gallate (EGCG) and quercetin, have been found to inhibit angiogenesis [114], with quercetin being proposed as a sensitizer of cisplatin in human head and neck cancer [115].

Black raspberry extracts [116,117], as well as a number of other principles extracted from garlic (Alliin) [118], mushroom [119], curcuminoid derivatives [120] and Ginkgo biloba [121], have been reported to exert antiangiogenic/anti-tumor activity by interfering with the NOS pathway.

However, the clinical benefits of these plant principles remain to be validated in large clinical studies.

Conclusions

After the discovery of NO, many functions have been attributed to this mediator. Starting from the observation that angiogenesis is accompanied by vasodilation and that many angiogenic molecules possess vasodilating properties, the existence of a close molecular/biochemical link among vasodilation, NO production and angiogenesis has been established. In human tumors, NOS expression and activity correlate with tumor growth and aggressiveness, through angiogenesis stimulation and regulation of angiogenic factor expression. However, NO actions are multifaceted and appear to be context-dependent being affected by the NO concentration, the duration of its release, the cell type, the presence of scavengers, or other reactive molecules impairing or amplifying NO effects. Recently, genetic variability of NOS isoforms has been proposed to interpret altered NO production and biological effects.

NO plays several and contradictory roles in tumors as its effects vary depending on the circumstances mentioned above, like the selective modification of various gene products and the complexity of cell responses. Exogenously or endogenously induced NO synthesis might be a promising adjuvant in radiation or chemotherapy. Inhibition of the NOS pathway may be foreseen as an anti-tumor strategy that may reduce edema, inhibit angiogenesis or facilitate the delivery of chemotherapeutic agents. However, the profound systemic effects produced by interfering with the NOS cascade limit the clinical use of this class of compounds, thereby restricting their use to be confined within the tumor volume.

The in-depth knowledge of the NOS pathway has been instrumental in defining the mechanisms of action of several protagonists of the angiogenic process, favoring the development of compounds controlling angiogenesis and tumor progression.

Acknowledgements. This work was supported by grants from the University of Siena, Italy (PAR project), Associazione Italiana Ricerca sul Cancro (AIRC) and EEC funding EICO-SANOX (project n. LSHM-CT-2004-005033). The authors are indebt to the late P.M. Gullino, a pioneer in the research of tumor associated angiogenesis, for his inspiring teaching in the research on tumor microenvironment.

References

1. Palmer RMJ, Ferrige AG, Moncada S. Nitric oxide release accounts for the biological activity of endothelium-derived relaxing factor. Nature 1987;327: 524–6.
2. Ignarro LJ, Buga GM, Wood KS, et al. Endothelial-derived relaxing factor produced and released from artery and vein is nitric oxide. Proc Natl Acad Sci USA 1987; 84: 9265–9.
3. Knowles RG, Moncada S. Nitric oxide synthases in mammals. Biochem J 1994; 298:249–58.
4. Huang PL, Huang Z, Mashimo H, et al. Hypertension in mice lacking the gene for endothelial nitric oxide synthase. Nature 1995; 377: 239–42.
5. Lee PC, Salyapongse AN, Bragdon GA, et al. Impaired wound healing and angiogenesis in eNOS-deficient mice. Am J Physiol 1999; 277: H1600–8.
6. MacMicking JD, Nathan C, Hom G, et al. Altered responses to bacterial infection and endotoxic shock in mice lacking inducible nitric oxide synthase. Cell 1995; 81: 641–50.
7. Huang PL, Dawson TM, Bredt DS, et al. Targeted disruption of the neuronal nitric oxide synthase gene. Cell 1993; 175: 1273–86.
8. Ziche M, Morbidelli L, Masini E, et al. Nitric oxide mediates angiogenesis in vivo and endothelial cell growth and migration in vitro promoted by substance P. J Clin Invest 1994; 94:2036–44.

9. Ziche M, Morbidelli L. Nitric oxide and angiogenesis. J Neuro-Oncology 2000; 50:139–48.

10. Ziche M, Morbidelli L, Masini E, et al. Nitric oxide promotes DNA synthesis and cyclic GMP formation in endothelial cells from postcapillary venules. Biochem Biophys Res Commun 1993; 192:1198–203.

11. Konturek SJ, Brzozowski T, Majka J, et al. Inhibition of nitric oxide synthase delays healing of chronic gastric ulcers. Eur J Pharmacol 1993; 239:215–7.

12. Yamasaki K, Edington HDJ, McClosky C, et al. Reversal of impaired wound repair in iNOS-deficient mice by topical adenoviral-mediated iNOS gene transfer. J Clin Invest 1998; 101:967–71.

13. Murohara T, Asahara T, Silver C, et al. Nitric oxide synthase modulates angiogenesis in response to tissue ischemia. J Clin Invest 1998; 101:2567–78.

14. Ziche M, Morbidelli L, Choudhuri R, et al. Nitric oxide synthase lies downstream from vascular endothelial growth factor-induced but not basic fibroblast growth factor-induced angiogenesis. J Clin Invest 1997; 99:2625–64.

15. Morbidelli L, Chang C-H, Douglas JG, et al. Nitric oxide mediates mitogenic effect of VEGF on coronary venular endothelium. Am J Physiol 1996; 270:H411–5.

16. Parenti A, Morbidelli L, Cui XL, et al. Nitric oxide is an upstream signal for vascular endothelial growth factor-induced extracellular signal-regulated kinases1/2 activation in postcapillary endothelium. J Biol Chem 1998; 273:4220–6.

17. Ku DD, Zaleski JK, Liu S, et al. Vascular endothelial growth factor induces EDRF-dependent relaxation in coronary arteries. Am J Physiol 1993; 265:H586–92.

18. Hood JD, Meininger CJ, Ziche M, et al. VEGF upregulates ecNOS message, protein, and NO production in human endothelial cells. Am J Physiol 1998; 43:H1054–8.

19. Garcia-Cardeña G, Fan R, Shah V, et al. Dynamic activation of endothelial nitric oxide synthase by Hsp90. Nature 1998; 392:821–4.

20. Radisavljevic Z, Avraham H, Avraham S. Vascular endothelial growth factor up-regulates ICAM-1 expression via phosphatidylinositol 3OH-kinase/AKT/nitric oxide pathway and modulates migration of brain microvascular endothelial cells. J Biol Chem 2000; 275:20770–4.

21. He H, Venema VJ, Gu X, et al. Vascular endothelial growth factor signals endothelial cell production of nitric oxide and prostacyclin through Flk-1/KDR activation of c-src. J Biol Chem 1999; 274:25130–5.

22. Venema RC, Venema VJ, Ju H, et al. Novel complexes of guanylate cyclase with heat shock protein 90 and nitric oxide synthase. Am J Physiol 2003; 285:H669–78.

23. Papapetropoulos, A, Zhou Z, Gerassimou C, et al. Interaction between the 90-kDa heat shock protein and soluble guanylyl cyclase: physiological significance and mapping of the domains mediating binding. Mol Pharmacol 2005; 68:1133–41.

24. Pyriochou A, Beis D, Koika V, et al. Soluble guanylyl cyclase activation promotes angiogenesis. J Pharmacol Exp Ther 2006; 319:663–71.

25. Parenti A, Morbidelli L, Ledda F, et al. The bradykinin/B1 receptor promotes angiogenesis by upregulation of endogenous FGF-2 in endothelium via the nitric oxide synthase pathway. FASEB J 2001; 15:1487–9.

26. Ziche M, Parenti A, Ledda F, et al. Nitric oxide promotes proliferation and plasminogen activator production by coronary venular endothelium through endogenous bFGF. Circ Res 1997; 80:845–52.

27. Donnini S, Solito R, Giachetti A, et al. Fibroblast growth factor-2 mediates ACE inhibitor-induced angiogenesis in coronary endothelium. J Pharm Exp Therap 2006; 319:515–22.

28. Pipili-Synetos E, Sakkoula E, Haralabopoulos G, et al. Evidence that nitric oxide is an endogenous antiangiogenic mediator. Br J Pharmacol 1994; 111.894–902.

29. Lala PK, Chakraborty C. Role of nitric oxide in carcinogenesis and tumor progression. Lancet Oncol 2001; 3:149–56.

30. Fukumura D, Kashiwagi S, Jain RK. The role of nitric oxide in tumor progression. Nat Rev Cancer 2006; 6:521–34.

31. Romagnani P, Lazzeri E, Lasagni L, et al. IP10 and Mig production by glomerular cells in human proliferative glomerulonephritis and regulation by nitric oxide. J Am Soc Nephrol 2002;13:53–64.

32. Thomsen LL, Lawton FG, Knowles RG, et al. Nitric oxide synthase activity in human gynaecological cancer. Cancer Res 1994; 54:1352–4.

33. Thomsen LL, Miles DW, Happerfield L, et al. Nitric oxide synthase activity in human breast cancer. Br J Cancer 1995; 72:41–4.

34. Cobbs CS, Brenman JE, Aldape KD, et al. Expression of nitric oxide synthase in human central nervous system tumours. Cancer Res 1995; 55:727–30.

35. Yamaguche K, Saito H, Oro S, et al. Expression of inducible nitric oxide synthase is significantly correlated with expression of vascular endothelial growth facto rand dendritic cell infiltration in patients with advanced gastric carcinoma. Oncology 2005;68:471–8.

36. Klotz T, Bloch W, Volberg C, et al. Selective expression of inducible nitric oxide synthase in human prostatic carcinoma. Cancer 1998; 82:1897–903.

37. Fujimoto H, Ando Y, Yamashita T, et al. Nitric oxide synthase activity in human lung cancer. Jpn. J Cancer Res 1997; 88:1190–8.

38. Gallo O, Masini E, Morbidelli L, et al. Role of nitric oxide in angiogenesis and tumor progression in head and neck cancer. J Natl Cancer Inst 1998; 90:587–96.

39. Jenkins DC, Charles IG, Thomsen LL, et al. Roles of nitric oxide in tumor growth. Proc Natl Acad Sci USA 1995; 92:4392–6.

40. Thomsen LL, Scott JM, Topley P, et al. Selective inhibition of inducible nitric oxide synthase inhibits tumor growth in vivo: studies with 1400W, a novel inhibitor. Cancer Res 1997; 57: 3300–4.

41. Jadeski LC, Hum KO, Chakraborty C, et al. Nitric oxide promotes murine mammary tumor growth and metastasis by stimulating tumor cell migration, invasiveness and angiogenesis, Int J Cancer 2000; 86:30–9.

42. Malone JM, Saed GM, Diamond MP, et al. The effects of the inhibition of inducible nitric oxide synthase on angiogenesis of epithelial ovarian cancer. Am J Obstet Gynecol 2006;194(4):1110–8.

43. Kostourou V, Robinson SP, Cartwright JE, et al. Dimethylarginine dimethylaminohydrolase I enhances tumor growth and angiogenesis. Br J Cancer. 2002;87(6):673–80.

44. Kostourou V, Robinson SP, Whitley GS, et al. Effects of overexpression of dimethylarginine dimethylaminohydrolase on tumor angiogenesis assessed by susceptibility magnetic resonance imaging. Cancer Res. 2003;63(16):4960–6.

45. Mukhopadhyay D, Tsiokas L, Sukhatme VP. Wild type p53 and v-src exert opposing influences on human vascular endothelial growth factor gene expression. Cancer Res 1995; 55:6161–5.

46. Ambs S, Merriam WG, Ogunfusika MO, et al. P53 and vascular endothelial growth factor regulate tumor growth of NOS2 expressing human carcinoma cells. Nat Med 1998; 4:1371–6.

47. Ambs S, Ogunfusika MO, Merriam WP, et al. Up-regulation of inducible nitric oxide synthase expression in cancer-prone p53 knockout mice. Proc Natl Acad Sci. USA 1998; 95:8823–8.

48. Kasuno K, Takabuchi S, Fukuda K, et al. Nitric oxide induces hypoxia –inducible factor 1 activation that is dependent on MAPK and phosphatidylinositol 3-kinase signaling. J Biol Chem 2004;279(4):2550–8.

49. Zhang J, Peng B, Chen X. Expressions of nuclear factor kappaB, inducible nitric oxide synthase, and vascular endothelial growth factor in adenoid cystic carcinoma of salivary glands: correlations with the angiogenesis and clinical outcome. Clin Cancer Res 2005;11(20):7334–43.

50. Gansauge S, Nussler AK, Beger HG, et al. Nitric oxide-induced apoptosis in human pancreatic carcinoma cell lines is associated with a G1-arrest and an increase of the cyclin-dependent kinase inhibitor p21WAF1/CIP1. Cell Growth Differ 1998; 9:611–7.

51. Mortensen K, Skouv J, Hougaard DM, et al. Endogenous endothelial cell nitric-oxide synthase modulates apoptosis in cultured breast cancer cells and is transcriptionally regulated by p53. J Biol Chem 1999; 274:37679–84.

52. Mann JR, Backlund MG, DuBois RN. Mechanisms of disease: Inflammatory mediators and cancer prevention. Nat Clin Pract Oncol 2005;4:202–10.

53. Kim SF, Huri DA, Snyder SH. Inducible nitric oxide synthase binds, S-nitrosilates and activates cycloxygenase-2. Science 2005; 310:1966–70.

54. Park SW, Lee SG, Song SH, et al. The effect of nitric oxide on cycloxygenase-2 (COX-2) overexpression in head and neck cancer cell lines. Int J Cancer 2003; 107:729–38.

55. Timoshenko AV, Lala PK, Chakraborty C. PGE2-mediated upregulation of iNOS in murine breast cancer cells through the activation of EP4 receptors. Int J Cancer 2004; 108(3):384–9.

56. Marrogi AJ, Travis WD, Welsh JA, et al. Nitric oxide synthase, cyclooxygenase 2, and vascular endothelial growth factor in the angiogenesis of non-small cell lung carcinoma. Clin Cancer Res 2000; 6:4739–44.

57. Bing RJ, Miyataka M, Rich KA, et al. Nitric oxide, prostanoids, cyclooxygenase, and angiogenesis in colon and breast cancer. Clin Cancer Res 2001; 7:3385–92.

58. Lasagna N, Fantappiè O, Solazzo M, et al. Hepatocyte growth factor and inducible nitric oxide synthase are involved in multidrug resistant induced angiogenesis in hepatocellular carcinoma cell lines. Cancer Res 2006; 66:2673–82.

59. Hata AN, Breyer RM. Pharmacology and signaling of prostaglandin receptors: multiple roles in inflammation and immune modulation. Pharmacol Ther 2004; 103:147–66.

60. Majima M, Amano H, Hayashi I. Prostanoid receptor signaling relevant to tumor growth and angiogenesis. Trends Pharmacol Sci 2003; 24:524–9.

61. Pai R, Soreghan B, Szabo IL, et al. Prostaglandin E2 transactivates EGF receptor: A novel mechanism for promoting colon cancer growth and gastrointestinal hypertrophy. Nat Med 2002; 8:289–93.

62. Buchanan FG, Wang D, Bargiacchi F, et al. Prostaglandin E2 regulates cell migration via the intracellular activation of the epidermal growth factor receptor. J Biol Chem 2003; 278:35451–7.

63. Donnini S, Finetti F, Solito R. et al. EP2 prostanoid receptor promotes A431 cell growth and invasion by iNOS and ERK1/2 pathway. Proceedings of the American Association for Cancer Research (AACR), Washington, 1–5 April, 2006.

64. Donnini S, Finetti F, Solito R, et al., EP2 prostanoid receptor promotes squamous cell carcinoma growth through epidermal growth factor receptor transactivation and iNOS and ERK1/2 pathways. FASEB J 2007;21(10):2418–30.

65. Hefler LA, Ludwig E, Lampe D, et al. Polymorphisms of the endothelial nitric oxide synthase gene in ovarian cancer. Gynecol Oncol. 2002;86(2):134–7.

66. Lu J, Wei Q, Bondy ML, et al. Promoter polymorphism (−786t>C) in the endothelial nitric oxide synthase gene is associated with risk of sporadic breast cancer in non-Hispanic white women age younger than 55 years. Cancer 2006;107(9):2245–53.

67. Medeiros R, Morais A, Vasconcelos A, et al. Endothelial nitric oxide synthase gene polymorphisms and the shedding of circulating tumor cells in the blood of prostate cancer patients. Cancer Lett 2003;189(1):85–90.

68. Marangoni K, Neves AF, Cardoso AM, et al. The endothelial nitric oxide synthase Glu-298-Asp polymorphism and its mRNA expression in the peripheral blood of patients with prostate cancer and benign prostatic hyperplasia. Cancer Detect Prev 2006;30(1):7–13.

69. Choi JY, Lee KM, Noh DY, et al. Genetic polymorphisms of eNOS, hormone receptor status, and survival of breast cancer. Breast Cancer Res Treat 2006;100(2):213–8.

70. Riener EK, Hefler LA, Grimm C, et al. Polymorphisms of the endothelial nitric oxide synthase gene in women with vulvar cancer. Gynecol Oncol 2004;93(3):686–90.

71. Conde MC, Ramirez-Lorca R, Lopez-Jamar JM, et al. Genetic analysis of caveolin-1 and eNOS genes in colorectal cancer. Oncol Rep 2006;16(2):353–9.

72. Shen J, Wang RT, Wang LW, et al. A novel genetic polymorphism of inducible nitric oxide synthase is associated with an increased risk of gastric cancer. World J Gastroenterol 2004;10(22): 3278–83.

73. Fransen K, Elander N, Soderkvist P. Nitric oxide synthase 2 (NOS2) promoter polymorphisms in colorectal cancer. Cancer Lett 2005;225(1):99–103.

74. Jadeski LC, Lala PK. Nitric oxide synthase inhibition by NG-nitro-L-arginine methyl ester inhibits tumor-induced angiogenesis in mammary tumours. Am J Pathol 1999; 155:1381–90.

75. Garvey EP, Oplinger JA, Furfine ES, et al. 1400W is a slow, tight binding, and highly selective inhibitor of inducible nitric-oxide synthase in vitro and in vivo. J Biol Chem 1997; 272:4959–63.

76. Cheshire DR. Use of nitric oxide synthase inhibitors for the treatment of inflammatory disease and pain. Drugs 2001; 4:795–802.

77. Gratton JP, Lin MI, Yu J, et al. Selective inhibition of tumor microvascular permeability by cavtratin blocks tumor progression in mice. Cancer Cell 2003; 4:31–9.

78. Broet A, DeWever J, Martinive P, et al. Antitumor effects of in vivo caveolin gene delivery are associated with the inhibition of the proangiogenic and vasodilatatory effects of nitric oxide. FASEB J 2005; 19:602–4.

79. Felley-Bosco E, Bender FC, Courjault-Gautier F, et al. Caveolin-1 downregulates inducible nitric oxide synathase via the proteasome pathway in human colon carcinoma cells. Proc Natl Acad Sci USA 2000; 97:14334–9.

80. Gratton JP, Bernatchez P, Sessa WC. Caveolae and caveolins in the cardiovascular system. Circ Res 2004; 94:1408–17.

81. Rao CV, Indranie C, Simi B, et al. Chemopreventive properties of a selective inducible nitric oxide synthase inhibitor in colon carcinogenesis, administered alone or in combination with celecoxib, a selective cycloxygenase-2 inhibitor. Cancer Res 2002; 62:165–70.

82. Cahlin C, Gelin J, Delbro D, et al. Effect of cycloxygenase and nitric oxide synthase inhibitors on tumor growth in mouse tumor models with and without cancer cachexia related to prostanoids. Cancer Res 2000; 60:1742–9.

83. Pritchard R, Flitney FW, Darkes MA, et al. Ruthenium-based nitric oxide scavengers inhibit tumour growth by reducing tumour vasculature. Clin Exp Metast 1999; 17:776.

84. Morbidelli L, Donnini S, Filippi S, et al. Antiangiogenic properties of selected ruthenium(III) complexes that are nitric oxide scavengers. Br J Cancer 2003; 88:1484–91.

85. Park IS, Kang SW, Shin YJ, et al. Arginine deiminase: a potential inhibitor of angiogenesis and tumour growth. Br J Cancer 2003;89(5):907–14.

86. Ascierto PA, Scala S, Castello G, et al. Pegylated arginine deiminase treatment of patients with metastatic melanoma: results from phase I and II studies. J Clin Oncol 2005;23(30):7660–8.

87. Lechner M, Lirk P, Rieder J. Inducible nitric oxide synthase (iNOS) in tumor biology: the two sides of the same coin. Semin Cancer Biol 2005; 15:277–89.

88. Juang SH, Xie K, Xu L, et al. Suppression of tumorigenicity and metastasis of human renal carcinoma cells by infection with retroviral vectors harboring the murine inducible nitric oxide synthase gene. Hum Gene Ther 1998; 9: 845–54.

89. Xie K, Huang S, Dong Z, et al. Transfection with the inducible nitric oxide synthase gene suppresses tumorigenicity and abrogates metastasis by K-1735 murine melanoma cells. J Exp Med 1995; 181:1333–43.

90. Wang B, Xiong Q, Shi Q, et al. Intact nitric oxide synthase II gene is required for interferon-beta-mediated suppression of growth and metastasis of pancreatic adenocarcinoma. Cancer Res 2001; 61:71–5.

91. Windbichler GH, Hausmaninger H, Stummvoll W, et al. Interferon-gamma in the first-line therapy of ovarian cancer: a randomized phase III trial. Br J Cancer 2000; 82:1138–44.

92. Serafini P, Meckel K, Kelso M, et al. Phosphodiesterase-5 inhibition augments endogenous antitumor immunity by reducing myeloid-derived suppressor cell function. J Exp Med 2006;203(12):2691–702.

93. Jain RK. Normalization of tumor vasculature: an emerging concept in antiangiogenic therapy. Science 2005; 307:58–62.

94. Kashiwagi S, Izumi Y, Gohongi T, et al. NO mediates mural cell recruitment and vessel morphogenesis in murine melanoma and tissue engineered blood vessels. J Clin Invest 2005; 115:1816–27.

95. Jordan BF, Sonveaux P, Feron O, et al. Nitric oxide as a radiosensitizer: Evidence for an intrinsic role in addition to its effect on oxygen delivery and consumption. Int J Cancer 2004; 109:768–73.

96. Sonveaux P, Dessy C, Brouet A, et al. Modulation of the tumor vasculature functionality by ionizing radiation accounts for tumor radiosensitization and promotes gene delivery. FASEB J 2002; 16:1979–81.

97. Oshikawa T, Okamoto M, Ahmed SU, et al. TX-1877, a bifunctional hypoxic cell radiosensitizer, enhances anticancer host response: immune cell migration and nitric oxide production. Int J Cancer 2005;116(4):571–8.

98. Wang D, and DuBois RN. Cyclooxygenase 2-derived prostaglandin E2 regulates the angiogenic switch. Proc Natl Acad Sci USA 2004;101(2):415–6.

99. Rao CV, Reddy BS. NSAIDs and chemoprevention. Curr Cancer Drug Targets 2004;4(1):29–42.

100. Huguenin S, Fleury-Feith J, Kheuang L, et al. Nitrosulindac (NCX 1102): a new nitric oxide-donating non-steroidal anti-inflammatory drug (NO-NSAID), inhibits proliferation and induces apoptosis in human prostatic epithelial cell lines. Prostate 2004;61(2):132–41.

101. Leonetti C, Scarsella M, Zupi G, et al. Efficacy of a nitric oxide-releasing nonsteroidal anti-inflammatory drug and cytotoxic drugs in human colon cancer cell lines in vitro and xenografts. Mol Cancer Ther 2006;5(4):919–26.

102. Bratasz A, Weir NM, Parinandi NL, et al. Reversal to cisplatin sensitivity in recurrent human ovarian cancer cells by NCX-4016, a nitro derivative of aspirin. Proc Natl Acad Sci USA 2006;103(10):3914–9.

103. Shibuya M. Differential roles of vascular endothelial growth factor receptor-1 and receptor-2 in angiogenesis. J Biochem Mol Biol 2006,(39):469–78.

104. Brugarolas J. Renal-cell carcinoma—molecular pathways and therapies. N Engl J Med 2007;356(2):185–7.

105. Lee SH, Nishino M, Mazumdar T, et al. 16-kDa prolactin down-regulates inducible nitric oxide synthase expression through inhibition of the signal transducer and activator of transcription 1/IFN regulatory factor-1 pathway. Cancer Res 2005;65(17):7984–92.

106. Mauriz JL, Linares P, Macias RI, et al. TNP-470 inhibits oxidative stress, nitric oxide production and nuclear factor kappa B activation in a rat model of hepatocellular carcinoma. Free Radic Res 2003;37(8):841–8.

107. Chiou WF, Chou CJ, Chen CF. Camptothecin suppresses nitric oxide biosynthesis in RAW 264.7 macrophages. Life Sci 2001; 69:625–35.

108. Yin MB, Li ZR, Toth K, et al. Potentiation of irinotecan sensitivity by Se-methylselenocysteine in an in vivo tumor model is associated with downregulation of cyclooxygenase-2, inducible nitric oxide synthase, and hypoxia-inducible factor 1alpha expression, resulting in reduced angiogenesis. Oncogene 2006;25(17):2509–19.

109. Tozer GM, Prise VE, Wilson J, et al. Mechanisms associated with tumor vascular shut-down induced by combretastatin A-4 phosphate: intravital microscopy and measurement of vascular permeability. Cancer Res 2001;61(17):6413–22.

110. de Wilt JH, Manusama ER, van Etten B, et al. Nitric oxide synthase inhibition results in synergistic anti-tumour activity with melphalan and tumour necrosis factor alpha-based isolated limb perfusions. Br J Cancer 2000; 83:1176–82.

111. Chanvorachote P, Nimmannit U, Stehlik C, et al. Nitric oxide regulates cell sensitivity to cisplatin-induced apoptosis through S-nitrosylation and inhibition of Bcl-2 ubiquitination. Cancer Res 2006;66(12):6353–60.

112. Cullis ER, Kalber TL, Ashton SE, et al. Tumour overexpression of inducible nitric oxide synthase (iNOS) increases angiogenesis and may modulate the anti-tumour effects of the vascular disrupting agent ZD6126. Microvasc Res 2006;71(2):76–84.

113. Beltz LA, Bayer DK, Moss AL, et al. Mechanisms of cancer prevention by green and black tea polyphenols. Anticancer Agents Med Chem 2006;6(5):389–406.

114. Donnini S, Finetti F, Lusini L, et al. Divergent effects of quercetin conjugates on angiogenesis Br J Nutrition 2006; 95(5):1016–23

115. Sharma H, Sen S, Singh N. Molecular pathways in the chemo-sensitization of cisplatin by quercetin in human head and neck cancer. Cancer Biol Ther 2005;4(9):949–55.

116. Rodrigo KA, Rawal Y, Renner RJ, et al. Suppression of the tumorigenic phenotype in human oral squamous cell carcinoma cells by an ethanol extract derived from freeze-dried black raspberries. Nutr Cancer 2006;54(1):58–68.

117. Chen T, Hwang H, Rose ME, et al. Chemopreventive properties of black raspberries in N-nitrosomethylbenzylamine-induced rat esophageal tumorigenesis: down-regulation of cyclooxy-genase-2, inducible nitric oxide synthase, and c-Jun. Cancer Res 2006;66(5):2853–9.

118. Mousa AS, Mousa SA. Anti-angiogenesis efficacy of the garlic ingredient alliin and antioxidants: role of nitric oxide and p53. Nutr Cancer 2005;53(1):104–10.

119. Song YS, Kim SH, Sa JH, et al. Anti-angiogenic and inhibitory activity on inducible nitric oxide production of the mushroom Ganoderma lucidum. J Ethnopharmacol 2004;90(1):17–20.

120. Leyon PV, Kuttan G. Studies on the role of some synthetic cur-cuminoid derivatives in the inhibition of tumour specific angio-genesis. J Exp Clin Cancer Res. 2003 Mar;22(1):77–83.

121. DeFeudis FV, Papadopoulos V, Drieu K. Ginkgo biloba extracts and cancer: a research area in its infancy. Fundam Clin Pharmacol 2003;17(4):405–17.

122. Hobbs AJ, Higgs A, Moncada S. Inhibition of nitric oxide syn-thase as a potential therapeutic target. Annu Rev Pharmacol Toxicol 1999;39:191–220.

123. Domenico R. Pharmacology of nitric oxide:molecular mechanisms and therapeutic strategies. Curr Pharm Des 2004;10(14):1667–76.

124. Alderton WK, Cooper CE, Knowles RG. Nitric oxide synthases: structure, function and inhibition. Biochem J 2001;357(Pt3):593–615.

125. Redington AE. Modulation of nitric oxide pathways: therapeu-tic potential in asthma and chronic obstructive pulmonary dis-ease. Eur J Pharmacol 2006;533(1–3):263–76.

Chapter 18
VEGF Signal Tranduction in Angiogenesis

Harukiyo Kawamura[1], Xiujuan Li[1], Michael Welsh[2], and Lena Claesson-Welsh[1,*]

Keywords: VEGF, VEGF receptors, signal transduction, PLC, PI3K, Src, Shb, TSAd, VE-cadherin, vascular permeability

Abstract: The development of a number of novel tumor therapies targeting the function of vascular endothelial growth factors (VEGFs) and their receptors has promoted an interest in understanding signal transduction regulating angiogenesis, i.e. formation of new blood vessels. The VEGFRs regulate many if not all aspects of endothelial cell function during active angiogenesis, and mediate survival signals during endothelial cell quiescence. Most tumors produce VEGF as a consequence of the hypoxic tumor microenvironment, leading to persistent stimulation of angiogenesis necessary for an expansion of the tumor as well as tumor spread through the circulation. Increased understanding of VEGFR signal transduction properties may allow development of fine-tuned therapy, targeting pathways critical in formation of new tumor vessels while preserving pathways required for survival of endothelial cells in normal vessels.

Introduction to Vascular Endothelial Growth Factor (VEGF) Signal Transduction

The term "VEGF" denotes both the prototype family member now named VEGF-A, and the family of five structurally related, homodimeric polypeptides of 40 kDa; VEGF-A, -B, -C, -D and placenta growth factor, PlGF. VEGF-A is alterna-

tively spliced to generate VEGF-A121, VEGF-A145, VEGF-A165 and VEGF-A189 (indicating the number of amino acid residues in the human splice variants; mouse variants are each one amino acid shorter) endowed with different biological properties [1]. A newly discovered splice variant, denoted VEGF-A165b, contains exon 8b, encoding a unique stretch of 5-amino acid residues. VEGF-A165b binds to VEGFR2 with high affinity but fails to transduce biological responses and may be an antagonist of VEGF-A165 [2]. VEGF-like proteins from the ORF virus family, denoted VEGF-E, cause contagious pustular dermatitis in sheep and goats and is transmissible to humans by direct contact [3, 4]. Snake venom-derived VEGF-like proteins (denoted VEGF-F) have unique structural features [5]. The mammalian VEGFs bind to different extents to three receptor tyrosine kinases, VEGF receptor-1, -2 and -3 [6].

The VEGF receptors are transmembrane glycoproteins with an extracellular ligand-binding domain, which in VEGFR1 and VEGFR2 is organized in 7 immunoglobulin-like loops. In VEGFR3, one of the loops is replaced by a disulfide bridge. The intracellular domain of each receptor is endowed with a ligand-activated kinase domain, which is split in two parts by the insertion of a "kinase insert" sequence of 70 amino acid residues.

In addition to the full-length receptor tyrosine kinase, VEGFR1 occurs as a soluble splice variant composed of the extracellular domain only [7]. The full-length form is expressed on a number of different cell types, including monocytes/macrophages and vascular endothelial cells (ECs) [8]. The soluble splice variant is highly expressed during gestation and has been associated with pre-eclampsia [9]. Deletion of the *vegfr1* gene leads to embryonic lethality at embryonic day (E) 11.5 due to excessive proliferation of ECs [10]. Thus, VEGFR1 is thought to serve as a negative regulator of VEGFR2, in part through the soluble variant which acts as a trap for VEGF-A. It cannot be excluded that the activated VEGFR1 kinase domain induces negative regulatory signaling in the target endothelial cell [11, 12]. However, priming cells by activation of VEGFR1 has been shown to enhance subsequent signal transduction via VEGFR2 [13]. The full length VEGFR1 mediates migration of hematopoietic precursors and monocytes (for a review, see

[1] Deptartment of Genetics and Pathology, Uppsala University, Dag Hammarskjöldsv. 20, 751 85 Uppsala, Sweden

[2] Deptartment of Medical Cell Biology, Uppsala University, Biomedical Center, Box 571,751 23 Uppsala, Sweden

*Corresponding Author:
E-mail: Lena.Welsh@genpat.uu.se

[14]). Altogether, the contribution of VEGFR1 to signal transduction in endothelial cells either directly, or indirectly via VEGFR2, remains unclear.

VEGFR2 is implicated in most if not all aspects of vascular endothelial cell biology. A number of signal transduction pathways induced downstream of VEGFR2 have been identified (see below). During development, VEGFR2 is the first specific endothelial marker to be expressed on hematopoietic/endothelial progenitors [15]. Subsequently, expression of VEGFR2 is turned off in hematopoeitic cells. Expression of VEGFR2 on ECs declines during the third trimester, but is induced again in conjunction with active angiogenesis [16, 17]. Inactivation of the *vegfr2* gene leads to embryonic death at mouse E 8.5–9.5, due to lack of proper differentiation and/or migration of ECs [18].

VEGFR3 is found primarily on lymphatic endothelial cells, and is critical for lymphatic EC development and function [19]. VEGFR3 may also be expressed on fenestrated capillaries, tumor ECs and on monocytes/macrophages [20, 21]. Mice deficient for VEGFR3 die at E9.5 due to defective remodeling and maturation of the primitive blood vascular plexus into larger vessels [22]. VEGFR3 is the only VEGFR for which naturally occurring mutations have been described [23].

Regulation of VEGF/VEGFR Expression

Hypoxia, i.e. low oxygen tension, is an important regulator of physiological and pathological angiogenesis. In hypoxia, the transcription factor hypoxia-inducible factor (HIF)-1 accumulates, allowing increased transcription of a multitude of genes through binding of HIF-1 to the hypoxia-responsive element (HRE). HIF-1 may also be induced by a number of other stimuli, such as growth factors, under normoxic conditions [24]. Such activation involves several different signal transduction pathways, including the phosphoinositide 3′ kinase (PI3K), extracellular regulated kinase (Erk) 1/2 and PKC pathways, which act through increasing HIF-1 translation or through regulatory phosphorylation [24, 25] (see chapters 15 & 16 for detailed discussion). The VEGFR2 promoter appears to lack a classical HRE, but has been shown to be regulated by the related HIF-2 [26].

Members of the Ets (E26 transforming sequence in avian erythroblastosis virus) family of transcription factors are expressed in endothelial cells and modify expression of several genes implicated in angiogenesis and inflammation; for example, Ets-1 regulates the expression of VEGFR1 and VEGFR2. Transcriptional activity by Ets is regulated e.g. by Erk1/2-mediated serine phosphorylation as well as through a number of other mechanisms (for a review, see (27)).

Activation of VEGFRs

Binding of VEGF leads to dimerization of receptor molecules followed by activation of the intrinsic tyrosine kinase.

The VEGFRs have been shown to form both homo- and heterodimers in vitro [28–30]. The activated receptor molecules in the dimers transphosphorylate each other, on tyrosine residues. An initial phosphorylation on positive regulatory tyrosine residue(s) in the kinase activation loop precedes full activation of the kinase. This is followed by phosphorylation on other tyrosine residues in the intracellular domain of the receptor to create binding sites for signaling intermediates, thereby initiating signaling cascades.

Numerous tyrosine phosphorylation sites have been identified on VEGFR1 [31], VEGFR2 [32–34], and VEGFR3 [28] (Fig. 18.1). It is noteworthy that VEGFR1 lacks phosphorylation on positive regulatory tyrosine residues [31], due to replacement of a conserved residue in the activation loop, from Asp to Asn at position 1050 [35]. This may explain why VEGFR1 kinase activity is difficult to induce. Positive regulatory tyrosine phosphorylation is found both in VEGFR2 and VEGFR3. Interestingly, for all three VEGF receptors, certain tyrosine phosphorylation sites are used selectively. Thus, for VEGFR1, phosphorylation site usage is dictated by the particular activating VEGF ligand, such as PlGF [13] and VEGF-A [31], which have been shown to induce different phosphorylation site patterns. For VEGFR2, the Y951 phosphorylation site, located in the insert region between the two parts of the kinase domain, is used primarily when the receptor is expressed in endothelial cells engaged in active angiogenesis [34]. Thirdly, C-terminal sites in VEGFR3 are phosphorylated in VEGFR3 homodimers, but not when VEGFR3 is heterodimerized with VEGFR2 [28]. The implication of these findings is that VEGFR tyrosine phosphorylation is both highly dynamic and tightly regulated, in agreement with the versatility of these receptors in endothelial biology.

Trimeric Gαq/Gα11 proteins have been implicated as important regulators of VEGFR2 signaling. Thus, antisense-mediated suppression of Gq/11 expression completely attenuated VEGFR2 tyrosine phosphorylation and signal transduction through a mechanism involving direct association between the receptor and the trimeric G-proteins [36].

Down-regulation of VEGFR Activity

How are VEGFRs turned off in order to halt signal transduction? One important mechanism involves dephosphorylation by phosphotyrosine phosphatases (PTPs; [37]). One interesting example is the receptor type PTP denoted vascular endothelial (VE)-PTP (also denoted PTP receptor type B; PTPRB), which is required for maintenance and remodeling of blood vessels during development [38]. VEGFRs do not appear to be direct substrates for VE-PTP, however. The broadly expressed transmembrane PTP denoted density-enhanced phosphatase-1 (DEP1)/CD148, has been implicated in regulation of endothelial cell junctional integrity and silencing of VEGFR2 in dense cells [39]. Moreover, the Src Homology-2 (SH2) domain-containing PTP SHP-2, which is a ubiquitously expressed

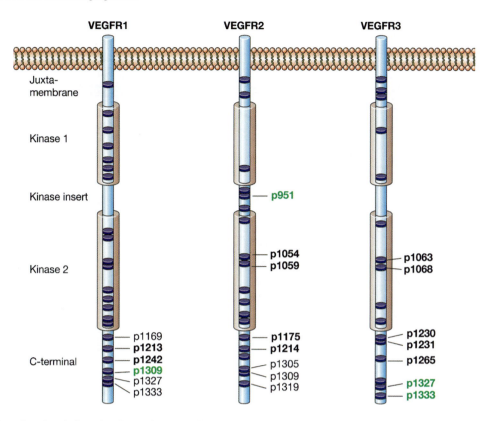

FIG. 18.1. Tyrosine phosphorylation sites in VEGFRs. Activated VEGFRs are represented as single intracellular domains (excised from dimerized receptor complexes). Tyrosine residues are indicated as *blue ovals*. Tyrosine phosphorylation sites are indicated as *p* (phospho) followed by *numbers* which show the position of the residue in the amino acid sequence. *Figures* in *green* indicate that phosphorylation sites are used selectively, i.e. p1309 is phosphorylated when VEGFR1 is activated by PlGF but not by VEGF, 951 is phosphorylated when VEGFR2 is expressed in endothelial cells devoid of a pericyte coat and 1327/1333 are phosphorylated when VEGFR3 is homodimerized and not when it is heterodimerized with VEGFR2 (see text description).

cytoplasmic PTP, has been shown to antagonize VEGF-function, possibly by direct dephosphorylation of VEGFR2 [40].

Another mechanism for clearance of activated VEGFRs involves rapid internalization and/or degradation, which have been studied in different in vitro EC models. Internalization of VEGFR2 expressed in human umbilical vein endothelial cells involves protein kinase C (PKC)-dependent serine phosphorylation on a C-terminal residue [41]. There are several reports that VEGFR2 is recycled to the cell surface from an endosomal compartment when endothelial cells are exposed to VEGF [42, 43]. Ubiquitination of VEGFR2 which at least in vitro involves activation of the ubiquitin ligase Cbl [44], leads to efficient degradation of receptors in the lysosome [42].

VEGF Co-receptors

A co-receptor is here defined as a molecular entity that binds to the ligand as well as to the receptor and thereby modulates the downstream signal transduction. Co-receptors may lack intrinsic enzymatic activity and do not necessarily signal independently of the receptor tyrosine kinase. There are at least two co-receptors for the VEGF/VEGFRs, heparan sulfate (HS) proteoglycans (HSPGs) and neuropilins (NRPs) that fit this definition (Fig. 18.2). HSPGs are composed of a protein backbone modified by attachment of repeated units of sulfated glucosaminoglycans (GAGs) that form long linear sugar chains. GAG sulfation confers a net negative charge, which allows binding to many different growth modulatory factors [45]. In cells lacking heparan sulfate completely, or express defective heparan sulfate with reduced degree of sulfation, there is no response to VEGF even though VEGF receptors are expressed [46]. The mode of presentation of HSPGs (on both endothelial cells and perivascular cells, or only on perivascular cells) determines the level and longevity of receptor activation [47].

Neuropilins (1 and 2) are transmembrane molecules with a short cytoplasmic tail, which lacks intrinsic enzymatic activity [48]. Neuropilins were first identified as negative regulators of neuronal axon guidance through binding of members of the class 3 Semaphorin (Sema) family [49, 50]. Binding of Sema to NRPs allows coupling to plexins, which have established roles in regulating Rho-family GTPases [51]. Neuropilin-1

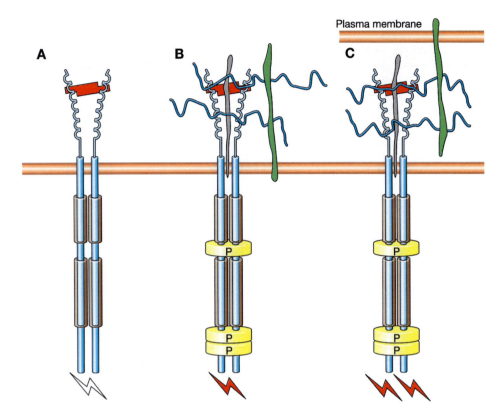

FIG. 18.2. VEGF co-receptor function. **A** VEGFR2 is unable to transduce signals leading to establishment of cellular responses in cells devoid of coreceptors HSPG and NRP1. **B** HRPG (GAG side chains indicate in *blue* and protein part in *green*) or NRP1 (indicated in *grey*) expressed on endothelial cells are engaged in the VEGF/VEGFR signaling complex and may affect signaling quantitatively (by stabilizing the complex) and qualitatively (by allowing transduction of signaling pathway not induced in the absence of coreceptors. **C** Presentation of HSPGs in trans, e.g., on pericytes, leads to further stabilization of the VEGF/VEGFR signaling complex, and prolonged signal transduction (*red activity arrows*).

was subsequently shown to bind exon-7-containing VEGF-A isoforms such as VEGF-A165 [52]. Neuropilin-1 -/- mice die at embryonic day (E)10.5–12.5 due to defects in vascular and neuronal development [53]. It is noteworthy that NRP-1 may be modified by chondroitin and heparan sulfation [54], potentially allowing binding of VEGF to the protein core as well as to the HS-side chains. Through these interactions, NRPs become an integral part of the VEGF/VEGFR signaling complex and may thereby enhance the activity of the VEGFR kinase. Whereas neuropilin-1 is engaged in VEGFR2 signaling, neuropilin-2 interacts with the VEGF-C/VEGFR3 signaling complex in lymphendothelial cells [55]. Gene targeting of neuropilin-2 leads to severe reduction of small lymphatic vessels and capillaries [56, 57].

VEGFR Signal Transduction Pathways

The Phospholipase Cγ (PLCγ) Pathway

Within the family of phospholipases (PLCs), PLCγ1 and 2 are equipped with Src homology 2 (SH2) domains, which confer binding to activated growth factor receptors [58]. PLCγ2 is preferentially expressed in hematopoietic cells, whereas PLCγ1 is ubiquitously expressed and will henceforth be referred to as "PLCγ". PLCγ is a substrate for all three VEGF receptors. Upon binding to phosphorylated Y1173 (1175 in the human) in the VEGFR2 C-terminal tail, PLCγ becomes tyrosine phosphorylated and thereby activated [59] (Fig. 18.3). PLCγ hydrolyzes phosphatidylinositol 4,5 bisphosphate (PI-4,5-P2), a plasma membrane lipid. The hydrolysis results in generation of inositol 1,4,5-P3 and diacylglycerol (DAG), which leads to release of Ca^{2+} from intracellular stores and activation of PKC, respectively. VEGF-induced PLCγ activation has been shown to lead to Ras-independent activation of Erk1/2, via PKC [59]. This pathway has been implicated in VEGF-driven proliferation of endothelial cells in vitro [60]. A knockin mutation replacing Tyr1173 with Phe, thereby removing the PLCγ-binding site, leads to embryonic lethality [61], and overall features that are similar to those observed for the complete receptor knockout [18]. The Y1175 phosphorylation site on VEGFR2 is also a binding site for the adaptor molecules Shb [62] and Sck [63]. It is noteworthy that inactivation of the

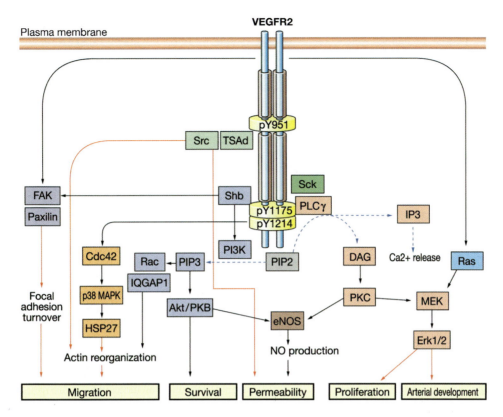

FIG. 18.3. Signal tranduction pathways induced by VEGFR2. Binding of signaling molecules to certain phosphorylation sites (*yellow ovals*) initiate signaling cascades leading to establishment of biological responses (*migration, survival*, etc). *Dashed arrows* indicate enzyme/substrate reactions involving PI-4,5-P2 (*PIP2*), PI 3,4,5-P3 (*PIP3*) or inositol 1,4,5-P3 (*IP3*). Both *Shb, Sck* and *PLCγ1* associate with *pY1175* whereas *TSAd* associates with *pY951*. *DAG* Diacylglycerol; *eNOS* endothelial nitric oxide synthase; *Erk* extracellular regulated kinase; *FAK* focal adhesion kinase; *HSP27* heat shock protein 27; *MAPK* mitogen activated protein kinase; *PI3K* phosphoinositide 3′kinase; *PKC* protein kinase C; *PLCγ*; phospholipase Cγ, SH2 and β-cells; *TSAd* T cell-specific adapter.

murine PLCγ1 gene is accompanied by defective vasculogenesis and erythropoiesis and early embryonic death [64]. Inactivation of the zebra fish PLCγ gene leads to loss in arterial specification [65].

The PI3K Pathway

VEGF is known to be important for endothelial cell survival. Under conditions where VEGF function is interrupted (e.g., by neutralization using VEGF-reactive antibodies, or VEGF receptor kinase inhibitors), fenestrated capillaries in endocrine organs regress [66]. One important mediator of VEGF-dependent survival is the PI3K pathway [67]. The PI3K family of lipid kinases phosphorylate the 3′-hydroxyl group of phosphatidylinositol and phosphoinositides (for a review, see [68]). PI3K is activated downstream of VEGFR2 and VEGFR3, but only indirectly and not through binding of the regulatory p85 subunit directly to tyrosine phosphorylation sites in the receptors (for a review, see [6]). Interestingly, Gille and colleagues identified a repressor sequence in the juxtamembrane of VEGFR1 which, when exchanged for the corresponding

domain from VEGFR2, allowed activation of the PI3K pathway [69]. VEGFR2-dependent activation of PI3K occurs downstream of focal adhesion kinase (FAK) [70] or vascular endothelial (VE)-cadherin [71]. Furthermore, VEGF-induced activation of PI3K is inhibited by siRNA-mediated knockdown of the adaptor molecule Shb [62]. Accumulation of PI 3,4,5-P3 as a result of PI3K activation, in turn allows membrane recruitment of molecules equipped with a pleckstrin homology (PH) domain, notably the serine/threonine (Ser/Thr) kinase Akt (also known as protein kinase B; PKB). The activity of the membrane-localized Akt is stimulated through phosphorylation on Ser/Thr by phosphoinositide-dependent kinases (PDKs), and Target of rapamycin (Tor). Akt promotes endothelial cell survival via regulation of several downstream effectors, such as the the pro-apoptotic Bcl-2 related protein, BAD [72], and the forkhead family of transcription factors (for a review, see [68]).

PI3K-dependent activation of Akt leading to phosphorylation of endothelial nitric oxide synthase (eNOS) and generation of NO, has been implicated in VEGF-induced vascular permeability [73]. Expression in endothelial cells of constitutively active Akt, modified by attachment of a myristyl group that confers constitutive plasma membrane-association, leads to formation of enlarged, hyperpermeable vessels resembling tumor blood vessels [74]. However, although these vessels expressing constitutive active Akt show features of chronic vascular permeability, they still respond to VEGF with increased acute permeability, indicating the complexity of this response.

PI3K also regulates the cytoskeleton by modulating the activity of small GTPases belonging to the Rho family. According to consensus, these are activated via PI-3,4, 5-P3-dependent recruitment of guanine nucleotide exchange factors (GEFs;[75]). However, certain aspects of this regulatory mechanism remain elusive, since the sequence of the PH domain of several of these GEFs do not conform to the consensus sequence for binding of PI3K products. Regulation of the actin cytoskeleton is critically involved in endothelial cell migration and formation of the vascular tube, by controlling filopodia, lamellipodia and actin stress fiber formation.

Interestingly, different aspects of endothelial cell biology seem to require a balanced activity of PLCγ1 versus PI3K. This includes arterial/vein specification (for a review, see [76]) and the formation and stability of the vascular tube [77]. Moreover, activation of eNOS by VEGFR1 and -2, have been shown to involve activation of PI3K and PLCγ, respectively [78]. These lipid modifiers compete for the same substrate, PI-4,5-P2, which may constitute a limiting factor, guiding the relative strengths of downstream signaling pathways.

The SRC Family of Cytoplasmic Tyrosine Kinases

Cytoplasmic Ser/Thr kinases of the Src family are known to be vital for EC function and involved in regulation of the cytoskeletal architecture, which is critical in EC migration and formation of the three-dimensional aspect of the vessel. It is not clear how Src kinases (here primarily indicating the ubiquitously expressed Src, Yes and Fyn kinases) are activated in response to VEGF treatment, e.g., via VEGFR2. One possible mechanism involves the adaptor VEGF receptor associated protein (VRAP; [79]), also denoted TSAd (T cell specific adapter), that binds to phosphorylated Y949/951 in VEGFR2, and which has been shown to associate with Src in a VEGF-regulated manner [34]. An important role for Src appears to be in regulation of endothelial cell junctions and, consequently, endothelial permeability [80]. Interestingly, although endothelial cells express the highly related Src, Fyn and Yes, only Src and Yes appear to be involved in regulation of VEGF-mediated permeability. Thus, gene inactivation of Src or Yes attenuates VEGF-mediated permeability without apparent blockade in other aspects of endothelial cell function [80, 81]. Src appears in complex with VEGFR2 and VE-cadherin, a main component of endothelial cell adherens junctions, and the stability of the complex is an important aspect in vascular leakage [71, 81, 82]. In vitro, VE-cadherin is a substrate for Src [83, 84]. It appears that VEGF causes disruption of the VEGFR2/VE-cadherin complex in a Src-dependent manner, causing release

of VE-cadherin-associated β-catenin. Thereby, β-catenin signaling becomes integrated in VEGF-induced responses [81]. For further discussion on signal transduction in vascular permeability, see below.

Endothelial Cell Biology and Signal Transduction

Endothelial Cell Survival and Proliferation

VEGF is an important survival factor for endothelial cells, through activation of the Akt pathway (see above). To what extent VEGF induces EC proliferation in primary cells or in vivo, is not clear, although several different pathways for VEGF-regulated endothelial cell DNA synthesis have been suggested. By mutation of the PLCγ-binding site in VEGFR2, and by use of PLCγ neutralizing antibodies, it has been demonstrated that the PLCγ-MEK/Erk pathways is critical for VEGF-induced DNA synthesis [60]. These data have been corroborated by exchange of the VEGFR2 Y1173 residue in vivo, which leads to early embryonic lethality [61]. It is noteworthy that Y1173/1175 is a binding site also for other signal transduction molecules, moreover, at least in certain cell types, VEGF-induced DNA synthesis appears to involve activation of Ras [85, 86].

Actin Cytoskeleton, Migration and Formation of the Vascular Tube

Several different signal transduction pathways have been implicated in regulation of endothelial cell actin cytoskeleton and migration. The migratory response is a complex series of coordinated events involving mechanisms for orientation of the movement, as well as release and formation of new contacts with the underlying substrate. It is conceivable that different signaling pathways regulate different stages in cell migration. It is furthermore likely that pathways that have been implicated in regulation of the cell cytoskeleton and migration in many different cell types are important also in ECs. Thus, FAK and its substrate paxillin which are involved in focal adhesion turnover during cell migration [87] are known to be activated in VEGF-stimulated endothelial cells. Other pathways implicated in cell migration such as PI3K, Src and small GTPases (Rac and cdc42) are also regulated by VEGF during EC migration (see below).

Removal of the binding site for the adaptor molecule TSAd by mutation of the Y949/951 phosphorylation site in VEGFR2 to Phe, or siRNA-mediated down-regulation of TSAd, attenuates VEGF-induced actin reorganization and migration of endothelial cells [34]. Another pathway implicated in VEGF-induced actin reorganization and EC migration is dependent on binding of the adaptor molecule Shb to Y1173/1175 in VEGFR2. VEGF induces tyrosine phosphorylation of Shb in a Src-dependent manner, allowing further downstream coupling to PI3K and FAK activation [62]. Endothelial cell motility appears furthermore to be regulated via a newly identified binding partner of phosphorylated VEGFR2, IQGAP1, which binds to and activates Rac1 [88]. Phosphorylation of Y1212/1214 in the C-terminal tail of VEGFR2 has been implicated in VEGF-induced actin remodelling via triggering of a signaling complex involving the adaptor Nck, the cytoplasmic tyrosine kinase Fyn and p21-activated kinase (PAK)-2, which in turn promotes activation of Cdc42 and p38 mitrogen associated protein kinase (MAPK) [89]. Cdc42 is a Rho-family GTPase regulating cytoskeletal remodelling [90], and p38MAPK has been implicated in neoangiogenesis [91], by phosphorylating Hsp27, a heat-shock protein involved in VEGF-induced actin reorganization and migration. Nck may also regulate the formation of focal adhesions [92]. Thus, VEGF induces a plethora of responses that all have the ability to cause cytoskeletal rearrangements and cell migration.

Arterial-Vein Specification

Arteries and veins are structurally and functionally distinct. Arteries, which have to tolerate large changes in blood pressure generated by the contractions of the heart muscle, develop a thicker outer coat with several layers of smooth muscle cells. In contrast, veins have a thin smooth muscle cell coat, which locally may fail to cover the vessel. The blood pressure as well the blood flow have been considered important in establishing such differences and, consequently, in arterial/vein specification [93]. Arteries and veins also differ with regard to gene expression patterns, such as the transmembrane ligand ephrin B2 and its receptor tyrosine kinase EphB4, which are expressed on arteries and veins, respectively [94]. Interestingly, a balance in the PI3K and PLCγ pathways appears to direct the development of arteries. Activation of Erk1/2 downstream of PLCγ is required for arterial development, and this pathway is opposed by activation of PI3K [95]. Erk activity may be essential in VEGF-mediated induction of the Notch signaling pathways, in an as yet unidentified circuit. Notch ligands and receptors are required for arterial development downstream of VEGF/VEGFR2 (for reviews, see refs. [76, 96]).

Signal Transduction in Vascular Permeability

VEGF-A was originally discovered as a vascular permeability factor (VPF) [97, 98]. VEGF-A is unique in its ability to induce permeability with unusual potency and kinetics without involvement of mast cell degranulation or endothelial cell damage. Pathological angiogenesis, such as in cancer, is often accompanied by vascular permeability, leading to formation of edema and ascites, and allowing the distant spread of metastases. These serious clinical complications have led to an intense interest in determining the signal transduction pathways regulating vascular permeability. Thus far, many different pathways

have been implicated, possibly due to the involvement of more than one mechanism, or because different endothelial cell types are differently regulated. One principal mechanism appears to involve assembly of vesiculo-vacuolar organelles (VVOs) to form trans-endothelial pores, allowing passage of large molecules [99]. Another mechanism involves loosening of endothelial cell-cell junctions. Endothelial adherens junctions are crucial for the maintenance and regulation of normal microvascular function [100]. The major cell–cell adhesion molecule in adherens junctions (also denoted zonula adherens) is VE-cadherin. VE-cadherin gene inactivation allows formation of a primitive vascular plexus, but vascular remodeling is deficient leading to an early lethal phenotype [71]. Adherens junctions are destabilized by VEGFR2-induced activation of the cytoplasmic kinases Src and Yes, leading to dissociation of a VEGFR2/VE-cadherin complex (Fig. 18.4). This complex is central to the integrity of the junction. It has also been implicated in regulation of the signaling strength of VEGFR2, via receptor dephosphorylation by junctional phosphatases, which would be an underlying mechanism in contact-inhibition of growth [39]. A critical role for Src in permeability is indicated by the fact that Src blockade stabilizes the VEGFR2/VE-cadherin complex, reducing edema and tissue injury following myocardial infarction [81]. However, although VE-cadherin indeed is a substrate for Src, loosening of adherens junctions may not require tyrosine phosphorylation of VE-cadherin. Instead, VEGF has been shown to induce rapid endocytosis of VE-cadherin in a pathway involving Src-mediated tyrosine phosphorylation and activation of Vav2, which is a GEF for Rac. Via activation of the p21-activated kinase (PAK), Rac in turn induces serine phosphorylation of VE-cadherin, which triggers endocytosis [101].

VEGF-induced vascular permeability also depends on nitric oxide (NO) production, which requires activation of eNOS either as a consequence of PLCγ activation and calcium influx, or through phosphorylation of eNOS by PKB/Akt [73, 102]. In agreement, targeted deletion of eNOS abrogates VEGF-induced vascular permeability [103].

Studies on different VEGF family members that preferentially bind to one or both of VEGFR1 and −2 indicate that vascular permeability involves both VEGFR receptors. A member of the snake venom VEGF-Fs, (*T. flavoviridis* svVEGF), is a potent inducer of vascular permeability [104]. *T.f.* svVEGF preferentially binds and activates VEGFR1, but also to some extent, VEGFR2. These results may suggest that cooperative action of VEGFR1 and VEGFR2 may be critical for generation of vascular permeability.

Changes in VEGF Signaling as a Consequence of Changes in the Endothelial Microenvironment

Targeted inactivation of the *vegf-a* gene has shown that the level of VEGF-A expression, and thus the amplitude or duration of signaling during vasculogenesis is critical, as deletion of only one allele leads to embryonic death [105, 106]. Interestingly,

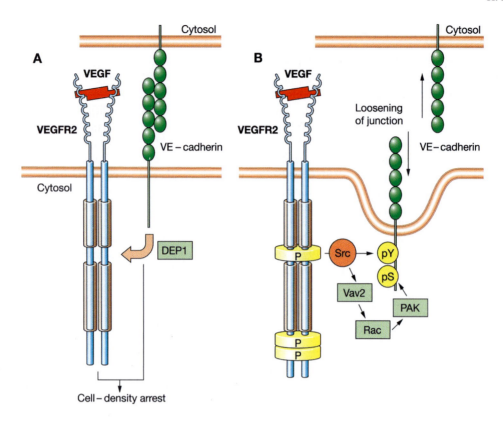

FIG. 18.4. Role of VE-cadherin and VEGFR2 interactions. **A** In dense cells, the density enhanced phosphatase (*DEP*)-1 is regulated via VE-cadherin, leading to dephosphorylation of VEGFR2 and attenuates endothelial cell proliferation. **B** Vascular permeability involves loosening of adherens junctions through disruption of the VEGFR2/VE-cadherin complex followed by Src-dependent tyrosine (*Y*) as well as *PAK* (p21 activated kinase)-dependent serine (*S*) phosphorylation of VE-cadherin.

different levels of VEGF may be required for induction of different VEGFR2-dependent processes [107]. Thus, tissues vascularized through vasculogenesis, such as spleen and lung, express higher levels of VEGF-A than tissues vascularized through angiogenesis such as brain [107]. Whether different levels of VEGF induce qualitatively distinct signals remains to be shown. Local changes in the microenvironment may lead to altered VEGF expression/signaling, possibly as part of an adaptive response. Such examples have been delineated by knockout studies. During embryonic development of mice lacking expression of either the platelet-derived growth factor-B or the PDGF-β receptor as a result of gene targeting, endothelial cells fail to attract pericytes in several organs, including the brain [108]. VEGF-A becomes up-regulated in these embryos, leading to the appearance of endothelial cells with an altered ultrastructure. VEGF-A gene expression is also increased in β3-integrin null mice in the male heart, with abnormal endothelial ultrastructure as a consequence [108].

Blood flow and shear stress may induce VEGFR activation and signal transduction independently of VEGF-binding. Thus, a mechanosensory complex consisting of VEGFR2, platelet-endothelial cell adhesion molecule-1 (PECAM-1) and VE-cadherin has been described [109]. Formation of this complex induces integrin activation and signal transduction.

It is quite well established that many aspects of endothelial cell biology involve convergence of integrin and VEGFR signaling pathways (for a review, see [110]). Other signaling systems also converge with that of the VEGFRs, such as the angiopoietin/Tie receptor complex, which plays an essential role in angiogenic remodeling. A characteristic feature of the angiopoietin ligands is their opposing effects on Tie2 receptor activation, where the eventual endothelial cell response is influenced by concurrent VEGFR signaling (for a review, see ref. [111]). Furthermore, signal transduction by TGF-β family of growth factors and receptors strongly modulates VEGF-induced angiogenesis [112]. Thus, VEGF-signaling is context-dependent and is influenced by cell–cell and cell–matrix interactions and by a wealth of soluble and cell-associated mediators of different kinds. One recent example is the Notch family of ligands and receptors which are critical in VEGF-dependent formation of tip cells in angiogenic sprouts [113].

Conclusions

The recent advances in development of VEGF-targeted therapy to arrest deregulated vascularization in conjunction with chronic inflammation and cancer have clearly demonstrated the critical in vivo contribution of VEGFR signaling in angiogenesis. Moreover, such therapy may be further developed, as combined neuropilin-1 and VEGF-neutralization appears superior in arresting tumor vascularization compared to VEGF-neutralization alone [114]. From studies on the consequence of VEGF-neutralization on normal vasculature, however, it seems that VEGF has a prominent and perhaps unique role in mediating endothelial cell survival in quiescent vessels [66]. Therefore, current VEGF-targeted therapy needs to be refined to preserve VEGF-induced survival, while suppressing other VEGF-induced responses such as permeability and formation of the three-dimensional vascular tube. Clearly, future antiangiogenic therapy will benefit from the steady accumulation of data on signaling in endothelial cells.

Acknowledgements. The authors acknowledge funding for their work from the Swedish Cancer foundation, the Swedish Research Council and the 6th frame work EU programs Angiotargeting (LSHG-CT-2004-504743) and Lymphangiogenomics (LSHG-CT-2004-503573).

References

1. Ferrara N, Gerber HP, LeCouter J. The biology of VEGF and its receptors. Nat Med 2003;9(6):669–76.

2. Woolard J, Wang WY, Bevan HS, et al. VEGF165b, an inhibitory vascular endothelial growth factor splice variant: mechanism of action, in vivo effect on angiogenesis and endogenous protein expression. Cancer Res 2004;64(21):7822–35.

3. Ogawa S, Oku A, Sawano A, Yamaguchi S, Yazaki Y, Shibuya M. A novel type of vascular endothelial growth factor, VEGF-E (NZ-7 VEGF), preferentially utilizes KDR/Flk-1 receptor and carries a potent mitotic activity without heparin-binding domain. J Biol Chem 1998;273(47):31273–82.

4. Wise LM, Veikkola T, Mercer AA, et al. Vascular endothelial growth factor (VEGF)-like protein from orf virus NZ2 binds to VEGFR2 and neuropilin-1. Proc Natl Acad Sci USA 1999;96(6):3071–6.

5. Yamazaki Y, Tokunaga Y, Takani K, Morita T. C-terminal heparin-binding peptide of snake venom VEGF specifically blocks VEGF-stimulated endothelial cell proliferation. Pathophysiol Haemost Thromb 2005;34(4–5):197–9.

6. Olsson AK, Dimberg A, Kreuger J, Claesson-Welsh L. VEGF receptor signalling - in control of vascular function. Nat Rev Mol Cell Biol 2006;7(5):359–71.

7. Kendall RL, Thomas KA. Inhibition of vascular endothelial cell growth factor activity by an endogenously encoded soluble receptor. Proc Natl Acad Sci USA 1993;90(22):10705–9.

8. Shibuya M. Structure and dual function of vascular endothelial growth factor receptor-1 (Flt-1). Int J Biochem Cell Biol 2001;33(4):409–20.

9. Tsatsaris V, Goffin F, Munaut C, et al. Overexpression of the soluble vascular endothelial growth factor receptor in preeclamptic patients: pathophysiological consequences. J Clin Endocrinol Metab 2003;88(11):5555–63.

10. Fong GH, Rossant J, Gertsenstein M, Breitman ML. Role of the Flt-1 receptor tyrosine kinase in regulating the assembly of vascular endothelium. Nature 1995;376(6535):66–70.

11. Zeng H, Dvorak HF, Mukhopadhyay D. Vascular permeability factor (VPF)/vascular endothelial growth factor (VEGF) peceptor-1 down-modulates VPF/VEGF receptor-2-mediated endothelial cell proliferation, but not migration, through phosphatidylinositol 3-kinase-dependent pathways. J Biol Chem 2001;276(29):26969–79.

12. Rahimi N, Dayanir V, Lashkari K. Receptor chimeras indicate that the vascular endothelial growth factor receptor-1 (VEGFR1) modulates mitogenic activity of VEGFR2 in endothelial cells. J Biol Chem 2000;275(22):16986–92.

13. Autiero M, Waltenberger J, Communi D, et al. Role of PlGF in the intra- and intermolecular cross talk between the VEGF receptors Flt1 and Flk1. Nat Med 2003;9(7):936–43.

14. Takahashi H, Shibuya M. The vascular endothelial growth factor (VEGF)/VEGF receptor system and its role under physiological and pathological conditions. Clin Sci (Lond) 2005;109(3):227–41.

15. Kabrun N, Buhring HJ, Choi K, Ullrich A, Risau W, Keller G. Flk-1 expression defines a population of early embryonic hematopoietic precursors. Development 1997;124(10):2039–48.

16. Kaipainen A, Korhonen J, Pajusola K, et al. The related FLT4, FLT1, and KDR receptor tyrosine kinases show distinct expression patterns in human fetal endothelial cells. J Exp Med 1993;178(6):2077–88.

17. Hatva E, Kaipainen A, Mentula P, et al. Expression of endothelial cell-specific receptor tyrosine kinases and growth factors in human brain tumors. Am J Pathol 1995;146(2):368–78.

18. Shalaby F, Rossant J, Yamaguchi TP, et al. Failure of blood-island formation and vasculogenesis in Flk-1-deficient mice. Nature 1995;376(6535):62–6.

19. Tammela T, Petrova TV, Alitalo K. Molecular lymphangiogenesis: new players. Trends Cell Biol 2005;15(8):434–41.

20. Partanen TA, Arola J, Saaristo A, et al. VEGF-C and VEGF-D expression in neuroendocrine cells and their receptor, VEGFR-3, in fenestrated blood vessels in human tissues. Faseb J 2000;14(13):2087–96.

21. Hamrah P, Chen L, Zhang Q, Dana MR. Novel expression of vascular endothelial growth factor receptor (VEGFR)-3 and VEGF-C on corneal dendritic cells. Am J Pathol 2003;163(1):57–68.

22. Dumont DJ, Jussila L, Taipale J, et al. Cardiovascular failure in mouse embryos deficient in VEGF receptor-3. Science 1998;282(5390):946–9.

23. Karkkainen MJ, Ferrell RE, Lawrence EC, et al. Missense mutations interfere with VEGFR-3 signalling in primary lymphoedema. Nat Genet 2000;25(2):153–9.

24. Dery MA, Michaud MD, Richard DE. Hypoxia-inducible factor 1: regulation by hypoxic and non-hypoxic activators. Int J Biochem Cell Biol 2005;37(3):535–40.

25. Pages G, Pouyssegur J. Transcriptional regulation of the Vascular Endothelial Growth Factor gene–a concert of activating factors. Cardiovasc Res 2005;65(3):564–73.

26. Elvert G, Kappel A, Heidenreich R, et al. Cooperative interaction of hypoxia-inducible factor-2alpha (HIF-2alpha) and Ets-1 in the transcriptional activation of vascular endothelial growth factor receptor-2 (Flk-1). J Biol Chem 2003;278(9):7520–30.

27. Oettgen P. Regulation of vascular inflammation and remodeling by ETS factors. Circ Res 2006;99(11):1159–66.

28. Dixelius J, Makinen T, Wirzenius M, et al. Ligand-induced vascular endothelial growth factor receptor-3 (VEGFR-3) heterodimerization with VEGFR2 in primary lymphatic endothelial

cells regulates tyrosine phosphorylation sites. J Biol Chem 2003;278(42):40973–9.

29. Ito N, Claesson-Welsh L. Dual effects of heparin on VEGF binding to VEGF receptor-1 and transduction of biological responses. Angiogenesis 1999;3(2):159–66.

30. Huang K, Andersson C, Roomans GM, Ito N, Claesson-Welsh L. Signaling properties of VEGF receptor-1 and -2 homo- and heterodimers. Int J Biochem Cell Biol 2001;33(4):315–24.

31. Ito N, Wernstedt C, Engstrom U, Claesson-Welsh L. Identification of vascular endothelial growth factor receptor-1 tyrosine phosphorylation sites and binding of SH2 domain-containing molecules. J Biol Chem 1998;273(36):23410–8.

32. Cunningham SA, Arrate MP, Brock TA, Waxham MN. Interactions of FLT-1 and KDR with phospholipase C gamma: identification of the phosphotyrosine binding sites. Biochem Biophys Res Commun 1997;240(3):635–9.

33. Dougher-Vermazen M, Hulmes JD, Bohlen P, Terman BI. Biological activity and phosphorylation sites of the bacterially expressed cytosolic domain of the KDR VEGF-receptor. Biochem Biophys Res Commun 1994;205(1):728–38.

34. Matsumoto T, Bohman S, Dixelius J, et al. VEGF receptor-2 Y951 signaling and a role for the adapter molecule TSAd in tumor angiogenesis. Embo J 2005;24(13):2342–53.

35. Meyer RD, Mohammadi M, Rahimi N. A single amino acid substitution in the activation loop defines the decoy characteristic of VEGFR1/FLT-1. J Biol Chem 2006;281(2):867–75.

36. Zeng H, Zhao D, Yang S, Datta K, Mukhopadhyay D. Heterotrimeric G alpha q/G alpha 11 proteins function upstream of vascular endothelial growth factor (VEGF) receptor-2 (KDR) phosphorylation in vascular permeability factor/VEGF signaling. J Biol Chem 2003;278(23):20738–45.

37. Kappert K, Peters KG, Bohmer FD, Ostman A. Tyrosine phosphatases in vessel wall signaling. Cardiovasc Res 2005;65(3):587–98.

38. Baumer S, Keller L, Holtmann A, et al. Vascular endothelial cell-specific phosphotyrosine phosphatase (VE-PTP) activity is required for blood vessel development. Blood 2006;107(12):4754–62.

39. Lampugnani MG, Orsenigo F, Gagliani MC, Tacchetti C, Dejana E. Vascular endothelial cadherin controls VEGFR2 internalization and signaling from intracellular compartments. J Cell Biol 2006;174(4):593–604.

40. Gallicchio M, Mitola S, Valdembri D, et al. Inhibition of vascular endothelial growth factor receptor 2-mediated endothelial cell activation by Axl tyrosine kinase receptor. Blood 2005;105(5):1970–6.

41. Singh AJ, Meyer RD, Band H, Rahimi N. The carboxyl terminus of VEGFR2 is required for PKC-mediated down-regulation. Mol Biol Cell 2005;16(4):2106–18.

42. Ewan LC, Jopling HM, Jia H, et al. Intrinsic tyrosine kinase activity is required for vascular endothelial growth factor receptor 2 ubiquitination, sorting and degradation in endothelial cells. Traffic 2006;7(9):1270–82.

43. Gampel A, Moss L, Jones MC, Brunton V, Norman JC, Mellor H. VEGF regulates the mobilization of VEGFR2/KDR from an intracellular endothelial storage compartment. Blood 2006;108(8):2624–31.

44. Duval M, Bedard-Goulet S, Delisle C, Gratton JP. Vascular endothelial growth factor-dependent down-regulation of Flk-1/KDR involves Cbl-mediated ubiquitination. Consequences on nitric oxide production from endothelial cells. J Biol Chem 2003;278(22):20091–7.

45. Esko JD, Selleck SB. Order out of chaos: assembly of ligand binding sites in heparan sulfate. Annu Rev Biochem 2002;71:435–71.

46. Dougher AM, Wasserstrom H, Torley L, et al. Identification of a heparin binding peptide on the extracellular domain of the KDR VEGF receptor. Growth Factors 1997;14(4):257–68.

47. Jakobsson L, J. K, K. H, et al. Heparan sulfate in trans potentiates VEGFR-mediated angiogenesis. Dev Cell 2006.

48. Fujisawa H, Kitsukawa T, Kawakami A, Takagi S, Shimizu M, Hirata T. Roles of a neuronal cell-surface molecule, neuropilin, in nerve fiber fasciculation and guidance. Cell Tissue Res 1997;290(2):465–70.

49. Kolodkin AL, Levengood DV, Rowe EG, Tai YT, Giger RJ, Ginty DD. Neuropilin is a semaphorin III receptor. Cell 1997;90(4):753–62.

50. He Z, Tessier-Lavigne M. Neuropilin is a receptor for the axonal chemorepellent Semaphorin III. Cell 1997;90(4):739–51.

51. Kruger RP, Aurandt J, Guan KL. Semaphorins command cells to move. Nat Rev Mol Cell Biol 2005;6(10):789–800.

52. Soker S, Takashima S, Miao HQ, Neufeld G, Klagsbrun M. Neuropilin-1 is expressed by endothelial and tumor cells as an isoform-specific receptor for vascular endothelial growth factor. Cell 1998;92(6):735–45.

53. Kitsukawa T, Shimizu M, Sanbo M, et al. Neuropilin-semaphorin III/D-mediated chemorepulsive signals play a crucial role in peripheral nerve projection in mice. Neuron 1997;19(5):995–1005.

54. Shintani Y, Takashima S, Asano Y, et al. Glycosaminoglycan modification of neuropilin-1 modulates VEGFR2 signaling. Embo J 2006;25(13):3045–55.

55. Karpanen T, Heckman CA, Keskitalo S, et al. Functional interaction of VEGF-C and VEGF-D with neuropilin receptors. Faseb J 2006;20(9):1462–72.

56. Giger RJ, Cloutier JF, Sahay A, et al. Neuropilin-2 is required in vivo for selective axon guidance responses to secreted semaphorins. Neuron 2000;25(1):29–41.

57. Yuan L, Moyon D, Pardanaud L, et al. Abnormal lymphatic vessel development in neuropilin 2 mutant mice. Development 2002;129(20):4797–806.

58. Carpenter G, Ji Q. Phospholipase C-gamma as a signal-transducing element. Exp Cell Res 1999;253(1):15–24.

59. Takahashi T, Ueno H, Shibuya M. VEGF activates protein kinase C-dependent, but Ras-independent Raf-MEK-MAP kinase pathway for DNA synthesis in primary endothelial cells. Oncogene 1999;18(13):2221–30.

60. Takahashi T, Yamaguchi S, Chida K, Shibuya M. A single autophosphorylation site on KDR/Flk-1 is essential for VEGF-A-dependent activation of PLC-gamma and DNA synthesis in vascular endothelial cells. Embo J 2001;20(11):2768–78.

61. Sakurai Y, Ohgimoto K, Kataoka Y, Yoshida N, Shibuya M. Essential role of Flk-1 (VEGF receptor 2) tyrosine residue 1173 in vasculogenesis in mice. Proc Natl Acad Sci U S A 2005;102(4):1076–81.

62. Holmqvist K, Cross MJ, Rolny C, et al. The adaptor protein shb binds to tyrosine 1175 in vascular endothelial growth factor (VEGF) receptor-2 and regulates VEGF-dependent cellular migration. J Biol Chem 2004;279(21):22267–75.

63. Warner AJ, Lopez-Dee J, Knight EL, Feramisco JR, Prigent SA. The Shc-related adaptor protein, Sck, forms a complex with the vascular-endothelial-growth-factor receptor KDR in transfected cells. Biochem J 2000;347(Pt 2):501–9.

64. Liao HJ, Kume T, McKay C, Xu MJ, Ihle JN, Carpenter G. Absence of erythrogenesis and vasculogenesis in Plcg1-deficient mice. J Biol Chem 2002;277(11):9335–41.

65. Lawson ND, Mugford JW, Diamond BA, Weinstein BM. phospholipase C gamma-1 is required downstream of vascular endothelial growth factor during arterial development. Genes Dev 2003;17(11):1346–51.

66. Kamba T, Tam BY, Hashizume H, et al. VEGF-dependent plasticity of fenestrated capillaries in the normal adult microvasculature. Am J Physiol Heart Circ Physiol 2006;290(2):H560–76.

67. Brader S, Eccles SA. Phosphoinositide 3-kinase signalling pathways in tumor progression, invasion and angiogenesis. Tumori 2004;90(1):2–8.

68. Engelman JA, Luo J, Cantley LC. The evolution of phosphatidylinositol 3-kinases as regulators of growth and metabolism. Nat Rev Genet 2006;7(8):606–19.

69. Gille H, Kowalski J, Yu L, et al. A repressor sequence in the juxtamembrane domain of Flt-1 (VEGFR1) constitutively inhibits vascular endothelial growth factor-dependent phosphatidylinositol 3′-kinase activation and endothelial cell migration. Embo J 2000;19(15):4064–73.

70. Qi JH, Claesson-Welsh L. VEGF-induced activation of phosphoinositide 3-kinase is dependent on focal adhesion kinase. Exp Cell Res 2001;263(1):173–82.

71. Carmeliet P, Lampugnani MG, Moons L, et al. Targeted deficiency or cytosolic truncation of the VE-cadherin gene in mice impairs VEGF-mediated endothelial survival and angiogenesis. Cell 1999;98(2):147–57.

72. Downward J. PI 3-kinase, Akt and cell survival. Semin Cell Dev Biol 2004;15(2):177–82.

73. Fulton D, Gratton JP, McCabe TJ, et al. Regulation of endothelium-derived nitric oxide production by the protein kinase Akt. Nature 1999;399(6736):597–601.

74. Phung TL, Ziv K, Dabydeen D, et al. Pathological angiogenesis is induced by sustained Akt signaling and inhibited by rapamycin. Cancer Cell 2006;10(2):159–70.

75. Welch HC, Coadwell WJ, Stephens LR, Hawkins PT. Phosphoinositide 3-kinase-dependent activation of Rac. FEBS Lett 2003;546(1):93–7.

76. Lamont RE, Childs S. MAPping out arteries and veins. Sci STKE 2006;2006(355):pe39.

77. Im E, Kazlauskas A. Regulating angiogenesis at the level of PtdIns-4,5-P2. Embo J 2006;25(10):2075–82.

78. Ahmad S, Hewett PW, Wang P, et al. Direct evidence for endothelial vascular endothelial growth factor receptor-1 function in nitric oxide-mediated angiogenesis. Circ Res 2006;99(7):715–22.

79. Wu LW, Mayo LD, Dunbar JD, et al. VRAP is an adaptor protein that binds KDR, a receptor for vascular endothelial cell growth factor. J Biol Chem 2000;275(9):6059–62.

80. Eliceiri BP, Paul R, Schwartzberg PL, Hood JD, Leng J, Cheresh DA. Selective requirement for Src kinases during VEGF-induced angiogenesis and vascular permeability. Mol Cell 1999;4(6):915–24.

81. Weis S, Shintani S, Weber A, et al. Src blockade stabilizes a Flk/cadherin complex, reducing edema and tissue injury following myocardial infarction. J Clin Invest 2004;113(6):885–94.

82. Dejana E, Spagnuolo R, Bazzoni G. Interendothelial junctions and their role in the control of angiogenesis, vascular permeability and leukocyte transmigration. Thromb Haemost 2001;86(1):308–15.

83. Wallez Y, Cand F, Cruzalegui F, et al. Src kinase phosphorylates vascular endothelial-cadherin in response to vascular endothelial growth factor: identification of tyrosine 685 as the unique target site. Oncogene 2007;26(7):1067–77.

84. Lambeng N, Wallez Y, Rampon C, et al. Vascular endothelial-cadherin tyrosine phosphorylation in angiogenic and quiescent adult tissues. Circ Res 2005;96(3):384–91.

85. Shu X, Wu W, Mosteller RD, Broek D. Sphingosine kinase mediates vascular endothelial growth factor-induced activation of ras and mitogen-activated protein kinases. Mol Cell Biol 2002;22(22):7758–68.

86. Meadows KN, Bryant P, Pumiglia K. Vascular endothelial growth factor induction of the angiogenic phenotype requires Ras activation. J Biol Chem 2001;276(52):49289–98.

87. Parsons JT, Martin KH, Slack JK, Taylor JM, Weed SA. Focal adhesion kinase: a regulator of focal adhesion dynamics and cell movement. Oncogene 2000;19(49):5606–13.

88. Yamaoka-Tojo M, Ushio-Fukai M, Hilenski L, et al. IQGAP1, a novel vascular endothelial growth factor receptor binding protein, is involved in reactive oxygen species–dependent endothelial migration and proliferation. Circ Res 2004;95(3):276–83.

89. Lamalice L, Houle F, Huot J. Phosphorylation of Tyr1214 within VEGFR2 triggers the recruitment of Nck and activation of Fyn leading to SAPK2/p38 activation and endothelial cell migration in response to VEGF. J Biol Chem 2006;281(45):34009–20.

90. Ridley AJ. Rho GTPases and actin dynamics in membrane protrusions and vesicle trafficking. Trends Cell Biol 2006;16(10):522–9.

91. Matsumoto T, Turesson I, Book M, Gerwins P, Claesson-Welsh L. p38 MAP kinase negatively regulates endothelial cell survival, proliferation, and differentiation in FGF-2-stimulated angiogenesis. J Cell Biol 2002;156(1):149–60.

92. Stoletov KV, Gong C, Terman BI. Nck and Crk mediate distinct VEGF-induced signaling pathways that serve overlapping functions in focal adhesion turnover and integrin activation. Exp Cell Res 2004;295(1):258–68.

93. le Noble F, Fleury V, Pries A, Corvol P, Eichmann A, Reneman RS. Control of arterial branching morphogenesis in embryogenesis: go with the flow. Cardiovasc Res 2005;65(3):619–28.

94. Shin D, Garcia-Cardena G, Hayashi S, et al. Expression of ephrinB2 identifies a stable genetic difference between arterial and venous vascular smooth muscle as well as endothelial cells, and marks subsets of microvessels at sites of adult neovascularization. Dev Biol 2001;230(2):139–50.

95. Hong CC, Peterson QP, Hong JY, Peterson RT. Artery/vein specification is governed by opposing phosphatidylinositol-3 kinase and MAP kinase/ERK signaling. Curr Biol 2006;16(13):1366–72.

96. Weinstein BM, Lawson ND. Arteries, veins, Notch, and VEGF. Cold Spring Harb Symp Quant Biol 2002;67:155–62.

97. Dvorak HF. Discovery of vascular permeability factor (VPF). Exp Cell Res 2006;312(5):522–6.

98. Senger DR, Galli SJ, Dvorak AM, Perruzzi CA, Harvey VS, Dvorak HF. Tumor cells secrete a vascular permeability factor that promotes accumulation of ascites fluid. Science 1983;219(4587):983–5.

99. Dvorak AM, Feng D. The vesiculo-vacuolar organelle (VVO). A new endothelial cell permeability organelle. J Histochem Cytochem 2001;49(4):419–32.

100. Bazzoni G, Dejana E. Endothelial cell-to-cell junctions: molecular organization and role in vascular homeostasis. Physiol Rev 2004;84(3):869–901.

101. Gavard J, Gutkind JS. VEGF controls endothelial-cell permeability by promoting the beta-arrestin-dependent endocytosis of VE-cadherin. Nat Cell Biol 2006;8(11):1223–34.

102. Bates DO, Harper SJ. Regulation of vascular permeability by vascular endothelial growth factors. Vascul Pharmacol 2003;39(4–5):225–37.

103. Fukumura D, Gohongi T, Kadambi A, et al. Predominant role of endothelial nitric oxide synthase in vascular endothelial growth factor-induced angiogenesis and vascular permeability. Proc Natl Acad Sci U S A 2001;98(5):2604–9.

104. Takahashi H, Hattori S, Iwamatsu A, Takizawa H, Shibuya M. A novel snake venom vascular endothelial growth factor (VEGF) predominantly induces vascular permeability through preferential signaling via VEGF receptor-1. J Biol Chem 2004;279(44):46304–14.

105. Carmeliet P, Ferreira V, Breier G, et al. Abnormal blood vessel development and lethality in embryos lacking a single VEGF allele. Nature 1996;380(6573):435–9.

106. Ferrara N, Carver-Moore K, Chen H, et al. Heterozygous embryonic lethality induced by targeted inactivation of the VEGF gene. Nature 1996;380(6573):439–42.

107. Miquerol L, Gertsenstein M, Harpal K, Rossant J, Nagy A. Multiple developmental roles of VEGF suggested by a LacZ-tagged allele. Dev Biol 1999;212(2):307–22.

108. Hellstrom M, Gerhardt H, Kalen M, et al. Lack of pericytes leads to endothelial hyperplasia and abnormal vascular morphogenesis. J Cell Biol 2001;153(3):543–53.

109. Tzima E, Irani-Tehrani M, Kiosses WB, et al. A mechanosensory complex that mediates the endothelial cell response to fluid shear stress. Nature 2005;437(7057):426–31.

110. Serini G, Valdembri D, Bussolino F. Integrins and angiogenesis: A sticky business. Exp Cell Res 2006;312(5):651–8.

111. Eklund L, Olsen BR. Tie receptors and their angiopoietin ligands are context-dependent regulators of vascular remodeling. Exp Cell Res 2006;312(5):630–41.

112. Scharpfenecker M, van Dinther M, Liu Z, et al. BMP-9 signals via ALK1 and inhibits bFGF-induced endothelial cell proliferation and VEGF-stimulated angiogenesis. J Cell Sci 2007.

113. Hellstrom M, Phng LK, Hofmann JJ, et al. Dll4 signalling through Notch1 regulates formation of tip cells during angiogenesis. Nature 2007;445(7129):776–80.

114. Pan Q, Chanthery Y, Liang WC, et al. Blocking neuropilin-1 function has an additive effect with anti-VEGF to inhibit tumor growth. Cancer Cell 2007;11(1):53–67.

Chapter 19
Delta-Like Ligand 4/Notch Pathway in Tumor Angiogenesis

Gavin Thurston, Irene Noguera-Troise, Ivan B. Lobov, Christopher Daly, John S. Rudge, Nicholas W. Gale, Stanley J. Wiegand, and George D. Yancopoulos

Keywords: Delta-like ligand 4 (Dll4), notch signaling, developmental angiogenesis

Abstract: The process of angiogenesis, be it physiological or pathological, requires the coordinated interplay of a variety of vascular growth factor systems. Many preclinical models, and more recently several clinical trials, have shown that vascular endothelial growth factor (VEGF) is an essential mediator of developmental and pathological angiogenesis. Recent evidence suggests that another pathway—the Delta/ Notch pathway, and the Delta-like ligand 4 (Dll4) in particular—also plays a specific and critical role in angiogenesis, acting in part to restrain VEGF-mediated angiogenesis. Perturbation of this Dll4-mediated restraint can result in excessive non-productive vessel growth. For example, during embryogenesis, genetic deletion of even one allele of Dll4 in mice results in profound vascular defects and significant embryonic lethality at approx E10.5. The vascular defects include abnormal vascular remodeling in the yolk sac and reduced vascular invasion of the placental labyrinth, poor formation of the major arteries in the embryo, and excessive sprouting/branching in certain vessel beds. In genetic backgrounds that permit survival of Dll4 heterozygous mice, other vascular defects are found, including abnormal maturation of the vascular bed in the developing post natal retina. Dll4 also plays a fundamental role in pathological angiogenesis, as blockers of the Dll4 / Notch pathway result in decreased tumor growth, even for tumors resistant to anti-VEGF therapies. This reduced tumor growth is associated with markedly increased tumor vascularity, enhanced angiogenic sprouting, and more vessel branching. However, the increased vascularity is disorganized and non-productive, as evidenced by poor perfusion and increased tumor hypoxia. The current model is that VEGF induces Dll4 as a negative feedback regulator of angiogenesis, thus helping to coordinate VEGF-induced sprouting and promoting the functional specialization of the endothelial cells (ECs) in a network. Although Dll4 is clearly induced by VEGF and helps regulate VEGF-mediated vascular growth, it appears to also have functions that are independent of VEGF, as blockade of both VEGF and Dll4 can show more potent anti-tumor effects than blockade of either pathway alone. Thus, blockade of Dll4 in tumors presents a novel therapeutic approach, even for tumors resistant to anti-VEGF therapies.

Introduction

Most solid tumors require the induction of new vessels (angiogenesis) for continued growth, thus numerous approaches to block or disrupt tumor angiogenesis have been explored as potential tumor therapy [1]. The best validated of these approaches involves blockade of the vascular endothelial growth factor (VEGF) signaling pathway, either by blocking VEGF itself or by blocking the primary VEGF angiogenesis receptor, VEGFR2. Blockade of VEGF is a potent inhibitor of tumor growth in numerous preclinical models [2, 3] and has recently demonstrated efficacy in human cancer trials [4, 5]. The importance of VEGF as an angiogenic factor emerged in early studies with blocking antibodies [6] and recombinant soluble receptors of VEGF [7], which showed broad inhibition of tumor growth. More dramatically, gene targeting of VEGF showed that loss of even a single allele of VEGF was embryonic lethal in mice [8, 9], with severe defects in vascular and blood cell development. Recently, potent inhibitors of VEGF signaling have shown antiangiogenic effects in models of solid tumors [10–12], retinal and choroidal neovascularization [13], and ovarian cycling [14,15]. Thus, VEGF plays a critical initiating role in the

Regeneron Pharmaceuticals, 777 Old Saw Mill River Road, Tarrytown, NY 10591, USA
E-mail: Gavin.Thurston@regeneron.com

formation of a vascular network during normal development, as well as in pathologic settings such tumor angiogenesis.

However, not all tumors are equally sensitive to anti-VEGF therapies, and some tumors that are initially sensitive may become resistant with prolonged treatments. Analogously, other signaling pathways are clearly required for vascular development. Gene targeting studies have shown vascular defects in many homozygous gene-targeted mice, including targeted deletion of ephrinB2, EphB4, neuropilin-1, Unc5b, VE-PTP, edg1, angiopoietin-1, angiopoietin-2, Tie1, and Tie2 (for reviews, see [16–18]). However, not all of these pathways have been verified as being required for tumor angiogenesis or to be amenable as targets for antiangiogenic therapy.

One signaling pathway implicated in vascular development by a variety of experimental approaches is the Delta/ Notch signaling pathway [19, 20]. While initial studies had shown a key role of the Notch signaling pathway during vascular development in zebrafish [21], recent studies have also provided evidence for a key role of this pathway in tumor angiogenesis. Indeed, blockade of this pathway has been identified as a potentially useful target for antiangiogenic therapy. Unlike VEGF blockade, which results in a loss of many tumor vessels and an apparent 'normalization' of the remaining tumor vessels, blockade of Dll4/Notch results in an increased number of tumor vessels, which are more pathological. Strikingly, blockade of Dll4/Notch suppresses tumor growth in some tumors that are highly resistant to VEGF blockade.

While certain components of the Delta/Notch pathway are necessary for angiogenesis, other members are critical for development of various other tissues. Perhaps not surprisingly, defects in the Notch pathway can lead to increased cell proliferation and cancers of other cell types (for recent reviews see [22–25]). However, rather than discussing all aspects of Notch signaling in cancer, this chapter will focus on recent results defining a role for Dll4/Notch signaling in tumor angiogenesis. First, we will briefly summarize the molecular components of the Delta/Notch signaling pathway; second, we will describe results from Dll4 expression analysis and gene targeting experiments in mice; third, we will describe the effects of the manipulation of Dll4/Notch signaling in tumor angiogenesis; fourth, we will describe the role of Dll4/Notch signaling in another model of postnatal developmental angiogenesis, the retinal model; and finally, we will present a model of how Dll4/Notch regulates VEGF-induced angiogenesis.

Molecular Components of the Delta/Notch Pathway

The Delta/Notch system is an evolutionarily conserved signaling pathway that enables cell-cell communication in multicellular organisms [26]. The pathway utilizes a distinct molecular mechanism to transduce a signal from the cell surface to the nucleus, and thus to regulate expression of particular target genes. In general, Delta/Notch signaling acts to regulate cell fate decisions, thereby influencing cell proliferation, differentiation, and apoptosis.

Delta and Jagged proteins are single-pass transmembrane ligands for Notch receptors. The two Notch ligands in Drosophila (Delta and Serrate) are represented by five ligands in mammals (Delta1, 3 & 4 and Jagged1 & 2). Notch receptors (one family member in Drosophila, four family members in mice and humans, called Notch1, 2, 3 & 4) are single-pass transmembrane proteins that signal upon ligand binding via proteolytic release of the intra-cellular domain (NICD). Upon ligand-induced proteolytic cleavage of Notch receptor near the plasma membrane by γ-secretase, NICD translocates to the cell nucleus [26, 27]. Once in the nucleus, NICD acts as a potent transcriptional co-activator through its interaction with the CSL family of transcription factors [28], converting CSL from a transcriptional repressor to a transcriptional activator and inducing transcription of a panel of target genes. Target genes of this transcriptional complex include several Helix-loop-helix type transcription factors known as hairy and enhancers of split (HES – e.g., HES1, 5 and 7), and HES-related repressor proteins (HERP – e.g., HERP1, 2 and 3), among other genes.

While the Delta/Notch system has been implicated in embryonic development of a variety of organ systems, it is clearly and absolutely required at early stages of development of the cardiovascular system. For example, genetic deletion of Notch1 in mice results in severe vascular defects and embryonic lethality associated with cardiovascular defects [29]. Genetic deletion of Notch4 in mice, while not producing an overt phenotype on its own, appears to exacerbate the defects associated with Notch1 [19, 30]. Mutations in Notch3 in mice result in marked structural and functional defects in arteries [31].

Genetic deletions in mice of several Notch ligands and downstream target genes have also produced vascular defects. For example, homozygous deletion of the Jagged1 gene in mice results in a failure of vascular remodeling in the yolk sac and embryo proper, frequent hemorrhage in the cranial region, and embryonic lethality at approximately E10. Deletion of Delta-like ligand 1 (Dll1) in mice also results in vascular defects and lethality at approx E12, although other significant defects also occur, such as defects in somitogenesis and left-right asymmetry [32, 33]. In contrast, deletion of Jagged2 results in skeletal and other defects [34], but no overt vascular defects.

Delta-like ligand 4 (Dll4), the last mammalian Delta family member to be cloned, was identified in a cDNA library from adipose tissue [35, 36]. Dll4 exhibits the hallmark structure of other Notch ligands, in particular a DSL domain followed by a series of EGF-like repeats (eight, in the case of Dll4), as well as a transmembrane domain and short cytoplasmic tail. Dll4 was found to be specifically expressed in the vasculature [36–38]. Of note, Dll4 was found to be strongly expressed in tumor vessels [37, 39]. While the cloning of Dll4 may have languished behind that of other components of the Notch system, its specific and potent actions in the vasculature have prompted much interest in recent years, leading to a rush to create gene targeted Dll4 mice.

Gene Targeting Studies Show that Dll4 is Required for Developmental Angiogenesis

Recent gene targeting approaches in mice have shown that Dll4 is absolutely required for normal vascular development and proper arterial formation. The genetic requirement for Dll4 is remarkably dose-dependent, as it was discovered that heterozygous Dll4 gene targeted embryos exhibited haploid embryonic lethality. Three separate groups targeted one allele of the Dll4 gene in mouse embryonic stem cells, and then used these cells to produce mice that are genetically deficient for Dll4 [39–41]. Two of the groups replaced the Dll4 gene with a β-galactosidase reporter gene, which was useful in subsequent studies of Dll4 expression. Following generation of chimeric mice, fewer than expected numbers of heterozygous Dll4 mice were born, indicating a partially penetrant embryonic lethal phenotype. Indeed, depending upon the inbred strain of mice, this heterozygous lethal phenotype is between 100% (fully lethal) and approximately 60%. Interestingly, VEGF is the only other gene reported to cause strongly penetrant embryonic lethality with vascular defects.

When examined in utero, Dll4 gene targeted mice appear normal at E8.5, but exhibit increasingly severe defects at E9.5, including obvious pericardial edema and a lack of large vessels (arterioles) in the yolk sac. Embryonic lethality in the heterozygous mice occurred at approximately E10.5. Analysis of the vasculature in the Dll4 heterozygous embryos showed distinctive defects, including: lack of well-defined major arteries such as the internal carotid artery, severe atrophy of the aorta, more primitive plexus of undifferentiated vessels in the head and yolk sac, and an increased number of vessel branches and vascular sprouts associated with the leading front of the growing vascular plexus in the yolk sac [39]. In contrast, the venous side of the circulation appeared relatively normal. The phenotype of the gene targeted Dll4 mice resembles that of the Notch1 gene targeted mice [29], suggesting that the primary signaling pathway in vascular development is through Notch1 via Dll4. Further studies will be needed to determine whether other receptors (in particular Notch4) are involved in specific aspects or types of Dll4-mediated vessel remodeling.

While deletion of a single allele of Dll4 results in partially penetrant embryonic lethality and vascular defects, deletion of both alleles of Dll4 results in a more severe and uniform vascular phenotype. Due to the high degree of lethality after deletion of even a single allele of Dll4 in the mouse strains that are commonly used for gene targeting, generating homozygous null Dll4 embryos was a major challenge. To overcome this challenge, Dll4 null embryonic stem (ES) cells were generated by sequential targeting in vitro, and then fully ES cell-derived Dll4 deficient embryos were produced using a new technology ("VelociMouse"; [42]). In embryos null for Dll4, the somewhat variable defects observed in heterozygous embryos were now highly consistent and detectable at earlier stages of embryonic development. By E9.5, all knockout (KO) embryos had severely perturbed vascular remodeling defects in the yolk sacs, were dramatically developmentally delayed and had severely malformed arterial systems [42]. Similar findings have been observed in Dll4 null embryos generated by conventional mating as well (N. Gale, unpublished observations).

Recently, a hypomorphic mutation in the zebrafish homolog of Dll4, supplemented with elegant studies using the morpholino approach to reduce Dll4 expression, was reported to produce a strong and specific defect in vascular development [43, 44]. Unlike some previous studies on Notch pathway defects in Zebrafish, this study of Dll4 mutants did not report defects in arterio-venous fate; the most notable defects in the Dll4 mutants were in vessel branching and endothelial cell proliferation [43]. The defects in vessel branching were attributed to a failure of the growing vessels to cease sprouting activity at the appropriate time/location. These results were confirmed using pharmacologic inhibition of Notch as well as inhibiting downstream effectors of Notch function [43, 44]. Further experiments indicate that Notch1b is the primary receptor for these angiogenic responses in zebrafish [45]. The phenotype of Dll4 reduction in Zebrafish is very consistent with that found in embryonic and postnatal angiogenesis in mice, supporting a common role for Dll4/Notch in vascular development in these widely different species.

The β-galactosidase gene inserted into the Dll4 locus gene served as a useful reporter of Dll4 expression in developing and adult mice, confirming and extending the previous analyses that had been done with in situ hybridization. Expression analysis was done in chimeric mice, in embryonic Dll4 heterozygous mice that were embryonic lethal, and in the surviving Dll4 heterozygous mice [39]. During embryonic development, Dll4 was found to be expressed almost exclusively in ECs, but expression was also noted in a band along the ventral portion of the neural tube. Early in development, up to approximately E10.5, Dll4 expression is restricted to the major arteries, including the aorta, the internal carotid artery, the umbilical artery, and the vitelline artery, and their major branches. Expression was very weak or absent in veins. Later in development, expression of Dll4 begins to shift toward smaller arteries and arterioles and becomes less prominent in the larger vessels. In adult mice, Dll4 is expressed in many microvascular beds and in small arterioles, and is absent or barely detectable in large arteries and veins [39]. An important site of extravascular Dll4 expression in adults is in the thymus.

Dll4 Plays a Key Role in Regulating Tumor Angiogenesis

Several studies have shown that tumor endothelial cells express Dll4, both in preclinical tumor models and in various human tumors. In human tumors grown in mice, RNase protection assays showed expression of Dll4 by the host (mouse) tissue and not by the tumor tissue [37]. In these tumors, Dll4 expression was detected in most, but not all, of the tumor blood vessels by in situ hybridization [37]. Notably, Dll4 expression appeared to be much higher in the tumor vessels than in adjacent normal vessels. Similarly, analysis of tumors grown in gene-targeted mice, in which a β-galactosidase reporter was knocked-in to the Dll4 gene locus, showed specific reporter gene expression in most tumor blood vessels [39, 45]. Dll4 was strongly expressed in small tumor vessels, but was weak or absent in most of the larger tumor vessels. Again, Dll4 expression was stronger in the tumor vessels than in the adjacent normal vessels [39, 45].

In human clear cell-renal tumors, Dll4 expression was localized to the blood vessels [37] and was found to be at 9-fold higher levels than in normal kidney tissue [46]. As assessed by quantitative PCR, the levels of Dll4 expression in the renal tumors show a correlation with the levels of VEGF expression [46]. In human bladder cancer, Dll4 expression was found to be increased approximately 2-fold in superficial tumors, and approximately 1.5-fold in invasive tumors, compared to normal bladder tissue [47]. In the superficial bladder tumors, the levels of Dll4 expression again showed a correlation with the levels of VEGF expression [47]. In a transgenic hepatocarcinoma model in mice, both Notch4 and Dll4 were significantly up-regulated during tumor progression from hyperplasia to diffuse carcinoma [48].

The expression of Dll4 in tumor vessels appears to be directly regulated by VEGF. For example, blockade of VEGF in tumor bearing mice results in a rapid and profound reduction of Dll4 expression by the tumor blood vessels [45]. Conversely, stimulation of cultured ECs with VEGF results in an increase in Dll4 expression [46]. Thus, the relatively high levels of Dll4 expression on tumor vessels may be a result of relatively high levels of VEGF signaling in these vessels compared to most normal vessels.

As a result of the suggestive expression studies, several groups sought to determine the functional role of Dll4/Notch signaling in tumor angiogenesis by developing soluble reagents that could specifically inhibit Dll4-mediated signaling [45, 49]. As of this writing, one group used soluble forms of Dll4 as well as antibodies to Dll4 that block the binding of mouse Dll4 to Notch1 receptor [45], while the other group generated a humanized phage antibody that binds with high affinity to Dll4 and blocks the binding of mouse and human Dll4 to Notch1 receptor [49].

Systemic treatment of mice with these Dll4/Notch inhibitors results in significantly reduced tumor growth, with little or no observed systemic toxicity [45, 49]. Blockade of Dll4/Notch was tested in a variety of established human and rodent tumors grown subcutaneously in mice, with reduction of tumor growth from 50% to more than 90%, depending on the tumor model (Fig. 19.1A) [45, 49]. Remarkably, the reduced tumor growth was associated with an increase in the vessel density in the tumors (Fig. 19.1B), with increased sprouts and vessel branches (Fig. 19.1C, D). In contrast, blockade of VEGF results in a decrease of tumor vessel density. The increased vascularity following Dll4 blockade is non-functional, as the tumor vessels were poorly perfused and the tumors showed increased areas of hypoxia [45].

In addition to the studies with systemic blockers of Dll4, one group also used a tumor overexpression system to locally produce either a blocker or an activator of the Dll4/Notch pathway [45]. Using this system, activation or blockade of Dll4/Notch caused reciprocal effects on the tumor vessels. Similar to what was observed with systemic treatment, local overexpression of blocker of Dll4/Notch signaling (soluble Dll4) resulted in decreased tumor growth and increased tumor vessel density. Reciprocally, overexpressing the full-length membrane bound form of Dll4 caused Notch activation in the tumor blood vessels, and resulted in a reduction of tumor vessel density, with straighter vessels, less branches and reduced sprouting. These unbranched vessels were functional, as they had flow and the tumors were not hypoxic [45]. Thus, the tumor growth rate was similar to control tumors in this model.

The effectiveness of blocking Dll4/Notch has been tested in tumor models that are resistant to anti-VEGF treatments [45, 49]. Importantly, Dll4 blockade can be effective even in tumors that are quite resistant to VEGF blockade. Treatment of mice bearing HT-1080-RM tumors with VEGF inhibitors results in very little inhibition of tumor growth, yet treatment of these tumors with inhibitors of Dll4/Notch results in approximately 90% reduction in growth [45]. Although the reduced tumor growth was associated with a dramatic increase in the vessel density [45], the vessels appear disorganized and the tumors show evidence of increased hypoxia, such as higher levels of VEGF expression by the tumor cells (I. Noguera-Troise, G. Thurston, unpublished results). Similarly, the growth of WEHI3 tumors in mice is scarcely slowed by a VEGF inhibitor, yet is slowed by more than 50% with Dll4 blockers [49].

While having anti-tumor effects on their own, the combination of both Dll4 blockers and VEGF blockers appears to be more potent at controlling tumor growth than inhibiting either

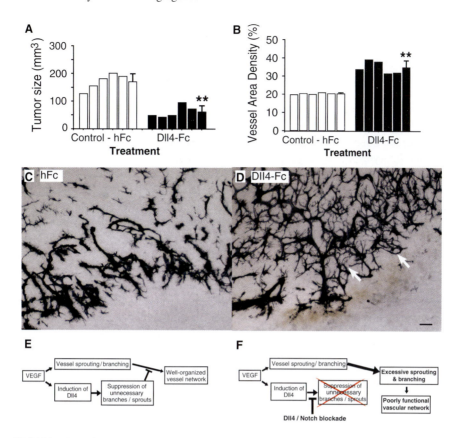

FIG. 19.1. Blockade of Dll4/Notch pathway results in reduced tumor growth, but increased tumor vessel density. C6 rat glioma tumor cells were implanted subcutaneously in mice, which were then treated systemically with Dll4-Fc (a Notch pathway blocker) or hFc as a control. ** denotes significantly different from the control tumor group (P < 0.01). A Tumors treated with Dll4-Fc were smaller than control tumors. Data show size of individual tumors, last bar in each group show mean ± SD. B Vessel area density is increased in Dll4-Fc treated tumors compared to controls. Vessel area density was measured morphometrically in thick sections of tumors stained with antibody to CD31/Pecam. C,D Microscographs of tumor sections immunostained for blood vessels (CD31). Control tumor (C) has relatively sparse network of tumor vessels, whereas tumor treated with systemic Dll4-Fc (D) has a dense network of vessels, including many fine branches and vascular sprouts (arrows). Scale bar: 50 μm. E,F Schematic views showing effect of D114/Notch blockade. Figure adapted from Noguera-Troise et al., [45].

pathway alone. Although experimental results are somewhat limited to date, the treatment combination of a VEGF antibody and a Dll4 antibody results in better suppression of the growth of MV-522 tumors [49]. Similarly, in resistant HT1080-RM tumors, the combination of VEGF Trap with Dll4 inhibitors results in tumor stasis for several weeks (I. Noguera-Troise, G. Thurston, unpublished results). It will be important to extend these findings to other tumor models, and to document the effect of the combination therapy on tumor vessel morphology and function.

Dll4 Plays a Key Role in Postnatal Vascular Development: Studies of Retinal Angiogenesis

The retina can serve as a useful model system in which to study angiogenesis, in part because the retinal vasculature in mice develops in a stereotypic manner during the early post-natal period. Several groups have used the retinal model in mice to evaluate the role of Dll4 in normal post-embryonic angiogenesis. By whole mount in situ hybridization in the mouse retina [50–52] or by using the Dll4 β-gal reporter mice [53], Dll4 expression is prominent in the most active regions of vascular growth during normal postnatal retinal development. For example, at postnatal day 7 (P7), Dll4 is strongly expressed by the endothelial cells of actively growing capillaries at the leading front of the superficial vascular plexus. Lower levels of expression were found in maturing capillaries and in newly forming arteries and veins. By immunostaining [51, 53], Dll4 protein was found in a subset of the leading endothelial cells and in some adjacent cells in the vascular plexus. Dll4 was localized in the cell body and perinuclear region of ECs, suggesting that it had been internalized following recent Notch engagement. As in tumor vessels, the expression of Dll4 was found to be regulated by VEGF, because inhibition of VEGF in the retina resulted in a reduced expression of Dll4 in the growing vessels [52, 53].

A variety of approaches was used to inhibit Dll4 in the retinal model, including genetic insufficiency (heterozygous mice), soluble Dll4 protein, Dll4 blocking antibodies, and chemical inhibitors of the enzymatic activation of Notch receptor (γ-secretase inhibitors) [49, 51–53]. Despite this wide range of approaches to inhibit Dll4/Notch, these studies reported very similar findings: in particular, inhibition of Dll4 in the early postnatal retina results in markedly enhanced angiogenic sprouting and a denser network of primary capillaries at the growing front of retinal vessels. The increased vascularity is associated with an increase in the number of endothelial cells and a small increase in EC proliferation in the growing capillaries. Despite the increase in vascular sprouting, inhibition of Dll4 causes a moderate delay in the rate of outgrowth of the capillary plexus towards the retinal periphery, as well as causing a delay in the maturation of the retinal arteries. The inhibition of Dll4 resulted in increased expression of VEGFR2 in the retinal endothelial cells [49, 52, 53], perhaps leading to an amplification of the VEGF signal and accounting for some of the increased sprouting activity of the vessels.

Proposed Model for Function of Dll4 in Regulating Angiogenesis

The in vivo and in vitro studies to date on the role of Dll4/Notch in angiogenesis suggest a model in which Dll4 helps to regulate the cellular actions of VEGF. In particular, VEGF-mediated signaling in the leading endothelial cells (so-called tip cells) induces the expression of Dll4 in these cells. The induced Dll4 then provides a signal to adjacent Notch receptor-bearing endothelial cells to down-regulate VEGF-induced sprouting and branching [51]. The down-regulation of sprouting and branching, which is also accompanied by other functional specializations of the downstream ECs (so-called stalk cells) including the formation of a patent vessel lumen and interaction with pericytes, may be a result of decreased expression of VEGF receptors. These combined actions of VEGF and Dll4/Notch help to generate a functional specialization of the endothelial cells in a network, and thereby promote a coordinated vascular response to VEGF.

Although a number of direct and indirect target genes for Notch have been identified in ECs, it is not yet known how activation of Notch and subsequent expression/repression of these target genes in angiogenic endothelial cells would lead to their functional specialization. As mentioned above, down-regulation of VEGFR2 may be one component of this specialization, but other genes are also likely to play a role. Other potential effector genes that appear to be regulated by Notch activation in endothelial cells include platelet-derived growth factor-B (PDGF-B) and Unc5b (both down-regulated), and ephrinB2 (up-regulated), all of which are required for normal vascular development in mice [54–56]. However, it is not known precisely how these genes regulate

endothelial specialization, or whether they indeed play a role in Dll4/Notch-mediated actions.

Blockade of Dll4/Notch signaling, either by blocking Dll4-Notch interaction or by blocking Notch signaling, thus results in a failure of the endothelial cells to specialize within the vascular network, associated with a failure to down-regulate VEGF-induced sprouting and branching. In most normal adult tissues, where VEGF signaling is low and vessels are quiescent, this blockade has few consequences. In the tumor setting, where VEGF signaling is strong and the angiogenic process is already abnormal, the decreased specialization can lead to increased vessel density but a less functional vascular network. Although tumor vessels are abnormal in a number of aspects [57–58], the recent studies show that they can be made even more abnormal and less efficient. Blockade of Dll4/Notch in tumors results in decreased tumor perfusion and increased tumor hypoxia. Thus, even tumor vasculature requires a regulated balance of angiogenic growth factors to form a hierarchy of functioning vessels. Ongoing studies are aimed at characterizing in more detail the morphologic and functional changes in tumor vessels following Dll4/Notch blockade. Future studies will help to determine the relative importance of the different Notch receptors in tumor angiogenesis (in particular, Notch1 and Notch4), as well as identifying other angiogenic pathways that are modulated by Dll4/Notch signaling. Another important challenge will be to resolve at the cellular level which endothelial cells in the vascular network of a growing tumor are expressing Dll4 and which cells are receiving Notch signals from Dll4 ligand.

Dll4 as a Target for Antiangiogenesis Therapy

One of the challenges for developing an antiangiogenic therapy based on blocking Dll4/Notch signaling will be to specifically block angiogenic signaling while limiting interaction of the therapeutic agent with other members of the Delta/Notch pathway that may play a role in normal tissue homeostasis. Protein-based reagents appear to be the most promising approach to meet this challenge. For example, therapeutic antibodies that bind to Dll4 may provide a means to inhibit the interaction of Dll4 with Notch receptors, and thus specifically block Dll4, without affecting signaling to Notch receptors via other Delta or Jagged ligands. Although chemical inhibitors of Notch signaling are available in the form of γ-secretase inhibitors, these are likely to block all forms of Notch signaling, as well as other events mediated by γ-secretase, and thus have a variety of undesirable effects that are not found with specific Dll4 inhibitors, such as over-proliferation of goblet cells in the intestinal epithelium [49].

Most antiangiogenesis approaches to date have attempted to eliminate tumor blood vessels and thereby starve the tumors. For example, effective blockade of VEGF in tumors generally results in regression of some vessels and inhibition or reduction in the growth of new vessels. The tumor vessels that remain after VEGF blockade are apparently more 'normal' [59]. The recent studies with blockers of Dll4/Notch seem to offer a counter-intuitive approach; that Dll4 blockade may further compromise tumor vascular function by causing excessive non-productive 'pathological' angiogenesis, which can in turn restrict tumor growth. Recent studies have shown that blockade of Dll4 can have potent vascular effects on tumors that are resistant to blockers of VEGF. Furthermore, and importantly for the field of antiangiogenesis, simultaneously blocking both VEGF and Dll4 has more potent anti-tumor effects than blockade of either pathway alone. While blockade of VEGF clearly has an important role in antiangiogenic therapy in a wide variety of tumors, the Dll4/Notch pathway represents a new therapeutic target in antiangiogenesis. Specific blockade of Dll4, alone or in combination with other angiogenesis inhibitors, might be beneficial for patients with tumors that are resistant to anti-VEGF therapies.

References

1. Folkman, J. (1992). The role of angiogenesis in tumor growth. Semin Cancer Biol *3*, 65–71.
2. Ferrara, N. (2004). Vascular endothelial growth factor as a target for anticancer therapy. Oncologist *9 Suppl 1*, 2–10.
3. Rudge, J. S., Thurston, G., Davis, S., Papadopoulos, N., Gale, N., Wiegand, S. J., and Yancopoulos, G. D. (2005). VEGF trap as a novel antiangiogenic treatment currently in clinical trials for cancer and eye diseases, and VelociGene- based discovery of the next generation of angiogenesis targets. Cold Spring Harb Symp Quant Biol *70*, 411–418.
4. Hurwitz, H., Fehrenbacher, L., Novotny, W., Cartwright, T., Hainsworth, J., Heim, W., Berlin, J., Baron, A., Griffing, S., Holmgren, E., *et al*. (2004). Bevacizumab plus irinotecan, fluorouracil, and leucovorin for metastatic colorectal cancer. N Engl J Med *350*, 2335–2342.
5. Laskin, J. J., and Sandler, A. B. (2005). First-line treatment for advanced non-small-cell lung cancer. Oncology (Williston Park) *19*, 1671–1676; discussion 1678–1680.
6. Kim, K. J., Li, B., Winer, J., Armanini, M., Gillett, N., Phillips, H. S., and Ferrara, N. (1993). Inhibition of vascular endothelial growth factor-induced angiogenesis suppresses tumour growth in vivo. Nature *362*, 841–844.
7. Kong, H. L., Hecht, D., Song, W., Kovesdi, I., Hackett, N. R., Yayon, A., and Crystal, R. G. (1998). Regional suppression of tumor growth by in vivo transfer of a cDNA encoding a secreted form of the extracellular domain of the flt-1 vascular endothelial growth factor receptor. Hum Gene Ther *9*, 823–833.
8. Carmeliet, P., Ferreira, V., Breier, G., Pollefeyt, S., Kieckens, L., Gertsenstein, M., Fahrig, M., Vandenhoeck, A., Harpal, K., Eberhardt, C., *et al*. (1996). Abnormal blood vessel development and lethality in embryos lacking a single VEGF allele. Nature *380*, 435–439.
9. Ferrara, N., Carver-Moore, K., Chen, H., Dowd, M., Lu, L., O'Shea, K. S., Powell-Braxton, L., Hillan, K. J., and Moore, M. W. (1996). Heterozygous embryonic lethality induced by targeted inactivation of the VEGF gene. Nature *380*, 439–442.
10. Beebe, J. S., Jani, J. P., Knauth, E., Goodwin, P., Higdon, C., Rossi, A. M., Emerson, E., Finkelstein, M., Floyd, E., Harriman, S., *et al*. (2003). Pharmacological characterization of CP-547,632, a novel vascular endothelial growth factor receptor-2 tyrosine kinase inhibitor for cancer therapy. Cancer Res *63*, 7301–7309.
11. Hicklin, D. J., and Ellis, L. M. (2005). Role of the vascular endothelial growth factor pathway in tumor growth and angiogenesis. J Clin Oncol *23*, 1011–1027.
12. Holash, J., Davis, S., Papadopoulos, N., Croll, S. D., Ho, L., Russell, M., Boland, P., Leidich, R., Hylton, D., Burova, E., *et al*. (2002). VEGF-Trap: a VEGF blocker with potent antitumor effects. Proc Natl Acad Sci U S A *99*, 11393–11398.
13. Kim, I. K., Husain, D., Michaud, N., Connolly, E., Lane, A. M., Durrani, K., Hafezi-Moghadam, A., Gragoudas, E. S., O'Neill, C. A., Beyer, J. C., and Miller, J. W. (2006). Effect of intravitreal injection of ranibizumab in combination with verteporfin PDT on normal primate retina and choroid. Invest Ophthalmol Vis Sci *47*, 357–363.
14. Ferrara, N., Chen, H., Davis-Smyth, T., Gerber, H. P., Nguyen, T. N., Peers, D., Chisholm, V., Hillan, K. J., and Schwall, R. H. (1998). Vascular endothelial growth factor is essential for corpus luteum angiogenesis. Nat Med *4*, 336–340.
15. Fraser, H. M., Wilson, H., Morris, K. D., Swanston, I., and Wiegand, S. J. (2005). Vascular endothelial growth factor Trap suppresses ovarian function at all stages of the luteal phase in the macaque. J Clin Endocrinol Metab *90*, 5811–5818.
16. Carmeliet, P. (2005). Angiogenesis in life, disease and medicine. Nature *438*, 932–936.
17. Jain, R. K. (2005a). Antiangiogenic therapy for cancer: current and emerging concepts. Oncology (Williston Park) *19*, 7–16.
18. Yancopoulos, G. D., Davis, S., Gale, N. W., Rudge, J. S., Wiegand, S. J., and Holash, J. (2000). Vascular-specific growth factors and blood vessel formation. Nature *407*, 242–248.
19. Gridley, T. (2001). Notch signaling during vascular development. Proc Natl Acad Sci U S A *98*, 5377–5378.
20. Shawber, C. J., and Kitajewski, J. (2004). Notch function in the vasculature: insights from zebrafish, mouse and man. Bioessays *26*, 225–234.
21. Lawson, N. D., Scheer, N., Pham, V. N., Kim, C. H., Chitnis, A. B., Campos-Ortega, J. A., and Weinstein, B. M. (2001). Notch signaling is required for arterial-venous differentiation during embryonic vascular development. Development *128*, 3675–3683.
22. Grabher, C., von Boehmer, H., and Look, A. T. (2006). Notch 1 activation in the molecular pathogenesis of T-cell acute lymphoblastic leukaemia. Nat Rev Cancer *6*, 347–359.
23. Miele, L., Golde, T., and Osborne, B. (2006). Notch signaling in cancer. Curr Mol Med *6*, 905–918.
24. Radtke, F., Clevers, H., and Riccio, O. (2006). From gut homeostasis to cancer. Curr Mol Med *6*, 275–289.
25. Wilson, A., and Radtke, F. (2006). Multiple functions of Notch signaling in self-renewing organs and cancer. FEBS Lett *580*, 2860–2868.
26. Artavanis-Tsakonas, S., Rand, M. D., and Lake, R. J. (1999). Notch signaling: cell fate control and signal integration in development. Science *284*, 770–776.
27. Iso, T., Hamamori, Y., and Kedes, L. (2003). Notch signaling in vascular development. Arterioscler Thromb Vasc Biol *23*, 543–553.
28. Kopan, R., Schroeter, E. H., Weintraub, H., and Nye, J. S. (1996). Signal transduction by activated mNotch: importance of proteo-

lytic processing and its regulation by the extracellular domain. Proc Natl Acad Sci U S A *93*, 1683–1688.

29. Swiatek, P. J., Lindsell, C. E., del Amo, F. F., Weinmaster, G., and Gridley, T. (1994). Notch1 is essential for postimplantation development in mice. Genes Dev *8*, 707–719.

30. Krebs, L. T., Xue, Y., Norton, C. R., Shutter, J. R., Maguire, M., Sundberg, J. P., Gallahan, D., Closson, V., Kitajewski, J., Callahan, R., *et al.* (2000). Notch signaling is essential for vascular morphogenesis in mice. Genes Dev *14*, 1343–1352.

31. Domenga, V., Fardoux, P., Lacombe, P., Monet, M., Maciazek, J., Krebs, L. T., Klonjkowski, B., Berrou, E., Mericskay, M., Li, Z., *et al.* (2004). Notch3 is required for arterial identity and maturation of vascular smooth muscle cells. Genes Dev *18*, 2730–2735.

32. Hrabe de Angelis, M., McIntyre, J., 2nd, and Gossler, A. (1997). Maintenance of somite borders in mice requires the Delta homologue Dll1. Nature *386*, 717–721.

33. Przemeck, G. K., Heinzmann, U., Beckers, J., and Hrabe de Angelis, M. (2003). Node and midline defects are associated with left-right development in Delta1 mutant embryos. Development *130*, 3–13.

34. Jiang, R., Lan, Y., Chapman, H. D., Shawber, C., Norton, C. R., Serreze, D. V., Weinmaster, G., and Gridley, T. (1998). Defects in limb, craniofacial, and thymic development in Jagged2 mutant mice. Genes Dev *12*, 1046–1057.

35. Rao, P. K., Dorsch, M., Chickering, T., Zheng, G., Jiang, C., Goodearl, A., Kadesch, T., and McCarthy, S. (2000). Isolation and characterization of the notch ligand delta4. Exp Cell Res *260*, 379–386.

36. Shutter, J. R., Scully, S., Fan, W., Richards, W. G., Kitajewski, J., Deblandre, G. A., Kintner, C. R., and Stark, K. L. (2000). Dll4, a novel Notch ligand expressed in arterial endothelium. Genes Dev *14*, 1313–1318.

37. Mailhos, C., Modlich, U., Lewis, J., Harris, A., Bicknell, R., and Ish-Horowicz, D. (2001). Delta4, an endothelial specific notch ligand expressed at sites of physiological and tumor angiogenesis. Differentiation *69*, 135–144.

38. Yoneya, T., Tahara, T., Nagao, K., Yamada, Y., Yamamoto, T., Osawa, M., Miyatani, S., and Nishikawa, M. (2001). Molecular cloning of delta-4, a new mouse and human Notch ligand. J Biochem (Tokyo) *129*, 27–34.

39. Gale, N. W., Dominguez, M. G., Noguera, I., Pan, L., Hughes, V., Valenzuela, D. M., Murphy, A. J., Adams, N. C., Lin, H. C., Holash, J., *et al.* (2004). Haploinsufficiency of delta-like 4 ligand results in embryonic lethality due to major defects in arterial and vascular development. Proc Natl Acad Sci U S A *101*, 15949–15954.

40. Duarte, A., Hirashima, M., Benedito, R., Trindade, A., Diniz, P., Bekman, E., Costa, L., Henrique, D., and Rossant, J. (2004). Dosage-sensitive requirement for mouse Dll4 in artery development. Genes Dev *18*, 2474–2478.

41. Krebs, L. T., Shutter, J. R., Tanigaki, K., Honjo, T., Stark, K. L., and Gridley, T. (2004). Haploinsufficient lethality and formation of arteriovenous malformations in Notch pathway mutants. Genes Dev *18*, 2469–2473.

42. Poueymirou, W. T., Auerbach, W., Frendewey, D., Hickey, J. F., Escaravage, J. M., Esau, L., Dore, A. T., Stevens, S., Adams, N. C., Dominguez, M. G., *et al.* (2007). F0 generation mice fully derived from gene-targeted embryonic stem cells allowing immediate phenotypic analyses. Nat Biotechnol *25*, 91–99.

43. Leslie, J. D., Ariza-McNaughton, L., Bermange, A. L., McAdow, R., Johnson, S. L., and Lewis, J. (2007). Endothelial

signalling by the Notch ligand Delta-like 4 restricts angiogenesis. Development.

44. Siekmann, A. F., and Lawson, N. D. (2007). Notch signalling limits angiogenic cell behaviour in developing zebrafish arteries. Nature.

45. Noguera-Troise, I., Daly, C., Papadopoulos, N. J., Coetzee, S., Boland, P., Gale, N. W., Lin, H. C., Yancopoulos, G. D., and Thurston, G. (2006). Blockade of Dll4 inhibits tumour growth by promoting non-productive angiogenesis. Nature *444*, 1032–1037.

46. Patel, N. S., Li, J. L., Generali, D., Poulsom, R., Cranston, D. W., and Harris, A. L. (2005). Up-regulation of delta-like 4 ligand in human tumor vasculature and the role of basal expression in endothelial cell function. Cancer Res *65*, 8690–8697.

47. Patel, N. S., Dobbie, M. S., Rochester, M., Steers, G., Poulsom, R., Le Monnier, K., Cranston, D. W., Li, J. L., and Harris, A. L. (2006). Up-regulation of endothelial delta-like 4 expression correlates with vessel maturation in bladder cancer. Clin Cancer Res *12*, 4836–4844.

48. Hainaud, P., Contreres, J. O., Villemain, A., Liu, L. X., Plouet, J., Tobelem, G., and Dupuy, E. (2006). The Role of the Vascular Endothelial Growth Factor-Delta-like 4 Ligand/Notch4-Ephrin B2 Cascade in Tumor Vessel Remodeling and Endothelial Cell Functions. Cancer Res *66*, 8501–8510.

49. Ridgway, J., Zhang, G., Wu, Y., Stawicki, S., Liang, W. C., Chanthery, Y., Kowalski, J., Watts, R. J., Callahan, C., Kasman, I., *et al.* (2006). Inhibition of Dll4 signalling inhibits tumour growth by deregulating angiogenesis. Nature *444*, 1083–1087.

50. Claxton, S., and Fruttiger, M. (2004). Periodic Delta-like 4 expression in developing retinal arteries. Gene Expr Patterns *5*, 123–127.

51. Hellstrom, M., Phng, L. K., Hofmann, J. J., Wallgard, E., Coultas, L., Lindblom, P., Alva, J., Nilsson, A. K., Karlsson, L., Gaiano, N., *et al.* (2007). Dll4 signalling through Notch1 regulates formation of tip cells during angiogenesis. Nature.

52. Suchting, S., Freitas, C., le Noble, F., Benedito, R., Breant, C., Duarte, A., and Eichmann, A. (2007). The Notch ligand Delta-like 4 negatively regulates endothelial tip cell formation and vessel branching. PNAS *in press*.

53. Lobov, I. B., Renard, R. A., Papadopoulos, N. J., Gale, N. W., Thurston, G., Yancopoulos, G. D., and Wiegand, S. J. (2007). Delta-like ligand 4 (dll4) is induced by VEGF as a negative regulator of angiogenic sprouting. PNAS *in press*.

54. Lindahl, P., Johansson, B. R., Leveen, P., and Betsholtz, C. (1997). Pericyte loss and microaneurysm formation in PDGF-B-deficient mice. Science *277*, 242–245.

55. Lu, X., Le Noble, F., Yuan, L., Jiang, Q., De Lafarge, B., Sugiyama, D., Breant, C., Claes, F., De Smet, F., Thomas, J. L., *et al.* (2004). The netrin receptor UNC5B mediates guidance events controlling morphogenesis of the vascular system. Nature *432*, 179–186.

56. Wang, H. U., Chen, Z. F., and Anderson, D. J. (1998). Molecular distinction and angiogenic interaction between embryonic arteries and veins revealed by ephrin-B2 and its receptor Eph-B4. Cell *93*, 741–753.

57. Baluk, P., Hashizume, H., and McDonald, D. M. (2005). Cellular abnormalities of blood vessels as targets in cancer. Curr Opin Genet Dev *15*, 102–111.

58. McDonald, D. M., and Foss, A. J. (2000). Endothelial cells of tumor vessels: abnormal but not absent. Cancer Metastasis Rev *19*, 109–120.

59. Jain, R. K. (2005b). Normalization of tumor vasculature: an emerging concept in antiangiogenic therapy. Science *307*, 58–62.

Chapter 20
Immune Cells and Inflammatory Mediators as Regulators of Tumor Angiogenesis

Michele De Palma[1] and Lisa M. Coussens[2]

Keywords: angiogenesis, cancer, inflammation, immune cells, leukocytes, monocytes.

Abstract: Tumors conscript immune cells to support neoplastic progression. One important mechanism by which immune cells contribute to tumor growth is the promotion of angiogenesis. In this chapter, we discuss the role of immune cells and inflammatory mediators in this process. Targeting immune cells and their proangiogenic programs in tumors may represent a new frontier of antiangiogenic therapy.

Innate and Adaptive Immune Cells in the Initiation and Promotion of Tumorigenesis

During the past 25 years, cancer research has primarily focused on the role of activating and/or inactivating mutations in genes involved in cell proliferation, growth and death, and the ever-growing knowledge of genetic and epigenetic changes that occur in tumor cells, that together have led us to view cancer as a genetic disease. Moreover, the sequential activation/inactivation of a relatively low number of molecular pathways, common to many cancer types, has put forward the 'multi-step model of tumorigenesis', that regards the progressive accumulation of genetic changes in somatic cells as the major process in the etiology of cancer [1]. However, although generally applicable, the multi-step model of tumorigenesis overlooks the role of the microenvironment in the development of cancer. Both experimental and clinical studies have recently highlighted the causal role of host-derived, extrinsic factors—such as the extra-cellular matrix (ECM), soluble molecules and tumor-associated cells—in the initiation and/or progression of cancer [2–9]. Solid tumors contain both neoplastic and non-neoplastic stromal cells, the latter of which not only passively support neoplastic cells by providing a scaffold for their growth, but also promote neoplastic development and regulate progression to malignancy. Cellular components of tumor stroma include (myo)fibroblasts, vascular cells and infiltrating leukocytes. The fact that some leukocytes promote—rather than restrict—tumor growth may be viewed as an apparent paradox [10, 11]. Historically, leukocytes found in and around developing tumors were thought to represent an attempt by the host to eradicate transformed cells. Undeniably, certain leukocytes, such as some T lymphocyte subsets and natural killer (NK) cells, play a vital function in constraining tumor development [12], and it has been postulated that many more tumors arise than those that eventually develop to fully malignant disease thanks to such activity. However, a growing body of research has recently implicated tumor-infiltrating leukocytes as causal players in cancer development [2, 3, 7, 10, 13, 14].

Leukocytes, represent a diverse assortment of immune cells that can be divided into innate (myeloid) and adaptive (lymphoid) lineages. Innate immune cells, including macrophages, granulocytes, mast cells, dendritic cells (DCs) and NK cells, represent the first line of defence against pathogens and foreign agents. When tissue homeostasis is perturbed, tissue-resident macrophages and mast cells locally secrete soluble factors, such as bioactive mediators, matrix-remodelling proteins, cytokines and chemokines, that recruit additional leukocytes from the circulation into the damaged tissue, a process known as inflammation. The recruited innate immune cells (also referred to as inflammatory cells) can directly eliminate pathogenic agents in situ. DCs, on the other hand, take up foreign antigens (including tumor antigens) and migrate to lymphoid organs where they present their antigens to adaptive immune cells. Upon recognition of a

[1] Angiogenesis and Tumor Targeting Research Unit and Telethon Institute for Gene Therapy, San Raffaele Scientific Institute, via Olgettina, 58, 20132 Milan, Italy

[2] Department of Pathology and Comprehensive Cancer Center, University of California, San Francisco, 2340 Sutter St, Rm N221, San Francisco, CA 94143, USA

foreign antigen presented by DCs or other professional antigen-presenting cells, adaptive immune cells, e.g., CD4+ T lymphocytes and B lymphocytes, undergo clonal expansion and mount an "adaptive" response targeted to the foreign agent. Thus, acute activation of innate immunity sets the stage for the activation of a more sophisticated, antigenically committed adaptive immune response. Once the foreign agent has been eliminated, inflammation resolves and tissue homeostasis is restored.

The inflammatory response required for commencing immune responses may also set the ground for promoting neoplastic disease. As early as 1863, Virchow first postulated that cancer originates at sites of chronic inflammation, in part based on his hypothesis that some classes of irritants that cause inflammation also enhance cell proliferation. When tissues are injured or exposed to chemical irritants, damaged cells are removed by induction of cell death pathways, while cell proliferation is enhanced to facilitate tissue regeneration or wound healing, in the attempt to re-establish tissue homeostasis. Proliferation and inflammation resolve only after the insulting agent is removed or tissue repair completed. By contrast, when the insulting agent persists over time, sustained cycles of cell proliferation and death in environments rich in inflammatory cells and their bioactive products may increase neoplastic risk and/or foster tumor progression [15]. Thus, while sporadic or inherited genetic mutations in critical genes regulating cell cycle, differentiation, metabolism and cell adhesion may represent the initiating events in tumorigenesis ("initiation"), chronic inflammation then favors the selection of additional features in initiated cells and may promote their full malignant transition ("promotion").

It has been estimated that more than 15% of malignancies worldwide can be attributed to chronic inflammatory disease [7, 16]. Perhaps the most compelling clinical evidence for a causative link between chronic inflammation and cancer comes from epidemiological studies reporting that inhibiting chronic inflammation in patients with pre-malignant disease, or who are predisposed to cancer development, has chemo-preventative potential [17]. These studies revealed that long-term usage of anti-inflammatory drugs, such as aspirin and selective cyclooxygenase-2 (COX-2) inhibitors, significantly reduces cancer risk, indicating that COX-2 or other key molecules involved in prostaglandin biosynthesis might be effective anti-cancer targets.

Tumor microenvironments are rich in immune cell-derived cytokines, and growth factors, including tumor necrosis factor-α (TNF-α), transforming growth factor-β (TGF-β), vascular endothelial growth factor (VEGF), and interleukin-1 (IL-1) and -6 (IL-6). These molecules not only foster tumor development via the modulation of gene expression programs in initiated neoplastic cells – culminating in altered cell cycle progression and enhanced survival – but also influence tumor-associated stromal cells and tissue remodeling. The availability of immune-competent mouse models of de novo carcinogenesis has facilitated

a mechanistic evaluation of links between pre-malignant conditions, chronic inflammation and pro-inflammatory factors, with tumor progression. The transcription factor nuclear factor κB (NFκB), a mediator of cell survival, proliferation, and growth arrest, has been identified as an important molecule linking chronic inflammation to cancer [14, 18]. Two studies [19, 20] revealed that the NFκB pathway promoted tumor development by a dual mechanism, by preventing apoptosis of cells with malignant potential, and by stimulating production of pro-inflammatory cytokines—including TNF-α—by innate immune cells. Pro-inflammatory cytokines then contributed to neoplastic cell proliferation in a paracrine fashion and increased survival and progression of initiated tumor cells. More recently, insight into the pathways linking innate and adaptive immunity in tumorigenesis have been provided, and have revealed that B lymphocytes and factors present in serum are essential for establishing chronic inflammatory states associated with pre-malignant progression in skin [21, 22]. Antigens that are present in early neoplastic tissues, possibly derived from initiated neoplastic cells or stromal tissue components, may be transported to lymphoid organs by DCs, where antigen presentation triggers adaptive immune responses and B cell activation. Following B cell activation, immunoglobulins (Igs) are released in the circulation and accumulate at the site where antigens are expressed. Interstitial Ig deposition, that has been observed both in experimental and human pre-malignant tissues, can trigger the local activation/recruitment of innate immune cells (via the cross-linking of Ig receptors) and establish inflammation [23]. These observations linking Ig deposition to chronic inflammation not only establish a connection between initiation, adaptive and innate immune responses in cancer development, but also suggest that activating humoral immune responses in patients predisposed to cancer development might enhance neoplastic programming of tissue rather than preventing it.

Interplay between Immune Cells and Tumor Angiogenesis

In adulthood, most blood vessels are quiescent and angiogenesis—the growth of new blood vessels from pre-existing ones—only occurs during the female reproductive cycle and under certain pathophysiological conditions, such as tissue remodelling and wound healing. During angiogenesis, neo-vessels are formed through a well-orchestrated series of events, encompassing endothelial cell (EC) proliferation, as well as directional migration of ECs through remodelled basement membrane and toward angiogenic stimuli [24]. Once a primitive endothelial layer is formed, recruitment of perivascular support cells enables stabilization of nascent vessels, functional lumen formation and blood flow [25]. In normal tissues, activation of proangiogenic molecular and cellular programs is regulated at many levels and controlled by a diverse assortment of positively and negatively acting soluble and insoluble mediators, whose balanced equilibrium is kept tightly in check under homeostatic conditions. However, under conditions of tissue stress, such as those that occur during the onset of incipient neoplasia, this balance may be rapidly upset, favoring proangiogenic programs [26, 27]. Whereas the cellular and molecular programs are common to both physiological and tumor angiogenesis, constitutively activated proangiogenic signaling in tumors make the tumor-associated vasculature distinctly irregular and chaotic in organization and inherently unstable, poorly functional and leaky [28, 29].

There is a tight interplay between innate immune cells and the vascular system. ECs mediate immune cell recruitment to extravascular tissues by expressing a repertoire of leukocyte adhesion molecules. On the other hand, innate immune cells produce a number of soluble factors that influence EC behavior. In many physiological conditions, recruitment of an inflammatory infiltrate functionally supports angiogenesis and tissue remodelling. During endometrial, decidual or retinal angiogenesis, inflammatory cells regulate vascular proliferation and pattern-

ing by producing both proangiogenic and antiangiogenic factors [24, 30–34]. Innate immune cells, e.g., granulocytes (neutrophils, basophils and eosinophils), DCs, macrophages, NK and mast cells, are also prominent components of pre-malignant and malignant tissues. They functionally contribute to cancer development by releasing of a myriad of cytokines, chemokines, matrix metalloproteinases (MMPs), serine proteases, DNA-damaging molecules (reactive oxygen species), histamine and other bioactive mediators that regulate cell survival, proliferation and motility, along with tissue remodeling and angiogenesis (Fig. 20.1).

Several immune cell-secreted factors exert an important role in supporting tumor angiogenesis. As early as 1971, Judah Folkman postulated that inhibition of angiogenesis would be an effective strategy to treat human cancer [35]. An active search for angiogenesis inducers and inhibitors began thereafter. Extensive research has identified several regulators of angiogenesis, some of which may represent therapeutic targets [24, 27, 36]. The appreciation that immune cells produce many proangiogenic factors has put forward the concept that targeting these cells in tumors may represent a valuable anticancer therapy.

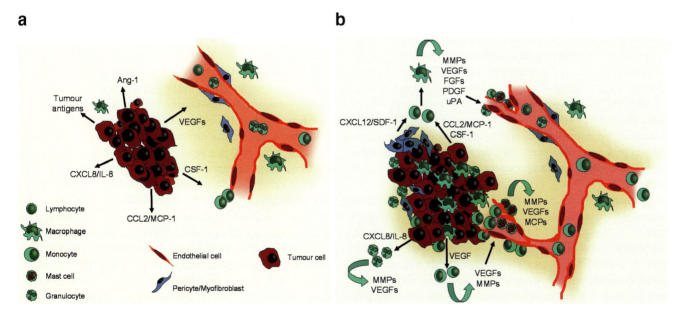

a   **b**

FIG. 20.1. Tumor-derived factors promote angiogenesis by recruiting proangiogenic hematopoietic cells. **A** Initiated neoplastic cells secrete factors that recruit different leukocytic populations, mostly myeloid-lineage cells, including *mast cells*, *neutrophils* and *monocytes*. These tumor-secreted factors include monocyte (colony-stimulating factor-1, *CSF-1*; monocyte chemoattractant protein-1, *MCP-1*; vascular endothelial growth factors, *VEGFs*) and neutrophil (inerleukin-8, *IL-8*) chemoattractants. Other tumor-secreted factors that likely recruit leukocytes are the Angiopoietins (*Ang-1*). Some of these factors also recruit vascular endothelial cells and thus directly stimulate angiogenesis (such as *VEGFs*). **B** The influx of leukocytes into premalignant tissues provides additional factors that foster tumor progression and angiogenesis. Some monocyte chemoattractants stimulate the differentiation of monocytes into macrophages. These cells, together with tumor-infiltrating neutrophils and mast cells, produce a wide array of proangiogenic factors, such as *VEGFs*, fibroblast growth factors (*FGFs*), matrix metalloproteinases (*MMPs*), platelet-derived growth factor (*PDGF*), mast cell-specific serine proteases (*MCPs*) and urokinase-type plasminogen activator (*uPA*). In addition, tumor-infiltrating mesenchymal cells, such as myofibroblasts, produce high levels of stromal cell-derived factor-1 (*SDF-1*), which further recruits CXCR4+ leukocytes to tumors and stimulates the migration of endothelial cells engaged in the angiogenic process.

The Extra-cellular Matrix and Proteases in Tumor Angiogenesis

ECM proteins are produced by both epithelial and stromal cells, with epithelial cells typically contributing basement membrane components (including type IV collagen and laminin) and stromal cells, mainly fibroblasts, supplying connective tissue components, such as type I collagens [37]. ECM strongly influences tumor development. Perturbation of ECM, either through proteolysis and degradation or altered deposition, can positively contribute to tumor progression in a significant way [4, 9, 38–40]. On the other hand, restoration of 'normal' cell-matrix interactions would supply "normalizing" signals to transformed epithelial cells. The matrix acts as a depot for a plethora of growth and proangiogenic factors, including basic fibroblast growth factor (bFGF), VEGF, transforming growth factor-β (TGF-β), and proteolysis of the matrix can unleash a powerful burst of frequently pro-tumorigenic signaling events. Proteolytically generated fragments of matrix molecules are themselves potent soluble signaling molecules, as is particularly evident with the anti-angiogenic fragments angiostatin, endostatin, tumstatin, and others. Thus, the ECM may provide both pro- and antiangiogenic signals to incipient tumors.

ECM-degrading enzymes can be produced within tumor microenvironments by many cell types. Typically, these are proteases of the metallo-, serine and cysteine classes, but other enzymes, for example, heparanase, which is a glucuronidase, are also important [40]. MMPs are a family of highly homologous, secreted or plasma membrane-associated zinc-binding proteinases [41, 42]. MMPs are produced by multiple cell types, regulate many developmental processes and participate in a variety of pathological conditions, including cancer. Initially, MMPs were thought to facilitate neoplastic progression by merely degrading ECM structural components, thereby allowing migration of tumor or ECs. Indeed, cleavage of collagen type I is required for EC invasion of ECM and vessel formation. Several proangiogenic growth factors, most notably VEGF, bFGF and TNF-α, are highly expressed in developing tumors. However, their bioavailability is limited, as they are either sequestered to ECM molecules or tethered to the cell membrane. MMPs (and other extracellular proteases) regulate release of these factors, rendering them available for interaction with cognate receptors on vascular cells and thus activating development of tumor-associated vasculature [43]. This function of MMPs was highlighted in mouse models of multi-step carcinogenesis. In the RIP1-Tag2 mouse model of pancreatic islet carcinogenesis [44], the angiogenic switch that promotes islet angiogenesis and progression to malignancy specifically occurs upon activation of VEGF receptor-2 (VEGFR-2) on ECs. Interestingly, Bergers and colleagues observed that VEGF and acidic FGF were both constitutively expressed in normal islet β-cells of control mice and in all stages of the RIP1-Tag2 islet carcinogenesis. However, these authors demonstrated that, during the angiogenic switch, cells expressing MMP-9/gelatinase-B infiltrated dysplastic islets and MMP-9 produced by tumor-infiltrating immune cells released matrix-sequestered VEGF, making it available for interaction with its receptor on ECs [45]. Coussens and colleagues similarly demonstrated that MMPs and serine proteases that promoted tumor angiogenesis and progression in the K14-HPV16 model of skin carcinogenesis [46] were produced by bone marrow (BM)-derived myeloid cells [6, 47]. In one study [47], infiltration by mast cells and activation of MMP-9 coincided with the angiogenic switch in pre-malignant skin lesions. Mast cells infiltrating hyperplasias, dysplasias, and invasive fronts of carcinomas were shown to degranulate in close apposition to capillaries and epithelial basement membranes, releasing MMP-9 and the mast cell-specific serine proteases MCP-6 (tryptase) and MCP-4 (chymase). MCP-6 is a mitogen for dermal fibroblasts that proliferate in the reactive stroma, whereas MCP-4 activates pro-MMP-9 and induces hyperplastic skin to become angiogenic. In this model, MMP-9 increased the rate and broadened the distribution of hyperproliferation of oncogene-expressing keratinocytes, enhancing malignant conversion of dysplasias into frank carcinomas, and affecting differentiation characteristics of emergent tumors. Notably, mast cell deficiency in K14-HPV16/KIT^{mut} mice resulted in a severe attenuation of early neoplasia, strengthening the notion that mast cells are functionally important in the angiogenic switch. In another study [6], the transplantation of wild-type BM cells into K14-HPV16/MMP-9^{null} mice, which have decreased incidence of skin tumors as compared to K14-HPV16 mice, restored angiogenesis and full neoplastic progression in the mutant mice. In this study, MMP-9 was shown to be predominantly expressed by BM-derived mast cells, monocytes/macrophages and neutrophils infiltrating both pre-malignant and malignant stages of skin carcinogenesis. Together, these studies have uncovered important regulatory capabilities for MMP-9, both during pancreatic and skin carcinogenesis. Importantly, MMP-9 imparted these regulatory capabilities on oncogene-positive neoplastic cells as a paracrine factor, originating from inflammatory cells conscripted to support neoplastic growth and progression.

The processing of pro-growth factors is not a unique property of MMP-9. In fact, several MMP family members are known to possess this property, and some of them (such as MMP-2, MMP-7, MMP-12) also regulate inflammation, tissue remodeling and angiogenesis through their ability to process ECM and ECM-embedded chemokines [42]. A major role for cysteine cathepsin proteases as important mediators of angiogenesis and cancer development has also been recently appreciated [8, 48, 49]. Cysteine cathepsins are lysosomal proteases produced by epithelial cells and leukocytes. These enzymes are involved in many physiological and pathological processes, including tissue remodeling, epithelial homeostasis, degradation of ECM, cell migration and invasion, regulation of inflammatory and immune responses and activation of angiogenesis. Joyce and colleagues recently demonstrated association of increased cathepsin activity with angiogenic vasculature and invasive fronts of carcinomas during tumorigenesis in transgenic mouse models of pancreatic and cervical carcinogenesis [8, 48, 49]. In particular, cysteine cathepsins C and H (among 11 members analyzed) were found to be specifically expressed by immune cells. Of note, the pharmacological inhibition of cysteine cathepsin activity impaired angiogenic switching, tumor growth, and invasion in a pancreatic tumor model, suggesting that broad-spectrum cysteine cathepsin inhibitors may effectively block multiple biological aspects of tumor development, offering new therapeutic opportunities in anti-cancer therapy.

As mentioned above, many matrix-remodeling proteases are stored within and rapidly released from tumor-infiltrating myeloid cells. It has been recently shown that amino-bisphosphonate-mediated blockade of MMP-9 production by macrophages significantly reduced cervical cancer development in K14-HPV16 mice [50]. Thus, drugs that inhibit selected proangiogenic MMPs or other ECM-modifying enzymes should hold promise for effective anti-cancer therapies [51]. However, whereas remodeling of ECM or certain matrix-bound molecules confers a proangiogeneic phenotype (e.g., VEGF), remodeling of others may confer antiangiogenic properties [52].

Embedded within some ECM molecules are bioactive cryptic protein fragments released by proteolytic cleavage. The first example of the release of a bioactive ECM fragment was the isolation of angiostatin from the urine of mice with Lewis lung cell carcinoma [53]. Angiostatin is a plasminogen cleavage product that inhibits EC proliferation and is thought responsible for maintaining Lewis lung cell metastases in a dormant state. Several MMPs, including MMP-2, -7, -9, and -12, can generate angiostatin. Another ECM fragment with antiangiogenic properties is endostatin, produced by cleavage of collagen type XVIII by MMP-3, -7, -9, -12, -13, and -20 and acts by inhibiting EC proliferation [54].

In conclusion, MMP-generated cleavage products of ECM and soluble molecules act either as activators or suppressors of angiogenesis, often in tissue-dependent and stage-dependent manners, and implicate MMP-producing cells—immune cells in particular—as important mediators of tumor-associated angiogenesis by both pro-tumor and anti-tumor mechanisms. Thus, the bi-functional activity of MMPs and of other ECM remodelling enzymes in the context of tumor angiogenesis should be taken into account when designing anti-cancer therapies that selectively target their functions [51].

Growth Factors, Chemokines and Cytokines in Tumor Angiogenesis

Several tumor-infiltrating immune cells produce growth factors and chemokines that are potent proangiogenic mediators (Fig. 20.1). Some of these are described in greater detail below.

VEGF family members and the MMP connection

VEGF family molecules are protypical proangiogenic factors that are produced both by tumor cells and tumor-infiltrating leukocytes, macrophages in particular [55–61]. VEGF-A gene expression is up-regulated by hypoxia, a common feature of the tumor microenvironment. VEGF-A binds to two EC receptor tyrosine kinases, VEGFR1 (Flt-1) and VEGFR2 (Flk-1 or KDR). VEGFR-2 is the major mediator of the mitogenic, angiogenic and permeability-inducing effects of VEGF-A on ECs. The role of VEGFR1 in the regulation of EC biology and angiogenesis is more complex. The VEGFR1, which is not uniquely expressed by ECs, but also by hematopoietic progenitors, monocytes/macrophages and BM stromal cells, can also function as a 'decoy' receptor that sequesters VEGF-A and prevents its interaction with VEGFR2 [62, 63]. In addition to VEGF-A, the VEGF homologue placental growth factor (PlGF) also binds to VEGFR1 and stimulates angiogenesis [59].

There is growing evidence that VEGFR1 has important roles in hematopoiesis and in recruitment of monocytes and other BM-derived cells to tumors and ischemic tissues [64, 59, 60]. Tissue injury and tumor growth can indeed induce plasma elevation of both VEGF-A and PlGF, which promote chemotaxis of hematopoietic progenitors, their mobilization from the BM and recruitment to neo-angiogenic niches [65, 66]. Interestingly, macrophages from VEGFR1 deficient mice display significantly reduced migration in response to VEGF in a mouse model of embryonic angiogenesis [67]. By using a genetic approach to switch expression of VEGF on and off in the heart or liver, Keshet and colleagues showed that an increase in VEGF expression in these organs induced robust angiogenesis [68]. The authors found that locally expressed VEGF efficiently mobilized VEGFR1+ myeloid cells from BM and recruited them to the target organs, where they stimulated angiogenesis by producing proangiogenic factors. Although this study did not investigate tumor angiogenesis, it contributes to the emerging view that proangiogenic factors (such as VEGF) not only directly stimulate the local proliferation of ECs, but also recruit proangiogenic BM-derived hematopoietic cells at sites of angiogenesis [69].

Neuropilin-1 (NP-1) is a VEGF-A co-receptor expressed by ECs and hematopoietic cells that enhances VEGF signaling through VEGFR2 [70]. It has been recently proposed that the NP-1–VEGF complex on hematopoietic cells exogenously stimulates VEGFR2 activation and induces brisk proliferation of ECs. According to this model, leukocytes expressing NP-1, such as monocytes/macrophages that often acquire a peri-endothelial position in tumors, may cluster the proangiogenic factor VEGF in the immediate vicinity of ECs, thus enhancing angiogenesis [71,72].

Besides a direct chemoattractant function, VEGF and PlGF can up-regulate MMP-9 expression in tissue stromal cells, including mesenchymal cells, ECs and leukocytes. Locally expressed MMP-9 then facilitates remodeling of the ECM, enhances cell migration (e.g., in tumors and BM microenvironment), and increases bio-availability of VEGF for ECs [43, 73]. Rafii and colleagues have shown that BM suppression or plasma elevation of VEGF-A and PlGF results in a timely up-regulation of MMP-9 in the BM, that facilitates proteolytical release of bioactive soluble Kit ligand (sKitL). sKitL (also known as stem cell factor, SCF) then promotes proliferation and mobilization of BM c-Kit+ hematopoietic progenitors [74]. These studies indicate that BM activation of MMPs may serve as a checkpoint for mobilization of myeloid progenitors from the BM to the peripheral circulation. Mobilized myeloid-lineage cells, expressing VEGFR1, are then recruited to VEGF-expressing tissues and tumors, where they promote angiogenesis in a paracrine fashion.

In summary, VEGF family members not only directly activate VEGFRs on ECs (thus enabling EC proliferation and angiogenesis), but also indirectly stimulate angiogenesis by (1) recruiting proangiogenic (VEGFR1+) inflammatory cells to tumors and other angiogenic tissues, and by (2) activating MMPs and facilitating tissue remodelling.

Stromal Cell-derived Factor-1

BM-derived and mesenchymal stromal cells, constituting a large proportion of the non-neoplastic cells found in tumors, secrete the chemokine stromal cell-derived factor-1 (SDF-1/CXCL12). SDF-1 secreted by stromal cells functions as a chemoattractant for cells expressing its cognate receptor, CXCR4, that is expressed by a broad range of cell types, including cancer cells, ECs, hematopoietic cells and their progenitors [75]. It has been recently highlighted that the SDF-1/CXCR4 axis promotes tumor progression by both direct and indirect mechanisms. In addition to stimulating the survival and migration of CXCR4+ cancer cells, SDF-1 promotes tumor angiogenesis by attracting CXCR4+ ECs and myeloid cells to the tumor microenvironment [76, 77]. Weinberg and colleagues reported that carcinoma-associated (myo)fibroblasts isolated from breast cancers, but not normal tissue fibroblasts, secrete high levels of SDF-1, that enhanced tumor angiogenesis by functioning as a potent chemoattractant for locally derived ECs and BM-derived hematopoietic cells [78]. In addition, it has been shown that proangiogenic VEGFR1+ myeloid cells also express CXCR4 and that the up-regulation of SDF-1 by perivascular cells is a major determinant for recruitment and retention of proangiogenic VEGFR1+CXCR4+ myeloid cells around nascent vessels [68, 79]. Accordingly, inhibition of the SDF-1/CXCR4

axis decreased the growth of experimental tumors through the suppression of tumor angiogenesis [76]. As such, CXCR4 antagonists, although initially developed for the treatment of non-neoplastic diseases, may actually become effective agents for the treatment of cancer.

Colony Stimulating Factor-1 and Monocyte Chemoattractant Protein-1

The recruitment of monocytes/macrophages to pre-malignant stages has been shown to promote angiogenesis and malignant progression [80, 81]. Colony-stimulating factor-1 (CSF-1) is a key macrophage growth factor, responsible for the survival, proliferation, differentiation and chemotaxis of monocytes/macrophages. CSF-1 is broadly expressed by tumors of the reproductive system, and its expression was found to correlate with the extent of leukocyte infiltration and a poor prognosis in these tumors [81]. Pollard and colleagues have reported that in a Polyoma Middle T antigen-induced mouse model of mammary carcinogenesis (MMTV-PyMT), an increase of macrophage infiltration at the primary tumor site occurred immediately before the angiogenic switch and the onset of malignancy [80, 82]. By using PyMT mice carrying a *Csf-1* null mutation (*Csf-1op/op*), these authors further demonstrated that depletion of CSF-1 markedly decreased infiltration of macrophages at tumor sites, and this inhibited the angiogenic switch and significantly delayed tumor progression. The knock-down of CSF-1 in transplanted tumor cells (by using antisense oligonucleotides) also resulted in inhibition of tumor growth, with tumors exhibiting extensive necrosis and poor vascularization, phenotypes that could be reversed by treatment of the mice with CSF-1. The premature macrophage infiltration in the mammary gland of MMTV LTR-CSF-1 transgenic mice induced robust angiogenesis even at early pre-malignant stages, providing evidence for a direct link between macrophage infiltration and angiogenesis, independent of tumor stage [82]. These studies demonstrate that CSF-1 is a major regulator of macrophage recruitment to tumors, and have shed light on the important roles of macrophages in tumor progression and in particular with tumor-associated angiogenesis.

Similar to CSF-1, several CC chemokines, particularly CCL2 (formally monocyte chemoattractant protein-1, or MCP-1) and CCL5 (RANTES, or regulated on activation normal T cells expressed and secreted), have been implicated in the recruitment of monocytes to tumors [81, 83]. CCL2/MCP-1 over-expression by genetically modified tumor cells implanted in mice promoted monocyte uptake by the tumor mass. In human tumors, CCL2/MCP-1 and CCL5/RANTES are mainly produced by tumor cells and fibroblasts, and their expression has been shown to correlate with macrophage infiltration in many tumors, including bladder, cervix, ovary, breast, lung and brain cancers. Furthermore, both CCL2/MCP-1 and CCL5/RANTES were shown to stimulate monocyte/macrophage-lineage cells to secrete MMP-9 and urokinase-type plasminogen activator (uPA), which through their

ECM-remodeling functions are potent activators of angiogenesis [83]. Thus, thanks to their ability to attract proangiogenic monocytes/macrophages to tumors, both CSF-1 and CCL2/MCP-1 can be regarded as major players in the orchestration of the angiogenic process in tumors.

Interleukin-8

Apart from cell-autonomous effects in tumor cells, Ras oncogene expression regulates tumor-host interactions that are essential for neoplastic progression. Constitutive Ras activity promotes tumor cell invasiveness and angiogenesis by enhancing the expression of MMPs and VEGF, and by down-regulating expression of the antiangiogenic factor thrombospondin-1 (TSP-1). Bar-Sagi and colleagues have recently implicated Ras in inflammation-dependent angiogenesis [84]. These authors demonstrated that activation of Ras proto-oncogenes in cancer cells resulted in up-regulation of the inflammatory chemokine interleukin-8 (IL-8, also known as CXCL8), which in turn promoted tumor growth by enhancing leukocyte infiltration and angiogenesis. In tumors, IL-8 is a potent chemo-attractant for neutrophils that express the IL-8 receptors CXCR-1 and CXCR-2. The authors found that neutralization of IL-8 with antibodies inhibited tumorigenic growth of cancer cells xenografted in immunocompromised mice. Indeed, anti-IL-8 antibodies reduced recruitment of host innate immune cells, decreased tumor vascularization and slowed tumor growth. Although the precise identity of cell type targeted by the tumor-derived IL-8 was not identified in this study, and it is formally possible that IL-8 may directly activate ECs in the angiogenic process, one likely scenario is that angiogenesis was promoted by a product of inflammatory cells recruited to the tumor via IL-8-induced chemotaxis. Thus, IL-8 establishes a direct link between oncogene activation and inflammation in cancer [13]. Moreover, IL-8 adds to the list of tumor-derived factors (VEGF, SDF-1, CSF-1, CCL2/MCP-1, CCL5/RANTES, and others) that indirectly promote angiogenesis by recruiting proangiogenic innate immune cells—mostly neutrophils and macrophages—to tumors.

Immune Cells that Promote Tumor Angiogenesis

Myeloid-lineage immune cells, such as mast cells, macrophages and neutrophils, have been demonstrated to promote tumor progression by exerting a number of pro-tumoral activities, e.g., by stimulating angiogenesis [47, 71, 85–87], suppressing anti-tumor immunity [5, 11, 88] and enhancing tumor cell migration and metastasis [89, 90]. Nucleated hematopoietic cells that have been directly implicated in tumor angiogenesis include mast cells [47], tumor-associated macrophages [2, 91, 92], Tie2-expressing monocytes [85, 93] neutrophils [86], dendritic cell precursors [94] and myeloid-derived suppressor cells [95, 96]. Other hematopoietic cell types, such as platelets [97], eosinophils [98] and hematopoietic progenitors [99], also participate in angiogenic processes, but it remains to be established whether they can directly promote the growth of tumor blood vessels.

Mast Cells

Mast cells accumulate during the premalignant stages of tumor progression and at the periphery of invasive tumors, consistent with a role in activation of angiogenesis. Mast cells have direct proangiogenic activity owing to their production of MMPs, particularly MMP-9, and secretion of angiogenic factors, including basic FGF, VEGF, and IL-8. In addition, mast cells indirectly stimulate angiogenesis by secreting mast cell-specific serine proteases (MCP-4 and MCP-6) that activate pro-MMPs and stimulate stromal fibroblasts to synthesize collagens [47, 100]. Mast cell-secreted VEGF can also up-regulate MMP-9 expression in ECs and other stromal cells and enhance the angiogenic response in the tumor microenvironment. In one study, the genetic deficiency of mast cells was shown to prevent the angiogenic switch and abate premalignant progression in a K14-HPV16/KIT-mutant mouse [47]. In addition to promoting angiogenesis, mast cells are a rich source of cytokines and chemokines, such as IL-1, IL-3, IL-4, granulocyte-macrophage colony-stimulating factor (GM-CSF), CCL-2/MCP-1, macrophage inflammatory protein (MIP-1)-α and β, TNF-α and interferon-γ. Many of these molecules contribute to the tumor microenvironment by enhancing tumor cell growth and invasion either directly or through cell intermediaries such as macrophages.

Tumor-Associated Macrophages (TAMs)

Many human tumors are infiltrated by macrophages, that may derive both from locally recruited tissue macrophages or, more likely, from circulating monocytes [92]. As discussed above, CCL2/MCP-1 and CSF-1 are well known chemoattractants for monocytes/macrophages in tumors. Despite the importance of macrophage infiltrates found in human tumors, the biological significance and prognostic value of these infiltrates has been obscure for many years. Historically, activated macrophages were merely thought of as effector cells that phagocyte microorganisms and kill tumor cells. However, during the past 15 years many clinical reports have highlighted a correlation between the macrophage infiltration in tumors and a poor clinical prognosis [81, 92].

It is known that TAMs heavily infiltrate necrotic areas in tumors, where they scavenge cellular debris. In addition, TAMs accumulate at hypoxic tumor areas, where they may cooperate with tumor cells to promote angiogenesis [83, 101]. In fact, hypoxia stimulates expression of several pro-angiogenic molecules by activating hypoxia-inducible factors (HIFs) in TAMs. Expression of the monocyte chemoattracts VEGF, endothelin 2, and endothelial monocyte-activating polypeptide II (EMAP II) by hypoxic tumor cells can attract TAMs into hypoxic areas within tumors. It is believed that TAMs are then retained in hypoxic tumor areas due to abrogation of chemotactic signal transduction and the down-regulation of chemoattractant receptors. Once in hypoxic areas, TAMs produce a wide array of proangiogenic molecules and matrix-remodeling factors, including IL-8/CXCL8, VEGF, FGF, platelet-derived growth factor (PDGF), MMPs and uPA, but it remains to be clarified how crucial these TAM-secreted factors are in the economy of tumor angiogenesis, since many proangiogenic molecules are also produced by other components of the tumor stroma and by the tumor cells themselves. In addition, TAMs are characterized by a high degree of heterogeneity, and it may be difficult to establish whether TAMs in general, rather than specific subsets of these cells, play a critical role in tumor angiogenesis [2, 91].

In addition to the aforementioned proangiogenic factors, TAMs release other molecules that can influence angiogenesis [92]. TAMs are key producers of TNF-α, which can up-regulate expression of thymidine phosphorylase (TP) and MMP-9. TAMs also produce IL-1, which may increase VEGF transcription by up-regulating expression of HIF-1α through COX 2. In addition, TAMs also release nitric oxyde (NO), a molecule that provokes vasodilation and increases vascular flow, through the activity of inducible NO synthase (iNOS).

It has been proposed that many tumor-secreted factors, such as IL-4, IL-13, IL-10, CSF-1, TGF-β and prostaglandin E2, can blunt the tumoricidal activity of macrophages and activate them to acquire a growth-promoting and proangiogenic function [2]. However, further studies are required to better understand macrophage heterogeneity and function in tumors.

Tie2-Expressing Monocytes (TEMs)

TEMs are a subset of circulating and tumor-infiltrating monocytes characterized by expression of the angiopoietin receptor Tie2 [102], a molecule previously known to be restricted to ECs and haematopoietic stem cells [103]. TEMs have been observed in several mouse tumor models—including subcutaneous tumor grafts, orthotopically growing gliomas and spontaneous pancreatic tumors—where they represent 1–15% of the total CD11b$^+$ myeloid cells. A peculiar feature of TEMs is that they preferentially localize around angiogenic blood vessels in tumors, a figure that is consistent with their marked proangiogenic activity in transplantation assays [85]. It appears that TEMs are a sub-population of TAMs, possibly overlapping with alternatively activated macrophages [2]. However, TEMs can be distinguished from other TAM populations by their surface marker profile, their preferential localization around angiogenic tumor vessels, their absence from necrotic tumor regions, and their marked proangiogenic activity.

TEMs can substantially accelerate vascularization of tumor grafts and are required for angiogenesis in certain tumors, thus they may represent pivotal triggers of the angiogenic switch during tumor growth. Targeted elimination of TEMs by means of a suicide gene impaired neovascularization of

human gliomas grafted in the mouse brain and induced substantial tumor regression [85]. Because TEM elimination did not affect recruitment of TAMs to necrotic tumor areas, it is unlikely that TEMs comprise precursors of other TAM populations. Rather, it would appear that TEMs represent a distinct monocyte/macrophage subset with inherent proangiogenic activity; indeed, TEMs possess proangiogenic activity already when they circulate in the peripheral blood, before reaching the tumor site. In this regard, identification of proangiogenic TEMs among the heterogeneous TAM population may challenge the notion that transition of tumor macrophage phenotype between growth-inhibitory and growth-promoting activity is exclusively and contextually modulated by the tumor microenvironment [2, 85].

Neutrophils

The role of neutrophils in tumor progression has been controversial. During inflammatory responses, neutrophils are among the first cells to arrive at inflamed sites, where they phagocytize cellular debris and microorganisms and release chemokines and proteases that in turn recruit additional immune effector cells. Moreover, neutrophils are involved in graft rejection, indicating that they might also be tumoricidal. However, whereas some reports have shown that increased neutrophil infiltration was linked to poor outcome, others suggested that neutrophil infiltration correlated with favorable prognosis [81]. Despite contradictory clinical studies, experimental studies that using mouse tumor models have found that tumor-associated neutrophils are involved in tumor angiogenesis and therefore can be pro-tumoral. The proangiogenic activity of neutrophils may derive from their production of canonical proangiogenic factors, such as VEGF, IL-8, MMPs and elastases [104]. Recently, Hanahan and colleagues investigated the putative role of neutrophils in multi-stage pancreatic carcinigenesis [86]. These authors observed that scarce MMP-9-expressing neutrophils were located inside angiogenic islet dysplasias and tumors, whereas more abundant MMP-9-expressing macrophages were mainly distributed along the periphery of such lesions. Interestingly, the transient ablation of neutrophils by using anti-Gr-1 antibodies reduced the frequency of angiogenic switching in pre-neoplastic lesions, but did not inhibit angiogenesis and tumor progression in late-stage tumors. These data are consistent with the proposition that neutrophils, that are 10-fold less abundant than macrophages in both angiogenic islets and tumors, may provide an initial, non-redundant source of MMP-9 for catalyzing the angiogenic switch in pre-angiogenic lesions.

Dendritic Cell Precursors, Myeloid-derived Suppressor Cells and Endothelial-like Monocyte-derived Cells

Dendritic cells (DCs) play important roles both in activation and suppression of anti-tumor immunity. Recently, DC precursors have been implicated in tumor angiogenesis

[94, 105]. Coukos and colleagues described a population of CD45+CD11c+MHC-II+ DC precursors that infiltrated human ovarian carcinomas. These cells were termed "vascular leukocytes" because, in addition to the aforementioned hematopoietic markers, they expressed EC-specific markers, including vascular endothelial cadherin (VE-Cad). CD45+VE-Cad+ cells isolated from human ovarian cancers by cell sorting formed perfused vascular channels in matrigel *in vivo*, suggesting that vascular leukocytes can directly participate in angiogenesis. The mixed hematopoietic/EC phenotype of vascular leukocytes is consistent with other reports demonstrating that monocytes and immature myeloid cells can acquire an endothelial-like phenotype under angiogenic conditions [106, 107].

Another myeloid population recently implicated in tumor angiogenesis are the so-called myeloid-derived suppressor cells [95, 108]. Myeloid-derived suppressor cells are a heterogeneous population, comprising myeloid progenitors, monocytes and neutrophils, that express low to undetectable levels of MHC-II and co-stimulatory molecules, therefore they cannot induce anti-tumor responses. Rather, these cells promote tumor development by exerting a profound inhibitory activity on both tumor-specific and non-specific T lymphocytes and, as recently described, by providing factors essential for tumor growth and neovascularization [96]. The frequency of myeloid-derived suppressor cells is significantly increased in the BM and spleen of cancer patients and mice carrying large tumors. Lin and colleagues [96] found that Gr-1+CD11b+ myeloid-derived suppressor cells obtained from spleens of tumor-bearing mice promoted angiogenesis and tumor growth when co-injected with tumor cells. Myeloid-derived suppressor cells produced high levels of MMP-9, and deletion of MMP-9 in these cells completely abolished their tumor-promoting activity. Similar to DC precursors described above, Gr-1+CD11b+ cells were also found to occasionally incorporate into tumor endothelium as endothelial-like cells [96].

The aforementioned studies highlight the ability of myeloid-lineage cells to promote tumor angiogenesis and sustain tumor progression. Although it appears that the proangiogenic function of myeloid cells in tumors mostly consists of the production of growth factors and matrix-remodeling proteins that stimulate angiogenic processes in a paracrine manner, the occasional luminal incorporation of myeloid cells in vascular endothelium has also been documented as a rare phenomenon often regarded as evidence for post-natal vasculogenesis [94, 96].

Monocytes/macrophages and ECs share phenotypical and functional features, including the expression of common metabolic and surface markers, as well as an ability to form vascular-like structures [106, 107, 110]. Surface markers co-expressed by ECs and hematopoietic subsets include VEGFR1, Sca-1, Tie2, AC133, CD31 (PECAM-1), von Willebrand Factor and CD146 (S-endo-1 or P1H12). This may have led some monocyte-derived populations to be incorrectly regarded as bona fide BM-derived "endothelial progenitor cells" [93, 109, 111].

Monocytes are a highly plastic cell types that can modulate their phenotype according to local conditions, and increasing reports suggest that in vitro cultured monocytes can be differentiated into endothelial-like cells [106, 107, 112, 113]. These ex-vivo expanded endothelial-like cells have been shown to participate in vascular healing and angiogenesis under certain experimental conditions. This function is likely enhanced by the release of several monocyte-derived proangiogenic factors that may promote the local recruitment and proliferation of ECs [68, 112].

In summary, several types of immune cells may play important functions in tumor angiogenesis. The ability of certain hematopoietic cells—and in particular monocytes/macrophages—to migrate within tissues, even in hypoxic conditions, and remodel the extracellular environment would account for their critical role in angiogenesis. This concept may provide an alternative explanation for the contribution of BM-derived hematopoietic cells to post-natal vasculogenesis.

Anti-inflammatory Drugs Meet Antiangiogenic Therapy

Hematopoietic cells of the innate immune system have a requisite role in tumor development, and increasing evidence suggests that one of the mechanisms by which they foster tumorigenesis is the promotion of angiogenesis [93, 114]. Inflammatory immune cells that are thought to promote angiogenesis in a paracrine manner (e.g., by releasing proangiogenic factors or ECM-modifying enzymes) include several myeloid cell populations. From the data discussed here, it follows that antiangiogenic therapy—such as treatments using anti-VEGF antibodies, VEGFR inhibitors or vascular disrupting agents—may be improved by drugs that concomitantly target proangiogenic inflammatory cells [115]. Clinical studies have recently shown that regular use of non-steroidal anti-inflammatory drugs (NSAIDs) is associated with reduced risk of some cancers [17]. Given the intimate association between inflammation and angiogenesis, NSAIDs may also function as angiopreventive molecules. On the other hand, the growing body of evidence that canonical regulators of angiogenesis, such as VEGF and SDF-1, also stimulate mobilization and recruitment of proangiogenic myeloid cells to tumors, has broadened the variety of cell types that might be concomitantly targeted by conventional antiangiogenic agents [114, 116]. Assessing the specific contribution of different BM-derived hematopoietic cell types to tumor angiogenesis, together with the identification of selective targets, may have important implications for the design of improved anti-cancer therapies. Macrophages and other innate immune cells are genetically stable cells that are less likely to develop drug resistance than cancer cells, thus drugs that inhibit selected macrophage functions should hold promise for effective anticancer therapies [50, 89]. The outstanding challenge is, however, to understand which targets, if any, could distinguish immune cells implicated in tumor growth from those that regulate important physiological processes, such as immunity and tissue homeostasis.

Acknowledgements. We acknowledge all the scientists who made contributions to the areas of research reviewed here that were not cited due to space constraints. M.D.P. wishes to thank Luigi Naldini for his mentorship and financial support. The authors were supported by grants from the Associazione Italiana per la Ricerca sul Cancro (M.D.P.), the National Institutes of Health, Sandler Program in Basic Sciences, National Technology Center for Networks and Pathways and a Department of Defense Era of Hope Scholar Award (L.M.C.).

References

1. Hanahan, D., and Weinberg, R. A. (2000). The hallmarks of cancer. Cell *100*, 57–70.
2. Balkwill, F., Charles, K. A., and Mantovani, A. (2005). Smoldering and polarized inflammation in the initiation and promotion of malignant disease. Cancer Cell *7*, 211–217.
3. Balkwill, F., and Mantovani, A. (2001). Inflammation and cancer: back to Virchow? Lancet *357*, 539–545.
4 Bissell, M. J., and Radisky, D. (2001). Putting tumors in context. Nat Rev Cancer *1*, 46–54.
5 Blankenstein, T. (2005). The role of tumor stroma in the interaction between tumor and immune system. Curr Opin Immunol *17*, 180–186.
6. Coussens, L. M., Tinkle, C. L., Hanahan, D., and Werb, Z. (2000). MMP-9 supplied by bone marrow-derived cells contributes to skin carcinogenesis. Cell *103*, 481–490.
7. Coussens, L. M., and Werb, Z. (2002). Inflammation and cancer. Nature *420*, 860–867.
8. Joyce, J. A. (2005). Therapeutic targeting of the tumor microenvironment. Cancer Cell *7*, 513–520.
9. Mueller, M. M., and Fusenig, N. E. (2004). Friends or foes - bipolar effects of the tumor stroma in cancer. Nat Rev Cancer *4*, 839–849.
10. de Visser, K. E., Eichten, A., and Coussens, L. M. (2006). Paradoxical roles of the immune system during cancer development. Nat Rev Cancer *6*, 24–37.
11. Zou, W. (2005). Immunosuppressive networks in the tumor environment and their therapeutic relevance. Nat Rev Cancer *5*, 263–274.
12. Dunn, G. P., Old, L. J., and Schreiber, R. D. (2004). The immunobiology of cancer immunosurveillance and immunoediting. Immunity *21*, 137–148.
13. Karin, M. (2005). Inflammation and cancer: the long reach of Ras. Nat Med *11*, 20–21.
14. Karin, M. (2006). Nuclear factor-kappaB in cancer development and progression. Nature *441*, 431–436.
15. Vakkila, J., and Lotze, M. T. (2004). Inflammation and necrosis promote tumor growth. Nat Rev Immunol *4*, 641–648.
16. Finch, C. E., and Crimmins, E. M. (2004). Inflammatory exposure and historical changes in human life-spans. Science *305*, 1736–1739.
17. Dannenberg, A. J., and Subbaramaiah, K. (2003). Targeting cyclooxygenase-2 in human neoplasia: rationale and promise. Cancer Cell *4*, 431–436.
18. Balkwill, F., and Coussens, L. M. (2004). Cancer: an inflammatory link. Nature *431*, 405–406.
19. Greten, F. R., Eckmann, L., Greten, T. F., Park, J. M., Li, Z. W., Egan, L. J., Kagnoff, M. F., and Karin, M. (2004). IKKbeta links Inflammation and tumorigenesis in mouse model for colitis-associated cancer. Cell *118*, 285–296.
20. Pikarsky, E., Porat, R. M., Stein, I., Abramovitch, R., Amit, S., Kasem, S., Gutkovich-Pyest, E., Urieli-Shoval, S., Galun, E., and Ben-Neriah, Y. (2004). NF-kappaB functions as a tumor promoter in inflammation-associated cancer. Nature *431*, 461–466.
21. de Visser, K. E., Korets, L. V., and Coussens, L. M. (2005). De novo carcinogenesis promoted by chronic inflammation is B lymphocyte dependent. Cancer Cell *7*, 411–423.
22. Mantovani, A. (2005). Cancer: inflammation by remote control. Nature *435*, 752–753.
23. Tan, Tingting, and Coussens, L.M., (2007). Humoral immunity, inflammation and cancer. Curr Opin Immunol, *in press*.
24 Carmeliet, P. (2005). Angiogenesis in life, disease and medicine. Nature *438*, 932–936.
25. Jain, R. K. (2003). Molecular regulation of vessel maturation. Nat Med *9*, 685–693.
26. Bergers, G., and Benjamin, L. E. (2003). Tumorigenesis and the angiogenic switch. Nat Rev Cancer *3*, 401–410.
27. Hanahan, D., and Folkman, J. (1996). Patterns and emerging mechanisms of the angiogenic switch during tumorigenesis. Cell *86*, 353–364.
28. McDonald, D. M., and Choyke, P. L. (2003). Imaging of angiogenesis: from microscope to clinic. Nat Med *9*, 713–725.
29. Morikawa, S., Baluk, P., Kaidoh, T., Haskell, A., Jain, R. K., and McDonald, D. M. (2002). Abnormalities in pericytes on blood vessels and endothelial sprouts in tumors. Am J Pathol *160*, 985–1000.
30. Gariano, R. F., and Gardner, T. W. (2005). Retinal angiogenesis in development and disease. Nature *438*, 960–966.
31. Girling, J. E., and Rogers, P. A. (2005). Recent advances in endometrial angiogenesis research. Angiogenesis 8, 89–99.
32. Hanna, J., Goldman-Wohl, D., Hamani, Y., Avraham, I., Greenfield, C., Natanson-Yaron, S., Prus, D., Cohen-Daniel, L., Arnon, T. I., Manaster, I., *et al.* (2006). Decidual NK cells regulate key developmental processes at the human fetal-maternal interface. Nat Med *12*, 1065–1074.
33. Ishida, S., Usui, T., Yamashiro, K., Kaji, Y., Amano, S., Ogura, Y., Hida, T., Oguchi, Y., Ambati, J., Miller, J. W., *et al.* (2003a). VEGF164-mediated inflammation is required for pathological, but not physiological, ischemia-induced retinal neovascularization. J Exp Med *198*, 483–489.
34. Ishida, S., Yamashiro, K., Usui, T., Kaji, Y., Ogura, Y., Hida, T., Honda, Y., Oguchi, Y., and Adamis, A. P. (2003b). Leukocytes mediate retinal vascular remodeling during development and vaso-obliteration in disease. Nat Med *9*, 781–788.
35. Folkman, J. (1971). Tumor angiogenesis: therapeutic implications. N Engl J Med *285*, 1182–1186.
36. Carmeliet, P., and Jain, R. K. (2000). Angiogenesis in cancer and other diseases. Nature *407*, 249–257.
37. Kalluri, R. (2003). Basement membranes: structure, assembly and role in tumor angiogenesis. Nat Rev Cancer *3*, 422–433.
38. Littlepage, L. E., Egeblad, M., and Werb, Z. (2005). Coevolution of cancer and stromal cellular responses. Cancer Cell *7*, 499–500.
39. Tlsty, T. D. (2001). Stromal cells can contribute oncogenic signals. Semin Cancer Biol *11*, 97–104.

40. van Kempen, L. C., de Visser, K. E., and Coussens, L. M. (2006). Inflammation, proteases and cancer. Eur J Cancer *42*, 728–734.

41. Bergers, G., and Coussens, L. M. (2000). Extrinsic regulators of epithelial tumor progression: metalloproteinases. Curr Opin Genet Dev *10*, 120–127.

42. Egeblad, M., and Werb, Z. (2002). New functions for the matrix metalloproteinases in cancer progression. Nat Rev Cancer *2*, 161–174.

43. Heissig, B., Hattori, K., Friedrich, M., Rafii, S., and Werb, Z. (2003). Angiogenesis: vascular remodeling of the extracellular matrix involves metalloproteinases. Curr Opin Hematol *10*, 136–141.

44. Hanahan, D. (1985). Heritable formation of pancreatic beta-cell tumors in transgenic mice expressing recombinant insulin/simian virus 40 oncogenes. Nature *315*, 115–122.

45. Bergers, G., Brekken, R., McMahon, G., Vu, T. H., Itoh, T., Tamaki, K., Tanzawa, K., Thorpe, P., Itohara, S., Werb, Z., and Hanahan, D. (2000). Matrix metalloproteinase-9 triggers the angiogenic switch during carcinogenesis. Nat Cell Biol *2*, 737–744.

46. Arbeit, J. M., Munger, K., Howley, P. M., and Hanahan, D. (1994). Progressive squamous epithelial neoplasia in K14-human papillomavirus type 16 transgenic mice. J Virol *68*, 4358–4368.

47. Coussens, L. M., Raymond, W. W., Bergers, G., Laig-Webster, M., Behrendtsen, O., Werb, Z., Caughey, G. H., and Hanahan, D. (1999). Inflammatory mast cells up-regulate angiogenesis during squamous epithelial carcinogenesis. Genes Dev *13*, 1382–1397.

48. Gocheva, V., Zeng, W., Ke, D., Klimstra, D., Reinheckel, T., Peters, C., Hanahan, D., and Joyce, J. A. (2006). Distinct roles for cysteine cathepsin genes in multistage tumorigenesis. Genes Dev *20*, 543–556.

49. Joyce, J. A., Baruch, A., Chehade, K., Meyer-Morse, N., Giraudo, E., Tsai, F. Y., Greenbaum, D. C., Hager, J. H., Bogyo, M., and Hanahan, D. (2004). Cathepsin cysteine proteases are effectors of invasive growth and angiogenesis during multistage tumorigenesis. Cancer Cell *5*, 443–453.

50. Giraudo, E., Inoue, M., and Hanahan, D. (2004). An aminobisphosphonate targets MMP-9-expressing macrophages and angiogenesis to impair cervical carcinogenesis. J Clin Invest *114*, 623–633.

51. Coussens, L. M., Fingleton, B., and Matrisian, L. M. (2002). Matrix metalloproteinase inhibitors and cancer: trials and tribulations. Science *295*, 2387–2392.

52. Sottile, J. (2004). Regulation of angiogenesis by extracellular matrix. Biochim Biophys Acta *1654*, 13–22.

53. O'Reilly, M. S., Holmgren, L., Shing, Y., Chen, C., Rosenthal, R. A., Moses, M., Lane, W. S., Cao, Y., Sage, E. H., and Folkman, J. (1994). Angiostatin: a novel angiogenesis inhibitor that mediates the suppression of metastases by a Lewis lung carcinoma. Cell *79*, 315–328.

54. O'Reilly, M. S., Boehm, T., Shing, Y., Fukai, N., Vasios, G., Lane, W. S., Flynn, E., Birkhead, J. R., Olsen, B. R., and Folkman, J. (1997). Endostatin: an endogenous inhibitor of angiogenesis and tumor growth. Cell *88*, 277–285.

55. Alitalo, K., and Carmeliet, P. (2002). Molecular mechanisms of lymphangiogenesis in health and disease. Cancer Cell *1*, 219–227.

56. Ferrara, N., Gerber, H. P., and LeCouter, J. (2003). The biology of VEGF and its receptors. Nat Med *9*, 669–676.

57. Hicklin, D. J., and Ellis, L. M. (2005). Role of the vascular endothelial growth factor pathway in tumor growth and angiogenesis. J Clin Oncol *23*, 1011–1027.

58. Hiratsuka, S., Maru, Y., Okada, A., Seiki, M., Noda, T., and Shibuya, M. (2001). Involvement of Flt-1 tyrosine kinase (vascular endothelial growth factor receptor-1) in pathological angiogenesis. Cancer Res *61*, 1207–1213.

59. Luttun, A., Tjwa, M., and Carmeliet, P. (2002a). Placental growth factor (PlGF) and its receptor Flt-1 (VEGFR-1): novel therapeutic Targest for angiogenic disorders. Ann NY Acad Sci *979*, 80–93.

60. Shibuya, M. (2006). Differential roles of vascular endothelial growth factor receptor-1 and receptor-2 in angiogenesis. J Biochem Mol Biol *39*, 469–478.

61. Shibuya, M., and Claesson-Welsh, L. (2006). Signal transduction by VEGF receptors in regulation of angiogenesis and lymphangiogenesis. Exp Cell Res *312*, 549–560.

62. Eubank, T. D., Roberts, R., Galloway, M., Wang, Y., Cohn, D. E., and Marsh, C. B. (2004). GM-CSF induces expression of soluble VEGF receptor-1 from human monocytes and inhibits angiogenesis in mice. Immunity *21*, 831–842.

63. Kendall, R. L., Wang, G., and Thomas, K. A. (1996). Identification of a natural soluble form of the vascular endothelial growth factor receptor, FLT-1, and its heterodimerization with KDR. Biochem Biophys Res Commun *226*, 324–328.

64. Kopp, H. G., Ramos, C. A., and Rafii, S. (2006). Contribution of endothelial progenitors and proangiogenic hematopoietic cells to vascularization of tumor and ischemic tissue. Curr Opin Hematol *13*, 175–181.

65. Hattori, K., Heissig, B., Wu, Y., Dias, S., Tejada, R., Ferris, B., Hicklin, D. J., Zhu, Z., Bohlen, P., Witte, L., *et al.* (2002). Placental growth factor reconstitutes hematopoiesis by recruiting VEGFR1(+) stem cells from bone-marrow microenvironment. Nat Med *8*, 841–849.

66. Luttun, A., Tjwa, M., Moons, L., Wu, Y., Angelillo-Scherrer, A., Liao, F., Nagy, J. A., Hooper, A., Priller, J., De Klerck, B., *et al.* (2002b). Revascularization of ischemic tissues by PlGF treatment, and inhibition of tumor angiogenesis, arthritis and atherosclerosis by anti-Flt1. Nat Med *8*, 831–840.

67. Hiratsuka, S., Minowa, O., Kuno, J., Noda, T., and Shibuya, M. (1998). Flt-1 lacking the tyrosine kinase domain is sufficient for normal development and angiogenesis in mice. Proc Natl Acad Sci U S A *95*, 9349–9354.

68. Grunewald, M., Avraham, I., Dor, Y., Bachar-Lustig, E., Itin, A., Yung, S., Chimenti, S., Landsman, L., Abramovitch, R., and Keshet, E. (2006). VEGF-induced adult neovascularization: recruitment, retention, and role of accessory cells. Cell *124*, 175–189.

69. Ruiz de Almodovar, C., Luttun, A., and Carmeliet, P. (2006). An SDF-1 trap for myeloid cells stimulates angiogenesis. Cell *124*, 18–21.

70. Lee, P., Goishi, K., Davidson, A. J., Mannix, R., Zon, L., and Klagsbrun, M. (2002). Neuropilin-1 is required for vascular development and is a mediator of VEGF-dependent angiogenesis in zebrafish. Proc Natl Acad Sci USA *99*, 10470–10475.

71. Takakura, N. (2006). Role of hematopoietic lineage cells as accessory components in blood vessel formation. Cancer Sci *97*, 568–574.

72. Yamada, Y., Oike, Y., Ogawa, H., Ito, Y., Fujisawa, H., Suda, T., and Takakura, N. (2003). Neuropilin-1 on hematopoietic cells as a source of vascular development. Blood *101*, 1801–1809.

73. Hiratsuka, S., Nakamura, K., Iwai, S., Murakami, M., Itoh, T., Kijima, H., Shipley, J. M., Senior, R. M., and Shibuya, M.

(2002). MMP9 induction by vascular endothelial growth factor receptor-1 is involved in lung-specific metastasis. Cancer Cell 2, 289–300.

74. Heissig, B., Hattori, K., Dias, S., Friedrich, M., Ferris, B., Hackett, N. R., Crystal, R. G., Besmer, P., Lyden, D., Moore, M. A., et al. (2002). Recruitment of stem and progenitor cells from the bone marrow niche requires MMP-9 mediated release of kit-ligand. Cell 109, 625–637.

75 Burger, J. A., and Kipps, T. J. (2006). CXCR4: a key receptor in the crosstalk between tumor cells and their microenvironment. Blood 107, 1761–1767.

76. Guleng, B., Tateishi, K., Ohta, M., Kanai, F., Jazag, A., Ijichi, H., Tanaka, Y., Washida, M., Morikane, K., Fukushima, Y., et al. (2005). Blockade of the stromal cell-derived factor-1/CXCR4 axis attenuates in vivo tumor growth by inhibiting angiogenesis in a vascular endothelial growth factor-independent manner. Cancer Res 65, 5864–5871.

77. Kryczek, I., Lange, A., Mottram, P., Alvarez, X., Cheng, P., Hogan, M., Moons, L., Wei, S., Zou, L., Machelon, V., et al. (2005). CXCL12 and vascular endothelial growth factor synergistically induce neoangiogenesis in human ovarian cancers. Cancer Res 65, 465–472.

78. Orimo, A., Gupta, P. B., Sgroi, D. C., Arenzana-Seisdedos, F., Delaunay, T., Naeem, R., Carey, V. J., Richardson, A. L., and Weinberg, R. A. (2005). Stromal fibroblasts present in invasive human breast carcinomas promote tumor growth and angiogenesis through elevated SDF-1/CXCL12 secretion. Cell 121, 335–348.

79. Jin, D. K., Shido, K., Kopp, H. G., Petit, I., Shmelkov, S. V., Young, L. M., Hooper, A. T., Amano, H., Avecilla, S. T., Heissig, B., et al. (2006). Cytokine-mediated deployment of SDF-1 induces revascularization through recruitment of CXCR4(+) hemangiocytes. Nat Med 12, 557–567.

80. Lin, E. Y., Nguyen, A. V., Russell, R. G., and Pollard, J. W. (2001). Colony-stimulating factor 1 promotes progression of mammary tumors to malignancy. J Exp Med 193, 727–740.

81. Lin, E. Y., and Pollard, J. W. (2004). Role of infiltrated leucocytes in tumor growth and spread. Br J Cancer 90, 2053–2058.

82. Lin, E. Y., Li, J. F., Gnatovskiy, L., Deng, Y., Zhu, L., Grzesik, D. A., Qian, H., Xue, X. N., and Pollard, J. W. (2006). Macrophages regulate the angiogenic switch in a mouse model of breast cancer. Cancer Res 66, 11238–11246.

83. Murdoch, C., Giannoudis, A., and Lewis, C. E. (2004). Mechanisms regulating the recruitment of macrophages into hypoxic areas of tumors and other ischemic tissues. Blood 104, 2224–2234.

84. Sparmann, A., and Bar-Sagi, D. (2004). Ras-induced interleukin-8 expression plays a critical role in tumor growth and angiogenesis. Cancer Cell 6, 447–458.

85. De Palma, M., Venneri, M. A., Galli, R., Sergi, L. S., Politi, L. S., Sampaolesi, M., and Naldini, L. (2005). Tie2 identifies a hematopoietic lineage of proangiogenic monocytes required for tumor vessel formation and a mesenchymal population of pericyte progenitors. Cancer Cell 8, 211–226.

86. Nozawa, H., Chiu, C., and Hanahan, D. (2006). Infiltrating neutrophils mediate the initial angiogenic switch in a mouse model of multistage carcinogenesis. Proc Natl Acad Sci USA 103, 12493–12498.

87. Okamoto, R., Ueno, M., Yamada, Y., Takahashi, N., Sano, H., Suda, T., and Takakura, N. (2005). Hematopoietic cells regulate the angiogenic switch during tumorigenesis. Blood 105, 2757–2763.

88. Bronte, V., Cingarlini, S., Marigo, I., De Santo, C., Gallina, G., Dolcetti, L., Ugel, S., Peranzoni, E., Mandruzzato, S., and Zanovello, P. (2006). Leukocyte infiltration in cancer creates an unfavorable environment for antitumor immune responses: a novel target for therapeutic intervention. Immunol Invest 35, 327–357.

89. Condeelis, J., and Pollard, J. W. (2006). Macrophages: obligate partners for tumor cell migration, invasion, and metastasis. Cell 124, 263–266.

90. Wyckoff, J., Wang, W., Lin, E. Y., Wang, Y., Pixley, F., Stanley, E. R., Graf, T., Pollard, J. W., Segall, J., and Condeelis, J. (2004). A paracrine loop between tumor cells and macrophages is required for tumor cell migration in mammary tumors. Cancer Res 64, 7022–7029.

91. Lewis, C. E., and Pollard, J. W. (2006). Distinct role of macrophages in different tumor microenvironments. Cancer Res 66, 605–612.

92. Pollard, J. W. (2004). Tumor-educated macrophages promote tumor progression and metastasis. Nat Rev Cancer 4, 71–78.

93. De Palma, M., and Naldini, L. (2006). Role of haematopoietic cells and endothelial progenitors in tumor angiogenesis. Biochim Biophys Acta 1766, 159–166.

94. Coukos, G., Benencia, F., Buckanovich, R. J., and Conejo-Garcia, J. R. (2005). The role of dendritic cell precursors in tumor vasculogenesis. Br J Cancer 92, 1182–1187.

95. Serafini, P., Borrello, I., and Bronte, V. (2006). Myeloid suppressor cells in cancer: recruitment, phenotype, properties, and mechanisms of immune suppression. Semin Cancer Biol 16, 53–65.

96. Yang, L., DeBusk, L. M., Fukuda, K., Fingleton, B., Green-Jarvis, B., Shyr, Y., Matrisian, L. M., Carbone, D. P., and Lin, P. C. (2004). Expansion of myeloid immune suppressor Gr+CD11b+ cells in tumor-bearing host directly promotes tumor angiogenesis. Cancer Cell 6, 409–421.

97. Kisucka, J., Butterfield, C. E., Duda, D. G., Eichenberger, S. C., Saffaripour, S., Ware, J., Ruggeri, Z. M., Jain, R. K., Folkman, J., and Wagner, D. D. (2006). Platelets and platelet adhesion support angiogenesis while preventing excessive hemorrhage. Proc Natl Acad Sci USA 103, 855–860.

98. Puxeddu, I., Alian, A., Piliponsky, A. M., Ribatti, D., Panet, A., and Levi-Schaffer, F. (2005). Human peripheral blood eosinophils induce angiogenesis. Int J Biochem Cell Biol 37, 628–636.

99. Takakura, N., Watanabe, T., Suenobu, S., Yamada, Y., Noda, T., Ito, Y., Satake, M., and Suda, T. (2000). A role for hematopoietic stem cells in promoting angiogenesis. Cell 102, 199–209.

100. Ribatti, D., Vacca, A., Nico, B., Crivellato, E., Roncali, L., and Dammacco, F. (2001). The role of mast cells in tumor angiogenesis. Br J Haematol 115, 514–521.

101. Lewis, C., and Murdoch, C. (2005). Macrophage responses to hypoxia: implications for tumor progression and anti-cancer therapies. Am J Pathol 167, 627–635.

102. De Palma, M., Venneri, M. A., Roca, C., and Naldini, L. (2003). Targeting exogenous genes to tumor angiogenesis by transplantation of genetically modified hematopoietic stem cells. Nat Med 9, 789–795.

103. Jones, N., Iljin, K., Dumont, D. J., and Alitalo, K. (2001). Tie receptors: new modulators of angiogenic and lymphangiogenic responses. Nat Rev Mol Cell Biol 2, 257–267.

104. Benelli, R., Albini, A., and Noonan, D. (2003). Neutrophils and angiogenesis: potential initiators of the angiogenic cascade. Chem Immunol Allergy *83*, 167–181.

105. Conejo-Garcia, J. R., Buckanovich, R. J., Benencia, F., Courreges, M. C., Rubin, S. C., Carroll, R. G., and Coukos, G. (2005). Vascular leukocytes contribute to tumor vascularization. Blood *105*, 679–681.

106. Schmeisser, A., Garlichs, C. D., Zhang, H., Eskafi, S., Graffy, C., Ludwig, J., Strasser, R. H., and Daniel, W. G. (2001). Monocytes coexpress endothelial and macrophagocytic lineage markers and form cord-like structures in Matrigel under angiogenic conditions. Cardiovasc Res *49*, 671–680.

107. Urbich, C., Heeschen, C., Aicher, A., Dernbach, E., Zeiher, A. M., and Dimmeler, S. (2003). Relevance of monocytic features for neovascularization capacity of circulating endothelial progenitor cells. Circulation *108*, 2511–2516.

108. Gallina, G., Dolcetti, L., Serafini, P., De Santo, C., Marigo, I., Colombo, M. P., Basso, G., Brombacher, F., Borrello, I., Zanovello, P., *et al.* (2006). Tumors induce a subset of inflammatory monocytes with immunosuppressive activity on CD8+ T cells. J Clin Invest *116*, 2777–2790.

109. Barber, C. L., and Iruela-Arispe, M. L. (2006). The ever-elusive endothelial progenitor cell: identities, functions and clinical implications. Pediatr Res *59*, 26R-32R.

110. Moldovan, N. I., Goldschmidt-Clermont, P. J., Parker-Thornburg, J., Shapiro, S. D., and Kolattukudy, P. E. (2000). Contribution of monocytes/macrophages to compensatory neovascularization: the drilling of metalloelastase-positive tunnels in ischemic myocardium. Circ Res *87*, 378–384.

111. Ingram, D. A., Caplice, N. M., and Yoder, M. C. (2005). Unresolved questions, changing definitions, and novel paradigms for defining endothelial progenitor cells. Blood *106*, 1525–1531.

112. Rehman, J., Li, J., Orschell, C. M., and March, K. L. (2003). Peripheral blood "endothelial progenitor cells" are derived from monocyte/macrophages and secrete angiogenic growth factors. Circulation *107*, 1164–1169.

113. Schmeisser, A., Graffy, C., Daniel, W. G., and Strasser, R. H. (2003). Phenotypic overlap between monocytes and vascular endothelial cells. Adv Exp Med Biol *522*, 59–74.

114. Ferrara, N., and Kerbel, R. S. (2005). Angiogenesis as a therapeutic target. Nature *438*, 967–974.

115. Albini, A., Tosetti, F., Benelli, R., and Noonan, D. M. (2005). Tumor inflammatory angiogenesis and its chemoprevention. Cancer Res *65*, 10637–10641.

116. Kerbel, R. S. (2006). Antiangiogenic therapy: a universal chemosensitization strategy for cancer? Science *312*, 1171–1175.

Chapter 21
Contribution of Endothelial Progenitor Cells to the Angiogenic Process

Marco Seandel[1,2], Andrea T. Hooper[1,3], and Shahin Rafii[1,3*]

Keywords: endothelial progenitor cells, hematopoietic cells, bone marrow-derived cells, stem cells

Abstract: Among the molecular and cellular processes that orchestrate construction of new blood vessels, the distinct contribution of circulating cells has recently become appreciated. The endothelial progenitor cell (EPC) was one of the first identified circulating cell types found to directly contribute to neo-vessels walls. Subsequently, a complex network of signals between EPCs and other circulating bone marrow-derived cell types has been described. Significant temporal and spatial heterogeneity in utilization of circulating progenitor cells exists between tumor types, complicating analysis in both animal models and in patients. A lack of standardized cell surface markers and techniques for quantitation of such rare cells has further complicated such studies. Nonetheless, levels of EPCs and other bone marrow- derived progenitors may hold prognostic significance for cancer patients or may be used in the future to guide therapy and EPCs may themselves represent a valid therapeutic target.

Introduction and Historical Significance

Vascular architecture in vertebrates results from two related yet distinct processes: vasculogenesis and angiogenesis. Vasculogenesis refers to the de novo formation of blood vessels by the differentiation, migration and patterned assembly of angioblasts, whereas angiogenesis is the sprouting, maturation and growth of endothelial cells (EC) from the preexisting vasculature [1]. However, in the adult, the bone marrow is a vast reservoir of stem cells which are capable of generating an array of lineage-specific cell types, in particular hematopoietic stem cells (HSCs) and endothelial progenitor cells (EPCs), which contribute to vasculogenesis and endothelial cell diversification [2–5].

Although it had been shown that there exists a population of circulating endothelial cells (CECs) in the peripheral blood [6], it was not until a landmark study in 1997, in which Asahara and colleagues challenged the traditional dogma of blood vessel formation in adult tissues, when they isolated and characterized a subset of CD34+ human endothelial progenitor cells that could differentiate into cells with endothelial cell–like characteristics [3]. These cells were termed "endothelial progenitor cells" (EPCs). The Asahara study was the first of its kind that suggested circulating endothelial progenitor cells in adult peripheral blood might contribute to new vessel formation, demonstrating vasculogenesis in adult vertebrates. Since that time, it has been established that EPCs are derived from the bone marrow (although other sites of origin cannot be excluded as discussed below), and that a wide variety of bone marrow-derived cells can differentiate into endothelium and/or support the process of angiogenesis, not only during development but also in the adult, particularly during pathogenesis and even pregnancy [7–10, 5].

Bone marrow-derived cells not only incorporate into functional blood vessels, but may also supply a gamut of angiogenic factors that facilitate angiogenesis and vasculogenesis simultaneously, demonstrating an additional means by which bone marrow-derived progenitors contribute to the process of neovascularization [11]. An ever-increasing body of evidence suggests that EPCs circulate in the blood and play an important role in the formation of new blood vessels during pathogenesis and also contribute to vascular homeostasis at steady state, contributing to endothelial diversity [12]. The

[1]Department of Genetic Medicine, Howard Hughes Medical Institute, Weill Medical College of Cornell University, New York, NY 10021, USA

[2]Division of Medical Oncology, Department of Medicine, Memorial Sloan-Kettering Cancer Center, New York, NY 10021, USA

[3]Department of Physiology, Biophysics and Systems Biology, Weill Cornell Graduate School of Medical Sciences, New York, NY 10021, USA

*Corresponding Author:
Shahin Rafii, Weill Cornell Medical College, HHMI, 1300 York Avenue, Room A-863, New York, NY 10021, USA
E-mail: srafii@med.cornell.edu

clinical significance of the mechanisms of vessel homeostasis and function in pathogenesis by EPCs is readily apparent. Endothelial abnormalities and dysfunction are fundamental to the pathogenesis of atherosclerosis, diabetes, and tumor growth, as well as during therapeutic myeloablation and irradiation. Therefore, a more profound understanding of these mechanisms will lead to new therapeutic approaches for both pro- and antiangiogenesis [13].

Endothelial Progenitor Cells

Endothelial progenitor cells (EPCs) may be considered as embryonic angioblasts that proliferate and migrate within the bone marrow, subsequently circulate in the periphery, and have the ability to promote tissue neovascularization via heterogeneous modes of action (Figure 21.1). EPCs were initially identified through their expression of CD34 (a surface marker common to hematopoietic stem cells and mature endothelial cells) and vascular endothelial cell growth-factor receptor 2 (VEGFR-2, kinase-domain-related/KDR receptor), antigens shared by the angioblast and the hematopoietic

progenitor but not by fully differentiated endothelial cells [3]. In steady-state conditions, VEGFR-2$^+$ EPCs represent less than 0.01% of circulating mononuclear cells in the peripheral circulation, and studies have shown that elevation of VEGF-A in adult mice mobilizes EPCs into circulation from the bone marrow [14]. EPCs were subsequently shown to express VE-cadherin (CD144) and AC133 antigen, which is currently believed to be expressed on immature endothelial progenitors but not mature endothelium [4].

In mice, EPCs are defined as CD45$^-$/c-Kit(CD117)$^+$/VEGFR2$^+$ (as reviewed by [15]). Importantly, functional assays, such as the ability to form endothelial colonies (CFU-EC) and the high proliferation rate of EPCs which distinguish EPCs from mature circulating endothelial cells (CECs) that are sloughed off walls of preexisting vasculature during physiological shear, have provided another method of differential characterization of EPCs, from CECs and from HSCs [16]. Still others have isolated various populations of EPCs using various combinations of markers from porcine, rat, murine and human bone marrow [11, 17–23]. It is of note that monocytes have been reported by several groups to differentiate along the endothelial lineage and even express what were previously

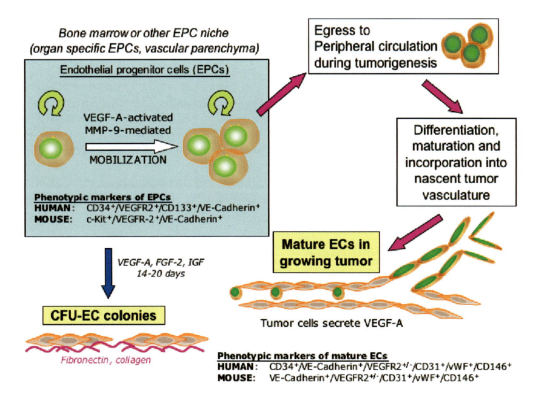

FIG. 21.1. Phenotypic and functional identification of human and mouse EPCs during tumorigenesis. EPCs originate from the bone marrow niche or alternatively from putative resident progenitors within organ niches or the vascular parenchyma. EPCs within these niches can be identified and isolated phenotypically by cell surface markers such as CD34, VEGFR-2, VE-Cadherin & CD133 for human and c-Kit, VEGFR-2 and VE-Cadherin for mouse. In vivo, once the bone marrow is stimulated by VEGF-A, such as that secreted by an actively growing tumor, stem cells are mobilized via release and activation of MMP-9. Subsequently EPCs are released into the periphery and undergo differentiation, maturation and ultimate incorporation into nascent tumor vasculature via a vasculogenic process. In addition, EPCs can be isolated and functionally characterized in vitro via CFU-EC assays as shown. Either after incorporating into the vasculature of the tumor or alternatively, in vitro at the endpoint of the CFU-EC assay, the endothelial cells will express mature pan-endothelial markers such as CD31, VE-Cadherin, vWF and VEGFR-2.

known as endothelial-specific markers [24–27]. EPCs defined in the above manner probably represent a heterogeneous population, which, in combination with the lack of a consensual definition, complicates the interpretation of work in this field. The current lack of unified definition underscores the importance of combining cell surface phenotypic markers with functional assays for EPC identification in future studies.

The mechanism by which EPCs are recruited from the bone marrow to sites of neoangiogenesis remains an active area of study. It was demonstrated that recruitment of EPCs is impaired in endothelial nitric oxide synthase (eNOS) deficient mice, resulting in a disruption of neoangiogenesis [28]. Interestingly, nitric oxide production was linked to activation of matrix metalloproteinase-9 (MMP-9) and release of soluble Kit-Ligand (sKitL), previously demonstrated to be the mechanism of HSC release from the bone marrow [28, 29]. These data suggest that mobilization of EPCs may be regulated by activation of specific matrix metalloproteinases, such as MMP-9. Even so, the mechanism of EPC egress from the bone marrow and subsequent recruitment to sites of neoangiogenesis warrants further investigation.

EPCs and Hematopoietic Cells – a two way street

Endothelial cells and hematopoietic cells are believed to share the same common progenitor [2]. HSCs have been shown to stimulate the assembly of endothelium into blood vessels. As mentioned previously, VEGFR2+ EPCs are mobilized in response to VEGF-A in adult mice. In addition, VEGF-A concomitantly induces the mobilization of VEGFR1/FLT-1+ HSCs, raising the possibility that co-mobilization of EPCs and HSCs is essential for neo-vascularization and even perhaps organ-specific incorporation of mobilized EPCs contributing to endothelial heterogeneity in the adult [14]. Indeed, hematopoietic lineage cells are present at sites of active neoangiogenesis and not only contribute in a paracrine fashion but can also incorporate and stabilize the neovasculature itself [30, 31].

During development, bone marrow-derived cells function in the homeostasis of the vasculature. In particular, the contribution of hematopoietic cells in the maintenance of vascular diversity was emphasized in several studies that demonstrated that the adaptor protein Slp-76 and the tyrosine kinase Syk, which are expressed primarily in hematopoietic cells, contribute to the separation of the lymphatic and blood vascular endothelial cell border [32]. The authors found that mice lacking either of these two factors exhibited a profound disorganization of blood and lymphatic vessels, resulting in arteriovenous-lymphatic shunts, and that the phenotype is rescued by expression of Slp-76 in a subset of hematopoietic cells [12, 32]. These studies suggest that a subset of bone marrow-derived cells, likely to include Slp-76/Syk-expressing EPCs, are the major determinants of separation of the lymphatic and

blood vascular system. This highlights the important role of hematopoietic cells in vascular homeostasis [33, 12].

The contribution of vasculogenesis, as mediated by angioblasts/EPCs, to the postnatal vasculature was first elucidated in a landmark study by Young et al. [34]. Previously, it had been thought that neovasculaturization in the early postnatal developmental period was exclusively mediated via VEGF-independent angiogenic sprouting. However, the study by Young et al. demonstrated for the first time that bone marrow-derived EPCs incorporate into the vasculature of newborns in the absence of myeloablation and that this process is enhanced and potentially regulated by VEGF. This process is reduced as the animals mature, thereby exhibiting a vasculogenic-to-angiogenic switch [34]. Taken together, these data not only demonstrate that EPCs/angioblasts contribute at least in part to postnatal vasculogenesis, but critically, the data accentuate the potential for utilization of EPCs (enhanced endothelialization) as a therapeutic approach for congenital vascular illness [34].

In addition to support of growth, development and maintenance of blood vessels by bone marrow-derived cells, it is now understood that the opposite is also true: the vasculature of the bone marrow not only coexists alongside, but also supports hematopoietic precursors. In the adult, endothelial cells, particularly those of the bone marrow, are known to support hematopoietic precursors nurturing them until they are mature and ready for egress from the marrow into circulation [35, 36]. In addition, hematopoietic cells function as proangiogenic agents, accelerating hemangiogenesis within the marrow and revascularization of ischemic hind limbs [37, 30]. Additionally, bone marrow endothelial cells support multilineage hematopoiesis by elaboration of soluble cytokines [36]. Taken together, these reports emphasize the functional role of endothelial cells as a major component of the bone marrow microenvironment. This complex interaction regulates the trafficking and homing of hematopoietic progenitor and stem cells, suggesting that hematopoietic tissue can provide a source of vascular endothelial progenitor cells throughout life (reviewed in [38]).

EPCs and Disease

Although during embryonic development angioblasts are the primary contributor to blood vessel formation, the adult myocardial endothelium—and in fact most of the endothelium in the body—remains in a quiescent state until provoked by injury. During a coronary event, for example, the vasculature proliferates and attempts to compensate for the ischemic insult. Even so, the capillary network of the heart is ultimately unable to support the greater demands of the hypertrophied myocardium, which results in increasing myocyte loss, subsequent infarct extension, and an irreversible deficit in cardiac function. Until recently, it was thought that the damaged adult vasculature could only repair itself through angiogenic processes, and this was critically limited in the heart. The pioneering work of Jeffrey Isner, Takayuki Asahara, and others has had a major implication in this regard: putative EPCs or angioblasts were isolated from human peripheral blood, differentiated into ECs, and incorporated into sites of active angiogenesis—such as that found in ischemic conditions [3]. These findings suggested for the first time that EPCs could augment collateral vessel growth to ischemic tissues—in a process of adult vasculogenesis—and could therefore be harnessed for therapeutic angiogenesis [3, 8, 39]. Indeed, an inverse correlation between the number of cardiovascular risk factors and the number and migratory potential of EPCs has been demonstrated in human patients [40], indicating that EPCs could be used as surrogates or biomarkers of cardiovascular or other ischemic illnesses.

Since these pioneering studies in cardiac ischemia, EPCs have been implicated both in the pathogenesis of and as a treatment for many conditions, including post-myocardial infarction, atherosclerosis, diabetes, hemophilia, smoking cessation, wound healing, limb ischemia, retinal and lymphoid organ neovascularization, inflammation and tumor growth [7–9, 15, 17, 18, 39, 41–45]. As of today, the most exciting goal is to dissect the anatomy of the tumor and the associated EPCs. In this chapter, the mechanism whereby recruitment of EPCs may contribute to tumor angiogenesis will be further explored along with perspectives on current therapeutic strategies.

Utilization of EPCs by Tumors: Animal Models

The paradigm for utilization of EPCs by tumors was originally established in mice and later confirmed in humans, although the magnitude of the effect is known to be context-dependent. Using *Id*-mutant mice, Lyden et al. [46] first observed that deficient recruitment of bone marrow-derived cells led to resistance to tumor growth, an effect that is abrogated upon transplantation of wild type bone marrow [46]. In that system, B6RV2 lymphomas and Lewis lung carcinomas growth required mobilization of EPCs in a VEGFR2-dependent manner. Notably, EPCs appeared to play a particularly important role in the early phase of tumor growth, since the majority of tumor neovessels were bone marrow-derived in the first several days but this proportion declined to ~50% by two weeks after inoculation. Later, it was found that MCA/129 fibrosarcomas exhibited ~50% BM-derived endothelial cells [47]. Furthermore Ruzinova et al. found, in spontaneous tumor models, that in PTEN$^{+/-}$ or PTEN$^{+/-}$Id1$^{-/-}$Id3$^{+/-}$ uterine carcinomas 15–20% of vessels were bone marrow-derived, in contrast to lymphoid hyperplasias in the same animals in which the vessels were not bone marrow-derived [48].

In spontaneous prostate adenocarcinomas in TRAMP mice, which express SV40 T antigen under control of a prostate specific probasin promoter, incorporation of EPCs varied with tumor grade, such that ~14% of tumor vessels in poorly differentiated tumors contained BM-derived EC, while well-differentiated tumors contained none [49]. More recently, Li and colleagues [50] evaluated endothelial progenitor incorporation in either high-grade mouse sarcomas or B16 melanomas and found high rates (up to 30%) of incorporation at day 5 and much lower rates at day 25 [50]. Duda et al. examined incorporation of labeled bone marrow-derived ECs following transplant in Lewis lung carcinomas and B16 melanomas and found 1–60% of vessels positive depending on the tumor type and site [51].

Even in light of these data, several studies, using transgenic mice expressing reporter genes driven by endothelial specific promoters, have failed to find evidence for a role for EPCs in murine tumor models [52–55]. However, none of these studies clearly demonstrated that the reporter genes were faithfully expressed, either in a bone marrow-derived repopulating endothelial progenitor or in the organ specific EPCs originally described by Ingram et al. [56]). Therefore, it is essential to discover an improved battalion of true repopulating EPC-specific promoters, in order to interrogate the contribution of organ-specific EPCs to tumor growth.

Evidence for Tissue Resident EPCs

It has been suggested that EPCs may reside within the vasculature of peripheral organs, in addition to the bone marrow. The luminal walls of large adult human arteries have been shown to contain cells with endothelial colony forming potential [57], the ability to mimic angiogenesis in vitro [58], and the ability to proliferate in vivo [59]. Recently, Bruno et al. identified a population of bipotential CD34$^-$CD133$^+$ cells in human renal cell carcinomas [60]. These stem-like cells could form both endothelial cells and epithelial cells and supported development of xenograft tumors in mice. Therefore, tumor angiogenesis may be directly supported by local, regional, and distal (bone marrow-derived) sources of endothelial progenitors.

Systemic Signals that Regulate EPCs

The identity and temporal sequence of signals required for EPCs mobilization, homing and incorporation into tumors has been the subject of much investigation [61]. It was initially found that this process is highly dependent upon specific chemokines and cytokines in mice, in particular VEGF-A and stromal cell-derived factor-1 (SDF-1) [14, 29, 62, 63]. Proof of principle for the role of VEGF-A in human EPC mobilization was provided in a study of newly diagnosed rectal cancer patients undergoing specific anti-VEGF-A therapy, which demonstrated decreased EPC counts after two weeks of treatment [64]. Placental-derived growth factor (PlGF) has also been strongly implicated in EPC mobilization [50].

Recently, CC chemokines were found to mediate EPC mobilization [65]. In the latter study, employing the transgenic (RIP-Tag) model of pancreatic islet cell tumors, it is notable that EPC incorporation was a relatively late and progressive process in tumorigenesis [65], in contrast to other model systems in which the reverse pattern was seen [46]. In the Spring et al. study, the authors also suggested that the tumor endothelium itself represents a key source of signals for EPC mobilization [65]. In contrast, a recent study by the Weinberg group demonstrated stromal fibroblasts within breast carcinoma are crucial for generating SDF-1, which then mobilized EPCs to the tumor [66]. Interestingly, neither normal (non-tumor associated) fibroblasts nor even tumor cells alone were able to mediate the same effect, suggesting an epigenetic disposition of tumor-associated fibroblasts to enhance angiogenesis through mobilization of EPCs. Therefore, when taken together, these studies establish that various host constituent cell types are usurped by the tumor to produce signals directing EPCs into the neoplastic sites undergoing neoangiogenesis.

While endogenous inhibitors of angiogenesis, such as angiostatin, endostatin and thrombospondin are well characterized (for a comprehensive review see chapters 11–13), only recently has it become apparent that the same natural products may directly modulate levels of endothelial progenitors. Endostatin, for example, was able to suppress both baseline EPC levels in mice and tumor-mobilized EPCs [67]. Shaked et al. demonstrated 5-fold increased EPC levels in thrombospondin-1 deficient mice compared to controls and that a TSP-1 peptide mimetic could abrogate this effect [68]. Therefore, both positive and negative regulators of angiogenesis appear to exert their effects, in part, through modulation of EPCs.

Relationship Between Tumor Anatomy and Circulating BM-derived cells

While it has been known for some time that tumor vasculature is both heterogeneous and structurally distinct from normal vessels [69], the significant spatial variation in bone marrow-derived vascular cells within tumors has only more recently

been appreciated. Of note, a mathematical model generated to describe the function of EPCs in tumor angiogenesis suggested that the most prominent effects of EPCs would be found at the dense peripheral edge of the tumor and especially at early time points in tumor growth [70]. Consistent with this, Li and colleagues noted that BM-derived vessels were located mainly at the periphery of the tumor, which could contribute to the widely reported variation in the magnitude of contribution of BM-derived cells to tumor vessels [50]. The relationship of EPC function to tumor microanatomy was recently elucidated in a study of the mechanisms of action of vascular disrupting agents (VDAs) [63]. It was found first that single doses of VDAs, which cause marked central necrosis in tumors, induce a prompt rise in the levels of EPCs. The authors found that the viable tumor border is a key site of action of EPCs, since pretreatment with anti-VEGFR2 antibody abrogates the residual angiogenic activity present following VDA administration, attributable to the mobilized EPC [63]. Thus, future studies need to control for intratumoral heterogeneity in assessing the EPC-specific contribution to tumor angiogenesis.

Pathophysiology of EPCs in Human Cancer

What, then, is the significance of EPCs in human cancer? In contrast to the plethora of clinical studies demonstrating incorporation of circulating cells into vessel walls in benign diseases in humans, as recently reviewed by [71], it has been more challenging to determine whether the tumor endothelium is constructed from circulating endothelial progenitors. Several groups have studied EPCs and CECs in the post-bone marrow transplant setting. For example, long-term endothelial proliferative capability (attributed to EPCs) in vitro was restricted to donor-derived cells, while short term proliferation resided in host-derived endothelial cells [16]. A seminal study by Peters et al. took advantage of cancers occurring

TABLE 21.1. Selected clinical studies of EPCs in cancer patients employing flow cytometry or culture based assays.

Reference	Population	EPC definition	Conclusions
[73]	Newly diagnosed NSCLC	VEGFR2+CD133+ *or* VEGFR2+CD34+	EPCs higher in NSCLC patients than HC; EPCs go down in responders to therapy; higher pre-treatment EPCs correlates with poor outcome
[74]	Primary breast cancer	CD45loCD34+VEGFR2+	If low baseline EPCs, then mobilization on chemotherapy (otherwise no change)
[75]	Myelodysplastic syndromes	CD133+	Higher EPCs in patients vs HC
[76]	Various primary cancers	CD133+	Subset of patients exhibit higher EPCs after chemotherapy
[77]	Metastatic breast cancer	CD133+	No correlation between EPCs and outcome
[78]	Hepatocellular carcinoma	Culture based	CFU correlate with unresectable vs resectable disease
[64]	Primary rectal cancer	CD133+	EPCs go down on anti-VEGF therapy
[79]	Newly dx'ed AML	CD45+CD31+ CD34+CD133+	Higher EPCs in patients vs HC
[80]	Advanced heme cancer	CD45-CD146+CD133+	Higher EPCs in patients vs HC
[81]	Multiple myeloma	culture based	CFU higher in myeloma vs HC
[82]	Various solid tumors	CD133 mRNA	Elevated CD133 in metastatic patients esp. bone metastasis

Key: *NSCLC* Non-small cell lung cancer; *HC* healthy controls; *AML* acute myeloid leukemia after sex-mismatched BMT, in which host vs. donor-derived EC could be counted, following fluorescence in situ hybridization using sex-chromosome specific probes [72]. While the patient population was small and heterogeneous by necessity, it was found that 1–12% of tumor endothelial cells were donor derived. Certain tumors such as lymphomas had the highest incorporation of EPCs into their vessel wall.

Human solid tumor patients, especially breast cancer patients, have been the subject of several studies of human EPCs (Table 21.1). Employing flow cytometry, Mancuso et al. found that CD133$^+$ EPCs were elevated in a fraction of patients during recovery from cytotoxic chemotherapy [76]. Subsequently, the same group found higher levels of apoptotic CECs but no difference in EPCs, in the subset of patients found to have a clinical benefit to therapy at two months [77]. Using real-time quantitative polymerase chain reaction, another study examined the peripheral blood of newly diagnosed breast cancer patients and found a trend toward increased CD133 mRNA and significantly elevated Tie-2 mRNA expression, with the highest levels associated with patients harboring infiltrating carcinoma, as compared to controls [83]. Similarly, a recent study of mixed solid tumor patients, including a sizable proportion of metastatic prostate patients, revealed that a high CD133 mRNA level correlates with the presence of bone metastasis and is an independent poor prognostic factor for survival, implicating EPC mobilization in tumor progression, although the authors could not formally rule out the contribution of circulating CD133$^+$ tumor cells [82].

An alternative approach to assessing circulating cells in peripheral blood was taken by Hilbe et al. [84]. The authors interrogated non-small cell lung cancer (NSCLC) tissue versus adjacent uninvolved areas for the presence of CD133$^+$ vascular cells, which were more abundant in tumor tissue. More recently, Dome et al., also focusing on NSCLC, found an increased number of EPCs (defined as either CD34$^+$VEGFR2$^+$ or VEGFR2$^+$CD133$^+$) in patients compared to controls and a correlation between high pretreatment levels and poor survival (Table 21.1) [73].

Flow cytometric analyses of peripheral blood from patients with hematologic malignancies have also yielded provocative results to support the significance of EPCs in human cancer (Table 21.1). In patients with myelodysplastic syndrome, CD133$^+$ EPCs were higher than controls but did not correlate with particular MDS subclasses or survival [75]. Yee et al. also found increased CECs in patients with various advanced hematologic malignancies [80], although the use of the immunophenotype CD133$^+$CD146$^+$ to define EPCs as has been criticized in other studies [85]. Using CD34$^+$CD31$^+$CD133$^+$ to denote EPCs, a study of acute myelogenous leukemia patients found increased EPCs in newly diagnosed patients vs. controls and a decrease upon achievement of complete remission [79].

Multiple studies, validating the quantitation of EPCs in vitro indirectly following isolation either from the blood or from the bone marrow, have laid the foundation for in vitro-based clinical studies on cancer patients [16, 86, 87]. In studies pooling patients with various tumor types, variable results have been obtained, with some studies demonstrating increased EPCs [88] and some studies demonstrating no difference between patients and healthy controls (Table 21.1) [89]. In multiple myeloma patients, however, it was recently found that endothelial cell colony-forming EPCs were increased compared to healthy controls and that these levels were lower in responders to therapy [81]. Similarly, higher peripheral blood endothelial colony forming units were found in hepatocellular carcinoma patients compared to controls and were also relatively higher in patients with recurrent disease following surgery [78]. Therefore, in vitro colony formation provides complementary supporting evidence for EPC function in humans.

Novel Stem-cell like Cell Populations: CXCR4$^+$VEGFR1$^+$ hemangiocytes

A subpopulation of marrow-derived CXCR4$^+$VEGFR1$^+$ proangiogenic hematopoietic cells ("hemangiocytes") that contribute to revascularization has recently been identified [30]. These proangiogenic myelomonocytic cells are co-mobilized with the endothelial progenitors from the bone marrow to peripheral sites of neo-angiogenesis upon stimulation by hematopoietic cytokines such as thrombopoietin, stem cell factor (kit ligand), and to a lesser extent by erythropoietin, G-CSF and GM-CSF. Such hematopoietic cytokine-driven mobilization and recruitment of hemangiocytes is modulated by the SDF-1/CXCR4 chemokine signaling axis. Release of SDF-1 from circulating platelets contributes to this mobilization process. Hemangiocytes localize to the perivascular zone of neovasculature and stabilize nascent vessels. The SDF-1/CXCR4 signaling pathway also mediates the retention of hemangiocytes within the neovascular niche. Hemangiocytes functionally contribute to the neovascularization by releasing proangiogenic factors such as angiopoietin-2. However, it is not yet known the extent to which these putative progenitors undergo in situ proliferation and/or differentiation within the neovascular niche.

Other Vascular Progenitor Populations in Tumors

While the concept that immature vascular cells are delivered to the site of tumor vessels was originally developed based on the endothelial progenitor paradigm, it has become apparent that other classes of vascular cells also differentiate from progenitors in situ. Among these, a class of Tie2-expressing mesenchymal progenitors was recently identified that are capable of expansion in vitro and of generating Tie2-negative, α smooth muscle actin-postive cells when re-inoculated into tumors [31]. In the same study, while not finding evidence for endothelial progenitors, the authors did describe bone marrow-derived Tie2-expressing monocytes, which are crucial for tumor angiogenesis. Another distinct but related class of immature vascular cells crucial for tumor angiogenesis was found by Song et al., who isolated PDGFRβ- perivascular progenitor cells in tumors, that give rise to different pericyte subtypes [90]. The study also found the bone marrow Sca1+ population to be a potential source of these cells.

Stem Cells in Tumor Angiogenesis: Future Perspectives

While there is abundant evidence for the existence EPCs, proof of a functional requirement for EPC mobilization and vascular recruitment in human cancer necessitates further experimentation. Larger scale clinical studies, utilizing homogenous patient populations will be a prerequisite for such a conclusion. In addition, the temporal and spatial contribution of EPCs to a specific tumor require the generation of novel tumor models, where the contribution of EPCs could be evaluated during the initial phases of the neo-angiogenic switch and oncogenic transformation. For example, many studies suggest that EPCs contribute to the early phases of tumor growth and may function as molecular hubs dictating the specification of the venous, lymphatic and arterial systems. The existence of specific phenotypes of bone marrow-derived EPCs (e.g., Slp-76+/Syk+) and their function in defining lymphatic versus vascular endothelial cells with verifiable biological roles in the adult vertebrate is testament to the therapeutic potential of endothelial progenitors [12, 32]. Sophisticated animal models will be needed to explore the significance of these Slp-76/Syk EPCs at various stages of tumorigenesis in order to bring the therapies to the clinic.

While extremely sensitive methodological tools now exist to measure proangiogenic stem and progenitor cells, there is no consensus on the best approach in clinical studies [91]. Automated systems, such as those combining immunomagnetic beads and automated fluorescent microscopy could greatly facilitate standardization of such studies [92]. Apart from the use of vascular stem and progenitor cells as biomarkers in patients, a more profound understanding of their regulation will allow an informed selection of drugs from the multitude of targeted agents now under development. It is likely that future antiangiogenic treatment protocols will employ one or more components to specifically block the contribution of rare but potent immature vascular cells in tumors. Alternatively, the dosing and scheduling of cytotoxic agents may be optimized to block mobilization of endothelial progenitors from the bone marrow and even peripheral niches [15].

Despite these potential hurdles, the novel therapeutic strategy to modulate the presence and function of EPCs during tumor growth should be approached with much enthusiasm as it remains a realistic and promising strategy for controlling blood vessel growth and ultimately tumor growth in human patients.

References

1. Risau W (1997) Mechanisms of angiogenesis. Nature 386:671–674
2. Choi K, Kennedy M, Kazarov A, Papadimitriou JC, Keller G (1998) A common precursor for hematopoietic and endothelial cells. Development 125:725–732
3. Asahara T, Murohara T, Sullivan A, Silver M, van der ZR, Li T, Witzenbichler B, Schatteman G, Isner JM (1997) Isolation of putative progenitor endothelial cells for angiogenesis. Science 275:964–967
4. Peichev M, Naiyer AJ, Pereira D, Zhu Z, Lane WJ, Williams M, Oz MC, Hicklin DJ, Witte L, Moore MA, Rafii S (2000) Expression of VEGFR-2 and AC133 by circulating human CD34(+) cells identifies a population of functional endothelial precursors. Blood 95:952–958
5. Rafii S (2000) Circulating endothelial precursors: mystery, reality, and promise. J Clin Invest 105:17–19
6. Hladovec J, Rossman P (1973) Circulating endothelial cells isolated together with platelets and the experimental modification of their counts in rats. Thombosis Research 3:665–674
7. Schatteman GC, Hanlon HD, Jiao C, Dodds SG, Christy BA (2000) Blood-derived angioblasts accelerate blood-flow restoration in diabetic mice. J Clin Invest 106:571–578
8. Kalka C, Masuda H, Takahashi T, Kalka-Moll WM, Silver M, Kearney M, Li T, Isner JM, Asahara T (2000) Transplantation of ex vivo expanded endothelial progenitor cells for therapeutic neovascularization. Proc Natl Acad Sci USA 97:3422–3427
9. Takahashi T, Kalka C, Masuda H, Chen D, Silver M, Kearney M, Magner M, Isner JM, Asahara T (1999) Ischemia- and cytokine-induced mobilization of bone marrow-derived endothelial progenitor cells for neovascularization. Nat Med 5:434–438
10. Sugawara J, Mitsui-Saito M, Hoshiai T, Hayashi C, Kimura Y, Okamura K (2005) Circulating endothelial progenitor cells during human pregnancy. J Clin Endocrinol Metab 90:1845–1848
11. Kamihata H, Matsubara H, Nishiue T, Fujiyama S, Tsutsumi Y, Ozono R, Masaki H, Mori Y, Iba O, Tateishi E, Kosaki A, Shintani S, Murohara T, Imaizumi T, Iwasaka T (2001) Implantation of bone marrow mononuclear cells into ischemic myocardium enhances collateral perfusion and regional function via side supply of angioblasts, angiogenic ligands, and cytokines. Circulation 104:1046–1052
12. Sebzda E, Hibbard C, Sweeney S, Abtahian F, Bezman N, Clemens G, Maltzman JS, Cheng L, Liu F, Turner M, Tybulewicz V, Koretzky GA, Kahn ML (2006) Syk and Slp-76 mutant mice reveal a cell-autonomous hematopoietic cell contribution to vascular development. Dev Cell 11:349–361
13. Rafii S, Lyden D, Benezra R, Hattori K, Heissig B (2002) Vascular and haematopoietic stem cells: novel targets for anti-angiogenesis therapy? Nat Rev Cancer 2:826–835
14. Hattori K, Dias S, Heissig B, Hackett NR, Lyden D, Tateno M, Hicklin DJ, Zhu Z, Witte L, Crystal RG, Moore MA, Rafii S (2001) Vascular endothelial growth factor and angiopoietin-1 stimulate postnatal hematopoiesis by recruitment of vasculogenic and hematopoietic stem cells. J Exp Med 193:1005–1014
15. Bertolini F, Shaked Y, Mancuso P, Kerbel RS (2006) The multifaceted circulating endothelial cell in cancer: towards marker and target identification. Nat Rev Cancer 6:835–845

16. Lin Y, Weisdorf DJ, Solovey A, Hebbel RP (2000) Origins of circulating endothelial cells and endothelial outgrowth from blood. J Clin Invest 105:71–77

17. Orlic D, Kajstura J, Chimenti S, Bodine DM, Leri A, Anversa P (2003) Bone marrow stem cells regenerate infarcted myocardium. Pediatr Transplant 7 Suppl 3:86–88

18. Kocher AA, Schuster MD, Szabolcs MJ, Takuma S, Burkhoff D, Wang J, Homma S, Edwards NM, Itescu S (2001) Neovascularization of ischemic myocardium by human bone-marrow-derived angioblasts prevents cardiomyocyte apoptosis, reduces remodeling and improves cardiac function. Nat Med 7:430–436

19. Jackson KA, Majka SM, Wang H, Pocius J, Hartley CJ, Majesky MW, Entman ML, Michael LH, Hirschi KK, Goodell MA (2001) Regeneration of ischemic cardiac muscle and vascular endothelium by adult stem cells. J Clin Invest 107:1395–1402

20. Fuchs S, Baffour R, Zhou YF, Shou M, Pierre A, Tio FO, Weissman NJ, Leon MB, Epstein SE, Kornowski R (2001) Transendocardial delivery of autologous bone marrow enhances collateral perfusion and regional function in pigs with chronic experimental myocardial ischemia. J Am Coll Cardiol 37:1726–1732

21. Kawamoto A, Gwon HC, Iwaguro H, Yamaguchi JI, Uchida S, Masuda H, Silver M, Ma H, Kearney M, Isner JM, Asahara T (2001) Therapeutic potential of ex vivo expanded endothelial progenitor cells for myocardial ischemia. Circulation 103:634–637

22. Kobayashi T, Hamano K, Li TS, Katoh T, Kobayashi S, Matsuzaki M, Esato K (2000) Enhancement of angiogenesis by the implantation of self bone marrow cells in a rat ischemic heart model. J Surg Res 89:189–195

23. Murohara T, Ikeda H, Duan J, Shintani S, Sasaki K, Eguchi H, Onitsuka I, Matsui K, Imaizumi T (2000) Transplanted cord blood-derived endothelial precursor cells augment postnatal neovascularization. J Clin Invest 105:1527–1536

24. Fernandez PB, Lucibello FC, Gehling UM, Lindemann K, Weidner N, Zuzarte ML, Adamkiewicz J, Elsasser HP, Muller R, Havemann K (2000) Endothelial-like cells derived from human CD14 positive monocytes. Differentiation 65:287–300

25. Schmeisser A, Garlichs CD, Zhang H, Eskafi S, Graffy C, Ludwig J, Strasser RH, Daniel WG (2001) Monocytes coexpress endothelial and macrophagocytic lineage markers and form cord-like structures in Matrigel under angiogenic conditions. Cardiovasc Res 49:671–680

26. Rohde E, Malischnik C, Thaler D, Maierhofer T, Linkesch W, Lanzer G, Guelly C, Strunk D (2006) Blood monocytes mimic endothelial progenitor cells. Stem Cells 24:357–367

27. Walenta K, Friedrich EB, Sehnert F, Werner N, Nickenig G (2005) In vitro differentiation characteristics of cultured human mononuclear cells-implications for endothelial progenitor cell biology. Biochem Biophys Res Commun 333:476–482

28. Aicher A, Heeschen C, Mildner-Rihm C, Urbich C, Ihling C, Technau-Ihling K, Zeiher AM, Dimmeler S (2003) Essential role of endothelial nitric oxide synthase for mobilization of stem and progenitor cells. Nat Med 9:1370–1376

29. Heissig B, Hattori K, Dias S, Friedrich M, Ferris B, Hackett NR, Crystal RG, Besmer P, Lyden D, Moore MA, Werb Z, Rafii S (2002) Recruitment of stem and progenitor cells from the bone marrow niche requires MMP-9 mediated release of kit-ligand. Cell 109:625–637

30. Jin DK, Shido K, Kopp HG, Petit I, Shmelkov SV, Young LM, Hooper AT, Amano H, Avecilla ST, Heissig B, Hattori K, Zhang F, Hicklin DJ, Wu Y, Zhu Z, Dunn A, Salari H, Werb Z, Hackett NR, Crystal RG, Lyden D, Rafii S (2006) Cytokine-mediated deployment of SDF-1 induces revascularization through recruitment of CXCR4+ hemangiocytes. Nat Med 12:557–567

31. De Palma M, Venneri MA, Galli R, Sergi SL, Politi LS, Sampaolesi M, Naldini L (2005) Tie2 identifies a hematopoietic lineage of proangiogenic monocytes required for tumor vessel formation and a mesenchymal population of pericyte progenitors. Cancer Cell 8:211–226

32. Abtahian F, Guerriero A, Sebzda E, Lu MM, Zhou R, Mocsai A, Myers EE, Huang B, Jackson DG, Ferrari VA, Tybulewicz V, Lowell CA, Lepore JJ, Koretzky GA, Kahn ML (2003) Regulation of blood and lymphatic vascular separation by signaling proteins Slp-76 and Syk. Science 299:247–251

33. Rafii S, Skobe M (2003) Splitting vessels: keeping lymph apart from blood. Nat Med 9:166–168

34. Young PP, Hofling AA, Sands MS (2002) VEGF increases engraftment of bone marrow-derived endothelial progenitor cells (EPCs) into vasculature of newborn murine recipients. Proc Natl Acad Sci U S A 99:11951–11956

35. Rafii S, Shapiro F, Rimarachin J, Nachman RL, Ferris B, Weksler B, Moore MA, Asch AS (1994) Isolation and characterization of human bone marrow microvascular endothelial cells: hematopoietic progenitor cell adhesion. Blood 84:10–19

36. Rafii S, Shapiro F, Pettengell R, Ferris B, Nachman RL, Moore MA, Asch AS (1995) Human bone marrow microvascular endothelial cells support long-term proliferation and differentiation of myeloid and megakaryocytic progenitors. Blood 86:3353–3363

37. Kopp HG, Hooper AT, Broekman MJ, Avecilla ST, Petit I, Luo M, Milde T, Ramos CA, Zhang F, Kopp T, Bornstein P, Jin DK, Marcus AJ, Rafii S (2006) Thrombospondins deployed by thrombopoietic cells determine angiogenic switch and extent of revascularization. J Clin Invest 116:3277–3291

38. Kopp HG, Avecilla ST, Hooper AT, Rafii S (2005) The bone marrow vascular niche: home of HSC differentiation and mobilization. Physiology (Bethesda) 20:349–356

39. Asahara T, Masuda H, Takahashi T, Kalka C, Pastore C, Silver M, Kearne M, Magner M, Isner JM (1999a) Bone marrow origin of endothelial progenitor cells responsible for postnatal vasculogenesis in physiological and pathological neovascularization. Circ Res 85:221–228

40. Vasa M, Fichtlscherer S, Aicher A, Adler K, Urbich C, Martin H, Zeiher AM, Dimmeler S (2001) Number and migratory activity of circulating endothelial progenitor cells inversely correlate with risk factors for coronary artery disease. Circ Res 89:E1–E7

41. Lin Y, Chang L, Solovey A, Healey JF, Lollar P, Hebbel RP (2002) Use of blood outgrowth endothelial cells for gene therapy for hemophilia A. Blood 99:457–462

42. Kondo T, Hayashi M, Takeshita K, Numaguchi Y, Kobayashi K, Iino S, Inden Y, Murohara T (2004) Smoking cessation rapidly increases circulating progenitor cells in peripheral blood in chronic smokers. Arterioscler Thromb Vasc Biol 24:1442–1447

43. Sata M, Saiura A, Kunisato A, Tojo A, Okada S, Tokuhisa T, Hirai H, Makuuchi M, Hirata Y, Nagai R (2002) Hematopoietic stem cells differentiate into vascular cells that participate in the pathogenesis of atherosclerosis. Nat Med 8:403–409

44. Otani A, Kinder K, Ewalt K, Otero FJ, Schimmel P, Friedlander M (2002) Bone marrow-derived stem cells target retinal astrocytes and can promote or inhibit retinal angiogenesis. Nat Med 8:1004–1010

45. Grant MB, May WS, Caballero S, Brown GA, Guthrie SM, Mames RN, Byrne BJ, Vaught T, Spoerri PE, Peck AB, Scott EW (2002)

Adult hematopoietic stem cells provide functional hemangioblast activity during retinal neovascularization. Nat Med 8:607–612

46. Lyden D, Hattori K, Dias S, Costa C, Blaikie P, Butros L, Chadburn A, Heissig B, Marks W, Witte L, Wu Y, Hicklin D, Zhu Z, Hackett NR, Crystal RG, Moore MA, Hajjar KA, Manova K, Benezra R, Rafii S (2001) Impaired recruitment of bone-marrow-derived endothelial and hematopoietic precursor cells blocks tumor angiogenesis and growth. Nat Med 7:1194–1201

47. Garcia-Barros M, Paris F, Cordon-Cardo C, Lyden D, Rafii S, Haimovitz-Friedman A, Fuks Z, Kolesnick R (2003) Tumor response to radiotherapy regulated by endothelial cell apoptosis. Science 300:1155–1159

48. Ruzinova MB, Schoer RA, Gerald W, Egan JE, Pandolfi PP, Rafii S, Manova K, Mittal V, Benezra R (2003) Effect of angiogenesis inhibition by Id loss and the contribution of bone-marrow-derived endothelial cells in spontaneous murine tumors. Cancer Cell 4:277–289

49. Li H, Gerald WL, Benezra R (2004) Utilization of bone marrow-derived endothelial cell precursors in spontaneous prostate tumors varies with tumor grade. Cancer Res 64:6137–6143

50. Li B, Sharpe EE, Maupin AB, Teleron AA, Pyle AL, Carmeliet P, Young PP (2006) VEGF and PlGF promote adult vasculogenesis by enhancing EPC recruitment and vessel formation at the site of tumor neovascularization. FASEB J 20:1495–1497

51. Duda DG, Cohen KS, Kozin SV, Perentes JY, Fukumura D, Scadden DT, Jain RK (2006b) Evidence for incorporation of bone marrow-derived endothelial cells into perfused blood vessels in tumors. Blood 107:2774–2776

52. De Palma M, Venneri MA, Roca C, Naldini L (2003) Targeting exogenous genes to tumor angiogenesis by transplantation of genetically modified hematopoietic stem cells. Nat Med 9:789–795

53. Gothert JR, Gustin SE, van Eekelen JA, Schmidt U, Hall MA, Jane SM, Green AR, Gottgens B, Izon DJ, Begley CG (2004) Genetically tagging endothelial cells in vivo: bone marrow-derived cells do not contribute to tumor endothelium. Blood 104:1769–1777

54. Rajantie I, Ilmonen M, Alminaite A, Ozerdem U, Alitalo K, Salven P (2004) Adult bone marrow-derived cells recruited during angiogenesis comprise precursors for periendothelial vascular mural cells. Blood 104:2084–2086

55. Udagawa T, Puder M, Wood M, Schaefer BC, D'Amato RJ (2006) Analysis of tumor-associated stromal cells using SCID GFP transgenic mice: contribution of local and bone marrow-derived host cells. FASEB J 20:95–102

56. Ingram DA, Mead LE, Tanaka H, Meade V, Fenoglio A, Mortell K, Pollok K, Ferkowicz MJ, Gilley D, Yoder MC (2004) Identification of a novel hierarchy of endothelial progenitor cells using human peripheral and umbilical cord blood. Blood 104:2752–2760

57. Ingram DA, Mead LE, Moore DB, Woodard W, Fenoglio A, Yoder MC (2005) Vessel wall-derived endothelial cells rapidly proliferate because they contain a complete hierarchy of endothelial progenitor cells. Blood 105:2783–2786

58. Zengin E, Chalajour F, Gehling UM, Ito WD, Treede H, Lauke H, Weil J, Reichenspurner H, Kilic N, Ergun S (2006) Vascular wall resident progenitor cells: a source for postnatal vasculogenesis. Development 133:1543–1551

59. Koizumi K, Tsutsumi Y, Kamada H, Yoshioka Y, Watanabe M, Yamamoto Y, Okamoto T, Mukai Y, Nakagawa S, Tani Y, Mayumi T (2003) Incorporation of adult organ-derived endothelial cells into tumor blood vessel. Biochem Biophys Res Commun 306:219–224

60. Bruno S, Bussolati B, Grange C, Collino F, Graziano ME, Ferrando U, Camussi G (2006) CD133+ Renal Progenitor Cells Contribute to Tumor Angiogenesis. Am J Pathol 169:2223–2235

61. Rafii S, Lyden D (2003) Therapeutic stem and progenitor cell transplantation for organ vascularization and regeneration. Nat Med 9:702–712

62. Asahara T, Takahashi T, Masuda H, Kalka C, Chen D, Iwaguro H, Inai Y, Silver M, Isner JM (1999b) VEGF contributes to postnatal neovascularization by mobilizing bone marrow-derived endothelial progenitor cells. EMBO J 18:3964–3972

63. Shaked Y, Ciarrocchi A, Franco M, Lee CR, Man S, Cheung AM, Hicklin DJ, Chaplin D, Foster FS, Benezra R, Kerbel RS (2006) Therapy-induced acute recruitment of circulating endothelial progenitor cells to tumors. Science 313:1785–1787

64. Willett CG, Boucher Y, di Tomaso E, Duda DG, Munn LL, Tong RT, Chung DC, Sahani DV, Kalva SP, Kozin SV, Mino M, Cohen KS, Scadden DT, Hartford AC, Fischman AJ, Clark JW, Ryan DP, Zhu AX, Blaszkowsky LS, Chen HX, Shellito PC, Lauwers GY, Jain RK (2004) Direct evidence that the VEGF-specific antibody bevacizumab has antivascular effects in human rectal cancer. Nat Med 10:145–147

65. Spring H, Schuler T, Arnold B, Hammerling GJ, Ganss R (2005) Chemokines direct endothelial progenitors into tumor neovessels. Proc Natl Acad Sci U S A 102:18111–18116

66. Orimo A, Gupta PB, Sgroi DC, Arenzana-Seisdedos F, Delaunay T, Naeem R, Carey VJ, Richardson AL, Weinberg RA (2005) Stromal fibroblasts present in invasive human breast carcinomas promote tumor growth and angiogenesis through elevated SDF-1/CXCL12 secretion. Cell 121:335–348

67. Capillo M, Mancuso P, Gobbi A, Monestiroli S, Pruneri G, Dell'Agnola C, Martinelli G, Shultz L, Bertolini F (2003) Continuous infusion of endostatin inhibits differentiation, mobilization, and clonogenic potential of endothelial cell progenitors. Clin Cancer Res 9:377–382

68. Shaked Y, Bertolini F, Man S, Rogers MS, Cervi D, Foutz T, Rawn K, Voskas D, Dumont DJ, Ben David Y, Lawler J, Henkin J, Huber J, Hicklin DJ, D'Amato RJ, Kerbel RS (2005) Genetic heterogeneity of the vasculogenic phenotype parallels angiogenesis; Implications for cellular surrogate marker analysis of antiangiogenesis. Cancer Cell 7:101–111

69. Jain RK (2005) Normalization of tumor vasculature: an emerging concept in antiangiogenic therapy. Science 307:58–62

70. Stoll BR, Migliorini C, Kadambi A, Munn LL, Jain RK (2003) A mathematical model of the contribution of endothelial progenitor cells to angiogenesis in tumors: implications for antiangiogenic therapy. Blood 102:2555–2561

71. Liew A, Barry F, O'Brien T (2006) Endothelial progenitor cells: diagnostic and therapeutic considerations. Bioessays 28:261–270

72. Peters BA, Diaz LA, Polyak K, Meszler L, Romans K, Guinan EC, Antin JH, Myerson D, Hamilton SR, Vogelstein B, Kinzler KW, Lengauer C (2005) Contribution of bone marrow-derived endothelial cells to human tumor vasculature. Nat Med JID - 9502015 11:261–262

73. Dome B, Timar J, Dobos J, Meszaros L, Raso E, Paku S, Kenessey I, Ostoros G, Magyar M, Ladanyi A, Bogos K, Tovari J (2006) Identification and clinical significance of circulating endothelial progenitor cells in human non-small cell lung cancer. Cancer Res 66:7341–7347

74. Furstenberger G, von Moos R, Lucas R, Thurlimann B, Senn HJ, Hamacher J, Boneberg EM (2006) Circulating endothelial cells

and angiogenic serum factors during neoadjuvant chemotherapy of primary breast cancer. Br J Cancer 94:524–531

75. Cortelezzi A, Fracchiolla NS, Mazzeo LM, Silvestris I, Pomati M, Somalvico F, Bertolini F, Mancuso P, Pruneri GC, Gianelli U, Pasquini MC, Cortiana M, Deliliers GL (2005) Endothelial precursors and mature endothelial cells are increased in the peripheral blood of myelodysplastic syndromes. Leuk Lymphoma 46:1345–1351

76. Mancuso P, Burlini A, Pruneri G, Goldhirsch A, Martinelli G, Bertolini F (2001) Resting and activated endothelial cells are increased in the peripheral blood of cancer patients. Blood 97:3658–3661

77. Mancuso P, Colleoni M, Calleri A, Orlando L, Maisonneuve P, Pruneri G, Agliano A, Goldhirsch A, Shaked Y, Kerbel RS, Bertolini F (2006) Circulating endothelial-cell kinetics and viability predict survival in breast cancer patients receiving metronomic chemotherapy. Blood 108:452–459

78. Ho JW, Pang RW, Lau C, Sun CK, Yu WC, Fan ST, Poon RT (2006) Significance of circulating endothelial progenitor cells in hepatocellular carcinoma. Hepatology 44:836–843

79. Wierzbowska A, Robak T, Krawczynska A, Wrzesien-Kus A, Pluta A, Cebula B, Smolewski P (2005) Circulating endothelial cells in patients with acute myeloid leukemia. Eur J Haematol 75:492–497

80. Yee KW, Hagey A, Verstovsek S, Cortes J, Garcia-Manero G, O'Brien SM, Faderl S, Thomas D, Wierda W, Kornblau S, Ferrajoli A, Albitar M, McKeegan E, Grimm DR, Mueller T, Holley-Shanks RR, Sahelijo L, Gordon GB, Kantarjian HM, Giles FJ (2005) Phase 1 study of ABT-751, a novel microtubule inhibitor, in patients with refractory hematologic malignancies. Clin Cancer Res 11:6615–6624

81. Zhang H, Vakil V, Braunstein M, Smith EL, Maroney J, Chen L, Dai K, Berenson JR, Hussain MM, Klueppelberg U, Norin AJ, Akman HO, Ozcelik T, Batuman OA (2005) Circulating endothelial progenitor cells in multiple myeloma: implications and significance. Blood 105:3286–3294

82. Mehra N, Penning M, Maas J, Beerepoot LV, van Daal N, van Gils CH, Giles RH, Voest EE (2006) Progenitor marker CD133 mRNA is elevated in peripheral blood of cancer patients with bone metastases. Clin Cancer Res 12:4859–4866

83. Sussman LK, Upalakalin JN, Roberts MJ, Kocher O, Benjamin LE (2003) Blood markers for vasculogenesis increase with tumor progression in patients with breast carcinoma. Cancer Biol Ther 2:255–256

84. Hilbe W, Dirnhofer S, Oberwasserlechner F, Schmid T, Gunsilius E, Hilbe G, Woll E, Kahler CM (2004) CD133 positive endothelial progenitor cells contribute to the tumour vasculature in non-small cell lung cancer. J Clin Pathol 57:965–969

85. Duda DG, Cohen KS, di Tomaso E, Au P, Klein RJ, Scadden DT, Willett CG, Jain RK (2006a) Differential CD146 expression on circulating versus tissue endothelial cells in rectal cancer patients: implications for circulating endothelial and progenitor cells as biomarkers for antiangiogenic therapy. J Clin Oncol 24:1449–1453

86. Gehling UM, Ergun S, Schumacher U, Wagener C, Pantel K, Otte M, Schuch G, Schafhausen P, Mende T, Kilic N, Kluge K, Schafer B, Hossfeld DK, Fiedler W (2000) In vitro differentiation of endothelial cells from AC133-positive progenitor cells. Blood 95:3106–3112

87. Reyes M, Dudek A, Jahagirdar B, Koodie L, Marker PH, Verfaillie CM (2002) Origin of endothelial progenitors in human postnatal bone marrow. J Clin Invest 109:337–346

88. Rabascio C, Muratori E, Mancuso P, Calleri A, Raia V, Foutz T, Cinieri S, Veronesi G, Pruneri G, Lampertico P, Iavarone M, Martinelli G, Goldhirsch A, Bertolini F (2004) Assessing tumor angiogenesis: increased circulating VE-cadherin RNA in patients with cancer indicates viability of circulating endothelial cells. Cancer Res 64:4373–4377

89. Kim HK, Song KS, Kim HO, Chung JH, Lee KR, Lee YJ, Lee DH, Lee ES, Kim HK, Ryu KW, Bae JM (2003) Circulating numbers of endothelial progenitor cells in patients with gastric and breast cancer. Cancer Lett 198:83–88

90. Song S, Ewald AJ, Stallcup W, Werb Z, Bergers G (2005) PDGFRbeta+ perivascular progenitor cells in tumours regulate pericyte differentiation and vascular survival. Nat Cell Biol 7:870–879

91. Goon PK, Lip GY, Boos CJ, Stonelake PS, Blann AD (2006) Circulating endothelial cells, endothelial progenitor cells, and endothelial microparticles in cancer. Neoplasia 8:79–88

92. Smirnov DA, Foulk BW, Doyle GV, Connelly MC, Terstappen LW, O'Hara SM (2006) Global gene expression profiling of circulating endothelial cells in patients with metastatic carcinomas. Cancer Res 66:2918–2922

Chapter 22
Tumor Angiogenesis and the Cancer Stem Cell Model

Chris Folkins[1,2] and Robert S. Kerbel[1,2]

Keywords: cancer stem cell, tumor vasculature, CD133

Abstract: In recent years, research and interest in the area of cancer stem cells has grown tremendously. An increasing number of studies are finding that many different cancers contain a subpopulation of tumor cells that display several defining characteristics of adult tissue stem cells, including multipotent differentiation potential, long-term self-renewal capacity, and the expression of various molecular markers of stemness. Most importantly, these stem-like cancer cells also appear to possess the strongest tumor-initiating potential of all the cells in the tumor, a finding that has led to the development of the cancer stem cell model for tumor progression. This model suggests that tumors are organized in a developmental hierarchy (similar to a healthy tissue), with long-term tumor progression being driven by a self-renewing tumor stem cell at the top of the hierarchy. As this new model for tumor progression takes shape, researchers are beginning to investigate how cancer stem cells fit into various other aspects of cancer biology. In this regard, several recent studies are uncovering an intriguing relationship between tumor angiogenesis and cancer stem cells. This chapter reviews recent data suggesting that cancer stem cells may play an important role in promoting tumor angiogenesis, and that tumor vasculature may in turn have a role in supporting and maintaining cancer stem cells. Related work suggesting that antiangiogenic therapy may be used as a strategy to eliminate the critical CSC population is also discussed.

The Cancer Stem Cell Hypothesis: a New Model for Tumor Progression

More than a decade ago, pioneering studies by John Dick and colleagues identified a subpopulation of leukemic cells with stem-like properties in patients with acute myelogenous leukemia [1, 2]. These leukemic stem cells (LSCs) were found exclusively in the CD34+CD38– leukemic cell fraction, a surface immunophenotype also found on hematopoietic stem cells [3]. Only the CD34+CD38– cells were able to transplant disease into immune deficient mice. The resulting disease displayed a phenotypic heterogeneity resembling that of the original human donor, and CD34+CD38– cells isolated from primary engrafted mice and serially transplanted into secondary recipient mice produced disease with comparable efficiency, indicating that LSCs are capable of differentiation and self-renewal, two key properties of stem cells. These findings provided some of the first significant evidence supporting the 'cancer stem cell' model for tumor progression [4]. This model posits that a tumor is organized as a hierarchy—a distorted mirror image of its normal tissue counterpart. At the top of the hierarchy is a cancer stem cell (CSC), a self-renewing cancer cell that expresses surface markers of primitive cells and possesses differentiation potential and limitless proliferative potential. The CSC can divide asymmetrically, maintaining its proliferative potential through self-renewal, or giving rise to a transit amplifying daughter cell that can divide rapidly for a limited period of time, differentiating (aberrantly) and giving rise to the bulk of cells in the tumor mass, which are non-CSCs. Since only the CSC population can divide repeatedly without loss of proliferative potential, it is the CSC population that is responsible for tumor initiation and driving tumor progression in the long term.

Several years after the initial discovery of the leukemic stem cell, similar studies in breast cancer by Clarke and colleagues revealed that experimental tumor initiating capacity resides largely in a minority subpopulation of self-renewing breast tumor cells with primitive surface immunophenotype and differentiation potential [5], demonstrating for the first time that the CSC model may also be applicable to solid tumors. In the years since these early groundbreaking studies, interest in the CSC model has grown tremendously, and CSCs have now been implicated and characterized to varying degrees in many different cancers, including brain [6–12], prostate [13–15], melanoma [16], colon [17, 18], lung [19], ovarian [20, 21], gastric [22], pancreatic [23], retinoblastoma [24], bone sarcoma [25], hepatocellular carcinoma [26–28], head and neck squamous cell carcinoma [29,30], multiple myeloma [31], and chronic myelogenous leukemia [32–34]. While more extensive work will be required to determine how

[1] Deparment of Molecular and Cellular Biology Research, Sunnybrook Health Sciences Centre, Toronto, Ontario, Canada

[2] Department of Medical Biophysics, University of Toronto, Toronto, Ontario, Canada

closely the current CSC model fits with reality, it is clear at this time that a wide array of cancers contain a subset of stem-like tumor cells, and that tumor-initiating capacity exists largely or exclusively within this stem-like fraction. These findings have the potential to fundamentally shift our understanding of tumor progression, and will likely have implications that will alter our perception and study of many different aspects of tumor biology. In this regard, one of the most intriguing possibilities raised by the continually expanding body of literature supporting the CSC model is that we may gain a new level of understanding of key processes in tumor progression, such as angiogenesis, by re-examining them in the novel context provided by this model.

Considering Angiogenesis in a New Light

Individual cells within a tumor are known to be heterogeneous in many respects, and angiogenic potential is no exception. For example, it is common to observe unevenly distributed regions of hypoxia, perfusion, and microvessel density within a tumor, which may be reflective of heterogeneity in angiogenic phenotype in cells throughout the tumor. A direct demonstration of heterogeneity in angiogenic potential was provided in a study by Folkman in 2001 [35], which showed that subclones generated from single cells of a human liposarcoma cell line differed widely in proangiogenic capacity, giving rise to either aggressive, highly angiogenic tumors, poorly angiogenic tumors, or nonangiogenic, 'dormant' tumors when implanted in immune deficient mice. Most interestingly, when pieces of the resulting tumors were transplanted into new recipient mice, secondary tumors were generated with phenotypes that generally matched that of the transplanted tissue, suggesting that some cells within a tumor cell population are inherently more strongly angiogenic than others. Conventional wisdom in tumor biology would attribute this angiogenic heterogeneity to, for example, genetic differences due to ongoing mutagenesis, or variations in local cellular microenvironments throughout the tumor. Although these are certainly contributing factors, the CSC model predicts that tumor heterogeneity results largely from the fact that a tumor is made up of cells from all stages of the tumor proliferative hierarchy, i.e. tumors include cells in various states of (abnormal) differentiation, which are phenotypically very different from one another. Is it possible, then, that angiogenic heterogeneity is a product of the variety of differentiation states present in a tumor initiated by a CSC? In other words, considering the results of Folkman's liposarcoma tumor dormancy study in the context of the cancer stem cell model, is it possible to attribute angiogenic potential to a particular stage (or stages) of the tumor proliferative hierarchy? Before it is possible to address such questions, significant work will be required to validate the existence and elucidate the composition of a true, multi-stage tumor proliferative hierarchy. In the meantime, however, a number of studies have begun to address a related but more

basic question: what is the contribution of the stem-like tumor cell fraction to tumor angiogenesis?

Contribution of Cancer Stem Cells to Tumor Angiogenesis

As studies characterizing cancer stem cells continue to accumulate, a number of researchers are beginning to address the role of CSCs in tumor angiogenesis. The rationale for undertaking such studies is easy to understand. Since angiogenesis has long been recognized as a limiting factor for significant tumor progression [36], and since CSCs are extraordinarily tumorigenic, it stands to reason that CSCs may be particularly strong promoters of tumor angiogenesis. A recent report by Bao et al. [37] suggests that this may indeed be the case. The authors show that the CD133+ fraction of human glioma specimens is enriched in stem-like cells and expresses 10–20 fold more vascular endothelial growth factor (VEGF) than the CD133– fraction. More modestly increased expression of factors, such as angiogenin, interleukin 6, interleukin 8, and basic fibroblast growth factor (bFGF), were also noted in the CD133+ cell fraction. Conditioned media from CD133+ cells induced higher levels of migration and tubule formation by human microvascular endothelial cells in vitro compared to media from CD133– cells, and CD133+ cells consistently gave rise to aggressive, strongly angiogenic tumors in immune deficient mice, whereas CD133– cells, which were rarely tumorigenic, yielded only tiny, poorly angiogenic tumors. Significantly, blockade of VEGF by the humanized monoclonal antibody bevacizumab reduced angiogenesis-associated endothelial cell behaviors induced by CD133+ cell conditioned media in vitro, and vascularity of CD133+ tumors in vivo, to levels comparable to those induced by CD133– cells, seemingly completely negating the proangiogenic advantage of CD133+ tumor cells. These results suggest that CSCs may be the most strongly angiogenic cells in the tumor, and that their angiogenic advantage, at least in some cases, is linked to a significantly increased expression of VEGF. Indeed, other studies are consistent with this hypothesis, and hint that it may be applicable to CSCs from other cancers as well. For example, brain tumor sphere cultures enriched in stem-like cells have been found to express more VEGF than matched monolayer cultures with lower stem-like cell content [38]. Microarray analysis revealed that these brain tumor spheres also expressed increased levels of hypoxia inducible factor 1-α and angiopoietin 2, and decreased levels of the angiogenesis inhibitor thrombospondin 1. Similar studies in breast cancer revealed that stem-like cell-enriched tumor sphere cultures of the MCF7 breast cancer cell line expressed more VEGF than non-CSC-enriched monolayer MCF7 cultures [39].

Despite evidence supporting a proangiogenic role for CSCs, the notion that CSCs are major contributors to tumor angiogenesis may seem counterintuitive. CSCs by definition represents only a minority subfraction of all the cells in a tumor. Indeed, in colon cancer, for example, limiting dilution transplant assays in immune deficient mice estimate that one in every 5.7×10^4 tumor cells is capable of initiating tumor formation [17]. How could such a small cell population have any significant impact on tumor angiogenesis? Even if 0.1% of tumor cells were CSCs, and CSCs expressed 10 or even 100 times more proangiogenic factors than non-CSC tumor cells, the proangiogenic signaling output from CSCs would still comprise only 1–10% of the total proangiogenic output of the tumor as a whole—hardly a major contribution. Contrary to what this scenario implies, however, an increased angiogenic output of CSCs *is* likely to be relevant, at times when CSCs comprise a more significant proportion of the tumor mass. These times would include (as predicted by the CSC model [4]): early during tumor initiation, early during initiation of a metastatic lesion, and during tumor regrowth following therapy (because CSCs are predicted to be inherently resistant to many conventional anticancer therapies, and thus are expected to remain and re-seed tumor growth following treatment with agents that shrink the tumor by killing, presumably, non-CSC tumor cells—indeed, brain tumor stem cells are found to be enriched following radiotherapy [40]). In these cases, early proangiogenic output by CSCs may play an important role in initiating an angiogenic phenotype permissive for tumor growth. In this regard, a recent study by Indraccolo et al. found that transient, strong proangiogenic signals provided by short-term exogenous infusion of VEGF and bFGF or co-implantation of short-term surviving highly angiogenic feeder cells are able to convert a dormant, poorly angiogenic tumor into a progressively growing, actively angiogenic tumor [41]. These findings imply that short-term proangiogenic signals early during tumor initiation may be sufficient to turn on angiogenesis in the long term to drive tumor progression. Therefore, it is conceivable that strong proangiogenic output from CSCs is critical at the earliest stages of tumor growth (or regrowth after anticancer therapy), providing the "angiogenic burst" required to initiate active angiogenesis and tumor expansion.

In interpreting the results reported by Bao et al., as well as others, another important point to note is that these studies characterized the angiogenic potential of a population in which CSCs are enriched, but not necessarily purified. For example, Bao et al. demonstrate increased angiogenic potential of the CD133+ fraction of brain tumors, but even in the best reported example, at least 100 CD133+ cells are required for tumor initiation [10], while tumor initiation should theoretically be achievable by a much smaller number of cells from a pure CSC population. Increased angiogenic capacity in a CSC-enriched fraction may reflect, as discussed above, increased angiogenic capacity of CSCs, or it may reflect a more generalized increased angiogenic capacity in the more primitive stages of the tumor proliferative hierarchy. In the latter scenario, the strongly angiogenic cells would represent a more significant proportion of the tumor mass, and thus could potentially make a significant contribution to driving angio-

genesis throughout tumor progression, rather than exclusively in the earliest stages of tumor growth. This example illustrates that, although current evidence suggests that stem-like tumor cells preferentially contribute to angiogenesis in some cases, a clear picture of the timing and extent of contribution of CSCs, and possibly primitive but non-CSC tumor cells, to angiogenesis will require a more thorough understanding of the putative tumor proliferative hierarchy and better methods for enriching (and ideally purifying) the CSC population.

The need for additional work in characterizing the roles of CSCs in tumor angiogenesis is further highlighted by reports indicating that, in some cases, the more mature cells in the tumor may be more strongly angiogenic than the CSC-enriched, primitive cell fraction. For example, it was recently reported that CSC-enriched brain tumor sphere cultures induced less microvessel outgrowth in a rat aortic ring assay and expressed lower levels of VEGF and CXCL12 mRNA than matched adherent brain tumor cell cultures not enriched in CSCs [42]. Another study shows that primary cultures derived from human brain tumor biopsies form nonangiogenic tumors when transplanted in immune deficient mice [43]. These tumors expressed neural precursor markers such as nestin and musashi-1, but expression of primitive neural cell markers decreased when tumors switched to an actively angiogenic phenotype. These findings, which seemingly conflict with the previously discussed studies describing a proangiogenic role for CSCs, suggest that the way in which cells from various stages of the putative tumor proliferative hierarchy contribute to tumor angiogenesis may be complex and context dependant. For example, CSC contribution to angiogenesis may vary among different types of cancer or among individual patients with similar disease type (implied by the seemingly contradictory findings in the various studies in brain cancer discussed above), and may differ depending on the stage of tumor progression (for example, as discussed above, more primitive cells may play a more significant role at an early stage when the tumor is smaller, whereas more mature cells may become more important as the tumor increases in size). Another related possibility is that the angiogenic potential of CSCs is dynamic, and changes in response to cues from the tumor microenvironment. Interestingly, the results of the previously discussed study on tumor dormancy in liposarcoma [35] seem to be consistent with this notion. This study showed that some cells within a tumor are initially strongly angiogenic and can seed the growth of aggressive tumors, while others are initially poorly angiogenic but can eventually switch to an angiogenic phenotype and drive tumor progression. The fact that both the strongly angiogenic and initially angiogenesis-dormant cell populations can initiate progressively growing tumors implies that both populations contain CSCs. This in turn implies that CSCs may exist in either a pre-angiogenic (angiogenesis-dormant) or actively angiogenic state. Following this reasoning, the shift of the CSCs from the pre-angiogenic to actively angiogenic state may represent a key event in the initiation of tumor angiogenesis

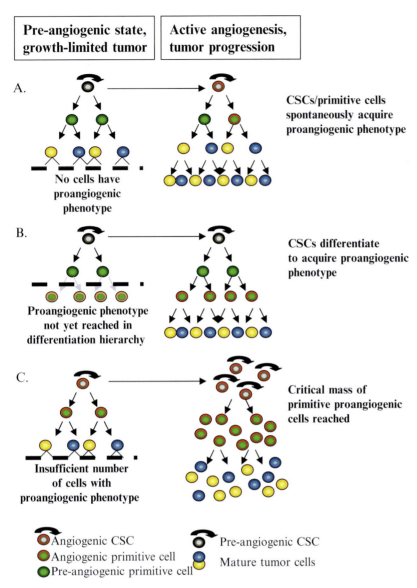

| Pre-angiogenic state, growth-limited tumor | Active angiogenesis, tumor progression |

A. CSCs/primitive cells spontaneously acquire proangiogenic phenotype

No cells have proangiogenic phenotype

B. CSCs differentiate to acquire proangiogenic phenotype

Proangiogenic phenotype not yet reached in differentiation hierarchy

C. Critical mass of primitive proangiogenic cells reached

Insufficient number of cells with proangiogenic phenotype

Angiogenic CSC / Pre-angiogenic CSC
Angiogenic primitive cell / Mature tumor cells
Pre-angiogenic primitive cell

FIG. 22.1. Models of the angiogenic switch in the context of the cancer stem cell model. Evidence suggests that stem-like and/or primitive tumor cells are strong promoters of tumor angiogenesis, and these cells may figure into the angiogenic switch in several different ways. **A** Pre-switch, all cells in the tumor proliferative hierarchy are nonangiogenic. Switch occurs when CSCs and/or primitive (but non-CSC) tumor cells spontaneously acquire a proangiogenic phenotype. **B** Pre-switch, primitive tumor cells have not yet reached a proangiogenic stage of differentiation. Switch occurs when differentiation gives rise to a population of proangiogenic cells (that still retain a primitive phenotype). **C** Pre-switch, CSCs and/or primitive tumor cells have proangiogenic capacity, but sufficient numbers aren't present to initiate an angiogenic state in the tumor. Switch occurs when CSCs/primitive tumor cells expand to reach a critical mass with combined proangiogenic capacity sufficient to initiate tumor angiogenesis.

– the so-called "angiogenic switch" (Fig. 22.1). Alternatively, the angiogenic switch may mark the point at which non-stem-like progeny of CSCs reach a stage of differentiation conducive to a strong proangiogenic phenotype. Clearly, many intriguing speculations can be made as to how the CSC model may fit into, and potentially modify, our current conceptions of tumor angiogenesis, and as our knowledge of cancer stem cells improves, a new level of understanding of tumor angiogenesis may be possible.

Contribution of Cancer Stem Cells to Tumor Vasculature by Alternative Mechanisms

In addition to a possible role in angiogenesis, it should be noted that CSCs may also contribute to tumor vascularization by other mechanisms. One possibility is that these primitive, undifferentiated cells are able to differentiate into functional endothelial cells and incorporate into the vessel walls of nascent tumor capillaries. In this regard, human chronic myelogenous leukemia progenitor cells are able to generate endothelial cells and contribute to neovascularization at sites of injury in the mouse gastrointestinal tract in an in vivo assay [34]. These results indicate that primitive tumor cells may retain endothelial cell differentiation potential possessed by stem/progenitor cells in their normal tissue counterpart. This raises the possibility that CSCs from tissues whose normal stem/progenitor cells have demonstrated plasticity toward the endothelial lineage, such as kidney [44,45], brain [46,47], skin [48], and muscle [49], may also have the ability to contribute to tumor vasculature by differentiating into vessel-incorporating endothelial cells. Another, related possibility, is that, rather than differentiation into frank endothelial cells, CSCs may preferentially participate in vasculogenic mimicry—a phenomenon characterized by the acquisition of endothelial cell markers and formation of fluid-conducting, vessel-like channels by tumor cells. Vasculogenic mimicry has been observed in a wide variety of malignancies, and is associated with tumor cells displaying an undifferentiated phenotype [50], suggesting that CSCs may be more likely than other tumor cells to participate in this process.

Role of Tumor Vasculature in Supporting Cancer Stem Cells

Evidence supporting a proangiogenic role for CSCs suggests that stem-like tumor cells and tumor vascular biology may be intimately intertwined. Interestingly, another recent study extends this notion with evidence suggesting that CSCs may in turn rely on nearby tumor vasculature for maintenance of their stem-like, tumorigenic state, therefore implying an interdependent relationship between stem-like tumor cells and tumor vasculature. This work, performed by Calabrese et al. [51], showed that stem-like brain tumor cells exist closer to tumor vasculature than non-stem-like tumor cells, and interact physically with vascular endothelial cells. Co-culture with primary human endothelial cells (PHECs) supported maintenance and expansion of stem-like tumor cells in vitro, and co-implantation of PHECs with tumor cells in immune deficient mice yielded tumors with enhanced growth rate and an increased fraction of stem-like cells, suggesting that factors secreted by endothelial cells promote self-renewal and expansion of CSCs. In another set of experiments, a medulloblastoma cell line overexpressing ERBB2 (which results in increased VEGF expression) was found to give rise to tumors with increased microvessel density and higher stem-like cell fraction. Significantly, inhibition of angiogenesis by either VEGF blockade using bevacizumab, or inhibition of ERBB2 activity with the tyrosine kinase inhibitor erlotinib, selectively eradicated stem-like cells from the tumor in the absence of any direct effects on tumor cell proliferation, apoptosis, or necrosis.

Taken together, the results of the Calabrese et al. study show that the brain tumor vasculature may be essential in supporting and preserving the survival, stem-like properties, and functions of CSCs which are critical for their ability to drive tumor progression in the long term. In other words, it appears that the vasculature may be providing a niche effect for brain cancer stem cells. A niche is a specialized microenvironment that, through contact-mediated as well as paracrine signaling interactions, maintains stem cells, provides protective effects, and controls stem cell proliferation and fate determination [52]. The study by Calabrese et al. is noteworthy in that it demonstrates for the first time that niche effects may also be at play, at least to a certain extent, in the maintenance of cancer stem cells. The fact that the putative brain CSC niche consists of a vascular microenvironment is particularly interesting, since stem-like cells from many other types of cancer can be maintained and expanded in culture in the presence of proangiogenic growth factors, such as bFGF and epidermal growth factor (EGF) (Table 22.1), which would be found in higher concentrations in highly angiogenic areas of the tumor. Since the CSC-supportive effects of these factors are seemingly broadly applicable, it is conceivable that CSCs in many other cancers may also be maintained by a vascular niche.

The notion of a vascular niche for CSCs has a number of potentially exciting implications. For example, it would add a new dimension to the role of angiogenesis in tumor progression, as continuous blood vessel growth would be required not only to provide a blood supply, but also to keep pace with the niche demands of increasing numbers of CSCs in the growing tumor. In this regard, it is interesting to consider the proangiogenic capacity of CSCs discussed in the previous section, which may play an additional role as a built-in fail-safe to ensure that nascent or disseminating CSCs can induce the formation of a vascular microenvironment to support their growth and survival. Another interesting point to consider is the potential effect that tumor oxygenation has on the CSC fraction since, contrary to what may be expected assuming a vascular niche for CSCs, hypoxia has been found to promote a dedifferentiated state in neuroblastoma [53] and breast cancer cells [54], and hypoxic culture conditions have been reported to expand the fraction of cells expressing the neural stem cell marker CD133 in a medulloblastoma cell line [55]. Perhaps the most pertinent impact of the CSC vascular niche concept, however, is illustrated by the results of Calabrese et al. showing that bevacizumab or erlotinib can reduce the stem-like cell fraction in brain tumors. A CSC vascular niche may represent an important novel therapeutic target, and accordingly suggests a possible new role for antiangiogenic therapy as a means of eradicating the critical cancer stem cell population.

Cancer stem cells and antiangiogenic therapy

One of the most compelling aspects of the CSC model is the potential implication it may hold for anti-cancer therapy. As (theoretically) the only cells capable of proliferating in the long term, and hence the one cell population that is absolutely critical for continued tumor progression and metastasis, the CSC

TABLE 22.1. Proangiogenic factors supporting maintenance and expansion of stem-like cancer cells in culture.

Cancer	Proangiogenic factors (associated references in superscript)
Brain	bFGF[6–12], EGF[7,8–12], PDGF[6]
Breast	bFGF[39], EGF[39]
Prostate	EGF[14], SCF[14]
Melanoma	bFGF[16]
Colon	bFGF[18], EGF[18]
Bone	bFGF[25], EGF[25]
Head and neck	bFGF[30], EGF[30]
Chronic myelogenous leukemia	EGF[34], PDGF[34], IL6[34], SCF[33], GMCSF[33], EPO[33]

bFGF Basic fibroblast growth factor; *EGF* epidermal growth factor; *PDGF* platelet-derived growth factor; *SCF* stem cell factor; *IL6* interleukin 6; *GMCSF* granulocyte-macrophage colony stimulating factor; *EPO* erythropoietin population must by definition be eradicated in order to completely and permanently arrest disease progression. The unfortunate irony, however, is that CSCs are also by definition likely to be the most difficult cell population to target. Among the many other properties they share with normal tissue stem cells, CSCs are also expected to possess a number of characteristics that make them inherently resistant to many conventional forms of anti-cancer therapy, including a reduced proliferation rate, enhanced DNA damage repair, and expression of anti-apoptotic proteins and multidrug resistance transporters [56]. Indeed, enhanced radioresistance [40,55], chemoresistance [57], and increased expression of multidrug resistance-associated proteins [42] have been observed in stem-like brain cancer cells. Add to these factors the protective effects provided by the niche microenvironment, and it becomes evident that targeting CSCs will likely be a very challenging prospect. While a number of strategies for targeting CSCs are currently under investigation, including CSC-directed immunotherapy [38,58], antibodies targeting CSC-specific surface markers [59], and agents interrupting signaling pathways important for CSC maintenance and function [60–62], targeting a CSC vascular niche may represent an opportunity to indirectly hamper the function and/or reduce the viability of CSCs while circumventing many aspects of CSC-associated intrinsic drug resistance.

While current antiangiogenic therapies have shown promising results in the clinic, in many cases they have yet to produce the dramatic effects that might be anticipated from a therapy capable of eliminating CSCs [63]. This highlights the fact that, in order to take full advantage of antiangiogenic therapies for the purpose of targeting the putative CSC vascular niche, it will first be necessary to gain a better understanding of the mechanisms by which the tumor vasculature supports and protects CSCs, which will in turn inform the development of rational anti-CSC therapies incorporating an antiangiogenic component. An example of the potential value of such an approach is provided by the bevacizumab/erlotinib results from the Calabrese et al. study. Both of these therapies effectively remove VEGF from the system. While VEGF is obviously important for angiogenesis, it has also been found to promote the survival of neural stem cells [64]. This suggests that by choosing the right antiangiogenic therapy (i.e. targeting VEGF in brain cancer, in this example), it may be possible to simultaneously impact both the vascular niche and the CSC population directly. Indeed, this type of two-pronged effect may account for the recently reported efficacy of bevacizumab (in combination with irinotecan) in the treatment of patients with recurrent grade III-IV glioma [65]. Since EGF has been implicated in supporting CSC maintenance and expansion in vitro in a wide variety of cancers (Table 22.1), antiangiogenic therapies targeting this system are also likely candidates for a dual anti-niche/anti-CSC effect.

Another important step in adapting antiangiogenic therapies for use in eradicating CSCs will be to investigate how different antiangiogenic therapeutic approaches impact the CSC population. In this regard, a recent study [66] shows that VEGF blockade in tumor xenografts of the rat C6 glioma cell line in nude mice using the anti-mouse VEGF receptor 2 monoclonal antibody DC101 (which would only prevent VEGF signaling

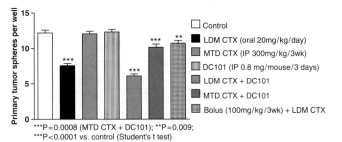

*** P = 0.0008 (MTD CTX + DC101); ** P = 0.009;
*** P < 0.0001 vs. control (Student's t test)

Fig. 22.2. Impact of cytotoxic, antiangiogenic, and combination cytotoxic plus antiangiogenic therapeutic strategies on the tumor stem-like cell fraction in experimental brain tumors. Established subcutaneous tumor xenografts of the rat C6 glioma cell line in nude mice were treated for 2 weeks (n = 7 mice per group) with either continuous low-dose "metronomic" (LDM) cyclophosphamide (CTX), conventional maximum tolerated dose (MTD) CTX, the anti-VEGF receptor 2 monoclonal antibody DC101, combinations of DC101 with LDM or MTD CTX, or LDM CTX combined with an upfront bolus dose of CTX. To evaluate the relative size of the stem-like tumor cell fraction after treatment, the fraction of cells capable of forming non-adherent tumor spheroid colonies at clonal density in serum-free conditions was determined. Sphere-forming capacity in these conditions is a behavior associated with stem-like tumor cells. Therapies that were exclusively antiangiogenic (DC101) or cytotoxic (MTD CTX) did not affect the size of the stem-like cell fraction, whereas all other therapies tested, all of which combine antiangiogenic plus cytotoxic effects, selectively eradicated stem-like tumor cells. Figure reprinted with permission from [66].

in host cells and, hence, unlike bevacizumab/erlotinib in the Calabrese et al. study, would be unable to potentially impact possible pro-survival effects of VEGF on CSCs) is on its own insufficient to selectively eradicate stem-like cancer cells. When DC101 was combined with either continuous low-dose "metronomic" or conventional maximum tolerated dose schedules of cyclophosphamide, however, significant eradication of stem-like tumor cells was achieved (Fig. 22.2). Furthermore, metronomic cyclophosphamide, which is known to produce antiangiogenic effects [67], was sufficient on its own to selectively eliminate stem-like cancer cells (although not to the same extent as when it was combined with DC101), whereas cytotoxic, maximum tolerated dose cyclophosphamide had no effect on the size of the stem-like cell fraction. Taken together, these results suggest that, at least in some cases, inhibition of angiogenesis alone may not be sufficient to eradicate CSCs, but antiangiogenic therapy may sensitize CSCs to the cytotoxic effects of chemotherapy. This CSC-specific chemosensitizing effect may be due to disruption of a CSC vascular niche, resulting in loss of CSC protective/pro-survival signals and/or loss of CSC maintenance and consequent loss or reduction of CSC-associated mechanisms of intrinsic drug resistance.

The possibility that antiangiogenic agents can chemosensitize CSCs is particularly relevant since, in a clinical setting, targeted antiangiogenic agents are rarely effective on their own in the treatment of advanced stage cancers, and must be combined with chemotherapy to show any significant therapeutic benefit [68]. The commonly observed chemosensitizing effects of antiangiogenic drugs may seem counterintuitive,

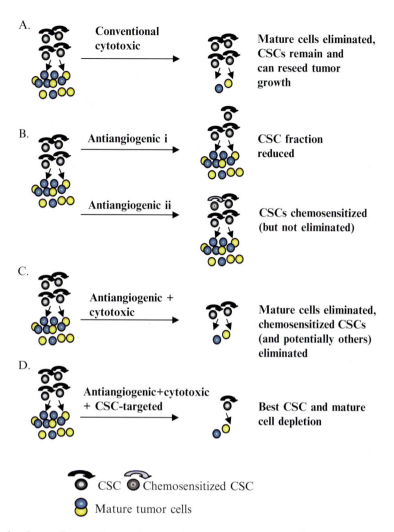

FIG. 22.3. Predicted impact of various anticancer therapeutic strategies on cancer stem cell and mature tumor cell populations. **A** Conventional cytotoxic chemotherapy eliminates rapidly dividing mature tumor cells but spares drug resistant CSCs, which can reseed tumor growth. **B** Antiangiogenic therapy, (i) as reported by Calabrese et al. [51], eliminates CSCs due to disruption of vascular niche, or (ii) as reported by Folkins et al. [66], chemosensitizes but does not eliminate CSCs. **C** Cytotoxic chemotherapy eliminates mature cells and sensitized CSCs; antiangiogenic therapy may also eliminate non-sensitized CSCs as in (i). **D** CSC and mature cell eradication as in **C**, plus CSC-targeted therapy eliminates residual CSCs.

as one might expect that disruption of angiogenesis would impede the delivery of chemotherapeutic agents to tumor tissue. Consequently, several hypotheses have been proposed to explain this seemingly paradoxical effect, including transient normalization of tortuous tumor vasculature resulting in a temporary "window" of improved drug delivery [69], slowing of tumor cell repopulation between successive cycles of chemotherapy [70], and enhanced toxicity of chemotherapeutic agents to proliferating endothelial cells in the tumor vasculature and bone marrow-derived circulating endothelial progenitor cells [67,68,71,72]. Disruption of a CSC vascular niche by antiangiogenic therapy may represent yet another mechanism for the chemosensitizing effects of antiangiogenic drugs, thereby extending our understanding of the action of these agents. Interestingly, this hypothesis would also help explain why antiangiogenic therapies in combination with chemotherapy often produce a survival benefit that is independent of tumor response (shrinkage) [73]: This is precisely what would

be expected from a therapy that targets the CSC population, since these cells are critical for tumor progression yet represent only a minority fraction of the overall tumor mass.

Considering that a vascular niche may provide signals that promote survival and inhibit differentiation of CSCs, another possible impact of niche disruption by antiangiogenic therapy is an increased sensitization of CSCs to therapies that selectively kill CSCs or induce their differentiation. Although such therapies are only in the earliest stages of preclinical development (as illustrated in the examples referenced above), it is interesting to consider how they might be applied. Combining CSC-targeting agents with what we know of conventional cytotoxic chemotherapy, antiangiogenic therapy, and the CSC model, it is intriguing to speculate how an optimal combination of these therapeutic approaches might come together for maximum therapeutic benefit (Fig. 22.3). For example, a regimen of bolus plus low-dose chemotherapy [72] would combine antiangiogenic effects (continuous low doses) to slow tumor growth, disrupt the vascular niche and chemosensitize and/or eliminate CSCs, and cytotoxic effects (bolus doses) to eliminate non-CSC tumor cells as well as sensitized CSCs. Combining this with a targeted antiangiogenic agent whose target is also a survival factor for CSCs would enhance the antiangiogenic, niche-disruptive effects, improving CSC chemosensitization, and would also directly reduce CSC viability. Finally, the addition of a CSC-targeted agent to this combination would help eradicate sensitized as well as non-sensitized CSCs. Certainly, as these speculations illustrate, the possibility that CSCs are dependant on the tumor vasculature presents a number of novel, potentially exciting therapeutic approaches for further investigation, and future work in this area may play a role in shaping how we understand and apply antiangiogenic therapy in years to come.

References

1. Bonnet D and Dick JE Human acute myeloid leukemia is organized as a hierarchy that originates from a primitive hematopoietic cell, Nature Med, *3:* 730–737, 1997.

2. Lapidot T, Sirard C, Vormoor J, Murdoch B, Hoang T, Caceres-Cortes J, Minden M, Paterson B, Caligiuri MA and Dick JE A cell initiating human acute myeloid leukemia after transplantation into SCID mice, Nature, *367:* 645–648, 1994.

3. Bhatia M, Wang JC, Kapp U, Bonnet D and Dick JE Purification of primitive human hematopoietic cells capable of repopulating immune-deficient mice, Proc Natl Acad Sci USA, *94:* 5320–5325, 1997.

4. Reya T, Morrison SJ, Clarke MF and Weissman IL Stem cells, cancer, and cancer stem cells, Nature, *414:* 105–111, 2001.

5. Al Hajj M, Wicha MS, Benito-Hernandez A, Morrison SJ and Clarke MF Prospective identification of tumorigenic breast cancer cells,Proc Natl Acad Sci USA, *100:* 3983–3988, 2003.

6. Kondo T, Setoguchi T and Taga T Persistence of a small subpopulation of cancer stem-like cells in the C6 glioma cell line, PNAS, *101:* 781–786, 2004.

7. Galli R, Binda E, Orfanelli U, Cipelletti B, Gritti A, Vitis SD, Fiocco R, Foroni C, Dimeco F and Vescovi A Isolation and characterization of tumorigenic, stem-like neural precursors from human glioblastoma, Cancer Res, *64:* 7011–7021, 2004.

8. Singh SK, Clarke ID, Terasaki M, Bonn VE, Hawkins C, Squire J and Dirks PB Identification of a cancer stem cell in human brain tumors, Cancer Res, *63:* 5821–5828, 2003.

9. Yuan X, Curtin J, Xiong Y, Liu G, Waschsmann-Hogiu S, Farkas DL, Black KL and Yu JS Isolation of cancer stem cells from adult glioblastoma multiforme, Oncogene, *23:* 9392–9400, 2004.

10. Singh SK, Hawkins C, Clarke ID, Squire JA, Bayani J, Hide D, Henkelman RM, Cusimano MD and Dirks PB Identification of human brain tumor initiating cells, Nature, *432:* 396–401, 2004.

11. Hemmati HD, Nakano I, Lazareff JA, Masterman-Smith M, Geschwind DH, Bronner-Fraser M and Kornblum HI Cancerous stem cells can arise from pediatric brain tumors,Proc Natl Acad Sci USAS, *100:* 15178–15183, 2003.

12. Taylor MD, Poppleton H, Fuller C, Su X, Liu Y, Jenesen P, Magdaleno S, Dalton J, Calabrese C, Board J, MacDonald T, Rutka J, Guha A, Gajjar A, Curran T and Gilbertson RJ Radial glial cells are candidate stem cells of ependymoma, Cancer Cell, *8:* 323–335, 2005.

13. Patrawala L, Calhoun T, Schneider-Broussard R, Li H, Bhatia B, Tang S, Reilly JG, Chandra D, Zhou J, Claypool K, Coghlan L and Tang DG Highly purified CD44+ prostate cancer cells from xenograft human tumors are enriched in tumorigenic and metastatic progenitor cells, Oncogene, *25:* 1696–1708, 2006.

14. Collins AT, Berry PA, Hyde C, Stower MJ and Maitland NJ Prospective identification of tumorigenic prostate cancer stem cells, Cancer Res, *65:* 10946–10951, 2005.

15. Xin L, Lawson DA and Witte ON The Sca-1 cell surface marker enriches for a prostate-regenerating cell subpopulation that can initiate prostate tumorigenesis,Proc Natl Acad Sci USA, *102:* 6942–6947, 2005.

16. Fang D, Nguyen TK, Leishear K, Finko R, Kulp AN, Hotz S, Van Belle PA, Xu X, Elder DE and Herlyn M A tumorigenic subpopulation with stem cell properties in melanomas, Cancer Res, *65:* 9328–9337, 2005.

17. O'Brien CA, Pollett A, Gallinger S and Dick JE A human colon cancer cell capable of initiating tumor growth in immunodeficient mice, Nature, *445:* 106–110, 2007.

18. Ricci-Vitiani L, Lombardi DG, Pilozzi E, Biffoni M, Todaro M, Peschle C and De Maria R Identification and expansion of human colon-cancer-initiating cells, Nature, *445:* 111–115, 2007.

19. Kim CF, Jackson EL, Woolfenden AE, Lawrence S, Babar I, Vogel S, Crowley D, Bronson RT and Jacks T Identification of bronchioalveolar stem cells in normal lung and lung cancer, Cell, *121:* 823–835, 2005.

20. Bapat SA, Mali AM, Koppikar CB and Kurrey NK Stem and progenitor-like cells contribute to the aggressive behaviour of human epithelial ovarian cancer, Cancer Res, *65:* 3025–3029, 2005.

21. Szotek PP, Pieretti-Vanmarcke R, Masiakos PT, Dinulescu DM, Connolly D, Foster R, Dombkowski D, Preffer F, MacLaughlin DT and Donahoe PK Ovarian cancer side population defines cells with stem cell-like characteristics and Mullerian Inhibiting Substance responsiveness,Proc Natl Acad Sci USA, *103:* 11154–11159, 2006.

22. Houghton JM, Stoicov C, Nomura S, Rogers AB, Carlson J, Li H, Cai X, Fox JG, Goldenring JR and Wang TC Gastric cancer originating from bone marrow-derived cells, Science, *306:* 1568–1571, 2004.

23. Li C, Heidt DG, Dalerba P, Burant CF, Zhang L, Adsay V, Wicha M, Clarke MF and Simeone DM Identification of pancreatic cancer stem cells, Cancer Res, *67:* 1030–1037, 2007.

24. Seigel GM, Campbell LM, Narayan M and Gonzalez-Fernandez F Cancer stem cell characteristics in retinoblastoma, Mol Vis, *11:* 729–737, 2005.

25. Gibbs CP, Kukekov VG, Reith JD, Tchigrinova O, Suslov ON, Scott EW, Ghivizzani SC, Ignatova TN and Steindler DA Stem-like cells in bone sarcomas: implications for tumorigenesis, Neoplasia, *7:* 967–976, 2005.

26. Suetsugu A, Nagaki M, Aoki H, Motohashi T, Kunisada T and Moriwaki H Characterization of CD133+ hepatocellular carcinoma cells as cancer stem/progenitor cells, BBRC, 824, 2006.

27. Chiba T, Kita K, Zheng YW, Yokosuka O, Saisho H, Iwama A, Nakauchi H and Taniguchi H Side population purified from hepatocellular carcinoma cells harbors cancer stem cell-like properties, Hepatology, *44:* 240–251, 2006.

28. Yin S, Li J, Hu C, Chen X, Yao M, Yan M, Jiang G, Ge C, Xie H, Wan D, Yang S, Zheng S and Gu J CD133 positive hepatocellular carcinoma cells possess high capacity for tumorigenicity, Int J Cancer, *120:* 1444–1450, 2007.

29. Prince ME, Sivanandan R, Kaczorowski A, Wolf GT, Kaplan MJ, Dalerba P, Weissman IL, Clarke MF and Ailles LE Identification of a subpopulation of cells with cancer stem cell properties in head and neck squamous cell carcinoma,Proc Natl Acad Sci USA, *104:* 973–978, 2007.

30. Zhou L, Wei X, Cheng L, Tian J and Jiang JJ CD133, one of the markers of cancer stem cells in Hep-2 cell line, Laryngoscope, *117:* 455–460, 2007.

31. Matsui W, Huff CA, Wang Q, Malehorn MT, Barber J, Tanhehco Y, Smith BD, Civin CI and Jones RJ Characterization of clonogenic multiple myeloma cells, Blood, *103:* 2332–2336, 2004.

32. Eisterer W, Jiang X, Christ O, Glimm H, Lee KH, Pang E, Lambie K, Shaw G, Holyoake TL, Petzer AL, Auewarakul C, Barnett MJ, Eaves CJ and Eaves AC Different subsets of primary chronic myeloid leukemia stem cells engraft immunodeficient mice and produce a model of the human disease, Leukemia, *19:* 435–441, 2005.

33. Jamieson CHM, Ailles LE, Dylla SJ, Muijtjens M, Jones C, Zehnder JL, Gotlib J, Li K, Manz MG, Keating A, Sawyers CL and Weissman IL Granulocyte-macrophage progenitors as candidate leukemic stem cells in blast-crisis CML, N Eng J Med, *351:* 657–667, 2004.

34. Fang B, Zheng C, Liao L, Han Q, Sun Z, Jiang X and Zhao RCH Identification of human chronic myelogenous leukemia progenitor cells with hemangioblastic characteristics, Blood, *105:* 2733–2740, 2005.

35. Achilles EG, Fernanadez A, Allred EN, Kisker O, Udagawa T, Beecken WD, Flynn E and Folkman J Heterogeneity of angiogenic activity in a human liposarcoma : a proposed mechanism for "no take" of human tumors in mice, J Natl Cancer Inst, *93:* 1075–1081, 2001.

36. Hannahan D and Weinberg RA The hallmarks of cancer, Cell, *100:* 57–70, 2000.

37. Bao S, Wu Q, Sathornsumetee S, Hao Y, Li Z, Hjelmeland AB, Shi Q, McLendon RE, Bigner DD and Rich JN Stem cell-like glioma cells promote tumor angiogenesis through vascular endothelial growth factor, Cancer Res, *66:* 7843–7848, 2006.

38. Pellegatta S, Polianai PL, Corno D, Menghi F, Ghielmetti F, Suarez-Merino B, Caldera V, Nava S, Ravanini M, Facchetti F, Bruzzone MG and Finocchiaro G Neurospheres enriched in cancer stem-like cells are highly effective in eliciting a dendritic cell-mediated immune response against malignant gliomas, Cancer Res, *66:* 10247–10252, 2006.

39. Ponti D, Costa A, Zaffaroni N, Pratesi G, Petrangolini G, Coradini D, Pilotti S, Pierotti MA and Daidone MG Isolation and in vitro propagation of tumorigenic breast cancer cells with stem/progenitor cell properties, Cancer Res, *65:* 5506–5511, 2005.

40. Bao S, Wu Q, McLendon RE, Hao Y, Shi Q, Hjelmeland AB, Dewhirst MW, Bigner DD and Rich JN Glioma stem cells promote radioresistance by preferential activation of the DNA damage response, Nature, *444:* 756–760, 2006.

41. Indraccolo S, Stievano L, Minuzzo S, Tosello V, Esposito G, Piovan E, Zamarchi R, Chieco-Bianchi L and Amadori A Interruption of tumor dormancy by a transient angiogenic burst within the tumor microenvironment, Proc Natl Acad Sci USA, *103:* 4216–4221, 2006.

42. Salmaggi A, Boiardi A, Gelati M, Russo A, Calatozzolo C, Ciusani E, Sciacca FL, Ottolina A, Parati EA, La Porta C, Alessandri G, Marras C, Croci D and De Rossi M Glioblastoma-derived tumorspheres identify a population of tumor stem-like cells with angiogenic potential and enhanced multidrug resistance phenotype, Glia, *54:* 850–860, 2006.

43. Sakariassen PO, Prestegarden L, Wang J, Skaftnesmo KO, Mahesparan R, Molthoff C, Sminia P, Sundlisaeter E, Misra A,

Tysnes BB, Chekenya M, Peters H, Lende G, Kalland KH, Oyan AM, Petersen K, Jonassen I, van der Kogel A, Feuerstein BG, Terzis AJA, Bjerkvig R and Enger PO Angiogenesis-independant tumor growth mediated by stem-like cancer cells,Proc Natl Acad Sci USA, *103:* 16466–16471, 2006.

44. Bussolati B, Bruno S, Grange C, Buttiglieri S, Deregibus MC, Cantino D and Camussi G Isolation of renal progenitor cells from adult human kidney, Am J Pathol, *166:* 545–555, 2005.

45. Bruno S, Bussolati B, Grange C, Collino F, Graziano ME, Ferrando U and Camussi G CD133+ renal progenitor cells contribute to tumor angiogenesis, Am J Pathol, *169:* 2223–2235, 2006.

46. Wurmser AE, Nakashima K, Summers RG, Toni N, D'Amour KA, Lie DC and Gage FH Cell fusion-independant differentiation of neural stem cells to the endothelial lineage, Nature, *430:* 350–356, 2004.

47. Oishi K, Kobayashi A, Fujii K, Kanehira D, Ito Y and Uchida MK Angiogenesis in vitro: vascular tube formation from the differentiation of neural stem cells, J Pharmacol Sci, *96:* 208–218, 2004.

48. Belicchi M, Pisati F, Lopa R, Porretti L, Fortunato F, Sironi M, Scalamogna M, Parati EA, Bresolin N and Torrente Y Human skin-derived stem cells migrate throughout forebrain and differentiate into astrocytes after injection into adult mouse brain, J Neurosci Res, *77:* 475–486, 2004.

49. Qu-Petersen Z, Deasy B, Jankowski R, Ikezawa M, Cummins J, Pruchnic R, Mytinger J, Cao B, Gates C, Wernig A and Huard J Identification of a novel population of muscle stem cells in mice: potential for muscle regeneration, J Cell Biol, *157:* 851–864, 2002.

50. Hendrix MJ, Seftor EA, Hess AR and Seftor RE Vasculogenic mimicry and tumor cell plasticity: lessons from melanoma, Nat Rev Cancer, *3:* 411–421, 2003.

51. Calabrese C, Poppleton H, Kocak M, Hogg TL, Fuller C, Hamner B, Oh EY, Gaber MW, Finklestein D, Allen M, Frank A, Bayazitov IT, Zakharenko SS, Gajjar A, Davidoff A and Gilbertson RJ A perivascular niche for brain tumor stem cells, Cancer Cell, *11:* 69–82, 2007.

52. Li L and Neaves WB Normal stem cells and cancer stem cells: The niche matters, Cancer Res, *66:* 4553–4557, 2006.

53. Jogi A, Ora I, Nilsson H, Lindeheim A, Makino Y, Poellinger L, Axelson H and Pahlman S Hypoxia alters gene expression in human neuroblastoma cells toward an immature and neural crest-like phenotype,Proc Natl Acad Sci USA, *99:* 7021–7026, 2002.

54. Helczynska K, Kronblad A, Jogi A, Nilsson E, Beckman S, Landberg G and Pahlman S Hypoxia promotes a dedifferentiated phenotype in ductal breast carcinoma in situ, Cancer Res, *63:* 1441–1444, 2003.

55. Blazek ER, Foutch JL and Maki G DAOY medulloblastoma cells that express CD133 are radioresistant relative to CD133– cells, and the CD133+ sector is enlarged by hypoxia, Int J Radiation Oncology Biol Phys, *67:* 1–5, 2007.

56. Dean M, Fojo T and Bates S Tumor stem cells and drug resistance, Nat Rev Cancer, *5:* 275–284, 2005.

57. Liu G, Yuan X, Zeng Z, Tunici P, Ng H, Abdulkadir IR, Lu L, Irvin D, Black KL and Yu JS Analysis of gene expression and chemoresistance of CD133+ cancer stem cells in glioblastoma, Mol Cancer, *5:* 67–78, 2006.

58. Wu A, Wiesner S, Xiao J, Ericson K, Chen W, Hall WA, Low WC and Ohlfest JR Expression of MHC I and NK ligands on human CD133+ glioma cells: possible targets of immunotherapy, J Neurooncol, 2006.

59. Jin L, Hope KJ, Zhai Q, Smadja-Joffe F and Dick JE Targeting of CD44 eradicates human acute myeloid leukemic stem cells, Nature Med, *12:* 1167–1174, 2006.

60. Piccirillo SGM, Reynolds BA, Zanetti N, Lamorte G, Binda E, Broggi G, Brem H, Olivi A, Dimeco F and Vescovi AL Bone morphogenetic proteins inhibit the tumorigenic potential of human brain tumor-initiating cells, Nature, *444:* 761–765, 2006.

61. Fan X, Matsui W, Khaki L, Stearns D, Chun J, Li YM and Eberhart CG Notch pathway inhibition depletes stem-like cells and blocks engraftment in embryonal brain tumors, Cancer Res, *66:* 7445–7452, 2007.

62. Yilmaz OH, Valdez R, Theisen BK, Guo W, Ferguson DO, Wu H and Morrison SJ PTEN dependence distinguishes haematopoietic stem cells from leukaemia-initiating cells, Nature, *441:* 475–482, 2006.

63. Shih T and Lindley C Bevacizumab: An angiogenesis inhibitor for the treatment of solid malignancies, Clin Ther, *28:* 1779–1802, 2006.

64. Wada T, Haigh JJ, Ema M, Hitoshi S, Chaddah R, Rossant J, Nagy A and van der Kooy D Vascular endothelial growth factor directly inhibits primitive neural stem cell survival but promotes definitive neural stem cell survival, J Neurosci, *26:* 6803–6812, 2006.

65. Vredenburgh JJ, Desjardins A, Herndon JE 2nd, Dowell JM, Reardon DA, Quinn JA, Rich JN, Sathornsumetee S, Gururangan S, Wagner M, Bigner DD, Friedman AH and Friedman HS Phase II trial of bevacizumab and irinotecan in recurrent malignant glioma, Clin Cancer Res, *13:* 1253–1259, 2007.

66. Folkins C, Man S, Xu P, Shaked Y, Hicklin DJ and Kerbel RS Anticancer therapies combining antiangiogenic and tumor cell cytotoxic effects reduce the tumor stem-like cell fraction in glioma xenograft tumors, Cancer Res, *In press.:*2007.

67. Kerbel RS and Kamen BA The anti-angiogenic basis of metronomic chemotherapy, Nat Rev Cancer, *4:* 423–436, 2004.

68. Kerbel RS Antiangiogenic therapy: a universal chemosensitization strategy for cancer?, Science, *312:* 1171–1175, 2006.

69. Jain RK Normalization of tumor vasculature: an emerging concept in antiangiogenic therapy, Science, *307:* 58–62, 2005.

70. Hudis CA Clinical implications of antiangiogenic therapies, Oncology, *19:* 26–31, 2005.

71. Bertolini F, Paul S, Mancuso P, Monestiroli S, Gobbi A, Shaked Y and Kerbel RS Maximum tolerable dose and low-dose metronomic chemotherapy have opposite effects on the mobilization and viability of circulating endothelial progenitor cells, Cancer Res, *63:* 4342–4346, 2003.

72. Shaked Y, Emmenegger U, Francia G, Chen L, Lee CR, Man S, Paraghamian A, Ben-David Y and Kerbel RS Low-dose metronomic combined with intermittent bolus-dose cyclophosphamide is an effective long-term chemotherapy treatment strategy, Cancer Res, *65:* 7045–7051, 2005.

73. Jubb AM, Oates AJ, Holden S and Koeppen H Predicting benefit from anti-angiogenic agents in malignancy, Nat Rev Cancer, *6:* 626–635, 2006.

Chapter 23
Targeting the Tumor Microenvironment (Stroma) for Treatment of Metastasis

Isaiah J. Fidler, Cheryl Hunt Baker, Kenji Yokoi, Toshio Kuwai, Toru Nakamura, Monique Nilsson, J. Erik Busby, Robert R. Langley, and Sun-Jin Kim

Keywords: tumor microenvironment; organ-specific metastasis; pathogenesis of metastasis

Abstract: The major cause of death from cancer is metastases that are resistant to conventional therapies. Metastases can be located in different organs and in different regions of the same organ that influence the response to therapy. Primary tumor in general and metastatic lesions in particular are biologically heterogeneous and contain multiple cell populations with diverse characteristics of growth rate, karyotype, cell surface receptors, antigenicity, immunogenicity, enzymes, hormone receptors, sensitivity to different cytotoxic drugs, production of extracellular matrix proteins, adhesion molecules, angiogenic potential, invasiveness, and metastatic potential. The outcome of metastasis depends on multiple interactions of metastatic cells with homeostatic mechanisms, which tumor cells often usurp. At the primary metastatic sites, tumor cells interact with host cells, such as endothelial cells, pericytes, epithelial cells, fibroblasts, myoepithelial cells, and leukocytes. The tissues composed by these normal cells are biologically unique, and the organ microenvironments they provide for tumor cells are also unique.

The pathogenesis of a metastasis consists of many sequential steps that must be completed to produce clinically relevant lesions. Preferential metastasis of tumor cells to certain organs is independent of vascular anatomy, rate of blood flow, and number of tumor cells delivered to each organ. The outcome of metastasis depends on multiple continuous interactions between unique subpopulations of tumor cells ("seed") and specific host factors within the organ microenvironment, such as vasculature ("soil").

Understanding the mechanisms responsible for the development of biological heterogeneity in primary cancers and metastases, and the processes that regulate tumor cell dissemination to and proliferation in distant tissues, is a major goal

Department of Cancer Biology, The University of Texas M. D. Anderson Cancer Center, Houston, TX, USA

of research. For many years, all of our efforts to treat cancer metastases have concentrated on the inhibition or destruction of tumor cells. Because all cells in the body depend on an adequate supply of oxygen and nutrients, and on the ability of the circulatory system to remove toxic molecules, therapeutic regimens directed against tumor-associated endothelial cell can destroy tumor cells regardless of their biologic heterogeneity. New strategies to treat tumor cells by modulating their interaction with the organ microenvironment and by targeting receptors expressed on tumor-associated endothelial cells present unprecedented possibilities for the treatment of cancer metastasis.

Introduction

The major cause of death from cancer is metastases that are resistant to conventional therapy. Metastases can be located in different organs or in different regions of the same organ, and the anatomic location of metastases influences response to therapy. Tumor cells in general and metastatic cells in particular are genetically unstable [1], leading to the generation of biological heterogeneity in primary tumors and metastases. These lesions therefore contain multiple cell populations with diverse growth rates, karyotypes, cell surface receptors, antigenicities, immunogenicities, enzyme profiles, hormone receptor compositions, production of extracellular matrix proteins, adhesion molecule profiles, angiogenic, invasive, and metastatic potentials, and sensitivities to different cytotoxic drugs [reviewed in Refs. 2,3].

The pathogenesis of metastasis is highly selective and consists of a series of interrelated and sequential steps. After the initial growth to about 1 mm in diameter, further growth requires the development of an additional blood supply. This is accomplished through the synthesis and secretion of several proangiogenic factors by tumor and infiltrating host cells, which in turn leads to the extension of a capillary network from the surrounding host tissues. Next, tumor cells produce

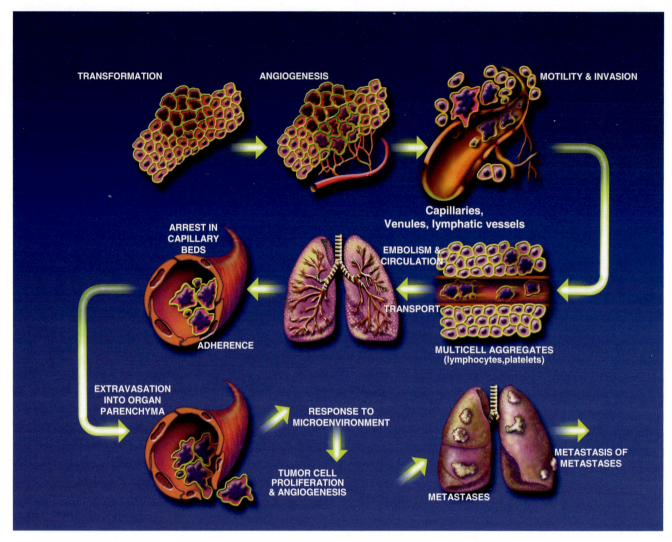

FIG. 23.1. The pathogenesis of cancer metastasis. To produce a metastasis, tumor cells must complete a series of selective events that few cells can complete. Once metastases are established, the process can be repeated.

a series of degradative enzymes (i.e., collagenases) to invade lymphatic or small blood vessels through which they can reach distal tissues. Many tumor cells form homotypic or heterotypic aggregates with lymphocytes and platelets, thereby enhancing their survival in the turbulent circulation. The surviving tumor emboli arrest in the microcirculation of organs to proliferate within the lumen of vessels or tumor cells or extravasate into the organ parenchyma, and then proliferate to produce micrometastases. The growth of these microscopic lesions likewise requires the development of a vascular supply and evasion of host defenses. Metastatic lesions can shed tumor cells into the circulation to produce metastasis of metastases. During the pathogenesis of a metastasis, tumor cells continuously interact with host cells such as endothelial cells, pericytes, epithelial cells, fibroblasts, myoepithelial cells, and leukocytes [2–4]. The outcome of metastasis is governed by multiple interactions of metastatic cells with homeostatic mechanisms, which tumor cells often usurp (Fig. 23.1).

Paget 1889: The "Seed and Soil" Hypothesis

Clinical observations of cancer patients and studies with experimental rodent tumors have led cancer biologists to conclude that the metastatic pattern of certain tumors is organ-specific and independent of vascular anatomy, rate of blood flow, and number of tumor cells delivered to each organ [reviewed in 2,5]. Indeed, the distribution and fate of hematogenously disseminated, radiolabeled melanoma cells in syngeneic mice conclusively demonstrated that tumor cells can reach the microvasculature of many organs, but proliferate to produce metastasis only in specific organs [6]. These findings, however, were not new.

In 1889, Stephen Paget noted the discrepancy between considerations of blood flow and the frequency of metastases in different organs. He studied the autopsy records of 735 women who died of breast cancer, and of many other patients with different neoplasms, and noticed the high frequency of breast cancer metastasis to the ovaries and the variations in incidence of skeletal metastases produced by different primary tumors. These findings were not compatible with the then-prevalent view that metastatic spread was due to "a matter of chance" or that tissues "played a passive role" in the process. Paget concluded that metastasis occurred only when certain tumor cells (that he labeled the "seed") had a special affinity for the growth milieu provided by certain specific organs (that he labeled the "soil"). Metastasis occurred only when the right cells interacted with the compatible organ environment [7].

Experimental data supporting Paget's 1889 seed and soil hypothesis were provided 90 years later by Hart and Fidler [8], who studied the preferential growth of B16 melanoma metastases in specific organs. Following the intravenous injection of B16 melanoma cells into syngeneic C57BL/6 mice, tumor growths developed in the natural lungs and in grafts of pulmonary or ovarian tissue implanted either subcutaneously or intramuscularly. In contrast, neoplastic lesions failed to develop in control grafts of similarly implanted renal tissue or at the site of surgical trauma. Parabiosis experiments suggested that the growth of the B16 melanoma in ectopic lung or ovarian tissue was due to the immediate arrest of circulating neoplastic cells and not to shedding of malignant cells from foci growing in the natural lungs. Quantitative analysis of tumor cell arrest and distribution using cells labeled with [^{125}I]-5-iodo-2'-deoxyuridine indicated that the tumor cells reached all organs but produced metastases only in the lungs [8]. These data demonstrated that the outcome of metastasis is dependent on both tumor cell properties and host factors, and supported the "seed and soil" hypothesis as an explanation of the nonrandom pattern of cancer metastasis.

The introduction of peritoneovenous shunts for palliation of malignant ascites produced by ovarian carcinoma provided an opportunity to study some of the factors affecting metastatic spread in humans. Tarin and colleagues described the outcome in patients with malignant ascites draining into the venous circulation, with the resulting entry of viable tumor cells into the jugular veins [9]. Beneficial results with minimal complications were reported for 29 patients with ovarian cancer. The autopsy findings in 15 patients substantiated the clinical observations that the shunts did not increase the risk of metastasis. The shunts introduced millions of cells into the circulation, and over many months, the number of circulating tumor cells likely reached hundreds of millions. Upon examination, the cells were found to be viable. Nevertheless, metastases in the lung (the first capillary bed encountered) were rare [9].

Another clear demonstration of organ-site specific metastasis comes from studies of experimental brain metastasis. Two murine melanomas were injected into the carotid artery of mice to simulate the hematogenous spread of tumor emboli to the brain. The K-1735 melanoma syngeneic to the C3H/HeN mouse produced lesions only in the brain parenchyma, whereas the B16 melanoma syngeneic to the C57BL/6 mouse produced only meningeal growths [10]. Similarly, different human melanomas [11] injected into the internal carotid artery of nude mice produced unique parenchymal or meningeal brain metastasis. Distribution analysis of radiolabeled melanoma cells injected into the internal carotid artery ruled out the possibility that the patterns of initial cell arrest in the microvasculature of the brain predicted the eventual sites of growth. Rather, the different sites of tumor growth in the brain involved the interaction between metastatic melanoma cells and brain endothelial cells, and the response of tumor cells to local astrocyte growth factors. In other words, site-specific metastases were produced by tumor cells that were receptive to their new environment [1–4].

The current understanding of the interaction between the tumor cells and the organ microenvironment consists of two principles. First, all malignant neoplasms are biologically heterogeneous and consist of multiple subpopulations of cells with distinct properties. Second, the establishment and progressive growth of metastases depend on multiple interactions ("cross talk") between metastatic cells and homeostatic mechanisms that tumor cells usurp [1–3]. The microenvironments of different organs are biologically unique. For example, endothelial cells in the vasculature of different organs express different cell surface receptors [12,13], and parenchymal cells in different organs can release different factors that influence establishment and growth of metastases [14]. Therapy of metastases can therefore be directed against homeostatic factors that promote tumor cell growth, survival, and metastasis. A primary example is the vasculature system.

The survival and growth of all cells in the body are dependent on an adequate supply of oxygen and, hence, on the vasculature. Oxygen can diffuse from capillaries only to a limited distance. Analysis of clinical specimens of lung cancer brain metastasis demonstrated that all proliferating tumor cells are located at a distance of less than 100 μm from the nearest capillary, whereas apoptotic tumor cells are located at a distance exceeding 150 μm from the nearest blood vessel [15]. This finding supports the original conclusion of Judah Folkman [16] that the progressive growth and survival of tumor cells requires the induction and maintenance of angiogenesis. The extent of angiogenesis is dependent on the balance between proangiogenic and antiangiogenic molecules released by tumor cells and host cells into the tumor microenvironment (Fig. 23.2).

Regulation of Angiogenesis by the Microenvironment

The production of vascular endothelial growth factor (VEGF), basic fibroblast growth factor (bFGF), interleukin-8 (IL-8), and IL-6 by tumor cells or host cells or the release of latent angiogenic molecules from the extracellular matrix induces the growth of endothelial cells and the formation of blood vessels. The organ environment can directly contribute to the induction

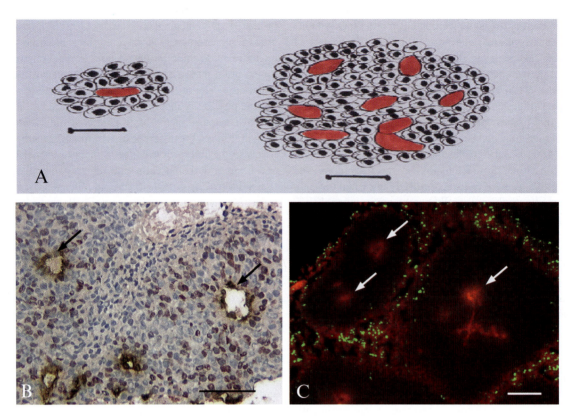

FIG. 23.2. Distance of dividing and apoptotic tumor cells from blood vessels in brain metastasis. **A** Diffusion distance of oxygen from a capillary (*bar* = 100 μm). **B** Dividing tumor cells within brain metastasis are identified by BrdU-positive nuclei staining red. The average distance from the nearest blood vessel is 74 μm. **C** TUNEL-positive tumor cells in brain metastasis are identified by *green staining* nuclei. Endothelial cells are identified by *red staining* CD31. The average distance of TUNEL-positive tumor cells from the nearest capillary is 166 μm (*bar* = 100 μm).

and maintenance of the angiogenic factors bFGF [17,18], IL-8 [19], and IL-6 [20]. For example, in renal cell carcinoma patients, the level of bFGF in the serum or urine inversely correlates with survival [21,22]. Human renal cancer cells implanted into different organs of nude mice had different metastatic potentials: those implanted into the kidney produced a high incidence of lung metastasis, whereas those implanted subcutaneously did not produce metastasis [17]. The subcutaneous-intramuscular tumors had a lower level of bFGF mRNA transcripts than did cells cultured in vitro. In contrast, tumors growing in the kidney of nude mice had 20-fold the levels of bFGF mRNA and protein as compared to cultured cells. Histopathological examination of the tumors revealed that subcutaneous tumors had few blood vessels, whereas the tumors in the kidney of nude mice had extensive vascularization [17,18].

Constitutive expression of IL-8 has been shown to directly correlate with the metastatic potential of the human melanoma cells [19]. IL-8 contributes to angiogenesis by inducing proliferation, migration, and invasion of endothelial cells [23]. Several organ-derived cytokines (produced by inflammatory cells) can upregulate expression of IL-8 in normal and tumorigenic cells [24]. IL-8 expression was upregulated in coculture of melanoma cells with keratinocytes (skin), whereas it was inhibited in cells cocultured with hepatocytes (liver). Similar results obtained with conditioned media from keratinocyte and hepatocyte cultures

suggested that organ-derived factors, e.g., IL-1 and TGF-β, can modulate the expression of IL-8 in human melanoma cells [22].

The influence of the microenvironment on the expression of VEGF/VPF, angiogenesis, tumor cell proliferation, and metastasis was investigated using human gastric cancer cells implanted in orthotopic (stomach) and ectopic (subcutaneous) organs of nude mice. Tumors growing in the stomach wall were highly vascularized and expressed higher levels of VEGF/VPF than did subcutaneous tumors [24]. Moreover, only tumors implanted in the stomach produced metastasis, suggesting that the biology of vascularization and metastasis of human gastric cancer cells is regulated by the organ microenvironment [24].

IL-6 was originally identified as a B-lymphocyte differentiation factor [25]. It has since been recognized as a critical mediator of physiologic processes, such as hematopoiesis, platelet production, osteoclast activation, and production of acute-phase proteins [26]. Several recent reports have implicated IL-6 as an important modulator of tumor progression [27,28]. The serum level of IL-6 is frequently elevated in women with ovarian carcinoma and is predictive of poor clinical outcome [29,30]. Our recent studies concluded that IL-6 enhanced endothelial cell migration and that endothelial cells of the skin (but few other organs) express IL-6 receptor [20]. Because IL-6 was capable of inducing a robust angiogenic response in the cutaneous microenvironment, it is possible that IL-6 contributes to the vascularization of skin tumors. Our finding correlates with published reports that serum levels of IL-6 are elevated in patients with metastatic melanoma [31] and that overexpression of IL-6 in basal cell carcinoma is associated with enhanced angiogenesis and tumor growth [32]. Additional evidence for the role of IL-6 in angiogenesis comes from a most recent report that a peptide specifically binding to the IL-6 receptor can inhibit vessel formation and growth of tumors in the subcutis of severe combined immunodeficiency mice [33].

Antivascular Therapy by Inhibition of Phosphorylation of Cell Surface Tyrosine Kinase Receptors on Tumor-Associated Endothelial Cells

Preclinical studies of antiangiogenesis therapy against experimental tumors have shown great promise, whereas clinical trials of similar agents have yielded mixed results [34]. Several reasons can account for this discrepancy. First, in preclinical models, tumors are not always grown in the orthotopic microenvironment, yet the site of tumor growth is important in determining the cross talk between tumor cells and tumor-associated endothelial cells. Second, most preclinical tumors grow at a rapid rate, and their vasculature is more akin to that in wound healing rather than the vasculature of a tumor that has existed for months or years. Third, the challenge in the clinic is to eradicate an existing disease rather than inhibit the growth of small lesions as is the case in preclinical studies. In other words, the challenge in the clinic is to destroy existing vasculature rather than inhibit the formation of new vasculature. The former is referred to as antivascular therapy, whereas the latter is referred to as antiangiogenic therapy [34].

The antivascular therapy of cancer metastasis is based on the hypothesis that induction of apoptosis in tumor-associated endothelial cells will trigger a second wave of apoptosis in stromal and tumor cells deprived of oxygen. Specifically, as all cells require oxygen for their survival, apoptosis induced in endothelial cells will lead to apoptosis of all tumor cells regardless of their biological properties. The induction of apoptosis in endothelial cells can be accomplished by multiple means.

Pancreatic Carcinoma

Pancreatic cancer is the fourth leading cause of cancer death in the United States with 28,000 new cases diagnosed annually [35]. The difficulty in detecting pancreatic cancer at an early stage, the aggressive nature of the disease, and the lack of effective therapy are all responsible for the high mortality from this disease [36]. Most patients develop local recurrence and metastatic disease. Although gemcitabine can prolong survival of patients, only less than 3% survive 5 years after the initial diagnosis, and the median survival duration is less than 6 months [37,38]. As discussed above, a different approach to therapy can be provided by preventing tumor cells from manipulating the organ microenvironment to their gain. Growth factors produced by tumor cells or host cells activating specific growth factor receptors is a primary example.

The epidermal growth factor (EGF) binds and phosphorylates its receptor (EGFR) and subsequently stimulates multiple signaling pathways that are involved in cell proliferation [39–42]. The overexpression of EGF and EGFR by cancer cells has been shown to correlate with metastasis, resistance to apoptosis, resistance to chemotherapy and, hence, poor prognosis [43–45]. These findings indicate that inhibiting EGFR signaling is a good strategy for therapy.

Production of another growth modulator, VEGF, or vascular endothelial growth factor/vascular permeability factor (VPF), is increased in many malignant tumors and is associated with angiogenesis and poor prognosis [46]. VEGF/VPF is not only a permeability and proliferating factor but also an anti-apoptotic survival factor for vascular endothelial cells [47,48]. Inhibiting VEGF signaling could lead to regression of vascular endothelial cells in the tumor microenvironment.

Platelet-derived growth factor (PDGF) and its receptor (PDGFR) are expressed in many types of cancers, including prostate, lung, gastric, and pancreatic [49,50]. In one study, 29 of 31 human pancreatic cancer specimens expressed phosphorylated PDGFR [51]. PDGFR signaling has been reported to increase proliferation of tumor cells in an autocrine manner [52,53] to stimulate angiogenesis and recruit and activate pericytes [52,54]. PDGFR signaling also controls the interstitial fluid pressure in the stroma and influences transvascular transport of chemotherapeutic agents in a paracrine manner [55,56]. Inhibition of PDGFR activity by tyrosine kinase inhibitor STI571 (imatinib/Gleevec) [57] in an orthotopic nude mouse model of pancreatic cancer decreased the growth of primary pancreatic tumors and decreased the incidence of peritoneal metastases when combined with gemcitabine [51].

Most recent data indicate that the biologic heterogeneity of neoplasms also includes the expression of cell surface tyrosine kinase receptors [52], and human pancreatic cancer cells growing in the pancreas of nude mice expressed either EGFR or PDGFR or both. These findings suggest that inhibition of one receptor's signaling may not be sufficient to inhibit the progressive growth and spread of the cancer. To overcome this receptor heterogeneity, we treated human pancreatic cancer growing in the pancreas of nude mice by simultaneous administration of AEE788 , a synthesized small molecule inhibitor of both EGFR and VEGFR tyrosine kinases [58], and STI571, an inhibitor of PDGFR, Bcr-abl, and c-kit tyrosine kinase [57].

Human pancreatic cancer cells growing in the pancreas of nude mice expressed high levels of EGF, VEGF, PDGF-BB, and their receptors. All of the receptors were phosphorylated. Tumor-associated endothelial cells (but not endothelial cells in tumor-free tissue) also expressed these receptors. Oral treatment with AEE788 and STI571 inhibited the phosphorylation of EGFR, VEGFR, and PDGFR on both the human pancreatic cancer cells and the tumor-associated endothelial cells of the recipient mice [59].

The human pancreatic cancer cells growing in the pancreas of nude mice were resistant to treatment with gemcitabine. Therapy with AEE788 directed at EGFR and VEGFR, and STI571 directed against PDGFR in combination with gemcitabine led to the fewest PCNA-positive tumor cells, the lowest microvessel density (MVD), the highest number of apoptotic cells, a significant decrease in tumor size ($P < 0.001$) and, most impressive, prolonged survival ($P < 0.0001$) [59].

The decrease in MVD was most dramatic. Tumor-associated endothelial cells expressed not only EGFR and VEGFR, but also PDGFR, which would provide another target for inhibition of its signaling by STI571. PDGFR as well as EGFR and VEGFR signaling, which activates the anti-apoptotic protein Akt and bcl-2, acts like a survival factor for endothelial cells [13,60]. With the inhibition of survival mechanisms by AEE788 and STI571, tumor-associated endothelial cells, whose proliferating frequency is 20–2,000 times higher than that of endothelial cells in normal organs [61,62], are more sensitive to anticycling chemotherapeutic treatment.

Collectively, these data demonstrate that pancreatic cancer cells produce EGF, VEGF, and PDGF. These ligands can activate their receptors on tumor cells in an autocrine manner and on tumor-associated endothelial cells by a paracrine manner. As a consequence, both tumor cells and tumor-associated endothelial cells have increased survival and resistance to chemotherapeutic agents [13]. Inhibiting these signaling pathways by tyrosine kinase inhibitors combined with conventional chemotherapy can induce a significant apoptosis in tumor-associated endothelial cells and tumor cells, resulting in decreased tumor size and significant prolongation of survival [59].

Renal Cancer

Human renal cell carcinoma (HRCC) cells growing in the kidneys or bones of nude mice express EGFR and phosphorylated EGFR (pEGFR) [63,64]. Systemic therapy with the EGFR tyrosine kinase inhibitor, PKI166 [65] (in combination with cytotoxic chemotherapy), inhibiting the activation of EGFR leads to therapy of the orthotopically growing HRCC cells [63,64]. Immunohistochemical analyses of tumors growing in the kidney (primary) or bone (metastasis) reveal that tumor-associated endothelial cells (but not endothelial cells in tumor-free kidney or bone) express activated EGFR. The activation of EGFR on endothelial cells, however, was only found in tumors that produce TGF-α/EGF [66]. Since both tumor cells and endothelial cells expressed activated EGFR, whether the primary target of PKI166 was the EGFR on tumor cells or tumor-associated endothelial cells (or both) was not clear [66]. We therefore determined whether the primary target for therapy could be the tumor-associated endothelial cells [67].

First, we confirmed that the activation of EGFR on tumor-associated endothelial cells (but not endothelial cells in kidney free of tumor) is dependent upon the production of TGF-α by tumor cells. Second, we found that kidney tumors produced by human renal cancer cells that do not produce EGF or TGF-α were relatively resistant to treatment with the protein tyrosine kinase inhibitor PKI166. In sharp contrast, PKI166 treatment against kidney tumors produced by human renal cancer cells that express TGF-α significantly decreased tumor volume. PKI166 inhibited the EGFR phosphorylation on endothelial cells only in the TGF-α-positive kidney tumors [67].

Clinical trials of the EGFR protein tyrosine kinase inhibitor Iressa in lung cancer were originally based on the rationale that many such cancers have high levels of EGFR protein. In the early trials, however, little correlation was found between the response of individual cancers and levels of EGFR protein in the lung cancer [68,68]. In other studies of Iressa [70,71], response in human tumors and in human cell lines have been associated with EGFR levels ranging from very low to very high [72–77]. One implication of these data is that EGFR content does not reflect the level of receptor activation. Most recent studies have suggested that many lung cancers responding to Iressa have somatic activating mutations in the EGFR gene [78,79], whereas in lung cancers that did not respond to Iressa, the frequency of such EGFR mutations was much lower. In our study with renal cancer, DNA sequence analysis identified no EGFR mutations in the HRCC cells (Fig. 23.3).

The studies cited above have focused on the content and level of EGFR (total and activated) in tumor cells. The possibility that a target for protein tyrosine kinase inhibitors of the EGFR is on the tumor-associated endothelial cells has been overlooked. Our data recommend that the clinical use of tyrosine kinase inhibitors specific to EGFR may be more effective against neoplasms that express high levels of TGF-α because both the tumor cells and the tumor-associated endothelial cells are the targets for therapy. The destruction of tumor-vasculature will lead to apoptosis of surrounding tumor cells [67].

Prostate Cancer

Prostate cancer is the second greatest cause of cancer-related death among men in North America [80,81]. Bone metastases that do not respond to conventional therapies cause devastating symptoms such as intractable bone pain, nerve compression, and pathologic fractures [82,83]. Despite progress in the detection of primary prostate cancer, approximately 25% of patients have metastatic disease at initial diagnosis, and in an additional 30% of the patients, metastases to lymph node and bone are discovered during surgical staging. A significant number of patients with clinically localized disease who are treated with radical prostatectomy may also develop metastasis. The present day therapy for advanced prostate cancer is anti-androgen therapy and chemotherapy [84]. The unfortunate and unavoidable emergence of tumor cells resistant to therapy

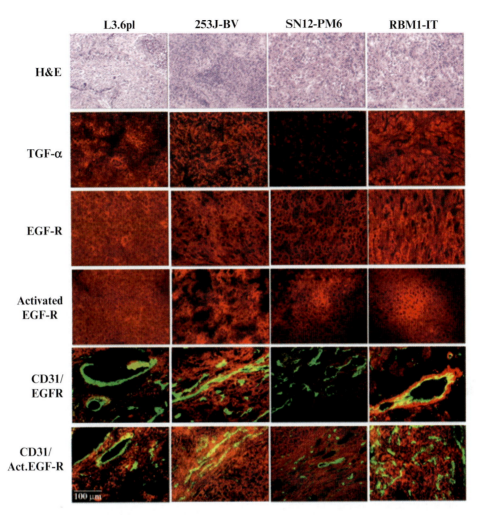

FIG. 23.3. Expression of *EGFR* and phosphorylated *EGFR* on tumor cells and tumor-associated endothelial cells depends on expression of the ligand, *TGF-α*. Human pancreatic cancer *L3.6pl*, human bladder cancer *253J-BV*, and human renal cancer *SN-12PM6, RBM1-IT* cells were implanted into their respective orthotopic organs in nude mice. Only tumors expressing the ligand *TGF-α* expressed phosphorylated *EGFR* on tumor cells and on tumor-associated endothelial cells. EGFR = *green*; CD31 = *red*; EGFR-positive endothelial cells = *yellow*.

is responsible for an average patient survival of 12 months [85–87]. The failure to eradicate these metastases by conventional therapies dictates that we search for a different approach. Targeting endothelial cells in tumor-associated blood vessels is an attractive possibility [88,89].

An excellent example of the adaptation of tumor cells to the organ microenvironment is the finding that prostate cancer cells growing adjacent to bone, e.g., bone metastases, express and release increased levels of PDGF and its receptor, PDGFR [88]. PDGF binding activates the PDGFR, leading to autophosphorylation of specific tyrosines in the cytoplasmic domain of PDGFR and bringing about the stimulation of cell division and survival [88].

Growth factors and their receptors play a pivotal role in the regulation of cancer progression and neovascularization. Overexpression of PDGF and PDGFR by many neoplasms has been shown to correlate with cancer metastasis, resistance to chemotherapy and, hence, poor prognosis. Inhibiting sig-

naling pathways through PDGFR represents a good strategy for therapeutic intervention. Recent studies have shown that endothelial cells in normal tissues survive 3–5 years, whereas in tumors, 2–9% of endothelial cells divide daily [61,62]. These findings suggest that tumor-associated endothelial cells are likely to be sensitive to chemotherapeutic agents [90]. Preclinical studies revealed that in tumors expressing a high level of PDGF, tumor-associated endothelial cells express phosphorylated PDGFR that up-regulates expression of several antiapoptotic molecules, thus rendering these endothelial cells resistant to anticycling drugs. Hence, the inhibition of phosphorylation of the PDGFR in dividing endothelial cells should lead to an increase in susceptibility to anticycling chemotherapeutics.

Imatinib mesylate (Gleevec), a derivative of 2-phenylamino-pyrimidine, was originally developed as a competitor for an adenosine triphosphate-binding site of the Ab1 protein tyrosine kinase. Imatinib is also a potent inhibitor of the tyrosine kinase activities of c-KIT and PDGFR, and it inhibits cell proliferation and induces apoptosis [91]. The oral administration of imatinib with or without intraperitoneal injections of paclitaxel was associated with a significant decrease in the incidence and size of human prostate cancer bone metastases in athymic nude mice. Mice treated with imatinib plus paclitaxel had fewer dividing tumor cells. Increased apoptosis in tumor-associated endothelial cells was associated with a significant decrease in vasculature and a significant decrease in the incidence of lymph node metastasis [92,93].

Blockade of the PDGFR can alter phosphatidylinositol 3' kinase activity and lead to a change in pericytes that surround blood vessels and to a change in interstitial fluid homeostasis, which results in an increase in transcapillary transport. The induction of apoptosis in tumor-associated endothelial cells suggests that the therapy is directed against tumor vasculature [93]. Most recent experiments with multidrug-resistant human prostate cancer cells growing in the bone of nude mice provide a confirmation. Administration of paclitaxel alone did not reduce experimental bone tumors, whereas administration of imatinib plus paclitaxel produced significant apoptosis in tumor-associated endothelial cells that was followed by apoptosis of tumor cells, i.e., necrosis and a reduction in production of lymph node metastasis [93].

We reported that human prostate cancer cells growing adjacent to bone tissue express high levels of EGF and PDGF and that the tumor cells and tumor-associated endothelial cells, but not endothelial cells in uninvolved bone, expressed EGFR and PDGFR [92,93].

Detailed analysis demonstrated that the expression of EGFR and PDGFR on the surface of human prostate cancer cells directly correlates with their proximity to bone tissue. Tumor cells that proliferate in bone tissue induce inflammatory changes. Inflammation or damage to bone tissue is associated with expression of transforming growth factor-β (TGF-β) which, in turn, induces expression of EGF [94–97] and PDGF [98,99] in osteoblasts and osteoclasts. The production of EGF and PDGF can up-regulate and activate the expression of their respective receptors in an autocrine manner (tumor cells) and paracrine manner (tumor-associated endothelial cells). Indeed, endothelial cells expressing EGFR or PDGFR on their surface are located within 20–40 μm of tumor cells expressing EGF/EGFR and/or PDGF/PDGFR [93].

The expression of EGF or PDGF and their receptors has been shown to correlate with progressive growth of human carcinomas of the prostate [100,101], and therefore, blockade of their signaling pathways has been developed as a new therapeutic strategy [102–104]. Specifically, blockade of EGFR signaling has been shown to produce arrest at the G1 restriction point [105,106] and to increase apoptosis [107,108]. Imatinib inhibits phosphorylation of PDGFR, which in turn inhibits depolymerization of microtubules [109,110].

Activation of the PDGFR has been shown to increase expression of Bcl-2 and P13K/Akt and to decrease the level of caspase in cells, i.e., it increases resistance to apoptosis [111].

Since treatment of cells with imatinib increases their sensitivity to an anticycling drug, e.g., taxol, the results support the hypothesis that PDGF is a survival factor for tumor cells and endothelial cells. Since the activation of the PDGFR on tumor-associated endothelial cells can enhance their resistance to anticycling drugs [13], the administration of PTK inhibitors such as imatinib should inhibit the resistance of dividing endothelial cells to anticycling drugs, such as paclitaxel. Indeed, the oral administration of PKI166 and imatinib without paclitaxel completely blocked the phosphorylation of EGFR and PDGFR, but did not produce therapeutic effects, whereas the combination of PKI166, imatinib, and paclitaxel produced significant therapy ($P < 0.001$) by targeting the tumor cells and tumor-associated endothelial cells [93].

A modular phase I trial in men with androgen-independent prostate cancer and bone metastases was designed to translate the preclinical findings to clinical reality. Twenty-eight men, most of whom had already been given chemotherapy, received a lead-in treatment of 30 days consisting of imatinib alone at 600 mg daily. This was followed by weekly docetaxel for 4 weeks in 6-week cycles. During the lead-in period with imatinib alone, 7% of the patients had a decline in prostate-specific antigen (PSA) levels of <50%. Twenty-one patients were then treated with the imatinib/docetaxel combination therapy. Of these, 38% had a >50% decline in PSA levels, 29% had a <50% decline in PSA levels, and 33% had a PSA level increase. To date, this trial has produced long-term survival in 50% of the patients [112]. A multi-institutional phase II trial is currently ongoing.

Antivascular Therapy of Multidrug Resistant Prostate Cancer

The success of current treatments for prostate cancer is limited by the emergence of cells that are resistant to chemotherapy. These resistant cells often express the multidrug resistance gene (MDR1) and its product, P-glycoprotein (P-gp) [113,114]. Targeting tumor-associated endothelial cells should destroy tumor cells, regardless of their biological heterogeneity, including ones that express MDR1. In other words, tumor-associated endothelial cells can serve as a primary target for tyrosine kinase inhibitors, and killing of these endothelial cells should lead to the death of all surrounding cells from nutrient deprivation. Regardless of sensitivity, resistance to hormones or chemotherapeutic drugs, all tumor cells are dependent on a viable vasculature for growth and survival [3,115,116]. As discussed above, the combination of imatinib and paclitaxel induces apoptosis of tumor-associated endothelial cells as well as PDGFR-expressing tumor cells. These data suggest that a major target for imatinib and paclitaxel therapy could well be the tumor-associated endothelial cell.

We tested this possibility by using MDR human prostate cancer cells. Like parental PC-3MM2 cells [92,93], multidrug resistant PC-3MM2-MDR cells growing in the bone (but not in the muscle) expressed IL-8, bFGF, EGF, EGFR, PDGF, and PDGFR. Endothelial cells of tumor-associated vessels in bone lesions also expressed PDGFR on their surface, and treatment with imatinib and paclitaxel inhibited the phosphorylation of the PDGFR on both tumor cells and endothelial cells [117].

In vitro, the PC-3MM2-MDR cells were highly resistant (67-fold) to paclitaxel. Furthermore, imatinib did not sensitize the multidrug resistant cells to paclitaxel. In vivo, however, the PC-3MM2-MDR bone lesions responded to systemic administration of imatinib and paclitaxel (but not to paclitaxel administered alone), raising the possibility that imatinib could have sensitized the cancer stroma to paclitaxel [117].

As stated previously, endothelial cells in normal tissues rarely divide, whereas 2–3% of endothelial cells in prostate cancer divide daily [61,62]. These dividing endothelial cells should be sensitive to anticycling drugs such as paclitaxel. Nevertheless, in the present experiment, treatment with only paclitaxel did not decrease the MVD in tumors.

After 14 daily treatments with imatinib and paclitaxel (once per week), apoptosis (as reflected by TUNEL-positive cells) was mostly limited to tumor-associated endothelial cells, i.e., after 14 consecutive days of treatment, the MDR tumor cells were treatment-resistant. In contrast, PC-3MM2 parental cells underwent extensive apoptosis in mice treated with imatinib and paclitaxel. Thus, the first wave of apoptosis in bone tumors from mice treated with imatinib and paclitaxel for only 2 weeks occurred in tumor-associated endothelial cells, followed by apoptosis of tumor cells and ultimately necrosis. By the fourth week of treatment with imatinib and paclitaxel or imatinib alone, concurrent apoptosis of tumor cells and tumor-associated endothelial cells was observed. Thus, the imatinib-induced blockade of PDGFR combined with paclitaxel targeted the tumor-associated endothelial cells [117].

Conclusions

The primary cause of death from cancer is the progressive growth of metastases that are resistant to conventional therapies. Metastases develop when specialized tumor cells interact with specific organ microenvironments. Neoplasms are heterogeneous: they consist of cells with different biologic properties. Cancer cells are genetically unstable, and by the time of diagnosis, malignant neoplasms contain multiple subpopulations of cells with different biologic properties, including varying potential for invasion and metastasis. The outcome of the growth and spread of cancer depends on multiple interactions ("cross-talk") of tumor cells with homeostatic mechanisms, which tumor cells usurp. The induction and maintenance of the vasculature in neoplasms is a good example of the way in which tumor cells manipulate a physiologic process for their advantage.

Growth factors and their receptors play a major role in the regulation of tumor neovascularization. Overexpression of EGF, EGFR, VEGF, VEGFR, PDGF, and PDGFR by many neoplasms correlates with progressive growth. Inhibition of tyrosine kinase activity of these receptors in combination with chemotherapy can be used to induce apoptosis in tumor-associated endothelial cells, leading to therapy of primary tumors and metastasis. This anti-vascular therapy of cancer is most encouraging but demands adherence to several principles. First, targeted therapy requires a target and, therefore, prior to treatment, it is required to verify that a tumor expresses an activated growth factor receptor. Second, tumors are heterogeneous for expression of phosphorylated growth factor receptors. A combination of several therapeutic agents is mandatory. Third, cancer is a chronic disease and, as such, anti-vascular therapy must be administered chronically, i.e., the disease is 'managed' rather than cured.

References

1. Fidler IJ. Critical factors in the biology of human cancer metastasis: twenty-eighth GHA Clowes memorial award lecture. Cancer Res 1990; 50:6130–38.
2. Fidler IJ. The pathogenesis of cancer metastasis: the 'seed and soil' hypothesis revisited (Timeline). Nat. Rev. Cancer 2003; 3:453–8.
3. Fidler IJ. Biology of cancer metastasis. In: Abeloff MD, Armitage JO, Niederhuber JE, Kastan MB, McKenna WG, editors. Clinical oncology, 3rd edition. Philadelphia, PA: Elsevier Science, 2004:59–79.
4. Liotta LA, Steeg PS, Stetler-Stevenson WG. Cancer metastasis and angiogenesis: an imbalance of positive and negative regulation. Cell 1991; 64:327–36.
5. Weiss L. Metastasis of cancer: a conceptual history from antiquity to the 1990s. Cancer Metastasis Rev 2000; 19:193–400.
6. Fidler IJ. Metastasis: quantitative analysis of distribution and fate of tumor emboli labeled with ^{125}I-5-iodo-2'-deoxyuridine. J Natl Cancer Inst 1970; 45:773–82.
7. Paget S. The distribution of secondary growths in cancer of the breast. Lancet 1889; 1:571–3.
8. Hart IR, Fidler IJ. Role of organ selectivity in the determination of metastatic patterns of the B16 melanoma. Cancer Res 1980; 40:2281–7.
9. Tarin D, Price JE, Kettlewell MG, et al. Mechanisms of human tumor metastasis studied in patients with peritoneovenous shunts. Cancer Res 1984; 44:3584–92.
10. Schackert G, Fidler IJ. Site-specific metastasis of moue melanomas and a fibrosarcoma in the brain or meninges of syngeneic animals. Cancer Res 1988; 48:3478–84.
11. Schackert G, Price JE, Zhang RD, et al. Regional growth of different human melanoma as metastasis in the brain of nude mice. Am J Pathol 1990; 136:95–102.
12. Langley RR, Ramirez KM, Tsan RZ, et al. Tissue-specific microvascular endothelial cell lines from H-$2k^b$-tsA58 mice for studies of angiogenesis and metastasis. Cancer Res 2003; 63:2971–6.
13. Langley RR, Fan D, Tsan RZ, et al. 2004. Activation of the platelet-derived growth factor receptor enhances survival of murine bone endothelial cells. Cancer Res (Adv in Brief) 2004; 64: 3727–30.

14. Fidler IJ. The organ microenvironment and cancer metastasis (*Review*). Differentiation 2002; 70:498–505.

15. Fidler IJ, Yano S, Zhang RD, et al. The seed and soil hypothesis: vascularization and brain metastases (*Personal View*). Lancet Oncol 2002; 3:53–7.

16. Folkman J. Angiogenesis in cancer, vascular, rheumatoid, and other diseases. Nat Med 1995; 1:27–31.

17. Singh RK, Bucana CD, Gutman M, et al. Organ site-dependent expression of basic fibroblast growth factor in human renal cell carcinoma cells. Am J Pathol 1994; 145:365–74.

18. Singh RK, Gutman M, Bucana CD, et al. Interferons alpha and beta downregulate the expression of basic fibroblast growth factor in human carcinomas. Proc Natl Acad Sci USA 1995; 92: 4562–6.

19. Singh RK, Gutman M, Radinsky R, et al. Expression of interleukin 8 correlates with the metastatic potential of human melanoma cells in nude mice. Cancer Res 1994; 54:3242–7.

20. Nilsson MB, Langley RR, Fidler IJ. Interleukin-6, secreted by human ovarian carcinoma cells, is a potent proangiogenic cytokine. Cancer Res 2005; 65:10794–800.

21. Nanus DM, Schmitz-Drager BJ, Motzer RJ, et al. Expression of basic fibroblast growth factor in primary human renal tumors: correlation with poor survival. J Natl Cancer Inst 1994; 85: 1587–99.

22. Nguyen M, Watanabe H, Budson AE, et al. Elevated levels of an angiogenic peptide, basic fibroblast growth factor, in the urine of patients with a wide spectrum of cancers. J Natl Cancer Inst 1994; 86:356–61.

23. Leek RD, Harris AL, Lewis CE. Cytokine networks in solid human tumors: regulation of angiogenesis. J Leukoc Biol 1994; 56:423–35.

24. Gutman M, Singh RK, Bucana CD, et al. Regulation of IL-8 expression in human melanoma cells by the organ environment. Cancer Res 1995; 55:2470–5.

25. Hirano T, Yasukawa K, Harada H, et al. Complementary DNA for a novel human interleukin (BSF-2) that induces B lymphocytes to produce immunoglobulin. Nature 1986; 324:73–6.

26. Hirano T. Interleukin 6 and its receptor: ten years later. Int Rev Immunol 1998;16:249–84.

27. Salgado R, Junius S, Benoy I, et al. Circulating interleukin-6 predicts survival in patients with metastatic breast cancer. Int J Cancer 2003; 103:642–6.

28. Culig Z, Bartsch G, Hobisch A. Interleukin-6 regulates androgen receptor activity and prostate cancer cell growth. Mol Cell Endocrinol 2002; 197:231–8.

29. Plante M, Rubin SC, Wong GY, et al. Interleukin-6 level in serum and ascites as a prognostic factor in patients with epithelial ovarian cancer. Cancer 1994; 73:1882–8.

30. Tempfer C, Zeisler H, Sliutz G, et al. Serum evaluation of interleukin 6 in ovarian cancer patients. Gynecol Oncol 1997; 66: 27–30.

31. Moretti S, Chiarugi A, Semplici F, et al. Serum imbalance of cytokines in melanoma patients. Melanoma Res 2001; 11: 395–9.

32. Jee SH, Shen SC, Chiu HC, et al. Overexpression of interleukin-6 in human basal cell carcinoma cell lines increases anti-apoptotic activity and tumorigenic potency. Oncogene 2001; 20:198–208.

33. Su J, Kai K, Chen C, et al. A novel peptide specifically binding to interleukin-6 receptor (gp80) inhibits angiogenesis and tumor growth. Cancer Res 2005; 65:4827–35.

34. Fidler IJ, Langley RR, Kerbel RS, et al. Biology of cancer: angiogenesis. In: DeVita VT Jr, Hellman S, Rosenberg SA, editors. Cancer: principles and practice of oncology, 7th edition. Philadelphia, PA: Lippincott & Wilkins, 2005: 129–137.

35. Jemal A, Tiwari RC, Murray T, et al. Cancer statistics, 2004. CA Cancer J Clin 2004; 54:8–29.

36. Li D, Xie K, Wolff R, et al. Pancreatic cancer. Lancet 2004; 363:1049–57.

37. Burris HA III, Moore MJ, Andersen J, et al. Improvements in survival and clinical benefit with gemcitabine as first-line therapy for patients with advanced pancreas cancer: a randomized trial. J Clin Oncol 1997; 15:2403–13.

38. Abbruzzese JL. New application of gemcitabine and future directions in the management of pancreatic cancer. Cancer 2002; 95:941–5.

39. Perugini RA, McDade TP, Vittimberga FJ, et al. Pancreatic cancer cell proliferation is phosphatidylinositol 3-kinase dependent. J Surg Res 2000; 90:29–44.

40. Nicholson KM, Anderson NG. The protein kinase B/Akt signaling pathway in human malignancy. Cell Signal 2002; 14: 381–95.

41. Wang W, Abbruzzese JL, Evans DB, et al. The nuclear factor-kappa B RelA transcription factor is constitutively activated in human pancreatic adenocarcinoma cells. Clin Cancer Res 1999; 5:119–27.

42. Douziech N, Calvo E, Laine J, et al. Activation of MAP kinases in growth responsive pancreatic cancer cells. Cell Signal 1999; 11:591–602.

43. Ghaneh P, Kawesha A, Evans JD, et al. Molecular prognostic markers in pancreatic cancer. J Hepatobiliary Pancreat Surg 2002; 9:1–11.

44. Kuwahara K, Sasaki T, Kuwada Y, et al. Expressions of angiogenic factors in pancreatic ductal carcinoma: a correlative study with clinico-pathologic parameters and patient survival. Pancreas 2003; 26:344–9.

45. Yamanaka Y, Friess H, Kobrin MS, et al. Coexpression of epidermal growth factor receptor and ligands in human pancreatic cancer is associated with enhanced tumor aggressiveness. Anticancer Res 1993; 13:565–9.

46. Ferrara N, Alitalo K. Clinical applications of angiogenic growth factors and their inhibitors. Nat Med 1999; 5:1359–64.

47. Gerber HP, Dixit V, Ferrara N. Vascular endothelial growth factor induces expression of the antiapoptotic proteins Bcl-2 and A1 in vascular endothelial cells. J Biol Chem 1998; 273:13313–6.

48. Tran J, Rak J, Sheehan C, et al. Marked induction of the IAP family antiapoptotic proteins survivin and XIAP by VEGF in vascular endothelial cells. Biochem Biophys Res Commun 1999; 264:781–8.

49. Kim SJ, Uehara H, Yazici S, et al. Modulation of bone microenvironment with Zoledronate enhances the therapeutic effects of STI571 and Paclitaxel against experimental bone metastasis of human prostate cancer. Cancer Res 2005; 65:3707–15.

50. Ebert M, Yokoyama M, Friess H, et al. Induction of platelet-derived growth factor A and B chains and over-expression of their receptors in human pancreatic cancer. Int J Cancer 1995; 62:529–35.

51. Hwang RF, Yokoi K, Bucana CD, et al. Inhibition of platelet-derived growth factor receptor phosphorylation by STI571 (Gleevec) reduces growth and metastasis of human pancreatic carcinoma in an orthotopic nude mouse model. Clin Cancer Res 2003; 9:6534–44.

52. Ostman A. PDGF receptors-mediators of autocrine tumor growth and regulators of tumor vasculature and stroma. Cytokine Growth Factor Rev 2004; 15(4):275–86.

53. Heldin C-H, Westermark B. Mechanism of action and *in vivo* role of platelet-derived growth factor. Physiol Rev 1999; 79: 1283–316.

54. Bergers G, Song S, Meyer-Morse N, et al. Benefits of targeting both pericytes and endothelial cells in the tumor vasculature with kinase inhibitors. J Clin Invest 2003; 111(9):1287–95.

55. Pietras K, Rubin K, Sjoblom T, et al. Inhibition of PDGF receptor signaling in tumor stroma enhances antitumor effect of chemotherapy. Cancer Res 2002; 62:5476–84.

56. Pietras K. Increasing tumor uptake of anticancer drugs with imatinib. Semin Oncol 2004; 31:18–23.

57. Buchdunger E, Cioffi CL, Law N, et al. Abl protein-tyrosine kinase inhibitor STI571 inhibits *in vitro* signal transduction mediated by c-Kit and platelet-derived growth factor receptors. J Pharmacol Exp Ther 2000; 295:139–45.

58. Traxler P, Allegrini PR, Brandt R, et al. AEE788: a dual family epidermal growth factor receptor/ErbB2 and vascular endothelial growth factor receptor tyrosine kinase inhibitor with antitumor and antiangiogenic activity. Cancer Res 2004; 64:4931–41.

59. Yokoi K, Sasaki T, Bucana CD, et al. Simultaneous inhibition of EGF-R, VEGF-R, and PDGF-R signaling combined with gemcitabine produces therapy of human pancreatic carcinoma and prolongs survival in an orthotopic nude mouse model. Cancer Res 2005; 65:10371–80.

60. Gerber HP, McMurtrey A, Kowalski J, et al. Vascular endothelial growth factor regulates endothelial cell survival through the phosphatidylinositol 3'-kinase/Akt signal transduction pathway. Requirement for Flk-1/KDR activation. J Biol Chem 1998; 273:30336–43.

61. Hobson B, Denekamp J. Endothelial proliferation in tumours and normal tissues: continuous labeling studies. Br J Cancer 1984;49:405–13.

62. Eberhard A, Kahlert S, Goede V, et al. Heterogeneity of angiogenesis and blood vessel maturation in human tumors: implications for antiangiogenic tumor therapies. Cancer Res 2000; 60:1388–93.

63. Weber KL, Doucet M, Price JE, et al. Blockade of epidermal growth factor-receptor signaling leads to inhibition of renal cell carcinoma growth in the bone of nude mice. Cancer Res 2003; 63: 2940–7.

64. Kedar D, Baker CH, Killion JJ, et al. Blockade of the epidermal growth factor receptor signaling inhibits angiogenesis leading to the regression of human renal cell carcinoma growing orthotopically in nude mice. Clin Cancer Res 2002; 8:3592–3600.

65. Traxler P, Buchdunger E, Furet P, et al. Preclinical profile of PKI166-a novel and potent EGF-R tyrosine kinase inhibitor for clinical development. Clin Cancer Res 1999; 5:3750–8.

66. Baker CH, Kedar D, McCarty MF, et al. Blockade of epidermal growth factor receptor signaling on tumor cells and tumor-associated endothelial cells for therapy of human carcinomas. Am J Pathol 2002; 161:929–38.

67. Baker CH, Pino MS, Fidler IJ. Phosphorylated epidermal growth factor receptor on tumor-associated endothelial cells in human renal cell carcinoma is a primary target for therapy by tyrosine kinase inhibitors. Neoplasia 2006; 8: 470–6.

68. Saltz L, Rubin M, Hochster H, et al. Cetuximab (IMC-225) plus irinotecan (CPT-11) is active in CPT-11-refractory colorectal cancer that expresses epidermal growth factor receptors. Proc Am Soc Clin Oncol 2001; 20:3–7.

69. Moasser MM, Basso A, Averbuch SD, et al. The tyrosine kinase inhibitor ZD1839 ("Iressa") inhibits HER2-driven signaling and suppresses the growth of HER2-overexpressing tumor cells. Cancer Res 2001; 61:7184–8.

70. Solorzano CC, Baker CH, Tsan R, et al. Optimization for the blockade of epidermal growth factor receptor signaling for therapy of human pancreatic carcinoma. Clin Cancer Res 2001; 7:2563–72.

71. Rak J, Filmus J, Kerbel RS. Reciprocal paracrine interactions between tumor cells and endothelial cells: the 'angiogenesis progression' hypothesis. Eur J Cancer 1996; 32A:2438–50.

72. Yoneda J, Kuniyasu H, Crispens MA, et al. Expression of angiogenesis-related genes and progression of human ovarian carcinomas in nude mice. J Natl Cancer Inst 1998; 90:447–54.

73. Radinsky R, Risin S, Fan D, et al. Level and function of epidermal growth factor receptor predict the metastatic potential of human colon carcinoma cells. Clin Cancer Res 1995; 1:19–31.

74. Albanell J, Rojo F, Averbuch S, et al. Pharmacodynamic studies of the epidermal growth factor receptor inhibitor ZD1839 in skin from cancer patients: histopathologic and molecular consequences of receptor inhibition. J Clin Oncol 2002; 20:110–24.

75. Baselga J, Rischin D, Ranson M, et al. Phase I safety, pharmacokinetic, and pharmacodynamic trial of ZD1839, a selective oral epidermal growth factor receptor tyrosine kinase inhibitor, in patients with five selected solid tumor types. J Clin Oncol 2002; 20:4292–302.

76. Sirotnak FM, Zakowski MF, Miller VA, et al. Efficacy of cytotoxic agents against human tumor xenografts is markedly enhanced by coadministration of ZD1839 (Iressa), an inhibitor of EGF-R tyrosine kinase. Clin Cancer Res 2000; 6:4885–92.

77. Ciardiello F, Caputo R, Bianco R, et al. Antitumor effect and potentiation of cytotoxic drugs activity in human cancer cells by ZD-1839 (Iressa), an epidermal growth factor receptor-selective tyrosine kinase inhibitor. Clin Cancer Res 2000; 6:2053–63.

78. Hirata A, Uehara H, Izumi K, et al. Direct inhibition of EGF receptor activation in vascular endothelial cells by gefitinib ("Iressa", ZD 1839). Cancer Sci 2004; 95:614–8.

79. Naito S, Walker SM, Fidler IJ. *In vivo* selection of human renal cell carcinoma cells with high metastatic potential in nude mice. Clin Exp Metastasis 1989; 7:381–9.

80. Landis SH, Murray T, Bolden-Wingo PA. Cancer statistics, 1999. Cancer 1999; 49:8–31.

81. Soh S, Kattan MW, Berkman S, et al. Has there been a recent shift in the pathological features and prognosis of patients treated with radical prostatectomy? J Urol 1997; 157:2212–8.

82. Barrettoni BA, Carter JR. Mechanisms of cancer metastasis to bone. J Bone Joint Surg Am 1986; 68A:308–12.

83. Jacobs SC. Spread of prostatic cancer to bone. Urology 1983; 21:337–44.

84. Huggins C, Hodges CV. Studies on prostate cancer: The effects of castration on advanced carcinoma of the prostate (abstr). Arch Surg 1941; 43:209.

85. Byar DP, Corle DK. Hormone therapy for prostate cancer: results of the Veterans Administration Cooperative Urological Research Group studies. NCI Monograph 1988; 7:165–70.

86. Millikan R, Thall PF, Lee SJ, et al. Randomized multicenter phase II trial of two multicomponent regimens in androgen-independent prostate cancer. J Clin Oncol 2003; 21:878–83.

87. Smith DC, Esper P, Strawderman M, et al. Phase II trial of oral estramustine, oral etoposide, and intravenous paclitaxel in hormone-refractory prostate cancer. J Clin Oncol 1999; 17: 1664–71.

88. Kim SJ, Uehara H, Yazici S, et al. Simultaneous blockade of platelet-derived growth factor-receptor and epidermal growth factor-receptor signaling and systemic administration of paclitaxel as therapy for human prostate cancer metastasis in bone of nude mice. Cancer Res 2004; 64:4201–4208.

89. Yazici S, Kim SJ, Busby JE, et al. Dual inhibition of the epidermal growth factor and vascular endothelial growth factor phosphorylation for antivascular therapy of human prostate cancer in the prostate of nude mice. Prostate 2005; 65:203–15.

90. Klement G, Baruchel S, Rak J, et al. Continuous low-dose therapy with vinblastine and VEGF receptor-2 antibody induces sustained tumor regression without overt toxicity. J Clin Invest 2000; 105:R15–R24.

91. Druker BJ, Tamura S, Buchdunger E, et al. Effects of a selective inhibitor of the Abl tyrosine kinase on the growth of Bcr-Abl-positive cells. Nat Med 1996; 2:561–5.

92. Uehara H, Kim SJ, Karashima T, et al. Effects of blocking platelet-derived growth factor-receptor signaling in a mouse model of prostate cancer bone metastasis. J Natl Cancer Inst 2003; 95:458–70.

93. Kim SJ, Uehara H, Karashima T, et al. Blockade of epidermal growth factor receptor signaling in tumor cells and tumor-associated endothelial cells for therapy of androgen-independent human prostate cancer growing in the bone of nude mice. Clin Cancer Res 2003; 9:1200–10.

94. Bruns CJ, Solorzano CC, Harbison MT, et al. Blockade of the epidermal growth factor receptor signaling by a novel tyrosine kinase inhibitor leads to apoptosis of endothelial cells and therapy of human pancreatic carcinoma. Cancer Res 2000;60:2926–35.

95. Hou X, Johnson AC, Rosner MR. Induction of epidermal growth factor receptor gene transcription by transforming growth factor beta 1: association with loss of protein binding to a negative regulatory element. Cell Growth Differ 1994; 5:801–9.

96. Saha D, Datta PK, Sheng H, et al. Synergistic induction of cyclo-oxygenase-2 by transforming growth factor-β1 and epidermal growth factor inhibits apoptosis in epithelial cells. Neoplasia 1999; 1:508–17.

97. Vinals F, Pouyssegur J. Transforming growth factor (TGF)-β1 promotes endothelial cell survival during *in vitro* angiogenesis via an autocrine mechanism implicating TGF-α signaling. Mol Cell Biol 2001; 21:7218–30.

98. Soory M, Virdi H. Implications of minocycline, platelet-derived growth factor, and transforming growth factor-β on inflammatory repair potential in the periodontium. J Periodontol 1999; 70:1136–43.

99. Seifert RA, Coats SA, Raines EW, et al. Platelet-derived growth factor (PDGF)-receptor α-subunit mutant and reconstituted cell lines demonstrate that transforming growth factor-β can be mitogenic through PDGF A-chain-dependent and -independent pathways. J Biol Chem 1994; 269:13951–5.

100. Lorenzo GD, Tortora G, D'Armiento FP, et al. Expression of epidermal growth factor receptor correlates with disease relapse and progression to androgen-independence in human prostate cancer. Clin. Cancer Res 2002; 8:3438–44.

101. Fudge K, Bostwick DG, Stearns ME. Platelet-derived growth factor A and B chains and α and β receptors in prostatic intraepithelial neoplasia. Prostate 1996; 29:282–6.

102. Levitzki A. Tyrosine kinases as targets for cancer therapy. Eur J Cancer 2002; 38(suppl 5): S11–8.

103. Capdeville R, Buchdunger E, Zimmermann J, et al. Glivec (STI571, imatinib), a rationally developed, targeted anticancer drug. Nat Rev Drug Discov 2002; 1:493–502.

104. Manley PW, Cowan-Jacob SW, Buchdunger E, et al. Imatinib: a selective tyrosine kinase inhibitor. Eur J Cancer 2002; 38(suppl 5):S19–27.

105. Ulrich A, Coussens L, Hayflick JS, et al. Human epidermal growth factor receptor cDNA sequence and aberrant expression of the amplified gene in A431 epidermoid carcinoma cells. Nature (*Lond*) 1984; 309:418–25.

106. Ciardiello F, Damiano V, Bianco R, et al. Antitumor activity of combined blockade of epidermal growth factor receptor and protein kinase A. J Natl Cancer Inst 1996; 88:1770–6.

107. Uckun FM, Narla RK, Zeren T, et al. *In vivo* toxicity, pharmacokinetics, and anticancer activity of Genistein linked to recombinant human epidermal growth factor. Clin Cancer Res 1988; 4:1125–34.

108. Uckun FM, Narla RK, Jun X, et al. Cytotoxic activity of epidermal growth factor-genistein against breast cancer cells. Clin. Cancer Res 1988; 4:901–12.

109. Thyberg J. The microtubular cytoskeleton and the initiation of DNA synthesis. Exp Cell Res 1984; 155:1–8.

110. Yoon SY, Tefferi A, Li CY. Cellular distribution of platelet-derived growth factor, transforming growth factor-β, basic fibroblast growth factor, and their receptors in normal bone marrow. Acta Haematol 2000; 104:151–7.

111. Saharinen P, Alitalio K. Double target for tumor mass destruction. J Clin Invest 2003; 111:1277–80.

112. Mathew P, Fidler IJ, Logothetis CJ. Combination Docetaxel and platelet derived growth factor receptor inhibition with imatinib mesylate in prostate cancer. Semin. Oncol 2004; 31:24–9.

113. Gottesman MM, Pastan I. Biochemistry of multidrug resistance mediated by the multidrug transporter. Annu Rev Biochem 1993; 62:385–427.

114. Van Brussel JP, Jan Van Steenbrugge G, Van Krimpen C, et al. Expression of multidrug resistance related proteins and proliferative activity is increased in advanced clinical prostate cancer. J Urol 2001; 165:130–5.

115. Kerbel RS, Folkman J. Clinical translation of angiogenesis inhibitors. Nat Rev Cancer 2002; 2:727–39.

116. Preise D, Mazor O, Koudinova N, et al. Bypass of tumor drug resistance by antivascular therapy. Neoplasia 2003; 5:475–80.

117. Kim SJ, Uehara H, Yazici S, et al. Targeting platelet-derived growth factor receptor on endothelial cells of multidrug resistant prostate cancer. J Natl Cancer Inst 2006; 98:783–93.

Section IV
Functional Assessments of Angiogenesis

Chapter 24
Normalization of Tumor Vasculature and Microenvironment

Rakesh K. Jain[1,*], Dan G. Duda[1], Tracy T. Batchelor[2], A. Gregory Sorensen[3], and Christopher G. Willett[4]

Keywords: vessel normalization, tumor vasculature, trastuzumab, bevacizumab, normalization window

Abstract: Solid tumors require blood vessels for growth, and many new cancer therapies are targeted against the tumor vasculature. The widely held view is that these antiangiogenic therapies destroy the tumor vasculature, thereby depriving the tumor of oxygen and nutrients. Indeed, that is the ultimate goal of antiangiogenic therapies. However, emerging preclinical and clinical evidence support an alternative hypothesis, that judicious application of agents that block angiogenesis directly (e.g., bevacizumab, AZD2171) and indirectly (e.g., trastuzumab) can also transiently "normalize" the abnormal structure and function of tumor vasculature. In addition to being more efficient for oxygen and drug delivery, the normalized vessels are fortified with pericytes, which can hinder intravasation of cancer cells, a necessary step in hematogenous metastasis. Drugs that induce vascular normalization can also normalize the tumor microenvironment—reduce hypoxia and interstitial fluid pressure—and thus increase the efficacy of many conventional therapies if both are carefully scheduled. Reduced interstitial fluid pressure can decrease tumor-associated edema as well as the probability of lymphatic dissemination. Independent of these effects, alleviation of hypoxia can decrease the selection pressure for a more malignant phenotype. Finally, the increase in proliferation of cancer cells during the "vascular normalization window" can potentially sensitize tumors to cytotoxic agents. Our recent Phase II clinical trial in glioblastoma patients shows that the normalization window—identified using advanced magnetic resonance imaging (MRI) techniques—can last one to four months, and the resulting changes in tumor vasculature correlate with circulating molecular and cellular biomarkers in these patients.

Introduction

After nearly four decades of basic research and clinical development of antiangiogenic therapy for cancer, two anti-VEGF approaches have yielded survival benefit in patients with metastatic cancer in randomized phase III trials [11,20]. In one approach, the addition of bevacizumab, a vascular endothelial growth factor (VEGF)-specific antibody, to standard therapy improved overall survival (OS) in colorectal and non-small cell lung cancer patients and progression-free survival (PFS) in breast cancer and renal cell cancer patients [14,31]. In the second approach, multi-targeted tyrosine kinase inhibitors (TKIs), that block not only VEGF receptor kinases but also other kinases in both endothelial and cancer cells, demonstrated survival benefit in gastrointestinal stromal tumor, renal cell carcinoma and hepatocellular carcinoma patients [7]. By contrast, bevacizumab failed to increase survival with chemotherapy in patients with previously treated and refractory metastatic breast cancer or pancreatic cancer. Furthermore, addition of vatalanib, a multi-targeted TKI developed as a VEGF receptor-selective agent, to conventional cytotoxic therapy did not show a similar benefit in metastatic colorectal

[1] Steele Laboratory, Department of Radiation Oncology, Massachusetts General Hospital, 100 Blossom Street, Cox-734, Boston, MA 02114, USA

[2] Department of Neuro-Oncology, Massachusetts General Hospital, 55 Fruit Street, Yawkey-9E, Boston, MA 02114, USA

[3] Department of Radiology, A.A. Martinos Imaging Center, Massachusetts General Hospital and Harvard Medical School, Charelestown, MA 02129, USA

[4] Department of Radiation Oncology, Box 3085, Duke University Medical Center, Durham, NC 27710, USA

*Corresponding Author:
Rakesh K. Jain, Andrew Werk Cook Professor of Tumor Biology, Director, Edwin L. Steele Laboratory for Tumor Biology, Department of Radiation Oncology, Massachusetts General Hospital and Harvard Medical School, 100 Blossom St, Cox 7, Boston, MA 02114, USA
E-mail: jain@steele.mgh.harvard.edu

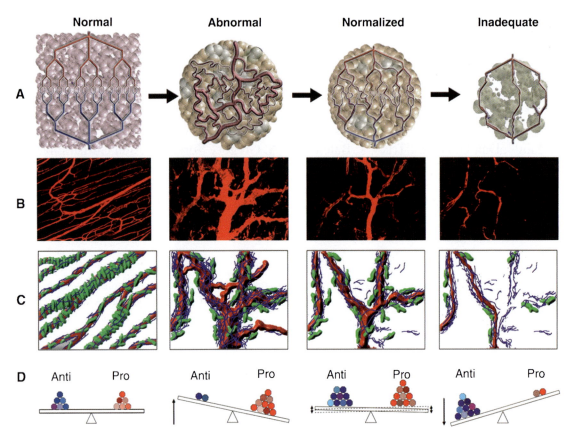

FIG. 24.1. Proposed role of vessel normalization in the response of tumors to antiangiogenic therapy. **A** Tumor vasculature is structurally and functionally abnormal. It is proposed that antiangiogenic therapies initially improve both the structure and the function of tumor vessels. However, sustained or aggressive antiangiogenic regimens may eventually prune away these vessels, resulting in a vasculature that is both resistant to further treatment and inadequate for delivery of drugs or oxygen [17]. **B** Dynamics of vascular normalization induced by VEGFR2 blockade. *Left* Two-photon image showing normal blood vessels in skeletal muscle; subsequent images show human colon carcinoma vasculature in mice at day 0, day 3 and day 5 after administration of VEGR2-specific antibody [33]. **C** Diagram depicting the concomitant changes in pericyte (*red*) and basement membrane (*blue*) coverage during vascular normalization [33,39]. **D** These phenotypic changes in the vasculature may reflect changes in the balance of pro- and antiangiogenic factors in the tissue. (Reproduced with permission from R. K. Jain, *Science*, 2005;307:58–62.)

cancer patients. Finally, several agents that target oncogenic signaling pathways (such as HER2 by trastuzumab) may indirectly inhibit angiogenesis and have yielded increased OS with chemotherapy in clinical trials.

These contrasting responses raise critical questions about how these agents work in patients and how to combine them optimally. Here, of the many potential mechanisms of action of antiangiogenic agents, we will focus on pruning and normalization of tumor vasculature for improved delivery and efficacy of therapeutics [4,20]. After summarizing preclinical evidence in support of vascular normalization, we will present clinical evidence from two trials: treatment of rectal carcinoma patients receiving bevacizumab [36,37] and recurrent glioblastoma patients receiving AZD2171 [1]. Then we will discuss the progress on the identification of potential biomarkers for anti-VEGF agent efficacy in humans: molecular and cellular parameters obtained from tissue biopsies, interstitial fluid pressure, circulating endothelial cells, protein levels in bodily fluids and physiological parameters measured with

various imaging techniques. We will end by outlining potential avenues of further investigation.

Normalization Hypothesis

Beginning with the seminal work of Dr. Beverly Teicher, several preclinical and clinical studies have shown that antiangiogenic therapy improves the outcome of cytotoxic therapies [19,20]. This is paradoxical. One would expect that destroying the vasculature would severely compromise the delivery of oxygen and therapeutics to the solid tumor, producing hypoxia that would render many chemotherapeutics, as well as radiation, less effective. To resolve this paradox, we hypothesized in 2001 that the judicious application of antiangiogenic agents can "normalize" the abnormal tumor vasculature, resulting in more efficient delivery of drugs and oxygen to the targeted cancer cells (Fig. 24.1) [17]. Increased penetration of drugs throughout the tumor would enhance the outcome of chemotherapy, while the

TABLE 24.1. Morphological and functional characteristics of the vasculature in normal tissue, an untreated tumor, a tumor during early stages of treatment with an antiangiogenic drug ("normalized"), and a tumor treated with high doses of an antiangiogenic drug over a long period ("regressing"). (Reproduced with permission from R. K. Jain, *Science*, 2005;307:58–62).

Properties	Vessel Type			
	Normal	Tumor (untreated)	Tumor (normalized)	Tumor (regressing)
Global organization	Normal	Abnormal	Normalized	Fragmented
Pericyte coverage	Normal	Absent or detached	Closer to normal	Missing
Basement membrane	Normal	Absent or too thick	Closer to normal, some ghost	Ghost
Vessel diameter	Normal distribution	Dilated	Closer to normal	Closer to or less than normal
Vascular density	Normal, homogeneous distribution	Abnormal, heterogeneous distribution	Closer to normal	Extremely low
Permeability to large molecules	Normal	High	Intermediate	N/A
MVP[a] and IFP	MVP > IFP	MVP ~ IFP	MVP > IFP	Low IFP
Plasma (P) and Interstitial (I) oncotic pressure[b]	P > I	P ~ I	P > I	N/A
pO_2	Normal	Hypoxia	Reduced hypoxia	Hypoxia
Drug penetration	N/A	Heterogeneous	More homogeneous	Inadequate

[a] *MVP* Microvascular pressure; *IFP* interstitial fluid pressure
[b] Osmotic pressure exerted by plasma proteins

ensuing increased level of oxygen would enhance the efficacy of radiation therapy and many chemotherapeutic agents.

Rationale for Normalizing the Tumor Vasculature

To obtain nutrients for their growth and to metastasize to distant organs, cancer cells co-opt host vessels, sprout new vessels from existing ones (angiogenesis), and/or recruit endothelial cells from the bone marrow (post-natal vasculogenesis) [5,11]. The resulting vasculature is structurally and functionally abnormal (Table 24.1) [18]. These structural abnormalities contribute to spatial and temporal heterogeneity in tumor blood flow. In addition, solid stress generated by proliferating cancer cells compresses intra-tumoral blood and lymphatic vessels, which further impairs not only the blood flow but also lymphatic flow [29]. Collectively, these vascular abnormalities lead to an abnormal tumor microenvironment characterized by interstitial hypertension (elevated hydrostatic pressure outside the blood vessels), hypoxia, and acidosis.

Impaired blood supply and interstitial hypertension interfere with the delivery of therapeutics to solid tumors. Hypoxia renders tumor cells resistant to both radiation and several cytotoxic drugs. And, independent of these effects, hypoxia also induces genetic instability and selects for more malignant cells with increased metastatic potential. Hypoxia and low pH also compromise the cytotoxic functions of immune cells that infiltrate a tumor. Unfortunately, cancer cells are able to survive in this abnormal microenvironment. Interstitial hypertension pushes the fluid from the tumor margin into the peri-tumoral tissue (or fluid) contributing to tumor-associated edema and lymphatic metastasis [22]. In essence, the abnormal vasculature of tumors and the resulting abnormal microenvironment together pose a formidable barrier to delivery and efficacy of cancer therapy. This suggests that if we knew how to correct the structure and function of tumor vessels, we would have a chance to normalize the tumor microenvironment and ultimately to improve cancer treatment. The fortified tumor vasculature may also inhibit the shedding of cancer cells into the circulation, a prerequisite for metastasis.

Blocking VEGF Signaling Normalizes Tumor Vessels in Transplanted Tumors

In normal tissues, the collective action of angiogenic stimulators (e.g., VEGF) is counter-balanced by the collective action of angiogenic inhibitors such as thrombospondin-1 (Fig. 24.1). This balance tips in favor of the stimulators in both pathological and physiological angiogenesis. However, in pathological angiogenesis, the imbalance persists. Therefore, restoring the balance may render the tumor vasculature close to normal. On the other hand, tipping this balance in favor of inhibitors may lead to vascular regression and, ultimately, to tumor regression.

Of all the known angiogenic molecules, VEGF (also referred to as VEGF-A) appears the most critical. VEGF promotes the survival and proliferation of endothelial cells, and increases vascular permeability [10]. VEGF is overexpressed in the majority of solid tumors. Thus, if one were to judiciously down-regulate VEGF signaling in tumors, then the vasculature might revert back to a more "normal" state. Indeed, blockade of VEGF signaling passively prunes the immature and leaky vessels of transplanted tumors in mice and actively remodels the remaining vasculature so that it more closely resembles the normal vasculature (Fig. 24.1). This "normalized" vasculature is characterized by less leaky, less dilated and less tortuous vessels with a more normal basement membrane and greater coverage by pericytes (Fig. 24.1). These morphological

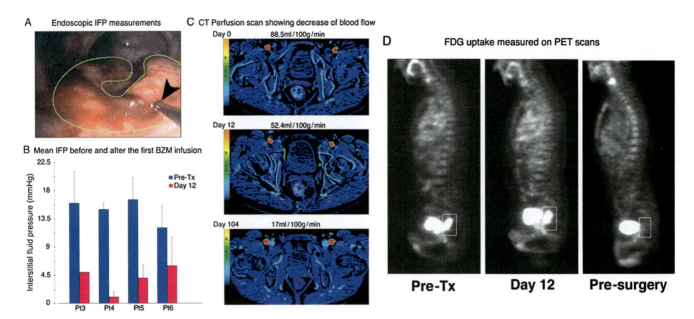

FIG. 24.2. Effect of treatment on tumors in rectal cancer patients who completed bevacizumab-chemoradiation treatment, and surgery. **A** Endoscope-guided tumor interstitial fluid pressure (IFP) measurements. **B** Bevacizumab decreases tumor IFP at day 12 after the first infusion. **C** Representative functional CT images of blood perfusion before treatment (day 0), after bevacizumab (day 12) and after completion of treatment (day 104). **D** Tumor FDG uptake before treatment (pretreatment), 12 days after bevacizumab treatment and 6–7 weeks after completion of all neoadjuvant therapy (presurgery). The images are sagittal projections of FDG-PET scans. Tumor is outlined in *box*, posterior to bladder. (Adapted with permission from C. G. Willett, Nature Medicine 2004; 10:145–147.)

changes are accompanied by functional changes: decreased interstitial fluid pressure, increased tumor oxygenation, and improved penetration of drugs in these tumors (Table 24.1) [12,13,15,24,26,32,33, 35,39,40].

First Clinical Evidence of Normalization in Human Tumors

In a Phase I/II clinical trial in rectal carcinoma patients receiving bevacizumab and chemoradiation, we recently examined the effect of bevacizumab on human tumors [36–38]. The treatment regimen consisted of a bevacizumab infusion followed by three cycles of bevacizumab, 5-fluorouracil (5-FU) and external beam local radiotherapy. Six to nine weeks after completion of the combined treatment, patients underwent resection of the tumor. Using multiple functional, cellular and molecular investigations, direct evidence was demonstrated of the antivascular effect of bevacizumab in these rectal carcinoma patients. In brief, bevacizumab reduced the tumor vascular density by approximately 50% 12 days after the first infusion, significantly reduced the tumor blood flow (evaluated by computed tomography (CT) scans) and the number of viable circulating endothelial cells (CECs) and progenitor cells (CPCs) [9]. The tumors in both the low-dose bevacizumab (5 mg/kg) and the high-dose bevacizumab (10 mg/kg)

infusion groups had a less hyperemic/hemorrhagic appearance. However, there was no significant tumor regression post-bevacizumab treatment by flexible sigmoidoscopy examination, which permitted macroscopic visualization of the rectal carcinoma. This reduction in the tumor vasculature led to a significant increase in cancer cell apoptosis. Interestingly, it also led to the emergence of a more mature (pericyte-covered) vasculature, and increased cancer cell proliferation. Twelve days after bevacizumab therapy alone, the tumor interstitial fluid pressure (IFP) was consistently decreased, particularly in patients with high baseline values (Fig. 24.2). This suggested that in human tumors, similar to mouse models, the tumor microenvironment was normalized by the reduction of the excess vasculature and was potentially sensitized to subsequent cytotoxic therapy.

Imaging studies have also provided supportive data for the normalization hypothesis. Fluorodeoxyglucose (FDG)-positron emission tomography (PET) detects metabolically active malignant lesions, and may be used to stage and monitor tumor response to treatment. Despite the significant reduction in vessel density and blood flow, FDG-uptake measured from PET scans did not significantly change 12 days after the initial bevacizumab infusion. Moreover, despite the decrease in vessel surface area (S) and blood flow, the permeability-surface area (P•S) product (proportional to the penetration of tracer in tumor) evaluated by perfusion CT scans did not significantly

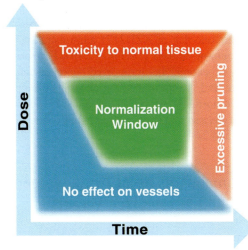

Proposed normalization window

However, whether the increase in tumor cell apoptosis was due to a direct effect of bevacizumab or an indirect effect due to insufficient perfusion in certain areas is unclear.

In summary, these results in patients mirrored those seen in transplanted tumors in mice. Two weeks after a single injection of bevacizumab, the global (mean) blood flow of tumors, as measured by contrast-enhanced CT, decreased by 30–50% in six consecutive patients. Tumor microvascular density, vascular volume, and interstitial fluid pressure were also found to be reduced. Surprisingly, however, there was no concurrent decrease in the uptake of FDG in tumors, suggesting that vessels in the residual "normalized" tumor vasculature were more efficient in supporting cancer cell metabolic activity than they were prior to bevacizumab treatment. Furthermore, whereas apoptosis of cancer cells increased as expected based on the decrease in vessel density, the proliferation rate in cancer cells increased supporting the normalization of microenvironment and potentially explaining the chemosensitizing effect of bevacizumab [37,38].

FIG. 24.3. Proposed effect of drug dose and schedule on tumor vascular normalization. The efficacy of cancer therapies that combine anti-angiogenic and cytotoxic drugs depends on the dose and delivery schedule of each drug. The vascular normalization model posits that a well-designed strategy should passively prune away immature, dysfunctional vessels and actively fortify those remaining, while incurring minimal damage to normal tissue vasculature. During this "normalization" window (*green*), cancer cells may be more vulnerable to traditional cytotoxic therapies and to novel targeted therapies. Note that the degree of normalization will be spatially and temporally dependent in a tumor. Vascular normalization will occur only in regions of the tumor where the imbalance of pro- and antiangiogenic molecules has been corrected. (Reproduced with permission from R. K. Jain, *Science*, 2005; 307:58–62.)

Identification of the Normalization Window in Mice and Patients

Optimal scheduling of antiangiogenic therapy with chemotherapy and/or radiation requires knowledge of the time window during which the vessels initially become normalized, as well as an understanding of how long they remain in that state (Fig. 24.3). Our recent studies, in which mice bearing human glioblastoma xenografts were treated with an antibody to VEGF receptor-2 (VEGFR2), have identified such a "normalization window"; that is, a period during which the addition of radiation therapy yields the best therapeutic outcome (Fig. 24.4A) [39]. This window was short-lived in the mice (about 6 days) and was characterized by an increase

change at day 12. The changes observed in these two parameters suggested that the tumor vasculature became more efficient after anti-VEGF treatment in some areas of the tumors.

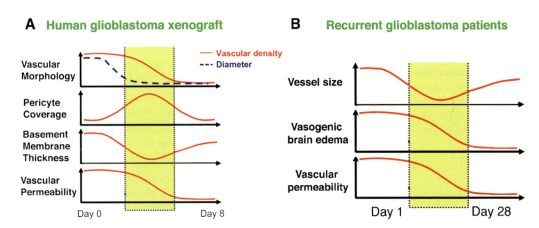

FIG. 24.4. Time course and mechanisms of vascular normalization, illustrated in schematic diagrams. **A** VEGFR2 blockade in mice produces a time window of morphological and functional normalization of glioblastoma xenograft vessels. **B** Similarly, blockade of VEGF receptors in recurrent glioblastoma patients transiently normalizes the tumor vasculature. (Adapted with permission from F. Winkler, *Cancer Cell*, 2004;6:553–563.)

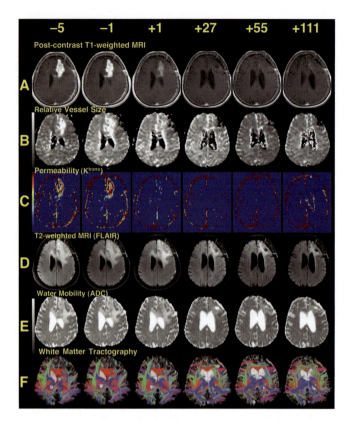

FIG. 24.5. Representative images from the best-responding recurrent glioblastoma patient to AZD2171 treatment. **A** T1-weighted anatomic images after intravenous administration of a contrast agent (gadolinium-DTPA), demonstrating a region of bright signal corresponding to the recurrent brain tumor in the left frontal lobe shrinking over time (all images are displayed per standard radiographic convention). Note also the decreased mass effect on the left lateral ventricle. **B** Map of relative microvessel size, also showing decrease over time. **C** Maps of Ktrans, a measure of blood-brain barrier permeability. Note the substantial change after the first dose. **D** T2-weighted images acquired with a fluid-attenuated inversion recovery sequence (FLAIR), where edema is seen surrounding the tumor enhancement evident in (**A**), also decreasing over time. **E** Images of apparent diffusion coefficient (*ADC*) demonstrating water mobility, which identifies areas of vasogenic edema as high (bright) signal surrounding the region of enhancing tumor; these also reduce over time. The displacement of the ventricle is also reduced over time. **F** Tractography. These images demonstrate directional water mobility suggesting the presence of white matter tracts. As the vasogenic edema decreases and the mass effect subsides, these white matter tracts become more evident. (Adapted with permission from T. T. Batchelor, Cancer Cell, 2007;11:83–95.)

in tumor oxygenation. During the normalization window, but not before or after it, VEGFR2 blockade was found to increase pericyte coverage of vessels in a human brain tumor xenograft in mice. Vessel normalization was causally related to upregulation of angiopoietin 1 and activation of matrix metalloproteinases (MMPs) [39].

More recently, we demonstrated that such a window exists in cancer patients. Using advanced MRI protocols in a Phase

II trial of AZD2171—an oral pan-VEGFR TKI—in recurrent glioblastoma patients, we showed that vascular normalization, as measured by reduction in vessel diameter and permeability, begins within one day of AZD2171 administration and lasts for a minimum of 28 days (1) (Fig. 24.4B). Furthermore, the tumor vessels become "de-normalized" upon discontinuation of treatment and "re-normalize" upon drug resumption [1]. The vascular normalization is accompanied by a reduction in brain edema and a steroid-sparing effect in this patient population (Figs. 24.4 and 24.5).

Trastuzumab Acts as an Antiangiogenic Cocktail and Normalizes Vessels

The constellation of angiogenic molecules expressed in a tumor increases with malignant progression. For example, early stages of breast tumors may require only VEGF for angiogenesis, whereas at later stages, angiogenesis in these tumors may be driven by additional factors. Thus, a late-stage breast tumor may escape anti-VEGF treatment by exploiting alternative angiogenic factors to generate/maintain its neovasculature. This may help explain in part why bevacizumab and chemotherapy did not prolong the survival of breast cancer patients in a recent Phase III trial [20]. In rectal carcinoma patients, we discovered that plasma VEGF and placental growth factor (PlGF) levels increase after VEGF blockade with bevacizumab [36,37] and reproduced these findings in recurrent glioblastoma patients treated with the pan-VEGF receptor inhibitor AZD2171 [1]. Moreover, plasma basic fibroblast growth factor (bFGF) and stromal cell-derived factor 1α (SDF1α) levels correlated with enlargement of vessel size and radiographic disease progression during AZD2171 treatment in recurrent glioblastoma patients [1]. Thus, optimal cancer treatment may require the targeting of multiple angiogenic pathways, and the challenge for the oncologist will be to formulate cocktails of antiangiogenic agents specifically tailored to the angiogenic profile of individual tumors.

We have recently shown in preclinical models that anti-cancer agents such as trastuzumab can "mimic" antiangiogenic cocktails [16]. Trastuzumab lowers expression of several pro-angiogenic molecules while increasing expression of the anti-angiogenic molecule thrombospondin-1. Interestingly, while trastuzumab lowered the expression of VEGF in tumor cells, the host cells within the tumor stroma produced compensatory VEGF; thus, additional anti-VEGF treatment could improve the efficacy of trastuzumab. These findings provide a powerful rationale for the clinical trial in which trastuzumab is combined with bevacizumab for treatment of HER2-positive breast cancer [20].

A further finding from this study was that trastuzumab "normalized" the vasculature of human breast cancer xenografts. Whereas vessels in the control antibody-treated tumors were dilated and leaky, those in the trastuzumab-treated tumors had

TABLE 24.2. Similarities between the effects of anti-VEGF therapy in tumor-bearing mice and cancer patients.

Properties	Preclinical data		Clinical data	
	Change after treatment	Refs.	Change after	Refs. treatment
Blood volume	↓	[33,39]	↓	[34,36,37]
Vascular density	↓	[33,39]	↓	[36,37]
Permeability (BSA)	↓	[33,39]	N/A	
Permeability-surface area (P S) product (small molecules)	N/A		↓	[36–38]
Interstitial fluid pressure	↓	[33,40]	↓	[36,37]
Perivascular cell coverage	↑	[33,39]	↑	[36,37]
Ang-2 level	↓	[33]	↓	[37]
Tumor apoptosis	↑	[26,30]	↑	[34,37]
Plasma VEGF level	↑	[27]	↑	[1,6,28,37]

TABLE 24.3. Dissimilarities between the effects of anti-VEGF therapy in tumor-bearing mice and cancer patients.

Properties	Preclinical data		Clinical data	
	Change after treatment	Refs.	Change after treatment	Refs.
Vascular density	↓	[33]	→	[34]
Cancer cell proliferation	↓	[30]	↑, →	[34,37]
Circulating endothelial cells (CECs)	↑	[2]	↓	[36,37]
Plasma VEGF	→[a]	[3]	↑[a]	[1,6,28,37]

[a]In mice, plasma VEGF was increased by DC101 and RAFL-1, two anti-VEGFR-2 antibodies, but not by PTK787 or SU5416, two VEGFR-selective tyrosine kinase inhibitors. In cancer patients, plasma VEGF was increased by anti-VEGF therapy with bevacizumab, PTK787, AZD2171, and SU11248

diameters and vascular permeability closer to those of normal vessels. Thus, trastuzumab and other drugs that target upstream receptors might act as mimics of antiangiogenic cocktails; that is, these drugs improve their own delivery as well as that of other therapeutics given in combination. This improvement in delivery and alleviation of hypoxia presumably contributes to their efficacy.

Perspective

At least three major challenges must be met before therapies based on the vascular normalization model can be successfully translated to the clinic. The first challenge is to determine which other direct or indirect antiangiogenic therapies lead to vascular normalization. In principle, any therapy that restores the balance between pro- and antiangiogenic molecules should induce normalization. Indeed, withdrawing hormones from a hormone-dependent tumor lowers VEGF levels and leads to vascular normalization [21]. Recently, metronomic therapy, a drug delivery method in which low doses of chemotherapeutic agents are given at frequent intervals, has also been shown to increase the expression of thrombospondin-1, which is a potent endogenous angiogenesis inhibitor [25]. Conceivably, this therapy might also induce normalization and improve oxygenation and drug penetration into tumors. Whether various synthetic kinase inhibitors, endogenous inhibitors, anti-vasocrine agents, conventional chemotherapeutic agents, and vascular targeting agents can do the same remains to be seen. Some of these agents may be particularly effective because they target both stromal and cancer cells. To date, most clinical trials in patients with solid tumors are designed primarily to measure changes in the size of the contrast-enhancing tumor. However, these response criteria may prove inadequate to ascertain "tumor response" when patients are treated with agents that reduce vascular permeability and the leakage of

contrast agents [1,36]. Alternative methods of assessment may be necessary to judge activity of these anti-VEGF therapies. Moreover, these clinical trials need to be tightly integrated with preclinical studies so that the similarities (Table 24.2) and differences (Table 24.3) could be understood and exploited.

The second challenge is to identify suitable surrogate markers of changes in the structure and function of the tumor vasculature, and to develop imaging technology that will help to identify the timing of the normalization window during antiangiogenic therapy. Measurement of blood vessel density requires tissue biopsy and provides little information on vessel function. Although imaging techniques are expensive and far from optimal, they can provide serial measures of vascular permeability, relative vessel size, blood volume, blood perfusion, and uptake of some drugs, and can therefore be used to monitor the window of normalization in patients [1]. The number of blood circulating mature endothelial cells and progenitor cells decreases after VEGF blockade, both in animals and patients, but whether this decline coincides with the normalization window is not known [7,8,23]. During the course of therapy, serial blood measurements of molecules involved in vessel maturation have the potential to identify surrogate markers. Positron emission tomography with [F-18] fluoromisonidazole (PET-MISO) and MRI can provide some indication of tumor oxygenation, and might be useful for tracking the normalization window. Finally, the measurement of the interstitial fluid pressure is minimally invasive, inexpensive, and easy to implement for anatomically accessible tumors [36].

The third challenge is to fill the gaps in our understanding of the molecular and cellular mechanisms of the vascular normalization process so that we may be able to extend this window in patients. With rapid advances in genomic and proteomic technology and access to tumor tissues during the course of therapy, we can begin to monitor tumor response to antiangiogenic therapies at the molecular level. Our recent trials have identified PlGF (in rectal cancer), and bFGF and SDF1α (in glioblastoma) as potential therapeutic targets to extend this window [1,36,37].

Addressing all of these challenges will not only benefit patients with cancer, but possibly patients with other diseases, e.g., age-related wet macular degeneration. These principles may also be useful for stabilizing plaques, controlling tumor-associated edema, and improving regenerative medicine, where the goal is to create and maintain a functionally normal vasculature.

Acknowledgments. We would like to thank the members of the Steele Lab, especially Y. Boucher, E. di Tomaso, D. Fukumura, K. R. Kozak and L. L. Munn for their collaborative support, and our many clinical collaborators. We also thank the National Cancer Institute for supporting our work. This chapter is adapted and updated from a recent review in Science (R. K. Jain, *Science*, 2005;307:58–62) and incorporates some elements of another recent review in Nature Clinical Practice Oncology (R. K. Jain, D. G. Duda, J. W. Clark and J. S. Loeffler, *Nat Clin Pract Oncol*, 2006;3:24–40).

References

1. Batchelor, T. T., Sorensen, A. G., di Tomaso, E., Zhang, W. T., Duda, D. G., Cohen, K. S., Kozak, K. R., Cahill, D. P., Chen, P. J., Zhu, M., Ancukiewicz, M., Mrugala, M. M., Plotkin, S., Drappatz, J., Louis, D. N., Ivy, P., Scadden, D. T., Benner, T., Loeffler, J. S., Wen, P. Y., and Jain, R. K. AZD2171, a pan-VEGF receptor tyrosine kinase inhibitor, normalizes tumor vasculature and alleviates edema in glioblastoma patients. Cancer Cell, *11:* 83–95, 2007.

2. Beaudry, P., Force, J., Naumov, G. N., Wang, A., Baker, C. H., Ryan, A., Soker, S., Johnson, B. E., Folkman, J., and Heymach, J. V. Differential effects of vascular endothelial growth factor receptor-2 inhibitor ZD6474 on circulating endothelial progenitors and mature circulating endothelial cells: implications for use as a surrogate marker of antiangiogenic activity. Clin Cancer Res, *11:* 3514–3522, 2005.

3. Bocci, G., Man, S., Green, S. K., Francia, G., Ebos, J. M., du Manoir, J. M., Weinerman, A., Emmenegger, U., Ma, L., Thorpe, P., Davidoff, A., Huber, J., Hicklin, D. J., and Kerbel, R. S. Increased plasma vascular endothelial growth factor (VEGF) as a surrogate marker for optimal therapeutic dosing of VEGF receptor-2 monoclonal antibodies. Cancer Res, *64:* 6616–6625, 2004.

4. Carmeliet, P. Angiogenesis in life, disease and medicine. Nature, *438:* 932–936, 2005.

5. Carmeliet, P. and Jain, R. K. Angiogenesis in cancer and other diseases. Nature, *407:* 249–257, 2000.

6. Drevs, J., Zirrgiebel, U., Schmidt-Gersbach, C. I., Mross, K., Medinger, M., Lee, L., Pinheiro, J., Wood, J., Thomas, A. L., Unger, C., Henry, A., Steward, W. P., Laurent, D., Lebwohl, D., Dugan, M., and Marme, D. Soluble markers for the assessment of biological activity with PTK787/ZK 222584 (PTK/ZK), a vascular endothelial growth factor receptor (VEGFR) tyrosine kinase inhibitor in patients with advanced colorectal cancer from two phase I trials. Ann Oncol, *16:* 558–565, 2005.

7. Duda, D. G., Batchelor, T. T., Willett, C. G., and Jain, R. K. VEGF-targeted cancer therapy strategies: current progress, hurdles and future prospects. Trends Mol Med, *13:* 223–230, 2007.

8. Duda, D. G., Cohen, K. S., di Tomaso, E., Au, P., Klein, R. J., Scadden, D. T., Willett, C. G., and Jain, R. K. Differential CD146 expression on circulating versus tissue endothelial cells in rectal cancer patients: implications for circulating endothelial and progenitor cells as biomarkers for antiangiogenic therapy. J Clin Oncol, *24:* 1449–1453, 2006.

9. Duda, D. G., Cohen, K. S., Scadden, D. T., and Jain, R. K. A protocol for phenotypic detection and enumeration of circulating endothelial cells and circulating progenitor cells in human blood. Nat Protoc, *2:* 805–810, 2007.

10. Dvorak, H. F. Vascular permeability factor/vascular endothelial growth factor: a critical cytokine in tumor angiogenesis and a potential target for diagnosis and therapy. J Clin Oncol, *20:* 4368–4380, 2002.

11. Folkman, J. Tumor angiogenesis: therapeutic implications. N Engl J Med, *285:* 1182–1186, 1971.

12. Gazit, Y., Baish, J. W., Safabakhsh, N., Leunig, M., Baxter, L. T., and Jain, R. K. Fractal characteristics of tumor vascular architecture during tumor growth and regression. Microcirculation, *4:* 395–402, 1997.

13. Hobbs, S. K., Monsky, W. L., Yuan, F., Roberts, W. G., Griffith, L., Torchilin, V. P., and Jain, R. K. Regulation of transport pathways in tumor vessels: role of tumor type and microenvironment. Proc Natl Acad Sci USA, *95:* 4607–4612, 1998.

14. Hurwitz, H., Fehrenbacher, L., Novotny, W., Cartwright, T., Hainsworth, J., Heim, W., Berlin, J., Baron, A., Griffing, S., Holmgren, E., Ferrara, N., Fyfe, G., Rogers, B., Ross, R., and Kabbinavar, F. Bevacizumab plus irinotecan, fluorouracil, and leucovorin for metastatic colorectal cancer. N Engl J Med, *350:* 2335–2342, 2004.

15. Inai, T., Mancuso, M., Hashizume, H., Baffert, F., Haskell, A., Baluk, P., Hu-Lowe, D. D., Shalinsky, D. R., Thurston, G., Yancopoulos, G. D., and McDonald, D. M. Inhibition of vascular endothelial growth factor (VEGF) signaling in cancer causes loss of endothelial fenestrations, regression of tumor vessels, and appearance of basement membrane ghosts. Am J Pathol, *165:* 35–52, 2004.

16. Izumi, Y., Xu, L., di Tomaso, E., Fukumura, D., and Jain, R. K. Tumor biology - Herceptin acts as an anti-angiogenic cocktail. Nature, *416:* 279–280, 2002.

17. Jain, R. K. Normalizing tumor vasculature with anti-angiogenic therapy: a new paradigm for combination therapy. Nat Med, *7:* 987–989, 2001.

18. Jain, R. K. Molecular regulation of vessel maturation. Nat Med, *9:* 685–693, 2003.

19. Jain, R. K. Normalization of tumor vasculature: an emerging concept in antiangiogenic therapy. Science, *307:* 58–62, 2005.

20. Jain, R. K., Duda, D. G., Clark, J. W., and Loeffler, J. S. Lessons from phase III clinical trials on anti-VEGF therapy for cancer. Nat Clin Pract Oncol, *3:* 24–40, 2006.

21. Jain, R. K., Safabakhsh, N., Sckell, A., Chen, Y., Jiang, P., Benjamin, L., Yuan, F., and Keshet, E. Endothelial cell death, angiogenesis, and microvascular function after castration in an androgen-dependent tumor: Role of vascular endothelial growth factor. Proc Natl Acad Sci U S A, *95:* 10820–10825, 1998.

22. Jain, R. K., Tong, R. T., and Munn, L. L. Effect of vascular normalization by antiangiogenic therapy on interstitial hypertension, peritumor edema, and lymphatic metastasis: insights from a mathematical model. Cancer Res, *67:* 2729–2735, 2007.

23. Jubb, A. M., Hurwitz, H. I., Bai, W., Holmgren, E. B., Tobin, P., Guerrero, A. S., Kabbinavar, F., Holden, S. N., Novotny,

W. F., Frantz, G. D., Hillan, K. J., and Koeppen, H. Impact of vascular endothelial growth factor-A expression, thrombospondin-2 expression, and microvessel density on the treatment effect of bevacizumab in metastatic colorectal cancer. J Clin Oncol, *24:* 217–227, 2006.

24. Kadambi, A., Carreira, C. M., Yun, C., Padera, T. P., Dolmans, D., Carmeliet, P., Fukumura, D., and Jain, R. K. Vascular endothelial growth factor (VEGF)-C differentially affects tumor vascular function and leukocyte recruitment: Role of VEGF-receptor 2 and host VEGF-A. Cancer Res, *61:* 2404–2408, 2001.

25. Kerbel, R. S. and Kamen, B. A. The anti-angiogenic basis of metronomic chemotherapy. Nat Rev Cancer, *4:* 423–436, 2004.

26. Lee, C. G., Heijn, M., di Tomaso, E., Griffon-Etienne, G., Ancukiewicz, M., Koike, C., Park, K. R., Ferrara, N., Jain, R. K., Suit, H. D., and Boucher, Y. Anti-vascular endothelial growth factor treatment augments tumor radiation response under normoxic or hypoxic conditions. Cancer Res, *60:* 5565–5570, 2000.

27. Lee, L., Sharma, S., Morgan, B., Allegrini, P., Schnell, C., Brueggen, J., Cozens, R., Horsfield, M., Guenther, C., Steward, W. P., Drevs, J., Lebwohl, D., Wood, J., and McSheehy, P. M. Biomarkers for assessment of pharmacologic activity for a vascular endothelial growth factor (VEGF) receptor inhibitor, PTK787/ZK 222584 (PTK/ZK): translation of biological activity in a mouse melanoma metastasis model to phase I studies in patients with advanced colorectal cancer with liver metastases. Cancer Chemother Pharmacol, *57:* 761–771, 2006.

28. Motzer, R. J., Michaelson, M. D., Redman, B. G., Hudes, G. R., Wilding, G., Figlin, R. A., Ginsberg, M. S., Kim, S. T., Baum, C. M., DePrimo, S. E., Li, J. Z., Bello, C. L., Theuer, C. P., George, D. J., and Rini, B. I. Activity of SU11248, a multitargeted inhibitor of vascular endothelial growth factor receptor and platelet-derived growth factor receptor, in patients with metastatic renal cell carcinoma. J Clin Oncol, *24:* 16–24, 2006.

29. Padera, T. P., Stoll, B. R., Tooredman, J. B., Capen, D., di Tomaso, E., and Jain, R. K. Pathology: cancer cells compress intratumour vessels. Nature, *427:* 695, 2004.

30. Prewett, M., Huber, J., Li, Y., Santiago, A., O'Connor, W., King, K., Overholser, J., Hooper, A., Pytowski, B., Witte, L., Bohlen, P., and Hicklin, D. J. Antivascular endothelial growth factor receptor (fetal liver kinase 1) monoclonal antibody inhibits tumor angiogenesis and growth of several mouse and human tumors. Cancer Res, *59:* 5209–5218, 1999.

31. Sandler, A., Gray, R., Perry, M. C., Brahmer, J., Schiller, J. H., Dowlati, A., Lilenbaum, R., and Johnson, D. H. Paclitaxel-carboplatin alone or with bevacizumab for non-small-cell lung cancer. N Engl J Med, *355:* 2542–2550, 2006.

32. Teicher, B. A. A systems approach to cancer therapy. (Antioncogenics + standard cytotoxics–>mechanism(s) of interaction). Cancer Metastasis Rev, *15:* 247–272, 1996.

33. Tong, R. T., Boucher, Y., Kozin, S. V., Winkler, F., Hicklin, D. J., and Jain, R. K. Vascular normalization by vascular endothelial growth factor receptor 2 blockade induces a pressure gradient across the vasculature and improves drug penetration in tumors. Cancer Res, *64:* 3731–3736, 2004.

34. Wedam, S. B., Low, J. A., Yang, S. X., Chow, C. K., Choyke, P., Danforth, D., Hewitt, S. M., Berman, A., Steinberg, S. M., Liewehr, D. J., Plehn, J., Doshi, A., Thomasson, D., McCarthy, N., Koeppen, H., Sherman, M., Zujewski, J., Camphausen, K., Chen, H., and Swain, S. M. Antiangiogenic and antitumor effects of bevacizumab in patients with inflammatory and locally advanced breast cancer. J Clin Oncol, *24:* 769–777, 2006.

35. Wildiers, H., Guetens, G., De Boeck, G., Verbeken, E., Landuyt, B., Landuyt, W., de Bruijn, E. A., and van Oosterom, A. T. Effect of antivascular endothelial growth factor treatment on the intratumoral uptake of CPT-11. Br J Cancer, *88:* 1979–1986, 2003.

36. Willett, C. G., Boucher, Y., di Tomaso, E., Duda, D. G., Munn, L. L., Tong, R. T., Chung, D. C., Sahani, D. V., Kalva, S. P., Kozin, S. V., Mino, M., Cohen, K. S., Scadden, D. T., Hartford, A. C., Fischman, A. J., Clark, J. W., Ryan, D. P., Zhu, A. X., Blaszkowsky, L. S., Chen, H. X., Shellito, P. C., Lauwers, G. Y., and Jain, R. K. Direct evidence that the VEGF-specific antibody bevacizumab has antivascular effects in human rectal cancer. Nat Med, *10:* 145–147, 2004.

37. Willett, C. G., Boucher, Y., Duda, D. G., di Tomaso, E., Munn, L. L., Tong, R. T., Kozin, S. V., Petit, L., Jain, R. K., Chung, D. C., Sahani, D. V., Kalva, S. P., Cohen, K. S., Scadden, D. T., Fischman, A. J., Clark, J. W., Ryan, D. P., Zhu, A. X., Blaszkowsky, L. S., Shellito, P. C., Mino-Kenudson, M., and Lauwers, G. Y. Surrogate markers for antiangiogenic therapy and dose-limiting toxicities for bevacizumab with radiation and chemotherapy: continued experience of a phase I trial in rectal cancer patients. J Clin Oncol, *23:* 8136–8139, 2005.

38. Willett, C. G., Duda, D. G., di Tomaso, E., Boucher, Y., Czito, B. G., Vujaskovic, Z., Vlahovic, G., Bendell, J., Cohen, K. S., Hurwitz, H. I., Bentley, R., Lauwers, G. Y., Poleski, M., Wong, T. Z., Paulson, E., Ludwig, K. A., and Jain, R. K. Complete pathological response to bevacizumab and chemoradiation in advanced rectal cancer. Nat Clin Pract Oncol, *4:* 316–321, 2007.

39. Winkler, F., Kozin, S. V., Tong, R. T., Chae, S. S., Booth, M. F., Garkavtsev, I., Xu, L., Hicklin, D. J., Fukumura, D., di Tomaso, E., Munn, L. L., and Jain, R. K. Kinetics of vascular normalization by VEGFR2 blockade governs brain tumor response to radiation: role of oxygenation, angiopoietin-1, and matrix metalloproteinases. Cancer Cell, *6:* 553–563, 2004.

40. Yuan, F., Chen, Y., Dellian, M., Safabakhsh, N., Ferrara, N., and Jain, R. K. Time-dependent vascular regression and permeability changes in established human tumor xenografts induced by an anti-vascular endothelial growth factor/vascular permeability factor antibody. Proc Natl Acad Sci USA, *93:* 14765–14770, 1996.

Chapter 25
Targeted Drug Delivery to the Tumor Neovasculature

Grietje Molema

Keywords: drug delivery, target epitopes, carrier systems

Abstract: The angiogenic process that accompanies solid tumor growth provides an excellent opportunity for the development of targeted drug delivery to selectively block the blood supply of a tumor. Several tumor endothelial associated molecules, preferentially expressed during the angiogenic phase, have been put forward as target epitopes for this purpose, and numerous strategies have now been investigated for therapeutic potential. This chapter provides a concise, though not complete, overview on delivery devices, target epitopes and targeting ligands, and pharmacological effector moieties studied for this purpose. As the final aim is to develop targeted drug delivery systems for the patient, the last part of the chapter focuses on issues that (still) need to be addressed to make successful clinical application a realistic aim.

Introduction

Targeted drug delivery aims to increase the efficacy, and at the same time reduce the toxicity, of drugs in the body. In addition, incorporation of unstable drugs in delivery systems can protect them from being degraded by the biological environment. The term 'drug' in this respect is defined as a molecule that exerts a pharmacological activity, and includes new chemical and new molecular entities, peptides and proteins, small interfering RNA, plasmids encoding therapeutic genes, and toxins. Upon formulating these drugs in a drug delivery system, the behavior of the carrier will largely determine the pharmacokinetics and cellular distribution of the drug. Selective delivery into target cells may facilitate the build-up of a high drug concentration at or in the target cells, or even in specific compartments of the target cells. As a result, drug efficacy can be enhanced, while toxic side-effects are abolished [1,2].

For cancer treatment, initial efforts endeavoured the development of drug delivery systems to selectively and actively destroy tumor cells. While being successful in the treatment of some hematological malignancies, established solid tumors are much less responsive. Among others, high interstitial fluid pressure in solid tumors significantly interferes with convection-based transport of macromolecular systems in tumor tissue. Moreover, the vascular wall of the tumor capillaries turns out to be a major barrier for adequate access of the blood-borne macromolecular delivery devices to the tumor cells [3,4]. The recognition that the tumor vasculature is 'different' from the vasculature elsewhere in the body, and the subsequent identification of molecular determinants that could be used to create tumor vascular specificity, ignited a new research field aimed to develop therapeutics that selectively deliver drugs at or into the tumor endothelial cells (ECs). Besides being more accessible than tumor cells, tumor ECs form one of the cornerstones of solid tumor growth. Hence, interfering with the tumor's blood supply is considered as a therapeutic approach that 'hits the tumor where it hurts' [5].

This chapter will introduce some basic principles on the development of vascular targeted drug delivery strategies for the treatment of solid tumors that have evolved in the last decades. Carrier systems, target epitopes on (angiogenic) tumor vascular endothelial cells, and pharmacological effector moieties will be summarized. It highlights a selection of key approaches yielding strong anti-tumor efficacy in preclinical tumor models, and discusses some important issues for further development of this new class of targeted therapeutics toward clinical application.

Passive Drug Delivery to Solid Tumors

Drug delivery systems for the treatment of solid tumors consist of a carrier molecule or vehicle, an entity with pharmacological activity, and where appropriate separate modalities that create

Departments of Pathology and Medical Biology, Medical Biology Section, Laboratory for Endothelial Biomedicine and Vascular Drug Targeting Research, University Medical Center Groningen (UMCG), University of Groningen, Groningen, The Netherlands

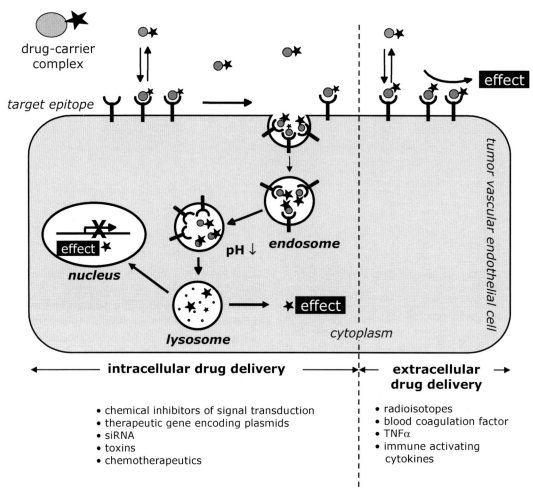

FIG. 25.1. Simplified scheme of cellular handling of targeted tumor endothelial specific drug delivery systems, bringing their pharmacologically active effector moiety either inside the cell (*left*), or at the outer membrane only (*right*). In the case of intracellular drug delivery, binding of the (homing ligand modified) drug-carrier complex to its target epitope is followed by receptor mediated endocytosis and intracellular trafficking via endosomal vesicles to the lysosomes. In the lysosomes, or en route to the lysosomes, the drug-carrier complexes are degraded, thereby releasing the drug. In the case of extracellular delivery, binding of the drug-carrier complex to its target epitope can be followed by direct pharmacological action of the drug.

cell specificity and facilitate drug release at or in the target cells. In drug delivery, two strategies can be distinguished that aim at controlled delivery and release of drugs in the body. In the case of *passive drug delivery*, the drug-carrier complex does not have to be equipped with a targeting ligand as it has an intrinsic quality of being recognized by cells of the monocyte–phagocytic system located in liver, spleen, and/or in the tumor tissue. Cellular uptake of the complex is followed by gradual degradation and (slow) release of the incorporated drugs from the cells into the systemic circulation or tumor microenvironment. Also, extracellular release can be achieved, as discussed below.

Active, or *targeted drug delivery* on the other hand, aims to selectively deliver a drug at or into a target cell by means of target cell specific recognition. In the case of targeted drug delivery to tumor neovasculature, the behavior of the molecular target upon binding will determine whether the complex remains at the outside of the endothelial cell or enters an internalization route to deliver the active component inside the cell (Fig. 25.1).

In solid tumors, the neovascularization process and concomitant production of growth factors, cytokines and chemokines temporarily interfere with vascular integrity, as a consequence of which an increase in vascular permeability arises. Both human tumors and preclinical tumor models present with deviant permeability, the extent being dependent on the tumor type and the site of tumor growth [6–8]. Enhanced permeability of the tumor vasculature facilitates local retention of macromolecular drug delivery systems, a process referred to as 'enhanced permeability and retention' (EPR) [9]. Many passive drug delivery approaches employ this enhanced permeability-facilitated accumulation of macromolecular systems, not only for therapy of cancer but also for therapy of chronic inflammatory diseases which present with a similar feature of increased vascular permeability [10].

Molecular weight and charge are two important physicochemical characteristics of a drug delivery device that determine its fate in the body, its pharmacokinetic behavior, and its passive accumulation in tumor tissue. Small peptides and antibody fragments often have a short plasma half-life due to

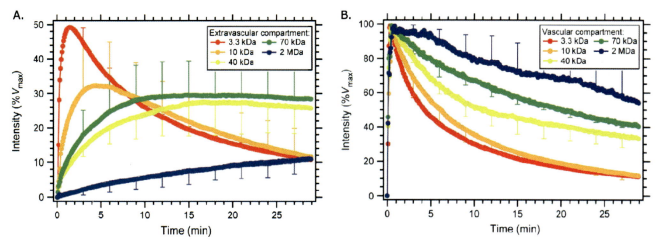

FIG. 25.2. Extravascular tumor accumulation (**A**) and intravascular pharmacokinetics (**B**) of dextrans with various molecular weights, exemplifying the behaviour of a non-targeted carrier system for use in passive drug delivery. 0.5 mg fluorescein labelled anionic dextran was intravenously injected into BALB/c *nu/nu* mice equipped with a dorsal skinfold window chamber in which FaDu human squamous cell carcinoma cells were grown. At the time of the experiment, tumors measured 2–3 mm in diameter. Fluorescein intensity was normalized to the maximum intensity V_{max} as described in the original publication. The data show that the rate of accumulation of dextrans in the extravascular compartment was most rapid for dextrans with low molecular weights, but that these detrans were also rapidly cleared from this compartment, mirroring the plasma pharmacokinetics. Despite the lower apparent permeability and rate of extravasation for dextrans with higher molecular weights, tumor accumulation of these dextrans was more durable and sustained. (From Dreher et al., Tumor vascular permeability, accumulation, and penetration of macromolecular drug carriers, J. Natl. Cancer Inst., 2006; 98(5): 335–344. Reprinted by permission of the authors and Oxford University Press).

rapid (renal) clearance, yet their penetration into the tumor tissue is superior to that of larger molecules. In an elegant study, Dreher and colleagues recently showed in a dorsal skinfold human squamous cell carcinoma xenograft model that upon systemic administration, small, 3.3- and 10-kDa dextrans deeply, rapidly (Fig. 25.2A), and homogeneously penetrated into tumor tissue. Dextrans of higher molecular weight, i.e., 40 and 70kDa, however, had a longer plasma half-life than their low molecular weight counterparts (Fig. 25.2B). As a consequence, the time available for extravasation and tumor accumulation was increased, yet they located more closely to the vascular surface compared to the smaller dextrans. Plasma half-life of 2,000-kDa dextran was longest, but tumor vascular permeability limited its accumulation. For these macromolecular devices, the optimal molecular weight for superior tumor accumulation was therefore assigned to be between 40 and 70kDa [11]. Since overall charge also affects tissue distribution and clearance [12], different carrier systems likely have their own combination of physicochemical characteristics to facilitate optimal access to the tumor.

Upon extravasation from the systemic circulation into the tumor tissue, the devices designed for passive drug delivery can be extracellularly modified or ingested by cells to release the drug into the tumor environment. To facilitate extracellular local release, prodrug strategies have been developed in which the linker between the drug and the carrier system is cleaved by enzymes that are overexpressed by tumor or tumor stroma-associated cells. An example is the D-Ala-Phe-Lys tripeptide linker between doxorubicin as effector molecule and cyclic Arg-Gly-Asp (RGD) containing peptide as an αvβ3 integrin specific homing ligand. Upon binding to the integrin on tumor and/or

tumor endothelial cells, tumor-produced plasmin enzymatically freed the doxorubicin in the vicinity of the tumor cells [13]. More recently, succinylated-Ala-Ala-Asn-Leu-doxorubicin prodrug was reported to have a prolonged plasma half-life and improved tumor accumulation characteristics, while being devoid of cardiac toxicity and myelosuppression. In a number of mouse tumor models, it showed superior anti-tumor activity by local catalytic conversion of the prodrug to doxorubicin by legumain, an asparaginyl endopeptidase overexpressed by tumor and tumor stromal cells including the endothelium [14].

Corticosteroids have been investigated extensively for use in cancer therapy after initial reports on their angiostatic effects in the 1980s. We recently showed that incorporation of prednisolone-phosphate in long circulating liposomes evoked 80–90% inhibition of s.c. B16.F10 melanoma and C26 colon carcinoma growth in mice at a single or weekly dose of 20mg/kg [15]. In contrast, free prednisolone-phosphate administered at even higher doses did not exert any effect. The liposomal formulation mainly localized around the tumor vasculature, and was predominantly present in endosomal compartments of tumor-associated macrophages. Tumor histology of liposomal corticosteroid treated mice revealed an appearance remarkably similar to that of dormant tumors at the start of tumor outgrowth. A number of possible scenarios may explain the strong anti-tumor effects, including inhibitory effects of the released drug on macrophage activity, as a consequence of which tumor neovascularization is halted [16]. Another scenario depicts a gradual degradation of the liposomes by the tumor-associated macrophages to release the drug, thereby exposing the tumor endothelium to the drug over a prolonged period of time. Such a long-term drug exposure profile would

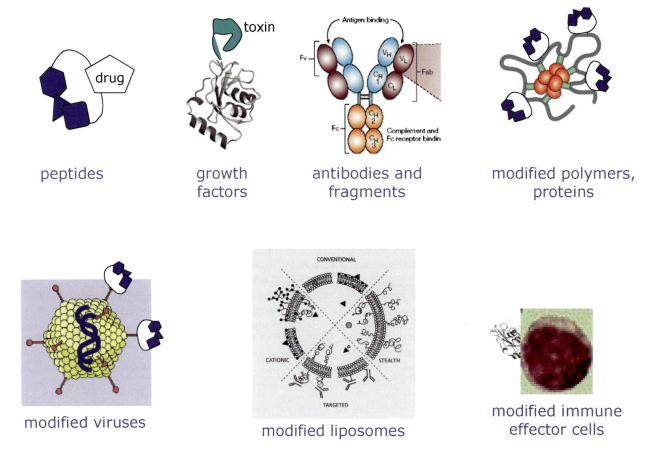

FIG. 25.3. The most frequently used carrier systems for tumor vasculature targeted drug delivery. For those carriers devoid of intrinsic cell specificity, tripeptides, antibodies and antibody fragment, and growth factors can be employed as homing ligands.

resemble exposure as created by metronomic dosing of anti-cancer drugs [17]. Superior effects of some antiangiogenic drugs when administered in a metronomic dosing schedule imply that passive delivery of these drugs by e.g., liposomal formulations are an attractive approach to increase efficacy while preventing systemic side effects.

Targeted Drug Delivery to the Tumor Neovasculature

Vasculature targeted drug delivery systems contain a core carrier molecule, which either contains an intrinsic specificity for target molecules on the tumor endothelial cells or can be harnessed with homing ligands that endow the system with the desired specificity. Similar to passive drug delivery, pharmacological entities, and where appropriate, modalities that facilitate drug release at or in the target cells are integral parts of the systems. All tumor-associated stromal cells, including ECs, pericytes, fibroblasts, and infiltrated immune cells, can be envisioned as targets for cell selective therapy [18]. Most studies on non-tumor cell-directed drug delivery strategies, however, specifically addressed the angiogenic ECs, which will be the focus of the remainder of this chapter.

Decades of research in the area of drug delivery yielded major progress in carrier development which now serves the development of tumor vasculature directed drug delivery strategies. The selection of appropriate homing ligands and pharmacological entities, on the other hand, is still depending on the ongoing generation of new knowledge regarding the cellular and molecular processes underlying tumor endothelial behavior during neovascularization.

Carrier Systems for Targeted Drug Delivery

Carrier systems for drug delivery are divided into particle type, soluble, and cellular carriers. Particle type carriers include liposomes, lipid particles, microspheres, nanoparticles, polymeric micelles, and (modified) viruses [19–22]. Soluble carriers comprise monoclonal antibodies and fragments thereof, (modified) proteins, peptides, polysaccharides, and biodegradable carriers consisting of polymers of various chemical composition [23–26]. Immune cells and stem cells have also been employed as vehicles to specifically deliver either intrinsic or engineered effector activities to the target tissue [27–29]. Figure 25.3 provides an overview of the carrier systems most frequently studied as part of tumor vasculature directed drug targeting strategies.

Target Epitopes on the Tumor Neovasculature

An important determinant for the therapeutic success of targeted drug delivery therapeutics is the selectivity of the delivery system for the target cells. This parameter is determined by selectivity of target molecule expression by the tumor ECs combined with selectivity of the homing ligand in the drug targeting construct. In theory, any protein expressed on the (apical) membrane of the tumor ECs can serve as a target epitope provided it is absent from, or expressed to a limited extent by, other easily accessible cells in the body.

The first molecular targets on tumor ECs were identified while studying endothelial cell behavior in angiogenesis. Later on, the search for targets moved toward more direct investigations regarding the differential molecular make-up of the angiogenic endothelium compared to quiescent cells. For this purpose, methods like in vivo phage display [30], serial analysis of gene expression (SAGE) [31], and microarray-based gene expression profiling of freshly isolated endothelial cells [32] were employed. More recently, Schnitzer and colleagues described an adapted phage library screening method to select vasculature targeting antibodies by using luminal EC plasma membranes isolated from tissue as a source of proteins exposed to the bloodstream [33]. Table 25.1 summarizes putative target epitopes on tumor ECs and tumor vasculature associated structures reported in the last two decades. When routine clinical procedure includes local tumor radiation therapy, radiation-induced neoantigens including P-and E-selectin may also be considered suitable target epitopes for targeted therapeutics [34].

TABLE 25.1. Summary of epitopes expressed by angiogenic endothelial cells and basement membrane components produced during angiogenesis that in theory can serve as targets for tumor vasculature selective drug targeting strategies. A selection of strategies employing targets in this list is detailed in the text.

Target	Reference
30.5 kD antigen	(35)
CD34	(36)
High molecular weight melanoma-associated antigen / NG2	(37)
Endosialin	(38)
VEGF-VEGFR complex	(39)
selectins	(40;41)
αv integrins	(42)
Endoglin	(43)
Tie-2	(44)
Basement membrane component	(45)
CD44 isoforms and related molecules	(46–48)
Prostate-specific membrane antigen (PSMA)	(49)
EDB domain of fibronectin	(50)
Angiostatin receptor	(51)
MMP-2 / MMP-9	(52)
Galectin-1	(53)
CD13 / Aminopeptidase N	(54)
Endostatin receptor	(55)
TEM 1 / 5 / 8	(31)
ROBO4	(56)
VE-cadherin cryptic epitope	(57)
IGFBP-3	(58)
Annexin A1	(59)
Plexin D1	(60)
Hela type caldesmon	(61)

Abbreviations: *IGFBP* Insulin growth factor binding protein; *MMP* matrix metalloproteinase; *TEM* tumor endothelial marker; *VEGF(R)* Vascular endothelial cell growth factor (receptor)

Homing Ligands for Tumor Vasculature Targeted Drug Delivery

To provide drug-carriers with tumor endothelial cell specificity, homing ligands consisting of antibodies or antibody fragments, such as recombinant single-chain variable fragments (ScFv), and peptides recognizing the above mentioned angiogenic target molecules, have been investigated [25,62]. In addition, aptamers selected by systemic evolution of ligands by exponential enrichment (SELEX) technology that have selectivity for angiogenesis and tumor growth-associated antigens have been reported for this purpose [63,64]. Antibodies and derivatives recognizing antigens that have inter-species sequence homology as well as some peptides have the advantage that they can be studied in different species and tumor models, which facilitates their development toward clinically applicable therapeutics.

A multitude of antibodies has been generated against the tumor vascular targets summarized in Table 25.1. A limited number of them have entered clinical studies by now, to block antigen function, to facilitate tumor and metastasis imaging, or to deliver an effector modality. Among the clinically tested antibodies are affinity matured, humanized antibody against αvβ3 integrin [65], and a recombinant, dimeric antibody fragment

recognizing EDB domain of fibronectin [66]. Clinical studies with antibodies that recognize angiogenesis markers that are also avidly expressed by the tumor cells comprise [186]Re-labeled humanized anti-CD44v6 antibody [67] and anti-prostate-specific membrane antigen PSMA conjugated to [90]Y [68]. For more details and updated information on the status of antibody-based tumor vasculature targeted therapeutics in clinical trials, the reader is referred to www.ClinicalTrials.gov.

Table 25.2 recapitulates peptide sequences recognizing known as well as unidentified tumor vascular targets that have been studied for application in vascular drug targeting strategies. Since internalization is an important feature when intracellular drug delivery is aimed for, the search for cell-penetrating peptides may in the near future yield new valuable homing ligands [62].

Pharmacological Agents for Targeted Delivery

In general, tumor vasculature directed therapies either aim at the specific interference with endothelial intracellular signaling during the neovascularization process, or at instant blockade of tumor blood flow by inducing local blood coagulation or tumor EC death. The therapeutic potential of vasculature targeted cytokines interleukin(IL)-2, IL-12, and Interferon

TABLE 25.2. Peptides specific for angiogenic endothelial cells and examples of their application in tumor neovasculature targeted drug delivery strategies.

Peptide sequence	Recognized target	Drug delivery application	Ref
RGD	αv integrins	RGD-doxorubicin	(69;70)
		Rec. RGD-truncated Tissue Factor	(71)
		RGD-PEG-liposomal doxorubicin	(72)
		RGD-PEG-polyethyleneimine siRNA	(73)
		Rec. RGD-TNFα	(74)
		RGD-anti-CD3 antibody	(28)
		RGD-HSA-PTK787	(75)
		RGD-based imaging	(76;77)
NGR	CD13	NGR-doxorubicin	(54;69)
		NGR-$_D$(KLAKLAK)$_2$	(78)
		NGR-TNFα	(79)
		NGR-PEG-liposomal doxorubicin	(80)
		NGR-IFNγ	(81)
		NGR-endostatin	(82)
ATWLPPR	VEGF-R	Naked peptide	(83)
		ATWLPPR-PVA-Verteporfin	(84)
APRPG	n.k.	APRPG-liposomal adriamycin	(85)
		APRPG-liposomal CNDAC	(86)
SMSIARL	n.k.	Peptide –$_D$(KLAKLAK)$_2$	(87)
TAASGVRSMH and LTLRWVGLMS	NG2	Naked peptide-phage	(88)
CDSDSDITWDQLWDLMK	E-selectin	Peptide-Cy5.5-labeled nanoparticles	(89;90)
GPLPLR	MT1-MMP	GPLPLR-liposomal CNDAC	(91)

Abbreviations/acronyms: *CNDAC* anti-tumor pentofuranosylcytosine derivative; $_D$*(KLAKLAK)$_2$*: pro-apoptotic, 14-amino acid synthetic peptide; *MT1-MMP* Membrane type-1 matrix metalloproteinase; *NG2* rat proteoglycan / human melanoma proteoglycan / high molecular weight melanoma-associated antigen ; *n.k.* not known; *PEG* polyethyleneglycol; *PVA* polyvinylalcohol; *Rec* recombinant

(IFN)-γ to enhance cellular immunity against tumors has also been investigated [81,92–94], but will not be discussed here.

Once selectivity of expression of target epitopes has been established, knowledge regarding cellular handling of the targets upon ligand binding and the choice of effector molecules to be delivered are intricately associated. In the case of using e.g., bacterial toxins, toxic drugs, plasmids encoding therapeutic proteins, or siRNA, intracellular delivery is a prerequisite, as their effects are exerted in the cells' interior. The target epitopes therefore need to be internalized after binding of the drug targeting construct. In contrast, for blockade of tumor blood flow by selective delivery of e.g., a blood coagulation inducing protein, or by cytotoxic T lymphocyte mediated killing of the endothelial cells, delivery of the effector at the outer membrane of the tumor endothelium suffices.

Inhibitors of Angiogenesis Associated Signal Transduction

The class of chemical receptor tyrosine kinase (RTK) inhibitors, including inhibitors of vascular endothelial growth factor (VEGF), platelet-derived growth factor (PDGF), and fibroblast growth factor-2 (FGF2) receptor signaling, act immediately downstream of receptor dimerization and target the kinase ATP sites. As a consequence, endothelial intracellular signal transduction is halted, which often leads to endothelial apoptosis (95). SU5416 and SU6668 represent members of the first generation of inhibitors that were shown to actively affect

VEGF and VEGF/FGF2/PDGF-induced signal transduction, respectively [96,97]. Other kinase inhibitors reported to be highly potent in inhibiting angiogenesis include SU11248, ZD6474, and the ZD6474 analogue PTK787 [98]. More general drugs, including phosphoinositol 3-kinase (PI3K) inhibitors wortmannin and LY294002, MEK inhibitor PD98059, cyclooxygenase (COX)-2 and mitogen activated protein (MAP) kinase inhibitors, the immunosuppressive drug rapamycin, and histone deacteylase inhibitors [99–104], can also efficiently restrain endothelial behavior during tumor neovascularization, and are therefore good candidates for incorporation in targeted drug delivery systems. Since rapamycin could inhibit VEGF-driven angiogenesis at relatively low doses [100], targeted drug delivery could be instrumental in dissociating the desired antiangiogenic from the more unfavorable immunosuppressive effects.

Although specificity for angiogenesis-associated receptor signaling implies that a number of these drugs are devoid of systemic toxicity, recent clinical studies suggest that for these RTK inhibitors, systemic toxicity will be a limiting factor when administered in free form [97,105]. Frequent absence of objective clinical responses upon systemic therapy reported so far also indicate that drug access into the target cells is limited. These observations definitely justify exploring incorporation of these drugs into targeted drug delivery formulations. Until now, however, the small chemical entities have not been extensively studied as effector moieties integrated in these systems. Strict delivery into the tumor endothelial cells may negatively

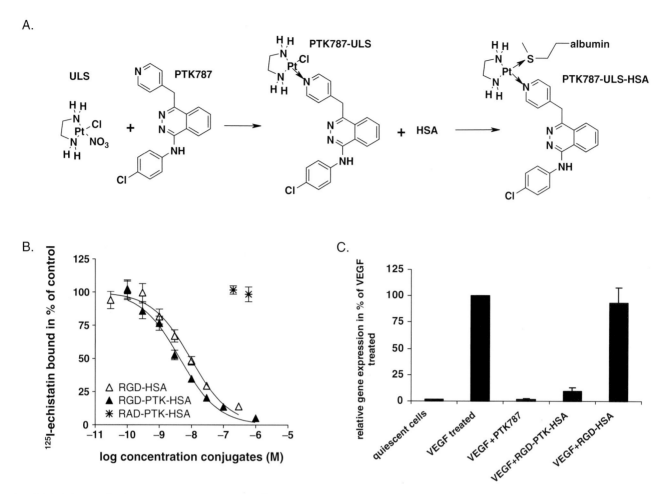

Fig. 25.4. Scheme of universal linkage system (ULS™) based conjugation of the VEGF-R kinase inhibitor PTK787 to the proteinaceous RGD-albumin carrier (**A**), its αvβ3 binding affinity as determined by competition of ^{125}I-echistatin pre-bound to human umbilical vein endothelial cells (**B**), and the in vitro pharmacological activity of the PTK-RGD-HSA conjugate exemplified by inhibitory effect on EGR3 gene expression (**C**). Adapted from [75].

influence their efficacy, as some of the drugs do not derive their potency only from interfering with the tumor vasculature, but also from affecting tumor and stromal cell behavior. In addition, the physicochemical characteristics of these small chemical drugs often do not favor their incorporation in drug delivery systems. For example, incorporation in polymeric or proteinaceous carriers requires that the drugs have chemical handle bars for covalent conjugation without interfering with drug activity upon local release. In a multidisciplinary approach combining (pharmaco)chemistry, pharmacology, and drug delivery expertise, we studied the applicability of the so-called Platinum-based universal linkage system (ULS™; Pt(cis-ethylenediamine)nitrate chloride) to protein conjugation of chemicals that lack reactive groups required for conventional chemical conjugation. Using the ULS™ system, we could covalently couple the VEGFR kinase inhibitor PTK787 to RGD-peptide modified albumins (Fig. 25.4). The conjugates exerted high binding affinity and specificity for αvβ3 integrin via the RGD-peptides, and internalization

by the proliferating ECs resulted in significant inhibition of VEGF-induced expression of zinc finger transcription factor EGR3, and nuclear receptors NR4A1 and NR4A3 [75]. In a similar approach, the p38 MAPK inhibitor SB202190 could be covalently incorporated in a protein carrier to exert pharmacological activity upon internalization and intracellular release [106].

DNAs encoding inhibitors of angiogenic signaling pathways that are instrumental in new blood vessel formation are chemically unstable, have inferior pharmacokinetics upon systemic administration, and are unable to cross cellular membranes. These features rationalize their formulation into targeted drug delivery systems. Examples explored include plasmids encoding mutant Raf gene *ATP$^\mu$-Raf*, and dominant negative NFκB, that were both delivered into angiogenic endothelial cells by RGD-αvβ3 integrin recognition using nanoparticles and chemically modified adenoviruses, respectively [107,108]. Also, genes encoding the antiangiogenic proteins platelet factor 4, thrombospondin, and Tie2 receptor were studied for their antiangiogenic effects, although specific delivery into the tumor endothelium was not pursued in this study [109].

Small interfering RNAs (siRNAs) blocking the expression of angiogenic proteins encoded by the tumor endothelium need to be formulated to deliver them functionally active into the cytoplasm of the target cells. By complexing negatively charged siRNA specific for VEGFR2 with positively charged PEGylated polyethyleneimine (PEI) equipped with RGD-peptides, we could alter whole body distribution of the siRNA. While, after i.v. administration, the majority of PEI nanoplexes distributed to tumor, lungs, liver, and spleen, RGD-targeted PEGylated PEI nanoplexes showed preferential redistribution to the tumor. The local VEGFR2 knock-down by the delivered siRNA caused strong inhibition of N2A neuroblastoma outgrowth [73]. Using PCR technology, McNamara and colleagues recently showed that chimeric RNAs consisting of SELEX-derived aptameric RNAs specific for a prostate associated antigen linked to pharmacologically active siRNA were effective in inhibiting tumor outgrowth when locally injected into the tumor [64]. The relatively small size of these aptameric siRNA chimeras (<15 kDa) would rationalize additional modifications by e.g., PEGylation or multimerization to increase plasma half-life upon systemic administration.

Instant Blockade of Tumor Blood Flow

Massive tumor load debulking can be achieved by instant blockade of tumor blood flow through tumor endothelial cell killing or local blood coagulation induction. One of the first tumor infarction approaches showing its potential consisted of an immunotoxin conjugate of ricin A chain and an antibody specifically recognizing a tumor endothelial epitope. It caused complete vascular thrombosis in the tumor, as a consequence of which extensive tumor cell necrosis and dramatic tumor regression occurred [110]. Other tumor vascular toxins reported afterwards include VEGF and VEGF isoforms as homing ligands chemically or recombinantly linked to diphteria toxin moieties [111,112].

Other components that actively induce tumor (endothelial) cell death are radioisotopes and pro-apoptotic peptides. Radioisotope ^{90}Y was covalently linked to an antibody against prostate membrane-specific antigen (PMSA) [68], and as effector moiety incorporated in $\alpha v \beta 3$ and VEGFR2 specific nanoparticles [113]. The apoptosis inducing peptide $_D$(KLAKLAK)$_2$ gained tumor endothelial binding preference and cell specific cytotoxicity upon modification with tumor vascular homing peptides [78,87].

Ever since its discovery, tumor necrosis factor (TNFα) is considered one of the most potent inducers of cell death. Its relatively small therapeutic window has limited its clinical application to local administration, e.g., in hyperthermic isolated limb perfusion protocols. Recombinant conjugation of the CD13-recognizing peptide CNGRCG to murine and human TNFα endowed the cytokine with specificity for tumor vasculature with concomitant 15- to 30-fold increased efficacy in decreasing tumor burden [79]. Similarly, RGD-TNFα exhibited strong in vitro endothelial death signaling and in vivo anti-tumor effects, the latter being even more pronounced when combined with the chemotherapeutic drug melphalan [74]. As in theory the cytokine-fused peptides can serve as a hapten to induce an immunological response, the recently reported low immunogenicity of the NGR motif favors development of these NGR-based tumor vasculature targeted therapeutics for clinical application [114]. At present, a phase I clinical study of NGR-TNFα is recruiting patients with advanced solid tumors of different origin (see www.ClinicalTrials.gov).

Also, the therapeutic window of conventional chemotherapeutics can be improved when incorporating the drugs into delivery systems. Doxorubicin has been investigated in depth for this purpose (see Table 25.2 for selected references) as the drug can be easily conjugated to protein or polymeric backbones, and quite efficiently incorporated into liposomal formulations. A novel pro-drug approach was recently reported in which the tumor selectivity of endostatin was combined with cytosine deaminase enzymatic activity to convert 5-fluorocytosine into 5-fluorouracil at the site of the tumor. Besides an endostatin related inhibitory effect on tumor endothelial cells, local production of the cytotoxic drug resulted in increased tumor cell apoptosis [115].

Studies investigating immunological approaches to selectively attack existing tumor neovasculature have also generated new therapeutics with high potency. In an elegant study by Niederman and colleagues, cytotoxic T lymphocytes were transduced with a chimeric T cell receptor consisting of VEGF sequences linked to the T cell receptor ζ chain. Multiple adoptive transfer of these cells into CT26, B16.F10, or LS174T tumor-bearing mice strongly inhibited tumor outgrowth [27]. A cytotoxic response toward proliferating endothelial cells could also be generated using a chemical strategy to produce bifunctional, RGD-peptide modified anti-CD3 monoclonal antibodies. The bifunctional antibody enabled cytotoxic T lymphocytes to cross link to the endothelium, resulting in major histocompatibility complex (MHC) independent cell killing [28].

The most effective approach to block blood flow in established tumors reported to date is the instantaneous local induction of blood coagulation by vascular delivery of a blood coagulation factor, Tissue Factor (TF). Bringing the truncated form of TF in close proximity to the endothelial cells in the tumor blood vessels resulted in substantial tumor mass reduction [116]. This first study showed the therapeutic potential of acute blockade of tumor blood supply. It was followed by a number of other reports that used a similar approach of tumor vasculature selective delivery of TF aimed at different target epitopes. Comparing these studies provides some important directions for further research, as will be discussed below.

Challenges for Further Research

Tumor vasculature targeted drug delivery strategies have been instrumental in demonstrating the therapeutic potential of attacking the tumor blood supply as a means to inhibit solid tumor outgrowth and to debulk established solid tumor masses. All strategies have been investigated in preclinical animal tumor models so far, and in the coming years the step towards the clinic has to be made. Important issues that need to be addressed to make this step a successful one include tumor vascular heterogeneity and the dynamics of tumor endothelial cell behavior under therapeutic pressure. Furthermore, we will have to put effort into devising a straightforward method to produce delivery systems with suitable shelf-lives to enable application in daily clinical care. Lastly, the design of rational treatment schedules for effective combination therapies aimed at multiple vascular targets as well as at different cells types within a tumor mass needs attention. Combining research focusing on these issues with the development of tools that enable molecular imaging of tumor angiogenesis status and identification of biomarkers representative of the pharmacological effects aimed for, will make successful clinical application a realistic goal.

Tumor Endothelial Heterogeneity

From the targeted Tissue Factor studies aimed at different molecular targets on the tumor endothelial cells some important lessons can be learned. In the first study reported on this tumor infarction strategy, we delivered TF to an artificially induced tumor endothelial target epitope, MHC class II, resulting in ~40% cure and long-lasting disease-free survival [116]. This study was followed by targeted delivery of TF to Hodgkin's tumor induced endothelial VCAM-1 [117]. Absence of cure was likely due to the fact that the target epitope expression levels as well as the number of tumor vessels positive for the target epitope were less than in the MHC class II model. Rapid tumor regrowth was also reported upon targeting TF to the EDB domain of fibronectin using a truncated TF-anti-EDB ScFv fusion protein [118], and although superior anti-tumor effects were observed when delivering truncated TF to a combination of target epitopes, neither cure nor disease-free survival were achieved [71].

Although not experimentally proven, these data, as well as the data from the other tumor vasculature targeting studies that did not show factual regression of extensive tumor load, imply that if one cannot block the large majority of tumor blood vessels within a short period of time, a clinically relevant effect is not likely to occur. Moreover, the extent of endogenously produced tumor endothelial target epitopes is probably too limited to facilitate sufficient extra- or intracellular delivery of pharmacological active compound to considerably affect tumor endothelial behavior or tumor blood flow. Strategies aimed at delivering drugs via multiple targets to affect all blood vessels in established tumors should therefore be investigated. Research efforts should focus on identifying the molecular meaning of tumor vascular heterogeneity in established (human) tumors instead of extending single target studies in angiogenesis-synchronized vasculature of quickly growing animal tumors. Based on the molecular signatures underlying tumor endothelial heterogeneity, a rational combination of target epitopes can be chosen that facilitate therapeutic intervention in all tumor blood vessels, or in a selection of blood vessels which can be discriminated from healthy vasculature. Systematic studies on the extent and level of expression of the angiogenesis-associated epitopes summarized in Table 25.1 in human and established mouse tumors are lacking, yet will be an essential basis for creating the conditions for targeted systems to become mainstay therapeutics for cancer therapy.

While in the late 1980s and early 1990s tumor angiogenesis was conceptually considered as a process with common features, we now know that tumor type, tumor growth stage, local microenvironment in the tumor, location of the tumor within the body, and the immunological status of the individual all influence the molecular and cellular fine tuning of tumor neovascularization, and hence tumor vascular responsiveness to therapeutic intervention [119–121]. Since endothelial cells quickly lose their environmentally controlled behavior when enzymatically released, the molecular meaning of tumor microvascular endothelial heterogeneity should be studied in an in vivo context. Our laboratory therefore developed an experimental protocol in which tumor ECs are microdissected from frozen tissue biopsies and their RNA subjected to gene expression analysis by quantitative reverse transcriptase (RT)-PCR. This protocol now enables us to investigate the molecular differences in behavior between morphologically or antigenically identified subsets of tumor ECs in vivo in their pathophysiological environment, and the local effects of (targeted) drugs on microvascular endothelial cell behavior [122,123]. The protocol was designed in such a way, that it can be applied to small biopsies of animal and human tumors alike. A major technological challenge will be to combine this in vivo endothelial gene expression profiling with analysis of complex kinase activities and proteomics (Fig. 25.5). Studies like these will not only provide a rational basis for multi-target strategies for targeted drug delivery, but also have the potential to provide detailed knowledge on pharmacological targets present in endothelial subsets. By combining the protocol of microdissection with in vivo phage display [124], new opportunities are created to select for endothelial subset specific targeting ligands.

Production of Drug Delivery Systems

Straightforward, reproducible production of targeted drug delivery systems which can be scaled up to yield sufficient material with homogenous chemical make-up and well-defined physicochemical characteristics is essential for taking these therapeutics to clinical development. Chemical conjugation of effector molecules to drug carriers often

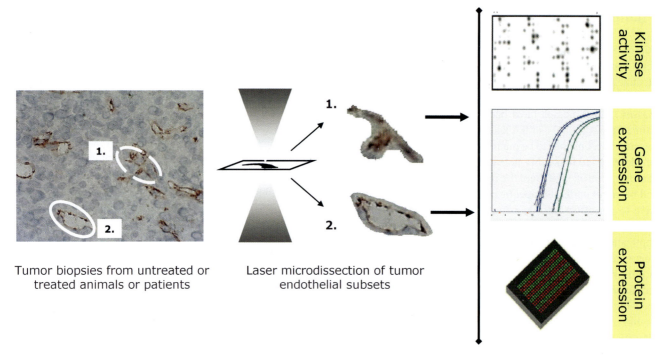

FIG. 25.5. To generate a molecular signature that underlies tumor endothelial heterogeneity, the microvascular endothelium needs to be investigated in its pathophysiological microenvironment. Laser microdissection of selected subsets of tumor endothelial followed by kinase activity arrays, and gene and protein expression analyses can provide an integrated view on their molecular behaviour. Proposed targets for targeted drug delivery as well as new pharmacological targets can be investigated using this experimental strategy. Furthermore, (lack of) effects of (targeted) therapeutics on selected vascular compartments within a tumor mass can be studied in greater detail.

results in heterogenous products, with low product yield at every synthesis step. Production of fusion molecules by recombinant DNA technology has in this respect a more favorable outcome. For vesicular delivery devices, drug incorporation efficacy and stability of incorporation are important features which are often not compatible with the physicochemical qualities of new chemical entities selected in the drug discovery process. The latter compounds are often chosen based on their pharmacological effects in cell systems, i.e., on their capacity to cross cell membranes and to interact with the molecular docking sites in target enzymes. The more lipophilic nature of the compounds often prohibits inclusion of the drugs into the hydrophilic compartment of vesicle-like carriers, while physical capture in the core ensures prevention of unwanted in vivo drug exchange with biological membranes prior to target cell specific delivery. Prodrug-based approaches in which new chemical entities are modified to produce more hydrophilic derivatives after lysosomal release in target cells are valuable advances for which collaboration between academia and pharmaceutical industry is vital. Incorporating such formulation strategies in the drug development process may provide highly potent chemical entities that exert intrinsic toxicity with a future as therapeutic modalities for the clinic.

Combination Therapies

Cancer is a complex molecular and cellular disease. It is highly unlikely that therapeutic strategies aimed at one molecular target or at one cellular compartment will lead to total eradication of the tumor mass. Therapeutic options combining tumor vasculature targeted devices with antiangiogenic or chemotherapeutic drugs as well as with immunotherapeutic approaches are considered feasible strategies.

Combining Tumor Vasculature Targeted Drug Delivery Therapeutics with Antiangiogenic Drugs

The majority of combination therapy studies reported to date combined antiangiogenic or vascular disrupting agents that exhibit a direct vascular occlusion effect with chemotherapeutics, as recently extensively reviewed [125]. Combining antiangiogenic drugs with tumor vasculature targeted drug delivery systems is investigated to a lesser extent, but may provide good opportunities to attack the tumor at different vascular compartments. The fact that antiangiogenic therapies can normalize the abnormal structure and function of tumor vasculature may moreover partly overcome the problem of tumor vascular heterogeneity. During the normalization process, e.g., induced by VEGFR2 blockade, immature vessels are pruned while the remaining vasculature show improved integrity and function [126] (see Chapter 24 on normalization of the tumor vasculature for a detailed description of the concept). Recently, it was shown that treatment with VEGF/PDGF receptor tyrosine kinase inhibitor AG-013736 resulted in a significantly diminished $\alpha v \beta 3$ integrin expression on pancreatic islet tumor blood vessels in the RIP-tag2 transgenic mice [127]. Since the surviving blood vessels remain under the influence of tumor produced factors, they likely still express molecular markers that discriminate them from normal vasculature. However, as they are in a more synchronized stage of (angiogenic) activation, they may exert a more homogenous expression of epitopes to be used for targeted drug delivery. Detailed analysis of the molecular make up of normalized vessels remains to be elucidated.

Treatment with antiangiogenic drugs can, on the other hand, shift the vascularization strategy of a solid tumor from sprouting angiogenesis toward vessel co-option [128]. As with vessel normalization, the consequence of such a shift for the target epitope expression to be used for targeted drug delivery following antiangiogenic therapy needs to be investigated. The dynamics of tumor vascularization in response to treatment were also illustrated in a study with the vascular disrupting agent OXi-4503, a derivative of combretastatin. As a consequence of OXi-4503 treatment, levels of circulating endothelial progenitor cells were induced, which was followed by increased progenitor cell association with the tumor vasculature and increased blood flow. Combining the vascular disrupting agent with the antiangiogenic VEGFR2-inhibiting monoclonal antibody DC101 enhanced anti-tumor activity [129]. Tumor neovascularization is noticeably flexible to adapt to changing conditions, a feature that needs to be taken into account when aiming to combine tumor vasculature targeted drug delivery systems with other therapeutics.

Combining Tumor Vasculature Targeted Drug Delivery Therapeutics with Chemotherapeutics

Besides the phase I clinical study on NGR-TNFα monotherapy, a phase IB clinical study on a combination of NGR-TNFα and doxorubicin is at present recruiting patients. The first preclinical data on combining tumor vasculature targeted TNFα with chemotherapeutics showed improved therapeutic outcome when combining the targeted TNFα with either melphalan or doxorubicin [79,130]. At high doses, NGR-TNFα itself inhibited tumor perfusion as a result of vascular damage leading to vessel occlusion. At low, subnanogram doses, synergism between NGR-TNFα and the chemotherapeutics was observed. This effect consisted of targeted TNFα interaction with TNF-R1 on endothelial cells and improved drug penetration into tumor tissue, the latter process possibly being facilitated by TNFα-mediated increase in vascular permeability and decrease in interstitial fluid pressure [130].

The occurrence of synergism between low dose NGR-TNFα and chemotherapy was dependent on dosing schedule, with the most pronounced effects at a 2-h time interval between treatments [131].

Sequential attack of the different functional compartments of a solid tumor was achieved by constructing a nanoparticle consisting of biodegradable poly-(lactic-co-glycolic) acid (PLGA) within an extranuclear PEGylated-lipid envelope. The antiangiogenic drug combretastatin was incorporated in the outer envelope, while doxorubicin was conjugated to the PLGA inner nanoparticle. After EPR-based accumulation in the tumor upon systemic administration, vascular shutdown was created by temporal release of combretastatin, after which doxorubicin release was confined to the tumor compartment [132].

Most of the studies discussed in this chapter were performed in animals with a relatively low tumor burden. In the clinical setting, however, tumor masses are often much more extensive, with heterogeneous neovascularization and tumor cell proliferation stages. It is highly likely that interference with tumor blood flow as a consequence of targeted delivery of drugs to the tumor neovasculature will only affect parts of the tumor. In these areas, vascular shutdown will lead to tumor cell death, which may positively or negatively affect tumor proliferative behavior, inflammatory cell recruitment, and neovascularization responses in close proximity to and at sites distant from the affected areas. Knowing the kinetics and dynamics of these processes will be of pivotal importance to design effective combination treatment schedules with superior performance.

Concluding Remarks

The vasculature of solid tumors and growing metastases is an attractive site for targeted drug delivery therapeutics. A barrier between the systemic circulation that transports the macromolecular devices and the target cells is non-existent; each target cell faces the systemic supply route and hence the therapeutics, and the vasculature is the cornerstone of the pathology of solid tumor growth. In the last decade, seminal studies in animal models have demonstrated the therapeutic potential of targeted drug delivery to the tumor neovasculature. The challenge for the coming decade will be to address a number of crucial issues described in this chapter to make these therapeutics successful in eradicating established tumor masses, and to fill the gaps in knowledge that need to be closed before clinical success becomes reality.

Acknowledgements. Dr. Astrid J Schraa and Dr. Kai Temming are acknowledged for their assistance in preparing Figures 1 and 4, respectively. Elise Langenkamp is acknowledged for critically reading the manuscript.

References

1. Molema G, Meijer DKF. Drug Targeting: Organ-Specific Strategies. Weinheim: Wiley-VCH, 2001.
2. Proost JH. Pharmacokinetic/Pharmacodynamic modelling in drug targeting. In: Molema G, Meijer DKF, editors. Drug Targeting - Organ Specific Strategies. Weinheim: Wiley-VCH, 2001: 333–370.
3. Heldin CH, Rubin K, Pietras K, Ostman A. High interstitial fluid pressure - an obstacle in cancer therapy. Nat Rev Cancer 2004; 4(10):806–813.
4. Jain RK. Transport of molecules across tumor vasculature. Cancer Metastasis Rev 1987; 6(4):559–593.
5. Sunassee K, Vile R. Tumour angiogenesis: hitting cancer where it hurts. Curr Biol 1997; 7:R282–R285.
6. Boucher Y, Salehi H, Witwer B, Harsh GR4, Jain RK. Interstitial fluid pressure in intracranial tumours in patients and in rodents. Br J Cancer 1997; 75(6):829–836.
7. Yuan F, Chen Y, Dellian M, Safabakhsh N, Ferrara N, Jain RK. Time-dependent vascular regression and permeability changes in established human tumor xenografts induced by an anti-vascular endothelial growth factor/vascular permeability factor antibody. Proc Natl Acad Sci USA 1996; 93(25):14765–14770.
8. Monsky WL, Mouta Carreira C, Tsuzuki Y, Gohongi T, Fukumura D, Jain RK. Role of host microenvironment in angiogenesis and microvascular functions in human breast cancer xenografts: mammary fat pad versus cranial tumors. Clin Cancer Res 2002; 8(4):1008–1013.
9. Maeda H, Wu J, Sawa T, Matsumura Y, Hori K. Tumor vascular permeability and the EPR effect in macromolecular therapeutics: a review. J Contr Rel 2000; 65(1–2):271–284.
10. Metselaar JM, Wauben MH, Wagenaar-Hilbers JP, Boerman OC, Storm G. Complete remission of experimental arthritis by joint targeting of glucocorticoids with long-circulating liposomes. Arthritis Rheum 2003; 48(7):2059–2066.
11. Dreher MR, Liu W, Michelich CR, Dewhirst MW, Yuan F, Chilkoti A. Tumor vascular permeability, accumulation, and penetration of macromolecular drug carriers. J Natl Cancer Inst 2006; 98(5):335–344.
12. Tabata Y, Kawai T, Murakami Y, Ikada Y. Electric charge influence of dextran derivatives on their tumor accumulation after intravenous injection. Drug Delivery 1997; 4:213–221.
13. de Groot FMH, Broxterman HJ, Adams HPHM et al. Design, synthesis and biological evaluation of a dual tumor-specific motive containing integrin-targeted plasmin-cleavable doxorubicin prodrug. Molc Cancer Ther 2002; 1:901–911.
14. Wu W, Luo Y, Sun C et al. Targeting cell-impermeable prodrug activation to tumor microenvironment eradicates multiple drug-resistant neoplasms. Cancer Res 2006; 66(2):970–980.
15. Schiffelers RM, Metselaar JM, Janssen AP et al. Liposome-encapsulated prednisolone phosphate inhibits growth of established tumors in mice. Neoplasia 2005; 7(2):118–127.
16. Lin EY, Li JF, Gnatovskiy L et al. Macrophages Regulate the Angiogenic Switch in a Mouse Model of Breast Cancer. Cancer Res 2006; 66(23):11238–11246.
17. Kerbel RS, Kamen BA. The anti-angiogenic basis of metronomic chemotherapy. Nat Rev Cancer 2004; 4(6):423–436.
18. Carmeliet P. Angiogenesis in life, disease and medicine. Nature 2005; 438(7070):932–936.
19. Torchilin VP. Recent advances with liposomes as pharmaceutical carriers. Nat Rev Drug Discov 2005; 4(2):145–160.
20. Torchilin VP, Lukyanov AN, Gao Z, Papahadjopoulos-Sternberg B. Immunomicelles: Targeted pharmaceutical carriers for poorly soluble drugs. Proc Natl Acad Sci USA 2003; 100(10):6039–6044.
21. Moghimi SM, Hunter AC, Murray JC. Nanomedicine: current status and future prospects. FASEB J 2005; 19(3):311–330.
22. Liu Y, Deisseroth A. Tumor vascular targeting therapy with viral vectors. Blood 2006; 107(8):3027–3033.
23. Wu AM, Senter PD. Arming antibodies: prospects and challenges for immunoconjugates. Nat Biotechnol 2005; 23(9):1137–1146.
24. Kok RJ, Ásgeirsdóttir SA, Verweij WR. Development of proteinaceous drug targeting constructs using chemical and recombinant DNA approaches. In: Molema G, Meijer DKF, editors. Drug Targeting - Organ specific strategies. Weinheim: Wiley-VCH, 2001: 275–308.
25. Temming K, Schiffelers RM, Molema G, Kok RJ. RGD-based strategies for selective delivery of therapeutics and imaging agents to the tumour vasculature. Drug Resist Update 2005; 8(6):381–402.
26. Duncan R. Polymer conjugates as anticancer nanomedicines. Nat Rev Cancer 2006; 6(9):688–701.
27. Niederman TM, Ghogawala Z, Carter BS, Tompkins HS, Russell MM, Mulligan RC. Antitumor activity of cytotoxic T lymphocytes engineered to target vascular endothelial growth factor receptors. Proc Natl Acad Sci USA 2002; 99(10):7009–7014.
28. Schraa AJ, Kok RJ, Botter SM et al. RGD-modified anti-CD3 antibodies redirect cytolytic capacity of cytotoxic T lymphocytes toward alphav-beta3 expressing endothelial cells. Int J Cancer 2004; 112(279):285.
29. De Palma M, Venneri MA, Roca C, Naldini L. Targeting exogenous genes to tumor angiogenesis by transplantation of genetically modified hematopoietic stem cells. Nat Med 2003; 9(6):789–795.
30. Rajotte D, Arap W, Hagedorn M, Koivunen E, Pasqualini R, Ruoslahti E. Molecular heterogeneity of the vascular endothelium revealed by in vivo phage display. J Clin Invest 1998; 102(2):430–437.
31. St Croix B, Rago C, Velculescu V et al. Genes expressed in human tumor endothelium. Science 2000; 289(5482):1197–1202.
32. Pai JT, Ruoslahti E. Identification of endothelial genes up-regulated in vivo. Gene 2005; 347(1):21–33.
33. Valadon P, Garnett JD, Testa JE, Bauerle M, Oh P, Schnitzer JE. Screening phage display libraries for organ-specific vascular immunotargeting in vivo. Proc Natl Acad Sci USA 2006; 103(2):407–412.
34. Stacy DR, Lu B, Hallahan DE. Radiation-guided drug delivery systems. Expert Rev Anticancer Ther 2004; 4(2):283–288.
35. Hagemeier HH, Vollmer E, Goerdt S, Schulze Osthoff K, Sorg C. A monoclonal antibody reacting with endothelial cells of budding vessels in tumors and inflammatory tissues, and non-reactive with normal adult tissues. Int J Cancer 1986; 38(4):481–488.
36. Schlingemann RO, Rietveld FJ, De Waal RM et al. Leukocyte antigen CD34 is expressed by a subset of cultured endothelial cells and on endothelial abluminal microprocesses in the tumor stroma. Lab Invest 1990; 62(6):690–696.
37. Schlingemann RO, Rietveld FJ, Kwaspen F, van de Kerkhof PC, De Waal RM, Ruiter DJ. Differential expression of markers for endothelial cells, pericytes, and basal lamina in the micro-

vasculature of tumors and granulation tissue. Am J Pathol 1991; 138(6):1335–1347.

38. Rettig WJ, Garin Chesa P, Healey JH, Su SL, Jaffe EA, Old LJ. Identification of endosialin, a cell surface glycoprotein of vascular endothelial cells in human cancer. Proc Natl Acad Sci USA 1992; 89(22):10832–10836.

39. Brown LF, Berse B, Jackman RW et al. Increased expression of vascular permeability factor (vascular endothelial growth factor) and its receptors in kidney and bladder carcinomas. Am J Pathol 1993; 143(5):1255–1262.

40. Nguyen M, Strubel NA, Bischoff J. A role for sialyl Lewis-X/A glycoconjugates in capillary morphogenesis. Nature 1993; 365(6443):267–269.

41. Hallahan DE, Staba Hogan MJ, Virudachalam S, Kolchinsky A. X-ray-induced P-selectin localization to the lumen of tumor blood vessels. Cancer Res 1998; 58(22):5216–5220.

42. Brooks PC, Montgomery AM, Rosenfeld M et al. Integrin alpha v beta 3 antagonists promote tumor regression by inducing apoptosis of angiogenic blood vessels. Cell 1994; 79(7):1157–1164.

43. Burrows FJ, Tazzari PL, Amlot P et al. Endoglin is an endothelial cell proliferation marker that is selectively expressed in tumor vasculature. Clin Cancer Res 1995; 1:1623–1634.

44. Sato TN, Tozawa Y, Deutsch U et al. Distinct roles of the receptor tyrosine kinases Tie-1 and Tie-2 in blood vessel formation. Nature 1995; 376(6535):70–74.

45. Epstein AL, Khawli LA, Hornick JL, Taylor CR. Identification of a monoclonal antibody, TV-1, directed against the basement membrane of tumor vessels, and its use to enhance the delivery of macromolecules to tumors after conjugation with interleukin 2. Cancer Res 1995; 55(June 15):2673–2680.

46. Henke CA, Roongta U, Mickelson DJ, Knutson JR, McCarthy JB. CD44-related chondroitin sulfate proteoglycan, a cell surface receptor implicated with tumor cell invasion, mediates endothelial cell migration on fibrinogen and invasion into a fibrin matrix. J Clin Invest 1996; 97(11):2541–2552.

47. Griffioen AW, Coenen MJH, Damen CA et al. CD44 is involved in tumor angiogenesis: an activation antigen on human endothelial cells. Blood 1997; 90(3):1150–1159.

48. Forster-Horvath C, Meszaros L, Raso E et al. Expression of CD44v3 protein in human endothelial cells in vitro and in tumoral microvessels in vivo. Microvasc Res 2004; 68(2):110–118.

49. Liu H, Moy P, Kim S et al. Monoclonal antibodies to the extracellular domain of prostate- specific membrane antigen also react with tumor vascular endothelium. Cancer Res 1997; 57(17):3629–3634.

50. Neri D, Carnemolla B, Nissim A et al. Targeting by affinity-matured recombinant antibody fragments of an angiogenesis associated fibronectin isoform. Nat Biotechnol 1997; 15(12):1271–1275.

51. Moser TL, Stack MS, Asplin I et al. Angiostatin binds ATP synthase on the surface of human endothelial cells. Proc Natl Acad Sci USA 1999; 96:2811–2816.

52. Koivunen E, Arap W, Valtanen H et al. Tumor targeting with a selective gelatinase inhibitor. Nat Biotechnol 1999; 17(8):768–774.

53. Clausse N, van den Brule F, Waltregny D, Garnier F, Castronova V. Galectin-1 expression in prostate tumor-associated capillary endothelial cells is increased by prostate carcinoma cells and modulates heterotypic cell-cell adhesion. Angiogenesis 1999; 3:317–325.

54. Pasqualini R, Koivunen E, Kain R et al. Aminopeptidase N is a receptor for tumor-homing peptides and a target for inhibiting angiogenesis. Cancer Res 2000; 60(3):722–727.

55. Karumanchi SA, Jha V, Ramchandran R et al. Cell surface glypicans are low-affinity endostatin receptors. Mol Cell 2001; 7(4):811–822.

56. Huminiecki L, Gorn M, Suchting S, Poulsom R, Bicknell R. Magic roundabout is a new member of the roundabout receptor family that is endothelial specific and expressed at sites of active angiogenesis. Genomics 2002; 79(4):547–552.

57. Corada M, Zanetta L, Orsenigo F et al. A monoclonal antibody to vascular endothelial-cadherin inhibits tumor angiogenesis without side effects on endothelial permeability. Blood 2002; 100(3):905–911.

58. Schmid MC, Bisoffi M, Wetterwald A et al. Insulin-like growth factor binding protein-3 is overexpressed in endothelial cells of mouse breast tumor vessels. Int J Cancer 2003; 103(5):577–586.

59. Oh P, Li Y, Yu J et al. Subtractive proteomic mapping of the endothelial surface in lung and solid tumours for tissue-specific therapy. Nature 2004; 429(6992):629–635.

60. Roodink I, Raats J, van der Zwaag B et al. Plexin D1 expression is induced on tumor vasculature and tumor cells: a novel target for diagnosis and therapy? Cancer Res 2005; 65(18):8317–8323.

61. Zheng PP, van der Weiden M, Kros JM. Differential expression of Hela-type caldesmon in tumour neovascularization: a new marker of angiogenic endothelial cells. J Pathol 2005; 205(3):408–414.

62. Ruoslahti E, Duza T, Zhang L. Vascular homing peptides with cell-penetrating properties. Curr Pharm Des 2005; 11(28):3655–3660.

63. Blank M, Weinschenk T, Priemer M, Schluesener H. Systematic evolution of a DNA aptamer binding to rat brain tumor microvessels. Selective targeting of endothelial regulatory protein pigpen. J Biol Chem 2001; 276(19):16464–16468.

64. McNamara JO, Andrechek ER, Wang Y et al. Cell type-specific delivery of siRNAs with aptamer-siRNA chimeras. Nat Biotechnol 2006; 24(8):1005–1015.

65. Posey JA, Khazaeli MB, DelGrosso A et al. A pilot trial of Vitaxin, a humanized anti-vitronectin receptor (anti alpha v beta 3) antibody in patients with metastatic cancer. Cancer Biother Radiopharm 2001; 16(2):125–132.

66. Santimaria M, Moscatelli G, Viale GL et al. Immunoscintigraphic detection of the ED-B domain of fibronectin, a marker of angiogenesis, in patients with cancer. Clin Cancer Res 2003; 9(2):571–579.

67. Borjesson PK, Postema EJ, Roos JC et al. Phase I therapy study with (186)Re-labeled humanized monoclonal antibody BIWA 4 (bivatuzumab) in patients with head and neck squamous cell carcinoma. Clin Cancer Res 2003; 9(10 Pt 2):3961S–3972S.

68. Milowsky MI, Nanus DM, Kostakoglu L, Vallabhajosula S, Goldsmith SJ, Bander NH. Phase I trial of yttrium-90-labeled anti-prostate-specific membrane antigen monoclonal antibody J591 for androgen-independent prostate cancer. J Clin Oncol 2004; 22(13):2522–2531.

69. Arap W, Pasqualini R, Ruoslahti E. Cancer treatment by targeted drug delivery to tumor vasculature in a mouse model. Science 1998; 279(January 16):377–380.

70. Kim JW, Lee HS. Tumor targeting by doxorubicin-RGD-4C peptide conjugate in an orthotopic mouse hepatoma model. Int J Mol Med 2004; 14(4):529–535.

71. Hu P, Yan J, Sharifi J, Bai T, Khawli LA, Epstein AL. Comparison of three different targeted tissue factor fusion proteins for inducing tumor vessel thrombosis. Cancer Res 2003; 63(16):5046–5053.

72. Schiffelers RM, Koning GA, ten Hagen TL et al. Anti-tumor efficacy of tumor vasculature-targeted liposomal doxorubicin. J Cont Rel 2003; 91:115–122.

73. Schiffelers RM, Ansari A, Xu J et al. Cancer siRNA therapy by tumor selective delivery with ligand-targeted sterically stabilized nanoparticle. Nucleic Acids Res 2004; 32(19):e149.

74. Curnis F, Gasparri A, Sacchi A, Longhi R, Corti A. Coupling tumor necrosis factor-alpha with alphaV integrin ligands improves its antineoplastic activity. Cancer Res 2004; 64(2):565–571.

75. Temming K, Lacombe M, Schaapveld RQJ et al. Rational design of RGD-albumin conjugates for targeted delivery of the VEGF-R kinase inhibitor PTK787 to angiogenic endothelium. Chem Med Chem 2007; 2(4):433–435.

76. Chen X, Park R, Tohme M, Shahinian AH, Bading JR, Conti PS. MicroPET and autoradiographic imaging of breast cancer alpha v-integrin expression using 18F- and 64Cu-labeled RGD peptide. Bioconjug Chem 2004; 15(1):41–49.

77. Decristoforo C, Faintuch-Linkowski B, Rey A et al. [(99 m)Tc]HYNIC-RGD for imaging integrin alpha(v)beta(3) expression. Nucl Med Biol 2006; 33(8):945–952.

78. Ellerby HM, Arap W, Ellerby LM et al. Anti-cancer activity of targeted pro-apoptotic peptides. Nat Med 1999; 5(9):1032–1038.

79. Curnis F, Sacchi A, Borgna L, Magni F, Gasparri A, Corti A. Enhancement of tumor necrosis factor alpha antitumor immunotherapeutic properties by targeted delivery to aminopeptidase N (CD13). Nat Biotechnol 2000; 18(11):1185–1190.

80. Pastorino F, Brignole C, Marimpietri D et al. Vascular damage and anti-angiogenic effects of tumor vessel-targeted liposomal chemotherapy. Cancer Res 2003; 63(21):7400–7409.

81. Curnis F, Gasparri A, Sacchi A, Cattaneo A, Magni F, Corti A. Targeted delivery of IFNgamma to tumor vessels uncouples antitumor from counterregulatory mechanisms. Cancer Res 2005; 65(7):2906–2913.

82. Yokoyama Y, Ramakrishnan S. Addition of an aminopeptidase N-binding sequence to human endostatin improves inhibition of ovarian carcinoma growth. Cancer 2005; 104(2):321–331.

83. Binetruy-Tournaire R, Demangel C, Malavaud B et al. Identification of a peptide blocking vascular endothelial growth factor (VEGF)-mediated angiogenesis. EMBO J 2000; 19(7): 1525–1533.

84. Renno RZ, Terada Y, Haddadin MJ, Michaud NA, Gragoudas ES, Miller JW. Selective photodynamic therapy by targeted verteporfin delivery to experimental choroidal neovascularization mediated by a homing peptide to vascular endothelial growth factor receptor-2. Arch Ophthalmol 2004; 122(7):1002–1011.

85. Oku N, Asai T, Watanabe K et al. Anti-neovascular therapy using novel peptides homing to angiogenic vessels. Oncogene 2002; 21(17):2662–2669.

86. Asai T, Shimizu K, Kondo M et al. Anti-neovascular therapy by liposomal DPP-CNDAC targeted to angiogenic vessels. FEBS Lett 2002; 520(1–3):167–170.

87. Arap W, Haedicke W, Bernasconi M et al. Targeting the prostate for destruction through a vascular address. Proc Natl Acad Sci USA 2002; 99(3):1527–1531.

88. Burg MA, Pasqualini R, Arap W, Ruoslahti E, Stallcup WB. NG2 proteoglycan-binding peptides target tumor neovasculature. Cancer Res 1999; 59(12):2869–2874.

89. Martens CL, Cwirla SE, Lee RY et al. Peptides which bind to E-selectin and block neutrophil adhesion. J Biol Chem 1995; 270(36):21129–21136.

90. Funovics M, Montet X, Reynolds F, Weissleder R, Josephson L. Nanoparticles for the optical imaging of tumor E-selectin. Neoplasia 2005; 7(10):904–911.

91. Kondo M, Asai T, Katanasaka Y et al. Anti-neovascular therapy by liposomal drug targeted to membrane type-1 matrix metalloproteinase. Int J Cancer 2004; 108(2):301–306.

92. Carnemolla B, Borsi L, Balza E et al. Enhancement of the antitumor properties of interleukin-2 by its targeted delivery to the tumor blood vessel extracellular matrix. Blood 2002; 99(5):1659–1665.

93. Dickerson EB, Akhtar N, Steinberg H et al. Enhancement of the antiangiogenic activity of interleukin-12 by peptide targeted delivery of the cytokine to alphavbeta3 integrin. Mol Cancer Res 2004; 2(12):663–673.

94. Gafner V, Trachsel E, Neri D. An engineered antibody-interleukin-12 fusion protein with enhanced tumor vascular targeting properties. Int J Cancer 2006; 119(9):2205–2212.

95. Gschwind A, Fischer OM, Ullrich A. The discovery of receptor tyrosine kinases: targets for cancer therapy. Nat Rev /Focus/ Therapeutic Proteins 2006; 9:S48–S57.

96. Kuenen BC, Giaccone G, Ruijter R et al. Dose-finding study of the multitargeted tyrosine kinase inhibitor SU6668 in patients with advanced malignancies. Clin Cancer Res 2005; 11(17):6240–6246.

97. Fury MG, Zahalsky A, Wong R et al. A Phase II study of SU5416 in patients with advanced or recurrent head and neck cancers. Invest New Drugs 2007; 25(2):165–172.

98. Morabito A, De Maio E, Di Maio M, Normanno N, Perrone F. Tyrosine kinase inhibitors of vascular endothelial growth factor receptors in clinical trials: current status and future directions. Oncologist 2006; 11(7):753–764.

99. Dormond O, Foletti A, Paroz C, Ruegg C. NSAIDs inhibit alpha V beta 3 integrin-mediated and Cdc42/Rac-dependent endothelial-cell spreading, migration and angiogenesis. Nat Med 2001; 7(9):1041–1047.

100. Guba M, von Breitenbuch P, Steinbauer M et al. Rapamycin inhibits primary and metastatic tumor growth by antiangiogenesis: involvement of vascular endothelial growth factor. Nat Med 2002; 8(2):128–135.

101. Sengupta S, Sellers LA, Li RC et al. Targeting of mitogen-activated protein kinases and phosphatidylinositol 3 kinase inhibits hepatocyte growth factor/scatter factor-induced angiogenesis. Circulation 2003; 107(23):2955–2961.

102. Yazawa K, Tsuno NH, Kitayama J et al. Selective inhibition of cyclooxygenase (COX)-2 inhibits endothelial cell proliferation by induction of cell cycle arrest. Int J Cancer 2005; 113(4):541–548.

103. Monnier Y, Zaric J, Ruegg C. Inhibition of angiogenesis by non-steroidal anti-inflammatory drugs: from the bench to the bedside and back. Curr Drug Targets Inflamm Allergy 2005; 4(1):31–38.

104. Qian DZ, Kato Y, Shabbeer S et al. Targeting tumor angiogenesis with histone deacetylase inhibitors: the hydroxamic acid derivative LBH589. Clin Cancer Res 2006; 12(2):634–642.

105. Millward MJ, House C, Bowtell D et al. The multikinase inhibitor midostaurin (PKC412A) lacks activity in metastatic melanoma: a phase IIA clinical and biologic study. Br J Cancer 2006; 95(7):829–834.

106. Temming K, Lacombe M, van der Hoeven P et al. Delivery of the p38 MAPKinase inhibitor SB202190 to angiogenic endothelial cells: development of novel RGD-equipped and pegylated drug-albumin conjugates using platinum(II)-based drug linker technology. Bioconjug Chem 2006; 17:1246–1255.

107. Hood JD, Bednarski M, Frausto R et al. Tumor regression by targeted gene delivery to the neovasculature. Science 2002; 296(5577):2404–2407.

108. Ogawara KI, Kuldo JM, Oosterhuis K et al. Functional inhibition of NF-êB signal transduction in αvβ3 integrin expressing endothelial cells by using RGD-PEG-modified adenovirus with a mutant IêB gene. Arthritis Res Ther 2006; 8(1):R32.

109. Kuo CJ, Farnebo F, Yu EY et al. Comparative evaluation of the antitumor activity of antiangiogenic proteins delivered by gene transfer. Proc Natl Acad Sci USA 2001; 98(8):4605–4610.

110. Burrows FJ, Thorpe PE. Eradication of large solid tumors in mice with an immunotoxin directed against tumor vasculature. Proc Natl Acad Sci USA 1993; 90:8996–9000.

111. Olson TA, Mohanraj D, Roy S, Ramakrishnan S. Targeting the tumor vasculature: inhibition of tumor growth by a vascular endothelial growth factor-toxin conjugate. Int J Cancer 1997; 73(6):865–870.

112. Arora N, Masood R, Zheng T, Cai J, Smith DL, Gill PS. Vascular endothelial growth factor chimeric toxin is highly active against endothelial cells. Cancer Res 1999; 59(1):183–188.

113. Li L, Wartchow CA, Danthi SN et al. A novel antiangiogenesis therapy using an integrin antagonist or anti-Flk-1 antibody coated 90Y-labeled nanoparticles. Int J Radiat Oncol Biol Phys 2004; 58(4):1215–1227.

114. Di Matteo P, Curnis F, Longhi R et al. Immunogenic and structural properties of the Asn-Gly-Arg (NGR) tumor neovasculature-homing motif. Mol Immunol 2006; 43(10):1509–1518.

115. Ou-Yang F, Lan KL, Chen CT et al. Endostatin-cytosine deaminase fusion protein suppresses tumor growth by targeting neovascular endothelial cells. Cancer Res 2006; 66(1):378–384.

116. Huang X, Molema G, King S, Watkins L, Edgington TS, Thorpe PE. Tumor infarction in mice by antibody-directed targeting of tissue factor to tumor vasculature. Science 1997; 275(24 January):547–550.

117. Ran S, Gao B, Duffy S, Watkins L, Rote N, Thorpe PE. Infarction of solid Hodgkin's tumors in mice by antibody-directed targeting of tissue factor to tumor vasculature. Cancer Res 1998; 58:4646–4653.

118. Nilsson F, Kosmehl H, Zardi L, Neri D. Targeted delivery of tissue factor to the ED-B domain of fibronectin, a marker of angiogenesis, mediates the infarction of solid tumors in mice. Cancer Res 2001; 61(2):711–716.

119. Bergers G, Javaherian K, Lo KM, Folkman J, Hanahan D. Effects of angiogenesis inhibitors on multistage carcinogenesis in mice. Science 1999; 284(5415):808–812.

120. Ozawa CR, Banfi A, Glazer NL et al. Microenvironmental VEGF concentration, not total dose, determines a threshold between normal and aberrant angiogenesis. J Clin Invest 2004; 113(4):516–527.

121. Tsujie M, Uneda S, Tsai H, Seon BK. Effective anti-angiogenic therapy of established tumors in mice by naked anti-human endoglin (CD105) antibody: differences in growth rate and therapeutic response between tumors growing at different sites. Int J Oncol 2006; 29(5):1087–1094.

122. Asgeirsdottir SA, Werner N, Harms G, van den Berg A, Molema G. Analysis of in vivo endothelial cell activation applying RT-PCR following endothelial cell isolation by laser dissection microscopy. Ann NY Acad Sci 2002; 973:586–589.

123. Asgeirsdottir SA, Kamps JAAM, Bakker HI et al. Site-specific inhibition of glomerulonephritis progression by targeted delivery of dexamethasone to glomerular endothelium. Molc Pharmacol 2007; 72:121–131.

124. Ruan W, Sassoon A, An F, Simko JP, Liu B. Identification of clinically significant tumor antigens by selecting phage antibody library on tumor cells in situ using laser capture microdissection. Mol Cell Proteomics 2006; 5(12):2364–2373.

125. Horsman MR, Siemann DW. Pathophysiologic effects of vascular-targeting agents and the implications for combination with conventional therapies. Cancer Res 2006; 66(24):11520–11539.

126. Tong RT, Boucher Y, Kozin SV, Winkler F, Hicklin DJ, Jain RK. Vascular normalization by vascular endothelial growth factor receptor 2 blockade induces a pressure gradient across the vasculature and improves drug penetration in tumors. Cancer Res 2004; 64(11):3731–3736.

127. Yao VJ, Ozawa MG, Varner AS et al. Antiangiogenic therapy decreases integrin expression in normalized tumor blood vessels. Cancer Res 2006; 66(5):2639–2649.

128. Leenders WP, Kusters B, Verrijp K et al. Antiangiogenic therapy of cerebral melanoma metastases results in sustained tumor progression via vessel co-option. Clin Cancer Res 2004; 10(18 Pt 1):6222–6230.

129. Shaked Y, Ciarrocchi A, Franco M et al. Therapy-induced acute recruitment of circulating endothelial progenitor cells to tumors. Science 2006; 313(5794):1785–1787.

130. Curnis F, Sacchi A, Corti A. Improving chemotherapeutic drug penetration in tumors by vascular targeting and barrier alteration. J Clin Invest 2002; 110(4):475–482.

131. Sacchi A, Gasparri A, Gallo-Stampino C, Toma S, Curnis F, Corti A. Synergistic antitumor activity of cisplatin, paclitaxel, and gemcitabine with tumor vasculature-targeted tumor necrosis factor-alpha. Clin Cancer Res 2006; 12(1):175–182.

132. Sengupta S, Eavarone D, Capila I et al. Temporal targeting of tumour cells and neovasculature with a nanoscale delivery system. Nature 2005; 436(7050):568–572.

Chapter 26
Models for Angiogenesis

Robert Auerbach

Keywords: in vitro and in vivo assays, cell proliferation, cell migration assays, tube formation, organ cultures, chorioallantoic membrane (CAM), Matrigel Plug Assay, animal models

Abstract: There are many systems which can serve as models for assessing angiogenic responses, both in vivo and in vivo. Each model has its advantages and disadvantages, its problems, technical difficulties and limitations, as well as its beneficial features. In this review we evaluate the major test systems currently in use, with special emphasis on comparing cell cultures, organ explants, whole embryo cultures and in vivo assay systems.

Introduction

The need for angiogenesis assays has never been greater. At the end of 2006, the PUBMED data base included 29,615 papers retrieved under "angiogenesis", of which 22,607 were published in 2006. Even more impressive is the fact that papers specifically indexed (MeSH terms) for angiogenesis numbered 16,121, of which 12,765 were published in 2006. Put another way, more than three-quarters of all publications indexed for angiogenesis were published in the last year. Yet, even 35 years after the publication of the classical paper of Judah Folkman in 1971 [1], alerting the scientific community about the importance of angiogenesis in the progression of cancers, there is still a largely unmet need for developing reliable, quantitative, assay methods.

There have been numerous reviews in the past few years, and major collections of papers, published as a separate volume, provide much information describing and evaluating assay methods [2–14]. In this chapter, we will review the most important types of assays, include their origin, and describe their major advantages and limitations. But, given the large influx of publications in the area, it is probable that

University of Wisconsin, Madison, WI 53706, USA
E-mail: rauerbac@wisc.edu

newer assays or, more likely, major modifications of existing assays, will need to be consulted for technical details and contemporary usage. For example, we published a paper on lymphocyte-induced angiogenesis in 1975 in which we quantitated the angiogenic response by counting the number of vascular divarications around an injection site [15,16]. Subsequently, image analysis methods became available providing more accurate measurement [17], followed by fractal analysis [18] and, most recently, by mathematical modeling [19].

What will remain contemporary for a longer period of time are the cautions that must be exercised in choosing and interpreting data obtained with the various angiogenesis assays. We will emphasize these cautions throughout the course of this chapter.

Features of Angiogenesis that are Amenable to Analysis

New blood vessels are derived from two major sources: the extension from pre-existing vessels and those derived from progenitor cells. Traditionally, these two modes of origin have been designated, respectively as "angiogenesis" and "vasculogenesis" [20]. However, as we have pointed out previously [21], the two modes are not mutually exclusive, both contributing to the formation of new blood vessels (neovascularization) in response to wounding, inflammation, tumor development, ischemia and the panoply of "angiogenic diseases" [22]. In this chapter we will use the term "angiogenesis" to describe new blood vessels regardless of the mode of origin.

The primary cells involved in angiogenesis are the endothelial cells, which line all blood and lymphatic vessels and constitute virtually the entirety of capillaries of both the blood- and lymphatic circulatory systems. Thus, analysis of angiogenesis largely involves the study of endothelial cells. Endothelial cells arising from preexisting vessels must be released from these vessels, breaking down or passing through the extracellular matrix. They must migrate, proliferate and reorganize. Those endothelial cells arising from progenitor cells must differentiate, passing through a sequence of developmental stages. They must adhere to the site where new vessels are being generated and they must be incorporated into the existing vascular architecture. Each of these processes—basement membrane disruption or penetration, migration, proliferation, and three-dimensional reorganization—can be tested in vitro using endothelial cell culture methods. However, throughout the process of angiogenesis there are interactions between the endothelial cells and the surrounding tissues, such as pericytes and smooth muscle cells. In part, these interactions can also be studied in vitro, using organ culture and tissue recombination methods. Even so, angiogenesis is in large part regulated by systemic factors ranging from cytokines and hormones to hypoxia and shear stress, and for this reason in vitro methods are never entirely adequate. Thus, although in vitro studies provide significant initial information, in vivo assays are ultimately almost imperative. Finally, there are numerous aspects of angiogenesis in patients that reflect pathologies that are only partially understood (e.g., Alzheimer's disease) and that cannot as yet be approached in model systems. For this reason, descriptive studies continue to be a mainstay of clinical research.

Implicit in these considerations is that no single angiogenesis assay is likely to be a reliable representative for all angiogenic reactions, and that assays chosen to evaluate a single process such as cell migration or vessel reorganization cannot accurately predict the more comprehensive and complicated process underlying the development of new blood vessels, be that in animal models or in patients. Furthermore, all angiogenesis reactions are not alike [23]. For example, vessels induced by activated lymphocytes are different from those induced by tumor cells [24]. Ultimately, meaningful evaluation of experimental results will have to rely on intelligent interpretation.

In Vitro Assays

Although the tissue culture literature on blood vessels goes back to the early 1900s [25], the major event leading to the use of in vitro assays to study angiogenesis was the description of reproducible cell culture techniques for the growth of vascular endothelial cells [26–28]. Originally described for large vessel cells such as those from the aortic and umbilical vein and dorsal aorta, methods for growth of microvascular endothelial from the adrenal gland were subsequently published [29]. These in particular have been expanded to encompass endothelial cells from various other organ sites [30–36,37–39]. Methods for the growth of lymphatic large vessel endothelial cells were reported almost simultaneously (33,34), but were only infrequently utilized. Methods for the isolation and growth of microvascular lymphatic endothelial cells have now also been described [40,41]. The extent of endothelial cell heterogeneity is only now being appreciated, largely through the analysis of tumor cell / endothelial cell selective adhesion and the use of phage display techniques [30,33,34,42–44].

It needs to be emphasized that endothelial cells in culture are by their very nature in an activated state, different from endothelial cells in the organism where, except under special normal (e.g., during the estrus cycle) or pathological (e.g., in response to tumors, as an accompaniment of inflammation, or during wound healing) conditions. Cells in vitro both gain and lose attributes found in vivo, and for technical reasons it is almost never possible to use primary endothelial cell cultures (i.e., not passaged following initial isolation) for the study of angiogenic reactions. As cells are propagated in vitro, they are likely to lose a variety of cell surface antigens, their various receptors may be activated, synthesis of products required for cell division may be initiated, karyotypes may be modified.

The fact that endothelial cells in vitro are generally studied in the absence of supporting cells should also be kept in mind. Coculture systems combining endothelial cells with non-endothelial cells overcome this problem in part, and organ cultures are a still more biologically relevant means for achieving a model representative of in vivo conditions. Even so, systemic factors will be missing as will modulations in the tissue environment in vivo that will not be duplicated once cells have been placed in culture.

Proliferation Assays

The most frequently used in vitro assay has been the monitoring of cell proliferation. The net number of cells in a culture can be determined by hemocytometer counts, Coulter counter or flow cytometric enumeration, colorimetric quantitation (tetrazolium salts), or total DNA content. A more direct assessment of proliferation rates is usually made by measuring DNA synthesis at given time points, as determined by radioisotope incorporation or by cell cycle analysis using flow cytometry.

There are several caveats to these measurements. One important consideration, often overlooked, is that inhibition of cell proliferation may reflect generalized toxicity and not be specific to endothelial cells. Frequently, these measurements are made without adequate experimental controls. Protocols need to include representative non-endothelial cells and determination of cell survival is essential [14] (Fig. 26.1).

Much more important, but far more subtle, is the fact that all endothelial cells are not alike [30,33,34,42–44]. Although organ-specific differences in endothelial cells have been well documented experimentally, although there are major structural differences between endothelial cells from large blood vessels such as the dorsal aorta or pulmonary vein and those comprising capillaries from different microvascular beds, and although there are major differences between endothelial cells from the blood vascular and those of the lymphatic vascular systems, there are still an almost overwhelming number of studies that assume that a test in vitro system using human umbilical vein cells is sufficient for establishing facts about angiogenesis. (Note: Even all HUVEC cells are not alike and there are marked differences along the length of individual large blood vessels such as the aorta.)

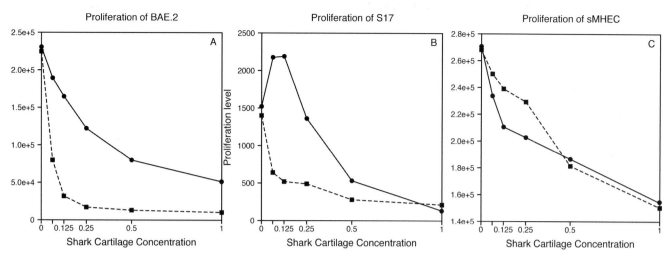

FIG. 26.1. This experiment was carried out in response to a published claim that a specific shark cartilage preparation was antiangiogenic. The preparation inhibited proliferation of both bovine aortic endothelial cells (**A**) and murine bone marrow-derived stromal cells (**B**). Note that, unlike the aortic endothelial cells, myocardium-derived microvascular endothelial cells (**C**) were not influenced by the shark extract. Microscopic examination of BAE and S17 cell cultures showed cell death within 4 h of explantation. *BAE* Bovine aortic endothelial cells; *S17* murine bone-marrow-derived stromal cells; *sMHEC* murine myocardium-derived microvascular endothelial cells.

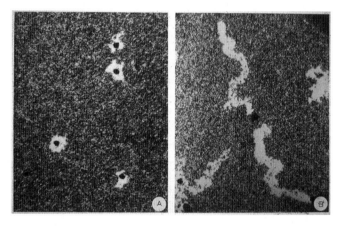

FIG. 26.2 Cell movement (phagokinetic track assay) of EOMA (murine endothelioma-derived endothelial) cells in (**A**) low (10 ml/l) versus (**B**) high (50 ml/l) concentrations of fetal bovine serum. Photographs were taken 24 h after explantation on a layer of 1-μM polystyrene beads.

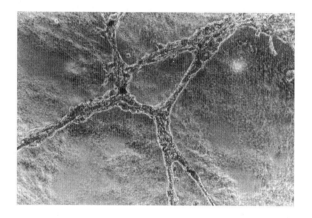

FIG. 26.3. Murine myocardium-derived endothelial cells were seeded on Matrigel. Tube formation was observed within 24 h.

Cell Migration Assays

Inasmuch as endothelial cells must migrate from preexisting vessels in order to form new blood vessels, cell migration assays represent a major in vitro approach for studying angiogenesis. Migration involves both chemokinesis and chemotaxis, and assays have been developed to measure each of these processes. One significant advantage of these assays is that the number of cells needed to measure chemotaxis or chemokinesis is orders of magnitude lower than the number needed for accurate cell proliferation tests [45,46] (Fig. 26.2).

Chemokinesis can be quantified by using a two-dimensional model in which endothelial cells are plated onto an indicator surface (colloidal gold, red blood cells or microspheres), and their

rate of cell movement assessed using computer imaging or time lapse photography [45]. Once again, however, the use of adequate controls is imperative, inasmuch as cell movement may be induced non-selectively or inhibited due to generalized toxicity.

Chemotaxis is best measured using various transwell or Boyden chamber-type culture systems. Cells using a commercially-available microwell assembly can be used to generate concentration gradients that permit measurement of rates of transwell migration of individual endothelial cells.

Endothelial cell migration can also be studied in an in vitro model of wound healing. A defined area in a monolayer culture of endothelial cells can be denuded, and the rate of re-endothelialization of the denuded area can be determined by quantitated by standard imaging techniques [47–49].

Tube Formation In Vitro

One of the best specific tests for angiogenesis in vitro measures the ability of endothelial cells to reorganize into three-dimensional structures (e.g., tube formation) [49–51]. Most endothelial cells will form tubes "spontaneously", given an appropriate medium and a reasonable length of time in culture. Tube formation can be enhanced by use of collagen or fibrin clots or by culture on complex matrix components such as those elaborated by tumor cells (e.g. Matrigel) (Fig. 26.3). Again, a word of caution seems necessary. Many non-endothelial cells will form tube-like structures when cultured on three-dimensional gels such as Matrigel. Careful assessment of histological preparations can resolve many of the questions arising from in situ observations, but confirmation by electron microscopy to confirm that true tight junctions have been established is not part of the usual protocol for tube formation in vitro [52,53].

Coculture Protocols

There are a seemingly endless number of coculture protocols, including the use of specific feeder layers or combinations between endothelial cells and tumor fragments,. Cocultures have also been designed to study interactions between different cell types separated by a nuclepore or millipore filter to prevent direct cell contact. For example, in our own laboratory, we have used murine melanoma cell aggregates or fragments placed on various microvascular endothelial cell monolayers to determine whether tube formation is augmented or endothelial cell migration is encouraged by the tumor cells. Transfilter assays have been used in which endothelial cells on one side of the filter were juxtaposed to tumor cells or mesenchymal cells on the opposing side, followed by assessment of endothelial cell reorganization. These systems, however, have been designed to answer highly specific questions and are therefore less generally applicable than are organ cultures that provide an already established three-dimensional organization [54].

Organ Culture Assays

The history of organ cultures that included endothelial cells goes back to the early twentieth century where explantation of embryonic rudiments, bone marrow and neonatal and adult organ fragments became prevalent in laboratory experimentation. More than 100 references to endothelium studied as a component of organ culture protocols were included in the historical overview of the literature of tissue culture published in the 1960s [25]. Organ cultures can more closely approximate the in vivo situation because complex cell interactions can take place among different tissues that comprise the organ. For example, spleen organ cultures, unlike spleen cells grown in suspension, can generate a primary immune response that sequentially gives rise first to IgM- and then to IgG-producing cells [55,56].

The most widely used organ culture system is the aortic ring assay [57–59]. In this assay, segments of the aorta, originally from rats but now from many other species including humans, are placed in culture, usually in a matrix-containing environment such as Matrigel. Over the next 7–14 days, explants are monitored for endothelial cell outgrowth, and the rate and pattern of this outgrowth permits the evaluation of angiogenesis-inducing and angiogenesis-inhibiting factors. This in vitro assay comes closer to simulating the in vivo condition than do endothelial cell cultures, but it must be kept in mind that the aorta is not representative of the microvasculature so frequently key to angiogenic responses. An alternate assay is the use of the carotid artery rather than the dorsal aorta, but this still involves large vessel endothelial cells rather than microvascular ones [60].

An only briefly reported modification of the aortic ring assay, using chick embryo aortic arch fragments, has found increased application [61]. Aortic arches are dissected from 12- to 14-day chick embryos, cut into rings similar to those of the rat aorta, and explanted on Matrigel. Unlike the adult aorta, the endothelial cells of the embryonic aortic arch share many properties with microvascular endothelial cells. Because the aortic arch is, from the time of explantation, in a highly proliferative state, outgrowth of endothelial cells can be observed within 24–48 h. Moreover, by everting the ring to exteriorize the endothelium, the outgrowing endothelial cells typically organize into three-dimensional tubes, providing yet another measure of neovascular differentiation [61] (Fig. 26.4).

Quantification of endothelial cell outgrowth both in the aortic ring and aortic arch cultures, can be achieved by using fluorescently labeled antibodies to CD31 (PECAM) or by staining with fluorescein-labeled lectins such as Bandeira simplifolica lectins (BSL-I and BSL-B4). (Note. Lectin staining is different for different species. For example, while rodent and pig endothelial cells are strongly and selectively bound by BSL-I and BSL-B4, human endothelial cells are bound preferentially by Ulex europaeus agglutinin I (UEA-I).)

In Vivo Assays

There is general agreement that assessment of angiogenesis-influencing factors ultimately requires whole animal/in vivo assessment. In vivo analysis of blood vessel development has been a mainstay of embryologists and pathologists for centuries going back into antiquity. A historical review of these studies, however, is beyond the scope of the present chapter. To this day, anatomical and histological observations provide detailed information concerning in vivo angiogenesis in patients, and are an essential component of virtually all in vivo experimental protocols.

The Chorioallantoic Membrane (CAM) Assay

The CAM has been used by embryologists for more than 60 years, primarily to study the growth and differentiation of grafted tissues [62]. The CAM was chosen because of its ready access and its propensity for rapidly providing the blood supply essential for survival of the grafts. A superb review by Richardson and Singh outlines the description of the CAM, its

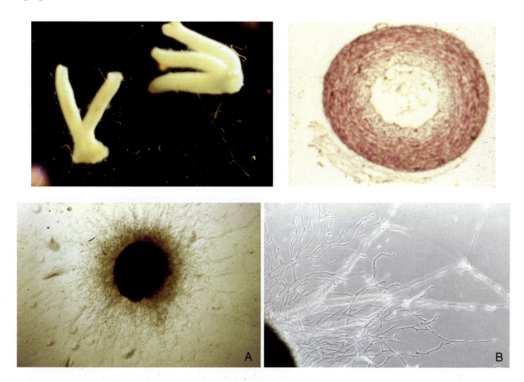

FIG. 26.4. *Top* Aortic arches obtained from day-13 embryos, shown following isolation from the embryo (*left*) and in cross-section (*right*). *Bottom* Outgrowth of cells from the explanted aortic arch. *Left panel* shows the total explant area; *right panel* shows more details of the outgrowth area, in which endothelial cells are seen forming tube-like structures.

major features, and many of the studies in which the CAM has been utilized for angiogenesis research [63].

In its original form, the CAM assay was performed on 7- to 9-day-old chicken embryos by making a window in the egg shell, and then placing tissue or organ grafts directly on the membrane [62]. The window was sealed, eggs were replaced in the incubator, and grafts were visualized at various time points and scored for growth and vascularization. More recently, a variety of total embryo culture methods have been developed, beginning with a simple eversion into a petri dish of the entire egg content at 3 days of incubation [64] (Fig. 26.5). The CAM would develop during the next few days of incubation. At this time, test materials or chemicals could be placed on the CAM and the subsequent angiogenic response could be monitored photographically. Various alternate methods for ex ovo cultivation of chicken embryos have been described [65–68]. On termination, the extent of vascularization could be further defined by histological examination. Although technically an "in vitro" approach, the assay is really as much "in vivo" as if the embryo and its membranes had been studied in ovo.

Among the most valuable aspects of CAM assays are the ease with which tissues or reagents can be administered, the ready availability and low expense of eggs, the fact that chick embryos do not require the extensive protocol approvals required (at least in the United States) for animal research, and the amenity of the assay to routine image analysis and other quantitative measures [17,69–71]. Grafts made using sponges or Matrigel plugs provide useful adjuncts of the CAM assay method [6,72].

A key point that needs to be remembered, however, is that the CAM is not a static membrane. CAM assays have been performed on chick embryos ranging from day 5 to day 14 of incubation, and there are marked differences in the vascular response, depending on the precise embryonic age at which an angiogenesis assay is initiated [73,74]. The CAM is highly proliferative until about day 11 of incubation: expansion of the CAM leads to alterations in its relationship to the underlying tissues, and there are marked changes in the physiology of the chicken embryo during the course of its normal development, changes which are reflected in the circulation of systemic factors that influence vascular responses [75–77].

Alternate assays include the use of quail embryos, which can be grown in whole embryo culture in a manner analogous to that used for chicken embryos [78,79]. The technique of intracoelomic grafts, which was technically exceedingly difficult to perform in ovo, can now be readily carried out in the whole embryo culture system. However, to date, this assay method has not been applied specifically to study neovascularization.

The Mesenteric Window Assay

Whereas the chorioallantoic membrane is an extraembryonic membrane and the assay uses avian species, the small gut mesenteric window of rats and mice is in many ways comparable to the CAM and has been used successfully for in vivo studies of mammalian angiogenesis [9,48,80,81]. Although technically not as simple as the CAM assay, the fact that it is of mammalian

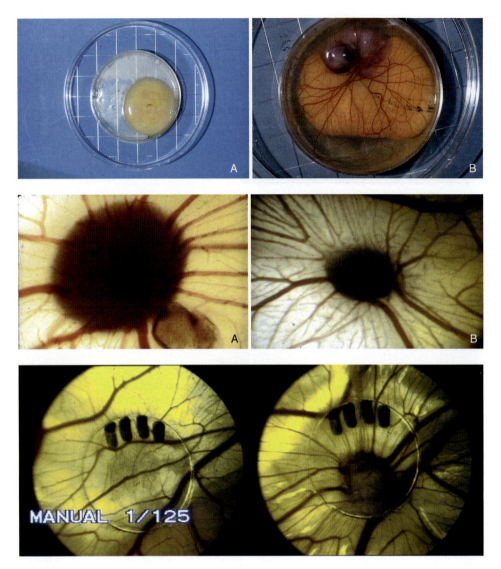

Fig. 26.5. CAM assay using explanted whole chick embryos. *Top left* Chick embryo at 72 h of incubation, shortly after being placed in a culture dish. *Top right* Chick embryo after 1 week of additional growth. Note that the chorio-allantoic membrane has now covered the entire surface of the culture dish. *Bottom right* FGF-2 was placed on the underside of plastic coverslips and the explant was photographed immediately after placing the disc on the CAM. *Bottom left* Same culture after a further 72 h of incubation. Note the massive influx of new blood vessels.

origin and can be carried out in normal adults makes it a valuable tool for studying in vivo interactions such as those mediated by inflammation. An important feature is that it permits simultaneous analysis of capillaries, small arterioles and small venules.

There is a long historically interesting surgical literature describing the use of the omentum in augmenting wound healing and in the treatment of burns [82,83]. Much of that literature is relevant to studies of angiogenesis.

The Corneal Angiogenesis Assay

This assay is considered to be one of the best in vivo models, largely because the assay monitors the penetration of new vessels into an essentially avascular cornea. The method was originally described for rabbit eyes [84], but has since been adapted to mice, where it has become one of the principal in vivo tests [24,85,86] (Fig. 26.6). In brief, a pocket is made in the cornea to test the response of tumor fragments, tissues or cells to test reagents. The use of osmotic pumps (Alzet) has provided a reliable means of administering test reagents systemically over prolonged periods of time [87]. Test substances can also be placed directly into the cornea, using various slow-release formulations or sponges covered by a diffusion-limiting barrier [88,89].

Among the advantages of the cornea assay is the ability to monitor progress of neovascularization over an extended time. In rabbits, this requires the use of slit lamp optics, but in mice, observations and recordings can be made under a standard ocular or dissecting microscope. Quantitation

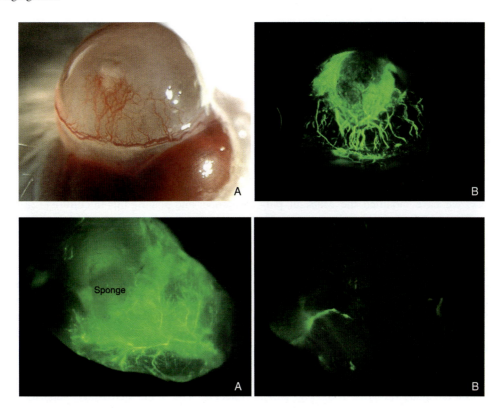

FIG. 26.6. *Top* FGF-2 induced angiogenesis in the mouse cornea. *Left* Numerous blood vessels have been induced by the implantation of a mini-sponge containing FGF-2 into the cornea. *Right* Fluoresceinated dextran can be used to generate quantitative data using imaging software. Photographs were taken after 7 days. *Bottom* The sponge-Matrigel assay was used to examine the effect of endostatin on neovascularization induced by 4T1 tumor fragments. Alzet pumps (7-day delivery) containing either PBS (*left*) or endostatin (*right*) were implanted subcutaneously, and replaced after 7 days for an additional 7 days. Photographs of the recovered plugs were made 14 days after the initiation of the experiment.

in living animals is best achieved by using fluorochrome-labeled high molecular weight dextran [90]. At termination of the experiment, corneas can be dissected, mounted, and analyzed using fluorescent markers, such as dextran or antibodies to lymphatic or blood vascular endothelial cells [90,91]. Quantitation is achieved by measuring the area of vessel penetration, progress of vessels toward the angiogenic stimulus, imaging of the vascularized cornea and by pixel counts above background.

Drawbacks to the assay include the fact that the surgical procedures are demanding and time consuming and, for mice, by the fact that the space available within the cornea is severely limited. The reaction includes a large number of circulating cells in addition to the endothelial cells comprising the neovasculature [92]. It seems also worth noting that there are distinctive differences between mice from different strains, sometimes making comparisons between studies difficult [93–95].

It is surprising that, given the prevalence of the corneal assay, there has been almost no use of the more spacious anterior eye chamber as a site for testing angiogenesis in mice. The anterior eye chamber has been utilized for tumor growth

and embryonic organ transplantation for more than half a century [96]. Tumors became rapidly vascularized when introduced into the anterior eye chamber. Implantation of mouse trophoblasts and of isolated ectoplacental cones yielded an explosive neovascular reaction [97], whereas embryonic rudiments vascularized in orderly fashion, vessels originating entirely from the choroidal rather than the limbal vasculature [98,99].

The Matrigel Plug Assay

Matrigel, extracted from Englebrth-Holm-Swarm tumors, is a complex mixture of laminin and other basement membrane proteins, and a variety of growth factors including FGF-2, NGF, IGF-1 and EGF [100]. It is liquid at 4°C and even at room temperature, but reconstitutes into a gelatinous mass (plug) at body temperature. In contrast to the corneal assay, the Matrigel plug assay is simple to carry out [101]. Cells, tissue fragments or test substances are combined with liquid Matrigel, which is then injected into animals subcutaneously, where it solidifies rapidly. Although the assay cannot be monitored during the course of the experiment, the Matrigel plug can be recovered intact, even after several weeks, and subjected to histological analysis and reactivity to immunological and lectin reagents. Once processed for histology, Matrigel plugs lend themselves well to imaging techniques [19]. A rapid assessment of angiogenesis is achieved by determining the hemoglobin content of Matrigel plugs. Although useful, this means of measuring angiogenesis is subject to undue influence by hemorrhaging and, by precluding subsequent histological analysis, becomes a less reliable means of quantifying the response.

In our laboratory, we have modified the Matrigel plug system to permit clearer delineation of the neovascular response [102]. In this assay, Matrigel is first injected subcutaneously to produce the plug. Small sponges containing cells or test substances are then implanted into the formed plug. New vessels can be visualized by injecting fluoresceinated dextran prior to removing the plugs for analysis. Again, Alzet pumps have been used to achieve chronic systemic administration of test reagents (Fig. 26.6).

One of the caveats for this and other in vivo assays is that there are marked regional differences so that, for example, a tumor implant, skin graft, or assay disc may differ as much as four-fold depending on the assay site [102–106] (Fig. 26.7). Where tumor grafts are placed, moreover, must also take into account the specific environment, as shown by the finding that orthotopic grafts have growth characteristics that differ significantly from heterotopic grafts [107–109]. Imaging techniques can help resolve some of the questions concerning orthotopic versus heterotopic neovascularization [108].

An earlier method which had many of the features of the Matrigel plug system involved the generation of sodium alginate beads which could incorporate test cells or substances and be injected in vivo [110,111]. Assay methods were similar to those for the Matrigel plug system, but bead generation was difficult and the assay has been almost entirely replaced by the use of Matrigel.

Sponge Implant Assays

Sponge implants provide a useful means of constraining test cells or substances (cf. the sponge/Matrigel assay described above). The basic sponge implant assay has been carried out in rats by placing a polyester sponge containing test substances subcutaneously into an air pocket [112,113] (cf. [114]), this air-pocket method ("Selye pouch") having been originally described by Selye [115]. The assay has been applied successfully to measure blood flow, using isotope washout methods (^{133}Xe clearance) [116].

The disc angiogenesis assay is a modification of the sponge implant, in which the sponge, cut in disc shape, is covered by impermeable material (e.g., sealed millipore discs) leaving only the rim exposed to the surrounding tissues [86,117–120]. Thus, endothelial cells can only penetrate the sponge from the edge. Progress of angiogen-

esis is scored on the basis of extent of penetration into the sponge, and this can be readily analyzed using computer imaging techniques (Fig. 26.8). Analysis can also be carried out histochemically or by extraction of hemoglobin. It is of historical interest that the initial covering by millipore filters recalls the early diffusion chamber experiments of Algire and his colleagues, who studied tissue interactions in vivo by introducing tumor cells into millipore-covered chambers which were then implanted into the peritoneal cavity [121]. Indeed, the millipore transwell culture system was originally designed by taking one half of the Algire chamber and placing this in culture [122].

Alternate Animal Models

In early studies of neovascularization, the hamster cheek pouch was a favored site [123]. However, the absence of good reagents for hamsters compared to mice and rats, as well as their limited availability, has reduced the importance of this technique. On the other hand, another older method, implanting assay tissues or reagents into the pinna of the ear of rabbits, is coming into more frequent use, both in rabbits and in mice, because of the increasingly sophisticated application of intravitral microscopy [124–126].

Among the most promising relatively recent additions to the panel of animal systems amenable to study of angiogenesis is the zebrafish [127–132]. Because of the transparency of early zebrafish embryos, vascular development is readily visible and can be monitored without dissection. Moreover, molecular genetic techniques have generated massive amounts of data on numerous genes relevant to vascular development, and unlike the generation of transgenic mice, manipulations and mutant screening of zebrafish embryos are easier, and generation of mutant zebrafish is rapid [127,133].

Amphibians such as *Rana* and *Xenopus* had long been used for making in vivo observations on the circulatory system, both during embryonic development and in adults. Exploitation of these older models for experiments that require visualization of the microcirculation has also begun to be appreciated [127,134,135].

We have not included in this chapter any discussion of clinical methods, although these are becoming ever more important as knowledge gained from experimental work in animal systems and in vitro find application to the clinic. There is no compelling evidence that in vitro studies carried out on human endothelial cells are more informative than those carried out on endothelial cells from other species, or that transplantation of human tumors into animals is more informative than analysis of compatible animal tumors.

Clinical studies of antiangiogenic therapies have sometimes included wound healing assays and various laboratory measurements, such as concentrations of circulating growth factors, to complement these procedures. To date, however,

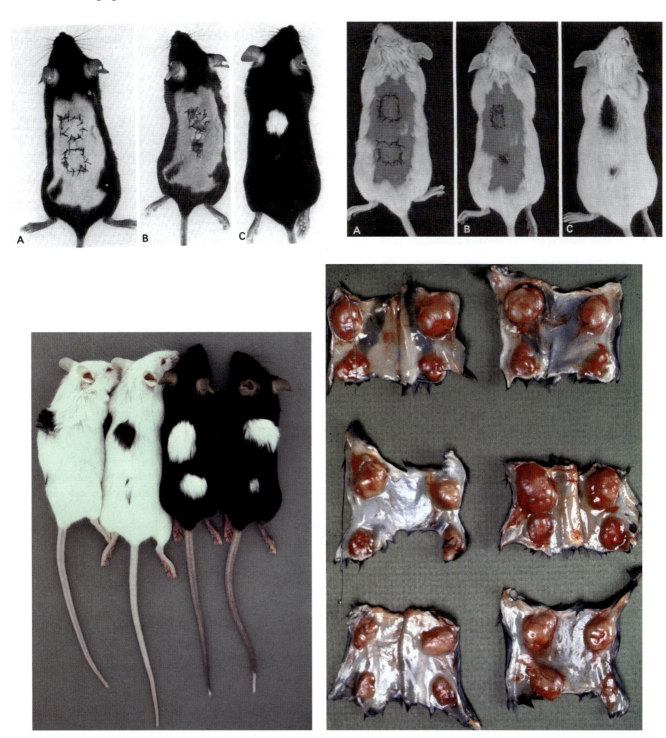

Fig. 26.7. Regional differences in growth of transplanted skin or tumor cells. *Top left* C57BL/6-c2 skin was grafted onto C57BL6J mice. **A** 1 day after grafting, **B** 1 week after grafting, **C** 1 month after grafting. *Bottom left* C57BL/6J skin grafted onto C57BL/6-c2 mice. **A–C** as above. *Top right* Reciprocal grafts involving exchanges between C57BL/6J and C57BL/6-c2J mice, 6 weeks after grafting. *Bottom right* C755 mammary carcinoma cells were injected intradermally into adult syngeneic (C57/BL6Au) mice. One complete set of six animals is shown to indicate range of growth differences. The anterior skin sites are at the top of each animal.

the prime methods employed in patients have been standard ones, such as determining the mean vessel density, assessing vascular patterns and blood flow using magnetic resonance imaging, positron emission tomography, and Doppler analysis, and the application of various other optical imaging methods to assess vessel growth or regression.

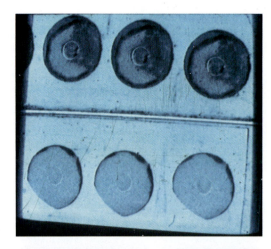

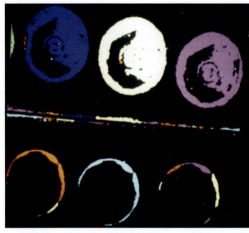

Fig. 26.8. The effect of a ribonuclease inhibitor on angiogenesis was tested in the disc angiogenesis assay. *Top panel, upper row* Discs from PBS control; *lower row* from RNasin-treated discs. *Bottom panel* Same discs following imaging to quantitate penetration of vessels into the discs.

Concluding Remarks

It is hard to underestimate the importance that in vitro studies have played and continue to play in furthering our knowledge about the process of angiogenesis. Most of these assays have used endothelial cells in isolation, as measured by migration, proliferation and reorganization into three-dimensional structures. They have been critical for experiments leading to the identification of angiogenesis-inducing factors such as VEGF and FGF2, angiogenesis-inhibiting substances such as endostatin and thrombospondin, and organizing factors such as the angiopoietins and matrix proteins.

Among the key problems arising from in vitro studies, however, are that all endothelial cells are not alike, that methods for growing endothelial cells in vitro do not replicate fully the cellular or humoral microenvironment found in in vivo, and that methods for replicating the rheological conditions seen within living organisms are limited. Multiple assays using different endothelial cells and examining different aspects of the angiogenic process, are recommended, as is the application of coculture or organ culture protocols. However, even then, in vitro methods cannot replace the need for in vivo assays.

In vivo tests such as the anterior eye chamber, corneal pocket grafts, Matrigel assays, and assays using various sponge and chamber formulations are all useful, but unfortunately are time-consuming and far more difficult to quantitate. Each of these tests has merit, and the choice of assay method(s) will depend on the specific needs of a given experimental investigation. Perhaps a useful dictum is that a model system is only a model.

Although reviews for many years have stressed the need for quantitation, improvements have been slow in coming, although numerous improved methods for quantifying the angiogenic response are now regularly appearing in the literature. They are largely based on radiological and fluorescence-assisted imaging techniques, combined with computer-assisted mathematical modeling [18,19,136–141]. Experimental systems that permit genetic manipulations, such as the use of morpholinos in zebrafish and gene targeting to generate transgenic mice, are gaining increased prominence. New media, new cell lines, in vitro/in vivo combination protocols, orthotopic transplantation procedures, and stem cell studies all may lead to new or improved methods.

Acknowledgments. I am indebted to Wanda Auerbach for extensive editorial assistance. The preparation of this manuscript was supported by grant # 1 G13 LM-08953-01 from the National Institutes of Health.

References

1. Folkman J. Tumor angiogenesis: therapeutic implications. N Engl J Med 1971;285:1182–6.
2. Auerbach R. An overview of current angiogenesis assays: Choice of assay, precautions in interpretation, future requirements and directions. In: Staton C, Bicknell R, Lewis C, eds. Angiogenesis Assays. Chichester, UK, 2007:410.
3. Auerbach R, Akhtar N, Lewis RL, Shinners BL. Angiogenesis assays: problems and pitfalls. Cancer Metastasis Rev 2000;19:167–72.
4. Auerbach R, Auerbach W. Assays to study angiogenesis. In: Voest E, D'Amore P, eds. Tumor angiogenesis and microcirculation. New York: Marcel Dekker, 2001:91–102.
5. Auerbach R, Auerbach W, Polakowski I. Assays for angiogenesis: a review. Pharmacol Ther 1991;51:1–11.
6. Auerbach R, Lewis R, Shinners B, Kubai L, Akhtar N. Angiogenesis assays: a critical overview. Clin Chem 2003;49:32–40.
7. Hasan J, Shnyder SD, Bibby M, Double JA, Bicknel R, Jayson GC. Quantitative angiogenesis assays *in vivo*–a review. Angiogenesis 2004;7:1–16.
8. Jain RK, Schlenger K, Hockel M, Yuan F. Quantitative angiogenesis assays: progress and problems. Nat Med 1997;3:1203–8.

9. Norrby K. *In vivo* models of angiogenesis. J Cell Mol Med 2006;10:588–612.

10. Ribatti D, Vacca A. Models for studying angiogenesis *in vivo*. Int J Biol Markers 1999;14:207–13.

11. Staton CA, Stribbling SM, Tazzyman S, Hughes R, Brown NJ, Lewis CE. Current methods for assaying angiogenesis *in vitro* and *in vivo*. Int J Exp Pathol 2004;85:233–48.

12. Tucker GC. Modèles expérimentaux d'angiogénèse (*in vitro* et chez l'animal) [Experimental models of angiogenesis (*in vitro* and *in vivo*)]. Therapie 2001;56:473–81.

13. Murray J. Angiogenesis Protocols. Totowa, NJ: Huamana Press, 2001.

14. Auerbach R, Popp B, Kiley L, Gilligan l. Angiogenesis assays: problems, pitfalls and potential. In: Hori W, ed. Therapeutic implications and mechanisms of angiogenesis – inhibitors and stimulators. Westborough, MA: IBC USA Conferences, Inc., 1996:1.3.1–1.3.9.

15. Sidky YA, Auerbach R. Lymphocyte-induced angiogenesis: a quantitative and sensitive assay of the graft-vs.-host reaction. J Exp Med 1975;141:1084–1100.

16. Sidky YA, Auerbach R. Lymphocyte-induced angiogenesis in tumor-bearing mice. Science 1976;192:1237–8.

17. Brooks PC, Montgomery AM, Cheresh DA. Use of the 10-day-old chick embryo model for studying angiogenesis. Methods Mol Biol 1999;129:257–69.

18. Gazit Y, Baish JW, Safabakhsh N, Leunig M, Baxter LT, Jain RK. Fractal characteristics of tumor vascular architecture during tumor growth and regression. Microcirculation 1997;4: 395–402.

19. Guidolin D, Vacca A, Nussdorfer GG, Ribatti D. A new image analysis method based on topological and fractal parameters to evaluate the angiostatic activity of docetaxel by using the Matrigel assay *in vitro*. Microvasc Res 2004;67:117–24.

20. Risau W. Mechanisms of angiogenesis. Nature 1997;386: 671–4.

21. Auerbach R, Auerbach W. Vasculogenesis and angiogenesis. In: Fan T-P, Kohn E, eds. The New Angiotherapy. Totowa, NJ: Humana Press, 2002:1–6.

22. Fan TP, Kohn E, eds. The New Angiotherapy. Totowa, NJ: Humana Press, 2002.

23. Auerbach R. Differential angiogenesis. In: Rifkin D, Klagsbrun M, eds. Angiogenesis. Mechanisms and Pathology. Cold Spring Harbor, NY: Cold Spring Harbor Laboratory, 1987:131–133.

24. Muthukkaruppan V, Auerbach R. Angiogenesis in the mouse cornea. Science 1979;205:1416–8.

25. Murray M, Kopech G. A bibliography of the research in tissue culture, 1884–1950. New York: Academic Press, 1953.

26. Nachman RL, Jaffe EA. Endothelial cell culture: beginnings of modern vascular biology. J Clin Invest 2004;114:1037–40.

27. Gimbrone MA, Jr., Cotran RS, Folkman J. Endothelial regeneration: studies with human endothelial cells in culture. Ser Haematol 1973;6:453–5.

28. Jaffe EA, Nachman RL, Becker CG, Minick CR. Culture of human endothelial cells derived from umbilical veins. Identification by morphologic and immunologic criteria. J Clin Invest 1973;52:2745–56.

29. Folkman J, Haudenschild CC, Zetter BR. Long-term culture of capillary endothelial cells. Proc Natl Acad Sci U S A 1979;76:5217–21.

30. Auerbach R. Vascular endothelial cell differentiation: organ-specificity and selective affinities as the basis for developing anti-cancer strategies. Int J Radiat Biol 1991;60:1–10.

31. Auerbach R, Alby L, Morrissey LW, Tu M, Joseph J. Expression of organ-specific antigens on capillary endothelial cells. Microvasc Res 1985;29:401–11.

32. Auerbach R, Joseph J. Cell surface markers on endothelial cells: A developmental perspective. In: Jaffe E, ed. The biology of endothelial cells. The Hague: Martinus Nijhoff, 1983: 393–400.

33. Auerbach R, Lu WC, Pardon E, Gumkowski F, Kaminska G, Kaminski M. Specificity of adhesion between murine tumor cells and capillary endothelium: an *in vitro* correlate of preferential metastasis *in vivo*. Cancer Res 1987;47:1492–6.

34. Gumkowski F, Kaminska G, Kaminski M, Morrissey LW, Auerbach R. Heterogeneity of mouse vascular endothelium. *In vitro* studies of lymphatic, large blood vessel and microvascular endothelial cells. Blood Vessels 1987;24:11–23.

35. Obeso J, Weber J, Auerbach R. A hemangioendothelioma-derived cell line: its use as a model for the study of endothelial cell biology. Lab Invest 1990;63:259–69.

36. Yu D, Auerbach R. Brain-specific differentiation of mouse yolk sac endothelial cells. Brain Res Dev Brain Res 1999;117: 159–69.

37. Wang SJ, Greer P, Auerbach R. Isolation and propagation of yolk-sac-derived endothelial cells from a hypervascular transgenic mouse expressing a gain-of-function fps/fes proto-oncogene. *In vitro* Cell Dev Biol Anim 1996;32:292–9.

38. Plendl J, Gilligan BJ, Wang SJ, et al. Primitive endothelial cell lines from the porcine embryonic yolk sac. *In vitro* Cell Dev Biol Anim 2002;38:334–42.

39. Jackson CL, Nguyen M. Human microvascular endothelial cells differ from macrovascular endothelial cells in their expression of matrix metalloproteinases. Int J Biochem Cell Biol 1997;29:1167–77.

40. Kriehuber E, Breiteneder-Geleff S, Groeger M, et al. Isolation and characterization of dermal lymphatic and blood endothelial cells reveal stable and functionally specialized cell lineages. J Exp Med 2001;194:797–808.

41. Helm CL, Zisch A, Swartz MA. Engineered blood and lymphatic capillaries in 3-D VEGF-fibrin-collagen matrices with interstitial flow. Biotechnol Bioeng 2007;96:167–76.

42. Rajotte D, Arap W, Hagedorn M, Koivunen E, Pasqualini R, Ruoslahti E. Molecular heterogeneity of the vascular endothelium revealed by *in vivo* phage display. J Clin Invest 1998;102:430–7.

43. Ruoslahti E, Rajotte D. An address system in the vasculature of normal tissues and tumors. Annu Rev Immunol 2000;18: 813–27.

44. Zetter BR. The cellular basis of site-specific tumor metastasis. N Engl J Med 1990;322:605–12.

45. Obeso JL, Auerbach R. A new microtechnique for quantitating cell movement *in vitro* using polystyrene bead monolayers. J Immunol Methods 1984;70:141–52.

46. Auerbach R, Bielich H, Obeso J, Weber J. Quantitation of endothelial cell movement: an *in vitro* approach to vasculogenesis and angiogenesis. Issues Biomed 1990;14:180–189.

47. Schor SL, Ellis IR, Harada K, et al. A novel 'sandwich' assay for quantifying chemo-regulated cell migration within 3-dimensional matrices: wound healing cytokines exhibit distinct

motogenic activities compared to the transmembrane assay. Cell Motil Cytoskeleton 2006;63:287–300.

48. Franzen L, Norrby K. A tissue model for quantitative studies on time course of healing, rate of healing, and cell proliferation after wounding. Acta Pathol Microbiol Immunol Scand [A] 1983;91:281–9.

49. Pepper MS, Belin D, Montesano R, Orci L, Vassalli JD. Transforming growth factor-beta 1 modulates basic fibroblast growth factor-induced proteolytic and angiogenic properties of endothelial cells in vitro. J Cell Biol 1990;111:743–55.

50. Montesano R, Orci L, Vassalli P. In vitro rapid organization of endothelial cells into capillary-like networks is promoted by collagen matrices. J Cell Biol 1983;97:1648–52.

51. Montesano R, Pepper MS. Three-dimensional in vitro assay of endothelial cell invasion and capillary tube morphogenesis. In: Little C, Mironov V, Sage E, eds. Vascular morphogenesis: In vivo, in vitro, in mente. Boston: Birkhausert, 1998:79–110.

52. Bahramsoltani M, Plendl J. [A new in vitro model to quantify angiogenesis]. Altex 2004;21:227–44.

53. Baker JH, Huxham LA, Kyle AH, Lam KK, Minchinton AI. Vascular-specific quantification in an in vivo Matrigel chamber angiogenesis assay. Microvasc Res 2006;71:69–75.

54. Korff T, Kimmina S, Martiny-Baron G, Augustin HG. Blood vessel maturation in a 3-dimensional spheroidal coculture model: direct contact with smooth muscle cells regulates endothelial cell quiescence and abrogates VEGF responsiveness. Faseb J 2001;15:447–57.

55. Globerson A, Auerbach R. Primary antibody response in organ cultures. J Exp Med 1966;124:1001–16.

56. Globerson A, Auerbach R. Primary immune reactions in organ cultures. Science 1965;149:991–3.

57. Blacher S, Devy L, Burbridge MF, et al. Improved quantification of angiogenesis in the rat aortic ring assay. Angiogenesis 2001;4:133–42.

58. Nicosia RF, Ottinetti A. Growth of microvessels in serum-free matrix culture of rat aorta. A quantitative assay of angiogenesis in vitro. Lab Invest 1990;63:115–22.

59. Masson VV, Devy L, Grignet-Debrus C, et al. Mouse Aortic Ring Assay: A New Approach of the Molecular Genetics of Angiogenesis. Biol Proced Online 2002;4:24–31.

60. Stiffey-Wilusz J, Boice JA, Ronan J, Fletcher AM, Anderson MS. An ex vivo angiogenesis assay utilizing commercial porcine carotid artery: modification of the rat aortic ring assay. Angiogenesis 2001;4:3–9.

61. Muthukkaruppan V, Shinners B, Lewis R, Park S-J, Baechler B, Auerbach R. The chick embryo aortic arch assay: a new, rapid, quantifiable in vitro method for testing the efficacy of angiogenic and anti-angiogenic factors in a three-dimensional, serum-free ogran culture system. Proc Am Assoc Cancer Res 2000;41:65.

62. Hamburger V. A manual of experimental embryology. Chicago: University of Chicago Press, 1942:212.

63. Richardson M, Singh G. Observations on the use of the avian chorioallantoic membrane (CAM) model in investigations into angiogenesis. Curr Drug Targets Cardiovasc Haematol Disord 2003;3:155–85.

64. Auerbach R, Kubai L, Knighton D, Folkman J. A simple procedure for the long-term cultivation of chicken embryos. Dev Biol 1974;41:391–4.

65. Dunn BE. Technique of shell-less culture of the 72-hour avian embryo. Poult Sci 1974;53:409–12.

66. Dunn BE, Boone MA. Growth of the chick embryo in vitro. Poult Sci 1976;55:1067–71.

67. Dunn BE, Boone MA. Photographic study of chick embryo development in vitro. Poult Sci 1978;57:370–7.

68. Dugan JD, Jr., Lawton MT, Glaser B, Brem H. A new technique for explantation and in vitro cultivation of chicken embryos. Anat Rec 1991;229:125–8.

69. Voss K, Jacob W, Roth K. A new image analysis method for the quantification of neovascularization. Exp Pathol 1984;26: 155–61.

70. Vu MT, Smith CF, Burger PC, Klintworth GK. An evaluation of methods to quantitate the chick chorioallantoic membrane assay in angiogenesis. Lab Invest 1985;53:499–508.

71. Vico PG, Kyriacos S, Heymans O, Louryan S, Cartilier L. Dynamic study of the extraembryonic vascular network of the chick embryo by fractal analysis. J Theor Biol 1998;195: 525–32.

72. Ribatti D, Gualandris A, Bastaki M, et al. New model for the study of angiogenesis and antiangiogenesis in the chick embryo chorioallantoic membrane: the gelatin sponge/chorioallantoic membrane assay. J Vasc Res 1997;34:455–63.

73. Storgard C, Mikolon D, Stupack DG. Angiogenesis assays in the chick CAM. Methods Mol Biol 2005;294:123–36.

74. Thompson WD, Reid A. Quantitative assays for the chick chorioallantoic membrane. Adv Exp Med Biol 2000;476:225–36.

75. Romanoff A. The avian embryo; structural and functional development. New York: Macmillan, 1960.

76. Ausprunk DH, Knighton DR, Folkman J. Vascularization of normal and neoplastic tissues grafted to the chick chorioallantois. Role of host and preexisting graft blood vessels. Am J Pathol 1975;79:597–628.

77. Ausprunk DH, Knighton DR, Folkman J. Differentiation of vascular endothelium in the chick chorioallantois: a structural and autoradiographic study. Dev Biol 1974;38:237–48.

78. Honda H, Yoshizato K. Formation of the branching pattern of blood vessels in the wall of the avian yolk sac studied by a computer simulation. Dev Growth Differ 1997;39:581–9.

79. Parsons-Wingerter P, Lwai B, Yang MC, et al. A novel assay of angiogenesis in the quail chorioallantoic membrane: stimulation by bFGF and inhibition by angiostatin according to fractal dimension and grid intersection. Microvasc Res 1998;55: 201–14.

80. Norrby K, Jakobsson A, Sorbo J. Quantitative angiogenesis in spreads of intact rat mesenteric windows. Microvasc Res 1990;39:341–8.

81. Norrby K, Jakobsson A, Sorbo J. Mast-cell-mediated angiogenesis: a novel experimental model using the rat mesentery. Virchows Arch B Cell Pathol Incl Mol Pathol 1986;52:195–206.

82. Gerber SA, Rybalko VY, Bigelow CE, et al. Preferential attachment of peritoneal tumor metastases to omental immune aggregates and possible role of a unique vascular microenvironment in metastatic survival and growth. Am J Pathol 2006;169:1739–52.

83. Kiricuta I, Popescu V. [Use of the omentum in the treatment of burns and severe hand trauma]. Ann Chir Plast 1976;21: 147–50.

84. Gimbrone MA, Jr., Cotran RS, Leapman SB, Folkman J. Tumor growth and neovascularization: an experimental model using the rabbit cornea. J Natl Cancer Inst 1974;52:413–27.

85. Muthukkaruppan VR, Kubai L, Auerbach R. Tumor-induced neovascularization in the mouse eye. J Natl Cancer Inst 1982;69:699–708.

86. Polakowski IJ, Lewis MK, Muthukkaruppan VR, Erdman B, Kubai L, Auerbach R. A ribonuclease inhibitor expresses anti-angiogenic properties and leads to reduced tumor growth in mice. Am J Pathol 1993;143:507–17.

87. Dickerson EB, Akhtar N, Steinberg H, et al. Enhancement of the antiangiogenic activity of interleukin-12 by peptide targeted delivery of the cytokine to alphavbeta3 integrin. Mol Cancer Res 2004;2:663–73.

88. Brem H, Folkman J. Inhibition of tumor angiogenesis mediated by cartilage. J Exp Med 1975;141:427–39.

89. Langer R, Folkman J. Polymers for the sustained release of proteins and other macromolecules. Nature 1976;263:797–800.

90. Kenyon BM, Voest EE, Chen CC, Flynn E, Folkman J, D'Amato RJ. A model of angiogenesis in the mouse cornea. Invest Ophthalmol Vis Sci 1996;37:1625–32.

91. Chang L, Kaipainen A, Folkman J. Lymphangiogenesis new mechanisms. Ann NY Acad Sci 2002;979:111–9.

92. Wilting J, Christ B. A morphological study of the rabbit corneal assay. Ann Anat 1992;174:549–56.

93. Rogers MS, D'Amato RJ. The effect of genetic diversity on angiogenesis. Exp Cell Res 2006;312:561–74.

94. Rohan RM, Fernandez A, Udagawa T, Yuan J, D'Amato RJ. Genetic heterogeneity of angiogenesis in mice. Faseb J 2000;14:871–6.

95. Shaked Y, Bertolini F, Man S, et al. Genetic heterogeneity of the vasculogenic phenotype parallels angiogenesis; Implications for cellular surrogate marker analysis of antiangiogenesis. Cancer Cell 2005;7:101–11.

96. Greene H. The heterologous transplantation of embryonic mammalian tissues. Cancer Res 1943;3:809–822.

97. Grobstein C. Intra-ocular growth and differentiation of clusters of mouse embryonic shields cultured with and without primitive endoderm and in the presence of possible inductors. J Exp Zool 1952;119:355–379.

98. Auerbach R. Analysis of the developmental effects of a lethal mutation in the house mouse. J Exp Zool 1954;127:305–330.

99. Auerbach R. Genetic control of thymus lymphoid differentiation. Proc Natl Acad Sci U S A 1961;47:1175–1181.

100. Vukicevic S, Kleinman HK, Luyten FP, Roberts AB, Roche NS, Reddi AH. Identification of multiple active growth factors in basement membrane Matrigel suggests caution in interpretation of cellular activity related to extracellular matrix components. Exp Cell Res 1992;202:1–8.

101. Passaniti A, Taylor RM, Pili R, et al. A simple, quantitative method for assessing angiogenesis and antiangiogenic agents using reconstituted basement membrane, heparin, and fibroblast growth factor. Lab Invest 1992;67:519–28.

102. Akhtar N, Dickerson EB, Auerbach R. The sponge/Matrigel angiogenesis assay. Angiogenesis 2002;5:75–80.

103. Auerbach R, Auerbach W. Regional differences in the growth of normal and neoplastic cells. Science 1982;215:127–34.

104. Auerbach R, Morrissey LW, Sidky YA. Regional differences in tumor growth: studies of the vascular system. Int J Cancer 1978;22:40–6.

105. Kubai L, Auerbach R. Regional differences in the growth of skin transplants. Transplantation 1980;30:128–31.

106. Auerbach R, Morrissey LW, Sidky YA. Regional differences in the incidence and growth of mouse tumors following intradermal or subcutaneous inoculation. Cancer Res 1978;38:1739–44.

107. Killion JJ, Radinsky R, Fidler IJ. Orthotopic models are necessary to predict therapy of transplantable tumors in mice. Cancer Metastasis Rev 1998;17:279–84.

108. Yang M, Baranov E, Wang JW, et al. Direct external imaging of nascent cancer, tumor progression, angiogenesis, and metastasis on internal organs in the fluorescent orthotopic model. Proc Natl Acad Sci USA 2002;99:3824–9.

109. Fidler IJ. Critical factors in the biology of human cancer metastasis: twenty-eighth G.H.A. Clowes memorial award lecture. Cancer Res 1990;50:6130–8.

110. Plunkett ML, Hailey JA. An in vivo quantitative angiogenesis model using tumor cells entrapped in alginate. Lab Invest 1990;62:510–7.

111. Robertson NE, Discafani CM, Downs EC, et al. A quantitative in vivo mouse model used to assay inhibitors of tumor-induced angiogenesis. Cancer Res 1991;51:1339–44.

112. Andrade SP, Fan TP, Lewis GP. Quantitative in-vivo studies on angiogenesis in a rat sponge model. Br J Exp Pathol 1987;68:755–66.

113. Andrade SP, Machado RD, Teixeira AS, Belo AV, Tarso AM, Beraldo WT. Sponge-induced angiogenesis in mice and the pharmacological reactivity of the neovasculature quantitated by a fluorimetric method. Microvasc Res 1997;54:253–61.

114. Ellis L, Gilston V, Soo CC, Morris CJ, Kidd BL, Winyard PG. Activation of the transcription factor NF-kappaB in the rat air pouch model of inflammation. Ann Rheum Dis 2000;59:303–7.

115. Selye H. Use of the "granuloma pouch technique" in the study of antiphagocytic corticoids. Proc Soc Exp Biol Med 1953;82:328–333.

116. Fan TP, Hu DE, Smither RL, Gresham GA. Further studies on angiogenesis in a rat sponge model. Exs 1992;61:308–14.

117. Fajardo LF, Kowalski J, Kwan HH, Prionas SD, Allison AC. The disc angiogenesis system. Lab Invest 1988;58:718–24.

118. Kowalski J, Kwan HH, Prionas SD, Allison AC, Fajardo LF. Characterization and applications of the disc angiogenesis system. Exp Mol Pathol 1992;56:1–19.

119. Nelson MJ, Conley FK, Fajardo LF. Application of the disc angiogenesis system to tumor-induced neovascularization. Exp Mol Pathol 1993;58:105–13.

120. Grant DS, Kinsella JL, Fridman R, et al. Interaction of endothelial cells with a laminin A chain peptide (SIKVAV) in vitro and induction of angiogenic behavior in vivo. J Cell Physiol 1992;153:614–25.

121. Algire G. An adaptation of the transparent chamber technique to the mouse. J Natl Cancer Inst 1943;4:1–11.

122. Grobstein C. Morphogenetic interaction between embryonic mouse tissues separated by a membrane filter. Nature 1953;172:869–871.

123. Sewell IA. Studies of the microcirculation using transparent tissue observation chambers inserted in the hamster cheek pouch. J Anat 1966;100:839–56.

124. Riley CM, Fuegy PW, Firpo MA, Shu XZ, Prestwich GD, Peattie RA. Stimulation of in vivo angiogenesis using dual growth factor-loaded crosslinked glycosaminoglycan hydrogels. Biomaterials 2006;27:5935–43.

125. Ichioka S, Shibata M, Kosaki K, Sato Y, Harii K, Kamiya A. Effects of shear stress on wound-healing angiogenesis in the rabbit ear chamber. J Surg Res 1997;72:29–35.

126. Cho CH, Sung HK, Kim KT, et al. COMP-angiopoietin-1 promotes wound healing through enhanced angiogenesis, lymphangiogenesis, and blood flow in a diabetic mouse model. Proc Natl Acad Sci USA 2006;103:4946–51.

127. Ny A, Autiero M, Carmeliet P. Zebrafish and Xenopus tadpoles: small animal models to study angiogenesis and lymphangiogenesis. Exp Cell Res 2006;312:684–93.

128. Childs S, Chen JN, Garrity DM, Fishman MC. Patterning of angiogenesis in the zebrafish embryo. Development 2002;129:973–82.

129. Lawson ND, Weinstein BM. *In vivo* imaging of embryonic vascular development using transgenic zebrafish. Dev Biol 2002;248:307–18.

130. Serbedzija GN, Flynn E, Willett CE. Zebrafish angiogenesis: a new model for drug screening. Angiogenesis 1999;3:353–9.

131. Lee P, Goishi K, Davidson AJ, Mannix R, Zon L, Klagsbrun M. Neuropilin-1 is required for vascular development and is a mediator of VEGF-dependent angiogenesis in zebrafish. Proc Natl Acad Sci USA 2002;99:10470–5.

132. Seng WL, Eng K, Lee J, McGrath P. Use of a monoclonal antibody specific for activated endothelial cells to quantitate angiogenesis *in vivo* in zebrafish after drug treatment. Angiogenesis 2004;7:243–53.

133. Kajimura S, Aida K, Duan C. Understanding hypoxia-induced gene expression in early development: *in vitro* and *in vivo* analysis of hypoxia-inducible factor 1-regulated zebra fish insulin-like growth factor binding protein 1 gene expression. Mol Cell Biol 2006;26:1142–55.

134. Levine AJ, Munoz-Sanjuan I, Bell E, North AJ, Brivanlou AH. Fluorescent labeling of endothelial cells allows *in vivo*, continuous characterization of the vascular development of Xenopus laevis. Dev Biol 2003;254:50–67.

135. Ny A, Koch M, Schneider M, et al. A genetic *Xenopus laevis* tadpole model to study lymphangiogenesis. Nat Med 2005;11:998–1004.

136. Chaplain MA, Graziano L, Preziosi L. Mathematical modelling of the loss of tissue compression responsiveness and its role in solid tumour development. Math Med Biol 2006;23:197–229.

137. Chaplain MA, McDougall SR, Anderson AR. Mathematical modeling of tumor-induced angiogenesis. Annu Rev Biomed Eng 2006;8:233–57.

138. Charalampidis D, Pascotto M, Kerut EK, Lindner JR. Anatomy and flow in normal and ischemic microvasculature based on a novel temporal fractal dimension analysis algorithm using contrast enhanced ultrasound. IEEE Trans Med Imaging 2006;25:1079–86.

139. Geninatti Crich S, Bussolati B, Tei L, et al. Magnetic resonance visualization of tumor angiogenesis by targeting neural cell adhesion molecules with the highly sensitive gadolinium-loaded apoferritin probe. Cancer Res 2006;66:9196–201.

140. Lin PC. Optical imaging and tumor angiogenesis. J Cell Biochem 2003;90:484–91.

141. Padhani AR, Neeman M. Challenges for imaging angiogenesis. Br J Radiol 2001;74:886–90.

Chapter 27
Surrogates for Clinical Development

Sylvia S. W. Ng[1,3] and Kim N. Chi[1,2,4]

Keywords: surrogate markers, non-invasive imaging, microvessel density

Abstract: Unlike cytotoxic drugs, antiangiogenic agents by themselves seldom induce rapid tumor shrinkage over short periods of time. Therefore, traditional clinical endpoints (e.g., complete or partial response) that are used to determine the anti-tumor activity of the former may not necessarily be applicable to the latter. With a myriad of angiogenesis inhibitors undergoing and entering clinical trials for cancer treatment, there is an urgent need to develop and validate novel surrogate biomarkers for defining optimal drug doses and schedules, monitoring drug efficacy, and predicting drug response. This chapter describes the molecular, cellular, and functional imaging candidates currently being explored as potential surrogate pharmacodynamic markers for antiangiogenic therapies. Their advantages and pitfalls are also discussed.

Introduction

Tumor growth is limited by the supply of oxygen and nutrients. Solid tumors cannot grow beyond $2–3\,mm^3$ without the formation of new blood vessels [1]. Since Judah Folkman [2] proposed that inhibition of tumor angiogenesis would be an effective strategy to treat human cancer, a multitude of novel agents that target tumor blood vessels have entered the drug development pipeline. Tumor vasculature can be targeted by antiangiogenic agents or vascular disrupting agents. While 'dedicated' antiangiogenic agents inhibit new microvessel growth and induce the regression of newly formed microvessels, vascular disrupting agents cause collapse of both newly formed and established vessels and massive intratumoral necrosis. The US Food and Drug Administration has recently approved bevacizumab (Avastin®; a recombinant humanized anti-vascular endothelial growth factor (VEGF) monoclonal antibody), sorafenib (Nexavar®; a Raf-kinase inhibitor that also targets the VEGF receptor 2, VEGFR2), and sunitinib (Sutent®; an inhibitor of multiple tyrosine kinases including VEGFR1, VEGFR2, VEGFR3) for the treatment of colorectal cancer, renal cell carcinoma, and gastrointestinal stromal tumor. While most antiangiongenic agents by themselves produce limited objective responses (with some exceptions, notably when used in renal cell carcinoma) and virtually no long-term survival benefits, their combination with chemotherapy often demonstrate promising anti-tumor activity. Interestingly, recent data indicate that tumor angiogenesis is also inhibited by "metronomic chemotherapy": the administration of comparatively low doses of conventional chemotherapeutic drugs on a frequent or continuous schedule without extended breaks over long periods of time [3]. Such regimens, in comparison to pulsatile maximum tolerated dose (MTD) chemotherapy, are generally much less toxic [4] and thus do not require supportive care measures. They have also demonstrated surprisingly effective antitumor activity that are sometimes even superior to the more cytotoxic—and toxic—regimens of the same drug in preclinical models and encouraging activity in several Phase I/II clinical trials [3,5–10]. More importantly, tumors that became resistant to the MTD regimen were shown to respond to the metronomic regimen of the same drug [5], suggesting a shift of treatment focus from drug-resistant tumor cells to drug-sensitive, genetically 'stable' endothelial cells in metronomic regimens [3,11–13]. The mechanistic basis of metronomic chemotherapy is believed to be primarily antiangiogenic, either as a result of direct killing of endothelial cells in the growing tumor vasculature [5,14], and/or destruction of bone marrow-derived endothelial progenitor cells [6,15].

Departments of Advanced Therapeutics[1] and Medical Oncology[2], British Columbia Cancer Agency; Faculty of Pharmaceutical Sciences[3] and Faculty of Medicine[4], University of British Columbia, Vancouver, Canada

Antiangiogenesis-based therapies are often cytostatic and do not cause immediate, direct toxicity to tumor cells. Rapid tumor shrinkage, as seen after MTD cytotoxic chemotherapy, seldom occurs following treatment with antiangiogenic agents. Therefore, traditional Phase II anti-tumor activity endpoints, defined by standard RECIST (response evaluation criteria in solid tumors) measurable disease criteria, may not be appropriate response markers for assessing the therapeutic potential of these agents. Two major issues arise in the initial clinical assessment of a particular agent. The first question being how to demonstrate clinical evidence of biologic activity (i.e., an antiangiogenic effect) and second, how to evaluate clinical anti-tumor activity in the potential absence of RECIST defined responses. The second question has largely been addressed by incorporating time to progression as primary endpoint, frequently necessitating larger, randomized phase II designs. This further emphasizes the importance of addressing the first question of establishing proof of principal evidence of clinical biologic activity and, therefore, defining a biologically effective dose and schedule of an antiangiogenic treatment modality to ensure the best chance of success in subsequent trials. Traditional phase I design selects an MTD based on toxicity for determining phase II dosing. Antiangiogenic agents, however, may be relatively non-toxic compared to cytotoxic drugs, and have shown biological activity at doses well under the MTD, if one can be defined at all. A suboptimal dose leads to treatment ineffectiveness, whereas too high a dose may hinder the delivery of concurrently administered chemotherapy by shutting down the entire tumor vasculature or, worse still, lead to toxicity (e.g., adverse cardiovascular events) by damaging the normal vasculature. While the severity of a skin rash and the expression/mutation of the epidermal growth factor receptor (EGFR) may be used to monitor and predict treatment response to the EGFR inhibitor erlotinib (Tarceva®), surrogate markers that can do the same for antiangiogenesis-based therapies are still in the early stage of clinical development.

To demonstrate clinical biologic activity of antiangiogenesis-based therapies, evidence of vascular (i.e., target) inhibition need to be sought. Determining tumor endothelial cell death in biopsies before and after antiangiogenesis-based therapies would appear to be the most direct and intuitive measurement of an antiangiogenic effect. However, repeated biopsies are invasive and sometimes impossible or unethical. Even when serial biopsies can be performed, tumor heterogeneity raises difficult questions such as whether an increase in endothelial cell apoptosis in a tumor biopsy section after treatment reflects a global antiangiogenic effect in the entire tumor or a lack thereof truly indicates treatment failure. Similar issues and others have plagued the use of intratumoral microvessel density (MVD). Although intratumoral MVD has proven to be a useful independent prognostic marker for various types of cancer [16,17], its clinical usefulness as a surrogate pharmacodynamic marker for antiangiogenic agents is contentious [17,18]. A decrease in MVD is suggestive of a drug's antiangiogenic activity. However, a lack thereof or even an increase in MVD does not necessarily reflect drug inactivity [17,18]. Clearly, alternative surrogate markers for assessing treatment response to antiangiogenic therapies need to be identified. Sensitivity, robustness, easy accessibility for repeated sampling, cost-effectiveness are all important characteristics of such markers.

Soluble Molecular Markers in Biological Fluids

Circulating Levels of Angiogenesis Regulators

Although most frequently used, measuring plasma, serum, and urine levels of soluble angiogenesis stimulators (e.g., VEGF, bFGF) and inhibitors (e.g., thrombospondin-1, endostatin) has yielded inconclusive and often conflicting results. Their levels have been reported to increase, decrease, or remain unchanged following administration of various antiangiogenic treatment modalities [10,19–27]. The discrepancies between studies can be partially attributed to the assessment of these factors in serum versus plasma. Accurate measurements in serum are precluded by the fact that platelets also produce and secrete many of these factors. In addition, binding of these factors to serum proteins render them unavailable for antibody detection by enzyme-linked immunosorbent assays. When changes in any one or two of these factors were indeed detected in plasma, serum, or urine, significant correlation with clinical outcome was only evident in some clinical studies but not others [10, 21,22,24,26–28]. Given that a tumor's angiogenicity is determined by the concerted action of at least 40 endogenous proangiogenic and antiangiogenic molecules whose expression can vary during tumor progression, the evaluation of one or two of them is unlikely to reflect the net effect of the complex interactions amongst all the 40 molecules [10]. It is therefore not surprising that detectable changes in one or two angiogenesis regulators provide little, if any, information on the activity of antiangiogenic agents.

Endothelial cell-associated molecules, such as the vascular cell adhesion molecule-1 (VCAM-1) and E-selectin, as well as soluble forms of VEGFR1, VEGFR2, and Tie-2, are shed into the circulation. Their levels have also been investigated in some clinical studies as potential surrogate markers for monitoring antiangiogenic drug activity. The levels of sE-selectin and sTie-2 were not altered by PTK787/ZK222584 which inhibits all three VEGFRs [24]. On the other hand, sE-selectin was increased by the angiogenesis inhibitors SU5416 [21,23] and CM-101 [29]. VCAM-1 levels was unchanged by SU5416 [23] but increased by CM-101 [29], or highly variable during celecoxib plus metronomic vinblastine or cyclophosphamide treatment [26]. These results were obtained from Phase I or pilot studies in which a small number of patients was involved and correlation between changes in the levels of the molecules and clinical outcome was not determined. However, in a Phase II trial ($n = 43$), Harris et al. [30] showed that a reduction in sTie-2 but not sVEGFR1 levels on treatment with the antiangiogenic agent razoxane was associated with stable disease ($P = 0.05$) and improved survival ($P = 0.04$). So far, the clinical usefulness of the aforementioned soluble factors or molecules as surrogate pharmacodynamic markers for antiangiogenic therapies has not been validated. Key issues remain unresolved. For example, to what extent, if at all, do levels of these molecules in biological fluids such as plasma, serum, and urine reflect those in the tumor microenvironment? Are circulating and/or tumor levels of these molecules good indicators of tumor angiogenesis? Do changes in circulating and/or tumor levels of these molecules indicate drug effect or progressive disease? These questions can only be addressed by conducting correlative studies in prospective Phase II and Phase III clinical trials.

Circulating Cellular Markers

Circulating Endothelial Cells and Circulating Endothelial Progenitors

It has been reported that tumor blood vessels shed endothelial cells into the circulation [31] and that endothelial progenitors are recruited from bone marrow to tumor sites to participate in angiogenesis [32–36]. Although controversies still exist on the relative contribution of bone marrow-derived endothelial precursors to tumor angiogenesis, they have not diminished the potential or prevented the development of circulating endothelial cells (CECs) and circulating endothelial progenitors (CEPs) as surrogate pharmacodynamic markers for antiangiogenic treatment modalities. Francesco Bertolini's group developed a multiparameter flow cytometry assay to enumerate CECs and CEPs in preclinical cancer models and in cancer patients using a combination of cell surface markers [6, 37–39]. Total CECs consist of CEPs and mature CECs. Murine CECs and CEPs are designated as CD45$^-$CD13$^+$VEGFR2$^+$CD117$^-$ and CD45$^-$CD13$^+$VEGFR2$^+$CD117$^+$, respectively. Human CECs and CEPs are defined as CD45$^-$CD31$^+$P1H12$^+$CD133$^-$ and CD45$^-$CD31$^+$P1H12$^+$CD133$^+$, respectively. Some studies have also used CD105 or CD106 to distinguish activated CECs from their resting counterparts, and 7AAD to differentiate viable versus apoptotic CECs [37,40]. Using this flow cytometry assay and associated data analysis criteria, the levels of resting and activated CECs in newly diagnosed breast cancer ($n = 46$) or lymphoma ($n = 30$) patients were shown to be five times higher than those in healthy controls ($n = 20$) [37]. It was speculated that such CEC increase in cancer patients might originate from "newly formed tumor vessels, ingress of proliferating endothelial cells from neighboring normal vessels, or distant uninvolved vessels activated by tumor-derived cytokines" [37]. Interestingly, there were marked reductions in CECs in lymphoma patients who achieved complete remission after chemotherapy [37], and in breast cancer patients who underwent

quadrantectomy [37] or anthracycline/taxane-based chemotherapy [41]. Using an animal model of human lymphoma, the same group subsequently showed that MTD cyclophosphamide increases predominantly the levels of apoptotic [7AAD$^+$) circulating hematopoietic cells and to a lesser extent those of apoptotic CECs [38]. In contrast, the angiogenesis inhibitor endostatin induces apoptosis only in CECs but not in circulating hematopoietic cells [38,39]. Endostatin treatment also decreased the total number of CEPs, the differentiation and clonogenic potential of these CEPs, and tumor growth [39]. Furthermore, low-dose metronomic cyclophosphamide was shown to induce a persistent decrease in CEP number and viability as well as a more durable tumor growth inhibition compared to MTD cyclophosphamide [6]. Consistent with earlier findings, Beerepoot et al. [42] evaluated 112 cancer patients with various tumor types and reported a 3.6-fold increase in CECs in patients with progressive disease but not in those with stable disease compared to healthy controls ($n = 46$). However, CEC number was shown to increase following cytotoxic taxane chemotherapy (in 7 out of 8 tested patients). In another study, thalidomide, a known inhibitor of angiogenesis, induced a 10-fold decrease in activated bone marrow endothelial cells in patients with myelodysplastic syndrome [43]. A notable decline in both CECs and CEPs was reported in rectal cancer patients treated with bevacizumab [44]. On the other hand, treatment of mice bearing Lewis lung carcinoma with the VEGFR2 inhibitor ZD6474 or the vascular disrupting agent ZD6126 induced an increase in CECs and had no effect on CEPs [45]. Similar CEC increases were also seen in patients treated with endostatin [46] or ZD6126 [47, 48]. It was suggested that the sloughing of fragile, mature endothelium from tumor vasculature and other sources may contribute to the CEC increases [45]. Mancuso et al. [49] recently analyzed CEC and CEP kinetics in 81 advanced breast cancer patients at baseline and after 2 months of metronomic cyclophosphamide/methotrexate chemotherapy. A significant improvement in progression-free survival ($P = 0.001$) and overall survival ($P = 0.005$) was observed in those patients with a post-treatment CEC count of greater than 11/μl [49]. The rise in CEC number in responders was attributed to the increase in the apoptotic fraction of these cells [49]. No correlation was found between CEP number and clinical outcome [49]. The origin of the apoptotic CECs was investigated in four preclinical cancer models [49]. The rise in apoptotic CECs was only evident in tumor-bearing but not in tumor-free mice during metronomic chemotherapy regimens, suggesting a release of these cells from tumors [49]. Whether this is the case in treated cancer patients remains to be determined. The concurrent assessment of CEC kinetics and endothelial cell apoptosis in serial sections of tumor biopsies obtained whenever possible may shed light on this question.

Robert Kerbel's group recently performed a series of elegant preclinical studies and demonstrated that viable CEP levels can be used to determine the optimal biologic doses for antiangiogenic drugs [50] and for metronomic chemotherapy [15], at least in mice. The dose of DC101 (800 µg), a rat monoclonal antibody against mouse VEGFR2, that induced the lowest levels of viable CEPs elicited the greatest reduction in tumor volume, whereas dose escalation up to 2,000 µg failed to decrease both parameters any further [50]. In addition, metronomic doses of cyclophosphamide, vinorelbine, and cisplatin that induced maximum reductions in viable CEPs were correlated with greatest tumor growth inhibition [15]. The levels of viable CEPs were also used to estimate the optimal biologic doses for metronomic UFT (a 5-FU prodrug) [51] and ABI-007 (Abraxane®; an albumin-bound nanoparticle paclitaxel) regimens [7], both of which demonstrated antitumor activity in various human tumor xenograft models. The clinical usefulness of viable CEP levels in determining optimal biologic doses of antiangiogenic therapies remains to be tested. Shaked et al. [52] showed that vascular disrupting agents, contrary to 'dedicated' antiangiogenic agents and metronomic chemotherapy, induce a prominent increase in CEPs. These CEPs subsequently home to the viable rim in the tumor periphery and contribute to tumor regrowth [52]. It should be noted that conventional MTD chemotherapy also increases CEP levels [6,41], possibly due to mobilization of bone marrow-derived hematopoietic progenitors in response to chemotherapy-induced myelosuppression [78,79]. This CEP surge, also seen after myocardial infarctions [53], is likely due to an acute reactive response to vascular damage resulting from exposure to vascular targeting agents or cytotoxic chemotherapy [52].

The above preclinical and clinical findings may seem confusing initially in that CECs and CEPs can increase or decrease in response to 'dedicated' antiangiogenic agents, metronomic chemotherapy, and vascular disrupting agents. However, different panels of cell surface markers and different techniques were sometimes used by different groups to define and quantify CECs and CEPs, making head-to-head comparisons between studies difficult. In some cases, the number of patients evaluated for CECs and CEPs is too small to draw definitive conclusions. Earlier studies focused on the total number and viability of CECs and CEPs, while latter ones also evaluated the apoptotic fraction of CECs, which is emerging as a useful surrogate marker for response to antiangiogenic treatments. Tumor vasculature, normal vasculature, and bone marrow all contribute to total CEC count [49]. It was suggested that antiangiogenic therapies may exert dual effects on these compartments by increasing CECs and reducing bone marrow-derived CEPs [49]. The low CEP counts in humans raise concerns about the sensitivity and robustness of flow cytometry assays to detect CEP changes in the clinical setting. Other diseases including cardiovascular disorders [54], not uncommon in elderly cancer patients, can also affect CEC and CEP counts and kinetics. The specificity of CEC and CEP changes as surrogate pharmacodynamic markers for antiangiogenic or antivascular agents is unclear under these circumstances. Despite these uncertainties, CECs and CEPs appear to be two of the most promising surrogate biomarkers at present and warrant further clinical development.

Circulating Tumor Cells

Tumor cells shed from solid tumors have been detected in the circulation of cancer patients [55–63]. The presence of 5 or more circulating tumor cells (CTCs) in 7.5 ml of blood at the time of diagnosis was associated with shorter progression-free ($P<0.001$) and overall survival ($P<0.001$) in patients with metastatic breast cancer ($n = 177$) [64,65]. Similar results were reported in prostate cancer [66,67]. However, very few studies have examined the use of CTCs as a surrogate marker for treatment response. A reduction of CTCs was reported in Her2/neu+ breast cancer patients treated with trastuzumab-based regimens [65], while the opposite was observed in colorectal cancer patients during palliative chemotherapy [68]. In another breast cancer study, Her2/neu overexpression was not detected on CTCs of three patients before initiation of trastuzumab therapy but was detected on their CTCs that underwent a concomitant increase in number during the course of treatment [69]. More studies are clearly required to validate the utility of CTCs in monitoring and predicting response to chemotherapy and molecularly targeted drugs including antiangiogenic agents. Considering that CTCs as well as CECs and CEPs can be measured from a blood sample, concurrent assessment of the kinetics and viability of all three cell types over time may reflect/predict response to antiangiogenic or antivascular agents more reliably than assessment of either cell type alone. Browder et al. [5] showed that tumor cell apoptosis occurs 3–4 days after endothelial cell apoptosis in response to antiangiogenic or metronomic cyclophosphamide treatment. We speculate that an effective antiangiogenic treatment modality may induce an initial increase in apoptotic CECs and a reduction in CEPs followed by an increase in apoptotic CTCs. Recent Phase I trials reported improvement in bone scans in prostate cancer patients after sorafenib treatment despite PSA increases [70], further highlighting how traditional measures of response, PSA in this case, may not be applicable to antiangiogenic agents, and why alternative surrogate pharmacodynamic markers such as CECs, CEPs, and CTCs are needed.

Functional Imaging Markers

Non-invasive imaging holds great promise in detecting early response to antiangiogenic or vascular disrupting therapies because it allows concurrent visualization of vascular architecture and assessment of vascular function in the entire tumor. Tumor blood flow, blood volume, vessel size, hypoxia, perfusion, and metabolism can be assessed by a combination of three-dimensional ultrasound, dynamic contrast enhanced-magnetic resonance imaging (DCE-MRI) or computed tomography (CT), and positron emission tomography (PET) (discussed in Chapter 28). Using PET imaging with [^{15}O]H$_2$O, tumor blood flow was shown to be reduced by endostatin in a Phase I trial [71]. A significant decline in tumor vascular permeability was detected by DCE-MRI and correlated with non-progressive disease in colorectal cancer patients after two cycles of PTK787 treatment [72]. DCE-MRI also revealed a decrease in tumor blood flow [73] and K^{trans} (a parameter that reflects vascular permeability, vessel surface area, and perfusion) [74] following infusion of the vascular disrupting agent combretastatin A-4. In another study [44], functional CT scans showed that bevacizumab reduces tumor blood perfusion and blood volume in rectal cancer patients after 12 days of treatment. Notably, these vascular functional changes were accompanied by concurrent increases in pericyte coverage of tumor vessels and decreases in interstitial pressure and viable CECs, providing evidence of an early response to bevacizumab despite the lack of tumor shrinkage in 5 out of 6 patients [44]. There are clearly no more direct ways to monitor the effects of antiangiogenic or antivascular therapies on the tumor vasculature than dynamic assessment of vascular function in situ. Concomitant evaluation of changes in one or more of the above circulating markers with those in functional imaging parameters will allow more accurate and reliable assessment of response to antiangiogenesis-based therapies. However, better computer software still needs to be developed for the accurate analyses of the latter. Expensive instrumentation and operating cost may preclude the widespread use of functional imaging.

Concluding Remarks and Future Perspectives

The number of patients receiving antiangiogenic therapies for cancer treatment will increase rapidly in the coming years. There is an urgent need to define optimal drug doses and schedules, monitor drug efficacy and toxicity, and predict drug response. Current surrogate pharmacodynamic markers being explored for antiangiogenic therapies include circulating molecular and cellular markers as well as functional imaging markers. As much potential as they hold, we have yet to refine our understanding of what the changes of these markers represent, identify more specific markers for different subpopulations of circulating cells and standardize the techniques used to measure them, as well as validate their clinical usefulness. Advances in flow cytometry will allow the use of more markers to better characterize CECs/CEPs and reflect drug–target interaction. For instance, it would be interesting to examine the fraction of phosphoVEGFR1$^-$phosphoVEGFR2$^-$ phosphoVEGFR3$^-$ CECs before and after treatment with SU11248, an inhibitor of all three VEGFRs. In addition to CEPs, other bone marrow-derived cell populations in the circulation such as Tie2-expressing monocytes [75], tumor-associated stromal cells [76], and tumor-associated dendritic cells [77] have recently been shown to contribute to tumor angiogenesis. The potential of these cells to serve as surrogate markers for antiangiogenic drug activity remains to be determined. As our understanding of the tumor vasculature deepens, we are also in the position to develop more reliable diagnostic methods for the identification of patients who will benefit most from specific antiangiogenic therapies. It may be possible to sort a patient's CECs/CEPs and CTCs at cancer diagnosis for gene expression/mutation and protein expression/phophorylation analyses. This may identify an angiogenic signature for the rational, tailored design of treatment protocols involving antiangiogenic agents with different mechanisms of action in combination with chemotherapy or radiotherapy. For patients with advanced disease who are on antiangiogenesis-based therapies, periodic evaluation of the CEC/CEP/CTC profile in peripheral blood using multiparameter flow cytometry and that of tumor vascular function using non-invasive imaging may be incorporated as part of long-term follow-up. As a tumor may depend on different angiogenic factors at different stages of growth, methods also need to be developed to identify the switch of dependency and modify treatment accordingly for maintenance of tumor growth inhibition. New platforms that are able to analyze a large panel of growth factors/cytokines using very small volumes of biological fluids will aid the monitoring of response to antiangiogenic therapies. The aforementioned surrogate markers should continue to be explored in the Phase I and II settings with the most promising ones to be incorporated into Phase III trials for validation.

References

1. Folkman, J. Tumor angiogenesis: a possible control point in tumor growth. Ann Intern Med, *82:* 96–100, 1975.
2. Folkman, J. Tumor angiogenesis: therapeutic implications. N Engl J Med, *285:* 1182–1186, 1971.
3. Kerbel, R. S. and Kamen, B. A. The anti-angiogenic basis of metronomic chemotherapy. Nat Rev Cancer, *4:* 423–436, 2004.
4. Emmenegger, U., Man, S., Shaked, Y., Francia, G., Wong, J. W., Hicklin, D. J., and Kerbel, R. S. A comparative analysis of low-dose metronomic cyclophosphamide reveals absent or low-grade toxicity on tissues highly sensitive to the toxic effects of maximum tolerated dose regimens. Cancer Res, *64:* 3994–4000, 2004.
5. Browder, T., Butterfield, C. E., Kraling, B. M., Shi, B., Marshall, B., O'Reilly, M. S., and Folkman, J. Antiangiogenic scheduling of chemotherapy improves efficacy against experimental drug-resistant cancer. Cancer Res, *60:* 1878–1886, 2000.
6. Bertolini, F., Paul, S., Mancuso, P., Monestiroli, S., Gobbi, A., Shaked, Y., and Kerbel, R. S. Maximum tolerable dose and low-dose metronomic chemotherapy have opposite effects on the mobilization and viability of circulating endothelial progenitor cells. Cancer Res, *63:* 4342–4346, 2003.
7. Ng, S. S., Sparreboom, A., Shaked, Y., Lee, C., Man, S., Desai, N., Soon-Shiong, P., Figg, W. D., and Kerbel, R. S. Influence of formulation vehicle on metronomic taxane chemotherapy: albumin-bound versus cremophor EL-based paclitaxel. Clin Cancer Res, *12:* 4331–4338, 2006.
8. Garcia, A. A., Oza, A. M., Hirte, H., Fleming, G., Tsao-Wei, D., Roman, L., Swenson, S., Gandara, D., Scudder, S., and Morgan, R. Interim report of a phase II clinical trial of bevacizumab (Bev) and low dose metronomic oral cyclophosphamide (mCTX) in recurrent ovarian (OC) and primary peritoneal carcinoma: A California Cancer Consortium Trial. Proc Am Soc Clin Oncol, *Abstract #5000,* 2005.
9. Colleoni, M., Rocca, A., Sandri, M. T., Zorzino, L., Masci, G., Nole, F., Peruzzotti, G., Robertson, C., Orlando, L., Cinieri, S.,

de, B. F., Viale, G., and Goldhirsch, A. Low-dose oral methotrex-
ate and cyclophosphamide in metastatic breast cancer: antitumor
activity and correlation with vascular endothelial growth factor
levels. Ann Oncol, 13: 73–80, 2002.

10. Kieran, M. W., Turner, C. D., Rubin, J. B., Chi, S. N., Zimmerman,
M. A., Chordas, C., Klement, G., Laforme, A., Gordon, A.,
Thomas, A., Neuberg, D., Browder, T., and Folkman, J. A fea-
sibility trial of antiangiogenic (metronomic) chemotherapy in
pediatric patients with recurrent or progressive cancer. J Pediatr
Hematol Oncol, 27: 573–581, 2005.

11. Baguley, B. C., Holdaway, K. M., Thomsen, L. L., Zhuang, L.,
and Zwi, L. J. Inhibition of growth of colon 38 adenocarcinoma
by vinblastine and colchicine: evidence for a vascular mechanism. Eur J Cancer, 27: 482–487, 1991.

12. Bocci, G., Nicolaou, K. C., and Kerbel, R. S. Protracted low-
dose effects on human endothelial cell proliferation and survival
in vitro reveal a selective antiangiogenic window for various chemotherapeutic drugs. Cancer Res, 62: 6938–6943, 2002.

13. Bocci, G., Francia, G., Man, S., Lawler, J., and Kerbel, R. S.
Thrombospondin 1, a mediator of the antiangiogenic effects of
low-dose metronomic chemotherapy. Proc Natl Acad Sci USA,
100: 12917–12922, 2003.

14. Klement, G., Baruchel, S., Rak, J., Man, S., Clark, K., Hicklin, D.
J., Bohlen, P., and Kerbel, R. S. Continuous low-dose therapy
with vinblastine and VEGF receptor-2 antibody induces sustained tumor regression without overt toxicity. J Clin Invest, 105:
15–24, 2000.

15. Shaked, Y., Emmenegger, U., Man, S., Cervi, D., Bertolini, F.,
Ben David, Y., and Kerbel, R. S. The optimal biological dose of
metronomic chemotherapy regimens is associated with maximum
antiangiogenic activity. Blood, 106: 3058–3061, 2005.

16. Gasparini, G. and Harris, A. L. Prognostic significance of
tumor vascularity, p. 317–399. Totowa, New Jersey: Human
Press, 1999.

17. Hlatky, L., Hahnfeldt, P., and Folkman, J. Clinical application
of antiangiogenic therapy: microvessel density, what it does and
doesn't tell us. J Natl Cancer Inst, 94: 883–893, 2002.

18. Kerbel, R. and Folkman, J. Clinical translation of angiogenesis
inhibitors. Nat Rev Cancer, 2: 727–739, 2002.

19. Braybrooke, J. P., O'Byrne, K. J., Propper, D. J., Blann, A.,
Saunders, M., Dobbs, N., Han, C., Woodhull, J., Mitchell, K.,
Crew, J., Smith, K., Stephens, R., Ganesan, T. S., Talbot, D. C.,
and Harris, A. L. A phase II study of razoxane, an antiangiogenic
topoisomerase II inhibitor, in renal cell cancer with assessment
of potential surrogate markers of angiogenesis. Clin Cancer Res,
6: 4697–4704, 2000.

20. Herbst, R. S., Hess, K. R., Tran, H. T., Tseng, J. E., Mullani, N.
A., Charnsangavej, C., Madden, T., Davis, D. W., McConkey, D. J.,
O'Reilly, M. S., Ellis, L. M., Pluda, J., Hong, W. K., and Abbruzzese,
J. L. Phase I study of recombinant human endostatin in patients with
advanced solid tumors. J Clin Oncol, 20: 3792–3803, 2002.

21. Kuenen, B. C., Levi, M., Meijers, J. C., van Hinsbergh, V. W.,
Berkhof, J., Kakkar, A. K., Hoekman, K., and Pinedo, H. M.
Potential role of platelets in endothelial damage observed during
treatment with cisplatin, gemcitabine, and the angiogenesis
inhibitor SU5416. J Clin Oncol, 21: 2192–2198, 2003.

22. Fiedler, W., Serve, H., Dohner, H., Schwittay, M., Ottmann, O. G.,
O'Farrell, A. M., Bello, C. L., Allred, R., Manning, W. C., Cher-
rington, J. M., Louie, S. G., Hong, W., Brega, N. M., Massimini, G.,
Scigalla, P., Berdel, W. E., and Hossfeld, D. K. A phase 1 study of

SU11248 in the treatment of patients with refractory or resistant
acute myeloid leukemia (AML) or not amenable to conventional
therapy for the disease. Blood, 105: 986–993, 2005.

23. Dowlati, A., Robertson, K., Radivoyevitch, T., Waas, J., Ziats, N. P.,
Hartman, P., Abdul-Karim, F. W., Wasman, J. K., Jesberger, J.,
Lewin, J., McCrae, K., Ivy, P., and Remick, S. C. Novel Phase
I dose de-escalation design trial to determine the biological
modulatory dose of the antiangiogenic agent SU5416. Clin Cancer
Res, 11: 7938–7944, 2005.

24. Drevs, J., Zirrgiebel, U., Schmidt-Gersbach, C. I., Mross, K.,
Medinger, M., Lee, L., Pinheiro, J., Wood, J., Thomas, A. L.,
Unger, C., Henry, A., Steward, W. P., Laurent, D., Lebwohl, D.,
Dugan, M., and Marme, D. Soluble markers for the assessment
of biological activity with PTK787/ZK 222584 (PTK/ZK), a
vascular endothelial growth factor receptor (VEGFR) tyrosine
kinase inhibitor in patients with advanced colorectal cancer from
two phase I trials. Ann Oncol, 16: 558–565, 2005.

25. Levine, A. M., Tulpule, A., Quinn, D. I., Gorospe, G., 3rd, Smith,
D. L., Hornor, L., Boswell, W. D., Espina, B. M., Groshen, S. G.,
Masood, R., and Gill, P. S. Phase I study of antisense oligonucleotide against vascular endothelial growth factor: decrease in
plasma vascular endothelial growth factor with potential clinical
efficacy. J Clin Oncol, 24: 1712–1719, 2006.

26. Stempak, D., Gammon, J., Halton, J., Moghrabi, A., Koren, G.,
and Baruchel, S. A pilot pharmacokinetic and antiangiogenic
biomarker study of celecoxib and low-dose metronomic vinblastine or cyclophosphamide in pediatric recurrent solid tumors.
J Pediatr Hematol Oncol, 28: 720–728, 2006.

27. Fury, M. G., Zahalsky, A., Wong, R., Venkatraman, E., Lis, E.,
Hann, L., Aliff, T., Gerald, W., Fleisher, M., and Pfister, D. G. A
Phase II study of SU5416 in patients with advanced or recurrent
head and neck cancers. Invest New Drugs, 25: 165–172, 2007.

28. Fine, H. A., Figg, W. D., Jaeckle, K., Wen, P. Y., Kyritsis, A. P.,
Loeffler, J. S., Levin, V. A., Black, P. M., Kaplan, R., Pluda, J. M.,
and Yung, W. K. Phase II trial of the antiangiogenic agent thalidomide in patients with recurrent high-grade gliomas. J Clin Oncol,
18: 708–715, 2000.

29. DeVore, R. F., Hellerqvist, C. G., Wakefield, G. B., Wamil, B. D.,
Thurman, G. B., Minton, P. A., Sundell, H. W., Yan, H. P., Carter,
C. E., Wang, Y. F., York, G. E., Zhang, M. H., and Johnson, D. H.
Phase I study of the antineovascularization drug CM101. Clin
Cancer Res, 3: 365–372, 1997.

30. Harris, A. L., Reusch, P., Barleon, B., Hang, C., Dobbs, N., and
Marme, D. Soluble Tie2 and Flt1 extracellular domains in serum
of patients with renal cancer and response to antiangiogenic therapy. Clin Cancer Res, 7: 1992–1997, 2001.

31. Chang, Y. S., di Tomaso, E., McDonald, D. M., Jones, R., Jain, R. K.,
and Munn, L. L. Mosaic blood vessels in tumors: frequency of
cancer cells in contact with flowing blood. Proc Natl Acad Sci
USA, 97: 14608–14613, 2000.

32. Lyden, D., Hattori, K., Dias, S., Costa, C., Blaikie, P., Butros, L.,
Chadburn, A., Heissig, B., Marks, W., Witte, L., Wu, Y., Hicklin,
D., Zhu, Z., Hackett, N. R., Crystal, R. G., Moore, M. A., Hajjar,
K. A., Manova, K., Benezra, R., and Rafii, S. Impaired recruitment of bone-marrow-derived endothelial and hematopoietic
precursor cells blocks tumor angiogenesis and growth. Nat Med,
7: 1194–1201, 2001.

33. Davidoff, A. M., Ng, C. Y., Brown, P., Leary, M. A., Spurbeck, W.
W., Zhou, J., Horwitz, E., Vanin, E. F., and Nienhuis, A. W. Bone
marrow-derived cells contribute to tumor neovasculature and,

when modified to express an angiogenesis inhibitor, can restrict tumor growth in mice. Clin Cancer Res, *7:* 2870–2879, 2001.

34. Peters, B. A., Diaz, L. A., Polyak, K., Meszler, L., Romans, K., Guinan, E. C., Antin, J. H., Myerson, D., Hamilton, S. R., Vogelstein, B., Kinzler, K. W., and Lengauer, C. Contribution of bone marrow-derived endothelial cells to human tumor vasculature. Nat Med, *11:* 261–262, 2005.

35. Duda, D. G., Cohen, K. S., Kozin, S. V., Perentes, J. Y., Fukumura, D., Scadden, D. T., and Jain, R. K. Evidence for incorporation of bone marrow-derived endothelial cells into perfused blood vessels in tumors. Blood, *107:* 2774–2776, 2006.

36. Aghi, M., Cohen, K. S., Klein, R. J., Scadden, D. T., and Chiocca, E. A. Tumor stromal-derived factor-1 recruits vascular progenitors to mitotic neovasculature, where microenvironment influences their differentiated phenotypes. Cancer Res, *66:* 9054–9064, 2006.

37. Mancuso, P., Burlini, A., Pruneri, G., Goldhirsch, A., Martinelli, G., and Bertolini, F. Resting and activated endothelial cells are increased in the peripheral blood of cancer patients. Blood, *97:* 3658–3661, 2001.

38. Monestiroli, S., Mancuso, P., Burlini, A., Pruneri, G., Dell'Agnola, C., Gobbi, A., Martinelli, G., and Bertolini, F. Kinetics and viability of circulating endothelial cells as surrogate angiogenesis marker in an animal model of human lymphoma. Cancer Res, *61:* 4341–4344, 2001.

39. Capillo, M., Mancuso, P., Gobbi, A., Monestiroli, S., Pruneri, G., Dell'Agnola, C., Martinelli, G., Shultz, L., and Bertolini, F. Continuous infusion of endostatin inhibits differentiation, mobilization, and clonogenic potential of endothelial cell progenitors. Clin Cancer Res, *9:* 377–382, 2003.

40. Bertolini, F., Shaked, Y., Mancuso, P., and Kerbel, R. S. The multifaceted circulating endothelial cell in cancer: towards marker and target identification. Nat Rev Cancer, *6:* 835–845, 2006.

41. Furstenberger, G., von Moos, R., Lucas, R., Thurlimann, B., Senn, H. J., Hamacher, J., and Boneberg, E. M. Circulating endothelial cells and angiogenic serum factors during neoadjuvant chemotherapy of primary breast cancer. Br J Cancer, *94:* 524–531, 2006.

42. Beerepoot, L. V., Mehra, N., Vermaat, J. S., Zonnenberg, B. A., Gebbink, M. F., and Voest, E. E. Increased levels of viable circulating endothelial cells are an indicator of progressive disease in cancer patients. Ann Oncol, *15:* 139–145, 2004.

43. Bertolini, F., Mingrone, W., Alietti, A., Ferrucci, P. F., Cocorocchio, E., Peccatori, F., Cinieri, S., Mancuso, P., Corsini, C., Burlini, A., Zucca, E., and Martinelli, G. Thalidomide in multiple myeloma, myelodysplastic syndromes and histiocytosis. Analysis of clinical results and of surrogate angiogenesis markers. Ann Oncol, *12:* 987–990, 2001.

44. Willett, C. G., Boucher, Y., di Tomaso, E., Duda, D. G., Munn, L. L., Tong, R. T., Chung, D. C., Sahani, D. V., Kalva, S. P., Kozin, S. V., Mino, M., Cohen, K. S., Scadden, D. T., Hartford, A. C., Fischman, A. J., Clark, J. W., Ryan, D. P., Zhu, A. X., Blaszkowsky, L. S., Chen, H. X., Shellito, P. C., Lauwers, G. Y., and Jain, R. K. Direct evidence that the VEGF-specific antibody bevacizumab has antivascular effects in human rectal cancer. Nat Med, *10:* 145–147, 2004.

45. Beaudry, P., Force, J., Naumov, G. N., Wang, A., Baker, C. H., Ryan, A., Soker, S., Johnson, B. E., Folkman, J., and Heymach, J. V. Differential effects of vascular endothelial growth factor receptor-2 inhibitor ZD6474 on circulating endothelial progenitors and mature circulating endothelial cells: implications for use as a surrogate marker of antiangiogenic activity. Clin Cancer Res, *11:* 3514–3522, 2005.

46. Davis, D. W., McConkey, D. J., Abbruzzese, J. L., and Herbst, R. S. Surrogate markers in antiangiogenesis clinical trials. Br J Cancer, *89:* 8–14, 2003.

47. Radema, S. A., Beerepoot, L. V., Witteveen, P. O., Gebbink, M. F., Wheeler, C., and Voest, E. E. Clinical evaluation or the novel-targeting agent, ZD6126: assessment of toxicity and surrogate markers of vascular damage. Proc Am Soc Clin Oncol, *21:* 110A, 2002.

48. Beerepoot, L. V., Radema, S. A., Witteveen, E. O., Thomas, T., Wheeler, C., Kempin, S., and Voest, E. E. Phase I clinical evaluation of weekly administration of the novel vascular-targeting agent, ZD6126, in patients with solid tumors. J Clin Oncol, *24:* 1491–1498, 2006.

49. Mancuso, P., Colleoni, M., Calleri, A., Orlando, L., Maisonneuve, P., Pruneri, G., Agliano, A., Goldhirsch, A., Shaked, Y., Kerbel, R. S., and Bertolini, F. Circulating endothelial-cell kinetics and viability predict survival in breast cancer patients receiving metronomic chemotherapy. Blood, *108:* 452–459, 2006.

50. Shaked, Y., Bertolini, F., Man, S., Rogers, M. S., Cervi, D., Foutz, T., Rawn, K., Voskas, D., Dumont, D. J., Ben David, Y., Lawler, J., Henkin, J., Huber, J., Hicklin, D. J., D'Amato, R. J., and Kerbel, R. S. Genetic heterogeneity of the vasculogenic phenotype parallels angiogenesis: Implications for cellular surrogate marker analysis of antiangiogenesis. Cancer Cell, *7:* 101–111, 2005.

51. Munoz, R., Man, S., Shaked, Y., Lee, C. R., Wong, J., Francia, G., and Kerbel, R. S. Highly efficacious nontoxic preclinical treatment for advanced metastatic breast cancer using combination oral UFT-cyclophosphamide metronomic chemotherapy. Cancer Res, *66:* 3386–3391, 2006.

52. Shaked, Y., Ciarrocchi, A., Franco, M., Lee, C. R., Man, S., Cheung, A. M., Hicklin, D. J., Chaplin, D., Foster, F. S., Benezra, R., and Kerbel, R. S. Therapy-induced acute recruitment of circulating endothelial progenitor cells to tumors. Science, *313:* 1785–1787, 2006.

53. Urbich, C. and Dimmeler, S. Endothelial progenitor cells: characterization and role in vascular biology. Circ Res, *95:* 343–353, 2004.

54. Hill, J. M., Zalos, G., Halcox, J. P., Schenke, W. H., Waclawiw, M. A., Quyyumi, A. A., and Finkel, T. Circulating endothelial progenitor cells, vascular function, and cardiovascular risk. N Engl J Med, *348:* 593–600, 2003.

55. Ashworth, T. R. A case of cancer in which cells similar to those in the tumors were seen in the blood after death. Aus Med J, *14:* 146, 1869.

56. Myerowitz, R. L., Edwards, P. A., and Sartiano, G. P. Carcinocythemia (carcinoma cell leukemia) due to metastatic carcinoma of the breast: report of a case. Cancer, *40:* 3107–3111, 1977.

57. Ross, A. A., Cooper, B. W., Lazarus, H. M., Mackay, W., Moss, T. J., Ciobanu, N., Tallman, M. S., Kennedy, M. J., Davidson, N. E., Sweet, D., and et al. Detection and viability of tumor cells in peripheral blood stem cell collections from breast cancer patients using immunocytochemical and clonogenic assay techniques. Blood, *82:* 2605–2610, 1993.

58. Brugger, W., Bross, K. J., Glatt, M., Weber, F., Mertelsmann, R., and Kanz, L. Mobilization of tumor cells and hematopoietic progenitor cells into peripheral blood of patients with solid tumors. Blood, *83:* 636–640, 1994.

59. Brandt, B., Junker, R., Griwatz, C., Heidl, S., Brinkmann, O., Semjonow, A., Assmann, G., and Zanker, K. S. Isolation of prostate-derived single cells and cell clusters from human peripheral blood. Cancer Res, 56: 4556–4561, 1996.

60. Racila, E., Euhus, D., Weiss, A. J., Rao, C., McConnell, J., Terstappen, L. W., and Uhr, J. W. Detection and characterization of carcinoma cells in the blood. Proc Natl Acad Sci USA, 95: 4589–4594, 1998.

61. Kraeft, S. K., Sutherland, R., Gravelin, L., Hu, G. H., Ferland, L. H., Richardson, P., Elias, A., and Chen, L. B. Detection and analysis of cancer cells in blood and bone marrow using a rare event imaging system. Clin Cancer Res, 6: 434–442, 2000.

62. Fehm, T., Sagalowsky, A., Clifford, E., Beitsch, P., Saboorian, H., Euhus, D., Meng, S., Morrison, L., Tucker, T., Lane, N., Ghadimi, B. M., Heselmeyer-Haddad, K., Ried, T., Rao, C., and Uhr, J. Cytogenetic evidence that circulating epithelial cells in patients with carcinoma are malignant. Clin Cancer Res, 8: 2073–2084, 2002.

63. Allard, W. J., Matera, J., Miller, M. C., Repollet, M., Connelly, M. C., Rao, C., Tibbe, A. G., Uhr, J. W., and Terstappen, L. W. Tumor cells circulate in the peripheral blood of all major carcinomas but not in healthy subjects or patients with nonmalignant diseases. Clin Cancer Res, 10: 6897–6904, 2004.

64. Cristofanilli, M., Budd, G. T., Ellis, M. J., Stopeck, A., Matera, J., Miller, M. C., Reuben, J. M., Doyle, G. V., Allard, W. J., Terstappen, L. W., and Hayes, D. F. Circulating tumor cells, disease progression, and survival in metastatic breast cancer. N Engl J Med, 351: 781–791, 2004.

65. Cristofanilli, M., Hayes, D. F., Budd, G. T., Ellis, M. J., Stopeck, A., Reuben, J. M., Doyle, G. V., Matera, J., Allard, W. J., Miller, M. C., Fritsche, H. A., Hortobagyi, G. N., and Terstappen, L. W. Circulating tumor cells: a novel prognostic factor for newly diagnosed metastatic breast cancer. J Clin Oncol, 23: 1420–1430, 2005.

66. Moreno, J. G., O'Hara, S. M., Gross, S., Doyle, G., Fritsche, H., Gomella, L. G., and Terstappen, L. W. Changes in circulating carcinoma cells in patients with metastatic prostate cancer correlate with disease status. Urology, 58: 386–392, 2001.

67. Moreno, J. G., Miller, M. C., Gross, S., Allard, W. J., Gomella, L. G., and Terstappen, L. W. Circulating tumor cells predict survival in patients with metastatic prostate cancer. Urology, 65: 713–718, 2005.

68. Staritz, P., Kienle, P., Koch, M., Benner, A., von Knebel Doeberitz, M., Rudi, J., and Weitz, J. Detection of disseminated tumour cells as a potential surrogate-marker for monitoring palliative chemotherapy in colorectal cancer patients. J Exp Clin Cancer Res, 23: 633–639, 2004.

69. Hayes, D. F., Walker, T. M., Singh, B., Vitetta, E. S., Uhr, J. W., Gross, S., Rao, C., Doyle, G. V., and Terstappen, L. W. Monitoring expression of HER-2 on circulating epithelial cells in patients with advanced breast cancer. Int J Oncol, 21: 1111–1117, 2002.

70. Dahut, W. L., Posadas, E. M., Weu, S., Arlen, P. M., Gulley, J. L., Wright, J., Chen, C. C., Jones, E., and Figg, W. D. Bony metastatic disease responses to sorafenib (BAY43-9006) independent of PSA in patients with metastatic androgen independent prostate cancer. J Clin Oncol ASCO Annual Meeting Proceedings, 24(18S): 4506, 2006.

71. Herbst, R. S., Mullani, N. A., Davis, D. W., Hess, K. R., McConkey, D. J., Charnsangavej, C., O'Reilly, M. S., Kim, H. W., Baker, C., Roach, J., Ellis, L. M., Rashid, A., Pluda, J., Bucana, C., Madden, T. L., Tran, H. T., and Abbruzzese, J. L. Development of biologic markers of response and assessment of anti-angiogenic activity in a clinical trial of human recombinant endostatin. J Clin Oncol, 20: 3804–3814, 2002.

72. Morgan, B., Thomas, A. L., Drevs, J., Hennig, J., Buchert, M., Jivan, A., Horsfield, M. A., Mross, K., Ball, H. A., Lee, L., Mietlowski, W., Fuxuis, S., Unger, C., O'Byrne, K., Henry, A., Cherryman, G. R., Laurent, D., Dugan, M., Marme, D., and Steward, W. P. Dynamic contrast-enhanced magnetic resonance imaging as a biomarker for the pharmacological response of PTK787/ZK 222584, an inhibitor of the vascular endothelial growth factor receptor tyrosine kinases, in patients with advanced colorectal cancer and liver metastases: results from two phase I studies. J Clin Oncol, 21: 3955–3964, 2003.

73. Dowlati, A., Robertson, K., Cooney, M., Petros, W. P., Stratford, M., Jesberger, J., Rafie, N., Overmoyer, B., Makkar, V., Stambler, B., Taylor, A., Waas, J., Lewin, J. S., McCrae, K. R., and Remick, S. C. A phase I pharmacokinetic and translational study of the novel vascular targeting agent combretastatin a-4 phosphate on a single-dose intravenous schedule in patients with advanced cancer. Cancer Res, 62: 3408–3416, 2002.

74. Galbraith, S. M., Maxwell, R. J., Lodge, M. A., Tozer, G. M., Wilson, J., Taylor, N. J., Stirling, J. J., Sena, L., Padhani, A. R., and Rustin, G. J. Combretastatin A4 phosphate has tumor antivascular activity in rat and man as demonstrated by dynamic magnetic resonance imaging. J Clin Oncol, 21: 2831–2842, 2003.

75. De Palma, M., Venneri, M. A., Galli, R., Sergi Sergi, L., Politi, L. S., Sampaolesi, M., and Naldini, L. Tie2 identifies a hematopoietic lineage of proangiogenic monocytes required for tumor vessel formation and a mesenchymal population of pericyte progenitors. Cancer Cell, 8: 211–226, 2005.

76. Udagawa, T., Puder, M., Wood, M., Schaefer, B. C., and D'Amato, R. J. Analysis of tumor-associated stromal cells using SCID GFP transgenic mice: contribution of local and bone marrow-derived host cells. Faseb J, 20: 95–102, 2006.

77. Conejo-Garcia, J. R., Benencia, F., Courreges, M. C., Kang, E., Mohamed-Hadley, A., Buckanovich, R. J., Holtz, D. O., Jenkins, A., Na, H., Zhang, L., Wagner, D. S., Katsaros, D., Caroll, R., and Coukos, G. Tumor-infiltrating dendritic cell precursors recruited by a beta-defensin contribute to vasculogenesis under the influence of Vegf-A. Nat Med, 10: 950–958, 2004.

78. Lipidot, T., and Petit, I. Current understanding of stem cell mobilization: the roles of chemokine. Exp Hematol, 20: 973–981, 2002.

79. Schwartzberg, L. S., Birch, R., Hazelton, B., Tauer, K. W., Lee, P., Altemose, M. D., George, C., Blanco, R., Wittlin, F., Cohen, J., Muscato, J., and West, W. H. Peripheral blood stem cell mobilization by chemotherapy with and without recombinant human granulocyte colony-stimulating factor. J Hematol, 1: 317–327, 1992.

Chapter 28
Imaging of Angiogenesis

Tristan Barrett and Peter L. Choyke

Keywords: magnetic resonance imaging (MRI), computed tomography (CT), positron emission tomography (PET),

Abstract: The imaging of angiogenesis has advanced from a scientific curiosity to an important tool in the assessment of new drugs. Virtually every imaging modality has a potential role to play in this arena. The best understood and widely available are dynamic contrast enhanced magnetic resonance imaging (MRI) and computed tomography (CT), but these modalities are by no means the only methods. Here, we discuss other MRI techniques, as well as ultrasound, optical and positron emission tomography (PET), as potential tools in assessing angiogenesis. We also consider the future role of molecular imaging methods that rely on targeting probes of biomarkers of angiogenesis. Finally, we review the considerations in implementing imaging methods into clinical trials of angiogenic inhibitors.

Introduction

The development of antiangiogenic and anti-vascular drugs has prompted a search for suitable biomarkers with which to properly select and monitor patients. An imaging assay is appealing because it could localize tumors as well as quantify an index of angiogenesis in vivo on repeated scans without requiring invasive procedures. The ideal imaging method would be entirely integrated with the decision to employ a particular therapy and then could be used to monitor its results.

The traditional 'gold standard' for cancer diagnosis is histology which is inherently invasive. A classic histological measure of angiogenesis is microvascular density (MVD). However, this method does not demonstrate vessel functionality and, moreover, tumors are heterogeneous and MVD estimates of angiogenesis will vary according to the tumor area sampled. Imaging modalities can potentially overcome these problems by depicting the entire tumor and demonstrating the functional microvasculature within it. Additionally, novel antiangiogenic drugs present new problems for physicians, particularly in the selection of an optimal dose. Unlike traditional chemotherapeutic drugs, antiangiogenic agents are not cytotoxic, thus the 'maximum tolerated dose' may not be the optimal one. Furthermore, antiangiogenic drug-induced reductions in tumor size or growth rate are likely to occur over a longer time period than with standard treatment [1]. Functional imaging has the advantage of being able to illustrate tumor treatment response quickly and independently of size changes. Thus, in theory, optimal dosage and drug efficacy can be established at an early stage of therapy, while non-responders may be more rapidly detected and their management plans altered.

There are two basic categories of angiogenesis imaging methods, non-targeted and targeted. Non-targeted imaging methods rely on the physiologic properties of tumor vessels, such as blood volume and vascular permeability, to imply the status of the neovasculature. Methods that exploit the hyperpermeable nature of tumor neo-vessels include scans that measure perfusion and permeability or changes in intravascular blood volume, with or without the use of contrast agents [2]. Various modalities have been used to demonstrate the physiology of angiogenesis including magnetic resonance imaging (MRI), computed tomography (CT), positron emission tomography (PET), single photon emission computed tomography (SPECT), ultrasound (US), and optical imaging (Table 28.1).

Targeted imaging agents can be designed to specifically target known endothelial cell surface markers associated with angiogenesis that are located on vessel walls. Targeted imaging has two basic problems to overcome; there are relatively low numbers of these targets in comparison to background signal from the unbound agent which remains in the lumen of the vessel, and

Molecular Imaging Program, National Cancer Institute, Building 10, Room 1B40, Bethesda, MD 20892-1088, USA

*Corresponding Author:
E-mail: pchoyke@nih.gov

TABLE 28.1. Comparison of the advantages and disadvantages of the different imaging modalities used for angiogenesis imaging, arbitrary units.

Modality	Resolution	Cost	Radiation exposure	Linearity	Quantification	Targetability
MRI	+++	+++	–	–	+	+/–
CT	++++	++	+++	++	+++	–
US	+	++	–	–	+	+
Optical	++	+	–	+	+	++
PET/SPECT	+	+++	++	++	+++	+++

blood vessels comprise only a small (<10%) percentage of the total mass of a typical tumor. Thus, binding needs to be highly specific, whilst washout of unbound conjugates must be rapid, and imaging must be sensitive enough to detect very low concentrations of the agent. Modalities with sufficiently high sensitivity to enable targeted imaging include optical, SPECT and PET. Optical imaging is primarily important in the research setting and in the discovery of new imaging agents, since they can be tested quickly and cheaply; although endovascular imaging may eventually make optical imaging clinically relevant as well. SPECT and PET have been used for targeted imaging of the VEGFR2, $\alpha_v\beta_3$ integrin and fibronectin in patients with metastatic solid tumors [3,4]. Targeting of these receptors have been successfully demonstrated across a number of modalities including MRI, US (using targeted microbubbles), SPECT, PET, and optical imaging, both in vitro and using animal studies [5].

The ideal method for imaging angiogenesis is yet to be agreed upon, however herein we describe the different approaches that have been employed to image angiogenesis, along with their inherent advantages and disadvantages.

Modalities Used for Imaging Angiogenesis

Magnetic Resonance Imaging

MRI does not expose the patient to ionizing radiation, an important consideration if multiple follow-up studies are required during a clinical trial where there are strict regulatory limits on radiation exposure in research subjects. MRI offers good spatial resolution and moderate sensitivity. Several different MRI techniques have been developed for imaging angiogenesis, with or without the use of external contrast agents. Gadolinium-diethylenetriamine pentaacetic acid (Gd-DTPA) and related gadolinium chelates are the most commonly used contrast agents and they all have a good toxicity profile. However, these agents are low in molecular weight and readily leak from non-tumor vessels, including inflammatory vessels. New macromolecular intravascular contrast agents demonstrate less non-specific leakage. They are under development for clinical use and are slowly gaining clinical approval. These may be more specific for neoplastic angiogenic vessels since the fenestra between endothelial cells in tumor angiogenesis tend to be large.

There are several limitations to MRI. It is sensitive to motion and susceptibility artifacts and it may be difficult to perform in very young children without sedation and adults

must be cooperative. Another limitation of MRI is that the relationship of contrast agent concentration to signal intensity is not linear and varies according to the T1 relaxivity of the tissue being evaluated, making quantification and inter-patient evaluation more problematic. Furthermore, because MRI is so flexible and terminology varies among manufacturers, standardization of acquisition protocols and analysis techniques is needed between centers to allow accurate comparisons.

Dynamic Contrast-enhanced (DCE) MRI

Dynamic contrast-enhanced imaging of angiogenesis has been performed with both CT and MRI, and these are probably the most commonly employed methods of assessing angiogenesis in clinical trials. DCE-MRI uses off-the-shelf agents and does not involve ionizing radiation. DCE methods provide a functional assessment of the tumor vasculature coupled with morphological depiction. Similar principles are applied to dynamic CT and MRI; baseline images are performed, then contrast is injected at a standardized rate and a series of scans are acquired to image the flow of contrast through the tumor region of interest (ROI) over a period of 5–30 min. Additionally, an arterial input function is selected from a nearby large artery in order to correct for ejection fraction. DCE-MRI is usually performed with a low-molecular weight contrast agent (i.e. Gd-DTPA). In DCE-MRI, the area that can be scanned is limited; thus, careful pre-selection of the target region is essential. Furthermore, signal intensity values are relative to the native T1 of the tissue, rather than absolute, which should be considered when acquiring and analyzing the data. During a typical DCE-MRI examination, the hyperpermeable tumor neovascularity allows the contrast agent to fill in the tumor area ('wash in') and then 'wash out' quickly in comparison to normal tissue. Dynamic contrast scans can be analyzed by semi-quantitative or quantitative means; these parameters can provide an assessment of angiogenesis by providing information on blood volume, blood flow, or permeability, and have been shown to correlate with the degree of angiogenesis within tumors [6].

Semi-quantitative measures include determining the initial area under the gadolinium concentration curve (IAUGC), the slope of the wash-in and washout concentration curves, or the time to maximal enhancement. Whilst these parameters are easy to obtain they are also more dependent on the exact injection protocol and acquisition variables used in the study, making inter-center comparisons more problematic. The total AUGC over a set time range can be used to derive indirect

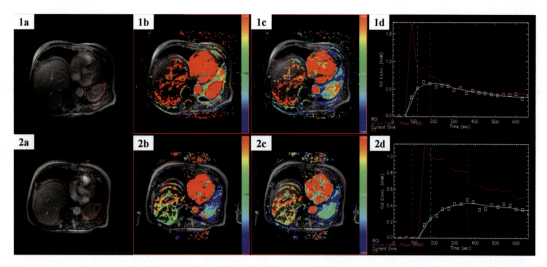

FIG. 28.1. Patient diagnosed with non-small cell lung cancer on sorafenib is followed with DCE-MRI. DCE-MRI at Baseline (*row 1*) and at 3 weeks following treatment with the anti-angiogenesis agent sorafenib (*row 2*). K^{trans} (**b**) and k_{ep} maps (**c**) show a reduced intensity despite little change in lesion size (**a**). Gadolinium concentration curves (**d**) also show reduced 'wash-in' and 'wash-out' over the treatment period.

measures of blood volume. DCE-MRI semi-quantitative methods of detecting blood flow have been validated, particularly for brain studies in relation to ischemic stroke [7]. Blood flow (BF) measures are typically derived from their relationship to blood volume (BV) and the mean transit time (MTT) of contrast agent within the tissue, whereby BF = BV/MTT.

Model-based quantitative analysis of dynamic-contrast acquired data is more directly related to changes in the physiological end-point of interest [8]. Pharmacokinetic measures of angiogenesis are derived from two compartmental models to fit the time-Gadolinium concentration curves, typically based on the General Kinetic Model as modified by Tofts et al. [9] The model assumes that the contrast agent transfers between the plasma and the extravascular, extracellular space (EES), but does not cross the cell membrane. Consensus has recently been agreed upon for the terminology of the derived parameters: K^{trans}, the forward transfer rate, a measure of blood flow and permeability; k_{ep}, the reverse rate constant; fpv, the fraction of plasma volume, related to whole tissue volume; and v_e, the extravascular, extracellular leakage volume. Gd-DTPA based DCE-MRI is rapidly becoming established as an imaging method of choice for quantifying treatment response and guiding drug dosage in clinical trials of antiangiogenic chemotherapeutic agents, vide infra (Fig. 28.1).

DCE-MRI can also be performed with the newer macromolecular contrast media (MMCM). MMCM, as their name implies, have a much higher molecular weight, typically >30,000 Da. A number of different MMCMs have been developed for use in DCE-MRI. The macromolecular size of these agents potentially allows for the incorporation of multiple targeting and payload molecules, including therapeutic 'payloads'. The signal will be amplified by the addition of multiple paramagnetic or superparamagnetic atoms into the MMCM, and is further increased via the slower molecular rotation of these larger molecules. Their increased size enables MMCMs to have a prolonged intravascular retention time, making them

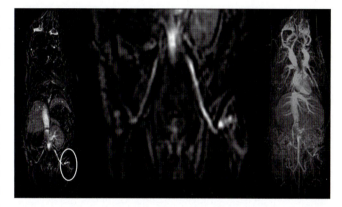

FIG. 28.2. Targeted MR imaging of αvβ3-expressing M21 melanoma in a murine model of cancer. A G4 dendrimer containing Gadolinium and with αvβ3-specific 'RGD' epitopes on the surface was injected (G4-[c(RGDfK)-Aminooxy]4[Gd-1B4M]59). T1 image taken 5 min after contrast agent injection (*left*) shows preferential uptake of the targeted agent within the tumor (*circled*), more clearly demonstrated on the magnified image (*middle*). For comparison, a G6-dendrimer MR angiogram within the same mouse depicts the vasculature (*right*). Image courtesy of Andrew Boswell (NCI, Bethesda, USA).

suitable as MR angiography agents (Fig. 28.2). However, this prolonged retention also compromises their potential as targeted imaging agents, because sufficient clearance of the unbound contrast may take hours, or days, before satisfactory detection of the vessel wall-bound agents is possible. MMCMs have another attractive property for imaging tumor angiogenesis: they may more accurately measure vascular permeability because their leakage is more specific for tumor vessels. The parameter K^{trans} reflects both blood flow and capillary permeability. In situations of high permeability (i.e. low molecular weight contrast agents, outside the brain) a flow limited state exists, and K^{trans} predominantly represents plasma flow. Conversely, in low permeability states (MMCMs) a permeability limited state exists. MMCMs have a reduced permeability due to their increased size, therefore when flow is adequate, permeability becomes the limiting factor and K^{trans} more closely reflects the permeability surface area product. However, despite these obvious advantages, currently the only MMCM that has potential for angiogenesis imaging and has gained clinical approval for human use is the Ultrasmall super-paramagnetic iron oxide molecule (USPIO). Due to restrictions on the rate of injection of USPIOs (they must be injected over 30 min in humans), much of the work with USPIOs has been limited to animal studies, wherein K^{trans} has been shown to correlate with both tumor grade[10] and histological MVD [11].

Stem Cell MR Imaging

Labeled endothelial progenitor stem cells potentially offer a safe and reliable way of targeting angiogenesis. These cells naturally track to regions of increased endothelial proliferation and areas of neovascularization. In order to track progress of the cells in vivo, they must be labeled with an imaging agent. The cells can potentially continue to be imaged during their life span, thus such agents must have low toxicity. Additionally, once the imaging agent is incorporated into the stem cell's cytoplasm, the harsh environment of the lysosome or endosome, may lead to dissociation of the paramagnetic atom (e.g. Gadolinium) from its chelate, thus rendering such compounds toxic. To this end, iron particles have emerged as the best stem cell labeling contrast agent. Iron particles are not toxic to the stem cells and do not affect their ability to function or differentiate [12]. Furthermore, even single stem cells loaded with iron particles can be detected by MRI [13]. However, iron is a negative contrast agent, making it difficult to obtain a range of signal intensities that could be used to quantify the number of stem cells, or differentiate negative contrast from areas of signal void due to artifact or tissue absence. Stem cells incorporating iron particles have been shown to be safe in cardiac patients, where they have been used to repair damaged myocardium or to induce vasculogenesis in ischemic areas [14,15]. Although yet to be tested in patients, the efficacy of stem cell angiogenesis imaging has been successfully demonstrated in a number of animal studies [16,17].

Arterial Spin Labeling (ASL) MRI

Arterial spin labeling (ASL) techniques do not require extrinsic contrast agents and can provide estimates of blood flow and blood volume within tumors. ASL MRI is a non-invasive imaging technique for measuring tissue perfusion. It uses radio frequency pulses to invert or "tag" the nuclear spins of protons in the arterial blood 'upstream' of the region of interest [18]. This 'magnetically labels' the blood and enables it to exchange with water in the tissue of interest, changing its magnetization. The effect is small, and is only visible after subtraction images are produced, *i.e.* the image performed without the ASL pulse is subtracted from the image with the ASL pulse (Fig. 28.3). Perfusion maps can be generated and ASL has proven to be

a reliable and reproducible measure of blood flow and blood volume. The non-invasive nature of ASL makes it appropriate when patients require multiple repeat scans [19].

Most ASL research has centered on brain imaging, particularly of ischemic stroke, due to the relative ease of 'tagging' blood in the carotid arteries. However, ASL studies of blood flow within brain tumors have shown a good correlation to the histologically derived measure MVD [20]. ASL MRI has also been investigated as a means of determining cerebral arterial blood volume (CBVa), and it is possible these techniques could be adapted to provide estimates of tumor blood volume. In experimental models, this has been achieved by modulating the tissue signal via magnetization transfer effects [21]. Herein, a long radio-frequency pulse is used to saturate the tissue and reduce its signal, whilst a separate ASL coil is used simultaneously to change the inflowing blood signal; this allows separation of the tissue and blood signals, and thus quantification of the blood volume. Results correlate well with blood volume estimates derived from monocrystalline iron oxide nanoparticles injected as a blood pool contrast agent [22]. A potential source of artifact with ASL is the length of transit time from the region of tagging to the region of interest wherein the 'tagged' spins lose their tagging due to T1 relaxation; this becomes more of an issue when scanning extracranial regions. Nevertheless, there has been some success in using ASL to image peripheral tumors [23].

Computed Tomography

CT has a number of advantages: it is quick to perform, widely available and has a very high spatial resolution. Quantification is reliable because there is a linear relation between the concentration of the iodine-based contrast agents and the attenuation units (expressed in Hounsfield units). Dynamic CT studies yield measurements of blood flow, blood volume and capillary permeability surface area product. The linear relationship between iodine concentration and signal intensity means that intra- and inter-patient comparisons can be made with confidence. These functional measures of angiogenesis complement the detailed morphologic information [24]. The major limitation of CT, however, is the repeated exposure to ionizing radiation over the area of interest. This is a particularly important consideration for angiogenesis imaging, where patients often require repeated examinations over time. Approval by radiation safety boards to conduct research in patients with ionizing radiation can be problematic. Additionally, iodine contrast agents can be nephrotoxic in selected high risk patients.

Dynamic CT

Dynamic CT is typically performed with a low molecular weight iodine contrast agent (Fig. 28.4). Research has begun into CT macromolecular contrast agents, for instance P743 [25], but these agents await clinical approval. As with MRI, a number of descriptive measures have been employed for analyzing dynamic CT data, including mean peak attenuation, maximum

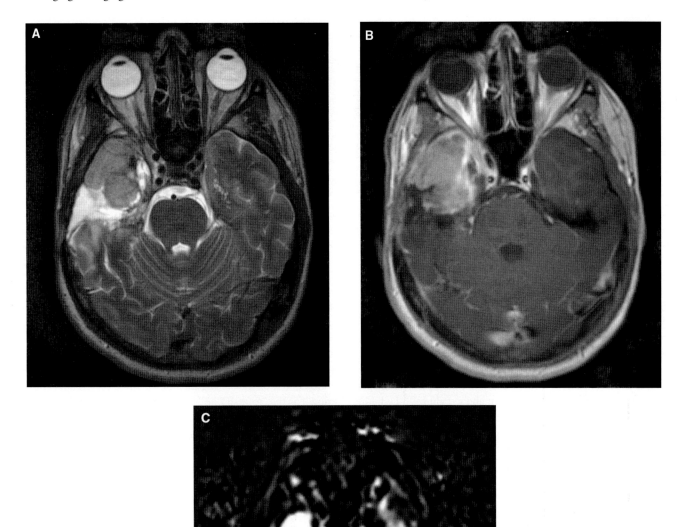

FIG. 28.3. Axial brain MRI images in a patient diagnosed with glioblastoma multiforme. T2 weighted (**a**), T1 weighted post Gadolinium contrast (**b**), and pulsed ASL (**c**) images. The primary lesion is clearly seen in the right temporal region on the post contrast image, and also demonstrates increased blood flow on ASL. Image courtesy of Neville Gai and John Butman (NIH, Bethesda, USA).

slope, enhancement rate, and peak enhancement. These measures have been validated by favorable comparisons to both MVD and VEGF levels in patients with adenocarcinoma of the lung [26]. In addition, semi-quantitative measures have shown good correlation with PET derived measures of blood flow in patients with gliomas or cerebral arterio-venous malformations [27]. However, because these measures have no intrinsic meaning they are less appealing than model-derived parameters that, at least in theory, measure physiologic phenomena. Model-derived pharmacokinetic analysis also allows better standardization

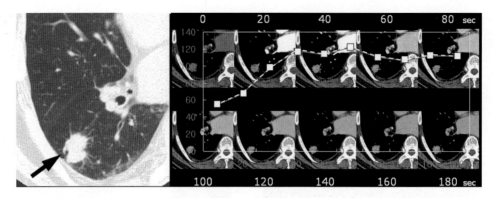

Fig. 28.4. Patient with adenocarcinoma of the lung. *Arrow* indicates lobulated lesion with a speculated margin on CT (*left*). Dynamic CT scans (*right*), with serial images obtained at 20-s intervals shows the enhancement dynamics of the nodule, with signal enhancement curve overlaid. Image reproduced with permission from Dr K.S. Lee (Sungkyunkwan University School of Medicine, Seoul, Korea)[31], by permission of Radiology.

and compensates for variations in patient hemodynamics. A two-compartmental model can be applied to derive parameters equivalent to those of DCE-MRI [28]. A more direct conversion of imaging data into a 'standardized perfusion value' has also been suggested [29]. This is analogous to the standardized uptake value of PET, and represents the ratio of the tumor perfusion to the patient's cardiac output and weight [30]. Dynamic CT has been less widely employed than DCE-MRI, nevertheless, there have been some encouraging results, with a recent study using dynamic CT for determining whether lung nodules are benign or malignant with an accuracy of 78% [31]. However, the ionizing radiation is a significant limitation. In the United States research subjects can only be exposed to 5cGy per year which could restrict the number of DCE-CT studies to two, or maximally three per year.

Ultrasound

Ultrasound is widely available within the clinic, portable, low in cost and has no adverse side effects. Disadvantages include its limited spatial resolution, limited depth penetration, poor imaging of the lungs, and the fact that it is highly operator-dependent. Doppler US is able to show blood flow, but it can only image flow within millimeter-diameter vessels and not the micron-sized vessels that make up the majority of angiogenic vessels. The introduction of microbubble-based contrast agents has encouraged the use of ultrasound to measure blood flow within tumors. Contrast-enhanced ultrasound is very sensitive and even a single microbubble can be resolved with high resolution equipment.

Contrast-enhanced Ultrasound (CEUS)

Gas-filled microbubbles can be employed as US contrast agents. They are administered intravenously and are very echogenic, producing excellent detection sensitivity even at a small dosage (Fig. 28.5). The combined use of Doppler ultrasound and harmonic encoding further increases sensitivity, enabling detection of single microbubbles and blood vessels with a diameter as small as 70μm [32]. Angiogenesis can be demonstrating within

tumors as an increase in the density of vessels, blood volume, increased blood flow, shunts between vessels, and increased vessel tortuosity [33]. CEUS can be used to derive measures of tumor blood flow and blood volume by using ultrasound with a high mechanical index to destroy microbubbles in a given region and then measuring the rate at which they reappear also known as the 'refresh rate' or replenishment kinetics [34]. The maximum US signal intensity measured after complete refilling is proportional to the blood volume in the field of view. The initial increase of microbubble replacement depends on the average blood velocity inside the vessels. The blood flow can be derived as a product of the velocity and blood volume, assuming there is a constant velocity of refilling [35]. More complex models have been applied to account for variability of blood flow velocities, which may be particularly apparent in tumors; these models have been shown to be more accurate and more consistent. Additionally, it is possible to functionalize a microbubble with targeting ligands to bind to cell surface markers on the endothelium, such as integrins. Thus, microbubbles offer potential as targeted imaging agents, as well as drug delivery vectors. Important limitations of CEUS is that it remains operator dependent, the relationship between signal intensity and microbubble concentration is non-linear and the microbubbles are micron sized making visualization of tiny vessels more difficult.

Optical Imaging

Optical imaging typically uses imaging in the near-infrared range of the optical spectra because it has better penetration (1–2 cm) than fluorophores that emit light at lower wavelengths. It has the advantage of being relatively inexpensive, portable and offers excellent spatial resolution. Optical modalities can offer real-time imaging and resolution at a cellular level, thus it has the potential to image tumor angiogenesis. The main drawback of optical imaging is the limited depth penetration. At present, imaging deeper than 1–2 cm below the skin surface is impractical, and has mainly limited its use to animal models. Intravital microscopy

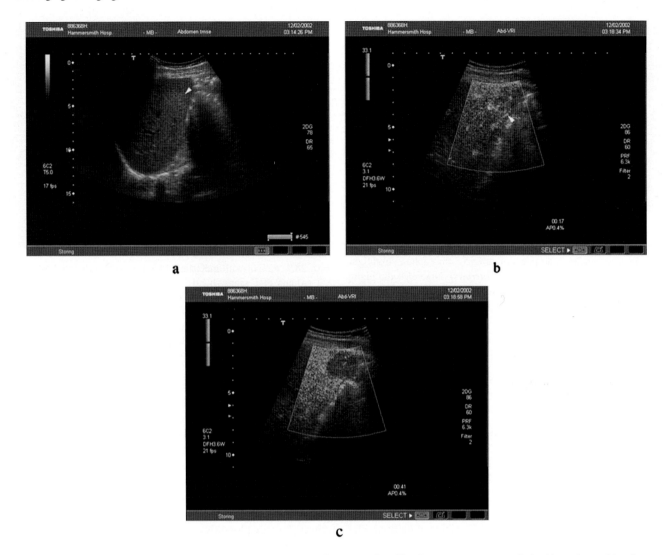

FIG. 28.5. Ultrasound microbubble vascular imaging. A subtle lesion (*arrowhead* in **a**) suggests a metastasis in this patient with a known colorectal carcinoma. Using a low power (MI ~0.1) microbubble-specific mode in which the contrast information is shown in the colour overlay, a peripheral feeding vessel (*arrowhead* in **b**) is seen the arterial phase at 17 s. Later (**c**, 41 s) the contrast medium has washed out of the tumor, but is continuing to accumulate in the large vascular volume of the liver so that the metastasis now appears as a prominent colour defect, leaving no doubt as to its presence. (In this mode motion of contrast medium in large vessels is shown as *red* or *blue*, according to their flow direction relative to the transducer, while stationary blood, i.e. the microcirculation, is depicted in *green*.). Image reproduced with permission from Dr D Cosgrove (Hammersmith Hospital, London, UK) [32], by permission of British Journal of Radiology.

can be used to image angiogenesis; it allows quantification of the vascular volume whilst concurrently showing vessel functionality and, furthermore, fluorescent contrast agents can be used to demonstrate permeability [36]. However, these methods are unlikely to translate to the clinic.

A number of instruments are available for optical imaging in vivo. However, experimental models have demonstrated the ability of optical instruments to image blood flow and hemoglobin concentration changes simultaneously, which would be ideal for the in vivo assessment of angiogenesis in superficial tumors [37]. The optical technique of orthogonal polarization spectral imaging has been used to demonstrate functional vessel density, diameter of microvessels and red blood cell velocity in an experimental model [38]. Response to the angiogenic inhibitor SU5416 was also demonstrated, and the data for functional vessel density correlated with that from fluorescence microscopy. Molecularly targeted optical fluorophores can be used to demonstrate angiogenic blood vessels, but unless the tumor is very superficial these changes are not visible outside the body. Endoscopic evaluation remains possible, but current technology does not permit endoscopic imaging of tumor vessels, nor is it clear such a method would ever be approved for humans. However, optical techniques do have the potential to image blood flow (Fig. 28.6) and demonstrate morphological and functional properties of the tumor microvasculature in more superficial tumors, for example breast cancer [39].

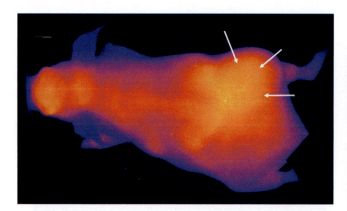

FIG. 28.6. Near infrared optical imaging of blood flow in a 'TUBO' tumor-bearing murine model. Frame from a dynamic optical imaging movie, taken 90 s after injection of Cy5.5 fluorophore, highlighting the extensive tumor vasculature (*arrows*). Note also the highly vascularized 'snout area'. Image courtesy of Andrew Boswell (NIH, Bethesda, USA).

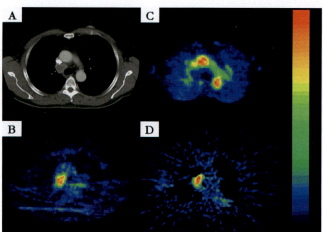

FIG. 28.7. Patient with metastatic colon cancer to the mediastinum. CT (**a**) and PET scans (**b,c,d**), taken on the same day. A metastatic mediastinal lymph node is indicated (*arrows*), color scale for PET images is shown on the *right* (*red* = maximum uptake, *blue* = minimum uptake). The lesion shows high uptake on [18]FDG-PET (**b**), reduced blood volume on [11]CO-PET (**c**), and increased blood flow on [15]H$_2$O-PET (**d**).

Positron Emission Tomography (PET) and Single Photon Emission Computed Tomography (SPECT)

These two types of nuclear medicine scan provide high sensitivity and excellent quantification. Once corrected for depth attenuation, the radioactivity counts are directly proportional to the concentration of the agent. Both techniques offer whole body imaging from a single scan and in combination with CT can provide both functional and anatomic information. As with any modality that depends on ionizing radiation, PET and SPECT are limited by how many scans can be performed during a research trial. Another disadvantage is the low spatial resolution (3–6 mm) that can be achieved with these modalities. PET has a 10-fold higher sensitivity than SPECT and is able to detect picomolar concentrations of a tracer, making it more quantitative then SPECT, and highly suited for molecularly targeted molecular imaging of cell surface receptors involved in angiogenesis [40]. Moreover, the spatial resolution of PET is better than SPECT. SPECT radionuclides are easier to prepare and are less expensive compared to their PET counterparts [5] and SPECT is the more widely available of the two modalities.

Nuclear medicine techniques were among the first to offer a means of quantifying blood flow and blood volume by imaging [41]. Historically, these techniques were developed to map the functions of the brain by monitoring changes in cerebral blood flow. These methods have now been adapted for cancer imaging. A number of perfusion tracers have been studied, including [15O]-water (Fig. 28.7d) and [11C]-carbon monoxide (Fig. 28.7c). In the case of the latter, following inhalation of tracer quantities of [11]CO, there is irreversible binding to red blood cells which distribute in the vascular space, providing accurate estimates of regional cerebral blood flow or blood volume [42,43]. [15O]-water is the most commonly used tracer for perfusion; [15]O has a short half-life (2.1 min), which means the tracer must be used almost immediately after production in a cyclotron and scanning must be completely quickly, however, the short half-life also means the study can be repeated within 20 min [44]. Following intravenous infusion, the total counts can be recorded per voxel during the buildup of radioactivity, allowing quantitation of regional tumor blood flow, which can be then be expressed as perfusion maps [45]. Image quality is often poor because the scans must be obtained dynamically before adequate counts are accumulated [46]. The PET agent [18F]-fluorodeoxyglucose (FDG) is most commonly used to detect tumors by their increased metabolic activity. However, it should be noted that this may not be a reliable means of measuring response to angiogenesis inhibition. Antiangiogenic agents may paradoxically lead to increased uptake of ^{18}F-FDG PET, because the hypoxia induced by antiangiogenic agents leads to less efficient glycolysis (anaerobic metabolism) which requires more molecules of glucose per molecule of ATP produced and hence increased tracer uptake. Eventually, however, FDG uptake will decrease as apoptosis and necrosis dominates the tumor. The high sensitivity of PET may make it feasible to directly image cell surface markers associated with angiogenesis. Indeed, new PET agents such as [18]Fluoro-Galacto-RGD have been developed to target the $\alpha_v\beta_3$ integrin, present on the endothelium of angiogenic vessels.

SPECT has been used to assess cerebral blood flow in ischemic stroke [47]. The scan must be performed within a few hours after intravenous administration of the tracer, and regions of interest are compared to an area on the contra-lateral side of the brain (which is assumed to be normal). Technetium-99 m

(^{99m}Tc) is the γ emitting tracer typically used in SPECT. For the assessment of blood flow, ^{99m}Tc is linked to a lipophilic compound to enable passage through the intact blood-brain barrier and uptake of Tc-99m by brain tissue in a manner proportional to blood flow [48]. Such compounds include hexamethylpropylene amine oxime (HMPAO) and ethyl cysteinate dimer. It is conceivable that SPECT scans can indirectly demonstrate angiogenesis within tumors by quantifying blood flow; studies evaluating the blood flow in cerebral tumors have already been reported [49]. SPECT measurements of blood flow have been shown to have a linear relationship with those obtained by PET[50]. SPECT is also a reliable method of quantifying regional cerebral blood volume, following administration of ^{99m}Tc-labeled red blood cells [51].

Imaging of Angiogenesis in Clinical Trials

It is instructive to review the modalities that have been used in angiogenesis-targeted drug trials. These studies tend to use either correlation to histological markers of angiogenesis (i.e. MVD), or blood levels of angiogenesis markers (e.g. VEGF) [52] as their gold standard. However, from a drug efficacy point of view, it has been proposed that factors such as pathological stage or grade, and disease progression are more appropriate as a 'gold standard' to which imaging techniques are compared [53]. Thus, imaging modalities that assess antiangiogenic agents should demonstrate early efficacy of the drug, be able to show dose-dependent responses, and predict long-term outcome and response to treatment. To date, the most popular method of monitoring these agents is DCE-MRI, followed by PET, and ultrasound.

DCE-MRI has been used to monitor treatment with Combretastatin A4 phosphate (C-A4-P), a microtubule inhibitor, with reductions in K^{trans} apparent at 4-h and 24-h time points [54]. A number of studies have also shown significant reductions in K^{trans} [55,56], or the semi-quantitative measure AUGC [57] two days after treatment with the VEGF tyrosine kinase inhibitor PTK787/ZK22584 (PTK/ZK). Most DCE-MRI studies have demonstrated that the degree, and persistence, of response to drug treatment is dose-dependent. Patients treated with the higher dose of C-A4-P (≥ 52 mg/m^2) showed more significant K^{trans} reductions on DCE-MRI [54]. For PTK/ZK, Mross et al. demonstrated that patients receiving doses >750 mg/day showed a statistically significant early reduction in K^{trans} and that disease progression was significantly correlated with the percentage change from baseline K^{trans}; furthermore, the optimal dose could be calculated to be 1,200 mg/day [56]. Patients with a best response to PTK/ZK have been shown to have a significantly greater reduction in K^{trans} after day 2 and at the end of treatment cycle 1 compared to those with progressive disease [55]. Treatment with this drug and a >40% reduction in the K^{trans} at day 2 has also been shown to be predictive for non-progression of disease [57]. It should be noted that reductions in K^{trans} may not correlate with actual tumor death but may simply indicate reductions in microvessel permeability.

Early efficacy of treatment with C-A4-P has also been demonstrated with PET. Patients treated with the higher dose (≥ 52 mg/m^2) had significant and persistent tumor-related reductions in blood flow (^{15}O-H$_2$O PET) and blood volume (^{15}O-CO PET) as early as 30 min after treatment, and still demonstrable at 24 h [58]. Color Doppler US can demonstrate tumor response to thalidomide treatment by measuring tumor volume and vascularity (number of visible vessels within the tumor) [59]. However, dose-dependent studies and early prediction of response have yet to be evaluated by US.

Ultimately, the aim of angiogenesis imaging is the accurate prediction of disease outcome and progression, not simply reductions in angiogenesis. Awareness of the effect of concurrent medications and non antiangiogenic effects of drugs is also essential. For instance anxiolytic agents, often administered to patients with 'mild' claustrophobia prior to MRI, may cause vasodilation, which can affect measures of blood volume. Glucocorticoids, which afford symptomatic relieve in patients with brain tumors, but have no effect on outcome, have been shown to cause a decrease in vascular permeability and regional blood volume [60]. Bevacizumab is a humanized monoclonal antibody to VEGF that has been trialed as an antiangiogenic agent. A known side effect of this drug is hypertension, thought to be caused by a decreased VEGF-mediated production of nitric oxide [61]. Nitric oxide is a vasodilator, therefore reduced levels cause a relative vasoconstriction. Furthermore, decreased levels lead to reduced renal sodium excretion which may further contribute to the hypertension [62]. Thus, early time-point reductions in blood flow and blood volume after drug administration may represent the vascular effects of nitric oxide depravation rather that treatment efficacy.

Summary

The increasing focus on angiogenesis and the imperfections of existing quantification assays has lead to a search for an imaging technique that can accurately quantify angiogenesis in vivo. Imaging modalities can potentially depict angiogenesis within entire tumors, demonstrating functionality within them, and help to select the optimal dose of antiangiogenic drugs. Techniques typically exploit physiological changes induced by the tumor neovasculature on blood flow, blood volume, tissue oxygenation, and/or permeability. Various imaging modalities have been used to image angiogenesis, including MRI, CT, PET, SPECT, US, and optical imaging. However, each modality brings with it disadvantages and no imaging modality in use today is wholly satisfactory. It may be that multi-modality imaging holds the key to angiogenesis imaging.

To date, the most popular method of monitoring antiangiogenic drugs is DCE-MRI, which can provide estimates of blood flow and blood volume, and quantitative measures of permeability within the tumor vasculature. Other clinically established modalities include PET and color Doppler US. ^{15}O-H$_2$O PET and ^{15}O-CO PET have been used to quantify blood flow and blood volume, respectively, but FDG-PET has potential inaccuracies when imaging angiogenesis. Another method of angiogenesis imaging is to directly target markers of angiogenesis on the endothelial surface. This method of imaging is more suited to the modalities with higher sensitivity, i.e. PET, SPECT. However, targeted imaging is also feasible with MRI and US; monoclonal antibodies bound to contrast agents can increase specificity and higher concentrations of agent may allow sufficiently increased sensitivity. The discovery and selection of targets is crucial to enable highly specific angiogenesis imaging. Presently, the $\alpha_v\beta_3$ integrin is the leading candidate as an angiogenesis target, because it is expressed almost exclusively on the activated endothelia of angiogenic vessels. However, many other promising cell surface markers could be targeted for imaging and therapy. Another means of targeting angiogenesis is the use of labeled endothelial progenitor stem cells. Stem cells incorporating iron particles have been shown to be safe in cardiac patients. Targeted agents are still to be trialed in patients, but they show great potential, and future agents may be able to combine specific imaging of angiogenesis along with the concurrent delivery of chemotherapeutic agents directly to tumors.

Low molecular weight Gd-DTPA enhanced dynamic MRI is establishing itself as the current imaging technique of choice for measuring tumor angiogenesis and clinical response in trials of antiangiogenesis agents. Whilst a number of trials have demonstrated an early response to treatment and/or provided an accurate predictor of clinical outcome, multicenter clinical trials are still lacking. Furthermore, there are still obstacles to overcome: DCE-MRI is still not routinely available in the clinic, and consensus needs to be reached on the exact kinetic model to be used for analysis. Moreover, it requires an analytical infrastructure not universally available in most imaging departments. Eventually, as the technique becomes more available, and acquisition protocols and analysis methods become standardized across centers, it could become the 'gold standard' method for imaging.

References

1. Hahnfeldt P, Panigrahy D, Folkman J, et al. Tumor development under angiogenic signaling: a dynamical theory of tumor growth, treatment response, and postvascular dormancy. Cancer Res 1999;59:4770–5.
2. Sinusas AJ. Imaging of angiogenesis. J Nucl Cardiol. 2004 Sep-Oct;11(5):617–33.
3. Li S, Peck-Radosavljevic M, Kienast O, et al. Imaging gastrointestinal tumours using vascular endothelial growth factor-165 (VEGF165) receptor scintigraphy. Ann Oncol. 2003 Aug;14(8):1274–7.
4. Santimaria M, Moscatelli G, Viale GL, et al. Immunoscintigraphic detection of the ED-B domain of fibronectin, a marker of angiogenesis, in patients with cancer. Clin Cancer Res. 2003 Feb;9(2):571–9.
5. Miller JC, Pien HH, Sahani D, et al. Imaging angiogenesis: applications and potential for drug development. J Natl Cancer Inst. 2005 Feb 2;97(3):172–87.
6. Padhani AR, Husband JE. Dynamic contrast-enhanced MRI studies in oncology with an emphasis on quantification, validation and human studies. Clin Radiol 2001; 56: 607–620.
7. Igarashi H, Hamamoto M, Yamaguchi H, et al. Cerebral blood flow index: dynamic perfusion MRI delivers a simple and good predictor for the outcome of acute-stage ischemic lesion. J Comput Assist Tomogr. 2003 Nov-Dec;27(6):874–81.
8. Galbraith SM, Lodge MA, Taylor NJ, et al. Reproducibility of dynamic contrast-enhanced MRI in human muscle and tumours: comparison of quantitative and semi-quantitative analysis. NMR Biomed. 2002 Apr;15(2):132–42.
9. Tofts PS, Brix G, Buckley DL, et al. Estimating kinetic parameters from dynamic contrast-enhanced T(1)-weighted MRI of a diffusable tracer: standardized quantities and symbols. J Magn Reson Imaging. 1999 Sep;10(3):223–32.
10. Turetschek K, Huber S, Floyd E, et al. MR imaging characterization of microvessels in experimental breast tumors by using a particulate contrast agent with histopathologic correlation. Radiology. 2001 Feb;218(2):562–9.
11. van Dijke CF, Brasch RC, Roberts TP, et al. Mammary carcinoma model: correlation of macromolecular contrast-enhanced MR imaging characterizations of tumor microvasculature and histologic capillary density.Radiology. 1996 Mar;198(3):813–8.
12. Arbab A, Bashaw L, Miller B, et al. Intracytoplasmic tagging of cells with ferumoxides and transfection agent for cellular magnetic resonance imaging after cell transplantation: methods and techniques. Transplantation. 2003;76: 1123–1130.
13. Arbab AS, Yocum GT, Wilson LB, et al. Comparison of transfection agents in forming complexes with ferumoxides, cell labeling efficiency, and cellular viability. Mol Imaging. 2004 Jan;3(1):24–32.
14. Britten MB, Abolmaali ND, Assmus B, et al. Infarct remodeling after intracoronary progenitor cell treatment in patients with acute myocardial infarction (TOPCARE-AMI): mechanistic insights from serial contrast-enhanced magnetic resonance imaging. Circulation. 2003 Nov 4;108(18):2212–8.
15. Strauer BE, Brehm M, Zeus T, et al. Repair of infarcted myocardium by autologous intracoronary mononuclear bone marrow cell transplantation in humans. Circulation. 2002 Oct 8;106(15):1913–8.
16. Anderson S, Glod J, Arbab A, et al. Noninvasive MR imaging of magnetically labeled stem cells to directly identify neovasculature in a glioma model. Blood. 2005 Jan 1;105(1):420–5.
17. Arbab AS, Frenkel V, Pandit SD, et al. Magnetic resonance imaging and confocal microscopy studies of magnetically labeled endothelial progenitor cells trafficking to sites of tumor angiogenesis. Stem Cells. 2006 Mar;24(3):671–8.
18. Detre JA, Alsop DC. Perfusion magnetic resonance imaging with continuous arterial spin labeling: methods and clinical applications in the central nervous system. Eur J Radiol 1999;30: 115–124.
19. Jahng GH, Song E, Zhu XP, et al. Human brain: reliability and reproducibility of pulsed arterial spin-labeling perfusion MR imaging. Radiology. 2005 Mar;234(3):909–16.
20. Kimura H, Takeuchi H, Koshimoto Y, et al. Perfusion imaging of meningioma by using continuous arterial spin-labeling: comparison with dynamic susceptibility-weighted contrast-enhanced MR images and histopathologic features. AJNR Am J Neuroradiol. 2006 Jan;27(1):85–93.
21. Kim T, Kim SG. Quantification of cerebral arterial blood volume and cerebral blood flow using MRI with modulation of

tissue and vessel (MOTIVE) signals. Magn Reson Med. 2005 Aug;54(2):333–42.

22. Kim T, Kim SG. Quantification of cerebral arterial blood volume using arterial spin labeling with intravoxel incoherent motion-sensitive gradients. Magn Reson Med. 2006 May;55(5):1047–57.

23. Boss A, Martirosian P, Schraml C, et al. Morphological, contrast-enhanced and spin labeling perfusion imaging for monitoring of relapse after RF ablation of renal cell carcinomas. Eur Radiol. 2006 Jan 27:1–11.

24. Kan Z, Phongkitkarun S, Kobayashi S, et al. Functional CT for quantifying tumor perfusion in antiangiogenic therapy in a rat model. Radiology. 2005 Oct;237(1):151–8.

25. Fournier LS, Cuenod CA, de Bazelaire C, et al. Early modifications of hepatic perfusion measured by functional CT in a rat model of hepatocellular carcinoma using a blood pool contrast agent. Eur Radiol. 2004 Nov;14(11):2125–33.

26. Tateishi U, Nishihara H, Watanabe S, et al. Tumor angiogenesis and dynamic CT in lung adenocarcinoma: radiologic-pathologic correlation. J Comput Assist Tomogr. 2001 Jan-Feb;25(1):23–7.

27. Gillard JH, Minhas PS, Hayball MP, et al. Assessment of quantitative computed tomographic cerebral perfusion imaging with H2(15)O positron emission tomography. Neurol Res. 2000 Jul;22(5):457–64.

28. Lee TY, Purdie TG, Stewart E. CT imaging of angiogenesis. Q J Nucl Med. 2003 Sep;47(3):171–87.

29. Miles KA, Griffiths MR, Fuentes MA. Standardized perfusion value: universal CT contrast enhancement scale that correlates with FDG PET in lung nodules. Radiology 2001;220:548–53.

30. Miles KA. Perfusion CT for the assessment of tumour vascularity: which protocol? Brit J of Radiology (2003) 76, S36–S42.

31. Yi CA, Lee KS, Kim EA, et al. Solitary pulmonary nodules: dynamic enhanced multi-detector row CT study and comparison with vascular endothelial growth factor and microvessel density. Radiology. 2004 Oct;233(1):191–9.

32. D Cosgrove. Angiogenesis imaging – ultrasound. Brit J of Radiology (2003) 76, S43–S49.

33. Kedar RP, Cosgrove D, McCready, et al. Microbubble contrast agent for color Doppler US: effect on breast masses. Work in progress. Radiology 1996;198:679–86.

34. Krix M, Kiessling F, Farhan N, et al. A multivessel model describing replenishment kinetics of ultrasound contrast agent for quantification of tissue perfusion. Ultrasound Med Biol. 2003 Oct;29(10):1421–30.

35. Wei K, Jayaweera AR, Firoozan S, et al. Quantification of myocardial blood flow with ultrasound-induced destruction of microbubbles administered as a constant venous infusion. Circulation. 1998 Feb 10;97(5):473–83.

36. Brown EB, Campbell RB, Tsuzuki Y, et al. In vivo measurement of gene expression, angiogenesis and physiological function in tumors using multiphoton laser scanning microscopy. Nat Med. 2001 Jul;7(7):864–8.

37. Dunn AK, Devor A, Bolay H, et al. Simultaneous imaging of total cerebral hemoglobin concentration, oxygenation, and blood flow during functional activation. Opt Lett 2003;28:28–30.

38. Pahernik S, Harris AG, Schmitt-Sody M, et al. Orthogonal polarisation spectral imaging as a new tool for the assessment of antivascular tumour treatment in vivo: a validation study. Br J Cancer 2002;86:1622–7.

39. Ntziachristos V, Yodh A, Schnall M, et al. Concurrent MRI and diffuse optical tomography of breast after indocyanine green enhancement. Proc Natl Acad Sci USA 2000;97:2767–72.

40. Rohren EM, Turkington TG, Coleman RE. Clinical applications of PET in oncology. Radiology. 2004 May;231(2):305–32.

41. Phelps ME, Huang SC, Hoffman EJ, et al. Validation of tomographic measurement of cerebral blood volume with C11-labeled carboxyhemoglobin. J Nucl Med 1979;20:328–34.

42. McDonald DM, PL Choyke. Imaging of angiogenesis: from microscope to clinic. Nat Med, 2003. 9(6): p. 713–25.

43. Saha GB, MacIntyre WJ, Go RT. Radiopharmaceuticals for brain imaging. Semin Nucl Med. 1994 Oct;24(4):324–49.

44. Raichle ME. Measurement of local cerebral blood flow and metabolism in man with positron emission tomography. Fed Proc. 1981 Jun;40(8):2331–4.

45. Frackowiak RS, Friston KJ. Functional neuroanatomy of the human brain: positron emission tomography–a new neuroanatomical technique. J Anat. 1994 Apr;184 (Pt 2):211–25.

46. Raichle ME. Positron emission tomography. Annu Rev Neurosci. 1983;6:249–67.

47. Latchaw RE. Cerebral perfusion imaging in acute stroke. J Vasc Interv Radiol. 2004;15(1 Pt 2):S29–S46.

48. Matsuda H, Tsuji S, Shuke N, et al. A quantitative approach to technetium-99 m hexamethylpropylene amine oxime. Eur J Nucl Med 1992. 19:195.

49. Suess E, Malessa S, Ungersbock K, et al. Technetium-99 m-d,1-hexamethylpropyleneamineoxime(HMPAO)uptakeandglutathione content in brain tumors. J Nucl Med. 1991 Sep;32(9):1675–81.

50. Kirkness CJ. Cerebral blood flow monitoring in clinical practice. AACN Clin Issues. 2005 Oct-Dec;16(4):476–87.

51. Sabatini U, Celsis P, Viallard G, et al. Quantitative assessment of cerebral blood volume by single-photon emission computed tomography. Stroke. 1991 Mar;22(3):324–30.

52. Schaefer JF, Schneider V, Vollmar J, et al. Solitary pulmonary nodules: association between signal characteristics in dynamic contrast enhanced MRI and tumor angiogenesis.Lung Cancer. 2006 Jul;53(1):39–49.

53. Mayr NA, Hawighorst H, Yuh WT, et al. MR microcirculation assessment in cervical cancer: correlations with histomorphological tumor markers and clinical outcome. J Magn Reson Imaging. 1999 Sep;10(3):267–76.

54. Galbraith SM, Maxwell RJ, Lodge MA, et al. Combretastatin A4 phosphate has tumor antivascular activity in rat and man as demonstrated by dynamic magnetic resonance imaging. J Clin Oncol. 2003 Aug 1;21(15):2831–42.

55. Morgan B, Thomas AL, Drevs J, et al. Dynamic contrast-enhanced magnetic resonance imaging as a biomarker for the pharmacological response of PTK787/ZK 222584, an inhibitor of the vascular endothelial growth factor receptor tyrosine kinases, in patients with advanced colorectal cancer and liver metastases: results from two phase I studies. J Clin Oncol. 2003 Nov 1;21(21):3955–64.

56. Mross K, Drevs J, Muller M, et al. Phase I clinical and pharmacokinetic study of PTK/ZK, a multiple VEGF receptor inhibitor, in patients with liver metastases from solid tumours. Eur J Cancer. 2005 Jun;41(9):1291–9.

57. Lee L, Sharma S, Morgan B, et al. Biomarkers for assessment of pharmacologic activity for a vascular endothelial growth factor (VEGF) receptor inhibitor, PTK787/ZK 222584 (PTK/ZK): translation of biological activity in a mouse melanoma metastasis model to phase I studies in patients with advanced colorectal cancer with liver metastases. Cancer Chemother Pharmacol. 2006 Jun;57(6):761–71.

58. Anderson HL, Yap JT, Miller MP, et al. Assessment of pharmacodynamic vascular response in a phase I trial of combretastatin A4 phosphate. J Clin Oncol. 2003 Aug 1;21(15): 2823–30.

59. Lassau N, Chawi I, Rouffiac V, et al. Interest of color Doppler ultrasonography to evaluate a new anti-angiogenic treatment with thalidomide in metastatic renal cell carcinoma. Bull Cancer. 2004 Jul-Aug;91(7–8):629–35.

60. Ostergaard L, Hochberg FH, Rabinov JD, et al. Early changes measured by magnetic resonance imaging in cerebral blood flow, blood volume, and blood-brain barrier permeability following dexamethasone treatment in patients with brain tumors. J Neurosurg. 1999 Feb;90(2):300–5.

61. Shen BQ, Lee DY, Zioncheck TF. Vascular endothelial growth factor governs endothelial nitric-oxide synthase expression via a KDR/Flk-1 receptor and protein kinase C signaling pathway. J Biol Chem 1999;274:3057–63.

62. Granger JP, Alexander BT. Abnormal pressure-natriuresis in hypertension: role of nitric oxide. Acta Physiol Scand 2000;168:161–68.

Chapter 29
Tumor Endothelial Markers

Janine Stevens and Brad St. Croix

Keywords: tumor endothelial markers; TEMs, angiogenesis; vasculature targeting agents, VTAs, antiangiogenic agents.

Abstract: Advances in the field of antiangiogenesis therapy have clearly demonstrated that this approach can augment more traditional methods of cancer treatment which aim to destroy the tumor cells directly. Although antiangiogenic agents such as neutralizing anti-VEGF antibodies have revealed clear clinical utility in certain cancer types, disease inevitably progresses, highlighting the continued need for the development of more effective therapeutic agents. The successful development of novel antiangiogenic and vascular disrupting agents that specifically target the tumor vasculature depends on a detailed knowledge of the molecular signatures of the tumor vasculature. Several laboratories have exploited recent advances in proteomics and global gene expression analysis to uncover a variety of cell surface tumor endothelial markers (TEMs) that may prove useful in the design of novel strategies for targeting tumor vasculature. A better understanding of these TEMs is key to uncovering their potential role in angiogenesis as well as their suitability as therapeutic targets.

Introduction

Angiogenesis is a complex process involving the growth factor-dependent activation of endothelial cells (ECs), extracellular matrix degradation and remodeling, endothelial migration and proliferation, and pericyte investment. In mammals, physiologic angiogenesis occurs during embryonic development and wound healing, and also occurs in females during menstruation, ovulation and pregnancy. Normal regulation of angiogenesis is governed by a fine balance between proangiogenic factors and antiangiogenic factors known as the "angiogenic switch" [1]. When this balance is shifted such that proangio-

genic factors predominate, angiogenesis ensues. Pathological angiogenesis, involving the abnormal proliferation of blood vessels, is associated with over 20 diseases, including rheumatoid arthritis, age-related macular degeneration, diabetes and cancer [2,3]. Aberrant vessel growth by itself is usually not responsible for initiating disease, but in the case of age-related macular degeneration and most cancers is now considered a rate-limiting step critical for disease progression. Consequently, many investigators have become interested in the possibility of inhibiting aberrant blood vessel growth as a means of halting disease progression. For example, treatment of age-related macular degeneration (AMD) with ranibizumab (Lucentis), a humanized antigen binding fragment (Fab) that neutralizes the vascular endothelial growth factor (VEGF), can prevent or delay the onset of blindness that occurs as a result of inappropriate neovascularization in the macula [4,5]. The growth, survival and metastasis of most solid tumors and certain leukemias are also dependent upon the nutrients provided by new blood vessel formation. Thus, the tumor vasculature has become an extremely attractive target for anti-cancer therapy [6]. The recent success of bevacizumab (Avastin), a humanized antibody against VEGF, in prolonging the lives of patients with metastatic colon and lung cancer establishes the validity of the antiangiogenic approach [7,8].

Historically, most therapies to combat cancer have been selected based on their ability to target rapidly dividing cells. However, cross-reactivity with normal proliferating cells and resistance to treatment remain major obstacles [9]. The recurrence of cancer following treatment with conventional anti-hormonal, radio- or chemotherapy is thought to be driven by the genetic instability of tumor cells, which ultimately allows for the selection of mutants that are able to resist treatment. An elegant example of this principle can be found in the treatment of chronic myelogenous leukemia (CML) patients with imatinib (Gleevec), a small molecular weight agent that targets the rearranged bcr-abl fusion product that is the underlying cause of the disease. Complete responses to imatinib were originally heralded as a great triumph in the treatment of this life-threatening disease. However, this initial excitement has

Tumor Angiogenesis Section, Mouse Cancer Genetics Program, National Cancer Institute at Frederick, Frederick, MD 21702, USA now been tempered by the eventual relapse and development of resistance in a large number of these patients due to multiple mechanisms including subtle point mutations in the bcr-abl gene itself [10]. As first proposed by Kerbel in 1991, targeting the genetically stable tumor vasculature instead of the tumor cells could, in theory, prevent the development of drug resistance [11]. Evidence supporting this idea was first provided by Boehem and colleages who were able to rid mice of large tumors using the antiangiogenic agent endostatin [12,13]. In more recent studies, however, Klagsbrun's group has found cytogenetic abnormalities in tumor ECs, suggesting that some host-derived stromal cells may not be as genetically stable as originally thought [14]. Also, clinical trials which include bevacizumab as part of the chemotherapeutic regimen in the treatment of metastatic lung and colorectal cancer have shown that the majority of tumors, even those initially responsive, eventually progress despite the continued use of therapy [7,8]. Although an increased genetic plasticity of tumor ECs could potentially explain these findings, an alternative, perhaps more likely, explanation has to do with the fact that multiple natural proangiogenic and antiangiogenic regulators exist. Although many of these angiogenesis modulators are likely to lie in the same pathway, if more than one angiogenic pathway exists then any antiangiogenic agent that targets a single pathway may ultimately select for the emergence of tumor cells that are able to evoke angiogenesis through an alternative pathway. In this model, the genetic instability of tumor cells is still ultimately responsible for resistance to the antiangiogenic agent given. The selection results from the interdependence of the tumor and stromal cells on each other, and an inability of the tumor to expand beyond 1–2 mm in size without new blood vessel growth. The model also allows for the presence of genetically stable ECs, which would always be expected at the onset of angiogenesis and near the borders where new vessels connect with pre-existing mature vessels. Clearly, the most effective antiangiogenic therapies will require a better knowledge of how many independent proangiogenic pathways exist. One recent study suggested that a fibroblast growth factor (FGF)-mediated pathway might be able to evoke angiogenesis when the classic VEGF/VEGFR2 pathway is blocked [15]. However, other studies have suggested that VEGF is a primary downstream mediator of FGF-mediated angiogenesis because inhibiting VEGF signaling blocked angiogenesis in vivo in response to FGF [16–18]. Thus, strategies that block VEGF could ulti-

mately select for cells with higher VEGF levels that are able to overwhelm the VEGF blockade, thereby rendering these cells resistant to the therapy. Regardless of the mechanisms involved, it seems likely that we will need to completely block every angiogenic pathway in order to prevent the development of resistance to antiangiogenic agents. It is currently unknown how many angiogenic mediators work through a VEGF-dependent pathway, and how many are VEGF-independent.

In addition to avoiding the problem of drug resistance, targeting tumor vessels instead of the tumor cells themselves offers several other potential advantages. First, ECs are directly exposed to circulating blood, thereby facilitating drug delivery and enabling the use of high molecular weight therapeutics. Second, there is likely to be a significant bystander effect when targeting the vascular endothelium because each capillary is thought to support hundreds of tumor cells. Third, because cancer is considered to be a large group of diseases, and 90% of cancers are solid tumors, a pharmaceutical agent that can effectively inhibit angiogenesis is likely to be effective against the majority of them. Fourth, because tumor metastasis may rely on an angiogenic phenotype, blocking angiogenesis may prevent the spread of cancerous cells. And finally, since most adult endothelium is quiescent, side-effects of antiangiogenesis therapy could be minimal.

"Antiangiogenic" Versus "Vascular Disrupting" Agents

There exist two major approaches for targeting the tumor vasculature (Table 29.1) [19]. The first involves using "antiangiogenic agents" which block the growth of new blood vessels. This approach, first proposed by Judah Folkman in 1971 [20], has been the most widely studied. In general, these agents are relatively non-toxic, cytostatic in nature and display an optimal biologic dose (OBD) that is well below the maximum tolerated dose (MTD). Examples include antibodies that block the function of molecules critical for angiogenesis as well as several natural or synthetic antiangiogenic agents. Most clinical drugs currently used to target tumor endothelium fall into this category, such as bevacizumab and sunitinib.

The second approach employs the use of vascular disrupting agents (VDAs), also referred to as vascular targeting agents. The idea of attacking new blood vessels with cytotoxic agents

TABLE 29.1. Differential characteristics of antiangiogenic and vascular disrupting agents.

Antiangiogenic agents	Vascular disrupting agents
Inhibits tumor growth	Causes rapid collapse of tumor blood flow and tumor necrosis
Cytostatic	Cytotoxic
OBD < MTD	OBD = MTD
Long treatment period	Rapid treatment period
Drug resistance is a concern	May circumvent the development of drug resistance
Specificity of target less important	Specificity of target very important
Targets must be functionally required for angiogenesis (preferably indispensable)	Target molecules need only be correlated with angiogenesis
Examples include: VEGF or VEGFR antagonists, thalidomide, endostatin	Examples include: immunotoxins, TF-coagulants, CA4P, DMXAA

was first put forward in the early 1980s by Julia Denekamp, who noticed that tumor blood vessels had different characteristics than normal blood vessels [21,22]. VDAs are designed to cause a rapid collapse in tumor blood flow by targeting all existing tumor blood vessels, including those which are already established. Compared to antiangiogenic agents, VDAs are much more toxic and, like conventional chemotherapeutics, they are usually most effective at their maximum tolerated dose. Unlike antiangiogenic agents, the targets of these agents need not be required for angiogenesis, but only correlated with the process. Hence, redundancy of target function is not a concern for VDAs. The use of a vascular disrupting approach is desirable because it could facilitate a more rapid treatment of disease and potentially circumvent the aforementioned problems associated with drug resistance/angiogenic pathway redundancy. Preclinical proof that a vascular targeting approach could be feasible was first provided by Thorpe's group, who were able to eradicate tumors in mice using an immunotoxin that targeted tumor vessels [23]. A subsequent study by the same group demonstrated the possibility of treating tumors by directing tissue factor to tumor vessels in order to induce tumor thrombosis [24]. However, these proof-of-concept studies relied on MHC class II molecules as the molecular target of tumor endothelium, which required artificial induction by challenging mice with tumors specifically engineered to overexpress interferon-γ. In subsequent work, tissue factor has been targeted to various proteins naturally expressed on tumor endothelium such as VCAM, the ED-B domain of fibronectin, and prostate-specific membrane antigen [25]. However, a thin rim of viable tumor cells at the tumor periphery usually survives, suggesting that this type of treatment may be best used in combination with chemotherapy that targets the proliferating tumor cells.

VDAs can be subdivided into two main classes: small-molecule and ligand-directed [25]. Ligand-based approaches involve the use of peptides, antibodies or growth factors to deliver toxins, pro-coagulants or apoptotic effectors to pathologic vessels. Small-molecules, on the other hand, do not specifically localize to such vessels, but rather exploit the differences between them and normal vessels in order to induce selective vascular disruption [26]. Small-molecules represent the most advanced group of VDAs, and several of them, such as combretastatin A4 phosphate (CA4P) and 5,6-dimethylxanthenone-4-acetic acid (DMXAA) are currently in clinical development. These drugs were developed based on their ability to cause a rapid collapse in tumor blood flow. Although the reasons for the specificity of CA4P for tumor vessels are likely to be multiple, their fast acting nature suggests that destabilization of the cytoskeleton combined with a high interstitial pressure of tumors may lead to vascular collapse. The molecules which enable these agents to enter cells and elicit their function may also be altered in tumor vessels.

For ligand-directed VDAs, it is imperative that the targets recognized by these agents are selectively expressed on tumor endothelium. Indeed, the large number of antiangiogenic agents in clinical trials compared to ligand-directed VDAs is likely a reflection of the fact that, until recently, markers with the necessary selectivity for tumor endothelium have been elusive. Only with advances in genomic and proteomic technologies over the last decade have we begun to understand the complexity of the EC surface. Below, we discuss some of the approaches that have been taken to identify markers of the tumor vasculature, and what is currently known about some of the key cell surface markers that have been identified. We have limited the discussion to some of the most well-characterized cell surface tumor endothelial markers with the hope that these markers may be more readily exploitable in the near future due to their immediate accessibility to pharmacologic intervention. A better understanding of the function of the genes expressed in tumor endothelium will undoubtedly aid in our understanding of the basic mechanisms underlying angiogenesis. Regardless of their function, however, it is hoped that some of the new targets either alone or in combination, will have the desired specificity needed to deliver new molecular therapeutics directly to tumor endothelium.

Methods Used to Uncover Tumor Endothelial Markers

The first systematic attempt to identify markers of tumor endothelium involved the use of peptide phage display. By injecting peptide libraries into tumor-bearing mice, followed by repeated rounds of phage isolation and re-injection, Ruoslahti and co-workers were able to identify several peptides which could selectively home to tumor vessels [27]. These pioneering studies supported the idea that the surface of tumor endothelium contained distinct molecular signatures that could potentially be exploited for therapeutic purposes. One advantage of this in vivo approach is that it selects for peptides which recognize antigens which are exposed at the cell surface. However, a relatively low binding affinity of these peptides for their cognate binding partners has hindered this approach and made receptor identification challenging. Nevertheless, some potential targets of the tumor vasculature have been identified, including aminopeptidase N, aminopeptidase A and nucleolin [28–30] (Table 29-2).

Another systematic proteomic approach was pioneered by Jan Schnitzer's laboratory [31]. This group labeled the luminal surface of blood vessels of tumor-bearing mice with silica, performed subcellular fractionation of the isolated tissues, then used mass spectrometry to identify specific proteins expressed at the EC surface. Like the phage display approach, the targets identified are immediately accessible to therapeutic intervention. So far, this approach has only been applied to rodent tissues. However, annexin A1, the most specific tumor endothelial target identified by this approach, was also found to be overexpressed in multiple human tumors. Importantly, targeting annexin A1 with radio-immunotherapy was able to destroy xenotransplanted tumors and increase survival in preclinical models [31].

TABLE 29.2. Cell surface molecules selectively overexpressed on tumor vasculature.

Tumor EC Target	Comment	References
αvβ3 integrin	Also expressed by a variety of normal non-endothelial cell types	[98]
Aminopeptidase A	Expressed in tumor-associated pericytes	[29,99]
Aminopeptidase N (CD13)	Expressed in tumor-associated pericytes	[28]
Annexin A1	Normally considered a cytosolic protein, annexin A1 was detected on the cell surface of tumor endothelium	[31]
CD44H	The TES-23 antibody recognizes a tumor associated form of CD44 that is post-translationally modified	[100,101]
Delta-like 4	Blocking Dll4 function inhibits tumor growth in preclinical mouse models	[96,97]
Nucleolin	Normally considered a nuclear protein, nucleolin was detected on the cell surface of tumor endothelium	[30]
Phosphatidylserine	mAB 3G4 has shown anti-tumor effects alone and in combination with chemotherapy in xenograft tumor models	[102,103]
PSMA-1	Elevated expression in human tumor vessels is not recapitulated in mice	[78,79,81]
TEM1 (Endosialin)	Expressed in tumor associated endothelial cells, pericytes and fibroblasts	[36,41,43,44]
TEM5	Putative G-protein coupled receptor	[41]
TEM7	Elevated expression in human tumor vessels is not recapitulated in mice	[35,41,59]
TEM8	Elevated in pathologic but not physiologic angiogenesis	[41,67]
Thy-1	Expression in hematopoietic cells could complicate a vascular targeting approach	[34,36,104]
Robo4	Expression in various adult mouse tissues could preclude its use as a vascular targeting agent	[86,87]
VEGF:receptor complexes	Expression of receptors in normal vessels could complicate its targeting	[105]
VE-Cadherin	Ubiquitous distribution throughout vasculature but still a possible target	[106,107]

A similar approach for identifying organ-specific antigens was recently described which involved the perfusion and biotinylation of accessible molecules on the luminal surface of blood vessels. Following purification on streptavidin resin, biotinylated proteins were identified by mass spectrometry [32]. This approach has now been applied ex vivo using surgically resected human kidneys with tumors [33]. So far, this approach seems to preferentially detect extracellular matrix molecules, presumably due to their high abundance.

Advances in global gene expression analysis have also enabled a systematic identification of genes upregulated in human tumor endothelium. In the first study of this kind, serial analysis of gene expression (SAGE) was used to delineate gene expression profiles in human ECs isolated from normal colonic mucosa or colorectal tumors [34]. Since then, a few independent groups have further characterized endothelial gene expression profiles from other tumor types such as human breast and brain cancers [35–38]. In each of these studies, ECs from both normal and tumor tissue of the same patients were isolated from fresh tissues following enzymatic dissociation. SAGE technology is uniquely suited for such studies because it can provide an unbiased highly-quantitative account of gene expression from as few as 50,000 cells. SAGE is also independent of pre-existing databases, allowing novel genes to be discovered and cloned [39]. In the original study on colorectal cancer, 46 tumor endothelial markers (TEMs) were identified with >10-fold overexpression in tumor- versus normal-derived endothelium [34]. While the majority of the SAGE tags identified in tumor endothelium corresponded to previously uncharacterized genes, several of those that had been characterized had already been implicated in angiogenesis, including some cell surface markers. For example,

integrin α1 had already been shown to be important for vascularization using mouse knockout (KO) models [40].

Based upon the methodology by which they were discovered, nine of the previously uncharacterized genes were sequenced and given the name Tumor Endothelial Marker (i.e. TEM1 to TEM9). Four of these, TEM1, TEM5, TEM7, and TEM8, reside at the cell surface. Aside from the commonality of their name and shared expression patterns, these genes are unrelated to one another, each belonging to a different gene family. Mouse orthologs of each of these four cell surface TEMs were also identified and found to be expressed in the endothelium of the developing embryo [41]. With the exception of TEM8, each of the TEMs analyzed were also found to be turned on in the corpus luteum of ovaries. Thus, most of the markers identified are presumably important for angiogenesis in general, and are not limited to tumor angiogenesis per say. The following section provides more detail for some of the best characterized cell surface tumor endothelial markers.

TEM1 (Endosialin)

TEM1 encodes a Type I transmembrane protein of 757 amino acids. TEM1 is currently an orphan receptor with a domain structure similar to that of thrombomodulin. Its extracellular domain contains three epidermal growth factor (EGF) repeats and a globular C-type lectin-like (CTLD) carbohydrate recognition domain [42]. Although a basal level of TEM1 mRNA has been observed in certain normal tissues, it appears to be markedly elevated in variety of cancer types and localizes to tumor neovasculature [41, 43–45].

TEM1 was originally discovered by Retig and colleagues as the antigen recognized by a monoclonal antibody called FB5, which was raised against human fetal fibroblasts during a search for novel molecular targets of the tumor stroma [43]. FB5 was found to bind to a heavily-sialylated glycoprotein on tumor ECs, which the authors named endosialin. By histology, the FB5 antibody selectively labeled tumor vessels in 84 immunoreactive cancers of various origins, but not in normal control tissues. Almost 10 years later, Christian et. al. identified TEM1 as the gene encoding the endosialin protein recognized by FB5 [46]. In 2001, Opavsky et. al. demonstrated a cell density-dependent upregulation of TEM1 mRNA and protein in vitro, suggesting a possible role in cell–cell adhesion [47]. TEM1 also contains a conserved WIGL motif within the CTLD domain, a unique feature found in other cell surface proteins that modulate endocytosis [48]. Interestingly, TEM1-expressing cells were able to endocytose antibodies to cell surface TEM1, a selective uptake that could prove extremely useful for targeting tumor endothelium [43].

TEM1 was recently re-discovered in a search for genes upregulated in tumor ECs derived from human malignant gliomas [36]. Overexpression in the vessels of highly malignant and invasive brain tumors was also found at the protein level by Brady and co-workers [45]. TEM1 was most highly expressed in high-grade primary and metastatic tumors, lower in benign tumors and virtually absent in normal tissue surrounding the tumor. This correlation with high tumor grade and aggressive behavior suggests that TEM1 expression may be a useful prognostic indicator.

Direct evidence that TEM1 is important for the growth and progression of tumors has come from recent studies of TEM1 knockout (KO) mice [49]. The viability of these mice suggests that TEM1 is not critical for early vascular development. TEM1 also appeared to be dispensable for wound healing and growth of subcutaneous tumors. However, following orthotopic implantation of colorectal cancer cells into the abdomen of KO and wild type (WT) mice, significant differences in tumor growth, invasiveness and metastasis were observed. Survival of KO mice was significantly enhanced compared to WT mice, and the latter had a much higher incidence of both peritoneal carcinomatosis and liver metastasis. Although the mechanisms regulating the reduced tumor aggressiveness in TEM1 KO mice are unknown, these results demonstrate the need for caution when interpreting results solely from conventional subcutaneous tumor models.

Subcellular localization studies have shown that expression of TEM1 in mouse embryos is predominantly on the abluminal surface of the endothelium [44]. In addition to ECs, several studies have documented an elevated expression of TEM1 in tumor fibroblasts and pericytes, both integral components of the tumor stroma [45,50–52]. Expression in these cells may not necessarily impede a vascular targeting strategy since the same cells in normal adult tissues display little or no TEM1 expression. Indeed, some studies suggest that targeting activated pericytes and ECs simultaneously could have added benefit [53].

TEM5

From a therapeutic standpoint, TEM5 is a particularly attractive target because it is predicted to encode a G-protein-coupled receptor (GPCR). The human genome contains over 800 GPCRs, and approximately 50% of all prescription drugs target these receptors [54]. GPCRs are commonly involved in cell signaling and are known to respond to a vast array of ligands, including photons, odorants, lipids and peptides. TEM5, a 1331 amino acid protein, is considered part of the adhesion family of GPCRs, also known as the EGF-TM7 or LN-TM7 family [55]. Like other members of this family, TEM5 has a long amino terminus and is heavily glycosylated (Stevens, unpublished data).

The physiologic function of TEM5 and its ligand(s) are currently unknown. Vallon et. al. have found a soluble form of TEM5 (sTEM5) that is shed into the medium of TEM5-transfected human umbilical vein endothelial cells during differentiation into capillary-like networks [56]. Furthermore, they found that a proteolytically processed version of sTEM5 (ppsTEM5) was able to inhibit apoptosis of ECs by binding to integrin $\alpha v\beta 3$ via a cryptic RGD domain located in the extracellular domain. Thus, TEM5 may play a role in EC survival during the formation of new blood vessels.

Whether or not TEM5 is a true signaling GPCR is unknown. One study, conducted by Yamamoto et al. in 2004, identified TEM5 as well as its closest homologue GPR125 (TEM5-like) in a yeast two-hybrid screen for binding partners of the human homologue of the Drosophila disc large tumor suppressor gene (hDlg) [57]. The authors proposed that hDlg is recruited to the plasma membrane by TEM5 where it functions as a scaffold for the receptor during angiogenesis. Interestingly, recent work by Gregorc et al. provides evidence suggesting that hDlg is a very early caspase-3 target in apoptosis, and that loss of hDlg membrane localization precedes the disruption of cell to cell contacts, leading to apoptosis [58].

TEM7

TEM7 is elevated in the endothelium of a variety of human tumor types, including colon, lung, breast, bladder, esophageal and brain cancer [34,37,59]. Gene expression data derived from the endothelium of breast and brain tumors has validated the overexpression of TEM7 in the endothelium of colorectal cancer [35,37]. One comprehensive study compared all SAGE libraries generated from the endothelium of breast, colon and brain tumors simultaneously, and then bioinformatically subtracted genes that were found to be expressed in SAGE libraries from corresponding normal ECs, along with a large number of libraries derived from normal whole tissues [37]. Following this stringent analysis, TEM7 was the only tumor endothelial gene identified encoding a cell surface receptor broadly expressed among various tumor types but absent in all of the normal ECs and tissues analyzed.

Following its identification, it was noted that the mouse orthologue of human TEM7, called mTEM7, differed from its human counterpart in that it did not appear to be elevated in the tumor endothelium of mice [41]. This initial result has now been confirmed in our laboratory using a large number of tumor models, and multiple independent approaches including in situ hybridization, SAGE, and RT-PCR of purified ECs (St. Croix, unpublished observations). Unexpectedly, in mice, TEM7 was found to be expressed in neuronal cells, particularly in the Purkinje cells of the adult cerebellum [41]. Lee et al. have confirmed this neuronal pattern of staining in rats [60]. An inability to detect elevated levels of TEM7 mRNA in mouse tumor ECs led us to search for TEM7 homologues that might compensate for the lack of TEM7 overexpression in tumor ECs. This search resulted in the identification of TEM7-related (TEM7R), which is also a transmembrane protein that shares a similar predicted structure to TEM7 and 57% amino acid identity. Importantly, both human and mouse TEM7R were found to be expressed by tumor endothelium. Based on the patterns of expression observed, we proposed that both TEM7 and TEM7R are important for angiogenesis in humans, whereas TEM7R may play a dominant role in mice [42]. However, in mice, a high expression of TEM7R was noted in the vessels of normal muscle and lung [41] which could complicate the targeting of this particular molecule. Thus, although TEM7 appears to be one of the most promising targets of tumor vessels in humans, an appropriate preclinical model for targeting TEM7 in vivo is still lacking.

The physiological ligand of TEM7 is currently unknown. One study showed that TEM7 is able to bind to the extracellular matrix protein nidogen [61]. Our previous work has shown that a 9 amino acid peptide derived from cortactin is able bind both TEM7 and TEM7R [59]. Although cortactin has been shown to be important for endothelial remodeling that occurs following an angiogenic stimulus [62], given its cytoplasmic location it is unlikely to be the physiological ligand of the membrane bound form of TEM7. Nevertheless, the nine amino acid region of cortactin that binds TEM7/TEM7R may allow for future construction of small molecular weight compounds that can target tumor endothelium.

TEM8

TEM8 has been studied extensively since its initial identification in 2000. In 2001, interest in TEM8 increased substantially with its identification as an anthrax toxin receptor (ATR) [63]. TEM8 encodes a type I transmembrane protein with a large cytoplasmic tail. Its extracellular domain includes a von Willebrand factor type A (vWA) domain containing a metal ion-dependent adhesion (MIDAS) motif. The MIDAS motif appears to be critical for anthrax toxin binding, since the toxin–TEM8 interaction was abolished in the presence of EDTA. Binding of TEM8 to anthrax toxin also requires palmitoylation and ubiquitination [64].

While not expressed in the angiogenic corpus luteum, TEM8 has been found to be consistently overexpressed in the vessels of a variety of tumor types and occasionally by the tumor cells themselves [65–67]. This observation, as well as its lack of expression during normal physiologic angiogenesis in the adult, makes TEM8 a particularly attractive target for treating solid tumors using antiangiogenic and vascular targeting approaches.

The function of TEM8 and its physiologic ligand are currently unknown. One potential ligand is Collagen VI α3, an extracellular matrix molecule recently found to bind TEM8 [67]. TEM8 was also found to bind type I collagen and gelatin and influence EC motility on these substrates, although it had no effect on proliferation, survival, or differentiation into tubes [68]. Subsequent research by Werner et. al. has expanded these findings by implicating an essential role for the TEM8 cytosolic domain in cell spreading [69].

Antibodies that have been generated against TEM8 demonstrate that the protein is overexpressed in the tumor endothelium of a variety of tumor types [67]. However, other studies have reported detectable TEM8 expression in the epithelial cells of certain normal tissues [70]. Furthermore, it is becoming apparent that expression of TEM8 protein at the cell surface requires the presence of other co-factors, such as LRP6 [71]. Thus, in order to most effectively exploit TEM8 as a vascular target it will be important to determine whether or not TEM8 protein is preferentially expressed on the cell surface of tumor ECs in vivo. Although the anti-vascular effects of targeting TEM8 (described below) suggest this may be the case, this has not yet been directly shown.

TEM8 has been used as a DNA vaccine target of the tumor endothelium in preclinical mouse tumor models [72]. Although minimally effective on its own, the TEM8 vaccine was able to prolong survival of mice when combined with either a Her2/neu DNA vaccine to treat breast tumors, or an hTRP1/hgp75 DNA vaccine to treat B16F10 melanoma. Although the immunoregulatory mechanism by which the TEM8 vaccine enhanced survival is unclear, further studies in this area are warranted [73].

The discovery of TEM8 as an anthrax toxin receptor may help to explain the potent anti-tumor responses that have been obtained using anthrax toxin in tumor-bearing mice [42]. Injections of the toxin at doses that elicited no overt toxicity have led to impressive anti-tumor responses, in some cases causing complete regression [74]. Indeed, an antiangiogenic effect has been proposed due to a paucity of vessels in the treated tumors. However, subsequent studies demonstrated microscopic signs of toxicity at the effective dose range [75], possibly due to the more ubiquitous expression of capillary morphogenesis protein 2 (CMG2), another anthrax toxin receptor that shares 40% amino acid identity with TEM8. The protective antigen subunit of anthrax toxin is responsible for binding to TEM8. Recently, Leppla's group has re-engineered protective antigen in order to achieve even greater specificity for tumors in vivo, taking advantage of the fact that PA requires cleavage by a ubiquitous cell surface furin-like protease in order to be taken up by cells. By changing the cleavage recognition site to one which is recognized by urokinase plasminogen activator (uPA), they have been able to achieve enhanced anti-tumor responses with fewer side effects [75]. Although uPA is frequently overexpressed by tumor cells, its expression by activated cells of the stroma [76] may also contribute to the improved specificity of this therapeutic agent. The recent development of uPA-dependent protective antigen variants that can preferentially target TEM8 instead of CMG2 offers an even more promising strategy for the selective targeting of tumor vasculature [77].

Additional Markers of Tumor Vasculature

The detailed study of individual genes has also led to the identification of several additional cell surface tumor endothelial markers that have potential utility as vascular targets. Prostate-specific membrane antigen (PSMA) was found to be overexpressed in the neovasculature of a large variety of solid tumors in humans, in addition to normal and malignant prostate cells [78–80]. Importantly, its expression in normal vessels appears to be absent or below detection levels. In mice, however, PSMA is undetectable in tumor endothelium hindering the preclinical translation of PSMA as a neovascular target [81]. Nevertheless, in humans, ^{111}In-labeled anti-PSMA antibodies have been shown to localize to tumors of non-prostatic origin, presumably by targeting the tumor vessels [82]. In addition to prostate epithelium, PMSA expression has also been found in a variety of other normal tissues, such as duodenal mucosa, renal tubules and normal bladder [78,79,83–85], suggesting the need for a watchful eye on normal tissue toxicity as novel PSMA-targeting agents are developed.

Robo4 is a transmembrane receptor expressed on vascular pericytes and ECs that may bind Slit family members, or other uncharacterized ligands. Its expression has been shown to be elevated at sites of active angiogenesis, including tumors [86,87]. Gene knock-down studies in zebrafish suggest that Robo4 may help guide ECs to their targets during angiogenic sprouting, analogous to related family members involved in axonal guidance of the nervous system [88]. Although overexpressed in endothelium, Robo4 has also been detected in normal tissues such as heart, lung, liver and kidney, which could potentially impair a vascular targeting strategy [89].

Delta-like 4 (Dll4) is a member of the delta family of notch ligands and appears to be critical for normal vascular development in mice [90,91]. Dll4 has been found to be overexpressed at sites of physiological and tumor angiogenesis, but is largely undetectable in normal adult endothelium in vivo [92,93] (see Chapter 19). Interestingly, in vitro studies have shown that either over-expression or down-regulation of Dll4 inhibits endothelial cell proliferation, suggesting that an optimal level of Dll4 is essential for tumor angiogenesis [92,94]. Consistent with the idea that Dll4 levels are carefully regulated, loss of a single Dll4 allele results in embryonic lethality due to severe vascular abnormalities [90,91]. Dll4 expression is turned on by VEGF stimulation in endothelial cells, and signaling through a VEGF-Dll4/Notch4-ephrin B2 pathway is thought to play a critical role in vessel remodeling during tumor angiogenesis [95]. Recent studies using mouse tumor models have shown that antagonists of Dll4 synergize with anti-VEGF therapy [96,97], making this a particularly attractive avenue for tumor vascular targeting.

Concluding Remarks

Most conventional chemotherapeutic drugs were designed to target rapidly dividing tumor cells. However, such drugs often suffer from poor selectivity because they also target many non-malignant proliferating cell types. Furthermore, even when selectivity is achieved, the genetic instability of tumor cells can results in the selection of clonal populations that are able to resist drug therapy. Indeed, our rapidly expanding knowledge of the gene expression profiles of tumors suggests that cancer actually encompasses hundreds of diseases, and gene signatures of tumor cells can change rapidly following therapy. Targeting the ECs of tumor vessels offers an alternative approach that could help alleviate many of the problems associated with tumor cell targeting. Accumulating knowledge of a number of cell surface tumor endothelial markers suggests that these targets may be more stably expressed across a broad variety of solid tumor types. However, it should also be noted that most of the currently known molecular targets of tumor endothelium are also overexpressed at active sites of physiological angiogenesis in the adult. Thus, development of the most effective antiangiogenic or vascular targeting agents may still require a better understanding of the more subtle differences between normal physiologic angiogenesis and pathologic angiogenesis. Advances in proteomics and gene expression technology are beginning to provide us with a plethora of potential new targets which should aid in the development of antiangiogenic agents or ligand-based VDAs that can selectively home to tumor vessels. The challenge facing us now is to determine which of the new targets will be the most useful, and the best means by which to exploit them.

References

1. Bergers G, Benjamin LE. Tumorigenesis and the angiogenic switch. Nat Rev Cancer 2003; 3:401–10.
2. Folkman J. Angiogenesis in cancer, vascular, rheumatoid and other disease. Nat Med 1995; 1:27–31.
3. Carmeliet P. Angiogenesis in life, disease and medicine. Nature 2005; 438:932–6.
4. Pieramici DJ, Avery RL. Ranibizumab: treatment in patients with neovascular age-related macular degeneration. Expert Opin Biol Ther 2006; 6:1237–45.
5. Ferrara N, Damico L, Shams N, et al. Development of ranibizumab, an anti-vascular endothelial growth factor antigen bind-

ing fragment, as therapy for neovascular age-related macular degeneration. Retina 2006; 26:859–70.

6. Folkman J. Tumor angiogenesis: a possible control point in tumor growth. Ann Intern Med 1975; 82:96–100.

7. Sandler A, Gray R, Perry MC, et al. Paclitaxel-carboplatin alone or with bevacizumab for non-small-cell lung cancer. N Engl J Med 2006; 355:2542–50.

8. Hurwitz H, Fehrenbacher L, Novotny W, et al. Bevacizumab plus irinotecan, fluorouracil, and leucovorin for metastatic colorectal cancer. N Engl J Med 2004; 350:2335–42.

9. Gottesman MM. Mechanisms of cancer drug resistance. Annu Rev Med 2002; 53:615–27.

10. Ritchie E, Nichols G. Mechanisms of resistance to imatinib in CML patients: a paradigm for the advantages and pitfalls of molecularly targeted therapy. Curr Cancer Drug Targets 2006; 6:645–57.

11. Kerbel RS. Inhibition of tumor angiogenesis as a strategy to circumvent acquired resistance to anti-cancer therapeutic agents. Bioessays 1991; 13:31–6.

12. Kerbel RS. A cancer therapy resistant to resistance. Nature 1997; 390:335–6.

13. Boehm T, Folkman J, Browder T, et al. Antiangiogenic therapy of experimental cancer does not induce acquired drug resistance. Nature 1997; 390:404–7.

14. Hida K, Klagsbrun M. A new perspective on tumor endothelial cells: unexpected chromosome and centrosome abnormalities. Cancer Res 2005; 65:2507–10.

15. Casanovas O, Hicklin DJ, Bergers G, et al. Drug resistance by evasion of antiangiogenic targeting of VEGF signaling in late-stage pancreatic islet tumors. Cancer Cell 2005; 8:299–309.

16. Seghezzi G, Patel S, Ren CJ, et al. Fibroblast growth factor-2 (FGF-2) induces vascular endothelial growth factor (VEGF) expression in the endothelial cells of forming capillaries: an autocrine mechanism contributing to angiogenesis. J Cell Biol 1998; 141:1659–73.

17. Tille JC, Wood J, Mandriota SJ, et al. Vascular endothelial growth factor (VEGF) receptor-2 antagonists inhibit VEGF- and basic fibroblast growth factor-induced angiogenesis in vivo and in vitro. J Pharmacol Exp Ther 2001; 299:1073–85.

18. Kanda S, Miyata Y, Kanetake H. Fibroblast growth factor-2-mediated capillary morphogenesis of endothelial cells requires signals via Flt-1/vascular endothelial growth factor receptor-1: possible involvement of c-Akt. J Biol Chem 2004; 279:4007–16.

19. Neri D, Bicknell R. Tumour vascular targeting. Nat Rev Cancer 2005; 5:436–46.

20. Folkman J. Tumor angiogenesis: therapeutic implications. N Engl J Med 1971; 285:1182–6.

21. Denekamp J. Endothelial cell proliferation as a novel approach to targeting tumour therapy. Br J Cancer 1982; 45:136–9.

22. Denekamp J. Review article: angiogenesis, neovascular proliferation and vascular pathophysiology as targets for cancer therapy. Br J Radiol 1993; 66:181–96.

23. Burrows FJ, Thorpe PE. Eradication of large solid tumors in mice with an immunotoxin directed against tumor vasculature. Proc Natl Acad Sci USA 1993; 90:8996–9000.

24. Huang X, Molema G, King S, et al. Tumor infarction in mice by antibody-directed targeting of tissue factor to tumor vasculature. Science 1997; 275:547–50.

25. Thorpe PE. Vascular targeting agents as cancer therapeutics. Clin Cancer Res 2004; 10:415–27.

26. Tozer GM, Kanthou C, Baguley BC. Disrupting tumour blood vessels. Nat Rev Cancer 2005; 5:423–35.

27. Arap W, Pasqualini R, Ruoslahti E. Cancer treatment by targeted drug delivery to tumor vasculature in a mouse model. Science 1998; 279:377–80.

28. Pasqualini R, Koivunen E, Kain R, et al. Aminopeptidase N is a receptor for tumor-homing peptides and a target for inhibiting angiogenesis. Cancer Res 2000; 60:722–7.

29. Marchio S, Lahdenranta J, Schlingemann RO, et al. Aminopeptidase A is a functional target in angiogenic blood vessels. Cancer Cell 2004; 5:151–62.

30. Christian S, Pilch J, Akerman ME, et al. Nucleolin expressed at the cell surface is a marker of endothelial cells in angiogenic blood vessels. J Cell Biol 2003; 163:871–8.

31. Oh P, Li Y, Yu J, et al. Subtractive proteomic mapping of the endothelial surface in lung and solid tumours for tissue-specific therapy. Nature 2004; 429:629–35.

32. Rybak JN, Ettorre A, Kaissling B, et al. In vivo protein biotinylation for identification of organ-specific antigens accessible from the vasculature. Nat Methods 2005; 2:291–8.

33. Castronovo V, Waltregny D, Kischel P, et al. A chemical proteomics approach for the identification of accessible antigens expressed in human kidney cancer. Mol Cell Proteomics 2006; 5:2083–91.

34. St Croix B, Rago C, Velculescu V, et al. Genes expressed in human tumor endothelium. Science 2000; 289:1197–202.

35. Parker BS, Argani P, Cook BP, et al. Alterations in vascular gene expression in invasive breast carcinoma. Cancer Res 2004; 64:7857–66.

36. Madden SL, Cook BP, Nacht M, et al. Vascular gene expression in nonneoplastic and malignant brain. Am J Pathol 2004; 165:601–8.

37. Beaty RM, Edwards JB, Boon K, et al. PLXDC1 (TEM7) is identified in a genome-wide expression screen of glioblastoma endothelium. J Neurooncol 2007; 81:241–8.

38. Allinen M, Beroukhim R, Cai L, et al. Molecular characterization of the tumor microenvironment in breast cancer. Cancer Cell 2004; 6:17–32.

39. Velculescu VE, Zhang L, Vogelstein B, et al. Serial analysis of gene expression. Science 1995; 270:484–7.

40. Pozzi A, Moberg PE, Miles LA, et al. Elevated matrix metalloprotease and angiostatin levels in integrin alpha 1 knockout mice cause reduced tumor vascularization. Proc Natl Acad Sci USA 2000; 97:2202–7.

41. Carson-Walter EB, Watkins DN, Nanda A, et al. Cell surface tumor endothelial markers are conserved in mice and humans. Cancer Res 2001; 61:6649–55.

42. Nanda A, St Croix B. Tumor endothelial markers: new targets for cancer therapy. Curr Opin Oncol 2004; 16:44–9.

43. Rettig WJ, Garin-Chesa P, Healey JH, et al. Identification of endosialin, a cell surface glycoprotein of vascular endothelial cells in human cancer. Proc Natl Acad Sci USA 1992; 89:10832–6.

44. Rupp C, Dolznig H, Puri C, et al. Mouse endosialin, a C-type lectin-like cell surface receptor: expression during embryonic development and induction in experimental cancer neoangiogenesis. Cancer Immun 2006; 6:10.

45. Brady J, Neal J, Sadakar N, et al. Human endosialin (tumor endothelial marker 1) is abundantly expressed in highly malignant and invasive brain tumors. J Neuropathol Exp Neurol 2004; 63:1274–83.

46. Christian S, Ahorn H, Koehler A, et al. Molecular cloning and characterization of endosialin, a C-type lectin-like cell surface receptor of tumor endothelium. J Biol Chem 2001; 276:7408–14.

47. Opavsky R, Haviernik P, Jurkovicova D, et al. Molecular characterization of the mouse Tem1/endosialin gene regulated by cell density in vitro and expressed in normal tissues in vivo. J Biol Chem 2001; 276:38795–807.

48. Dean YD, McGreal EP, Akatsu H, et al. Molecular and cellular properties of the rat AA4 antigen, a C-type lectin-like receptor with structural homology to thrombomodulin. J Biol Chem 2000; 275:34382–92.

49. Nanda A, Karim B, Peng Z, et al. Tumor endothelial marker 1 (Tem1) functions in the growth and progression of abdominal tumors. Proc Natl Acad Sci USA 2006; 103:3351–6.

50. MacFadyen J, Savage K, Wienke D, et al. Endosialin is expressed on stromal fibroblasts and CNS pericytes in mouse embryos and is downregulated during development. Gene Expr Patterns 2007; 7:363–9.

51. Virgintino D, Girolamo F, Errede M, et al. An intimate interplay between precocious, migrating pericytes and endothelial cells governs human fetal brain angiogenesis. Angiogenesis 2007.

52. Huber MA, Kraut N, Schweifer N, et al. Expression of stromal cell markers in distinct compartments of human skin cancers. J Cutan Pathol 2006; 33:145–55.

53. Bergers G, Song S, Meyer-Morse N, et al. Benefits of targeting both pericytes and endothelial cells in the tumor vasculature with kinase inhibitors. J Clin Invest 2003; 111:1287–95.

54. Tyndall JD, Sandilya R. GPCR agonists and antagonists in the clinic. Med Chem 2005; 1:405–21.

55. Bjarnadottir TK, Fredriksson R, Hoglund PJ, et al. The human and mouse repertoire of the adhesion family of G-protein-coupled receptors. Genomics 2004; 84:23–33.

56. Vallon M, Essler M. Proteolytically processed soluble tumor endothelial marker (TEM) 5 mediates endothelial cell survival during angiogenesis by linking integrin alpha(v)beta3 to glycosaminoglycans. J Biol Chem 2006; 281:34179–88.

57. Yamamoto Y, Irie K, Asada M, et al. Direct binding of the human homologue of the Drosophila disc large tumor suppressor gene to seven-pass transmembrane proteins, tumor endothelial marker 5 (TEM5), and a novel TEM5-like protein. Oncogene 2004; 23:3889–97.

58. Gregorc U, Ivanova S, Thomas M, et al. hDLG/SAP97, a member of the MAGUK protein family, is a novel caspase target during cell-cell detachment in apoptosis. Biol Chem 2005; 386:705–10.

59. Nanda A, Buckhaults P, Seaman S, et al. Identification of a binding partner for the endothelial cell surface proteins TEM7 and TEM7R. Cancer Res 2004; 64:8507–11.

60. Lee HK, Bae HR, Park HK, et al. Cloning, characterization and neuronal expression profiles of tumor endothelial marker 7 in the rat brain. Brain Res Mol Brain Res 2005; 136:189–98.

61. Lee HK, Seo IA, Park HK, et al. Identification of the basement membrane protein nidogen as a candidate ligand for tumor endothelial marker 7 in vitro and in vivo. FEBS Lett 2006; 580:2253–7.

62. Li Y, Uruno T, Haudenschild C, et al. Interaction of cortactin and Arp2/3 complex is required for sphingosine-1-phosphate-induced endothelial cell remodeling. Exp Cell Res 2004; 298:107–21.

63. Bradley KA, Mogridge J, Mourez M, et al. Identification of the cellular receptor for anthrax toxin. Nature 2001; 414:225–9.

64. Abrami L, Leppla SH, van der Goot FG. Receptor palmitoylation and ubiquitination regulate anthrax toxin endocytosis. J Cell Biol 2006; 172:309–20.

65. Koo HM, VanBrocklin M, McWilliams MJ, et al. Apoptosis and melanogenesis in human melanoma cells induced by anthrax lethal factor inactivation of mitogen-activated protein kinase kinase. Proc Natl Acad Sci USA 2002; 99:3052–7.

66. Oberthuer A, Skowron M, Spitz R, et al. Characterization of a complex genomic alteration on chromosome 2p that leads to four alternatively spliced fusion transcripts in the neuroblastoma cell lines IMR-5, IMR-5/75 and IMR-32. Gene 2005; 363:41–50.

67. Nanda A, Carson-Walter EB, Seaman S, et al. TEM8 interacts with the cleaved C5 domain of collagen alpha 3(VI). Cancer Res 2004; 64:817–20.

68. Hotchkiss KA, Basile CM, Spring SC, et al. TEM8 expression stimulates endothelial cell adhesion and migration by regulating cell-matrix interactions on collagen. Exp Cell Res 2005; 305:133–44.

69. Werner E, Kowalczyk AP, Faundez V. Anthrax toxin receptor 1/tumor endothelium marker 8 mediates cell spreading by coupling extracellular ligands to the actin cytoskeleton. J Biol Chem 2006; 281:23227–36.

70. Bonuccelli G, Sotgia F, Frank PG, et al. ATR/TEM8 is highly expressed in epithelial cells lining Bacillus anthracis' three sites of entry: implications for the pathogenesis of anthrax infection. Am J Physiol Cell Physiol 2005; 288:C1402–10.

71. Wei W, Lu Q, Chaudry GJ, et al. The LDL receptor-related protein LRP6 mediates internalization and lethality of anthrax toxin. Cell 2006; 124:1141–54.

72. Felicetti P, Mennecozzi M, Barucca A, et al. Tumor endothelial marker 8 enhances tumor immunity in conjunction with immunization against differentiation Ag. Cytotherapy 2007; 9:23–34.

73. St. Croix B. Vaccines targeting tumor vasculature: a new approach for cancer immunotherapy. Cytotherapy 2007; 9:1–3.

74. Duesbery NS, Resau J, Webb CP, et al. Suppression of ras-mediated transformation and inhibition of tumor growth and angiogenesis by anthrax lethal factor, a proteolytic inhibitor of multiple MEK pathways. Proc Natl Acad Sci USA 2001; 98:4089–94.

75. Liu S, Aaronson H, Mitola DJ, et al. Potent antitumor activity of a urokinase-activated engineered anthrax toxin. Proc Natl Acad Sci USA 2003; 100:657–62.

76. Gutierrez LS, Schulman A, Brito-Robinson T, et al. Tumor development is retarded in mice lacking the gene for urokinase-type plasminogen activator or its inhibitor, plasminogen activator inhibitor-1. Cancer Res 2000; 60:5839–47.

77. Chen KH, Liu S, Bankston LA, et al. Selection of anthrax toxin protective antigen variants that discriminate between the cellular receptors TEM8 and CMG2 and achieve targeting of tumor cells. J Biol Chem 2007.

78. Chang SS, Reuter VE, Heston WD, et al. Five different anti-prostate-specific membrane antigen (PSMA) antibodies confirm PSMA expression in tumor-associated neovasculature. Cancer Res 1999; 59:3192–8.

79. Chang SS, O'Keefe DS, Bacich DJ, et al. Prostate-specific membrane antigen is produced in tumor-associated neovasculature. Clin Cancer Res 1999; 5:2674–81.

80. Chang SS, Reuter VE, Heston WD, et al. Metastatic renal cell carcinoma neovasculature expresses prostate-specific membrane antigen. Urology 2001; 57:801–5.

81. Huang X, Bennett M, Thorpe PE. Anti-tumor effects and lack of side effects in mice of an immunotoxin directed against human and mouse prostate-specific membrane antigen. Prostate 2004; 61:1–11.

82. Milowsky MI, Nanus DM, Kostakoglu L, et al. Vascular targeted therapy with anti-prostate-specific membrane antigen monoclonal antibody J591 in advanced solid tumors. J Clin Oncol 2007; 25:540–7.

83. Silver DA, Pellicer I, Fair WR, et al. Prostate-specific membrane antigen expression in normal and malignant human tissues. Clin Cancer Res 1997; 3:81–5.

84. Gala JL, Loric S, Guiot Y, et al. Expression of prostate-specific membrane antigen in transitional cell carcinoma of the bladder: prognostic value? Clin Cancer Res 2000; 6:4049–54.

85. Kinoshita Y, Kuratsukuri K, Landas S, et al. Expression of prostate-specific membrane antigen in normal and malignant human tissues. World J Surg 2006; 30:628–36.

86. Huminiecki L, Gorn M, Suchting S, et al. Magic roundabout is a new member of the roundabout receptor family that is endothelial specific and expressed at sites of active angiogenesis. Genomics 2002; 79:547–52.

87. Seth P, Lin Y, Hanai J, et al. Magic roundabout, a tumor endothelial marker: expression and signaling. Biochem Biophys Res Commun 2005; 332:533–41.

88. Bedell VM, Yeo SY, Park KW, et al. roundabout4 is essential for angiogenesis in vivo. Proc Natl Acad Sci USA 2005; 102:6373–8.

89. Park KW, Morrison CM, Sorensen LK, et al. Robo4 is a vascular-specific receptor that inhibits endothelial migration. Dev Biol 2003; 261:251–67.

90. Gale NW, Dominguez MG, Noguera I, et al. Haploinsufficiency of delta-like 4 ligand results in embryonic lethality due to major defects in arterial and vascular development. Proc Natl Acad Sci USA 2004; 101:15949–54.

91. Krebs LT, Shutter JR, Tanigaki K, et al. Haploinsufficient lethality and formation of arteriovenous malformations in Notch pathway mutants. Genes Dev 2004; 18:2469–73.

92. Patel NS, Li JL, Generali D, et al. Up-regulation of delta-like 4 ligand in human tumor vasculature and the role of basal expression in endothelial cell function. Cancer Res 2005; 65:8690–7.

93. Mailhos C, Modlich U, Lewis J, et al. Delta4, an endothelial specific notch ligand expressed at sites of physiological and tumor angiogenesis. Differentiation 2001; 69:135–44.

94. Williams CK, Li JL, Murga M, et al. Up-regulation of the Notch ligand Delta-like 4 inhibits VEGF-induced endothelial cell function. Blood 2006; 107:931–9.

95. Hainaud P, Contreres JO, Villemain A, et al. The Role of the Vascular Endothelial Growth Factor-Delta-like 4 Ligand/Notch4-Ephrin B2 Cascade in Tumor Vessel Remodeling and Endothelial Cell Functions. Cancer Res 2006; 66:8501–10.

96. Noguera-Troise I, Daly C, Papadopoulos NJ, et al. Blockade of Dll4 inhibits tumour growth by promoting non-productive angiogenesis. Nature 2006; 444:1032–7.

97. Ridgway J, Zhang G, Wu Y, et al. Inhibition of Dll4 signalling inhibits tumour growth by deregulating angiogenesis. Nature 2006; 444:1083–7.

98. Brooks PC, Clark RA, Cheresh DA. Requirement of vascular integrin alpha v beta 3 for angiogenesis. Science 1994; 264: 569–71.

99. Schlingemann RO, Oosterwijk E, Wesseling P, et al. Aminopeptidase a is a constituent of activated pericytes in angiogenesis. J Pathol 1996; 179:436–42.

100. Taniguchi K, Harada N, Ohizumi I, et al. Recognition of human activated CD44 by tumor vasculature-targeted antibody. Biochem Biophys Res Commun 2000; 269:671–5.

101. Ohizumi I, Taniguchi K, Saito H, et al. Suppression of solid tumor growth by a monoclonal antibody against tumor vasculature in rats: involvement of intravascular thrombosis and fibrinogenesis. Int J Cancer 1999; 82:853–9.

102. Ran S, He J, Huang X, et al. Antitumor effects of a monoclonal antibody that binds anionic phospholipids on the surface of tumor blood vessels in mice. Clin Cancer Res 2005; 11:1551–62.

103. Huang X, Bennett M, Thorpe PE. A monoclonal antibody that binds anionic phospholipids on tumor blood vessels enhances the antitumor effect of docetaxel on human breast tumors in mice. Cancer Res 2005; 65:4408–16.

104. Lee WS, Jain MK, Arkonac BM, et al. Thy-1, a novel marker for angiogenesis upregulated by inflammatory cytokines. Circ Res 1998; 82:845–51.

105. Brekken RA, Thorpe PE. VEGF-VEGF receptor complexes as markers of tumor vascular endothelium. J Control Release 2001; 74:173–81.

106. Liao F, Doody JF, Overholser J, et al. Selective targeting of angiogenic tumor vasculature by vascular endothelial-cadherin antibody inhibits tumor growth without affecting vascular permeability. Cancer Res 2002; 62:2567–75.

107. Lamszus K, Brockmann MA, Eckerich C, et al. Inhibition of glioblastoma angiogenesis and invasion by combined treatments directed against vascular endothelial growth factor receptor-2, epidermal growth factor receptor, and vascular endothelial-cadherin. Clin Cancer Res 2005; 11:4934–40.

Section V
Clinical Translation of Angiogenesis Inhibitors

Chapter 30
Overview and Clinical Applications of VEGF-A

Napoleone Ferrara

Keywords: Vascular endothelial growth factor; endothelium; angiogenesis; tyrosine kinases; tumor growth

Abstract: Vascular endothelial growth factor (VEGF)-A is an endothelial cell-specific mitogen and an angiogenic inducer. Two tyrosine kinases, VEGFR1 and VEGFR-2, are VEGF receptors. Loss of a single VEGF-A allele results in defective vascularization and embryonic lethality. Anti-VEGF-A monoclonal antibodies or other VEGF inhibitors block growth and neovascularization in tumor models. We developed a humanized anti-VEGF-A monoclonal antibody (bevacizumab). Bevacizumab demonstrated clinical efficacy, including a survival advantage, in multiple tumor types. Bevacizumab has been approved by the FDA for the treatment of previously untreated and relapsed metastatic colorectal cancer, non-small-cell lung cancer, and previously untreated metastatic breast cancer, in combination with chemotherapy. Also, two small molecule inhibitors of VEGF receptor tyrosine kinase activity have been approved by the FDA for the treatment of metastatic renal cell carcinoma. Furthermore, VEGF-A is implicated in intraocular neovascularization associated with active proliferative retinopathies and the wet form of age-related macular degeneration (AMD). A humanized anti-VEGF-A Fab (ranibizumab) has been developed for the treatment of the neovascular form of AMD. Ranibizumab administration maintained and even improved visual acuity and was approved by the FDA for the treatment of AMD in June 2006.

Introduction

Angiogenesis is known to be fundamental to a variety of physiological processes including embryonic and postnatal development, reproductive functions and wound healing [1] Furthermore, neovascularization plays an important pathogenic role in tumorigenesis and in the vision loss associated with ischemic retinal disorders and the wet form of age-related macular degeneration (AMD). Research performed in recent decades has established that angiogenesis is a complex and coordinated process, which requires a series of signaling steps in endothelial and mural cells elicited by numerous families of ligands (reviewed in [2]). Moreover, a variety of endogenous inhibitors of angiogenesis have been identified, including endostatin, tumstatin and vasostatin [3]. However, despite such complexity and potential redundancy, vascular endothelial growth factor (VEGF)-A appears to be necessary for growth of blood vessels in a variety of normal and pathological circumstances [4]. *VEGF-A* is the prototype member of a gene family that includes placenta growth factor (PlGF), VEGF-B, VEGF-C, VEGF-D and the orf-virus encoded VEGF-E (reviewed in [5, 6]).

Definitive clinical studies, resulting in approval by the U.S. Food and Drug administration (FDA) of several drugs, have established that VEGF-A is an important therapeutic target for cancer and wet AMD (reviewed in [7]). This chapter summarizes the basic biology of VEGF-A and provides an update on the clinical progress in targeting VEGF.

History of VEGF

Independent lines of research contributed to the discovery of VEGF [4]. In 1983, Senger et al. reported the identification in the supernatant of a guinea pig tumor cell line of a permeability-enhancing protein, which was named "vascular permeability factor" (VPF) and was proposed to be involved in the high permeability of tumor vessels [8]. However, these efforts did not yield the full purification of the VPF protein. The lack of amino acid sequence information precluded cDNA cloning and elucidation of the identity of VPF. Therefore, very limited progress in understanding the role of this factor took place over the subsequent several years.

In 1989, we reported the isolation of an endothelial cell mitogen from the supernatant of bovine pituitary cells, which we named "vascular endothelial growth factor" (VEGF) [9]. The NH_2-terminal amino acid sequence of VEGF did not

Genentech, Inc., 1 DNA Way, South San Francisco, CA 94080, USA
E-mail: ferrara.napoleone@gene.com

match any known protein in available databases [9]. Subsequently, Connolly's group at Monsanto Co. reported the isolation and sequencing of VPF [10]. By the end of 1989, we isolated cDNA clones encoding bovine VEGF$_{164}$ and three human VEGF isoforms: VEGF$_{121}$, VEGF$_{165}$ and VEGF$_{189}$ [11]. The Monsanto group described a human VPF clone, which encoded a protein identical to VEGF$_{189}$ [12]. These studies indicated that, unexpectedly, a single molecule was responsible for both mitogenic and permeabiliy-enhancing activities.

Biological Effects of VEGF-A

VEGF-A stimulates the growth of vascular endothelial cells derived from arteries, veins and lymphatics [11,13]. VEGF-A induces angiogenesis in a variety of in vivo models [13]. Administration of VEGF also induces rapid and transient increases in microvascular permeability in several experimental model systems (reviewed in [14]).

Inactivation of a single VEGF-A allele results in embryonic lethality between day 11 and 12, indicating that during early development there is a critical VEGF-A gene-dosage requirement [15]. VEGF-A also plays an important role in early postnatal life. Administration of VEGF inhibitors, including monoclonal antibodies and soluble receptors, results in growth arrest and lethality in mice when the treatment is initiated at day 1 or day 8 postnatally [16,17]. VEGF is important for endochondral bone formation and growth plate angiogenesis and morphogenesis. VEGF-A blockade reversibly inhibits skeletal growth [18]. Another key function of VEGF-A is the regulation of the cyclical angiogenesis that occurs in the female reproductive tract [19]. VEGF-A is also a survival factor for endothelial cells, both in vitro and in vivo [20–23]. VEGF induces expression of the anti-apoptotic proteins Bcl-2, A1 [21] and survivin [24] in endothelial cells. In vivo, VEGF's pro-survival effects are developmentally regulated. Inhibition of VEGF results in apoptotic changes and regression of the vasculature of neonatal, but not adult mice [16]. Endothelial cells are the primary targets of VEGF-A, but several studies have reported mitogenic and non-mitogenic effects of VEGF-A on non-endothelial cell types including neurons (reviewed in [25]).

It is now well established that VEGF-A promotes monocyte chemotaxis [26,27]. VEGF deficient hematopoietic stem cells and bone marrow mononuclear cells fail to re-populate lethally irradiated hosts, despite co-administration of a large excess of wildtype cells [28].

VEGF-A Isoforms

Alternative exon splicing results in the generation of four main VEGF-A isoforms, which have respectively 121, 165, 189 and 206 amino acids after the signal sequence is cleaved (VEGF$_{121}$, VEGF$_{165}$, VEGF$_{189}$, VEGF$_{206}$) [29,30]. Less frequent splice variants have also been reported, including VEGF$_{145}$, VEGF$_{183}$, VEGF$_{162}$, and VEGF$_{165b}$ (reviewed in [13]).

Like VEGF$_{165}$, native VEGF is a heparin-binding homodimeric glycoprotein of 45 kDa [9, 31]. In contrast, VEGF$_{121}$ lacks heparin-binding properties [32]. VEGF$_{189}$ and VEGF$_{206}$ bind to heparin with affinity comparable to that of bFGF [32]. Whereas VEGF$_{121}$ is a freely diffusible protein, VEGF$_{189}$ and VEGF$_{206}$ are almost completely bound to heparin-like moieties in the cell surface or in the extracellular matrix. VEGF$_{165}$ has intermediate properties in terms of heparin-affinity and bioavailability [33]. The long isoforms may be released in a diffusible form by proteolytic cleavage. Early studies showed that plasmin is able to cleave VEGF$_{165}$ at the COOH terminus, generating VEGF$_{110}$, a bioactive fragment consisting of the first 110 NH$_2$-terminal amino acids [32,34]. Interestingly, recent studies have shown that various matrix metalloproteinases (MMPs), particularly MMP-3, may also cleave VEGF$_{165}$ to generate diffusible, non-heparin binding fragments [35]. Proteolytic processing of VEGF$_{165}$ by MMP-3 occurs in steps, with sequential cleavage at residues 135, 120, and finally at residue 113 [35]. Thus, the final product of MMP-3 processing, VEGF$_{113}$, is expected to be biologically and biochemically very similar to the plasmin-generated VEGF fragment.

VEGF Receptors

VEGF-A binds two highly related receptor tyrosine kinases (RTK), VEGFR1 [36] and VEGFR2 (Flk-1/KDR) [37]. VEGFR1 was the first RTK to be identified as a VEGF receptor more than a decade ago [38], but the precise function of this molecule is still debated in the field [13]. The functions and signaling properties of VEGFR1 appear to be vary with the developmental stage and the cell type, e.g., endothelial versus non-endothelial cells. VEGFR1 binds not only VEGF-A but also PlGF and VEGF-B, and fails to mediate a strong mitogenic signal in endothelial cells [39,40]. Non-mitogenic functions mediated by VEGFR1 in the vascular endothelium include the release of growth factors [41] and the induction of MMP-9 [42]. Furthermore, VEGFR1 mediates hematopoiesis [43] and monocyte chemotaxis [27] in response to VEGF-A or PlGF.

VEGFR2 also binds VEGF-A with high affinity [37, 44]. VEGF-C and VEGF-D may also bind and activate VEGFR2, following their proteolytic cleavage [6]. The key role of VEGFR2 in developmental angiogenesis and hematopoiesis is underscored by the lack of vasculogenesis and failure to develop blood islands and organized blood vessels in Flk-1 null mice [45]. There is now agreement that VEGFR2 is the major mediator of the angiogenic and permeability-enhancing effects of VEGF-A. VEGFR2 undergoes dimerization and strong ligand-dependent tyrosine phosphorylation in intact cells and results in a mitogenic, chemotactic and pro-survival signal. Several tyrosine residues have been shown to be phosphorylated (for review see [46]).

In 1998, Soker et al. identified a receptor for isoforms of VEGF-A containing the exon 7-encoded heparin-binding domain [47]. Surprisingly, this receptor proved identical to Neuropilin-1 (NRP1), a molecule that has been previously shown to be implicated in axon guidance as a receptor for members of the collapsin/semaphorin family [47]. NRP1 appears to present $VEGF_{165}$ to VEGFR2 in a configuration that enhances the effectiveness of VEGFR2-mediated signal transduction [47].

Regulation of VEGF Gene Expression

Oxygen Tension

Oxygen tension plays a key role in regulating the expression of a variety of genes [48]. VEGF mRNA expression is induced by exposure to low pO_2 [49, 50]. A 28-base sequence has been identified in the promoter of the VEGF gene, which represents a binding site for the hypoxia-inducible factor 1 (HIF-1) transcription factor [51]. HIF-1 is a basic, heterodimeric, helix-loop-helix protein consisting of two subunits, HIF-1α and aryl hydrocarbon receptor nuclear translocator (ARNT), also known as HIF-1β [52]. The critical role of the product of the von Hippel-Lindau (VHL) tumor suppressor gene in HIF-1-dependent hypoxic responses has been elucidated (for review see [53]). The VHL gene is inactivated in patients with von Hippel-Landau disease and in most sporadic clear-cell renal carcinomas [54]. One of the functions of VHL is to be part of a ubiquitin ligase complex that targets HIF subunits for proteasomal degradation [55,56]. Oxygen promotes the hydroxylation of HIF at specific proline residues [55,56]. A family of prolyl hydroxylases related to the *Egl-9 C. elegans* gene product was identified as HIF prolyl hydroxylases [48,57,58]. (Refer to chapters 15 & 16 on the hypoxic regulation of angiogenesis by HIF1 and HIF2.)

Growth Factors, Hormones and Oncogenes

Several growth factors, including endothelial growth factor (EGF), transforming growth factor (TGF)-α, TGF-β, keratinocyte growth factor (KGF), insulin-like growth factor (IGF)-1, FGF and PDGF, upregulate VEGF mRNA expression [59–61], suggesting that paracrine or autocrine release of such factors cooperates with local hypoxia in regulating VEGF release in the microenvironment. Also, inflammatory cytokines such as interleukin (IL)-1-α and IL-6 induce expression of VEGF in several cell types, including synovial fibroblasts [62,63].

A variety of transforming events also result in induction of VEGF gene expression. Oncogenic mutations or amplification of ras lead to VEGF upregulation [64,65]. Mutations in the wnt-signaling pathway, that are frequently associated with pre-malignant colonic adenomas, also result in upregulation of VEGF [66]. Interestingly, VEGF is upregulated in polyps from mouse models of human familial adenomatous, the $Apc^{\Delta716}$ [67] and the $Apc^{+/min}$ [68].

Role of VEGF-A in Tumor Angiogenesis

Many tumor cell lines secrete VEGF-A in vitro (reviewed in [13]). In situ hybridization studies have demonstrated that VEGF mRNA is expressed in many human tumors [14]. Renal cell carcinomas have a particularly high level of VEGF-A expression, consistent with the notion that inactivating mutation in the von Hippel-Lindau (VHL) tumor suppressor gene, resulting in high transcription of HIF-target genes under normoxic conditions, occur in ~50% of such tumors [54].

In 1993, monoclonal antibodies targeting VEGF-A were reported to inhibit the growth of several tumor cell lines in nude mice [69]. Inhibition of tumor growth has been achieved also with other anti-VEGF-A treatments, including small molecule inhibitors of VEGFR2 signaling (reviewed in [70]), anti-VEGFR2 antibodies [71] and soluble VEGF receptors [72,73].

Although tumor cells frequently represent the major source of VEGF-A, tumor-associated stroma is also an important site of VEGF production [72]. Recent studies have shown that tumor-derived PDGF-A may be especially important for the recruitment of an angiogenic stroma that produces VEGF-A and potentially other angiogenic factors [74,75].

Combining anti-VEGF treatment with chemotherapy [76] or radiation therapy [77] results in a greater anti-tumor effect than either of these therapies alone. The mechanism of such potentiation is under debate. One hypothesis is that antiangiogenic agents normalize the tumor vasculature, improving delivery of chemotherapy to tumor cells [78].

Clinical Trials in Cancer Patients with VEGF Inhibitors

Several VEGF inhibitors have been developed as anti-cancer agents. These include a humanized anti-VEGF-A monoclonal antibody (bevacizumab; Avastin™) [79,80], an anti-VEGFR2 antibody [71], various small molecules inhibiting VEGFR2 signal transduction [70], and a VEGF receptor chimeric protein [73]. For recent reviews, see [81–83] and the subsequent chapters which will provide an in-depth discussion of the clinical development of these antiangiogenic agents. The next section will briefly highlight a few of these compounds.

The clinical trial that resulted in the FDA approval of bevacizumab (February 2004) was a randomized, double-blind, phase III study in which bevacizumab was administered in combination with bolus-IFL (irinotecan, 5FU, leucovorin) chemotherapy as first-line therapy for previously untreated metastatic colorectal cancer [84]. Median survival was increased from 15.6 months in the bolus-IFL + placebo arm to 20.3 months in the bolus-IFL + bevacizumab arm. Similar increases were seen in progression-free survival, response rate, and duration of response. The clinical benefit of bevacizumab was seen in all subject subgroups [84]. Although bevacizumab was generally well tolerated, some serious and unusual toxicities were observed. Hypertension requiring medical intervention with standard anti-hypertensive therapy developed in 11% of bevacizumab treated patients. In addition, gastrointestinal perforation was noted in ~2% of patients. In a combined analysis of five randomized trials involving bevacizumab, including the pivotal trial in colorectal cancer, the incidence of arterial thromboembolic complications, including stroke, myocardial infarction, transient ischemic attacks and unstable angina was approximately doubled the incidence seen with chemotherapy alone.

The clinical benefit of bevacizumab is being evaluated in a broad variety of tumor types and lines of therapy. Several combination studies with biologicals are also ongoing, which include inhibitors of tyrosine kinases (sorafenib, bay 43-9006), the proteosome (bortezomib), and the mammalian target of rapamycin (mTOR) (CCI-779). Bevacizumab combined with weekly paclitaxel in women with previously untreated metastatic breast cancer provided a substantial improvement in the primary endpoint of progression free survival (11.0 versus 6.1 months, $P < 0.001$) relative to paclitaxel alone (reviewed in [7]. Combining bevacizumab with paclitaxel and carboplatin in patients with previously untreated, nonsquamous, non-small cell lung cancer (NSCLC) provided a significant improvement in the primary endpoint of overall survival (12.5 versus 10.2 months, $P = 0.007$) [85]. An earlier, phase II study of bevacizumab in NSCLC had identified pulmonary bleeding as a significant adverse event in this tumor type [86]. Squamous cell histology was identified as a major risk factor for bleeding and these patients were excluded from the phase III study. This markedly reduced the rate of serious bleeding associated with bevacizumab use [85]. Also, combining bevacizumab with 5-fluorouricil, leucovorin, and oxaliplatin (FOLFOX) in patients with previously treated metastatatic colorectal cancers provided a significant improvement in the primary endpoint of survival (12.9 versus 10.8 months, $P = 0.002$) [87].

Preliminary, encouraging, data with bevacizumab have been reported in a number of other cancers including renal cell [88], ovarian and prostate (reviewed in [7]). Taken together, these findings suggest that targeting VEGF-A with a neutralizing antibody may be a broadly applicable approach to the treatment of human cancer.

Besides bevacizumab, several other types of VEGF inhibitors are being developed (reviewed in [8,70,83]). Among these, a variety of small molecule RTK inhibitors targeting the VEGF receptors are at different stages of clinical development. The most advanced are SU11248 (sunitinib) and Bay 43-9006. SU11248 inhibits tyrosine phosphorylation of VEGFRs, platelet-derived growth factor receptors (PDGFRs), c-kit and Flt-3 [89] and has been reported to have efficacy in imatinib-resistant gastrointestinal stromal tumor (GIST) [90]. AG-013736 (axitinib), which has a similar spectrum of kinase inhibition as SU11248, has also shown therapeutic promise in metastatic renal cell carcinoma in a Phase II monotherapy study [91]. SU11248 is FDA-approved for the treatment of imatinib-resistant GIST and metastatic renal cell carcinoma [92]. Phase III data indicates that Bay 43-9006 monotherapy results in a significant increase in progression-free survival in patients with advanced renal cell carcinoma [93]. In 2006, Bay 43-9006 and SU-11248 were approved by the FDA for the treatment of metastatic renal cell carcinoma.

For an updated overview of cancer clinical trials with VEGF inhibitors and other antiangiogenic agents, see http://www.cancer.gov/clinicaltrials/developments/anti-angio-table.

Role of VEGF-A in Intraocular Neovascular Syndromes

The expression of VEGF-A mRNA is spatially and temporally correlated with neovascularization in several animal models of retinal ischemia [20,94]. This is consistent with the fact that VEGF-A gene expression is up-regulated by hypoxia, via HIF-dependent transcriptional activation [48]. In 1994, it was reported that the levels of VEGF-A are elevated in the aqueous and vitreous humor of human eyes with proliferative retinopathy secondary to diabetes and other conditions [95,96]. Subsequently, animal studies using various VEGF inhibitors, including soluble VEGF receptor chimeric proteins [97], anti-VEGF-A monoclonal antibodies [98] and small molecule VEGF RTK inhibitors [99], have directly demonstrated the role of VEGF as a mediator of ischemia-induced intraocular neovascularization.

Age-related macular degeneration (AMD) is the most common cause of severe, irreversible vision loss in the elderly [100]. AMD is classified as nonexudative (dry) or exudative (wet or neovascular) disease. Although the exudative form accounts for ~10–20% of cases, it is responsible for 80–90% of the visual loss associated with AMD [101]. Pharmacological therapies for neovascular AMD have been approved by the FDA. One is verteporfin (Visudyne®) photodynamic therapy (PDT) [102] for only predominantly classic lesions, in which 50% or more of the lesion consists of classic choroidal neovascularization (CNV). The other is pegaptanib sodium (Macugen®) [103] approved in December 2004 for all angiographic subtypes of neovascular AMD. Although both treatments can slow the progression of vision loss, only a small percentage of treated patients experience any improvement in visual acuity. Most recently (June 2006), ranibizumab (Lucentis®) was approved by the FDA for the treatment of all subtypes of neovascular AMD [104].

Clinical Studies of Anti-VEGF Therapy for Neovascular AMD: Pegaptanib and Ranibizumab

Pegaptanib and ranibizumab are the first ocular anti-VEGF treatments evaluated in large, randomized, controlled clinical trials for the treatment of neovascular AMD. Both are administered locally by intravitreal injection into the back of the eye. Pegaptanib sodium is a pegylated oligonucelotide aptamer that binds to and inactivates $VEGF_{165}$ [105]. In a combined analysis of the VISION trials—two identical, large, controlled, double-masked, randomized, multicenter clinical trials involving patients with all CNV lesion types—pegaptanib prevented moderate vision loss (the primary endpoint, which was defined as loss <15 letters of vision) in 70% of subjects compared with 55% for the control group at 1 year ($P < 0.001$) [103]. However, on average, patients in the pegaptinib group lost ~ 8 letters at 1 year, compared with a loss of ~15 letters in the sham injection group ($P < 0.002$). The proportion of subjects who experienced a moderate gain in vision (defined as a change of >= 15 letters at 1 year from baseline) was −6% in the pegaptinib group versus 2% in the sham injection group ($P = 0.04$). Key adverse events observed in the pegaptinib groups were uncommon and included endophthalmitis in 1.3%, traumatic lens injury in 0.7%, and retinal detachment in 0.6% of patients.

Ranibizumab is a recombinant, humanized Fab that binds to and potently neutralizes the biological activities of all known human VEGF-A isoforms, as well as the proteolytic cleavage product VEGF$_{110}$ [104,106]. Ranibizumab is currently being evaluated in two large, phase III, multicenter, randomized, double-masked, controlled pivotal trials in different neovascular AMD patient populations.

The MARINA trial randomized subjects with minimally classic (less than 50% of the lesion consisting of classic CNV) or occult without classic CNV to monthly sham injections or monthly intravitreal injections of one of two doses of ranibizumab [107]. In the primary analysis at 1 year, the study met its primary endpoint, with a significantly greater proportion of ranibizumab subjects avoiding moderate vision loss than sham-injected subjects. Moreover, on average, ranibizumab-treated subjects gained vision at 1 year compared with baseline while sham-injection subjects lost vision. A significantly larger percentage of subjects treated with ranibizumab gained ≥15 letters at 1 year than did the sham-injection group. Key serious ocular adverse events occurring in ranibizumab-treated subjects included uveitis and endophthalmitis and were uncommon. Analysis of complete 2-year follow-up show that the visual acuity benefits observed at 1 year were maintained through the second year and that the cumulative 2-year safety profile was similar to that observed at 1 year [107].

The ANCHOR trial randomized subjects with predominantly classic CNV to verteporfin PDT with monthly sham ocular injections or to monthly intravitreal injections of one of two doses of ranibizumab with a sham PDT procedure. In the primary analysis at 1 year, the study met its primary endpoint, with a significantly greater proportion of ranibizumab subjects avoiding moderate vision loss compared with subjects treated with verteporfin PDT [108]. In addition, on average, ranibizumab-treated subjects gained vision at 1 year compared with baseline while verterporfin PDT subjects lost vision, and a significantly larger percentage of subjects treated with ranibizumab gained ≥15 letters at 1 year than did the verteporfin PDT group. The ocular and non-ocular safety profile observed in ANCHOR was similar to that observed in MARINA.

Perspectives

Research conducted for almost two decades has established that VEGF-A is important for regulation of the normal angiogenesis processes. Moreover, VEGF inhibition has been shown to suppress pathological angiogenesis in a variety of cancer models, leading to the clinical development of a variety of VEGF inhibitors. Definitive clinical studies have proven that VEGF inhibition, by means of bevacizumab in combination with chemotherapy, provides a significant clinical benefit, including increased survival, in patients with previously untreated metastatic colorectal cancer [84]. Furthermore, SU11248 and Bay 43-9006 have been recently approved by the FDA for metastatic renal cell carcinoma and their mechanism of tumor suppression consists, at least partly, of inhibition of VEGF signaling [81–83].

A particularly active area of research concerns the elucidation of the mechanisms of refractoriness or resistance to anti-VEGF therapy. Tumor cell-intrinsic or treatment-induced expression of angiogenic factors may be implicated [109,110]. Very recent studies have provided evidence that, at least is some murine models, refractoriness to anti-VEGF therapy is related to the ability of the tumor to recruit CD11b+Gr1+ myeloid cells, which promote angiogenesis [111]. It remains to be established whether these findings also apply to human tumors.

Most clinical studies with VEGF inhibitors have been conducted in patients with advanced malignancies. Preclinical studies suggested that such agents may be particularly effective when the tumor burden is low [112]. Thus, the clinical benefit of these therapies may be greater if the treatment were initiated at earlier stages of malignancy. Adjuvant clinical trials with bevacizumab in breast, colorectal and NSCLC patients are presently ongoing and the results should be known within the next few years.

Reliable markers are needed to monitor the activity of anti-angiogenic drugs. Circulating endothelial cells and their progenitor subsets are a potential candidate, as is MRI dynamic measurement of vascular permeability/flow in response to angiogenesis inhibitors, but neither has been clinically validated [82]. Emphasizing the difficulty of identifying predictive markers, a recent study found that VEGF-A and thrombospondin expression or microvessel density in tumor sections do not correlate with clinical response to bevacizumab in patients with metastatic colorectal cancer and patients showed a survival benefit from the treatment, irrespective of these parameters [113].

VEGF inhibitors have also demonstrated a marked clinical benefit in wet AMD. Blockade of all VEGF-A isoforms and bioactive fragments with ranibizumab not only slowed down vision loss, but unexpectedly, appears to have the potential to enable many AMD patients to obtain a meaningful and sustained gain of vision. Further research is needed to determine whether the vision gain conferred by ranibizumab extends beyond 24 months and whether additional intraocular neovascular syndromes may benefit from this treatment.

References

1. Folkman J, Klagsbrun M. Angiogenic factors. Science 1987;235:442–7.
2. Yancopoulos GD, Davis S, Gale NW, Rudge JS, Wiegand SJ, Holash J. Vascular-specific growth factors and blood vessel formation. Nature 2000;407(6801):242–8.

3. Sund M, Hamano Y, Sugimoto H, *et al.* Function of endogenous inhibitors of angiogenesis as endothelium-specific tumor suppressors. Proc Natl Acad Sci USA 2005;102(8):2934–9.

4. Ferrara N. VEGF and the quest for tumour angiogenesis factors. Nat Rev Cancer 2002;2(10):795–803.

5. Ferrara N, Gerber HP, LeCouter J. The biology of VEGF and its receptors. Nature Med 2003;9:669–76.

6. Alitalo K, Tammela T, Petrova TV. Lymphangiogenesis in development and human disease. Nature 2005;438:946–53.

7. Ferrara N, Mass RD, Campa C, Kim R. Targeting VEGF-A to treat cancer and age-related macular degeneration. Annu Rev Med 2007;58:491–504.

8. Senger DR, Galli SJ, Dvorak AM, Perruzzi CA, Harvey VS, Dvorak HF. Tumor cells secrete a vascular permeability factor that promotes accumulation of ascites fluid. Science 1983;219:983–5.

9. Ferrara N, Henzel WJ. Pituitary follicular cells secrete a novel heparin-binding growth factor specific for vascular endothelial cells. Biochem Biophys Res Commun 1989;161(2):851–8.

10. Connolly DT, Olander JV, Heuvelman D, *et al.* Human vascular permeability factor. Isolation from U937 cells. J Biol Chem 1989;264:20017–24.

11. Leung DW, Cachianes G, Kuang WJ, Goeddel DV, Ferrara N. Vascular endothelial growth factor is a secreted angiogenic mitogen. Science 1989;246(4935):1306–9.

12. Keck PJ, Hauser SD, Krivi G, *et al.* Vascular permeability factor, an endothelial cell mitogen related to PDGF. Science 1989;246(4935):1309–12.

13. Ferrara N. Vascular endothelial growth factor: basic science and clinical progress. Endocr Rev 2004;25:581–611.

14. Dvorak HF. Vascular permeability factor/vascular endothelial growth factor: a critical cytokine in tumor angiogenesis and a potential target for diagnosis and therapy. J Clin Oncol 2002;20(21):4368–80.

15. Ferrara N, Carver Moore K, Chen H, *et al.* Heterozygous embryonic lethality induced by targeted inactivation of the VEGF gene. Nature 1996;380(6573):439–42 issn: 0028-836.

16. Gerber HP, Hillan KJ, Ryan AM, *et al.* VEGF is required for growth and survival in neonatal mice. Development 1999;126:1149–59.

17. Malik AK, Baldwin ME, Peale F, *et al.* Redundant roles of VEGF-B and PlGF during selective VEGF-A blockade in mice. Blood 2006;107(550–557).

18. Gerber HP, Vu TH, Ryan AM, Kowalski J, Werb Z, Ferrara N. VEGF couples hypertrophic cartilage remodeling, ossification and angiogenesis during endochondral bone formation. Nat Med 1999;5:623–8.

19. Ferrara N, Chen H, Davis-Smyth T, *et al.* Vascular endothelial growth factor is essential for corpus luteum angiogenesis. Nat Med 1998;4:336–40.

20. Alon T, Hemo I, Itin A, Pe'er J, Stone J, Keshet E. Vascular endothelial growth factor acts as a survival factor for newly formed retinal vessels and has implications for retinopathy of prematurity. Nat Med 1995;1(10):1024–8 i

21. Gerber HP, Dixit V, Ferrara N. Vascular Endothelial Growth Factor Induces Expression of the Antiapoptotic Proteins Bcl-2 and A1 in Vascular Endothelial Cells. J Biol Chem 1998;273:13313–6.

22. Gerber HP, McMurtrey A, Kowalski J, *et al.* VEGF Regulates Endothelial Cell Survival by the PI3-kinase/Akt Signal Transduction Pathway. Requirement for Flk-1/KDR Activation. J Biol Chem 1998;273:30336–43.

23. Benjamin LE, Golijanin D, Itin A, Pode D, Keshet E. Selective ablation of immature blood vessels in established human tumors follows vascular endothelial growth factor withdrawal [see comments]. J Clin Invest 1999;103(2):159–65.

24. Tran J, Master Z, Yu JL, Rak J, Dumont DJ, Kerbel RS. A role for survivin in chemoresistance of endothelial cells mediated by VEGF. Proc Natl Acad Sci USA 2002;99(7):4349–54.

25. Carmeliet P. Angiogenesis in life, disease and medicine. Nature 2005;438(7070):932–6.

26. Clauss M, Gerlach M, Gerlach H, *et al.* Vascular permeability factor: a tumor-derived polypeptide that induces endothelial cell and monocyte procoagulant activity, and promotes monocyte migration. J Exp Med 1990;172(6):1535–45.

27. Barleon B, Sozzani S, Zhou D, Weich HA, Mantovani A, Marme D. Migration of human monocytes in response to vascular endothelial growth factor (VEGF) is mediated via the VEGF receptor flt-1. Blood 1996;87(8):3336–43.

28. Gerber H-P, Malik A, Solar GP, *et al.* Vascular endothelial growth factor regulates hematopoietic stem cell survival by an internal autocrine loop mechanism. Nature 2002;417:954–8.

29. Houck KA, Ferrara N, Winer J, Cachianes G, Li B, Leung DW. The vascular endothelial growth factor family: identification of a fourth molecular species and characterization of alternative splicing of RNA. Mol Endocrinol 1991;5(12):1806–14.

30. Tischer E, Mitchell R, Hartman T, *et al.* The human gene for vascular endothelial growth factor. Multiple protein forms are encoded through alternative exon splicing. J Biol Chem 1991;266(18):11947–54.

31. Plouet J, Schilling J, Gospodarowicz D. Isolation and characterization of a newly identified endothelial cell mitogen produced by AtT20 cells. EMBO J 1989;8:3801–8.

32. Houck KA, Leung DW, Rowland AM, Winer J, Ferrara N. Dual regulation of vascular endothelial growth factor bioavailability by genetic and proteolytic mechanisms. J Biol Chem 1992;267(36):26031–7.

33. Park JE, Keller G-A, Ferrara N. The vascular endothelial growth factor isoforms (VEGF): Differential deposition into the subepithelial extracellular matrix and bioactivity of extracellular matrix-bound VEGF. Mol Biol Cell 1993;4:1317–26.

34. Keyt BA, Berleau LT, Nguyen HV, *et al.* The carboxyl-terminal domain (111–165) of vascular endothelial growth factor is critical for its mitogenic potency. J Biol Chem 1996;271(13):7788–95.

35. Lee S, Jilani SM, Nikolova GV, Carpizo D, Iruela-Arispe ML. Processing of VEGF-A by matrix metalloproteinases regulates bioavailability and vascular patterning in tumors. J Cell Biol 2005;169(4):681–91.

36. Shibuya M, Yamaguchi S, Yamane A, *et al.* Nucleotide sequence and expression of a novel human receptor-type tyrosine kinase (flt) closely related to the fms family. Oncogene 1990;8:519–27.

37. Terman BI, Dougher-Vermazen M, Carrion ME, *et al.* Identification of the KDR tyrosine kinase as a receptor for vascular endothelial cell growth factor. Biochem Biophys Res Commun 1992;187(3):1579–86.

38. de Vries C, Escobedo JA, Ueno H, Houck K, Ferrara N, Williams LT. The fms-like tyrosine kinase, a receptor for vascular endothelial growth factor. Science 1992;255(5047):989–91.

39. Park JE, Chen HH, Winer J, Houck KA, Ferrara N. Placenta growth factor. Potentiation of vascular endothelial growth factor bioactivity, in vitro and in vivo, and high affinity binding to Flt-1 but not to Flk-1/KDR. J Biol Chem 1994;269(41):25646–54 issn: 0021-9258.

40. Olofsson B, Korpelainen E, Pepper MS, *et al.* Vascular endothelial growth factor B (VEGF-B) binds to VEGF receptor-1 and regulates plasminogen activator activity in endothelial cells. Proc Natl Acad Sci USA 1998;95(20):11709–14.

41. LeCouter J, Moritz DR, Li B, *et al.* Angiogenesis-independent endothelial protection of liver: role of VEGFR-1. Science 2003;299:890–3.

42. Hiratsuka S, Nakamura K, Iwai S, *et al.* MMP9 induction by vascular endothelial growth factor receptor-1 is involved in lung-specific metastasis. Cancer Cell 2002;2(4):289–300.

43. Hattori K, Heissig B, Wu Y, *et al.* Placental growth factor reconstitutes hematopoiesis by recruiting VEGFR1(+) stem cells from bone-marrow microenvironment. Nat Med 2002;8(8):841–9.

44. Quinn TP, Peters KG, De Vries C, Ferrara N, Williams LT. Fetal liver kinase 1 is a receptor for vascular endothelial growth factor and is selectively expressed in vascular endothelium. Proc Natl Acad Sci USA 1993;90(16):7533–7.

45. Shalaby F, Rossant J, Yamaguchi TP, *et al.* Failure of blood-island formation and vasculogenesis in Flk-1-deficient mice. Nature 1995;376(6535):62–6.

46. Olsson AK, Dimberg A, Kreuger J, Claesson-Welsh L. VEGF receptor signalling - in control of vascular function. Nat Rev Mol Cell Biol 2006;7(5):359–71.

47. Soker S, Takashima S, Miao HQ, Neufeld G, Klagsbrun M. Neuropilin-1 is expressed by endothelial and tumor cells as an isoform-specific receptor for vascular endothelial growth factor. Cell 1998;92(6):735–45.

48. Safran M, Kaelin WJ, Jr. HIF hydroxylation and the mammalian oxygen-sensing pathway. J Clin Invest 2003;111:779–83.

49. Dor Y, Porat R, Keshet E. Vascular endothelial growth factor and vascular adjustments to perturbations in oxygen homeostasis. Am J Physiol 2001;280:C1367–74.

50. Semenza GL. Angiogenesis in ischemic and neoplastic disorders. Annu Rev Med 2003;54:17–28.

51. Madan A, Curtin PT. A 24-base-pair sequence 3' to the human erythropoietin gene contains a hypoxia-responsive transcriptional enhancer. Proc Natl Acad Sci USA 1993;90(9):3928–32.

52. Wang GL, Semenza GL. Purification and characterization of hypoxia-inducible factor 1. J-Biol-Chem 1995;270(3):1230–7 issn: 0021–9258.

53. Mole DR, Maxwell PH, Pugh CW, Ratcliffe PJ. Regulation of HIF by the von Hippel-Lindau tumour suppressor: implications for cellular oxygen sensing. IUBMB Life 2001;52(1–2):43–7.

54. Lonser RR, Glenn GM, Walther M, *et al.* von Hippel-Lindau disease. Lancet 2003;361(9374):2059–67.

55. Jaakkola P, Mole DR, Tian YM, *et al.* Targeting of HIF-alpha to the von Hippel-Lindau ubiquitylation complex by O2-regulated prolyl hydroxylation. Science 2001;292(5516):468–72.

56. Ivan M, Kondo K, Yang H, *et al.* HIF-alpha targeted for VHL-mediated destruction by proline hydroxylation: implications for O2 sensing. Science 2001;292(5516):464–8.

57. Maxwell PH, Ratcliffe PJ. Oxygen sensors and angiogenesis. Semin Cell Dev Biol 2002;13(1):29–37.

58. Pugh CW, Ratcliffe PJ. Regulation of angiogenesis by hypoxia: role of the HIF system. Nat Med 2003;9(6):677–84.

59. Frank S, Hubner G, Breier G, Longaker MT, Greenhalg DG, Werner S. Regulation of VEGF expression in cultured keratinocytes. Implications for normal and impaited wound healing. J Biol Chem 1995;270:12607–13.

60. Pertovaara L, Kaipainen A, Mustonen T, *et al.* Vascular endothelial growth factor is induced in response to transforming growth factor-beta in fibroblastic and epithelial cells. J Biol Chem 1994;269(9):6271–4.

61. Warren RS, Yuan H, Matli MR, Ferrara N, Donner DB. Induction of vascular endothelial growth factor by insulin-like growth factor 1 in colorectal carcinoma. J Biol Chem 1996;271(46):29483–8.

62. Ben-Av P, Crofford LJ, Wilder RL, Hla T. Induction of vascular endothelial growth factor expression in synovial fibroblasts. FEBS Lett 1995;372:83–7.

63. Cohen T, Nahari D, Cerem LW, Neufeld G, Levi BZ. Interleukin 6 induces the expression of vascular endothelial growth factor. J Biol Chem 1996;271(2):736–41 issn: 0021–9258.

64. Grugel S, Finkenzeller G, Weindel K, Barleon B, Marme D. Both v-Ha-Ras and v-Raf stimulate expression of the vascular endothelial growth factor in NIH 3T3 cells. J Biol Chem 1995;270(43):25915–9 issn: 0021–9258.

65. Okada F, Rak JW, Croix BS, *et al.* Impact of oncogenes in tumor angiogenesis: mutant K-ras up-regulation of vascular endothelial growth factor/vascular permeability factor is necessary, but not sufficient for tumorigenicity of human colorectal carcinoma cells. Proc Natl Acad Sci USA 1998;95(7):3609–14.

66. Zhang X, Gaspard JP, Chung DC. Regulation of vascular endothelial growth factor by the Wnt and K-ras pathways in colonic neoplasia. Cancer Res 2001;61(16):6050–4.

67. Seno H, Oshima M, Ishikawa TO, *et al.* Cyclooxygenase 2- and prostaglandin E(2) receptor EP(2)-dependent angiogenesis in Apc(Delta716) mouse intestinal polyps. Cancer Res 2002;62(2):506–11.

68. Korsisaari N, Kasman IM, Forrest WF, *et al.* Inhibition of VEGF-A prevents the angiogenic switch and results in increased survival of Apc+/min mice. Proc Natl Acad Sci USA 2007;104(25):10625–30

69. Kim KJ, Li B, Winer J, *et al.* Inhibition of vascular endothelial growth factor-induced angiogenesis suppresses tumor growth in vivo. Nature 1993;362:841–4.

70. Manley PW, Martiny-Baron G, Schlaeppi JM, Wood JM. Therapies directed at vascular endothelial growth factor. Expert Opin Investig Drugs 2002;11(12):1715–36.

71. Prewett M, Huber J, Li Y, *et al.* Antivascular endothelial growth factor receptor (fetal liver kinase 1) monoclonal antibody inhibits tumor angiogenesis. Cancer Res 1999;59:5209–18.

72. Gerber HP, Kowalski J, Sherman D, Eberhard DA, Ferrara N. Complete inhibition of rhabdomyosarcoma xenograft growth and neovascularization requires blockade of both tumor and host vascular endothelial growth factor. Cancer Res 2000;60:6253–8.

73. Holash J, Davis S, Papadopoulos N, *et al.* VEGF-Trap: a VEGF blocker with potent antitumor effects. Proc Natl Acad Sci USA 2002;99(17):11393–8.

74. Dong J, Grunstein J, Tejada M, *et al.* VEGF-null cells require PDGFR alpha signaling-mediated stromal fibroblast recruitment for tumorigenesis. EMBO J 2004;23:2800–10.

75. Tejada M, Yu L, Dong J, *et al.* Tumor-driven paracrine PDGF receptor -a signaling is a key determinent of stromal cell recruitment in a model of human lung carcinoma. Clin Cancer Res 2006;12:2676–88.

76. Klement G, Baruchel S, Rak J, *et al.* Continuous low-dose therapy with vinblastine and VEGF receptor-2 antibody induces sustained tumor regression without overt toxicity [see comments]. J Clin Invest 2000;105(8):R15–24.

77. Lee CG, Heijn M, di Tomaso E, *et al.* Anti-Vascular endothelial growth factor treatment augments tumor radiation response under normoxic or hypoxic conditions. Cancer Res 2000;60:5565–70.

78. Jain RK. Normalization of tumor vasculature: an emerging concept in antiangiogenic therapy. Science 2005;307(5706):58–62.

79. Presta LG, Chen H, O'Connor SJ, *et al.* Humanization of an anti-VEGF monoclonal antibody for the therapy of solid tumors and other disorders. Cancer Res 1997;57:4593–9.

80. Ferrara N, Hillan KJ, Gerber HP, Novotny W. Discovery and development of bevacizumab, an anti-VEGF antibody for treating cancer. Nat Rev Drug Discov 2004;3(5):391–400.

81. Gasparini G, Longo R, Toi M, Ferrara N. Angiogenic inhibitors: a new therapeutic strategy in oncology. Nat Clin Pract 2005;2(11):562–77.

82. Ferrara N, Kerbel RS. Angiogenesis as a therapeutic target. Nature 2005;438:967–74.

83. Jain RK, Duda DG, Clark JW, Loeffler JS. Lessons from phase III clinical trials on anti-VEGF therapy for cancer. Nat Clin Pract 2006;3(1):24–40.

84. Hurwitz H, Fehrenbacher L, Novotny W, et al. Bevacizumab plus irinotecan, fluorouracil, and leucovorin for metastatic colorectal cancer. N Engl J Med 2004;350:2335–42.

85. Sandler A, Gray R, Perry MC, et al. Paclitaxel-Carboplatin Alone or with Bevacizumab for Non-Small-Cell-Lung Cancer. N Engl J Med 2006;355:2542–50.

86. Johnson DH, Fehrenbacher L, Novotny WF, et al. Randomized phase II trial comparing bevacizumab plus carboplatin and paclitaxel with carboplatin and paclitaxel alone in previously untreated locally advanced or metastatic non-small-cell lung cancer. J Clin Oncol 2004;22(11):2184–91.

87. Giantonio BJ, Catalano PJ, Meropol NJ, et al. Bevacizumab in combination with oxaliplatin, fluorouracil, and leucovorin (FOLFOX4) for previously treated metastatic colorectal cancer: results from the Eastern Cooperative Oncology Group Study E3200. J Clin Oncol 2007;25(12):1539–44.

88. Yang JC, Haworth L, Sherry RM, et al. A randomized trial of bevacizumab, an anti-VEGF antibody, for metastatic renal cancer. N Engl J Med 2003;349:427–34.

89. Smith JK, Mamoon NM, Duhe RJ. Emerging roles of targeted small molecule protein-tyrosine kinase inhibitors in cancer therapy. Oncol Res 2004;14(4–5):175–225.

90. Maki RG, Fletcher JA, Heinrich MC, et al. Results from a continuation trial of SU11248 in patients (pts) with imatinib (IM)-resistant gastrointestinal stromal tumor (GIST). ASCO Annual Meeting Proceedings 2005;Abstract 9011.

91. Rini B, Rixe O, Bukowski R, et al. AG-013736, a multi-target tyrosine kinase receptor inhibitor, demonstrates anti-tumor activity in a Phase 2 study of cytokine-refractory, metastatic renal cell cancer (RCC). ASCO Annual Meeting Proceedings 2005;Abstract 4509.

92. Motzer RJ, Hutson TE, Tomczak P, et al. Sunitinib versus Interferon Alfa in Metastatic Renal-Cell Carcinoma. N Engl J Med 2007;356(2):115–24.

93. Escudier B, Eisen T, Stadler WM, et al. Sorafenib in advanced clear-cell renal-cell carcinoma. N Engl J Med 2007;356(2):125–34.

94. Miller JW, Adamis AP, Shima DT, et al. Vascular endothelial growth factor/vascular permeability factor is temporally and spatially correlated with ocular angiogenesis in a primate model. Am J Pathol 1994;145(3):574–84.

95. Aiello LP, Avery RL, Arrigg PG, et al. Vascular endothelial growth factor in ocular fluid of patients with diabetic retinopathy and other retinal disorders. [see comments]. N Eng J Med 1994;331(22):1480–7.

96. Malecaze F, Clemens S, Simorer-Pinotel V, et al. Detection of vascular endothelial growth factor mRNA and vascular endothelial growth factor-like activity in proliferative diabetic retinopathy. Arch Ophthalmol 1994;112:1476–82.

97. Aiello LP, Pierce EA, Foley ED, et al. Suppression of retinal neovascularization in vivo by inhibition of vascular endothelial growth factor (VEGF) using soluble VEGF-receptor chimeric proteins. Proc Natl Acad Sci USA 1995;92(23):10457–61

98. Adamis AP, Shima DT, Tolentino MJ, et al. Inhibition of vascular endothelial growth factor prevents retinal ischemia-associated iris neovascularization in a nonhuman primate. Arch Ophthalmol 1996;114(1):66–71.

99. Ozaki H, Seo MS, Ozaki K, et al. Blockade of vascular endothelial cell growth factor receptor signaling is sufficient to completely prevent retinal neovascularization. Am J Pathol 2000;156(2):697–707.

100. Congdon N, O'Colmain B, Klaver CC, et al. Causes and prevalence of visual impairment among adults in the United States. Arch Ophthalmol 2004;122(4):477–85.

101. Ferris FL, 3rd, Fine SL, Hyman L. Age-related macular degeneration and blindness due to neovascular maculopathy. Arch Ophthalmol 1984;102(11):1640–2.

102. Photodynamic Therapy of Subfoveal Choroidal Neovascularization in Age-related Macular Degeneration With Verteporfin. One-Year Results of 2 Randomized Clinical Trials—TAP Report 1. Arch Ophthalmol 1999;117:1329–45.

103. Gragoudas ES, Adamis AP, Cunningham ET, Jr., Feinsod M, Guyer DR. Pegaptanib for neovascular age-related macular degeneration. N Engl J Med 2004;351(27):2805–16.

104. Ferrara N, Damico L, Shams N, Lowman H, Kim R. Developmemt of Ranibizumab, an anti-vascular endothelial growth factor antigen binding fragment, as therapy for neovascular age-related macular degeneration. Retina 2006;26:859–70.

105. Ng EW, Sima DT, Calias P, Cunningham ET, Jr., Guyer DR, Adamis AP. Pegaptanib, a targeted anti-VEGF aptamer for ocular vascular disease. Nat Rev Drug Discov 2006;5:123–32.

106. Chen Y, Wiesmann C, Fuh G, et al. Selection and analysis of an optimized anti-VEGF antibody: crystal structure of an affinity-matured Fab in complex with antigen. J Mol Biol 1999;293(4):865–81.

107. Rosenfeld PJ, Brown DM, Heier JS, et al. Ranibizumab for neovascular age-related macular degeneration. N Engl J Med 2006;355:1419–31.

108. Brown DM, Kaiser PK, Michels M, et al. Ranibizumab versus verteporfin for neovascular age-related macular degeneration. N Engl J Med 2006;355(14):1432–44.

109. Casanovas O, Hicklin DJ, Bergers G, Hanahan D. Drug resistance by evasion of antiangiogenic targeting of VEGF signaling in late-stage pancreatic islet tumors. Cancer Cell 2005;8(4):299–309.

110. Kerbel RS, Yu J, Tran J, et al. Possible mechanisms of acquired resistance to anti-angiogenic drugs: implications for the use of combination therapy approaches. Cancer Metastasis Rev 2001;20(1–2):79–86.

111. Shojaei F, Wu X, Malik AK, et al. Tumor refrectoriness to anti-VEGF treatment is mediated by CD11b+Gr1+ myeloid cells. Nat Biotechnol 2007;25:911–20.

112. Gerber HP, Ferrara N. Pharmacology and pharmacodynamics of bevacizumab as monotherapy or in combination with cytotoxic therapy in preclinical studies. Cancer research 2005;65(3):671–80.

113. Jubb AM, Hurwitz HI, Bai W, et al. Impact of Vascular Endothelial Growth Factor-A Expression, Thrombospondin-2 Expression, and Microvessel Density on the Treatment Effect of Bevacizumab in Metastatic Colorectal Cancer. J Clin Oncol 2006;24:217–27.

Chapter 31
Protein Tyrosine Kinase Inhibitors as Antiangiogenic Agents

Alexander Levitzki

Keywords: signal transduction therapy, protein tyrosine kinase, resistance, signaling inhibitors

Abstract: An essential element in cancer therapy is to inhibit the blood supply to the tumor. In this chapter, we discuss the few tyrosine phosphorylation inhibitors (tyrphostins/PTK inhibitors) that have been developed to achieve that goal. The inhibitors that block VEGFR2, PDGFR and EGFR and not solely VEGFR2 seem to do a better job towards this end. It is, however, not entirely clear that the anti-tumor effects observed with the agents described are entirely due to their antiangiogenic effects. This is because of the many other targets within the kinome they hit, which may be part of the oncogenic network, driving the cancers for which they were approved. None of the agents is life saving.

Signal Transduction Therapy

Principles of Signal Transduction

Understanding the patterns of signal transduction pathways in normal cells and diseased cells has led to the identification of elements in these pathways that have gone awry in many diseases. The most visible examples are in the field of cancer biology where numerous chromosomal changes occur. Most notable among these changes are mutations in proto-oncogenes converting them to oncogenes, the loss-of-function of tumor suppressor genes and aneuploidy. Cumulatively, these genetic changes lead to cell immortalization and the eventual full blossomed transformation to a metastatic cancer cell. Over the past two decades, it has become apparent that the manifestation of the many changes in the genetic make-up of the cell upon transformation results in a complex, yet biochemically fully defined, cellular network. The biochemical definition of signal transduction pathways gone awry and the identification of specific signaling proteins involved in cancer has re-directed cancer therapy in the past decade. Currently, we are in the middle of a paradigm shift from a shot gun approach to targeted therapy, or more precisely: signal transduction therapy as originally defined [1]. Efforts are being made to generate agents that target key signaling proteins that play a pivotal role in driving the autonomous tumor.

Initially, we were trained to think that signal transduction pathways are linear since the first ones to be elucidated, such as the hormone-dependent activation of adenylyl cyclase, were indeed linear. In that case, the end product of the pathway cAMP activates numerous elements through the cAMP activated protein kinase (PKA). Unlike G protein coupled receptors, which signal to a single biochemical machinery like adenylyl cyclase or phospholipase Cβ (PLCβ), protein tyrosine kinases (PTKs) that play a key role in cancer signal to a number of elements and each of these downstream elements in turn signals to a set of more signaling elements. Thus, in the case of PTKs, one actually deals with a network of signaling events emanating from the initiator PTK. Moreover, many of the downstream elements of the PTK signaling receptor are themselves sometimes mutated and/or amplified leading to oncogenesis. Furthermore, we now also know that receptor tyrosine kinase pathways can be transactivaed by G protein coupled receptors (for review, see [2]).

There are very few cancers that can be linked to one major signaling event. For example, chronic myelogenous leukemia (CML) at its chronic (early) phase is driven by Bcr-Abl tyrosine kinase and therefore can be successfully treated by a selective Bcr-Abl inhibitor [2]. Another example is polycythemia vera, which is caused by a single mutation in the Jak2 peudokinase domain that drives the proliferative condition [3]. Similarly, the non-malignant development of restenosis subsequent to balloon angioplasty is largely driven by platelet-derived growth factor (PDGF) and its receptor (PDGFR), and therefore the process can be effectively halted by a stent eluting a PDGFR kinase inhibitor (tyrphostin AG 2043) [4]. In inflammatory conditions, tumor necrosis factor-α (TNFα) has been identified as a major player, and therefore antibodies to

Unit of Cellular Signaling, Department of Biological Chemistry, The Hebrew University of Jerusalem, Jerusalem, Israel

TABLE 31.1. Signal transduction inhibitors.

In the clinic and in development	
Agent	**Targeted disease(s)**
Herceptin (anti-Her-2 antibody)	Breast cancer in women
Gleevec/STI 571, a tyrosine kinase inhibitor aimed at Bcr-Abl, Kit	Early CML, GIST (gastrointestinal stromal tumor)
Iressa, a tyrosine kinase inhibitor aimed at EGFR	Non small cell lung carcinoma, colon cancer, glioblastoma multiforme (GBM)
Tarceva, atyrosine kinase inhibitor aimed at EGFR	Non small cell lung carcinoma, glioblastoma multiforme (GBM)
Lapatinib (GW 2016) aimed at EGFR and Her-2	Inflammatory or Herceptin resistant breast cancer
Erbitux, antibody aimed at EGFR	Head and neck cancer, colon cancer
Anti-VEGF antibody	Colon cancer
SU 11248 (SUTENT) inhibits PDGFR, FGFR, VEGFR2	Gleevec resistant GIST, renal cell carcinoma
BAY 43-9006 (B-Raf and VEGFR2 inhibitor)	Renal cell carcinoma, metastatic melanoma
Aimed at non-cancer targets	
Antibodies against TNFα	Rheumatoid arthritis
Rapamycin (inhibits mTor)	Inhibition of restenosis using drug eluting stent
AGL 2043 (inhibits PDGFR)	Inhibition of restenosis using drug eluting stent (in development)

this cytokine or a soluble TNFα receptor have a profound clinical effect. Table 31.1 summarizes the most advanced signal transduction inhibitors in the clinic and the ones in advanced preclinical development.

Are Signal Transduction Inhibitors Successful?

The most dramatic success of signal transduction inhibitors was achieved with imatinib (Gleevec), but this is actually not surprising. Bcr-Abl is a survival element in CML at the early (chronic) phase of the disease, so depriving the cells from their survival signal induced their apoptotic death. However, this is not the case in the more advanced stages of the disease for which imatinib has indeed a temporary effect and the patients recur within 2–6 months. Patients with gastrointestinal stromal tumors (GIST) who harbor certain mutations in exon 11 of c-Kit do better on imatinib than others whose c-Kit harbors the mutations elsewhere [5]. As it turns out, the persistently active, mutated c-Kit at exon 11 acts as a survival protein similar to Bcr-Abl for CML cells at the chronic (early) phase of the disease. It should be noted that both CML patients and GIST patients who respond to imatinib have never been taken off the drug. In fact, the imatinib treatment has converted the disease from a life-threatening terminal disease to a chronic disease. A few percent of the treated patients develop resistance over time and hence the treatment needs to be further modified. One of the agents that has recently been introduced quite successfully is the Bcr-Abl/Src dual inhibitor BMS 354825, which targets the activated state of Bcr-Abl and of Src, and thus may become promising for imatinib-resistant patients. It seems then that we should seriously consider the prospect that cancer will never be actually "cured", but rather tamed to become a chronic condition that can be managed with the appropriate regimen of signal transduction therapy. This is actually a grim prospect and a rather expensive way of life, since the cost of treatment with these agents is very high. For example, one year of treatment with imatinib costs between US$24,000 and US$35,000, depending on the country where treatment is being received.

Unlike CML at the chronic phase, most cancers are driven by an aberrant signaling network, which consists of a number of oncoproteins. Thus, there is the potential need to simultaneously inhibit a number of signaling elements in order to elicit a pro-apoptotic effect and perhaps to probably combine these inhibitors with stimulators of apoptosis. Furthermore, since the genome of the cancer cell is highly unstable and continuously changes, the signaling network of the evolving cancer cell is also continuously changing, presenting the clinician and the scientist with a highly complex situation. The scientist needs to develop methods to identify quickly the ever-changing signaling network of the evolving cancer, which presents the clinician with a moving target with a changing cocktail. It seems that because we are pretty far from that scenario, our success in the clinic in curing cancer is rather non-existent and will take a long while to achieve. Indeed, all the agents listed in Table 1 perform rather weakly in the clinic, offering life extension from weeks to months and, more rarely, for years. In the meantime, we should seek to develop agents that block the robust anti-apoptotic shield the cancer cell develops, as well as anti-proliferative and pro-apoptotic agents aimed at well-defined targets. These will then have to be combined to treat the tumor and be modified according to the progression of the disease. Since cancer is a heterogeneous entity it is clear that we will have to utilize "smart cocktails", and that their composition may change according to the state of the disease. It should be emphasized that the cocktail is likely to include agents that target the microenvironment of the tumor, in order to cut the blood supply and interfere with the unique metabolic features of the tumor. Agents that immediately come to mind are anti-VEGF antibodies, VEGFR2/PDGFR/FGFR kinase inhibitors, hypoxia-inducible factor (HIF)-1α inhibitors and agents that target the unique metabolic features of the cancer cell such as dichloroacetic acid.

Types of Signaling Inhibitors

Recognizing the families of proteins involved with the initiation, progression and spread of cancer one can rationally think about the pathways that need to be manipulated in order to suppress the disease.

Targeting the Protein Kinase Signaling Network

The human genome comprises about 600 protein kinases, among them ~90 protein tyrosine kinases (PTKs). All of the PTKs and a subset of Ser/Thr kinases play key roles in signaling and their enhanced activities are pivotal in cancer as well as in certain non-malignant conditions. The activity network emanating from protein tyrosine kinases is activated in many cancers and therefore, key branch points are excellent targets for inhibition [2].

Targeting the Anti-apoptotic Pathways

Anti-apoptotic proteins, except for the protein kinase B (PKB) (also known as Akt) module, are devoid of catalytic activity and operate by virtue of their interaction and inhibition of pro-apoptotic proteins. In these cases, one can develop small molecules that can inhibit proteins such as Bcl2, IAP and Mdm2, as well as agonistic antibodies against TRAIL receptors. These agents as well as PKB inhibitors can be utilized to dismantle the cancer cell from its anti-apoptotic shield and re-instate its sensitivity to pro-apoptotic or pro-necrotic regimens. In the case of PKB, it has been shown that the substrate competitive inhibitor PTR6164 can obtain a 10- to 20-fold sensitization to pro-apoptotic agents with no adverse effect on the tumor bearing animal. Similarly, the synergy between antagonists of Bcl2, Mdm2 and IAP and cytotoxic agents may prove to be a useful therapeutic approach. Although promoting apoptosis in cancer is a promising strategy, research and development in this area is still in its early stage.

Resistance

The relatively short experience with imatinib already shows that signal transduction inhibitors will not be spared the drug resistance problems. In the case of imatinib, a number of mechanisms have already been identified: mutations in the active site and beyond that diminish the affinity towards imatinib, overexpression of the Bcr-Abl protein, and the enhancement of Bcr-Abl independent oncogenic pathways. Interestingly, the existence of point mutations rendering the Bcr-Abl protein refractory to imatinib inhibition was found to be present prior to treatment. Thus, one would actually be able to predict whether a patient will develop resistance even before treatment and design an a priori regimen that will include the use of imatinib with another anti-cancer agent. Resistance to gefitinib (Iressa) has also been seen in the clinic and correlated with a decreased sensitivity to the drug concomitantly with increased activities of FGF and PDGF receptors as well as of PKB/Akt. It has also been observed that resistance to gefitinib is accompanied with the deletion of the tumor suppressor gene PTEN and that re-introduction of PTEN to the resistant cells re-sensitizes them to the EGFR kinase inhibitor. It seems, therefore, that combining an EGFR kinase inhibitor with a PKB inhibitor may achieve the desired goal of inflicting an apoptotic blow to the EGFR kinase inhibitor-resistant

cell. In most cases, unlike CML at its early phase, there will not be one "Achilles heel" which identifies a single oncogene to which the cancer cell is addicted. Rather we will have to identify the particular signaling *network* of oncogenic elements which will have to be simultaneously manipulated in order to push the cancer cell "over the edge".

More strategies?

It is apparent that the main obstacle to "tame" the cancer and convert it to a manageable chronic condition, if not annihilating it altogether, is the continuous emergence of anti-apoptotic elements. Therefore, understanding these processes will enhance our efforts to design strategies to enhance pro-apoptotic signals in the cancer cell without harnessing them in normal cells. These strategies, combined with signal transduction inhibitors and chemotherapeutic agents, can become the basis of effective treatments.

Inhibiting Angiogenic Signaling

The rationale

Angiogenesis play an important role in the growth of most solid tumors and the progression of metastasis. Recently, it has been reported that the specific inhibition of tumor-induced angiogenesis suppresses the growth of many types of solid tumors. For this reason, it is considered that the inhibition of angiogenesis is a novel therapeutic approach against such tumors. The vascular endothelial growth factor (VEGF) is a key angiogenic factor and is secreted by malignant tumors; it induces the proliferation and the migration of vascular endothelial cells. From the accumulating knowledge, it seems that the VEGF receptor, VEGFR2, is not the only growth factor receptor involved in angiogenesis. PDGF and FGF receptors seem to play important roles too. Furthermore, it is known that overexpression of EGFR can lead to the production of VEGFR by the tumor. In fact, since the expression of HIF-1α, the master transcription factor regulating the expression of VEGF, is overexpressed in many tumors, even under normoxia, it is considered as a target for cancer therapy. Since HIF-1 also upregulates other pro-oncogenic pathways, such as metabolic pathways, it may even be a better target than the VEGF/VEGFR2 system.

Combination of Protein Kinase Inhibitors with Cytotoxic Agents and Targeted Agents

The synergistic effect of adding Avastin/bevacizumab to traditional chemotherapy has been explained by the concept of vascular normalization in which antiangiogenic therapy improves delivery of cytotoxic agents by "normalizing" the morphology and function of aberrant tumor vasculature. Vascular changes induced during normalization include decreased vessel density, diameter, and tortuosity, increased pericyte coverage, decreased permeability, reduced interstitial fluid pressure (IFP), increased oxygen tension (pO$_2$), and improved extravasation of intravenously administered molecules [6, 7]. It is therefore believed that kinase inhibitors aimed at receptors promoting angiogenesis, like VEGFR2, FGFR, and PDGFR when used in combination with other agents will also produce a desired therapeutic benefit. Indeed, current clinical studies are focused on the combination of protein kinase inhibitors with cytotoxic agents. An emerging strategy is to use the anti-angiogenic agents either in combination with other targeted agents or in combination with another antiangiogenic compound of a different mechanism; therefore, producing a less toxic cumulative effect.

Chemistry of the Inhibitors

The first proof of principle that one can generate the VEGFR2 kinase inhibitor as an antiangiogenic agent was reported in 1996 [8]. A number of tyrphostins exhibited reasonable antiangio-genic activity. Based on this study, Sugen scientists began to develop antiangiogenic inhibitors aimed to inhibit VEGFR2 and later on also FGFR and PDGFR. Numerous scaffolds have been developed in order to target VEGFR2. Initially only VEGFR2 was targeted, but quickly multi-targeted kinase inhib-itors entered the search and development. This is mostly notable in the development of indolinones by Sugen (now Pfizer). Initially, SU 5416 was developed as a specific VEGFR2 kinase inhibitor, but SU 6668 and SU 11248 (sunitinib or SUTENT) were developed to broadly target VEGFR2, PDGFR and FGFR. Subsequently, BAY 43-9006 (sorafenib) was discovered by Onyx as a c-Raf inhibitor, and was later found to exhibit antian-giogenic efficacy, most probably due to its inhibitory potency against VEGFR2. A more comprehensive discussion of the various chemical scaffolds is given in a recent publication [9]. Below, we will briefly discuss some of the inhibitors that went into clinical development (a detailed discussion of these compounds is provided in later chapters of this volume).

Indolinones

Sugen developed a number of compounds aimed to be antian-giogenic. Indol-2-ones or 2-oxoindole-based tyrosine kinase inhibitors were originally identified in the late 1980s by researchers in Japan and Italy. Sugen identified this pharma-cophore as a potent inhibitor of VEGFR2. Since it is unlikely that antiangiogenic therapy would work by itself, it was assumed that a combination of an antiangiogenic agent with cytotoxic agents could benefit the patient. As a single agent, SU 5416 proved to extend no benefit to patients. However, it was quickly found that combining SU 5416 with cytotoxic agents has adverse effects that led to episodes of bleeding. This was also found in clinical trials in which bevacizumab was combined with cytotoxic agents [10]. Since SU 5416 had limited efficacy and elicited undesirable side effects, further development of the drug was abandoned. Studies with SU6668, a second generation Sugen compound, have also been disap-pointing. The effects of SU6668 on receptor phosphorylation

and biological surrogates of angiogenesis inhibition in paired tumor biopsies obtained in patients with advanced solid malig-nancies were weak. Microvessel densities were also quite variable and relatively weak. In summary, the surrogate mark-ers accounted for the failure of both SU 5416 and SU 6668 [11]. SU11248 (sunitinib), a third generation indolinone, proved to be more successful. SU11248 is a broad-spectrum, orally available tyrosine kinase inhibitor, inhibiting VEGFR, PDGFR, c-Kit and Flt-3 kinase activity [12]. The compounds has shown efficacy against renal cell carcinoma [13] and in the treatment of imatinib-resistant, GIST patients [14].

PTK 787 (Vatalanib)

Following on the success of the combination treatment with bevacizumab, attention turned to PTK787 (vatalanib), an anti-angiogenic tyrosine kinase inhibitor that blocks signaling from VEGF receptors, as well as receptors for platelet-derived growth factor (PDGF) and fibroblast-derived growth factor (FGF). PTK787 demonstrated impressive efficacy in preclinical and Phase I/II trials where it significantly reduced tumor vessel den-sity and in some cases induced tumor regression [15]. However, Phase III trials examining the efficacy of PTK787 in combina-tion with chemotherapy showed no significant improvement in progression free survival (PFS), prompting the discontinuation of one study and leaving the drug's future in doubt.

ZD6474 (Vandetanib)

ZD6474 is a multi-targeted kinase inhibitor, which inhibits VEGFR2, PDGFR EGFR and RET. Preclinical studies of van-detanib have demonstrated antitumor efficacy against multiple human cancer xenografts in subcutaneous, orthotopic and metastatic models. Phase I clinical trials have demonstrated that vandetanib is well tolerated. Phase II clinical studies in patients with non-small-cell lung cancer have shown promis-ing results, employing vandetanib as both monotherapy and in combination with docetaxel. Phase II studies in other cancers have likewise been initiated.

BAY 43-9006 (Sorafenib)

BAY 43-9006 is a novel bi-aryl urea developed by Bayer and which inhibits a broad spectrum of kinases including VEGFR2 and VEGFR3, and, with even greater potency, B-Raf, an onco-gene that contributes to the development of many cancers. In fact, this compound was initially discovered as a c-Raf inhibitor and subsequently later as a VEGFR2 inhibitor. BAY 43-9006 did not succeed in clinical trials on melanoma patients, 66% of whom express mutated B-Raf, which is strongly inhibited by the BAY 43-9006. Its high potency on VEGFR2 re-focused the efforts to develop the compound as an antiangiogenic agent. Indeed, sorafenib has been approved for the treatment of renal cell carcinoma patients most likely due to its antiangiogenic activity [16]. The compound has been used in combination with doxorubicin in clinical trials on other solid tumors [17].

Conclusions

The development of protein kinase inhibitors aimed at the inhibition of angiogenesis is quite successful. Recognizing that angiogenesis depends on a number of PTKs leaves room for further development of yet more potent multi-targeted protein kinase inhibitors.

References

1. A. Levitzki. in *Faseb J* 1992;3275–82
2. A. Levitzki and E. Mishani. in *Annu Rev Biochem* 2006; 93–109
3. R. L. Levine and D. G. Gilliland. in *Curr Opin Hematol* 2007;43–7
4. S. Banai, S. D. Gertz, L. Gavish, M. Chorny, L. S. Perez, G. Lazarovichi, M. Ianculuvich, M. Hoffmann, M. Orlowski, G. Golomb and A. Levitzki. in *Cardiovasc Res* 2004;165–71
5. M. C. Heinrich, C. L. Corless, C. D. Blanke, G. D. Demetri, H. Joensuu, P. J. Roberts, B. L. Eisenberg, M. von Mehren, C. D. Fletcher, K. Sandau, K. McDougall, W. B. Ou, C. J. Chen and J. A. Fletcher. in *J Clin Oncol* 2006;4764–74
6. F. Winkler, S. V. Kozin, R. T. Tong, S. S. Chae, M. F. Booth, I. Garkavtsev, L. Xu, D. J. Hicklin, D. Fukumura, E. di Tomaso, L. L. Munn and R. K. Jain. in *Cancer Cell* 2004;553–63
7. R. T. Tong, Y. Boucher, S. V. Kozin, F. Winkler, D. J. Hicklin and R. K. Jain. in *Cancer Res* 2004;3731–6
8. L. M. Strawn, G. McMahon, H. App, R. Schreck, W. R. Kuchler, M. P. Longhi, T. H. Hui, C. Tang, A. Levitzki, A. Gazit, I. Chen, G. Keri, L. Orfi, W. Risau, I. Flamme, A. Ullrich, K. P. Hirth and L. K. Shawver. in *Cancer Res* 1996;3540–5
9. S. J. Boyer. in *Curr Top Med Chem* 2002;973–1000
10. P. W. Manley, G. Martiny-Baron, J. M. Schlaeppi and J. M. Wood. in *Expert Opin Investig Drugs* 2002;1715–36
11. D. W. Davis, H. Q. Xiong, R. S. Herbst, J. L. Abbruzzese, J. V. Heymach, G. D. Demetri, W. Stadler and D. J. McConkey. in *Proc Am Assoc Cancer Res* 2004;Abstract #3988 P
12. D. B. Mendel, A. D. Laird, X. Xin, S. G. Louie, J. G. Christensen, G. Li, R. E. Schreck, T. J. Abrams, T. J. Ngai, L. B. Lee, L. J. Murray, J. Carver, E. Chan, K. G. Moss, J. O. Haznedar, J. Sukbuntherng, R. A. Blake, L. Sun, C. Tang, T. Miller, S. Shirazian, G. McMahon and J. M. Cherrington. in *Clin Cancer Res* 2003;327–37
13. J. Motzer, B. I. Rini, M. D. Michaelson, B. G. Redman, G. R. Hudes, G. Wilding, R. M. Bukowski, D. J. George, S. T. Kim and B.C.M. in *ASCO Annual Meeting* 2005;16 S, 4508R
14. C. Blanke. in *Clin Adv Hematol Oncol* 2006;582–3
15. W. P. Steward, A.Thomas, B. Morgan, B. Wiedenmann, C. Bartel, U. Vanhoefer, T. Trarbach, U. Junker, D. Laurent and D. Lebwohl. in *J Clin Oncol* 2004;3556
16. B. Escudier, C. Szczylik and T. Eisen. in *Proc Am Soc Clin Oncol* 2005;380s Abstract LBA4510
17. H. Richly, B. F. Henning, P. Kupsch, K. Passarge, M. Grubert, R. A. Hilger, O. Christensen, E. Brendel, B. Schwartz, M. Ludwig, C. Flashar, R. Voigtmann, M. E. Scheulen, S. Seeber and D. Strumberg. in *Ann Oncol* 2006;866–73

Chapter 32
Therapeutic Strategies that Target the HIF System

Kristina M. Cook and Christopher J. Schofield

Keywords: HIF, drug development, HIF inhibitors

Abstract: Human cells respond to conditions of limiting oxygen by a switching on a response involving a gene array regulated by the transcription factor hypoxia inducible factor (HIF). This hypoxic response works both to limit the damage of hypoxia and improve oxygen supply to tissues including by upregulating angiogenesis and erythropoiesis. Studies on the molecular mechanism of the HIF system have identified a mechanism by which the HIF system senses changes in oxygen levels. Degradation of the HIF-α subunit by the ubiquitin proteasome pathway is signalled for by oxygenase catalyzed post-translational prolyl-hydroxylation. HIF-α asparaginyl hydroxylation reduces by the transcriptional activity of HIF. The dependence of the HIF hydroxylase on oxygen for catalysis enables an oxygen dependent response. Modulation of the activity of HIF is a potential therapeutic avenue for the treatment of both ischemic disease and cancer. This chapter summarises the molecular components of the HIF system and ongoing therapeutic strategies.

Introduction

Hypoxia inducible factor (HIF) is an α,β-heterodimeric transcription factor that regulates cellular adaptations to hypoxia in metazoans [1–5]. HIF is composed of two basic helix–loop–helix (bHLH) proteins—HIF-α and HIF-β—of the PAS family [PER (period circadian protein), ARNT (aryl-hydrocarbon-receptor nuclear translocator), SIM (single-minded protein)]. The α,β-HIF dimer binds to a core DNA motif (G/ACGTG) in hypoxia-response elements (HREs) that are associated with a very broad range of genes involved in the hypoxic response. To date, about 40 human genes have been shown to be directly

regulated by HIF and it is predicted that there may be up to 200–300 human HIF target genes. HIF-regulated genes play central roles in both systemic responses to hypoxia, such as in cell proliferation, angiogenesis, erythropoiesis, and in intracellular responses, such as regulation of glycolysis. HIF target genes include proteins that have already been medicinally targeted such as erythropoietin (EPO), vascular endothelial growth factor (VEGF) and nitric oxide synthase. The medicinal importance of HIF-regulated genes coupled to breakthroughs in our knowledge of the biochemistry of hypoxia, have led to significant interest in manipulating the response for therapeutic benefit. Medicinal stimulation of the natural HIF-mediated response might be used to treat ischemic diseases; in contrast, inhibition of the HIF-mediated hypoxic response might be used for the treatment of tumors via inhibition of angiogenesis. In this chapter, we review the biochemistry of the HIF system, identify possible points for therapeutic intervention and summarize reported work (for reviews, see [6–10]).

The Biochemistry of HIF

The HIF-β subunit (ca. 95 kDa), which is identical to the aryl carbon receptor nuclear translocator (ARNT), is a constitutive nuclear protein that also has roles in transcription not associated with HIF-α. In contrast to HIF-β, the levels of the HIF-α subunits and their transcriptional activity are regulated by oxygen availability. There are three related forms of human HIF-α (HIF-1α, HIF-2α, and HIF-3α), each of which is encoded by a distinct genetic locus. HIF-1α and HIF-2α have very similar functional domain structures and are regulated in a similar manner by oxygen [11–14]. The domain structure of HIF-3α is less closely related and its regulation is not as well studied. The expression of HIF-3α is tissue-specific and appears to exist in a number of differently spliced variants; one truncated form of murine HIF-3α is termed inhibitory PAS domain protein (IPAS) and has been shown to inhibit the induction of HIF target genes via heterodimerization with HIF-1α [15,16]. High levels of IPAS in the corneal epithelium

The Chemistry Research Laboratory, University of Oxford, Mansfield Road Oxford, OX1 3TA, UK of the eye may reduce the HIF-mediated expression of genes involved in angiogenesis [16].

Both HIF-1α and HIF-2α subunits (ca. 90kDa) are rapidly degraded by the proteasome in the presence of sufficient oxygen, (Fig. 32.2). Mutagenesis studies have demonstrated that both HIF-1α and HIF-2α possess a central oxygen-dependent degradation domain (ODD), which is absent in HIF-β [17–19]. The ODD is composed of two sub-domains, denoted NODD and CODD (for N- and C- terminal ODD), each of which can independently enable degradation of the HIF-α subunit. The von Hippel-Lindau tumor suppressor protein (pVHL) enables binding of the ODD of HIF-α to a multicomponent ubiquitin E3 ligase complex (pVHL–elongin B–elongin C–Cul2–Rbx) that catalyses polyubiquitinylation of HIF-α, thereby targeting it for hydrolysis by the ubiquitin–proteasome pathway [20,21].

Two transcriptional activation domains are present in HIF-1α and HIF-2α: one internal (N-terminal transactivation domain, NTAD) that overlaps with the ODD, and one at the carboxyl-terminus (C-terminal transactivation domain, CTAD). The transcriptional activity of the CTAD is reduced in the presence of oxygen in a process independent of HIF-α proteasomal degradation and which involves oxygen-dependent ablation of the binding of the CTAD with the CH-1 (cysteine/histidine rich) domain of the transcriptional co-activator p300 [22].

Splice variants of HIF-1α have been identified, some of which lack one or both of the NTAD or CTAD [23–25].

Binding of HIF to DNA and HIF-α,β- dimerization

Dimerization of HIF-α/β and binding of necessary coactivators, including p300, CBP, SRC-1, and TIF2, forms the active transcription complex that binds to the HRE associated with HIF target genes [26]. Dimerization of HIF-1α/β is mediated by the bHLH PAS domain present in all HIF subunits (Fig. 32.1).

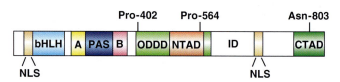

Fig. 32.1. Domain structure of HIF-1α. Major structural domains of HIF-1α are shown. The subunit HIF-1α contains defined domains including a *bHLH*: basic helix-loop-helix domain; *PAS*: Per/ARNT/Sim domain; *ODD*: oxygen dependent degradation domain; *NTAD*: N-terminal transactivation domain; *ID*: inhibitor domain; *NLS*: nuclear localization signal; and *CTAD*: C-terminal transactivation domain. Pro-402 and Pro-564 are hydroxylated in a reaction catalyzed by the PHDs, while Asn-803 hydroxylation is catalyzed by FIH.

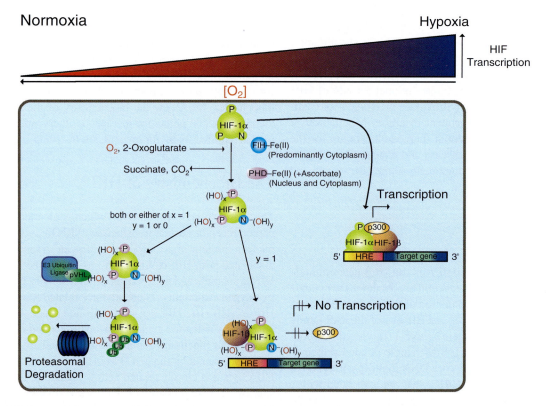

Fig. 32.2. HIF Transactivation or Degradation. Under normoxic conditions, HIF-1α is hydroxylated in reactions catalyzed by both the PHDs (nucleus and cytoplasm) and FIH (predominantly cytoplasm). Hydroxylation by the PHDs enables pVHL, a part of an E3 ubiquitin ligase complex to bind and ubiquitinate HIF-1α, signaling for proteasomal degradation of HIF-1α. Hydroxylation by FIH inhibits HIF mediated transactivation, by blocking HIF-α binding to p300. Under hypoxic conditions, hydroxylation by the PHDs or FIH is prevented, and HIF-1α is able to enter the nucleus, where it dimerizes with HIF-1β and binds p300 to activate the transcription of target genes.

The bHLH domain determines the specificity of the complex formation and bind to DNA. The PAS domain mediates interactions with other proteins and contributes to dimerization specificity [27]. Truncation studies involving mutation have shown that the bHLH domains can dimerize alone, but the presence of the PAS domains are required for stability and DNA binding [27]. NMR studies on the PAS domain of HIF-2α have demonstrated that its interaction with HIF-β is through solvent-exposed residues on the β-sheet face of HIF-2α [27,28].

Regulation of HIF-1α Activity Through Hydroxylation

Both the HIF-α and HIF-β subunits are produced constitutively, but in normoxia HIF-1α and HIF-2α undergo oxygen-dependent, post-translational hydroxylations that deactivate them and signal for their proteasomal degradation. To date, these are the only direct interfaces (i.e., involving covalent modification) between HIF, dioxygen, and the HIF hydroxylases. Consequently, the enzymes that catalyze hydroxylation have been termed oxygen sensors (for reviews, see [29–32]).

HIF Prolyl Hydroxylation

Hydroxylation of HIF-α at two conserved prolyl residues in the ODD (Pro-402 and Pro-564 in HIF-1α) regulates separate interactions with pVHL [33–36]. *trans*-4-Prolyl hydroxylation increases the affinity of HIF-α peptides for the pVHL–elonginB–elonginC (VBC) complex by at least three orders of magnitude in vitro. X-ray crystallographic analyses of a hydroxylated HIF-1α CODD peptide fragment complexed to VBC revealed that the discrimination between hydroxylated and nonhydroxylated sequences is mediated, at least in part, by two optimized hydrogen bonds, which are formed between the alcohol of the hydroxylated proline and two residues of pVHL (Ser111 and His115) [37,38] (Fig. 32.3). In hypoxia, prolyl-4-hydroxylation is reduced or absent, thus enabling the HIF-α to avoid proteasomal hydrolysis and to accumulate. To date, there are no reports that HIF-α hydroxylation is reversible.

von Hippel-Lindau disease affects ca. 1 in 35,000 humans and is associated with tumor growth (particularly in the brain/spinal cord and kidney); it is caused by mutations to the *VHL* gene resulting in increased HIF levels (for review, see [39–43]). Inherited mutations in the *VHL* gene can also cause polycythemia which can occur as part of the von Hippel-Lindau syndrome or independently. The functional effects of some mutations in pVHL leading to von Hippel-Lindau disease have been rationalized by the structures of pVHL in complex with HIF$_{(hyp564)}$ peptide [37,38]. The pVHL α domain enables binding to elongin C whilst the β domain is responsible for HIF-α binding. pVHL disease-associated mutations are associated with surface residues of both α and β pVHL domains, revealing that they are both of functional importance [21].

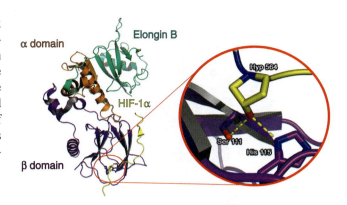

FIG. 32.3. View from a crystal structure of HIF-1α and von Hippel Lindau protein. The HIF-1α CODD segment (in *yellow*) binds to the β domain of VHL (in *purple*). The α domain of VHL is shown in *orange* and Elongin B is shown in *turquoise*. The interaction between the alcohol of the hydroxylated proline (Hyp) in HIF-1α and the two residues of pVHL (Ser111 and His115) is shown in enlarged view [37,38].

Three closely related human ferrous iron- and 2-oxoglutarate (2OG)-dependent oxygenases catalyze HIF prolyl hydroxylation. Prolyl hydroxylase domain PHD1, PHD2 and PHD3 (or EGLN 1–3) enzymes have been identified in humans [44,45]; one of the PHDs (PHD3) is itself upregulated by hypoxia in MCF7 cells [46]. There is evidence that the hydroxylation at the two sites is linked (at least modification at the CODD sites affects hydroxylation at the NODD), and that PHD2 and PHD3 are selective for the CODD over the NODD of HIF-1α [44,47]. Since the catalytic domains of all three PHDs are highly conserved, it is possible that they are all selective for the CODD over the NODD sequences, but vary in the degree of selectivity. However, further experiments are required to fully define the selectivity of the PHDs.

Structural and mechanistic studies on the PHDs are ongoing. They are members of a ubiquitous family of oxygenases that utilize a single ferrous iron as a cofactor and 2OG as a cosubstrate [48–52]. The 2OG oxygenases catalyze the oxidative decarboxylation of 2OG to give succinate (into which one of the atmospheric oxygen atoms is incorporated), carbon dioxide and an enzyme bound Fe(IV)=O ferryl intermediate that effects the two electron oxidation (hydroxylation) of the substrate (e.g., HIF) [48,50,53] (Fig. 32.7). Crystallographic analyses on PHD2, the form of the enzyme thought to be most involved in the hypoxic response in most normal tissues [54], has revealed a double-stranded β-helix core common to the 2OG oxygenase family and a triad of conserved iron binding residues [55] (Fig. 32.9a). The structure also suggests why PHD2 appears to have an unusually high affinity for iron and 2OG compared to other studied family members. The structure of PHD2 bound to an inhibitor is shown in Fig. 32.9a,b. Kinetic analyses have been carried out on the HIF hydroxylases and the oxygen binding capabilities of PHD2 appear to be within the normal range for 2OG oxygenases [47,56,57]. Recent kinetic studies also indicate

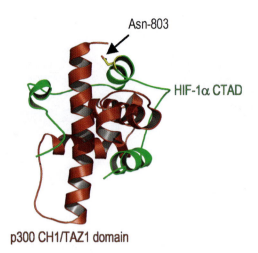

FIG. 32.4. View from a crystal structure of HIF-1α C-terminal transactivation domain and p300 CH1 domain. HIF-1α CTAD domain (shown in *green*) bound to the p300 CH1/TAZ1 domain (in *red*). The amino acid hydroxylated by FIH (Asn-803) is indicated in *yellow*.

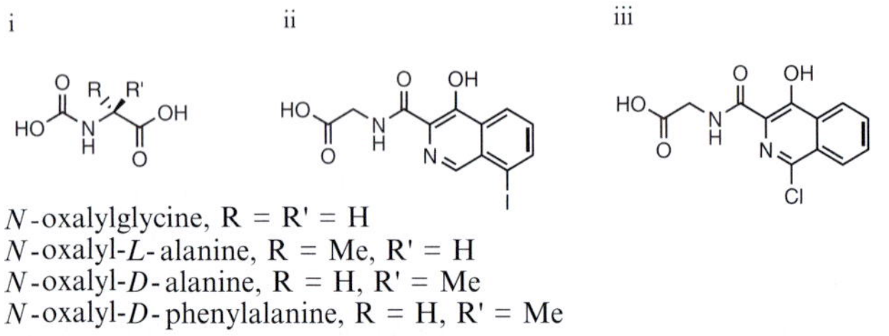

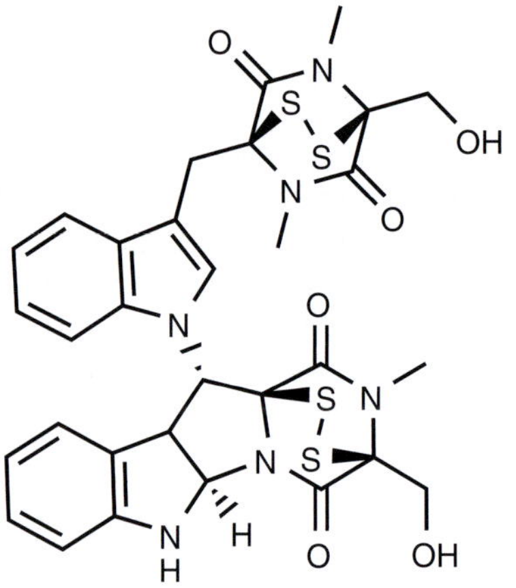

FIG. 32.6. Structure of chetomin. Chetomin disrupts the interaction between HIF-1α CTAD and p300 CH1.

i   ii   iii

N-oxalylglycine, R = R' = H
N-oxalyl-*L*-alanine, R = Me, R' = H
N-oxalyl-*D*-alanine, R = H, R' = Me
N-oxalyl-*D*-phenylalanine, R = H, R' = Me

FIG. 32.5 HIF hydroxylase inhibitors. (*i*) *N*-oxalylglycine, R = R' = H; *N*-oxalyl-*L*-alanine, R = Me, R' = H; *N*-oxalyl-*D*-alanine, R = H, R' = Me; *N*-oxalyl-*D*-phenylalanine, R = H, R' = CH₂Ph (NOFD) [129]; (*ii*) PHD2 inhibitor, {[(4-hydroxy-8-iodoisoquinolin-3-yl)carbonyl]amino}acetic acid [55]. (*iii*) PHD2 inhibitor, an isoquinoline derivative [139].

that the preferred substrates for the PHDs are significantly longer than the relatively short peptides (<20–40 residues) used in some analyses [57,58]. Coupled with the structural analyses it seems possible that the substrate specificity of the PHDs, in part, is determined by regions relatively remote from the iron center and may involve the variable N- and C-terminal regions.

An inherited mutation in PHD2, Pro-317-Arg, linked with a familial erythrocytosis has been identified [59]. As shown in Fig. 32.9b, Pro-317 of PHD2 is located two residues from the iron binding aspartyl residue (Asp-315) in a β-turn and is located close to the active site entrance suggesting that mutation to arginine at this position may alter iron and/or substrate binding.

HIF Asparaginyl Hydroxylation

A second type of HIF-α hydroxylation event occurs at Asn-803, wherein its β-carbon is hydroxylated by the factor inhibiting HIF-1 (FIH) [22,60,61]. Hydroxylation of Asn-803 ablates the binding of HIF-1α to p300/CBP (CREB Binding Protein), a transcriptional coactivator protein [22] (Fig. 32.4).

NMR studies of an unhydroxylated CTAD polypeptide complexed to CH-1 reveal that HIF-1α Asn-803 is part of an α-helix buried at the complex interface, and suggest that β-hydroxylation of Asn-803 may prevent HIF-1α binding to CH-1 by disrupting both the hydrophobic interactions between the molecules and/or formation of the α-helix adopted by CTAD in this complex [62,63]. The structure of HIF-1α CTAD bound to p300 CH1 is shown in Fig. 32.4.

Like prolyl hydroxylation, HIF asparaginyl hydroxylation is also catalyzed by a member of the 2OG dependent-oxygenase superfamily, but by an enzyme that is significantly different from the three closely related PHDs, FIH, which was originally identified as a protein that binds HIF-α [64] (Fig. 32.9a,b). Mutation experiments with HIF-1α have shown that, with the exception of Val-802, individually conserved residues around Asn-803 are not essential for efficient hydroxylation in cell-based assays [65]; similar conclusions have been drawn for PHD-catalyzed prolyl hydroxylation [66,67]. Although the general mechanisms of FIH and PHDs are likely to be very similar, as are the ways in which they bind the active site iron, there are significant differences between both their active sites and overall folds which are both based on a double-strand β-helix core. Unlike the PHDs, FIH is homodimeric [68] and is also a member of the JumonjiC (JmjC) family of human transcriptional regulator proteins [69]. Many of these were originally proposed to be zinc-dependent transcription factors, but in the light of hydroxylation activity and crystal structure of FIH, were proposed to be iron- and 2OG- dependent oxygenases [61]. This proposal was supported by the recent assignment of various members of the JmjC family as histone demethylases operating via initial hydroxylation of the methyl groups on methylated lysyl residues [70,71].

To date, the only PHD structures reported are complexed with iron and an inhibitor but without substrate [55] (Fig. 32.9). FIH has been crystallized in the presence of a CTAD peptide and the resultant structures revealed the existence of two distinct FIH-CTAD interaction sites, one involving the hydroxylation

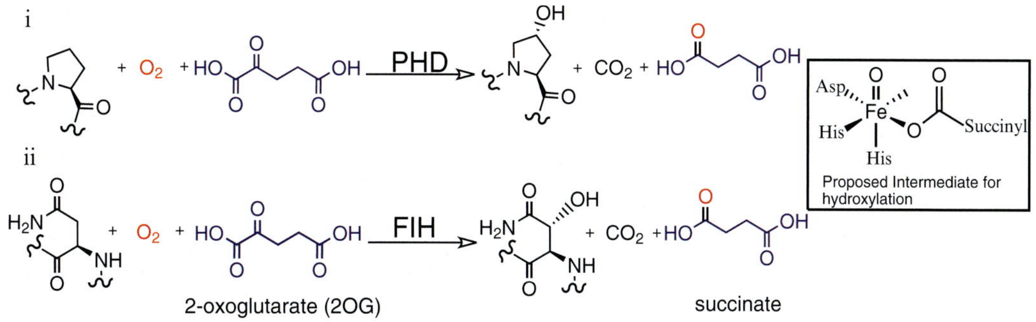

FIG. 32.7. Reactions catalyzed by the HIF hydroxylases. (*i*) Reaction catalyzed by the PHDs (HIF-α prolyl hydroxylases). (*ii*) Reaction catalyzed by FIH (HIF-α asparaginyl hydroxylase).

site (CTAD795–806) and a second lying to the C-terminal side of this site (CTAD813–822) [72]. At the hydroxylation site, the CTAD795–803 residues are bound in a groove and adopt an extended conformation linked to FIH by hydrogen bonds. Asn-803 of the CTAD is completely buried at the active site and lies directly adjacent to the Fe(II). CTAD Asn-803 and Ala-804 form a tight turn, stabilized by a hydrogen bond between the backbone carbonyl of Val-802 and NH of Ala-804, which projects the side chain of Asn-803 towards the Fe(II) [68,72,73].

Recently, FIH has been shown to also catalyze hydroxylation of ankyrin repeat domain proteins from the NF-κβ family at highly conserved asparaginyl residues [74]. However, the functional significance of this hydroxylation is unclear. Various alternative substrates have been proposed for the PHDs, including RNA polymerase II [75] and IκB kinase-β [76] (which is negatively regulated by PHD1). However, unlike ankyrin hydroxylation by FIH, to date none has been verified by direct demonstration of hydroxylation. The discovery that 2OG oxygenases can repair certain forms of methylated nucleic acids by hydroxylation of the methyl groups has also raised the possibility that signaling involving oxygen might also occur via the methylation/demethylation of nucleic acids [77,78].

Endogenous Hydroxylase Regulation by Small Molecules

As well as the availability of their common cofactor ferrous iron, it is possible that small molecules can regulate the activities of the HIF hydroxylases. Interestingly, the activity of the PHDs appears to be more dependent on the presence of ascorbate than does FIH, and PHD2 binds Fe(II) and 2OG unusually tight [79]. Ascorbate is commonly used with in vitro studies on 2OG oxygenases, possibly to maintain the iron in the ferrous form, and is particularly important to catalysis by collagen prolyl-4-hydroxylase [80]. Impaired activity of collagen prolyl-4-hydroxylase due to lack of ascorbate leads to scurvy. Two tricarboxylic acid cycle intermediates, 2OG and succinate, are involved in HIF hydroxylase

catalysis, one as a substrate and one as a product respectively. This provides a potential link between respiration and the HIF system. Several studies have examined tricarboxylic acid cycle intermediates as HIF hydroxylase inhibitors [69,81,82]. The HIF hydroxylases are inhibited, though not potently, by fumarate (and to a lesser extent by their product, succinate) leading to the proposal that elevated levels of fumarate, due to impaired succinate dehydrogenase activity, in some cancers may lead to activation of the HIF system (and may possibly be a contributory factor to the Warburg effect) [81]. The relationship between nitric oxide (NO) and the HIF system is complex, with apparent activation of HIF transcriptional activity by different NO donors in normoxia [83,84] (possibly mediated by inhibition of the HIF hydroxylases) and reduction of HIF transcriptional activity in hypoxia [85–88] (for reviews, see [89–91]).

Other Modifications to and Regulators of HIF-α

In addition to the oxygen-regulated mechanisms of HIF-1α activation, non-hypoxia-regulated mechanisms may also regulate HIF-1α. HIF-α is phosphorylated, but the specific sites of phosphorylation have yet to be fully identified. Both the phosphatidylinositol 3-kinase (PI3K)-Akt-FKBP12-rapamycin associated protein (FRAP, also known as mammalian target of rapamycin or mTOR) and the mitogen-activated protein kinase (MAPK) cascades are proposed to modulate HIF-α activity through different modes of action (for review, see [92–94]). Stimulation of the PI3K-AKT-FRAP pathway occurs via growth factors and cytokines including epidermal growth factor, fibroblast growth factor 2, heregulin, insulin, insulin-like growth factor 1 and 2 and interleukin-1b that bind to receptors which activate the PI3K pathway through receptor tyrosine kinases.

Although kinase pathways clearly modulate HIF activity, the effect of the PI3K/MAPK pathways may be mediated through modulation of the phosphorylation status of HIF binding partners, e.g., p300. Consequently, until the exact mechanism by which phosphorylation enhances HIF activation is elucidated, rational targeting of kinase pathways as a means of HIF regulation remains challenging.

Acetylation of the HIF-1α subunit at Lys352, mediated by ARD1 (arrest defective protein 1), has been reported to enhance the interaction of HIF-1α with pVHL [95]. However, more recent studies have failed to reproduce the earlier work [96–98], and although it seems that ARD1 does interact with HIF-α there is presently no solid evidence for HIF-α acetylation.

There is also evidence for a p53 dependent pathway for the degradation of HIF-1 that is oxygen-independent [99,100]. Binding of HIF-1α peptides to p53 was shown to occur independent of prolyl-4-hydroxylation [100]. The relative significance of the p53 pathway in comparison with the pVHL pathway in terms of the hypoxic response is unknown.

Modulating the HIF System for Therapeutic Benefit: Overview

Broadly, medicinal manipulation of HIF can be divided into HIF inactivation and HIF activation strategies. Medicines that inactivate HIF could be useful in the treatment of tumors or diabetic retinopathy, whereas those that activate HIF could be useful for the treatment of ischemic diseases including myocardial infarction, stroke, and ischemic limb disease, or as alternatives to the use of recombinant human erythropoietin [101–103]. There is good evidence that treatments targeting HIF, pVHL or the HIF hydroxylases will result in a physiological effect in humans. There is also evidence, primarily from genetic diseases, but also from emerging data on different types of therapies targeting the HIF system, that it will be possible to develop approaches that are sufficiently safe for use in some clinical applications. In the case of prospective treatments that up-regulate HIF, a concern is that they may cause tumorigenesis. Further data is required, but the available evidence suggest that this may not be of concern—at least for shorter term treatments as might be envisaged for the treatment of some types of heart disease or stroke [104].

The central role of HIF-1 in the transcriptional activation of genes under hypoxic conditions, including those involved in angiogenesis and cell proliferation, makes it a promising target for cancer treatment, especially since HIF is reported to be overexpressed in the majority of tumors. Tumor hypoxia and overexpression of HIF-1 has been shown to be associated with resistance to traditional radiation and chemotherapeutic treatments, increased aggressiveness of tumors, increased risk of metastasis and poor outcome [105–110]. However, several studies in embryonic stem cell tumors found that inhibition of HIF increased tumor growth [111,112] and activation of HIF led to a slower growth rate than the wildtype cells [113]. It was also found in mouse tumor models that HIF-2α overexpression, while increasing vascularity, slowed the growth of the tumor, in part by increasing tumor cell apoptosis. In this same model, siRNA or dominant negative HIF-2α was used to knock-down HIF-2α, leading to a decrease in angio-genesis, but an increase in the rate of tumor growth [114]. Since HIF regulates multiple cellular activities, which may either promote or hinder carcinogenesis, caution as to the role of the HIF system in cancer, needs to be exercised in drawing conclusions from results on experiments from a limited number of cell types.

Early evidence that modulating the HIF system by small molecules might be possible came from work with peptides. Under normoxic conditions, introduction of HIF-α CODD or NODD peptides fused to cell-penetrating *tat*-sequences into cells, led to the induction of HIF-1α [115]. In addition, appropriately treated sponges implanted subcutaneously into mice assayed the in vivo effect of the peptides. An increase in vascularity and gene expression of VEGF and Glut-1 was observed in the peptide-treated sponges. The peptides probably up-regulate HIF-α either by competing with HIF-α for binding to pVHL and/or to the PHDs. In an alternative approach, PR39 (proline-rich peptide 39), a macrophage delivered antimicrobial peptide known to stabilize HIF via an undefined mechanism, has been shown to increase vascularization in mouse heart tissue [116]. Further evidence for efficacy of targeting the HIF system has been accrued over the last five years or so.

The Question of Selectivity

Transcription factors have not been extensively targeted as drug targets, probably because many of the potential targets are protein–protein/protein–DNA interactions and probably in part due to the concern of whether this approach is selective enough, at least for non-cancer applications. Since there are now good precedents for small molecules that block protein–protein interactions and there are 'classical' enzyme targets for small molecules in the HIF hydroxylases, the question of selectivity is perhaps more pressing in the case of the HIF system. For some applications involving modulation of the HIF system, it might be optimal to up-regulate a particular HIF target gene or set of genes, but in other cases it may be advantageous to augment or induce the natural hypoxic response (though the type of natural response itself may vary depending on the context) in which an array of genes is activated.

Although advances are being made, little is understood at the molecular level about how selectivity is achieved for different target genes within the HIF system. Given the apparent plethora of HIF-regulated genes, it seems likely that there are unidentified mechanisms for the context-dependent up- or down-regulation of HIF target genes. There is already evidence that some genes are more strongly regulated by either HIF-1α or HIF-2α [114,117,118]. There may also be some selectivity at the HIF-α:PHD interface, and the PHDs may have different physiological roles. However, further evidence is required, particularly in whole organisms, before different HIF hydroxylases/forms of HIF can be definitively targeted for any particular disease. Another important question is how the HIF signaling system interfaces with other signaling and homeostasis systems.

As outlined above, molecular connections have been made between the HIF and the NF-κβ system via the HIF hydroxylase FIH though the functional significance, if any, of the connections have yet to be reported. Links between the HIF system and Notch have also been reported [119]. HIF also likely plays an important role in development since knockout mice experiments (pVHL, HIF-α, HIF-2α, HIF-1α) are lethal to the embryo and/or result in vascular and other defects [39,120–122].

Targets Within the HIF System

HIF/ HIF target gene *activation* targets include: (1) HIF hydroxylase inhibition, (2) blocking the HIF-α:pVHL interaction, (3) proteasome inhibition, and (4) blocking non-oxygen dependent pathways for HIF degradation.

HIF/ HIF target gene *inactivation* targets include (1) blocking the HIF-α:p300 interaction or interactions between HIF and other coactivators, (2) activating/up-regulating/increasing the efficiency of the HIF hydroxylases, (3) blocking HIF-α:HIF-β dimerization, (4) blocking HIF: HRE/DNA binding, and (5) promoting non-oxygen-dependent pathways for HIF degradation.

Below, we discuss work on some of these different possible approaches focusing on the HIF axis itself rather than HIF target genes (e.g., VEGF, EPO etc.) and summarize work to date including further evidence for efficacy.

HIF/HIF Target Gene Activation Inactivation with the HIF Hydroxylases as Targets

Since the hypoxic response is mediated by reduction in oxygen levels, it may be argued that the proposed natural oxygen sensing mechanism, i.e., the HIF hydroxylases, are a preferred target for modulation of the response. Indeed, in terms of small molecular therapies, most reported effort has been directed towards HIF hydroxylase inhibitors. Given that HIF-α prolyl hydroxylation signals for proteasomal degradation, HIF prolyl hydroxylase inhibitors might be regarded as being related to inhibitors that actually target the proteasome. One proteasome inhibitor, bortezomib (Velcade) is already in clinical use for cancer (multiple myeloma) giving further support to the validation of the PHDs as targets [123]. It seems possible that targeting the mechanisms that signal for proteasomal degradation might be more selective than targeting the proteasome itself.

Since tumors cells are often hypoxic, a potential therapy directed towards HIF inactivation involves optimization or increase of HIF hydroxylase activity. This will be difficult if the hydroxylase activity is solely limited by oxygen availability, but if other factors are limiting (e.g., ferrous iron, 2OG, or ascorbate, enzyme) it may be possible to increase hydroxylase activity. While little work has been reported on this effect, addition of ascorbate to cell lines have been shown to stimulate hydroxylase activity [124]. It should also be noted that the HIF system is linked with iron metabolism and reactive oxidizing species; thus, a simple relationship between hydroxylase activity and iron concentration or redox potential seems unlikely in humans. One strategy may be to up-regulate the HIF hydroxylases, either by gene/protein therapy or by small molecules; PHD3 is hypoxically regulated but little is reported on the non-oxygen-dependent regulation of the HIF hydroxylases.

Considerably more efforts have been made in evaluating HIF activation via HIF hydroxylase inhibition. Given the importance of 2OG oxygenases in human metabolism, DNA repair and regulation, it is probably important that any clinically used HIF hydroxylase inhibitors are selective for their target(s). Further, given the potentially different roles of the PHDs and the observation that FIH has non-HIF substrates, it may be desirable that particular HIF hydroxylases are targeted. The significant structural differences between FIH and the PHDs imply that it should be possible to achieve selective inhibition in this regard. It is anticipated that selective inhibition of PHD isoforms will be more difficult, but precedent with selective inhibition of isoforms of other metalloenzymes, e.g., the cyclooxygenases and metallo-proteases, suggests that this should be achievable. The apparent differences in the predominantly intracellular localization of the HIF hydroxylases might also be exploited for medicinal chemistry (e.g., FIH is predominantly cytoplasmic whereas PHD2 is nuclear).

Historic observations that cobalt and other transition metal ions stimulate EPO production [125] can be rationalized by inhibition of the HIF hydroxylases by competition with the ferrous iron essential for catalysis. Similarly, iron chelators such as desferrioxamine can induce the hypoxic response by depriving the hydroxylases of iron [126–128]. However, potent chelators of free iron are unlikely to be of widespread therapeutic utility as they are non-selective. It seems likely that the HIF hydroxylase inhibitors that will be developed for clinical use will preferentially bind to the iron at the active site. Most of the reported inhibitors also likely compete with 2OG for binding to the active site iron.

Early studies on the inhibition of the PHDs exploited compounds that had been developed for procollagen prolyl-4-hydroxylase inhibition with the aim of developing anti-fibrotic agents. *N*-Oxalylglycine, or its cell penetrating form dimethyloxalylglycine, has been extensively used as hypoxia mimics (Fig. 32.5). *N*-Oxalylglycine is a close analogue of 2OG and crystallographic analyses have shown it binds in an almost identical manner to 2OG at the active site of FIH [72]. The substitution of an NH for a methylene group at the 3-position of 2OG either renders *N*-Oxalylglycine stable to attack by activated dioxygen or hinders dioxygen binding [129]. Derivatives of *N*-oxalylglycine have been used to demonstrate that selective inhibition of HIF hydroxylases is possible. Thus, *N*-oxalyl-*L*-alanine was a more potent PHD2 inhibitor than *N*-oxalyl-*D*-alanine whereas the opposite was observed for FIH [129]. *N*-Oxalyl-*D*-phenylalanine inhibited FIH but not PHD2. A crystal structure of this inhibitor complexed to FIH (Fig. 32.8) reveals that it binds to the 2OG and, likely, in the dioxygen binding site [129]. Several of the HIF hydroxylase inhibitors are shown in Fig. 32.5. Other analogues of 2OG

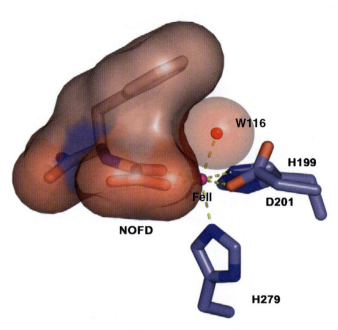

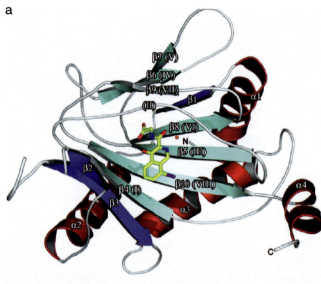

FIG. 32.8. View from a crystal structure of FIH bound to an inhibitor. Selected residues of FIH are shown bound to the inhibitor, *N*-oxalyl-*D*-phenylalanine (NOFD)[129].

have also been explored as PHD inhibitors, including thiols and analogues of dealanylalahopcin, a naturally occurring procollagen prolyl hydroxylase inhibitor [130,131]. Quercetin, a dietary flavonoid, has also been shown to inhibit FIH [132]. FG-0041, developed as a procollagen prolyl hydroxylase inhibitor, was shown to improve ventricular function in a rodent model of myocardial infarction [133,134]. Although the effect was proposed to be due to impairment of fibrosis via procollagen prolyl hydroxylase inhibition, it may in fact be due, at least in part, to HIF activation through inhibition of the HIF hydroxylases.

Various inhibitors developed for procollagen prolyl hydroxylase have been screened for HIF hydroxylase inhibition and in some cases selectivity was demonstrated for the PHDs over procollagen prolyl hydroxylase or vice versa [133]. Pyridine 2,5-dicarboxylate inhibited procollagen prolyl hydroxylase but not the PHDs. However, other compounds, such as pyridine 2,4-dicarboxylate, have been shown to inhibit both procollagen prolyl hydroxylase and the PHDs. Following these earlier studies, further inhibitors have been developed, including some from employing structure-based designs that focus on the PHDs [135]. Most, if not all of these, bind to the active site iron and compete with 2OG [136–139]. At least two HIF hydroxylase inhibitors, FG2216 and FG4592, have been progressed into human clinical trials for anemia [140,141].

HIF-1α and the Transcriptional Coactivator p300

Disrupting the interaction between the HIF-1α CTAD domain and the CH1 domain of the coactivator p300 is a target for HIF inactivation. A high-throughput screen led to the identification

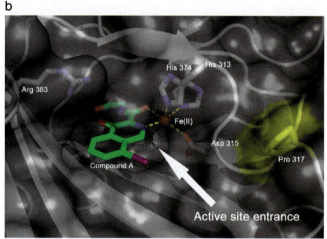

FIG. 32.9 **a** View from a crystal structure of PHD2 bound to an inhibitor PHD2 is shown bound to a 2OG-competitive isoquinoline inhibitor {[(4-hydroxy-8-iodoisoquinolin-3-yl)carbonyl]amino}acetic acid [55]. Roman numerals in parenthesis indicate the eight strands of the double strand β-helix core fold. (From McDonough et al. [55] by permission of M.A. McDonough and PNAS). **b** Closeup view of PHD2 and inhibitor. Compound 1{[(4-hydroxy-8-iodoisoquinolin-3-yl)carbonyl]amino}acetic acid shown bound to PHD2[55]. Selected residues of PHD2 are indicated including Pro-317, which is mutated in a form of familial erythropoiesis [59].

of the compound. Chetomin (Fig. 32.4) that blocked the binding of p300 to either HIF-1α or HIF-2α in both in vitro and in vivo assays [142]. Chetomin is a cyclic disulfide diketopiperazine metabolite of the fungus *Chaetomium* species, and was previously known to have antimicrobial activity. Studies demonstrated that systemic administration of Chetomin attenuates HIF-1-mediated gene expression mice, causing a reduction in tumor size [142]. Unfortunately, high levels of necrosis in tumor tissues were observed, rendering its use as a chemotherapeutic drug unlikely. Further work aimed at understanding the structural basis of the CTAD:p300 disruption by Chetomin may provide clues to a less toxic derivative.

HIF Complex and DNA Binding

Blocking the HIF complex from binding to the hypoxia response element (HRE) DNA sequence will prevent transcription of angiogenesis and hypoxia-related genes. The viability of this target was demonstrated through experiments using polyamides that bound specifically to the minimal HRE sequence 5'-WTWCGW-3' from the VEGF promoter. The polyamides successfully reduced hypoxia induced VEGF transcription by 60% in cellular assays, bringing VEGF transcription levels close to those seen in normoxic cells. This demonstrated that inhibition of this target site had an effect on HIF transactivation [143].

Echinomycin, another small molecule inhibitor, was found to bind to a portion of the HRE (5'-ACGT-3') where the transcription complex binds [144]. The DNA sequence is also found within the consensus sequence for c-Myc, another transcription factor involved in cancer biology [144,145]. Echinomycin was thought to serve as an inhibitor of transcription complexes that bind to this sequence by preventing the complex from binding to its target genes. However, clinical trials for the use of Echinomycin in cancer treatment were disappointing [146]. Since Echinomycin binds to a relatively short sequence of DNA itself, and not the transcription complex, a possible problem is the lack of specificity [145]. While Echinomycin failed to show the results that were hoped for, a small molecule that is able to bind the HIF-HRE complex still holds promise as a selective HIF transactivation inhibitor.

In a recent study, Chinese hamster ovary cells, which stably expressed a luciferase reporter construct under the control of a hypoxia response element, were used to screen 15,000 compounds. One of the compounds, DJ12, was shown to significantly inhibited HIF-1 transactivation activity by blocking HIF-1 HRE DNA binding thereby downregulating the transcription of HIF-α downstream targets [147].

Dimerization of HIF-1α and HIF-1β

Blocking the dimerization of HIF-1α and HIF-1β PAS domains could serve as a specific method of inhibition of HIF. By targeting this site, it is possible that only HIF transactivation would be affected, unlike some of the other methods discussed for targeting HIF. However, it may be a difficult challenge to select for the specific PAS domains of HIF. Mutational studies of the PAS domains in both subunits demonstrated that disrupting dimerization impairs the hypoxic response in cells, indicating the importance of dimerization for transactivation ability [148]. Even though the PAS domains have the extensive protein–protein interaction surface, other studies have shown that PAS domains are predisposed to binding small molecule ligands within their core that lead to structural and conformational changes. With this knowledge, it may be possible to screen compounds for the ability to disrupt their interaction [148,149].

Modulating Alternative Pathways for HIF-α Degradation, e.g., Preventing the HIF-1α:Hsp90 Interaction

Heat shock protein 90 (Hsp90) is proposed to mediate folding of HIF-1α and to protect it from proteasomal degradation [150]; blocking the HIF-α:Hsp90 interaction should therefore destabilize HIF-1α, leading to an increased rate of degradation. Indeed, inhibition of Hsp90 by known inhibitors, e.g., Geldanamycin, a natural antibiotic, is reported to promote HIF-α degradation via a pVHL-independent pathway [150]. A derivative of geldanamycin is in very early stage clinical trials as an anti-tumor agent, and the effect of this compound and others targeting Hsp90 on the HIF system is of interest [151]. Another chaperone, TRiC, has been associated with PHD3 and is another potential target [152]. However, given that most chaperones probably have roles outside of HIF, their inhibition seems unlikely to result in specific HIF targeting. Selectivity will also likely be an issue for some of the other potential targets discussed below.

Topoisomerase Inhibitors

Inhibitors of topoisomerases I and II are reported to decrease HIF-1α levels in tumor cells [153,154]. Topotecan, currently used in the clinic to treat small cell lung cancer and ovarian cancer, was found to inhibit HIF-1α translation through a topoisomerase I-dependent mechanism, but one separate from the replication-mediated DNA damage [153]. The topoisomerase II inhibitor GL331 appears to regulate HIF-1α through transcriptional repression, lowering levels of HIF-1α protein and mRNA [155].

Microtubule Stabilization and Destabilization

Compounds that either stabilize or destabilize microtubules have been found to inhibit HIF-1α. They include some anti-tumor agents that are widely used in the clinic, such as taxol and vincristine. These compounds act by affecting the stability of microtubules, which is believed to lead to a reduction in HIF-1α protein levels and HIF-1 transactivating activity; the mechanism by which this occurs is not understood [156].

2-Methoxyestradiol (2ME2) is a human metabolite of estradiol that functions by destabilizing microtubules [157]. 2ME2 inhibits tubulin polymerization, resulting in cell cycle arrest and has shown activity in several tumor models, as well as in the NCI 60 tumor cell line assay (reviewed in [158]). 2ME2 inhibits HIF-1α protein levels by acting at the translational level, but it does not affect the rates of HIF-1α gene transcription or HIF-1α proteasomal degradation [157]. A greater understanding of the relationship between microtubules and HIF-1α may provide new therapeutic avenues.

PI3K Pathway Inhibitors

Over 70% of human cancer cell lines contain high levels of HIF-1 under normoxic conditions [157,159]. Clearly, hypoxic regulation is not responsible for these high levels. The PI3K/

Akt/mTOR signaling pathway causes HIF-1 upregulation by increasing the rate of HIF-1α translation. Several inhibitors of the PI3K pathway have been shown to decrease levels of HIF-1α including LY294002, wortmannin, 9-β-D-arabinofurano-syl-2-fluoroadenine (FARA-A), SU5416 (Semaxanib), and Rapamycin and its derivatives [102,160,161]. FARA-A acts by inhibiting Akt activation and SU5416 inhibits the VEGF receptor, which leads to a decrease in HIF-1α protein levels through the PI3K/Akt pathway [160]. Rapamycin and derivatives are inhibitors of mTOR [160]. While the P13K pathway inhibitors have all demonstrated varying levels of inhibition of HIF-1α and/or tumor growth, it is unknown to what degree the decrease in HIF-1α actually plays in the anti-tumor activity. This is because the PI3K/Akt/mTOR pathway affects many cellular processes and could be acting upon multiple downstream targets that lead to the anti-tumor activity. Further, an approach involving the P13K pathway and HIF would only work in tumors that show up-regulation of HIF through the PI3K/Akt/mTOR pathway.

Farnesyl Transferase Inhibitors

Another method for inhibiting HIF-1α activity is by blocking the farnesylation of the upstream MAPK pathway activators, Rho and Ras. Ras is involved in stabilizing HIF-1α and targeting it could lead to destabilization of HIF-1α [162]. Two compounds that have shown promise under this mechanism are tipifarnib (R115777) and lonafarnib (SCH66336) [160].

Thioredoxin Inhibitors

Thioredoxin (Trx-1) is a small, ubiquitously expressed redox protein that is often found in increased levels in human tumors [163,164]. It participates in the regulation of transcription factor activity for HIF-1α, among other proteins [165]. An increase in Trx-1 expression was also shown to have an increase in HIF-1α protein and enhanced tumor angiogenesis [166]. Potent inhibitors of Trx-1, e.g., PX-12, have been reported to inhibit HIF activation [167,168]. However, a recent study using other inhibitors reported that treatment of cancer cell lines with novel Trx inhibitors (AJM290 or AW464) reduced HIF-1 CTAD transcription activity and DNA binding, but also inhibited degradation of HIF, in contrast to other Trx inhibitors [169].

Other Inhibitors

YC-1, a drug initially developed for circulatory disorders, is an activator of soluble guanyl cyclase (sGC)[170]. It has been shown to decrease hypoxia-induced HIF-1α levels in vitro and the expression of HIF-1 downstream target genes [171]. YC-1 also possessed anti-tumor activity in animal models [172]. However, experiments have shown that YC-1 functions independently of the sGC pathway in its effects on HIF-1α [171], so its actual mechanism of action is unknown.

103D5R reduced HIF-1α levels under hypoxia in a cell-based screening assay. In addition, it also reduced the amount of phosphorylation of SAPK/JNK, Akt and Erk1/2. It appears to cause diminished synthesis of HIF-1α [173].

Genetic Approaches

The HIF system presents opportunities for gene therapy including the use of sense or antisense vectors for HIF-α, pVHL and the HIF hydroxylases. The most developed avenue has involved fusion of the DNA binding and dimerization domains of HIF-1 (without the ODD) with the transactivation domain from the herpes simplex virus VP16 protein to treat a rabbit model of hindlimb ischemia via HIF activation [174]. Similar effects have been reported in an acute myocardial infarction model in the rat [175]. Human clinical trials are presently ongoing for the treatment of peripheral vascular disease, though there is the possibility of treating other ischemic diseases, particularly heart disease. Experiments have demonstrated that overexpression of pVHL can slow the growth of tumor cell lines and C6 cells overexpressing pVHL displayed a reduced growth rate compared to the parental cell line when subcutaneously implanted in mice [176]. Additional studies have inhibited HIF-1α using an antisense HIF-1α plasmid in murine tumors with smaller tumors regressing in size after treatment and larger tumors showing little to no HIF-1α protein [177].

Future Prospects

The past decade has led to major advances in the molecular understanding of the hypoxic response. This basic research has led to the identification of new targets associated with both activation and inactivation of the master transcriptional regulator HIF. These targets include the oxygen sensing enzymes of the hypoxic response, and protein–protein, and protein–DNA interactions involved in the HIF system. It is likely that other targets associated with HIF will be discovered, either by small molecule screening for effects on HIF or by biochemical studies on the system. Ongoing studies involving both small molecule and gene therapy approaches are evaluating some of the molecular targets—in some cases in a clinical environment—and data regarding efficacy and safety is currently emerging.

Acknowledgements. We thank the Biotechnology and Biological Research Council, The Wellcome Trust, and National Institutes of Health for funding and all members of our Laboratory for their contributions to our work on HIF. We thank Michael McDonough for his help with the illustrations. We apologize for lack of full citations due to space constraints.

References

1. Semenza GL, Wang GL. A nuclear factor induced by hypoxia via de novo protein synthesis binds to the human erythropoietin gene enhancer at a site required for transcriptional activation. Mol Cell Biol 1992; 12:5447–5454.

2. Wang GL, Semenza GL. General involvement of hypoxia-inducible factor 1 in transcriptional response to hypoxia. Proc Natl Acad Sci USA 1993; 90:4304–4308.

3. Maxwell PH, Pugh CW, Ratcliffe PJ. Inducible operation of the erythropoietin 3' enhancer in multiple cell lines: Evidence for a widespread oxygen-sensing mechanism. Proc Natl Acad Sci USA 1993; 90:2423–2427.

4. Wang GL, Semenza GL. Purification and characterization of hypoxia-inducible factor 1. J Biol Chem 1995; 270:1230–1237.

5. Wang GL, Semenza GL. Characterization of hypoxia-inducible factor 1 and regulation of DNA binding activity by hypoxia. J Biol Chem 1993; 268:21513–21518.

6. Semenza GL. Hypoxia-inducible factor 1: oxygen homeostasis and disease pathophysiology. Trends Mol Med 2001; 7:345–350.

7. Wenger RH. Cellular adaptation to hypoxia: O_2-sensing protein hydroxylases, hypoxia-inducible transcription factors, and O_2-regulated gene expression. FASEB J 2002; 16:1151–1162.

8. Pugh CW, Ratcliffe PJ. Regulation of angiogenesis by hypoxia: role of the HIF system. Nat Med 2003; 9:677–684.

9. Semenza GL. Targeting HIF-1 for cancer therapy. Nat Rev Cancer 2003; 3:721–732.

10. Huang LE, Bunn HF. Hypoxia-inducible Factor and Its Biomedical Relevance. J Biol Chem 2003; 278:19575–19578.

11. Ema M, Taya S, Yokotani N, et al. A novel bHLH-PAS factor with close sequence similarity to hypoxia-inducible factor 1 alpha regulates the VEGF expression and is potentially involved in lung and vascular development. Proc Natl Acad Sci USA 1997; 94:4273–4278.

12. Hogenesch JB, Chan WK, Jackiw VH, et al. Characterization of a subset of the basic-helix-loop-helix-PAS superfamily that interacts with components of the dioxin signaling pathway. J Biol Chem 1997; 272:8581–8593.

13. Tian H, McKnight SL, Russell DW. Endothelial PAS domain protein 1 (EPAS1), a transcription factor selectively expressed in endothelial cells. Genes Dev 1997; 11:72–82.

14. Flamme I, Frohlich T, von Reutern M, et al. HRF, a putative basic helix-loop-helix-PAS-domain transcription factor is closely related to hypoxia-inducible factor-1α and developmentally expressed in blood vessels. Mech Develop 1997; 63:51–60.

15. Gu YZ, Moran SM, Hogenesch JB, et al. Molecular characterization and chromosomal localization of a third alpha-class hypoxia inducible factor subunit, HIF-3alpha. Gene Expression 1998; 7:205–213.

16. Makino Y, Cao R, Svensson K, et al. Inhibitory PAS domain protein is a negative regulator of hypoxia-inducible gene expression. Nature 2001; 414:550–554.

17. Pugh CW, O'Rourke JF, Nagao M, et al. Activation of hypoxia-inducible factor-1; Definition of regulatory domains within the alpha subunit. J Biol Chem 1997; 272:11205–11214.

18. Jiang BH, Zheng JZ, Leung SW, et al. Transactivation and inhibitory domains of hypoxia-inducible factor 1 alpha. Modulation of transcriptional activity by oxygen tension. J Biol Chem 1997; 272:19253–19260.

19. Huang LE, Gu J, Schau M, et al. Regulation of hypoxia-inducible factor 1alpha is mediated by an O2-dependent degradation domain via the ubiquitin-proteasome pathway. Proc Natl Acad Sci USA 1998; 95:7987–7992.

20. Maxwell PH, Wiesener MS, Chang GW, et al. The tumour suppressor protein VHL targets hypoxia-inducible factors for oxygen-dependent proteolysis. Nature 1999; 399:271–275.

21. Ohh M, Park CW, Ivan M, et al. Ubiquitination of hypoxia-inducible factor requires direct binding to the [bgr]-domain of the von Hippel-Lindau protein. Nat Cell Biol 2000; 2:423–427.

22. Lando D, Peet DJ, Whelan DA, Gorman JJ, Whitelaw ML. Asparagine hydroxylation of the HIF transactivation domain: A hypoxic switch. Science 2002; 295:858–861.

23. Chun YS, Choi E, Yeo EJ, et al. A new HIF-1 alpha variant induced by zinc ion suppresses HIF-1-mediated hypoxic responses. J Cell Sci 2001; 114:4051–4061.

24. Thrash-Bingham CA, Tartof KD. aHIF: a natural antisense transcript overexpressed in human renal cancer and during hypoxia. J Natl Cancer Inst 1999; 91:143–151.

25. Chun YS, Choi E, Kim TY, et al. A dominant-negative isoform lacking exons 11 and 12 of the human hypoxia-inducible factor-1alpha gene. Biochem J 2002; 362:71–79.

26. Semenza GL. HIF-1 and mechanisms of hypoxia sensing. Curr Opin Cell Biol 2001; 13:167–171.

27. Erbel PJA, Card PB, Karakuzu O, et al. Structural basis for PAS domain heterodimerization in the basic helix-loop-helix-PAS transcription factor hypoxia-inducible factor. Proc Natl Acad Sci USA 2003; 100:15504–15509.

28. Card PB, Erbel PJA, Gardner KH. Structural basis of ARNT PAS-B dimerization: Use of a common beta-sheet interface for hetero- and homodimerization. J Mol Bio 2005; 353:664–677.

29. Schofield CJ, Ratcliffe PJ. Oxygen sensing by HIF hydroxylases. Nat Rev Mol Cell Biol 2004; 5:343–354.

30. Schofield CJ, Ratcliffe PJ. Signalling hypoxia by HIF hydroxylases. Biochem Bioph Res Co 2005; 338:617–626.

31. Kaelin WG. Proline hydroxylation and gene expression. Annu Rev Biochem 2005; 74:115–128.

32. Dann III CE, Bruick RK. Dioxygenases as O_2-dependent regulators of the hypoxic response pathway. Biochem Bioph Res Commun 2005; 338:639–647.

33. Yu F, White SB, Zhao Q, et al. HIF-1alpha binding to VHL is regulated by stimulus-sensitive proline hydroxylation. Proc Natl Acad Sci USA 2001; 98:9630–9635.

34. Ivan M, Kondo K, Yang H, et al. HIFα Targeted for VHL-Mediated Destruction by Proline Hydroxylation: Implications for O_2 Sensing. Science 2001; 292:464–468.

35. Jaakkola P, Mole DR, Tian YM, et al. Targeting of HIF-alpha to the von Hippel-Lindau ubiquitylation complex by O_2-regulated prolyl hydroxylation. Science 2001; 292:468–472.

36. Masson N, William C, Maxwell PH, et al. Independent function of two destruction domains in hypoxia-inducible factor-alpha chains activated by prolyl hydroxylation. EMBO 2001; 20:5197–5206.

37. Hon WC, Wilson MI, Harlos K, et al. Structural basis for the recognition of hydroxyproline in HIF-1α by pVHL. Nature 2002; 417:975–978.

38. Min JH, Yang H, Ivan M, et al. Structure of an HIF-1alpha - pVHL complex: Hydroxyproline recognition in signaling. Science 2002; 296:1886–1889.

39. Kaelin WG. The von Hippel-Lindau tumor suppressor protein: Roles in cancer and oxygen sensing. Cold Spring Harb Sym 2005; 70:159–166.

40. Semenza GL. VHL and p53: Tumor suppressors team up to prevent cancer. Mol Cell 2006; 22:437–439.

41. Haase VH. The VHL//HIF oxygen-sensing pathway and its relevance to kidney disease. Kidney Int 2006; 69:1302–1307.

42. Kaelin J. The von Hippel-Lindau protein, HIF hydroxylation, and oxygen sensing. Biochem Bioph Res Commun 2005; 338: 627–638.

43. Kaelin WG. Molecular basis of the VHL hereditary cancer syndrome. Nat Rev Cancer 2002; 2:673–682.

44. Epstein ACR, Gleadle JM, McNeill LA, et al. C. elegans EGL-9 and mammalian homologs define a family of dioxygenases that regulate HIF by prolyl hydroxylation. Cell 2001; 107:43–54.

45. Bruick RK, McKnight SL. A conserved family of prolyl-4-hydroxylases that modify HIF. Science 2001; 294:1337–1340.

46. Elvidge GP, Glenny L, Appelhoff RJ, et al. Concordant regulation of gene expression by hypoxia and 2-oxoglutarate-dependent dioxygenase inhibition: the role of HIF-1α, HIF-2α, and other pathways. J Biol Chem 2006; 281:15215–15226.

47. Hirsila M, Koivunen P, Gunzler V, et al. Characterization of the human prolyl 4-hydroxylases that modify the hypoxia-inducible factor. J Biol Chem 2003; 278:30772–30780.

48. Schofield CJ, Zhang Z. Structural and mechanistic studies on 2-oxoglutarate-dependent oxygenases and related enzymes. Curr Opin Struc Biol 1999; 9:722–731.

49. Clifton IJ, McDonough MA, Ehrismann D, et al. Structural studies on 2-oxoglutarate oxygenases and related double-stranded β-helix fold proteins. J Inorg Biochem 2006; 100:644–669.

50. Costas M, Mehn MP, Jensen MP, et al. Dioxygen activation at mononuclear nonheme iron active sites: Enzymes, models, and intermediates. Chem Rev 2004; 104:939–986.

51. Hausinger RP. Fe(II)/{alpha}-ketoglutarate-dependent hydroxy-lases and related enzymes. Crit Rev Biochem Mol Biol 2004; 39:21–68.

52. Hegg EL, Que Jr L. The 2-His-1-carboxylate facial triad - An emerging structural motif in mononuclear non-heme iron(II) enzymes. Eur J Biochem 1997; 250:625–629.

53. Ryle MJ, Hausinger RP. Non-heme iron oxygenases. Curr Opin Chem Biol 2002; 6:193–201.

54. Berra E, Benizri E, Ginouves A, et al. HIF prolyl-hydroxylase 2 is the key oxygen sensor setting low steady-state levels of HIF-1α in normoxia. EMBO J 2003; 22:4082–4090.

55. McDonough MA, Li V, Flashman E, et al. Cellular oxygen sensing: Crystal structure of hypoxia-inducible factor prolyl hydroxylase (PHD2). Proc Natl Acad Sci USA 2006; 103:9814–9819.

56. Jiang BH, Semenza GL, Bauer C, et al. Hypoxia-inducible factor 1 levels vary exponentially over a physiologically relevant range of O_2 tension. Am J Physiol Cell Physiol 1996; 271: C1172–C1180.

57. Ehrismann D, Flashman E, Genn DN, Mathioudakis N, et al. Studies on the activity of the hypoxia-inducible-factor hydroxylases using an oxygen consumption assay. Biochem J 2007; 401:227–234.

58. Koivunen P, Hirsila M, Kivirikko KI, et al. The length of peptide substrates has a marked effect on hydroxylation by the hypoxia-inducible factor prolyl 4-hydroxylases. J Biol Chem 2006; 281:28712–28720.

59. Percy MJ, Zhao Q, Flores A, et al. A family with erythrocytosis establishes a role for prolyl hydroxylase domain protein 2 in oxygen homeostasis. Proc Natl Acad Sci USA 2006; 103:654–659.

60. McNeill LA, Hewitson KS, Claridge TD, et al. Hypoxia-inducible factor asparaginyl hydroxylase (FIH-1) catalyses hydroxylation at the beta-carbon of asparagine-803. Biochem J 2002; 367:571–575.

61. Hewitson KS, McNeill LA, Riordan MV, et al. Hypoxia-inducible factor (HIF) asparagine hydroxylase is identical to factor inhibiting HIF (FIH) and is related to the cupin structural family. J Biol Chem 2002; 277:26351–26355.

62. Freedman SJ, Sun ZY, Poy F, et al. Structural basis for recruitment of CBP/p300 by hypoxia-inducible factor-1alpha. Proc Natl Acad Sci USA 2002; 99:5367–5372.

63. Dames SA, Martinez-Yamout M, De Guzman RN, et al. Structural basis for Hif-1alpha /CBP recognition in the cellular hypoxic response. Proc Natl Acad Sci USA 2002; 99: 5271–5276.

64. Mahon PC, Hirota K, Semenza GL. FIH-1: a novel protein that interacts with HIF-1alpha and VHL to mediate repression of HIF-1 transcriptional activity. Genes Dev 2001; 15:2675–2686.

65. Linke S, Stojkoski C, Kewley RJ, et al. Substrate requirements of the oxygen-sensing asparaginyl hydroxylase factor-inhibiting hypoxia-inducible factor. J Biol Chem 2004; 279:14391–14397.

66. Huang J, Zhao Q, Mooney SM, et al. Sequence determinants in hypoxia-inducible factor-1 alpha for hydroxylation by the prolyl hydroxylases PHD1, PHD2, and PHD3. J Biol Chem 2002; 277:39792–39800.

67. Li D, Hirsila M, Koivunen P, et al. Many amino acid substitutions in a hypoxia-inducible transcription factor (HIF)-1{alpha}-like peptide cause only minor changes in its hydroxylation by the HIF prolyl 4-hydroxylases: Substitution of 3,4-dehydroproline or azetidine-2-carboxylic acid for the proline leads to a high rate of uncoupled 2-oxoglutarate decarboxylation. J Biol Chem 2004; 279:55051–55059.

68. Dann CE, III, Bruick RK, Deisenhofer J. Structure of factor-inhibiting hypoxia-inducible factor 1: An asparaginyl hydroxylase involved in the hypoxic response pathway. Proc Natl Acad Sci USA 2002; 99:15351–15356.

69. Hewitson KS, Lienard BMR, McDonough MA, et al. Structural and mechanistic studies on the inhibition of the HIF hydroxylases by tricarboxylic acid cycle intermediates. J Biol Chem 2006; in press.

70. Chen Z, Zang J, Whetstine J, et al. Structural insights into histone demethylation by JMJD2 family members. Cell 2006; 125: 691–702.

71. Trewick SC, McLaughlin PJ, Allshire RC. Methylation: lost in hydroxylation? EMBO Rep 2005; 6:315–320.

72. Elkins JM, Hewitson KS, McNeill LA, et al. Structure of factor-inhibiting hypoxia-inducible factor (HIF) reveals mechanism of oxidative modification of HIF-1 alpha. J Biol Chem 2003; 278:1802–1806.

73. Lee C, Kim SJ, Jeong DG, et al. Structure of human FIH-1 reveals a unique active site pocket and interaction sites for HIF-1 and von Hippel-Lindau. J Biol Chem 2003; 278:7558–7563.

74. Cockman ME, Lancaster DE, Stolze IP, et al. Posttranslational hydroxylation of ankyrin repeats in I{kappa}B proteins by the hypoxia-inducible factor (HIF) asparaginyl hydroxylase, fac-

tor inhibiting HIF (FIH). Proc Natl Acad Sci USA 2006; 103: 14767–14772.

75. Kuznetsova AV, Meller J, Schnell PO, et al. von Hippel-Lindau protein binds hyperphosphorylated large subunit of RNA polymerase II through a proline hydroxylation motif and targets it for ubiquitination. Proc Natl Acad Sci USA 2003; 100:2706–2711.

76. Cummins EP, Berra E, Comerford KM, et al. Prolyl hydroxylase-1 negatively regulates IκB kinase-beta, giving insight into hypoxia-induced NFκB activity. Proc Natl Acad Sci USA 2006; 103:18154–18159.

77. Falnes PO, Johansen RF, Seeberg E. AlkB-mediated oxidative demethylation reverses DNA damage in Escherichia coli. Nature 2002; 419:178–182.

78. Trewick SC, Henshaw TF, Hausinger RP, et al. Oxidative demethylation by Escherichia coli AlkB directly reverts DNA base damage. Nature 2002; 419:174–178.

79. McNeill LA, Flashman E, Buck MRG, et al. Hypoxia-inducible factor prolyl hydroxylase 2 has a high affinity for ferrous iron and 2-oxoglutarate. Mol Biosyst 2005; 1:321–324.

80. Prockop JD. Collagens: Molecular biology, diseases, and potentials for therapy. Annu Rev Biochem 1995; 64:403–434.

81. King A, Selak MA, Gottlieb E. Succinate dehydrogenase and fumarate hydratase: linking mitochondrial dysfunction and cancer. Oncogene; 25:4675–4682.

82. Koivunen P, Hirsila M, Remes AM, et bal. Inhibition of HIF hydroxylases by citric acid cycle intermediates: Possible links between cell metabolism and stabilization of HIF. J Biol Chem 2006; in press.

83. Metzen E, Zhou J, Jelkmann W, et al. Nitric oxide impairs normoxic degradation of HIF-1α by inhibition of prolyl hydroxylases. Mol Biol Cell 2003; 14:3470–3481.

84. Kimura H, Weisz A, Kurashima Y, et al. Hypoxia response element of the human vascular endothelial growth factor gene mediates transcriptional regulation by nitric oxide: control of hypoxia-inducible factor-1 activity by nitric oxide. Blood 2000; 95:189–197.

85. Wang F, Sekine H, Kikuchi Y, et al. HIF-1α-prolyl hydroxylase: molecular target of nitric oxide in the hypoxic signal transduction pathway. Biochem Bioph Res Commun 2002; 295: 657–662.

86. Agani FH, Puchowicz M, Chavez JC, et al. Role of nitric oxide in the regulation of HIF-1alpha expression during hypoxia. Am J Physiol Cell Physiol 2002; 283:C178–C186.

87. Sogawa K, Numayama-Tsuruta K, Ema M, et al. Inhibition of hypoxia-inducible factor 1 activity by nitric oxide donors in hypoxia. Proc Natl Acad Sci USA 1998; 95:7368–7373.

88. Huang LE, Willmore WG, Gu J, et al. Inhibition of Hypoxia-inducible Factor 1 Activation by Carbon Monoxide and Nitric Oxide. Implications for oxygen sensing and signaling. J Biol Chem 1999; 274:9038–9044.

89. Brune B, von Knethen A, Sandau KB. Transcription factors p53 and HIF-1α as targets of nitric oxide. Cell Signal 2001; 13:525–533.

90. Brune B, Zhou J. The role of nitric oxide (NO) in stability regulation of hypoxia inducible factor α (HIF-1α). Curr Med Chem 2003; 10:845–855.

91. Zhou J, Schmid T, Brune B. HIF-1α and p53 as targets of NO in affecting cell proliferation, death and adaptation. Curr Mol Med 2004; 4:741–751.

92. Sansal I, Sellers WR. The biology and clinical relevance of the PTEN tumor suppressor pathway. J Clin Oncol 2004; 22: 2954–2963.

93. Vivanco I, Sawyers CL. The phosphatidylinositol 3-kinase-AKT pathway in human cancer. Nat Rev Cancer 2002; 2:489–501.

94. Vara JAF, Casado E, de Castro J, et al. PI3K/Akt signaling pathway and cancer. Cancer Treat Rev 2004; 30:193–204.

95. Jeong JW, Bae MK, Ahn MY, et al. Regulation and destabilization of HIF-1α by ARD1-Mediated Acetylation. Cell 2002; 111:709–720.

96. Murray-Rust TA, Oldham NJ, Hewitson KS, et al. Purified recombinant hARD1 does not catalyse acetylation of Lys532 of HIF-1[alpha] fragments in vitro. FEBS Lett 2006; 580: 1911–1918.

97. Bilton R, Mazure N, Trottier E, et al. Arrest-defective-1 protein, an acetyltransferase, does not alter stability of hypoxia-inducible factor (HIF)-1α and is not induced by hypoxia or HIF. J Biol Chem 2005; 280:31132–31140.

98. Fisher TS, Des Etages S, Hayes L, et al. Analysis of ARD1 function in hypoxia response using retroviral RNA interference. J Biol Chem 2005; 280:17749–17757.

99. Ravi R, Mookerjee B, Bhujwalla ZM, Sutter CH, Artemov D, Zeng Q et al. Regulation of tumor angiogenesis by p53-induced degradation of hypoxia-inducible factor 1alpha. Genes Dev 2000; 14:34–44.

100. Hansson LO, Friedler A, Freund S, et al. Two sequence motifs from HIF-1alpha bind to the DNA-binding site of p53. Proc Natl Acad Sci USA 2002; 99:10305–10309.

101. Hewitson KS, Schofield CJ. The HIF pathway as a therapeutic target. Drug Disc Today 2004; 9:704–711.

102. Giaccia A, Siim BG, Johnson RS. HIF-1 as a target for drug development. Nat Rev Drug Discov 2003; 2:803–811.

103. Paul SAM, Simons JW, Mabjeesh NJ. HIF at the crossroads between ischemia and carcinogenesis. J Cell Physiol 2004; 200:20–30.

104. Safran M, Kaelin WG, Jr. HIF hydroxylation and the mammalian oxygen-sensing pathway. J Clin Invest 2003; 111: 779–783.

105. Hockel M, Vaupel P. Tumor hypoxia: Definitions and current clinical, biologic, and molecular aspects. J Natl Cancer Inst 2001; 93:266–276.

106. Brizel DM, Dodge RK, Clough RW, et al. Oxygenation of head and neck cancer: changes during radiotherapy and impact on treatment outcome. Radiother Oncol 1999; 53:113–117.

107. Brown JM, Le QT. Tumor hypoxia is important in radiotherapy, but how should we measure it? Int J Radiat Oncol 2002; 54:1299–1301.

108. Harrison LB, Chadha M, Hill RJ, et al. Impact of tumor hypoxia and anemia on radiation therapy outcomes. Oncologist 2002; 7:492–508.

109. Janssen HLK, Haustermans KMG, Sprong D, et al. HIF-1α, pimonidazole, and iododeoxyuridine to estimate hypoxia and perfusion in human head-and-neck tumors. Int J Radiat Oncol 2002; 54:1537–1549.

110. Brown JM, Giaccia AJ. The unique physiology of solid tumors: Opportunities (and problems) for cancer therapy. Cancer Res 1998; 58:1408–1416.

111. Blouw B, Song H, Tihan T, et al. The hypoxic response of tumors is dependent on their microenvironment. Cancer Cell 2003; 4:133–146.

112. Carmeliet P, Dor Y, Herbert JM, et al. Role of HIF-1α in hypoxia-mediated apoptosis, cell proliferation and tumour angiogenesis. Nature 1998; 394:485–490.

113. Mack FA, Rathmell WK, Arsham AM, et al. Loss of pVHL is sufficient to cause HIF dysregulation in primary cells but does not promote tumor growth. Cancer Cell 2003; 3:75–88.

114. Acker T, Diez-Juan A, Aragones J, et al. Genetic evidence for a tumor suppressor role of HIF-2α. Cancer Cell 2005; 8: 131–141.

115. Willam C, Masson N, Tian YM, et al. Peptide blockade of HIF alpha degradation modulates cellular metabolism and angiogenesis. Proc Natl Acad Sci USA 2002; 99:10423–10428.

116. Li J, Post M, Volk R, et al. PR39, a peptide regulator of angiogenesis. Nat Med 2000; 6:49–55.

117. Sowter HM, Raval R, Moore J, et al. Predominant role of hypoxia-inducible transcription factor (HIF)-1α versus HIF-2α in regulation of the transcriptional response to hypoxia. Cancer Res 2003; 63:6130–6134.

118. Wiesener MS, Jurgensen JS, Rosenberger C, et al. Widespread, hypoxia-inducible expression of HIF-2 alpha in distinct cell populations of different organs. FASEB J 2002; 17:271–273.

119. Cejudo-Martin P, Johnson RS. A new notch in the HIF belt: How hypoxia impacts differentiation. Dev Cell 2005; 9: 575–576.

120. Kotch LE, Iyer NV, Laughner E, et al. Defective vascularization of HIF-1a-null embryos is not associated with VEGF deficiency but with mesenchymal cell death. Dev Biol 1999; 209: 254–267.

121. Tian H, Hammer RE, Matsumoto AM, et al. The hypoxia-responsive transcription factor EPAS1 is essential for catecholamine homeostasis and protection against heart failure during embryonic development. Genes Dev 1998; 12:3320–3324.

122. Adelman DM, Gertsenstein M, Nagy A, et al. Placental cell fates are regulated in vivo by HIF-mediated hypoxia responses. Genes Dev 2000; 14:3191–3203.

123. Roccaro AM, Hideshima T, Richardson PG, et al. Bortezomib as an antitumor agent. Curr Pharm Biotechno 2006; 7:441–448.

124. Knowles HJ, Raval RR, Harris AL, et al. Effect of ascorbate on the activity of hypoxia-inducible factor in cancer cells. Cancer Res 2003; 63:1764–1768.

125. Goldberg MA, Dunning SP, Bunn HF. Regulation of the erythropoietin gene: evidence that the oxygen sensor is a heme protein. Science 1988; 242:1412–1415.

126. Wang GL, Semenza GL. Desferrioxamine induces erythropoietin gene expression and hypoxia- inducible factor 1 DNA-binding activity: implications for models of hypoxia signal transduction. Blood 1993; 82:3610–3615.

127. Ran R, Xu H, Lu A, et al. Hypoxia preconditioning in the brain. Dev Neurosci Basel 2005; 27:87–92.

128. Richardson DR. Molecular mechanisms of iron uptake by cells and the use of iron chelators for the treatment of cancer. Curr Med Chem 2005; 12:2711–2729.

129. McDonough MA, McNeill LA, Tilliet M, et al. Selective inhibition of factor inhibiting hypoxia-inducible factor. J Am Chem Soc 2005; 127:7680–7681.

130. Mole DR, Schlemminger I, McNeill LA, et al. 2-Oxoglutarate analogue inhibitors of HIF prolyl hydroxylase. Bioorgan Med Chem Lett 2003; 13:2677–2680.

131. Schlemminger I, Mole DR, McNeill LA, et al. Analogues of dealanylalahopcin are inhibitors of human HIF prolyl hydroxylases. Bioorgan Med Chem Lett 2003; 13:1451–1454.

132. Welford RWD, Schlemminger I, McNeill LA, et al. The selectivity and inhibition of AlkB. J Biol Chem 2003; 278: 10157–10161.

133. Ivan M, Haberberger T, Gervasi DC, et al. Biochemical purification and pharmacological inhibition of a mammalian prolyl hydroxylase acting on hypoxia-inducible factor. Proc Natl Acad Sci USA 2002; 99:13459–13464.

134. Nwogu JI, Geenen D, Bean M, et al. Inhibition of collagen synthesis with prolyl 4-hydroxylase inhibitor improves left ventricular function and alters the pattern of left ventricular dilatation after myocardial infarction. Circulation 2001; 104:2216–2221.

135. Banerji B, Conejo-Garcia A, McNeill LA, et al. The inhibition of factor inhibiting hypoxia-inducible factor (FIH) by β-oxocarboxylic acids. Chem Commun 2005; 5438–5440.

136. Warshakoon NC, Wu S, Boyer A, et al. A novel series of imidazo[1,2-a]pyridine derivatives as HIF-1α prolyl hydroxylase inhibitors. Bioorgan Med Chem Lett 2006; 16:5598–5601.

137. Warshakoon NC, Wu S, Boyer A, et al. Structure-based design, synthesis, and SAR evaluation of a new series of 8-hydroxy-quinolines as HIF-1α prolyl hydroxylase inhibitors. Bioorgan Med Chem Lett 2006; 16:5517–5522.

138. Warshakoon NC, Wu S, Boyer A, et al. Design and synthesis of substituted pyridine derivatives as HIF-1α prolyl hydroxylase inhibitors. Bioorgan Med Chem Lett 2006; 16:5616–5620.

139. Warshakoon NC, Wu S, Boyer A, et al. Design and synthesis of a series of novel pyrazolopyridines as HIF 1-α prolyl hydroxylase inhibitors. Bioorgan Med Chem Lett 2006; 16:5687–5690.

140. Klaus S, Arend M, Fourney P, et al. Induction of erythropoiesis and iron utilization by the HIF prolyl hydroxylase inhibitor FG-4592. American Society of Nephrology (ASN) Renal Week 2005 Session: Complications of ESRD: Bone Disease, Malnutrition, and Anemia, Abstract F-FC050. 11-11–2005.

141. Gunzler V, Muthukrishnan E, Neumayer HH, et al. FG-2216 increases hemoglobin concentration in anemic patients with chronic kidney disease. American Society of Nephrology (ASN) Renal Week 2005 Session: Anemia and Bone Disease, Abstract SA-PO924. 11-12–2005.

142. Kung AL, Zabludoff SD, France DS, et al. Small molecule blockade of transcriptional coactivation of the hypoxia-inducible factor pathway. Cancer Cell 2004; 6:33–43.

143. Olenyuk BZ, Zhang GJ, Klco JM, et al. Inhibition of vascular endothelial growth factor with a sequence-specific hypoxia response element antagonist. Proc Natl Acad Sci USA 2004; 101:16768–16773.

144. Kong D, Park EJ, Stephen AG, Calvani M, Cardellina JH, Monks A et al. Echinomycin, a small-molecule inhibitor of hypoxia-inducible factor-1 DNA-binding activity. Cancer Res 2005; 65:9047–9055.

145. Semenza GL. Development of novel therapeutic strategies that target HIF-1. Expert Opin Ther Tar 2006; 10:267–280.

146. William JG, Nicholas JV, Lary JK, et al. A phase II clinical trial of echinomycin in metastatic soft tissue sarcoma. Invest New Drug 1995; 13:171–174.

147. Jones DT, Harris AL. Identification of novel small-molecule inhibitors of hypoxia-inducible factor-1 transactivation and DNA binding. Mol Cancer Ther 2006; 5:2193–2202.

148. Yang J, Zhang L, Erbel PJA, et al. Functions of the Per/ARNT/Sim domains of the hypoxia-inducible factor. J Biol Chem 2005; 280:36047–36054.

149. Amezcua CA, Harper SM, Rutter J, et al. Structure and interactions of PAS kinase N-terminal PAS domain: Model for intramolecular kinase regulation. Structure 2002; 10:1349–1361.

150. Isaacs JS, Jung YJ, Mimnaugh EG, et al. Hsp90 regulates a von Hippel Lindau-independent hypoxia-inducible factor-1alpha - degradative pathway. J Biol Chem 2002; 277:29936–29944.

151. Neckers L, Neckers K. Heat shock protein 90 inhibitors as novel cancer chemotherapeutic agents. Exp Opin Emerg Drugs 2002; 7:277–288.

152. Masson N, Appelhoff RJ, Tuckerman JR, et al. The HIF prolyl hydroxylase PHD3 is a potential substrate of the TRiC chaperonin. FEBS Lett 2004; 570:166–170.

153. Rapisarda A, Uranchimeg B, Sordet O, et al. Topoisomerase I-mediated inhibition of hypoxia-inducible factor 1: Mechanism and therapeutic implications. Cancer Res 2004; 64: 1475–1482.

154. Li TK, Liu LF. Tumor cell death induced by topoisomerase-targeting drugs. Annu Rev Pharmacol 2001; 41(1):53–77.

155. Chang H, Shyu KG, Lee CC, et al. GL331 inhibits HIF-1α expression in a lung cancer model. Biochem Bioph Res Co 2003; 302:95–100.

156. Powis G, Kirkpatrick L. Hypoxia inducible factor-1{alpha} as a cancer drug target. Mol Cancer Ther 2004; 3(5):647–654.

157. Mabjeesh NJ, Escuin D, LaVallee TM, et al. 2ME2 inhibits tumor growth and angiogenesis by disrupting microtubules and dysregulating HIF. Cancer Cell 2003; 3:363–375.

158. William DF, Erwin AK, Douglas KP, et al. Inhibition of angiogenesis: Treatment options for patients with metastatic prostate cancer. Invest New Drug 2002; 20:183–194.

159. Maxwell PH. The HIF pathway in cancer. Semin Cell Dev Biol 2005; 16:523–530.

160. Diaz-Gonzalez JA, Russell J, Rouzaut A, et al. Targeting hypoxia and angiogenesis through HIF-1alpha inhibition. Cancer Biol Ther 2005; 4:1055–1062.

161. Zhong H, Chiles K, Feldser D, et al. Modulation of hypoxia-inducible factor 1α expression by the epidermal growth factor/phosphatidylinositol 3-kinase/PTEN/AKT/FRAP pathway in human prostate cancer cells: Implications for tumor angiogenesis and therapeutics. Cancer Res 2000; 60:1541–1545.

162. Chen C, Pore N, Behrooz A, et al. Regulation of glut1 mRNA by hypoxia-inducible Factor-1. Interaction between H-ras and hypoxia. J Biol Chem 2001; 276:9519–9525.

163. Powis G, Montfort WR. Properties and biological activities of thioredoxins. Annu Rev Pharmacol 2001; 41(1):261–295.

164. Berggren M, Gallegos A, Gasdaska JR, et al. Thioredoxin and thioredoxin reductase gene expression in human tumors and cell lines, and the effects of serum stimulation and hypoxia. Anticancer Research 1996; 16:3459–3466.

165. Huang LE, Arany Z, Livingston DM, et al. Activation of hypoxia-inducible transcription factor depends primarily upon redox-sensitive stabilization of its alpha subunit. J Biol Chem 1996; 271:32253–32259.

166. Welsh SJ, Bellamy WT, Briehl MM, et al. The redox protein thioredoxin-1 (Trx-1) increases hypoxia-inducible factor 1α protein expression: Trx-1 overexpression results in increased vascular endothelial growth factor production and enhanced tumor angiogenesis. Cancer Res 2002; 62:5089–5095.

167. Kirkpatrick DL, Kuperus M, Dowdeswell M, et al. Mechanisms of inhibition of the thioredoxin growth factor system by antitumor 2-imidazolyl disulfides. Biochem Pharmacol 1998; 55(7):987–994.

168. Kirkpatrick L, Dragovich T, Ramanathan R, et al. Results from Phase I study of PX-12, a thioredoxin inhibitor in patients with advanced solid malignancies. J Clin Oncol (Meeting Abstracts) 2004; 22:3089.

169. Jones DT, Pugh CW, Wigfield S, et al. Novel thioredoxin inhibitors paradoxically increase hypoxia-inducible factor-α expression but decrease functional transcriptional activity, DNA binding, and degradation. Clin Cancer Res 2006; 12:5384–5394.

170. Ko FN, Wu CC, Kuo SC, et al. YC-1, a novel activator of platelet guanylate cyclase. Blood 1994; 84:4226–4233.

171. Chun YS, Yeo EJ, Choi E, et al. Inhibitory effect of YC-1 on the hypoxic induction of erythropoietin and vascular endothelial growth factor in Hep3B cells. Biochem Pharmacol 2001; 61(8):947–954.

172. Yeo EJ, Chun YS, Cho YS, et al. YC-1: A potential anticancer drug targeting hypoxia-inducible factor 1. J Natl Cancer Inst 2003; 95:516–525.

173. Belozerov VE, Van Meir EG. Hypoxia inducible factor-1: a novel target for cancer therapy. Anti-Cancer Drugs 2005; 16:901–909.

174. Vincent KA, Shyu KG, Luo Y, et al. Angiogenesis is induced in a rabbit model of hindlimb ischemia by naked DNA encoding an HIF-1α/VP16 hybrid transcription factor. Circulation 2000; 102:2255–2261.

175. Shyu KG, Wang MT, Wang BW, et al. Intramyocardial injection of naked DNA encoding HIF-1α/VP16 hybrid to enhance angiogenesis in an acute myocardial infarction model in the rat. Cardiovasc Res 2002; 54:576–583.

176. Sun X, Liu M, Wei Y, et al. Overexpression of von Hippel-Lindau tumor suppressor protein and antisense HIF-1α eradicates gliomas. Cancer Gene Ther 2005; 13:428–435.

177. Sun X, Kanwar JR, Leung E, et al. Gene transfer of antisense hypoxia inducible factor-1α enhances the therapeutic efficacy of cancer immunotherapy. Nature Gene Therapy 2001; 8: 638–645.

Chapter 33
The Clinical Utility of Bevacizumab

Jeanny B. Aragon-Ching[1], Ravi A. Madan[2], and James L. Gulley[2]

Keywords: Clinical Trials, FDA approval, Survival benefit, Toxicities, Colorectal Cancer, Lung Cancer, Breast Cancer, Prostate Cancer, Ovarian Cancer, Renal Cancer

Abstract: Targeting angiogenesis is a rapidly emerging field of cancer research. Bevacizumab is a humanized monoclonal antibody against vascular endothelial growth factor (VEGF) that is at the forefront of clinical investigations and is FDA-approved for several neoplasms, including advanced colon, lung, and renal cancers. This chapter discusses key clinical trials that led to the approval of bevacizumab, its promising use in a variety of other cancers such as ovarian, prostate and breast cancer, as well as safety and toxicity monitoring in the use of bevacizumab.

Introduction

Bevacizumab (Avastin®; Genentech, Inc., South San Francisco, CA), a humanized monoclonal antibody IgG1, was developed from a murine antihuman vascular endothelial growth factor (VEGF) monoclonal antibody. It is composed predominantly of human protein sequences (93%) and a small murine protein sequence (7%) in the complementarity-determining region [1–3]. Bevacizumab specifically recognizes and functions as an inhibitor of all major isoforms of human VEGF-A, with biological and pharmacokinetic properties similar to its parent murine antibody, except for its longer terminal elimination half-life [4]. The promising neutralizing effects on several cancer cell lines in vivo [1, 5, 6] by a mouse monoclonal antibody targeting VEGF was first demonstrated in 1993, leading to the subsequent development of bevacizumab. Since then, bevacizumab in combination with chemotherapy has been shown to have clinical activity in a variety of cancers. It was approved by the Food and Drug Administration (FDA) in 2004 [7, 8]. This chapter reviews the major clinical trials evaluating the use of bevacizumab.

Mechanism of Action

Numerous studies have shown that angiogenesis plays a significant role in the progression, growth, and metastases of various tumors [9–13]. Bevacizumab has been shown to inhibit the VEGF pathway in various in vivo and in vitro models [6, 14, 15]. The drug's effects are manifested in endothelial cells and tumor vasculature, where it causes regression of existing vessels, [16] inhibition of new vessel formation, and normalization of tumor vasculature [17, 18]. Normalizing tumor vasculature may reduce the risk of tumor cells permeating vessel walls, and thus reduce the risk of metastasis [19]. Abnormal tumor microvasculature may lead to heterogeneous areas of high vessel density, large-diameter vessels, and leaky vessels, each of which can steal blood away from other areas of the tumor. Normalization of blood flow within the tumor may also result from decreased intratumoral pressure or interstitial fluid pressure following administration of bevacizumab, allowing for more efficient delivery of cytotoxic chemotherapy drugs administered in conjunction with bevacizumab [14, 17, 20–22]. Improved blood flow brought about by bevacizumab may also enhance the antitumor effects of other therapeutic modalities, such as radiation (improved oxygen delivery to tumor microenvironment), targeted small molecule therapy, and immunotherapy (more oxygen and nutrients in the tumor microenvironment, which immune cells need in order to function).

Early Clinical Trials

Preclinical models used trough serum concentrations of 10–30 µg/ml to achieve maximal tumor growth inhibition [7]. Based on these early trials, a phase I dose-finding clinical

[1] Medical Oncology Branch, [2] Laboratory of Tumor Immunology and Biology, National Cancer Institute, National Institutes of Health, Bethesda, MD, USA

study of bevacizumab treated 25 patients with dosages ranging from 0.1 to 10 mg/kg as 90-min intravenous infusions on days 0, 28, 35, and 42 [23]. Bevacizumab appeared to have a linear pharmacokinetic profile at doses of 0.3–10 mg/kg, with a half-life of approximately 20 days (range: 11 to 50 days). Steady-state levels were predicted to be reached within 100 days, with an accumulation ratio of 2.8 following a biweekly dose of 10 mg/kg [24]. Clearance appeared to be higher in men than in women and in patients with greater tumor burden [25]. In a phase Ib trial in 12 patients examining the safety and pharmacokinetics of weekly intravenous bevacizumab at 3 mg/kg plus with 3 different chemotherapy combinations, no apparent increase in toxicity was noted over that expected with chemotherapy alone [3].

Bevacizumab at doses of 0.3 mg/kg or more has been shown to suppress free serum VEGF levels. [23] Although different trials used variable dosing schedules, a population pharmacokinetic noncompartmental analysis that included different doses and frequencies of bevacizumab every 2 or 3 weeks showed very similar overall exposure, [26] probably because of the drug's long terminal half-life. This finding supports the use of bevacizumab at the same dosing frequency as the concomitant chemotherapy. However, a clear dose-response relationship was not consistently observed across tumor types [7]. For instance, low-dose bevacizumab (5 mg/kg) initially showed superior response rates compared to high-dose bevacizumab (10 mg/kg) in patients with metastatic colorectal cancer, [27] but the opposite was true in studies of non-small cell lung cancer [28] and renal cell cancer, [29] where higher doses corresponded to better overall response and time to progression (TTP). Escalating doses of bevacizumab were also associated with better overall response rates, but no overall survival advantage, in metastatic breast cancer. [30] These findings suggest that an optimal dosing regimen of bevacizumab may depend on the tumor type and chemotherapy combination [31].

Definitive Clinical Trials in Colorectal Cancer

Colorectal cancer is the second leading cause of cancer deaths in Western countries. In 2007 alone, it is estimated that colorectal cancer will be diagnosed in 11% (79,130) of men and 10% (74,630) of women in this country [32]. Approximately 20–50% of patients will be diagnosed with metastatic disease. Between 1990 and 2003, death rates for colon cancer declined by approximately 25.4% in men and 21% in women. Historically, best supportive care of metastatic colorectal cancer patients has yielded median survival of 5–9 months [33–35]. These numbers have steadily improved since the use of various chemotherapeutic regimens that included the fluoropyrimidine 5-fluorouracil (5-FU) as the backbone, with modulation by leucovorin (LV) [36]. and recently approved agents in the past decade, including irinotecan and oxaliplatin.

The use of bevacizumab in combination with 5-FU/LV and either irinotecan or oxaliplatin has resulted in median survival times in excess of 20 months (Table 33.1). In a phase II trial of 104 previously untreated patients with metastatic colorectal cancer, the addition of 5 mg/kg bevacizumab to 5-FU/LV every 2 weeks resulted in a median overall survival (OS) of 21.5 months (95% confidence interval, CI, 17.3 to undetermined). In comparison, a control arm receiving 5-FU/LV alone had a median OS of 13.8 months (95% CI, 9.1 to 23 months) and a third arm receiving 5-FU/LV with 10 mg/kg bevacizumab had a median OS of 16.1 months (95% CI, 11–20.7 months) [27]. The low-dose bevacizumab arm also had higher response rates and longer median TTP compared to the control or high-dose bevacizumab arm. Although it was not clear why high-dose bevacizumab was less effective than low-dose bevacizumab, it was postulated that differences in randomization resulted in more patients with poor prognostic features being placed in the high-dose arm, resulting in more favorable

TABLE 33.1. Selected bevacizumab trials in colorectal cancer leading to FDA approval.

Authors	Phase	Chemotherapy regimen	Bevacizumab dose	Randomization arm, number of patients	Outcomes
First-line metastatic					
Kabbinavar et al. [27]	II	Weekly bolus 5 FU 500 mg/m^2 and LV 500 mg/m^2 × 6 of q 8-wk cycles	5 mg/kg q 2 wks; 10 mg/kg q 2 wks	5 FU/LV alone, n=36 5 FU/LV + Bev 5 mg/kg, n=35 5 FU/LV + Bev 10 mg/kg, n=33	Improved OS for 5 FU + Bev 5 mg/kg (21.5 mos) and time to disease progression (5.2 mos) and RR (40%)
Hurwitz et al. [40]	III	Weekly irinotecan 125 mg/m^2 and 5FU 500 mg/m^2 and LV 20 mg/m2 (IFL) × 4 q 6-wk cycles; weekly 5 FU and LV: 500 mg/m^2 × 6 q 8-wk cycles	5 mg/kg q 2 wks	IFL + placebo, n=411 IFL + Bev, n = 402 Discontinued arm: 5 FU/LV + Bev, n=110	Improved OS for IFL + Bev (20.3 mos), PFS (10.6 mos), and RR (44.8%)
Second-line metastatic					
Giantonio et al. [52]	III	FOLFOX4: oxaliplatin 85 mg/m^2 D1, 5-FU 400 mg/m^2 bolus D1 and infusional 600 mg/m^2 D1 and D2, LV 200 mg/m^2 D1 and D2	10 mg/kg q 2 wks	FOLFOX4 + Bev, n=286 FOLFOX4 alone, n=291 Bevacizumab alone, n=243	Improved OS for FOLFOX4 + Bev arm (12.9 mos) and PFS (7.3 mos) and RR (22.7%)

OS Overall survival; *PFS* progression-free survival; *5 FU* 5 fluorouracil; Bev bevacizumab; *LV* leucovorin; *RR* response rate; *q 2 wks* every 2 weeks; *mos* months outcomes for patients randomized to the low-dose bevacizumab arm. This study did not definitively establish the superiority of low-dose over high-dose bevacizumab, and subsequent trials have continued to use high-dose bevacizumab in chemotherapy combination regimens. Nevertheless, the encouraging results from this phase II trial paved the way for the phase III trial that ultimately gained FDA approval in 2004 for the use of bevacizumab in combination with chemotherapy as first-line treatment for metastatic colorectal cancer in the United States.

Bevacizumab as First-line Treatment for Colorectal Cancer

At the time that these first clinical trials of bevacizumab were taking place, irinotecan with 5-FU was being studied as first-line treatment of colorectal cancer in North America, [37, 38] while oxaliplatin-containing regimens were being developed in Europe [39]. Thus, based on the benefits observed in the phase II trial described above, the same low-dose regimen of 5 mg/kg bevacizumab was employed in phase III trials, in combination with 5-FU/irinotecan. The pivotal phase III trial consisted of 813 patients with metastatic colorectal cancer [40]. The chemotherapy regimen consisted of irinotecan, 5-fluorouracil, and leucovorin (IFL), with irinotecan at 125 mg/m², 5-FU at 500 mg/m², and leucovorin at 20 mg/m², given once weekly for 4 weeks, with cycles repeating every 6 weeks. Patients were initially randomized to 3 groups: IFL with placebo ($n = 100$), IFL with bevacizumab 5 mg/kg ($n = 103$), and 5-FU/LV with bevacizumab 5 mg/kg ($n = 110$). The latter arm was felt to be necessary since the safety of the IFL regimen with bevacizumab had not been previously tested. However, when the first interim analysis showed no difference in toxicity between the 2 IFL groups, the 5-FU/LV/bevacizumab arm was halted and accrual continued with a total of 411 patients in the IFL/placebo group and 402 patients in the IFL/bevacizumab group. A statistically significant difference in the OS (primary endpoint) was observed between the IFL/bevacizumab arm (20.3 months) and the IFL/placebo arm (15.6 months; hazard ratio, HR = 0.66, $P < 0.001$) (Fig. 33.1). Furthermore, improvements in median progression-free survival (PFS) (10.6 vs 6.2 months, $P < 0.001$) and response rates (44.8% vs 34.8%, $P = 0.004$) were noted in the IFL/bevacizumab arm. These positive findings formed the basis for the first anti-VEGF targeted agent that was FDA-approved for cancer patients.

Subsequent trials with multiple chemotherapy agents in combination with bevacizumab in different populations of colorectal cancer patients showed persistent activity. A separately reported analysis of the halted 5 FU/LV/bevacizumab cohort ($n = 110$) in the previous phase III trial [40], and a comparison between this arm and the IFL/placebo arm ($n = 100$), showed median OS of 18.3 and 15.1 months, respectively [41]. The stratified HR for death in the 5-FU/LV/bevacizumab arm compared with the IFL/placebo arm was 0.82, but the 95% CI was 0.59–1.15 and did not reach statistical significance ($P = 0.25$). Nevertheless, the improved efficacy seen with bevacizumab-containing regimens offered a promising alternative

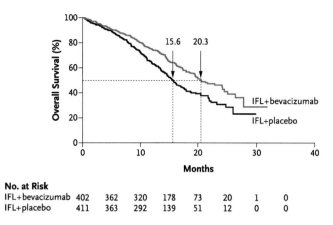

Fig. 33.1. Kaplan-Meier curve of median overall survival in a phase III trial comparing IFL/bevacizumab versus IFL/placebo as first-line therapy for metastatic colorectal cancer.
Source: Hurwitz, H., L. Fehrenbacher, et al. (2004). Bevacizumab plus irinotecan, fluorouracil, and leucovorin for metastatic colorectal cancer. N Engl J Med 350(23): 2335–42.

for patients who otherwise could not tolerate irinotecan [42]. A further analysis used primary efficacy data from 3 independent studies [27, 40, 42] that showed improvements in OS (17.9 months vs 14.6 months, $P = 0.008$), PFS (8.8 months vs 5.6 months, $P \leq 0.0001$), and response rates (34.1% vs 24.5%, $P = 0.019$) with the combination of 5-FU/LV/bevacizumab compared to either 5-FU/LV or IFL alone [43].

High-dose bevacizumab (10 mg/kg) in combination with IFL was further studied in an Eastern Cooperative Oncology Group (ECOG) single-arm phase II trial (E2200) with 81 evaluable patients [44]. Again, meaningful responses were seen, including an overall response rate of 49.4%, with 6.2% complete responses, and median OS of 26.3 months, median PFS of 10.7 months, and 1-year survival of 85%, all substantially better than historical controls.

In 2004, the North Central Cancer Treatment Group demonstrated improvements in OS in a trial that randomized 795 patients to either IFL, FOLFOX (infusional 5-FU/LV/oxaliplatin), or IROX (irinotecan/oxaliplatin) [45, 46]. Subsequent trials utilizing bevacizumab in combination with FOLFOX showed improved clinical activity over FOLFOX alone [47, 48]. Bevacizumab used in combination with any intravenous 5-FU-based chemotherapy is currently FDA-indicated as first-line treatment of metastatic colorectal cancer [49]. The addition of bevacizumab to any of the 5-FU-based first-line chemotherapy regimens—FOLFOX, FOLFIRI, 5-FU/LV, or CapeOx (capecitabine, an oral fluoropyrimidine, with oxaliplatin)—is considered an acceptable standard of care in the United States [50, 51].

Bevacizumab as Second-line Treatment for Metastatic Colorectal Cancer

In June 2006, bevacizumab was once again approved by the FDA for the treatment of colorectal cancer in the United States. This second approval was based on an ECOG study (E3200) that was a randomized, open-labeled, 3-arm trial of 829 patients

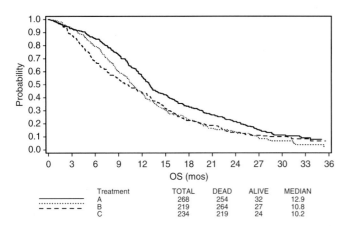

FIG. 33.2. Kaplan-Meier estimates of overall survival in a phase III trial comparing oxaliplatin, fluorouracil, and leucovorin (FOLFOX4) in combination with bevacizumab (A) compared to FOLFOX4 (B) or bevacizumab (C) alone.
Source: Giantonio BJ, Catalano PJ, Meropol NJ, et al. Bevacizumab in combination with oxaliplatin, fluorouracil, and leucovorin (FOLFOX4) for previously treated metastatic colorectal cancer: results from the Eastern Cooperative Oncology Group Study E3200. J Clin Oncol 2007;25(12):1539–44.

(9 of whom were deemed ineligible) with metastatic colorectal cancer. Patients were assigned to 1 of 3 groups: FOLFOX4 with bevacizumab 10 mg/kg (n = 286), FOLFOX4 without bevacizumab (n = 291), or bevacizumab alone (n = 243) [52]. Patients had previously failed treatment with fluoropyrimidine and irinotecan. The primary endpoint was OS, with secondary evaluation of PFS, response, and toxicity. However, the bevacizumab-alone arm discontinued accrual midway through the trial when an early efficacy study conducted by the Data Monitoring Committee revealed inferior survival in the bevacizumab-alone arm compared to the chemotherapy-containing arms. Randomization continued in a 1:1 fashion for both FOLFOX4 and FOLFOX4 with bevacizumab. Median OS was 12.9 months (FOLFOX4/bevacizumab) and 10.8 months (FOLFOX4 alone) (P = 0.0011, stratified log-rank test), with median PFS of 7.3 months and 4.7 months respectively (log-rank P < 0.0001) (Fig. 33.2) [52,53]. The corresponding median OS and PFS for the terminated bevacizumab-alone arm were 10.2 months and 2.7 months, suggesting inferior outcomes for bevacizumab as a monotherapy.

There may still be a rationale for continuing bevacizumab in patients who fail front-line chemotherapy with bevacizumab, although this has not been studied extensively. Some data suggest that the combination of bevacizumab, irinotecan, and cetuximab, a chimeric antibody targeting epidermal growth factor receptor (EGFR), may lead to higher response rates than bevacizumab and cetuximab alone in patients with known irinotecan-refractory disease [54]. Other studies are ongoing evaluating the use of bevacizumab as third-line treatment for patients who have failed standard chemotherapy [55].

Bevacizumab as Adjuvant Treatment for Colorectal Cancer

Adjuvant treatment of stage III colorectal cancer with fluoropyrimidine 5-FU backbone, [56–58] in combination with leucovorin, [59, 60] and with oxaliplatin-containing regimens, [61–63] has shown promising results. Based on these findings, the addition of bevacizumab to chemotherapy regimens is being studied as adjuvant treatment for colorectal cancer [64–66]. The AVANT study is a randomized, 3-arm, multinational phase III study with planned accrual of 3,450 patients (1,150 patients/treatment arm) evaluating adjuvant chemotherapy with or without bevacizumab for stage II or III colorectal cancer. The study includes the following regimens: bevacizumab every 2 or 3 weeks in combination with either XELOX (intermittent capecitabine and oxaliplatin every 3 weeks) or FOLFOX4 (infusional 5-FU/LV with oxaliplatin) versus FOLFOX4 alone [64]. There is also an ECOG study (E5202) of adjuvant therapy for stage II colorectal cancer in high-risk patients (microsatellite stable; 18 q loss of heterozygosity) using FOLFOX with or without bevacizumab versus observation, with a planned accrual of 3,610 patients [65].

Emerging Role of Bevacizumab in Treatment of Other Malignancies

Lung Cancer

Lung cancer is the leading cause of cancer deaths for both men and women in the United States. In 2007 alone, there will be an estimated 213,380 new diagnoses of lung cancer and 160,390 deaths from this disease [32]. Over 85% of patients diagnosed with lung cancer have the non-small cell lung cancer (NSCLC) histological subtype [67]. A recent study of standard chemotherapy consisting of a platinum-based doublet showed a 19% response rate, with 33% and 11% 1- and 2-year survival, respectively [68].

Although previous attempts to add a third chemotherapy agent to standard regimens showed no increased survival, emerging data on the role of angiogenesis in lung cancer development and progression have led to several clinical trials involving bevacizumab [69–75]. Initial clinical studies evaluated bevacizumab in combination with the standard chemotherapy regimen of carboplatin/paclitaxel (C/P) [3, 23]. One study randomized 99 patients with metastatic NSCLC to C/P alone, C/P and bevacizumab 7.5 mg/kg, or C/P and bevacizumab 15 mg/kg. All regimens were given every 3 weeks in the first-line treatment setting, and patients who did not receive bevacizumab were able to cross over at progression. Nonsignificant higher response rates were seen in the 2 bevacizumab arms (28.1% and 31.5%, respectively) compared to the C/P-alone arm (18.8%). The higher-dose bevacizumab arm had a significant increase in TTP of 7.4 months compared to 4.2 months in the C/P-alone arm (P = 0.023). There was also a trend toward improved OS in this arm (17.7 months vs 14.9 months in the C/P-alone arm). A further analysis of the data demonstrated increased morbidity in the subgroup of patients treated with bevacizumab who also had squamous cell carcinoma. Because these toxicities could potentially require exclusion of such patients, investigators reanalyzed the data. When this group of 17 patients was excluded from the analysis, there were still trends toward increased response and OS in both groups treated with C/P and bevacizumab [28]. This led to the initiation of an ECOG study in patients with advanced-stage (IIIB/IV) disease, excluding patients with squamous cell histology. In this study, patients were treated with C/P (*n* = 444) or C/P and bevacizumab 15 mg/kg (*n* = 434). Patients who received bevacizumab had statistically significant increases in response rate (35% vs 15%), TTP (6.2 vs 4.5 months) and OS (12.3 vs 10.3 months) (Fig. 33.3). There were, however, increases in morbidity and mortality in patients treated with bevacizumab who had baseline hemoptysis and tumor cavitation, suggesting that this subset of patients might not benefit from bevacizumab [76]. Because of improvements

in survival, and in spite of increased adverse events, the FDA approved C/P and bevacizumab as a first-line treatment for nonsquamous NSCLC in October 2006 [7, 77].

Another regimen commonly used to treat NSCLC is cisplatin/gemcitabine (C/G). A phase III trial compared C/G alone to C/G plus 2 doses of bevacizumab (7.5 mg/kg and 15 mg/kg) every 3 weeks in 1,043 chemotherapy-naïve patients with no squamous cell subtype or significant hemoptysis. Statistically significant increases in response rates were seen at both dose levels (34% for 7.5 mg and 30% for 15 mg, compared to 20% with C/G alone), and TTP in both bevacizumab arms was 6.1 months compared to 4.7 months for C/G alone. When these data were presented at the American Society of Clinical Oncology's 2007 Annual Meeting, evaluation of OS benefit with the addition of bevacizumab was awaiting additional follow-up [78].

Bevacizumab as part of second-line therapy also showed promise in combination with erlotinib, an EGFR inhibitor. Patients with NSCLC who were previously treated with at least one previous chemotherapy regimen (55% had 2 or more) tolerated bevacizumab at 15 mg/kg every 3 weeks in combination with daily erlotinib at 150 mg. Median TTP was 6.2 months and OS was 12.6 months, with 20% of patients having partial responses [79]. Another study randomized 120 patients to either: (1) second-line chemotherapy (docetaxel or pemetrexed) alone, (2) second-line chemotherapy combined with bevacizumab, or (3) erlotinib and bevacizumab. There were trends to improved TTP and 6-month survival in both arms containing bevacizumab compared to chemotherapy alone, further suggesting the clinical benefit of bevacizumab as a second-line agent in NSCLC [80].

Breast Cancer

Breast cancer is the most commonly diagnosed cancer in women in the United States, with 178,480 new diagnoses expected in 2007, and the second leading cause of cancer death, with 40,460 expected in 2007 [32]. In 2003, clinical studies demonstrated that bevacizumab was safe and effective in patients with metastatic breast cancer. As a single agent, however, the benefits were limited, with only a 9.3% response rate and median TTP of 2.4 months [30]. A phase III study evaluated bevacizumab at 15 mg/kg every 3 weeks combined with capecitabine 2,500 mg/m^2/day for days 1 to 14 of each 21-day cycle. The study included 462 patients with metastatic breast cancer who were previously treated with adjuvant anthracycline and taxane therapy, and then either relapsed within 12 months or received 1 to 2 chemotherapy regimens after recurrence beyond 12 months. There was a significantly higher response rate for patients who received capecitabine and bevacizumab compared to capecitabine alone (19.8% vs 9.1%), although increases in TTP and OS were minimal [81].

ECOG has compared paclitaxel 90 mg/m^2 weekly for 3 of 4 weeks in 350 patients with the same chemotherapy and bevacizumab 10 mg/kg every 2 weeks in 365 patients with

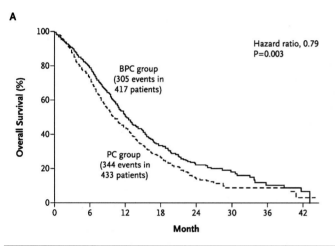

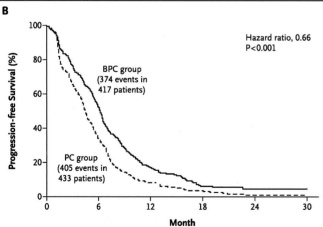

FIG. 33.3. Benefits of adding bevacizumab to standard-of-care therapy with carboplatin and paclitaxel in non-small cell lung cancer. Source: Sandler A, Gray R, Perry MC, et al. Paclitaxel-carboplatin alone or with bevacizumab for non-small-cell lung cancer. N Engl J Med 2006;355(24):2542–50.

metastatic breast cancer. Preliminary findings indicated that patients treated with bevacizumab had statistically significant increases in responses (29.9% vs 13.8%) and TTP (11.4 vs 6.11 months); there was also a trend to increased survival (28.4 vs 25.2 months; $P = 0.12$) [82]. Based on this study, the National Comprehensive Cancer Network now lists paclitaxel in combination with bevacizumab as a treatment option for patients with metastatic breast cancer [83]. Smaller studies have also investigated bevacizumab combined with vinorelbine, docetaxel, and albumin-bound paclitaxel, with promising results [84–86]. Other ongoing studies are evaluating bevacizumab in the neoadjuvant and adjuvant setting and in combination with anti-estrogen therapy for metastatic breast cancer [87–90].

Ovarian Cancer

Although the estimated 22,430 new cases of ovarian cancer in 2007 represent only 3% of new cancer diagnoses in American women, the 15,280 deaths will make up 6% of cancer deaths in this population [32]. Early studies indicate that bevacizumab could also play a role in the treatment of ovarian cancer [91]. A study by the Gynecological Oncology Group (GOG) evaluated 62 patients with epithelial ovarian cancer or primary peritoneal cancer who had measurable disease after up to 2 previous platinum-based chemotherapeutic regimens. Bevacizumab was given as a single agent every 21 days at 15 mg/kg. The overall response rate was 17.7%, with 8 patients having a partial response and 3 patients having a complete response. Stable disease was seen in 54.8% of these patients and median duration of response was 10.25 months. The 38.7% 6-month PFS compared favorably to historical controls [92].

A second study focused on patients with ovarian cancer who progressed on a platinum-based regimen within 6 months, and progressed within 3 months on a second-line chemotherapy regimen. The 44 patients enrolled were treated with bevacizumab alone at 15 mg/kg every 21 days. Despite their chemotherapy-refractory disease, there was an overall response rate of 15.9%, with a 4.2-month duration of response. In addition, 27.4% of patients had more than 6 months of PFS [93]. Other preliminary studies combining bevacizumab with agents such as low-dose (metronomic) oral cyclophosphamide and sorafenib have also yielded promising results [7, 94].

There are currently 2 ongoing phase III trials involving bevacizumab as frontline therapy. GOG 0218 began enrolling in September 2005, and International Collaborative Ovarian Neoplasm (ICON) 7 opened in October 2006. Both studies will evaluate a standard-of-care platinum-taxane (P/T) pair for 6 cycles in combination with bevacizumab. GOG 0218 will have 3 arms: P/T alone, P/T and bevacizumab (15 mg/kg every 3 weeks) and P/T with a longer course of bevacizumab (concurrent with and for 16 cycles after the 6 cycles of chemotherapy). GOG 0218 will enroll 2,000 patients with stage III and IV disease, with a primary endpoint of OS. ICON 7 will have 2 arms: P/T alone or bevacizumab (7.5 mg/kg every 3 weeks) concurrently with P/T and then for 12 cycles beyond the initial

6 cycles of chemotherapy. ICON 7 will enroll 1,520 patients with either high-risk early-stage disease or late-stage disease. The primary endpoint of this trial will be PFS [91].

Renal Cell Cancer

Renal cell cancer has long been of interest to investigators using VEGF inhibition. One of the disease's prominent characteristics, a defective von Hippel-Lindau tumor suppressor gene, leads to overproduction of VEGF [95, 96]. An initial study done at the National Cancer Institute (NCI) treated 116 patients with metastatic renal cell cancer who either could not tolerate or progressed on interleukin-2, the standard of care at that time. Patients were randomized to 1 of 2 biweekly doses of bevacizumab (3 mg/kg or 10 mg/kg) or a placebo. The trial was stopped prematurely because of a significant increase in TTP of 4.8 months vs 2.5 months ($P < 0.001$) in the 10 mg/kg group compared to placebo (Fig. 33.4). (There was also a trend to improved TTP at the smaller dose.) At the time these data were published, however, the bevacizumab groups

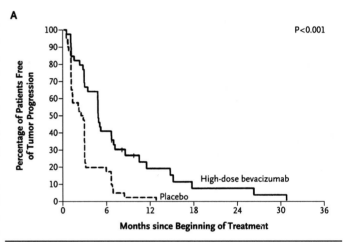

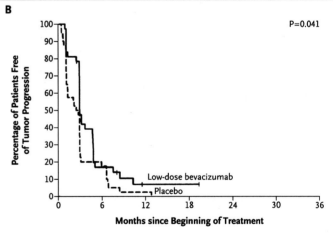

FIG. 33.4. Progression-free survival in patients receiving high-dose (**A**) or low-dose bevacizumab (**B**), compared with placebo.
Source: Yang JC, Haworth L, Sherry RM, et al. A randomized trial of bevacizumab, an anti-vascular endothelial growth factor antibody, for metastatic renal cancer. N Engl J Med 2003;349(5):427–34.

showed no survival benefit [29]. Results of a large randomized phase III trial comparing interferon-α 3 times a week with and without bevacizumab 10 mg/kg biweekly as a first-line treatment in locally advanced renal cell cancer were recently published. The international study enrolled 649 patients and demonstrated a significantly higher response rate in the interferon-α/bevacizumab group (30.6%) compared to interferon-α alone (12.4%; $P < 0.0001$). There was also a significant difference in PFS favoring the bevacizumab group of 10.2 months vs 5.4 months (P< 0.0001). There was a trend to improved OS ($P = 0.067$) among all patients in the combination arm, but the data suggest that an improved survival benefit may manifest if higher-risk patients are excluded [97]. Thus, these data suggest that bevacizumab has activity in renal cell cancer.

Prostate Cancer

In a major breakthrough in 2004, chemotherapy was first shown to extend survival in metastatic castrate-resistant prostate cancer (mCRPC) [98]. Many agents, including bevacizumab, have been added to this docetaxel-based regimen in an attempt to enhance this increased OS in mCRPC. A Cancer and Leukemia Group B (CALGB) study first showed promise in 2003 when bevacizumab (15 mg/kg every 21 days) was added to docetaxel and estramustine. This study demonstrated that of 17 patients with evaluable disease, 9 had partial responses. Furthermore, 65% of the 20 patients evaluable for prostate-specific antigen (PSA) had a >50% decline in PSA [99]. More recently, a trial at the NCI combined bevacizumab (15 mg/kg every 21 days) with docetaxel, prednisone, and daily thalidomide. Of 39 evaluable patients, 87% had a >50% decline in PSA and 67% had a PSA decline >80%. Of 17 patients with evaluable disease, there was a 59% response rate, with 1 complete response and 9 partial responses [100]. Currently, a large CALGB-initiated intergroup phase III trial is completing enrollment of mCRPC patients to compare standard-of-care docetaxel and prednisone with and without bevacizumab [101].

Toxicity and Safety Monitoring

In the initial phase I study, bevacizumab was well tolerated with no definitively attributable grade 3 or 4 toxicities, as measured by the NCI's Common Toxicity Criteria [23]. The true frequency and spectrum of bevacizumab toxicities became more apparent in subsequent larger clinical trials, and when bevacizumab was combined with other agents (Table 33.2). The most frequent high-grade (3 or 4) toxicity is hypertension, occurring to some degree in up to 32% of patients. This was seen in several of the earlier studies in colon cancer and was usually treated successfully with oral agents such as angiotensin-converting enzyme inhibitors or calcium channel blockers [27, 28, 40, 102]. For a minority of patients, the resulting hypertension could not be controlled with oral agents and up

TABLE 33.2. Serious toxicities associated with bevacizumab.

Toxicity	Estimated incidence	Additional information
Bleeding	20–53%	Most commonly manifests as self-limiting epistaxis. Appears to be associated with cavitating tumors and squamous cell histology in lung cancer.
Proteinuria	27–38%	Urine dipsticks are required before bevacizumab can be administered. If 2+ or greater, a 24-h urine must yield less than 2 g for treatment to be administered.
Hypertension	22–32%	Hypertension can usually be managed with oral agents (primarily angiotensin converting enzyme inhibitors or calcium channel lockers). Blood pressure should be checked prior to each treatment with bevacizumab.
Gastrointestinal perforation	1.4–11.4%	Possible risk factors may include diverticulitis, intestinal obstruction by tumor, abdominal carcinomatosis and a history of abdominal radiation.
Poor wound healing	6.7–10%	Surgery should be avoided within 1–2 months of the last dose of bevacizumab.
Arterial thrombosis	3.8%	Increased risk for patients over 65 years old and for patients with a history of atherosclerois.
Reversible posterior leukoencephalopathy syndrome	<0.1%	Central nervous system symptoms resolve when bevacizumab is discontinued.

Adapted from: Panares RL, Garcia AA. Bevacizumab in the management of solid tumors. Expert Rev Anticancer Ther 2007;7(4):433–45

to 1.7% of patients had to discontinue bevacizumab due to hypertension. [7] Based on these data, guidelines suggest that blood pressure should be monitored prior to each administration of the drug, or at least every 2–3 weeks. Any patient who has a hypertensive crisis or hypertension uncontrolled with oral medications should discontinue bevacizumab [102].

Patients on bevacizumab have an increase in asymptomatic proteinuria. In colon cancer patients, the incidence of proteinuria increased from 11.4% to 21.7% in patients on chemotherapy alone, compared to 22.8–38% in patients who received chemotherapy and bevacizumab [27, 40, 28]. Less than 2% of patients reportedly have grade 3 or 4 proteinuria resulting in more than 3.5 grams of protein in the urine over 24 h, and the risk of developing nephritic syndrome is 0.5% [7]. Patients should have a urine dipstick analysis prior to each administration of the drug. If results indicate ≥ +2 protein by this crude evaluation, a 24-h urine collection should be done. If the urine protein is >2 g in the 24-h urine collection, bevacizumab should be discontinued until levels decrease. Patients who develop nephrotic symptoms should be taken off bevacizumab [102].

Bleeding as a complication of bevacizumab occurs in up to 53% of patients and most commonly manifests as self-limiting epistaxis [7]. The most serious complications were seen in initial trials in patients with NSCLC, where carboplatin and paclitaxel were combined with bevacizumab. Of 99 patients,

66 received bevacizumab, and of those, 6 (9%) had life-threatening hemoptysis or hematemesis. Four of these patients had fatal complications. A further analysis indicated that 4 of 13 patients with squamous cell NSCLC had a significant bleed, and 5 of the 6 patients had cavitation of their tumor at baseline [28]. Because of this, squamous cell pathology tumors, which are often centrally located, have been excluded from subsequent NSCLC trials involving bevacizumab. Tumors in proximity to major blood vessels are thought to convey an increased risk of hemorrhage as well, and while no studies have been done in patients with central nervous system metastasis, until more data becomes available these patients should also be excluded from treatment. For patients on anticoagulation therapy, or who have a coagulopathy or an underlying bleeding diathesis, caution and close monitoring is advised [102].

Venous thrombosis can also be an issue for 6–9% of patients taking bevacizumab, but this does not seem to differ significantly from patients on chemotherapy alone, based on the studies of colorectal cancer patients [27, 28, 40, 102]. An increased risk of arterial thrombosis, however, does appear to present in patients on bevacizumab relative to those on chemotherapy alone across different tumor types. A review of 1,745 patients with breast, colorectal, and NSCLC pathology showed that the incidence of arterial thrombi increased from 1.7% to 3.8% ($P < 0.01$). A multivariate analysis indicated that an age of >65 years and a history of atherosclerosis were independent predictors of arterial thrombi. The authors of this analysis caution that, in spite of the increased risk in this subgroup of patients, many of these patients may still benefit from therapy with bevacizumab [103]. Patients who have arterial thrombi while on therapy should, however, immediately discontinue use of bevacizumab [102].

Wound healing can be delayed or altered by bevacizumab. This was demonstrated in a phase III study that randomized patients to 1 of 2 chemotherapy regimens including bevacizumab or one without. These patients were then further randomized to surgery 1–2 months prior to starting therapy or while undergoing treatment. There were significant increases in wound healing complications in the groups of patients who had surgery while on bevacizumab [104]. Based on these results and the 21-day half-life of bevacizumab, it was determined that elective surgeries should be postponed until at least 30 days after the last dose of bevacizumab [102].

Gastrointestinal perforation is another complication seen in patients on bevacizumab. This adverse event was most evident in the ovarian cancer study using bevacizumab alone (15 mg/kg every 3 weeks) in patients who progressed on, or within 6 months of, a platinum-based regimen, and then progressed within 3 months of starting a second-line treatment with either liposomal doxorubicin or topotecan. In these patients with relatively advanced, chemoresistant disease, 5 of the first 44 patients enrolled had perforations, which led to premature termination of the trial [93]. The overall risk of perforation in patients treated with bevacizumab across several trials was only 2.4%, but almost one third of these events resulted in death [7]. Although the risk factors for gastrointestinal perforation are not clear, precipitating factors may include diverticulitis, bowel obstruction (especially obstruction caused by tumor), peritoneal carcinomatosis, history of abdominal radiation, and colitis [7, 93, 102]. Presentation of these risk factors is cause for caution, although therapy that includes bevacizumab is not absolutely contraindicated [102].

Other rare toxicities in patients on bevacizumab include reversible posterior leukoencephalopathy syndrome (RPLS) and nasal septum perforation. RPLS occurs in <0.1% of patients treated with bevacizumab and manifests in the form of headaches, visual changes, nausea, and altered mental status. This syndrome is thought to result from a capillary leak in the intracranial vasculature, possibly associated with fluid retention, hypertension, or increased permeability of the vascular endothelium due to the drug. Diagnosis is confirmed by specific findings on MRI. As the name implies, RPLS resolves after the offending agent is discontinued [85, 105]. Nasal septum perforation has also been reported and is thought to be due to the same factors that cause poor wound healing and gastrointestinal perforation. The authors who first reported nasal septum perforation do not recommend discontinuing bevacizumab due to this complication [106].

References

1. Kim KJ, Li B, Houck K, Winer J, Ferrara N. The vascular endothelial growth factor proteins: identification of biologically relevant regions by neutralizing monoclonal antibodies. Growth Factors 1992;7(1):53–64.
2. Presta LG, Chen H, O'Connor SJ, et al. Humanization of an anti-vascular endothelial growth factor monoclonal antibody for the therapy of solid tumors and other disorders. Cancer Res 1997;57(20):4593–9.
3. Margolin K, Gordon MS, Holmgren E, et al. Phase Ib trial of intravenous recombinant humanized monoclonal antibody to vascular endothelial growth factor in combination with chemotherapy in patients with advanced cancer: pharmacologic and long-term safety data. J Clin Oncol 2001;19(3):851–6.
4. Lin YS, Nguyen C, Mendoza JL, et al. Preclinical pharmacokinetics, interspecies scaling, and tissue distribution of a humanized monoclonal antibody against vascular endothelial growth factor. J Pharmacol Exp Ther 1999;288(1):371–8.
5. Borgstrom P, Hillan KJ, Sriramarao P, Ferrara N. Complete inhibition of angiogenesis and growth of microtumors by anti-vascular endothelial growth factor neutralizing antibody: novel concepts of angiostatic therapy from intravital videomicroscopy. Cancer Res 1996;56(17):4032–9.
6. Kim KJ, Li B, Winer J, et al. Inhibition of vascular endothelial growth factor-induced angiogenesis suppresses tumour growth in vivo. Nature 1993;362(6423):841–4.
7. Panares RL, Garcia AA. Bevacizumab in the management of solid tumors. Expert Rev Anticancer Ther 2007;7(4):433–45.
8. Ferrara N, Hillan KJ, Gerber HP, Novotny W. Discovery and development of bevacizumab, an anti-VEGF antibody for treating cancer. Nat Rev Drug Discov 2004;3(5):391–400.
9. Folkman J. Tumor angiogenesis: therapeutic implications. N Engl J Med 1971;285(21):1182–6.

10. Folkman J. Role of angiogenesis in tumor growth and metastasis. Semin Oncol 2002;29(6 Suppl 16):15–8.

11. Weidner N, Folkman J, Pozza F, et al. Tumor angiogenesis: a new significant and independent prognostic indicator in early-stage breast carcinoma. J Natl Cancer Inst 1992;84(24):1875–87.

12. Weidner N, Carroll PR, Flax J, Blumenfeld W, Folkman J. Tumor angiogenesis correlates with metastasis in invasive prostate carcinoma. Am J Pathol 1993;143(2):401–9.

13. Weidner N, Semple JP, Welch WR, Folkman J. Tumor angiogenesis and metastasis–correlation in invasive breast carcinoma. N Engl J Med 1991;324(1):1–8.

14. Gerber HP, Ferrara N. Pharmacology and pharmacodynamics of bevacizumab as monotherapy or in combination with cytotoxic therapy in preclinical studies. Cancer Res 2005;65(3):671–80.

15. Warren RS, Yuan H, Matli MR, Gillett NA, Ferrara N. Regulation by vascular endothelial growth factor of human colon cancer tumorigenesis in a mouse model of experimental liver metastasis. J Clin Invest 1995;95(4):1789–97.

16. Ferrara N, Gerber HP, LeCouter J. The biology of VEGF and its receptors. Nat Med 2003;9(6):669–76.

17. Jain RK. Normalization of tumor vasculature: an emerging concept in antiangiogenic therapy. Science 2005;307(5706):58–62.

18. Jain RK, Duda DG, Clark JW, Loeffler JS. Lessons from phase III clinical trials on anti-VEGF therapy for cancer. Nat Clin Pract Oncol 2006;3(1):24–40.

19. Lee TH, Avraham HK, Jiang S, Avraham S. Vascular endothelial growth factor modulates the transendothelial migration of MDA-MB-231 breast cancer cells through regulation of brain microvascular endothelial cell permeability. J Biol Chem 2003;278(7):5277–84.

20. Wildiers H, Guetens G, De Boeck G, et al. Effect of antivascular endothelial growth factor treatment on the intratumoral uptake of CPT-11. Br J Cancer 2003;88(12):1979–86.

21. Tong RT, Boucher Y, Kozin SV, Winkler F, Hicklin DJ, Jain RK. Vascular normalization by vascular endothelial growth factor receptor 2 blockade induces a pressure gradient across the vasculature and improves drug penetration in tumors. Cancer Res 2004;64(11):3731–6.

22. Lee CG, Heijn M, di Tomaso E, et al. Anti-Vascular endothelial growth factor treatment augments tumor radiation response under normoxic or hypoxic conditions. Cancer Res 2000;60(19):5565–70.

23. Gordon MS, Margolin K, Talpaz M, et al. Phase I safety and pharmacokinetic study of recombinant human anti-vascular endothelial growth factor in patients with advanced cancer. J Clin Oncol 2001;19(3):843–50.

24. Kramer I, Lipp HP. Bevacizumab, a humanized anti-angiogenic monoclonal antibody for the treatment of colorectal cancer. J Clin Pharm Ther 2007;32(1):1–14.

25. Genentech. US prescribing information. In: South San Francisco, CA: Genentech Inc.; 2005.

26. Gaudreault J, Bruno R, Kabbinavar F, Sing A, Johnson DH, Lu J. Clinical pharmacokinetics of bevacizumab following every 2- or every 3-week dosing. J Clin Oncol (Meeting Abstracts) 2004;22(14_suppl):3041.

27. Kabbinavar F, Hurwitz HI, Fehrenbacher L, et al. Phase II, randomized trial comparing bevacizumab plus fluorouracil (FU)/leucovorin (LV) with FU/LV alone in patients with metastatic colorectal cancer. J Clin Oncol 2003;21(1):60–5.

28. Johnson DH, Fehrenbacher L, Novotny WF, et al. Randomized phase II trial comparing bevacizumab plus carboplatin and paclitaxel with carboplatin and paclitaxel alone in previously untreated locally advanced or metastatic non-small-cell lung cancer. J Clin Oncol 2004;22(11):2184–91.

29. Yang JC, Haworth L, Sherry RM, et al. A randomized trial of bevacizumab, an anti-vascular endothelial growth factor antibody, for metastatic renal cancer. N Engl J Med 2003;349(5):427–34.

30. Cobleigh MA, Langmuir VK, Sledge GW, et al. A phase I/II dose-escalation trial of bevacizumab in previously treated metastatic breast cancer. Semin Oncol 2003;30(5 Suppl 16):117–24.

31. Bergsland E, Dickler MN. Maximizing the potential of bevacizumab in cancer treatment. Oncologist 2004;9 Suppl 1:36–42.

32. Jemal A, Siegel R, Ward E, Murray T, Xu J, Thun MJ. Cancer statistics, 2007. CA Cancer J Clin 2007;57(1):43–66.

33. Scheithauer W, Rosen H, Kornek GV, Sebesta C, Depisch D. Randomised comparison of combination chemotherapy plus supportive care with supportive care alone in patients with metastatic colorectal cancer. Bmj 1993;306(6880):752–5.

34. Expectancy or primary chemotherapy in patients with advanced asymptomatic colorectal cancer: a randomized trial. Nordic Gastrointestinal Tumor Adjuvant Therapy Group. J Clin Oncol 1992;10(6):904–11.

35. Simmonds PC. Palliative chemotherapy for advanced colorectal cancer: systematic review and meta-analysis. Colorectal Cancer Collaborative Group. Bmj 2000;321(7260):531–5.

36. Thirion P, Michiels S, Pignon JP, et al. Modulation of fluorouracil by leucovorin in patients with advanced colorectal cancer: an updated meta-analysis. J Clin Oncol 2004;22(18):3766–75.

37. Saltz LB, Cox JV, Blanke C, et al. Irinotecan plus fluorouracil and leucovorin for metastatic colorectal cancer. Irinotecan Study Group. N Engl J Med 2000;343(13):905–14.

38. Douillard JY, Cunningham D, Roth AD, et al. Irinotecan combined with fluorouracil compared with fluorouracil alone as first-line treatment for metastatic colorectal cancer: a multicentre randomised trial. Lancet 2000;355(9209):1041–7.

39. de Gramont A, Figer A, Seymour M, et al. Leucovorin and fluorouracil with or without oxaliplatin as first-line treatment in advanced colorectal cancer. J Clin Oncol 2000;18(16):2938–47.

40. Hurwitz H, Fehrenbacher L, Novotny W, et al. Bevacizumab plus irinotecan, fluorouracil, and leucovorin for metastatic colorectal cancer. N Engl J Med 2004;350(23):2335–42.

41. Hurwitz HI, Fehrenbacher L, Hainsworth JD, et al. Bevacizumab in combination with fluorouracil and leucovorin: an active regimen for first-line metastatic colorectal cancer. J Clin Oncol 2005;23(15):3502–8.

42. Kabbinavar FF, Schulz J, McCleod M, et al. Addition of bevacizumab to bolus fluorouracil and leucovorin in first-line metastatic colorectal cancer: results of a randomized phase II trial. J Clin Oncol 2005;23(16):3697–705.

43. Kabbinavar FF, Hambleton J, Mass RD, Hurwitz HI, Bergsland E, Sarkar S. Combined analysis of efficacy: the addition of bevacizumab to fluorouracil/leucovorin improves survival for patients with metastatic colorectal cancer. J Clin Oncol 2005;23(16):3706–12.

44. Giantonio BJ, Levy DE, O'Dwyer P J, Meropol NJ, Catalano PJ, Benson AB, 3rd. A phase II study of high-dose bevacizumab in combination with irinotecan, 5-fluorouracil, leucovorin, as initial therapy for advanced colorectal cancer: results from the Eastern Cooperative Oncology Group study E2200. Ann Oncol 2006;17(9):1399–403.

45. Goldberg RM, Sargent DJ, Morton RF, et al. A randomized controlled trial of fluorouracil plus leucovorin, irinotecan, and

oxaliplatin combinations in patients with previously untreated metastatic colorectal cancer. J Clin Oncol 2004;22(1):23–30.

46. Thomas M, Hoff P, Wolff R. The MD Anderson Manual of Medical Oncology. New York: McGraw-Hill; 2006.

47. Emmanouilides C, Sfakiotaki G, Androulakis N, et al. Front-line bevacizumab in combination with oxaliplatin, leucovorin and 5-fluorouracil (FOLFOX) in patients with metastatic colorectal cancer: a multicenter phase II study. BMC Cancer 2007;7:91.

48. Hochster HS, Hart LL, Ramanathan RK, Hainsworth JD, Hedrick EE, Childs BH. Safety and efficacy of oxaliplatin/fluoropyrimidine regimens with or without bevacizumab as first-line treatment of metastatic colorectal cancer (mCRC): Final analysis of the TREE-Study. J Clin Oncol (Meeting Abstracts) 2006;24(18_suppl):3510.

49. Bevacizumab (Avastin) Drug Label. (Accessed August 15, 2007, at http://www.fda.gov/cder/drug/infopage/avastin/default.htm.)

50. NCCN. National Comprehensive Cancer Network (NCCN) Practice Guidelines version 2.2007. In: http://www.nccn.org/professionals/physician_gls/PDF/colon.pdf; Accessed on August 11, 2007.

51. Ellenhorn J, Cullinane C, Coia L, Alberts SR. Cancer Management: A Multidisciplinay Approach. 10th ed. Lawrence, KS: CMP Medica; 2007.

52. Giantonio BJ, Catalano PJ, Meropol NJ, et al. Bevacizumab in combination with oxaliplatin, fluorouracil, and leucovorin (FOLFOX4) for previously treated metastatic colorectal cancer: results from the Eastern Cooperative Oncology Group Study E3200. J Clin Oncol 2007;25(12):1539–44.

53. Cohen MH, Gootenberg J, Keegan P, Pazdur R. FDA drug approval summary: bevacizumab plus FOLFOX4 as second-line treatment of colorectal cancer. Oncologist 2007;12(3):356–61.

54. Saltz LB, Lenz HJ, Hochster H, et al. Randomized phase II trial of cetuximab/bevacizumab/irinotecan (CBI) versus cetuximab/bevacizumab (CB) in irinotecan-refractory colorectal cancer. J Clin Oncol (Meeting Abstracts) 2005;23(16_suppl):3508.

55. Emmanouilides C, Pegram M, Robinson R, Hecht R, Kabbinavar F, Isacoff W. Anti-VEGF antibody bevacizumab (Avastin) with 5FU/LV as third line treatment for colorectal cancer. Tech Coloproctol 2004;8 Suppl 1:s50–2.

56. Laurie JA, Moertel CG, Fleming TR, et al. Surgical adjuvant therapy of large-bowel carcinoma: an evaluation of levamisole and the combination of levamisole and fluorouracil. The North Central Cancer Treatment Group and the Mayo Clinic. J Clin Oncol 1989;7(10):1447–56.

57. Moertel CG, Fleming TR, Macdonald JS, et al. Levamisole and fluorouracil for adjuvant therapy of resected colon carcinoma. N Engl J Med 1990;322(6):352–8.

58. Moertel CG, Fleming TR, Macdonald JS, et al. Fluorouracil plus levamisole as effective adjuvant therapy after resection of stage III colon carcinoma: a final report. Ann Intern Med 1995;122(5):321–6.

59. O'Connell MJ, Mailliard JA, Kahn MJ, et al. Controlled trial of fluorouracil and low-dose leucovorin given for 6 months as postoperative adjuvant therapy for colon cancer. J Clin Oncol 1997;15(1):246–50.

60. Wolmark N, Rockette H, Fisher B, et al. The benefit of leucovorin-modulated fluorouracil as postoperative adjuvant therapy for primary colon cancer: results from National Surgical Adjuvant Breast and Bowel Project protocol C-03. J Clin Oncol 1993;11(10):1879–87.

61. Andre T, Boni C, Mounedji-Boudiaf L, et al. Oxaliplatin, fluorouracil, and leucovorin as adjuvant treatment for colon cancer. N Engl J Med 2004;350(23):2343–51.

62. Kuebler JP, Wieand HS, O'Connell MJ, et al. Oxaliplatin combined with weekly bolus fluorouracil and leucovorin as surgical adjuvant chemotherapy for stage II and III colon cancer: results from NSABP C-07. J Clin Oncol 2007;25(16):2198–204.

63. Wolpin BM, Meyerhardt JA, Mamon HJ, Mayer RJ. Adjuvant treatment of colorectal cancer. CA Cancer J Clin 2007;57(3):168–85.

64. NCT00112918. http://clinicaltrials.gov/ct/show/NCT00112918 Accessed on August 11, 2007.

65. NCT00217737. http://clinicaltrials.gov/ct/show/NCT00217737. Accessed on August 11, 2007.

66. NCT00096278. http://clinicaltrials.gov/ct/show/NCT00096278. Accessed on August 11, 2007.

67. Govindan R, Page N, Morgensztern D, et al. Changing epidemiology of small-cell lung cancer in the United States over the last 30 years: analysis of the surveillance, epidemiologic, and end results database. J Clin Oncol 2006;24(28):4539–44.

68. Schiller JH, Harrington D, Belani CP, et al. Comparison of four chemotherapy regimens for advanced non-small-cell lung cancer. N Engl J Med 2002;346(2):92–8.

69. Delbaldo C, Michiels S, Syz N, Soria JC, Le Chevalier T, Pignon JP. Benefits of adding a drug to a single-agent or a 2-agent chemotherapy regimen in advanced non-small-cell lung cancer: a meta-analysis. Jama 2004;292(4):470–84.

70. Folkman J. The role of angiogenesis in tumor growth. Semin Cancer Biol 1992;3(2):65–71.

71. Fontanini G, Bigini D, Vignati S, et al. Microvessel count predicts metastatic disease and survival in non-small cell lung cancer. J Pathol 1995;177(1):57–63.

72. Lucchi M, Fontanini G, Mussi A, et al. Tumor angiogenesis and biologic markers in resected stage I NSCLC. Eur J Cardiothorac Surg 1997;12(4):535–41.

73. D'Amico TA, Massey M, Herndon JE, 2nd, Moore MB, Harpole DH, Jr. A biologic risk model for stage I lung cancer: immunohistochemical analysis of 408 patients with the use of ten molecular markers. J Thorac Cardiovasc Surg 1999;117(4):736–43.

74. Ushijima C, Tsukamoto S, Yamazaki K, Yoshino I, Sugio K, Sugimachi K. High vascularity in the peripheral region of non-small cell lung cancer tissue is associated with tumor progression. Lung Cancer 2001;34(2):233–41.

75. Sledge GW, Jr., Miller KD. Angiogenesis and antiangiogenic therapy. Curr Probl Cancer 2002;26(1):1–60.

76. Sandler A, Gray R, Perry MC, et al. Paclitaxel-carboplatin alone or with bevacizumab for non-small-cell lung cancer. N Engl J Med 2006;355(24):2542–50.

77. FDA Approves New Combination Therapy for Lung Cancer. (Accessed August 11, 2007, at http://www.fda.gov/bbs/topics/news/2006/new01488.html.)

78. Manegold C, von Pawel, J, Zatloukal, P, Ramlau, R, Gorbounova, V,Hirsh, V, Leighl, N, Mezger, J, Archer, V, Reck M. Randomised, double-blind multicentre phase III study of bevacizumab in combination with cisplatin and gemcitabine in chemotherapy-naïve patients with advanced or recurrent non-squamous non-small cell lung cancer (NSCLC): BO17704. J Clin Onc 25 (18S), 2007: LBA7514.

79. Herbst RS, Johnson DH, Mininberg E, et al. Phase I/II trial evaluating the anti-vascular endothelial growth factor monoclonal

antibody bevacizumab in combination with the HER-1/epidermal growth factor receptor tyrosine kinase inhibitor erlotinib for patients with recurrent non-small-cell lung cancer. J Clin Oncol 2005;23(11):2544–55.

80. Fehrenbacher L, O'Neill, V., Belani, CP et al. A Phase II, multicenter, randomized clinical trial to evaluate the efficacy and safety of bevacizumab in combination with either chemotherapy of (docetaxel or pemetrexed) or erlotinib hydrochloride compared with chemotherapy alone for the treatment of recurrent or refractory non-small cell lung cancer. In: 42nd ASCO Annual Meeting; 2006; 2006.

81. Miller KD, Chap LI, Holmes FA, et al. Randomized phase III trial of capecitabine compared with bevacizumab plus capecitabine in patients with previously treated metastatic breast cancer. J Clin Oncol 2005;23(4):792–9.

82. Miller K, Wang, M., Gralow, J. et al. A randomized phase III trial of paclitaxel versus paclitaxel plus bevacizumab plus paclitaxel as first line therapy for recurrent or metastatic breast cancer: a trial coordinated by the Eastern Cooperative Oncology Group (E2100). In: 28th Annual San Antonio Breast Cancer Symposium, San Antonio (TX): Breast Cancer Research and Treatment December, 2005.

83. NCCN Practice Guidelines in Oncology, Breast Cancer, V.2.2007. (Accessed August 11, 2007, at http://www.nccn.org/professionals/physician_gls/PDF/breast.pdf.)

84. Burstein H, Parker L, Savoie, J Phase II trial of the anti-VEGF antibody bevacizumab in combination with vinorelbine for refractory advamced breast cancer. Breast Cancer Res Treat 2002;76:2115:446.

85. Ramaswamy B, Elias AD, Kelbick NT, et al. Phase II trial of bevacizumab in combination with weekly docetaxel in metastatic breast cancer patients. Clin Cancer Res 2006;12(10):3124–9.

86. Link J, Waisman, JR, Nguyen, B, Jacobs, CI. Bevacizumab and albumin-bound paclitaxel treatment in metastatic breast cancer. J Clin Onc, Vol 25, No 18S, 2007: 1101;.

87. Traina TA, Rugo HS, Dickler M. Bevacizumab for advanced breast cancer. Hematol Oncol Clin North Am 2007;21(2):303–19.

88. Traina T, Rugo, H, Caravelli, J, Yeh, B, Panageas, K, Bruckner, J, Norton, L, Park, J, Hudis, C, Dickler, M. Letrozole with bevacizumab is feasible in patients with hormone receptor-positive metastatic breast cancer J Clin Onc, 24 (18S), 2006: 3050.

89. Forero-Torres A, Percent, I, Galleshaw, J, Nabell, L,Carpenter, J,Falkson, C, Jones, C, Krontriras, H, De Los Santos, J, Saleh, M A study of pre-operative (neoadjuvant) letrozole in combination with bevacizumab in post-menopausal women with newly diagnosed operable breast cancer: A preliminary safety report. J of Clin Onc 25(18S), 2007: 11020.

90. Mayer E, Miller, KD, Rugo, HS, Peppercorn, JM, Carey, LA, Ryabin, N, Winer, EP, Burstein, HJ. A pilot study of adjuvant bevacizumab after neoadjuvant chemotherapy for high-risk breast cancer. J of Clin Onc 25(18S), 2007: 561.

91. Burger RA. Experience with bevacizumab in the management of epithelial ovarian cancer. J Clin Oncol 2007;25(20): 2902–8.

92. Burger R, Sill, M, Monk, BJ,Greer, B.,Sorosk, J. Phase II trial of bevacizumab in persistent or recurrent epithelial ovarian cancer or primary peritoneal cancer: a Gynecologic Oncology Group study. J Clin Onc 23(16S) 2005: 5009.

93. Cannistra S, Matulonis, U, Penson, R, Wenham, R, Armstrong, D, Burger, RA, Mackey, H, Douglas, J, Hambleton, J, McGuire, W. Bevacizumab in patients with advanced platinum-resistant ovarian cancer. J Clin Onc 24(18S) 2005: 5006.

94. Azad N, Posadas, EM, Kwitkowski, VE, Annunziata, CM, Barrett, T, Premkumar, A, Kotz, HL, Sarosy, GA,Minasian, LM, Kohn, EC. Increased efficacy and toxicity with combination anti-VEGF therapy using sorafenib and bevacizumab. J Clin Onc 24(18S) 2006: 3004.

95. Mukhopadhyay D, Knebelmann B, Cohen HT, Ananth S, Sukhatme VP. The von Hippel-Lindau tumor suppressor gene product interacts with Sp1 to repress vascular endothelial growth factor promoter activity. Mol Cell Biol 1997;17(9):5629–39.

96. Gnarra JR, Zhou S, Merrill MJ, et al. Post-transcriptional regulation of vascular endothelial growth factor mRNA by the product of the VHL tumor suppressor gene. Proc Natl Acad Sci USA 1996;93(20):10589–94.

97. Escudier B, Koralewski, P, Pluzanska, A. Ravaud, A, Bracarda, S, Szczylik, C, Chevreau, C, Filipek, M, Melichar, B, Moore, N. A randomized, controlled, double-blind phase III study (AVOREN) of bevacizumab/interferon-α2a vs placebo/interferon-α2a as first-line therapy in metastatic renal cell carcinoma. J Clin Onc 25(18S) 2007: 3.

98. Tannock IF, de Wit R, Berry WR, et al. Docetaxel plus prednisone or mitoxantrone plus prednisone for advanced prostate cancer. N Engl J Med 2004;351(15):1502–12.

99. Picus J, Halabi, S, Rini, B, Vogelzang, N, Whang, Y, Kaplan, E, Kelly, W, Small, E. The use of bevacizumab with docetaxel and estramustine in hormone refractory prostate cancer: Initial results of CALGB 90006. Proc Am Soc Clin Oncol 22: 2003 (abstr 1578).

100. Ning Y, Arlen, P, Gulley, J, Latham, L, Jones, E, Chen, C, Parnes, H, Wright, J, Figg, WD, Dahut, WL. A phase II trial of thalidomide, bevacizumab, and docetaxel in patients with metastatic androgen-independent prostate cancer. J Clin Onc 25(16S) 2007: 5114;.

101. Phase III Randomized Study of Docetaxel and Prednisone With Versus Without Bevacizumab in Patients With Hormone-Refractory Metastatic Adenocarcinoma of the Prostate (Accessed August 11, 2007, at http://www.cancer.gov/clinicaltrials/CALGB-90401.)

102. Gordon MS, Cunningham D. Managing patients treated with bevacizumab combination therapy. Oncology 2005;69 Suppl 3:25–33.

103. Skillings J, Johnson, DH, Miller, K, Kabbinavar, F, Bergsland, E, Holmgren, E, Holden, SN,Hurwitz, F, Scappaticci, F. Arterial thromboembolic events in a pooled analysis of 5 randomized, controlled trials (RCTs) of bevacizumab with chemotherapy. J Clin Onc 23(16S), 2005: 3019.

104. Scappaticci FA, Fehrenbacher L, Cartwright T, et al. Surgical wound healing complications in metastatic colorectal cancer patients treated with bevacizumab. J Surg Oncol 2005;91(3):173–80.

105. Glusker P, Recht L, Lane B. Reversible posterior leukoencephalopathy syndrome and bevacizumab. N Engl J Med 2006;354(9):980–2; discussion –2.

106. Fakih MG, Lombardo JC. Bevacizumab-induced nasal septum perforation. Oncologist 2006;11(1):85–6.

Chapter 34
Development of Thalidomide and Its IMiD Derivatives

Cindy H. Chau[1,2], William Dahut[2], and William D. Figg[1,2,*]

Keywords: thalidomide, lenalidomide, immunomodulatory drugs, multiple myeloma

Abstract: Despite its controversial past, the establishment of thalidomide as an anti-inflammatory, immunomodulatory, and antiangiogenic agent bolstered intense research into its mechanism of action and therapeutic range. The precise pharmacologic mechanism through which thalidomide exerts its activity is complex and not fully understood. The enhancement of thalidomide's immunomodulatory effects while minimizing the adverse reactions brought about a class of novel analogues termed the Immunomodulatory drugs (IMiDs). This chapter reviews and highlights some of the clinical activities and development of thalidomide and its IMiDs derivatives in the treatment of hematological cancers and various solid tumors.

Introduction

Thalidomide was synthesized in Germany in 1954 from the glutamic acid derivative α-phthaloylisoglutamine. It was originally prescribed outside the United States (US) as a sedative and anti-emetic for morning sickness. Bearing structural resemblance to barbiturates, thalidomide soon became a popular sedative and was subsequently marketed worldwide in the late 1950 (see [1] and [2] for a review of the history of thalidomide). However, the drug never received approval by the US Food and Drug Administration (FDA) due in part to concerns raised about potentially irreversible peripheral neuritis and the drug's safety profile [3]. Initial reports of limb abnormalities and other congenital defects were observed in women who took as little as a single dose of thalidomide during gestation [4]. The highest risk for teratogenicity occurred when the drug was taken between weeks 3 and 8 after conception. The use of thalidomide in pregnant women resulted in the birth of over 10,000 infants worldwide with severe malformations, and led to its withdrawal from Europe and other countries in the early 1960s. The drug resurfaced in 1965 when it was found to be effective in the treatment of erythema nodosum leprosum (ENL) lesions, a painful inflammatory dermatological reaction of lepromatous leprosy (also known as Hansen's Disease). Thalidomide was subsequently granted approval by the US FDA in 1998 for this indication (Sheskin 1965; [5]).

In 1994, the laboratory of Judah Folkman uncovered another clinical potential of thalidomide that redefined the drug. They hypothesized that the thalidomide-associated malformations were the result of the drug's interference with vasculogenesis, and that a similar mechanism might prevent the growth of blood vessels recruited by solid tumors [6]. A new role had been discovered for thalidomide based on its inhibitory effect on angiogenesis. The establishment of thalidomide as an anti-inflammatory, immuno-modulatory, and antiangiogenic agent soon inspired researchers to investigate its mechanism of action and clinical range. The precise pharmacologic mechanism through which thalidomide exerts its activity has not been conclusively determined and remains controversial. This chapter reviews and highlights some of the therapeutic activities and clinical development of thalidomide and its immunomodulatory analogues (IMiDs) in the treatment of hematological cancers and various solid tumors.

Pharmacological Mechanisms

Extensive studies have been conducted in an attempt to elucidate the pharmacological activity thalidomide and its IMiD derivatives. The mechanism of action of thalidomide and its analogues is complex and not yet fully understood, as a clear molecular target remains to be identified to account for its clinical activity, teratogenicity, or side effects. This class of compounds possesses unique therapeutic properties, which probably act in concert to

[1]Molecular Pharmacology Section, [2]Medical Oncology Branch, Center for Cancer Research, National Cancer Institute, Bethesda, MD, USA

Corresponding Author:
William D. Figg, National Cancer Institute, Building 10, Room 5A01, 9000 Rockville Pike, Bethesda, MD, USA
E-mail: wdfigg@helix.nith.gov

produce the clinical efficacy observed in various disease states. Indeed, the response of certain diseases to thalidomide therapy may occur via modulation of multiple pathways simultaneously in order to elicit an additive or synergistic effect. Alternatively, these diverse mechanisms of action may all be downstream effects of thalidomide's interaction with an unknown single target.

Antiangiogenic Activity

Thalidomide was originally shown to inhibit angiogenesis by D'Amato et al. in 1994. Using a rabbit cornea micropocket assay, thalidomide inhibited bFGF-induced angiogenesis in vivo [6]. This was subsequently confirmed in several different in vitro and ex vivo assays [6–10]. The antiangiogenic activity of thalidomide is believed to require enzymatic activation via cytochrome P450 (CYP450)-mediated metabolism. Thalidomide alone did not inhibit angiogenesis in the mouse or rat aorta ex vivo assays, but the addition of human liver microsomes induced activity [7,8,11]. The species specificity of thalidomide may be due to differences in metabolism, since activity was not observed with rat liver microsomes [9]. Therefore, it is hypothesized that a metabolite of thalidomide, formed by the human CYP450 system, is responsible for the inhibition of angiogenesis. Moreover, Fujita et al. found an inverse correlation between TNF-α concentration and antiangiogenic effect, postulating that suppression of TNF-α is the mechanism by which thalidomide is exerting its effect [8]. This is in contrast to earlier studies that demonstrated TNF-α inhibition is not responsible for the antian-

giogenic activity in bFGF- or VEGF-driven corneal neovascularization [10].

The extent to which the antiangiogenic properties of thalidomide and the IMiDs play a role in its anti-cancer activity, particularly in multiple myeloma (MM), is not clearly understood. However, several mechanisms have been proposed involving the down-regulation of cytokines in the endothelial cell (EC), the inhibition of EC proliferation, or a decrease in the level of circulating endothelial cells, as well as the modulation of adhesion molecules between the MM cells and the endogenous bone marrow (BM) stromal cells, thereby decreasing the production of VEGF and interleukin-6 (IL-6) [12–15] (Fig. 34.1). VEGF expression is known to cause proliferation of MM cell lines and promote the self-renewal of leukemia progenitors [15,16]. Secretion of both VEGF and bFGF from tumor and BM stromal cells is suppressed in the presence of thalidomide and its analogues, thereby reducing EC migration and adhesion [6,12,17]. Moreover, the thalidomide-induced antiangiogenic action may also occur via depletion of the VEGF receptors through modulation of ceramide and sphingosine-1-phosphaste signaling in zebrafish studies [18]. A recent report demonstrated that thalidomide exerts inhibitory effects on nitric oxide-mediated angiogenesis by altering subcellular actin polymerization pattern, resulting in inhibition of endothelial cell migration [19]. In particular, the thalidomide analogue lenalidomide was found to attenuate growth factor-induced angiogenesis and endothelial cell migration via inhibiting the phosphorylation of Akt and its subsequent

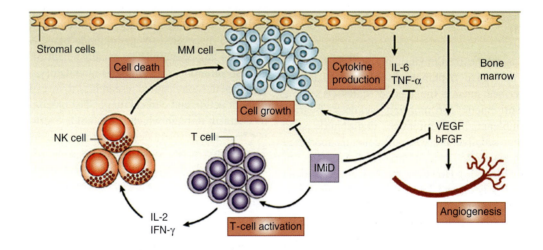

FIG. 34.1. The pharmacological mechanisms of thalidomide and the IMiDs in multiple myeloma. Immunomodulatory drugs (IMiDs) induce growth arrest and/or apoptosis in multiple myeloma (MM) cells and inhibit adhesion of MM cells to bone-marrow stromal cells. Stromal-cell expression of vascular endothelial growth factor (VEGF) and basic fibroblast growth factor (bFGF) is reduced by IMiDs, which decreases angiogenesis. Expression of interleukin-6 (IL-6) and tumor-necrosis factor-α (TNF-α) by the stromal cells is also reduced, which inhibits growth of MM cells. The IMiDs also enhance T-cell stimulation and proliferation. The activated T cells release IL-2 and interferon-γ (IFN-γ), which activate natural-killer (NK) cells (which might also be activated directly) and causes MM-cell death. Reprinted by permission from Macmillan Publishers Ltd: Nature Reviews Cancer [39], copyright 2004.

signaling [20]. Furthermore, it has been suggested that thalidomide and its analogues display antiangiogenic activity that is independent of their immunomodulatory effects [6,12].

Immunomodulatory and Anti-inflammatory Activity

The immunomodulatory and anti-inflammatory properties of thalidomide are closely related in that they exhibit biological effects on both cytokines and cell-mediated immunity. One of the key mediators of the anti-inflammatory response observed with the use of thalidomide is tumor necrosis factor-α (TNF-α) [21]. Thalidomide has also been shown to modulate a range of cytokines, including IL-2, IL-4, IL-6, IL-12 and interferon-γ [22]. The effect of thalidomide on TNF-α is dependent on the type of cell being treated, decreasing the synthesis of TNF-α in LPS-stimulated PBMC or increasing TNF-α synthesis in adherent cells [23]. This inhibition of TNF-α in monocytes is due to increased degradation of TNF-α mRNA [24]. It is possible that the high affinity binding of α1-acid glycoprotein by thalidomide may also affect the activity, since this protein is known to potentiate TNF-α secretion [25,26]. However, the inhibitory activity of TNF-α has previously been shown to be enantioselective towards the (-)-S-enantiomer of thalidomide and other chirally fixed analogs exhibiting better inhibition, suggesting specificity of the cellular target-protein(s) to which it is binding [27]. The rapid reduction in circulating TNF-α in patients with erythema nodosum leprosum (ENL) is believed to be the mechanism underlying the healing of the inflammatory skin lesions [28].

In addition to its effect on cytokine production, thalidomide has been shown to induce changes in multiple cell types. Depending on the type of immune cell that is activated and the type of stimulus that the cell receives, the immunomodulatory effects of thalidomide are variable. Thalidomide affects both T-cell proliferation and cytokine production by acting as a costimulator [29]. This effect is more pronounced for cytotoxic T-cells (CD8+) than for helper T-cells (CD4+). Thalidomide significantly increased IL-2 production by CD8+ T-cells, and interferon-γ by both CD8+ and CD4+ T-cells, while the production of TNF-α was not significantly increased in either T cell type [29]. Moreover, thalidomide affects the balance between T-helper (Th)-1 and -2 subsets, at least in part through cytokine modulation of IL-4, IL-5, and IFN-γ [30]. Thalidomide also appears to increase production of natural killer cells, which subsequently induces tumor cell lysis [31]. Furthermore, thalidomide blocks the activation of the transcription factor nuclear factor κ B (NF-κ B), a key regulator of TNF-α and IL-8 production, through inhibition of the inhibitor of kappa B kinase (IKK) [32].

Other Mechanisms of Action

Thalidomide and its analogues exhibit anti-proliferative and pro-apoptotic activity in hematologic malignancies. This class of agents inhibited the proliferation of MM cell lines that are resistant to standard chemotherapy through inhibition of IL-6 production [33]. The effects on apoptosis in MM cells involve cell cycle arrest at the G1 phase and activation of multiple apoptotic signaling pathways including the potentiation of TNF-related apoptosis-inducing ligand (TRAIL), increased sensitivity to Fas-induced apoptosis, and initiation of caspase-8 activation as well as down-regulation of NF-κB activity and expression of apoptosis-inhibitory protein [33,34]. In particular, lenalidomide has specific anti-proliferative actions in chromosome 5 deleted hematopoietic tumors in vitro by inducing cell cycle arrest, inhibiting Akt and Gab1 phosphorylation as well as interfering with the adaptor protein complex assembly, and thus providing insights into its clinical efficacy in myelodysplastic syndromes [35].

Thalidomide inhibits the growth of tumors through cyclooxygenase-2 (COX-2) degradation. While COX-2 is believed to play a role in tumor angiogenesis, the antitumor effects of thalidomide that is dependent on COX-2 expression appears to occur independent of its antiangiogenic activity [36]. Treatment of MCF-7 and HL-60 cell lines with thalidomide resulted in reduced expression of COX-2, coupled with a decrease in bcl-2, TNF-α, VEGF, GSH and an increase in cytochrome c, but exhibited no effect on COX-1. Likewise, cells that did not express COX-2 were not sensitive to thalidomide. The inhibition of COX-2 expression in LPS-, TNF-α-, and IL-1β-stimulated peripheral blood mononuclear cells partially involved elevated levels of IL-10 production [37]. Another study demonstrated that thalidomide inhibited the interleukin-1β (IL-1β)-mediated induction of COX-2 protein and mRNA expression in Caco-2 cells. Thalidomide destabilized the COX-2 mRNA through inhibition of cytoplasmic shuttling of p38 mitogen-activated protein kinase (MAPK) and HuR, an mRNA stabilizing protein [38]. These recent findings elucidating the molecular mechanisms of this class of compounds further expanded the functional roles of thalidomide and its analogues in attempt to answer the therapeutic implications for their clinical efficacy.

Development of Thalidomide Analogues

Although the clinical effects of thalidomide as an antiangiogenic agent were beneficial for multiple myeloma and other cancer types, its safety profile remains a significant concern to the scientific community. The next wave of drug discovery efforts was aimed at searching for derivatives of thalidomide with improved therapeutic efficacy and less adverse effects began. Using the thalidomide structural backbone as a template, the first-generation compounds were synthesized with increased immunological and anti-cancer properties and decreased toxicity compared with the parent compound [39,40]. The second-generation analogues were focused on improved inhibition of TNF-α, generating a series of amino-phthaloyl-substituted thalidomide analogues

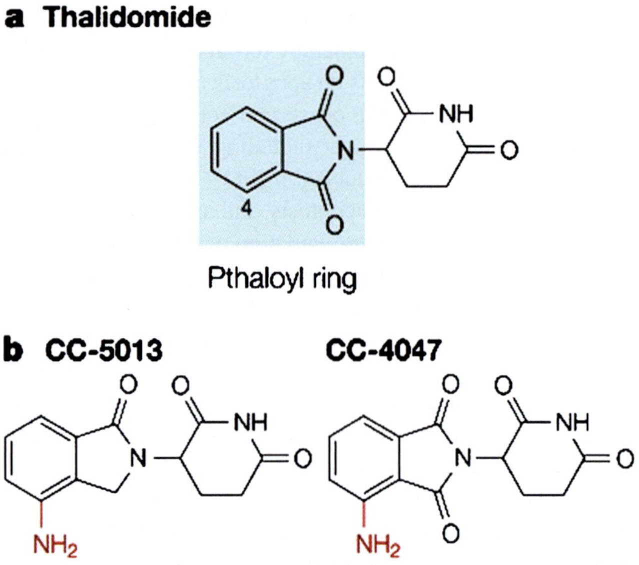

a **Thalidomide**

Pthaloyl ring

b **CC-5013**   **CC-4047**

FIG. 34.2. Structure of thalidomide and its IMiD derivatives. The thalidomide structure (**a**) was modified by adding an amino (NH$_2$-) group at the 4 position of the phthaloyl ring to generate the IMiDs CC-5013 and CC-4047 (**b**). For CC-5013, one of the carbonyls (C = O) of the 4-amino-substituted phthaloyl ring has been removed. Reprinted by permission from Nature Publishing Group: Nature Reviews Cancer [39], copyright 2004.

[41–43] (Fig. 34.2). The addition of an amino group to the fourth carbon of the pthaloyl ring of thalidomide enhanced the anti-TNF-α activity of these 4-amino analogues up to 50,000 times more potent than the parent compound in vitro. Lenalidomide (Revlimid) and CC-4047 (Actimid) are among the lead compounds in clinical development derived from this second-generation series. Third-generation IMiDs may offer even greater benefits with improved anti-cancer and antiangiogenic activities as well as enhanced immunomodulatory effects.

Role of Thalidomide in Cancer Therapy

The discovery of thalidomide's antiangiogenic potential came at an opportune time with the quest of validating the emerging concept of angiogenesis inhibition as a possible therapeutic option for cancer treatment. The role of thalidomide in cancer therapy is based on evidence of support stemming from its anti-TNF-α activity. Since TNF-α appears to be involved in angiogenesis by upregulating the expression of endothelial integrin [44] and increased serum levels of TNF-α in cancer patients is often associated with advanced disease, the anti-TNF-α effects of thalidomide may proved beneficial in this treatment setting. The immunomodulatory and antiangiogenic effects coupled with the diversity of its pharmacological action make thalidomide and its derivatives a unique class of anti-cancer agents with broad implications for the treatment of hematologic malignancies and solid tumors.

FDA Approved Indications for Thalidomide and Lenalidomide

In May 2006, thalidomide (Thalmid, Celgene) was granted accelerated approval for use in combination with dexamethasone for the treatment of newly diagnosed multiple myeloma (MM) patients. The effectiveness of thalidomide is based on objective improved response rates and shorter time to response in a phase 3 trial of 207 diagnosed patients randomly assigned to receive thalidomide plus dexamethasone or dexamethasone alone [45]. The response rate with thalidomide plus dexamethasone was significantly higher than with dexamethasone alone (63% vs 41%, respectively; $P = .0017$]. There are no controlled trials thus far demonstrating a clinical benefit. Because of its known teratogenicity, the marketing and use of thalidomide in the United States is restricted through the System for Thalidomide Education and Prescribing Safety (S.T.E.P.S.) program. This mandatory registry includes authorized patients, prescribers and pharmacies, extensive patient education regarding thalidomide's safety, and is designed to prevent fetal exposure to thalidomide during pregnancy. Women of childbearing age must undergo pregnancy testing before initiation therapy with repeat testing every 2–4 weeks during therapy. They must use two highly effective contraceptive methods. Male patients, even those with successful vasectomy, must use a condom while under treatment since thalidomide was detected in the semen. All patients must continue these measures for at least 1 month after discontinuation of the drug. Thus the S.T.E.P.S. program is designed to prevent fetal exposure to thalidomide during pregnancy. The most common toxicities associated with thalidomide were grade 3 or higher venous thromboembolism (VTE), neuropathy, somnolence, constipation, and rash.

Lenalidomide (Revlimid, Celgene) has undergone rapid clinical development for the treatment of multiple myeloma [46]. It received fast-track designation from the US FDA and was designated an orphan drug by the European Commission for Multiple Myeloma. Efficacy and safety of lenalidomide were demonstrated in two randomized, double-blind, multi-center, multi-national, placebo-controlled phase 3 studies comparing the combination of lenalidomide plus oral dexamethasone versus dexamethasone alone in multiple myeloma patients who had received at least one prior therapy. Interim analysis of these studies showed that treatment with the combination of lenalidomide and dexamethasone significantly increased overall response (OR) and time to progression (TTP). Based on the superior response rates and prolonged TTP, both studies were stopped by the independent data monitoring committees. The US FDA approved lenalidomide in June 2006 for use in combination with dexamethasone in patients with multiple myeloma who have received one prior therapy. Retrospective data analysis of the two trials demonstrated that the overall response rate (58.9% vs 20.9%; $P < 0.001$) and median TTP (60 vs 20 weeks, $P < 0.001$) was significantly better in elderly patients on the combination treatment arm versus dexamethasone alone, respectively; and the median OS was 79 weeks and has not been reached ($P \leq 0.001$) in the combination group [47]. Thus, lenalidomide in combination with dexamethasone improves OR and prolongs TTP and OS. The marketing and use of lenalidomide is made available under a special restricted distribution program called RevAssist. In December 2005, lenalidomide was also granted US FDA approval for patients with myelodysplastic syndromes associated with a deletion 5q cytogenetic abnormality with or without additional chromosomal abnormalities. Approval was based on trial results that demonstrated lenalidomide reduced transfusion requirements and reversed cytologic and cytogenetic abnormalities in patients who have the myelodysplastic syndrome with the 5q deletion [48]. The dose-limiting toxicity of lenalidomide is myelosuppression, which is manageable with dose interruption and/or reduction. Lenalidomide causes less neuropathy, sedation or constipation compared to thalidomide. Treatment with lenalidomide is generally well tolerated with a manageable safety profile that is more favorable than thalidomide.

A major concern in the use of these immunomodulatory agents is the incidence of venous thromboembolism (VTE). Randomized trials have reported significantly higher VTE rates in multiple myeloma patients receiving thalidomide- or lenalidomide-based therapy compared with standard therapies [49]. Thus, prophylactic anticoagulation is recommended for patients receiving thalidomide or lenalidomide in combination with dexamethasone or other chemotherapies. Future studies are needed to establish optimal anticoagulant regimens for the prevention of thrombosis, which may include low molecular weight heparin, therapeutic doses of warfarin, or aspirin.

Clinical Development of Thalidomide Andlenalidomide

Thalidomide and its analogs have therapeutic potential in a wide spectrum of diseases given their range of pharmacologic effects. Thalidomide is currently being investigated in over 90 clinical trials involving hematologic malignancies or solid tumors as monotherapy, in combination with conventional chemotherapy, radiation therapy, or immunotherapy. Over 25 of the trials are specifically for further evaluation of thalidomide with different treatment regimens in multiple myeloma. Early phase trials of thalidomide in solid tumors have shown activity in prostate cancer, Kaposi's sarcoma, renal cell cancinoma (RCC), melanoma, neuroendocrine tumors, hepatocellular carcinoma (HCC), lung cancer, and gliomas [50,51]. Phase II trials of thalidomide in androgen-independent prostate cancer (AIPC) have demonstrated promising acitivity [52–54]. Specifically, a phase II weekly docetaxel and thalidomide trial of 75 patients with metastatic AIPC demonstrated that the addition of thalidomide to docetaxel resulted in an encouraging PSA decline and overall median survival rate for patients in the combination arm (docetaxel plus thalidomide - median survival, 25.9 months versus doxetaxel alone - median survival, 14.7 months; $P = 0.0407$) [52,55]. Since both the

docetaxel/thalidomide and the docetaxel/bevacizumab combinations exhibit significant activity in prostate cancer [56], presumably through targeting different angiogenic factors, a phase II trial of a four-drug combination consisting of docetaxel, prednisone, thalidomide, and bevacizumab is underway in men with chemotherapy-naïve progressive AIPC and early results to this study are extremely encouraging [57]. The mechanism for thalidomide activity in prostate cancer has been proposed to involve early modulation of the tumor microenvironment [58]. While phase II trials have shown promising results with thalidomide in combination with cytotoxic chemotherapy, conducting randomized phase III trials is essential in order to validate these findings.

The use of thalidomide in RCC has demonstrated variable responses. A Phase II study evaluating the combination of interferon-α and thalidomide in metastatic RCC has minimal efficacy and considerable toxicity [59,60]. In a separate Phase II trial, thalidomide was found not to be superior to medroxyprogesterone acetate in patients with metastatic RCC, who had either progressed after first-line immunotherapy or who were not suitable for immunotherapy [61]. However, a Phase I/II study of thalidomide in combination with IL-2 found this regimen to be tolerable and to produce durable, active responses in patients with RCC, but a larger study is needed to confirm these initial results [62]. The ongoing clinical trials of thalidomide in various solid tumors and hematologic malignanices should prove useful in identifying disease states that would benefit from a thalidomide-based treatment regimen.

A substantial number of trials are currently underway evaluating lenalidomide as monotherapy or in combinations with conventional chemotherapy for not only relapsed and refractory MM but also a broad range of solid tumors. Several ongoing Phase III trials are evaluating lenalidomide alone or combined with dexamethasone in MM populations. The activity of IMiDs has been reported in both myeloid and lymphoid malignancies, in particular non-Hodgkin's lymphoma, acute myeloid leukemia, and myelofibrosis with myeloid metaplasia [51]. A Phase II, open-label study of lenalidomide in patients with metastatic RCC demonstrated an anti-tumor effect as evidenced by durable partial responses. The overall activity of lenalidomide in this patient population will need to be addressed in larger Phase III studies [63]. A Phase II trial of lenalidomide in patients with relapsed or refractory B-cell chronic lymphocytic leukemia (B-CLL) demonstrated that lenalidomide is a clinically active regimen with a 47% overall response rate and 9% of the patients attaining complete remission [64]. Moreover, preliminary results from a phase II study of patients with relapsed or refractory aggressive non-Hodgkin's lymphoma demonstrate that lenalidomide oral monotherapy is an active regimen with manageable side effects in this patient population [65]. These encouraging results will need to be confirmed in future clinical studies but demonstrate thus far that lenalidomide is an effective agent in a broad range of cancers. Phase I/II trials evaluating lenalidomide in combination with the proteasome inhibitor bortezomib and dexamethasone are currently underway for patients with newly diagnosed MM, with relapsed MM, or relapsed and refractory MM. Treatment combinations of lenalidomide with dexamethasone and/or targeted biological agents such as bortezomib represent attractive regimens and while initial study results demonstrate promising response rates, further studies are needed to elucidate the therapeutic benefit of this treatment approach [66].

Conclusions

Thalidomide and its IMiD derivatives represent a class of compound that is clinically effective and that possesses a broad spectrum of pharmacological mechanisms including antiangiogenic, anti-tumor, immunomodulatory, and anti-inflammatory properties. Although the increased rate of thromboembolic complications warrants caution and prophylactic anticoagulation is recommended, the overall toxicity profile appears to be manageable for the IMiDs. Ongoing and future studies are underway to further determine the clinical efficacy of thalidomide and the IMiDs, either in combination with conventional chemotherapy, other targeted therapy, or as single agents, and define their role in solid tumors and other hematologic malignancies.

References

1. Lenz W. A short history of thalidomide embryopathy. Teratology 1988;38:203–15.
2. Stephens T, Brynner R. Dark Remedy: The Impact of Thalidomide and Its Revival as a Vital Medicine. Cambridge, MA: Perseus Publishing, 2001.
3. Kelsey FO. Thalidomide update: regulatory aspects. Teratology 1988;38:221–6.
4. Mellin GW, Katzenstein M. The saga of thalidomide. Neuropathy to embryopathy, with case reports of congenital anomalies. N Engl J Med 1962;267:1238–44 concl.
5. Teo SK, Resztak KE, Scheffler MA, et al. Thalidomide in the treatment of leprosy. Microbes Infect 2002;4:1193–202.
6. D'Amato RJ, Loughnan MS, Flynn E, et al. Thalidomide is an inhibitor of angiogenesis. Proc Natl Acad Sci U S A 1994;91:4082–5.
7. Bauer KS, Dixon SC, Figg WD. Inhibition of angiogenesis by thalidomide requires metabolic activation, which is species-dependent. Biochem Pharmacol 1998;55:1827–34.
8. Fujita K, Asami Y, Murata E, et al. Effects of thalidomide, cytochrome P-450 and TNF-α on angiogenesis in a three-dimensional collagen gel-culture. Okajimas Folia Anat Jpn 2002;79:101–6.
9. Marks MG, Shi J, Fry MO, et al. Effects of putative hydroxylated thalidomide metabolites on blood vessel density in the chorioallantoic membrane (CAM) assay and on tumor and endothelial cell proliferation. Biol Pharm Bull 2002;25:597–604.
10. Kenyon BM, Browne F, D'Amato RJ. Effects of thalidomide and related metabolites in a mouse corneal model of neovascularization. Exp Eye Res 1997;64:971–8.
11. Price DK, Ando Y, Kruger EA, et al. 5'-OH-thalidomide, a metabolite of thalidomide, inhibits angiogenesis. Ther Drug Monit 2002;24:104–10.
12. Dredge K, Marriott JB, Macdonald CD, et al. Novel thalidomide analogues display anti-angiogenic activity independently of immunomodulatory effects. Br J Cancer 2002;87:1166–72.
13. Zhang H, Vakil V, Braunstein M, et al. Circulating endothelial progenitor cells in multiple myeloma: implications and significance. Blood 2005;105:3286–94.
14. Ng SS, Gutschow M, Weiss M, et al. Antiangiogenic activity of N-substituted and tetrafluorinated thalidomide analogues. Cancer Res 2003;63:3189–94.
15. Gupta D, Treon SP, Shima Y, et al. Adherence of multiple myeloma cells to bone marrow stromal cells upregulates vascular endothelial growth factor secretion: therapeutic applications. Leukemia 2001;15:1950–61.
16. Bellamy WT, Richter L, Sirjani D, et al. Vascular endothelial cell growth factor is an autocrine promoter of abnormal localized immature myeloid precursors and leukemia progenitor formation in myelodysplastic syndromes. Blood 2001;97:1427–34.
17. Lentzsch S, LeBlanc R, Podar K, et al. Immunomodulatory analogs of thalidomide inhibit growth of Hs Sultan cells and angiogenesis in vivo. Leukemia 2003;17:41–4.
18. Yabu T, Tomimoto H, Taguchi Y, et al. Thalidomide-induced antiangiogenic action is mediated by ceramide through depletion of VEGF receptors, and is antagonized by sphingosine-1-phosphate. Blood 2005;106:125–34.
19. Tamilarasan KP, Kolluru GK, Rajaram M, et al. Thalidomide attenuates nitric oxide mediated angiogenesis by blocking migration of endothelial cells. BMC Cell Biol 2006;7:17.
20. Dredge K, Horsfall R, Robinson SP, et al. Orally administered lenalidomide (CC-5013) is anti-angiogenic in vivo and inhibits endothelial cell migration and Akt phosphorylation in vitro. Microvasc Res 2005;69:56–63.
21. Sampaio EP, Sarno EN, Galilly R, et al. Thalidomide selectively inhibits tumor necrosis factor α production by stimulated human monocytes. J Exp Med 1991;173:699–703.
22. Meierhofer C, Dunzendorfer S, Wiedermann CJ. Theoretical basis for the activity of thalidomide. BioDrugs 2001;15:681–703.
23. Shannon EJ, Sandoval F. Thalidomide can be either agonistic or antagonistic to LPS evoked synthesis of TNF-α by mononuclear cells. Immunopharmacol Immunotoxicol 1996;18:59–72.
24. Moreira AL, Sampaio EP, Zmuidzinas A, et al. Thalidomide exerts its inhibitory action on tumor necrosis factor α by enhancing mRNA degradation. J Exp Med 1993;177:1675–80.
25. Boutten A, Dehoux M, Deschenes M, et al. Alpha 1-acid glycoprotein potentiates lipopolysaccharide-induced secretion of interleukin-1 β, interleukin-6 and tumor necrosis factor-α by human monocytes and alveolar and peritoneal macrophages. Eur J Immunol 1992;22:2687–95.
26. Turk BE, Jiang H, Liu JO. Binding of thalidomide to α1-acid glycoprotein may be involved in its inhibition of tumor necrosis factor α production. Proc Natl Acad Sci U S A 1996;93:7552–6.
27. Wnendt S, Finkam M, Winter W, et al. Enantioselective inhibition of TNF-α release by thalidomide and thalidomide-analogues. Chirality 1996;8:390–6.

28. Sampaio EP, Kaplan G, Miranda A, et al. The influence of thalidomide on the clinical and immunologic manifestation of erythema nodosum leprosum. J Infect Dis 1993;168:408–14.

29. Haslett PA, Corral LG, Albert M, et al. Thalidomide costimulates primary human T lymphocytes, preferentially inducing proliferation, cytokine production, and cytotoxic responses in the CD8+ subset. J Exp Med 1998;187:1885–92.

30. McHugh SM, Rifkin IR, Deighton J, et al. The immunosuppressive drug thalidomide induces T helper cell type 2 (Th2) and concomitantly inhibits Th1 cytokine production in mitogen- and antigen-stimulated human peripheral blood mononuclear cell cultures. Clin Exp Immunol 1995;99:160–7.

31. Davies FE, Raje N, Hideshima T, et al. Thalidomide and immunomodulatory derivatives augment natural killer cell cytotoxicity in multiple myeloma. Blood 2001;98:210–6.

32. Keifer JA, Guttridge DC, Ashburner BP, et al. Inhibition of NF-κB activity by thalidomide through suppression of IκB kinase activity. J Biol Chem 2001;276:22382–7.

33. Hideshima T, Chauhan D, Shima Y, et al. Thalidomide and its analogs overcome drug resistance of human multiple myeloma cells to conventional therapy. Blood 2000;96:2943–50.

34. Mitsiades N, Mitsiades CS, Poulaki V, et al. Apoptotic signaling induced by immunomodulatory thalidomide analogs in human multiple myeloma cells: therapeutic implications. Blood 2002;99:4525–30.

35. Gandhi AK, Kang J, Naziruddin S, et al. Lenalidomide inhibits proliferation of Namalwa CSN.70 cells and interferes with Gab1 phosphorylation and adaptor protein complex assembly. Leuk Res 2006;30:849–58.

36. Du GJ, Lin HH, Xu QT, et al. Thalidomide inhibits growth of tumors through COX-2 degradation independent of antiangiogenesis. Vascul Pharmacol 2005;43:112–9.

37. Payvandi F, Wu L, Haley M, et al. Immunomodulatory drugs inhibit expression of cyclooxygenase-2 from TNF-α, IL-1β, and LPS-stimulated human PBMC in a partially IL-10-dependent manner. Cell Immunol 2004;230:81–8.

38. Jin SH, Kim TI, Yang KM, et al. Thalidomide destabilizes cyclooxygenase-2 mRNA by inhibiting p38 mitogen-activated protein kinase and cytoplasmic shuttling of HuR. Eur J Pharmacol 2007;558:14–20.

39. Bartlett JB, Dredge K, Dalgleish AG. The evolution of thalidomide and its IMiD derivatives as anticancer agents. Nature Rev Cancer 2004;4:314–322.

40. Marriott JB, Muller G, Stirling D, et al. Immunotherapeutic and antitumour potential of thalidomide analogues. Expert Opin Biol Ther 2001;1:675–82.

41. Marriott JB, Westby M, Cookson S, et al. CC-3052: a water-soluble analog of thalidomide and potent inhibitor of activation-induced TNF-α production. J Immunol 1998;161:4236–43.

42. Muller GW, Corral LG, Shire MG, et al. Structural modifications of thalidomide produce analogs with enhanced tumor necrosis factor inhibitory activity. J Med Chem 1996;39:3238–40.

43. Muller GW, Chen R, Huang SY, et al. Amino-substituted thalidomide analogs: potent inhibitors of TNF-α production. Bioorg Med Chem Lett 1999;9:1625–30.

44. Ruegg C, Yilmaz A, Bieler G, et al. Evidence for the involvement of endothelial cell integrin αVβ3 in the disruption of the tumor vasculature induced by TNF and IFN-γ. Nat Med 1998;4:408–14.

45. Rajkumar SV, Blood E, Vesole D, et al. Phase III clinical trial of thalidomide plus dexamethasone compared with dexamethasone alone in newly diagnosed multiple myeloma: a clinical trial coordinated by the Eastern Cooperative Oncology Group. J Clin Oncol 2006;24:431–6.

46. Richardson PG, Mitsiades C, Hideshima T, et al. Lenalidomide in multiple myeloma. Expert Rev Anticancer Ther 2006;6:1165–73.

47. Chanan-Khan AA, Weber D, Dimopoulos M, et al. Lenalidomide in combination with dexamethasone improves survival and time to progression in elderly patients with relapsed or refractory multiple myeloma. Blood 2006;108:Abstract 3551.

48. List A, Dewald G, Bennett J, et al. Lenalidomide in the myelodysplastic syndrome with chromosome 5q deletion. N Engl J Med 2006;355:1456–65.

49. Bennett CL, Angelotta C, Yarnold PR, et al. Thalidomide- and lenalidomide-associated thromboembolism among patients with cancer. Jama 2006;296:2558–60.

50. Kulke MH, Stuart K, Enzinger PC, et al. Phase II study of temozolomide and thalidomide in patients with metastatic neuroendocrine tumors. J Clin Oncol 2006;24:401–6.

51. Kumar S, Witzig TE, Rajkumar SV. Thalidomid: current role in the treatment of non-plasma cell malignancies. J Clin Oncol 2004;22:2477–88.

52. Dahut WL, Gulley JL, Arlen PM, et al. Randomized phase II trial of docetaxel plus thalidomide in androgen-independent prostate cancer. J Clin Oncol 2004;22:2532–9.

53. Mathew P, Logothetis CJ, Dieringer PY, et al. Thalidomide/estramustine/paclitaxel in metastatic androgen-independent prostate cancer. Clin Genitourin Cancer 2006;5:144–9.

54. Figg WD, Li H, Sissung T, et al. Pre-clinical and clinical evaluation of estramustine, docetaxel and thalidomide combination in androgen-independent prostate cancer. BJU Int 2007;99:1047–55.

55. Figg WD, Retter A, Steinberg SM, et al. In reply, Inhibition of Angiogenesis: Thalidomide or low-molecular-weight heparin? J Clin Oncol 2005;23:2113–4.

56. Figg WD. The 2005 Leon I. Goldberg Young Investigator Award Lecture: Development of thalidomide as an angiogenesis inhibitor for the treatment of androgen-independent prostate cancer. Clin Pharmacol Ther 2006;79:1–8.

57. Ning YM, Arlen P, Gulley JL, et al. A phase II trial of thalidomide, bevacizumab, and docetaxel in patients with metastatic androgen-independent prostate cancer. J Clin Oncol 2007;25:Abstract 5114.

58. Efstathiou E, Troncoso P, Wen S, et al. Initial modulation of the tumor microenvironment accounts for thalidomide activity in prostate cancer. Clin Cancer Res 2007;13:1224–31.

59. Clark PE, Hall MC, Miller A, et al. Phase II trial of combination interferon-α and thalidomide as first-line therapy in metastatic renal cell carcinoma. Urology 2004;63:1061–5.

60. Vaishampayan UN, Heilbrun LK, Shields AF, et al. Phase II trial of interferon and thalidomide in metastatic renal cell carcinoma. Invest New Drugs 2007;25:69–75.

61. Lee CP, Patel PM, Selby PJ, et al. Randomized phase II study comparing thalidomide with medroxyprogesterone acetate in patients with metastatic renal cell carcinoma. J Clin Oncol 2006;24:898–903.

62. Amato RJ, Morgan M, Rawat A. Phase I/II study of thalidomide in combination with interleukin-2 in patients with metastatic renal cell carcinoma. Cancer 2006;106:1498–506.

63. Choueiri TK, Dreicer R, Rini BI, et al. Phase II study of lenalidomide in patients with metastatic renal cell carcinoma. Cancer 2006;107:2609–16.

64. Chanan-Khan A, Miller KC, Musial L, et al. Clinical efficacy of lenalidomide in patients with relapsed or refractory chronic lymphocytic leukemia: results of a phase II study. J Clin Oncol 2006;24:5343–9.

65. Wiernik PH, Lossos IS, Tuscano J, et al. Preliminary results from a phase II study of lenalidomide oral monotherapy in relapsed/ refractory aggressive non-Hodgkin lymphoma. J Clin Oncol 2007;25:Abstract 8052.

66. Wang M, Delasalle K, S. G, et al. Rapid control of previously untreated multiple myeloma with bortezomib-thalidomide-dexamethasone followed by early intensive therapy. Blood 2005;106:231a (abstract 784).

Chapter 35
TNP-470: The Resurrection of the First Synthetic Angiogenesis Inhibitor

Hagit Mann-Steinberg and Ronit Satchi-Fainaro*

Keywords: TNP-470, fumagillin, angiogenesis, vascular permeability, polymer therapeutics, clinical trials, first angiogenesis inhibitor

Abstract: TNP-470, an analogue of a naturally secreted antibiotic of Aspergillus fumigatus fresenius – named fumagillin, has been shown to inhibit angiogenesis in vitro and in vivo. TNP-470 was one of the first antiangiogenic compounds to enter clinical trials for cancer, making it a valuable prototype for future trials of angiogenesis inhibitors in oncology. It was halted in clinical Phase II trials following its evaluation in Kaposi's sarcoma, renal cell carcinoma, brain cancer, breast cancer, cervical cancer and prostate cancer. In early clinical reports, TNP-470 was tolerated up to 177 mg/m2 with neurotoxic effects being the principal dose limiting toxicity. To date, there are 500 papers published on TNP-470. They describe the antiangiogenic activity of TNP-470 including its anti-inflammatory, anti-infective and anti-obesity effects. Oxidative stress reduction and hyperpermeabilty decrease by TNP-470 account for yet another anti-cancer effect of this molecule. Mechanistic studies have identified cell cycle mediators and the protein methionine aminopeptidase-2 (MetAP-2) as molecular targets of TNP-470. Cell-cycle inhibition by TNP-470 is mediated at least in part by an activation of p21WAF1/CIP1 because of a p53-dependent mechanism, with reduction of the cyclin D-Cdk4 and cyclin E-Cdk 2 expression. Furthermore, it has been recently shown that FGFR1/PI3K/AKT signaling pathway is a novel target for the antiangiogenic effects of TNP-470. Co-administration of TNP-470 with chemotherapy increased drug delivery to the tumor followed by increased tumor response to cytotoxic therapy, implicating a potential role for these agents in combination treatment of solid tumors. Although TNP-470 is not approved for systemic use, this compound has one of the highest efficacies for the treatment of the broadest spectrum of tumor types. TNP-470 does exhibit side effects such as neurotoxicity that have deterred its acceptance as a viable treatment, but the current body of research on the synthesis of novel analogs and formulations of this molecule shows an exciting and promising "resurrection" of this drug as a potent antiangiogenic agent.

Introduction: The Discovery of TNP-470

Angiogenesis, the formation of new blood vessels from pre-existing vasculature is considered one of the most promising therapeutic targets to starve cancers and prevent metastases [1]. TNP-470, (O-(chloroacetylcarbamoyl) fumagillol), is a synthetic analog of fumagillin, a compound secreted by the fungus *Aspergillus fumigatus fresenius* which has antiangiogenic properties. The initial studies demonstrating its angio-inhibitory effects were done in the mid-1980s by Donald Ingber, a postdoctoral fellow in the laboratory of Judah Folkman at Harvard Medical School and Children's Hospital, Boston, at the time [2]. An accidental fungal contamination of an endothelial cell culture caught Ingber's attention when he noted that the adjacent monolayer of endothelial cells (ECs) near the contaminant had rounded to an abnormal morphology, suggesting that their ability to proliferate was impaired. With the industrial support of Takeda Chemical Industries Ltd., the fungus was identified as *Aspergillus fumagatus fresenius* and the active compound was isolated and determined to be fumagillin (dicyclohexylamine), a known polyene macrolide antibiotic used historically to treat amebiasis in humans and Nosema bombi disease in bees (marketed under the trade names of Fumidil B and/or Nosem-X). Fumagillin was originally patented by Upjohn, a pharmaceutical company, in 1953. In the rush to find new antibiotics to replace penicillin for humans, many antibiotics were developed and tested. Fumagillin was found to have no obvious potential for humans and Upjohn did not keep it. In 1957, Abbott Laboratories patented the product Fumidil B® for the treatment of Nosema apis in honeybees. About the same time, a Hungarian company,

Department of Physiology and Pharmacology, Sackler School of Medicine, Tel Aviv University, Ramat Aviv, Tel Aviv 69978, Israel
E-mail: ronitsf@post.tau.ac.il

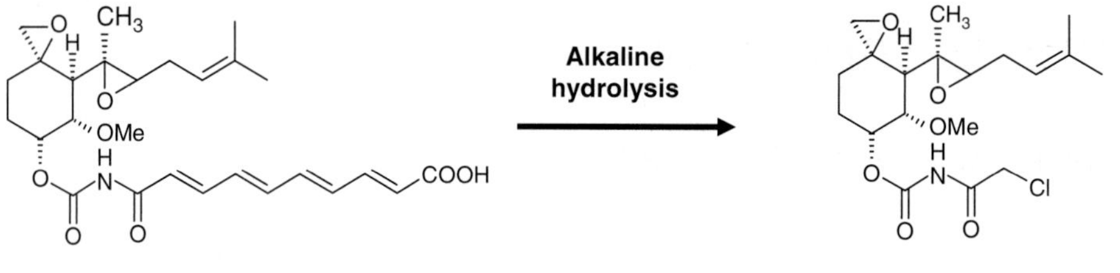

FIG. 35.1. Chemical structure of fumagillin and its synthetic derivative TNP-470.

called Chinoin, infringed on the rights to fumagillin and began marketing in Eastern Europe. Since fumagillin was a minor product in Abbott's huge range of pharmaceuticals they found it more convenient to use Chinoin as the source of fumagillin than to pursue patent infringement. A long list of drugs has been tested for treatment of nosema but so far nothing works as well as fumagillin. There has been no reported drug resistance of Nosema apis.

Subsequently, fumagillin was demonstrated to inhibit endothelial cell proliferation *in vitro* and to inhibit vessel growth *in vivo* [2]. Due to the induction of severe weight loss in mice, the compound was not used in models of *in vivo* tumor growth and instead the search began for synthetic fumagillin analogues that would be more potent than fumagillin with fewer limiting side-effects. Among the over 100 compounds synthesized, the most potent was O-(chloroacetylcarbamoyl) fumagillol, named TNP-470 (Takeda Neoplastic Product-470, also known as Angiogenesis Modulator; AGM-1470), in which the deca-tetraenoyl group of fumagillin was replaced with the chloroacetylcarbamoyl moiety (Fig. 35.1). Indeed, TNP-470 showed increased potency for angiogenesis inhibition and its reduced side effects have permitted its use in animal models of tumorigenesis as well as in human clinical trials.

In Vitro Characterization of TNP-470

The fact that endothelial cells go through several processes when they form new blood vessels raised the basis to investigate the antiangiogenic properties of TNP-470 in most of the common *in vitro* angiogenesis assays including EC proliferation assays, EC migration assays, capillary-like tube formation assays and organ culture assays.

TNP-470 Inhibits Endothelial Cell Proliferation

The inhibitory effect of TNP-470 on endothelial cell proliferation was evaluated by several methods, including measuring ^{3}H-thymidine incorporation [3], reduction of tetrazolium salts such as MTT or XTT in the mitochondria of living endothelial cells or by direct cell count [4]. The *in vitro* proliferation experiments have shown that TNP-470 inhibits endothelial cell

proliferation dose-dependently with an IC_{50} in the low picomolar range, approximately 50-fold more potent than its natural analog fumagillin [2, 5, 6]. Kusaka *et al.* reported that picomolar concentrations of TNP-470 has a cytostatic effect, arresting ECs in the G1 phase of the cell cycle, while micromolar concentrations of the drug are cytotoxic [5]. The cytostatic inhibition of the cells was reversible and endothelial cell proliferation resumed 4 days after withdrawal of TNP-470 from the culture medium [5].

TNP-470 Inhibits Endothelial Cell Migration

The effect of TNP-470 on growth factor-induced endothelial cell migration was evaluated by a modified Boyden chamber assay, in which ECs are placed on the upper layer of a cell-permeable filter and permitted to migrate in response to a chemoattractant factor placed in the cultured medium below the filter [7]. Migration was assessed by counting the number of cells that migrated through the filter toward a chemoattractant, such as vascular endothelial growth factor (VEGF) or basic fibroblast growth factor (bFGF) [8, 9]. Satchi-Fainaro *et al.* showed that treatment with TNP-470 dramatically inhibited the chemotactic migration of human microvascular endothelial cells (HMVECs) in response to VEGF by 68% [8]. TNP-470 also inhibited basal migration of HMVEC in the absence of VEGF by 70% [8].

TNP-470 Inhibits Endothelial Sprouting

The recognition that angiogenesis *in vivo* involves not only endothelial cells but also their surrounding cells has recently led to a shift to assess angiogenesis by organ culture methods [7]. An *ex vivo* model of chick aortic ring implanted in Matrigel was used to determine the effect of TNP-470 on endothelial sprouting. TNP-470 was found to reduce the number and length of vascular sprouts growing from the chick aortic ring at 50 pg/ml and completely prevent outgrowth at 100 pg/ml [10].

TNP-470 Disrupts Capillary-like Tube Formation

The effect of antiangiogenic compounds on endothelial tube formation is usually assessed by Matrigel assay or collagen I cultures in which endothelial cells are able to form three-dimensional structures. Using this method, Vander Schaft *et al.* demonstrated that TNP-470 could inhibit vascular cord and tube formation by HMVECs and human umbilical vein endothelial cells (HUVECs) [11]. Tube formation on the Matrigel matrix is not specific for ECs, since primary human fibroblasts and a number of tumor cell lines also form tubes when propagated on the Matrigel matrix [12]. TNP-470 was found not to affect tubular network formation by human melanoma MUM-2B and C8161 cells [11]. Kusaka *et al.* found that AGM-1470 (1–1,000 ng/ml) selectively inhibited the capillary-like tube formation of endothelial cells with a minimal effect on the non-endothelial cell growth [13]. Similar results have been reported by Rybak *et al.* in the B16F10 murine melanoma model underlining again the role of TNP-470 as an antiangiogenic agent [14]. As opposed to Matrigel assay where cancer and ECs form tubes, co-culture with fibroblasts allows only endothelial cells to form capillary-like networks [15]. Using this method, Friis *et al.* has demonstrated that network formation by HUVECs was inhibited by 1 μM of TNP-470 [15]. This result further underlines the specific antiangiogenic effects of TNP-470.

TNP-470 Affects Endothelial Cell Morphology

The effect of TNP-470 on the morphology of endothelial cells was evaluated by staining the cytoskeleton of the cells with fluorescently-labeled phallacidin that labels F-actin [16]. HUVECs treated with TNP-470 were reported to possess enhanced actin stress fibers and reduced number of cellular projections ending in their lamellipodia. Focal adhesions, the sites where stress fibers connect with the extracellular matrix, were also enhanced in the TNP-470-treated cells compared to the untreated ones [16]. These results suggest that TNP-470 alters endothelial cell morphology via its effect on cytoskeletal organization.

The effect of TNP-470 on EC morphology was also demonstrated on endothelial cells propagated in coculture with fibroblasts. The untreated cells were shown to form capillary-like networks of tubes and to retain important ultrastructural and physiological properties of endothelial cells [15]. Electron microscopy experiments revealed that the cells contained Weibel-Palade bodies, transport vesicles, diluted rough endoplasmic reticulum and mitochondria with homogeneous mitochondrial matrix, as well as regular arrangement of cytoskeletal filaments and microtubules. Treatment with 1 μM TNP-470 inhibited network formation, and resulted in EC morphology of short cords with only a few interconnections, suggesting that TNP-470 inhibits endothelial cell cord assembly and canalization [15]. Nevertheless, it should be noted that no morphological changes in microvascular structures were observed when tissues of mice treated with TNP-470 were examined with transmission electron microscope [8]. Compared to untreated tissue, vesiculo-vacuolar organelles (VVOs)

were normal with respect to frequency and size, although trancytosis of ferritin through VVOs was impaired, interendothelial cell junctions were normally closed, and fenestrae remained rare. In a different study, Nahari *et al.* showed that TNP-470 prevented VEGF-induced endothelial permeability, intercellular gap formation and ruffle formation by preventing Rac1 activation [17].

High-dose TNP-470 Inhibits the Proliferation of Non-Endothelial Cells

In addition to its effect on endothelial cells, TNP-470 was also shown to have an inhibitory effect on non-endothelial cells at higher concentrations. The literature includes several studies that demonstrate the inhibitory effect of TNP-470 on the proliferation of cancer cell lines. While EC proliferation was inhibited by picograms of TNP-470, inhibition of most cancer cell lines by TNP-470 required 3–4 logs higher concentrations of the drug [18, 19]. Differences in the effective doses of TNP-470 have been observed between monolayer cultures and soft agar cultures of PC-3 human prostate carcinoma cells and MDA-MB-231 human breast carcinoma cells [20]. Yamaoka *et al.* reported that the IC_{50} values were significantly lower when cells were grown in soft agar than in monolayer culture. As far as we know, the only cancer cell lines reported to be inhibited by TNP-470 at doses similar to its antiangiogenic range were U-87 MG human glioblastoma [21] and FU-MMT-1 human cells derived from a primary carcinosarcoma of the uterus [22], although the effects on the U-87 MG cell line are controversial [8]. Several normal non-endothelial cells were also reported to be suppressed by higher concentrations of the drug [23, 24]. Nahari *et al.* showed that TNP-470 had direct Cytotoxic Effects on Anaplastic thyroid Carcinoma Cells (DRO'90) [17] *in vitro*. Paradoxically, TNP-470 Increased VEGF secretion from Tumor Cells *in vitro* [17].

In Vivo Characterization of the Activity of TNP-470

The *in vivo* antiangiogenic activity of TNP-470 was evaluated in a wide variety of animal models. TNP-470 was administered in a variety of doses and schedule regimens, and its effects on tumor growth, metastases and intratumor neovascularization were determined. TNP-470 was reported to inhibit the growth of a broad spectrum of tumors and was found to be most effective when it was administered subcutaneously every other day (s.c. q.o.d.) at 30 mg/kg [18].

In Vivo Angiogenesis Models

Kusaka *et al.* have evaluated the antiangiogenic effect of TNP-470 in several *in vivo* angiogenesis models [13]. Locally administered TNP-470 inhibited angiogenesis in the chick embryo chorioallantoic membrane (CAM) assay and in the rat corneal micropocket assay. In the CAM assay, TNP-470 inhibited capillary growth by local administration and induced avascular zones in the CAM in a dose-dependent manner. In the rat corneal micropocket assay, pellets containing TNP-470 at 20 μg per cornea suppressed the number and length of new blood vessels induced by 250-ng pellets of bFGF, a potent angiogenic factor. The rat sponge implantation assay demonstrated that systemic administration of 30 mg/kg of TNP-470 was able to inhibit angiogenesis induced by bFGF and unknown factors involved in inflammation.

TNP-470's antiangiogenic activity was also evaluated on the partial hepatectomy model. This non-neoplastic model is a relatively rapid (8 days) *in vivo* angiogenesis-dependent process [25]. Liver regeneration post partial hepatectomy is angiogenesis-dependent and thus highly regulated by endothelial cells, similar to tumor growth [25]. On day 0 following partial hepatectomy, liver weight was approximately 0.4 g. Following partial hepatectomy, control mice regenerated their resected livers to their pre-operative mass (~1.2 g) by post-operative day 8. In mice treated with TNP-470 at 30 mg/kg s.c. q.o.d., the regeneration of the liver was inhibited and livers reached the average size of 0.7 g on post-operative day 8 [10, 25].

In Vivo Primary Murine Tumor Models

The effect of TNP-470 on tumor growth was evaluated in a wide variety of tumor models, using different animal species such as mice, rats, rabbits and hamsters. As a monotherapy, TNP-470 was shown to impair and decrease tumor growth rate [2]. Systemic administration of TNP-470 to mice bearing different types of solid tumors was found to inhibit the growth of both blood vessels and tumor size [26]. Systemic administration of TNP-470 to mice bearing B16BL6 melanoma, M5076 reticulum cell sarcoma, Lewis lung carcinoma, and Walker 256 carcinoma, implanted subcutaneously in C57Bl/6 mice, inhibited tumor size in a dose-dependent manner [27]. In another study, TNP-470 significantly reduced primary tumor volumes in mice injected with either TBJ or C1300 murine neuroblastoma cell lines [28]. TNP-470 was also shown to inhibit other primary tumor types as presented in Tables 35.1 and 35.2. In most of these tumor models, TNP-470 treatment resulted in tumor growth inhibition and in some cases, the antitumor effect of TNP-470 was accompanied by an increase in tumor necrosis and an increase in mean survival [27–29]. Furthermore, some studies show that inhibition of tumor growth by TNP-470 was correlated with decrease in vascularization of the tumor indicating that the antiangiogenic effect of TNP-470 was responsible for the observed inhibition [2, 27].

In Vivo Human Tumor Models

TNP-470 was effective in decreasing the growth of many human xenograft models. Administration of TNP-470 to severe combined immunodeficiency (SCID) mice bearing slowly growing (RT-4) and rapidly growing (MGH-U1) human bladder carcinoma cell lines decreased the tumor volume and

TABLE 35.1. *In vivo* studies of TNP-470 as monotherapy on primary and metastatic tumor models.

Tumor model	Percent Inhibition	Location of metastases	Reference
Primary murine tumors			
TBJ neuroblastoma	60		[28]
Retinoblastoma	60		[115]
Lewis lung carcinoma	67–81		[2, 70, 83, 116, 10, 117]
drug-resistant Lewis lung carcinoma	100 (in combination with cytoxan)		[118]
C-1300 neuroblastoma	67		[28]
Mammary carcinoma	70		[119]
B16 melanoma	71–74		[2, 117]
Colon 38 carcinoma	75		[120]
Renal cortical adenocarcinoma	77		[121]
MCA-105 fibrosarcoma	83		[120]
Hemangioendothelioma	90		[122]
M5076 reticulum cell sarcoma	91		[26]
B16BL6	70		[123]
Murine angiosarcoma cell line (ISOS-1)	55		[124]
Murine neuroblastoma cell line, C-1300	80		[125]
CT-26 colonic adenocarcinoma cell line	80		[45]
Sarcoma 180	77		[117]
SVR Murine Angiosarcoma	90		[126]
Human Tumors			
Gastric carcinoma	43		[39]
Embryonal rhabdomyosarcoma	47		[127]
Ovarian carcinoma	60		[18]

(continued)

TABLE 35.1. (continued)

Tumor model	Percent Inhibition	Location of metastases	Reference
Choriocarcinoma	60		[18]
Colon carcinoma	61		[128]
Meningioma (benign)	63		[129]
Medulloblastoma	66		[130]
Prostate carcinoma	67		[20]
MDA-MB-231 breast carcinoma	72		[20]
T98G glioblastoma	72		[131]
Meningioma (malignant)	77		[129]
MCF-7 breast carcinoma	80		[132]
U87 glioblastoma	95		[21, 133]
Breast carcinoma	96		[134]
Neurofibtosarcoma	97		[135]
Neurofibroma	100		[135]
Acoustic neuroma	100		[135]
Angiosarcoma	84		[136]
Neuroblastoma	47 (prevention)		[137]
Human squamous cell nasopharyngeal carcinoma NPC/HK1	41.3		[138]
Superficial transitional cell carcinoma (TCC) cell line (KK-47)	67		[139]
Invasive TCC cell line (MGH-U1)	68		[139]
Neuroblastoma	60		[140]
Tamoxifen-stimulated MCF-7 breast tumors	50		[141]
Pancreas adenocarcinoma PCI-43	prominent suppression of the establishment of peritoneal nodules		[142]
MYCN-amplified, human neuroblastoma cell line NBL-W-N	Significantly improved tumor-free survival at 12 weeks, and overall survival at 45 weeks, if it is administered in the setting of microscopic disease		[143]
Orthotopic bladder carcinoma	90		[144]
Human bladder cancer cell line 639 v	delayed 639 V tumor appearance but did not reduce its growth		[145]
Human bladder cancer cell line T24	TNP-470 did not significantly reduce T24 tumor incidence but it significantly delayed the time of tumor appearance and significantly slowed the tumor growth rate 80		[145]
A2058 human melanoma	66		[10]
Human uterine carcinosarcoma cell line, FU-MMT-1	64		[36]
Human pancreatic cell line PC-3	64		[146]
Human colon adenocarcinoma cell line Lovo	54–56		[147, 148]
Tumors in other animal species			
Walker 256 carcinosarcoma (rat)	64		[149]
Pancreatic adenocarcinoma (hamster)	67		[150]
Fibrosarcoma A5653HM (rat)	69		[151]
Osteosarcoma (rat)	92		[152]
VX-2 carcinoma (rabbit)	96		[153, 154]
Rat urinary bladder tumors	67% decrease in foci number, 86% decrease in tumor burden by treatment of TNP-470 at 30 mg/kg i.p.		[32]
Rhabdomyosarcomas (rat)	significant reduction of the growth rate, even for relatively large-sized (> 7 cm^3) tumors		[155]
Ascites hepatoma AH-130 cell line (rat)	73		[156]
Neuroectodermic tumor (rat)	92		[35]
Metastatic tumors			
H59 carcinoma	50	Lung	[157]
MCA-105 fibrosarcoma (mouse)	67	Lung	[46]

(continued)

TABLE 35.1. (continued)

Tumor model	Percent Inhibition	Location of metastases	Reference
Lewis lung carcinoma-L1 (mouse)	69	Lung	[46, 75, 157]
Fibrosarcoma A5653HM (rat)	69	Lymph nodes	[151]
GCH-1 choriocarcinoma (human)	73	Lung	[18]
TBJ neuroblastoma (mouse)	81	Lymph nodes	[28]
C-1300 neuroblastoma (mouse)	82	Lymph nodes	[28]
B16 B16 melanoma (mouse)	87	Lung	[74]
AH-130 hepatoma (rat)	89	Liver	[158, 159]
Bomirski Ab melanoma (hamster)	89	Lung	[160, 161]
B16F10 melanoma (mouse)	90	Lung	[157]
M27 lung carcinoma (mouse)	90	Lung	[157]
Hepatocellular carcinoma (human)	90	Liver	[38]
VX-2 carcinoma (rabbit)	92	Liver	[162, 163]
Renal adenocarcinoma (mouse)	92	Lung and liver	[41]
Colon adenocarcinoma (human)	93	Liver	[40, 164]
LM8 osteosarcoma (mouse)	97	Lung	[165]
M5076 reticulum sarcoma (mouse)	98	Liver	[26]
Breast carcinoma (human)	100	Lymph nodes	[134]
NUC-1 choriocarcinoma (human)	100	Lung	[18]
M5076 reticulum sarcoma (mouse)	100	Lung	[26]
Az-H5c gastric carcinoma (human)	100	Liver	[166]
MT-5 gastric carcinoma	100	Liver	[39]
MT-2 and MT-5 gastric cancer.	~36% inhibition of MT-2 and MT-5 tumor growth.	Liver	[159]
Hepatic metastatic model of rat hepatoma, AH-130	Prolonged survival for 4 months. Better survival rate at 15 mg/kg than 30 mg/kg.	No evidence of mac-roscopic metastatic foci after 4 months of treatment.	
B16BL6	85	Lung	[123]
DHD K12 colon carcinoma (rat)	77	Liver	[167]
KM12SM human colon cancer in nude rat	50–90	Lung	[168]
Colon 26/TC-1	60–64 in combination with fibronectin-binding domain	Liver	[169]
Mouse pancreatic adenocarcinoma cell line panc02	40	Lymph nodes and Liver	[89]
Human transitional cell carcinoma 253J B-V cells transplanted orthotopically in the urinary bladder	45–76 non-established 35–81 established	Lymph nodes	[31]
Human anaplastic thyroid carcinoma cells (DRO'90) into thyroid glands of nude mice.	TNP-470 Inhibits tumor growth, reduces metastases, and prolongs survival in Anaplastic thyroid cancer TNP-470 inhibits proliferation, increases apop-tosis, and increases VEGF secretion in thyroid tumors *in vivo*.	Liver (16.6% in TNP-470-treated mice ver-sus 50% in control mice)	[17]
Transgenic mice models			
Transgenic mouse model of pancreatic islet cell carcinogenesis (RIP1-Tag2).	11.7% reduction in angiogenic islets and 82% reduction in tumor burden in an intervention trial. 58.8% reduction in tumor burden in a regression trial.		[42]
Neuroblastoma TH-*MYCN* in transgenic mice created by targeted expression of *MYCN* to the neural-crest (under control of the rat tyrosine hydroxylase promotor).	Near complete ablation (~100%), with reduced proliferation, enhanced apoptosis, and vasculature disruption.		[43]

TABLE 35.2. In vivo studies of TNP-470 combined with cytotoxic agents.

Combined agent		TNP-470	Tumor type	Effect	References
Name	Dosing schedule	Dosing schedule			
Adriamycin	15–17.5 mg/kg i.p. 6–7 fractionated dosing schedule	120–140 mg/kg, s.c. 6–7 fractionated dosing schedule	B16BL6 melanoma	Reduction of tumor size.	[74]
			Lewis lung carcinoma		
BCNU +	15 mg/kg i.p days, 7,9,11	30 mg/kg q.o.d. s.c	EMT-6 mammary carcinoma	Increased tumor growth delay.	[80]
			Lewis lung carcinoma		
Minocyclin	10 mg/kg i.p. every day		9 L gliosarcoma		
Bromocriptine	300 μg/kg q.o.d. s.c.	50 mg/kg q.o.d. s.c.	Estrogen-induced pituitary tumors in rats	Suppressive effect on tumor vasuculture.	[170]
CHS 828 (Pyridyl cyanoguanidine)	20 mg/kg orally	15–30 mg/kg q.o.d. s.c	SH-SY5Y human neuroblastoma	Reduction of tumor volume.	[171]
Cisplatin (CDDP)	1.25 mg/kg, i.v. on days 21 and 24	2.5 mg/kg /week by Alzet osmotic pumps on days 7–21	Rat osteosarcoma	Increased anti-metastatic effect on lung metastases.	[172]
	4 mg/kg i.p. 6–7 fractionated dosing schedule	120–140 mg/kg, s.c 6–7 fractionated dosing schedule	B16BL6 melanoma	Reduction of tumor size	[74]
			Lewis lung carcinoma		
	0.25 mg/kg twice a week i.p.	30 mg/kg, twice a week i.p.	Rat model of bladder cancer	Decrease in the tumor vessel density with no added effect on tumor cell proliferation	[173]
Cisplatin (CDDP) + Minocyclin	10 mg/kg i.p. day 7	30 mg/kg q.o.d. s.c	EMT-6 mammary carcinoma	Increased tumor growth delay	[80]
	10 mg/kg i.p. every day		Lewis lung carcinoma 9 L gliosarcoma		
Cyclophosphamide	170 mg/kg, every 6 days s.c.	12.5 mg/kg every 6 days s.c.	Drug resistant Lewis lung carcinoma	84% eradication of cancer	[118]
Cyclophosphamide + Minocyclin	150 mg/kg i.p days, 7,9,11. 10 mg/kg i.p. every day	30 mg/kg q.o.d. s.c	EMT-6 mammary carcinoma Lewis lung carcinoma 9 L gliosarcoma	Increased tumor growth delay	[80]
Cyclophosphamide +	500 mg/kg i.p.	30 mg/kg q.o.d. s.c	Lewis lung carcinoma	Increase in tumor growth delay and 40–50% long term survivors animals	[75]
Minocyclin	10 mg/kg i.p. every day		FSaIIC fibrosarcoma	Increase in tumor cell killing	
Docetaxel	20 mg/kg/week, i.p.	105 mg/kg/week s.c.	Human metastatic transitional cell carcinoma (TCC) 253 J B-V	Increased anti-tumor and anti- metastatic effect	[73]
			PC-3 human prostate cancer	Increased tumor growth delay	[91]
Doxorubicin (ADR)	Single i.a. injection of sesame oil solution containing 1 mg ADR	Single i.a. injection of sesame oil solution containing 1 mg or 5 mg TNP-470	Rabbit VX-2 carcinoma	Increased anti-tumor effect	[153]
5-Fluorouracil	140 mg/kg i.p. days 3,5	150 mg/kg, s.c. days 3,5	B16BL6 melanoma	Reduction in tumor size	[74]
Gemcitabine	50 mg/kg, q.o.d. i.p.	30 mg/kg q.o.d. s.c	SW1990 human pancreatic carcinoma	Increased anti-tumor and anti-metastatic effect and improvement in survival rate.	[174]

(continued)

TABLE 35.2. In vivo studies of TNP-470 combined with cytotoxic agents.

Combined agent		TNP-470			
Name	Dosing schedule	Dosing schedule	Tumor type	Effect	References
	60 mg/kg i.p. once a week	15 mg/kg s.c. daily	Human bladder cancer KoTCC-1	Increased anti-tumor and anti-metastatic effect	[175]
Mitomycin C (MMC)	5 mg/kg i.p. days 3,5,7,9,11	150 mg/kg, s.c. days 3,5,7,9,11	B16BL6 melanoma (B16M)	Reduction in tumor size and pulmonary metastasis	[74]
			Lewis lung carcinoma		
Paclitaxel + Carboplatin + Minocyclin	36 mg/kg i.v. days 7–11 50 mg/kg i.p. day 7 10 mg/kg i.p. every day	30 mg/kg q.o.d. s.c	Lewis lung carcinoma	Increased tumor growth delay and decreased the number of lung metastases	[176]
rhTNF	3 or 8 microgram, by single i.v bolus injection	30 or 60 mg/kg s.c. 2 days pretreatment	MC38 mouse colon adenocarcinoma	80% reduction of tumor volume	[72]
Temozolomide (TMZ)	40 mg/kg intra-arterial on day 15	30 mg/kg s.c on days 6, 8, 10, 12, and 14	Rat C6 glioma	Reduction in efficacy against primary tumor cells	[177]

the blood vessel density in both tumor types [30]. Frequent administration of TNP-470 at an optimal biological dose provided maximal anti-tumor and anti-metastatic effects of human transitional cell carcinoma (TCC) of the urinary bladder [31].

The effect of the antiangiogenic agent TNP-470 on tumor growth, vascular area, vascular density and tumor perfusion was evaluated in a variety of human tumor models in mice. In the majority of the *in vivo* models treated with TNP-470, a reduction in number of microvessels surrounding tumor tissues was observed, indicating that the observed anti-tumor effect was mainly by the inhibition of angiogenesis [32–34]. However, there are some reports showing that the inhibition of tumor growth *in vivo* was not associated with corresponding decrease in tumor vasculature [35, 36]. For example, two subcutaneously implanted human glioma xenografts (E98 and E106) in nu/nu mice showed a small but significant tumor growth suppression. However, no differences in vascular parameters between TNP-470-treated tumors and controls could be found after 6 weeks of treatment [37]. Nahari *et al.* tested the effect of TNP-470 at 30 mg/kg, subcutaneously for 6 weeks on nu/nu mice bearing orthotopic human anaplastic thyroid carcinoma cells (DRO'90) [17]. TNP-470 prolonged survival and reduced liver metastases. TNP-470 had direct cytotoxic effects on anaplastic thyroid carcinoma cells *in vivo* without associated increase in tumor microvessel density [17].

A detailed list of the broad spectrum of tumor types inhibited by TNP-470 as monotherapy is described in Table 35.1 and as combination therapy with other chemotherapeutic agents in Table 35.2.

In Vivo Metastatic Models

In vivo therapy with TNP-470 was found to decrease the number and volume of metastases in several metastatic models. Xia *et al.* demonstrated the ability of TNP-470 to inhibit lung metastases in the highly metastatic model of human hepatocellular

carcinoma-LCI-D20 [38]. Kanai *et al.* showed that early treatment of MT-2 gastric cancer with TNP-470 completely inhibited the development of macroscopic foci in the liver and was significantly more effective than late treatment [39]. Tanaka *et al.* reported that intra-arterial injection of TNP-470 is the most effective method for preventing liver metastases in VX2 carcinoma model in rabbits [40]. TNP-470 was also shown to inhibit lung metastases of choriocarcinoma [18], lung and liver metastases of renal cell carcinoma [41] and lymph node metastases of human transitional cell carcinoma [31]. These findings support the idea that the induction of new blood vessels is crucial for the metastatic process and that potent antiangiogenic agents can prevent metastases formation.

In Vivo Transgenic Models

The genetically engineered mouse model of pancreatic islet cell carcinogenesis, Rat Insulin Promoter, large T antigen (RIP1-Tag2), in which insulinoma tumors arise spontaneously in the pancreatic islets, was used in order to assess the effect of TNP-470 on distinct stages of the carcinogenesis process [42]. In this transgenic mouse model, an angiogenic switch occurs in pre-malignant lesions, and angiogenesis persists during progression to expanded solid tumors and invasive carcinomas. Using this model, Bergers *et al.* demonstrated that TNP-470 intervened in the rapid expansion of small tumors to induce the regression of large end-stage cancers and even to extend the life-span of the mice [42]. However, early treatment with TNP-470 at the hyperplastic stage could not block the angiogenic switch before the initial formation of solid tumors, indicating that TNP-470 was effective only after the tumors have recruited blood vessels. Furthermore, this study shows that TNP-470 can also affect tumor progression in animal models where the cancer arises from normal cells in their natural microenvironment and progresses through multiple stages, as human cancer does.

In a different study, TNP-470 was also shown to inhibit the growth of a transgenic neuroblastoma tumor [43]. It is known that the expression of *MYCN* is associated with enhanced neo-vascularization and poor clinical outcome in neuroblastoma. Weiss *et al.* developed a mouse model of neuroblastoma targeting overexpression of *MYCN* to the neural-crest (TH-*MYCN*) [44]. Chesler *et al.* have serially characterized malignant progression, neovascularization, and sensitivity to angiogenic blockade in these Mycn-driven murine neuroblastomas. Tumors were highly proliferative, displayed a complex vasculature, and expressed a number of biomarkers found in human neuroblastoma, including the angiogenic ligand VEGF, the angiogenesis related proteins VEGFR2 and α-SMA, and matrix metalloproteinases MMP-2 and MMP-9. Treatment of established murine tumors with TNP-470, and Caplostatin [see section "HPMA copolymer-Gly-Phe-Leu-Gly-TNP-470 (Caplostatin)"] caused near complete tumor ablation, with reduced proliferation, enhanced apoptosis, and vascular disruption. This study highlights the importance of angiogenic blockade in an immuno-competent mouse model for neuroblastoma with native tumor microenvironment interactions [43].

Preclinical Toxicity of TNP-470

With few exceptions, weight loss (around 10–20% in body weight) was the most common side effect of treatment with TNP-470. High doses of the drug were also associated with local skin lesions at the site of injection and increased lethality [45]. There were also two reports in which decrease in splenic weight was noted [28,46], and another study that mentioned diminished extramedullary erythropoiesis and disrupted liver morphology [47]. TNP-470 was also reported to cause ataxia [48] and to affect mice neurological functions such as motor coordination and balance [10]. In general, all preclinical toxicities were reversible as soon as treatment was withdrawn. These reports are in agreement with the neurotoxicity observed in clinical trials.

Mechanism of Action

The mechanism of action by which TNP-470 affects endothelial cell proliferation is not yet clearly understood even though much effort has been put in this direction. Studies exploring TNP-470's mechanism of action revealed its effect on methionine aminopeptidase type 2 (MetAP-2) [49, 50], on cell cycle [5] and on vessel hyperpermeability [8] and are described below.

The Effects of TNP-470 on MetAP-2

The effort to understand the mechanism of angiogenesis inhibition by TNP-470 had led to the isolation and identification of the molecular target of fumagillin/ TNP-470. This target was later identified as MetAP-2 [49, 50]. MetAP-2, also known

as eukaryotic initiation factor 2 (eIF2)-associated protein, is one of two enzymes that catalyze the removal of the initiator methionine during protein translation [51]. TNP-470 was found to bind MetAP-2 covalently leading to specific inhibition of its methionine aminopeptidase activity. This inhibition was found to be very specific since it did not affect the activity of its closely related isozyme MetAP-1 [49, 50]. The covalent bond formed between fumagillin and the active site of MetAP-2 was further confirmed by chemical modifications of fumagillin and site-directed mutagenesis [52], and by a high resolution crystal structure of free and inhibited human MetAP-2 [53]. Several lines of evidence strongly support the notion that MetAP-2 is not only inhibited by TNP-470 but also plays a critical role in the effect of TNP-470 on endothelial cell proliferation [54–56]. First, by using a series of analogs of fumagilln, a strong correlation has been found between inhibition of MetAP-2 enzymatic activity and inhibition of EC proliferation [55]. Second, by using biotin-fumagillin conjugate as a probe, a correlation was observed between the drug concentrations required to inhibit EC proliferation and the concentrations needed to inactivate the endogenous MetAP-2 [56]. Also, the structural information from MetAP-2 crystal structure has been used to rationally design fumagillin analogs with increased potency in endothelial cell culture [54]. Thus, inhibition of MetAP-2 is strongly connected to EC inhibition by TNP-470. However, a report showing that MetAP2-depleted endothelial cells remain responsive to inhibition by either fumagillin or by LAF389, a newly identified MetAP-2 enzyme inhibitor, has raised some doubts about the role of MetAP-2 in mediating the antiangiogenic effect of TNP-470 [57]. In order to solve the discrepancy between the results an *in vivo* approach was used [58]. Since loss of MetAP-2 resulted in embryonic lethality, a conditional MetAP-2 knockout mouse strain was generated. Ubiquitous deletion of the MetAP-2 gene (*MAP*2) resulted in an early gastrulation defect indicating that MetAP-2 function is essential for embryonic development at gastrulation. This finding is consistent with another report, which demonstrated that TNP-470-mediated MetAP-2 inhibition blocks non-canonical Wnt signaling, which plays a critical role in development, cell differentiation and tumorigenesis [59]. In addition to ubiquitous knockout of MetAP-2 gene, targeted deletion of *MAP2* specifically in the hemangioblast lineage was generated. This deletion resulted in abnormal vascular development indicating that MetAP-2 is required for murine vasculogenesis[58]. However, in order to define the *in vivo* physiological specific role of MetAP-2 in vascular biology, it would be interesting to examine whether deleting MetAP-2 gene in other lineages would also affect the generation of blood vessels or just affect the proliferation of any lineage affected by the deletion.

The Effects of TNP-470 on Cell Cycle

The action of TNP-470 on the cell cycle of endothelial cells is considered complex and most likely affects several key points of the process. The early studies showed that TNP-470 specifically inhibited DNA synthesis in ECs, indicating that this drug affects one or several steps of the cell cycle [5]. Other studies had conferred this notion and reported that TNP-470 inhibited neovascularization via endothelial cell cycle arrest in the late G_1 phase [60, 61]. TNP-470 was shown not to inhibit early mitogenic events, but potently inhibited the phosphorylation of retinoblastoma (Rb) protein, a tumor suppressor retinoblastoma gene product, and abolished the growth factor-induced mRNA expression of CDC2, cyclin A and cyclin E [60]. The same study mentioned little or no effect on the mRNA level of cyclin dependent kinase (CDK) 2, 4 and cyclin D1. In contrast to this study, Hori et al. reported that a 4 h-treatment with TNP-470 suppressed cyclin D1 mRNA expression in mid G1 phase [3]. This discrepancy with respect to the effect of TNP-470 on cyclin D1 mRNA was assessed by Lien et al., who reported that TNP-470 actually induced cyclin D1 in endothelial cells due to the enhanced expression of its mRNA [6]. The induced cyclin D1 was shown to form a complex with cyclin-dependent kinase 4 (CDK4) and with p21. The protein p21 (WAF1/CIP1) belongs to the Cip/Kip family of CDK inhibitors (CKIs) that antagonizes the activation of the complex cyclin D1 with CDK4 which facilitates exit from G1 by phosphorylating the Rb protein [62]. Indeed, the ability of cyclin D1-associated CDK4 to phosphorylate the Rb protein was reduced in TNP-470 treated cells [6]. Other studies demonstrated that TNP-470 requires the induction of p21(WAF1/CIP1) and p53 for endothelial cell growth arrest [63, 64]. The requirement of p53 and p21(WAF1/CIP1) for the cell cycle inhibition by TNP-470 was underscored by the observation that cells deficient in p53 and p21(WAF1/CIP1) are resistant to TNP-470 [63, 64]. The concurrent increase of cyclin D1 protein with the increase of p53 and p21 protein in the TNP-470-treated endothelial cells was not surprising since a proper G1 arrest mediated by p53 is often associated with up-regulation of the latter two proteins. TNP-470 was also shown to promote HUVECs to undergo senescence, possibly via increased formation of the complex cyclin D1 with CDK2, reminiscent of an inactive CDK2 complex found in senescent fibroblasts. This observation suggests that the cytostatic effect of TNP-470 on endothelial cells is in part mediated by induction of senescence and that cyclin D1 is a key molecule participating in this event. In addition, Mauriz et al. showed that the increase in in vivo levels of CDK2, CDK4, cyclin D and cyclin E in a rat hepatocelular carcinoma was prevented [65]. In the same model, TNP-470 inhibited oxidative stress, nitric oxide production and NF-κB activation induced by experimental hepatocarcinogenesis [66].

Effect of TNP-470 on Other Molecular Targets Affecting Endothelial Cells

There are single reports showing the effect of TNP-470 on VEGF secretion, on cytoskeletal regulatory proteins cofilin and hsp27, and on E-selectin. Miura et al. showed that TNP-470 decreases VEGF secretion from FU-MMT-1 uterine carcinosarcoma cells, thus inhibiting the VEGFR2 pathways and affecting the FU-MMT-1 induced human arterial endothelial

cell (HAEC) tube formation [22]. Keezer et al. demonstrated that TNP-470 affected cytoskeletal organization and phosphorylation of the cytoskeletal regulatory proteins cofilin and hsp27. Both of these proteins are involved in actin dynamics and are regulated through differential phosphorylation and subcellular localization. HUVECs treated with TNP-470 were also shown to own more actin stress fibers and focal adhesions compared to untreated cells [16]. Budson et al. reported that the mRNA and protein levels of E-selectin, an endothelial leukocyte adhesion molecule, were specifically increased in bovine capillary cells treated with TNP-470. The increase in E-selectin mRNA and protein was always greater with subconfluent growing cells than with confluent cells. This observation might be a clue as to how TNP-470 is able to both reversibly inhibit endothelial cell proliferation in vitro and inhibit tumor growth in vivo without apparent effects to quiescent endothelium [67].

The Effects of TNP-470 on Vessel Hyperpermeability

TNP-470 and caplostatin [see section "HPMA copolymer-Gly-Phe-Leu-Gly-TNP-470 (Caplostatin)"] inhibit vascular hyperpermeability of tumor blood vessels as well as that induced in mouse skin by different mediators, such as VEGF, platelet activating factor (PAF), histamine, substance-P and serotonin [8]. Treatment with TNP-470 for 3 days was sufficient to reduce permeability of tumor blood vessels, delayed-type hypersensitivity, and pulmonary edema induced by IL-2. TNP-470 also inhibited VEGF-induced phosphorylation of VEGFR2, calcium influx, and RhoA activation in endothelial cells. These results identified an activity of TNP-470, that of inhibiting vessel hyperpermeability. This activity likely contributes to TNP-470's antiangiogenic effect and suggests that in its improved non-toxic forms (see section "Improvements and modifications of TNP-470"), it can be used in the treatment of cancer and inflammation.

The Effect of TNP-470 on Non-endothelial Cells

Although TNP-470 inhibits mostly the proliferation of endothelial cells, higher concentrations of the drug were found to affect tumor cell proliferation [17]. B16F10 murine melanoma cells were found to be especially vulnerable to TNP-470 cytotoxicity [68]. Addition of n-acetylcystein, a free radical scavenger, inhibited an increase in the generation of reactive oxygen species (ROS) and protected B16F10 cells from TNP-470-induced death, indicating that ROS generation is responsible for the toxic effect of TNP-470 on B16F10 cells [68]. Another cell line that is especially sensitive to the inhibitory effect of TNP-470 is the FU-MMT-1 cell line, which was derived from a human primary carcinosarcoma of the uterus. Emoto et al. has shown that TNP-470 inhibited the proliferation of FU-MMT-1 and their ability to secrete VEGF to the culture medium [69]. TNP-470 was also shown to inhibit VEGF synthesis in platelet derived growth factor-BB (PDGF-BB) activated hepatic stellate cells (HSCs). This blockage of VEGF synthesis was shown to be dependent on the mitogen-activated protein kinase (MAPK)/cyclooxygenase-2 (Cox-2) signaling pathway [23]. By reducing the secretion of VEGF from tumor cells, TNP-470 affects indirectly the proliferation, migration, survival and hyperpermeability of endothelial cells in the tumor microenvironment.

Combination Therapy Studies

The efficacy of TNP-470 with different anti-cancer modalities such as chemotherapeutic agents, cytokines or radiotherapy was evaluated in several combination therapy studies. Theses studies had shown that TNP-470 acts synergistically with most conventional cytotoxic agents with variability in the different models as presented in Table 35.2. In some cases, the additional effect was enhancement of anti-tumor efficacy [70], reduction of tumor volume [71, 72], and inhibition of metastases [73], while in some animal models, the combined effect was more than additive, leading to long-term survival [74–76].

The combination of TNP-470 with radiation has also been studied extensively. A synergistic effect was observed in most of the studies that have evaluated the effect of combining TNP-470 with radiation [77–79]. Teicher et al. has shown that combined application of TNP-470 with fractionated radiotherapy significantly increased tumor growth delay as compared to irradiation alone in Lewis lung carcinoma [80]. Concurrent administration of TNP-470 and minocycline with fractionated irradiation produced a greater growth delay of Lewis lung carcinoma than angiogenesis inhibitor with radiotherapy alone [80]. The combination of TNP-470 and minocycline and fractionated radiotherapy was also moderately effective in EMT-6 mammary carcinoma [81]. Shintani et al. observed that TNP-470 significantly enhanced the effect of radiation especially on cells with high neovascularization [78]. Lund et al. demonstrated that TNP-470 treatment given before the radiation treatment can prevent the typical acute microvascular damage after ionizing radiation [79].

TNP-470 was also shown to synergize with photodynamic therapy (PDT), a method that consists of the systemic or local administration of a photosensitizer and its subsequent activation by visible light [82]. PDT is approved for the treatment of several cancers. However, its induction of VEGF creates conditions favorable to enhanced tumor growth and metastases, therefore mitigating its cytotoxic and anti-vascular effects. Kosharskyy et al. demonstrated that combined treatment of PDT with TNP-470 decreased tumor growth, lymph node metastasis, and disease-related toxicity in an orthotopic model of prostate cancer [82].

TNP-470 was also evaluated in several combination studies with other antiangiogenic agents. Brem et al. demonstrated that the combination of TNP-470 together with α/β-interferon suppressed Lewis lung carcinoma growth by 80% compared with controls, and inhibited the growth of pulmonary metastases [83]. Parangi et al. reported reduction in tumor size and capillary density in RIP-Tag transgenic mice model when TNP-470 was administered together with minocyclin and α/β-interferon [84]. Administration of TNP-470, together with the antiangiogenic agent IL-12, resulted in augmented anti-tumor activity in Colon-26 and B16F10 melanoma models [85].

Other combination strategies that were examined together with TNP-470 included hyperthermia [86], antibodies against endoglin [87], antibodies against IL-8 [88], agents that increase tumor oxygenation [77], the immune stimulator α-galactosylceramide (KRN7000) [76], tumor-lysate pulsed dendritic cells [89] and even gene therapy [90]. In all of these strategies, TNP-470 was found to enhance the anti-tumor effect.

Much effort has been directed toward elucidating the mechanism by which TNP-470 synergizes with other agents. Several studies have found that TNP-470 is able to enhance the effectiveness of biologic or cytotoxic therapy by shifting the equilibrium between tumor cell proliferation and death [73, 84,91]. Other studies report that the addition of TNP-470 to cytotoxic drugs had increased both their uptake by the tumor and their cytotoxic action, probably through the effect of TNP-470 on the vasculature [81]. These reports are in agreement with the notion that antiangiogenic therapy can normalize the structure and function of the tumor vessels before their destruction, thereby improving drug delivery to the tumor [92]. This normalization effect may underlie the therapeutic benefit of combined antiangiogenic and cytotoxic therapies.

Pharmacokinetics and Metabolism of TNP-470

The pharmacokinetic of TNP-470 was initially studied using reverse phase high-performance liquid chromatography (HPLC). This study described very rapid clearance of TNP-470 in human plasma and identified one of its major active metabolites [93]. The biotransformation of the drug was further examined in primary cultures of human hepatocytes and in microsomal fractions of various human tissues [94]. TNP-470 was found to undergo rapid and extensive transformation into at least six metabolic derivatives (M-I to M-VI) within 30 min of administration [94]. The two predominant extracellular metabolites are M-II and M-IV, also known as AGM-1883. Metabolite M-IV formation involves ester hydrolysis of TNP-470 that is followed by conversion to metabolite M-II by microsomal epoxide hydrolase. The same pattern of TNP-470 metabolism was observed in rhesus monkey model, which is highly predictive of the pharmacokinetics of various compounds in humans [95]. The structures of TNP-470 and its two major metabolites in humans are shown in Fig. 35.2.

Since then, several methods for quantification of TNP-470 and its metabolites have been established in order to improve the limits of detection of quantification, to shorten the analysis time and enable higher throughput in measurement of the drug [96, 97]. Quantification of TNP-470 and its metabolites was done by several techniques including HPLC alone or HPLC combined with mass spectrometry (LC/MS/MS) [93, 96, 97].

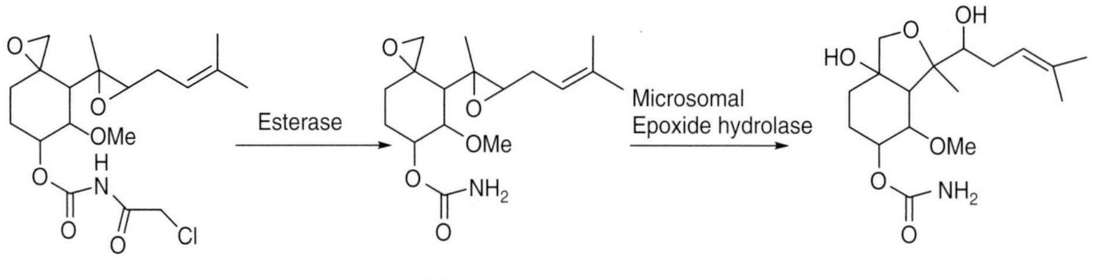

Fig. 35.2. Chemical structure of TNP-470 and its two major metabolites.

Overview of Clinical Trials of TNP-470

Clinical Trials Phase I with TNP-470

TNP-470 entered Phase I clinical trials in 1992 for HIV-associated Kaposi's sarcoma [98], only a few years after Ingber's discovery and before its full characterization. In the following years, other Phase I trials of TNP-470 were conducted for squamous cell carcinoma of the cervix [99], some advanced cancers [100] and androgen-independent prostate cancer [101]. These Phase I clinical trials with TNP-470 were conducted in order to determine the safety profile, pharmacokinetics and maximum tolerated dose (MTD) of this drug. The simultaneously performed Phase I trials including over 200 patients concluded that TNP-470 was most effective at 1 h-intravenous infusion of $60 \, mg/m^2$, three times per week and that neurotoxicity was the dose-limiting toxicity [102]. The principal toxicities of TNP-470 were dizziness, lightheadedness, vertigo, ataxia, decrease in concentration and short-term memory, confusion, anxiety, and depression, which occurred at doses of 133, 177, and $235 \, mg/m^2$. Overall, these neurological symptoms were dose-related, had an insidious onset, progressively worsened with treatment, and resolved completely within 2 weeks of stopping the drug [100]. The documented clinical responses to TNP-470 are summarized in Table 35.3 (as monotherapy) and Table 35.4 (as combination therapy trials) and include one complete response that was seen in a woman with diagnosis of recurrent squamous cell carcinoma of the cervix metastatic to lungs [103,99] and many cases of patients with partial response or stable disease. Other documented complete and durable responses to TNP-470 treatment are: high-grade sarcoma of the kidney [100], renal cell carcinoma (Stadler *et al.*, Proc AACR 17:310A, 1998), androgen-independent prostate cancer (Zukiwiski *et al.*, Proc AACR 13:A795, 1994) and Kaposi's sarcoma [98].

Clinical Pharmacokinetics

The pharmacokinetics of TNP-470 and its major metabolites were investigated in AIDS enrolled patients in two Phase I dose-escalation trials for the treatment of Kaposi's sarcoma [104,105]. In one study, the drug was administered to the patients by 1h-intravenous infusion in dose cohorts ranging from of 10, 20, 30, 40, 50 and $70 \, mg/m^2$ [105], and in the other study, TNP-470

dosage was increased in 13 sequential cohorts using a modified Fibonacci escalation scheme (4.6, 9.3, 15.4, 23.2, and $43.1 \, mg/m^2$) [104]. The plasma concentrations of TNP-470 and its two major metabolites were assayed either by reverse-phase HPLC or by HPLC/MS. Moore *et al.* mentioned that all patients had both metabolites in their plasma while the parent drug was undetectable at time points as early as 5 min after the end of the infusion for some patients. TNP-470 was also undetectable in the urine samples collected from the patients. Figg *et al.* reported that there was a linear relationship between the dose of TNP-470 and both area under the curve to infinity (AUC[inf]) and time to maximum concentration (Cmax). The Cmax ranged between $6.6 \, ng/ml$ at the lowest dosage ($4.6 \, mg/m^2$) and $597.1 \, ng/ml$ at the highest dosage ($43.1 \, mg/m^2$) indicating that concentrations of TNP-470 that have *in vitro* activity were achievable *in vivo*. The drug was rapidly cleared from the circulation after a single 1h-infusion and this observation is consistent with the pharmacokinetic preclinical data.

Clinical Toxicity of TNP-470

The toxic effects of TNP-470 observed in the clinical trials were moderate and reversible, and included asthenia, ataxia, fatigue, vertigo, dizziness, loss of concentration and short term memory, confusion, anxiety and depression [100], as well as nausea, bone pain and constipation [101]. The major dose-limiting side-effect was a reversible toxic reaction. Overall the neurological symptoms were dose-related, had an insidious onset, progressively worsened with treatment and resolved completely within 2 weeks of stopping the administration of the drug [100].

Phase II Studies with TNP-470

Based on the Phase I results, a number of Phase II studies have been initiated in renal cell, breast, cervical, pancreatic and brain cancers. One multi-institutional Phase II study targeting metastatic renal carcinomas has been completed, concluding that long-term treatment with TNP-470 was in general adequately tolerated but did not lead to any significant objective responses [106]. The observed lack of response could be due to the possibility that TNP-470 only prevents further tumor progression but does not lead to marked tumor shrinkage.

TABLE 35.3. TNP-470 as monotherapy in clinical trials.

Phase	Type of cancer	Dosing schedule of TNP-470	Results	References
Phase-I	**Kaposi's sarcoma**	Weekly 1-h i.v. infusion, 10–70 mg/m^2 up to 24 weeks	7 of 38 (18%) patients achieved partial response. One partial clinical response ≥ 50% flattening of all raised baseline lesions.	[98, 105]
Phase-I	**Inoperable recurring or metastatic squamous cell carcinoma of the cervix**	1-h i.v. infusion, 9.3–71.2 mg/m^2, q.o.d. for 28 d followed by a 14 d rest period.	1 of 18 patients achieved complete response, which continues for 26 months, and three patients with initially progressive disease stage had stable disease for 5, 7.7, and 19+ months.	[99, 103]
Phase-I	**Androgen-independent prostate cancer**	1-h i.v. infusion, 9.3–106 mg/m^2, q.o.d. for 28 d followed by a 14 d rest period	No definite antitumor activity was observed.	[101]
Phase-I	**Malignant melanoma**	4-h i.v. infusion, 25–235 mg/m^2, once a week	Stable disease.	[100]
Phase-I	**Adenocarcinoma of the colon**	4-h i.v. infusion, 25–235 mg/m^2, once a week	No clinically detectable disease for 13 months.	[100]
Phase-I	**Soft tissue sarcoma**	4-h i.v. infusion, 25–235 mg/m^2, once a week	No clinically detectable disease for > 3 years	[100]
Phase-I	**Metastatic lesions from breast cancer**		Partial response at the three-month follow-up and stabilization at the five-month follow-up Metastases at four sites: eye, lung, liver, bone.	[178]
Phase-II	**Metastatic renal cell carcinoma**	1-h i.v. infusion, 60 mg/m^2, 3 times per week.	One of 33 patients achieved partial response of short duration. Six patients (18%) remained on study for 6 or more months without toxicity or disease progression.	[106]

TABLE 35.4. TNP-470 in combination with chemotherapeutic agents in clinical trials.

Phase	Type of cancer	Dosing schedule of TNP-470	Dosing schedule of combined agent	Results	References
Phase-I	**Non-small-cell lung cancer**	Arm A: 1-h i.v. infusion, 60 mg/m^2, 3 times per week Arm B: 1-h i.v. infusion, 20–45 mg/m^2, 3 times per week	Arm A: Escalated doses of paclitaxel 135–225 mg/m^2 administered by 3-h i.v. on day 1 of 3 week cycle. Arm B: Constant dose of 225 mg/m^2 of paclitaxel administered by 3-h i.v. on day 1 of 3 week cycle.	8 of 32 patients (25%) achieved partial response, and 6 of 16 patients (38%) with NSCLS achieved stable disease. Median survival for all patients was 14.1 months.	[107]
Phase-I	**Solid tumors**	1-h i.v. infusion, 60 mg/m^2, 3 times per week	3-h i.v. of paclitaxel 225 mg/m^2 and 1-h i.v. of carboplatin 6 mg/ml x min on day 1 of 3-week cycle.	4 of 17 patients (24%) achieved partial response, and 8 patients (47%) had stable disease.	[108]

Combination Phase I Studies for TNP-470 and Cytotoxic Agents

TNP-470 was one of the first antiangiogenic agents examined with chemotherapy and shown to have synergy [75]. Therefore, it is not surprising that TNP-470 has been clinically evaluated with other cytotoxic agents. Paclitaxel, a potent anticancer drug, was the first to enter Phase I study in combination with TNP-470. The combination of TNP-470 administered at 60 mg/m^2 three times per week and paclitaxel dose of 225 mg/m^2 infused every 3 weeks was defined as both the maximum-tolerated dose and optimal dose [107]. The addition of TNP-470 to paclitaxel did not result in additive toxic effects compared with the literature toxic reports for paclitaxel alone.

Partial responses were reported in 8 (25%) of 32 patients with solid tumors and in 6 (38%) of 16 patients with non-small cell lung carcinoma (NSCLC). Another combination study that has recently been completed assessed the pharmacokinetics of TNP-470 in combination with paclitaxel and carboplatin [108]. The optimal dose of 60 mg/m^2 of TNP-470 was administered three times a week and paclitaxel and carboplatin were infused every 3 weeks at 225 mg/m^2 and AUC 5 mg/ml per min, respectively. The addition of TNP-470 to this paclitaxel and carboplatin doublet did not result in significant additional adverse events compared to those reported with carboplatin and paclitaxel combinations [107]. The results for this study included 4 (24%) of 17 patients with partial response and 8 (47%) with stable disease. To conclude, the combination of

TNP-470 and paclitaxel with or without carboplatin was well-tolerated and encouraging, but further studies are needed. The documented clinical responses to TNP-470 in combination with other cytotoxic agents are summarized in Table 35.4. Combination therapies with other cytotoxic agents may prove more beneficial to the patients.

Improvements and Modifications of TNP-470

HPMA Copolymer-Gly-Phe-Leu-Gly-TNP-470 (Caplostatin)

Satchi-Fainaro *et al.* synthesized and characterized a water-soluble N-(2-hydroxypropyl)methacrylamide (HPMA) copolymer-Gly-Phe-Leu-Gly-TNP-470 conjugate [10], named caplostatin (Fig. 35.3), which accumulated selectively in tumor vessels by the enhanced permeability and retention (EPR) effect [109, 110]. HPMA copolymer-TNP-470 conjugate substantially enhanced and prolonged the activity of TNP-470 *in vivo* in tumor and hepatectomy models. Caplostatin significantly inhibited the growth of Lewis lung carcinoma, U87 glioblastoma, A2058 melanoma, PC3 prostate carcinoma, COLO-205 colon carcinoma and Mycn-driven murine neuroblastomas in transgenic mice [8, 10, 43, 111]. Polymer conjugation prevented TNP-470 from crossing the blood-brain barrier and decreased its accumulation in normal organs thereby avoiding drug-related toxicities. Treatment with TNP-470 caused weight loss and neurotoxic effects in mice, while treatment with caplostatin did not. This novel approach for targeting angiogenesis inhibitors specifically to the tumor vasculature may provide a new strategy for the rational design of cancer therapies.

While investigating the mechanism of extravasation from leaky vessels, tumor accumulation and trafficking of caplostatin, Satchi-Fainaro *et al.* found in a different study that both TNP-470 and caplostatin inhibit vascular hyperpermeability of tumor blood vessels as well as that induced in mouse skin by different mediators (VEGF, PAF, histamine, substance P and serotonin) [8]. Treatment with TNP-470 or caplostatin for 3 days was sufficient to reduce permeability of tumor blood vessels, delayed-type hypersensitivity, and pulmonary edema induced by IL-2. TNP-470 and caplostatin also inhibited VEGF-induced phosphorylation of VEGFR2, calcium influx, and RhoA activation in endothelial cells. These results identified an additional activity of TNP-470, that of inhibiting vessel hyperpermeability and may explain the selectivity of TNP-470 to endothelial cells. Since MetAp-2 is present in most cells, TNP-470's selectivity to endothelial cells remained unclear by its blockage activity of MetAp-2 alone. This activity likely contributes to TNP-470's antiangiogenic effect and suggests that caplostatin can be used in the treatment of cancer and inflammation. Caplostatin was also shown to inhibit endometriosis

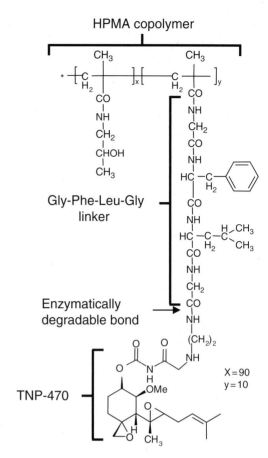

FIG. 35.3. Chemical structure of caplostatin.

and suppress the growth of endometriotic lesions by 59% compared with controls [112].

TNP-470 in a Medium Chain Triglyceride and Poly-Lactic Acid Microspheres.

As TNP-470 is very unstable both *in vitro* and *in vivo*, it has been difficult to verify its pharmacological efficacy in clinical medicine. The preparation of a drug delivery system (DDS) in a microsphere form was studied for the stable inclusion and controlled release of TNP-470 [113]. Medium-chain triglyceride (MCTG) as an effective stabilizer and poly-lactic acid (PLA) as a biodegradable carrier were used for this purpose. The release of TNP-470 from the MCTG containing DDS continued for approximately 2 weeks, while the release of TNP-470 from the one without MCTG stopped after only 5 days. It was proved that TNP-470 could be released much more stable for a much longer period from the MCTG-containing DDS compared to the one without DDS.

Poly(vinyl alcohol)-TNP-470 Conjugate

TNP-470 was conjugated in poly(vinyl alcohol) PVA by a dimethylaminopyridine-catalyzed reaction [114]. The TNP-470-

PVA inhibited the growth of HUVECs, similar to that of free TNP-470. The bovine retinal pigment epithelial cells (BRPECs) were less sensitive to TNP-470-PVA than were the HUVECs. TNP-470-PVA significantly inhibited the progression of CNV induced by subretinal injection of gelatin microspheres containing bFGF in rabbits. Histologic studies at 4 weeks after treatment demonstrated that the degree of vascular formation and the number of vascular endothelial cells in the subretinal membrane of the eyes treated with TNP-470-PVA were less than those of the control eyes. Yasukawa *et al.* concluded that TNP-470-PVA inhibited the proliferation of HUVECs more sensitively than that of BRPECs, and the targeted delivery of TNP-470-PVA may have potential as a treatment modality for CNV.

Summary

TNP-470 was one of the first antiangiogenic agents to enter clinical trials in cancer patients, which makes its clinical development a prototype for future antiangiogenic agents. TNP-470 clinical trials provided important information regarding the end points and parameters that were most informative to assess following trials of antiangiogenic agents. Some of the important lessons learnt from TNP-470's trial are probably the fact that one should bear in mind that realistic and positive goals of an antiangiogenic agent trial as a monotherapy are (1) the achievement of stable disease or regression to avascular state and not total eradication of tumor cells as one should expect from a chemotherapy trial, otherwise to be considered as treatment failure; (2) the duration of the treatment is long, and therefore (3) the expected regression is slow, hence antiangiogenic agents should be relatively non-toxic and be used for long-term periods of time. The clinical experience suggested that TNP-470 clinical trials were halted partially due to the dose-limiting effect, neurotoxicity. TNP-470's efficacy on a broad spectrum of tumor types was supported by documented tumor regression and a few complete remissions in human cancers. This was a clinical proof that the principle of antiangiogenesis as an anticancer strategy was a promising approach. TNP-470's poor pharmacokinetics with its short half-life in the body together with its clinical benefit makes it the perfect candidate for drug delivery systems. Three such systems were described here and other approaches are currently being investigated.

Since angiogenesis is a multi-step process, it is possible that TNP-470 alone will not induce maximal antiangiogenesis in all cancers. Combination therapies with other antiangiogenic agents, such as the FDA-approved monoclonal antibody to VEGF, bevacizumab [111], or low-dose metronomic scheduling of chemotherapeutics agents will prove more beneficial to the patient.

Future clinical trials for improved formulations of modified TNP-470 should include other parameters such as time to progression, disease-free survival, and ultimately- survival. Therefore, these new chemical entities (NCEs) should have few or no side effects and allow extended duration of patient care and follow-up time to demonstrate benefit. If tumor dormancy or regression to avascular disease will be an acceptable end-point for these trials, these novel modified TNP-470-NCEs may hopefully convert cancer to a "chronic manageable disease" by minimal negative impact on the patient's quality of life.

References

1. Hanahan D, Folkman J. Patterns and emerging mechanisms of the angiogenic switch during tumorigenesis. Cell 1996;86(3):353–64.
2. Ingber D, Fujita T, Kishimoto S, et al. Synthetic analogues of fumagillin that inhibit angiogenesis and suppress tumour growth. Nature 1990;348(6301):555–7.
3. Hori A, Ikeyama S, Sudo K. Suppression of cyclin D1 mRNA expression by the angiogenesis inhibitor TNP-470 (AGM-1470) in vascular endothelial cells. Biochem Biophys Res Commun 1994;204(3):1067–73.
4. Hotz HG, Reber HA, Hotz B, et al. Angiogenesis inhibitor TNP-470 reduces human pancreatic cancer growth. J Gastrointest Surg 2001;5(2):131–8.
5. Kusaka M, Sudo K, Matsutani E, et al. Cytostatic inhibition of endothelial cell growth by the angiogenesis inhibitor TNP-470 (AGM-1470). Br J Cancer 1994;69(2):212–6.
6. Lien WH, Chen CK, Lai LY, Chen YH, Wu MP, Wu LW. Participation of cyclin D1 deregulation in TNP-470-mediated cytostatic effect: involvement of senescence. Biochem Pharmacol 2004;68(4):729–38.
7. Auerbach R, Lewis R, Shinners B, Kubai L, Akhtar N. Angiogenesis assays: a critical overview. Clin Chem 2003;49(1):32–40.
8. Satchi-Fainaro R, Mamluk R, Wang L, et al. Inhibition of vessel permeability by TNP-470 and its polymer conjugate, caplostatin. Cancer Cell 2005;7(3):251–61.
9. Yoshida A, Anand-Apte B, Zetter BR. Differential endothelial migration and proliferation to basic fibroblast growth factor and vascular endothelial growth factor. Growth Factors 1996;13(1–2):57–64.
10. Satchi-Fainaro R, Puder M, Davies JW, et al. Targeting angiogenesis with a conjugate of HPMA copolymer and TNP-470. Nat Med 2004;10(3):255–61.
11. van der Schaft DW, Seftor RE, Seftor EA, et al. Effects of angiogenesis inhibitors on vascular network formation by human endothelial and melanoma cells. J Natl Cancer Inst 2004;96(19):1473–7.
12. Donovan D, Brown NJ, Bishop ET, Lewis CE. Comparison of three in vitro human 'angiogenesis' assays with capillaries formed in vivo. Angiogenesis 2001;4(2):113–21.
13. Kusaka M, Sudo K, Fujita T, et al. Potent anti-angiogenic action of AGM-1470: comparison to the fumagillin parent. Biochem Biophys Res Commun 1991;174(3):1070–6.
14. Rybak SM, Sanovich E, Hollingshead MG, et al. "Vasocrine" formation of tumor cell-lined vascular spaces: implications for rational design of antiangiogenic therapies. Cancer Res 2003;63(11):2812–9.
15. Friis T, Hansen AB, Houen G, Engel AM. Influence of angiogenesis inhibitors on endothelial cell morphology in vitro. Apmis 2006;114(3):211–24.
16. Keezer SM, Ivie SE, Krutzsch HC, Tandle A, Libutti SK, Roberts DD. Angiogenesis inhibitors target the endothelial cell cytoskeleton through altered regulation of heat shock protein 27 and cofilin. Cancer Res 2003;63(19):6405–12.

17. Nahari D, Satchi-Fainaro R, Chen M, et al. Tumor cytotoxicity and endothelial Rac inhibition induced by TNP-470 in anaplastic thyroid cancer. Mol Cancer Ther 2007;6(4):1329–37.

18. Yanase T, Tamura M, Fujita K, Kodama S, Tanaka K. Inhibitory effect of angiogenesis inhibitor TNP-470 on tumor growth and metastasis of human cell lines in vitro and in vivo. Cancer Res 1993;53(11):2566–70.

19. Sedlakova O, Sedlak J, Hunakova L, et al. Angiogenesis inhibitor TNP-470: cytotoxic effects on human neoplastic cell lines. Neoplasma 1999;46(5):283–9.

20. Yamaoka M, Yamamoto T, Ikeyama S, Sudo K, Fujita T. Angiogenesis inhibitor TNP-470 (AGM-1470) potently inhibits the tumor growth of hormone-independent human breast and prostate carcinoma cell lines. Cancer Res 1993;53(21):5233–6.

21. Takamiya Y, Brem H, Ojeifo J, Mineta T, Martuza RL. AGM-1470 inhibits the growth of human glioblastoma cells in vitro and in vivo. Neurosurgery 1994;34(5):869–75; discussion 75.

22. Miura S, Emoto M, Matsuo Y, Kawarabayashi T, Saku K. Carcinosarcoma-induced endothelial cells tube formation through KDR/Flk-1 is blocked by TNP-470. Cancer Lett 2004;203(1):45–50.

23. Wang YQ, Luk JM, Chu AC, et al. TNP-470 blockage of VEGF synthesis is dependent on MAPK/COX-2 signaling pathway in PDGF-BB-activated hepatic stellate cells. Biochem Biophys Res Commun 2006;341(1):239–44.

24. Koyama H, Nishizawa Y, Hosoi M, et al. The fumagillin analogue TNP-470 inhibits DNA synthesis of vascular smooth muscle cells stimulated by platelet-derived growth factor and insulin-like growth factor-I. Possible involvement of cyclin-dependent kinase 2. Circ Res 1996;79(4):757–64.

25. Greene AK, Wiener S, Puder M, et al. Endothelial-directed hepatic regeneration after partial hepatectomy. Ann Surg 2003;237(4):530–5.

26. Yamamoto T, Sudo K, Fujita T. Significant inhibition of endothelial cell growth in tumor vasculature by an angiogenesis inhibitor, TNP-470 (AGM-1470). Anticancer Res 1994;14(1A):1–3.

27. Yamaoka M, Yamamoto T, Masaki T, Ikeyama S, Sudo K, Fujita T. Inhibition of tumor growth and metastasis of rodent tumors by the angiogenesis inhibitor O-(chloroacetyl-carbamoyl)fumagillol (TNP-470; AGM-1470). Cancer Res 1993;53(18):4262–7.

28. Nagabuchi E, VanderKolk WE, Une Y, Ziegler MM. TNP-470 antiangiogenic therapy for advanced murine neuroblastoma. J Pediatr Surg 1997;32(2):287–93.

29. Isobe N, Uozumi T, Kurisu K, Kawamoto K. Antitumor effect of TNP-470 on glial tumors transplanted in rats. Anticancer Res 1996;16(1):71–6.

30. Beecken WD, Fernandez A, Joussen AM, et al. Effect of antiangiogenic therapy on slowly growing, poorly vascularized tumors in mice. J Natl Cancer Inst 2001;93(5):382–7.

31. Inoue K, Chikazawa M, Fukata S, Yoshikawa C, Shuin T. Frequent administration of angiogenesis inhibitor TNP-470 (AGM-1470) at an optimal biological dose inhibits tumor growth and metastasis of metastatic human transitional cell carcinoma in the urinary bladder. Clin Cancer Res 2002;8(7):2389–98.

32. Osawa S, Terashima Y, Kimura G, Akimoto M. Antitumour effects of the angiogenesis inhibitor AGM-1470 on rat urinary bladder tumours induced by N-butyl-N-(4-hydroxybutyl) nitrosamine. BJU Int 1999;83(1):123–8.

33. Ueda N, Kamata N, Hayashi E, Yokoyama K, Hoteiya T, Nagayama M. Effects of an anti-angiogenic agent, TNP-470, on the growth of oral squamous cell carcinomas. Oral Oncol 1999;35(6):554–60.

34. Kurebayashi J, Kurosumi M, Dickson RB, Sonoo H. Angiogenesis Inhibitor O-(Chloroacetyl-carbamoyl) fumagillol (TNP-470) Inhibits Tumor Angiogenesis, Growth and Spontaneous Metastasis of MKL-4 Human Breast Cancer Cells in Female Athymic Nude Mice. Breast Cancer 1994;1(2):109–15.

35. Morales C, Zurita M, Vaquero J. Antitumor effect of TNP-470 is not associated to decrease of angiogenesis in an experimental malignant neuroectodermic tumor. J Neurooncol 2002; 58(2):131–6.

36. Emoto M, Ishiguro M, Iwasaki H, Kikuchi M, Kawarabayashi T. Effect of angiogenesis inhibitor TNP-470 on the growth, blood flow, and microvessel density in xenografts of human uterine carcinosarcoma in nude mice. Gynecol Oncol 2003;89(1):88–94.

37. Bernsen HJ, Rijken PF, Peters H, Bakker H, van der Kogel AJ. The effect of the anti-angiogenic agent TNP-470 on tumor growth and vascularity in low passaged xenografts of human gliomas in nude mice. J Neurooncol 1998;38(1):51–7.

38. Xia JL, Yang BH, Tang ZY, Sun FX, Xue Q, Gao DM. Inhibitory effect of the angiogenesis inhibitor TNP-470 on tumor growth and metastasis in nude mice bearing human hepatocellular carcinoma. J Cancer Res Clin Oncol 1997;123(7):383–7.

39. Kanai T, Konno H, Tanaka T, et al. Effect of angiogenesis inhibitor TNP-470 on the progression of human gastric cancer xenotransplanted into nude mice. Int J Cancer 1997;71(5):838–41.

40. Tanaka T, Konno H, Matsuda I, Nakamura S, Baba S. Prevention of hepatic metastasis of human colon cancer by angiogenesis inhibitor TNP-470. Cancer Res 1995;55(4):836–9.

41. Fujioka T, Hasegawa M, Ogiu K, Matsushita Y, Sato M, Kubo T. Antitumor effects of angiogenesis inhibitor 0-(chloroacetyl-carbamoyl) fumagillol (TNP-470) against murine renal cell carcinoma. J Urol 1996;155(5):1775–8.

42. Bergers G, Javaherian K, Lo KM, Folkman J, Hanahan D. Effects of angiogenesis inhibitors on multistage carcinogenesis in mice. Science 1999;284(5415):808–12.

43. Chesler L, Goldenberg DD, Seales IT, et al. Angiogenesis inhibition in a murine transgenic model of neuroblastoma. Cancer Res 2007;in press.

44. Weiss WA, Aldape K, Mohapatra G, Feuerstein BG, Bishop JM. Targeted expression of MYCN causes neuroblastoma in transgenic mice. Embo J 1997;16(11):2985–95.

45. Cernaianu G, Frank S, Erbstosser K, et al. TNP-470 fails to block the onset of angiogenesis and early tumor establishment in an intravital minimal disease model. Int J Colorectal Dis 2006;21(2):143–54.

46. Schoof DD, Obando JA, Cusack JC, Jr., Goedegebuure PS, Brem H, Eberlein TJ. The influence of angiogenesis inhibitor AGM-1470 on immune system status and tumor growth in vitro. Int J Cancer 1993;55(4):630–5.

47. Noren-Nystrom U, Eriksson M, Eriksson B, Roos G, Bergh A, Holmberg D. Antitumor activity of the angiogenesis inhibitor TNP-470 on murine lymphoma/leukemia cells in vivo and in vitro. Exp Hematol 2003;31(2):143–9.

48. Drevs J, Hofmann I, Hugenschmidt H, et al. Effects of PTK787/ZK 222584, a specific inhibitor of vascular endothelial growth factor receptor tyrosine kinases, on primary tumor, metastasis, vessel density, and blood flow in a murine renal cell carcinoma model. Cancer Res 2000;60(17):4819–24.

49. Sin N, Meng L, Wang MQ, Wen JJ, Bornmann WG, Crews CM. The anti-angiogenic agent fumagillin covalently binds and inhib-

its the methionine aminopeptidase, MetAP-2. Proc Natl Acad Sci U S A 1997;94(12):6099–103.

50. Griffith EC, Su Z, Turk BE, et al. Methionine aminopeptidase (type 2) is the common target for angiogenesis inhibitors AGM-1470 and ovalicin. Chem Biol 1997;4(6):461–71.

51. Bradshaw RA, Brickey WW, Walker KW. N-terminal processing: the methionine aminopeptidase and N alpha-acetyl transferase families. Trends Biochem Sci 1998;23(7):263–7.

52. Griffith EC, Su Z, Niwayama S, Ramsay CA, Chang YH, Liu JO. Molecular recognition of angiogenesis inhibitors fumagillin and ovalicin by methionine aminopeptidase 2. Proc Natl Acad Sci U S A 1998;95(26):15183–8.

53. Liu S, Widom J, Kemp CW, Crews CM, Clardy J. Structure of human methionine aminopeptidase-2 complexed with fumagillin. Science 1998;282(5392):1324–7.

54. Han CK, Ahn SK, Choi NS, et al. Design and synthesis of highly potent fumagillin analogues from homology modeling for a human MetAP-2. Bioorg Med Chem Lett 2000;10(1):39–43.

55. Turk BE, Su Z, Liu JO. Synthetic analogues of TNP-470 and ovalicin reveal a common molecular basis for inhibition of angiogenesis and immunosuppression. Bioorg Med Chem 1998;6(8):1163–9.

56. Turk BE, Griffith EC, Wolf S, Biemann K, Chang YH, Liu JO. Selective inhibition of amino-terminal methionine processing by TNP-470 and ovalicin in endothelial cells. Chem Biol 1999;6(11):823–33.

57. Kim S, LaMontagne K, Sabio M, et al. Depletion of methionine aminopeptidase 2 does not alter cell response to fumagillin or bengamides. Cancer Res 2004;64(9):2984–7.

58. Yeh JR, Ju R, Brdlik CM, et al. Targeted gene disruption of methionine aminopeptidase 2 results in an embryonic gastrulation defect and endothelial cell growth arrest. Proc Natl Acad Sci USA 2006;103(27):10379–84.

59. Zhang Y, Yeh JR, Mara A, et al. A chemical and genetic approach to the mode of action of fumagillin. Chem Biol 2006;13(9):1001–9.

60. Abe J, Zhou W, Takuwa N, et al. A fumagillin derivative angiogenesis inhibitor, AGM-1470, inhibits activation of cyclin-dependent kinases and phosphorylation of retinoblastoma gene product but not protein tyrosyl phosphorylation or protooncogene expression in vascular endothelial cells. Cancer Res 1994;54(13):3407–12.

61. Antoine N, Greimers R, De Roanne C, et al. AGM-1470, a potent angiogenesis inhibitor, prevents the entry of normal but not transformed endothelial cells into the G1 phase of the cell cycle. Cancer Res 1994;54(8):2073–6.

62. Xiong Y, Hannon GJ, Zhang H, Casso D, Kobayashi R, Beach D. p21 is a universal inhibitor of cyclin kinases. Nature 1993;366(6456):701–4.

63. Zhang Y, Griffith EC, Sage J, Jacks T, Liu JO. Cell cycle inhibition by the anti-angiogenic agent TNP-470 is mediated by p53 and p21WAF1/CIP1. Proc Natl Acad Sci USA 2000;97(12):6427–32.

64. Yeh JR, Mohan R, Crews CM. The antiangiogenic agent TNP-470 requires p53 and p21CIP/WAF for endothelial cell growth arrest. Proc Natl Acad Sci U S A 2000;97(23):12782–7.

65. Mauriz JL, Gonzalez P, Duran MC, Molpeceres V, Culebras JM, Gonzalez-Gallego J. Cell-cycle inhibition by TNP-470 in an in vivo model of hepatocarcinoma is mediated by a p53 and p21(WAF1/CIP1) mechanism. Transl Res 2007;149(1):46–53.

66. Mauriz JL, Linares P, Macias RI, et al. TNP-470 inhibits oxidative stress, nitric oxide production and nuclear factor kappa B activation in a rat model of hepatocellular carcinoma. Free Radic Res 2003;37(8):841–8.

67. Budson AE, Ko L, Brasel C, Bischoff J. The angiogenesis inhibitor AGM-1470 selectively increases E-selectin. Biochem Biophys Res Commun 1996;225(1):141–5.

68. Okroj M, Kamysz W, Slominska EM, Mysliwski A, Bigda J. A novel mechanism of action of the fumagillin analog, TNP-470, in the B16F10 murine melanoma cell line. Anticancer Drugs 2005;16(8):817–23.

69. Emoto M, Ishiguro M, Iwasaki H, Kikuchi M, Kawarabayashi T. TNP-470 inhibits growth and the production of vascular endothelial growth factor of uterine carcinosarcoma cells in vitro. Anticancer Res 2000;20(1 C):601–4.

70. Voest EE, Kenyon BM, O'Reilly MS, Truitt G, D'Amato RJ, Folkman J. Inhibition of angiogenesis in vivo by interleukin 12. J Natl Cancer Inst 1995;87(8):581–6.

71. Satoh H, Ishikawa H, Fujimoto M, et al. Combined effects of TNP-470 and taxol in human non-small cell lung cancer cell lines. Anticancer Res 1998;18(2A):1027–30.

72. Amikura K, Matsuno S, Egawa S. Synergistic Antitumor Effect of an Angiogenesis Inhibitor (TNP-470) and Tumor Necrosis Factor in Mice. Surg Today 2006;36(12):1069–74.

73. Inoue K, Chikazawa M, Fukata S, Yoshikawa C, Shuin T. Docetaxel enhances the therapeutic effect of the angiogenesis inhibitor TNP-470 (AGM-1470) in metastatic human transitional cell carcinoma. Clin Cancer Res 2003;9(2):886–99.

74. Kato T, Sato K, Kakinuma H, Matsuda Y. Enhanced suppression of tumor growth by combination of angiogenesis inhibitor O-(chloroacetyl-carbamoyl)fumagillol (TNP-470) and cytotoxic agents in mice. Cancer Res 1994;54(19):5143–7.

75. Teicher BA, Holden SA, Ara G, et al. Potentiation of cytotoxic cancer therapies by TNP-470 alone and with other anti-angiogenic agents. Int J Cancer 1994;57(6):920–5.

76. Matsumoto G, Nagai S, Muta M, Tsuruta K, Okamoto A, Toi M. Survival benefit of KRN7000 immune therapy in combination with TNP470 in hamster liver metastasis model of pancreatic cancer. Oncol Rep 2003;10(5):1201–6.

77. Teicher BA, Dupuis NP, Emi Y, Ikebe M, Kakeji Y, Menon K. Increased efficacy of chemo- and radio-therapy by a hemoglobin solution in the 9 L gliosarcoma. In Vivo 1995;9(1):11–8.

78. Shintani S, Li C, Mihara M, et al. Anti-tumor effect of radiation response by combined treatment with angiogenesis inhibitor, TNP-470, in oral squamous cell carcinoma. Oral Oncol 2006;42(1):66–72.

79. Lund EL, Bastholm L, Kristjansen PE. Therapeutic synergy of TNP-470 and ionizing radiation: effects on tumor growth, vessel morphology, and angiogenesis in human glioblastoma multiforme xenografts. Clin Cancer Res 2000;6(3):971–8.

80. Teicher BA, Emi Y, Kakeji Y, Northey D. TNP-470/minocycline/cytotoxic therapy: a systems approach to cancer therapy. Eur J Cancer 1996;32A(14):2461–6.

81. Teicher BA, Holden SA, Dupuis NP, et al. Potentiation of cytotoxic therapies by TNP-470 and minocycline in mice bearing EMT-6 mammary carcinoma. Breast Cancer Res Treat 1995;36(2): 227–36.

82. Kosharskyy B, Solban N, Chang SK, Rizvi I, Chang Y, Hasan T. A mechanism-based combination therapy reduces local tumor growth and metastasis in an orthotopic model of prostate cancer. Cancer Res 2006;66(22):10953–8.

83. Brem H, Gresser I, Grosfeld J, Folkman J. The combination of antiangiogenic agents to inhibit primary tumor growth and metastasis. J Pediatr Surg 1993;28(10):1253–7.

84. Parangi S, O'Reilly M, Christofori G, et al. Antiangiogenic therapy of transgenic mice impairs de novo tumor growth. Proc Natl Acad Sci U S A 1996;93(5):2002–7.

85. Dabrowska-Iwanicka A, Olszewska D, Jalili A, et al. Augmented antitumour effects of combination therapy with TNP-470 and chemoimmunotherapy in mice. J Cancer Res Clin Oncol 2002;128(8):433–42.

86. Nishimura Y, Murata R, Hiraoka M. Combined effects of an angiogenesis inhibitor (TNP-470) and hyperthermia. Br J Cancer 1996;73(3):270–4.

87. Maier JA, Delia D, Thorpe PE, Gasparini G. In vitro inhibition of endothelial cell growth by the antiangiogenic drug AGM-1470 (TNP-470) and the anti-endoglin antibody TEC-11. Anticancer Drugs 1997;8(3):238–44.

88. Kawano T, Yanoma S, Nishimura G, Tsukuda M. The inhibitory effects of TNP470 on tumour growth of head and neck carcinoma cell producing interleukin-8. J Laryngol Otol 2001;115(10):802–7.

89. Miyazaki J, Tsuzuki Y, Matsuzaki K, et al. Combination therapy with tumor-lysate pulsed dendritic cells and antiangiogenic drug TNP-470 for mouse pancreatic cancer. Int J Cancer 2005;117(3):499–505.

90. Pu YS, Do KA, Luo W, Logothetis CJ, Lin SH. Enhanced suppression of prostate tumor growth by combining C-CAM1 gene therapy and angiogenesis inhibitor TNP-470. Anticancer Drugs 2002;13(7):743–9.

91. Muramaki M, Miyake H, Hara I, Kamidono S. Synergistic inhibition of tumor growth and metastasis by combined treatment with TNP-470 and docetaxel in a human prostate cancer PC-3 model. Int J Oncol 2005;26(3):623–8.

92. Jain RK. Antiangiogenic therapy for cancer: current and emerging concepts. Oncology (Williston Park) 2005;19(4 Suppl 3):7–16.

93. Figg WD, Yeh HJ, Thibault A, et al. Assay of the antiangiogenic compound TNP-470, and one of its metabolites, AGM-1883, by reversed-phase high-performance liquid chromatography in plasma. J Chromatogr 1994;652(2):187–94.

94. Placidi L, Cretton-Scott E, de Sousa G, Rahmani R, Placidi M, Sommadossi JP. Disposition and metabolism of the angiogenic moderator O-(chloroacetyl-carbamoyl) fumagillol (TNP-470; AGM-1470) in human hepatocytes and tissue microsomes. Cancer Res 1995;55(14):3036–42.

95. Cretton-Scott E, Placidi L, McClure H, Anderson DC, Sommadossi JP. Pharmacokinetics and metabolism of O-(chloroacetyl-carbamoyl) fumagillol (TNP-470, AGM-1470) in rhesus monkeys. Cancer Chemother Pharmacol 1996;38(2):117–22.

96. Ong VS, Stamm GE, Menacherry S, Chu S. Quantitation of TNP-470 and its metabolites in human plasma: sample handling, assay performance and stability. J Chromatogr B Biomed Sci Appl 1998;710(1–2):173–82.

97. Whalen CT, Hanson GD, Putzer KJ, Mayer MD, Mulford DJ. Assay of TNP-470 and its two major metabolites in human plasma by high-performance liquid chromatography-mass spectrometry. J Chromatogr Sci 2002;40(4):214–8.

98. Dezube BJ, Von Roenn JH, Holden-Wiltse J, et al. Fumagillin analog in the treatment of Kaposi's sarcoma: a phase I AIDS Clinical Trial Group study. AIDS Clinical Trial Group No. 215 Team. J Clin Oncol 1998;16(4):1444–9.

99. Kudelka AP, Levy T, Verschraegen CF, et al. A phase I study of TNP-470 administered to patients with advanced squamous cell cancer of the cervix. Clin Cancer Res 1997;3(9):1501–5.

100. Bhargava P, Marshall JL, Rizvi N, et al. A Phase I and pharmacokinetic study of TNP-470 administered weekly to patients with advanced cancer. Clin Cancer Res 1999;5(8):1989–95.

101. Logothetis CJ, Wu KK, Finn LD, et al. Phase I trial of the angiogenesis inhibitor TNP-470 for progressive androgen-independent prostate cancer. Clin Cancer Res 2001;7(5):1198–203.

102. Owa T, Yoshino H, Yoshimatsu K, Nagasu T. Cell cycle regulation in the G1 phase: a promising target for the development of new chemotherapeutic anticancer agents. Curr Med Chem 2001;8(12):1487–503.

103. Kudelka AP, Verschraegen CF, Loyer E. Complete remission of metastatic cervical cancer with the angiogenesis inhibitor TNP-470. N Engl J Med 1998;338(14):991–2.

104. Figg WD, Pluda JM, Lush RM, et al. The pharmacokinetics of TNP-470, a new angiogenesis inhibitor. Pharmacotherapy 1997;17(1):91–7.

105. Moore JD, Dezube BJ, Gill P, Zhou XJ, Acosta EP, Sommadossi JP. Phase I dose escalation pharmacokinetics of O-(chloroacetylcarbamoyl) fumagillol (TNP-470) and its metabolites in AIDS patients with Kaposi's sarcoma. Cancer Chemother Pharmacol 2000;46(3):173–9.

106. Stadler WM, Kuzel T, Shapiro C, Sosman J, Clark J, Vogelzang NJ. Multi-institutional study of the angiogenesis inhibitor TNP-470 in metastatic renal carcinoma. J Clin Oncol 1999;17(8):2541–5.

107. Herbst RS, Madden TL, Tran HT, et al. Safety and pharmacokinetic effects of TNP-470, an angiogenesis inhibitor, combined with paclitaxel in patients with solid tumors: evidence for activity in non-small-cell lung cancer. J Clin Oncol 2002;20(22):4440–7.

108. Tran HT, Blumenschein GR, Jr., Lu C, et al. Clinical and pharmacokinetic study of TNP-470, an angiogenesis inhibitor, in combination with paclitaxel and carboplatin in patients with solid tumors. Cancer Chemother Pharmacol 2004;54(4):308–14.

109. Matsumura Y, Maeda H. A new concept for macromolecular therapeutics in cancer chemotherapy: mechanism of tumoritropic accumulation of proteins and the antitumor agent smancs. Cancer Res 1986;46(12 Pt 1):6387–92.

110. Satchi-Fainaro R. Targeting tumor vasculature: reality or a dream? J Drug Target 2002;10(7):529–33.

111. Satchi-Fainaro R, Birsner A, Butterfield C, Akslen L, Short S, Folkman J. HPMA copolymer-TNP-470 (caplostatin) and Avastin show synergistic inhibition of human tumor growth in mice,. European Journal of Cancer Supplements 2005;3(2):15.

112. Becker CM, Wright RD, Satchi-Fainaro R, et al. A novel non-invasive model of endometriosis for monitoring the efficacy of antiangiogenic therapy. Am J Pathol 2006;168(6):2074–84.

113. Kakinoki S, Yasuda C, Kaetsu I, et al. Preparation of poly-lactic acid microspheres containing the angiogenesis inhibitor TNP-470 with medium-chain triglyceride and the in vitro evaluation of release profiles. Eur J Pharm Biopharm 2003;55(2):155–60.

114. Yasukawa T, Kimura H, Tabata Y, et al. Targeted delivery of anti-angiogenic agent TNP-470 using water-soluble polymer in the treatment of choroidal neovascularization. Invest Ophthalmol Vis Sci 1999;40(11):2690–6.

115. Marcus D, Kim S, Brem H. The effect of an antiangiogenic compound (AGM-1470) on transgenic murine retinoblastoma [abstract]. Invest Ophtalmol Vis Sci 1993;34:226.

116. Brem H, Gotto F, Budson A. Minimal drug resistance after prolonged antiangiogenic therapy with AGM-1470. Surg Forum 1994;45:674–7.

117. Koyanagi S, Nakagawa H, Kuramoto Y, Ohdo S, Soeda S, Shimeno H. Optimizing the dosing schedule of TNP-470 [O-(chloroacetyl-carbamoyl) fumagillol] enhances its antitumor and antiangiogenic efficacies. J Pharmacol Exp Ther 2003;304(2):669–74.

118. Browder T, Butterfield CE, Kraling BM, et al. Antiangiogenic scheduling of chemotherapy improves efficacy against experimental drug-resistant cancer. Cancer Res 2000;60(7):1878–86.

119. Murata R, Nishimura Y, Hiraoka M. An antiangiogenic agent (TNP-470) inhibited reoxygenation during fractionated radiotherapy of murine mammary carcinoma. Int J Radiat Oncol Biol Phys 1997;37(5):1107–13.

120. Brem H, Folkman J. Analysis of experimental antiangiogenic therapy. J Pediatr Surg 1993;28(3):445–50; discussion 50–1.

121. Morita T, Shinohara N, Tokue A. Antitumour effect of a synthetic analogue of fumagillin on murine renal carcinoma. Br J Urol 1994;74(4):416–21.

122. O'Reilly MS, Brem H, Folkman J. Treatment of murine hemangioendotheliomas with the angiogenesis inhibitor AGM-1470. J Pediatr Surg 1995;30(2):325–9; discussion 9–30.

123. Yatsunami J, Tsuruta N, Fukuno Y, Kawashima M, Taniguchi S, Hayashi S. Inhibitory effects of roxithromycin on tumor angiogenesis, growth and metastasis of mouse B16 melanoma cells. Clin Exp Metastasis 1999;17(2):119–24.

124. Ma G, Masuzawa M, Hamada Y, et al. Treatment of murine angiosarcoma with etoposide, TNP-470 and prednisolone. J Dermatol Sci 2000;24(2):126–33.

125. Yoshizawa J, Mizuno R, Yoshida T, et al. Inhibitory effect of TNP-470 on hepatic metastasis of mouse neuroblastoma. J Surg Res 2000;93(1):82–7.

126. Ruggeri BA, Robinson C, Angeles T, Wilkinson Jt, Clapper ML. The chemopreventive agent oltipraz possesses potent antiangiogenic activity in vitro, ex vivo, and in vivo and inhibits tumor xenograft growth. Clin Cancer Res 2002;8(1):267–74.

127. Kalebic T, Tsokos M, Helman LJ. Suppression of rhabdomyosarcoma growth by fumagillin analog TNP-470. Int J Cancer 1996;68(5):596–9.

128. Konno H, Tanaka T, Kanai T, Maruyama K, Nakamura S, Baba S. Efficacy of an angiogenesis inhibitor, TNP-470, in xenotransplanted human colorectal cancer with high metastatic potential. Cancer 1996;77(8 Suppl):1736–40.

129. Yazaki T, Takamiya Y, Costello PC, et al. Inhibition of angiogenesis and growth of human non-malignant and malignant meningiomas by TNP-470. J Neurooncol 1995;23(1):23–9.

130. Isobe N, Uozumi T, Kurisu K, Kawamoto K. Experimental studies of the antitumor effect of TNP-470 on malignant brain tumors. Antitumor effect of TNP-470 on a human medulloblastoma xenograft line. Neuropediatrics 1996;27(3):136–42.

131. Taki T, Ohnishi T, Arita N, et al. Anti-proliferative effects of TNP-470 on human malignant glioma in vivo: potent inhibition of tumor angiogenesis. J Neurooncol 1994;19(3):251–8.

132. McLeskey SW, Zhang L, Trock BJ, et al. Effects of AGM-1470 and pentosan polysulphate on tumorigenicity and metastasis of FGF-transfected MCF-7 cells. Br J Cancer 1996;73(9):1053–62.

133. Kragh M, Spang-Thomsen M, Kristjansen PE. Time until initiation of tumor growth is an effective measure of the antiangiogenic effect of TNP-470 on human glioblastoma in nude mice. Oncol Rep 1999;6(4):759–62.

134. Singh Y, Shikata N, Kiyozuka Y, et al. Inhibition of tumor growth and metastasis by angiogenesis inhibitor TNP-470 on breast cancer cell lines in vitro and in vivo. Breast Cancer Res Treat 1997;45(1):15–27.

135. Takamiya Y, Friedlander RM, Brem H, Malick A, Martuza RL. Inhibition of angiogenesis and growth of human nerve-sheath tumors by AGM-1470. J Neurosurg 1993;78(3):470–6.

136. Arbiser JL, Panigrathy D, Klauber N, et al. The antiangiogenic agents TNP-470 and 2-methoxyestradiol inhibit the growth of angiosarcoma in mice. J Am Acad Dermatol 1999;40(6 Pt 1):925–9.

137. Katzenstein HM, Rademaker AW, Senger C, et al. Effectiveness of the angiogenesis inhibitor TNP-470 in reducing the growth of human neuroblastoma in nude mice inversely correlates with tumor burden. Clin Cancer Res 1999;5(12):4273–8.

138. Qian CN, Min HQ, Lin HL, et al. Anti-tumor effect of angiogenesis inhibitor TNP-470 on the human nasopharyngeal carcinoma cell line NPC/HK1. Oncology 1999;57(1):36–41.

139. Beecken WD, Fernandez A, Panigrahy D, et al. Efficacy of antiangiogenic therapy with TNP-470 in superficial and invasive bladder cancer models in mice. Urology 2000;56(3):521–6.

140. Shusterman S, Grupp SA, Maris JM. Inhibition of tumor growth in a human neuroblastoma xenograft model with TNP-470. Med Pediatr Oncol 2000;35(6):673–6.

141. Takei H, Lee ES, Cisneros A, Jordan VC. Effects of angiogenesis inhibitor TNP-470 on tamoxifen-stimulated MCF-7 breast tumors in nude mice. Cancer Lett 2000;155(2):129–35.

142. Kato H, Ishikura H, Kawarada Y, Furuya M, Kondo S, Yoshiki T. Anti-angiogenic treatment for peritoneal dissemination of pancreas adenocarcinoma: a study using TNP-470. Jpn J Cancer Res 2001;92(1):67–73.

143. Katzenstein HM, Salwen HR, Nguyen NN, Meitar D, Cohn SL. Antiangiogenic therapy inhibits human neuroblastoma growth. Med Pediatr Oncol 2001;36(1):190–3.

144. Rooks V, Beecken WD, Iordanescu I, Taylor GA. Sonographic evaluation of orthotopic bladder tumors in mice treated with TNP-470, an angiogenic inhibitor. Acad Radiol 2001;8(2):121–7.

145. Rocchetti R, Talevi S, Margiotta C, Calza R, Corallini A, Possati L. Antiangiogenic drugs for chemotherapy of bladder tumours. Chemotherapy 2005;51(6):291–9.

146. Prox D, Becker C, Pirie-Shepherd SR, Celik I, Folkman J, Kisker O. Treatment of human pancreatic cancer in mice with angiogenic inhibitors. World J Surg 2003;27(4):405–11.

147. Huang ZH, Fan YF, Xia H, Feng HM, Tang FX. Effects of TNP-470 on proliferation and apoptosis in human colon cancer xenografts in nude mice. World J Gastroenterol 2003;9(2):281–3.

148. Fan YF, Huang ZH. Angiogenesis inhibitor TNP-470 suppresses growth of peritoneal disseminating foci of human colon cancer line Lovo. World J Gastroenterol 2002;8(5):853–6.

149. Yanai S, Okada H, Misaki M, et al. Antitumor activity of a medium-chain triglyceride solution of the angiogenesis inhibitor TNP-470 (AGM-1470) when administered via the hepatic artery to rats bearing Walker 256 carcinosarcoma in the liver. J Pharmacol Exp Ther 1994;271(3):1267–73.

150. Egawa S, Tsutsumi M, Konishi Y, et al. The role of angiogenesis in the tumor growth of Syrian hamster pancreatic cancer cell line HPD-NR. Gastroenterology 1995;108(5):1526–33.

151. Futami H, Iseki H, Egawa S, Koyama K, Yamaguchi K. Inhibition of lymphatic metastasis in a syngeneic rat fibrosarcoma model by an angiogenesis inhibitor, AGM-1470. Invasion Metastasis 1996;16(2):73–82.

152. Morishita T, Mii Y, Miyauchi Y, et al. Efficacy of the angiogenesis inhibitor O-(chloroacetyl-carbamoyl)fumagillol (AGM-1470) on osteosarcoma growth and lung metastasis in rats. Jpn J Clin Oncol 1995;25(2):25–31.

153. Yanai S, Okada H, Saito K, et al. Antitumor effect of arterial administration of a medium-chain triglyceride solution of an angiogenesis inhibitor, TNP-470, in rabbits bearing VX-2 carcinoma. Pharm Res 1995;12(5):653–7.

154. Kamei S, Okada H, Inoue Y, Yoshioka T, Ogawa Y, Toguchi H. Antitumor effects of angiogenesis inhibitor TNP-470 in rabbits bearing VX-2 carcinoma by arterial administration of microspheres and oil solution. J Pharmacol Exp Ther 1993;264(1):469–74.

155. Landuyt W, Theys J, Nuyts S, et al. Effect of TNP-470 (AGM-1470) on the growth of rat rhabdomyosarcoma tumors of different sizes. Cancer Invest 2001;19(1):35–40.

156. Kinoshita S, Hirai R, Yamano T, Yuasa I, Tsukuda K, Shimizu N. Angiogenesis inhibitor TNP-470 can suppress hepatocellular carcinoma growth without retarding liver regeneration after partial hepatectomy. Surg Today 2004;34(1):40–6.

157. Brem H, Ingber D, Blood C. Suppression of tumor metastasis by angiogenesis inhibition. Surg Forum 1991;42:439–41.

158. Ahmed MH, Konno H, Nahar L, et al. The angiogenesis inhibitor TNP-470 (AGM-1470) improves long-term survival of rats with liver metastasis. J Surg Res 1996;64(1):35–41.

159. Konno H. Antitumor effect of the angiogenesis inhibitor TNP-470 on human digestive organ malignancy. Cancer Chemother Pharmacol 1999;43 Suppl:S85–9.

160. Mysliwski A, Szmit E, Szatkowski D, Sosnowska D. Suppression of growth of Bomirski Ab melanoma and its metastasis in hamsters by angiogenesis inhibitor TNP-470. Anticancer Res 1998;18(1A):441–3.

161. Mysliwski A, Bigda J, Koszalka P, Szmit E. Synergistic effect of the angiogenesis inhibitor TNP-470 and tumor necrosis factor (TNF) on Bomirski Ab melanoma in hamsters. Anticancer Res 2000;20(6B):4643–7.

162. Tanaka H, Taniguchi H, Mugitani T, et al. Angiogenesis inhibitor TNP-470 prevents implanted liver metastases after partial hepatectomy in an experimental model without impairing wound healing. Br J Surg 1996;83(10):1444–7.

163. Matsumoto K, Ninomiya Y, Inoue M, Tomioka T. Intra-tumor injection of an angiogenesis inhibitor, TNP-470, in rabbits bearing VX2 carcinoma of the tongue. Int J Oral Maxillofac Surg 1999;28(2):118–24.

164. Konno H, Tanaka T, Matsuda I, et al. Comparison of the inhibitory effect of the angiogenesis inhibitor, TNP-470, and mitomycin C on the growth and liver metastasis of human colon cancer. Int J Cancer 1995;61(2):268–71.

165. Mori S, Ueda T, Kuratsu S, Hosono N, Izawa K, Uchida A. Suppression of pulmonary metastasis by angiogenesis inhibitor TNP-470 in murine osteosarcoma. Int J Cancer 1995;61(1):148–52.

166. Shishido T, Yasoshima T, Denno R, Sato N, Hirata K. Inhibition of liver metastasis of human gastric carcinoma by angiogenesis inhibitor TNP-470. Jpn J Cancer Res 1996;87(9):958–62.

167. Gervaz P, Scholl B, Padrun V, Gillet M. Growth inhibition of liver metastases by the anti-angiogenic drug TNP-470. Liver 2000;20(2):108–13.

168. Oda H, Ogata Y, Shirouzu K. The effect of angiogenesis inhibitor TNP-470 against postoperative lung metastasis following removal of orthotopic transplanted human colon cancer: an experimental study. Kurume Med J 2001;48(4):285–93.

169. Saito N, Mitsuhashi M, Hayashi T, et al. Inhibition of hepatic metastasis in mice treated with cell-binding domain of human fibronectin and angiogenesis inhibitor TNP-470. Int J Clin Oncol 2001;6(5):215–20.

170. Takechi A. Effect of angiogenesis inhibitor TNP-470 on vascular formation in pituitary tumors induced by estrogen in rats. Neurol Med Chir (Tokyo) 1994;34(11):729–33.

171. Svensson A, Backman U, Jonsson E, Larsson R, Christofferson R. CHS 828 inhibits neuroblastoma growth in mice alone and in combination with antiangiogenic drugs. Pediatr Res 2002;51(5):607–11.

172. Morishita T, Miyauchi Y, Mii Y, et al. Delay in administration of CDDP until completion of AGM-1470 treatment enhances antimetastatic and antitumor effects. Clin Exp Metastasis 1999;17(1):15–8.

173. Kong C, Zhu Y, Sun C, et al. Inhibition of tumor angiogenesis during cisplatin chemotherapy for bladder cancer improves treatment outcome. Urology 2005;65(2):395–9.

174. Jia L, Zhang MH, Yuan SZ, Huang WG. Antiangiogenic therapy for human pancreatic carcinoma xenografts in nude mice. World J Gastroenterol 2005;11(3):447–50.

175. Muramaki M, Miyake H, Hara I, Kawabata G, Kamidono S. Synergistic inhibition of tumor growth and metastasis by combined treatment with TNP-470 and gemcitabine in a human bladder cancer KoTCC-1 model. J Urol 2004;172(4 Pt 1):1485–9.

176. Herbst RS, Takeuchi H, Teicher BA. Paclitaxel/carboplatin administration along with antiangiogenic therapy in non-small-cell lung and breast carcinoma models. Cancer Chemother Pharmacol 1998;41(6):497–504.

177. Devineni D, Klein-Szanto A, Gallo JM. Uptake of temozolomide in a rat glioma model in the presence and absence of the angiogenesis inhibitor TNP-470. Cancer Res 1996;56(9):1983–7.

178. Offodile R, Walton T, Lee M, Stiles A, Nguyen M. Regression of metastatic breast cancer in a patient treated with the anti-angiogenic drug TNP-470. Tumori 1999;85(1):51–3.

Chapter 36
Clincal Development of VEGF Trap

John S. Rudge, Ella Ioffe, Jingtai Cao, Nick Papadopoulos, Gavin Thurston, Stanley J. Wiegand, and George D. Yancopoulos

Keywords: VEGF Trap, vascular eye disease, kinase inhibitors

Abstract: The inhibition of angiogenesis is proving to be an effective strategy in treating diseases involving pathological angiogenesis such as cancer and ocular vascular diseases. Since its discovery in the 1980s, vascular endothelial cell growth factor (VEGF) has been shown to play a vital role in both physiological and pathological angiogenesis, resulting in the development of numerous approaches to block VEGF and VEGF signaling, ranging from small molecule tyrosine kinase inhibitors to protein-based and RNA-based therapeutic candidates. VEGF Trap is one such protein-based agent that has been engineered to bind and sequester VEGF, as well as placental growth factor (PlGF), with high affinity. VEGF Trap has been shown to effectively inhibit pathological angiogenesis in numerous preclinical models of cancer and eye disease, and is now being evaluated in clinical trials in several types of cancer, as well as the 'wet' or neovascular form of age-related macular degeneration (AMD). This chapter will summarize the basic biology of VEGF and the progress of the VEGF Trap from the bench to the clinic.

Introduction

Angiogenesis is a vital process not only during development, but also in the adult in settings of wound repair and reproduction [1, 2]. However, in diseases characterized by uncontrolled, pathological angiogenesis, such as solid tumors or vascular diseases of the eye, inhibition of angiogenesis would be of therapeutic benefit. While diverse factors might be expected to regulate angiogenesis in different settings, a clear consensus is emerging that a single growth factor, VEGF is the

critical requisite driver of both physiological and pathological angiogenesis in most settings [3]. Thus, while blocking VEGF signaling during early development can lead to severe growth retardation, it can also produce highly beneficial effects when applied to disease states characterized by pathological neovascularization [3]. Recent clinical studies have validated VEGF as a bona fide target for therapeutic intervention in cancer as well as in vascular diseases affecting the eye, such as wet AMD. These studies have led to approval by the U.S. Food and Drug Administration (FDA) of drugs that target the VEGF pathway, specifically antibodies directed against VEGF, or kinase inhibitors which block activation of the VEGF receptors [4–7]. Emerging therapeutic candidates that otherwise target VEGF signaling have also reported encouraging results [8].

Biology of VEGF and Its Receptors

VEGF is widely known to promote angiogenesis, by stimulating vascular endothelial cell proliferation, migration and tube formation, and it can also markedly increase vascular permeability [9]. VEGF-A is the prototypical member of a family of factors that also consists of VEGF-B, VEGF-C, VEGF-D and PlGF, which bind differentially to VEGF receptors 1, 2 and 3 and neuropilin with different specificities [10,11]. In addition, alternative exon splicing results in the production of four major isoforms of human VEGF-A – $VEGF_{121}$, $VEGF_{165}$, $VEGF_{189}$, $VEGF_{206}$ differentiated by their heparin binding affinity [12]. $VEGF_{121}$ differs from the other isoforms in that it does not bind heparin and thus is freely diffusible. As the isoforms increase in molecular weight, they show increasing heparin binding affinity and consequently bind extracellular matrix such that they tend to be sequestered near their site of production [13–15]. Matrix metalloproteases (MMPs) appear to play an important role in regulating the release of these isoforms from matrix [16]. Although endothelial cells are the primary targets of VEGF actions, more recently it has been shown that other cell types, including monocytes, hematopoietic stem cells and neurons, can also respond to VEGF [17–19].

Regeneron Pharmaceuticals, 777 Old Saw Mill River Road
Tarrytown, NY 10591, USA
E-mail: John.Rudge@regeneron.com

The VEGF receptor 2 (VEGFR2) appears to be the major mediator of the mitogenic, angiogenic and pro-permeability actions of VEGF-A. Indeed, genetic deletion of VEGFR2 results in profound defects in embryonic vasculogenesis, including a failure to develop blood islands and organized blood vessels [20]. In contrast, the role of VEGFR1, which was the first VEGF receptor identified, is more difficult to discern [21]. To date, VEGFR1 has been implicated in MMP9 induction, hematopoiesis, and monocyte chemotaxis, and also appears to act as a 'biological sink' which sequesters VEGF from binding to the lower affinity receptor, VEGFR2 [11]. A further level of complexity was revealed when VEGF was shown to bind neuropilin, a molecule previously identified as a receptor for the collapsin/semaphorin family of ligands involved in axon guidance. The binding of VEGF to neuropilin is thought to result in presentation of VEGF to VEGFR2, augmenting VEGFR2 signal transduction [22].

The Evidence for VEGF as a Key Player in Tumor Angiogenesis

In situ hybridization studies have revealed that VEGF is highly expressed in a number of human tumors [23–27]. This is most apparent in renal cell carcinoma where mutations in the von Hippel-Lindau (VHL) tumor suppressor gene result in increased transcription of hypoxia-inducible factor (HIF1) genes [28,29].

Numerous preclinical studies have shown that the growth of many different tumor types can be inhibited using agents that variously inhibit VEGF signaling (small molecule inhibitors, VEGFR2 antibodies, soluble VEGF receptors including the VEGF Trap) [30–35]. These studies have also shown that VEGF derived from the stromal compartment, as well as the tumor itself, plays an important role in mediating the angiogenesis which supports tumor growth [32,36].

Not surprisingly, antiangiogenic therapy results in a cytostatic effect in many tumor types rather than frank regression. Thus, in many models, combination of VEGF blockade with chemotherapy or radiation therapy results in greater efficacy than either approach alone [37,38].

VEGF Pathway Inhibitors Approved for the Treatment of Cancer

Bevacizumab (Avastin®)

The positive results from preclinical studies have led to the testing of several VEGF inhibitors in clinical trials. These include the humanized anti-VEGF-A monoclonal antibody (bevacizumab, Avastin®), an anti-VEGFR antibody [39], small molecules that inhibit VEGFR signaling and VEGF Trap [4,40–42]. The first clinical validation of the anti-VEGF approach came with FDA approval of bevacizumab in 2004, based on the results of a randomized double-blind phase III trial in which bevacizumab was combined with bolus IFL (irinotecan, 5FU, leucovorin) chemotherapy as first line therapy for metatstatic colorectal cancer. Median survival increased from 15.6 months in the IFL alone arm to 20.3 months in the IFL + bevacizumab arm. Severe hypertension was observed in about 10% of bevacizumab treated patients, and gastrointestinal perforation was noted in ~2% of patients. In addition, the incidence of arterial thromboembolic complications (stroke, myocardial infarction, transient ischemic attacks, unstable angina) was double the incidence seen with chemotherapy alone [43].

Kinase Inhibitors

An alternative approach to using antibodies that bind and neutralize VEGF is the use of tyrosine kinase inhibitors that target the VEGF and other receptors. To date, two such kinase inhibitors have been approved by the FDA: SU11248 (sunitinib; Sutent®) and Bay 43–9006 (sorafenib; Nexavar®). Sunitinib inhibits tyrosine phosphorylation of VEGFR1, VEGFR2, platelet derived growth factor receptor (PDGFR), c-kit and Flt3, and has been approved for the treatment of Gleevec-resistant gastrointestinal stromal tumors (GIST) and metastatic renal cell carcinoma [7]. Sorafenib, which inhibits tyrosine phosphorylation of raf, VEGFR2/3, PDGFR, kit and Flt3, was approved in 2006 for the treatment of metastatic renal cell carcinoma [5,6]. It should be pointed out that these kinase inhibitors do not block VEGF signaling selectively or completely, and much of their clinical benefit could derive from the inhibition of other kinase pathways.

Development and Application of VEGF Trap in Preclinical Animal Models

Despite the important benefits in patient care provided by the currently approved VEGF pathway blockers, current evidence suggests that optimal VEGF blockade may not have yet been achieved (e.g., dose response studies do not indicate that saturation of benefit has been reached) [44,45], raising the possibility that more potent VEGF blockade could provide even more benefit for patients. The VEGF Trap may provide the opportunity to test this hypothesis, as it was designed to bind VEGF with exceedingly high-affinity. The VEGF Trap is a soluble chimeric receptor in which key domains of VEGFR1 (domain 2) and VEGFR2 (domain 3) are fused to the constant region (Fc portion) of human IgG1 [34]. This fully human protein is capable of binding all isoforms of VEGF-A with very high affinity (KD <0.5 pM for $hVEGF_{165}$). Moreover, and in contrast to antibodies directed against VEGF-A, the VEGF Trap also binds PlGF, a VEGF family member also implicated in pathological angiogenesis [46] (KD ~25 pM for hPlGF2). In addition, the VEGF Trap was engineered to exhibit excellent pharmacokinetic properties, and has a circulating half-life in humans of approximately 2–3 weeks, allowing for bi-weekly or even less frequent dosing.

Once VEGF Trap had been optimized, it was tested and shown to have significant antiangiogenic and anti-tumor efficacy in numerous preclinical tumor models [34,37,38,47–50]. When administered in combination with chemotherapy or radiation, VEGF Trap also produced a significant additive impact on tumor growth. Moreover, in an ovarian cancer model, VEGF Trap potently inhibited ascites formation [37,38,50]. The impressive efficacy profile of VEGF Trap in preclinical animal models justified its progression into clinical trials.

Interestingly, VEGF Trap also has an additional property that distinguishes it from antibodies: while antibodies form multimeric complexes with their antigens that are rapidly cleared from the systemic circulation, VEGF Trap forms an inert 1:1 complex with VEGF that remains in the circulation and can thus be readily measured. Thus, when given in doses that saturate VEGF binding, this unique property of the VEGF:VEGF Trap complex allows for the accurate determination of tumor and whole body VEGF production rates [51]. Somewhat surprisingly, when non-tumor bearing animals and humans are given the VEGF Trap, we find that total body production rates of VEGF are quite high, challenging previous claims that systemic VEGF levels can be used as a sensitive measure of tumor burden [51]. This finding has the important corollary that agents designed to bind and inactivate tumor-derived VEGF must be provided in amounts sufficient to ensure that they are not effectively consumed by the large amounts of VEGF normally produced by the body. Building on this finding, we have determined that measurement of the levels of free and bound VEGF Trap provides a clear index of the doses required to capture all available VEGF, offering a useful guide for dosing of this antiangiogenic agent. Based on these assays, the doses currently being used in the clinic for cancer appear to be in the therapeutic range.

VEGF Trap in Clinical Trials for Cancer

Multiple Phase 1 trials have been carried out using the VEGF Trap as a single agent, or in combination with various chemotherapeutic regimens, in patients with advanced cancers, with more than 300 patients treated to date. Numerous objective responses as well as cases of prolonged stable disease have been noted in these trials. Interestingly, in addition to anti-tumor responses, improvements in tumor-associated co-morbidities have been observed: in particular, substantial or complete resolution of tumor-associated ascites. The major adverse events noted in these trials are consistent with those seen with other anti-VEGF agents, and include hypertension and proteinuria. Importantly, no antibodies to VEGF Trap were observed in any patients.

VEGF Trap is currently in Phase II clinical trials and is about to initiate multiple large Phase 3 trials. Preliminary results from two of these studies were recently reported at the American Society of Clinical Oncology 2007 annual meeting [41,42] Results were reported from an interim analysis of a randomized, double-blind, multi-center Phase 2 trial comparing two doses of VEGF Trap in patients with recurrent, platinum-resistant epithelial ovarian cancer. The patients selected for this study were heavily pre-treated and had failed several other treatment regimens. While the study remains blinded with respect to dose, the pooled preliminary results from both the high and low dose groups demonstrated anti-tumor activity, as evidenced by an 8% objective tumor response rate, with a 13% response rate as measured by a 50% reduction in circulating levels of the tumor marker CA-125. Disease was judged to be stable in 77% of patients at 4 weeks, and in 41% of patients at 14 weeks. In addition, of 24 patients in the study who had tumor-associated ascites, 29% experienced complete resolution of their ascites while 54% experienced no increase in ascites during treatment. Tolerability was similar to other molecules in this class with hypertension being the most common grade 3/4 event (16%). Two patients (1.2%) experienced bowel perforation, both of whom recovered. In a collaboration involving Sanofi-Aventis and Regeneron Pharmaceuticals, the ongoing single-agent studies of VEGF Trap will be complemented by a large Phase 3 program combining VEGF Trap with standard chemotherapy regimens in at least 5 different advanced solid tumors: colorectal, non-small cell lung, prostate, pancreas and gastric cancer, with the first of these studies initiating in 2007.

Preclinical Studies with VEGF Trap in Vascular Eye Diseases

The finding that VEGF-A was up-regulated in the aqueous and vitreous humor of patients with proliferative diabetic retinopathy extended the focus of antiangiogenic agents from cancer to vascular eye disease [52,53]. Wet age-related macular degeneration (AMD) is the most common cause of vision loss in the elderly [53,54]. In this disease, VEGF is believed to mediate the abnormal growth of choroidal vessels that, together with the associated vascular leak and edema, disrupts the normal retinal architecture. VEGF Trap has also been tested in a number of rodent models of ocular vascular disease, where it has been shown to inhibit choroidal [55] and corneal [56] neovascularization in addition to suppressing vascular leak in the retina [57]. VEGF Trap has also been shown to improve the survival of corneal transplants [58]. In a primate model of AMD, in which a laser was used to induce choroidal vascular lesions, intravitreal administration of VEGF Trap was found not only to prevent the development of vascular leak and neovascularization when administered before the time of injury but also completely resolved vascular leak when administration was delayed until after the lesions had fully developed [59].

VEGF Trap in Clinical Trials for Vascular Eye Disease

The promising results in preclinical models supported the introduction of VEGF Trap into the clinic for treatment of both wet AMD and diabetic macular edema, using a version of VEGF Trap specifically formulated for intra-ocular administration, termed VEGF Trap-Eye. Following successful completion of Phase I studies in wet AMD, VEGF Trap-Eye has progressed to a Phase 2 trial, for which interim results were recently presented at the Association for Research in Vision and Ophthalmology (ARVO) 2007 annual meeting [60,61]. In this trial, VEGF Trap-Eye was given at monthly intervals at doses of 0.5 or 2.0 mg, or as a single injection of 0.5, 2.0 or 4.0 mg. In the interim analysis, VEGF Trap-Eye met the pre-specified primary endpoint of reducing retinal thickness as measured by ocular coherence tomography (OCT) at 12 weeks compared with baseline (all groups combined, decrease of 135 μm, $P <$ 0.0001). Mean change in visual acuity, a key secondary endpoint of the study, also demonstrated a statistically significant improvement (all groups combined, increase of 5.9 letters, $P < 0.0001$). Interestingly, patients in the highest monthly dose group achieved an average vision gain of more than 10 letters at 12 weeks, and even a single injection of VEGF Trap-Eye resulted in improved mean visual acuity at 12 weeks (all dose levels combined). Thus, the increased potency of VEGF Trap-Eye may offer the potential for improved efficacy and/or longer dosing intervals compared to the current standard of care [62–64]. In the VEGF Trap-Eye Phase 2 trial, there were no drug-related serious adverse events, and treatment with the VEGF Trap-Eye was generally well-tolerated. The most common adverse events were those typically associated with intra-vitreal injections. A Phase III program for VEGF Trap-Eye in wet AMD will be initiated in 2007.

In addition, as reported at ARVO 2007 [60,65], a Phase 1 study of VEGF Trap-Eye in diabetic macular edema (DME) showed that a single intravitreal injection in patients with longstanding diabetes resulted in a marked decrease in mean central retinal thickness and mean macular volume throughout the 6 week observation period. Additional trials with VEGF Trap-Eye in DME are anticipated in the future.

Conclusion

The VEGF Trap is an all human, chimeric receptor-based protein engineered to potently bind all forms of VEGF-A and PIGF, and to exhibit superior pharmacokinetic properties. The promise of the VEGF Trap in preclinical models of cancer and vascular eye diseases is being maintained during its initial evaluation in Phase 1 and 2 human trials, offering the hope that this therapeutic candidate may offer additional benefit to patients suffering from cancer or blinding vascular eye diseases.

Bibliography

1. Folkman J, Klagsbrun M. Angiogenic factors. Science 1987;235(4787):442–7.
2. Carmeliet P. Angiogenesis in life, disease and medicine. Nature 2005;438(7070):932–6.
3. Ferrara N. Role of vascular endothelial growth factor in physiologic and pathologic angiogenesis: therapeutic implications. Semin Oncol 2002;29(6 Suppl 16):10-
4. Ferrara N, Hillan KJ, Novotny W. Bevacizumab (Avastin), a humanized anti-VEGF monoclonal antibody for cancer therapy. Biochem Biophys Res Commun 2005;333(2):328–35.
5. Ratain MJ, Eisen T, Stadler WM, et al. Phase II placebo-controlled randomized discontinuation trial of sorafenib in patients with metastatic renal cell carcinoma. J Clin Oncol 2006;24(16):2505–12.
6. Escudier B, Eisen T, Stadler WM, et al. Sorafenib in advanced clear-cell renal-cell carcinoma. N Engl J Med 2007;356(2):125–34.
7. Motzer RJ, Hoosen S, Bello CL, Christensen JG. Sunitinib malate for the treatment of solid tumours: a review of current clinical data. Expert Opin Investig Drugs 2006;15(5):553–61.
8. Ho QT, Kuo CJ. Vascular endothelial growth factor: Biology and therapeutic applications. Int J Biochem Cell Biol 2007.
9. Dvorak HF. Vascular permeability factor/vascular endothelial growth factor: a critical cytokine in tumor angiogenesis and a potential target for diagnosis and therapy. J Clin Oncol 2002;20(21):4368–80.
10. Ferrara N, Gerber HP, LeCouter J. The biology of VEGF and its receptors. Nat Med 2003;9(6):669–76.
11. Yancopoulos GD, Davis S, Gale NW, Rudge JS, Wiegand SJ, Holash J. Vascular-specific growth factors and blood vessel formation. Nature 2000;407(6801):242–8.
12. Tischer E, Mitchell R, Hartman T, et al. The human gene for vascular endothelial growth factor. Multiple protein forms are encoded through alternative exon splicing. J Biol Chem 1991;266(18):11947–54.
13. Park JE, Keller GA, Ferrara N. The vascular endothelial growth factor (VEGF) isoforms: differential deposition into the subepithelial extracellular matrix and bioactivity of extracellular matrix-bound VEGF. Mol Biol Cell 1993;4(12):1317–26.
14. Houck KA, Leung DW, Rowland AM, Winer J, Ferrara N. Dual regulation of vascular endothelial growth factor bioavailability by genetic and proteolytic mechanisms. J Biol Chem 1992;267(36):26031–7.
15. Keyt BA, Berleau LT, Nguyen HV, et al. The carboxyl-terminal domain (111–165) of vascular endothelial growth factor is critical for its mitogenic potency. J Biol Chem 1996;271(13):7788–95.
16. Lee S, Jilani SM, Nikolova GV, Carpizo D, Iruela-Arispe ML. Processing of VEGF-A by matrix metalloproteinases regulates bioavailability and vascular patterning in tumors. J Cell Biol 2005;169(4):681–91.
17. Barleon B, Sozzani S, Zhou D, Weich HA, Mantovani A, Marme D. Migration of human monocytes in response to vascular endothelial growth factor (VEGF) is mediated via the VEGF receptor flt-1. Blood 1996;87(8):3336–43.
18. Clauss M, Gerlach M, Gerlach H, et al. Vascular permeability factor: a tumor-derived polypeptide that induces endothelial cell and monocyte procoagulant activity, and promotes monocyte migration. J Exp Med 1990;172(6):1535–45.
19. Gerber HP, Malik AK, Solar GP, et al. VEGF regulates haematopoietic stem cell survival by an internal autocrine loop mechanism. Nature 2002;417(6892):954–8.
20. Shalaby F, Rossant J, Yamaguchi TP, et al. Failure of blood-island formation and vasculogenesis in Flk-1-deficient mice. Nature 1995;376(6535):62–6.
21. de Vries C, Escobedo JA, Ueno H, Houck K, Ferrara N, Williams LT. The fms-like tyrosine kinase, a receptor for vascular endothelial growth factor. Science 1992;255(5047):989–91.
22. Soker S, Takashima S, Miao HQ, Neufeld G, Klagsbrun M. Neuropilin-1 is expressed by endothelial and tumor cells as an

isoform-specific receptor for vascular endothelial growth factor. Cell 1998;92(6):735–45.

23. Yoshiji H, Gomez DE, Shibuya M, Thorgeirsson UP. Expression of vascular endothelial growth factor, its receptor, and other angiogenic factors in human breast cancer. Cancer Res 1996;56(9):2013–6.

24. Sowter HM, Corps AN, Evans AL, Clark DE, Charnock-Jones DS, Smith SK. Expression and localization of the vascular endothelial growth factor family in ovarian epithelial tumors. Lab Invest 1997;77(6):607–14.

25. Volm M, Koomagi R, Mattern J. Prognostic value of vascular endothelial growth factor and its receptor Flt-1 in squamous cell lung cancer. Int J Cancer 1997;74(1):64–8.

26. Ellis LM, Takahashi Y, Fenoglio CJ, Cleary KR, Bucana CD, Evans DB. Vessel counts and vascular endothelial growth factor expression in pancreatic adenocarcinoma. Eur J Cancer 1998;34(3):337–40.

27. Tomisawa M, Tokunaga T, Oshika Y, et al. Expression pattern of vascular endothelial growth factor isoform is closely correlated with tumour stage and vascularisation in renal cell carcinoma. Eur J Cancer 1999;35(1):133–7.

28. Iliopoulos O, Levy AP, Jiang C, Kaelin WG, Jr., Goldberg MA. Negative regulation of hypoxia-inducible genes by the von Hippel-Lindau protein. Proc Natl Acad Sci USA 1996;93(20):10595–9.

29. Lonser RR, Glenn GM, Walther M, et al. von Hippel-Lindau disease. Lancet 2003;361(9374):2059–67.

30. Kim KJ, Li B, Winer J, et al. Inhibition of vascular endothelial growth factor-induced angiogenesis suppresses tumour growth in vivo. Nature 1993;362(6423):841–4.

31. Prewett M, Huber J, Li Y, et al. Antivascular endothelial growth factor receptor (fetal liver kinase 1) monoclonal antibody inhibits tumor angiogenesis and growth of several mouse and human tumors. Cancer Res 1999;59(20):5209–18.

32. Gerber HP, Kowalski J, Sherman D, Eberhard DA, Ferrara N. Complete inhibition of rhabdomyosarcoma xenograft growth and neovascularization requires blockade of both tumor and host vascular endothelial growth factor. Cancer Res 2000;60(22):6253–8.

33. Smith JK, Mamoon NM, Duhe RJ. Emerging roles of targeted small molecule protein-tyrosine kinase inhibitors in cancer therapy. Oncol Res 2004;14(4–5):175–225.

34. Holash J, Davis S, Papadopoulos N, et al. VEGF-Trap: a VEGF blocker with potent antitumor effects. Proc Natl Acad Sci U S A 2002;99(17):11393–8.

35. Rudge JS, Thurston G, Davis S, et al. VEGF trap as a novel antiangiogenic treatment currently in clinical trials for cancer and eye diseases, and VelociGene- based discovery of the next generation of angiogenesis targets. Cold Spring Harb Symp Quant Biol 2005;70:411–8.

36. Liang WC, Wu X, Peale FV, et al. Cross-species vascular endothelial growth factor (VEGF)-blocking antibodies completely inhibit the growth of human tumor xenografts and measure the contribution of stromal VEGF. J Biol Chem 2006;281(2):951–61.

37. Byrne AT, Ross L, Holash J, et al. Vascular endothelial growth factor-trap decreases tumor burden, inhibits ascites, and causes dramatic vascular remodeling in an ovarian cancer model. Clin Cancer Res 2003;9(15):5721–8.

38. Wachsberger PR, Burd R, Cardi C, et al. VEGF Trap in combination with radiotherapy improves tumor control in U87 glioblastoma. Int J Radiat Oncol Biol Phys 2007.

39. Shaheen RM, Ahmad SA, Liu W, et al. Inhibited growth of colon cancer carcinomatosis by antibodies to vascular endothelial

and epidermal growth factor receptors. Br J Cancer 2001;85(4):584–9.

40. Jain RK, Duda DG, Clark JW, Loeffler JS. Lessons from phase III clinical trials on anti-VEGF therapy for cancer. Nat Clin Pract Oncol 2006;3(1):24–40.

41. Massarelli E, Miller VA, Leighl N, et al. Phase II Study of the Efficacy and Safety of Intravenous (IV) AVE0005 (VEGF Trap) Given Every 2 Weeks in Patients (Pts) with Platinum- and Erlotinib-Resistant Adenocarcinoma of the Lung (NSCLA). ASCO Abs 2007.

42. Tew WP, Colombo N, Ray-Coquard I, et al. VEGF-Trap for patients (pts) with recurrent platinum-resistant epithelial ovarian cancer (EOC): Preliminary results of a randomized, multicenter phase III study. ASCO Abs 2007;In Press.

43. Hurwitz H, Fehrenbacher L, Novotny W, et al. Bevacizumab plus irinotecan, fluorouracil, and leucovorin for metastatic colorectal cancer. N Engl J Med 2004;350(23):2335–42.

44. Yang JC. Bevacizumab for patients with metastatic renal cancer: an update. Clin Cancer Res 2004;10(18 Pt 2):6367S-70S.

45. Yang JC, Haworth L, Sherry RM, et al. A randomized trial of bevacizumab, an anti-vascular endothelial growth factor antibody, for metastatic renal cancer. N Engl J Med 2003;349(5):427–34.

46. Luttun A, Tiwa M, Carmeliet P. Placental growth factor (PlGF) and its receptor Flt-1 (VEGFR1): novel therapeutic targets for angiogenic disorders. Ann NY Acad Sci 2002;979:80–93.

47. Fukasawa M, Korc M. Vascular endothelial growth factor-trap suppresses tumorigenicity of multiple pancreatic cancer cell lines. Clin Cancer Res 2004;10(10):3327–32.

48. Huang J, Frischer JS, Serur A, et al. Regression of established tumors and metastases by potent vascular endothelial growth factor blockade. Proc Natl Acad Sci USA 2003;100(13):7785–90.

49. Dalal S, Berry AM, Cullinane CJ, et al. Vascular endothelial growth factor: a therapeutic target for tumors of the Ewing's sarcoma family. Clin Cancer Res 2005;11(6):2364–78.

50. Hu L, Hofmann J, Holash J, Yancopoulos GD, Sood AK, Jaffe RB. Vascular endothelial growth factor trap combined with paclitaxel strikingly inhibits tumor and ascites, prolonging survival in a human ovarian cancer model. Clin Cancer Res 2005;11(19 Pt 1):6966–71.

51. Rudge JS, Holash J, Hylton D, et al. VEGF Trap Complex Formation measures production rates of VEGF, providing a new biomarker for predicting efficacious angiogenic blockade. PNAS 2007; Proc Natl Acad Sci USA 2007;104(47):18363–70.

52. Aiello LM, Cavallerano J. Diabetic retinopathy. Curr Ther Endocrinol Metab 1994;5:436–46.

53. Malecaze F, Clamens S, Simorre-Pinatel V, et al. Detection of vascular endothelial growth factor messenger RNA and vascular endothelial growth factor-like activity in proliferative diabetic retinopathy. Arch Ophthalmol 1994;112(11):1476–82.

54. Congdon N, O'Colmain B, Klaver CC, et al. Causes and prevalence of visual impairment among adults in the United States. Arch Ophthalmol 2004;122(4):477–85.

55. Saishin Y, Saishin Y, Takahashi K, et al. VEGF-TRAP(R1R2) suppresses choroidal neovascularization and VEGF-induced breakdown of the blood-retinal barrier. J Cell Physiol 2003;195(2):241–8.

56. Wiegand S, Cao J, Renard R, Rudge J, Yancopoulos G. Long-lasting inhibition of corneal neovascularization following systemic administration of the VEGF Trap. Invest Ophthalmol Vis Sci 2003;44(Abstract):829.

57. Qaum T, Xu Q, Joussen AM, et al. VEGF-initiated blood-retinal barrier breakdown in early diabetes. Invest Ophthalmol Vis Sci 2001;42(10):2408–13.

58. Cursiefen C, Cao J, Chen L, et al. Inhibition of hemangiogenesis and lymphangiogenesis after normal-risk corneal transplantation by neutralizing VEGF promotes graft survival. Invest Ophthalmol Vis Sci 2004;45(8):2666–73.

59. Wiegand SJ, Zimmer E, Nork TM, et al. VEGF Trap Both Prevents Experimental Choroidal Neovascularization and Causes Regression of Established Lesions in Non–Human Primates. Invest Ophthalmol Vis Sci 2005;46:E-Abstract 1411.

60. Ciulla TA. VEGF Trap, Potential Therapeutic Role for Neovascular AMD and DME. ARVO 2007.

61. Benzl MS, Nguyen QD, Chu K, et al. CLEAR-IT-2: Interim Results Of The Phase II, Randomized, Controlled Dose-and Interval-ranging Study Of Repeated Intravitreal VEGF Trap Administration In Patients With Neovascular Age-related Macular Degeneration (AMD). 2007.

62. Brown DM, Kaiser PK, Michels M, et al. Ranibizumab versus verteporfin for neovascular age-related macular degeneration. N Engl J Med 2006;355(14):1432–44.

63. Rosenfeld PJ, Brown DM, Heier JS, et al. Ranibizumab for neovascular age-related macular degeneration. N Engl J Med 2006;355(14):1419–31.

64. Rosenfeld PJ, Rich RM, Lalwani GA. Ranibizumab: Phase III clinical trial results. Ophthalmol Clin North Am 2006;19(3):361–72.

65. Dol DV, Nguyen QD, Browning DJ, et al. An Exploratory Study of the Safety, Tolerability and Biological Effect of a Single Intravitreal Administration of VEGF Trap in Patients with Diabetic Macular Edema ARVO 2007.

Chapter 37
Recent Advances in Angiogenesis Drug Development

Cindy H. Chau and William D. Figg*

Keywords: classification of antiangiogenic agents, small-molecule receptor tyrosine kinase inhibitors, investigational agents

Abstract: Early cancer researchers investigating the conditions necessary for cancer metastasis observed that one of the critical events required for tumor growth is an increased vascularization and the formation of a new network of blood vessels called *angiogenesis*. Indeed it was nearly 70 years ago that the existence of tumor-derived factors responsible for promoting new vessel growth was postulated [1] and that tumor growth is essentially dependent on vascular induction and the development of a neovascular supply [2]. By the late 1960s, Dr. Judah Folkman and colleagues had begun the search for a tumor angiogenesis factor [3]. In his 1971 landmark report, Folkman proposed that inhibition of angiogenesis by means of holding tumors in a nonvascularized dormant state would be an effective strategy to treat human cancer, and hence laid the groundwork for the concept behind the development of "antiangiogenic" drugs [4]. This fostered the search for angiogenic factors, regulators of angiogenesis, and antiangiogenic molecules over the next three decades, shedding light on angiogenesis as an important therapeutic target for the treatment of cancer and other angiogenesis-dependent disease states.

Successful development and clinical translation of antiangiogenic agents depends on the complete understanding of the biology of angiogenesis and the regulatory proteins that govern this angiogenic process, topics which have been covered in greater detail in another section of this book. This chapter will discuss the recent advances in the angiogenesis drug development arena, highlighting the FDA approved drugs and investigational agents in the preclinical and clinical stage of development.

Identification of Antiangiogenic Targets within the Angiogenic Process

Understanding the angiogenic process will facilitate the identification of cellular antiangiogenic targets as potential candidates for drug development. Cancer cells promote angiogenesis at an early stage of tumorigenesis, beginning with the release of molecules that send signals to the surrounding normal host tissue and stimulating the migration of microvascular endothelial cells (ECs) in the direction of the angiogenic stimulus. These angiogenic factors not only mediate EC migration, but also EC proliferation and microvessel formation in tumors undergoing the switch to the angiogenic phenotype [5]. Thus, targeting these angiogenic factors and/or the endothelial cells represent attractive sites for drug development in inhibiting the process of angiogenesis.

Dormant tumors have been discovered during autopsies of individuals who died of causes other than cancer [6]. These autopsy studies suggest that the vast majority of microscopic, in situ cancers never switch to the angiogenic phenotype during a normal lifetime, are usually not neovascularized, and can remain harmless to the host for long periods of time [7,8]. These incipient tumors cannot expand beyond the initial microscopic size and become clinically detectable, lethal, tumors until they have switched to the angiogenic phenotype [9–11] through neovascularization and/or blood vessel cooption [12]. Depending on the tumor type and the environment, this switch can occur at different stages of the tumor progression pathway and ultimately depends on a net balance of positive and negative regulators. Thus, the angiogenic phenotype may result from the production of growth factors by tumor cells and/or the downregulation of negative modulators. Researchers are thus largely interested in searching for triggers to the angiogenic

Molecular Pharmacology Section, Medical Oncology Branch, Center for Cancer Research, National Cancer Institute, Bethesda, MD, USA

Corresponding Author:
William D. Figg, National Cancer Institute, Building 10, Room 5A01, 9000 Rockville Pike, Bethesda, MD 20892, USA
E-mail: wdfigg@helix.nih.gov

switch, and drug discovery efforts are aimed at targeting the components of the angiogenic balance.

Proangiogenic growth factors and their receptors represent attractive therapeutic targets These positive regulators of angiogenesis include vascular endothelial growth factor (VEGF), basic fibroblast growth factor, (bFGF), platelet-derived growth factor (PDGF), placental growth factor, transforming growth factor-β, pleiotrophins and others [13]. Activation of the hypoxia-inducible factor-1-α via tumor-associated hypoxic conditions is also involved in the upregulation of several angiogenic factors [14]. The angiogenic switch also involves down-regulation of angiogenesis suppressor proteins that include endostatin, angiostatin, thrombospondin and others (reviewed in [15,16]). Moreover, oncogenes play a significant role in driving the angiogenic switch by not only inducing the expression of stimulators but also downregulating inhibitors of angiogenesis (reviewed in [17]).

Changes in the angiogenic balance affecting the levels of activator and inhibitor molecules dictate whether an endothelial cell will be in a quiescent or an angiogenic state. The development of new blood vessels shifts the balance in favor of the angiogenic state with an increase in the amount of activators and a decrease in inhibitors. This prompts the activation, growth and division of vascular ECs resulting in the formation of new blood vessels. The activated ECs produce and release matrix metalloproteinases (MMPs) into the surrounding tissue. These degradative enzymes breakdown the extracellular matrix to allow the ECs to migrate into the surrounding tissues and organize themselves into hollow tubes that eventually evolve into a mature network of blood vessels. Understanding the angiogenic process is critical as many antiangiogenic drugs have been designed to target not only the endothelial cells but also MMPs and the surrounding matrix.

Rationale for Antiangiogenic Therapy

Antiangiogenic therapy stems from the fundamental concept that tumor growth, invasion, and metastatsis are angiogenesis-dependent. The microvascular endothelial cell recruited by a tumor has become an important second target in cancer therapy. Unlike the cancer cell (the primary target of cytotoxic chemotherapy) that is genetically unstable with unpredictable mutations, the genetic stability of endothelial cells may make them less susceptible to acquired drug resistance [18]. Moreover, endothelial cells in the microvascular bed of a tumor may support 50–100 tumor cells. Coupling this amplification potential, together with the lower toxicity of most angiogenesis inhibitors, results in the use of antiangiogenic therapy that should be significantly less toxic than conventional chemotherapy. Therefore, treating both the cancer cell and the endothelial cell in a tumor may be more effective than treating the cancer cell alone.

Classification of Antiangiogenic Agents

The timeline reflecting the discovery of angiogenesis inhibitors spans across almost three decades and includes both synthetic angiogenesis inhibitors and endogenous angiogenesis inhibitors. The first angiogenesis inhibitor was reported in 1980 and involved low dose administration of interferon-α [19–21]. Over the next decade, several compounds were discovered to have potent antiangiogenic activity and included protamine and platelet-factor 4 [22], trahydrocortisol [23], and the fumagillin analogue TNP-470 [24]. At least 28 endogenous angiogenesis inhibitors have been identified to date [15,16]. There are currently seven US US Food and Drug Admistration (FDA)-approved antiangiogenic agents and over 40 investigational drugs in preclinical and/or clinical testing.

The proof of the concept that targeting angiogenesis is an effective strategy for treating cancer came with the approval of the first angiogenesis inhibitor, bevacizumab, by the US FDA following a phase III study showing a survival benefit. Since then, several anti-angiogenesis agents have received FDA approval for cancer treatment, and two additional agents (pegatanib and ranibizumab) are approved for the treatment of age-related macular degeneration (Table 37.1). There are numerous investigational angiogenesis inhibitors currently being tested in clinical trials (Table 37.2). The reader is referred to reference [25] for an excellent review of the current state of drug development of angiogenesis inhibitors.

The inhibitors being investigated fall into several different categories, depending on their mechanism of action. Some inhibit endothelial cells directly, while others inhibit the angiogenesis signaling cascade or block the ability of endothelial cells to break down the extracellular matrix. Some antiangiogenic agents may target VEGF directly through neutralizing the protein, blocking the tumor expression of the angiogenic factor, or blocking the receptor for the angiogenic factor on the endothelial cells. Finally, these inhibitors may also be characterized by the degree of their blocking potential: drugs that block one main angiogenic protein, drugs that block two or three main angiogenic proteins, or drugs that have a broad spectrum effect, blocking a range of angiogenic regulators [25]. These broad spectrum inhibitors may target the angiogenic regulators and/or signaling pathways in both the tumor and endothelial cells.

Angiogenesis inhibitors may also be referred to as either being *exclusive* or *inclusive*. Drugs that are *exclusively* antiangiogenic only have one known function, which is to exhibit antiangiogenic activity. Examples of these drugs include bevacizumab, VEGF-Trap, etc. For other angiogenesis inhibitors, the antiangiogenic activity is *included* with other functions of the drug. Among them are certain cancer agents that exhibit dual roles. In many cases, the antiangiogenic activity is discovered as a secondary function after the drug has received FDA approval for a different primary function. For example, bortezomib is a proteasome inhibitor that is approved for multiple

TABLE 37.1. Antiangiogenesis agents that have received US FDA approval.

Drug	Indication (year of approval)	Class	Mechanism (cellular targets)
Bevacizumab (Avastin)	Colorectal cancer (2004); Non-small cell lung cancer (2006)	Anti-VEGF monoclonal antibody	VEGF
Sorafenib (Nexavar, BAY439006)	Renal cell cancer (2005)	Small-molecule raf kinase & tyrosine kinase inhibitor	VEGFR2, VEGFR3, PDGFR, FLT3, c-Kit
Sunitinib (Sutent, SU11248)	Gastrointestinal stromal tumor & Renal cell cancer (2006)	Small-molecule tyrosine kinase receptor inhibitor	VEGFR1, VEGFR2, VEGFR3, PDGFR, FLT3, c-Kit, RET
Thalidomid (Thalomid)	Multiple myeloma (2006)	Immunomodulatory agent	Unknown
Lenalidomide (Revlimid, CC-5013)	Myelodysplastic syndrome (2005); Multiple myeloma (2006)	Immunomodulatory agent	Unknown
Pegaptanib (Macugen)	Age-related macular degeneration (2004)	Anti-VEGF aptamer	VEGF
Ranibizumab (Lucentis)	Age-related macular degeneration (2006)	Anti-VEGF monoclonal antibody fragment	VEGF

TABLE 37.2. Examples of antiangiogenic agents in clinical development.

Drug	Cellular targets	Clinical development
Anti-VEGF agents		
VEGF-AS (antisense oligonucleotide, Veglin)	VEGF, VEGF-C, VEGF-D	RCC, mesothelioma, leukemia, lymphoma
VEGF Trap	VEGF, PlGF	NSCLC, RCC, ovarian
(Multi)targeted agents: small molecule TKIs		
Axitinib (AG13736)	VEGFR1, VEGFR2, PDGFR	Breast, NSCLC, melanoma, thyroid, pancreatic, RCC
AZD2171	VEGFR1, VEGFR2, VEGFR3, PDGFR, c-Kit	NSCLC, CRC, breast, HRPC, HCC, GIST, RCC, ovarian, glioblastoma, melanoma, mesothelioma,
BMS582664	VEGFR2, FGFR	HCC
Motesanib (AMG706)	VEGFR1, VEGFR2, VEGFR3, PDGFR, c-Kit	Solid tumors, NSCLC, GIST, breast, thyroid, CRC
OSI930	VEGFR2, c-Kit	Solid tumors
Pazopanib (GW786034)	VEGFR1–3, PDGFR, c-Kit	Solid tumors, breast, cervical, NSCLC, glioma, RCC, glioblastoma, CRC, mesothelioma
Tandutinib (MLN518/CT53518)	FLT3, PDGFR, c-Kit	HRPC, glioblastoma, RCC
Vatalanib (PTK787/ZK222584)	VEGFR1, VEGFR2, VEGFR3, PDGFR, c-Kit	NSCLC, neuroendocrine, RCC, pancreatic, breast, hematological malignancies
Vandetanib (ZD6474)	VEGFR2, VEGFR3, EGFR, Ret	CRC, NSCLC, HNC, thyroid, solid tumors
XL184	VEGFR2, MET, c-Kit, FLT3, Tie2	Solid tumors
XL999	VEGFRs, FGFR, PDGFR, FLT3	Solid tumors, CRC, RCC, NSCLC, MM
Agents that target other proteins or pathways		
AMG386	Angiopoietins	Solid tumors
AMG479	Insulin-like growth factor-1 receptor	Solid tumors
ATN161	α5beta1 integrin receptor	RCC, malignant glioma
CP751871	Insulin-like growth factor-1 receptor	MM, HRPC, breast, lung
Volociximab (M200)	Alpha5beta1 integrin receptor	RCC, melanoma

CRC Colorectal cancer; *HCC* hepatocellular carcinoma; *HNC* head and neck cancer; *HRPC* hormone-refractory prostate cancer; *MM* multiple myeloma; *NSCLC* non-small cell lung cancer; *RCC* renal cell carcinoma

myeloma and was later found to possess antiangiogenic acitivity via inhibiting VEGF. Certain orally available small molecule drugs display their antiangiogenic activity through inducing the expression of endogenous angiogenesis inhibitors such as celecoxib, a cox-2 inhibitor which inhibits angiogenesis by increasing levels of endostatin [26]. Examples of these inclusive angiogenesis inhibitors are highlighted in Table 37.3.

Direct Angiogenesis Inhibitors

There are two general classes of angiogenesis inhibitors. A direct angiogenesis inhibitor blocks vascular endothelial cells from proliferating, migrating, or increasing their survival in response to proangiogenic proteins. Direct angiogenesis inhibitors are presumed to be less likely to induce acquired drug resistance because they target the genetically stable

TABLE 37.3. *Inclusive* angiogenesis inhibitors: examples of drugs with antiangiogenic activity as a secondary function.

Drug class	Example(s)
Antibiotic	Doxycycline
Bisphosphonate	Zoledronic acid
COX-2 inhibitor	Celecoxib
EGFR small-molecule receptor tyrosine kinase inhibitors	Gefitinib, Erlotinib
EGFR/HER2 monoclonal antibodies	Cetuximab, Panitumumab, Trastuzumab
HDAC inhibitors	Belinostat (PXD101), LBH589, Vorinostat (SAHA)
mTOR inhibitors	Everolimus, Temsirolimus (CCI-779)
PPAR-γ agonist	Rosiglitazone
Proteasome inhibitor	Bortezomib

endothelial cells. Examples in this category include angiostatin, thrombospondin-1 (TSP-1), etc. ABT-510, a promising new agent, is a TSP-1 analogue. A phase I trial in patients with advanced solid malignancies showed a favorable toxicity profile, and stability of disease was observed in some patients [27]. A phase II study evaluating ABT-510 in head and neck cancer is currently underway.

Other drugs, which interact with the integrin protein, can also promote the destruction of proliferating endothelial cells. Integrins are cell surface adhesion molecules that play an essential role in cell–cell and cell–matrix adhesion. They are responsible for transmitting signals important for cell migration, invasion, proliferation and survival. One member of the integrin family, α5β3-integrin, is expressed on tumor and endothelial cells. The involvement of integrin in tumor angiogenesis was demonstrated in studies that show the β4 subunit of integrin promoting endothelial migration and invasion [28]. Therefore, agents that target integrins are currently being evaluated as potential therapeutic options and include abegrin, a monoclonal antibody, and cilengitide (EMD-121974), a cyclic peptide. Both interfere with the α5β3-integrin and are under investigation in Phase I/II studies.

Indirect Angiogenesis Inhibitors

Indirect angiogenesis inhibitors decrease or block expression of a tumor cell product, neutralize the tumor product itself, or block its receptor on endothelial cells. Examples of drugs that interfere with the angiogenesis signaling pathway include bevacizumab, sorafenib, and sunitinib. These drugs target the major signaling pathways in tumor angiogenesis: VEGF, PDGF, and their respective receptors as well as other growth factors and/or signaling pathways. Most indirect angiogenesis inhibitors are designed to target growth factor signaling pathways and can block the activity of one, two, or a broad spectrum of proangiogenic proteins and/or their receptors. These compounds are known as small-molecule, receptor tyrosine kinase inhibitors (TKI). The limitation to indirect inhibitors is that, over time, tumor cells may acquire mutations that lead

to increased expression of other proangiogenic proteins not blocked by the indirect inhibitor. This may give the appearance of drug resistance and warrants the addition of a second antiangiogenic agent, one that would target the expression of these upregulated proangiogenic proteins.

Drugs that Block Extracellular Matrix Breakdown

Another group of angiogenesis inhibitors are directed against the MMPs, enzymes that catalyze the breakdown of the extracellular matrix (ECM). Because breakdown of the matrix is required to allow ECs to migrate into surrounding tissues and proliferate into new blood vessels, drugs that target MMPs also can inhibit angiogenesis. While inhibiting the activity of MMPs has been considered a potential target for tumor therapy, phase I/II clinical trials have yielded disappointing results. Incyclinide (COL-3), an oral MMP inhibitor (MMPI), showed biological activity in AIDS-related Kaposi's sarcoma [29]. A broad spectrum MMPI, BMS-275291, failed to achieve partial or complete tumor responses in patients with advanced or metastatic cancer [30]. In a recent study, this agent also demonstrated limited clinical activity in hormone-refractory prostate cancer with bone metastases [31]. Other MMPIs, such as BAY-129566 or BB-2516, have failed to show therapeutic efficacy in human malignancies despite preclinical antimetastatic and antiangiogenic activity. The reasons for the disappointing results observed with MMPIs in cancer therapy remain to be established. These negative results of MMPIs have definitely raised serious concerns and doubts about pursuing future developments of MMPIs as a therapeutic target.

FDA Approved Antiangiogenic Agents

Bevacizumab

Bevacizumab (Avastin, Genentech) is a recombinant humanized anti-VEGF-A monoclonal antibody that received US FDA approval in February 2004 for use in combination therapy with fluorouracil-based regimens for metastatic colorectal cancer (mCRC). Bevacizumab binds VEGF and prevents the interaction of VEGF to its receptors (Flt-1 and KDR) on the surface of endothelial cells. It is the first antiangiogenic agent clinically proven to extend survival, following a large, randomized, double-blind, phase III study in which bevacizumab was administered in combination with bolus irinotecan, 5-FU, and leucovorin (IFL) as first-line therapy for mCRC [32]. In 2006, its approval extended to first- or second-line treatment of patients with metastatic carcinoma of the colon or rectum. In 2006, it received an additional approval for use in combination with carboplatin and paclitaxel, and is indicated for first-line treatment of patients with unresectable, locally advanced, recurrent or metastatic non-squamous, non-small cell lung cancer [33]. Trials are currently underway to evaluate the clinical activity of bevacizumab alone or in combination regimens in metastatic breast cancer, renal cell carcinoma, pancreatic cancer, ovarian cancer, and hormone-refractory prostate cancer as well as other solid tumors [34]. Several other trials studying the clinical benefits of bevacizumab combined with different anti-EGFR agents, such as erlotinib, panitumumab or cetuximab, are ongoing. The reader is referred to chapter 33 for a comprehensive review of this agent.

Thalidomide & its Analogue Lenalidomide

Off the market for several decades, thalidomide (Thalomid, Celgene Corporation) re-emerged in recent years as a somewhat effective treatment for various cancers, neurological, and inflammatory diseases [35]. Although its exact mechanism of action is still unclear, in addition to its antiangiogenic activity, thalidomide and its analogues possess antiangiogenic, anti-inflammatory and immunomodulatory properties. Thalidomide analogues, referred to as "immunomodulatory drugs" (IMiDs), were developed to enhance the immunomodulatory effects while minimizing the toxic effects of the parent drug. In May 2006, thalidomide was granted accelerated approval for use in combination with dexamethasone for the treatment of newly diagnosed multiple myeloma (MM) patients. Because of thalidomide's known teratogenicity, the FDA is controlling thalidomide's marketing in the United States via the System for Thalidomide Education and Prescribing Safety (S.T.E.P.S.) program. Lenalidomide (Revlimid, Celgene Corporation), a thalidomide analogue, has undergone rapid clinical development for the treatment of multiple myeloma [36]. In December 2005, lenalidomide was approved for patients with myelodysplastic syndromes associated with a deletion 5q cytogenetic abnormality with or without additional chromosomal abnormalities. In June 2006, the US FDA approved lenalidomide for use in combination with dexamethasone in patients with MM who have received one prior therapy. Lenalidomide is also available under a special restricted distribution program called RevAssist. Thalidomide and its immunomodulatory analogues are being investigated in several Phase II/III trials for treating various tumors including RCC [37], prostate cancer and hepatocellular cancer. The reader is referred to chapter 34 for an extensive review of this class of agents.

Sorafenib

Sorafenib (Nexavar, Onyx Pharmaceuticals) is a small molecule Raf kinase and VEGF receptor kinase (VEGFR2 and VEGFR3) inhibitor. It exhibits broad spectrum effects on multiple targets (PDGFR, c-KIT, p38) that affect the maintenance of the tumor vasculature and angiogenesis [38]. In December 2005 the US FDA granted approval for sorafenib, considered the first multikinase inhibitor, for the treatment of patients with advanced renal cell carcinoma (RCC). The efficacy and safety of sorafenib was proven in the largest phase 3, multicenter, randomized, double-blind, placebo controlled study conducted

in advanced RCC. Treatment of sorafenib was shown to prolong progression-free survival as compared with placebo [39].

Sorafenib was generally well tolerated with a predictable safety profile. The most common adverse events include diarrhea, rash/desquamation, hand-foot skin reaction, alopecia, and nausea/vomiting. Grade 3/4 adverse events were 38% for sorafenib vs. 28% for placebo. Sorafenib induced hypertension in patients with metastatic RCC. The treatment-related hypertension was noted to be a class effect observed not only with VEGFR inhibitors but also with the VEGF monoclonal antibody as well. No significant relationship between previously described mediators of blood pressure and the magnitude of increase was found in a study evaluating the mechanism of sorafenib-induced hypertension in patients [40].

One therapeutic combination of particular interest involves simultaneous blockade of VEGFRs with depletion of secreted VEGF. The rationale behind this approach is based on the observation that serum VEGF levels increase following administration of VEGFR inhibitors, including sorafenib. Thus, the combination of sorafenib with bevacizumab is currently underway in phase II studies. In addition to bevacizumab, temsirolimus (CCI-779) is another agent that affects angiogenesis by targeting the mammalian target of rapamycin (mTOR). mTOR regulates the expression of HIF-1α, which is upregulated by the loss of the *von Hippel Lindau* gene in RCC. Thus, by downregulating HIF-α in the tumor cell, temsirolimus may complement the effects of sorafenib at the level of the endothelial cell and thus the combination of sorafenib and temsirolimus is also being evaluated in phase I/II studies. Sorafenib is also being investigated in other tumor types such as melanoma, hepatocellular and prostate cancer.

Sunitinib

Sunitinib (Sutent, Pfizer) is a novel, oral small molecule multitargeted receptor tyrosine kinase (RTK) inhibitor of angiogenesis. It exhibits potent antitumor and antiangiogenic activity. Sunitinib is a small-molecule inhibitor of multiple RTKs, including: VEGF receptors 1, 2, and 3; stem cell factor receptor (c-KIT); PDGF receptors; Fms-like tyrosine kinase 3 (FLT-3); colony stimulating factor receptor type 1 receptor; and the glial cell line–derived neurotrophic factor receptor (RET). Previous tyrosine kinase inhibitors such as SU6668 and SU5416 (semaxanib) had little success in the clinic due to poor pharmacologic properties and limited efficacy. Therefore, sunitinib was rationally designed and chosen for its high bioavailability and its nanomolar-range potency against the antiangiogenic RTKs.

In January 2006, sunitinib was granted approval by the US FDA for the treatment of gastrointestinal stromal tumor (GIST) after disease progression on, or intolerance to, imatinib and accelerated approval for the treatment of advanced renal cell cancer (RCC) [41]. The accelerated approval for RCC was based on durable partial responses, with a response rate of 26–37%, and a median duration of response of 54 weeks from two phase II, single-arm, multicenter trials of patients with cytokine-refractory RCC [42]. In February 2007, the FDA converted the accelerated approval of sunitinib for advanced RCC to regular approval following confirmation of an improvement in progression-free survival (PFS). Efficacy data, based on an interim PFS analysis, was determined in a phase III, multi-center, international randomized trial enrolling 750 patients with treatment-naïve metastatic RCC to receive either sunitinib or interferon-α [43]. Sunitinib demonstrated significant efficacy (prolonged median time to progression) in imatinib-resistant or -intolerant GIST in a randomized, double-blind, placebo-controlled phase III trial [44]. Adverse effects (grade 1 or 2 in severity), including diarrhea, mucositis, asthenia, skin abnormalities, and altered taste, were more common in patients receiving sunitinib. In addition, a decrease in left ventricular ejection fraction and severe hypertension were also more commonly reported in the sunitinib arm. Grade 3 or 4 treatment-emergent adverse events were reported in 56% versus 51% of patients on sunitinib versus placebo, respectively.

Sunitinib has demonstrated robust antitumor activity in preclinical studies resulting not only in tumor growth inhibition, but tumor regression in models of colon cancer, non-small-cell lung cancer, melanoma, renal carcinoma, and squamous cell carcinoma, which were associated with inhibition of VEGFR and PDGFR phosphorylation. Clinical activity was evaluated in neuroendocrine, colon, and breast cancers in phase II studies. Studies investigating sunitinib alone in various tumor types and in combination with chemotherapy are ongoing.

Pegaptanib and Ranibizumab for Macular Degeneration

Age-related macular degeneration (AMD) is the leading cause of adult blindness among individuals 55 years of age and older. The neovascular (wet) form of AMD is responsible for the most severe and rapid loss of central vision. Wet AMD occurs when new, abnormal blood vessels form to improve the blood supply to oxygen-deprived retinal tissue. These fragile blood vessels break easily, causing bleeding and damage to the surrounding tissue. The reader is referred to the chapter (Ch. 44) on ocular neovascularization for an in-depth discussion of this disease state. VEGF induces vascular leakage and retinal neovascularization, thus playing a critical role in the pathogenesis of neovascular AMD. Recent advances in the treatment of wet AMD have focused on therapy that target this angiogenic protein.

Pagaptanib (Macugen, OSI Pharmaceuticals) is the first anti-VEGF therapy shown to help reduce the risk of vision loss in wet AMD. It is a 28-nucleotide RNA aptamer, a pegylated modified oligonucleotide, which adopts a three-dimensional conformation that enables it to selectively bind and inactivate the major soluble VEGF isoform (VEGF165), thereby blocking its interaction with VEGFR2. Pegaptanib, administered by intravitreal injection, was approved to treat all patients with wet AMD in the US in late 2004, and in Europe in early 2006. The worldwide marketing approval of pegaptanib for wet AMD was based on two concurrent prospective, multicenter, double-blind, dose-ranging, randomized controlled trials of intravitreal injections of pegaptanib given every 6 weeks versus placebo [45]. The 2-year results of patients who have remained on treatment with pegaptanib demonstrated continuing visual benefits of treatment compared with placebo [46]. The 2-year safety profile of pegaptanib is favorable, with the most common ocular adverse events being transient, mild to moderate in intensity, and attributable to the injection preparation and procedure [47].

Ranibizumab (Lucentis, Genentech) is an anti-VEGF antibody that received FDA approval for the treatment of wet AMD in June 2006. It is a humanized monoclonal antibody fragment specifically designed to bind all forms of VEGF. Two prospective, multicenter, double-blind, controlled pivotal trials are the basis of US marketing approval for ranibizumab in wet AMD. The Marina study randomized patients to receive 24 monthly intravitreal injections of ranibizumab or sham injections. Results from this study showed that administration of ranibizumab for 2 years prevented vision loss and improved mean visual acuity, with low rates of serious adverse events [48]. Results from the ANCHOR study, which compared ranibizumab versus photodynamic therapy with verteporfin, demonstrated that ranibizumab is superior to verteporfin as intravitreal treatment with improvement in visual acuity and low rates of serious ocular side effects [49].

While pegaptanib targets the VEGF165 isoform, ranibizumab targets all of the biologically active isoforms of VEGF. This may explain the effectiveness results that are reported to be much better for ranibizumab than pegaptanib even though both are anti-VEGF therapies. As such, it is not surprising that bevacizumab is also used as an intravitreal injection as a low cost off-label alternative to ranibizumab to treat all patients with wet AMD, but it is not licensed for this indication. Bevacizumab is derived from the same source antibody as ranibizumab and is suspected, but not proven, to have effects similar to ranibizumab. Future controlled trials will need to address the clinical efficacy and safety of intravitreal bevacizumab for wet AMD, and a head-to-head comparison is necessary to determine whether bevacizumab and ranibizumab are interchangeable. The promise of antiangiogenic agents in the treatment of patients with wet AMD is limited by the requirement for continued and frequent intravitreal injections to sustain vision benefit. Studies are currently underway to examine alternative routes of delivery of these products to improve outcome.

Tyrosine Kinase Inhibitors in Development

There are numerous small-molecule, tyrosine kinase inhibitors (TKIs) in development (Table 37.2). Examples of those with demonstrated activity in preclinical and clinical models are highlighted below.

Vatalanib

Vatalanib (PTK-787/ZK222584, Novartis) is an oral TKI with a selective range of molecular targets that consists of PDGFR, c-KIT, and all VEGFRs with a higher potency toward VEGFR2. Preclinical studies demonstrated antitumor activity against a broad range of cancer types, including colorectal, prostate, renal, hepatocellular, myeloma, recurrent glioblastoma multiform, and ovarian. Vatalanib has been investigated in more than 30 Phase I/II studies, alone and in combination with chemotherapy. However, clinical evaluation of this agent has yielded mixed results. A Phase I clinical trial, evaluating the safety and efficacy of vatalanib in patients with liver metastases from solid tumors, showed a safe toxicity profile and promising activity [50]. Other early phase clinical trials, investigating the therapeutic benefits of vatalanib for NSCLC, metastatic neuroendocrine tumors, renal and breast cancers as well as hematological malignancies, are currently underway. While vatalanib has shown encouraging results in Phase I/II studies, results from two Phase III trials (CONFIRM-1 AND CONFIRM-2 studies) of vatalanib in combination with FOLFOX4 in metastatic CRC have been disappointing, showing no significant improvement in overall survival and failing to reach statistical significance in PFS [51]. Despite this, both studies have independently provided consistent evidence that vatalanib may be beneficial to a subgroup of patients with high baseline serum lactate dehydrogenase levels although the precise mechanism of this observation remains to be fully elucidated. The most frequently reported adverse events associated with vatalanib were nausea, fatigue, vomiting and dizziness [50]. The clinical benefits of vatalanib alone or in combination with chemotherapy in various tumors remains to be determined.

Motesanib

Motesanib (AMG706, Amgen) is a highly selective, oral TKI directed against VEGFR1-3, PDGFR, and the c-KIT receptor. It blocks HUVEC proliferation and vascularization of tumor xenografts in animal models [52]. Preliminary findings have shown this agent to be well tolerated, with acceptable safety, promising activity, and disease stabilization in early clinical trials in patients with advanced solid malignancies [53]. The main treatment-related adverse events that were most frequently reported included hypertension, fatigue, diarrhea, headache and nausea. Motesanib is currently undergoing several Phase I/II clinical trials alone or in combination with chemotherapy in thyroid and breast cancer, CRC, imatinib-resistant GIST, NSCLC and other solid tumors.

Vandetanib

Vandetanib (ZD-6474, AstraZeneca) is a orally available small molecule TKI that possesses potent inhibitory activity against VEGFR2, VEGFR3, EGFR, and the oncogenic RET kinases (reviewed in [54]). Preclinical studies have yielded data consistent with inhibition of VEGF-dependent angiogenesis as well as producing antitumor and antimetastatic effects across a broad range of cancer types including lung, thyroid, breast and multiple myeloma. While Phase I clinical trials have demonstrated that vandetanib is well tolerated [55], Phase II studies evaluating the drug in several cancers elicited variable response. Results of Phase II trials, examining the clinical benefits of vandetanib in patients with previously treated metastatic breast cancer or relapsed multiple myeloma, revealed the drug displayed limited monotherapy activity. In contrast, phase II studies of vandetanib in NSCLC have shown promising evidence of clinical activity both as monotherapy (versus gefitinib) and in combination with conventional chemotherapy (vandetanib plus docetaxel versus docetaxel alone), demonstrating a significant prolongation of PFS in the vandetanib treatment group [54]. In addition, preliminary data from the Phase II medullary thyroid cancer study suggest vandetanib may also have clinical activity in this tumor type, warranting further investigation. Common drug-related adverse events include rash, diarrhea, and asymptomatic Qtc prolongation. The positive outcomes of these trials have led to the initiation of Phase III studies of vandetanib in patients with NSCLC, and clinical development continues in early phase trials to investigate efficacy in other types of cancer such as CRC and head and neck cancer.

AZD2171

AZD2171 (cediranib, AstraZeneca) is a highly potent, and selective, oral pan-VEGF receptor (inhibiting VEGFR1-3) TKI with additional inhibitory activity against PDGFR and c-Kit. Preclinical data demonstrated this compound to be effective in a range of cancer types including colon, lung, breast, ovarian, and prostate [56]. Results from a Phase I study of AZD2171 in hormone refractory prostate cancer showed good tolerability with the most common reported adverse effects being fatigue, diarrhea, hypertension, headache, and hoarseness [57]. Preliminary results from an ongoing Phase II trial in recurrent glioblastomas demonstrate promising tumor responses with AZD2171 inducing vascular normalization and circulating biomarkers correlating with disease progression [58]. AZD2171 also has the clinical benefit of alleviating edema, a major cause of morbidity in glioblastoma. Future studies are needed to validate these important findings. AZD2171 is currently in Phase II/III trials for the treatment of NSCLC, CRC, breast and ovarian cancer, imatinib-resistant GIST, RCC, glioblastoma, melanoma, mesothelioma, and other solid tumors.

Developmental Issues in the Clinical Translational of Angiogenesis Inhibitors

The successful development of angiogenesis inhibitors relies on understanding how to strategically move the compounds from benchside into the clinic. Several key issues that remain to be addressed involve determining the most effective combination therapy and optimal dosing schedule, assessing the clinical response and safety of long-term administration of antiangiogenic therapy, and overcoming the resistance mechanism that is surfacing in this treatment paradigm.

Tumor angiogenesis is a highly complex process involving multiple growth factors and their receptor signaling pathways. Based on current evidence, effective therapy will probably rely on a combinatorial approach that involves targeting multiple pathways simultaneously. A number of studies have shown that antiangiogenic agents in combination with chemotherapy or radiotherapy result in additive or synergistic effects. Several models have been proposed to explain the mechanism responsible for this potentiation, keying in on the chemosensitizing effects of antiangiogenic therapy [59]. One hypothesis is that antiangiogenic therapy may normalize the tumor vasculature, thus resulting in improved oxygenation, better blood perfusion and, consequently, improved delivery of chemotherapeutic drugs [60]. A second model suggests that chemotherapy delivered at low doses, and at close, regular, intervals with no extended drug-free break periods, preferentially damages endothelial cells in the tumor neovasculature [61,62] and suppresses circulating endothelial progenitor cells [63,64]. This regimen, also called metronomic chemotherapy, sustains antiangiogenic activity and reduces acute toxicity [65]. Thus, the efficacy of metronomic chemotherapy may increase when administered in combination with specific antiangiogenic drugs. Finally, the third model addresses the use of antiangiogenic drugs to slow down tumor cell repopulation between successive cycles of cytotoxic chemotherapy [66]. This model underscores the importance of timing and sequence in achieving the maximal therapeutic benefit from combination therapies. Nevertheless, it remains a challenge to establish the most effective combination of antiangiogenic agents, other targeted therapies and conventional therapies in order to improve clinical outcomes.

Another important issue to consider in the development of angiogenesis inhibitors is the identification of effective biomarkers to measure the clinical activity of these inhibitors. Surrogate markers of tumor angiogenesis activity are important to guide clinical development of these agents and to select patients most likely to benefit from this approach. Several avenues are currently being investigated and include tumor biopsy analysis, microvessel density, non-invasive vascular imaging (positron emission tomography, MRI) and measuring circulating biomarkers (levels of angiogenic factors in serum, plasma, urine or circulating endothelial cells) [67]. Thus, reliable surrogate markers of activity are desparately needed to monitor and evaluate the clinical efficacy of these drugs. Researching new clinical end-points or methodological approaches are critical to optimally develop these agents in the clinic.

The development of resistance to VEGF inhibitors is a growing concern in antiangiogenic therapy and presents a challenge to researchers to develop more novel drugs to cir-

cumvent this issue. Resistance may be observed in late stage tumors when tumors re-grow during treatment, after an initial period of growth suppression from these antiangiogenic agents. The resistance to VEGF inhibitors involves reactivation of tumor angiogenesis and increased expression of other proangiogenic factors. As the disease progresses, it is possible that redundant pathways might be implicated, with VEGF being replaced by other angiogenic pathways, warranting the addition of a second antiangiogenic agent that would target these secondary growth factors and/or their activated receptor signaling pathways. Perhaps the administration of angiogenic drugs at earlier stages of the disease may be a more effective and beneficial approach. In addition, tumor cells bearing genetic alterations of the p53 gene may display a lower apoptosis rate under hypoxic conditions, which might reduce their reliance on vascular supply and thereby their responsiveness to antiangiogenic therapy [68]. Therefore, the selection and overgrowth of tumor-variant cells that are hypoxia resistant and thus less dependent [68] on angiogenesis and vasculature remodeling resulting in vessel stabilization [69] could also explain the resistance to antiangiogenic drugs. Finally, among the possible mechanisms for acquired resistance to antiangiogenic drugs [70,71] perhaps the most intriguing finding is that, although endothelial cells are assumed to be genetically stable, they may under some circumstances harbor genetic abnormalities and thus acquire resistance as well [72,73].

Conclusion

Angiogenesis inhibition has been shown to suppress tumor growth and metastasis in both preclinical models and clinical studies. These benefits have translated to the clinic with both marketed and investigational antiangiogenic agents. The most prominent target of these compounds is VEGF and its receptors, although other potential therapeutic targets can include integrins, multiple growth factors in the angiogenic signaling cascade, and endogenous angiogenesis inhibitors. Antiangiogenic therapy has been established as a fourth modality in cancer treatment validating that angiogenesis as an important target for cancer. Future research will focus on determining the tumor types and stages that will benefit most from antiangiogenic therapy. Well-designed clinical trials will help develop surrogate markers of tumor angiogenesis, determine the optimal dosing strategy, and identify the most effective combination therapy to ensure that patients will fully benefit from these new agents. Much progress is needed in understanding the emergence of targeted therapy resistance and assessing the potential cumulative toxicities that arise from combination therapies. Nonetheless, antiangiogenic therapy represents a promising strategy for the treatment of cancer with the overall goal of the clinical use of angiogenesis inhibitors being to convert cancer to a chronic manageable disease.

References

1. Ide AG, Baker NH, Warren SL. Vascularization of the Brown Pearce rabbit epithelioma transplant as seen in the transparent ear chamger. Am J Roentgenol 1939;42:891–899.

2. Algire GH, Chalkley HW, Legallais FY, et al. Vascular reactions of normal and malignant tissues in vivo. I. Vascular reactions of mice to wounds and to normal and neoplastic transplants. J Natl Cancer Inst 1945;6:73–85.

3. Folkman J, Merler E, Abernathy C, et al. Isolation of a tumor factor responsible for angiogenesis. J Exp Med 1971;133:275–88.

4. Folkman J. Tumor angiogenesis: therapeutic implications. N Engl J Med 1971;285:1182–6.

5. Hanahan D, Folkman J. Patterns and emerging mechanisms of the angiogenic switch during tumorigenesis. Cell 1996;86:353–64.

6. Black WC, Welch HG. Advances in diagnostic imaging and over-estimations of disease prevalence and the benefits of therapy. N Engl J Med 1993;328:1237–43.

7. Folkman J, Kalluri R. Cancer without disease. Nature 2004;427:787.

8. Weidner N, Semple JP, Welch WR, et al. Tumor angiogenesis and metastasis–correlation in invasive breast carcinoma. N Engl J Med 1991;324:1–8.

9. Udagawa T, Fernandez A, Achilles EG, et al. Persistence of microscopic human cancers in mice: alterations in the angiogenic balance accompanies loss of tumor dormancy. Faseb J 2002;16:1361–70.

10. Naumov GN, Bender E, Zurakowski D, et al. A model of human tumor dormancy: an angiogenic switch from the nonangiogenic phenotype. J Natl Cancer Inst 2006;98:316–25.

11. Holmgren L, O'Reilly MS, Folkman J. Dormancy of micrometastases: balanced proliferation and apoptosis in the presence of angiogenesis suppression. Nat Med 1995;1:149–53.

12. Holash J, Maisonpierre PC, Compton D, et al. Vessel cooption, regression, and growth in tumors mediated by angiopoietins and VEGF. Science 1999;284:1994–8.

13. Relf M, LeJeune S, Scott PA, et al. Expression of the angiogenic factors vascular endothelial cell growth factor, acidic and basic fibroblast growth factor, tumor growth factor beta-1, platelet-derived endothelial cell growth factor, placenta growth factor, and pleiotrophin in human primary breast cancer and its relation to angiogenesis. Cancer Res 1997;57:963–9.

14. Carmeliet P, Dor Y, Herbert JM, et al. Role of HIF-1alpha in hypoxia-mediated apoptosis, cell proliferation and tumour angiogenesis. Nature 1998;394:485–90.

15. Folkman J. Endogenous angiogenesis inhibitors. Apmis 2004;112:496–507.

16. Nyberg P, Xie L, Kalluri R. Endogenous inhibitors of angiogenesis. Cancer Res 2005;65:3967–79.

17. Rak J, Yu JL, Klement G, et al. Oncogenes and angiogenesis: signaling three-dimensional tumor growth. J Investig Dermatol Symp Proc 2000;5:24–33.

18. Kerbel RS. Inhibition of tumor angiogenesis as a strategy to circumvent acquired resistance to anti-cancer therapeutic agents. Bioessays 1991;13:31–6.

19. Brouty-Boye D, Zetter BR. Inhibition of cell motility by interferon. Science 1980;208:516–8.

20. Dvorak HF, Gresser I. Microvascular injury in pathogenesis of interferon-induced necrosis of subcutaneous tumors in mice. J Natl Cancer Inst 1989;81:497–502.

21. Sidky YA, Borden EC. Inhibition of angiogenesis by interferons: effects on tumor- and lymphocyte-induced vascular responses. Cancer Res 1987;47:5155–61.

22. Taylor S, Folkman J. Protamine is an inhibitor of angiogenesis. Nature 1982;297:307–12.

23. Crum R, Szabo S, Folkman J. A new class of steroids inhibits angiogenesis in the presence of heparin or a heparin fragment. Science 1985;230:1375–8.

24. Ingber D, Fujita T, Kishimoto S, et al. Synthetic analogues of fumagillin that inhibit angiogenesis and suppress tumour growth. Nature 1990;348:555–7.

25. Folkman J. Angiogenesis: an organizing principle for drug discovery? Nat Rev Drug Discov 2007;6:273–86.

26. Folkman J. Antiangiogenesis in cancer therapy–endostatin and its mechanisms of action. Exp Cell Res 2006;312:594–607.

27. Hoekstra R, de Vos FY, Eskens FA, et al. Phase I safety, pharmacokinetic, and pharmacodynamic study of the thrombospondin-1-mimetic angiogenesis inhibitor ABT-510 in patients with advanced cancer. J Clin Oncol 2005;23:5188–97.

28. Nikolopoulos SN, Blaikie P, Yoshioka T, et al. Integrin beta4 signaling promotes tumor angiogenesis. Cancer Cell 2004;6:471–83.

29. Dezube BJ, Krown SE, Lee JY, et al. Randomized phase II trial of matrix metalloproteinase inhibitor COL-3 in AIDS-related Kaposi's sarcoma: an AIDS Malignancy Consortium Study. J Clin Oncol 2006;24:1389–94.

30. Rizvi NA, Humphrey JS, Ness EA, et al. A phase I study of oral BMS-275291, a novel nonhydroxamate sheddase-sparing matrix metalloproteinase inhibitor, in patients with advanced or metastatic cancer. Clin Cancer Res 2004;10:1963–70.

31. Lara PN, Jr., Stadler WM, Longmate J, et al. A randomized phase II trial of the matrix metalloproteinase inhibitor BMS-275291 in hormone-refractory prostate cancer patients with bone metastases. Clin Cancer Res 2006;12:1556–63.

32. Hurwitz H, Fehrenbacher L, Novotny W, et al. Bevacizumab plus irinotecan, fluorouracil, and leucovorin for metastatic colorectal cancer. N Engl J Med 2004;350:2335–42.

33. Sandler A, Gray R, Perry MC, et al. Paclitaxel-carboplatin alone or with bevacizumab for non-small-cell lung cancer. N Engl J Med 2006;355:2542–50.

34. Shih T, Lindley C. Bevacizumab: an angiogenesis inhibitor for the treatment of solid malignancies. Clin Ther 2006;28:1779–802.

35. Teo SK, Stirling DI, Zeldis JB. Thalidomide as a novel therapeutic agent: new uses for an old product. Drug Discov Today 2005;10:107–14.

36. Richardson PG, Mitsiades C, Hideshima T, et al. Lenalidomide in multiple myeloma. Expert Rev Anticancer Ther 2006;6:1165–73.

37. Choueiri TK, Dreicer R, Rini BI, et al. Phase II study of lenalidomide in patients with metastatic renal cell carcinoma. Cancer 2006;107:2609–16.

38. Wilhelm SM, Carter C, Tang L, et al. BAY 43-9006 exhibits broad spectrum oral antitumor activity and targets the RAF/MEK/ERK pathway and receptor tyrosine kinases involved in tumor progression and angiogenesis. Cancer Res 2004;64:7099–109.

39. Escudier B, Eisen T, Stadler WM, et al. Sorafenib in advanced clear-cell renal-cell carcinoma. N Engl J Med 2007;356:125–34.

40. Veronese ML, Mosenkis A, Flaherty KT, et al. Mechanisms of hypertension associated with BAY 43-9006. J Clin Oncol 2006;24:1363–9.

41. Goodman VL, Rock EP, Dagher R, et al. Approval summary: sunitinib for the treatment of imatinib refractory or intolerant gastrointestinal stromal tumors and advanced renal cell carcinoma. Clin Cancer Res 2007;13:1367–73.

42. Motzer RJ, Michaelson MD, Redman BG, et al. Activity of SU11248, a multitargeted inhibitor of vascular endothelial growth factor receptor and platelet-derived growth factor receptor, in patients with metastatic renal cell carcinoma. J Clin Oncol 2006;24:16–24.

43. Motzer RJ, Hutson TE, Tomczak P, et al. Sunitinib versus interferon alfa in metastatic renal-cell carcinoma. N Engl J Med 2007;356:115–24.

44. Demetri GD, van Oosterom AT, Garrett CR, et al. Efficacy and safety of sunitinib in patients with advanced gastrointestinal stromal tumour after failure of imatinib: a randomised controlled trial. Lancet 2006;368:1329–38.

45. Gragoudas ES, Adamis AP, Cunningham ET, Jr., et al. Pegaptanib for neovascular age-related macular degeneration. N Engl J Med 2004;351:2805–16.

46. Chakravarthy U, Adamis AP, Cunningham ET, Jr., et al. Year 2 efficacy results of 2 randomized controlled clinical trials of pegaptanib for neovascular age-related macular degeneration. Ophthalmology 2006;113:1508 e1–25.

47. D'Amico DJ, Masonson HN, Patel M, et al. Pegaptanib sodium for neovascular age-related macular degeneration: two-year safety results of the two prospective, multicenter, controlled clinical trials. Ophthalmology 2006;113:992–1001 e6.

48. Rosenfeld PJ, Brown DM, Heier JS, et al. Ranibizumab for neovascular age-related macular degeneration. N Engl J Med 2006;355:1419–31.

49. Brown DM, Kaiser PK, Michels M, et al. Ranibizumab versus verteporfin for neovascular age-related macular degeneration. N Engl J Med 2006;355:1432–44.

50. Mross K, Drevs J, Muller M, et al. Phase I clinical and pharmacokinetic study of PTK/ZK, a multiple VEGF receptor inhibitor, in patients with liver metastases from solid tumours. Eur J Cancer 2005;41:1291–9.

51. Scott EN, Meinhardt G, Jacques C, et al. Vatalanib: the clinical development of a tyrosine kinase inhibitor of angiogenesis in solid tumours. Expert Opin Investig Drugs 2007;16:367–79.

52. Polverino A, Coxon A, Starnes C, et al. AMG 706, an Oral, Multikinase Inhibitor that Selectively Targets Vascular Endothelial Growth Factor, Platelet-Derived Growth Factor, and Kit Receptors, Potently Inhibits Angiogenesis and Induces Regression in Tumor Xenografts. Cancer Res 2006;66:8715–21.

53. Rosen L, Kurzrock R, Jackson E, et al. Safety and pharmacokinetics of AMG 706 in patients with advanced solid tumors. Proc. Am. Soc. Clin. Oncol 2005;23:3013.

54. Herbst RS, Heymach JV, O'Reilly MS, et al. Vandetanib (ZD6474): an orally available receptor tyrosine kinase inhibitor that selectively targets pathways critical for tumor growth and angiogenesis. Expert Opin Investig Drugs 2007;16:239–49.

55. Holden SN, Eckhardt SG, Basser R, et al. Clinical evaluation of ZD6474, an orally active inhibitor of VEGF and EGF receptor signaling, in patients with solid, malignant tumors. Ann Oncol 2005;16:1391–7.

56. Wedge SR, Kendrew J, Hennequin LF, et al. AZD2171: a highly potent, orally bioavailable, vascular endothelial growth factor receptor-2 tyrosine kinase inhibitor for the treatment of cancer. Cancer Res 2005;65:4389–400.

57. Ryan CJ, Stadler WM, Roth B, et al. Phase I dose escalation and pharmacokinetic study of AZD2171, an inhibitor of the vascular endothelial growth factor receptor tyrosine kinase, in patients with hormone refractory prostate cancer (HRPC). Invest New Drugs 2007.

58. Batchelor TT, Sorensen AG, di Tomaso E, et al. AZD2171, a pan-VEGF receptor tyrosine kinase inhibitor, normalizes tumor vasculature and alleviates edema in glioblastoma patients. Cancer Cell 2007;11:83–95.

59. Kerbel RS. Antiangiogenic therapy: a universal chemosensitization strategy for cancer? Science 2006;312:1171–5.

60. Jain RK. Normalization of tumor vasculature: an emerging concept in antiangiogenic therapy. Science 2005;307:58–62.

61. Browder T, Butterfield CE, Kraling BM, et al. Antiangiogenic scheduling of chemotherapy improves efficacy against experimental drug-resistant cancer. Cancer Res 2000;60:1878–86.

62. Klement G, Baruchel S, Rak J, et al. Continuous low-dose therapy with vinblastine and VEGF receptor-2 antibody induces sustained tumor regression without overt toxicity. J Clin Invest 2000;105:R15–24.

63. Bertolini F, Paul S, Mancuso P, et al. Maximum tolerable dose and low-dose metronomic chemotherapy have opposite effects on the mobilization and viability of circulating endothelial progenitor cells. Cancer Res 2003;63:4342–6.

64. Mancuso P, Colleoni M, Calleri A, et al. Circulating endothelial-cell kinetics and viability predict survival in breast cancer patients receiving metronomic chemotherapy. Blood 2006;108:452–9.

65. Kerbel RS, Kamen BA. The anti-angiogenic basis of metronomic chemotherapy. Nat Rev Cancer 2004;4:423–36.

66. Hudis CA. Clinical implications of antiangiogenic therapies. Oncology (Williston Park) 2005;19:26–31.

67. Davis DW, McConkey DJ, Abbruzzese JL, et al. Surrogate markers in antiangiogenesis clinical trials. Br J Cancer 2003;89:8–14.

68. Yu JL, Rak JW, Coomber BL, et al. Effect of p53 status on tumor response to antiangiogenic therapy. Science 2002;295:1526–8.

69. Glade Bender J, Cooney EM, Kandel JJ, et al. Vascular remodeling and clinical resistance to antiangiogenic cancer therapy. Drug Resist Updat 2004;7:289–300.

70. Sweeney CJ, Miller KD, Sledge GW, Jr. Resistance in the anti-angiogenic era: nay-saying or a word of caution? Trends Mol Med 2003;9:24–9.

71. Kerbel RS, Yu J, Tran J, et al. Possible mechanisms of acquired resistance to anti-angiogenic drugs: implications for the use of combination therapy approaches. Cancer Metastasis Rev 2001;20:79–86.

72. Streubel B, Chott A, Huber D, et al. Lymphoma-specific genetic aberrations in microvascular endothelial cells in B-cell lymphomas. N Engl J Med 2004;351:250–9.

73. Hida K, Hida Y, Amin DN, et al. Tumor-associated endothelial cells with cytogenetic abnormalities. Cancer Res 2004;64:8249–55.

Chapter 38
Combination of Antiangiogenic Therapy with Other Anticancer Therapies

Beverly A. Teicher

Keywords: Antiangiogenic therapy, anticancer therapy, combined treatments

Abstract: The study of angiogenesis in malignant tumors has a long history of more than 200 years. Formal modern studies of tumor angiogenesis and the search for therapeutic antiangiogenic agents have progressed very rapidly over the past 30 years. First generation antiangiogenic agents were very important because they allowed the opportunity to validate the concepts that blocking angiogeogenesis could be therapeutically important in cancer and that a role could be found for such molecules in combination therapeutic regimens for cancer. Through the 1970s and 1980s, investigators faced a dilemma regarding how to incorporate agents that "choked-off" the tumor blood supply into multi-component therapeutic regimens that had been developed empirically over many years. TNP470, a synthetic molecule derived from a fungal toxin, and minocycline, a tetracycline, were very useful in the preclinical studies that demonstrated that antiangiogenic therapy could enhance the activity to cytotoxic chemostherapy and radiation therapy. These fisrt generation molecules failed to reach clinical approval, but very quickly antiangiogenic kinase inhibitor and antiangiogenic antibodies, especially bevacizumab, demonstrated very similar effects in preclinical models and entered clinical trial. Returning to early observations, cytotoxic therapies administered in continuous low dose regimens can take best advantage of the toxicity of these molecules toward endothelial cells in 'metronomic' regimens. Most recently, large genomics efforts have been devoted to identifying differences between normal endothelium and tumor endothelium with a goal to define highly selective tumor endothelial targets for therapeutic attack. The antiangiogenic therapies that have become approved drugs are administered to patients as important components for combination tereatment regimens.

Introduction

Tumors are dynamic, complex, living tissues undergoing the varied processes of tissue growth under the guidance of aberrant malignant cells. Cytotoxic anticancer therapies have focused on the eradication of malignant cells, which is a necessity in cancer therapy However, even the most heroic therapeutic strategies rarely achieve cure of many types of tumors. While it is clear that cytotoxic therapeutics first contact tumor vasculature, the manner in which many of these drugs have been administered may not take full advantage of the antiangiogenic activity of these agents. The growth processes of tumors are normal processes, and the invasive processes of tumors are normal processes; it is the inappropriate activation of these processes that comprises the morbidity of malignant disease [1, 2].

The importance of normal cells and tissues to support growth of tumors has been recognized for centuries. The observations of Van der Kolk [3], Jones [4] and Paget [5] more than 100 years ago document this knowledge in the clinical science literature. Sixty years ago, Algire and Chalkey [6] reported that host vascular reactions could be elicited by growing tumors, and described in exquisite detail the extent and tumor-specific nature of the induction of host capillaries by transplanted tumors. The central hypothesis of Algire and Chalkey was that vascular induction by solid tumors may be the major, and possibly, the only distinguishing factor leading to tumor growth beyond normal tissue control levels. By the late 1960s, Folkman and colleagues had begun the search for a tumor angiogenesis factor (TAF), and in 1971 in his landmark report, Folkman proposed "antiangiogenesis" as a means of holding tumors in a nonvascularized dormant state [7–13]. Among the analogies used to characterize growth and invasion of solid tumors that of Dr. Stephen Paget known as the "seed-and-soil" hypothesis has continued to provoke thought for

Genzyme Corporation, 1 Mountain Road, Framingham, MA 01701, USA

E-mail: Beverly.Teicher@Genzyme.com

nearly 120 years [1, 5]. Mainly through observations made by the study of autopsy data, Paget realized that specific tumors had a predictable pattern of metastasis. Thus, breast cancer frequently metastasized to liver and certain bones, while many other organs were never sites of breast cancer growth. Paget concluded that the "soil" of a specific highly limited spectrum of tissues provided favourable conditions for mestastic tumor growth but that most tissues did not [1, 5]. Paget further concluded that the "remote organs cannot be altogether passive or indifferent" to the process of tumor growth and that the malignant cell has a "seminal influence on the tissues in which it lodges". Thus, the normal cells of the tissue under the influence of the malignant cells must be proactively involved for a tumor to grow. Paget's concepts apply to both primary and metastatic disease. The active involvement of normal cells in the vicinity of a malignant colony is required for a tumor mass to grow, and these normal cells become a major component of the malignant disease. Recent investigations have begun to elucidate the gene and protein expression changes in the "normal" cells in malignant disease, such as carcinoma-associated fibroblasts and tumor endotheial cells [1, 14–16].

The corollary is that both the normal and the cancer cells that comprise malignant disease are valid targets for therapeutic intervention. Broadly, the normal compartment includes vascular components, extracellular matrix, stromal and infiltrating cells. The ratio of these components can vary greatly, so that some tumors appear to be masses of malignant cells while for others, such as Hodgkin's disease, it is difficult to find a malignant cell [1]. Despite our enormous progress in understanding cell types and plasticity, and genomic alterations both mutations and epigenetic changes in the malignant and 'normal' cells, our progress with therapeutic improvements remains incremental. Still, clinical exploration of new concepts and new molecules is moving very rapidly. In the last 25–30 years, the field of antiangiogenic therapy has progressed to third generation molecules. The current chapter will describe preclinical studies from first, second and third generation antiangiogenic agents that have contributed to understanding how to most effectively apply these agents in clinical combination regimens.

The early modern search for antiangiogenic substances led to the discovery of proteins and small molecules that inhibit various steps in the breakdown of vascular basement membrane [1]. These included: naturally occurring proteins such as protamine [17], interferon-α [18], interferon-γ [19], platelet factor-4 [17, 20], tissue inhibitors of metalloproteinases (TIMPs) [21], interleukin-12 [22, 23], angiostatin [24] and later endostatin [25–29]; peptides derived from cartilage [30], vitreous humor [31], smooth muscle [32] and aorta [32]; as well as synthetic peptides such as synthetic laminin peptide (CDPG) YIGSR-NH2 [33], and somatostatin analogs such as somatuline [34]. Early antiangiogenic small molecules included naturally occurring heparins [35], a variety of steroids [36, 37], several retinoids and carotenoids [38–45], warfarin [46], genistein [47–52] and fumagillin [53–62], as well as

synthetic agents such as sulphated chitin derivatives [63], sulphated cyclodextrins [64, 65], linomide [66] and derivatives of fumagillin [53–62]. It was also highlighted during this time that cytotoxic chemotherapy and radiation therapy inhibited blood vessel growth [2, 67].

One of the first antiangiogenic antibodies was LM609, which is directed towards human integrin αvβ3 [68–71]. Extracellular matrix metalloproteinases (MMPs), enzymes active in growth and invasion, have been targeted with small molecules, largely hydroxaminc acid peptide analogs [72–75]. The most extensively studied were batimastat (BB94) and marimastat (BB256), which went into clinical trial. Marimastat, a synthetic inhibitor of MMP-1, -2, -3, -7 and -9, was the first orally bioavailable MMPI tested in the clinic [74, 75]. A major strategy to inhibit tumor angiogenesis is blocking the vascular endothelial growth factor (VEGF) signaling pathway with VEGF neutralizing monoclonal antibodies [76–85]. Bevacizumab (Avastin), a recombinant humanized anti-VEGF monoclonal antibody, is now an approved drug.

Findings with Early Antiangiogenic Agents in Combination Regimens

First generation antiangiogenic agents were useful to establish some principles related to the use of antiangiogenic agents in combination regimens with cytotoxic therapies. Minocycline is a tetracycline and a collagenase inhibitor which has demonstrated antiangiogenic activity [64, 65, 86–88]. The characteristics of minocycline as a modulator of cytotoxic therapies in the Lewis lung carcinoma have been described [87, 88]. TNP-470 is a synthetic analog of a fungal secretion product called fumagillin (see Chapter 35) [53–62, 89–95]. TNP-470 is a potent inhibitor of endothelial cell (EC) migration, proliferation and capillary tube formation. TNP-470 also inhibits angiogenesis as demonstrated in the chick chorio-allantoic membrane and the rabbit and rodent cornea assays. Hoechst 33342 is a DNA-binding fluorescent dye which diffuses from the vasculature through cell layers [90]. Fig. 38.1 shows typical fluorescence distributions for FSaIIC tumor cells from disaggregated tumors after tumor-bearing mice were injected intravenously with a tracer amount of the dye. The control tumor shows a fluorescence distribution typical of untreated or cytotoxic therapy treated tumors [57, 90]. The distribution of Hoechst 33342 into tumors grown during treatment with TNP-470 and minocycline was increased as compared with untreated control tumors, indicating that the treated tumors were more easily penetrated by this lipophilic dye. The vasculature forms the first barrier to penetration of molecules into tumors. Although the modulator treatments administered did not completely inhibit angiogenesis in these tumors, the vasculature of the treated tumors may be impaired. Overall, therefore, it is probable that the main targets for the antiangiogenic modulators are extracellular matrix processes and/or tumor endothelial cells,

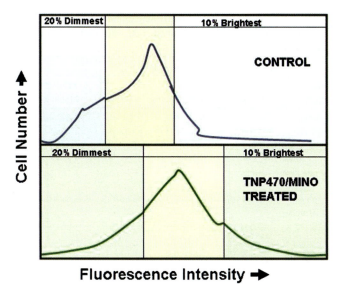

FIG. 38.1. Fluorescence intensity distributions of Hoechst 33342 in FSaIIC murine fibrosacoma cells after intravenous injection of tumor-bearing animals with Hoechst 33342 (2 mg/kg). Data are for a tumor from an untreated control mouse and for a tumor from a mouse treated with TNP470 (3 × 30 mg/kg, sc) and minocycline (5 × 10 mg/kg, ip). Tumors were 250–300 mm³ in volume and were excised 20 min post Hoechst 33342 injection. Fluorescence intensity is on a log scale with 3 logs shown.

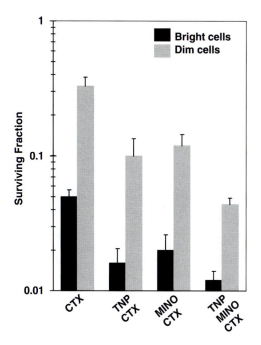

FIG. 38.2. Survival of FSaII murine fibrosarcoma cell sub-populations based upon Hoechst 33342 fluorescence intensity from mice bearing FSaIIC tumors. The tumor-bearing mice were treated with a single dose of cyclophosphamide (150 mg/kg, ip) on day 7 post tumor cell implantation alone or in combination with TNP470 (3 × 30 mg/kg, sc) on days 3, 5 and 7 and/or minocycline (5 × 10 mg/kg, ip) on days 3 through 7. The *bars* show the surviving fraction of fluorescently bright cells and fluorescently dim cells. Data are means of 3 independent experiments ± SEM.

and that inhibition and/or impairment of these non-malignant functions may improve therapeutic responses when used in combination with cytotoxic therapies [57].

When tumors were treated with TNP-470 and minocycline, there was a marked shift toward greater brightness of the entire tumor-cell population, so that the 10% brightest and the 20% dimmest cell sub-populations were composed containing much more dye than the same sub-populations in the control tumor. Cyclophosphamide was about 6-fold more toxic toward bright cells than toward dim cells (Fig. 38.2). Treatment of mice with TNP-470 (3 × 30 mg/kg) or minocycline (5 × 10 mg/kg) along with cyclophosphamide resulted in a 2.5- to 3.0-fold increase in the killing of the bright cells and the dim cell sub-populations compared with cyclophosphamide alone. Treatment with the combination was most effective resulting in a 3.6-fold increase in the killing of bright cells and a 6.8-fold increase in the killing of dim cells compared with cyclophosphamide alone [57–59].

The Lewis lung carcinoma growing subcutaneously in a hind leg of male C57BL mice is very hypoxic, having 92% of the pO_2 measurements ≤5 mm Hg as determined with a polarographic oxygen electrode [60]. Administration of a perflubron emulsion (8 ml/kg), a oxygen-delivery agent, along with carbogen (95% oxygen/5% carbon dioxide) breathing increased the tumor oxygen level so that 82% of the pO_2 readings were ≤5 mm Hg. Treating tumor-bearing mice with TNP-470 and minocycline daily beginning on day 4 after tumor cell implantation resulted in decreased hypoxia in the tumors on day 9 when pO_2 measurements were made.

The percent of pO_2 readings ≤5 mm Hg in the tumors of the TNP-470/minocycline-treated mice was 75%, which upon administration of the perflubron emulsion along with carbogen breathing was reduced to 45%. Therapeutically daily fractionated radiation (2, 3 or 4 Gray × 5) was used as an oxygen-dependent cytotoxic modality. The radiation response of the tumors in TNP-470/minocycline-treated mice was greater than that in the untreated tumors (Fig. 38.3). The addition of carbogen breathing for 1 h prior to and during radiation delivery further increased the raidtion response so that overall there was a 2.2-fold increase in the tumor growth delay produced by the fractionated radiation in the mice treated with TNP-470/minocycline compared with untreated mice. Administration of the perflubron emulsion along with carbogen breathing prior to and during radiation delivery resulted in a 3.4-fold increase in tumor growth delay by the fractionated radiation regimens in the TNP-470/minocycline-treated mice compared with the tumor growth delay obtained with radiation alone. There was a linear relationship between decrease in the percent of pO_2 readings ≤5 mm Hg and tumor growth delay at each radiation dose indicating that the diminution in tumor hypoxia produced by these treatments may be directly responsible for the increase in the effectiveness of the radiation therapy [60, 61].

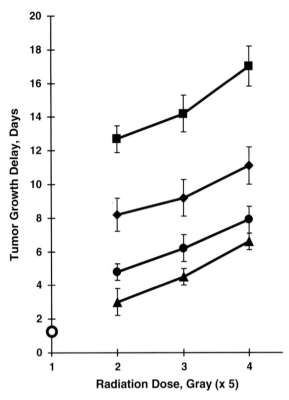

FIG. 38.3. Growth delay of the murine Lewis lung carcinoma produced by daily fractionated radiation delivered in fractions of 2, 3 or 4 (Gray) locally to the tumor-bearing limb for 5 days on days 7 through 11 post tumor cell implant alone (▲), in mice treated with TNP-470 (30 mg/kg, sc) on alternate days and minocycline (10 mg/kg, ip) daily on days 4 through 18 (●), in mice treated with TNP-470/minocycline (as above) and allowed to breathe carbogen (95% oxygen/5% carbon dioxide) for 1 h prior to and during radiation delivery (◆), and in animals treated with TNP-470/minocycline (as above) and injected intravenously with perflubron emulsion (8 ml/kg), an oxygen-carrying agent, and then allowed to breathe carbogen (95% oxygen/5% carbon dioxide) for 1 h prior to and during radiation delivery (■). The points are means of 15 mice and *bars* are SEM.

TABLE 38.1. Growth delay of the Lewis lung tumor produced by various anti-cancer treatment alone or in combination with potential antiangiogenic modulators.

Treatment Group	Tumor growth delay (days)			
	Alone	+Minocycline	+TNP470	+TNP470/ Minocycline
Controls	-	1.2+/−0.4	2.1+/−0.4	1.8+/−0.4
Cisplatin (10 mg/kg)	4.5+/−0.3	5.0+/−0.3	6.0+/−0.5	10.9+/−0.8
Cyclophosphamide (3 × 150 mg/kg)	21.5+/−1.7	32.4+/−1.8	25.3+/−2.2	44.8+/−2.8
Melphalan (10 mg/kg)	2.7+/−0.3	4.3+/−0.3	6.0+/−0.5	8.5+/−0.6
BCNU (3 × 15 mg/kg)	3.6+/−0.4	5.2+/−0.4	6.3+/−0.5	14.6+/−1.0
Radiation (5 × 3 Gray)	4.4+/−0.3	7.8+/−0.6	10.6+/−1.1	15.3+/−1.2

TABLE 38.2. Number of lung metastases on day 20 from subcutaneous Lewis lung tumors after various anti-cancer therapies alone or in combination with potential antiangiogenic modulators.

Treatment Group	Mean number of lung metastases (% large)			
	Alone	+Minocycline	+TNP470	+TNP470/ Minocycline
Controls	20 (62)	20 (50)	21 (51)	18 (54)
Cisplatin (10 mg/kg)	13 (58)	11 (48)	14.5 (34)	14 (50)
Cyclophosphamide (3 × 150 mg/kg)	12 (40)	6 (33)	6 (30)	2 (25)
Melphalan (10 mg/kg)	13 (48)	11 (50)	15 (47)	15 (45)
BCNU (3 × 15 mg/kg)	16 (53)	15 (38)	15 (47)	13 (38)
Radiation (5 × 3 Gray)	15 (40)	13 (30)	10 (40)	12 (42)

The Lewis lung carcinoma growing in C57BL mice was chosen for tumor growth delay studies because this tumor is relatively resistant to many cancer therapies and because it metastasizes avidly to the lungs from subcutaneous implants. In order to study tumor growth delay, each of the cytotoxic therapies was administered at full standard dose and schedule. TNP-470, administered subcutaneously on the back on alternate days beginning on day 4 and continuing until day 18, was a moderately effective modulator of the cytotoxic therapies (Table 38.1). TNP-470 as a single modulator was most effective in combination with melphalan, carmustine (BCNU) and radiation therapy, increasing the tumor growth delay produced by these treatments 1.8- to 2.4-fold. TNP-470 along with minocycline (10 mg/kg) administered intraperitonmeally daily on days 4 through 18 was a highly effective modulator combination. The increases in tumor growth delay produced by the

modulator combination TNP-470/minocycline along with the cytotoxic therapies ranged from 2- to 4-fold. In the treatment groups receiving TNP-470/minocycline and cyclophosphamide, approximately 40% of the mice were long-term (>120 days) survivors. Each of the cytotoxic therapies (including radiation therapy delivered locally to the tumor-bearing limb) produced a reduction in the number of lung metastases found on day 20 (Table 38.2). None of the modulators altered the number of lung metastases or the percentage of large (vascularized) lung metastases on day 20 nor was the number of lung metastases different from those obtained with the cytotoxic therapies, except in the case of cyclophosphamide; many mice treated with this drug and modulator combination had very few metastases on day 20, most of them very small [57–61].

Two conclusions may be drawn. First, combinations of antiangiogenic and/or antimetastatic agents can evoke a greater effect on tumor response to cytotoxic therapy than treatment with single agents of these classes. Second, treatment with antiangiogenic agents and/or antimetastatic agents can interact in a positive way with cytotoxic therapies.

These combined regimens produced additive or synergistic antitumor activity[57–61, 64, 65, 86–99]. Potentiation of the therapeutic effects with combined regimens may be related to increased access into the tumor mass of cytotoxic drugs or to enhanced oxygen pressure as a result of the enhanced permeability induced by antiangiogenic agents [57–61, 95, 97]. The greater than additive therapeutic effects may result from direct effects on tumor endothelial cells in addition to direct effects on tumor cells. In animals bearing Lewis lung carcinoma, TNP-470, minocycline, suramin and genstein, alone or in two agent combinations, with cytotoxic agents and radiation therapy, enhanced the regression of primary subcutaneous tumors and reduced the number and size of lung metastases. In an orthotopic animal model of transitional cell carcinoma, docetaxel, administered before TNP-470, significantly increased the complete response rate of non-established and established tumors compared with either compound alone [91]. The combined treatment inhibited angiogenesis by upregulation of basic fibroblast growth factor (bFGF) and MMP-9 and enhanced apoptosis, without altering the expression of VEGF, interleukin-8 (IL-8), MMP-2 and E-cadherin. Combinations of TNP-470 with various cytotoxic chemotherapeutic agents, such as paclitaxel and carboplatin in non-small cell lung cancer (NSCLC) and breast cancer models, paclitaxel in NSCLC, cisplatin in liver metastasis of human pancreatic cancer, and 5-fluorouracil in liver metastasis of colorectal cancer produced additive or synergistic antitumor activity [92–99].

'Antiangiogenic' Kinase Inhibitors in Combination Regimens

SU5416 was under development as a selective inhibitor of VEGFR2 (Flk-1, KDR) kinase activity. SU6668 and SU11248 are under development as broad-spectrum receptor kinase inhibitors being able to block VEGFR2, bFGF receptor (bFGFR) and platelet-derived growth factor receptor (PDGFR) kinase activities [100–109]. Early in vivo work with SU5416 suffered from the use of DMSO as a vehicle for the compound administered by intraperitoneal injection in mice once daily beginning 1 day after tumor cell implantation [103]. Geng et al. [101] found that SU5416 increases the sensitivity of the murine B16 melanoma and murine GL261 glioma to radiation therapy. SU5416 and SU6668 have been tested as single agents or in combination with fractionated radiation therapy in C3H mice bearing SCC VII squamous carcinomas [102]. Like STI571, SU5416, SU6668 and SU11248 inhibit c-kit (KIT), the stem cell factor receptor tyrosine kinase and FLT3 [108, 109]. C-kit is a key factor for development of normal hematopoietic cells and has a functional role in acute myeloid leukemia. The multi-targeted kinase inhibitor SU11248 blocks the activity of receptor tyrosine kinases located on both ECs and malignant cells. The potential therapeutic utility of SU11248 was evaluated in the MMTV-v-Ha-ras transgenic mouse model of mammary cancer and in the DMBA-carcinogen-induced rat mammary cancer model, as well as in the MX-1 human breast carcinoma

subcutaneously implanted xenograft and the MDA-MB-435 human breast carcinoma subline, 435/HAL-Luc, selected from a bone metastases alone and along with docetaxel, 5-fluorouracil or doxorubicin. The combined regimens resulted in longer survival times than either single chemotherapeutic agent or SU11248 alone [108–110]. Other small molecule tyrosine kinase inhibitors showing promise activity in early clinical trial include PTK787/ZK222584 and ZD6474. PTK787/ZK 222584 has shown activity in several experimental models [111–114]. Daily oral treatment with PTK787/ZK222584 resulted in a significant decrease in primary RENCA murine renal cell carcinoma grown in the subrenal capsule of Balb/c mice. The occurrence of lung metastases was reduced 98% and 78% on days 14 and 21, respectively; and development of lymph node metastases were delayed [114].

Small molecule approaches to inhibiting protein kinase C (PKC) have been directed toward the ATP-binding site or the diacylglycerol binding site. Conformationally constrained analogs of diacylglycerol-lactones with approximately the same binding affinity for PKC as phorbol esters were explored [115–118]. Safingol, the L-threo enantiomer of dihydrosphingosine, acts as a competitive inhibitor at the diacylglycerol-binding domain of the enzyme. Balanol analogs with benzophenone subunits were inhibitors of PKC and PKA [119]. The advent of stuarosporine-like bisindolylmaleimide structures opened up the field of inhibitors such as CGP41251, Go-6850, Ro-31-8220 and UCN-01 at the ATP-binding site of the enzyme [120, 121]. The National Cancer Institute 60-cell line identified the protein kinase C inhibitor UCN-01, 7-hydroxy-staurosporine. UCN-01 was shown to inhibit the growth of many tumor types in vitro and in vivo [122]. UCN-01 has undergone Phase I and II clinical trial alone and in combination regimens [123, 124].

The N-benzoylated staurosporine analog midostaurin was originally identified as a PKC inhibitor, but has subsequently been found to inhibit both tyrosine kinase and serine-threonine kinases [125]. Midostaurin is a potent radiation sensitizer both in cell culture and human tumor xenograft studies through a phosphoinositol 3-kinase (PI3K)/protein kinase B (PKB or Akt) pathway [127]. In B16F10 murine melanoma, midostaurin was anti-metastatic by inhibition of the invasive and/or platelet-aggregating activities of the melanoma cells [127].

In culture, midostaurin was more cytotoxic toward B-cell chronic lymphocytic leukemia than toward normal B-cells [128]. Exposure to midostaurin sensitized human U1810 NSCLC cells to DNA damage by etoposide [129]. Midostaurin suppressed Akt kinase activation and induced apoptosis in human multiple myeloma cells [130].

Midostaurin is a potent inhibitor of mutant FLT3 and of the abnormal fusion protein ZNF198-fibroblast growth factor receptor (FGFR)1 as well as mutant KIT tyrosine kinase activities [131, 132]. Gleevec (imatinib) resistant disease remains sensitive to midostaurin [133]. In culture, exposure of human leukemia cells to rapamycin along with midostaurin resulted in synergistic inhibition of cells expressing midostaurin-sensitive or

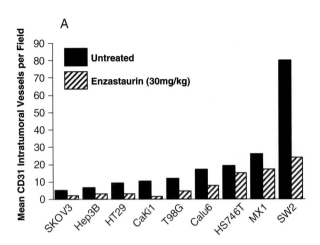
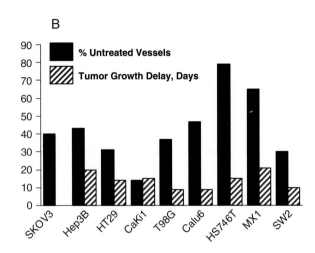

FIG. 38.4. **A** Countable intratumoral vessels by CD31 staining in human tumor xenografts either untreated or after treatment of the tumor-bearing animals with enzastaurin (30 mg/kg) p.o. twice per day for 10–14 days after tumor implantation. Data are the means of 10 determinations. **B** Percent of untreated intratumoral vessels after treatment with enzastaurin (30 mg/kg) and tumor growth delay produced in a series of human tumor xenografts by single agent treatment with enzastaurin (30 mg/kg) p.o. twice per day for 10–14 days after tumor cell implantation.

resistant leukemogenic FLT3 mutants. Similarly, exposure of human acute myelogenous leukemia cells with mutant FLT3 to 17-allylamino-demethoxygeldanamycin (17-AAG) or histone decaetylase inhibitor LAQ824 and midostaurin resulted in increased cell killing compared with either agent alone, and relatively increased cell killing compared with acute myelogenous leukemia cells expressing normal FLT3 [134, 135].

Midostaurin has undergone several Phase I clinical trials as a PKC inhibitor and as a tyrosine kinase inhibitor. Phase I single agent studies showed that midostaurin can be safely administered by chronic oral therapy. Phase I studies of midostaurin in combination with protracted continuous infusion 5-fluorouracil in advanced solid tumors, in combination with paclitaxel and carboplatin or gemcitabine and cisplatin in advanced NSCLC, established midostaurin doses that could be safely administered in these regimens [136, 137]. While Phase II trials of midostaurin alone and along with imatinib in mutant FLT3 acute myeloid leukemia or gastrointestinal stromal tumor (GIST) have shown activity, a mutation in the FLT3 tyrosine kinase domain which confers resistance to midostaurin has been identified [138].

Enzastaurin is a potent inhibitor of PKCβ [139]. PKCβ promoter activity and PKCβII mRNA expression in HCT116 colon cancer cells were inhibited enzastaurin demonstrating autoregulation of PKCβII expression. Endogenous PKCβII mRNA levels in HCT116 cells were significantly reduced by enzastaurin, consistent with the effect on PKCβ promoter activity [140]. Exposure to enzastaurin (600 nM, 72 h) profoundly inhibited proliferation of VEGF (20 ng/ml)-stimulated human umbilical vein endothelial cells (HUVECs). When human SW2 small cell lung carcinoma cells were exposed to enzastaurin for 72 h, a potency differential on the malignant cells versus HUVECs was apparent. Cell culture studies indicate that exposure to enzastaurin can cause growth inhibition

and apoptosis in human multiple myeloma cells, diffuse large cell lymphoma and mantle cell lymphoma [141–144].

Enzastaurin at 10 mg/kg orally twice per day for 10 days post-surgical implant of VEGF impregnated filters resulted in markedly decreased vascular growth to about one-half of the VEGF-stimulated controls, while 30 mg/kg decreased vascular growth to the level of surgical control [**139**]. Enzastaurin at 30 mg/kg orally twice per day for 10 days post-surgical implantation of bFGF resulted in decreased vascular growth to 26% of the bFGF control. When nude mice bearing human tumor xenografts were treated with enzastaurin orally twice daily for 2 weeks, the number of intratumoral vessels was decreased to one-half to one-quarter of the controls in treated animals (Fig. 38.4A) [139, 145–148]. Although some tumors responded to enzastaurin as an antiangiogenic agent in no case was angiogenesis completely blocked as in the corneal micropocket neoangiogenesis model. The tumor growth delay did not correlate with intratumoral vessel decrease (Fig. 38.4B). In most tumor models, the tumor growth delay produced by enzastaurin as a single agent was not sufficient to predict single agent activity in the clinic. However, the combination regimens suggested high activity. VEGF plasma levels in mice bearing the human SW2 SCLC and Caki-1 renal cell carcinomas treated or untreated with enzastaurin were obtained every 3 days starting on day 7 post-implantation, through treatment, and after the termination of treatment. Plasma VEGF levels were similar between the treated and untreated groups through day 20 [149, 150]. Plasma VEGF levels in the control groups continued to increase throughout the study. However, even after termination of enzastaurin treatment, plasma VEGF levels in enzastaurin treated mice were significantly decreased [151].

The secretion of VEGF by 12 human tumor cell lines grown in monolayer was compared with plasma VEGF levels

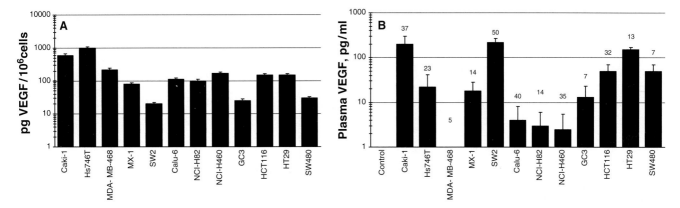

FIG. 38.5. **A** Mean levels of VEGF secretion by human tumor cell lines into cell culture medium. Cells were grown to 80% confluency then the medium was replaced with serum-free medium for 24 h. After 24 h, the medium was collected and stored at −80°C until analysis for VEGF. **B** Mean plasma levels of human VEGF from nude mice bearing human tumors. The *numbers* indicate the sample size analyzed for each tumor line. The *bars* represent the means +/− SEM. Nude mice of the same gender as the origin of the tumor were irradiated with 450 rad TBI. Human tumor cells (5 × 106) were suspended in 1:1 medium:Matrigel™, and injected subcutaneously into a hind leg of the mouse. Tumor measurements were recorded twice weekly. To obtain plasma, mice were killed with CO_2 and blood collected by cardiac puncture into EDTA tubes on ice. The blood samples were centrifuged at 800 g for 30 min at 4°C, followed by 10 min at 3,000 g. The plasmas were collected and stored at −80°C until analysis for VEGF.

when the same cell were grown as xenograft tumor in nude mice [150]. VEGF secreted into culture medium by human tumor cell lines and by normal cells found within tumor tissue was measured. VEGF was secreted by all of the tumor cell line cultures (Fig. 38.5). In culture, VEGF levels were in the range 0–1,064 pg/ml/10⁶ cells/24 h, with the highest levels secreted by HS746T gastric carcinoma cells (1,064 pg/10⁶ cells) and Caki-1 renal cell cvarcinoma cells (452 pg/10⁶ cells). Those lines secreting the lowest amounts of VEGF into culture medium were GC3 colon carcinoma cells, SW480 colon carcinoma cells and SW2 small cell lung carcinoma cells (25 pg/10⁶ cells). The same human tumor cell lines were grown as xenograft tumors in nude mice. Non-tumor-bearing mice had no detectable plasma levels of human VEGF. In tumor-bearing mice plasma VEGF levels ranged between 5 and 200 pg/ml, except MDA-MB-468 breast carcinoma for which no plasma VEGF was detectable. The highest plasma VEGF levels were obtained in mice bearing Caki-1 renal carcinoma, SW2 small cell lung carcinoma HT29 colon carcinoma and HCT116 colon carcinoma tumors. The data suggest that there is no correlation between in vitro and in vivo tumor cell VEGF secretion. There was a linear correlation between tumor volume and plasma VEGF levels in mice bearing SW2 small cell lung carcinoma, HCT116 colon carcinoma, Caki-1 renal cell carcinoma, HS746T gastric carcinoma and Calu6 non-small cell lung carcinoma tumors. In most cases, VEGF was not detectable when tumor volumes were below 800 mm³. Plasma VEGF levels continued to increase as the tumors grew [150].

Based upon these findings, the effect of enzastaurin, twice daily oral administration, on tumor growth, VEGF plasma levels and intratumoral vessel counts was tested in nude mice bearing human SW2 small cell lung carcinoma, HCT116 colon carcinoma, and Caki-1 renal cell carcinoma tumors [151].

For the faster growing SW2 small cell lung carcinoma and HCT116 colon carcinoma xenografts, enzastaurin treatment was initiated on day 14 after tumor cell implantation and was continued through day 30. For the slower growing CaKi-1 renal cell carcinoma xenograft, enzastaurin treatment was initiated on day 21 and continued through day 39 (Fig. 38.6). There was no observable toxicity to the mice from enzastaurin treatment as determined by body weight changes.

Tumor dimensions were measured twice weekly to assess tumor response to the single-agent enzastaurin treatment. Because the blood collections for plasma were terminal events, it was not possible to follow the progress of individual tumors over the course of the experiment. Tumor dimensions were measured from the three mice that were bled for each time-point. Therefore, each tumor volume data point and VEGF level is the mean of nine mice. With this experimental design, single-agent enzastaurin administration produced no tumor growth delay in mice bearing the SW2 small cell lung carcinoma or HCT116 colon carcinoma tumors. However, a tumor growth delay of about 15 days was observed in mice bearing the Caki-1 renal cell carcinoma (Fig. 6) [151].

The plasma concentrations of VEGF165 in mice bearing human SW2 small cell lung carcinoma, HCT116 colon carcinoma, and Caki-1 renal cell carcinoma tumors and in untreated control were measured over the time course of tumor growth. In the SW2 tumor control group, the plasma VEGF165 concentration continued to increase as the tumors grew reaching 400 pg/ml on day 40 after tumor cell implantation (Fig. 38.7). In the enzastaurin-treated group, the plasma VEGF levels reached 75 pg/ml by day 20 and remained at that concentration for the duration of the treatment course. Upon completion of the treatment regimen on day 30, plasma VEGF165 concentration in the enzastaurin-treated mice remained markedly decreased compared with the untreated control mice [151].

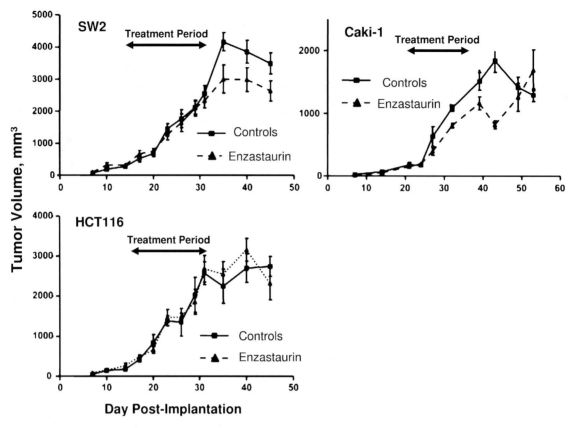

FIG. 38.6. Tumor volumes in nude mice bearing subcutaneously implanted human SW2 small cell lung carcinoma, human CaKi1 renal cell carcinoma or human HCT116 colon carcinoma, either untreated controls or treated with LY317615 (enzastaurin) orally twice daily on days 14–30 (days 21–39 for CaKi1-bearing mice). The data are the means from three independent experiments with each point derived from 9 individual tumors. *Bars* are SEM.

The plasma VEGF165 concentrations in the Caki-1 tumor-bearing control mice continued to increase throughout the experiment and peaked at 225 pg/ml on day 49 after tumor cell implantation. In the enzastaurin-treated group, the VEGF165 plasma concentrations remained <30 pg/ml throughout the treatment period (days 21–39) and until termination of the experiment on day 53. Plasma VEGF165 concentrations in mice bearing HCT116 colon carcinoma were lower than those measured in the plasma of SW2 and Caki-1 bearing mice. The plasma VEGF165 concentrations in the untreated control HCT116 colon carcinoma-bearing mice reached 50 pg/ml and remained at that level for the duration of the experiment. In contrast to the results obtained with the SW2 and Caki-1 bearing mice, treatment with enzastaurin did not result in a decrease in plasma VEGF165 concentrations in mice bearing HCT116 tumors. Both the Caki-1 tumors and the HCT116 tumors responded to enzastaurin treatment as determined by intratumoral vessel count with a 40% reduction in countable vessels. The kinetics of response to the antiangiogenic effects of enzastaurin appear to be slow. That is, although the decrease in plasma VEGF was evident shortly after administration of enzastaurin began, slowing of Caki-1 tumor growth was not evident until 1 week later. It is possible that the faster growing SW2 tumors were beginning to respond to the enzastaurin when

the experiment was terminated. Thus, the slower growth rate of the Caki-1 tumors may have made detection of a response to the single-agent enzastaurin therapy more likely [151].

Enzastaurin completed Phase I clinical studies in 2003 [152]. Phase II studies have been conducted in patients with relapsed diffuse large B-cell lymphoma and in patients with recurrent high grade gliomas. Several patients with multiple relapsed diffuse large B-cell lymphomas achieved prolonged periods of stable disease following enzastaurin treatment, although the objective tumor response rate was low [153]. In high grade glioma patients, enzastaurin was well tolerated and appeared to have promising antitumor activity in a significant percentage of highly pretreated patients [154, 155]. A Phase III randomized, open label registration trial of enzastaurin in lymphoma has been initiated.

Antiangiogenic Protein Therapeutics with Antitumor Agents

Inoue et al. [156, 157] found that combined regimens, including the monoclonal antibody C225 that blocks the epidermal growth factor receptor (EGFR) function, or the rat monoclonal antibody DC101 that blocks VEGFR2 function, with paclitaxel, had enhanced antitumor activity through inhibition

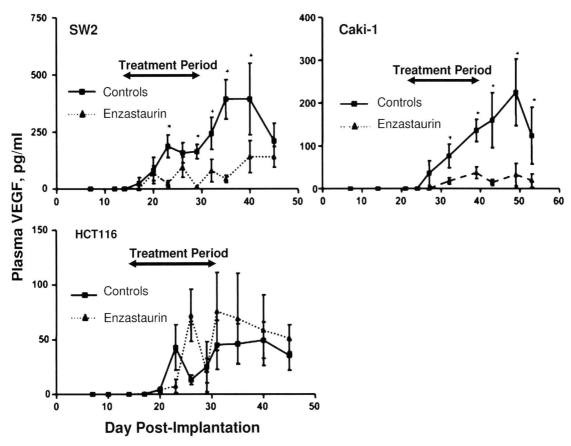

FIG. 38.7. Plasma VEGF concentrations in nude mice bearing subcutaneously implanted human SW2 small cell lung carcinoma, human CaKi1 renal cell carcinoma or human HCT116 colon carcinoma, either untreated controls or treated with LY317615 (enzastaurin) orally twice daily on days 14–30 (days 21–39 for CaKi1-bearing mice). The data are the means from three independent experiments with each point derived from 9 individual tumors. *Bars* are SEM.

of both angiogenesis and induction of apoptosis. Using male nude mice implanted with PC3-MM2 prostate cancer cells in the tibia, Kim et al. [158] found that the combination of oral administration of PKI 166, a selective EGFR tyrosine kinase inhibitor, and that weekly paclitaxel reduced the incidence and size of bone metastasis from prostate cancer and inhibited EGFR phosphorylation on tumor cells and endothelial cells with enhanced apoptosis. Administration of angiostatin or endostatin along with cytotoxic chemotherapeutic agents produced marked antitumor effects in an ovarian carcinoma model and the RIPTag transgenic mouse pancreatic adenocarcinoma model [159, 160].

Similar results have been reported by Klement et al. [161] who combined continuous low dose of vinblastine with a rat VEGFR2 neutralizing monoclonal antibody. In vitro synergistic antiangiogenic activity was reported for docetaxel, and a recombinant humanized monoclonal antibody directed toward VEGF or 2-methoxyestradiol. Docetaxel inhibited endothelial cell migration and proliferation with an IC_{50} of 10pM, which is similar to its cytotoxic IC_{50} against cancer cells in culture [162]. Since then, several additional preclinical studies have supported the notion that prolonged lower dose antiangiogenic therapies

plus cytotoxic therapies or combinations of cytotoxic therapies can have substantial antitumor effects. Further, these studies have shown that circulating endothelial precursor cells and circulating endothelial cells may be useful markers of biologcal effects for these regimens [163–170].

Hanahan et al [162] proposed the term of "metronomic" chemotherapy for schedules of cytotoxic agents given regularly at sub-cytotoxic doses and having the "activated" endothelium as principal target (i.e., the antiangiogenesis chemotherapy paradigm). Several studies, using subcutaneously implanted tumors in mice, documented the antiangiogenic activity of cytotoxic chemotherapeutic agents when administered continuously at low doses [160, 161]. Browder et al. [160] administering a combination regimen including cyclophosphamide in the drinking water on low dose metronomic schedule and TNP-470 eradicated Lewis lung carcinoma, a cyclophosphamide exquisitely sensitive tumor model, in the majority of treated mice. Some cytotoxic chemotherapeutic agents, including camptothecin analogs, vinca alkaloids and taxanes, are cytotoxic on normal mature endothelial cells at lower concentrations than those required to kill malignant cells ("metronomic" chemotherapy). Based upon the work of Steiner [2] and Browder et al. [162] the most promising antiangiogenic chemotherapeutic agents are cyclophosphamide, vinblastine, paclitaxel, and docetaxel.

Preclinical studies examined the feasibility of combining anti-VEGF therapy in the form of bevacizumab and/or its murine equivalent A4.6.1, with cytotoxic or biologic agents [82–85]. Combining bevacizumab/A4.6.1 with doxorubicin, topotecan, paclitaxel, docetaxel or radiotherapy resulted in additive or synergistic tumor growth inhibition. Changes in vascular fucntions were frequently reported in response to treatment. In some studies, these improvements resulted in an increase in intratumoral uptake of chemotherapy as was observed with earlier antiangiogenic agents [82, 85]. As was seen with earlier agents, bevacizumab/A4.6.1 treatment in combination with radiation therapy increased tumor oxugenation andf tumor growth delay. Bevacizumab/A4.6.1 treatment also reduced the development of ascites in ovarian cancer models [82, 85]. Bevacizumab (Avastin), a humanized anti-VEGF has now been approved as a treatment for several major malignancies in combination with standard of care chemotheraputic regimens [171, 172].

Other Small Molecules Antiangiogenic Agents in Combination Regimens

Numerous clinical trials have been initiated to test the efficacy of nonsteroidal anti-inflammatory cyclooxygenase-2 (COX-2) inhibitors, especially celecoxib, in combination therapy regimens in advanced solid tumors [173, 174]. These compounds exhibit anti-inflammatory, analgesic and antipyretic acitivities as well as block angiogenesis in animal models. Phase II clinical studies have combined celecoxib with a taxane, either docetaxel or paclitaxel, for treatment of NSCLC [174]. The combination was well tolerated with response rates trending toward improved activity with the celecoxib, without additional toxicity. In breast cancer, celecoxib in combination with exemestane was well tolerated with a trend towards more efficacy for the combination. Celecoxib combinations have been studied for therapy of esophageal cancer with irinotecan/cisplatin/concurrent radiation therapy, pancreatic cancer with gemcitabine, renal cell carcinoma with low dose cyclophosphamide, and malignant glioma with temozolomide [176]. In a Phase II study, Altorki et al. evaluated the combination of celecoxib with paclitaxel/carboplatin regimen as preoperative chemotherapy in early-stage NSCLC [175]. Compared with historical data, the addition of celecoxib enhanced response rate and normalized the prostaglandin E2 (PGE2) tissue levels.

Thalidomide is a compound with potential antiangiogenic, immunomodulatory and antitumor effects. The more promising clinical results with thalidomide have been observed in plasma cell malignancies, particularly multiple myeloma [176–179]. The clinical data with thalidomide in multiple myeloma have been confirmed with a dose-dependent therapeutic effect [177–179]. The mechanisms of thalidomide activity in multiple myeloma have not been well defined. (see Chapter 34).

New Target Discovery

The field of antiangiogenic therapies has moved very quickly from laboratory discoveries into the clinic. As with other areas of science the rapidity of the development of the antiangiogenic field was fueled by the availability of models and the identification of therapeutic targets. The field was also fueled by the early hypothesis which held that angiogenesis was the same no matter where it occurred. Therefore, angiogenesis during embryo development or wound healing was the same as angiogenesis during the growth of malignant disease [11–13, 180]. The corollary to this hypothesis was that models of normal embryo development and models working with mature well-differentiated ECs in culture would be sufficient and satisfactory models for tumor endothelial cells. This hypothesis also held that because endothelial cells involved in malignant disease were normal, these cells would be less susceptible to developing drug resistance because they were genetically stable [181, 182].

The current hypothesis is that angiogenesis occurring during malignant disease is abnormal, and that therapeutic targets identified by studying endothelial cells isolated from fresh samples of human cancers will be most relevant for developing therapeutic agents to treat human malignant disease [183–186].

Early studies of gene expression were carried out primarily with cell lines. As the importance of the tissue microenvironment and the easy plasticity with which cells alter gene expression in response to the microenvironment became evident, the severe limitations, indeed, inaccuracies in disease representation by monolayer cell culture were recognized. "Drug-target hunters" realized the need to get as close to the human disease as possible to identify critical molecular targets. To accomplish this, fresh samples of human malignant tumors and corresponding normal tissues were used as starting materials [187–201]. Gene expression profiling techniques, such as microarray analysis [187–196] and serial analysis of gene expression (SAGE) [198–201], have provided global views of the levels of mRNAs in malignant tissues compared with normal tissues, and allowed identification of genes and pathways involved in the malignant process. Specific diseases, including ovarian cancer, breast cancer, gastric cancer, multiple myeloma, lung adenocarcinoma, Wilm's tumor and neuroblastoma, have been analyzed for diagnostic and prognostic gene expression characteristics and for identification of potential drug targets [190–196]. Chief among the issues being faced by these studies is developing data analysis methods that allow investigators to draw biologically meaningful conclusions from very large datasets [188, 189].

One of the challenges for gene expression studies is to translate research findings of multigene expression signature classifiers/ genomic signatures of disease into applications in diagnostics and therapeutics [202–206]. Integrative computational and analytical data analysis approaches, including meta-analysis, functional enrichment analysis, interactome analysis, transcriptional network analysis and integrative model system analysis, are being applied to gene expression data. Some studies focus on the expression of mRNAs that code for enzymes as potential drug targets, some search for functional regulators driving large-scale transcriptional signatures and others focus on epigenetic alterations that regulate gene expression [203–208].

SAGE is a gene expression profiling method that allows global unbiased, quantitative determination the transcriptome of the sample at the time of RNA collection [197–202]. SAGE expression profiling depends upon the notions that a short (10–27 base-pair sequence) fragment of mRNA cut by a restriction enzyme is sufficient to uniquely identify a transcript, and that concatemerization of these fragments (tags) increases the efficiency of sequence-based transcriptome analyses [197]. Approximately 90% of genes are represented by SAGE tags (S. Madden, personal communication). Because SAGE does not depend upon a priori knowledge of the genes of interest, it can identify novel, un-named and unexpected transcripts. For these reasons SAGE methodology has been selected as the method of choice to examine gene expression from subpopulations of cells isolated from fresh clinical specimens [198–202].

Fresh specimens of colon carcinoma, normal colon mucosa, breast carcinoma, normal breast tissue, brain tumors and normal brain were obtained for analysis of cellular subpopulations by SAGE analysis (Fig. 38.8) [201, 202, 209, 210]. The tissues were disaggregated and the endothelial cells were isolated using selection with an antibody to P1H12 linked to a magnetic bead [210–212]. The RNA from the endothelial cells isolated from tumor and normal tissues was collected and subjected to SAGE analysis. This methodology allows elucidation of the RNA transcripts in the cells at the time of RNA isolation providing the identity of the transcript and the relative abundance of each transcript. Thus far, SAGE-derived transcript libraries have been generated for endothelial cells isolated from 7 fresh human tumor specimens and 5 specimens of corresponding normal tissues.

The first bioinformatics analysis was to compare the genes/ mRNA expressed in each of the three tumor types with the genes/mRNA expressed in each corresponding normal tissue. In each case, a similar pattern emerged. The vast majority of the genes/mRNA expressed by the tumor endothelial cells was very similar to the genes/mRNA expressed by the ECs from the corresponding normal tissue. However, there was a small subpopulation of genes/mRNA that was expressed at much higher levels by the tumor endothelial cells than by normal ECs and a different small subpopulation of genes/mRNA that was expressed at much higher levels by the normal ECs than by the tumor endothelial cells. Generally, the tumor endothelial cells appeared to be expressing at least a partial 'malignant phenotype'. The tumor endothelial cells appeared to be

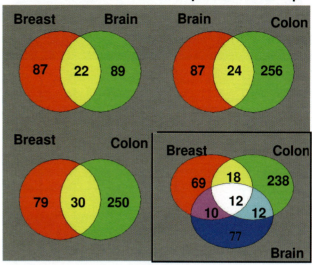

Tumor Endothelial Gene Expression Overlap

FIG. 38.8. Venn diagrams depicting the overlap in the number of genes expressed at higher levels in tumor endothelial cells derived from breast, brain and colon tumors compared with endothelial cells from the corresponding normal tissues. The selected genes were overexpressed in tumor endothelial cells with >98% confidence by Chi-square analysis. The data include known and un-named genes.

relatively de-differentiated or immature relative to the corresponding normal ECs.

The second bioinformatics analysis was to compare the genes/mRNA that were expressed at high levels by the tumor endothelial cells from the colon carcinoma, breast cancer and brain tumors with each other. Venn diagrams were developed for the subpopulations of genes that by the Chi-square test had >99% confidence of being over expressed in the tumor endothelial cells compared with the corresponding normal endothelial cells (Fig. 38.9). The genes/mRNA that fulfilled these criteria included 280 genes from the colon carcinoma, 109 genes from the breast carcinomas and 111 genes from the brain tumors. The number of genes that were overexpressed in endothelial cells from both breast cancer and brain cancers was 22, from brain cancers and colon cancer was 24, and from breast cancer and colon cancer was 30. Thus, there is a high degree of organ/tissue specificity in the endothelium and there is a high degree of heterogeneity among tumor endothelium. When the highly overexpressed genes from the endothelial cell libraries for each of the three tumor types were compared, there were only 12 genes that were highly overexpressed in all three tumor types. Based on these findings, it may be less likely that therapeutic antiangiogenesis targets can be identified that are universally applicable. It may be more likely that antiangiogenic therapeutic targets can be found that will apply to major tumor categories.

Hierarchical clustering analysis using GeneSpring™ software was applied to the SAGE data from the normal and tumor brain and breast endothelial cell libraries. Each SAGE library included 30—40,000 SAGE tags. When the complete gene

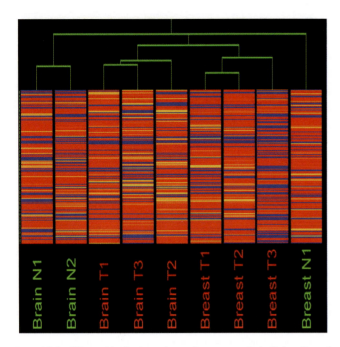

Fig. 38.9. Hierarchical clustering of tumor endothelial cell and normal tissue endothelial cells SAGE libraries by GeneSpring™ is shown for breast tumor and normal breast and brain tumors and normal brain. SAGE tags from statistical confidence filtering (90% confidence) were used. At the 90% confidence level shown, the tumor endothelial cells libraries formed distinctive sub-clusters from the normal endothelial cells libraries indicating that there is a group of genes expressed by endothelial cells involved in the switch from normal to tumor independent of tissue of origin.

expression libraries were analyzed the libraries formed two sub-clusters based upon the tissue of origin of the ECs, that is the normal and tumor breast endothelial cells clustered apart from the normal and tumor brain endothelial cells. Therefore genes that distinguish tumor from normal endothelial cells did not dictate the general gene expression profiles. Statistical confidence filtering was then applied to all the libraries to isolate genes upregulated with 90%, 95% and 99% confidence. When hierarchical clustering was applied to the gene population upregulated in these libraries with 90% and 95% confidence, tumor endothelial cells libraries formed a distinctive sub-cluster from the normal ECs libraries. Thus, a group of genes could be identified that were involved in the switch from normal tissue endothelium to malignant disease tissue endothelium without tissue type distinction (Fig. 38.9). Interestingly, when hierarchical clustering analysis was performed with genes up-regulated at the 99% confidence level, the libraries from different tissues formed distinctive sub-clusters. Thus, at this high level of statistical stringency, genes expressed by the endothelial cells were dominated by the tissue of origin of the cells and not by the normalcy or malignancy of tissue.

The final bioinformatics analysis was to examine the expression of the genes/mRNA that were highly up-regulated in tumor endothelial cells with genes/mRNA expressed in cells commonly used as a model systems in the angiogenesis

and antiangiogenesis fields. The cells whose gene/mRNA expression was examined included HUVECs, HMVECs (human microvascular endothelial cells) and EPCs (human endothelial precursor cells) [16]. SAGE libraries were available for several cell culture conditions including stimulated with VEGF and unstimulated cells. Many of the genes/mRNA expressed at high levels in the tumor endothelial cells isolated from fresh human tumor specimens were either not expressed or were expressed at very low levels in HUVEC and HMVEC under stimulated and unstimulated conditions.

Conclusions

Several anticancer therapeutics which have antiangiogenic activity as a major component of their clinical benefit have been approved by the FDA. Outstanding amongst these agents is the anti-VEGF antibody bevacizumab, and several kinase inhibitors, including sunitinib, erlotinib, sorafenib, and imatinib. Other antibodies and kinase inhibitors are in late clinical trials and likely to be approved shortly. When the clinical focus is on the antiangiogenic activity of these agents they are used in combination regimens. The first generation antiangiogenic agents allowed the principle to be established that antiangiogenic agents can contribute importantly to the anticancer activity of existing standard of care cytotoxic therapy regimens. The second generation antiangiogenic agents have now proven the value of the vasculature as a therapeutic target. The next generation of antianigenic agents will have improved selectivity for tumor versus normal vasculature and thus improved safety and tolerability of treatment regimens for patients.

References

1. Teicher BA: A systems approach to cancer therapy (antiangiogenics + standard cytotoxics mechanism(s) of interaction. Cancer Met Rev 199; 15: 247–72.
2. Steiner R. Angiostatic activity of anticancer agents in the chick embryo chorioallantoic membrane (CHE-CAM) assay. In: Steiner R, Weisz PB, Langer R (eds), Angiogenesis: key principles – science, technology, medicine; Birkhauser Verlaga, Basel, 1992; EXS 61: 449–54.
3. Van der Kolk S. In: Blood Supply of Tumors, vol 2; Montagna W and Ellis R, eds. 1826; pp123–149.
4. Jones T. Guy's Hospital Reports, 2nd Ser. 1850; 7: 1–94.
5. Paget S. Lancet 1889; March 23: 571–573.
6. Algire G, Chalkey H. J Natl Cancer Inst 1945; 6: 73–95.
7. Folkman MJ, Merler E, Abernathy C, Williams G. Isolation of a tumor factor responsible for angiogenesis. J Exp Med 1971; 133: 275–88.
8. Folkman MJ. Tumor Angiogenesis. Adv Cancer Res 1974; 19: 331–58.
9. Folkman MJ, Cotran R. Relation of vascular proliferation to tumor growth. Int Rev Exp Pathol 1976; 16: 207–48.
10. Folkman MJ. Tumor angiogenesis: therapeutic implications. New Engl J Med 1971; 285: 1182–6.
11. Folkman J. Angiogenesis. Annu Rev Med 2006; 57: 1–18.

12. Folkman J. Antiangiogenesis in cancer therapy – endostatin and its mechanisms of action. Exp Cell Res 2006; 312: 594–607.

13. Folkman J. Angiogenesis and apoptosis. Semin Cancer Biol 2003; 13: 159–167.

14. Weinberg CAFsBergers G, Benjamin LE: Tumorigenesis and the angiogenic switch. Nature Rev Cancer 2003; 3: 401–10.

15. St. Croix B, Rago C, Velculescu V, et al: Genes expressed in human tumor and endothelium. Science 2000; 289: 1197–1202.

16. Bagley R, Walter-Yohrling J, Cao X, et al: Endothelial precursor cells as a model of tumor endothelium: characterization and comparison to mature endothelial cells. Cancer Res 2003; 63: 5866–73.

17. Taylor S, Folkman J. Protamine is an inhibitor of angiogenesis. Nature 1982; 297: 307–12.

18. Groopman JE, Scadden DT. Interferon therapy for Kaposi sarcoma associated with the acquired immunodeficiency syndrome (AIDS). Ann Int Med 1989; 110: 335–7.

19. Strieter RM, Kunkel SL, Arenberg DA, Burdick MD, Polverini PJ. Interferon-g-inducible protein 10 (IP-10), a member of the C-X-C chemokine family, is an inhibitor of angiogenesis. Biochem Biophys Res Commun 1995; 210: 51–7.

20. Kolber DL, Kniselt TL, Maione TE. Inhibition of development of murine melanoma lung metastases by systemic administration of recombinant platelet factor 4. J Natl Cancer Inst 1995; 87: 304–9.

21. Stetler-StevensonWG, Krutzsch HC, Liotta LA. Tissue inhibitor of metalloproteinase (TIMP-2). A new member of the metalloproteinase inhibitor family. J Biol Chem 1989; 264: 17374–8.

22. Voest EE, Kenyon BM, O'Reilly MS, Truitt G, D'Amato RJ, Folkman J. Inhibition of angiogenesis in vivo by interleukin 12. J Natl Cancer Inst 1995; 87: 581–6.

23. Teicher BA, Ara G, Menon K, Schaub RG. In vivo studies with interleukin-12 alone and in combination with monocyte colony-stimulating factor and/or fractionated radiation treatment. Int J Cancer 1995; 65: 80–84.

24. O'Reilly MS, Holmgren L, Shing Y, Chen C, Rosenthal RA, Mosses M, Lane WS, Cao Y, Sage EH, Folkman J. Angiostatin: a novel angiogenesis inhibitor that mediates the suppression of metastases by a Lewis lung carcinoma. Cell 1994; 79: 315–28.

25. Mauceri HJ, Hanna NN, Beckett MA, et al: Combined effects of angiostatin and ionizing radiation in antitumour therapy. Nature 1998; 394: 287–291.

26. Capillo M, Mancuso P, Gobbi A, et al: Continuous infusion of endostatin inhibits differentiation, mobilization, and clonogenic potential of endothelial cell progenitors. Clin Cancer Res 2003; 9: 377–82.

27. Sudhakar A, Sugimoto H, Yang C, et al: Human tumstatin and human endostatin exhibit distinct antiangiogenic activities mediated by $\alpha v\beta 3$ and $\alpha 5\beta 1$ integrins. Proc Natl Acad Sci USA 2003; 100: 4766–71.

28. Morbidelli L, Donnini S, Chillemi F, et al: Angiosuppressive and angiostimulatory effects exerted by synthetic partial sequences of endostatin. Clin Cancer Res 2003; 9: 5358–69.

29. Yokoyama Y, Dhanabal M, Griffioen AW, et al: Synergy between angiostatin and endostatin: inhibition of ovarian cancer growth. Cancer Res 2000; 60: 2190–6.

30. Moses MA, Sudhalter J, Langer R. Identification of an inhibitor of neovascularization from cartilage. Science 1990; 248: 1408–10.

31. Taylor CM, Weiss JB. Partial purification of a 5.7K glycoprotein from bovine vitreous which inhibits both angiogenesis and collagenase activity. Biochem Biophys Res Commun 1985; 133: 911–6.

32. DeClerck YA. Purifcation and characterization of a collagenase inhibitor produced by bovine vascular smooth muscle cells. Arch Biochem Biophys 1988; 265: 28–37.

33. Sakamoto N, Iwahana M, Tanaka NG, Osada Y. Inhibition of angiogenesis and tumor growth by a synthetic laminin peptide CDPGYIGSR-NH2. Cancer Res 1991; 51: 903–6.

34. Bogden AE, Taylor JE, Moreau JP, Coy DH, LePage DJ. Response of human lung tumor xenografts to treatment with a somatostatin analog (somatuline). Cancer Res 1990; 50: 4360–5.

35. Folkman J, Weisz PB, Joullie MM, Li WW, Ewing WR. Control of angiogenesis with synthetic heparin substitutes. Science 1989; 243: 1490–3.

36. Crum R, Szabo S, Folkman J. A new class of steroids inhibits angiogenesis in the presence of heparin or a heparin fragment. Science 1985; 230: 1375–8.

37. Lee K-E, Iwamura M, Cockett ATK. Cortisone inhibition of tumor angiogenesis measured by a quantitative colorimetric assay in mice. Cancer Chemother Pharmacol 1990; 26: 461–3.

38. Ingber D, Folkman J. Inhibition of angiogenesis through modulation of collagen metabolism. Lab Invest 1988; 59: 44–51.

39. Oikawa T, Hirotani K, Nakamura O, Shudo K, Hiragun A, Iwaguchi T. A highly potent antiangiogenic activity of retionoids. Cancer Lett 1989; 48: 157–62.

40. Schwartz JL, Flynn E, Shklar G. The effects of carotenoids on the antitumot immune response in vivo and in vitro with hamster and mouse immune effectors. In: Bendich A, Chandra RK (ed) Symposium on micronutrients and immune functions. NY Acad Sci, NY 1990; pp 92–6.

41. Schwartz JL, Singh R, Teicher BA, Wright JE, Trites DH, Shklar G. Induction of a 70-kDa protein associated with the selective cytotoxicity of beta carotene in human epidermal carcinoma. Biocem Biophys Res Commun 1990; 169: 941–6.

42. Lippman SM, Kavanagh JJ, Paredes-Espinoza M, Delgadillo-Madrueno F, Paredes-Casillas P, Hong WK. Masimini G, Holdener EE, Krakoff IH. 13-cis-retinoic acid plus interferon-a2a in locally advanced squsamous cell carcinoma of the cervix. J Natl Cancer Inst 1993; 85: 499–500.

43. Schwartz JL, Tanaka J, Khandekar V, Herman Teicher BA. B-carotene and/or vitamin EW as modulators of alkylating agents in SCC-25 human squsamous carcinoma cells. Cancer Chemother Pharmacol 1992; 29: 207–13.

44. Denekamp J. Angiogenesis, neovascular proliferation and vascular pathophysiology as targets for cancer therapy. Br J Radiol 1993; 66: 181–96.

45. Teicher BA, Schwartz JL, Holden SA, Ara G, Northey D. In vivo modulation of several anticancer agents by b-carotene. Cancer Chemother. Pharmacol. 1994; 34: 235–41.

46. Majewski S, Szmurlo A, Marczak M, Jablonska S, Bollag W. Inhibition of tumor cell-induced angiogenesis by retinoinds, 1,25-dihydroxyvitamin D3 and their combination. Cancer Lett 1993; 75: 35–9.

47. Barnes S, Grubbs C, Setchell KDR, Carlson J. Soybeans inhibit mammary tumor in models of breast cancer. In: Pariza M, Liss AR (ed) Mutagens and carcinogens in the diet. Wiley-Liss, NY 1990; pp239–53.

48. Zwiller J, Sassone-Corsi P, Kakazu K, Boyton AL. Inhibition of PDGF-induced c-jun and c-fos expression by a tyrosine protein kinase inhibitor. Oncogene 1991; 6: 219–21.

49. McCabe Jr MJ, Orrenius S. Genistein induces apoptosis in imma-ture human thymocytes by inhibiting topoisomerase II. Biocehm Bipohys Res bCommun 1993; 194: 944–50.

50. KanataniY, Kasukabe T, Hozumi M, Motoyoshi K. Genistein exhibits preferential cytotoxicity to a leukemogenic variant but induces differentiation of a non-leukemogenic variant of the mouse monocytic leukaemia Mm cell line. Leuk Res 1993; 17: 847–53.

51. Fotsis T, Pepper M, Adlercreutz H, Fleischmann G, Hase T, Montesano R, Schweigerer L. Genistein, a dietary-derived inhibitor of in vitro angiogenesis. Proc Natl Acad Sci USA 1993; 90: 2690–4.

52. Uckun FM, Evans WE, Forsyth CJ, Waddick KG, Ahlgren LT, Chelstrom LM, Burkhardt A, Bolen J, Myers DE. Biotherapy of B-cell precursor leukaemia by targeting genistein to CD19-asso-ciated tyrosine kinases. Science 1995; 267: 886–9.

53. Yanase T, Tamura M, Fujita K, Kodama S, Tanaka K. Inhibitory effect of angiogenesis inhibitor TNP-470 in rabbits bearing VX-2 carcinoma by arterial administration of microspheres and oiul solution. J Pharmacol Exp Ther 1993; 264: 469–74.

54. Yamaoka M, Yamamoto T, Masaki T, Ikeyama S, Sudo K, Fujita T. Inhibition of tumor growth and metastasis of rodent tumors by the angiogenesis inhibitor O-(chloroacetyl-carbamoyl)fumagillin (TNP-470); AGM-1470). Cancer Res 1993; 53: 4262–7.

55. Toi M, Tamamoto Y, Imazawa T, Takayanagi T, Akutsu K, Tom-inga T. Antitumor effect of the angiogenesis inhibitor AGM-1470 and its combination effect with tamoxifen in DMBA induced mammary tumors in rats. Int J Oncol 1993; 3: 525–8.

56. Yamaoka ZM, Yamamoto T, Ikeyama S, Sudo K, Fujita T. Angiogenesis inhibitor TNP-470 (AGM-1470) potently inhib-its the tumor growth of hormone-independent human breast and prostate carcinoma cell lines. Cancer Res 1993; 53: 5233–6.

57. Teicher BA, Holden SA, Ara G, Alvarez Sotomayor E, Huang ZD, Chen Y-N, Brem H. Potentiation of cytotoxic cancer thera-pies by TNP-470 alone and with other anti-angiogenic agents. Int. J. Cancer 1994; 57 (6):920–5.

58. Teicher BA, Dupuis N, Kusumoto T, Robinson MF, Liu F, Menon K, Coleman CN. Antiangiogenic agents can increase tumor oxy-genation and response to radiation therapy. Radiat Oncol Invest 1995; 2: 269–76.

59. Teicher BA, Holden SA, Dupuis NP, Kakeji Y, Ikebe M, Emi Y, Goff D. Potentiation of cytotoxic therapies by TNP-470 and minocycline in mice bearing EMT-6 mammary carcinoma. Breast Cancer Res Treatment 1995; 36(2): 227–36.

60. Teicher BA, Dupuis NP, Robinson MF, Emi Y, Goff DA. (TNP-470/minocycline) increases tissue levels of anticancer drugs in mice bearing Lewis lung carcinoma. Oncology Research 1995; 7(5): 237–43.

61. Teicher BA, Holden SA, Ara G, Dupuis N, Liu F, Yuan J, Ikebe M, Kakeji Y. Influence of an anti-angiogenic treatment on 9L gliosarcoma: oxygenation and response to cytotoxic therapy. Int J Cancer 1995; 61(5): 732–7.

62. Tanaka T, Konno H, Matsuda I, Nakamura S, Baba S. Prevention of hepatic metastasis of human colon cancer angiogenesis inhibi-tor TNP-470. Cancer Res 1995; 55: 836–9.

63. Murata J, Saiki I, Makabe T, Tsuta Y, Tokura S, Azuma I. Inhibi-tion of tumor-induced angiogenesis by sulphated chitin deriva-tives. Cancer Res 1991; 51: 22–6.

64. Teicher BA, Alvarez Sotomayor E, Huang ZD, Ara G, Holden S, Khandekar V, Chen Y-N. ß-Cyclodextrin tetradecasulfate/tet-

rahydrocortisol ± minocycline as modulators of cancer therapies in vitro and in vivo against primary and metastatic Lewis lung carcinoma. Cancer Chemother. Pharmacol. 1993; 33:229–38.

65. Teicher BA, Holden SA, Ara G, Northey D. Response of the FSaII fibrosarcoma to antiangiogenic modulators plus cytotoxic agents. Anticancer Res. 1993; 13:2101–6.

66. Vukanovic J, Isaacs JT. Linomide inhibits angiogenesis, growth, metastasis and macrophage infiltration within rat prostatic can-cers. Cancer Res 1995; 55: 1499–1504.

67. Prionas SD, Kowalski J, Fajardo LF, Kaplan I, Kwan HH, Allison AC. Effects of x-irradiation on angiogenesis. Radiat Res 1990; 124: 43–9.

68. Brooks PC, Clark RA Cheresh DA. Requirement of vascular integrin avb3 for angiogenesis. Science 1994; 264: 569–73.

69. Brooks PC, Montgomery AMP, Rosenfeld M, Reisfeld RA, Hu T, Klier G, Cheresh DA. Integrin avb3 antagonists promote tumor regression by inducing apoptosis of angiogenic blood ves-sels. Cell 1994; 79: 1157–64.

70. Eliceiri BP, Cheresh DA: The role of α-v integrin during angio-genesis: insights into potential mechanisms of action and clinical development. J Clin Invest 1999; 103: 1227–30.

71. Hynes RO: A reevaluation of integrins as regulators of angiogen-esis. Nature Med 2002; 8: 918–21.

72. Anderson IC, Shipp MA, Docherty AJP, Teicher BA. Combina-tion therapy including a gelatinase inhibitor and cytotoxic agent reduces local invasion and metastasis of murine Lewis lung car-cinoma. Cancer Res 1996; 56(4): 715–718.

73. Bonomi P: Matrix metalloproteinases and matrix inhibitors in lung cancer. Sem Oncol 2002; 29: 78–86.

74. Coussens LM, Fingleton B, Matrisian LM: Matrix metallopro-teinse inhibitors and cancer trials and tribulations. Science 2002; 295: 2387–92.

75. Hidalgo M, Eckhardt SG: Development of matrix metalloproteinase inhibitors in cancer therapy. J Natl Cancer Inst 2001; 7: 178–93.

76. Ferrara N: Role of vascular endothelial growth factor in physi-ologic and pathologic angiogenesis: therapeutic implications. Semin Oncol 2002; 29: 10–14.

77. Ferrara N, Gerber HP, LeCouter J. The biology of VEGF and it receptors. Nat Med 2003; 9: 669–76.

78. Kim KJ, Li B, Winer J, et al. Inhibtion of vascular endothelial growth factor-induced angiogenesis suppresses tumor growth in vivo. Nature 1993; 362: 841–4.

79. Kabbinavar FF, Wong JT, Ayala RE, Wintroub AB, Kim KJ, Fer-rara N. The effect of antibody to vascular endothelial growth fac-tor and cisplatin on the growth of lung tumors in nude mice. Proc Am Assoc Cancer Res 1995; 36: 488.

80. Borgstrom P, Gold DP, Hillan KJ, Ferrara N. Importance of VEGF for breast cancer angiogenesis in vivo: implications from intra-vital microscopy of combination treatments with an anti-VEGF neutralizing antibody and doxorubicin. Anticancer res 1999; 19: 4203–14.

81. Liang W-C, Wu X, Peale FV, Lee CV, Meng YG, Gutierrez J, Fu L, Malik AK, Gerber H-P, Ferrara N, Fuh G. Cross-species vascular endothelial growth factor (VEGF)-blocking antibodies completely inhibit the growth of human xenografts and measure the contribu-tion of stromal VEGF. J Biol Chem 2006; 281: 951–61.

82. Gerber H-P, Ferrara N. Pharmacology and pharmacodynamics of bevacizumab as monotherapy or in combination with cytotoxic therapy in preclinical studies. Cancer Res 2005; 65: 671–80.

83. Dong J, Grunstein J, Tejada M, peale F, Frantz G, Liang W-C, Bai W, Yu L, Kowalski HJ, Liang X, Fuh G, Gerber H-P, Ferrara N. VEGF-null

cells require PDGFRa signaling-mediated stromal fibroblast recruitment for tumorigenesis. The EMBO J 2004; 23: 2800–10.

84. Ferrara N, Hillan KJ, Gerber H-P, Novotny W. Discovery and development of bevacizumab, an anti-VEGF antibody for treating cancer. Nat Rev Drug Disc 2004; 3: 391–400.

85. Gerber H-P, Kowalski J, Sherman DA, Eberhard DA, Ferrara N. Complete inhibition of rhabdomyosarcoma xenograft growth and neovascularization requires blockade of both tumor and host vascular endothelial growth factor. Cancer 2000; 60: 6253–8.

86. Kakeji Y, Teicher BA: Preclinical studies of the combination of angiogenic inhibitors with cytotoxic agents. Invest New Drugs 1997; 15: 39–48.

87. Teicher BA, Alvarez E, Huang ZD: Antiangiogenic agents potentiate cytotoxic cancer therapies against primary and metastatic disease. Cancer Res 1992; 52: 6702–4.

88. Alvarez Sotomayor E, Teicher BA Schwartz GN, Holden SA, Menon K, Herman TS, Frei E. Minocycline in combination with chemotherapy or radiation therapy in vitro and in vivo. Cancer Chemother Pharmacol 1992; 30: 377–84.

89. Ingber D, Fujita T, Kishimoto S, Sudo K, Kanamaru T, Brem H, Folkman J. Synthetic analogs of fumigillin that inhibit angiogenesis and suppress tumor growth. Nature 1990; 348: 555–7.

90. Teicher BA, Holden SA, Al-Achi A, Herman TS. Classification of antineoplastic treatments by their differential toxicity toward putative oxygenated and hypoxic tumor subpopulations in vivo in FSaIIC murine fibrosarcoma. Cancer Res 1990; 50: 3339–44.

91. Inoue K, Chikazawa M, Fukata S, et al: Docetaxel enhances the therapeutic effect of the angiogenesis inhibitor TNP-470 (AGM-1470) in metastatic human transitional cell carcinoma. Clin Cancer Res 2003; 9: 886–99.

92. Satoh H, Ishikawa H, Fujimoto M, et al: Combined effects of TNP-470 and taxol in human non-small cell lung cancer cell lines. Anticancer Res 1998; 18: 1027–30.

93. Shishido T, Yasoshima T, Denno R, et al: Inhibition of liver metastasis of human pancreatic carcinoma by angiogenesis inhibitor TNP-470 in combination with cisplatin. Jpn J Cancer Res 1998; 89: 963–9.

94. Ogawa H, Sato Y, Kondo M, et al: Combined treatment with TNP-470 and 5-fluorouracil effectively inhibits growth of murine colon cancer cells in vitro and liver metastasis in vivo. Oncol Rep 2000; 7: 467–72.

95. Herbst RS, Takeuchi H, Teicher BA: Paclitaxel/carboplatin administration along with antiangiogenic therapy in non-small cell lung and breast carcinoma models. Cancer Chemo Pharmacol 1998; 41: 497–504.

96. Teicher BA, Holden SA, Ara G, Korbut T, Menon K. Comparison of several antiangiogenic regimens alone and with cytotoxic therapies in the Lewis lung carcinoma. Cancer Chemother Pharmacol 1996; 38: 169–77.

97. Teicher BA, Emi Y, Kakeji Y, Northey D. TNP-470/minocycline/cytotoxic therapy: A systems approach to cancer therapy. Europ J Cancer 1996; 32A: 2461–66.

98. Teicher BA. The role of angiogenesis in the response to anticancer therapies. Drug Resistance Updates 1998; 1: 59–62.

99. Murata R, Nishimura Y, Hiraoka M. An antiangiogenic agent (TNP-470) inhibited reoxygenation during fractionated radiotherapy of murine mammary carcinoma. Int J Radiat Oncol Buiol Phys 1997; 37: 1107–13.

100. Laird AD, Cherrington JM: Small molecule tyrosine kinase inhibitors: clinical development of anticancer agents. Exp Opin Invest Drugs 2003; 12: 51–64.

101. Geng L, Donnelly E, McMahon G, et al: Inhibition of vascular endothelial growth factor receptor signaling leads to reversal of tumor resistance to radiotherapy. Cancer Res 2001; 61: 2413–9.

102. Ning S, Laird D, Cherrington JM, et al: The antiangiogenic agents SU5416 and SU6668 increase the antitumor effects of fractionated irradiation. Radiat Res 2002; 157: 45–51.

103 Fong TA, Shawver LK, Sun L, et al: SU5416 is a potent and selective inhibitor of the vascular endothelial growth factor (Flk-1/KDR) that inhibitors tyrosine kinase catalysis, tumor vascularization and growth of multiple tumor types. Cancer Res 1999; 59: 99–106.

104. Schuuring J, Bussink J, Bernsen H et al. Irradiation combined with SU5416: microvascular changes and growth delay in a human xenograft glioblastoma tumor line. Int J Radiat Oncol Biol Phys 2005; 61: 529–34.

105. Ning S, Laird D, Cherrington JM, Knox SJ. The antiangiogenic agents SU5416 and SU6668 increase the antitumor effects of fractionated irradiation. Radiat Res 2002; 157: 45–51.

106. Lund EL, Olsen MW, Lipson KE et al. Imporived effect of an antiangiogenic tyrosine kinase inhibitor (SU5416) by combinations with fractionated radiotherapy or low molecular weight heparin. Neoplasia 2003; 5: 11–6.

107. Lu B, Geng L, Musiek A, et al. Broad Sectrum receptor tyrosine kinase inhibitor, SU6668, sensitizes radiation via targeting survival pathway of vascular endothelium. Int J Radiat Oncol Biol Phys 2004; 58: 844–50.

108. Mendel DB, Laird AD, Xin X, et al: In vivo antitumor activity of SU11248, a novel tyrosine kinase inhibitor targeting vascular endothelial growth factor and platelet-derived growth factor receptors: determination of a pharmacokinetic/pharmacodynamic relationship. Clin Cancer Res 2003; 9: 327–37.

109. O'Farrell AM, Abrams TJ, Yuen HA, et al: SU11248 is a novel FLT3 tyrosine kinase inhibitor with potent activity in vitro and in vivo. Blood 2003; 101: 3597–605.

110. Schueneman AJ, Himmelfarb E, Geng L et al. SU11248 maintenance therapy prevents tumor regrowth after fractionated irradiation of murine tumor models. Cancer Res 2003; 63: 4009–16.

111. Hess C, Vuong V, Hegyi I et al.n Effect of VEGF receptor inhibitor PTK787/ZK222584 correction of ZK222548 combined with ionizing radiation on endothelial cells and tumor growth. Br J cancer 2001; 85: 2010–6.

112. Wood JM, Bold G, Buchdunger E, et al: PTK787/ZK 222584, a novel and potent inhibitor of vascular endothelial growth factor receptor tyrosine kinases, impairs vascular endothelial growth factor-induced responses and tumor growth after oral administration. Cancer Res 2000; 60: 2178–89.

113. Drevs J, Hofmann I, Hugenschmidt H, et al: Effects of PTK787/ZK 222584, a specific inhibitor of vascular endothelial growth factor receptor tyrosine kinases, on primary tumor, metastasis, vessel density and blood flow in a murine renal cell carcinoma model. Cancer Res 2000; 60: 4819–24.

114. Morgan B, Thomas AL, Drevs J, et al: Dynamic contrast-enhanced magnetic resonance imaging as a biomarker for the pharmacological response of PTK787/ZK222584, an inhibitor of the Vascular Endothelial growth Factor Receptor tyrosine kinases, in patients with advanced colorectal cancer and liver metastases: results from two phase I studies J Clin Oncol 2003; 21: 3955–64.

115. Goekjian PG, Jirousek MR: Protein kinase C inhibitors as novel anticancer drugs. Exp Opin Invest Drugs 2001; 10: 2117–40.

116. Swannie HC, Kaye SB: Protein kinase C inhibitors. Curr Oncol Rep 2002; 4: 37–46.

117. Michie AM, Nakagawa R. The link between PKCalpha regulation and cellular transformation. Immunol Lett 2005; 96: 155–62.

118. Tamamura H, Signano DM, Lewin NE, Peach ML, Nicklaus MC, Blumberg PM, Marquez VE. Conformationally constrained analogs of diacylglycerol (DAG). 23. hydrophobic ligand-protein interactions versus ligand-lipid interactions of DAG-lactones with protein kinase C (PKC). J Med Chem 2004; 47: 4858–64.

119. Lampe JW, Biggers CK, Defauw JM, Fogleson RJ, Hall SE, Heerding JM, Hollinshead SP, Hong H, Hughes PF, Jagdmann GE, Johnson MG et al J Med Chem 2002; 45: 2624–43.

120. Dieter P, Fitzke E. RO 31-8220 and RO 31-7549 show improved selectivity for protein kinase C over staurosporine in macrophages. Biochem Biophys Res Commun 1991; 181: 396–401.

121. Newton RC, Decicco CP. Therapeutic potential and strategies for inhibiting tumor necrosis factor-alpha. J Med Chem 1999; 42: 2295–314.

122. Sausville EA, Arbuck SG, Messmann R, Headlee D, Bauer KS, Lush RM, Murgo A, Figg WD, Lahusen T, Jaken S, Jing X, Roberge M, Fuse E, Kuwabara T, Senderowicz AM. Phase I trial of 72-hour continuous infusion UCN-01 in patients with refractory neoplasms. J Clin Oncol 2001; 19: 2319–33.

123. Mack PC, Lara PN, Longmate J, Gumerlack PH, Synold TW, Doroshow JH, Gandara DR. Phase I and correlative science trial of UCN-01 plus cisplatin (CDDP) in advanced solid tumors: a California cancer consortium study. Proc Amer Soc Clin Oncol 2004: Abstr 9591.

124. Hirte HW. A phase II study of UCN-01 in combination with topotecan in patients with advanced recurrent ovarian cancer: a Princess Margaret phase II consortium trial. Proc Amer Soc Clin Oncol 2005: Abstr 3127.

125 Si MS, Reitz BA, Borie DC. Effects of the kinase inhibitor CGP41251 (PKC 412) on lymphocyte activation and TNF-alpha production. Int Immunopharmacol 2005; 5: 1141–9.

126. Tenzer A, Zingg D, Rocha S, Hemmings B, Fabbro D, Glanzmann C, Schubiger PA, Bodis S, Pruschy M. The phosphatidylinositide 3'-kinase/Akt survival pathway is a target for the anticancer and radiosensitizing agent PKC412, an inhibitor of protein kinase C. Cancer Res 2001; 61: 8203–10.

127. Nakamura K, Yoshikawa N, Yamaguchi Y, Kagota S, Shinozuka K, Kunitomo M. Effect of PKC412, an inhibitor of protein kinase C, on spontaneous metastatic model mice. Anticancer Res 2003; 23: 1395–9.

128. Ganeshaguru K, Wickremasinghe RG, Jones DT, Gordon M, Hart SM, Virchis AE, Prentice HG, Hoffbrand AV, Man A, Champain K, Csermak K, Mehta AB. Actions of the selective protein kinase C inhibitor PKC412 on B-chronic lymphocytic leukemia cells in vitro. Haematologica 2002; 87: 167–76.

129. Hemstrom TH, Joseph B, Schulte G, Lewensohn R, Zhivotovsky B. PKC 412 sensitizes U1810 non-small cell lung cancer cells to DNA damage. Exp Cell Res 2005; 305: 200–13.

130. Bahlis NJ, Miao Y, Koc ON, Gerson S. N-benzoylstaurosporine (PKC412) inhibits AKT kinase inducing apoptosis in multiple myeloma cells. Proc Amer Soc Clin Oncol 2005: Abstr 6503.

131. Growney JD, Clark JJ, Adelsperger J, Stone R, Fabbro D, Griffin JD, Gilliland DG. Activation mutations of human c-KIT resistant to imatinib mesylate are sensitive to the tyrosine kinase inhibitor PKC412. Blood 2005; 106: 721–4.

132. Gotlib J, Berube C, Growney JD, Chen CC, George TI, Williams C, Kajiguchi T, Ruan J, Lilleberg SL, Durocher JA, Lichy JH, Wang Y, Cohen PS, Arber DA, Heinrich MC, Neckers L, Galli SJ, Gilliland DG, Coutre SE. Activity of the tyrosine kinase inhibitor PKC412 in a patient with mast cell leukemia with the D816V KIT mutation. Blood 2005; 106: 2865–70.

133. Cools J, Mentens N, Furet P, Fabbro D, Clark JJ, Griffin JD, Marynen P, Gilliland DG. Prediction of resistance to small molecule FLT3 inhibitors: implications for molecularly targeted therapy of acute leukemia. Cancer Res 2004; 64: 6385–9.

134. George P, Bali P, Cohen P, Tao J, Guo F, Sigua C, Vishvanath A, Fiskus W, Scuto A, Annavarapu S, Moscinski L, Bhalla K. Cotreatment with 17-allylamino-demethoxygeldaamycin and FLT-3 kinase inhibitor PKC412 is highly effective against human acute myelogenous leukemia cells with mutant FLT-3. Cancer Res 2004; 64: 3645–52.

135. Bali P, George P, Cohen P, Tao J, Guo F, Sigua C, Vishvanath A, Scuto A, Annavarapu S, Fiskus W, Moscinski L, Atadja P, Bhalla K. Superior activity of the combination of histone deacetylase inhibitor LAQ824 and the FLT-3 kinase inhibitor PKC412 against human acute myelogenous leukemia cells with mutant FLT-3. Clin Cancer Res 2004; 10: 4991–7.

136. Eder JP, Garcia-Carbonero R, Clark JW, Supko JG, Puchalski TA, Ryan DP, Deluca P, Wozniak A, Campbell A, Rothermel J, LoRusso P. A phase I trial of daily oral 4'-N-benzoyl-staurosporine in combination with protracted continuous infusion 5-fluorouracil in patients with advanced solid malignancies. Invest New Drugs 2004; 22: 139–50.

137. Monnerat C, Henriksson R, Le Chevalier T, Novello S, Berthaud P, Faivre S, Raymond E. Phase I study of PKC412 (N-benzoyl-staurosporine), a novel oral protein kinase C inhibitor, combined with gemcitabine and cisplatin in patients with non-small cell lung cancer. Ann Oncol 2004; 15: 316–23.

138. Heidel F, Solem FK, Breitenbuecher F, Lipka DB, Kasper S, Thiede MH, Brandts C, Serve H, Roesel J, Giles F, Feldman E, Ehninger G, Schiller GJ, Nimer S, Stone RM, Wang Y, Kindler T, Cohen PS, Huber C, Fischer T. Clinical resistance to the kinase inhibitor PKC412 in acute myeloid leukemia by mutation of Asn-676 in the FLT3 tyrosine kinase domain. Blood 2006; 107: 293–300.

139. Teicher BA, Alvarez E, Menon K, Esterman MA, Considine E, Shih C, Faul MM. Antiangiogenic effects of a protein kinase C beta-selective small molecule. Cancer Chemo Pharmacol 2002; 49: 69–77.

140. Liu Y, Su W, Thompson EA, Leitges M, Murray NR, Fields AP. Protein kinase CbetaII regulates its own expression in rat intestinal epithelial cells and the colonic epithelium in vivo. J Biol Chem 2004; 279: 45556–63.

141. Rizvi MA, Ghias K, Davies KM, Ma C, Krett NL, Rosen ST. Enzastaurin (LY317615), an oral protein kinase C b inhibitor, induces apoptosis in multiple myeloma cell lines. Proc Amer Soc Hematol 2005: Abstr 1577.

142. Podar K, Raab MS, Abtahi D, Tai Y-T, Lin B, Munshi NC, Hideshima T, Chauhan D, Anderson KC. The PKC-inhibitor enzastaurin inhibits MM cell growth, survival and migration in the bone marrow microenvironment. Proc Am Soc Hematol 2005: Abstr 1584.

143. Rossi RM, Henn AD, Conkling R, Guzmann ML, Bushnell T, Harvey J, Fisher RI, Jordan CT. The PKCβ selective inhibitor, enzastaurin (LY317615), inhibits growth of human lymphoma cells. Proc Am Soc Hematol 2005: Abstr 1483.

144. Rieken M, Weigert O, Pastore A, Hutter G, Zimmermann Y, Weinkauf M, Hiddemann W, Dreyling M. Inhibition of protein

kinase C beta by enzastaurin (LY317615) induces alterations of key regulators of cell cycle and apoptosis in mantle cell lymphoma and synergizes with chemotherapeutic agents in a sequence dependent manner. Proc Amer Soc Hematol 2005: Abstr 2416.

145. Teicher BA, Menon K, Alvarez E, Galbreath E, Shih C, Faul MM. Antiangiogenic and antitumor effects of a protein kinase C beta inhibitor in human HT-29 colon carcinoma and human Caki-1 renal cell carcinoma xenografts. Anticancer Res 2001; 21: 3175–84.

146. Teicher BA, Menon K, Alvarrez E, Galbreath E, Shih C, Faul MM. Antiangiogenic and antitumor effects of a protein kinase C beta inhibitor in murine Lewis lung carcinoma and human Calu-6 non-small cell lung carcinoma xenografts. Cancer Chemother Pharmacol 2004; 48: 473–80.

147. Teicher BA, Menon K, Alvarez E, Liu P, Shih C, Faul MM. Antiangiogenic and antitumor effects of a protein kinase C beta inhibitor in human hepatocellular and gastric cancer xenografts. In Vivo 2001; 15: 185–93.

148. Teicher BA, Menon K, Alvarez E, Galbreath E, Shih C, Faul MM. Antiangiogenic and antitumor effects of a protein kinase C Beta inhibitor in human T98G glioblastoma multiforme xenografts. Clin Cancer Res 2001; 7: 634–40.

149. Keyes K, Cox K, Treadway P, Mann L, Shih C, Faul MM, Teicher BA. An In vitro tumor model: analysis of angiogenic factor expression after chemotherapy. Cancer Res 2002; 62: 5597–602.

150. Keyes K, Mann L, Cox K, Treadway P, Iversen P, Chen Y-F, Teicher BA. Circulating angiogenic growth factor levels in mice bearing human tumors using Luminex multiplex technology. Cancer Chemo Pharmacol 2003; 51: 321–7.

151. Keyes KA, Mann L, Sherman M, Galbreath E, Schirtzinger L, Ballard D, Chen YF, Iversen P, Teicher BA. LY317615 decreases plasma VEGF levels in human tumor xenograft-bearing mice. Cancer Chemother Pharmacol 2004; 53: 133–40.

152. Herbst RS, Thornton DE, Kies MS, Sinha V, Flanagan S, Cassidy CA, Carducci MA. Phase 1 study of LY317615, a protein kinase Cβ inhibitor. Am Soc Clin Oncol 2002, abstr 326.

153. Robertson M, Kahl B, Vose J, de Vos S, Laughlin M, Flynn P, Rowland K, Cruz J, Goldberg S, Darnstein C, Enas N, Neuberg D, Savage K, Thornton D, Slapak C, Shipp M. A phase II study of enzastaurin, a protein kinase C-b (PKCb) inhibitor, in the treatment of relapsed diffuse large B-cell lymphoma (DLBCL). Proc Am Soc Hematol 2005: Abstr 934.

154. Fine HA, Kim L, Royce C, Mitchell S, Duic JP, Albert P, Musib L, Thornton D. A phase II trial of LY317615 in patients with recurrent high grade gliomas. Proc Am Soc Clin Oncol 2004: Abstr 1511.

155. Fine HA, Kim L, Royce C, Draper D, Haggarty I, Ellinzano H, Albert P, Kinney P, Musib L, Thornton D. Results from phase II trial of enzastaurin (LY317615) in patients with recurrent high grade gliomas. Proc Am Soc Clin Oncol 2004: Abstr 1504.

156. Inoue K, Slaton JW, Perrotte P, et al: Paclitaxel enhances the effects of the anti-epidermal growth factor receptor monoclonal antibody ImClone C225 in mice with metastatic human bladder transitional cell carcinoma. Clin Cancer Res 2000; 6: 4874–84.

157. Inoue K, Slaton JW, Davis DW, et al: Treatment of human metatstatic transitional cell carcinoma of the bladder in a murine model with tha anti-vascular endothelial growth factor receptor monoclonal antibody DC 101 and paclitaxel. Clin Cancer Res 2000; 6: 2635–43.

158. Kim SJ, Uehara H, Karashima T, et al: Blockage of epidermal growth factor receptor signalling in tumor cells and tumor-associated endothelial cells for therapy of androgen-independent human prostate cancer growing in the bone of nude mice. Clin Cancer Res 2003; 9: 1200–10.

159. Bergers G, Javaherian K, Lo KM, et al: Effects of angiogenesis inhibitors on multistage carcinogenesis in mice. Science 1999; 284: 808–12.

160. Browder T, Butterfield CE, Kraling BM, et al: Antiangiogenic cheduling of chemotherapy improves efficacy against experimental drug-resistant cancer. Cancer Res 2000; 60: 1878–86.

161. Klement G, Baruchel S, Rak J, et al: Continuous low-dose therapy with vinblastine and VEGF receptor-2 antibody induces sustained tumor regression without overt toxicity. J Clin Invest 2000; 105: R15–R24.

162. Hanahan D, Bergers G, Bergsland E: Less is more, regularly: metronomic dosing of cytotoxic drugs can target tumor angiogenesis in mice. J Clin Invest 2000; 105: 1045–7.

163. Gately S, Kerbel RS: Antiangiogenic scheduling of lower dose cancer chemotherapy. Cancer J 2001; 7: 427–36.

164. Bertolini F, Paul S, Mancuso P, Monestiroli S, Gobbi A, Shaked Y, Kerbel RS. Maximum tolerable dose and low-dose metronomic chemotherapy have opposite effects on the mobilization and viability of circulating endothelail progenitor cells. Cancer Res 2003; 63: 4342–6.

165. Pietras K, Hanahan D. A multitargeted, metronomic, and maximum-tolerated dose "chemo-switch" regimen is antiangiogenic, producing objective responses and survival benefit in a mouse model of cancer. J Clin Oncol 2005; 23: 939–52.

166. Shaked Y, Ciarrocchi A, Franco M, Lee CR, Man S, Cheung AM, Hicklin DJ, Chaplin D, Foster FS, Benezra R, Kerbel RS. Therapy-induced acute recruitment of circulating endothelial progenitor cells to tumors. Science 2006; 313: 1785–7.

167. Ferrara N, Kerbel RS. Angiogenesis as a therapeutic target. Nature 2005; 438: 967–74.

168. Bertolini F, Shaked Y, Mancuso P, Kerbel RS. The multifaceted circulating endothelial cell in cancer: towards marker and target identification. Nat Rev Cancer 2006; 6: 835–45.

169. Kerbel RS, Kamen BA. The anti-angiogenic basis of metromoic chemostherapy. Nat Rev Cancer 2004; 4: 423–36.

170. Shaked Y, Bertolini F, Man S, Rogers MS, Cervi D, Foutz T, Rawn K, Voskas D, Dumont DJ, Ben-David Y, Lawler J, Henkin J, Huber J, Hicklin DJ, D'Amato RJ, Kerbel RS. Genetic heterogeneity of the vasculogenic phenotype parallels angiogenesis: implications for cellular surrogate marker analysis of antiangiogenesis. Cancer Cell 2005; 7: 101–11.

171. Willett CG, Boucher Y, di Tomaso E, et al. Direct evidence that the VEGF-specific antibody bevacizumab has antivascular effects in human rectal cancer. Nat Med 2004; 10: 145–7.

172. Sandler A, Gray R, Perry MC, Brahmer J, Schiller JC, Dowlati A, Lilenbaum R, Johnson DH. Paclitaxel-carboplatin alone or with bevacizumab for non-small-cell lung cancer. New Eng J Med 2006; 355: 2542–50.

173. Hida T, Kozaki K, Muramatsu H, et al: Cyclo-oxygenase-2 inhibitor induces apoptosis and enhances cytotoxity of various anticancer agents in non-small cell lung cancer cell lines. Clin Cancer Res 2000; 6: 2006–11.

174. Gasparini G, Longo R, Sarmiento R, et al: COX-2 inhibitors (Coxibs): A new class of anticancer agents? Lancet Oncol 2003; 4: 605–15.

175. Altorki NK, Keresztes RS, Port JL, et al: Celecoxib, a selective cyclo-oxygenase-2 inhibitor, enhances the response to preoperative paclitaxel and carboplatin in early-stage non-small-cell lung cancer. J Clin Oncol 2003; 21: 2645–50.

176. Gasparini G, Morabito A, Magnani E, et al: Thalidomide: an old sedative-hypnotic with anticancer activity? Current Opin Invest Drugs 2001; 2: 1302–8.

177. Neben K, Moehler T, Benner A, et al: Dose-dependent effect of thalidomide on overall survival in relapsed multiple myeloma. Clin Cancer Res 2002; 8: 3377–82.

178. Barlogie B, Desikan R, Eddlemon P, et al: Extended survival in advenced and refractory multiple myeloma after single-agent thalidomide: identification of prognostic factors in a phase 2 study of 169 patients. Blood 2001; 98: 492–94.

179. Weber D, Rankin K, Gavino M, et al: Thalidomide alone or with dexamethasone for previously untreated multiple myeloma. J Clin Oncol 2003; 21: 16–19.

180. Kerbel R, Folkman J. Clinical translation of angiogenesis inhibitors. Nature Rev Cancer 2002; 2: 727–39.

181. Herbst RS, Onn A, Sandler A. Angiogenesis and lung cancer: prognostic and therapeutic implications. J Clin Oncol 2005; 23: 3243–3256.

182. Riesterer O, Honer M, Jochum W, Oehler C, Ametamey S, Pruschy M. Ionizing radiation antagonizes tumor hypoxia induced by antiangiogenic treatment. Clin Cancer Res 2006; 12: 3518–24.

183. Hida K, Hida Y, Amin DN, Flint AF, Panigrahy D, Morton CC, Klagsbrun M. Tumor-associated endothelial cells with cytogenetic abnormalities. Cancer Res 2004; 64: 8249–8255.

184. Gasparini G, Longo R, Fanelli M, Teicher BA. Combination of antiangiogenic therapy with other anticancer therapies: results, challenges and open questions. J Clin Oncol 2005; 23: 1295–1311.

185. Teicher BA. Hypoxia, tumor endothelium and targets for therapy. In: Advances in Experimental Medicine & Biology; eds. P Okunieff, J Williams, Y Chen. Springer: New York NY; 2005; 566: 31–8.

186. Fiegl H, Millinger S, Goebel G, Muller-Holzner E, Marth C, Laird PW, Widschwendter M. Breast cancer DNA methylation profiles in cancer ccells and tumor stroma: association with HER-2/neu status in primary breast cancer. Cancer Res 2006; 66: 29–33.

187. Weeraratna AT. Discovering causes and cures for cancer from gene expression analysis. Ageing Res Rev 2005; 4: 548–563.

188. Brentani RR, Carrari DM, Verjovski-Almeida S, Reis EM, Neves EJ, de Souza SJ, Carvalho AF, Brentani H, Reis LFL. Gene expression arrays in cancer research: methods and applications. Crit Rev Oncol/Hematol 2005; 54: 95–105.

189. Segal E, Friedman N, Kaminski N, Regev A, Koller D. From signatures to models: understanding cancer microarrays. Nature Gen 2005; 37: S38–45.

190. Sieben NL, Oosting J, Flanagan AM, Prat J, Roemen G MJM, Kolkman-Uljee SM, van Eijk R, Cornelisse CJ, Fleuren GJ, van Engeland M. Differential Gene Expression in Ovarian Tumors Reveals Dusp 4 and Serpina 5 As Key Regulators for Benign Behavior of Serous Borderline Tumors. J Clin Oncol 2005; 23: 7257–64

191. Espinosa E, Vara JAF, Redondo A, Sanchez JJ, Hardisson D, Zamora P, Pastrana F, Gomez CP, Martinez B, Suarez A, Calero F, Baron MG. Breast Cancer Prognosis Determined by Gene Expression Profiling: A Quantitative Reverse Transcriptase Polymerase Chain Reaction Study. J Clin Oncol 2005; 23: 7278–85.

192. Chen C-N, Lin J-J, Chen J JW, Lee P-H, Yang C-Y, Kuo M-L, Chang K-J, Hsieh F-J. Gene Expression Profile Predicts Patient Survival of Gastric Cancer After Surgical Resection. J Clin Oncol 2005; 23: 7286–95.

193. Agnelli L, Bicciato S, Mattioli M, Fabri S, Intini D, Verdelli D, Baldini L, Morabito F, Callea V, Lombardi L, Neri A. Molecular Classification of Multiple Myeloma: A Distinct Transcriptional Profile Characterizes Patients Expressing CCND1 and Negative for 14q32 Translocations. J Clin Oncol 2005; 23: 7296–7306.

194. Spinola M, Leoni V, Pignatiello C, Conti B, Ravagnani F, Pastorino U, Dragani TA. Functional FGFR4 Gly388Arg Polymorphism Predicts Prognosis in Lung Adenocarcinoma Patients. J Clin Oncol 2005; 23: 7307–11.

195. Grundy PE, Breslow NE, Li S, Perlman E, Beckwith JB, Ritchey ML, Shamberger RC, Haase GM, D'Angio GJ, Donaldson M, Coppes MJ, Malogolowkin M, Shearer P, Thomas PRM, Macklis R, Tomlinson G, Huff V, Green DM.Loss of Heterozygosity for Chromosomes 1p and 16q Is an Adverse Prognostic Factor in Favorable-Histology Wilms Tumor: A Report From the National Wilms Tumor Study Group. J Clin Oncol 2005; 23: 7312–21.

196. Bilke S, Chen Q-R, Westerman F, Schwab M, Catchpoole D, Khan J. Inferring a Tumor Progression Model for Neuroblastoma From Genomic Data. J Clin Oncol 2005; 23: 7322–31.

197. Velculescu VE, Zhang L, Vogelstein B, Kinzler KW. Serial analysis of gene expression. Science 1995; 270: 484–487.

198. Porter D, Yao J, Polyak K. SAGE and related approaches for cancer target identification. Drug Discov Today 2006; 11: 110–18.

199. Beauchamp NJ, van Achterberg TAE, Engelse MA, Pannekoek H, de Vries CJM. Gene expression profiling of resting and activated vascular smooth muscle cells by serial analysis of gene expression and clustering analysis. Genomics 2003; 82: 288–299.

200. Polyak K, Riggins GJ. Gene discovery using the serial analysis of gene expression technique: implications for cancer research. J Clin Oncol 2001; 19: 2948–2958.

201. Madden SL, Cook BP, Nacht M, Weber WD, Callahan MR, Jiang Y, Dufault MR, Zhang X, Zhang W, Walter-Yohrling J, Rouleau C, Akmaev VR, Wang CJ, Cao X, St. Martin TB, Roberts BL, Teicher BA, Klinger KW, Stan R-V, Lucey B, Carson-Walter EB, Laterra J, Walter KA. Vascular gene expression in non-neoplastic and malignant brain. Am J Pathol 2004; 165: 601–608.

202. Parker BS, Argani P, Cook BP, Liangfen H, Chartrand SD, Zhang M, Saha S, Bardelli A, Jiang Y, St. Martin TB, Nacht M, Teicher BA, Klinger KW, Sukumar S, Madden SL. Alerations in vascular gene expression in invasive breast carcinoma. Cancer Res 2004; 64: 7857–7866.

203. Simon R. Roadmap for developing and validating therapeutically relevant genomic classifiers. J Clin Oncol 2005; 23: 7332–41.

204. Bono H, Okazaki Y. The study of metabolic pathways in tumors based on the transcriptome. Semin Cancer Biol 2005; 15: 290–99.

205. Adler AS, Lin M, Horlings H, Nuyten DSA, van de Vijver MJ, Chang HY. Genetic regulators of large-scale transcritptional signatures in cancer. Nature Gen 2006; 38: 421–30.

206. Rhodes DR, Chinnaiyan AM. Integrative analysis of the cancer transcriptome. Nature Genetics 2005; 37: S31–37.

207. Yoo CB, Jones PA. Epigenetic therapy of cancer: past, present and future. Nature Rev Drug Discovery 2006; 5: 37–50.

208. B. A. Teicher, G. Ara, S. R. Keyes, R. S. Herbst, and E. Frei III, Acute in vivo resistance in high-dose therapy. Clin Cancer Res 1998; 4: 483–91.

209. St. Croix B, Rago C, Velculescu V, Traverso G, Romans KE, Montgomery E, Lal A, Riggins GJ, Lengauer C, Vogelstein B, Kinzler KW. Genes expressed in human tumor endothelium. Science 2000; 289: 1197–1202.

210. Bagley RG, Rouleau C, Morgenbesser SD, Weber W, Cook BP, Shankara S, Madden SL, Teicher BA. Pericytes from human non-small cell lung carcinomas: an attractive target for anti-angiogenic therapy. Vascular Res 2006; 71: 163–74.

211. Nanda A, St. Croix B. Tumor endothelial markers: new targets for cancer therapy. Curr Opin Oncol 2004; 16: 44–49.

212. Carson-Walter EB, Watkins DN, Nanda A, Vogelstein B, Kinzler K, St. Croix B. Cell surface tumor endothelial markers are conserved in mice and humans. Cancer Res 2001; 61: 6649–6655.

Chapter 39
Immunotherapy of Angiogenesis with DNA Vaccines

Chien-Fu Hung[1], Archana Monie[1], and T.-C. Wu[1–4,*]

Keywords: angiogenesis, immunotherapy, DNA vaccines, endothelial cells

Abstract: The understanding of the molecular mechanisms of tumor neovascularization has identified several important molecular targets that are specifically expressed on tumor neovasculature but not on physiological neovasculature. These molecular targets can potentially be used in the development of novel therapeutic approaches against tumor angiogenesis. Immunotherapy targeting angiogenesis has emerged as a potentially promising approach compared to the use of angiogenesis inhibitors due to its ability to afford long-term therapeutic protection. This presents remarkable opportunities for the development of innovative cancer therapies. Immunotherapy using DNA vaccines has gained momentum for antiangiogenesis therapy due to their stability, simplicity and excellent safety profile, and may prove to be a potentially useful strategy for targeting angiogenesis. In the current review, we discuss the various strategies and molecular targets employed in the form of DNA vaccines to target: (1) the endothelial cells within the tumor; (2) biological factors important for angiogenesis; and (3) the extracellular matrix and stromal cells associated with the tumor in order to control tumor angiogenesis in preclinical models.

Departments of Pathology[1], Oncology[2], Molecular Microbiology and Immunology[3], and Obstetrics and Gynecology[4], Johns Hopkins School of Medicine, Baltimore, MD, USA

*Corresponding Author:
T.-C. Wu, Department of Pathology, Johns Hopkins School of Medicine, Cancer Research Building II, Room 309, 1550 Orleans Street, Baltimore, MD, USA
E-mail: wutc@jhmi.edu

Introduction

Alternative Treatment is Necessary for Effective Control of Cancer

Cancer is one of the major causes of death in developed countries and is extremely difficult to cure. Current efforts to reduce this high mortality rate, including improvements in early detection and treatment, have been relatively unsuccessful. Thus, there is an urgent need to develop an innovative therapeutic approach to control cancer. Existing therapies, such as chemotherapy and radiation therapy, rarely result in long-term benefits for patients with locally advanced or metastatic disease. This is because of their inability to destroy tumors without harming normal cells. The ideal cancer therapy should have the potency to eradicate systemic tumors at multiple sites in the body, as well as the specificity to discriminate between malignant and normal cells. In both of these respects, anti-angiogenesis therapy is an attractive approach for combating cancer.

Antiangiogenesis as a Therapeutic Approach for Blocking Tumor Growth

Antiangiogenesis represents a potentially ideal therapeutic approach for the control of cancer. The natural remodeling of the body's vasculature is referred to as angiogenesis. In the normal adult body, a steady balance between activators and inhibitors of endothelial cell (EC) growth tightly controls the process of neovascularization. This occurs only during wound repair, pregnancy, and in the female reproductive cycle (for review see [1,2]). Antiangiogenic strategies for the control of cancer should aim to destroy the blood vessels that supply cancer cells the nutrients and oxygen necessary for their survival, without interfering with the normal neovascularization. As a result, they would lead to starvation of cancer cells and ultimate shrinkage of the tumor [3]. Since tumor neovasculature is different from the physiological neovasculature, antiangiogenesis strategies have the specificity to discriminate between malignant and normal cells.

Antiangiogenic therapy has a number of advantages over conventional cancer therapies. In general, most cancer therapies target the tumor cells, which undergo high rates of mutation. In contrast, antiangiogenic therapy targets endothelial cells and/or the surrounding stroma and matrix, which have a low rate of mutation, thus making drug resistance highly unlikely [4]. Since one endothelial cell supports ca. 10–100 tumor cells, targeting ECs may lead to amplification of tumor cell death. These properties make inhibition of angiogenesis advantageous to chemotherapy or radiotherapy in the long-term treatment of cancer [5]. In addition, antiangiogenic therapy could be applied universally to all kinds of cancers since it targets a process that is essential for the growth of all malignant tumors. Therefore, there is a significant interest in developing strategies to induce an antiangiogenic effect to control tumors.

Immunotherapy Represents a Promising Approach to Generate Sustained Anti-Angiogenesis Effect

Immunotherapy targeting angiogenesis has emerged as an attractive approach for the generation of sustained antiangiogenic effect to control tumors. Compared to other angiogenesis-based approaches, such as employment of angiogenesis inhibitors, the immunotherapy approach has a number of advantages. This approach overcomes the inherent difficulties of repeated administration of angiogenesis inhibitor drugs over a long course of treatment because immunotherapy can lead to immunological memory, which can provide long-term therapeutic effects. This feature may make immunotherapy more cost-effective compared to conventional treatment. In addition, immunotherapy may also be potentially less toxic than angiogenesis inhibitors derived from synthetic chemical compounds due to the specificity of the immune system.

Immunotherapy targeting angiogenesis also has considerable advantages over immunotherapy of cancer, where patients are immunized against antigens presented by tumor cells. Firstly, resistance is less likely to develop in immunotherapy targeting ECs because they are genetically more stable compared to tumor cells. Another major advantage of immunologic targeting of angiogenesis-associated products is that one vaccine can be applicable to many different types of tumors as long as the angiogenic process is similar among different tumors. In contrast, tumor immunotherapy may be specific to each cancer type. Furthermore, ECs are easily accessible in the bloodstream, and effector T cells or antibodies specific for the tumor neovasculature can effectively reach the target tissue. In comparison, immunotherapy targeting tumors may have to cross multiple cell layers in order to reach the tumor cells that are far away from the vasculature (For review, see [6]). Table 39.1 summarizes the comparison of immunotherapy targeting angiogenesis versus immunotherapy targeting tumor cells.

TABLE 39.1. Comparison between immunotherapy of angiogenesis and immunotherapy targeting tumor cells.

Immunotherapy of angiogenesis	Immunotherapy targeting tumor cells
Targets endothelial cells, stromal cells or extracellular matrix	Targets tumor cells
Genetically stable	Genetically unstable
Resistance to immunotherapy unlikely	Resistance to immunotherapy may occur
Applicable to all tumors	Tumor-specific
Readily accessible target	Difficult to penetrate tumor mass
May affect wound healing and fertility	No harmful side effects

DNA Vaccines as an Attractive Approach for Immunotherapy Targeting Angiogenesis

There are various forms of vaccines that may be employed in antiangiogenic immunotherapy, such as DNA-based, peptide-based, protein-based, viral and bacterial vector-based, and cell-based vaccines. Each of these vaccines has their advantages and disadvantages. Table 39.2 summarizes the advantages and disadvantages of the various forms of vaccine technology.

Among these various forms of vaccines, DNA-based vaccines have emerged as an attractive approach for anti-angiogenesis therapy (for review, see [7]). Naked DNA vaccines are advantageous because of their purity, simplicity of preparation, stability and excellent safety profile. Moreover, they are relatively inexpensive and easy to prepare. DNA vaccines allow for long-term sustained expression of antigen on MHC-peptide complexes compared to peptide or protein vaccines. In addition, various DNA vaccines targeting different angiogenesis-associated antigens can be combined together and repeatedly administered. However, a major challenge of DNA vaccines targeting angiogenesis is the breaking of immune tolerance to endogenous angiogenesis-associated molecular targets.

Strategies to Break Immune Tolerance to Endogenous Molecular Targets

Molecular targets related to tumor angiogenesis are mainly endogenous and are thus tolerated by the host's immune system. This 'self'-tolerance may be responsible for the weak immunogenicity of many angiogenesis-associated antigens. It is therefore important to develop methods for successfully breaking immunologic tolerance to endogenous angiogenesis-associated antigens. Several different methods for breaking tolerance have been employed with varying degrees of success. For example, DNA vaccines encoding xenogenic proteins have been found to break tolerance in mouse models by generating autoantibody and/or antigen-specific T cell-mediated immune responses against the target endogenous

TABLE 39.2. Comparison of the various forms of vaccines used in immunotherapy.

	Advantages	Drawbacks
Protein-based	Ease of production; multiple known adjuvants	Induces Ab production, but weak for CTL induction
Peptide-based	Ease of production; safety	Weakly immunogenic, patient's HLA must be matched
Vector-based	*Viral:* Highly immunogenic; multiple different viruses with immunological properties; versatility of construction *Bacterial:* Highly immunogenic; can deliver engineered plasmids; Abs mediate toxicity	Safety concerns, pre-existing immunity
DNA-based	Ease of production, storage and transport; versatility of engineering by adding targeting and/or co-stimulatory genes; stability of expression; capable of multiple immunizations; safety	Intrinsically weakly immunogenic;
RNA-based	Ease of production, storage and transport; versatility of engineering by adding targeting and/or co-stimulatory genes; no risks of integration or cellular transformation; capable of multiple immunizations	Intrinsically weakly immunogenic; unstable to store and handle
Dendritic cell-based	Highly immunogenic; multiple methods of Ag loading; enhanced potency with transduced DC-activating genes	Labor intensive individualized cell processing; variable quality control; do not home to draining lymph nodes
Gene-modified tumor cell-based	Useful when no antigen has been identified	Safety concerns regarding the administration of modified tumor cells

Ab Antibody; *CTL* cytotoxic T lymphocytes; *HLA* human leukocytes antigen; *Ag* antigen; *DC* dendritic cell

TABLE 39.3. Summary of DNA vaccines targeting angiogenesis.

Molecular targets	Origin of antigen	Mechanisms	References
Molecules expressed on endothelial cells of tumor			
VEGFR2	autologous	CTL	[13, 24, 28, 29]
Tie2	autologous	CTL	[35]
FGFR1	xenogeneic	Ab	[14]
Endoglin	autologous	CTL	[15]
Endoglin	xenogeneic	CTL, Ab	[16]
Integrin	xenogeneic	Ab	[17]
Angiomotin	xenogeneic	Ab	[44]
Survivin	autologous	CTL	[18]
Biological factors important for angiogenesis			
VEGF	xenogeneic	Ab	[11]
Extracellular matrix & tumor stromal cells			
MMP-2	xenogeneic	Ab	[9]
PDGFRbeta	autologous	CTL	[19]
TAM	autologous	CTL	[66]

VEGFR2 Vascular endothelial growth factor receptor-2; *Ab* antibody; *CTL* cytotoxic T lymphocyte

antigen [8–20]. Suicidal DNA vectors have also been shown to break tolerance against endogenous antigens [21]. DNA-based αviral RNA replicon vectors, also called suicidal DNA vectors, alleviate the concerns for potential chromosomal integration and cell transformation generated by the use of naked DNA vaccines because these vectors eventually cause apoptotic cell death of transfected cells. Furthermore, the use of bacterial vectors is another technique that can break tolerance to endogenous antigenic targets [22–26]. While the mechanism for breaking of tolerance by a bacterial vector carrying a DNA vaccine remains unclear, an attenuated strain of *Salmonella typhimurium* has been shown to efficiently deliver DNA encoding the antigen of interest to professional antigen-presenting cells (APCs) in the gastrointestinal tract via an oral route of administration. The Salmonella vector allows the use of a natural route of administration for delivery of DNA to cell types that have specifically evolved to induce an immune response [27]. These features make Salmonella an attractive means for delivering DNA vaccines and breaking immune tolerance to endogenous angiogenesis-associated targets.

In the following sections, we will discuss immunotherapy using DNA vaccines directed against the molecularly defined targets related to tumor angiogenesis. Conceptually, there are three different kinds of molecular targets for inducing antiangiogenic effects in tumors. These targets include

(1) molecules expressed on endothelial cells of the tumor
(2) biological factors important for angiogenesis and
(3) tumor extracellular matrix and tumor stromal cells.

DNA vaccines have been employed to target molecules important for tumor angiogenesis in all three categories. Table 39.3 provides a summary of DNA vaccines targeting angiogenesis that have been discussed in this chapter. Figure 39.1 illustrates the different antiangiogenic mechanisms of attack against the three kinds of molecular targets in the tumor vasculature.

DNA Vaccines Targeting Molecules Expressed on Endothelial Cells of the Tumor

An increased understanding of tumor angiogenesis has led to the identification of several potential molecular targets expressed on the surface of endothelial cells of the tumor.

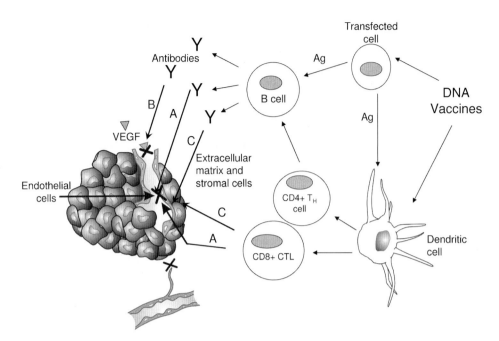

FIG. 39.1. Schematic figure depicting the various mechanisms of immunotherapy targeting angiogenesis through DNA vaccination. (A) Anti-angiogenic effect through vaccines targeting of molecules expressed on endothelial cells of the tumor (e.g., VEGFR2). This is mediated both by T cell-mediated and humoral responses. (B) Antiangiogenic effect through vaccines blocking biological factors important for angiogenesis (e.g., VEGF). This is mediated by antibody responses. (C) Antiangiogenic effect through vaccines targeting the extracellular matrix (e.g., MMPs), and stromal cells (e.g., PDGFRβ associated with the tumor). This is mediated either by antibody responses or by T cell-mediated immune responses. *VEGF* Vascular endothelial growth factor; *VEGFR2* vascular endothelial growth factor receptor 2; *MMP* matrix metalloproteinase; *PDGFRβ* platelet-derived growth factor receptor beta; *CTL* cytotoxic T lymphocyte; T_H *cell* T helper cell; *Ag* antigen.

These molecules serve as potential targets for the development of DNA vaccines.In this section, we will discuss the various DNA vaccination strategies targeting these molecules for antiangiogenesis therapy.

Vascular Endothelial Growth Factor Receptor-2 (VEGFR2)

Vascular endothelial growth factor (VEGF) is an important signaling protein involved in both vasculogenesis and angiogenesis of tumors. All members of the VEGF family stimulate cellular responses by binding to tyrosine kinase receptors, including the vascular endothelial growth factor receptors (VEGFRs) on the cell surface. The binding of VEGF to VEGFR2 causes the receptors to dimerize and become activated through transphosphorylation. VEGFR2 is overexpressed on activated ECs within the tumor and regulates crucial steps of angiogenesis such as EC proliferation, migration, survival or permeability in many cancer types. (See Chapter 8).

DNA vaccines targeting VEGFR2 have been employed in several preclinical models. For example, Niethammer et al. showed that oral admininstration of murine VEGFR2 express-ing DNA vaccines delivered using attenuated *S. typhimurium* led to VEGFR2-specific CD8+ T cell immune responses. This

resulted in effective protection from lethal challenges with several murine tumors, reducing the growth of established metastases in a therapeutic setting and prolonging the lifespan of tumor-challenged mice [24].

Oral administration of attenuated *S. typhimurium* containing DNA encoding minigenes derived from murine VEGFR2 has been shown to induce VEGFR2 peptide-specific CD8+ T cell immune responses, particularly the H-2Db-restricted VEGFR2 peptide (aa 400–408)-specific CD8 + T cell immune response [28]. The VEGFR2 peptide (aa 400–408)-specific cytotoxic T cells were capable of killing VEGFR2+ endothelial cells, resulting in suppression of angiogenesis and long-lived protection in different tumor models [28]. Oral administration of attenuated *S. typhimurium* containing murine VEGFR2 DNA vaccines has also been used in conjunction with intratumoral administration of murine interleukin-12 (IL-12) DNA constructs to generate a stronger cytotoxic T lymphocyte (CTL) response compared to VEGFR2 DNA or IL-12 DNA alone in a murine glioblastoma model [13]. This combination therapy resulted in reduced angiogenesis, apoptosis of tumor cells and significant inhibition of tumor growth [13]. In addition, naked DNA vaccines encoding the extracellular domains (domains 1–3) of the murine VEGFR2 have been used to generate endothelial cell-specific CTL activity and shown to reduce microvessel density and slow the growth of murine hepatoma tumors [29].

Tie2

Like VEGFR2, Tie2 is a tyrosine kinase receptor that is expressed on proliferating endothelial cells [30]. The interaction of Tie2 with its ligand angiopoietin-1 plays an important role in angiogenesis by stabilizing interactions between endothelial cells and pericytes [31]. Tie2 overexpression has been shown in the supporting vasculature of various carcinomas including breast [30], melanomas [32], non-small cell lung carcinomas [33] and gliomas/glioblastomas [34]. In addition, vaccination with DNA vaccine encoding a modified human Tie2 protein has been shown to generate HLA-A2-restricted Tie2-specific CD8+ T cell immune responses in HLA-A2 transgenic mice [35]. Furthermore, Tie2-specific CD8+ T cells were capable of killing HLA-A2+ endothelial cells overexpressing Tie2 [35]. Thus, Tie2 may serve as a feasible target for DNA vaccine development.

Fibroblast Growth Factor Receptor-1(FGFR1)

Fibroblast growth factors (FGFs) are a family of growth factors involved in wound healing and embryonic development. One of the most important functions of basic FGF (bFGF) is the promotion of angiogenesis and endothelial cell proliferation [36]. (see Chapter 7) Fibroblast growth factor receptor-1(FGFR1), also called CD331, is a high-affinity receptor for bFGF. Both bFGF and FGFR1 are expressed on endothelial cells within the tumor but barely expressed on the normal vasculature [1,37]. Thus, FGFR1 has the potential to be used as a molecular target for the development of DNA vaccine.

DNA vaccines encoding Xenopus FGFR1 have been shown to be effective at inducing protective and therapeutic anti-tumor immunity against H22 hepatoma cells, MMT mammary cancer cells and Meth A fibrosarcoma cell models in vaccinated mice [14]. The FGFR1 specific autoantibodies generated by DNA vaccination has been shown to be important for the anti-tumor effect since adoptive transfer of sera or immunoglobulins from Xenopus FGFR1 DNA vaccinated mice were capable of inhibiting tumor growth [14]. Depletion of CD4+ T cells, but not CD8+ T cells or NK cells, led to loss of autoantibody production as well as anti-tumor effect generated by the DNA vaccination [14]. Taken together, CD4+ T cell-dependent FGFR1 specific autoantibodies were responsible for the anti-tumor effect and suppression of angiogenesis in vaccinated mice.

Endoglin

Endoglin is a type I membrane glycoprotein located on the cell surface and is overexpressed on the tumor endothelium. It is part of the TGF-β1/TGF-β3 receptor complex and it modulates the effects of TGF-β which is a soluble stimulator of angiogenesis. Thus, endoglin serves as an ideal target for antiangiogenesis therapy (For review, see [38, 39]).

Oral administration of DNA vaccines encoding murine endoglin delivered by attenuated *S. typhimurium* to secondary lymphoid organs like Peyer's patches have been shown to induce a robust endoglin-specific CD8+ T cell mediated immune response in vaccinated mice. Vaccination with the DNA vaccine inhibited angiogenesis, resulting in suppression of tumor metastasis in breast cancer models and increased life span of tumor bearing mice [15].

In addition, intramuscular administration of naked DNA vaccines encoding the extracellular domain of porcine endoglin has also been shown to generate significant protective and therapeutic anti-tumor immunity and inhibition of angiogenesis in vaccinated mice in H22 hepatoma models [16]. The anti-tumor effect generated by the DNA vaccine appeared to be contributed by both murine endoglin-specific autoantibodies as well as CD4+ and CD8+ T cells [16].

Integrin αvβ3

Integrin αvβ3 (CD51/61) is highly expressed on tumor endothelium and serves as an important regulator for tumor angiogenesis. Integrin is a receptor for extracellular matrix proteins with exposed RGD sequence such as fibronectin. They play a role in the attachment of a cell to the extracellular matrix (ECM) and in signal transduction from the ECM to the cell [40]. (For review, see Chapter 6.)

DNA vaccines encoding the ligand-binding domain of chicken integrin β3 have been shown to be effective in both protective and therapeutic anti-tumor immunity in Meth A-fibrosarcoma cells and MMT mammary carcinoma cell models in vaccinated mice [17]. The integrin β3-specific autoantibodies generated by DNA vaccination were shown to be important for the anti-tumor effect since adoptive transfer of immunoglobulins from DNA vaccinated mice was capable of inhibiting tumor growth [17]. Furthermore, depletion of CD4+ T cells led to loss of the anti-tumor effect generated by the DNA vaccination [17]. Taken together, CD4+ T cells and integrin β3-specific autoantibodies are responsible for the anti-tumor effect generated by the DNA vaccination.

Angiomotin

Angiomotin is a molecule that is highly expressed in the tumor endothelium. Angiomotin is the receptor for angiostatin, an inhibitor of angiogenesis. It is a membrane-associated protein that mediates angiostatin inhibition of endothelial migration in vitro [41–43]. The expression of angiomotin in angiogenic vessels and its functional role as an angiostatin receptor makes it a potential target for antiangiogenesis therapy.

Intramuscular administration of DNA vaccines encoding the human p80 isoform of angiomotin followed by in vivo electroporation has been shown to generate significant anti-tumor effect against the TUBO murine breast cancer model [44]. The anti-tumor effect by the DNA vaccination was mainly contributed by CD4+ T cell-dependant antibody responses since depletion of CD4+ T cells or the employment of B cell knockout mice led to abrogation of the anti-tumor effect [44].

Other Potential Targets

There are several molecules that are expressed on tumor endothelial cells as well as on tumor cells, such as survivin and the epidermal growth factor receptor (EGFR). Thus, these molecules could also serve as feasible targets for antiangiogenesis therapy.

Survivin is a unique member of the IAP (inhibitor of apoptosis) family that is overexpressed in almost all human tumors, whereas expression is undetectable in most normal adult tissues. It can also be found in the angiogenic endothelium, thus making it a rational target for antiangiogenic therapy (For review, see [45]). For example, oral administration of attenuated *S. typhimurium* containing DNA vaccines co-expressing murine survivin and the secretary chemokine CCL21 has been shown to generate a potent antiangiogenesis, tumor cell-specific CTL-mediate immune responses and an anti-tumor effect in vaccinated mice [18]. This led to significant inhibition of pulmonary metastases of murine D121 lung cancer model in both prophylactic and therapeutic settings in vaccinated mice [18]. Thus, DNA vaccines targeting survivin may be able to directly inhibit tumor cell growth as well as induce antiangiogenic effects to generate a potent anti-tumor effect.

The epidermal growth factor receptor may be another potential target for the development of DNA vaccines to control tumor growth through direct tumor cell inhibition and angiogenesis inhibition. EGFR is the tyrosine kinase receptor for epidermal growth factor (EGF), which is abnormally elevated in most solid tumors and thus promotes the secretion of angiogenic stimulators. It is also expressed on endothelial cells in some tumors, thus directly regulating angiogenesis (For review, see [46]). Thus, DNA vaccines targeting EGFR may lead to direct inhibition of tumor cell growth resulting in significant anti-tumor and antiangiogenic effects.

DNA Vaccines Blocking Biological Factors Important for Angiogenesis

While tumor endothelial cells are crucial for tumor angiogenesis, other soluble proangiogenic molecules are important for tumor angiogenesis and therefore serve as potential targets for DNA vaccine development. These proangiogenic molecules can be derived from tumors, tumor endothelial cells and other kinds of cells. In this section, we will discuss some of the important proangiogenic molecules including vascular endothelial growth factor (VEGF) and basic fibroblast growth factor (bFGF).

Vascular endothelial growth factor is an important signaling protein involved in angiogenesis and vasculogenesis. VEGF activity is mainly restricted to cells of the vascular endothelium. VEGF levels are elevated in a majority of cancers and thus blocking VEGF activity results in inhibition of tumor angiogenesis and consequently inhibition of tumor growth. Thus, the importance of VEGF in tumor angiogenesis is highlighted as an excellent target for antiangiogenic therapy.

DNA vaccines encoding xenogeneic VEGF have been explored in preclinical models [11]. Over the past decades, monoclonal antibodies (mAb) against proangiogenic molecules have emerged as promising therapeutic agents for antiangiogenesis therapy [47–49]. Specifically, bevacizumab (Avastin), a humanized anti-VEGF mAb, has been approved by the FDA as first-line treatment for patients with metastatic colorectal cancer [50]. While humanized monoclonal antibodies represent a significantly promising approach for antiangiogenic therapies, a major drawback of this approach is the short biological half-life of these mAbs; thus, high doses are usually administered repeatedly for a long period of time to achieve clinical efficacy. Furthermore, one possible concern is the development of antibodies against the mAb in patients, subjecting the mAb to be cleared by the patient's immune system. As such, active immunization using DNA vaccines may potentially resolve these concerns.

Intramuscular administration of DNA vaccines encoding Xenopus VEGF has been shown to induce protective and therapeutic anti-tumor immunity in Meth A fibrosarcoma, H22 hepatoma and MA782/5S mammary cancer cell models [11]. The VEGF-specific autoantibodies generated by DNA vaccination were shown to be important for the suppression of angiogenesis since adoptive transfer of immunoglobulins from DNA vaccinated mice resulted in inhibition of VEGF-mediated endothelial cell proliferation [11]. Furthermore, depletion of CD4+ T cells, but not CD8+ T cells, led to loss of autoantibody production as well as the anti-tumor effect generated by DNA vaccination [11]. Taken together, CD4+ T cell-dependent VEGF-specific autoantibodies are responsible for the anti-tumor effect generated by the DNA vaccination.

Fibroblast growth factors may also serve as a potential target for DNA vaccine development. FGFs are a family of growth factors involved in cancer angiogenesis, wound healing and embryonic development. One of the most important functions of basic FGF (bFGF), also called FGF2, is the promotion of cancer angiogenesis and EC migration [36]. It has been shown that intramuscular administration of liposome-based peptide vaccines containing the heparin-binding domain of bFGF inhibit bFGF-mediated angiogenesis, tumor development and metastasis in vaccinated mice using several preclinical tumor models [51]. Thus, further exploration of DNA vaccination strategies targeting proangiogenic factors should be encouraged.

DNA Vaccines Targeting the Molecules Associated with the Extracellular Matrix and Tumor Stromal Cells

Extracellular matrix

It is well known that the extracellular matrix (ECM) plays an important role in the regulation of angiogenesis (reviewed in [52,53]). Angiogenesis is an invasive process requiring proteolysis of the ECM and proliferation and migration of endothelial cells. Inappropriate breaking down of the ECM is thought to play a role in invasive tumor growth and metastasis. Matrix metalloproteinases (MMPs) are zinc-dependent endopeptidases that are capable of degrading and remodeling the extracellular matrix in order to promote endothelial cell migration. Increased MMP activity allows the tumor to grow in its surrounding microenvironment and promotes angiogenesis. Numerous studies have shown that MMPs, such as MMP-2 and MMP-9, are frequently overexpressed in various solid tumor cells and peritumoral stromal cells and are associated with higher tumor grade and stage [54]. Thus, DNA vaccination strategies targeting MMPs serve as a potentially feasible approach to control angiogenesis of tumors.

Tumor Stromal Cells

The cells located in the tumor stroma play an essential role in the regulation of angiogenesis. The tumor stroma contains several cells including reactive fibroblasts, activated pericytes and tumor-associated macrophages. These cells play a vital role in physiologic processes such as embryonic development and wound healing, as well as in tumor angiogenesis. Thus, molecules expressed on stromal cells may potentially serve as vital targets for antiangiogenic therapies (for review, see [56]).

Platelet derived growth factor receptor β (PDGFRβ) is a receptor tyrosine kinase (RTK) that is overexpressed on tumor stromal fibroblasts and pericytes [57,58]. PDGFRβ expression is highly up-regulated in pathologic conditions such as cancer development [59–61]. PDGFRβ-expressing pericytes migrate towards the tumor vasculature on a chemotactic gradient of PDGF and are involved in the regulation of angiogenesis [62]. Oral administration of DNA vaccines encoding murine PDGFRβ delivered using attenuated *S. typhimurium* has been shown to induce PDGFRβ-specific CD8+ T cell immune responses in vaccinated mice [19]. In addition, the DNA vaccination was also capable of suppressing angiogenesis and inhibiting tumor growth in vaccinated mice in several murine tumor models [19].

Tumor-associated macrophages (TAMs) are a major inflammatory component of the tumor stroma. TAMs consist primarily of a polarized M2 (F4/80+/CD206+) macrophage population with little cytotoxicity for tumor cells because of their limited production of nitric oxide and proinflammatory cytokines [63]. These macrophages promote angiogenesis and tumor cell proliferation and metastasis by secreting a wide range of growth and proangiogenic factors as well as metalloproteinases, and by regulating the functions of fibroblasts in the tumor stroma (for review, see [64,65]). Thus, immunization against molecules overexpressed on TAMs has emerged as a novel strategy against tumor angiogenesis.

Legumain, a stress protein that is overexpressed by TAMs has been used as a target for DNA vaccines to control tumor angiogenesis. DNA vaccines encoding murine legumain have been shown to induce a robust CD8+ T cell immune response against TAMs in vaccinated mice. The DNA vaccination resulted in a dramatic reduction in the number of TAMs in

the tumor tissues and a decrease in the proangiogenic growth factors released by TAMs, such as TGF-β, TNF-α and VEGF [66]. Furthermore, the DNA vaccination resulted in suppression of tumor progression, metastasis, and reduction in tumor angiogenesis in murine preclinical models [66].

Future Directions

While antiangiogenic therapy using DNA vaccines has generated encouraging results in several preclinical models, there a number of challenges for future endeavors, including: (1) identification of novel molecular targets; (2) applications of innovative strategies to enhance DNA vaccine potency to lead to antiangiogenic effect; (3) overcoming the limitations of anti-angiogenesis therapy; and (4) employment of DNA for gene therapy to directly inhibit angiogenesis.

Identification of Novel Molecular Targets

The improvement of new technology has paved the way for the discovery of innovative molecular targets that are uniquely expressed on the tumor neovasculature compared to the physiological neovasculature. This may become possible with the availability of laser microdissection techniques in conjunction with techniques that allow us to characterize the gene expression profile of the tumor neovasculature using modern microarray analysis methods or serial analysis of gene expression (SAGE) with minimal amounts of RNA isolated from the specific tissues or cells [67,68]. The candidate molecules should be further characterized by in situ hybridization or immunocytochemical staining in order to confirm that they are uniquely expressed on the tumor neovasculature. Thus, comparison of the physiological neovasculature with tumor vasculature will be key to the identification of novel angiogenesis-associated antigens that can be used as potential targets for DNA vaccine development in the future.

Applications of Innovative Strategies to Enhance DNA Vaccine Potency to Lead to Antiangiogenic Effect

While DNA vaccines have the advantage of simplicity, stability and safety, one major limitation is their low immunogenicity. Several strategies have been developed in order to enhance DNA vaccine potency and overcome the issue of low immunogenicity. Since dendritic cells (DCs) play an integral role in DNA vaccine-mediated immune responses, modification of DCs represents an excellent strategy to improve DNA vaccine potency. There are several strategies to enhance DNA vaccine potency by modification of the properties of DCs which include: (1) increasing the number of antigen-expressing DCs; (2) improving antigen expression, processing and presentation in DCs; (3) promoting DC activation and function; and (4) enhancing DC and T cell interaction, to enhance T cell immune responses (for review see [69,70]). Table 39.4 summarizes the various strategies to enhance DNA vaccine potency

TABLE 39.4. Strategies to enhance DNA vaccine potency by modification of properties of APCs. For review, see [69, 70].

Strategies	Methods
Strategies to increase the number of antigen-expressing DCs	• Epidermal administration of DNA vaccines via gene gun • Intranodal injection • Electroporation • Intercellular antigen spreading • Linkage of antigen to molecules capable of binding to DCs • Employment of chemotherapy-induced apoptotic cell death • Attenuated *S. typhimurium* vector • Codon optimization
Strategies to improve antigen expression, processing, and presentation in DCs	• Employment of intracellular targeting strategies to enhance MHC class I and class II antigen presentation in DCs • Bypassing antigen processing – MHC class I single chain trimer (SCT)
Strategies to promote DC activation and function	• Employment of toll-like receptor ligands and other genetic adjuvants • Inhibition of immunosuppressive factors
Strategies to enhance DC and T cell interaction	• Employment of cytokines and co-stimulatory molecules • Prolonging DC survival • Induction of CD4+ T cell help • Eliminating immunosuppressive T regulatory cells • Promoting in vivo DC expansion

DC Dendritic cell; *MHC* major histocompatibility complex

by modification of the properties of DCs. For example, the employment of attenuated *S. typhimurium* vector for the delivery of DNA vaccines represents an important strategy to increase the number of antigen-expressing DCs [22–26]. This is due to efficient delivery of DNA encoding the antigen of interest to professional APCs in the gastrointestinal tract via an oral route of administration. Thus, these strategies can be potentially employed to increase the potency of DNA vaccines targeting tumor angiogenesis.

Overcoming the Limitations of Antiangiogenic Therapy

While active immunization using DNA vaccines can lead to sustained antiangiogenic effects, it is important to address the issues relating to the safety of these vaccines before they become widely used in the clinic [6]. Active immunization against angiogenesis-associated antigens could potentially induce autoimmunity that may affect physiological processes like wound healing and female reproduction. Moreover, there is a possibility that active immunization against proangiogenic molecules may elicit an unwanted response against the normal vasculature that may exhibit low level of expression of the same molecules. Thus, it is vital to obtain a better understanding of the effect of antiangiogenic therapies on the normal physiological processes in order to address the issue of toxicity of these vaccines. Furthermore,

in order to further improve the efficacy of antiangiogenic therapy using DNA vaccines, it is important to address the issues regarding the heterogeneity of the tumor microenvironment that could potentially affect the efficacy of this therapeutic approach.

Employment of DNA for Gene Therapy to Directly Inhibit Angiogenesis

An important future direction for antiangiogenic therapy using DNA constructs is the use of DNA encoding angiogenesis inhibitors as gene therapy to directly inhibit angiogenesis. The employment of DNA constructs for the expression of angiogenesis inhibitors may lead to persistent expression and release of these inhibitors over a long period of time, resulting in the control of tumor growth. For example, several studies have employed calreticulin (CRT) in DNA constructs, which have resulted in antiangiogenic effects mediating the control of tumor growth in DNA treated mice [71–74]. CRT and its protein fragment (aa 1–180) vasostatin have been shown to directly target endothelial cells, inhibit angiogenesis and suppress tumor growth [75,76]. Thus, DNA constructs targeting inhibitors of angiogenesis, such as CRT may potentially generate antiangiogenic effects.

Conclusions

With the identification of novel angiogenesis-associated antigens that can potentially be used as targets, a number of remarkable opportunities have opened up for the development of innovative immunization strategies for antiangiogenic therapy using DNA vaccines. However, it is important to seek other strategies to break immune tolerance as well as to further enhance vaccine potency due to the low immunogenicity of DNA vaccines. The encouraging results from preclinical models using DNA vaccines targeting tumor angiogenesis serve as an important foundation for future clinical translation. Clinical trials using DNA vaccines have provided a unique opportunity to identify the mechanisms of the immune response that correlate with DNA vaccine potency in humans. Thus, a better understanding of the immune mechanisms involved in DNA vaccines targeting angiogenesis will provide insights into the development of more effective antiangiogenesis strategies in the future. In addition, the combination of DNA vaccination targeting tumor angiogenesis with other cancer therapeutic regimens such as chemotherapy and radiation may further improve the therapeutic effects for the control of cancer.

Acknowledgements. This review is not intended to be encyclopedic and we apologize to any authors not cited. We would like to thank Dr. Richard Roden and Ping Mao for their critical review of the manuscript.

References

1. Carmeliet P. Mechanisms of angiogenesis and arteriogenesis. Nat Med 2000; 6(4):389–95.

2. Papetti M, Herman IM. Mechanisms of normal and tumor-derived angiogenesis. Am J Physiol Cell Physiol 2002; 282(5): C947–70.

3. Scappaticci FA. The therapeutic potential of novel antiangiogenic therapies. Expert Opin Investig Drugs 2003; 12(6):923–32.

4. Boehm T, Folkman J, Browder T, et al. Antiangiogenic therapy of experimental cancer does not induce acquired drug resistance. Nature 1997; 390(6658):404–7.

5. Hanahan D, Folkman J. Patterns and emerging mechanisms of the angiogenic switch during tumorigenesis. Cell 1996; 86(3): 353–64.

6. Li Y, Bohlen P, Hicklin DJ. Vaccination against angiogenesis-associated antigens: a novel cancer immunotherapy strategy. Current molecular medicine 2003; 3(8):773–9.

7. Boyd D, Hung CF, Wu TC. DNA vaccines for cancer. IDrugs 2003; 6(12):1155–64.

8. Liu JY, Wei YQ, Yang L, et al. Immunotherapy of tumors with vaccine based on quail homologous vascular endothelial growth factor receptor-2. Blood 2003; 102(5):1815–23.

9. Su JM, Wei YQ, Tian L, et al. Active immunogene therapy of cancer with vaccine on the basis of chicken homologous matrix metalloproteinase-2. Cancer Res 2003; 63(3):600–7.

10. Lu Y, Wei YQ, Tian L, et al. Immunogene therapy of tumors with vaccine based on xenogeneic epidermal growth factor receptor. J Immunol 2003; 170(6):3162–70.

11. Wei YQ, Huang MJ, Yang L, et al. Immunogene therapy of tumors with vaccine based on Xenopus homologous vascular endothelial growth factor as a model antigen. Proc Natl Acad Sci USA 2001; 98(20):11545–50.

12. Luo Y, Markowitz D, Xiang R, et al. FLK-1-based minigene vaccines induce T cell-mediated suppression of angiogenesis and tumor protective immunity in syngeneic BALB/c mice. Vaccine 2007; 25(8):1409–15.

13. Keke F, Hongyang Z, Hui Q, et al. A combination of flk1-based DNA vaccine and an immunomodulatory gene (IL-12) in the treatment of murine cancer. Cancer Biother Radiopharm 2004; 19(5):649–57.

14. He QM, Wei YQ, Tian L, et al. Inhibition of tumor growth with a vaccine based on xenogeneic homologous fibroblast growth factor receptor-1 in mice. J Biol Chem 2003; 278(24):21831–6.

15. Lee SH, Mizutani N, Mizutani M, et al. Endoglin (CD105) is a target for an oral DNA vaccine against breast cancer. Cancer Immunol Immunother 2006; 55(12):1565–74.

16. Jiao JG, Li YN, Wang H, et al. A plasmid DNA vaccine encoding the extracellular domain of porcine endoglin induces anti-tumour immune response against self-endoglin-related angiogenesis in two liver cancer models. Dig Liver Dis 2006; 38(8):578–87.

17. Lou YY, Wei YQ, Yang L, et al. Immunogene therapy of tumors with a vaccine based on the ligand-binding domain of chicken homologous integrin beta3. Immunological investigations 2002; 31(1):51–69.

18. Xiang R, Mizutani N, Luo Y, et al. A DNA vaccine targeting survivin combines apoptosis with suppression of angiogenesis in lung tumor eradication. Cancer Res 2005; 65(2):553–61.

19. Kaplan CD, Kruger JA, Zhou H, et al. A novel DNA vaccine encoding PDGFRbeta suppresses growth and dissemination of murine colon, lung and breast carcinoma. Vaccine 2006; 24 (47–48):6994–7002.

20. Perales MA, Blachere NE, Engelhorn ME, et al. Strategies to overcome immune ignorance and tolerance. Semin Cancer Biol 2002; 12(1):63–71.

21. tuLeitner WW, Hwang LN, deVeer MJ, et al. Alphavirus-based DNA vaccine breaks immunological tolerance by activating innate antiviral pathways. Nat Med 2003; 9(1):33–9.

22. Xiang R, Lode HN, Chao TH, et al. An autologous oral DNA vaccine protects against murine melanoma. Proc Natl Acad Sci USA 2000; 97(10):5492–7.

23. Niethammer AG, Primus FJ, Xiang R, et al. An oral DNA vaccine against human carcinoembryonic antigen (CEA) prevents growth and dissemination of Lewis lung carcinoma in CEA transgenic mice. Vaccine 2001; 20(3–4):421–9.

24. Niethammer AG, Xiang R, Becker JC, et al. A DNA vaccine against VEGF receptor 2 prevents effective angiogenesis and inhibits tumor growth. Nat Med 2002; 8(12):1369–75.

25. Niethammer AG, Xiang R, Ruehlmann JM, et al. Targeted interleukin 2 therapy enhances protective immunity induced by an autologous oral DNA vaccine against murine melanoma. Cancer Res 2001; 61(16):6178–84.

26. Luo Y, Zhou H, Mizutani M, et al. Transcription factor Fos-related antigen 1 is an effective target for a breast cancer vaccine. Proc Natl Acad Sci USA 2003; 100(15):8850–5.

27. Darji A, Guzman CA, Gerstel B, et al. Oral somatic transgene vaccination using attenuated S. typhimurium. Cell 1997; 91(6):765–75.

28. Zhou H, Luo Y, Mizutani M, et al. T cell-mediated suppression of angiogenesis results in tumor protective immunity. Blood 2005; 106(6):2026–32.

29. Lu F, Qin ZY, Yang WB, et al. A DNA vaccine against extracellular domains 1–3 of flk-1 and its immune preventive and therapeutic effects against H22 tumor cell in vivo. World J Gastroenterol 2004; 10(14):2039–44.

30. Peters KG, Coogan A, Berry D, et al. Expression of Tie2/Tek in breast tumour vasculature provides a new marker for evaluation of tumour angiogenesis. Br J Cancer 1998; 77(1):51–6.

31. Wong AL, Haroon ZA, Werner S, et al. Tie2 expression and phosphorylation in angiogenic and quiescent adult tissues. Circulation Res 1997; 81(4):567–74.

32. Kaipainen A, Vlaykova T, Hatva E, et al. Enhanced expression of the tie receptor tyrosine kinase mesenger RNA in the vascular endothelium of metastatic melanomas. Cancer Res 1994; 54(24):6571–7.

33. Takahama M, Tsutsumi M, Tsujiuchi T, et al. Enhanced expression of Tie2, its ligand angiopoietin-1, vascular endothelial growth factor, and CD31 in human non-small cell lung carcinomas. Clin Cancer Res 1999; 5(9):2506–10.

34. Koga K, Todaka T, Morioka M, et al. Expression of angiopoietin-2 in human glioma cells and its role for angiogenesis. Cancer Res 2001; 61(16):6248–54.

35. Ramage JM, Metheringham R, Conn A, et al. Identification of an HLA-A*0201 cytotoxic T lymphocyte epitope specific to the endothelial antigen Tie2. Int J Cancer 2004; 110(2):245–50.

36. Presta M, Dell'Era P, Mitola S, et al. Fibroblast growth factor/fibroblast growth factor receptor system in angiogenesis. Cytokine & growth factor reviews 2005; 16(2):159–78.

37. Folkman J. Tumor angiogenesis and tissue factor. Nat Med 1996; 2(2):167–8.

38. Fonsatti E, Altomonte M, Nicotra MR, et al. Endoglin (CD105): a powerful therapeutic target on tumor-associated angiogenetic blood vessels. Oncogene 2003; 22(42):6557–63.

39. Fonsatti E, Altomonte M, Arslan P, et al. Endoglin (CD105): a target for anti-angiogenetic cancer therapy. Current drug targets 2003; 4(4):291–6.

40. Eliceiri BP, Cheresh DA. Adhesion events in angiogenesis. Current opinion in cell biology 2001; 13(5):563–8.

41. Levchenko T, Bratt A, Arbiser JL, et al. Angiomotin expression promotes hemangioendothelioma invasion. Oncogene 2004; 23(7):1469–73.

42. Bratt A, Birot O, Sinha I, et al. Angiomotin regulates endothelial cell-cell junctions and cell motility. J Biol Chem 2005; 280(41):34859–69.

43. Bratt A, Wilson WJ, Troyanovsky B, et al. Angiomotin belongs to a novel protein family with conserved coiled-coil and PDZ binding domains. Gene 2002; 298(1):69–77.

44. Holmgren L, Ambrosino E, Birot O, et al. A DNA vaccine targeting angiomotin inhibits angiogenesis and suppresses tumor growth. Proc Natl Acad Sci USA 2006; 103(24):9208–13.

45. Altieri DC. Validating survivin as a cancer therapeutic target. Nat Rev Cancer 2003; 3(1):46–54.

46. van Cruijsen H, Giaccone G, Hoekman K. Epidermal growth factor receptor and angiogenesis: Opportunities for combined anticancer strategies. Int J Cancer 2005; 117(6):883–8.

47. Hou J, Tian L, Wei Y. Cancer immunotherapy of targeting angiogenesis. Cellular & molecular immunology 2004; 1(3):161–6.

48. Sanz L, Alvarez-Vallina L. Antibody-based antiangiogenic cancer therapy. Expert opinion on therapeutic targets 2005; 9(6):1235–45.

49. Herbst RS. Therapeutic options to target angiogenesis in human malignancies. Expert opinion on emerging drugs 2006; 11(4):635–50.

50. Willett CG, Boucher Y, di Tomaso E, et al. Direct evidence that the VEGF-specific antibody bevacizumab has antivascular effects in human rectal cancer. Nat Med 2004; 10(2):145–7.

51. Plum SM, Holaday JW, Ruiz A, et al. Administration of a liposomal FGF-2 peptide vaccine leads to abrogation of FGF-2-mediated angiogenesis and tumor development. Vaccine 2000; 19(9–10):1294–303.

52. Sottile J. Regulation of angiogenesis by extracellular matrix. Biochim Biophys Acta 2004; 1654(1):13–22.

53. Wang D, Anderson JC, Gladson CL. The role of the extracellular matrix in angiogenesis in malignant glioma tumors. Brain pathology (Zurich, Switzerland) 2005; 15(4):318–26.

54. John A, Tuszynski G. The role of matrix metalloproteinases in tumor angiogenesis and tumor metastasis. Pathol Oncol Res 2001; 7(1):14–23.

55. Vacca A, Moretti S, Ribatti D, et al. Progression of mycosis fungoides is associated with changes in angiogenesis and expression of the matrix metalloproteinases 2 and 9. Eur J Cancer 1997; 33(10):1685–92.

56. Kim JB, Stein R, O'Hare MJ. Tumour-stromal interactions in breast cancer: the role of stroma in tumourigenesis. Tumour Biol 2005; 26(4):173–85.

57. Bergsten E, Uutela M, Li X, et al. PDGF-D is a specific, protease-activated ligand for the PDGF beta-receptor. Nature cell biology 2001; 3(5):512–6.

58. Ostman A, Heldin CH. Involvement of platelet-derived growth factor in disease: development of specific antagonists. Adv Cancer Res 2001; 80:1–38.

59. Coltrera MD, Wang J, Porter PL, et al. Expression of platelet-derived growth factor B-chain and the platelet-derived growth factor receptor beta subunit in human breast tissue and breast carcinoma. Cancer Res 1995; 55(12):2703–8.

60. Kawai T, Hiroi S, Torikata C. Expression in lung carcinomas of platelet-derived growth factor and its receptors. Lab Invest 1997; 77(5):431–6.

61. Singer CF, Hudelist G, Lamm W, et al. Expression of tyrosine kinases in human malignancies as potential targets for kinase-specific inhibitors. Endocrine-related cancer 2004; 11(4):861–9.

62. Pietras K, Ostman A, Sjoquist M, et al. Inhibition of platelet-derived growth factor receptors reduces interstitial hypertension and increases transcapillary transport in tumors. Cancer Res 2001; 61(7):2929–34.

63. Mills CD, Kincaid K, Alt JM, et al. M-1/M-2 macrophages and the Th1/Th2 paradigm. J Immunol 2000; 164(12):6166–73.

64. Mantovani A, Sozzani S, Locati M, et al. Infiltration of tumours by macrophages and dendritic cells: tumour-associated macrophages as a paradigm for polarized M2 mononuclear phagocytes. Novartis Foundation symposium 2004; 256:137–45; discussion 46–8, 259–69.

65. Sica A, Schioppa T, Mantovani A, et al. Tumour-associated macrophages are a distinct M2 polarised population promoting tumour progression: potential targets of anti-cancer therapy. Eur J Cancer 2006; 42(6):717–27.

66. Luo Y, Zhou H, Krueger J, et al. Targeting tumor-associated macrophages as a novel strategy against breast cancer. J Clin Invest 2006; 116(8):2132–41.

67. Quinn DI, Henshall SM, Sutherland RL. Molecular markers of prostate cancer outcome. Eur J Cancer 2005; 41(6):858–87.

68. Peale FV, Jr., Gerritsen ME. Gene profiling techniques and their application in angiogenesis and vascular development. J Pathol 2001; 195(1):7–19.

69. Hung CF, Wu TC. Improving DNA vaccine potency via modification of professional antigen presenting cells. Curr Opin Mol Ther 2003; 5(1):20–4.

70. Tsen S-WD, Paik A, Hung C-F, et al. Enhancing DNA Vaccine Potency by Modifying the Properties of Antigen-Presenting Cells Expert Review of Vaccines 2007; (in press).

71. Cheng WF, Hung CF, Chai CY, et al. Tumor-specific immunity and antiangiogenesis generated by a DNA vaccine encoding calreticulin linked to a tumor antigen. J Clin Invest 2001; 108(5):669–78.

72. Xiao F, Wei Y, Yang L, et al. A gene therapy for cancer based on the angiogenesis inhibitor, vasostatin. Gene Ther 2002; 9(18):1207–13.

73. Cheng WF, Hung CF, Chen CA, et al. Characterization of DNA vaccines encoding the domains of calreticulin for their ability to elicit tumor-specific immunity and antiangiogenesis. Vaccine 2005; 23(29):3864–74.

74. Zhao KJ, Cheng H, Zhu KJ, et al. Recombined DNA vaccines encoding calreticulin linked to HPV6bE7 enhance immune response and inhibit angiogenic activity in B16 melanoma mouse model expressing HPV 6bE7 antigen. Arch Dermatol Res 2006; 298(2):64–72.

75. Pike SE, Yao L, Jones KD, et al. Vasostatin, a calreticulin fragment, inhibits angiogenesis and suppresses tumor growth. J Exp Med 1998; 188(12):2349–56.

76. Pike SE, Yao L, Setsuda J, et al. Calreticulin and calreticulin fragments are endothelial cell inhibitors that suppress tumor growth. Blood 1999; 94(7):2461–8.

Chapter 40
Challenges of Antiangiogenic Therapy of Tumors

Roberta Sarmiento, Raffaele Longo, and Giampietro Gasparini *

Keywords: endothelial cells, adverse events, clinical trial design, clinical study endpoints, vertical inhibiton, horizontal inhibition

Abstract: After the approval of the first antiangiogenic agent, bevacizumab, in February 2004, a humanized monoclonal antibody anti-vascular endothelial growth factor (VEGF), this category of new anticancer drugs expanded rapidly. At the end of 2006, the U.S. Food and Drug Administration (FDA) approved bevacizumab for therapy of advanced colorectal and non-small cell lung cancers (except squamous cell histotype), sunitinib for advanced-refractory renal cancer and for imatinib-resistant gastro-intestinal stromal tumors (GISTs), and sorafenib for recurrent advanced kidney cancer.

Twelve antiangiogenic drugs entered Phase III trials and at least another 15–20 are under evaluation in Phase I–II studies. Such a rapid clinical development of inhibitors of angiogenesis with different pharmacodynamic and pharmacokinetic characteristics opens a number of challenges, including: the identification of targets of choice; the optimal therapeutic strategy; the rational selection of the patients; proper study-design of clinical trials as well as the monitoring of drug efficacy; management of toxicity; and, finally, the determination of the disease stage to obtain the best benefit. All the above relevant issues are presented and discussed in this chapter.

Division of Medical Oncology, Azienda Ospedaliera "San Filippo Neri", Rome, Italy

*Correspondence Author:
Giampietro Gasparini, Division of Medical Oncology, Azienda Ospedaliera "San Filippo Neri", Via Martinotti, 20, 00135 Rome, Italy
E-mail: gasparini.oncology@tiscalinet.it

Introduction

In the early 1970s, Judah Folkman hypothesized for the first time the central role of angiogenesis in allowing tumor growth and metastasis [1]. The main scientific importance of research on angiogenesis is based on the knowledge of tumor biology involving: (i) the original paradigm that the microenvironment, namely the vascular endothelium, has an active role in promoting tumor proliferation; and (ii) "activated" endothelial cells (ECs) are viewed as a new target for anticancer therapy. Despite the modest clinical results of the first antiangiogenic agents tested in humans (SU-5416, TNP-470 and others), the approval in 2004 by the U.S. Food and Drug Administration (FDA) of bevacizumab, the first active selective inhibitor of angiogenesis, opened the door to this class of agents for therapy of a number of tumor types.

Carmeliet recently sustained that "angiogenesis research will probably change the face of medicine in the next decade, with more than 500 million people worldwide predicted to benefit from pro- or antiangiogenesis treatments" [2].

There is a great need for new effective anticancer treatments. In fact, as stated by Varmus, "cancer is minimally controlled by modern medicine" [3] as the age-adjusted mortality rate has only moderately improved in the past 25 years.

Angiogenesis research provided the evidence that it is feasible to attack cancer cells indirectly with the advantage of applying more "physiological" treatments based on natural (i.e., endogenous) angiogenesis inhibitors, and with the possibility of causing far fewer side-effects than standard cytotoxic agents. Based on the promising results of preclinical pharmacological studies, the combination of antiangiogenic drugs together with selective targeted therapy is now one of the more widely tested treatment paradigms in human studies, which include three "pure" antiangiogenic drugs (bevacizumab, sunitinib, sorafenib) and another five indirect inhibitors (bortezomib, temsirolimus, erlotinib, trastuzumab, thalidomide) already approved by the FDA (Table 40.1).

TABLE 40.1. Angiogenesis inhibitors approved for clinical use.

Date approved	Drug	Place	Disease
May 2003	Bortezomib	U.S. (FDA)	Multiple myeloma
December 2003	Thalidomide	Australia	Multiple myeloma
February 2004	Bevacizumab[a]	U.S. (FDA)	Colorectal cancer
November 2004	Erlotinib	U.S. (FDA)	Lung cancer
December 2004	Bevacizumab[a]	Switzerland	Colorectal cancer
December 2004	Macugen[a]	U.S. (FDA)	Macular degeneration
January 2005	Bevacizumab[a]	EMEA	Colorectal cancer
September 2005	Endostatin[a]	China	Lung cancer
December 2005	Sorafenib[a]	U.S. (FDA)	Kidney cancer
December 2005	Revlimid	U.S. (FDA)	Myelodysplastic sindrome
January 2006	Sunitinib[a]	U.S. (FDA)	Gastric (GIST), Kidney cancer
June 2006	Lucentis	U.S. (FDA)	Macular degeneration
October 2006	Bevacizumab[a]	U.S. (FDA)	Lung cancer

[a] "Pure" antiangiogenic agents

At least another 15–20 new angiogenesis inhibitors have entered early clinical studies (Table 40.2). As a consequence, the rapid development of this new class of anticancer agents has raised a number of relevant challenges for their optimal clinical use.

Possible Targets

More than 30 years of laboratory research have identified a number of potential targets for inhibition of angiogenesis and the discovery of more than 50 antiangiogenic agents acting with different mechanisms of action [4] (Fig. 40.1).

Targeting "Activated" Endothelial Cells

Angiomics research clearly demonstrated that the vascular endothelium within an invasive tumor is characterized by a distinct gene profile [5]. Endogenous antiangiogenic agents, such as angiostatin, endostatin, caplostatin and thrombospondin-1 (TSP1), selectively block the proliferation and migration of intratumoral vascular endothelium [6–9]. It is now recognized that recombinant human (rh)-endostatin blocks endothelial growth by complex mechanisms including: interaction with ATPase on the endothelial surface, angiomotion, binding the $\alpha_5\beta_1$ integrin, and inhibition of activity of metalloproteinases (MMPs) −2, −9 and −13 with an E-selectin-dependent action [10–13]. Finally, natural inhibitors of angiogenesis induce prolonged tumor dormancy in several experimental models [14, 15]. In September 2005, rh-endostatin was approved in China for the treatment of advanced non-small cell lung cancer (NSCLC) based on the positive results of the Phase III trial by Sun et al. [16]. Also, the TSP1 mimetic inhibitor ABT-510 entered early clinical evaluation as a single agent [17] or combined with fluoropyrimidines in patients with advanced solid malignancies [18], showing a good toxicity profile with a significant number of patients with prolonged stable disease.

TABLE 40.2. Drugs with antiangiogenic activity, December 2006. Not yet FDA approved.

Agent	Target
Phase II	
A6 (Angstrom Pharmaceuticals)	Binds to uPA cell surface receptor
ABT-510 (Abbott Laboratories)	Thrombospondin-1 receptor CD36
AG-013736 (Pfizer)	VEGF, PDGFR
AG3340 (Agouron Pharm.)	MMP inhibitor
AMG706 (Amgen)	VEGF, PDGFR, KITR, RetR
AS1404 (Antisoma)	Vascular distrupting; release TNF-α & vWF
Atiprimod (Callisto Pharm.)	VEGF, IL6
ATN-161 (Attenuon)	α 5 beta 1 antagonist
Combretastatin (Oxigene)	VE-Cadherin
E7820 (Elsai)	Inhibits integrin α 2 subunit on endothelium
EMD 121974 (EMD)	α v beta 3 and 5 antagonist
Genistein (McKesson Health Solutions)	Suppresses VEGF, neuropilin & MMP-9
INGN 241 (Intogen Therapeutics)	VEGF, MDA-7
Interleukin-12 (NCI)	Upregulates IP10
Panzem (2ME2) (EntreMed)	Inhibits tubulin polymerization
PI-88 (Progen Industries/Medigen)	bFGF, stimulates release of TSP1
PKC412 (Novartis)	VEGFR-2
Suramin (NCI)	IGF-I, EGFR, PDGFR, TGF-b, inhibits VEGF & bFGF
XL999 (Exelisis)	VEGFR, PDGFR, EGFR, Flt.3, Src
Phase III	
AZD2171 (AstraZeneca)	VEGFR1,-2,-3, PDGFR
Avastin (Genentech)	VEGF
BMS-275291 (Bristol Myers Squibb)	MMP inhibitor
CCI-779 (Wyeth)	MTOR inhibitor
CDP-791 (ImClone)	VEGFR-2/KDR
Celebrex (Pfizer)	Increases endostatin
LY317615 (Eli Lilly & Company)	VEGF
Neovastat (AE941) (Aeterna Zentaris)	MMP inhibitor
Sorafenib (BAY-43-9006) (Bayer/Onyx)	VEGFR-2 and PDGFR-beta
Vatalanib (PTK787) (Novartis)	VEGFR1, 2, PDGFR
RAD001 (Everolimus) (Novartis)	VEGFR/mTOR
Revlimid (CC5013) (Celgene)	VEGF, precursor endothelial cells
Sunitinib (SU11248) (Pfizer)	VEGFR1,2,3, PDGFR
Thalidomide (Celgene Corporation)	VEGF, precursor endothelial cells
VEGF Trap (Regeneron Pharm.)	VEGF
Bortezomib (PS341) (Millennium Pharm.)	VEGF
ZD6474 (AstraZeneca)	VEGFR-2, EGFR

The recent identification and characterization of the endothelial progenitor cell (EPC), and its capability to migrate from bone marrow into circulation and then into other tissues where it stimulates angiogenesis [19–21], may also open new selective anti-EPC therapeutic strategies.

Targeting Pericytes

Vessels in tumors are surrounded by pericytes that are mural cells differentiated from pools of c-kit+Sca-1+VEGFR1+perivascular progenitor cells mobilized from the bone marrow in response to the platelet-derived growth factor (PDGF)-BB [22, 23].

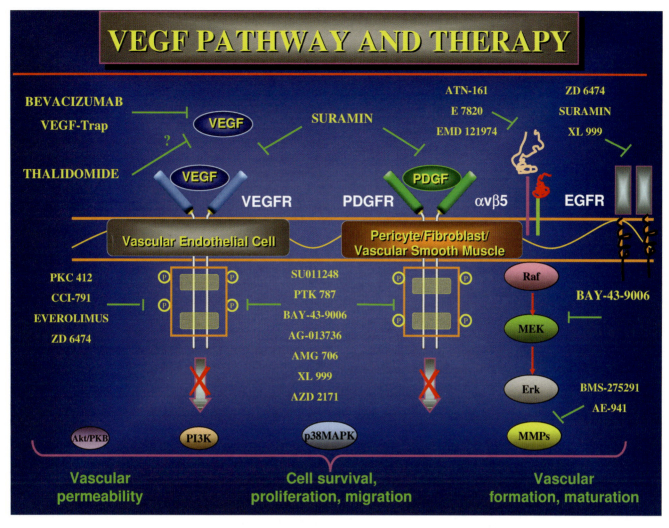

FIG. 40.1. VEGF pathway and therapy. Potential targets for inhibition of angiogenesis.

Pericytes under physiological conditions promote blood vessel stabilization by vascular endothelial growth factor (VEGF)-A, angiopoietin-1 (Ang1) and the Tie2 endothelial receptor tyrosine kinase (RTK). When PDGF is overexpressed, the tumor microvasculature is covered by a higher number of mural cells and tumor growth is accelerated [22]. Conversely, drugs targeting PDGFR-β inhibit the recruitment of pericytes, induce dilation of tumor vessels and stimulate ECs to enter apoptosis. The combined blockade of both VEGFR and PDGFR-β by multitargeted tyrosine kinase inhibitors, such as sunitinib, in advanced stage kidney cancer [24], increases the antiangiogenic effect, even in resistant late stage solid tumors [25].

Targeting Proangiogenic Factors

More than 20 angiogenesis promoting molecules have been identified and purified to date. Among these, VEGF is the most well-studied [26]. It is a key regulator of angiogenesis,

overexpressed in the majority of tumor types with prognostic significance, and more importantly, it is presently the selected target for most of the available antiangiogenic treatments. Among the anti-VEGF inhibitors, the most advanced in the clinic are the humanized monoclonal antibody bevacizumab and a number of selective VEGFR tyrosine kinase inhibitors, such as sunitinib, vatalanib, sorafenib, AZD-2171, CDP-791, LY-317615, AG-013736, PKC-412 and others (Table 40.2). Additional compounds targeting VEGF in clinical development include a VEGF trap and antibodies directed against VEGFR2 and/or VEGFR1. Anti-VEGF agents not only arrest proliferation and migration of endothelial cells, but also induce regression of existing vessels by inducing apoptosis and by suppressing the mobilization of EPCs from bone marrow [27–29]. The existence of three different mechanisms are also clinically relevant: (i) the improvement of cytotoxic drug delivery by normalizing tumor vasculature [30]; (ii) the reduction of vascular permeability and, consequently, of the interstitial fluid pressure; and (iii) the direct targeting of some malignant cells [31]. However, the focus on blocking only one proangiogenic factor is a potential drawback because tumors often make more than one type of angiogenesis-promoting molecule during progression [32]. Moreover, anti-VEGF drugs may increase the risk of vascular diseases, such as hypertension, heart attacks and coagulopathy [33, 34].

Targeting the Hypoxia Pathways and Nitric Oxide

Oxygen limitation is central in regulating angiogenesis, glucose metabolism, survival and tumor growth [35]. A major key factor is the hypoxia-inducible factor (HIF), which is a required transcriptional factor in nutrient stress signaling [36]. Briefly, HIF is a pleiotropic factor that controls the expression of two master angiogenic molecules: VEGF-A and angiopoietin-2 (Ang2). HIF binds to the *epo* gene and induces its transcription in hypoxia [37], and is activated at a cut-off point of approximately 5% oxygen (40 mmHg). Hypoxic stress also attenuates protein synthesis by the mammalian target of rapamycin (mTOR) pathway, the latter being a new anticancer target [38, 39]. Indeed, HIF also transcriptionally regulates the expression of the MET promoter that is a key regulator of invasive growth by driving cell motility and metastasis [40–44]. Somatic mutations of MET, or in the absence of mutations by the production of autocrine hepatocyte growth factor (HGF), are associated with increased tumor aggressiveness and malignant cell spread. Because MET is linked with cancer stem cells, its inhibition is an attractive target for anticancer therapy aimed to neutralize the growth of these cells [40]. An innovative approach coupling anti-viral-based gene therapy with engineered soluble decoy receptors of MET or HGF antagonists has offered proof of principle that MET inhibition is an effective anti-metastatic approach in mice [45, 46]. Additionally, HIF can be targeted by topotecan, and a new class of more selective agents such as PX-478 and decoy oligonucleotides are also directed against HIF-1α [47].

Nitric oxide (NO) is a multifunctional gaseous molecule and a highly reactive free radical that regulates several vascular functions, such as mediating VEGF- and Ang1-induced angiogenesis in vivo [48]. NO exposure directly stimulates the proliferation and migration of endothelial cells through the 5GC-cGMP pathway [49–51], transiently activates HIF-1α and inhibits the secretion of TSP1 [52]. NO also enhances vascular permeability and interstitial fluid pressure in tumors [53, 54]. Thus, inhibition of NO signaling is another potential antiangiogenic therapeutic strategy. A number of compounds that interfere with NO production include cavtratin, a caveolin-1 derived peptide, the nitric oxide synthase (NOS) inhibitor L-NNA, and NO-donating non-steroidal anti-inflammatory drugs (NSAIDs) [55–57]. Combinations of NOS inhibitors and cyclooxygenase (COX)-2 inhibitors have been shown to produce comprehensive chemopreventive effects against colon carcinogenesis in preclinical models [58].

Therapeutic Strategies

Monoclonal Antibodies (Moabs)

For many decades, it was recognized that antibodies that react with tumor-associated antigens could be selective agents for the treatment of cancer. In the past years, the difficulty of achieving their uniform production was a major obstacle to their clinical application. However, the ability to make humanized, chimeric monoclonal antibodies (MoAbs) has now reduced the limitations associated with the human anti-immunoglobulin (IG) immune response (e.g., human anti-mouse MoAb or anti-chimera MoAb responses) in humans, and thus has made possible repeated treatments over time [59].

The major limitation to the clinical application of MoAbs is the target antigen. Most MoAbs are not tumor-specific, so there is the potential for toxicity against normal tissues [60]. Because of their inability to pass through the cellular membrane, MoAbs can only act on molecules that are expressed on the cell surface or that are secreted [59]. Synergistic effect of the combination of MoAbs with conventional chemotherapy has been described, most probably due to some direct action on the apoptotic pathway [60, 61].

Bevacizumab is the first MoAb developed against the secreted proangiogenic protein VEGF-A, and it has been demonstrated to improve overall survival when combined with fluoropyrimidine-based chemotherapy in patients with metastatic colorectal cancer [62]. MoAbs against the extracellular domain of VEGF receptors have also been developed. DC101, a rat anti-mouse Flk-1 antibody in tumor xenografts inhibited the growth of human cancer cell lines such as ovarian, pancreatic, and renal cells However, clinical data are not yet available [63].

Small-molecule Inhibitors

Many small molecule agents targeting growth factor receptors and their signaling pathways have been developed and have entered clinical trials. Key signaling molecules, such as protein tyrosine kinases, have proven to be good targets for small molecule inhibitors that compete with ATP and that inhibit the kinase activity [64–66]. RTKs and non-RTKs are crucial mediators in signaling pathways of cell proliferation, differentiation, migration, angiogenesis, cell-cycle regulation and others. The biological role of tyrosine kinases differs, depending on the receptor and the cell type where they are expressed. Therefore, inhibition of tyrosine kinases is expected to elicit complex biological responses and thus elicit different effects.

Taking anti-VEGF strategies as a model for the development of such rationally-designed targeted therapies, a number of small molecule compounds inhibiting VEGFR tyrosine kinase have been developed. Semaxanib (SU-5416), the first competitive tyrosine kinase inhibitor targeting VEGFR2, VEGFR1, and c-kit, was investigated in Phase III studies. The negative findings in advanced colorectal cancer stopped

TABLE 40.3. Differences between MoAbs and SMs.

	MoAbs	TKIs
RR	18–24%	<20%
• Administration	I.V.	P.O.
• Molecular weigth	~150 kDa (large proteins)	~500 Da (sinth chemicals)
• CNS barriers	Inefficiency?	High efficiency
• **Tissues penetration, blood clearance**	**Low efficiency**	**High efficiency**
• **Half-life**	**Prolonged (3–21 days)**	**Short (26–48 h)**
Plasma variability	Low	High (oral administration)
• Degradation variability	Low (proteins)	High (chemicals)
• Sites of action	Cellular surface or secreted molecules	Regardless targets location
• **Immuno system**	**Yes (ADCC)**	**No**
• Synergic effect with CT	Yes	Variable

its further development [67]. Sunitinib is an oral, multitarget, antiangiogenic compound that inhibits the tyrosine kinase receptors of VEGF, basic fibroblast growth factor (bFGF), PDGF, and c-kit. Clinical results are positive for therapy in advanced renal cancer and resistant GISTs [68]. ZD6474 is an oral low molecular weight inhibitor of the VEGFR2 tyrosine kinase and it is currently in Phase III trials [69]. Vatalanib is another TKI that binds directly to the ATP-binding sites of VEGFR1 and VEGFR2 with good oral bioavailability, and is currently being investigated in ongoing Phase III studies [70, 71].

It should be emphasized that MoAbs and small molecule inhibitors differ in several pharmacological properties. First, the large molecular weight of MoAbs is the cause of their inefficient delivery into the brain because of the blood–brain barrier. Differences in terms of pharmacokinetics are also present such that MoAbs half-lives are much longer than those of small molecule compounds. Small molecule inhibitors diffuse into the cytoplasm, and can therefore be developed to target molecules regardless of their cellular location, as compared to MoAbs that can only target the cell surface or secreted molecules. Given the importance of the antitumor immunity for the destruction of tumor cells, another relevant difference between MoAbs and small molecule agents is the effect on the immune system, because small molecule agents do not directly act on it (Table 3). A final consideration concerns the huge monetary cost in the development of MoAbs compared with small molecule inhibitors, the latter being less expensive and more convenient to administer with similar activity [61].

Natural Peptide Inhibitors

A third approach for antiangiogenic therapeutic strategies is the administration of natural peptide inhibitors such as angiostatin and endostatin, internal fragments of larger proteins that specifically inhibit proliferation and migration of the "activated" endothelium, but not affecting resting endothelial cells,

nor tumor cells [72]. Angiostatin induces tumor dormancy in both the primary tumor and metastasis [6]. To be active, these compounds require continuous therapy for a lengthy period of time and up to now have been prevalently tested as monotherapy. Based on the promising results obtained by Sun et al. [16] with rh-endostatin, further studies are warranted to test the potential active combinations of endostatin with conventional chemotherapy or other antiangiogenic agents.

Prevention and Management of Adverse Events of Antiangiogenic Therapy

As compared to standard chemotherapy, most of the antiangiogenic agents are generally well tolerated, but may have a variety of adverse effects that are typically downstream events which arise from the suppression of signaling pathways involved in the regulation and maintenance of the microvasculature. Antiangiogenic agents may induce unusual toxicities that may be unexpected for oncologists and which may induce unnecessary treatment delays or interruptions. For example, bevacizumab therapy can be accompanied by hypertension, coagulatory alterations and, rarely, gastrointestinal (GI) perforation. To prevent the occurrence of the above toxicity, the selection of patients, the identification of potential risk factors and the careful monitoring of patients on treatment are of utmost importance. GI perforation is potentially fatal. A large community-based observational registry (BRiTE study) [73], which involved approximately 2,000 patients receiving bevacizumab and first-line chemotherapy for metastatic colorectal cancer, had an incidence of 1.7 %. The majority of cases occurred within the first 2 months of therapy, and at least one of the following risk factors have been recognized in two–thirds of the patients: unresected primary tumor, carcinomatosis, history of pelvic radiation therapy, or chronic bowel inflammatory diseases. Kamba et al. [74] found that anti-VEGF therapy causes regression of 34–46% of capillaries of the intestinal villi in mice, which in the presence of the above risk factors may be involved in the pathogenesis of GI perforation. A similar pathogenetic mechanism has been suggested to be responsible for the loss of thyroid capillaries and of the sunitinib-related hypothyroidism [74]. Sunitinib is also associated with fatigue, diarrhea, coagulopathies, hypothyroidism and hand–foot skin syndrome [24]. Sorafenib therapy can cause diarrhea, fatigue, hand-foot skin reaction and sensory neuropathy [75, 76].

The above adverse effects, although usually manageable with conventional medical approaches, differ from those usually found with chemotherapy. Therefore, it is important that recommendation guidelines be outlined for the prevention and treatment of antiangiogenic therapy-related toxicity. Also, patients that need major surgery require careful monitoring and adequate periods of treatment interruption.

The specific toxicological profile of inhibitors of angiogenesis also suggests that their combination with standard chemotherapy needs careful consideration and adequate study design to prevent additional potentially high-grade or irreversible toxicities [67].

Selection of the Patients

Clinical studies conducted with the administration of anti-VEGF compounds, such as bevacizumab and sunitinib, documented improvement in clinical outcomes even though the patient selection criteria to predict treatment responsiveness were not defined for the specific target. Thus, a rational strategy to improve the effectiveness and to minimize unnecessary side effects of antiangiogenic compounds is to determine the predictive indicators of response. Among potential predictive tools for patient selection, there is evidence for three distinct classes that include: (i) tumor and/or proteins (pro- and/or anti-angiogenic factors); (ii) circulating endothelial cells (CECs) and CEPs; and (iii) dynamic vascular imaging techniques.

Surrogate Biomarkers Predictive of Response

The potential role of tumor and/or circulating VEGF as definitive biomarkers were evaluated in clinical trials with bevacizumab and VEGF receptor tyrosine kinase inhibitors, with most of the published studies showing negative results or a lack of correlation [77]. The findings from these studies suggest that response to antiangiogenic therapy may not be related to pretreatment tumor VEGF expression or circulating VEGF levels. However, there are some limitations to the above studies, such as the small sample sizes, the low objective responses (for the breast cancer trial) and the measurement of VEGF in primary tumors of patients with metastatic disease. In addition, a number of methodological issues have to be considered that include the lack of a 'gold standard' detection method; the presence of VEGF isoforms and different VEGF non-tumor sources such as platelets and leukocytes, the binding of VEGF to serum proteins [28], and the variation in processing times and storage procedures. In fact, there is no definitive proof that circulating VEGF levels directly correlate with tumor angiogenesis or activation of the VEGF pathway [77, 28, 78], and quantification of VEGF can be confounded by a number of factors (Table 4). Ultimately the lack of a predefined, clinically-relevant, cut-off value is a significant hindrance to the clinical utility of VEGF measurements for therapy selection.

Given the diversity of the VEGF signaling network, it is important to consider VEGF expression in the context of other determinants of molecular activity, such as specific isoforms, other ligands, receptors and co-receptors, downstream components, and cross-talk with other molecular pathways. Recent data suggest that VEGF bioavailability, not total expression, could determine the response to VEGF inhibition [79]. The pattern of specific VEGF-A isoform expression might

TABLE 40.4. Summary of vascular endothelial growth factor (VEGF) possible detection methods.

Method and description	Comments
Immunohistochemistry (IHC) detects VEGF protein expression in whole tissue sections (usually formalin-fixed, paraffin-embedded tissue)	• Possible to differentiate between tumor and non-tumor VEGF expression • Easy and low cost • No standardized methodology of scoring procedure • Results variable and subjective
Enzyme-linked immunosorbent assay (ELISA) and chemiluminescence immunosorbent assay (ICMA) detect VEGF protein expression in tissue homogenate (fresh-frozen tissue), serum, or plasma	• Serum and plasma measurements convenient vs. tissue samples • Can be automated for high throughput • Cannot distinguish between tumor and non-tumor sources of VEGF • Serum measurements may be confounded by release of VEGF from platelets
Western blotting detects VEGF protein expression in tissue homogenate (fresh-frozen tissue)	• Cannot distinguish between tumor and non-tumor sources of VEGF • More complex to perform than IHC
In-situ hybridization (ISH) detects VEGF mRNA in whole tissue sections (ideally, fresh-frozen tissue)	• Can distinguish between tumor and non-tumor VEGF expression • May not relate to VEGF protein expression • More complex to perform than IHC
Northern blotting detects VEGF mRNA from tissue homogenates (fresh-frozen tissue)	• Cannot distinguish between tumor and non-tumor VEGF expression • May not relate to VEGF protein expression • More complex to perform than IHC
Reverse-transcription polymerase chain reaction (RT-PCR) detects VEGF mRNA in tissue homogenates (usually fresh-frozen)	• Quantitative method that can be automated for high-throughput • Cannot distinguish between tumor and non-tumor sources of VEGF • Sensitive to contamination • May not relate directly to VEGF protein expression
RNase protection assay detects VEGF mRNA in cellular extracts (tissue or circulating)	• Cannot distinguish between tumor and non-tumor VEGF expression • May not relate directly to VEGF protein expression • Relatively complex to perform

also influence the response to therapy without substantially changing total VEGF mRNA expression [80]. Finally, the identification of a valid predictive marker is likely to hinge on elucidating the precise mechanism by which anti-VEGF agents exert their effect in vivo.

Furthermore, studies by Yan et al. [81] demonstrated that metalloproteinase-9/neutrophil gelatinase activity is detectable in the urine from cancer patients and that it is correlated to disease status, stage and prognosis [82]. Because urinary detection is possible also in early tumor development, this assay may be potentially useful in predicting even non-invasive stages.

Circulating Levels of CECs and/or CEPs

In a preclinical model, ZD-6474 and ZD-6126 had differential effects, causing a dose-dependent increase in mature CECs but not CEPs, accompanied by a decrease in tumor microvessel density (MVD) and tumor volume after 3 days of treatment [83]. The apoptotic fraction of mobilized CEC was not significantly increased by treatment [83]. In locally advanced rectal carcinoma, one bevacizumab infusion reduced the percentage of viable CECs at day 3 in all patients with high CEC counts at baseline. However, patients treated with high bevacizumab doses maintained the same viable CEC count at day 12 after the first bevacizumab infusion despite the significant increase in levels of plasma VEGF and platelet-induced growth factor (PlGF) [84, 85].

In patients with advanced breast cancer receiving a metronomic schedule of chemotherapy, CECs decreased in those without clinical benefit (defined as a clinical response or stable disease) compared to those who had clinical benefit ($P = 0.015$). This difference was due to an increased fraction of apoptotic CECs. After a median follow-up of 17.4 months, univariate and multivariate analyses indicated that CEC values greater than $11/\mu l$ after 2 months of therapy were associated with significantly longer progression-free survival ($P = 0.001$) and overall survival ($P = 0.005$). In the same study, there was neither clinical benefit nor effect on CEC or on CEP count and viability in the chemotherapy plus thalidomide treatment group [86]. One proposed mechanism for the observed changes is that antiangiogenic treatments damage and/or kill endothelial cells, either in circulation or in tumor-associated blood vessels, with their subsequent release into the circulation [86]. In another study of 32 patients with advanced solid tumors treated with ZD-6126, CEC numbers increased to a maximum level over baseline after a median of 4 h after infusion. The ZD-6126 dose had no apparent correlation with the number of CECs [87].

In light of these findings, several questions remain to be addressed: (i) the sensitivity and reproducibility of the methods employed; (ii) the doubt if tumors may mobilize sufficient CEPs to be detected in clinical samples as surrogate predictive markers; (iii) the best antigen panel to characterize these cells; and (iv) the role of viable and non-viable cells [88].

Platelets

It has been recently recognized that blood platelets and megakaryocytes contain pro- and antiangiogenic factors in distinct α granules with important implications for regulating angiogenesis. Immunofluorescence microscopy technique revealed the presence of VEGF, bFGF, and PlGF as proangiogenic molecules and endostatin and TSP1 as natural inhibitors in separate α granules. Selective proteinase-activated receptor (PAR) agonists stimulated the release of endostatin-combining α granules [89]. These preliminary observations suggest the possibility that the analysis of angiogenic products from platelets may predict the prevalence of angiogenic pathways before and after antiangiogenic therapy.

Angiogenesis Genetic Profiling and Proteomics

Germ-line polymorphisms in genes critical to the regulation of angiogenesis include VEGF and endothelial nitric oxide synthase (eNOS). Schneider et al. [90] analyzed breast cancer samples and found that the host angiogenic genotype imprints the tumor genotype. Such determinations may be useful in the future to personalize treatment with antiangiogenic therapy.

Godl et al. [91] demonstrated for the first time that proteomic characterization of an angiogenesis inhibitor, namely SU-6668, has multiple impact on cellular kinase signaling. By employing immobilized SU-6668 for the affinity capture of cellular targets using mass spectrometry, the authors identified novel drug targets such as the Aurora kinase and TANK-binding kinase-1 (involved in antiviral and inflammatory responses). These results demonstrate the potential use of proteomics to identify more comprehensive panels of targets involved in the mechanisms of action of VEGFR2 and PDGFR tyrosine kinase inhibitors and to assess the pharmacodynamic properties of this class of compounds.

Vascular Imaging Techniques

These techniques are non-invasive and can assess a large volume of tumor to determine both tumor vascularity (a marker of antiangiogenesis efficacy) and size (a marker of cytotoxic efficacy). Evidence to support imaging as a surrogate marker for treatment response is based on three techniques: dynamic contrast-enhanced magnetic resonance imaging (DCE-MRI), PET scan, and dynamic CT scan [92, 93]. DCE-MRI yields parameters related to tissue perfusion by T2-methods and permeability by T1-methods and have been evaluated in a number of phase I/II studies [92, 93]. Morgan et al. [94] identified a significant inverse correlation between the reduced percentage of baseline bidirectional transfer constant (Ki), a measure of tumor permeability and vascularity, and increased vatalanib plasma levels in 26 patients with metastatic colorectal cancer. A dose-dependent reduction in contrast enhancement was evident at all dose levels. There was also a relationship between reduction in contrast enhancement and disease progression. The authors identified a dose associated with at least 40% reduction in contrast enhancement (60% baseline Ki) and with non-progressive disease. Another Phase I study evaluated the role of DCE-MRI in 31 patients with advanced solid tumors treated with AG-013736. This drug caused a significant and dose-dependent decreases in DCE-MRI vascular parameters by day 2 of treatment, but the correlation was not evaluated between vascularity and clinical response [95].

In the study by Wedam et al. [96], a decrease of the Ki transfer was observed after the first infusion of bevacizumab and continued during chemotherapy. However, no significant differences in any of vascular parameters were found between clinical responders and non-responders. In addition, the phosphorylation status of VEGFR2, at tyrosine residues 951 and 996, decreased after the first infusion of bevacizumab. There was no significant change in MVD, VEGF-A, and VEGFR2 expression in the experimental group of patients. Despite these interesting results, there are several issues that need to be taken into account, including: (i) the lack of 3 dimensional (3D) protocols; (ii) the poor standardization of the methods used; (iii) tumor heterogeneity; (iv) the reproducibility of the measurements; and (v) the need for validation in large prospective trials [92, 93].

PET is a sensitive and quantitative technique that has been used in clinical studies to assess tumor blood flow with oxygen-labeled water and tumor metabolism with fluoro-labeled fluorodeoxyglucose. The use of oxygen-labeled water offers several properties for the measurement of blood flow. It is freely diffusible, has a short half-life of 2 min, and has favorable dosimetric properties. However, there are potential limitations, such as the partial volume effects in small tumors, the "spill over" or "spill in" phenomenon from surrounding structures with high blood flow, the use of ionizing radiation limiting the number of studies that can be performed, and the short-life of many PET isotopes, requiring synthesis of the relevant compound. Tumors and, particularly, necrotic areas may not have a uniform exchange of water between blood and tissue with a lower volume of distribution of the tracer. The heterogeneity of delivery of drugs to solid tumors may lead to variability in the results obtained from PET scans and other imaging modalities [92, 93].

Dynamic or functional CT scans assess tumor perfusion, relative blood volume, capillary permeability, and leakage. This technique is simple, widely available, and reproducible. It has been validated against the oxygen-labeled water PET scan. However, the reduction in tumor perfusion by antiangiogenic compounds has yet to be demonstrated by dynamic CT in clinical studies. The reason for this is probably due to the lack of commercially available software to perform more precise quantitative analyses for calculating vascular parameters. This limitation is likely to be overcome by the rapid development of new CT software, such as the 3D assessment of spiral CT, and with the use of monoclonal anti-VEGF antibodies as possible tracers [92, 93].

Finally, although a number of surrogate markers for antiangiogenic therapy are currently being investigated, none has been clinically validated with standardized methods in prospective studies. Multiparametric studies evaluating more efficient methods of assessing angiogenic activity (i.e., biomarkers, CECs, CEPs and platelets, and imaging techniques) are warranted to improve the prediction of efficacy. A clearer understanding of the mechanisms of action of anti-VEGF agents and their relationship to tumor VEGF status, particularly in combination studies with conventional chemotherapy, should help define the optimal use of these new antiangiogenic compounds. There is also a need to determine different biomarkers for different agents, or to define a collective set of surrogate markers or "signature" that would be a better representative of predicting treatment response as compared to using only a single factor or parameter [77].

Study Design and End-points of Clinical Studies

The study design of most randomized clinical trials of first generation angiogenesis inhibitors consisted of the comparison of a standard chemotherapeutic regimen with or without the test compound. Clinical development of SU5416 was one of the first examples of this strategy, but the study was stopped because the interim analysis showed increased toxicity without any additional clinical benefit in colorectal cancer patients treated with this antiangiogenic agent.

Traditional phase I studies for chemotherapeutic agents are designed to find the maximum tolerated dose (MTD) and the dose-limiting toxicity (DLT), based on the assumption that there is a direct correlation among dose, activity and toxicity. With antiangiogenic agents, there is not a linear relationship between dose and activity, but toxicity may increase with the dose. In addition, MTD may not be defined if the drug has a wide therapeutic ratio. The phase I study design could be improved in several ways, including identifying the effects on the target, measuring "surrogates" for biological activity, and quantifying pharmacokinetic parameters. As proposed by Korn et al. [97], the identification of the target effect through pharmacodynamic studies is important for selecting the minimum target inhibiting dose (MTID) and the optimal schedule of administration. However, this methodology presents several issues, such as the difficulty in identifying the pivotal biological target and the lack of sensitive, specific and validated pharmacodynamic assays. Another issue is the lack of adequate tissue samples on which to measure the target, often requiring the use of "surrogate" tissues, such as skin or peripheral leukocytes. However, the biochemical effects of such target-based agents in the tumor may not be predicted by analysis of surrogate tissues. Two novel Phase I designs are based on the de-escalation method to determine the biological modulatory dose as tested for SU-5416 or on the patient-specific dosing paradigm proposed by Rogatko et al. [98], based on measurable characteristics of each single patient that may be predictive of adverse response to a specific therapy.

The standard end-point of phase II studies is to demonstrate the disease-oriented activity of a new drug and to evaluate the objective response rate by standard criteria (e.g., RECIST). In most cases, this approach is not adequate for the development of antiangiogenic agents as they are cytostatic and exhibit a low evidence of tumor shrinkage [99, 100]. Non-cytotoxic compounds need different end-points for assessment of activity since their action is mainly to control tumor growth. Therefore, considering response rate as the only end-point for a phase II study with antiangiogenic agents results in the potential risk of underestimating or rejecting potentially active drugs. A major challenge is to define alternative statistical designs or new end-points, such as time to progression (TTP), progression-free-survival (PFS), overall survival (OS), early progression rate and growth modulation index. Rubinstein et al. [101] suggested the possibility of developing "Phase II screening trials" consisting of non-definitive randomized comparisons of experimental regimens to standard treatments by adjusting the false positive error rates (type 1 error = α) and false negative error rates (type II error = β), so that the benefit of targeted therapy may be appropriate without the need of large sample sizes. A possible alternative methodological approach is the randomized discontinuation design that can distinguish the stabilization of disease due to natural history from that of treatment-related, but this requires a large number of cases [102].

Phase III trials with antiangiogenic agents have similar designs and the same end-points in terms of efficacy (OS, quality of life, etc.) to those with cytotoxic drugs. However, the key issue that remains is the selection of the patients. Molecular characterization of tumors may play a pivotal role for patient selection and may become an important stratification variable. Betensky et al. [103] showed that incorrect assumptions or lack of information regarding the molecular characteristics of tumors can lead to false negative results even in large randomized phase III trials, as observed with MMP inhibitors. Finally, patients with a larger tumor burden, pre-treated with several lines of chemotherapy and resistant to conventional therapies, may receive minimal benefit from antiangiogenic therapy. For this reason, the best option to validate antiangiogenic compounds is probably in the adjuvant setting.

In conclusion, there are several potential ways to improve study designs. First, the determination of validated surrogate biomarkers is necessary, such as the expression and/or the activity of the therapeutic target. Prospective multiparametric studies are needed to identify the method of choice that is feasible, reproducible and standardized prior to wide clinical use. Second, the direct effects of antiangiogenic therapy on mature CECs and EPCs and the value of new vascular imaging techniques can be evaluated as potential indicators for quantifying clinical and vascular responses to these agents. Third, pharmacodynamic indicators should confirm a therapeutic concentration of the drug at the level of the molecular target.

It may be difficult to demonstrate a conventional antitumor response (i.e., objective response) with antiangiogenic therapies in cohorts of patients with advanced disease resistant to conventional therapy. For many angiogenesis inhibitors perhaps the more appropriate clinical setting may be chemoprevention, adjuvant or maintenance therapy, once satisfactory

tolerability and activity has been proven by phase I studies in patients with advanced disease.

Presently, there are two different strategic assumptions regarding angiogenesis inhibitors. Should they be considered unselective anticancer agents, sustained by the rationale that all tumors need to acquire the angiogenic phenotype for their growth and metastasis? Or, on the contrary, are angiogenesis inhibitors selective molecularly targeted agents? The choice of the above therapeutic paradigm is important for the clinical study design (Table 5).

TABLE 40.5. Overall strategy of clinical development of angiogenesis inhibitors (AIs) [77].

	Hypothesis I	Hypothesis II
Assumption	AIs are unselective agents	AIs are selective targeted agents
Statement	AIs active in all cancers	AIs active only in presence of the specific angiogenic target
Strategy	Disease-oriented studies aimed to verify the efficacy and tolerability of AIs for each tumor type	Target-selected studies aimed to select the patients to be treated and the optimal schedule of administration with surrogate endpoints
Study Design	**Phase I** • Pharmacokinetics and definition of DLTs and MTD still relevant end-points	**Phase I** • Pharmacokinetic and pharmacodynamic studies • Toxicity may not be the appropriate end-point • Definition of the optimal biological dosage range • Identification of surrogate end-points (biomarkers, genetic tests, dynamic imaging techniques, etc)
	Phase II • Assessment of activity (RR, TTP) in unselected patients with advanced disease • End-point: RR	**Phase II-Randomized** • Randomized trials of AI alone or in combination *vs* standard therapy with the patients selected by the target • End-point: TTP • Validation of the determination of surrogate end-points
	Phase III • Large randomized prospective trials to compare efficacy (TTP and OS) and QoL of AIs alone or combined with chemotherapy *vs* standard chemotherapy • End-points: OS and QoL	**Phase III** • AI alone or in combination *vs* standard chemotherapy with patients selected for the target • Standardization with quality controls of the determination methods for surrogate biomarkers and/or imaging techniques • End-points: OS and QoL

RR Response rate; *TTP* time to tumor progression; *OS* overall survival; *DLT* dose-limiting toxicity; *MTD* maximum tolerated dose; *QoL* quality of life

New Strategies for Antiangiogenic Therapy Combinations

Broad-spectrum Targeting

Recently, Folkman proposed the paradigm to stop tumor growth indefinitely by the use of a broad-spectrum single multitarget agent capable simultaneously of blocking several angiogenesis pathways or, alternatively, by combinations of different selective drugs. The pros and cons of the above two therapeutic strategies are summarized in Table 6.

An example of the first strategy is sunitinib, a multitarget tyrosine kinase inhibitor acting on VEGFR, PDGFR, KIT and FLT3. The binding of sunitinib to the above receptors is an effective therapy for imatinib-resistant GISTs, blocking the proliferation of vascular endothelial cells via VEGFR, pericytes via PDGFR, and GIST tumor cells via PDGFR and KIT [24]. Sorafenib is also an orally active multitarget drug effective for therapy of metastatic renal cancer through inhibition of several receptors (VEGFR2, PDGFR, FLT-3 and KIT) and the B-Raf pathway [75, 76].

Examples of the second strategy, the use of polytreatments, include combinations of bevacizumab and the anti-epidermal growth factor receptor (EGFR) agents taking into account that angiogenesis is linked to other molecular pathways. Preclinical studies have demonstrated the important crosstalk between EGFR and VEGFR signaling pathways. They exert effects both directly and indirectly on tumor cells; thus, combining drugs directed against both targets may confer additional clinical benefit. EGFR has been detected on endothelial cells of the tumor vasculature [104]. Co-expression of EGFR and the transforming growth factor-α has been correlated with increased vascularity in invasive breast cancer [105]. VEGF is also down-regulated by EGFR inhibition [106, 107], and a recent study suggests that blockade of VEGF may also inhibit the EGFR autocrine signaling event [108]. A number of preclinical studies investigating the antitumor activity of combined anti-EGFR and anti-VEGF agents have shown promising results [109–113]. Furthermore, VEGF blockade is critical in preventing resistance to EGFR inhibition. The use of agents such as erlotinib and bevacizumab that target different signaling pathways and hence affect different cell types (tumor cells and endothelial cells) may abide by differ-

ent rules than standard cytotoxic chemotherapy because of the significant crosstalk that exists among the pathways in different cell types.

These encouraging data have led to the initiation of a number of clinical studies evaluating the combination of erlotinib with bevacizumab in a range of tumor types, including Phase II trials in renal cell carcinoma [114] and metastatic breast cancer [115], as well as a Phase I study in patients with head and neck squamous cell carcinoma [116].

Herbst et al. [117] evaluated the combination of bevacizumab and erlotinib in 40 patients (34 patients at the Phase II dose) with previously treated NSCLC. Eight patients (20.0%) had partial response and 26 (65.0%) had stable disease as their best treatment response. The median OS for the 34 patients treated at the Phase II dose was 12.6 months, with a PFS of 6.2 months. The most common adverse events were mild to moderate rash, diarrhea and proteinuria. In another multicenter Phase II trial, 63 patients with metastatic clear-cell renal carcinoma were treated with bevacizumab 10 mg/kg intravenously every 2 weeks and erlotinib 150 mg orally daily. Fifteen (25%) of 59 assessable patients had objective response to treatment and an additional 36 patients (61%) had stable disease after 8 weeks of treatment. The median and 1-year PFS was 11 months and 43%, respectively. After a median follow-up of 15 months, median OS has not been reached and survival at 18 months was 60%. Treatment was generally well tolerated with only two patients discontinued on treatment because of toxicity (skin rash). Grade 1/2 skin rash and diarrhea were the most frequent treatment-related toxicities [118].

Encouraging results have been reported with the combination of two different monoclonal antibodies with or without chemotherapy. The BOND II Trial evaluated the combination of cetuximab/bevacizumab +/− irinotecan in metastatic colorectal cancer patients after irinotecan failure. Preliminary data showed a better efficacy in the experimental arm in terms of response rate (37% vs 20%) and median TTP (7.9 months vs 5.6 months) [119]. Based on preclinical results showing co-upregulation of VEGF and HER-2/neu in breast cancer, a recent Phase I/II trial explored the combination of trastuzumab and bevacizumab in relapsed metastatic breast cancer HER-2 positive [120]. A further interesting approach is to combine TKIs targeting different molecular signaling pathways. For example, a recent European Organization for Research and Treatment of Cancer (EORTC) trial explored the combination of imatinib and sunitinib alone as second-line therapy in patients with resistant and/or refractory GISTs. In this study, the combination of different TKIs in the clinic needs to be carefully evaluated because of the risk of additional side effects without enhanced activity, which is probably due to competitive binding on the same targets.

Metronomic Chemotherapy

Another therapeutic strategy is based on the use of antiangiogenic schedules of cytotoxic agents (i.e., metronomic

TABLE 40.6. Single multitargeted agent (SMA) vs combination of highly selective compounds (CSCs).

	SMA	CSCs
Rationale	• Inhibition of multiple signaling pathways	• Additive or synergistic effects among the drugs
Pros	• Single drug administration	• High selectivity for the targets • Cross-resistance
Con	• Possible acquired resistance • Cross-reactivity with normal tissues	• Possibile Pharmacokinetics or pharmacodynamics interactions

chemotherapy) aimed to block endothelial cell proliferation. The most studied cytotoxic agent is oral cyclophosphamide, which is more active if combined with selective antiangiogenic agents such as the TSP1 peptide ABT-510, thalidomide, celecoxib, or anti-VEGF agents [121].

In a recent phase II study in pediatric cancer patients with recurrent or progressive poor prognosis tumors for which no curative therapy remained, Kieran et al, [122] tested the continuous administration of oral thalidomide and celecoxib with alternating etoposide and cyclophosphamide. This therapeutic approach was well tolerated and 40% of the patients had clinical benefit with 25% of all cases continuing to be progression-free for more than 123 weeks. Elevated circulating levels of TSP1 correlated with prolonged response.

Bidirectional Action of the Angiogenic Balance

The angiogenic activity of a tumor is the result of the net balance between the pro- and antiangiogenic molecules. It is therefore reasonable to hypothesize that the concurrent use of an anti proangiogenic molecule (i.e., anti-VEGF) with that of a naturally occurring angiogenesis inhibitor (i.e., endostatin) should be more effective than the use of either single class of agents alone. This strategy of bidirectional targeting of both pro- and antiangiogenic molecules attempts to restore the angiogenic balance. It is now feasible to investigate this approach in the clinic due to the approval of rh-endostatin in China for the treatment of NSCLC, a type of tumor that also responds to bevacizumab therapy [16]; thus, it remains to be determined whether the potential strategy of bidirectional targeting can be represented by the combination of rh-endostatin and bevacizumab to effectively treat NSCLC.

Conclusions

As outlined in the American Society of Clinical Oncology's report on the "Clinical cancer advances of 2006" [123], antiangiogenic drugs have been demonstrated to be effective new targeted therapies, especially for patients with GIST linked to mutation status and with advanced kidney cancer (sunitinib). Also, studies with temsirolimus, a targeted inhibitor of mTOR (a protein regulating both tumor cell growth and angiogenesis), improved survival in advanced, high-risk, kidney cancer.

Postulating that antiangiogenic therapy is considered a targeted, non-cytotoxic treatment approach, a situation that is of concern deals with the fact that most Phase I studies of targeted anticancer agents performed in the last 5 years used traditional end-points for selection of the recommended dosing schedules for subsequent clinical trials. Additionally, measures of molecular drug effects in tumors or surrogate tissues by functional imaging techniques are not routinely incorporated into the study design and are rarely the basis for dose selection [124]. These limitations represent the need for a more

rational approach in the clinical development of antiangiogenic agents. Genomic characterization of individual tumors must play a greater role in personalization of treatments and in the drug development process. A rational patient selection guideline beginning with Phase I/II studies may confer enormous advantages in trial design allowing for more efficient and cost-effective drug development. This strategy should be based on predictive factors for treatment-specific regimens capable of identifying the most appropriate target population for the subsequent larger Phase III trials.

In future years, the first step towards treatment will be to characterize the angiogenesis regulatory genes in patients in which the differential expression of these genes may determine the way an individual responds to treatment with selective antiangiogenic drugs. These analyses may help researchers understand, for example, why bevacizumab monotherapy is poorly effective in humans yet effective in rodents, and why the same drug in combination with chemotherapy provides overall survival benefit in certain but not all tumor types.

Assuming that inhibition of VEGF is necessary but not sufficient in advanced tumors to permanently halt angiogenesis, a multitargeted strategy should be applied and at least two modalities are suggested: first, *vertical inhibition* of the entire VEGF pathway cascade by using neutralizing antibodies for the ligands and the external VEGFR domains in combination with tyrosine kinase inhibitors to block the transduction of signals to the nucleus as well as removing the key VEGF-stimulating mechanisms via interfering with hypoxia, NO and related molecules;and second, *horizontal inhibition* by concurrently inhibiting several proangiogenic factors (i.e., VEGF, PDGF, bFGF and others) and administering natural antiangiogenic molecules such as endostatin or TSP1.

The history of chemotherapy for infectious diseases and tumors as well as anti-HIV therapy has clearly shown that appropriate combinations of targeted agents are more effective than the most active single-agent treatment. Similarly, a major challenge for the future is the development of combined "broad spectrum" antiangiogenic treatments that, probably, will be the winning strategy to fully exploit their therapeutic potential, particularly when applied in the adjuvant setting.

References

1. Folkman J. Tumor angiogenesis: Therapeutic implications. New Engl J Med 1971; 285; 1182–6
2. Carmeliet P. Angiogenesis in life, disease and medicine. Nature 2005:438; 932–36
3. Varmus H. The New Era in Cancer Research. Science 2006; 312; 1162–65
4. Kerbel R, and Folkman J. Clinical translation of angiogenesis inhibitors. Nature Rev Cancer 2002; 2; 727–39
5. St Croix B, Rago C, Velculescu V, et al. Genes expressed in human tumor endothelium. Science 2000; 289; 1197–202
6. O'Reilly MS, Holmgren L, Shing Y, et al. Angiostatin : a novel angiogenesis inhibitor that mediates the suppression of metastasis by a Lewis lung carcinoma. Cell 1994; 79; 315–28

7. O'Reilly MS, Holmgren L, Shing Y, et al. Endostatin : an endogenous inhibitor of angiogenesis and tumor growth. Cell 1997; 88; 277–85

8. O'Reilly MS, Pirie-Sphephered S, Lane WS, et al. Antiangiogenic activity of the cleaved conformation of the sepin antithrombin. Science 1999; 285; 1926–8

9. Satchi-Fainaro R, Mamluk R, Wang L, et al. Inhibition of vessel permeability by TNP-470 and its polymer conjugate, caplostatin. Cancer Cell 2005; 7; 251–61

10. Wickstrom SA, Alitalo K, Keski-Oja J. Endostatin associates with integrin alpha5beta1 and caveolin-1, and activates Src via a tyrosyl phasphatase-dependent pathway in human endothelial cells. Cancer Res; 2002; 62; 5580–89

11. Sudhakar A, Sugimoto H, Yang C, et al. Human tumstatin and human endostatin exhibit distinct antiangiogenic activities mediated by alpha v beta 3 and alpha 5 beta 1 integrins. Prooc Natl Acad Sci USA.; 2003; 100; 4766–4771

12. Yu Y, Moulton KS, Khan MK, et al. E-selectin is required for the antiangiogenic activity of endostatin. Prooc Natl Acad Sci USA.; 2004; 101; 8005–10

13. Nyberg P, Heikkila P, orsa T, et al. Endostatin inhibits human tongue carcinoma cell invasion and intravasation and blocks the activation of matrix metalloprotease-2, -9, and –13. J Biol Chem; 2003; 278: 22404–22411

14. Almog N, Henke V, Flores L, et al. Prolonged dormancy of human liposarcoma is associated with impaired tumor angiogenesis. FASEB J. 2006; 20: 1–10

15. Naumov GN, Bender E, Zurakowski D, et al. A model of human tumor dormancy : an angiogenic switch from the nonangiogenic phenotype. J Natl Cancer Inst 2006; 98: 316–25

16. Sun Y, Wang J, Liu Y, et al. Results of phase III trial of rh-endostatin (YH-16) in advanced non-small cell lung cancer (NSCLC) patients. J Clin Oncol 2005; 23

17. Hoekstra R, de Vos F.Y.F.L., Eskens F.A.L.M., et al. Phase I safety, pharmacodynamic study of the thrombospondin-1-mimetic angiogenesis inhibitor ABT-510 in patients with advanced cancer. J Clin Oncol 2005; 23: 5188–97

18. Hoekstra R, de Vos FYFL, Eskens FALM, et al. Phase I study of the thrombospondin-1-minetic angiogenesis inhibitor ABT-510 with 5-flururacil and leucovorin : A safe combination. Eur J Cancer 2006; 42 : 467–72

19. Asahara T, Murohara T, Sullivan A, et al. Isolation of putative progenitor endothelial cells for angiogenesis. Science 1997; 275; 964–7

20. Bagley RG, Weber W, Rouleau C, et al. Pericytes and endothelial precursor cells : cellular interactions and contributions to malignancy. Cancer Res 2005; 65 (21); 9741–50

21. Abramsson A, Lindblom P, Betsholtz C. Endothelial and non-endothelial sources of PDGF-B regulate pericyte recruitment and influence vascular pattern formation in tumors. J Clin Invest 2003; 112; 1142–51

22. Ostman A. PDGF receptors-mediators of autocrine tumor growth and regulators of tumor vasculatur and stroma. Cytokine Growth Factors Rev 2004; 15; 275–86

23. Song S, Ewald AJ, Stallcup W, et al. PDGFR b+ perivascular progenitor cells in tumours regulate pericyte differentiaion and vascular survival. Nature Cell Biol 2005; 7; 870–9

24. Motzer RJ, Hutson TE, Tomczak P, et al. Sunitinib versus interferon alfa in metastatic renal-cell carcinoma. N Engl J Med 2007; 356 (2): 115–24

25. Bergers G, Song S, Meyer-Morse, et al. Benefits of targeting both pericytes and endothelial cells in tumor vasculature with kinase inhibitors. J Clin Invest 2003; 111; 1287–95

26. Ferrara N, Houck K, Jakeman L, et al. Molecular and biological properties of the vascular endothelial growth factor family of proteins. Endocr Rev 1992; 13; 18–32

27. Carmeliet P, and Jain RK Angiogenesis in cancer and other diseases Nature; 2000; 407; 249–57

28. Dvorak HF. Vascular permeability factor/vascular endothelial growth factor: a critical cytokine in tumor angiogenesis and a potential target for diagnosis and therapy. J Clin Oncol 2002; 20; 4368–4380

29. Rafii S, et al. Vascular and haematopoietic stem cells: novel targets for anti-angiogenesis therapy? Nat Rev Cancer 2002; 2; 826–35

30. Jain RK. Normalization of tumor vasculature: an emerging concept in antiangiogenic therapy. Science 2005; 307; 58–62

31. Jain RK, Duda DG, Clark JW, et al. Lessons from phase III clinical trials on anti-VEGF therapy for cancer. Nature Clin Pract Oncol 2006; 3 (1); 24–40

32. Cooke R. Dr. Folkman's War: Angiogenesis and the struggle to defeat cancer. Random House 2001.

33. Genentech Inc. Avastin tm (bevacizumab): prescribing information (online). Available from URL: http://www.gene.com (Accessed 2005 Nov 1)

34. European Medicines Agency. Avastin 25 mg/mL: summary of product characteristics (online). Available from URL: http://www.roche.com/med-cor-2005-04-19 (Accessed 2005 Nov 1)

35. Pouyssegur J, Dayan F, and Mature NM. Hypoxia signaling in cancer and approaches to enforce tumour regression. Nature 2006; 441: 437–43

36. Hirota K, Semenza GL. Regulation of angiogenesis by hypoxia-inducible factor 1. Critical Rev Oncol Hematol 2006; 59: 15–26

37. Wang GL, and Semenza GL. Purification and characteriation of hypoxia-inducible factor 1. J Biol Chem 1995; 270; 1230–37

38. Guertin DA, and Sabatini DM. An expanding role for mTOR in cancer. Trends Mol Med 2005; 11: 353–61

39. Nobukini T, and Thomas G. The mTOR/S6K signalling pathway : the role of of the TSC1/2 tumour suppressor complex and the proto-oncogene Rheb. Novartis Found Symp 2004; 262: 148–54. Discussion 154–9, 265–8

40. Shook D, and Keller R. Mechanisms, mechanics and function of epithelial-mesenchimal transitions in early developmnt. Mech Dev 2003; 120: 1351–83

41. Huber MA, Kraut N, Beug H. Molecular requirements for epithelial-mesenchimal transiion during tumor progression. Curr Opin Cell Biol 2005; 17: 548–88

42. Thiery JP. Epithelial-mesenchimal transitions in tumour progression. Nature Rev Cancer 2002; 2: 442–54

43. Nakamura T, Teramoto H, Ichihara A. Purification and characterization of a growth factor from rat platelets for mature parenchymal hepatocytes in primary cultures. Proc Natl Acad Sci USA; 1986; 83; 6489–93

44. Boccaccio C, and Comoglio PM. Invasive growth: a MET-driven genetic programme for cancer and stem cells. Nature Rev Cancer 2006; 6; 637–45

45. Michieli P, et al. Targeting the tumor and its microenvironment by a dual-function decoy Met receptor. Cancer Cell 2004; &; 61–73

46. Mazzone M, et al. An uncleavable form of pro-scatter factor suppresses umor growth and dissemination in mice. J Clin Invest 2004; 114; 1418–32

47. Garber K. New drugs target hypoxia response in tumors. J Natl Cancer Inst 2005; 97(15); 1112–14

48. Fukumura D, Kashiwagi S, Jain RK. The role of nitric oxide in tumour progression. Nature Rev Cancer 2006; 6; 521–34

49. Oliveira CJR, et al. Nitric oxide and cGMP activate the Ras-MAP kinase pathway-stimulaing protein tyrosine phosphorylation in rabbit aortic endothelial cells. Free Radic Biol Med 2003; 35; 381–96

50. Kawasaki K, et al. Activation of the phosphatidylinositol 3-kinase/protein kinase Akt pathway mediates nitric oxide-induced endothelial cell migration and angiogenesis. Mol Cell Biol 2003; 23; 5726–37

51. Zaragoza C, et al. Activation of the mitogen activated protein kinase extracellular signal-regulated kinase 1 and 2 by the nitric oxide-cGMP –cGMP-dependent protein kinase axis regulates the expression of matrix metalloproteinase 13 in vascular endothelial cells. Mol Pharmacol 2002; 62; 927–35

52. Ridnour LA, et al. Nitric oxide regulates angiogenesis through a functional switch involving thrombospondin-1. Proc Natl Acad Sci USA 2005; 102; 13147–52

53. Fukumura D, Jain RK. Role of nitric oxide in angiogenesis and microcirculation in tumors. Cancer Metastasis Rev 1998; 17; 77–89

54. Kubes P. Nitric oxide affects microvascular permeability in the intact and inflamed vasculature. Microcirculation 1995; 2; 235–44

55. Gratton JP, et al. Selective inhibition of tumor microvascular permeability by cavtratin blocks tumor progression in mice. Cancer Cell 2003; 4; 31–39

56. Park SW, et al. The effect of nitric oxide on cyclooxygenase-2 (COX-2) overexpression in head and neck cancer cell lines. Int J Cancer 2003; 107; 729–38

57. Kim SF, Huri DA, Synder SH. Inducible nitric oxide synthase binds, S-nytrosylates, and activates cyclooxygenase-2. Science 2005; 310; 1966–70

58. Rao CV, et al. Chemopreventive properties of a selective inducible nitric oxide synthase inhibitor in colon carcinogenesis, administered alone or in combination with celecoxib, a selective cyclooxigenase-2 inhibitor. Cancer Res 2002; 62; 165–70

59. Dillman RO. Monoclonal antibodies for treating cancer. Ann Intern Med 1989; 11; 592–603

60. James K. Human monoclonal antibodies and engineered antibodies in the management of cancer. Cancer Biol 1990; 1; 243–53

61. Imai K, Takaoka A. Comparing antibody and small-molecule therapies for cancer. Nature 2006; 6; 714–25

62. Hurwitz H, Fehrenbacher L, Novotny W, et al. Bevacizumab plus irinotecan, fluoruracil, and leucovorin for metastatic colorectal cancer. N Engl J Med 2004; 350: 2335–42

63. Hicklin D, t al. Antitumor activity of anti-flk-1 monolonal antibodies. Proc Am Assoc Cancer Res 1997; 38; 266 (Abstr 1788)

64. Baselga J. targeting tyrosine kinases in cancer: the second wave. Science 2006; 312; 1175–78

65. Arora A, and Scholar EM: Role of tyrosine kinase in cancer therapy. J Pharmacol Exp Ther 2005; 315; 971–79

66. Krause DS, and Van Ettern RA. Tyrosine kinases as targets for cancer therapy. N Engl J Med; 2005; 353; 172–8

67. Kuenen BC, Rosen L, Smit EF, et al. Dose finding and pharmacokinetic study of cisplatin, gemcitabine, and SU5416 in patients with solid tumors. J Clin Oncol 2002; 20; 1657–67

68. Heinrich MC, Maki RG, Corless CL, et al. Sunitinib (SU) response in imatinib-resistant (IM-R) GIST correlates with KIT abd PDGFRA mutation status. J Clin Oncol 2006; 24 (Abstr 9502)

69. Wedge SR, Ogilvie DJ, Duks M, et al. ZD6474 inhibits vascular endothelial growth factor signaling, angiogenesis, and tumor growth following oral administration. Cancer Res 2002; 62 (16); 4645–55

70. Mross K, et al. Phase I clinical and pharmacokinetic study of PTK/ZK, a multiple VEGF receptor inhibitor, in patients with liver metastases from solid tumours. Eur J Cancer 2005; 41; 1291–99

71. Hecht JR, et al. A randomized, double-bind, placebo-controlled phase III study in patients with metastatic adenocarcinoma of the colon or rectum receiving first-line chemotherapy with oxaliplatin/5-fluoruracil/leucovorin and PTK787/ZK222584 or placebo. J Clin Oncol 2005;(Abstr 23); LBA3

72. Folkman J. Antiangiogenesis in cancer therapy-endostatin and its mechanims of action. Exp Cell Res 2006; 312: 594–607

73. Hedrick E, Kozloff M, Hainsworth J, et al. Safety of bevacizumab plus chemotherapy as first-line treatment of patients with metastatic colorectal cancer: Updated results from a large observational registry in the USA (BRiTE).Citation: J Clinical Oncol 2006; Proc Am Soc Clin Oncol 24 (18S) (Supplement) Abstr 3536

74. Kamba T, Tam BYY, Hashizume H, et al. VEGF-dependent plasticity of fenestrated capillaries in the normal adult microvasculature. Am J Physiol Heart Circ Physiol 2006; 290:H560–76

75. Clark JW, et al. Safety and pharmacokinetics of the dual action of raf kinase and vascular endothelial growth factor receptor inhibitor, BAY 43-9006, in patients with advanced, refractory solid tumors. Clin Cancer Res 2005; 11: 5472–80

76. Moore M, et al. Phase I study to determine the safety and pharmacokinetic of the novel Raf kinase and VEGFR inhibitor BAY 43-9006, administered for 28 days on/days off in patients with advanced, refractory solid tumors. Ann Oncol 2005; 16: 1688–94

77. Gasparini G, Longo R, Toi M, et al. Angiogenic inhibitors: a new therapeutic strategy in oncology. Nat Clin Pract Oncol 2005; 2(11): 562–77.

78. George ML, Eccles SA, Tutton MG, et al. Correlation of plasma and serum vascular endothelial growth factor levels with platelet count in colorectal cancer: clinical evidence of platelet scavenging? Clin Cancer Res 2000; 6:3147–3152

79. Davidoff AM, Ng CY, Zhang Y, et al, Careful decoy receptor titering is required to inhibit tumor angiogenesis while avoiding adversely altering VEGF bioavailability. Mol Ther 2005; 11: 300–10

80. Uthoff SM, Duchrow M, Schmidt MH, et al. VEGF isoforms and mutations in human colorectal cancer. Int J Cancer 2002;101: 32–6

81. Yan L, Borregaard N, Kjeldsen L, et al. The high molecular weight urinary matrix metalloproteinase (MMP) activity is a complex of gelatinase B/MMP-9 and neutrophil gelatinase-associated lipocalin (NGAL). J Biol Chem 2001; 276 (40): 37258–65

82. Roy R, Wewer UM, Zurakowsky D, et al. ADAM 12 cleaves extracellular matrix proteins and correlates with cancer status and stage. J Biol Chem 2004; 279 (49): 51323–330

83. Beaudry P, Force J, Naumov GN, et al. Differential effects of vascular endothelial growth factor receptor-2 inhibitor ZD6474

on circulating endothelial progenitors and mature circulating endothelial cells: implications for use as a surrogate marker of antiangiogenic activity. Clin Cancer Res 2005; 11: 3514–22

84. Willett CG, Boucher Y, di Tomaso E, et al. Direct evidence that the VEGF-specific antibody bevacizumab has antivascular effects in human rectal cancer. Nature Med 2004; 10: 145–47

85. Willet CG, Boucher Y, Duda DG, et al. Surrogate markers for antiangiogenic therapy and dose-limiting toxicities for bevacizumab with radiation and chemotherapy: continued experience of a phase I trial in rectal cancer patients. J Clin Oncol 2005; 23: 8136–9

86. Mancuso P, Colleoni M, Calleri A, et al. Circulating endothelial-cell kinetics and viability predict survival in breast cancer patients receiving metronomic chemotherapy. Blood 2006; 108: 452–9

87. Beerepoot LV, Radema SA, Witteveen EO, et al. Phase I clinical evaluation of weekly administration of the novel vascular-targeting agent, ZD6126, in patients with solid tumors. J Clin Oncol 2006; 24: 1491–8

88. Bertolini F, Shaked Y, Mancuso P, et al. The multifaceted circulating endothelial cells in cancer: towards marker and target identification. Nat Rev Cancer 2006; 6: 835–45

89. Italiano J, Richardson JL, Folkman J, et al. Blood platelets organize pro- and anti-angiogenic factors into separate, distinct Alpha granules: implications for the regulation of angiogenesis. Abstr 301 48th ASH Annual Meeting

90. Schneider BP, Skaar TC, Sledge GW, et al. Analysis of angiogenesis genes from paraffin-embedded breast tumor and lymph node. Breast Cancer Res Treatment 2006; 96:209–15

91. Godl K, Gruss OJ, Eickoff J, et al. Proteomic characterization of the angiogenesis inhibitor of SU6668 reveals multiple impacts on cellular kinase signaling. Cancer Res 2005; 65 (15): 6919–26

92. Rehman S and Jayson GC. Molecular imaging of antiangiogenic agents. Oncologist 2005; 10: 92–103

93. Miller JC, Pien HH, Sahani D, Sorensen AG. Imaging angiogenesis: applications and potential drug development. J Nat Cancer Inst 2005; 97: 172–87

94. Morgan B, Thomas AL, Drevs J, et al. Dynamic contrast-enhanced magnetic resonance imaging as a biomarker for the pharmacological response of PTK 787/ZD 222584, an inhibitor of the vascular endothelial growth factor receptor tyrosine kinases, in patients with advanced colorectal cancer and liver metastases: results from two phase I studies. J Clin Oncol 2003; 21: 3955–64

95. Liu G, Rugo HS, Wilding G, McShane TM, et al. Dynamic contrast-enhanced magnetic resonance imaging as a pharmacodynamic measure of response after acute dosing of AG-013736, an oral angiogenesis inhibitor, in patients with advanced solid tumors: results from a phase I study. J Clin Oncol 2005;23: 5464–73

96. Wedam SB, Low JA, Yang SX, et al. Antiangiogenic and antitumor effects of bevacizumab in patients with inflammatory and locally advanced breast cancer. J Clin Oncol 2006;24: 769–77

97. Korn EL, Arbuck SG, Pluda JM, et al. Clinical trial designs for cytostatic agents: Are new approaches needed? J Clin Oncol 2001; 19: 265–72

98. Rogatko A, Babb JS, Tighiouart M, et al. New paradigm in dose-finding trails : patient-specific dosing and beyond phase I. Clin Cancer Res 2005; 11; 5342–46

99. Therasse P, Eisenhauer EA, and Buyse M. Update in methodology and conduct of clinical cancer. Eur J Clin Oncol 2006; 42: 1322–30

100. Michaelis LC, Ratain M. Measuring response in a post-RECIST world : from black and white to shades of grey. Nature 2006; 6; 409–14

101. Rubinstein LV, Korn EL, Freidlin B, t al. Design issues randomized Phase II trials and a proposal for Phase II screening trials. J Clin Oncol 2005; 23; 7199–7206

102. Rosner GL, Stadler W, Ratain MJ. Randomized discontinuation design. Application to cytostatic antineoplastic agents. J Clin Oncol 2002; 20:4478–84

103. Betensky RA et al. influence of unrecognized molecular heterogeneity on randomized clinical trials. J Clin Oncol 2002; 20: 2495–2499

104. Kim SJ, Uehara H, Karashima T, et al. Blockade of epidermal growth factor receptor signaling in tumor cells and tumor-associated endothelial cells for therapy of androgen-independent human prostate cancer growing in the bone of nude mice. Clin Cancer Res 2003; 9:1200–1210

105. de Jong JS, van Diest PJ, van der Valk P, et al. Expression of growth factors, growth-inhibiting factors, and their receptors in invasive breast cancer. II: Correlations with proliferation and angiogenesis. J Pathol 1998; 184:53–57

106. Petit AM, Rak J, Hung MC, et al. Neutralizing antibodies against epidermal growth factor and ErbB2-neu receptor tyrosine kinase down-regulate vascular endothelial growth factor production by tumor cells in vitro and in vivo. Am J Pathol 1997; 151:1523–1530

107. Hirata A, Ogawa S, Kometani T, et al. ZD1839 (Iressa) induces antiangiogenic effects through inhibition of epidermal growth factor receptor tyrosine kinase. Cancer Res 2002; 62:2554–2560

108. Ciardiello F, Caputo R, Damiano V, et al. Antitumor effects of ZD6474, a small molecule vascular endothelial growth factor receptor tyrosine kinase inhibitor, with additional activity against epidermal growth factor receptor tyrosine kinase. Clin Cancer Res 2003; 9:1546–1556

109. Herbst RS, Mininberg E, Henderson T, et al. Phase I/II trial evaluating blockade of tumor blood supply and tumor cell proliferation with combined bevacizumab and erlotinib HCl as targeted cancer therapy in patients with recurrent non-small cell lung cancer. Eur J Cancer 2003;1:S293

110. Jung YD, Mansfield PF, Akagi M, et al. Effects of combination anti-vascular endothelial growth factor receptor and anti-epidermal growth factor receptor therapies on the growth of gastric cancer in a nude mouse model. Eur J Cancer 2002; 38:1133–1140

111. Ciardiello F, Bianco R, Damiano V, et al. Antiangiogenic and antitumor activity of anti-epidermal growth factor receptor C225 monoclonal antibody in combination with vascular endothelial growth factor antisense oligonucleotide in human GEO colon cancer cells. Clin Cancer Res 2000; 6:3739–3747

112. Shaheen RM, Ahmad SA, Liu W, et al. Inhibited growth of colon cancer carcinomatosis by antibodies to vascular endothelial and epidermal growth factor receptors. Br J Cancer 2001; 85:584–589

113. Guy SP, Ashton S, Hughes G, et al. Gefitinib (Iressa, ZD1839) enhances the activity of the novel vascular-targeting agent ZD6126 in human colorectal cancer and non-small cell lung can-

cer (NSCLC) xenograft models. Clin Cancer Res 2003;9:6142S (abstr; suppl B13)

114. Hainsworth JD, Jeffrey AS, Spigel DR, et al. Treatment of metastatic renal cell carcinoma with a combination of bevacizumab and erlotinib. J Clin Oncol 2005; 23(31): 7889–96

115. Dickler M, Rugo H, Caravelli J, et al. Phase II trial of erlotinib (OSI-774), an epidermal growth factor receptor (EGFR)-tyrosine kinase inhibitor, and bevacizumab, a recombinant humanized monoclonal antibody to vascular endothelial growth factor (VEGF), in patients (pts) with metastatic breast cancer (MBC). Proc Am Soc Clin Oncol 2004; 23:127, (abstr 2001)

116. Mauer AM, Cohen EEW, Wong SJ, et al. Phase I study of epidermal growth factor receptor (EGFR) inhibitor, erlotinib, and vascular endothelial growth factor monoclonal antibody, bevacizumab, in recurrent and/or metastatic squamous cell carcinoma of the head and neck (SCCHN). Proc Am Soc Clin Oncol 2004; 23:496(abstr 5539)

117. Herbst RS, Johnson DH, Mininberg E, et al. Phase I/II trial evaluating the anti-vascular endothelial growth factor monoclonal antibody bevacizumab in combination wit the HER-1/epidermal growth factor receptor tyrosine kinase inhibitor erlotinib for patients with recurrent non-small-cell lung cancer. J Clin Oncol 2005; 23:2544–2555

118. Hainsworth JD, Sosman JA, Spigel DR, et al. Treatment of metastatic renal cell carcinoma: a combination of bevacizumab and erlotinib. J Clin Oncol 2005; 23: 7889–96

119. Saltz LB, Chung KY. Antibody-based therapies for colorectal cancer. The Oncologist 2005; 10: 701–9

120. Konecny GE, Meng YG, Untch M, et al. Association between HER-2/neu and vascular endothelial growth factor expression predicts clinical outcome in primary breast cancer. Clin Cancer Res 2004; 10: 1706–16

121. Kerbel RS. Antiangiogenic therapy: a universal chemosensitization strategy for cancer? Science 2006; 312: 1171–78

122. Kieran MW, Turner CD, Rubin JB, et al. A feasibility trial of antiangiogenic(metronomic) chemotherapy in pediatric patients with recurrent or progressive cancer. J Pediatr Hematol Oncol 2005; 27: 573–81

123. Ozols RF, Herbst RS, Colson YL, et al. Clinical cancer advances 2006: major research advances in cancer treatment, prevention, and screening-A report from the American Society of Clinical Oncology. J Clin Oncol 2007; 25: 146–62

124. Parulekar WR, Eisenhauer EA. Phase I trial design for solid tumor studies of targeted, non- cytotoxic agents: theory and practice. J Natl Cancer Inst 2004;96: 990–997

Chapter 41
Pharmacogenetics of Antiangiogenic Therapy

Guido Bocci, Giuseppe Pasqualetti, Antonello Di Paolo,
Mario Del Tacca, and Romano Danesi*

Keywords: pharmacogenetics, VEGF, VEGFR, single nucleotide polymorphisms, antiangiogenic therapy, clinical trials

Abstract: The concept of targeting tumor angiogenesis has moved from a pioneering research field into clinical practice, and the novel drugs now available are defining a new and promising avenue of targeted cancer therapy. As with many other target-specific treatments, however, the need for a rational selection of patients to be administered anti-angiogenic treatments is emerging, since the clinical activity of these agents appears to be limited to specific patients and is not predictable on the basis of standard approaches. Candidate targets for treatment optimization are vascular endothelial growth factors (VEGF) and their receptors (VEGFR), and while the activity of drugs inhibiting VEGF and VEGFR signal transduction pathways are likely to also be influenced by their intrinsic biological activities and the regulation of the gene expression of VEGF and VEGFR as well. The pharmacogenetic approach to anti-angiogenic therapy should be considered a possible strategy for delivering the optimal treatment to specific groups of patients affected by tumors, as well as by other pathological conditions dominated by pathological angiogenesis, including age-related macular degeneration or endometriosis. While pharmacogenetic studies are building stronger foundations for the systematic investigations of phenotype–genotype relationships in many fields of medicine, pharmacogenetic data regarding anti-angiogenic drugs are still lacking. Here, we review preclinical and clinical genetic studies mainly focusing on VEGF-A and VEGFR-2. Many genetic variants are being discovered, and single nucleotide polymorphisms of VEGF and VEGFR genes appear to be able to affect VEGF transcription, affinity to its receptor, and biological activity of signal transduction pathway. We suggest that pharmacogenetic profiling of patients who are candidates for the currently available anti-angiogenic agents may help in the selection of subjects to be treated on the basis of their likelihood of responding to the drugs, while avoiding major toxicities, including hypertension and bleeding.

Division of Pharmacology and Chemotherapy, Department of Internal Medicine, University of Pisa, Pisa, Italy

*Corresponding Author:
Romano Danesi, Division of Pharmacology and Chemotherapy, Department of Internal Medicine, University of Pisa, Via Roma, 55, I-56126 Pisa, Italy
E-mail: r.danesi@med.unipi.it

Introduction

The systematic investigation of phenotype-genotype relationships in medicine [1, 2] is rapidly building new and important foundations for pharmacogenetics. Pharmacogenetic research on antiangiogenic drugs is still a poorly investigated area, although its potential to select patient candidates for antiangiogenic treatment is of obvious clinical importance. The pharmacological principles involved in the clinical use of antiangiogenic drugs could provide the starting basis for the design of pioneering genetic studies. Knowledge of the main mechanisms by which drugs decrease the neovascularization in solid tumors, or other pathological conditions such as age-related macular degeneration or endometriosis [3], could help prioritize candidate genes for pharmacogenetic studies.

Genetic polymorphisms could influence either the level of expression and secretion or the functional activity of proteins, depending on the variation involved. In this perspective, it may be possible to demonstrate clear relationships between specific genetic variants of target proteins (e.g., VEGF-A) for some antiangiogenic drugs (e.g., anti-VEGF antibodies) and clinical parameters such as the efficacy of the therapy (e.g., stabilization of the disease) and the optimal biological dose suggested by surrogate markers [4, 5]. However, selecting candidate genes for pharmacogenetic studies on antiangiogenic drugs, based essentially on a generally accepted primary action, may fail to identify some relevant gene variations. A promising approach to identify key genes primarily involved in the activity of antiangiogenic drugs could be the use of genetic animal models [6]. A strict correlation of antiangiogenic drug dose response (i.e., using antibodies targeting VEGFR-2) with respect to antitumor therapeutic activity and levels of viable circulating endothelial progenitors (CEPs) was demonstrated in genetically manipulated mouse strains such as TSP-1$^{-/-}$ C57 BL/6 J, VEGF-A$^{hi/+}$ and CD-1 *Tie-2* transgenic mice.

A strong expression of VEGF-A is present in a variety of tissues and conditions, either physiologically, including tissues of the female reproductive system, or pathologically, such as ischemic tissues, solid tumors and endometriosis [3, 7]. The most important inducer of VEGF-A gene expression is hypoxia both in vivo and in vitro, especially in the perinecrotic areas of solid tumors. Under hypoxic conditions, VEGF-A expression is mediated through the activation of the hypoxia-inducible factor 1 (HIF-1), a transcription factor that binds to a 28-bp enhancer in the 5′ upstream region of the VEGF-A gene [8]. Moreover, hypoxia determines a VEGF-A mRNA half-life increase by as much as 3- to 4-fold, as a result of the interplay of both stability and instability elements present in mRNA. The regulating elements are in the 3′ untranslated region (3′-UTR) of VEGF-A mRNA, where the binding of hypoxia-induced proteins, such as the RNA-binding protein HuR, results in a significantly increased half-life of the mRNA [9]. This post-transcriptional regulation affects not only the VEGF-A gene, but also other hypoxia-inducible genes like erythropoietin or tyrosine hydroxylase [10, 11]. In addition, somatic mutations of some oncogenes increase VEGF-A expression. It was indeed demonstrated that both levels of VEGF-A mRNA and of the secreted functional protein were increased in human and rodent tumor cell lines harboring mutant K-ras or H-ras oncogenes, respectively [12]. Moreover, K-ras oncogene mutation is associated with VEGF-A overexpression in patients with pancreatic carcinoma [13]. The squamous cell carcinoma related oncogene (SCCRO), a novel oncogene identified by positional cloning of a recurrent amplification at 3q26.3 and overexpressed in lung, head and neck, cervical and ovarian carcinomas, regulates angiogenesis through the increase of VEGF-A [14]. Instead, N-myc oncogene seems to enhance the angiogenic process by both directly regulating VEGF-A production by stimulating initiation of VEGF-A mRNA translation [15] and indirectly down-regulating interleukin (IL)-6, an inhibitor of VEGF-induced angiogenesis [16].

Neovascularization inhibition by antiangiogenic drugs has been demonstrated by numerous preclinical and, in some cases, clinical studies using anti-VEGF approaches such as monoclonal antibodies targeting VEGF-A and its high affinity receptor, VEGFR2, small molecule receptor tyrosine kinase inhibitors of VEGFR2 signaling, soluble receptors, antisense oligonucleotides and peptides targeting VEGF-A [17] (Table 41.1). Moreover, three anti-VEGF drugs have recently been made available on the market: bevacizumab, an anti-human VEGF-A monoclonal antibody; pegaptanib, a RNA aptamer directed against VEGF-A$_{165}$ [17]; and ranizumab, an anti-human VEGF-A monoclonal antibody, for ocular neovascularization.

Although the efficacy of these drugs is very promising in different pathologic conditions, such as colorectal cancer and non-small cell lung cancer (NSCLC) [18], the clinical application of antiangiogenic compounds is more complex than initially thought. As an example, bevacizumab monotherapy demonstrated a lack of increase in survival of cancer patients, in sharp contrast with the rate of efficacy of this drug combined with chemotherapeutic schedules, especially in the first-line treatment of patients with metastatic colorectal cancer. Indeed, resistance to angiangiogenic therapy due to tumor escape is emerging as an important clinical issue [19, 20]. Bevacizumab and some tyrosine kinase inhibitors may be associated with serious toxicities such hypertension, minor bleedings (e.g., epistaxis) or serious hemorrhages (e.g., gastrointestinal and subaracnoid hemorrhages), thromboembolic events, gastrointestinal perforations, wound healing complications (e.g., wound dehiscence), neuropathies and proteinuria with an etiology that is still poorly understood [18, 21]. Thus, although the causes of anti-VEGF therapy failure and of side-effects have not yet been elucidated, the hypothesis that some genetic mechanisms could be involved is reasonable. Indeed, germinal and somatic mutations could alter the VEGF-A gene expression because neoplastic cells usually represent the main source of secreted VEGF-A, with tumor-associated stroma and inflammatory cells also being important sites of growth factor production [22, 23]. Inflammatory angiogenesis and macrophage infiltration seem to be essential for the development of malignant tumors [24]. In various inflammatory responses, tumor-associated macrophages (TAMs) play an important role in providing a microenvironment that stimulates cell migration, survival and proliferation of various stromal cell types by producing angiogenic growth factors and cytokines [25]. The critical role of inflammatory cell infiltration in tumor

TABLE 41.1. Antiangiogenic drugs targeting VEGF-A and VEGFR-2 already approved or in advanced clinical development.

Drug	Target	Features	Status	Major Indication
Bevacizumab	VEGF-A	MoAb	Approved	CRC [58], NSCLC
Pegaptanib	VEGF-A	RNA aptamer	Approved	Retinopathy [79]
Vatalanib	VEGFR1, VEGFR-2, VEGFR-3, c-kit, PDGFR	TKI	Phase III	NSCLC, CRC
Semaxanib	VEGFR-2	TKI	Phase III	NSCLC, CRC
Sunitinib	VEGFR-2, PDGFR, c-kit	TKI	Approved	GIST [80], RCC [81, 82]
Sorafenib	VEGFR-2, B-RAF, c-kit, PDGFR	TKI	Approved	RCC [83]

MoAb Monoclonal antibody; *RCC* renal cell cancer; *NSCLC* non-small cell lung cancer; *GIST* gastro-intestinal stromal tumors; *CRC* colorectal cancer; *TKI* tyrosine kinase inhibitor angiogenesis has suggested the actual possibility to target this process for antitumor drug development.

Single Nucleotide Polymorphisms of VEGF-A gene

Strong interindividual variations of VEGF-A plasma levels and VEGF-A gene expression have been reported [26] in healthy and pathological populations. Numerous studies recorded different mean values of plasma VEGF-A, probably due to sample processing, age and gender of the subjects, methods of measurement [26] and genetic variants affecting protein production mainly represented by single nucleotide polymorphisms (SNPs) (Table 41.2) [27, 28]. Indeed, VEGF-A gene is highly polymorphic with more than 20 different polymorphisms [29]. Three polymorphisms (702 C/T, 936 C/T, 1612 G/A) were found in healthy subjects, and allele frequencies of 702 T, 936 T and 1612 A were 0.017, 0.160 and 0.471, respectively [30]. Subjects harboring the 936 T allele had VEGF-A plasma levels significantly lower than noncarriers, whereas the SNPs 702 C/T and the 1612 G/A revealed no associations with VEGF-A plasma concentrations. The possible mechanisms of the association between the 936 C/T polymorphism

and VEGF-A plasma levels could be (1) the loss of a potential transcription factor binding site for AP-4 due to the SNP 936 C/T, or (2) a linkage disequilibrium between this mutation and other yet unknown functional mutations in the VEGF-A gene sequence [30]. The role of 936 C/T polymorphism for breast cancer risk was investigated in a case-control study [26]. VEGF-A genotype was determined in 500 women with breast cancer and 500 sex- and age-matched healthy control subjects. Carriers of the VEGF-A 936 T allele were less frequent among cancer patients than in healthy subjects, indicating that this genetic variant may be protective against breast cancer. Therefore, it is reasonable to hypothesize that patients, harboring the 936 C allele, could have tumors more dependent on VEGF-A levels (higher than in 936 T patients) for their growth and thus more sensitive to anti-VEGF-A drugs such as bevacizumab. Additional findings provided evidence that 936 C is not clearly associated with invasive breast cancer but it is associated with reduced risk for in situ cancer [31.

VEGF-A promoter polymorphisms have been extensively studied for their possible influence on VEGF-A secretion. Indeed, peripheral-blood mononuclear cells from −1154 GG individuals produced significantly more VEGF-A than cells from −1154 AA subjects [32]. Similarly, VEGF-A production was significantly higher in cells from −2578 CC than those from −2578 AA individuals [32]. The −2578 C and −1154 G alleles were also associated with increased risk for invasive breast cancer. The evaluation of the correlation between VEGF-A SNPs and the risk for breast cancer among postmenopausal women in a cancer prevention study demonstrated no association between SNPs and breast cancer risk [31]. However, the −2578 C and −1154 G alleles, which are both described to increase expression of VEGF-A, were associated with increased risk for invasive breast cancer but not for in situ cancer [31]. A strong association of the −2578 AA genotype with larger and poorly differentiated gastric tumors was observed [33]. Moreover, the frequency of allele −1154 AA in 238 prostate cancer patients and 263 controls showed that it was significantly lower in patients than controls (6.3 vs 12.9%; $P = 0.01$) [34]. Thus, it was suggested that genotypes associated with low production of VEGF-A may confer protection to prostate cancer development, at variance with subjects that produce high amounts of VEGF-A [34]. Another case control study demonstrated that the −1154 GA and AA genotypes were negatively associated with prostate cancer risk compared with the GG genotype, and the presence of at least one A allele (GA + AA genotypes) was associated with a reduced incidence of cancer [35]. In NSCLC, low VEGF-A expression in tumor cells was significantly correlated with the presence of the −2578 CC, −634 GG and −1154 AA and GA alleles of the VEGF-A gene [36]. The vascular density measurement showed that tumors with −2578 CC had a significantly lower value if compared to the CA, whereas the AA cases had an intermediate vascularization. Furthermore, −634 GG and −1154 AA VEGF-A SNPs were also associated with significantly lower vascular density [36].

TABLE 41.2. SNPs of VEGF-A with associated clinical condition.

	Biologic effects	Associated clinical condition
936 T	Lower VEGF plasma levels [30]	Breast cancer [26]
936 C	Higher VEGF plasma levels [30]	*In situ* breast cancer [31]
−1154 G	High VEGF production [32]	Vascular dementia [39], melanoma [38], invasive breast cancer [32]
−1154 A	Reduced VEGF transcription [46]	Prostate cancer [34, 35]
−1154 AA/GA	Reduced VEGF transcription [36]	N/A
−2578 C	High VEGF production [32]	Acute rejection of renal transplant [32], invasive breast cancer [31]
−2578 CC/CA	Lower tumor vascular density [36]	N/A
−634 C	Reduced VEGF translation [46]	Henoch-Schönlein purpura [41]
−634 GG	Lower tumor vascular density [36]	N/A
−634 CC	Lower plasma VEGF levels [28]	Prognosis of chronic heart failure [40], gastric cancer [33]
−73 TT, −9228 TT, −8339 TT	N/A	Family history of bladder cancer [37]
9162 TT	N/A	Severe retinopathy [42]
674 CC	N/A	Macular degeneration [43]
936 TT	N/A	Protection against psoriatic arthritis [44]
−2578 CC, −460 CC	High expression	Heart transplant failure [45]

N/A Not available

A large scale evaluation of 1,433 SNPs in 386 genes involved in cancer-related pathways and their association with bladder cancer risk was conducted [37]. The SNPs that significantly correlated with cancer were in the 5'-UTR of VEGF-A. The analysis of 1,086 cases and 1,033 controls indicated that variants in the regulatory regions of VEGF-A could modify the risk for developing bladder cancer. Among the VEGF-A polymorphisms analyzed (−15648 A/C, −9228 G/T, −8339 A/T, −1497 C/T, 405 C/G, −73 C/T, and 1378 C/T), a strong association was found between two SNPs in the VEGF-A promoter (−9228 G/T and −8339 A/T) and family history of cancer.

Other VEGF-A SNPs have been correlated with melanoma [38], vascular dementia [39], chronic heart disease [40], nephritis in Henoch-Schonlein purpura [41], severe retinopathy [42], age-related macular degeneration [43], and psoriatic arthritis [44]. In order to determine if any common variants of VEGF-A are associated with long-term renal and retinal complications in type 1 diabetes, 1,369 diabetic Caucasian subjects from the DCCT/EDIC (Diabetes Control and Complications Trial/Epidemiology of Diabetes Interventions and Complications) genetic study were analyzed, and it was found that multiple VEGF-A variants are associated with the development of severe retinopathy, the most important of which being the 9162 TT [42]. Moreover, a case-control study of 258 patients affected by psoriatic arthritis demonstrated that 936 T/C allele of VEGF-A may be a protective allele in development of disease [44]. The implication of VEGF-A variants in heart transplant failure among racial and ethnic groups have been evaluated and it was found that −2578 CC and −460 CC VEGF-A genotypes were more frequent in African-Americans and Hispanics compared to Caucasians [45].

The −634 G/C was related to variations in circulating VEGF-A levels and VEGF-A gene expression and affected the activity of an internal ribosomal entry site B (IRES-B) involved in VEGF-A translation [46]. A significant correlation was found between VEGF-A production in peripheral blood mononuclear cells and genotype: the lowest VEGF-A protein production being observed in −634 CC, whereas the highest VEGF-A production was found in subjects with −634 GG genotype [28]. However, a significant association of the −634 CC genotype was found with an increased risk for gastric cancer together with a more advanced stage of disease [33]. These data may explain the possibility that advanced gastric tumors could be less dependent on VEGF-driven angiogenesis. Finally, the VEGF-A −634 CC genotype was associated with a poor clinical outcome in patients with chronic heart failure [40].

VEGF-A Promoter Haplotypes

The analysis of VEGF-A promoter haplotypes is important because the SNPs could not be associated with VEGF-A expression whereas a single haplotype could be significantly related to the VEGF-A phenotype such as lower VEGF-A pro-

duction [29]. Four different common haplotypes (including the location of SNPs −2578, −2549, −2489, −2447, −1498, −1198, −1190, −1154, −634, and −7 from the translation start point) have been identified in the promoter region, and at least one modulates VEGF-A production. Rogers and D'Amato [29] have compiled and numbered the VEGF-A gene promoter haplotypes from 1 to 4 based on the order in which they have been previously described [47]. Three out of four haplotypes have also been indicated with the capital letters A, B and C [48]. The haplotype frequencies in normal controls varied depending on the different studies: (1) the haplotype 1, otherwise named C (−2578 C, −2489 C, −2447 G, −1498 T, −1198 C, −1190 G, −1154 G, −634 C, and −7 C), displays a frequency between 42.8 [49] and 25% [50]; (2) the haplotype 2 frequency (−2578 C, −2489 C, −2447 G, −1498 T, −1198 C, −1190 G, −1154 G, −634 G, and −7 C) ranged from 19.6 [38] and 35.1% [51]; (3) the haplotype 3, also named A (−2578 A, −2489 T, −1498 C, −1198 T, −1190 A, −1154 G, −634 G, and −7 T), has a frequency from 12.5 [49] to 48% [52]; and finally, (4) the haplotype 4, also named B (−2578 A, −2489 T, −1498 C, −1198 C, −1190 A, −1154 A, −634 G, and −7 C), occurred between 12.5 [49] and 26.5% [38]. Genotyping of at least 3 polymorphisms (−2578, −1154, −634) is required to discriminate all common promoter haplotypes [29]. The importance in determining the promoter haplotype has been suggested by the evidence that no VEGF-A SNP confers susceptibility to or influences the prognosis of cutaneous malignant melanoma, whereas the VEGF −2578, −1154, −634 CAC haplotypes were significantly associated with less advanced disease [38]. Moreover, it was found that VEGF-A plasma levels were significantly reduced in amyotrophic lateral sclerosis (ALS) or healthy subjects carrying AAG/AAG or AGG/AGG genotypes [46]. With respect to gastric cancer susceptibility, however, Tzanakis et al. reported that none of the examined haplotypes had influence on it [33]. Among other chronic diseases related to pathological angiogenesis, the serum VEGF-A levels seemed to be influenced by the VEGF-A haplotype in patients with Crohn's disease (−2578/−1154/−634 AAG promoter haplotype) [53]. Moreover, an extensive haplotype analysis of VEGF-A promoter SNPs in age-related macular degeneration revealed that possession of the −1498 T, −1198 C, −1190 G, −1154 A, −634 C haplotypes was strongly associated with the onset of age-related macular degeneration and with the exudative form of the disease, thereby creating the possibility for predictive testing [43].

Role of Pharmacogenetics in Anti-VEGF Therapy

In this perspective, a pharmacogenetic approach might help in understanding some of the differences in clinical outcomes using anti-VEGF-A drugs. A number of polymorphisms of the VEGF-A gene have been detected and some of these seem to be closely related to the expression and plasma concentrations of this growth factor. Thus, it is conceivable that a cancer patient, bearing a VEGF-A SNP that affects gene expression, may have a tumor that is independent or poorly dependent on VEGF-A for its vascularization, and hence growth, and thus be resistant to anti-VEGF treatment. Indeed, many other growth factors can stimulate vascular growth and compensate for the lack of VEGF such as fibroblast growth factors (FGFs). The correlation of SNPs and plasma or tissue levels of VEGF-A may also explain, at least in part, the lack of response of tumors to anti-VEGF therapy as first-line treatment. Indeed, it is possible, for example, that placental-derived growth factor, expressed in many tumors, may compensate the relative deficiency of VEGF-A [54] without being a target of the anti-VEGF monoclonal antibody. Thus, a cancer patient who has low level of plasma VEGF-A or low expression of VEGF-A in tumor tissue may not respond to bevacizumab treatment.

The acquired resistance to antiangiogenic treatments is emerging as a forthcoming issue in the oncology field. Several mechanisms have been proposed to explain this phenomenon, and the influence of the genetic background of patients or individual tumors could have a primary role in some of them [55]. Moreover, other possible mechanisms directly related to the genetic mutations of tumor endothelial cells have been hypothesized, such as: (1) the somatic mutations in single endothelial cancer stem cells that may contribute to tumor angiogenesis, as they do in clonal hemangiomas; (2) the somatic mutations of PDGFRs and Tie2 that may arise in tumor vessels; and (3) the acquisition of novel mutations, gene amplification, reduced drug uptake or activation of downstream signaling pathways by tyrosine kinase inhibitors of VEGF receptors [55].

The sequence analysis of VEGF-A transcripts in 25 colorectal cancer specimens and in adjacent normal mucosa revealed at least six somatic polymorphisms in cancer cells [56]. A silent mutation was observed in exon 1 at position +70 relative to the amplification start site, whereas other mutations were a 1- and 2-base deletion (+172 and +171/172, respectively) in exon 3, a transition mutation (at position +248) in exon 3, and 2 transition mutations in exon 4 (+398 and +403).

VEGF-A SNPs could influence antiangiogenic therapy in many areas, because it has been proven that VEGF-A could have a relevant role in the pathogenesis of different diseases. Indeed, anti-VEGF-A therapy is currently being applied to a number of clinical trials. The tolerability of anti-VEGF therapy is becoming a key problem that may involve pharmacogenetic mechanisms. As an example, colorectal cancer patients treated with bevacizumab suffer from venous and arterial thromboembolism, pulmonary embolism, gastrointestinal perforations, bleeding, proteinuria and hypertension [18, 21]. Based on published data, the incidence of some side effects in anti-VEGF-A treatment seems not to be dose-related, as demonstrated in different clinical trials, and may suggest a mechanism due to the

genetic background of patients. Unfortunately, no reports are currently available in literature proving a link between the incidence of side-effects and the patient's genetic background. However, the risk of thromboembolic events increased with increasing age and prior history of arterial thrombotic events [18]. Epidemiological studies have identified several inherited states that may result in endothelial damage or altered hemostatic equilibrium, thereby predisposing patients to arterial thrombosis, including hyperhomocysteinemia, hereditary thrombophilias, and platelet hyper-reactivity [57]. In this perspective, a deeper investigation of VEGF-A expression in the above-mentioned diseases could be an initial platform for VEGF-A genetic profiling and identification of patients with higher risks of severe toxicities. Plasma concentrations of VEGF-A may also play a role in the development of hypertension in treated subjects. Hypertension is likely to be attributable to reduced vasodilation by nitric oxide and alteration of renal salt homeostasis [55]. However, only 22.4% of bevacizumab-treated patients developed this important side effect in a phase III clinical trial [58]. Further investigations will have to elucidate, for example, if subjects carrying the SNPs or the haplotype which is associated with low levels of plasma VEGF-A (e.g., 936 C/T allele or promoter haplotype 4) are more or less susceptible to develop hypertension during bevacizumab treatment.

Single Nucleotide Polymorphisms of VEGFRs Genes

The rising interest in angiogenic modulators has led to the design and synthesis of new small molecules that inhibit the signal transduction pathways triggered by VEGF-A. Some of these drugs, such as vatalanib or AE-941, are currently in advanced phase III clinical trials with promising results [17], but others have failed the expectations (i.e., semaxanib). Other agents, such as sunitinib or sorafenib, have a broad spectrum of activity that includes PDGFR-β, VEGFR2/KDR, c-Kit and FLT3 [17]. However, studies of VEGFR SNPs are few and not specifically designed to address drug response, given the recent introduction of antiangiogenic agents in clinical practice. As already described for EGFR [59], it is reasonable to predict that genetic variants of VEGFRs may alter the sensitivity to antiangiogenic drugs. However, the analogy between EGFR and VEGFR mutations may be limited, because the relevance of mutations could not be the same given that EGFR is expressed on genetically unstable tumor cells whereas VEGFRs are mainly expressed on endothelial cells. On the other hand, the recent description of the presence of VEGFRs on cancer cells and the direct antiproliferative activity of tyrosine kinase inhibitors [60–67] open new opportunities for research on somatic mutations in the VEGFR2 receptor. Indeed, somatic mutations in the VEGFR2 gene have been

TABLE 41.3. SNPs of VEGFRs with associated biologic effect or clinical condition.

	Receptor	Associated biologic effect or clinical condition
2399 G/A, 2455 C/T, 3217 G/A	VEGFR-2/KDR	CRC [68]
3439 TT	VEGFR-2/KDR	Hemangioma [69]
4422(AC)11–14[a]	VEGFR-2/KDR	Coronary arteriopathy in Kawasaki disease [70]
–519 C/T	VEGFR1/FLT1	P53-dependent responsiveness to genotoxic stress [71]
889 G/A, AGAG haplotype	VEGFR-2/KDR	Atopy/asthma [72]

[a]AC dinucleotide repeat at intron 2

CRC colorectal cancer

described in colorectal cancer. Three non-synonymous SNPs (2399 G/A, 2455 C/T, 3217 G/A) corresponding to amino acid substitutions in the kinase domain (G800D, R819Stop, and A1073T, respectively) of VEGFR2 have been detected [68]. Moreover, a 3439 C/T transition in exon 26 of VEGFR2 in a case of juvenile hemangioma, resulting in a proline-to-serine substitution at codon 1147 in tumor tissue, was also described (Table 41.3) [69].

The analysis of genetic polymorphisms of VEGFR2 in Japanese patients with Kawasaki disease (KD) and normal control subjects demonstrated that the frequency of the A1 allele with 11 AC dinucleotide repeats in intron 2 of VEGFR2 was significantly higher in KD patients with coronary artery lesions (CAL) than in those without CAL ($P = 0.013$) or control subjects ($P = 0.040$), thus suggesting a possible involvement of VEGF SNPs in the development of CAL in KD patients (Table 41.3) [70]. A C>T SNP upstream of the transcriptional start site of the VEGFR1 promoter within a putative p53 response element has been shown in approximately 6% of subjects, and it has been found that only the promoter with the T SNP was responsive to p53 (Table 41.3) [71]. In response to doxorubicin-induced DNA damage, there was an allelic discrimination based on p53 binding at the VEGFR1-T but not VEGFR1-C promoters, as well as p53-dependent induction of VEGFR1 mRNA, which required the presence of VEGFR1-T. This study provides evidence that p53 can differentially stimulate the transcription of a polymorphic variant of VEGFR1 promoter and directly places the VEGF system within the p53 stress-response network via VEGFR1 [71]. The VEGFR2 polymorphisms 54 A/G, 889 G/A, 1416 T/A and −92 G/A were investigated in atopic juvenile patients, and a study found that, while 889 G/A was marginally associated with the prevalence of asthma, the association was strong for the AGAG haplotype (Table 3) [72]. Finally, a report on the prevalence of SNPs located in the promoter, in the 5′-UTR and in the coding regions of VEGFR-2 in patients with amyotrophic lateral sclerosis, showed no significant association [73].

Pharmacogenetic Clinical Trials Involving Anti-VEGF Drugs

Pharmacogenetic clinical trials testing antiangiogenic drugs could be conducted as a "nested" study within large randomized phase II-III trials or as independent studies focused on the validation of specific genetic determinants. Both types of trials could be useful to gain different objectives: (1) to find out new pharmacogenetic determinants (i.e., sequencing numerous genes directly involved in the mechanism of action and toxicity of drugs such as TKI or bevacizumab) or study the epidemiology of polymorphisms and haplotypes in a large population of patients; and (2) to investigate the statistical correlations between selected SNPs or haplotypes and some hard or surrogate end-points. These trials can be conducted prospectively or retrospectively, such as in case-control studies.

Two types of approach have been defined in recent years in order to set up pharmacogenetic studies: the "candidate gene approach" and the "whole genome SNP approach". The first one involves a priori SNPs selection (maximum 3–5 per gene) and usually the study is set to confirm a fixed hypothesis (i.e., the "low production" VEGF SNPs 936 C/T may represent a suitable candidate gene for bevacizumab in order to investigate the possible resistance to the treatment). The second approach is much more costly but can investigate 100,000 SNPs per whole genome, and it may show unexpected correlations. However, both models can be integrated in the validation process of a SNP [74]. To move the pharmacogenetics of anti-VEGF-A therapy to clinical practice, the "pyramidal step model" approach should be followed [75], which includes: (1) sequencing of the candidate genes and in vitro studies to validate the functional role of VEGF polymorphisms; (2) proof of concept demonstration in clinical trials aimed also at documenting sufficient degree of variability; and (3) proof of superiority of selecting the treatment on the basis of pharmacogenetic versus the standard approach.

VEGF-A and VEGFR1/VEGFR2 gene polymorphisms have to be investigated in more detail, particularly with respect to somatic mutations in tissue specimens and especially in those tumors where the autocrine loop has been described. However, the candidate genes for drugs involved in blocking VEGF-induced angiogenesis should not be restricted to the VEGF-A pathway. Other genes could be studied, such as HIF-1, TIMP-3 and eNOS, because they are closely related to VEGF-A-dependent angiogenesis.

The selection of SNPs for a possible successful pharmacogenetic approach of bevacizumab-based therapy should not leave the four main functional VEGF-A polymorphisms (936 C/T, −634 G/C, −1154 G/A, −2578 C/A) out of consideration. Indeed, these SNPs have been directly correlated to the difference in production and secretion of the molecular target of drugs such as bevacizumab. Moreover, the VEGF-A gene promoter haplotypes need to be evaluated, at least the main ones (haplotype 1–4), because some of them have been shown to be related to altered VEGF-A production. The sample size of a pharmacogenetic clinical study including anti-VEGF drugs depends on the frequency of the selected SNPs or haplotypes in the selected patient population. Sample size determination remains a critical issue, at least in the oncology field. For example, the search for the 936 SNP should take into account that its frequency is around 16% [30]. Thus, to test the hypothesis that the SNP affects the clinical outcome (e.g., overall survival), we would need to enroll a statistically-powered number of patients.

Presently, few studies have evaluated the correlation between VEGF-A plasma levels and genetic background, and no studies have investigated the genetic impact in the clinical response of anti-VEGF drugs. Thus, there is an urgent need for pharmacogenetic studies on VEGF-A, evaluating both genotypes and phenotypes, and its correlation with clinical outcome. The VEGF-A phenotype could be analyzed from different point of views, such as VEGF-A immunohistochemistry of tumor specimens, free VEGF-A plasma levels and PDGF-stimulated PBMC VEGF-A production from blood samples. Among the proposed biomarkers [5] that may be applicable to anti-VEGF agents, circulating endothelial progenitors (CEPs) have attracted considerable interest [76], whereas measurement of serum VEGF-A levels could be a surrogate biomarker for bevacizumab [77]. Indeed, reduction of free serum VEGF-A in cancer patients treated with escalating doses of an anti-VEGF antibody when compared to basal serum concentrations has been described [78]. Based on these data, the ability to detect the biologically active free plasma VEGF-A would be a key step to uncover possible significant correlations with the genetic background of the patients. The end-points for clinical trials including antiangiogenic drugs are still being debated. However, the pivotal randomized phase III trial of first-line treatment in patients with metastatic colorectal cancer demonstrated, for the first time, that the addition of an antiangiogenic drug such as bevacizumab to conventional chemotherapy (irinotecan, fluorouracil and leucovorin) significantly improved the response rate, the median progression free survival, and the median duration of survival [58].

Conclusions

Molecular surrogate markers and pharmacogenetic profiling may improve the prediction of patients who will experience significant benefit or toxicity from the currently available antiangiogenic agents. Validation of these putative predictive factors in prospective clinical trials may represent the future approach to anti-VEGF therapy and the necessary step to allow a rational and systematic individualization of the antiangiogenic strategy.

References

1. Danesi R, de Braud F, Fogli S, de Pas TM, Di Paolo A, Curigliano G, et al. Pharmacogenetics of anticancer drug sensitivity in non-small cell lung cancer. Pharmacol Rev 2003;55:57–103.
2. Siest G, Marteau JB, Maumus S, Berrahmoune H, Jeannesson E, Samara A, et al. Pharmacogenomics and cardiovascular drugs: need for integrated biological system with phenotypes and proteomic markers. Eur J Pharmacol 2005;527:1–22.
3. Carmeliet P. Blood vessels and nerves: common signals, pathways and diseases. Nat Rev Genet 2003;4:710–20.
4. Bocci G, Man S, Green SK, Francia G, Ebos JM, du Manoir JM, et al. Increased plasma vascular endothelial growth factor (VEGF) as a surrogate marker for optimal therapeutic dosing of VEGF receptor-2 monoclonal antibodies. Cancer Res 2004;64:6616–25.
5. Shaked Y, Bocci G, Munoz R, Man S, Ebos JM, Hicklin DJ, et al. Cellular and molecular surrogate markers to monitor targeted and non-targeted antiangiogenic drug activity and determine optimal biologic dose. Curr Cancer Drug Targets 2005;5:551–9.
6. Shaked Y, Bertolini F, Man S, Rogers MS, Cervi D, Foutz T, et al. Genetic heterogeneity of the vasculogenic phenotype parallels angiogenesis: implications for cellular surrogate marker analysis of antiangiogenesis. Cancer Cell 2005;7:101–11.
7. Iruela-Arispe ML, Dvorak HF. Angiogenesis: a dynamic balance of stimulators and inhibitors. Thromb Haemost 1997;78: 672–7.
8. Mazure NM, Chen EY, Yeh P, Laderoute KR, Giaccia AJ. Oncogenic transformation and hypoxia synergistically act to modulate vascular endothelial growth factor expression. Cancer Res 1996;56:3436–40.
9. Levy NS, Chung S, Furneaux H, Levy AP. Hypoxic stabilization of vascular endothelial growth factor mRNA by the RNA-binding protein HuR. J Biol Chem 1998;273:6417–23.
10. Scandurro AB, Beckman BS. Common proteins bind mRNAs encoding erythropoietin, tyrosine hydroxylase, and vascular endothelial growth factor. Biochem Biophys Res Commun 1998;246:436–40.
11. Levy AP, Levy NS, Goldberg MA. Hypoxia-inducible protein binding to vascular endothelial growth factor mRNA and its modulation by the von Hippel-Lindau protein. J Biol Chem 1996;271:25492–7.
12. Rak J, Mitsuhashi Y, Bayko L, Filmus J, Shirasawa S, Sasazuki T, et al. Mutant ras oncogenes upregulate VEGF/VPF expression: implications for induction and inhibition of tumor angiogenesis. Cancer Res 1995;55:4575–80.
13. Ikeda N, Nakajima Y, Sho M, Adachi M, Huang CL, Iki K, et al. The association of K-ras gene mutation and vascular endothelial growth factor gene expression in pancreatic carcinoma. Cancer 2001;92:488–99.
14. Talbot SG, P Oc, Sarkaria IS, Ghossein R, Reddy P, Ngai I, et al. Squamous cell carcinoma related oncogene regulates angiogenesis through vascular endothelial growth factor-A. Ann Surg Oncol 2004;11:530–4.
15. Mezquita P, Parghi SS, Brandvold KA, Ruddell A. Myc regulates VEGF production in B cells by stimulating initiation of VEGF mRNA translation. Oncogene 2005;24:889–901.
16. Hatzi E, Murphy C, Zoephel A, Rasmussen H, Morbidelli L, Ahorn H, et al. N-myc oncogene overexpression down-regulates

IL-6; evidence that IL-6 inhibits angiogenesis and suppresses neuroblastoma tumor growth. Oncogene 2002;21:3552–61.

17. Jain RK, Duda DG, Clark JW, Loeffler JS. Lessons from phase III clinical trials on anti-VEGF therapy for cancer. Nat Clin Pract Oncol 2006;3:24–40.

18. Lyseng-Williamson KA, Robinson DM. Bevacizumab. A review of its use in advanced colorectal cancer, breast cancer and NSCLC. Am J Cancer 2006;5:43–60.

19. Kerbel RS. Antiangiogenic therapy: a universal chemosensitization strategy for cancer? Science 2006;312:1171–5.

20. Casanovas O, Hicklin DJ, Bergers G, Hanahan D. Drug resistance by evasion of antiangiogenic targeting of VEGF signaling in late-stage pancreatic islet tumors. Cancer Cell 2005;8:299–309.

21. Saif MW, Mehra R. Incidence and management of bevacizumab-related toxicities in colorectal cancer. Expert Opin Drug Saf 2006;5:553–66.

22. Luo Y, Zhou H, Krueger J, Kaplan C, Lee SH, Dolman C, et al. Targeting tumor-associated macrophages as a novel strategy against breast cancer. J Clin Invest 2006;116:2132–41.

23. Tang Y, Nakada MT, Kesavan P, McCabe F, Millar H, Rafferty P, et al. Extracellular matrix metalloproteinase inducer stimulates tumor angiogenesis by elevating vascular endothelial cell growth factor and matrix metalloproteinases. Cancer Res 2005;65:3193–9.

24. Pollard JW. Tumour-educated macrophages promote tumour progression and metastasis. Nat Rev Cancer 2004;4:71–8.

25. Coussens LM, Werb Z. Inflammation and cancer. Nature 2002;420:860–7.

26. Krippl P, Langsenlehner U, Renner W, Yazdani-Biuki B, Wolf G, Wascher TC, et al. A common 936 C/T gene polymorphism of vascular endothelial growth factor is associated with decreased breast cancer risk. Int J Cancer 2003;106:468–71.

27. Boiardi L, Casali B, Nicoli D, Farnetti E, Chen Q, Macchioni P, et al. Vascular endothelial growth factor gene polymorphisms in giant cell arteritis. J Rheumatol 2003;30:2160–4.

28. Watson CJ, Webb NJ, Bottomley MJ, Brenchley PE. Identification of polymorphisms within the vascular endothelial growth factor (VEGF) gene: correlation with variation in VEGF protein production. Cytokine 2000;12:1232–5.

29. Rogers MS, D'Amato RJ. The effect of genetic diversity on angiogenesis. Exp Cell Res 2006;312:561–74.

30. Renner W, Kotschan S, Hoffmann C, Obermayer-Pietsch B, Pilger E. A common 936 C/T mutation in the gene for vascular endothelial growth factor is associated with vascular endothelial growth factor plasma levels. J Vasc Res 2000;37:443–8.

31. Jacobs EJ, Feigelson HS, Bain EB, Brady KA, Rodriguez C, Stevens VL, et al. Polymorphisms in the vascular endothelial growth factor gene and breast cancer in the Cancer Prevention Study II cohort. Breast Cancer Res 2006;8:R22.

32. Shahbazi M, Fryer AA, Pravica V, Brogan IJ, Ramsay HM, Hutchinson IV, et al. Vascular endothelial growth factor gene polymorphisms are associated with acute renal allograft rejection. J Am Soc Nephrol 2002;13:260–4.

33. Tzanakis N, Gazouli M, Rallis G, Giannopoulos G, Papaconstantinou I, Theodoropoulos G, et al. Vascular endothelial growth factor polymorphisms in gastric cancer development, prognosis, and survival. J Surg Oncol 2006;94:624–30.

34. McCarron SL, Edwards S, Evans PR, Gibbs R, Dearnaley DP, Dowe A, et al. Influence of cytokine gene polymorphisms on the development of prostate cancer. Cancer Res 2002;62:3369–72.

35. Sfar S, Hassen E, Saad H, Mosbah F, Chouchane L. Association of VEGF genetic polymorphisms with prostate carcinoma risk and clinical outcome. Cytokine 2006;35:21–8.

36. Koukourakis MI, Papazoglou D, Giatromanolaki A, Bougioukas G, Maltezos E, Sivridis E. VEGF gene sequence variation defines VEGF gene expression status and angiogenic activity in non-small cell lung cancer. Lung Cancer 2004;46:293–8.

37. Garcia-Closas M, Malats N, Real FX, Yeager M, Welch R, Silverman D, et al. Large-scale evaluation of candidate genes identifies associations between VEGF polymorphisms and bladder cancer risk. PLoS Genet 2007;3:e29.

38. Howell WM, Bateman AC, Turner SJ, Collins A, Theaker JM. Influence of vascular endothelial growth factor single nucleotide polymorphisms on tumour development in cutaneous malignant melanoma. Genes Immun 2002;3:229–32.

39. Kim Y, Nam YJ, Lee C. Haplotype analysis of single nucleotide polymorphisms in VEGF gene for vascular dementia. Am J Med Genet B Neuropsychiatr Genet 2006;141:332–5.

40. van der Meer P, De Boer RA, White HL, van der Steege G, Hall AS, Voors AA, et al. The VEGF +405 CC promoter polymorphism is associated with an impaired prognosis in patients with chronic heart failure: a MERIT-HF substudy. J Card Fail 2005;11:279–84.

41. Rueda B, Perez-Armengol C, Lopez-Lopez S, Garcia-Porrua C, Martin J, Gonzalez-Gay MA. Association between functional haplotypes of vascular endothelial growth factor and renal complications in Henoch-Schonlein purpura. J Rheumatol 2006;33:69–73.

42. Al-Kateb H, Mirea L, Xie X, Sun L, Liu M, Chen H, et al. Multiple variants in Vascular Endothelial Growth Factor (VEGF) are risk factors for time to severe retinopathy in type 1 diabetes: The DCCT/EDIC genetics study. Diabetes 2007;56(8):2161–8.

43. Churchill AJ, Carter JG, Lovell HC, Ramsden C, Turner SJ, Yeung A, et al. VEGF polymorphisms are associated with neovascular age-related macular degeneration. Hum Mol Genet 2006;15:2955–61.

44. Butt C, Lim S, Greenwood C, Rahman P. VEGF, FGF1, FGF2 and EGF gene polymorphisms and psoriatic arthritis. BMC Musculoskelet Disord 2007;8:1.

45. Girnita DM, Webber SA, Ferrell R, Burckart GJ, Brooks MM, McDade KK, et al. Disparate distribution of 16 candidate single nucleotide polymorphisms among racial and ethnic groups of pediatric heart transplant patients. Transplantation 2006;82:1774–80.

46. Lambrechts D, Storkebaum E, Morimoto M, Del-Favero J, Desmet F, Marklund SL, et al. VEGF is a modifier of amyotrophic lateral sclerosis in mice and humans and protects motoneurons against ischemic death. Nat Genet 2003;34:383–94.

47. Awata T, Inoue K, Kurihara S, Ohkubo T, Watanabe M, Inukai K, et al. A common polymorphism in the 5'-untranslated region of the VEGF gene is associated with diabetic retinopathy in type 2 diabetes. Diabetes 2002;51:1635–9.

48. Stevens A, Soden J, Brenchley PE, Ralph S, Ray DW. Haplotype analysis of the polymorphic human vascular endothelial growth factor gene promoter. Cancer Res 2003;63:812–6.

49. Han SW, Kim GW, Seo JS, Kim SJ, Sa KH, Park JY, et al. VEGF gene polymorphisms and susceptibility to rheumatoid arthritis. Rheumatology (Oxford) 2004;43:1173–7.

50. Bhanoori M, Arvind Babu K, Pavankumar Reddy NG, Lakshmi Rao K, Zondervan K, Deenadayal M, et al. The vascular endothelial growth factor (VEGF) +405 G>C 5'-untranslated region polymorphism and increased risk of endometriosis in South Indian women: a case control study. Hum Reprod 2005;20: 1844–9.

51. Lu H, Shu XO, Cui Y, Kataoka N, Wen W, Cai Q, et al. Association of genetic polymorphisms in the VEGF gene with breast cancer survival. Cancer Res 2005;65:5015–9.

52. Jin Q, Hemminki K, Enquist K, Lenner P, Grzybowska E, Klaes R, et al. Vascular endothelial growth factor polymorphisms in relation to breast cancer development and prognosis. Clin Cancer Res 2005;11:3647–53.

53. Ferrante M, Pierik M, Henckaerts L, Joossens M, Claes K, Van Schuerbeek N, et al. The role of vascular endothelial growth factor (VEGF) in inflammatory bowel disease. Inflamm Bowel Dis 2006;12:870–8.

54. Autiero M, Luttun A, Tjwa M, Carmeliet P. Placental growth factor and its receptor, vascular endothelial growth factor receptor-1: novel targets for stimulation of ischemic tissue revascularization and inhibition of angiogenic and inflammatory disorders. J Thromb Haemost 2003;1:1356–70.

55. Carmeliet P. Angiogenesis in life, disease and medicine. Nature 2005;438:932–6.

56. Uthoff SM, Duchrow M, Schmidt MH, Broll R, Bruch HP, Strik MW, et al. VEGF isoforms and mutations in human colorectal cancer. Int J Cancer 2002;101:32–6.

57. Feinbloom D, Bauer KA. Assessment of hemostatic risk factors in predicting arterial thrombotic events. Arterioscler Thromb Vasc Biol 2005;25:2043–53.

58. Hurwitz H, Fehrenbacher L, Novotny W, Cartwright T, Hainsworth J, Heim W, et al. Bevacizumab plus irinotecan, fluorouracil, and leucovorin for metastatic colorectal cancer. N Engl J Med 2004;350:2335–42.

59. Lynch TJ, Bell DW, Sordella R, Gurubhagavatula S, Okimoto RA, Brannigan BW, et al. Activating mutations in the epidermal growth factor receptor underlying responsiveness of non-small-cell lung cancer to gefitinib. N Engl J Med 2004;350:2129–39.

60. Zhong X, Li X, Wang G, Zhu Y, Hu G, Zhao J, et al. Mechanisms underlying the synergistic effect of SU5416 and cisplatin on cytotoxicity in human ovarian tumor cells. Int J Oncol 2004;25:445–51.

61. Das B, Yeger H, Tsuchida R, Torkin R, Gee MF, Thorner PS, et al. A hypoxia-driven vascular endothelial growth factor/Flt1 autocrine loop interacts with hypoxia-inducible factor-1alpha through mitogen-activated protein kinase/extracellular signal-regulated kinase 1/2 pathway in neuroblastoma. Cancer Res 2005;65:7267–75.

62. Giatromanolaki A, Koukourakis MI, Turley H, Sivridis E, Harris AL, Gatter KC. Phosphorylated KDR expression in endometrial cancer cells relates to HIF1alpha/VEGF pathway and unfavourable prognosis. Mod Pathol 2006;19:701–7.

63. Hiramatsu A, Miwa H, Shikami M, Ikai T, Tajima E, Yamamoto H, et al. Disease-specific expression of VEGF and its receptors in AML cells: possible autocrine pathway of VEGF/type1 receptor of VEGF in t(15;17) AML and VEGF/type2 receptor of VEGF in t(8;21) AML. Leuk Lymphoma 2006;47:89–95.

64. Lacal PM, Ruffini F, Pagani E, D'Atri S. An autocrine loop directed by the vascular endothelial growth factor promotes invasiveness of human melanoma cells. Int J Oncol 2005;27:1625–32.

65. Vieira JM, Santos SC, Espadinha C, Correia I, Vag T, Casalou C, et al. Expression of vascular endothelial growth factor (VEGF) and its receptors in thyroid carcinomas of follicular origin: a potential autocrine loop. Eur J Endocrinol 2005;153:701–9.

66. Weigand M, Hantel P, Kreienberg R, Waltenberger J. Autocrine vascular endothelial growth factor signalling in breast cancer. Evidence from cell lines and primary breast cancer cultures in vitro. Angiogenesis 2005;8:197–204.

67. Wey JS, Fan F, Gray MJ, Bauer TW, McCarty MF, Somcio R, et al. Vascular endothelial growth factor receptor-1 promotes migration and invasion in pancreatic carcinoma cell lines. Cancer 2005;104:427–38.

68. Bardelli A, Parsons DW, Silliman N, Ptak J, Szabo S, Saha S, et al. Mutational analysis of the tyrosine kinome in colorectal cancers. Science 2003;300:949.

69. Walter JW, North PE, Waner M, Mizeracki A, Blei F, Walker JW, et al. Somatic mutation of vascular endothelial growth factor receptors in juvenile hemangioma. Genes Chromosomes Cancer 2002;33:295–303.

70. Kariyazono H, Ohno T, Khajoee V, Ihara K, Kusuhara K, Kinukawa N, et al. Association of vascular endothelial growth factor (VEGF) and VEGF receptor gene polymorphisms with coronary artery lesions of Kawasaki disease. Pediatr Res 2004;56: 953–9.

71. Menendez D, Krysiak O, Inga A, Krysiak B, Resnick MA, Schonfelder G. A SNP in the flt-1 promoter integrates the VEGF system into the p53 transcriptional network. Proc Natl Acad Sci USA 2006;103:1406–11.

72. Park HW, Lee JE, Shin ES, Lee JY, Bahn JW, Oh HB, et al. Association between genetic variations of vascular endothelial growth factor receptor 2 and atopy in the Korean population. J Allergy Clin Immunol 2006;117:774–9.

73. Brockington A, Wokke B, Nixon H, Hartley J, Shaw PJ. Screening of the transcriptional regulatory regions of vascular endothelial growth factor receptor 2 (VEGFR2) in amyotrophic lateral sclerosis. BMC Med Genet 2007;8:23.

74. Mooser V, Waterworth DM, Isenhour T, Middleton L. Cardiovascular pharmacogenetics in the SNP era. J Thromb Haemost 2003;1:1398–402.

75. Johnson JA, Cavallari LH. Cardiovascular pharmacogenomics. Exp Physiol 2005;90:283–9.

76. Bertolini F, Shaked Y, Mancuso P, Kerbel RS. The multifaceted circulating endothelial cell in cancer: towards marker and target identification. Nat Rev Cancer 2006;6:835–45.

77. Cristofanilli M, Charnsangavej C, Hortobagyi GN. Angiogenesis modulation in cancer research: novel clinical approaches. Nat Rev Drug Discov 2002;1:415–26.

78. Gordon MS, Margolin K, Talpaz M, Sledge GW, Jr., Holmgren E, Benjamin R, et al. Phase I safety and pharmacokinetic study of recombinant human anti-vascular endothelial growth factor in patients with advanced cancer. J Clin Oncol 2001;19: 843–50.

79. Chakravarthy U, Adamis AP, Cunningham ET, Jr., Goldbaum M, Guyer DR, Katz B, et al. Year 2 efficacy results of 2 randomized controlled clinical trials of pegaptanib for neovascular age-related macular degeneration. Ophthalmology 2006;113: 1508 e1–25.

80. Demetri GD, van Oosterom AT, Garrett CR, Blackstein ME, Shah MH, Verweij J, et al. Efficacy and safety of sunitinib in patients with advanced gastrointestinal stromal tumour after failure

of imatinib: a randomised controlled trial. Lancet 2006;368:
1329–38.

81. Motzer RJ, Michaelson MD, Redman BG, Hudes GR, Wilding
G, Figlin RA, et al. Activity of SU11248, a multitargeted inhibitor of vascular endothelial growth factor receptor and platelet-
derived growth factor receptor, in patients with metastatic renal
cell carcinoma. J Clin Oncol 2006;24:16–24.

82. Motzer RJ, Rini BI, Bukowski RM, Curti BD, George DJ, Hudes
GR, et al. Sunitinib in patients with metastatic renal cell carcinoma. JAMA 2006;295:2516–24.

83. Eisen T, Bukowski RM, Stehler M, Szczylik C. Randomized
phase III trial of sorafenib in advanced renal cell carcinoma
(RCC): impact crossover on survival. Proc Am Soc Clin Oncol
2006;24:4524.

Section VI
Angiogenesis in Health & Disease

Chapter 42
Angiogenesis in the Central Nervous System

Carmen Ruiz de Almodovar[1,2]*, Serena Zacchigna[1,2]*, and Peter Carmeliet**[1,2]

Keywords: CNS vascularization, ephrinB2, blood brain barrrier, CNS genetic diseasess, CNS disorders, CNS tumors

Abstract: Angiogenesis is critical for the development and repair, and contributes to disorders of the central nervous system (CNS). Identifying the signals that regulate CNS vascularization in health and disease offers novel opportunities to treat CNS disorders. We will review vascular development in the CNS (excluding in the retina, which is described in Chapter 44 on Ocular Neovascularization) in development and disease.

Angiogenesis in the Developing Nervous System

Development of any vascularized organ, and thus also of the CNS, relies on the formation of its vasculature. It is therefore not surprising that the vascularization and wiring of the CNS are tightly controlled and intertwined. In the first part of this chapter, we will review this process, with special attention to the role of vascular endothelial growth factor (VEGF). We will also discuss the formation and maintenance of the blood brain barrier (BBB), which is essential for maintaining and ensuring appropriate activities of neurons.

[1]Department for Transgene Technology and Gene Therapy, VIB, B-3000 Leuven, Belgium; [2]The Center for Transgene Technology and Gene Therapy (CTG), K.U. Leuven, B-3000 Leuven, Belgium

*Both authors contributed equally to this work

**Corresponding Author:

Peter Carmeliet, VIB Department of Transgene Technology and Gene Therapy, K.U. Leuven, Onderwijs en Navorsing 1, 9e verd, Campus Gasthuisberg, Herestraat 49, bus 912, 3000 Leuven, Belgium
E-mail: peter.carmeliet@med.kuleuven.be

Patterning of CNS Vascularization

Multiple cellular processes contribute to the formation of the embryonic vasculature: differentiation of angioblasts into endothelial cells (ECs) and their subsequent migration and assembly into vessels [1,2]. The two main structures of the CNS, the brain and the spinal cord, develop from the neural tube (NT). In higher organisms, vascularization of this primitive structure occurs at mid-gestation by the formation of a perineural vascular plexus (PNVP) around the neural tube (NT), symmetrically relative to the midline of the CNS (Fig. 42.1). This plexus branches off from primitive arterial tracts, located ventrally on either side of the floor plate, that originate from segmental branches of the aorta [3] and cardinal veins drain the deoxygenated blood back to the heart.

The NT itself does not contain endothelial progenitors, but is vascularized by invading exogenous endothelial precursors [4] derived from both the lateral plate mesoderm and the somites [5–8]. It remains poorly understood how angioblasts are instructed to migrate to the midline and form this vascular plexus. The NT, together with the notochord and endoderm, is a signaling center that organizes embryonic structures around the midline (Fig. 42.1). Coculture experiments of NT with presomitic mesoderm revealed that the NT instructs the formation of a vascular plexus [7]. Further evidence for a critical role of the NT is deduced from findings that, in *Tbx6* mutant embryos, two ectopic NTs form on each side of the midline [9], and that each of these ectopic NTs is surrounded by a vascular plexus [7].

Although angioblasts invade the dorsal regions of the NT [3], it is generally accepted that vascularization of the NT occurs via vessel sprouting from the PNVP [10]. Vessels sprout from the primitive arterial tracts lateral to the floor plate, with ECs extending long filopodia protruding between the neuroepithelium. Later, additional sprouts from the lateral PNVP invade the NT and grow in a medio-ventral direction to establish vascular connections with sprouts of the floor plate, thereby forming a loop around the motor column [3,7] (Fig. 42.1). Once vessels become lumenized, CNS circulation is initiated [3].

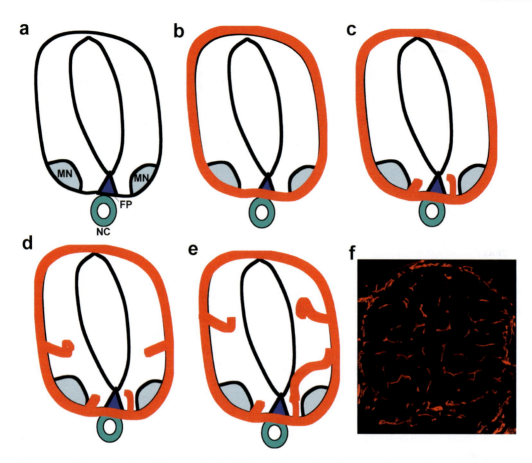

FIG. 42.1. Vascularization of the neural tube. **a** The neural tube (NT) itself is the signaling center that organizes the embryonic NT vasculature. **b** Lateral plate-derived and somite-derived angioblasts are attracted by VEGF, among other angiogenic factors, produce by motoneurons (MN) and by the floor plate (FP). **c–e** Vessel sprouting from the perineural vascular plexus (PNVP) into the NT. **f** Image of a E12.5 mouse NT. Section was immunostained with an EC marker (CD31) in order to visualize the developing vasculature.

As mentioned above, the main structure of the CNS, the brain, differentiates from the NT. Once the PNVP is formed, the cerebral vasculature develops by sprouting of vessels from the PNVP into the brain parenchyma, presumably in response to oxygen and metabolic demands [3]. Cerebral angiogenesis occurs in a ventral-to-dorsal temporal developmental gradient in the telencephalon [11]. The current established model of developmental brain angiogenesis proposes that ECs sprout from pial vessels into the brain and converge radially towards the ventricles. Once vessels penetrate the cortex, they extend second-order branches that surround the ventricles, or extend branches in reverse direction towards the pial surface [12] (Fig. 42.2).

Molecular Mechanisms Implicated in CNS Vascularization

Various signals have been proposed to regulate the vascularization of the NT: sonic hedgehog (Shh), VEGF and its receptor VEGFR2 (Flk-1), and, angiopoieitin-1 (Ang1) and its Tie2 receptor. Indeed, VEGF is expressed at the lateral edge of the embryonic NT precisely at the time of PNVP formation [7,13]. Furthermore, VEGF was shown to induce NT vascularization, while VEGF inhibitors inhibited PNVP formation in NT/presomitic mesoderm co-cultures [7]. Blocking

VEGF signaling genetically by using presomitic mesoderm explants from VEGFR2 deficient embryos also prevented neural tube-dependent vascular plexus formation [7]. The effects of Shh on vascular development appear to be indirect. For example, in zebrafish, Shh up-regulated VEGF expression in the adjacent somites, which in turn drove proper formation of the dorsal aorta [14]. In the CNS, although transgenic over-expression of Shh in the mouse NT results in hypervascularization [15], Shh failed to affect human umbilical vein endothelial cells (HUVECs) directly [16]. In the mouse NT, Shh, VEGF and Ang1 are expressed by neural progenitor cells in the ventricular zone, floor plate cells and motoneurons (only VEGF and Ang1) at stage E8.5-9.0 [7,13,17], precisely at the time of vessel sprouting. In addition, their receptors, VEGFR2 and Tie2, are expressed in ECs around and within the NT [13]. Interestingly, Shh, expressed in the floor plate and notochord, is essential for the induction of motoneuron identity [18]. Motoneuron induction, VEGF and Ang1 expression as well as vascularization of the NT are impaired in mouse embryos exposed to a Shh signaling inhibitor, stressing the importance of motoneurons, and of Ang1 and VEGF signaling for proper vascularization of the NT [13]. Consistent herewith, loss of Tie2 impairs sprouting from the PNVP [19]. Finally, ephrinB2 also regulates NT vascularization, as capil-

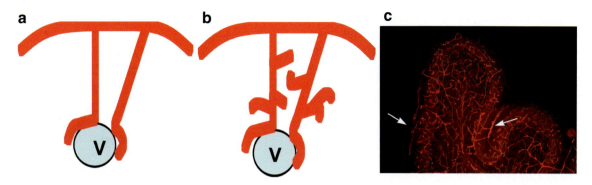

FIG. 42.2. Vascularization of the brain. **a** Vessels sprout from the perineural vascular plexus (PNVP) and begin to grow into the brain parenchyma in a radial direction towards the ventricle (*V*). **b** Once they reach the ventricle, some extend branches to surround the ventricle or extend them in a reverse direction towards the pia. VEGF is expressed in the ventricular zone and it is known to play an important role in the regulation of this angiogenic process. **c** Image of a 10-day-old mouse cerebellum. A 300-μm thick section was immunostained with an EC marker (Isolectin-B4) to observe the developing vasculature. *Arrows* point towards the pial vessels.

lary ingrowth into the NT fails to occur in eprhinB2 knockout mice and, instead, ehprinB2-expressing ECs stalled at the exterior surface of the NT [20].

Brain vascularization is similarly regulated by several angiogenic factors, the most prominent of these is again VEGF, which is expressed by neural progenitor cells in the ventricular zone of the developing mouse [21] and human [22] brain. Angiopoetins are also important for developmental brain angiogenesis [23]. For example, in the cerebellum, VEGF, Ang1/2 and their receptors are expressed in a temporal and spatial pattern consistent with blood vessel growth [24]. Indeed, VEGF and Ang1 are expressed in Purkinje cells and astrocytes during cerebellar development [24], while Ang2, VEGFR2, VEGFR1/Flt-1, Tie1 and Tie2 are expressed in ECs. These expression patterns suggest that VEGF and angiopoietins, together, may act in a coordinated manner to induce the formation of the complex and highly branched vascular network of the cerebellum [24].

Interestingly, a spatial VEGF gradient, established by the different affinity of distinct VEGF isoforms for heparin-rich matrix components, is necessary for proper vessel patterning in the brain. This was demonstrated in VEGF$^{120/120}$ mice (which lack the heparin-binding VEGF isoforms), in which ECs in the tip of growing sprouts failed to extend their filopodia towards the midline of the developing hindbrain [25]. Several studies investigated in more detail the function of VEGF in the developing brain by inactivating VEGF in neuronal cells. When VEGF was specifically eliminated from neural progenitors, both the invasion and directed ingrowth of capillaries in the developing brain were severely impaired [17,26]. Together, these findings indicate that VEGF, secreted by neural progenitor cells, induces brain angiogenesis and guides vessels towards the ventricular zone. Another VEGF co-receptor, Neuropilin-1 (NRP1), is critical for proper VEGF signaling in CNS angiogenesis, as brain angiogenesis was impaired in NRP1-deficient mice [27,28].

Vascular development in the CNS is also regulated by cell–matrix and cell–cell interaction. Integrins are heterodimeric extracellular matrix (ECM) receptors that control cell migration, adhesion and tissue organization. Genetic studies showed that integrins containing either the αv or β8 subunit are required for

proper capillary development in the CNS [29–31]. Indeed, both β8 or αv knockout mice show severe cerebral hemorrhage during embryogenesis [29,31,32]. Interestingly, however, specific deletion of β8 or αv in ECs did not cause vascular defects [30,33]. In contrast, deletion of β8 from the neuroepithelium by the use of Nestin-Cre mice resulted in aberrant capillary vessel morphology and abnormal clustering of ECs [33]. In β8 Nestin-Cre mutants, cerebral hemorrhages coincided with radial glial disorganization [33]. This is an intriguing observation, as radially oriented blood vessels in the brain often run parallel and in close apposition to radial glial processes, suggesting that defective association between ECs and glia might have caused the leaky vasculature [33]. Moreover, specific elimination of αv in glial cells (using GFAP-Cre transgene) resulted in embryonic and neonatal hemorrhages. Moreover, when αv was conditionally inactivated in glial cells and neurons, the resultant transgenic mice showed hemorrhages and severe neurological defects such as seizures and ataxia [30]. The fact that αv forms heterodimers with β1, β3, β5, β6 and β8 subunits [34], together with the finding that no vascular defects were observed in β6 mutants [35], β3/β5 double knockout mice [32] or in mice lacking β1 in neuroepithelial cells [36], suggests that α5 and β8 function together during proper vascular development in the CNS.

An intriguing, but thus far untested, question is whether spatio-temporal gradients of combinatorial codes of transcription factors determine the formation and identity of the distinct cerebral vascular beds in a similar way as neuroblasts differentiate into distinct neuronal cell types.

Blood Vessel Maturation in the CNS

Establishment of a functional vascular network requires that nascent vessels mature into durable, stable, non-leaky and functional vessels. This stabilization requires recruitment of mural periendothelial (pericytes) and smooth muscle cells (SMCs), deposition of an extracellular matrix and specialization of the vessel wall for structural support and regulation of vessel function. Pericytes and SMCs appear in the CNS shortly after the initial sprouting of ECs [3]. Recruitment of mural cells to nascent vessels relies on several regulatory pathways

(Ang/Tie2, PlGF/VEGFR1, TGF-β, PDGF/PDGFR). Among them, the platelet-derived growth factor (PDGF) family comprises four family members (e.g., PDGF-A to D) which bind, with distinct selectivity to the receptor tyrosine kinases PDGFR-α and -β, expressed on ECs and SMCs. PDGF-BB and its receptor PDGFR-β play essential roles in the stabilization of nascent blood vessels. By releasing PDGF-BB, ECs stimulate the growth and differentiation of PDGFR-ß-positive mesenchymal progenitors and recruit them around nascent vessels. In the CNS, PDGF-B/PDGFR-β signaling appears to be important as mice lacking either the ligand or the receptor develop cerebral hemorrhage as a consequence of the failure to recruit pericytes and SMCs [37,38].

The ECM is important to provide signals for proper vascular cell survival, proliferation, migration, differentiation and maturation. Several lines of evidence point to an important role for ECM-integrin interactions in regulating vessel maturation in the CNS. A study performed in the developing mouse CNS showed that maturation of blood vessels was accompanied by a marked up-regulation of integrin β1 expression together with a switch from expression of integrin α4 and α5 early in development to α1 and α6 later in development and in adulthood [39]. Interestingly, levels of fibronectin, an ECM-ligand that binds to α4/β1 and α5/β1 heterodimers, were timely down-regulated during blood vessel maturation in the CNS, while expression of laminin, a ligand for α1/β1 and α6/β1, increased with maturation of the CNS [39]. Thus, the developmental switch in ECM-integrin signaling may regulate cerebral vessel maturation—the precise molecular mechanisms of which still remain to be defined [40].

Blood Brain Barrier Formation and Maintenance

The BBB arises through a close interaction between ECs, perycites, astrocytes and neurons; they establish functional 'neurovascular units' (Fig. 42.3), which, among other functions, permit or facilitate the entry of essential nutrients and exclude potentially harmful compounds [41]. The BBB is established shortly after vascularization of the NT, already between E11 and E13, and relies, in part, on instructive signals from the surrounding neural environment (neurons and astrocytes) [42]. ECs of the BBB differ from ECs in other regions of the body in that they contain a low number of pinocytotic vesicles, lack fenestration and are connected by tight junctions [43]. The tight junction complex comprises several classes of transmembrane molecules, including occludins and claudins, which interact with transmembrane proteins of adjacent ECs [44]. These junctions significantly restrict the movement of even small ions such as Na+ or Cl− so that the transendothelial electrical resistance, which is typically in the range of 2–20 ohm. cm^2 in peripheral capillaries, can be more than 1,000 ohm.cm^2 in brain endothelium [45]. Mutual interactions between various neurovascular cellular components mediate the formation, maintenance and function of the BBB. Indeed, brain capillaries are surrounded or closely associated by perivascular endfeet of

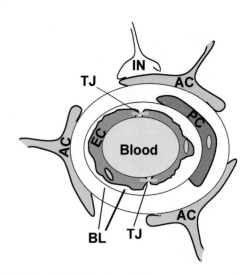

FIG. 42.3. Blood brain barrier (BBB) components. The BBB is formed by *ECs* with tight junctions (*TJ*) between them. ECs are then surrounded by basal lamina (*BL*) and perycites (*PC*). Astrocytes (*AC*) endfeets contact ECs through the basal lamina and also function as a link between interneurons (*IN*) and ECs. All together form the functional 'neurovascular unit'. Mutual interactions among all components contribute to the formation and maintenance of the BBB.

astrocytic glia, pericytes, microglia and neuronal processes [43] (Fig. 42.3); these close cell–cell associations (particularly between astrocytes-ECs) and contacts help to establish the BBB. For instance, astrocytes secrete factors such as basic fibroblast growth factor (bFGF) [46,47], transforming growth factor-ß (TGF-ß), glial-derived neurotrophic factor (GDNF) and Ang1 that induce typical BBB properties such as high electrical resistance and reduced permeability in ECs [45]. Moreover, the Src-suppressed C kinase substrate SSeCKS stimulates astrocytic secretion of Ang1 [48], and SSeCKS-conditioned medium increases the expression of tight junctional proteins and decreases vascular permeability of brain vessels [48]. Interestingly, this intercellular communication appears to be bi-directional, as brain ECs also enhance the growth and differentiation of perivascular astrocytes [49].

Other cell types also contribute to the BBB phenotype. For instance, pericytes, by releasing Ang1 [50] and TGF-β [51], induce and maintain critical BBB functions. Finally, neurons, which make indirect contact with ECs through astrocytes, participate synergistically with astrocytes in inducing the differentiation of brain ECs *in vitro* [52]. The blood retinal barrier (BRB) is formed by the retinal pigmented epithelium (RPE) and the blood vessels of the retina, which form the outer and inner barriers, respectively [53]. As in the BBB, in the inner barrier, ECs, astrocytes, pericytes and astrocytes/glia cells form functional 'vascular units'. Astrocytes also participate in the induction and maintenance of the BRB [43]. The BRB is similar to the BBB in that it serves as a selective barrier that regulates the local environment of the neural retina, specifically, it regulates osmotic balance, ionic concentration, and transport of nutrients, including sugars, lipids, and amino acids. Additionally, the BRB defines the retina as an immunologically-privileged site by acting as a barrier against immunoglobulins and circulating immune cells [54].

The Role of Angiogenesis in the CNS Diseases

Considering the interaction of blood vessels with neuronal cells in embryonic development and adult life, it is not surprising that angiogenesis also plays a role in a variety of CNS disorders—perhaps even more so than originally anticipated. In this section, we will first consider a series of inherited conditions in which blood vessel defects are associated and causally linked to neuronal pathology. Subsequently, we will highlight the role of angiogenesis in more common CNS disorders and discuss possible novel therapeutic strategies aimed at modifying the neurovascular link.

Genetic Diseases Affecting the CNS Vascular System

CADASIL and Other Hereditary Small Vessel Diseases

Hereditary forms of cerebrovascular disorders represent only a minor proportion (<10%) of the burden of neurological disease, but they are invariably associated with a high morbidity and mortality. To date, various autosomal dominant conditions and at least one recessive disorder causing either ischemic or hemorrhagic stroke have been characterized [55–57]. The most common familial cerebral angiopathy is CADASIL (cerebral autosomal dominant arteriopathy with subcortical infarcts and leukoencephalopathy), caused by mutations in the Notch3 gene [58,59]. Worldwide, about 400 families with CADASIL have been identified [60]. The morphological hallmark is a small-vessel non-amyloid, non-atherosclerotic arteriopathy, with accumulation of osmiophilic, granular electron-dense material (GOM) in the medial layer and degeneration of SMCs [60] (Fig. 42.4n, o). The endothelium remains grossly intact though functionally altered, as indicated by the elevation of the cerebrospinal fluid (CSF) protein content, likely due to breakdown of the BBB [61]. Despite a systemic involvement of the microvasculature, cerebral (and spinal) vessels are most affected, leading to recurrent ischemic strokes and progressive subcortical dementia in middle-aged patients [62]. Clinical hallmarks include migraine with aura and mood disorders [63]. Diagnosis often depends on magnetic resonance imaging (MRI), revealing diffuse hyperintensities in the white matter with frequent lacunar infarctions and microbleeds [64]. Symptoms are significantly associated with infarct lesion load, which may well underlie the stepwise decline in cognitive dysfunction [65]. So far, more than 100 mutations have been recorded, all restricted to the EGF repeat domain in the extracellular portion of the Notch3 receptor, and particularly concentrated in the third and fourth exons [66]. However, the precise pathogenesis of the disease remains largely unknown. The degeneration of arterial SMCs has been proposed to result from a toxic gain-of-function of Notch3. This is puzzling, as the presence of Notch3 protein in perivascular GOM deposits is a matter of debate, and the precise nature and composition of these deposits still remains largely unknown [67,68]. Further insights into the pathophysiology of this disease has been deduced from transgenic mice, expressing typical CADASIL mutations: they exhibit abnormal cerebrovascular autoregulation, suggesting that defective vessel reactivity may contribute to the pathogenesis of the disease [69].

Other rare genetic microangiopathies have also been reported: CARASIL (cerebral autosomal recessive arteriopathy with subcortical infarcts) [70], cerebroretinal vasculopathy [71], hereditary vascular retinopathy [72], HERNS (hereditary endotheliopathy with retinopathy, nephropathy and stroke syndrome) [73], retinal arteriolar tortuosity, and leucoencephalopathy [74] and familial amyloid angiopathies [75]. Very recently, distinct forms of hereditary multi-infarct dementia, originally suspected to be CADASIL, but with dissimilar clinical features and not associated to Notch3 mutations, have been reported, thus indicating that familial small vessel diseases are more frequent than initially thought and need further genetic characterization [56,57,76,77]. Thus far, the genetic basis of these neurovascular disorders remains entirely enigmatic.

Moya-Moya Disease

Moya-Moya disease is an idiopathic progressive arteriopathy, relatively common in Japan (prevalence 3 in 100,000) but almost nonexistent elsewhere, that is characterized by intimal hyperplasia and medial fibrosis of the distal internal carotid and basilar arteries, leading to progressive vascular stenosis [78]. The coincident development of profuse teleangiectatic neovessels, presumably induced by the associated ischemia, around the circle of Willis gives origin to the peculiar 'puff of smoke' picture visible on angiography (hence the name Moya-Moya, the Japanese translation of 'puff of smoke') [79] (Fig. 42.4k). The arterial narrowing leads to the appearance of ischemic symptoms in children, while later on, the development of fragile and dilated neovessels, prone to rupture often results in cerebral hemorrhage in adults. Familial forms of the disease show linkage to chromosomes 17q25 and 3p24.2-26, but the causative genes have not yet been identified. In addition, several sporadic cases also exist, and Moya-Moya-like clinical conditions are also found in other genetic diseases, such as neurofibromatosis type 1, tuberous sclerosis and pseudoxanthoma elasticum—the precise genetic causes remain, however, undefined to date [62]. Mitochondrial myopathy, encephalopathy, lactic acidosis and stroke-like episodes (MELAS)

MELAS is among the most frequently occurring maternally inherited mitochondrial diseases, mainly caused by a single nucleotide mutation in the gene for tRNA^{Leu}, which undergoes conformational changes impairing its ability to interact with leucyl-tRNA synthetase and eventually leading to a general reduction in mitochondrial protein synthesis. This in turn causes severe defects in the respiratory chain / oxidative

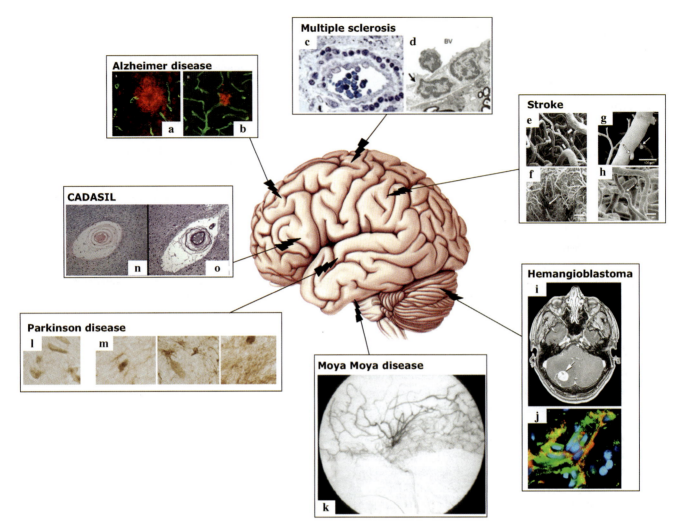

Fig. 42.4. Blood vessel abnormalities in CNS disorders. **a,b** Laser scanning micrograph of double immunofluorescent staining for both β-amyloid (*red*) and GLUT1 (an endothelial marker, in *green*) in a mouse model of AD, showing no direct contact of diffuse plaque and blood vessel (**b**) and reduced capillary density in the proximity of senile plaques (**a**) [201]. **c,d** in (**c**) (toluidine *blue*) a blood vessel, with red cells in the lumen of an acute MS lesion, is ringed by small lymphocytes and the surrounding parenchyma has undergone demyelination; in (**d**), an electron micrograph from an actively demyelinating lesion shows a small lymphocyte, possibly a T-cell, within the vessel lumen, adherent to the endothelium, another cell that has almost crossed the endothelium through a gap junction (*arrow*), and a third cell to the right already within the perivascular space. The perivascular cuff of cells is separated from the central nervous system parenchyma below (myelinated nerve fibers are seen) by a layer of astroglial cells [146]. **e–h** Scanning electron microscopy showing vascular casts from normal rat brain and following MCAO in a rat model of stroke: **e** leptomeningeal (*large arrows*) and small penetrating arterioles (*small arrows*) in normal brain; **f**: area of infarction where no blood vessels are visible (*arrow*); **g** vascular buds were visible 3 days after MCAO (*arrows*); **h** connections or 'nests' of small microvessels associating with surrounding vessels (*arrows*) [202]. **i,j** Contrast-enhanced MRI demonstrating an hemangioblastoma of the cerebellum (**i**) and immunofluorescent characterization of the neoplastic stromal cells in a hemangioblastoma tissue section, showing expression of the mesodermal marker brachyury (*green*), Flk-1 (*red*) and the DAPI nuclear staining (*blue*), indicating their origin from embryonic hemangioblasts (**j**) [175]. **k** The new vessels appear as a puff of smoke and hence the name Moya-Moya (Japanese for puff of smoke) [62]. **l, m** VEGF immunoreactivity in astrocyte-like cells in the substantia nigra compacta of patients with PD (**m**) is enhanced compared to control (**l**) [134]. **n, o** PAS stain highlights degeneration of a small vessel in the brain of a patient affected by CADASIL (**n**); Luxol Fast Blue stain further highlights vessel degeneration, with loss of smooth muscle cells (**o**) (reproduced from Digital Pathology @ Brown Medical School, contributed by Dr. Robert Bagdasaryan). The central brain picture is reproduced with permission from Bear, Connors and Paradiso, Neuroscience 2Ed, edited by Suzanne Katz, copyright © 2001 Lippincott Williams & Wilkins.

phosphorylation pathway, associated with reduced activity of complex I and cytochrome-c oxidase (COX) [80,81]. Neurological involvement is prominent in MELAS, with typical features of stroke-like episodes, hemiparesis and hemi-anopsia, encephalopathy with migraine, dementia and seizures, and lactic acidosis. Despite progressive characterization of the genetic defect, little is known about how a general mitochondrial dysfunction can cause infarct-like lesions, predominantly located in the occipital cortex [82]. Among the possible pathogenetic mechanisms, a mitochondrial angiopathy, with a striking accumulation of abnormal mitochondria within smooth muscle and endothelial cells has been proposed as the cause of the ischemic lesions [83,84]. Alternatively, mitochondrial energy failure in neurons might be responsible of non-ischemic neuronal loss and stroke-like symptoms [85–87]. In any case, no explanation for the preferential involvement of the posterior cortex has been provided so far.

Interestingly, MELAS, like other mitochondrial disorders, is characterized by a variation in the expression of mutated DNA between different tissues (heteroplasmy), thought to arise from random replicative segregation of mitochondria during germ-layer differentiation early in embryonic development. This is probably the cause of the phenotypic heterogeneity associated with mitochondrial mutations within the same family [88], and such heterogeneity may also result in the lack of a family history. Heteroplasmy might also explain the peculiar neurological feature of the syndrome. It has been shown that a cell-specific threshold of mitochondrial mutation load can be tolerated before mitochondrial respiratory chain dysfunction starts to appear [85]. It is therefore possible that neurons of the posterior cortex have a lower threshold for the mutation, resulting in a selective vulnerability of this brain region. However, a recent study challenged this hypothesis, by showing that the most severe COX deficiency is not associated to occipital neurons, but to leptomeningeal and cortical blood vessel wall, thus highlighting the importance of vascular mitochondrial dysfunction in the pathogenesis of the disease. Consistently, endothelial dysfunction, with impaired vasodilation capacity of small arteries, has been recently reported in MELAS patients [89].

The Role of Angiogenesis in Common CNS Disorders

Angiogenesis and Stroke

Apart from the rare genetic forms of ischemic brain disease described above, stroke generally occurs as a multifactorial acquired condition, for which several risk factors have been identified, including aging, hypertension, diabetes, smoking, obesity, atherosclerosis and poor physical activity. However, conventional risk factors account for only 40–50% of the overall risk, thus indicating that disregulation of other genes, perhaps associated with vessel formation or function, co-determine the risk of stroke [55]. For instance, a compensatory angiogenic response, occurring after an ischemic insult, affects the overall clinical outcome of stroke. Indeed, studies in the early 1990s reported that an increased vessel density in the peri-infarct region correlates with survival after stroke [90] (Fig. 42.4e–h). This finding can be explained by the fact that an increase in perfusion (either via dilatation or angiogenesis) might rescue an ischemic but potentially viable region, called the 'penumbra', around the infarcted core [91]. How-

ever, the observation that the functional recovery from a stroke may continue for several years after the ischemic insult raises the question whether angiogenesis may also not promote regeneration of the CNS, perhaps in part by also stimulating neurogenesis in the highly vascularized neurogenic niches in the adult human brain [92,93].

Which stimulus triggers new blood vessel formation in the border region of an infarct? At least in the early inflammatory phase of stroke, angiogenesis is activated by the release of a cocktail of growth factors and cytokines from infiltrating leucocytes, macrophages and platelets [94] including VEGF [95,96], PDGF [97], TGF-β [98], bFGF [99] and erythropoietin (EPO) [100,101]. Preclinical studies revealed that administration of VEGF (or conversely VEGF inhibitors) may improve or aggravate the histological appearance and clinical outcome of stroke, depending on the dose, timing and route of administration after the ischemic brain insult [102–105]. In particular, consistent with the up-regulation of VEGF receptors after middle cerebral artery occlusion [106], intravenous delivery of VEGF within 2 days after the ischemic event was effective in inducing angiogenesis in the penumbra, thus improving neural recovery [103]. The therapeutic potential of VEGF in cerebral ischemia is confirmed by the observation that intra-cerebroventricular (ICV) administration of VEGF 1 day after reperfusion reduces infarct size, stimulates angiogenesis and improves neurological performance [107]. Remarkably, ICV delivery of VEGF also enhances neurogenesis after stroke, by promoting the proliferation, migration an maturation of neural progenitors [108,109].

A concern about the use of VEGF for stroke treatment relates to its potent permeabilizing effect, which normally occurs spontaneously immediately after the onset of ischemia, when VEGF is up-regulated in the ischemic core [110]. It has been shown that intravenous infusion of VEGF immediately after embolic ischemic damage indeed enhances the leakage of the BBB with subsequent hemorrhagic transformation of the ischemic lesion [103], whereas a soluble receptor, trapping endogenous VEGF, reduces edema formation and results in a significant sparing of brain tissue [102]. On the other hand, other studies indicate that both intraventricular and intraparenchymal delivery of VEGF counteract edema formation and decrease infarct size [111,112]. Therefore, the effects of VEGF in the context of cerebral ischemia seem to be highly dependent on the dose, timing and route of administration, with side effects linked to vascular leakage mainly occurring after early (within 1 h of onset) and high dose intravascular delivery, where VEGF directly acts on endothelial cells. Despite extensive preclinical experimentation, only a limited number of trials for the induction of therapeutic angiogenesis (i.e., by administration of the proangiogenic factor erythropoietin) after stroke have been so far translated into the clinics [113] and no long-term follow-up is available to date. In conclusion, angiogenic therapy holds promise for the treatment of stroke, but careful optimization of the dose, timing, and route of administration of angiogenic will likely be required to obtain clinical success.

Angiogenesis and Alzheimer's Disease

Alzheimer's disease (AD) is a common and health-care demanding neurodegenerative disorder, affecting 10% of the population worldwide. Although the presence of senile plaques and fibrillary tangles are recognized as primary pathological lesions in AD, the etiologic mechanisms via which these deposits form and cause neuronal degeneration still remain largely unclear [114]. Recent epidemiological, clinical, pathological and functional studies point toward the vasculature as a possible player in AD. First, there is a strong association between cognitive decline in AD and cerebrovascular lesions, which also share a series of risk factors, including age, atherosclerosis, diabetes, hypertension, hyperlipidemia and the ε4 allele of the apolipoprotein E gene [115,116]. Second, cerebral amyloid angiopathy (CAA), consisting of the deposition of amyloid β peptide (Aβ) in small arterial vessels, is another cardinal component of some forms of AD, causing loss of smooth muscle cells, vessel dilation and rupture with hemorrhage [117]. CAA of pial and intracerebral arteries also impairs the physiological functional hyperemic response to increased synaptic activity, a phenomenon called 'neurovascular uncoupling' [118]. Third, significant alterations in the microvasculature, including reduced vessel density, vessel fragmentation and atrophy, irregular capillary surface, both enlarged and reduced vessel diameter, and collagen accumulation with thickening of capillary basement membrane, have all been described in AD [119,120] (Fig. 42.4a, b). Fourth, BBB dysfunction, with deficient clearance of Aβ has been documented in both AD animal models and patients [120]. Fifth, VEGF promoter polymorphisms lowering VEGF expression have been correlated with an increased risk of AD in some populations, though not in all [121,122]. On the other hand, VEGF and other angiogenic proteins have been documented to be up-regulated in the brain of AD patients, possible as a compensatory mechanism to hypoperfusion [123–125]. However, Aβ deposits avidly bind and sequester both bFGF and VEGF, thus locally lowering the soluble and active amount of angiogenic factors and reinforcing hypoperfusion [126,127]. Collectively, these observations challenge the classical neurocentric view of AD and open novel therapeutic options for AD, aimed at restoring proper blood flow, neurovascular coupling and BBB functional integrity.

Angiogenesis and Parkinson's Disease

Recent findings correlate abnormal angiogenesis to another common form of neurodegeneration, Parkinson's disease (PD). Traditionally considered a pure neurological syndrome, characterized by bradykinesia, tremor and rigidity, secondary to the specific loss of dopaminergic neurons in the substantia nigra, PD is currently viewed as a complex disorder, associated with vascular inflammation. Microglia activation and astrocytosis, with elevation of the levels of pro-inflammatory cytokines, are observed in PD patients and animal models [128]. Inflammation is often coupled to angiogenesis, and the vessel abnormalities, the increase in vascular density and angiogenesis in the subs-

tantia nigra of PD patients may be promoted by the local inflammatory process [129,130]. Interestingly, changes in angiogenesis and vascular remodelling also parallel neurogenesis [131]. This is interesting, as neuroblast activation, followed by migration and differentiation towards the dopaminergic lineage, has been recently reported in both PD patients and animal models [132,133], thus suggesting that nigral neurogenesis may compensate for the neuronal loss in the damaged brain. Interestingly, VEGF is overexpressed by astrocytes in the substantia nigra pars reticulata [134], a region often undergoing strong reactive gliosis and active neural reconstruction by migrating neural progenitor cells [132] (Fig. 42.4, m). The recent evidence that VEGF stimulates neurogenesis and exerts neurotrophic and anti-apoptotic activities on dopaminergic neurons [135–137] might provide a molecular link coupling neurogenesis and angiogenesis in PD brains. Similarly to what has been documented for stroke, a threshold dose of VEGF seems to separate beneficial from adverse effects, with lower VEGF doses being associated to neuroprotection and improved neuro-perfusion, while higher doses induce massive angiogenesis, glial proliferation, vessel leakage and edema, thereby further harming neuronal viability and disguising the neuroprotective effect [137,138].

Another hallmark of brain inflammation is BBB dysfunction [139]. Although the absence of brain edema and a normal cerebrospinal fluid composition suggest that the BBB is largely intact in PD patients [140], multiple and repetitive focal breakdowns of the BBB integrity, in association with periods of angiogenesis, may accompany hot spots of active inflammation [141]. Consistently, BBB dysfunction has been detected in PD patients [142] and has also been experimentally reproduced in a rat model of the disease [141,143]. Notably, L-dopa pharmacotherapy has been recently shown to stimulate angiogenesis in the affected brain regions. Since dopamine counteracts the mitogenic activity of VEGF on ECs directly, this angiogenic response likely represents an adaptation to the increased energy demands in brain nuclei that sustain pronounced changes in firing patterns, synaptic plasticity and macromolecular syntheses in response to the L-dopa pharmacotherapy [143]. Of note, since newly born microvessels often exhibit an increased leakiness and impaired BBB, the consequences of these changes on the pharmacokinetic profile of administered drugs (such as L-dopa, the access of which into the brain parenchyma is normally greatly limited by the BBB) should be taken into consideration.

Angiogenesis and Multiple Sclerosis

Substantial advances have been achieved during the last 10 years in the understanding of the mechanisms underlying the inflammation, demyelination and neurodegeneration characterizing multiple sclerosis (MS). In general, MS is considered an immune-mediated disease that arises in genetically susceptible people [144]. Although the sequence of events initiating the disease is not entirely understood, MS lesions invariably have a vascular component. Over 130 years ago, Rindfleish reported that MS lesions were associated with abnormal blood vessels [145]. It is generally believed that development of the inflammatory plaque, the histopathological hallmark of MS, originates from a breach in the BBB integrity, allowing leukocytes to home and cross vessel walls and to enter the normally immunologically privileged CNS [146] (Fig. 42.4c, d). Typically, MS lesions are centered on one or more veins in the periventricular white matter and extend finger-like projections (called Dawson's fingers) along the course of venules [144,147]. The abnormal vascularization pattern in MS is highlighted by the MRI finding of a 'ring enhancement' at the periphery of the lesion, indicating increased vascular permeability. Although this radiological finding has been traditionally assumed to represent the effect of inflammation-induced leakiness of pre-existing vessels [148], it might be equally indicative of new blood vessel formation and expanded blood volume [149]. In accordance, during relapsing-remitting and secondary-progressive phases of the disease, patients present an increase in vascular perfusion, which may correlate with an increased vessel formation in MS lesions [150]. Indeed, both acute and chronic MS lesions up-regulate VEGF and contain numerous vessels of irregular shape and size, an increased number in ECs and mitotic counts [149,151]. Notably, the process of neovascularization is not confined within the plaque, but often extends into the surrounding tissue, often corresponding to areas of peri-plaque active gliosis [149,151,152]. In contrast to the enhanced vascularization of MS plaques, perfusion of grey matter is reduced—possibly reflecting the decreased metabolism secondary to neuronal and axonal loss [150].

Several of the molecular players involved in the pathophysiology of MS are also associated with angiogenesis. Matrix metalloproteinases (MMPs) have a key role in the destabilization of the basement membrane during angiogenic sprouting, while endothelial specific adhesion molecules facilitate BBB crossing and homing of leukocytes into the MS brain [153] (Fig. 42.4c, d). Nitric oxide has been described as an 'actor', but also as a 'director' in angiogenesis, being capable of inducing both vasodilation and EC growth [154]. Interestingly, NO levels in MS patients correlate with clinical and MRI signs of disease progression [155]. The role of VEGF in MS remains largely undetermined. VEGF and its receptors have been found in the plaques of MS patients and animal models, and VEGF levels have been correlated with the extent of spinal cord lesions, suggesting a possible involvement of angiogenesis in longitudinal extension of the plaque [151,156]. However, as VEGF expression is highly influenced by inflammatory cytokines and ischemia, the accumulation of VEGF may be the result—not the cause—of MS. Also, VEGF has been proposed to aggravate immune-mediated inflammation via its activity to stimulate angiogenesis and vascular permeability (facilitating infiltration of inflammatory cells [157]) or via its immunostimulatory activity (by increasing Th1 cytokine production [158]). However, angiogenesis also promotes wound healing and may thus be critical for remyelination or axon repair. In addition, VEGF has also been implicated in immunosuppression (by inhibiting T-cells and dendritic cells or increasing Th2 cytokine production [159,160]) and in neuro-

protection (by stimulating neuron survival and axon integrity and outgrowth [161,162]). CNS infiltrating CD4+ T-cells can also rescue injured neurons by producing neurotrophic factors, such as VEGF [163].

In conclusion, although the specific role of angiogenesis has not yet been evaluated, it is worth mentioning that several of the drugs currently used for MS also display antiangiogenic properties. For instance, INFβ, which is exploited in MS to inhibit T cell proliferation, TNF production and antigen presentation, also interferes with the passage of leukocytes across the BBB and inhibits vessel growth in vitro and in vivo [164,165]. In addition, corticosteroids, beside their potent anti-inflammatory activity, also exert a stabilizing effect on the BBB and appear angiostatic in animal models of cancer and chronic inflammation [166–168]. Finally, immunosuppressive therapies, effective in slowing disease progression in about 10% of patients, are commonly exploited in cancer therapy, where they show significant antiangiogenic activity [169–171].

Angiogenesis and CNS Tumors

The essential role of angiogenesis in general tumor growth has been extensively reviewed elsewhere in this book. Here, we discuss the vascular aspects of two particular tumors, specifically affecting the CNS, i.e. hemangioblastomas and gliomas.

Hemangioblastomas

Hemangioblastomas are highly vascularized tumors, which occur sporadically or in the context of the von Hippel-Lindau disease (VHL), an autosomal dominant hereditary multitumor syndrome [172]. They account for 3% of all tumors of the CNS and specifically form in a limited subset of CNS regions, including the retina, the cerebellum, the brain stem and spinal cord [172] (Fig. 42.4i). Typical histological features of hemangioblastomas comprise a teleangiectasic capillary network and interstitial, polyhydric, lipid-laden, stromal cells [172]. The analysis of the familiar forms reveals that the stromal cells represent the VHL-deficient component of the tumor. However, the origin of these cells has remained enigmatic for a long time, and hemangioblastomas have been classified as "of uncertain histogenesis".

The observation that hemangioblastomas are able to support blood island formation and extramedullary hematopoiesis suggests that they might arise from developmentally arrested hemangioblasts [173,174]. Very recently, it has been shown that hemangioblstoma cells retain the ability to differentiate into both hematopoietic and vascular progenies [175]. In accordance, the same set of molecules driving hemangioblast commitment (including Scl, Csf-1R, Gata-1, VEGFR2 and Tie2 and Brachyury [175,176]) is also expressed by stromal cells in hemangioblastomas [177] (Fig. 42.4j). The expression of embryo-specific genes suggests that the progenitor cell for hemangioblastomas is already present during embryonic development, in accordance with the novel emerging concept of an early developmental origin of cancer, recently proposed for other tumor types [178,179]. Notably, Scl is transiently expressed by embryonic hemangioblasts in the retina, diencephalon, mesencephalon, metencephalon and spinal cord, but not in the telencephalon, thus matching the distribution pattern of hemangioblastomas [177,180]. This observation challenges the hypothesis that neoplastic cells migrate and home to specific sites in the CNS, and instead suggests that developmental processes govern the peculiar tissue-restricted distribution of these tumors. A better understanding of how hemangioblastomas arise is obviously of significant medical importance for the development of novel treatment paradigms. An intriguing question is also whether targeting specific molecular pathways regulating the differentiation of the hemangioblast into blood versus endothelium might offer any therapeutic potential [175]. The administration of antiangiogenic drugs, such as SU5416, a selective VEGFR-2 inhibitor [181], has recently been proposed for the treatment of hemangioblastoma, and resulted in clinical benefits, at least in the case of optic disc-associated lesions [182]. It still remains to be determined whether such an approach could be successfully proposed as a general therapeutic option for hemangioblastoma patients.

Gliomas

Glioma is the most common and vascularized tumor in the CNS [183]. Vessel density is an independent prognostic parameter for human astroglial tumors [184]. Analysis of several animal models has led to the identification of more than 25 angiogenic cytokines expressed by these tumor cells; these models also offered opportunities to assess the efficacy of antiangiogenic therapy in glioma growth [185]. Delivery of conventional cytotoxic drugs to the brain is significantly hampered by the existence of the BBB. In contrast, ECs are in direct contact with blood, and are thus more accessible to circulating antiangiogenic compounds. However, the first antiangiogenic therapy clinical trials in glioma patients, based on the use of thalidomide, interferons and metronomic chemotherapy, have not been as successful as initially expected. This might be attributable to several reasons.

First, antiangiogenic strategies rarely provide a definite cure for brain tumors, in part because tumors develop "antiangiogenic escape strategies" through activation of alternative angiogenic pathways [186]. This has been specifically observed as a consequence of VEGFR2 targeting in an experimental model of glioma, resulting in the compensatory up-regulation of both PDGF and Ang1 [187]. An alternative escape strategy of gliomas has recently been described in a mouse model, in which anti-VEGFR2 treatment lead to a massive invasion of tumor cells into healthy adjacent tissue, likely as a consequence of hypoxia-induced overexpression of invasive molecules (the hypoxia results from the inhibition of tumor vascularization by the anti-VEGFR2 antibody) [188]. The invading cells were not able to give rise to new tumor masses, possibly because of the ongoing antiangiogenic treatment, but they diffusely spread, co-opting host vessels, and eventually killed the host by diffuse gliomatosis [188]. Another limitation of antiangiogenic therapy

for glioma patients is the poor reliability of traditional imaging and laboratory parameters to monitor vascular changes in a clinical setting. Despite the high level of angiogenic cytokine expressed by glioma cells, their use as surrogate markers in the serum seems not reliable, likely because of the restricted permeability of the CNS conferred by the BBB [189]. In contrast, the level of the angiogenic factors, such as VEGF and bFGF, in the CSF has been shown to correlate with the degree of tumor vascularity and were adversely associated with patient survival [190]. This prompts the consideration of the measurement of angiogenic markers in the CSF for prognostic purposes, and also the monitoring of the efficacy of antiangiogenic therapy, but would require repetitive invasive lumbar punctions. An alternative promising approach involves determination of the number of circulating endothelial progenitor cells (CEPs) [191], although their predictive value still needs further investigation.

A thorough understanding of antiangiogenic resistance mechanisms, as well as the optimization of multimodality therapies, in which antiangiogenic compounds should be combined to conventional chemotherapy and/or novel molecular drugs targeting other features of glioma biology (cell invasion, drug resistance, resistance to apoptosis), will be of major interest for the near future.

Conclusions and Perspectives

Understanding the role and mechanisms of angiogenesis in the CNS will facilitate our comprehension of the importance of the neurovascular link in CNS development, health and disease. Many intriguing questions remain outstanding and require further study—to name a few: Recent studies documented that angiogenesis and neurogenesis are coordinated and tightly linked in the adult brain and that the vasculature is a source of neurotrophic factors such as brain-derived neurotrophic factor (BDNF) [192]. How general is this finding? Are blood vessels generally a source of neurotrophic, neuroprotective or neuro-active signals, and do they determine neuronal differentiation and patterning? Sensory nerves determine the pattern of arterial differentiation and blood vessel branching in the skin by the secretion of VEGF [193]. Do neural cells also determine vessel navigation in the CNS? What is the role of the lymphatic system in the CNS? Despite the fact that the CNS parenchyma does not contain lymphatics, extracranial lymphatic vessels along olfactory, cranial and spinal nerves are important for drainage of the cerebrospinal fluid (CSF) [194,195]. Do lymph vessels and neurons also communicate with each other through a "neuro-lymphatic" link, and will such studies help to resolve insight into the mechanisms involved in diseases characterized by abnormal CSF flow or drainage [196]? An increasing literature indicates that angiogenic factors, such as VEGF, also have important neuroprotective activity. Will it be possible to deliver VEGF for the treatment of neurodegenerative disorders (as we demonstrated for the motoneuron degenerative disease amyotrophic lateral sclerosis, or ALS [161,197–200])? These and many other intriguing questions beg for answers. Insight into these processes promises to broaden the medical armamentarium to treat neurovascular disorders.

References

1. Risau W. Mechanisms of angiogenesis. Nature 1997;386(6626): 671–4.
2. Carmeliet P. Angiogenesis in health and disease. Nat Med 2003;9(6):653–60.
3. Kurz H. Physiology of angiogenesis. J Neurooncol 2000;50 (1–2):17–35.
4. Noden DM. Embryonic origins and assembly of blood vessels. Am Rev Respir Dis 1989;140(4):1097–103.
5. Ambler CA, Nowicki JL, Burke AC, Bautch VL. Assembly of trunk and limb blood vessels involves extensive migration and vasculogenesis of somite-derived angioblasts. Dev Biol 2001;234(2):352–64.
6. Ambler CA, Schmunk GM, Bautch VL. Stem cell-derived endothelial cells/progenitors migrate and pattern in the embryo using the VEGF signaling pathway. Dev Biol 2003;257(1): 205–19.
7. Hogan KA, Ambler CA, Chapman DL, Bautch VL. The neural tube patterns vessels developmentally using the VEGF signaling pathway. Development 2004;131(7):1503–13.
8. Childs S, Chen JN, Garrity DM, Fishman MC. Patterning of angiogenesis in the zebrafish embryo. Development 2002;129(4):973–82.
9. Chapman DL, Papaioannou VE. Three neural tubes in mouse embryos with mutations in the T-box gene Tbx6. Nature 1998;391(6668):695–7.
10. Kurz H, Gartner T, Eggli PS, Christ B. First blood vessels in the avian neural tube are formed by a combination of dorsal angioblast immigration and ventral sprouting of endothelial cells. Dev Biol 1996;173(1):133–47.
11. Bar T. The vascular system of the cerebral cortex. Adv Anat Embryol Cell Biol 1980;59:I-VI,1–62.
12. Greenberg DA, Jin K. From angiogenesis to neuropathology. Nature 2005;438(7070):954–9.
13. Nagase T, Nagase M, Yoshimura K, Fujita T, Koshima I. Angiogenesis within the developing mouse neural tube is dependent on sonic hedgehog signaling: possible roles of motor neurons. Genes Cells 2005;10(6):595–604.
14. Lawson ND, Vogel AM, Weinstein BM. sonic hedgehog and vascular endothelial growth factor act upstream of the Notch pathway during arterial endothelial differentiation. Dev Cell 2002;3(1):127–36.
15. Rowitch DH, B SJ, Lee SM, Flax JD, Snyder EY, McMahon AP. Sonic hedgehog regulates proliferation and inhibits differentiation of CNS precursor cells. J Neurosci 1999;19(20):8954–65.
16. Pola R, Ling LE, Silver M, et al. The morphogen Sonic hedgehog is an indirect angiogenic agent upregulating two families of angiogenic growth factors. Nat Med 2001;7(6):706–11.
17. Raab S, Beck H, Gaumann A, et al. Impaired brain angiogenesis and neuronal apoptosis induced by conditional homozygous inactivation of vascular endothelial growth factor. Thromb Haemost 2004;91(3):595–605.
18. Yamada T, Pfaff SL, Edlund T, Jessell TM. Control of cell pattern in the neural tube: motor neuron induction by diffusible

19. Sato TN, Tozawa Y, Deutsch U, et al. Distinct roles of the receptor tyrosine kinases Tie-1 and Tie-2 in blood vessel formation. Nature 1995;376(6535):70–4.
20. Wang HU, Chen ZF, Anderson DJ. Molecular distinction and angiogenic interaction between embryonic arteries and veins revealed by ephrin-B2 and its receptor Eph-B4. Cell 1998;93(5):741–53.
21. Breier G, Albrecht U, Sterrer S, Risau W. Expression of vascular endothelial growth factor during embryonic angiogenesis and endothelial cell differentiation. Development 1992;114(2): 521–32.
22. Virgintino D, Errede M, Robertson D, Girolamo F, Masciandaro A, Bertossi M. VEGF expression is developmentally regulated during human brain angiogenesis. Histochem Cell Biol 2003;119(3):227–32.
23. Harrigan MR. Angiogenic factors in the central nervous system. Neurosurgery 2003;53(3):639–60; discussion 60–1.
24. Acker T, Beck H, Plate KH. Cell type specific expression of vascular endothelial growth factor and angiopoietin-1 and -2 suggests an important role of astrocytes in cerebellar vascularization. Mech Dev 2001;108(1–2):45–57.
25. Ruhrberg C, Gerhardt H, Golding M, et al. Spatially restricted patterning cues provided by heparin-binding VEGF-A control blood vessel branching morphogenesis. Genes Dev 2002;16(20): 2684–98.
26. Haigh JJ, Morelli PI, Gerhardt H, et al. Cortical and retinal defects caused by dosage-dependent reductions in VEGF-A paracrine signaling. Dev Biol 2003;262(2):225–41.
27. Kawasaki T, Kitsukawa T, Bekku Y, et al. A requirement for neuropilin-1 in embryonic vessel formation. Development 1999;126(21):4895–902.
28. Gerhardt H, Ruhrberg C, Abramsson A, Fujisawa H, Shima D, Betsholtz C. Neuropilin-1 is required for endothelial tip cell guidance in the developing central nervous system. Dev Dyn 2004;231(3):503–9.
29. Bader BL, Rayburn H, Crowley D, Hynes RO. Extensive vasculogenesis, angiogenesis, and organogenesis precede lethality in mice lacking all alpha v integrins. Cell 1998;95(4):507–19.
30. McCarty JH, Lacy-Hulbert A, Charest A, et al. Selective ablation of alphav integrins in the central nervous system leads to cerebral hemorrhage, seizures, axonal degeneration and premature death. Development 2005;132(1):165–76.
31. Zhu J, Motejlek K, Wang D, Zang K, Schmidt A, Reichardt LF. beta8 integrins are required for vascular morphogenesis in mouse embryos. Development 2002;129(12):2891–903.
32. McCarty JH, Monahan-Earley RA, Brown LF, et al. Defective associations between blood vessels and brain parenchyma lead to cerebral hemorrhage in mice lacking alphav integrins. Mol Cell Biol 2002;22(21):7667–77.
33. Proctor JM, Zang K, Wang D, Wang R, Reichardt LF. Vascular development of the brain requires beta8 integrin expression in the neuroepithelium. J Neurosci 2005;25(43):9940–8.
34. Calderwood DA. Integrin activation. J Cell Sci 2004;117 (Pt 5):657–66.
35. Huang XZ, Wu JF, Cass D, et al. Inactivation of the integrin beta 6 subunit gene reveals a role of epithelial integrins in regulating inflammation in the lung and skin. J Cell Biol 1996;133(4): 921–8.

factors from notochord and floor plate. Cell 1993;73(4): 673–86.

36. Graus-Porta D, Blaess S, Senften M, et al. Beta1-class integrins regulate the development of laminae and folia in the cerebral and cerebellar cortex. Neuron 2001;31(3):367–79.

37. Lindahl P, Hellstrom M, Kalen M, Betsholtz C. Endothelial-perivascular cell signaling in vascular development: lessons from knockout mice. Curr Opin Lipidol 1998;9(5):407–11.

38. Lindahl P, Johansson BR, Leveen P, Betsholtz C. Pericyte loss and microaneurysm formation in PDGF-B-deficient mice. Science 1997;277(5323):242–5.

39. Milner R, Campbell IL. Developmental regulation of beta1 integrins during angiogenesis in the central nervous system. Mol Cell Neurosci 2002;20(4):616–26.

40. Wang J, Milner R. Fibronectin promotes brain capillary endothelial cell survival and proliferation through alpha5beta1 and alphavbeta3 integrins via MAP kinase signalling. J Neurochem 2006;96(1):148–59.

41. Begley DJ, Brightman MW. Structural and functional aspects of the blood-brain barrier. Prog Drug Res 2003;61:39–78.

42. Bauer HC, Bauer H. Neural induction of the blood-brain barrier: still an enigma. Cell Mol Neurobiol 2000;20(1):13–28.

43. Kim JH, Kim JH, Park JA, et al. Blood-neural barrier: intercellular communication at glio-vascular interface. J Biochem Mol Biol 2006;39(4):339–45.

44. Wolburg H, Lippoldt A. Tight junctions of the blood-brain barrier: development, composition and regulation. Vascul Pharmacol 2002;38(6):323–37.

45. Abbott NJ, Ronnback L, Hansson E. Astrocyte-endothelial interactions at the blood-brain barrier. Nat Rev Neurosci 2006;7(1):41–53.

46. Reuss B, Dono R, Unsicker K. Functions of fibroblast growth factor (FGF)-2 and FGF-5 in astroglial differentiation and blood-brain barrier permeability: evidence from mouse mutants. J Neurosci 2003;23(16):6404–12.

47. Reuss B, von Bohlen und Halbach O. Fibroblast growth factors and their receptors in the central nervous system. Cell Tissue Res 2003;313(2):139–57.

48. Lee SW, Kim WJ, Choi YK, et al. SSeCKS regulates angiogenesis and tight junction formation in blood-brain barrier. Nat Med 2003;9(7):900–6.

49. Mi H, Haeberle H, Barres BA. Induction of astrocyte differentiation by endothelial cells. J Neurosci 2001;21(5):1538–47.

50. Hori S, Ohtsuki S, Hosoya K, Nakashima E, Terasaki T. A pericyte-derived angiopoietin-1 multimeric complex induces occludin gene expression in brain capillary endothelial cells through Tie-2 activation in vitro. J Neurochem 2004;89(2):503–13.

51. Dohgu S, Takata F, Yamauchi A, et al. Brain pericytes contribute to the induction and up-regulation of blood-brain barrier functions through transforming growth factor-beta production. Brain Res 2005;1038(2):208–15.

52. Schiera G, Bono E, Raffa MP, et al. Synergistic effects of neurons and astrocytes on the differentiation of brain capillary endothelial cells in culture. J Cell Mol Med 2003;7(2):165–70.

53. Cunha-Vaz JG. The blood-retinal barriers system. Basic concepts and clinical evaluation. Exp Eye Res 2004;78(3):715–21.

54. Erickson KK, Sundstrom JM, Antonetti DA. Vascular permeability in ocular disease and the role of tight junctions. Angiogenesis 2007;10(2):103–17.

55. Tournier-Lasserve E. New players in the genetics of stroke. N Engl J Med 2002;347(21):1711–2.

56. Verreault S, Joutel A, Riant F, et al. A novel hereditary small vessel disease of the brain. Ann Neurol 2006;59(2):353–7.

57. Low WC, Junna M, Borjesson-Hanson A, et al. Hereditary multi-infarct dementia of the Swedish type is a novel disorder different from NOTCH3 causing CADASIL. Brain 2007;130(Pt 2):357–67.

58. Tournier-Lasserve E, Joutel A, Melki J, et al. Cerebral autosomal dominant arteriopathy with subcortical infarcts and leukoencephalopathy maps to chromosome 19q12. Nat Genet 1993;3(3):256–9.

59. Joutel A, Corpechot C, Ducros A, et al. Notch3 mutations in CADASIL, a hereditary adult-onset condition causing stroke and dementia. Nature 1996;383(6602):707–10.

60. Vikelis M, Xifaras M, Mitsikostas DD. CADASIL: a short review of the literature and a description of the first family from Greece. Funct Neurol 2006;21(2):77–82.

61. Dichgans M, Wick M, Gasser T. Cerebrospinal fluid findings in CADASIL. Neurology 1999;53(1):233.

62. Razvi SS, Bone I. Single gene disorders causing ischaemic stroke. J Neurol 2006;253(6):685–700.

63. Chabriat H, Vahedi K, Iba-Zizen MT, et al. Clinical spectrum of CADASIL: a study of 7 families. Cerebral autosomal dominant arteriopathy with subcortical infarcts and leukoencephalopathy. Lancet 1995;346(8980):934–9.

64. van den Boom R, Lesnik Oberstein SA, Ferrari MD, Haan J, van Buchem MA. Cerebral autosomal dominant arteriopathy with subcortical infarcts and leukoencephalopathy: MR imaging findings at different ages–3rd-6th decades. Radiology 2003;229(3):683–90.

65. Liem MK, van der Grond J, Haan J, et al. Lacunar infarcts are the main correlate with cognitive dysfunction in CADASIL. Stroke 2007;38(3):923–8.

66. Federico A, Bianchi S, Dotti MT. The spectrum of mutations for CADASIL diagnosis. Neurol Sci 2005;26(2):117–24.

67. Ishiko A, Shimizu A, Nagata E, Takahashi K, Tabira T, Suzuki N. Notch3 ectodomain is a major component of granular osmiophilic material (GOM) in CADASIL. Acta Neuropathol (Berl) 2006;112(3):333–9.

68. Joutel A, Andreux F, Gaulis S, et al. The ectodomain of the Notch3 receptor accumulates within the cerebrovasculature of CADASIL patients. J Clin Invest 2000;105(5):597–605.

69. Lacombe P, Oligo C, Domenga V, Tournier-Lasserve E, Joutel A. Impaired cerebral vasoreactivity in a transgenic mouse model of cerebral autosomal dominant arteriopathy with subcortical infarcts and leukoencephalopathy arteriopathy. Stroke 2005;36(5):1053–8.

70. Yanagawa S, Ito N, Arima K, Ikeda S. Cerebral autosomal recessive arteriopathy with subcortical infarcts and leukoencephalopathy. Neurology 2002;58(5):817–20.

71. Grand MG, Kaine J, Fulling K, et al. Cerebroretinal vasculopathy. A new hereditary syndrome. Ophthalmology 1988;95(5):649–59.

72. Terwindt GM, Haan J, Ophoff RA, et al. Clinical and genetic analysis of a large Dutch family with autosomal dominant vascular retinopathy, migraine and Raynaud's phenomenon. Brain 1998;121 (Pt 2):303–16.

73. Jen J, Cohen AH, Yue Q, et al. Hereditary endotheliopathy with retinopathy, nephropathy, and stroke (HERNS). Neurology 1997;49(5):1322–30.

74. Vahedi K, Massin P, Guichard JP, et al. Hereditary infantile hemiparesis, retinal arteriolar tortuosity, and leukoencephalopathy. Neurology 2003;60(1):57–63.

75. Revesz T, Ghiso J, Lashley T, et al. Cerebral amyloid angiopathies: a pathologic, biochemical, and genetic view. J Neuropathol Exp Neurol 2003;62(9):885–98.

76. Hagel C, Groden C, Niemeyer R, Stavrou D, Colmant HJ. Subcortical angiopathic encephalopathy in a German kindred suggests an autosomal dominant disorder distinct from CADASIL. Acta Neuropathol (Berl) 2004;108(3):231–40.

77. Tomimoto H, Ohtani R, Wakita H, et al. Small artery dementia in Japan: radiological differences between CADASIL, leukoaraiosis and Binswanger's disease. Dement Geriatr Cogn Disord 2006;21(3):162–9.

78. Vilela P, Goulao A. Ischemic stroke: carotid and vertebral artery disease. Eur Radiol 2005;15(3):427–33.

79. Bendszus M, Koltzenburg M, Burger R, Warmuth-Metz M, Hofmann E, Solymosi L. Silent embolism in diagnostic cerebral angiography and neurointerventional procedures: a prospective study. Lancet 1999;354(9190):1594–7.

80. Schmiedel J, Jackson S, Schafer J, Reichmann H. Mitochondrial cytopathies. J Neurol 2003;250(3):267–77.

81. McKenzie M, Liolitsa D, Hanna MG. Mitochondrial disease: mutations and mechanisms. Neurochem Res 2004;29(3):589–600.

82. Betts J, Jaros E, Perry RH, et al. Molecular neuropathology of MELAS: level of heteroplasmy in individual neurones and evidence of extensive vascular involvement. Neuropathol Appl Neurobiol 2006;32(4):359–73.

83. Michelson DJ, Ashwal S. The pathophysiology of stroke in mitochondrial disorders. Mitochondrion 2004;4(5–6):665–74.

84. Ohama E, Ohara S, Ikuta F, Tanaka K, Nishizawa M, Miyatake T. Mitochondrial angiopathy in cerebral blood vessels of mitochondrial encephalomyopathy. Acta Neuropathol (Berl) 1987;74(3):226–33.

85. Chomyn A, Enriquez JA, Micol V, Fernandez-Silva P, Attardi G. The mitochondrial myopathy, encephalopathy, lactic acidosis, and stroke-like episode syndrome-associated human mitochondrial tRNALeu(UUR) mutation causes aminoacylation deficiency and concomitant reduced association of mRNA with ribosomes. J Biol Chem 2000;275(25):19198–209.

86. Iizuka T, Sakai F, Ide T, Miyakawa S, Sato M, Yoshii S. Regional cerebral blood flow and cerebrovascular reactivity during chronic stage of stroke-like episodes in MELAS - Implication of neurovascular cellular mechanism. J Neurol Sci 2007.

87. Iizuka T, Sakai F, Suzuki N, et al. Neuronal hyperexcitability in stroke-like episodes of MELAS syndrome. Neurology 2002;59(6):816–24.

88. de Vries D, de Wijs I, Ruitenbeek W, et al. Extreme variability of clinical symptoms among sibs in a MELAS family correlated with heteroplasmy for the mitochondrial A3243G mutation. J Neurol Sci 1994;124(1):77–82.

89. Koga Y, Akita Y, Junko N, et al. Endothelial dysfunction in MELAS improved by l-arginine supplementation. Neurology 2006;66(11):1766–9.

90. Krupinski J, Kaluza J, Kumar P, Wang M, Kumar S. Prognostic value of blood vessel density in ischaemic stroke. Lancet 1993;342(8873):742.

91. Heiss WD, Sobesky J, Hesselmann V. Identifying thresholds for penumbra and irreversible tissue damage. Stroke 2004;35(11 Suppl 1):2671–4.

92. Eriksson PS, Perfilieva E, Bjork-Eriksson T, et al. Neurogenesis in the adult human hippocampus. Nat Med 1998;4(11):1313–7.

93. Steindler DA, Pincus DW. Stem cells and neuropoiesis in the adult human brain. Lancet 2002;359(9311):1047–54.

94. Slevin M, Krupinski J, Kumar P, Gaffney J, Kumar S. Gene activation and protein expression following ischaemic stroke: strategies towards neuroprotection. J Cell Mol Med 2005;9(1):85–102.

95. Issa R, Krupinski J, Bujny T, Kumar S, Kaluza J, Kumar P. Vascular endothelial growth factor and its receptor, KDR, in human brain tissue after ischemic stroke. Lab Invest 1999;79(4):417–25.

96. Slevin M, Krupinski J, Slowik A, Kumar P, Szczudlik A, Gaffney J. Serial measurement of vascular endothelial growth factor and transforming growth factor-beta1 in serum of patients with acute ischemic stroke. Stroke 2000;31(8):1863–70.

97. Krupinski J, Issa R, Bujny T, et al. A putative role for platelet-derived growth factor in angiogenesis and neuroprotection after ischemic stroke in humans. Stroke 1997;28(3):564–73.

98. Krupinski J, Kumar P, Kumar S, Kaluza J. Increased expression of TGF-beta 1 in brain tissue after ischemic stroke in humans. Stroke 1996;27(5):852–7.

99. Issa R, AlQteishat A, Mitsios N, et al. Expression of basic fibroblast growth factor mRNA and protein in the human brain following ischaemic stroke. Angiogenesis 2005;8(1):53–62.

100. Mu D, Chang YS, Vexler ZS, Ferriero DM. Hypoxia-inducible factor 1alpha and erythropoietin upregulation with deferoxamine salvage after neonatal stroke. Exp Neurol 2005; 195(2):407–15.

101. Marti HJ, Bernaudin M, Bellail A, et al. Hypoxia-induced vascular endothelial growth factor expression precedes neovascularization after cerebral ischemia. Am J Pathol 2000;156(3):965–76.

102. van Bruggen N, Thibodeaux H, Palmer JT, et al. VEGF antagonism reduces edema formation and tissue damage after ischemia/reperfusion injury in the mouse brain. J Clin Invest 1999;104(11):1613–20.

103. Zhang ZG, Zhang L, Jiang Q, et al. VEGF enhances angiogenesis and promotes blood-brain barrier leakage in the ischemic brain. J Clin Invest 2000;106(7):829–38.

104. Wang Y, Kilic E, Kilic U, et al. VEGF overexpression induces post-ischaemic neuroprotection, but facilitates haemodynamic steal phenomena. Brain 2005;128(Pt 1):52–63.

105. Ozawa CR, Banfi A, Glazer NL, et al. Microenvironmental VEGF concentration, not total dose, determines a threshold between normal and aberrant angiogenesis. J Clin Invest 2004;113(4):516–27.

106. Lennmyr F, Ata KA, Funa K, Olsson Y, Terent A. Expression of vascular endothelial growth factor (VEGF) and its receptors (Flt-1 and Flk-1) following permanent and transient occlusion of the middle cerebral artery in the rat. J Neuropathol Exp Neurol 1998;57(9):874–82.

107. Sun Y, Jin K, Xie L, et al. VEGF-induced neuroprotection, neurogenesis, and angiogenesis after focal cerebral ischemia. J Clin Invest 2003;111(12):1843–51.

108. Sun FY, Guo X. Molecular and cellular mechanisms of neuroprotection by vascular endothelial growth factor. J Neurosci Res 2005;79(1–2):180–4.

109. Yano A, Shingo T, Takeuchi A, et al. Encapsulated vascular endothelial growth factor-secreting cell grafts have neuroprotective

and angiogenic effects on focal cerebral ischemia. J Neurosurg 2005;103(1):104–14.

110. Zhang ZG, Zhang L, Tsang W, et al. Correlation of VEGF and angiopoietin expression with disruption of blood-brain barrier and angiogenesis after focal cerebral ischemia. J Cereb Blood Flow Metab 2002;22(4):379–92.

111. Harrigan MR, Ennis SR, Sullivan SE, Keep RF. Effects of intraventricular infusion of vascular endothelial growth factor on cerebral blood flow, edema, and infarct volume. Acta Neurochir (Wien) 2003;145(1):49–53.

112. Hayashi T, Abe K, Itoyama Y. Reduction of ischemic damage by application of vascular endothelial growth factor in rat brain after transient ischemia. J Cereb Blood Flow Metab 1998;18(8):887–95.

113. Tonges L, Schlachetzki JC, Weishaupt JH, Bahr M. Hematopoietic cytokines–on the verge of conquering neurology. Curr Mol Med 2007;7(2):157–70.

114. Hardy J, Selkoe DJ. The amyloid hypothesis of Alzheimer's disease: progress and problems on the road to therapeutics. Science 2002;297(5580):353–6.

115. Casserly I, Topol E. Convergence of atherosclerosis and Alzheimer's disease: inflammation, cholesterol, and misfolded proteins. Lancet 2004;363(9415):1139–46.

116. Hofman A, Ott A, Breteler MM, et al. Atherosclerosis, apolipoprotein E, and prevalence of dementia and Alzheimer's disease in the Rotterdam Study. Lancet 1997;349(9046):151–4.

117. Greenberg SM, Gurol ME, Rosand J, Smith EE. Amyloid angiopathy-related vascular cognitive impairment. Stroke 2004;35(11 Suppl 1):2616–9.

118. Iadecola C. Neurovascular regulation in the normal brain and in Alzheimer's disease. Nat Rev Neurosci 2004;5(5):347–60.

119. Bailey TL, Rivara CB, Rocher AB, Hof PR. The nature and effects of cortical microvascular pathology in aging and Alzheimer's disease. Neurol Res 2004;26(5):573–8.

120. Zlokovic BV. Neurovascular mechanisms of Alzheimer's neurodegeneration. Trends Neurosci 2005;28(4):202–8.

121. Chapuis J, Tian J, Shi J, et al. Association study of the vascular endothelial growth factor gene with the risk of developing Alzheimer's disease. Neurobiol Aging 2006;27(9):1212–5.

122. Del Bo R, Scarlato M, Ghezzi S, et al. Vascular endothelial growth factor gene variability is associated with increased risk for AD. Ann Neurol 2005;57(3):373–80.

123. Kalaria RN, Cohen DL, Premkumar DR, Nag S, LaManna JC, Lust WD. Vascular endothelial growth factor in Alzheimer's disease and experimental cerebral ischemia. Brain Res Mol Brain Res 1998;62(1):101–5.

124. Vagnucci AH, Jr., Li WW. Alzheimer's disease and angiogenesis. Lancet 2003;361(9357):605–8.

125. Thirumangalakudi L, Samany PG, Owoso A, Wiskar B, Grammas P. Angiogenic proteins are expressed by brain blood vessels in Alzheimer's disease. J Alzheimers Dis 2006;10(1):111–8.

126. Siedlak SL, Cras P, Kawai M, Richey P, Perry G. Basic fibroblast growth factor binding is a marker for extracellular neurofibrillary tangles in Alzheimer disease. J Histochem Cytochem 1991;39(7):899–904.

127. Yang SP, Bae DG, Kang HJ, Gwag BJ, Gho YS, Chae CB. Co-accumulation of vascular endothelial growth factor with beta-amyloid in the brain of patients with Alzheimer's disease. Neurobiol Aging 2004;25(3):283–90.

128. McGeer PL, McGeer EG. Inflammation and neurodegeneration in Parkinson's disease. Parkinsonism Relat Disord 2004;10 Suppl 1:S3–7.

129. Barcia C, Emborg ME, Hirsch EC, Herrero MT. Blood vessels and parkinsonism. Front Biosci 2004;9:277–82.

130. Faucheux BA, Bonnet AM, Agid Y, Hirsch EC. Blood vessels change in the mesencephalon of patients with Parkinson's disease. Lancet 1999;353(9157):981–2.

131. Palmer TD, Willhoite AR, Gage FH. Vascular niche for adult hippocampal neurogenesis. J Comp Neurol 2000;425(4):479–94.

132. Yoshimi K, Ren YR, Seki T, et al. Possibility for neurogenesis in substantia nigra of parkinsonian brain. Ann Neurol 2005;58(1):31–40.

133. Zhao M, Momma S, Delfani K, et al. Evidence for neurogenesis in the adult mammalian substantia nigra. Proc Natl Acad Sci U S A 2003;100(13):7925–30.

134. Wada K, Arai H, Takanashi M, et al. Expression levels of vascular endothelial growth factor and its receptors in Parkinson's disease. Neuroreport 2006;17(7):705–9.

135. Jin K, Zhu Y, Sun Y, Mao XO, Xie L, Greenberg DA. Vascular endothelial growth factor (VEGF) stimulates neurogenesis in vitro and in vivo. Proc Natl Acad Sci USA 2002;99(18):11946–50.

136. Pitzer MR, Sortwell CE, Daley BF, et al. Angiogenic and neurotrophic effects of vascular endothelial growth factor (VEGF165): studies of grafted and cultured embryonic ventral mesencephalic cells. Exp Neurol 2003;182(2):435–45.

137. Yasuhara T, Shingo T, Kobayashi K, et al. Neuroprotective effects of vascular endothelial growth factor (VEGF) upon dopaminergic neurons in a rat model of Parkinson's disease. Eur J Neurosci 2004;19(6):1494–504.

138. Yasuhara T, Shingo T, Muraoka K, et al. The differences between high and low-dose administration of VEGF to dopaminergic neurons of in vitro and in vivo Parkinson's disease model. Brain Res 2005;1038(1):1–10.

139. Huber JD, Egleton RD, Davis TP. Molecular physiology and pathophysiology of tight junctions in the blood-brain barrier. Trends Neurosci 2001;24(12):719–25.

140. Haussermann P, Kuhn W, Przuntek H, Muller T. Integrity of the blood-cerebrospinal fluid barrier in early Parkinson's disease. Neurosci Lett 2001;300(3):182–4.

141. Carvey PM, Zhao CH, Hendey B, et al. 6-Hydroxydopamine-induced alterations in blood-brain barrier permeability. Eur J Neurosci 2005;22(5):1158–68.

142. Kortekaas R, Leenders KL, van Oostrom JC, et al. Blood-brain barrier dysfunction in parkinsonian midbrain in vivo. Ann Neurol 2005;57(2):176–9.

143. Westin JE, Lindgren HS, Gardi J, et al. Endothelial proliferation and increased blood-brain barrier permeability in the basal ganglia in a rat model of 3,4-dihydroxyphenyl-L-alanine-induced dyskinesia. J Neurosci 2006;26(37):9448–61.

144. Noseworthy JH, Lucchinetti C, Rodriguez M, Weinshenker BG. Multiple sclerosis. N Engl J Med 2000;343(13):938–52.

145. Rindfleisch E. Pathological Histology: An Introduction to the Study of Pathlogical Anatomy (translated by Kloman, W.C. and Miles F.T.): Trubner & Co.; 1872.

146. Frohman EM, Racke MK, Raine CS. Multiple sclerosis–the plaque and its pathogenesis. N Engl J Med 2006;354(9):942–55.

147. Tan IL, van Schijndel RA, Pouwels PJ, et al. MR venography of multiple sclerosis. AJNR Am J Neuroradiol 2000;21(6): 1039–42.

148. Plumb J, McQuaid S, Mirakhur M, Kirk J. Abnormal endothelial tight junctions in active lesions and normal-appearing white matter in multiple sclerosis. Brain Pathol 2002;12(2):154–69.

149. Kirk S, Frank JA, Karlik S. Angiogenesis in multiple sclerosis: is it good, bad or an epiphenomenon? J Neurol Sci 2004;217(2):125–30.

150. Rashid W, Parkes LM, Ingle GT, et al. Abnormalities of cerebral perfusion in multiple sclerosis. J Neurol Neurosurg Psychiatry 2004;75(9):1288–93.

151. Proescholdt MA, Jacobson S, Tresser N, Oldfield EH, Merrill MJ. Vascular endothelial growth factor is expressed in multiple sclerosis plaques and can induce inflammatory lesions in experimental allergic encephalomyelitis rats. J Neuropathol Exp Neurol 2002;61(10):914–25.

152. Wuerfel J, Bellmann-Strobl J, Brunecker P, et al. Changes in cerebral perfusion precede plaque formation in multiple sclerosis: a longitudinal perfusion MRI study. Brain 2004;127 (Pt 1):111–9.

153. Benveniste EN. Role of macrophages/microglia in multiple sclerosis and experimental allergic encephalomyelitis. J Mol Med 1997;75(3):165–73.

154. Ziche M, Morbidelli L. Nitric oxide and angiogenesis. J Neurooncol 2000;50(1–2):139–48.

155. Giovannoni G, Miller DH, Losseff NA, et al. Serum inflammatory markers and clinical/MRI markers of disease progression in multiple sclerosis. J Neurol 2001;248(6):487–95.

156. Su JJ, Osoegawa M, Matsuoka T, et al. Upregulation of vascular growth factors in multiple sclerosis: correlation with MRI findings. J Neurol Sci 2006;243(1–2):21–30.

157. Ishida S, Usui T, Yamashiro K, et al. VEGF164-mediated inflammation is required for pathological, but not physiological, ischemia-induced retinal neovascularization. J Exp Med 2003;198(3):483–9.

158. Mor F, Quintana FJ, Cohen IR. Angiogenesis-inflammation cross-talk: vascular endothelial growth factor is secreted by activated T cells and induces Th1 polarization. J Immunol 2004;172(7):4618–23.

159. Ohm JE, Carbone DP. VEGF as a mediator of tumor-associated immunodeficiency. Immunol Res 2001;23(2–3):263–72.

160. Ohm JE, Gabrilovich DI, Sempowski GD, et al. VEGF inhibits T-cell development and may contribute to tumor-induced immune suppression. Blood 2003;101(12):4878–86.

161. Storkebaum E, Lambrechts D, Carmeliet P. VEGF: once regarded as a specific angiogenic factor, now implicated in neuroprotection. Bioessays 2004;26(9):943–54.

162. Jin K, Mao XO, Greenberg DA. Vascular endothelial growth factor stimulates neurite outgrowth from cerebral cortical neurons via Rho kinase signaling. J Neurobiol 2006;66(3): 236–42.

163. Freeman MR, Schneck FX, Gagnon ML, et al. Peripheral blood T lymphocytes and lymphocytes infiltrating human cancers express vascular endothelial growth factor: a potential role for T cells in angiogenesis. Cancer Res 1995;55(18):4140–5.

164. Fidler IJ. Angiogenesis and cancer metastasis. Cancer J 2000;6 Suppl 2:S134–41.

165. Lindner DJ, Borden EC. Synergistic antitumor effects of a combination of interferon and tamoxifen on estrogen receptor-positive and receptor-negative human tumor cell lines in vivo and in vitro. J Interferon Cytokine Res 1997;17(11):681–93.

166. Colville-Nash PR, Alam CA, Appleton I, Brown JR, Seed MP, Willoughby DA. The pharmacological modulation of angiogenesis in chronic granulomatous inflammation. J Pharmacol Exp Ther 1995;274(3):1463–72.

167. Folkman J, Langer R, Linhardt RJ, Haudenschild C, Taylor S. Angiogenesis inhibition and tumor regression caused by heparin or a heparin fragment in the presence of cortisone. Science 1983;221(4612):719–25.

168. Nauck M, Karakiulakis G, Perruchoud AP, Papakonstantinou E, Roth M. Corticosteroids inhibit the expression of the vascular endothelial growth factor gene in human vascular smooth muscle cells. Eur J Pharmacol 1998;341(2–3):309–15.

169. Billington DC. Angiogenesis and its inhibition: potential new therapies in oncology and non-neoplastic diseases. Drug Des Discov 1991;8(1):3–35.

170. Hommes OR, Weiner HL. Results of an international questionnaire on immunosuppressive treatment of multiple sclerosis. Mult Scler 2002;8(2):139–41.

171. Polverini PJ, Novak RF. Inhibition of angiogenesis by the antineoplastic agents mitoxantrone and bisantrene. Biochem Biophys Res Commun 1986;140(3):901–7.

172. Lonser RR, Glenn GM, Walther M, et al. von Hippel-Lindau disease. Lancet 2003;361(9374):2059–67.

173. Vortmeyer AO, Frank S, Jeong SY, et al. Developmental arrest of angioblastic lineage initiates tumorigenesis in von Hippel-Lindau disease. Cancer Res 2003;63(21):7051–5.

174. Zec N, Cera P, Towfighi J. Extramedullary hematopoiesis in cerebellar hemangioblastoma. Neurosurgery 1991;29(1): 34–7.

175. Park DM, Zhuang Z, Chen L, et al. von Hippel-Lindau Disease-Associated Hemangioblastomas Are Derived from Embryologic Multipotent Cells. PLoS Med 2007;4(2):e60.

176. Huber TL, Kouskoff V, Fehling HJ, Palis J, Keller G. Haemangioblast commitment is initiated in the primitive streak of the mouse embryo. Nature 2004;432(7017):625–30.

177. Glasker S, Li J, Xia JB, et al. Hemangioblastomas share protein expression with embryonal hemangioblast progenitor cell. Cancer Res 2006;66(8):4167–72.

178. Samuelsen SO, Bakketeig LS, Tretli S, Johannesen TB, Magnus P. Head circumference at birth and risk of brain cancer in childhood: a population-based study. Lancet Oncol 2006;7(1): 39–42.

179. Trichopoulos D, Lagiou P, Adami HO. Towards an integrated model for breast cancer etiology: the crucial role of the number of mammary tissue-specific stem cells. Breast Cancer Res 2005;7(1):13–7.

180. van Eekelen JA, Bradley CK, Gothert JR, et al. Expression pattern of the stem cell leukaemia gene in the CNS of the embryonic and adult mouse. Neuroscience 2003;122(2):421–36.

181. Mendel DB, Laird AD, Smolich BD, et al. Development of SU5416, a selective small molecule inhibitor of VEGF receptor tyrosine kinase activity, as an anti-angiogenesis agent. Anticancer Drug Des 2000;15(1):29–41.

182. Aiello LP, George DJ, Cahill MT, et al. Rapid and durable recovery of visual function in a patient with von hippel-lindau syndrome after systemic therapy with vascular endothelial growth factor receptor inhibitor su5416. Ophthalmology 2002;109(9):1745–51.

183. Plate KH, Risau W. Angiogenesis in malignant gliomas. Glia 1995;15(3):339–47.

184. Leon SP, Folkerth RD, Black PM. Microvessel density is a prognostic indicator for patients with astroglial brain tumors. Cancer 1996;77(2):362–72.

185. Kieran MW. Anti-angiogenic chemotherapy in central nervous system tumors. Cancer Treat Res 2004;117:337–49.

186. Carmeliet P. Angiogenesis in life, disease and medicine. Nature 2005;438(7070):932–6.

187. Erber R, Thurnher A, Katsen AD, et al. Combined inhibition of VEGF and PDGF signaling enforces tumor vessel regression by interfering with pericyte-mediated endothelial cell survival mechanisms. Faseb J 2004;18(2):338–40.

188. Kunkel P, Ulbricht U, Bohlen P, et al. Inhibition of glioma angiogenesis and growth in vivo by systemic treatment with a monoclonal antibody against vascular endothelial growth factor receptor-2. Cancer Res 2001;61(18):6624–8.

189. Takano S, Yoshii Y, Kondo S, et al. Concentration of vascular endothelial growth factor in the serum and tumor tissue of brain tumor patients. Cancer Res 1996;56(9):2185–90.

190. Peles E, Lidar Z, Simon AJ, Grossman R, Nass D, Ram Z. Angiogenic factors in the cerebrospinal fluid of patients with astrocytic brain tumors. Neurosurgery 2004;55(3):562–7; discussion 7–8.

191. Shaked Y, Emmenegger U, Man S, et al. Optimal biologic dose of metronomic chemotherapy regimens is associated with maximum antiangiogenic activity. Blood 2005;106(9):3058–61.

192. Louissaint A, Jr., Rao S, Leventhal C, Goldman SA. Coordinated interaction of neurogenesis and angiogenesis in the adult songbird brain. Neuron 2002;34(6):945–60.

193. Mukouyama YS, Shin D, Britsch S, Taniguchi M, Anderson DJ. Sensory nerves determine the pattern of arterial differentiation and blood vessel branching in the skin. Cell 2002;109(6):693–705.

194. Johnston M. The importance of lymphatics in cerebrospinal fluid transport. Lymphat Res Biol 2003;1(1):41–4; discussion 5.

195. Zakharov A, Papaiconomou C, Johnston M. Lymphatic vessels gain access to cerebrospinal fluid through unique association with olfactory nerves. Lymphat Res Biol 2004;2(3):139–46.

196. Zhang J, Williams MA, Rigamonti D. Genetics of human hydrocephalus. J Neurol 2006;253(10):1255–66.

197. Lambrechts D, Devriendt K, Driscoll DA, et al. Low expression VEGF haplotype increases the risk for tetralogy of Fallot: a family based association study. J Med Genet 2005;42(6):519–22.

198. Lambrechts D, Storkebaum E, Morimoto M, et al. VEGF is a modifier of amyotrophic lateral sclerosis in mice and humans and protects motoneurons against ischemic death. Nat Genet 2003;34(4):383–94.

199. Oosthuyse B, Moons L, Storkebaum E, et al. Deletion of the hypoxia-response element in the vascular endothelial growth factor promoter causes motor neuron degeneration. Nat Genet 2001;28(2):131–8.

200. Storkebaum E, Lambrechts D, Dewerchin M, et al. Treatment of motoneuron degeneration by intracerebroventricular delivery of VEGF in a rat model of ALS. Nat Neurosci 2005;8(1):85–92.

201. Kouznetsova E, Klingner M, Sorger D, et al. Developmental and amyloid plaque-related changes in cerebral cortical capillaries in transgenic Tg2576 Alzheimer mice. Int J Dev Neurosci 2006;24(2–3):187–93.

202. Krupinski J, Stroemer P, Slevin M, Marti E, Kumar P, Rubio F. Three-dimensional structure and survival of newly formed blood vessels after focal cerebral ischemia. Neuroreport 2003;14(8):1171–6.

Chapter 43
Lymphatic Vascular System and Lymphangiogenesis

Leah N. Cueni and Michael Detmar*

Keywords: lymphangiogenesis, VEGF-C, VEGF, Prox1, podoplanin

Abstract: The lymphatic system plays an important role in the maintenance of tissue fluid homeostasis, in the afferent phase of the immune response, and in metastatic cancer spread. However, the scientific exploration of the lymphatic system has lagged behind that of the blood vascular system, largely due to the absence of specific markers for lymphatic endothelium and to the paucity of knowledge about the molecular regulators of its development and function. The recent identification of genes that specifically control lymphatic development and the growth of lymphatic vessels (lymphangiogenesis), as well as the discovery of novel lymphatic endothelium-specific markers, have now provided important insights into the molecular mechanisms that control lymphatic growth and function. Studies of genetic mouse models have led to a new molecular model for embryonic lymphatic vascular development, and they have identified molecular pathways whose mutational inactivation leads to human diseases associated with lymphedema. Moreover, recent evidence indicates that malignant tumors can directly promote lymphangiogenesis and lymphatic metastasis, and that lymphangiogenesis also has a major role in chronic inflammation.

Anatomical and Functional Features of the Lymphatic Vasculature

The Italian anatomist Gasparo Aselli, who identified lymphatic vessels as "milky veins" in the mesentery of a "well-fed" dog, was the first to describe the lymphatic vasculature back in the seventeenth century [1]. The lymphatic system consists of the lymphatic vessels and the lymphoid organs which include lymph nodes, thymus, tonsils, spleen and Peyer's patches. In contrast to the blood vasculature, a closed circle around which blood is pumped by the heart, the lymphatic system comprises a linear, open-ended network without a central driving force. Its branches extend into most tissues, except for avascular structures such as epidermis, hair, nails, cartilage and cornea, and some vascularized organs such as brain and retina. Lymph, the protein-rich exudate from blood vessels, is taken up by the lymphatic capillaries in the tissue. From there it is returned to the venous circulation via the collecting lymphatic vessels and the thoracic duct, which connects the lymphatic system to the inferior vena cava. The pressure gradients to move lymph through the vessels result from skeletal muscle action, respiratory movement and contraction of smooth muscle in vessel walls.

A single, non-fenestrated layer of overlapping endothelial cells lines the lymphatic capillaries, which, in contrast to blood vessels, lack a continuous basement membrane as well as pericyte or smooth muscle cell coverage. Instead, the lymphatic endothelial cells (LECs) are connected to the surrounding extracellular matrix by specialized anchoring filaments [2]. These thin, fibrillin-containing fibres exert tension on endothelial cells when interstitial fluid pressure rises, thereby widening the capillary lumen and pulling open the loose intercellular junctions. This enables uptake of fluid, large macromolecules and cells into the vessel. In contrast to the capillaries, the collecting lymphatic vessels contain a basement membrane and are coated by a layer of smooth muscle cells contributing to lymph propulsion. Unidirectional fluid transport is ensured by luminal valves.

The lymphatic system also contributes to the immune surveillance of the body. Lymphatic vessels serve as a route for the trafficking of immune cells, such as lymphocytes and antigen-presenting dendritic cells from the skin and other organs to regional lymph nodes, where specific immune responses are initiated. Furthermore, dietary fat and fat-soluble vitamins from the digestive system are taken up by the lacteal lymphatic vessels in the intestine and transported to the venous circulation.

Institute of Pharmaceutical Sciences, Swiss Federal Institute of Technology, ETH Zurich, Zurich, Switzerland

*Corresponding Author:
Michael Detmar, Institute of Pharmaceutical Sciences, Swiss Federal Institute of Technology ETH Zurich, Wolfgang-Pauli-Strasse 10, HCI H303, CH-8093 Zurich, Switzerland
E-mail: michael.detmar@pharma.ethz.ch

The lymphatic system is involved in a number of pathologic conditions, the most prominent of which is lymphedema, a disorder caused by failure of proper lymphatic drainage. Importantly, lymphatic vessels also have a role in several inflammatory and auto-immune diseases, and together with blood vessels they provide a major route for the metastatic spread of many malignant tumors.

Embryonic Development of the Lymphatic Vasculature

The first concept of lymphatic development was presented in 1902 by Florence Sabin, based upon ink-injection experiments in pigs. She proposed that the initial lymph sacs are formed by endothelial cells which bud from the veins during early embryonic development. The peripheral lymphatic system originates from these primary lymph sacs and spreads by endothelial sprouting into the surrounding tissues and organs, where local capillaries are formed [3]. In 1910, Huntington and McClure alternatively suggested a mesenchymal origin of the lymph sacs, arising independently of the veins from precursor cells (lymphangioblasts), with venous connections established only subsequently [4].

Evidence supporting a venous origin of LECs came from studies in Prox1 deficient mice, revealing the crucial role of this homeobox protein in lymphatic development [5,6] (Fig. 43.1A), and, only very recently, from studies in zebrafish, in which LECs of the thoracic duct arise from primitive veins [7]. There is, however, evidence for a dual origin of lymphatics in avians and *Xenopus* frogs, in which parts of the lymphatic system are derived from adjacent veins, while others likely originate from local lymphangioblasts [8–10]. It remains unknown, whether and how lymphangioblasts might also contribute to embryonic lymphangiogenesis in mammals.

Adult Lymphangiogenesis

Adult lymphangiogenesis takes place in the context of tissue repair, inflammation and tumor growth. At present, it is unclear whether lymphatic vessel growth in such conditions is exclusively due to proliferation of local endothelial cells and sprouting of pre-existing vessels, or whether it also involves incorporation of circulating endothelial progenitor cells at sites of active lymphangiogenesis. Evidence for an exclusive contribution of local endothelial cell sprouting comes from experiments using sublethally irradiated mice that were grafted with green fluorescent protein (GFP)-expressing bone marrow. When growth of lymphatic vasculature was induced by vascular endothelial growth factor-C (VEGF-C) application or by tumor implants in these mice, no GFP-positive donor cells were observed in the newly formed lymphatic vessels, suggesting that bone marrow-derived endothelial progenitor cells do not contribute to tumor- or VEGF-C-induced lymphangiogenesis [11].

On the other hand, putative lymphatic endothelial progenitor cells have been identified in human fetal liver and cord blood, a subset of cells co-expressing lymphatic endothelial and stem cell markers [12]. Furthermore, bone marrow-derived cells were incorporated into newly formed lymphatic vessels in the inflamed or fibroblast growth factor-2 (FGF2) treated corneas of GFP chimeric mice [13,14], as well as during inflammation-associated lymphangiogenesis in human renal transplants [15]. It has been suggested that these lymphatic endothelial progenitors may be macrophages, which transdifferentiate into LECs [14,15]. However, the relative contribution of these progenitor cells appears to be minor. In addition, macrophages and other inflammatory cells may play an indirect role in neovascularization through the secretion of lymphangiogenic factors such as VEGF-C or VEGF-D [16,17].

Experimental Models for Lymphatic Research

Apart from the early studies by Sabin and Huntington and McClure, the field of lymphatic research has been neglected for a long time, mainly because of the lack of specific lymphatic markers and growth factors, and of suitable in vitro and in vivo models to discover novel players in lymphangiogenesis. It was only in the last decade when major progress was made. The identification of several molecules specifically expressed on either lymphatic or blood vascular endothelium (Table 43.1) has enabled the isolation of those two cell types from human skin and the establishment of defined cultures of blood vascular (BECs) and lymphatic endothelial cells (LECs), which can be maintained for several passages in vitro without loss of their lineage-specific differentiation [18–21].

To study mechanisms of (early) lymphatic vessel development in vitro, murine embryoid bodies have recently been described as valuable tools. In these three-dimensional, embryonic stem cell-derived structures, LECs seem to bud off from pre-existing blood vessel-like structures, but were also found away from vascular areas, where they might either have migrated to, or developed locally from lymphatic progenitors [22,23].

Another valuable tool for studying the effect of potential lymphangiogenic factors is the chorioallantoic membrane (CAM) assay, since the differentiated avian CAM is drained by a regular network of lymphatic vessels with typical structural and molecular features [24].

Besides a number of genetically engineered mice displaying lymphatic phenotypes (Table 43.2), the *Xenopus laevis* tadpole and the zebrafish embryo have recently been established as novel vertebrate models to study lymphatic development and function [7,8,25]. These model organisms can be bred and maintained in large numbers in the laboratory, and are easily manipulated genetically and experimentally. Moreover, the early *Xenopus* embryo, and even more so the zebrafish embryo, are optically transparent, permitting a direct microscopic inspection of many developmental processes. Importantly, the

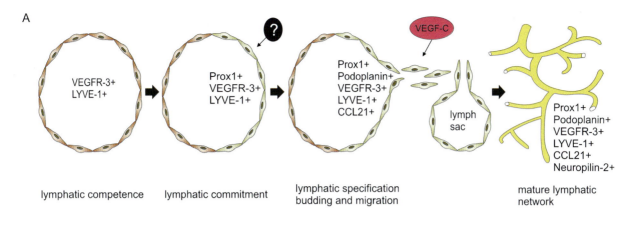

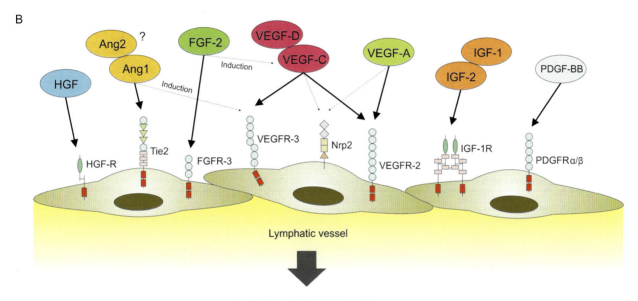

FIG. 43.1. **A** Current model of lymphatic vasculature development in the mammalian embryo. All endothelial cells of the embryonic cardinal vein express the lymphatic markers LYVE-1 and VEGFR3 and display lymphatic competence during early vascular development. A yet unidentified mesenchymal signal induces polarized expression of the transcription factor Prox1 in a subset of endothelial cells on one side of the cardinal vein, that become committed to the lymphatic lineage. Under the influence of VEGF-C, these cells bud off from the vein and migrate into the surrounding tissue to form primitive lymph sacs. During this process, they adopt the expression of additional lymphatic lineage markers. The formation of a mature lymphatic network is completed after the first postnatal days. **B** A variety of growth factors and their receptors expressed by lymphatic endothelium contribute to lymphangiogenesis. Lymphatic vessel growth is promoted by vascular endothelial growth factors (VEGF)-A, -C and -D, activating distinct VEGF receptors (VEGFR) and neuropilin-2 (Nrp2), angiopoietin-1 (Ang1) activating Tie2 and up-regulating VEGFR3 (*dashed line*), fibroblast growth factor-2 (FGF2) acting directly through FGF receptor 3 (FGFR-3) and via induction of VEGF-C (*dashed line*), and by hepatocyte growth factor (HGF), insulin-like growth factors (IGF) and platelet-derived growth factor-BB (PDGF-BB), acting directly through their respective receptors HGF-R, IGF-1R and PDGFR.

lymphatic vascular systems of the *Xenopus* tadpole and the zebrafish share a number of crucial functional, morphological and molecular features with the lymphatic vasculature of higher vertebrates [7,8,25].

Altogether, these novel experimental models provide excellent tools for future studies to improve the understanding of lymphatic development and function, as well as of several diseases associated with the lymphatic vasculature.

Lineage Markers of the Lymphatic Endothelium

One of the first markers of LECs to be identified was vascular endothelial growth factor receptor-3 (VEGFR3), also known as Flt4 [26]. It is a member of the fms-like tyrosine kinase (flt) family and serves as a specific receptor for the lymphangiogenic

TABLE 43.1. Specific markers for lymphatic (*LV*) versus blood vessels (*BV*).

Marker	LV	BV	Function
VEGFR3 [26]	+	−/(+)[a]	Growth factor receptor
Prox1 [6]	++	−	Transcription factor
LYVE-1 [4]	++	−	Hyaluronan receptor
Podoplanin [47,49]	++	−	Transmembrane glycoprotein
SLC/CCL21 [55]	+	−	CC-chemokine
CCL20/MIP-3α [18,19]	+ (++)[b]	− (++)[b]	CC-chemokine
Neuropilin-2 [80]	+	−/(+)[c]	Semaphorin and growth factor receptor
Desmoplakin [116]	+	−	Anchoring protein of adhering junctions
Macrophage mannose receptor 1 [117]	+	−	L-selectin receptor
Integrin α_9 [35,96]	+	−	Adhesion molecule, subunit of osteopontin and tenascin receptors, VEGFR3 coreceptor?
VEGFR1 [18]	−	+	Growth factor receptor
CD34 [118]	−/(+)[d]	++	L-selectin receptor
Neuropilin-1 [34,35]	−	+	Semaphorin and growth factor receptor
CD44 [19]	−	+	Hyaluronan receptor
VEGF-C [18,19]	−	+	Growth factor
PAL-E [119,120]	−	++	Caveolae-associated glycoprotein?
Laminin [35,121]	−/(+)[e]	++	Basement membrane molecule
Collagen IV [18]	−/(+)[e]	++	Extracellular matrix protein

[a] VEGFR3 is also present on blood capillaries during tumor neovascularization and in wound granulation tissue [28,29]
[b] Both BECs and LECs strongly express CCL20 after activation [19]
[c] Neuropilin-2 is also expressed in veins [80]
[d] CD34 expression has also been found on lymphatic endothelial cells [19,122,123]
[e] Large collecting lymphatic vessels have a complete basement membrane, some peripheral vessels may have an incomplete one

TABLE 43.2. Genetic mouse models with abnormalities of the lymphatic vasculature.

Gene	Function	Model	Phenotype
Prox1 [6,38]	Transcription factor	KO	No lymphatic vasculature (−/−), adult-onset obesity, chylous ascites (+/−)
SOX18 (ragged) [124]	Transcription factor	KO (spontaneous missense mutations)	Edema, chylous ascites, cardiovascular and hair follicle defects
Net (Elk3) [125]	Transcription factor	KO	Chylothorax, dilated lymphatic vessels
FOXC2 [126,127]	Transcription factor	KO	Abnormal patterning and pericyte investment of lymphatic vessels, absent valves, lymphatic dysfunction (−/−), lymphatic hyperplasia (+/−)
VEGF-C [66]	Growth factor	KO	No lymphatic vasculature (−/−), delayed lymphatic vascular development, lymphatic hypoplasia and lymphedema (+/−)
VEGF-C [64]	Growth factor	TG	Hyperplastic lymphatic vessels
VEGFR3 [70]	Growth factor receptor	KO	Cardiovascular failure, defective remodelling of vascular networks
VEGFR3 [72]	Growth factor receptor	Chy mice (inactivating mutation)	Lymphedema, chylous ascites, hypoplasia
Neuropilin-2 [80]	Growth factor receptor	KO	Severe reduction or absence of small lymphatic vessels and capillaries during development
Angiopoietin-1 [90]	Growth factor	TG	Hyperplastic lymphatic vessels
Angiopoietin-2 [92]	Growth factor	KO	Chylous ascites and peripheral edema, abnormal patterning of small lymphatic vessels
HGF [93]	Growth factor	TG	Enhanced formation and enlargement of lymphatic vessels
Integrin α_9 [96]	Adhesion receptor	KO	Chylothorax, lymphedema
LYVE-1 [45,46]	Hyaluronan receptor	KO	None or only subtle defects: increased interstitial-lymphatic flow, atypical shape of vessel lumen
Podoplanin [48]	Membrane glycoprotein	KO	Lymphedema, dilation of lymphatic vessels, diminished lymphatic transport
Ephrin B2 [42]	Ligand of EphB receptors	mutant lacking PDZ interaction site	Defective remodelling of lymphatic vascular network, hyperplasia, absent valves, chylothorax
Syk and SLP-76 [128,129]	Tyrosine kinase (Syk), adaptor protein (SLP-76)	KO	Abnormal blood-lymphatic connections, chylous ascites, defect in hematopoetic endothelial progenitors

FOXC2 Forkhead box C2; *HGF* hepatocyte growth factor; *KO* knockout; *LYVE-1* lymphatic vascular endothelial hyaluronan receptor-1; *PDZ* PSD-95, DISCS-large, and ZO-1; *SLP* Src homology 2-domain containing leukocyte protein; *SOX18* sex determining region Y-related high mobility group box 18; *TG* transgenic; *VEGF* vascular endothelial growth factor; *VEGFR* vascular endothelial growth factor receptor

vascular endothelial growth factors (VEGF)-C and -D. Before the initiation of lymphatic vasculature development in the early embryo, VEGFR3 is present both in venous and in presumptive lymphatic endothelia (Fig. 43.1A). Later, its expression is progressively down-regulated by venous endothelial cells, and becomes restricted to the lymphatic endothelium in normal adult tissues [26,27]. Nevertheless, VEGFR3 alone might not be a sufficiently specific marker for lymphatic vessels, since it is also expressed on some blood capillaries during tumor angiogenesis and in wound granulation tissue [28–31].

The most specific marker for lymphatic endothelium at present is the transcription factor Prox1, a homolog of the Drosophila homeobox gene *prospero* [32]. It is expressed in a variety of cell types, including LECs, but not blood vascular endothelial cells (BECs), in the embryo [6] as well as in lymphatic endothelia in adult tissues and tumors [5,33]. Prox1 acts as a master regulator of lymphatic development. Around mouse embryonic day (E) 9.5–E10.5, a presently unknown signal induces polarized Prox1-expression in the lymphatically competent endothelial cells on one side of the cardinal vein [5,6], which promotes lymphatic commitment and specification (Fig. 43.1A). In *Prox1* null mice, budding and sprouting of these cells from the veins is arrested prematurely at around E11.5–E12.0, resulting in a complete lack of lymphatic vasculature in such mice [6].

Deletion of *Prox1* prevents the expression of lymphatic endothelial markers in the budding endothelial cells [5], while differentiated BECs can be reprogrammed to adopt a lymphatic phenotype by ectopic expression of Prox1 [34,35]. A comparable effect was seen upon infection of cultured BECs with Kaposi's sarcoma-associated herpes virus (also known as human herpesvirus-8), which upregulates Prox1 [36,37]. *Prox1* +/− mice only survive in one genetic background [6], in which they develop chylous ascites and, interestingly, adult onset obesity, suggesting a link between impaired lymphatic development and function and adiposity [38].

Lymphatic vessel hyaluronan (HA) receptor-1 (LYVE-1) is a lymphatic endothelium-specific cell surface protein which is also expressed by some other cell types such as activated macrophages and sinusoidal endothelium of the liver and the spleen [39,40]. During development, LYVE-1 is the first marker of lympathic competence; it is expressed by endothelial cells of the anterior cardinal vein as early as mouse E9.0-9.5 [41] (Fig. 43.1A). In the mature vasculature, LYVE-1 expression remains high in lymphatic capillaries but is down-regulated in the collecting lymphatic vessels [42]. The biological function of LYVE-1 is as yet unknown. As a homolog of the blood vascular endothelium-specific HA receptor CD44 [43], it was proposed to be involved in HA metabolism and in leukocyte migration through the lymphatic vasculature [39,44]. Recently, it has also been suggested to mediate cell-surface retention of secreted growth factors such as platelet-derived growth factor-BB (PDGF-BB) [45]. LYVE-1 deficient mice, however, have no or only subtle lymphatic vascular defects [45,46].

The mucin-type transmembrane glycoprotein podoplanin is expressed by LECs, but not BECs, in vivo and in vitro [18,19,35,47–49]. During mouse embryonic development, it

is detectable on all endothelial cells of the cardinal vein at E11.5-E12.5 and later becomes specific for the budding lymphatically committed cells [48]. Although several studies indicated that podoplanin is involved in actin cytoskeleton reorganization of various cell types and thereby mediates cell motility [48,50,51], its exact biological function remains unknown. *Podoplanin* null mice die at birth of respiratory failure [52]. In addition, they display various lymphatic defects, suggesting that podoplanin is essential for the normal formation and function of the lymphatic vasculature [48]. Podoplanin might also play a role in tumor progression, because, in addition to LECs and several other cell types [49,53], podoplanin is expressed in a number of tumors, e.g., squamous cell carcinoma and certain germ cell tumors [53,54].

The chemokine CCL21, also known as secondary lymphoid chemokine, is specifically secreted by lymphatic endothelium [19] and plays an important role in immunoregulatory and inflammatory processes. Through its high-affinity binding to the CC chemokine receptor 7 (CCR-7), it mediates homing of lymphocytes and migration of antigen-stimulated dendritic cells from the tissues into the lymphatic vessels and the secondary lymphatic organs [55,56]. CCL21 can also attract CCR-7 expressing tumor cells towards the lymphatic endothelium, thereby contributing to metastasis [57,58].

Besides the few well established markers mentioned so far, comparative transcriptional profiling of LECs versus BECs meanwhile revealed a number of novel differentially expressed genes (Table 43.1), although the vast majority of all investigated genes (ca. 98%) are expressed at comparable levels in those genetically closely related cell types [18,35]. The exploration of the roles of these specifically expressed molecules will provide new insight into the functional regulation and physiological maintenance of the two types of vasculature.

Lymphangiogenic Growth Factors

VEGFs and Their Receptors

To date, several lymphangiogenic growth factors have been identified [59]. The most important and best characterized of these molecules are VEGF-C and VEGF-D. VEGF-C and -D are the only known ligands for VEGFR3 [60–63]. Overexpression of VEGF-C or VEGF-D in the skin of transgenic mice causes hyperplasia of the cutaneous lymphatic vessels [64,65], whereas deletion of VEGF-C results in a complete lack of lymphatic vasculature and prenatal death of mouse embryos [66]. Without the VEGF-C signal during embryonic development, the Prox1-positive lymphatically committed venous endothelial cells are unable to migrate out to form the initial lymph sacs (Fig. 43.1A).VEGF-D stimulates lymphangiogenesis in tissues and tumors [65,67], but is not essential for the development of the lymphatic vasculature since VEGF-D deficient mice have no lymphatic defects [66,68]. Probably VEGF-D is not expressed at the critical sites of lymph sac formation in the embryo, because only exogenous, but not endogenous,

VEGF-D can rescue the phenotype of the *Vegf-c^-/-* mice [66,69]. Assessing the role of VEGFR3 in lymphatic development is difficult, because VEGFR3 deficient mice die of cardiovascular failure at E9.5, before the emergence of lymphatic vessels [70]. An important role of VEGFR3 in lymphatic development and function is supported by the identification of VEGFR3 mutations in patients suffering from hereditary lymphedema [71], as well as in Chy mutant mice which are characterized by cutaneous lymphedema [72]. Indeed, amelioration of lymphatic function was observed after treatment of Chy mice with viral VEGF-C gene therapy [72] and, after administration of recombinant human VEGF-C protein, in a model of surgically induced lymphedema in the rabbit ear [73].

Since VEGF-C and VEGF-D, after enzymatic cleavage, also bind to VEGFR2, which is expressed on both LECs and BECs [18,19,74], VEGFR2 might also contribute to lymphangiogenesis. Exclusive activation of VEGFR3 signaling, however, is sufficient to promote lymphangiogenesis, as shown by overexpression of a VEGFR3-specific mutant of VEGF-C (VEGF-C156S [75]) in the skin of transgenic mice [65].

Another ligand for VEGFR2 is VEGF-A. Although VEGF-A cannot rescue the phenotype of VEGF-C deficient mice [66], several recent studies have established a role for VEGF-A/VEGFR2 signaling in adult lymphangiogenesis. Like VEGF-C [76], VEGF-A potently induces proliferation and survival of LECs in vitro [18]. Moreover, it strongly enhances lymphangiogenesis along with angiogenesis in vivo: in mouse ears after adenoviral gene delivery of murine VEGF-A [77], and during wound healing or skin inflammation in transgenic mice specifically overexpressing murine VEGF-A in the epidermis [74,78]. Signaling of VEGF-A through VEGFR2 is important for these effects, since they can be inhibited by a specific VEGFR2 blocking antibody. Additionally, attraction of inflammatory cells producing VEGF-C and -D might contribute to the lymphangiogenic activity of VEGF-A [17,79].

In addition to the three VEGFR tyrosine kinases, several VEGFs also interact with neuropilins. Two of these non-kinase transmembrane proteins, neuropilin 1 (Nrp1) and neuropilin 2 (Nrp2), have been described so far. In the vasculature, Nrp1 is primarily found in arterial endothelial cells, whereas Nrp2 expression is confined to veins and lymphatic capillaries [34,72,80]. Nrp2 binds several VEGF family members, including VEGF-C and -D [72,81], and interacts with VEGFR3 and VEGFR2 [82]. Thus, Nrp2 might enhance VEGF-C/-D signaling through VEGFR3, similar to the Nrp1-mediated promotion of VEGF-A signaling via VEGFR2 [72,83]. Indeed, Nrp2 deficient mice show a severe reduction of small lymphatic vessels during development [80], in addition to defects of the nervous system, which are explained by the role of neuropilins as semaphorin receptors in neuronal axon guidance [84,85].

Other Lymphangiogenic Factors

Along with the VEGF family, the angiopoietins are key players in vascular development. Although several mouse models have demonstrated the necessity of the angiopoietin signaling system for normal blood vessel development [86–88], little is known about its role in lymphangiogenesis. Angiopoietin-1 (Ang1) activates the endothelial-specific receptor tyrosine kinase Tie2 (tyrosine kinase with Ig-like loop and EGF homology domain-2) [89], which is expressed by LECs in vitro [19,35] and in vivo [90, 91]. When applied to the mouse cornea or to other adult mouse tissues (by viral gene delivery), Ang1 induces lymphangiogenesis [90,91], and its skin-specific overexpression in transgenic mice results in cutaneous lymphatic hyperplasia [90]. It remains unclear to what extent Ang1 exerts its effects on the lymphatic endothelium directly through Tie2, or whether it might also act indirectly via the VEGF-C/VEGFR3 pathway. The latter appears likely, since the effects of virally delivered *Ang1* in mice are inhibited by treatment with soluble VEGFR3, and VEGFR3 is up-regulated after Ang1 stimulation of LECs in vitro and in vivo [90].

Angiopoietin-2 (Ang2) is considered to act as an antagonist of Tie2 [88], and also seems to be required for the normal formation of the lymphatic vasculature. Ang2 deficient mice display chylous ascites, peripheral edema and abnormal patterning of small lymphatic vessels [92]. Interestingly, these defects, but not the blood vascular phenotype, can be rescued by Ang1 [92]. Thus, Ang1 and Ang2 might have redundant roles as agonists of Tie2 in lymphangiogenesis, while Ang2 counteracts Ang1 as a (context-dependent) antagonist of Tie2 in angiogenesis.

Hepatocyte growth factor (HGF, also known as scatter factor) was recently shown to promote proliferation, migration and tube formation of cultured LECs, and overexpression of HGF in transgenic mice as well as its subcutaneous delivery resulted in increased numbers and enlargement of lymphatic vessels [93]. These effects were not inhibited by a VEGFR3 blocking antibody, indicating that HGF can promote lymphangiogenesis independently of the VEGFR3 pathway via its receptor HGF-R [93]. In a mouse cornea assay, however, the lymphangiogenic effect of implanted HGF was attenuated after administration of a soluble VEGFR3, suggesting that HGF might also exert indirect effects on lymphangiogenesis [94]. The enhancement of LEC migration by HGF was found to be mediated by integrin $\alpha_9\beta_1$ [93], which is specifically expressed by LECs [35], binds directly to VEGF-C and VEGF-D [95] and is required for normal lymphatic function [96].

Further growth factors having lymphangiogenic activity include the insulin-like growth factors (IGF) 1 and 2 [97], fibroblast growth factor (FGF)-2 [98–100], and platelet derived growth factors (PDGF) [101]. IGFs and FGF2 are well known to promote angiogenesis [102–105], but also enhance proliferation and migration of LECs in vitro and stimulate lymphatic vessel growth in the mouse cornea [97–100]. IGFs and FGF2 both appear to at least partially act directly via their respective receptors IGF-1R and/or -2R and FGFR-3 expressed on lymphatic endothelium [97,100]. For FGF2, also an indirect mechanism involving the VEGF-C/VEGFR3 signaling pathway probably plays a role [98,99]. PDGF-BB was reported to induce lymphangiogenesis in the mouse cornea assay and within xenotransplanted tumors in mice, probably via activation of the PDGF receptors α and β [101].

During the process of lymphatic vessel formation and growth in physiological and/or pathological conditions, there is probably a complex interplay between the various growth factors

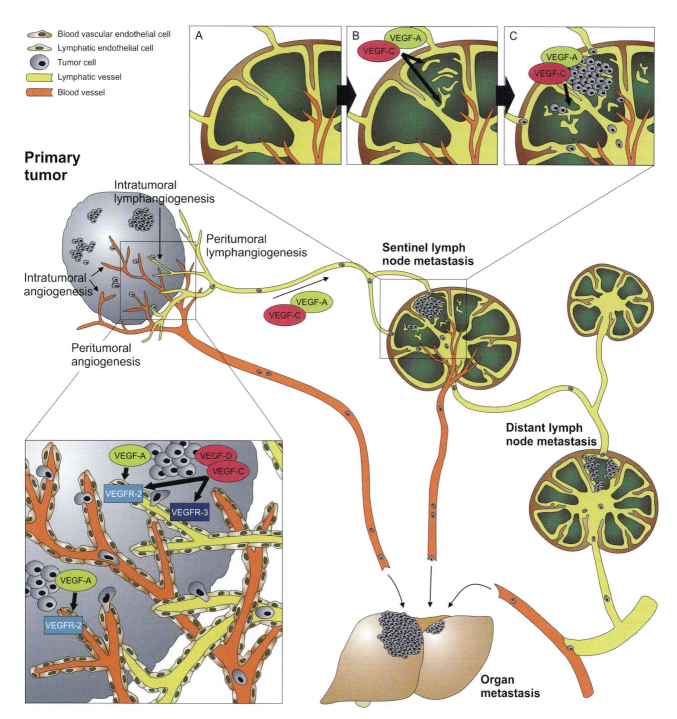

Fɪɢ. 43.2. Tumor and lymph node lymphangiogenesis: Impact on cancer metastasis. In normal adult tissues and their draining lymph nodes, there is no detectable lymphangiogenesis. Through secretion of VEGF-A and/or VEGF-C/-D, tumors can induce tumor-associated lymphatic vessel growth and also lymphangiogenesis within sentinel LNs, even before they metastasize, possibly preparing the LN for their later arrival. Metastatic tumor cells maintain their lymphangiogenic activity after metastasis to sentinel LNs, likely promoting further cancer spread to distant lymph nodes and organs. *VEGF* Vascular endothelial growth factor; *LN* lymph node.

(Fig. 43.1B). Several of them are likely also involved in tumor metastasis via the lymphatic vasculature (e.g., VEGF-A [106], HGF [93], IGF [97], FGF2 [99], PDGF-BB [101]) and, therefore, might serve as potential therapeutic targets.

Tumor Lymphangiogenesis and Metastasis

Tumor metastasis to regional (sentinel) lymph nodes frequently represents the first step of cancer dissemination and serves as an important indicator for cancer progression. The mechanisms of tumor cell entry into the lymphatic system have remained unclear, however. An increasing amount of evidence suggests that tumors can actively induce lymphangiogenesis, and that tumor-associated lymphangiogenesis actively promotes cancer metastasis. Studies in animal tumor models have revealed that increased levels of VEGF-C and/or VEGF-D promote active tumor lymphangiogenesis and lymphatic tumor spread to regional lymph nodes, and that these effects can be suppressed by blocking VEGFR3 signaling [67,107–109]. VEGF-A can also act as a tumor lymphangiogenesis factor and, importantly, tumor-secreted VEGF-A promotes expansion of the lymphatic network within draining, sentinel lymph nodes, even before cancer metastasis [106]. These results, together with the recent finding that VEGF-C-induced lymph node lymphangiogenesis promoted squamous cell carcinoma metastasis to distant lymph nodes and to the lung [110], suggest that lymph node lymphangiogenesis contributes to tumor metastasis beyond the sentinel lymph nodes (Fig. 43.2). Numerous studies have shown a direct correlation between expression of VEGF-C or VEGF-D by human cancers and tumor metastasis, suggesting that lymphangiogenesis also has an important role in promoting metastasis of human tumors [111]. Indeed, lymphangiogenesis is increased in primary melanomas of the skin that later metastasized, as compared to non-metastatic tumors, and the extent of tumor-associated lymphangiogenesis can serve as a potent predictor of lymph node metastasis and patient survival [112]. Tumor lymphangiogenesis was also shown to be the most significant parameter to predict the presence of sentinel lymph node metastases at the time of surgical excision of primary melanomas [113]. Overall, these studies indicate that tumor lymphangiogenesis is a novel, potent predictor of cancer metastasis and might be helpful for improved patient care, and that interference with lymphangiogenesis might reduce the incidence of cancer metastases.

Inflammation and Lymphangiogenesis

There is increasing evidence that lymphatic vessels have an active role in acute and chronic inflammation. Skin lesions in the chronic inflammatory disease psoriasis show lymphatic hyperplasia [78], and LEC proliferation and lymphatic hyperplasia are also observed in chronic skin inflammation in mice [78]. Lymphangiogenesis has also been found to be associated with kidney transplant rejection, where the LEC-derived chemokine CCL21 might further enhance the inflammatory process [114]. Moreover, inflammation induced by acute UVB irradiation of the skin results in edema and hyperpermeable, leaky lymphatic vessels that are functionally impaired [115]. It is of interest that inhibition of VEGFR3 signaling leads to prolonged edema and inflammation after UVB irradiation. Thus, in addition to removing inflammation-associated tissue edema, lymphatic vessels might also actively participate in the maintenance of chronic inflammatory diseases.

Outlook

The mechanisms controlling the normal and pathological development of the lymphatic vasculature are now being unraveled, mainly based on studies in genetic mouse models. However, distinct genetic defects have also been identified in lymphedema patients, and new insights into the molecular control of lymphatic vessel growth have led to the development of the first pro-lymphangiogenic therapies to treat lymphedema. The novel concept of tumor- and lymph node-associated lymphangiogenesis and its role in tumor metastasis is of particular importance for the understanding, and possibly inhibition, of cancer progression. These developments will hopefully lead to improved diagnosis and therapy of human cancers but also of chronic inflammatory diseases.

Acknowledgements. This work was supported by National Institutes of Health grants CA69184, CA86410, CA92644, American Cancer Society Research Project Grant 99-23901, Swiss National Fund grant 3100A0-108207, Austrian Science Foundation grant S9408-B11, Cancer League Zurich, and Commission of the European Communities grant LSHC-CT-2005-518178 (to M.D.).

References

1. Asellius G. *De lactibus sive lacteis venis*. Milan 1627:Mediolani.
2. Gerli R, Solito R, Weber E, et al. Specific adhesion molecules bind anchoring filaments and endothelial cells in human skin initial lymphatics. Lymphology 2000;33:148–57.
3. Sabin FR. On the origin of the lymphatic system from the veins and the development of the lymph hearts and thoracic duct in the pig. Am J Anat 1902;1:367–91.
4. Huntington GS, McClure CFW. The anatomy and development of the jugular lymph sac in the domestic cat (*Felis domestica*). Am J Anat 1910;10:177–311.
5. Wigle JT, Harvey N, Detmar M, et al. An essential role for Prox1 in the induction of the lymphatic endothelial cell phenotype. Embo J 2002;21:1505–13.
6. Wigle JT, Oliver G. Prox1 function is required for the development of the murine lymphatic system. Cell 1999;98:769–78.
7. Yaniv K, Isogai S, Castranova D, et al. Live imaging of lymphatic development in the zebrafish. Nat Med 2006;12:711–6.

8. Ny A, Koch M, Schneider M, et al. A genetic Xenopus laevis tadpole model to study lymphangiogenesis. Nat Med 2005;11:998–1004.

9. Schneider M, Othman-Hassan K, Christ B, et al. Lymphangioblasts in the avian wing bud. Dev Dyn 1999;216:311–9.

10. Wilting J, Aref Y, Huang R, et al. Dual origin of avian lymphatics. Dev Biol 2006;292:165–73.

11. He Y, Rajantie I, Ilmonen M, et al. Preexisting lymphatic endothelium but not endothelial progenitor cells are essential for tumor lymphangiogenesis and lymphatic metastasis. Cancer Res 2004;64:3737–40.

12. Salven P, Mustjoki S, Alitalo R, et al. VEGFR3 and CD133 identify a population of CD34+ lymphatic/vascular endothelial precursor cells. Blood 2003;101:168–72.

13. Religa P, Cao R, Bjorndahl M, et al. Presence of bone marrow-derived circulating progenitor endothelial cells in the newly formed lymphatic vessels. Blood 2005;106:4184–90.

14. Maruyama K, Ii M, Cursiefen C, et al. Inflammation-induced lymphangiogenesis in the cornea arises from CD11b-positive macrophages. J Clin Invest 2005;115:2363–72.

15. Kerjaschki D, Huttary N, Raab I, et al. Lymphatic endothelial progenitor cells contribute to *de novo* lymphangiogenesis in human renal transplants. Nat Med 2006;12:230–4.

16. Schoppmann SF, Birner P, Stockl J, et al. Tumor-associated macrophages express lymphatic endothelial growth factors and are related to peritumoral lymphangiogenesis. Am J Pathol 2002;161:947–56.

17. Cursiefen C, Chen L, Borges LP, et al. VEGF-A stimulates lymphangiogenesis and hemangiogenesis in inflammatory neovascularization via macrophage recruitment. J Clin Invest 2004;113:1040–50.

18. Hirakawa S, Hong YK, Harvey N, et al. Identification of vascular lineage-specific genes by transcriptional profiling of isolated blood vascular and lymphatic endothelial cells. Am J Pathol 2003;162:575–86.

19. Kriehuber E, Breiteneder-Geleff S, Groeger M, et al. Isolation and characterization of dermal lymphatic and blood endothelial cells reveal stable and functionally specialized cell lineages. J Exp Med 2001;194:797–808.

20. Makinen T, Veikkola T, Mustjoki S, et al. Isolated lymphatic endothelial cells transduce growth, survival and migratory signals via the VEGF-C/D receptor VEGFR-3. Embo J 2001;20:4762–73.

21. Podgrabinska S, Braun P, Velasco P, et al. Molecular characterization of lymphatic endothelial cells. Proc Natl Acad Sci USA 2002;99:16069–74.

22. Kreuger J, Nilsson I, Kerjaschki D, et al. Early lymph vessel development from embryonic stem cells. Arterioscler Thromb Vasc Biol 2006;26:1073–8.

23. Liersch R, Nay F, Lu L, et al. Induction of lymphatic endothelial cell differentiation in embryoid bodies. Blood 2006;107:1214–6.

24. Oh SJ, Jeltsch MM, Birkenhager R, et al. VEGF and VEGF-C: specific induction of angiogenesis and lymphangiogenesis in the differentiated avian chorioallantoic membrane. Dev Biol 1997;188:96–109.

25. Kuchler AM, Gjini E, Peterson-Maduro J, et al. Development of the zebrafish lymphatic system requires VEGFC signaling. Curr Biol 2006;16:1244–8.

26. Kaipainen A, Korhonen J, Mustonen T, et al. Expression of the fms-like tyrosine kinase 4 gene becomes restricted to lymphatic endothelium during development. Proc Natl Acad Sci USA 1995;92:3566–70.

27. Partanen TA, Arola J, Saaristo A, et al. VEGF-C and VEGF-D expression in neuroendocrine cells and their receptor, VEGFR3, in fenestrated blood vessels in human tissues. Faseb J 2000;14:2087–96.

28. Paavonen K, Puolakkainen P, Jussila L, et al. Vascular endothelial growth factor receptor-3 in lymphangiogenesis in wound healing. Am J Pathol 2000;156:1499–504.

29. Valtola R, Salven P, Heikkila P, et al. VEGFR3 and its ligand VEGF-C are associated with angiogenesis in breast cancer. Am J Pathol 1999;154:1381–90.

30. Kubo H, Fujiwara T, Jussila L, et al. Involvement of vascular endothelial growth factor receptor-3 in maintenance of integrity of endothelial cell lining during tumor angiogenesis. Blood 2000;96:546–53.

31. Partanen TA, Alitalo K, Miettinen M. Lack of lymphatic vascular specificity of vascular endothelial growth factor receptor 3 in 185 vascular tumors. Cancer 1999;86:2406–12.

32. Oliver G, Sosa-Pineda B, Geisendorf S, et al. Prox 1, a *prospero*-related homeobox gene expressed during mouse development. Mech Dev 1993;44:3–16.

33. Wilting J, Papoutsi M, Christ B, et al. The transcription factor Prox1 is a marker for lymphatic endothelial cells in normal and diseased human tissues. Faseb J 2002;16:1271–3.

34. Hong YK, Harvey N, Noh YH, et al. Prox1 is a master control gene in the program specifying lymphatic endothelial cell fate. Dev Dyn 2002;225:351–7.

35. Petrova TV, Makinen T, Makela TP, et al. Lymphatic endothelial reprogramming of vascular endothelial cells by the Prox-1 homeobox transcription factor. Embo J 2002;21:4593–9.

36. Wang HW, Trotter MW, Lagos D, et al. Kaposi sarcoma herpesvirus-induced cellular reprogramming contributes to the lymphatic endothelial gene expression in Kaposi sarcoma. Nat Genet 2004;36:687–93.

37. Hong YK, Foreman K, Shin JW, et al. Lymphatic reprogramming of blood vascular endothelium by Kaposi sarcoma-associated herpesvirus. Nat Genet 2004;36:683–5.

38. Harvey NL, Srinivasan RS, Dillard ME, et al. Lymphatic vascular defects promoted by *Prox1* haploinsufficiency cause adult-onset obesity. Nat Genet 2005;37:1072–81.

39. Jackson DG, Prevo R, Clasper S, et al. LYVE-1, the lymphatic system and tumor lymphangiogenesis. Trends Immunol 2001;22:317–21.

40. Jackson DG. The lymphatics revisited: new perspectives from the hyaluronan receptor LYVE-1. Trends Cardiovasc Med 2003;13:1–7.

41. Oliver G. Lymphatic vasculature development. Nat Rev Immunol 2004;4:35–45.

42. Makinen T, Adams RH, Bailey J, et al. PDZ interaction site in ephrinB2 is required for the remodeling of lymphatic vasculature. Genes Dev 2005;19:397–410.

43. Banerji S, Ni J, Wang SX, et al. LYVE-1, a new homologue of the CD44 glycoprotein, is a lymph-specific receptor for hyaluronan. J Cell Biol 1999;144:789–801.

44. Jackson DG. Biology of the lymphatic marker LYVE-1 and applications in research into lymphatic trafficking and lymphangiogenesis. Apmis 2004;112:526–38.

45. Huang SS, Liu IH, Smith T, et al. CRSBP-1/LYVE-1-null mice exhibit identifiable morphological and functional alterations of lymphatic capillary vessels. FEBS Lett 2006;580:6259–68.

46. Gale NW, Prevo R, Fematt JE, et al. Normal lymphatic development and function in mice deficient for the lymphatic hyaluronan receptor LYVE-1. Mol Cell Biol 2006.

47. Breiteneder-Geleff S, Soleiman A, Kowalski H, et al. Angiosarcomas express mixed endothelial phenotypes of blood and lymphatic capillaries: podoplanin as a specific marker for lymphatic endothelium. Am J Pathol 1999;154:385–94.

48. Schacht V, Ramirez MI, Hong YK, et al. T1alpha/podoplanin deficiency disrupts normal lymphatic vasculature formation and causes lymphedema. Embo J 2003;22:3546–56.

49. Wetterwald A, Hoffstetter W, Cecchini MG, et al. Characterization and cloning of the E11 antigen, a marker expressed by rat osteoblasts and osteocytes. Bone 1996;18:125–32.

50. Martin-Villar E, Megias D, Castel S, et al. Podoplanin binds ERM proteins to activate RhoA and promote epithelial-mesenchymal transition. J Cell Sci 2006;119:4541–53.

51. Wicki A, Lehembre F, Wick N, et al. Tumor invasion in the absence of epithelial-mesenchymal transition: podoplanin-mediated remodeling of the actin cytoskeleton. Cancer Cell 2006;9:261–72.

52. Ramirez MI, Millien G, Hinds A, et al. T1alpha, a lung type I cell differentiation gene, is required for normal lung cell proliferation and alveolus formation at birth. Dev Biol 2003;256:61–72.

53. Schacht V, Dadras SS, Johnson LA, et al. Up-regulation of the lymphatic marker podoplanin, a mucin-type transmembrane glycoprotein, in human squamous cell carcinomas and germ cell tumors. Am J Pathol 2005;166:913–21.

54. Martin-Villar E, Scholl FG, Gamallo C, et al. Characterization of human PA2.26 antigen (T1alpha-2, podoplanin), a small membrane mucin induced in oral squamous cell carcinomas. Int J Cancer 2005;113:899–910.

55. Gunn MD, Tangemann K, Tam C, et al. A chemokine expressed in lymphoid high endothelial venules promotes the adhesion and chemotaxis of naive T lymphocytes. Proc Natl Acad Sci USA 1998;95:258–63.

56. Tangemann K, Gunn MD, Giblin P, et al. A high endothelial cell-derived chemokine induces rapid, efficient, and subset-selective arrest of rolling T lymphocytes on a reconstituted endothelial substrate. J Immunol 1998;161:6330–7.

57. Shields JD, Emmett MS, Dunn DB, et al. Chemokine-mediated migration of melanoma cells towards lymphatics - a mechanism contributing to metastasis. Oncogene 2006.

58. Wiley HE, Gonzalez EB, Maki W, et al. Expression of CC chemokine receptor-7 and regional lymph node metastasis of B16 murine melanoma. J Natl Cancer Inst 2001;93:1638–43.

59. Cueni LN, Detmar M. New insights into the molecular control of the lymphatic vascular system and its role in disease. J Invest Dermatol 2006;126:2167–77.

60. Joukov V, Pajusola K, Kaipainen A, et al. A novel vascular endothelial growth factor, VEGF-C, is a ligand for the Flt4 (VEGFR-3) and KDR (VEGFR-2) receptor tyrosine kinases. Embo J 1996;15:1751.

61. Yamada Y, Nezu J, Shimane M, et al. Molecular cloning of a novel vascular endothelial growth factor, VEGF-D. Genomics 1997;42:483–8.

62. Achen MG, Jeltsch M, Kukk E, et al. Vascular endothelial growth factor D (VEGF-D) is a ligand for the tyrosine kinases VEGF receptor 2 (Flk1) and VEGF receptor 3 (Flt4). Proc Natl Acad Sci USA 1998;95:548–53.

63. Orlandini M, Marconcini L, Ferruzzi R, et al. Identification of a c-fos-induced gene that is related to the platelet- derived growth factor/vascular endothelial growth factor family. Proc Natl Acad Sci USA 1996;93:11675–80.

64. Jeltsch M, Kaipainen A, Joukov V, et al. Hyperplasia of lymphatic vessels in VEGF-C transgenic mice. Science 1997;276:1423–5.

65. Veikkola T, Jussila L, Makinen T, et al. Signalling via vascular endothelial growth factor receptor-3 is sufficient for lymphangiogenesis in transgenic mice. Embo J 2001;20:1223–31.

66. Karkkainen MJ, Haiko P, Sainio K, et al. Vascular endothelial growth factor C is required for sprouting of the first lymphatic vessels from embryonic veins. Nat Immunol 2004;5:74–80.

67. Stacker SA, Caesar C, Baldwin ME, et al. VEGF-D promotes the metastatic spread of tumor cells via the lymphatics. Nat Med 2001;7:186–91.

68. Baldwin ME, Halford MM, Roufail S, et al. Vascular endothelial growth factor D is dispensable for development of the lymphatic system. Mol Cell Biol 2005;25:2441–9.

69. Avantaggiato V, Orlandini M, Acampora D, et al. Embryonic expression pattern of the murine *figf* gene, a growth factor belonging to platelet-derived growth factor/vascular endothelial growth factor family. Mech Dev 1998;73:221–4.

70. Dumont DJ, Jussila L, Taipale J, et al. Cardiovascular failure in mouse embryos deficient in VEGF receptor-3. Science 1998;282:946–9.

71. Karkkainen MJ, Ferrell RE, Lawrence EC, et al. Missense mutations interfere with VEGFR-3 signalling in primary lymph-oedema. Nat Genet 2000;25:153–9.

72. Karkkainen MJ, Saaristo A, Jussila L, et al. A model for gene therapy of human hereditary lymphedema. Proc Natl Acad Sci USA 2001;98:12677–82.

73. Szuba A, Skobe M, Karkkainen MJ, et al. Therapeutic lymphangiogenesis with human recombinant VEGF-C. Faseb J 2002;16:1985–7.

74. Hong YK, Lange-Asschenfeldt B, Velasco P, et al. VEGF-A promotes tissue repair-associated lymphatic vessel formation via VEGFR-2 and the alpha1beta1 and alpha2beta1 integrins. Faseb J 2004;18:1111–3.

75. Joukov V, Kumar V, Sorsa T, et al. A recombinant mutant vascular endothelial growth factor-C that has lost vascular endothelial growth factor receptor-2 binding, activation, and vascular permeability activities. J Biol Chem 1998;273:6599–602.

76. Makinen T, Veikkola T, Mustjoki S, et al. Isolated lymphatic endothelial cells transduce growth, survival and migratory signals via the VEGF-C/D receptor VEGFR-3. Embo J 2001;20:4762–73.

77. Nagy JA, Vasile E, Feng D, et al. Vascular permeability factor/vascular endothelial growth factor induces lymphangiogenesis as well as angiogenesis. J Exp Med 2002;196:1497–506.

78. Kunstfeld R, Hirakawa S, Hong YK, et al. Induction of cutaneous delayed-type hypersensitivity reactions in VEGF-A transgenic mice results in chronic skin inflammation associated with persistent lymphatic hyperplasia. Blood 2004;104:1048–57.

79. Baluk P, Tammela T, Ator E, et al. Pathogenesis of persistent lymphatic vessel hyperplasia in chronic airway inflammation. J Clin Invest 2005;115:247–57.

80. Yuan L, Moyon D, Pardanaud L, et al. Abnormal lymphatic vessel development in neuropilin 2 mutant mice. Development 2002;129:4797–806.

81. Karpanen T, Heckman CA, Keskitalo S, et al. Functional interaction of VEGF-C and VEGF-D with neuropilin receptors. Faseb J 2006;20:1462–72.

82. Favier B, Alam A, Barron P, et al. Neuropilin-2 interacts with VEGFR-2 and VEGFR-3 and promotes human endothelial cell survival and migration. Blood 2006;108:1243–50.

83. Soker S, Takashima S, Miao HQ, et al. Neuropilin-1 is expressed by endothelial and tumor cells as an isoform-specific receptor for vascular endothelial growth factor. Cell 1998;92:735–45.

84. Chen H, Bagri A, Zupicich JA, et al. Neuropilin-2 regulates the development of selective cranial and sensory nerves and hippocampal mossy fiber projections. Neuron 2000;25:43–56.

85. Giger RJ, Cloutier JF, Sahay A, et al. Neuropilin-2 is required in vivo for selective axon guidance responses to secreted semaphorins. Neuron 2000;25:29–41.

86. Suri C, Jones PF, Patan S, et al. Requisite role of angiopoietin-1, a ligand for the Tie2 receptor, during embryonic angiogenesis. Cell 1996;87:1171–80.

87. Dumont DJ, Gradwohl G, Fong GH, et al. Dominant-negative and targeted null mutations in the endothelial receptor tyrosine kinase, tek, reveal a critical role in vasculogenesis of the embryo. Genes Dev 1994;8:1897–909.

88. Maisonpierre PC, Suri C, Jones PF, et al. Angiopoietin-2, a natural antagonist for Tie2 that disrupts in vivo angiogenesis. Science 1997;277:55–60.

89. Davis S, Aldrich TH, Jones PF, et al. Isolation of angiopoietin-1, a ligand for the Tie2 receptor, by secretion-trap expression cloning. Cell 1996;87:1161–9.

90. Tammela T, Saaristo A, Lohela M, et al. Angiopoietin-1 promotes lymphatic sprouting and hyperplasia. Blood 2005;105: 4642–8.

91. Morisada T, Oike Y, Yamada Y, et al. Angiopoietin-1 promotes LYVE-1-positive lymphatic vessel formation. Blood 2005;105:4649–56.

92. Gale NW, Thurston G, Hackett SF, et al. Angiopoietin-2 is required for postnatal angiogenesis and lymphatic patterning, and only the latter role is rescued by Angiopoietin-1. Dev Cell 2002;3:411–23.

93. Kajiya K, Hirakawa S, Ma B, et al. Hepatocyte growth factor promotes lymphatic vessel formation and function. Embo J 2005;24:2885–95.

94. Cao R, Bjorndahl MA, Gallego MI, et al. Hepatocyte growth factor is a lymphangiogenic factor with an indirect mechanism of action. Blood 2006;107:3531–6.

95. Vlahakis NE, Young BA, Atakilit A, et al. The lymphangiogenic vascular endothelial growth factors VEGF-C and -D are ligands for the integrin alpha9beta1. J Biol Chem 2005;280:4544–52.

96. Huang XZ, Wu JF, Ferrando R, et al. Fatal bilateral chylothorax in mice lacking the integrin alpha9beta1. Mol Cell Biol 2000;20:5208–15.

97. Bjorndahl M, Cao R, Nissen LJ, et al. Insulin-like growth factors 1 and 2 induce lymphangiogenesis in vivo. Proc Natl Acad Sci USA 2005;102:15593–8.

98. Chang LK, Garcia-Cardena G, Farnebo F, et al. Dose-dependent response of FGF-2 for lymphangiogenesis. Proc Natl Acad Sci USA 2004;101:11658–63.

99. Kubo H, Cao R, Brakenhielm E, et al. Blockade of vascular endothelial growth factor receptor-3 signaling inhibits fibroblast growth factor-2-induced lymphangiogenesis in mouse cornea. Proc Natl Acad Sci USA 2002;99:8868–73.

100. Shin JW, Min M, Larrieu-Lahargue F, et al. Prox1 promotes lineage-specific expression of FGF receptor-3 in lymphatic endothelium: a role for FGF signaling in lymphangiogenesis. Mol Biol Cell 2006;17:576–84.

101. Cao R, Bjorndahl MA, Religa P, et al. PDGF-BB induces intratumoral lymphangiogenesis and promotes lymphatic metastasis. Cancer Cell 2004;6:333–45.

102. Auguste P, Javerzat S, Bikfalvi A. Regulation of vascular development by fibroblast growth factors. Cell Tissue Res 2003;314:157–66.

103. Lee OH, Bae SK, Bae MH, et al. Identification of angiogenic properties of insulin-like growth factor II in in vitro angiogenesis models. Br J Cancer 2000;82:385–91.

104. Shigematsu S, Yamauchi K, Nakajima K, et al. IGF-1 regulates migration and angiogenesis of human endothelial cells. Endocr J 1999;46 Suppl:S59–62.

105. Shing Y, Folkman J, Sullivan R, et al. Heparin affinity: purification of a tumor-derived capillary endothelial cell growth factor. Science 1984;223:1296–9.

106. Hirakawa S, Kodama S, Kunstfeld R, et al. VEGF-A induces tumor and sentinel lymph node lymphangiogenesis and promotes lymphatic metastasis. J Exp Med 2005;201:1089–99.

107. Skobe M, Hawighorst T, Jackson DG, et al. Induction of tumor lymphangiogenesis by VEGF-C promotes breast cancer metastasis. Nat Med 2001;7:192–8.

108. Mandriota SJ, Jussila L, Jeltsch M, et al. Vascular endothelial growth factor-C-mediated lymphangiogenesis promotes tumour metastasis. Embo J 2001;20:672–82.

109. He Y, Kozaki K, Karpanen T, et al. Suppression of tumor lymphangiogenesis and lymph node metastasis by blocking vascular endothelial growth factor receptor 3 signaling. J Natl Cancer Inst 2002;94:819–25.

110. Hirakawa S, Brown LF, Kodama S, et al. VEGF-C-induced lymphangiogenesis in sentinel lymph nodes promotes tumor metastasis to distant sites. Blood 2007;109:1010–7.

111. Stacker SA, Achen MG, Jussila L, et al. Lymphangiogenesis and cancer metastasis. Nat Rev Cancer 2002;2:573–83.

112. Dadras SS, Paul T, Bertoncini J, et al. Tumor lymphangiogenesis: a novel prognostic indicator for cutaneous melanoma metastasis and survival. Am J Pathol 2003;162: 1951–60.

113. Dadras SS, Lange-Asschenfeldt B, Velasco P, et al. Tumor lymphangiogenesis predicts melanoma metastasis to sentinel lymph nodes. Mod Pathol 2005;18:1232–42.

114. Kerjaschki D, Regele HM, Moosberger I, et al. Lymphatic neoangiogenesis in human kidney transplants is associated with immunologically active lymphocytic infiltrates. J Am Soc Nephrol 2004;15:603–12.

115. Kajiya K, Detmar M. An important role of lymphatic vessels in the control of UVB-induced edema formation and inflammation. J Invest Dermatol 2006;126:919–21.

116. Ebata N, Nodasaka Y, Sawa Y, et al. Desmoplakin as a specific marker of lymphatic vessels. Microvasc Res 2001;61:40–8.

117. Irjala H, Johansson EL, Grenman R, et al. Mannose receptor is a novel ligand for L-selectin and mediates lymphocyte binding to lymphatic endothelium. J Exp Med 2001;194: 1033–42.

118. Young PE, Baumhueter S, Lasky LA. The sialomucin CD34 is expressed on hematopoietic cells and blood vessels during murine development. Blood 1995;85:96–105.

119. Niemela H, Elima K, Henttinen T, et al. Molecular identification of PAL-E, a widely used endothelial-cell marker. Blood 2005;106:3405–9.
120. Schlingemann RO, Dingjan GM, Emeis JJ, et al. Monoclonal antibody PAL-E specific for endothelium. Lab Invest 1985;52:71–6.
121. Barsky SH, Baker A, Siegal GP, et al. Use of anti-basement membrane antibodies to distinguish blood vessel capillaries from lymphatic capillaries. Am J Surg Pathol 1983;7:667–77.
122. Fiedler U, Christian S, Koidl S, et al. The sialomucin CD34 is a marker of lymphatic endothelial cells in human tumors. Am J Pathol 2006;168:1045–53.
123. Sauter B, Foedinger D, Sterniczky B, et al. Immunoelectron microscopic characterization of human dermal lymphatic microvascular endothelial cells. Differential expression of CD31, CD34, and type IV collagen with lymphatic endothelial cells vs blood capillary endothelial cells in normal human skin, lymphangioma, and hemangioma in situ. J Histochem Cytochem 1998;46:165–76.
124. Pennisi D, Gardner J, Chambers D, et al. Mutations in Sox18 underlie cardiovascular and hair follicle defects in ragged mice. Nat Genet 2000;24:434–7.
125. Ayadi A, Zheng H, Sobieszczuk P, et al. Net-targeted mutant mice develop a vascular phenotype and up-regulate egr-1. Embo J 2001;20:5139–52.
126. Kriederman BM, Myloyde TL, Witte MH, et al. FOXC2 haploinsufficient mice are a model for human autosomal dominant lymphedema-distichiasis syndrome. Hum Mol Genet 2003;12:1179–85.
127. Petrova TV, Karpanen T, Norrmen C, et al. Defective valves and abnormal mural cell recruitment underlie lymphatic vascular failure in lymphedema distichiasis. Nat Med 2004;10:974–81.
128. Sebzda E, Hibbard C, Sweeney S, et al. Syk and Slp-76 mutant mice reveal a cell-autonomous hematopoietic cell contribution to vascular development. Dev Cell 2006;11:349–61.
129. Abtahian F, Guerriero A, Sebzda E, et al. Regulation of blood and lymphatic vascular separation by signaling proteins SLP-76 and Syk. Science 2003;299:247–51.

Chapter 44
Ocular Neovascularization

Peter A. Campochiaro

Keywords: ocular neovascularization, retinal neovascularization, AMD, macular edema

Abstract: The molecular cascade leading to neovascularization is complex and can vary in different tissues and different disease processes. Each tissue has its own unique microenvironment with potential differences in blood vessels, surrounding cells, and extracellular matrix. These differences can result in differences in constitutive and induced-expression of neovascularization-related proteins or they can alter the effects of proteins; some proteins promote neovascularization in one setting and inhibit it in others. Proteins depend upon other molecules for their actions, and lack of expression of a binding partner in a tissue can have a major impact on the effect of a protein in that setting. In order to define the potential actions of a protein and its interactions with other proteins in neovascularization, it is useful to study its effects in several well-characterized vascular beds and different pathologic processes. The eye is an important organ in which to study neovascularization, because it has several useful features that facilitate such study, and also because neovascular diseases are prevalent causes of visual morbidity and blindness. The eye contains several vascular beds separated by avascular tissue. The vascular beds can be visualized in vivo, and the presence of neovascularization can be unequivocally identified and quantified because of the surrounding avascular tissue. Also, retina-specific promoters combined with inducible promoter systems provide a useful way to control expression of proteins of interest. By observing the different effects of proteins in different vascular beds in the eye, at different stages of development, and in differ-ent disease models, a more complete picture of the protein's actions and interactions with other proteins can emerge.

Vascular Beds in the Eye

The hyaloidal vasculature, consisting of the pupillary membrane, the tunica vasculosa lentis, and the hyaloid vessels, develops in the embryonic eye and extends from the optic nerve through the vitreous to surround the developing lens. These vessels are needed to provide oxygen and nutrients to the developing eye before other vascular beds have developed, but would interfere with vision if maintained in the adult eye. As the retinal vessels develop, the hyaloidal vessels are eliminated by programmed vascular regression. Understanding the molecular signals that regulate physiologic vascular regression may provide insights that will allow development of regressive treatments for pathologic neovascularization.

In rodents, retinal vascular development occurs entirely after birth facilitating the study of developmental angiogenesis. The central retinal artery enters the eye through the optic nerve, and retinal vessels develop from the optic nerve to the periphery of the retina. On the day of birth, postnatal day 0 (P0), there are no retinal vessels, but by P4, superficial retinal vessels extend from the optic nerve halfway to the peripheral edge of the retina. At P7, vessels cover the entire surface of the retina and sprouts from superficial vessels begin to grow into the retina to form the intermediate and deep capillary beds. By P18, all three vascular beds of the retina have been formed and remodeled, resulting in an adult retinal circulation, which is relatively stable thereafter. Retinal vascular endothelial cells (ECs) have tight junctions and modified vesicular transport, and are the site of the inner blood-retinal barrier. This is in part due to their interactions with pericytes and retinal glia [1].

The retinal circulation supplies the inner half of the retina. The outer nuclear layer, which is made up of the cell bodies of the photoreceptors, and the inner and outer segments of

Departments of Ophthalmology and Neuroscience, The Johns Hopkins University School of Medicine, Maumenee 719, 600 N. Wolfe Street, Baltimore, MD 21287-9277, USA
E-mail: pcampo@jhmi.edu

the photoreceptors, make up the outer, avascular half of the retina. It receives its oxygen and nutrients from the choroidal circulation, a high-flow system supplied by multiple long and short posterior ciliary arteries, which all feed into an extensive network of fenestrated capillaries, the choriocapillaris. The choriocapillaris allows plasma to pool beneath the retinal pigmented epithelium (RPE), which has tight junctions and specialized transport systems, and it constitutes the outer blood retinal barrier.

Developmental Retinal Neovascularization

Retinal astrocytes enter the retina from the optic nerve [2]. As they migrate away from the optic nerve into avascular retina they become hypoxic and express increased levels of vascular endothelial growth factor-A (VEGF-A) [3]. VEGF-A is one of the stimuli that guide the growth of blood vessels from the optic nerve to the periphery of the retina and, as the blood vessels become functional, they alleviate hypoxia, resulting in decreased expression of VEGF-A. The relative hypoxia of the deeper layers of the retina results in a VEGF-A gradient that favors sprouting from the superficial vessels resulting in penetrating branches that form the intermediate and deep capillary beds.

VEGF-A gradients appear to be needed for ordered retinal vascular development, and when such gradients are disrupted, pathologic sprouting occurs. VEGF-A gradients can be disrupted by an imbalance among heparin-binding and soluble VEGF-A isoforms or by excessive proteolytic activity that can cleave heparin-binding domains and result in excessive amounts of soluble VEGF-A [4] There is specialization within vascular sprouts so that endothelial cells at the tip of sprouts, "tip cells", respond to VEGF-A by migrating toward high levels, whereas other cells within sprouts that are not at the tip, "stalk cells", respond to VEGF-A by proliferating [5]. This specialization allows directed growth along VEGF-A gradients. The molecular basis of this differential responsiveness is not yet understood, although it has been noted that tip cells express high levels of the VEGF receptor, VEGFR2, relative to stalk cells. There are other differences as well, because tip cells produce PDGF-B and δ-like 4 notch ligand, while stalk cells do not.

VEGF is not the only guidance cue for developing retinal vessels. Norrin acts as a tissue-specific ligand for Frizzled 4 (Fz4) to help direct retinal vascular development [6]. Mutations in Norrin result in Norrie disease, in which retinal vascular development is incomplete, resulting in large areas of avascular retina, neovascularization, scarring, and retinal detachment. Heterozygosity for mutations in Fz4, a presumptive Wnt receptor, result in familial exudative vitreoretinopathy (FEVR), in which retinal vascular development is perturbed, but often not as severely as in Norrie disease, resulting in areas of avascular retina, neovascularization, and a spectrum that ranges from focal traction

retinal detachments to severe detachments and blindness. Norrin binds to Fz4 with high affinity and specificity, and induces downstream signaling, indicating that despite the lack of any structural homology to Wnts, Norrin functions as a ligand for Fz4. The mechanism by which the activation of Fz4 by Norrin influences retinal vascular development is not yet known, but null phenotypes suggest that it helps to orchestrate growth of retinal vessels. This signaling system is not completely unique to the eye, because it is also required for maintaining the vascularization of the inner ear, but is not known to function in any other organs.

In addition to stimulating retinal vascular development, VEGF-A also acts as a survival factor for the ECs of the newly formed retinal vessels [7]. Dependence on VEGF-A is transient and is probably eliminated by the development of cell-cell contacts, particularly association with pericytes that provide an alternative source of survival factors. Platelet-derived growth factor-B (PDGF-B) secreted by endothelial cells is critical for the recruitment of pericytes, and PDGF-deficient mice lack pericyte envelopment of retinal vessels [8].

Regression of the Hyaloidal Vasculature

In mice, the hyaloidal vasculature regresses between P3 and P9. There are a small number of resident macrophages in the pupillary membrane; 300–400 macrophages for a vessel network that contains 6,000–7,000 vascular endothelial cells. It is well established that macrophages play a passive role in recognizing and engulfing apoptotic cells. They will simply engulf cells after there has been a macrophage-independent signal for cell death. But there is also evidence suggesting that macrophages may play an active role in inducing apoptosis. Some of that evidence was obtained by observations in PU.1 mutant mice. PU.1 is a transcription factor that plays a role in hematopoiesis. Deleting the PU.1 gene eliminates mature tissue macrophages, and PU.1 mutant mice have persistent hyaloid vessels, indicating that macrophages are essential for hyaloid vessel regression [9].

The canonical Wnt pathway is also required for hyaloid vessel regression. The receptor complex for Wnts consists of Frizzled proteins and LRP5 or LRP6 co-receptors. After Wnt ligands bind to the receptor complex, signal transduction events result in stabilization of β-catenin, which then complexes with transcription factors of the LEF-TCF class to regulate gene expression. Wnt signaling has important roles in development and is activated in tumorigenesis. It is also important for hyaloid vessel regression, because mice with a mutation in LRP5 show no hyaloid vessel regression due to a failure of cell death pathways [10]. Mutations in LEF1 give a similar phenotype. These data suggest that the canonical Wnt pathway is required for hyaloid regression. Wnt signaling causes cell cycle entry, which enhances susceptibility to apoptosis. Angiopoietin-1 (Ang2) contributes by causing withdrawal of survival signals after cell cycle entry [11]. Macrophages contribute by producing Wnt7B and pericytes produce Ang2. At present, the working model for regression of the hyaloid vasculature is the following. Macrophages deliver Wnt7B locally to vascular endothelial cells by cell-cell contact and Wnt7B stimulates the ECs to enter the cell cycle. Expression of Ang2 by pericytes causes withdrawal of survival signals and without those signals, ECs that enter G1 undergo programmed cell death. This model is consistent with findings in models of ocular neovascularization [12]. Increased expression of Ang2 causes regression of retinal or choroidal neovascularization in which ECs are proliferating, but it has no effect on quiescent endothelial cells in mature vessels.

Pathologic Retinal Neovascularization

The ability to carefully examine the retina by ophthalmoscopy allowed clinicians to observe the strong correlation between occlusion of retinal vessels and retinal neovascularization. The development of fluorescein angiography, which uses intravascular injection of a fluorescent dye to document perfusion of the retina, helped to document the correlation of retinal neovascularization and nonperfusion [13]. There are several diseases in which closure of retinal vessels occurs including diabetic retinopathy, retinopathy of prematurity (ROP), central or branch retinal vein occlusions, or vasculitis. Each of these diseases differ with respect to the mechanism of damage to retinal vascular cells that leads to vessel closure, but once a sufficient area of retina becomes ischemic, they converge into a common mechanistic pathway that leads to retinal neovascularization; therefore, these diseases are grouped together and referred to as ischemic retinopathies. The molecular events that lead from retinal ischemia to retinal neovascularization can be studied using several different models in which closure of retinal vessels is achieved by different approaches and in different species. A commonly used model that requires use of neonatal animals is oxygen-induced ischemic retinopathy. It was first developed in kittens and dogs [14], but has been adapted to rats [15] and mice [16], which are less expensive and allow larger experimental groups. The murine model of oxygen-induced ischemic retinopathy has allowed use of genetically engineered mice and has led to widespread use of this model to explore the role of many different gene products in the pathogenesis of retinal neovascularization. One of the first proteins demonstrated to play an important stimulatory role was VEGF [17]. It was shown that VEGF is also critical for retinal vascular development, and that there is a narrow window of time during and after retinal vascular development in which retinal vascular endothelial cells are dependent upon VEGF for survival [7]. This provided insight into the molecular pathogenesis of models of oxygen-induced ischemic retinopathy. During a developmental stage when retinal ECs are still dependent upon VEGF for survival, neonates are placed in a high oxygen environment, which causes down-regulation of VEGF in the retina causing many ECs to die, which leads to closure of many vessels. When the neonates are returned to room air, the areas of retina in which vessels have closed become ischemic resulting in stabilization, reduced degradation, and increased levels of hypoxia-inducible factor-1 (HIF-1) transcription factor [18]. HIF-1 increases expression of genes with a *hypoxia response element* (HRE) in their promoter region. Proteins that have been shown to be increased in ischemic retina, or retina in which there is increased expression of HIF-1 from gene transfer, include VEGF-A, placental growth factor-A (PlGF-A), VEGF receptor 1 (VEGFR1), angiopoietin-2 (Ang2), angiopoietin -1 (Ang1), and PDGF-B [19]. The role of each of these proteins will be discussed below after mention of other models of retinal neovascularization, and the other major types of neovascularization affecting the retina, subretinal or choroidal neovascularization.

Models of Retinal Neovascularization

As noted above, models of oxygen-induced ischemic retinopathy are widely used and, although they mimic ROP most closely, they also provide insights into how retinal ischemia leads to neovascularization in other conditions, such as diabetic retinopathy. This is fortunate, because a period of several years of hyperglycemia is required to cause closure of retinal vessels, and, in general, unless an additional vascular insult (such as hypertension) is added, the life span of rodents is not long enough to achieve sufficient ischemia from hyperglycemia-induced vascular closure to generate retinal neovascularization. Retinal neovascularization is achievable after 5–10 years of diabetes in dogs [20], but the time and expense involved make this model impractical for most applications. In contrast, models of branch retinal vein occlusion are feasible, because retinal branch veins can be occluded by laser photocoagulation [21]. In monkeys, laser-induced retinal vein occlusions reproducibly cause iris neovascularization, but rarely cause retinal neovascularization, probably because monkeys have a liquid rather than a solid vitreous. Therefore, soluble factors released from ischemic retina are not sequestered at the retinal surface, but are concentrated near the anterior outflow channels of the eye adjacent to the iris [21, 22]. However, pigs have a solid vitreous, and laser-induced occlusions of retinal vessels cause retinal neovascularization if a sufficient area of retina is made ischemic [23]. Vessels may reopen, requiring additional laser to re-occlude them or use of photosensitizers to facilitate thrombosis, but the pig branch vein occlusion model has proven useful as a model of ischemia-induced retinal neovascularization in a relatively large eye [24–26].

Subretinal or Choroidal Neovascularization

Choroidal neovascularization (CNV) occurs in diseases in which there is compromise of the retinal pigmented epithelium (RPE) and/or Bruch's membrane, a complex 5-layered extracellular matrix (ECM) tissue that separates the RPE from the choroicapillaris. Diseases that result in cracks or breaks in Bruch's membrane have a high incidence of CNV, which can occur along the entire age spectrum. The most common entities in this category are pathologic myopia, ocular histoplasmosis, angioid streaks, and choroiditis (inflammatory diseases of the choroid).

Age-related macular degeneration (AMD) is a disease that accounts for more CNV than all of the other diseases combined [27]. It occurs in individuals over the age of 60 who develop deposits called drusen under the RPE and diffuse thickening of Bruch's membrane, followed by slow degeneration of RPE and photoreceptor cells resulting in gradual reduction in central vision. Although visible breaks are not seen, Bruch's membrane is compromised in AMD patients and they are prone to sudden development of CNV originating from the choroid [28] and/or retinal angiomatous proliferation (RAP) originating from the deep capillary bed of the retina [29]. Although RAP does not originate from choroidal vessels, it is usually lumped under the CNV umbrella, because it is often co-incident with neovascularization originating from the choroid and grows in the same location, the subretinal space. Technically speaking, RAP and CNV are subtypes of subretinal neovascularization, but from a practical standpoint CNV and subretinal neovascularization are used interchangeably.

Models of Subretinal Neovascularization

Rupture of Bruch's membrane with laser photocoagulation in monkeys was found to result in CNV at a high percentage of the rupture sites [30]. This model of laser-induced CNV has been adapted to rats [31], mice [32], and pigs [33]. Similar to the situation with retinal neovascularization, the ability to use the murine model of CNV in knockout and transgenic mice, and the relative inexpensiveness of mice, has led to its widespread use. However, the pig model of laser-induced CNV allows investigations in a near human-sized eye, which has important applications [34].

Transgenic mice with expression of VEGF in photoreceptors (rho/VEGF mice) develop subretinal neovascularization originating from the deep capillary bed of the retina and therefore provide a model for RAP [35, 36]. This model compliments the laser-induced CNV model and has proven extremely useful for genetic studies and testing potential therapeutic agents. In contrast to the laser-induced CNV model in which there is a burst of angiogenic activity immediately after the laser that rapidly declines [12], rho/VEGF mice show sustained expression of VEGF, the angiogenic stimulus, which is advantageous for some applications. Double transgenic mice with doxycycline-inducible expression of VEGF in photoreceptors (Tet/opsin/VEGF mice) secrete roughly 10-fold higher levels of VEGF than rho/VEGF mice and have rapid onset of severe subretinal neovascularization [37]; they are able to distinguish between treatments that are maximally or near-maximally effective in other models [38, 39].

Overexpression of VEGF in photoreceptors and RPE can also be achieved by subretinal injection of adeno-associated viral vectors (AAV) containing a VEGF expression cassette [40]. However, in contrast to the transgenic models, differences in the area of transduction, which can occur from a number of technical problems, results in variability that can complicate interpretation of effectiveness of therapeutic agents.

VEGF is a Major Stimulus for Retinal and Subretinal Neovascularization and Macular Edema

The initial strategy for identification of stimuli for retinal neovascularization was to isolate proteins responsible for angiogenic activity in retinal homogenates or extracts. Fibroblast growth factor-2 (FGF2) was a major candidate, because it is a potent angiogenic agent that was identified in brain and retinal extracts [41, 42]. However, compared to wild type mice, FGF2 knockout mice and mice that overexpress FGF2 in the retina showed no difference in retinal or choroidal neovascularization [32, 43]. While FGF2 is present in substantial levels in retina, it is sequestered in cells and the ECM and does not contribute to neovascularization unless there is substantial tissue disruption [44].

The focus shifted to VEGF when it was demonstrated to be up-regulated in ischemic tissue [45, 46], because of the correlation between ischemia and retinal neovascularization. In samples obtained from patients with ischemic retinopathies [47, 48] and in animal models [17, 22], an association was noted between increased levels of VEGF and retinal neovascularization, and the role of VEGF was confirmed by demonstration that blockade of VEGF signaling suppressed retinal neovascularization [49–51].

Since tissue hypoxia has not been demonstrated to be part of the pathogenesis of subretinal neovascularization, there was no smoking gun implicating VEGF as was the case for retinal neovascularization. VEGF was identified in surgically removed CNV membranes [52, 53], but so were many other growth factors, including FGF2 [54], which is now known to play little role. The demonstration that expression of VEGF in photoreceptors caused subretinal neovascularization suggested a possible role for VEGF [35, 36], and this was confirmed by demonstrations that blockade of VEGF signaling suppresses CNV [55, 56]. Thus, preclinical studies predicted that VEGF antagonists would provide benefit in patients with CNV, and this prediction has been shown to be correct by clinical trials. Phase III trials have demonstrated some benefit from intravitreous injections of pegaptanib [57], an aptamer that binds $VEGF_{165}$, and ranibizumab, an Fab fragment of an antibody that binds all isoforms of VEGF-A [58, 59]. The results with ranibizumab are impressive and mutually confirmatory in two trials investigating effects in slightly different patient populations and with different protocols. In the Marina trial, monthly intravitreous injections of 0.5 mg of ranibizumab in AMD patients with occult or minimally classic subfoveal CNV reduced the percentage of patients with moderate loss of vision (3 lines) over the course of a year from 38% in the sham injection group to 5%, and the percentage of patients who experienced substantial improvement in vision (3 lines) was increased from 4.6% to 34% [59]. In the Anchor trial, patients with predominantly classic CNV due to AMD were randomized to receive monthly intravitreous injections of ranibizumab or photodynamic therapy with Visudyne every 3 months [58]. At 12 months, only 4% of patients who had received monthly injections of 0.5 mg of ranibizumab lost 3 lines of vision, and 44% gained 3 lines compared to 36% who lost 3 lines and 6% who gained 3 lines in the photodynamic therapy group. These data suggest that antagonism of VEGF-A can result in stabilization of vision in more than 95% of AMD patients with CNV and substantial improvement in vision in 35–44%.

These results clearly demonstrate that VEGF-A as an important target in the treatment of CNV due to AMD, but what about CNV due to other causes? Systemic treatment with bevacizumab, a full-length humanized monoclonal antibody that binds all isoforms of VEGF-A, caused resorption of subretinal and intraretinal fluid, involution of subfoveal CNV, and improvement in visual acuity in two patients with pathologic myopia [60]. This strongly suggests that VEGF is an important stimulus for CNV in pathologic myopia as well as in AMD. An ongoing study in young patients with CNV due to several disease processes other than AMD implicates VEGF in their pathogenesis (Nguyen and Campochiaro, unpublished data). Several case series have suggested that intraocular injections of bevacizumab may also provide benefit in patients with CNV due to AMD [61–63].

There are no prospective clinical trials testing VEGF antagonists for proliferative diabetic retinopathy, but retrospective observations in patients treated for diabetic macular edema (DME), who also happened to have proliferative diabetic retinopathy, have suggested possible benefits [64]. It is unlikely that VEGF antagonists will replace scatter photocoagulation, which is an effective treatment for proliferative diabetic retinopathy [65], but they are likely to be used as adjuncts when laser photocoagulation cannot be delivered (e.g., when there is hemorrhage obscuring large parts of the retina) or when a rapid response is needed (when neovascularization is growing over the outflow channels of the eye causing glaucoma). Extensive active neovascularization within scar tissue causing diabetic traction retinal detachments is prone to bleeding during surgery, greatly complicating the surgery; preoperative administration of VEGF agonists may help to reduce bleeding during surgery and make it less complicated.

Recently, it has been demonstrated that retinal hypoxia contributes to DME [66], and this implicated VEGF in the pathogenesis since it is up-regulated by hypoxia and is a potent inducer of vascular permeability that has been shown to cause leakage from retinal vessels [67, 68]. In a small clinical trial, a nonspecific VEGF antagonist caused significant reduction in DME as measured by optical coherence tomography (OCT) [69]. In another small trial, 10 of 10 patients with DME treated with intraocular injections of 0.5 mg of ranibizumab showed improvement in visual acuity and foveal thickness measured by OCT [70]. This suggests that macular edema may prove to be the most important application for VEGF antagonists in patients with ischemic retinopathy.

VEGF Family Members

The VEGF family consists of several gene products, including VEGF-A, -B, -C, and -D, and placental growth factor (PlGF) 1 and 2. VEGF-B does not play an important role in retinal neovascularization, because mice deficient in VEGF-B have normal retinal vascular development and no difference in hypoxia-induced retinal neovascularization compared to wild type mice [71]. It is not known if VEGF-C or -D play any role in the eye, but PlGF acts synergistically with VEGF to promote retinal neovacularization, because PlGF-deficient mice have significantly less ischemia-induced retinal neovascularization than wild type mice [72], and anti-PlGF antibody substantially suppresses retinal neovascularization [73]. Can additional improvement be achieved by inhibiting PlGF as well as all isoforms of VEGF-A? Clinical trials investigating the efficacy of VEGF Trap should answer this question. VEGF Trap is a recombinant fusion protein consisting of the binding domains of VEGF receptors 1 and 2 and an Fc fragment of IgG that binds VEGF-A, PlGF, and other members of the VEGF family that bind VEGF receptors 1 or 2 [74]. VEGF Trap suppresses CNV in mice [75], and systemic administration of 1 mg/kg of VEGF Trap in patients with CNV due to AMD reduced leakage and retinal thickening, but 3 mg/kg caused substantial hypertension [76]. Intraocular injection of VEGF Trap is currently being tested and will help to determine if there are advantages to more generalized blockade of VEGF family members compared to specific blockade of VEGF-A.

The Effects of VEGF Family Members in the Eye are Mediated Through Several Receptors

There are 3 primary VEGF receptors that are designated VEGFR1, R2, and R3. VEGFR3 binds VEGF-C and -D and mediates lymphangiogenesis. VEGFR2 is the major mediator of mitogenesis of endothelial cells [77]. It binds VEGF-A and proteolytic fragments of VEGF-C and -D; therefore, while VEGF-C and -D function primarily in lymphangiogenesis, they can also contribute to angiogenesis. VEGFR1 binds VEGF-A, -B, and PlGF. The role of VEGFR1 is context-dependent; in embryos and some adult tissues it acts as a decoy receptor that suppresses angiogenesis [78, 79], and in some adult tissues it mediates VEGF signaling and is proangiogenic [73]. In the eye,

VEGFR1 is proangiogenic and its inhibition can suppress retinal or choroidal neovascularization [80]. VEGF also interacts with neuropilins (NRP), which were first identified as receptors for semaphorins that function in axon guidance (for reviews see [81, 82]). Absence or blockade of either NRP1 or NRP2 suppresses ocular neovascularization [83, 84].

There is substantial increase in VEGF receptors in endothelial cells participating in CNV and as a result, VEGF121, the most soluble isoform, can be used as a tool to direct destructive therapy to CNV [85]. After intravenous injection of 45 mg/kg of a chimeric protein consisting of VEGF121 coupled to the toxin gelonin (VEGF/rGel), but not uncoupled gelonin, there was immunofluorescent staining for gelonin within CNV in mice and regression of the CNV occurred. Intraocular injection of 5 ng of VEGF/rGel also caused significant regression of CNV or retinal neovascularization. Thus, VEGF is both a target and a homing device for treatment of CNV.

Tie Receptors and Angiopoietins

Tie1 and Tie2 receptors are selectively expressed on vascular endothelial cells and are required for embryonic vascular development [86,87]. The first binding partner identified for Tie2, angiopoietin-1 (Ang1), binds with high affinity and initiates Tie2 phosphorylation and downstream signaling [88]. Ang1 is also an agonist for Tie1 [89]. The second Tie2 binding partner identified, Ang2, binds with high affinity, but does not stimulate phosphorylation of Tie2 in cultured ECs, and is a competitive inhibitor of Ang1 at both Tie1 and Tie2 [89,90].

Several lines of evidence indicate that Ang2 is a developmentally- and hypoxia-regulated permissive factor for VEGF-induced neovascularization in the retina. Expression of Ang2 in the retina increases during the first week after birth, peaks around P8, during development of the deep capillary bed, and then decreases in adults [91]. Ang2/LacZ knockin mice show the spatial pattern of Ang2 expression in the retina. Between P0 and P7 there is Ang2 expression along the surface of the retina, and at P8 there is intense expression in the region of the deep capillary bed within horizontal cells, and this is maintained at a reduced level in adults [92]. In mice with oxygen-induced ischemic retinopathy, at P12 when the mice are removed from hyperoxia, the retina becomes hypoxic, and within hours ectopic expression of Ang2 occurs at the surface of the retina. Within a few days, new vessels sprout from superficial vessels and there is intense expression of Ang2 within and around the sprouts. Homozygous Ang2 knockouts show very poor development of the superficial capillary bed and almost no development of the deep capillary bed, indicating that Ang2 is required for normal retinal vascular development. They also show persistence of the hyaloid vessels. Double transgenic Tet/opsin/ang2 and Tet/IRBP/ang2 mice with inducible expression of Ang2 have also helped to elucidate the effects of Ang2 in the eye [12]. During retinal vascular development, increased expression of Ang2

causes an increase in the density of the deep capillary bed at P12 that normalizes by P18. Expression of Ang2 in adult mice has no effect on normal retinal vessels. In mice with ischemic retinopathy, increased expression of Ang2 during the hypoxic period, P12 to P17, results in a marked increase in the amount of retinal neovascularization. In this model, VEGF levels reach a peak, plateau between P17 and P19, and then decline. Onset of Ang2 expression at P20 results in rapid regression of neovascularization. In Rho/VEGF transgenic mice or mice with CNV due to laser-induced rupture of Bruch's membrane, high-level expression of Ang2 results in regression of neovascularization. Thus, it is not just the presence of VEGF that seems to modulate the effect of Ang2, but it is the ratio of Ang2 to VEGF. If the level of Ang2 is far above the level of VEGF, then new vessels regress, while mature vessels are unaffected. In Tet/opsin/ang2 mice, in the absence of doxycycline, injection of an adenoviral vector that expresses VEGF results in neovascularization of the cornea and iris, but no neovascularization of the retina except for a few sprouts in the area of needle penetration (due to a combination of VEGF and the injury) [93]. In the presence of doxycycline, which causes expression of Ang2, injection of the adenoviral vector expressing VEGF results in florid retinal neovascularization that originates from both the superficial and deep capillaries.

Thus, retinal vessels require expression of both VEGF and Ang2 for neovascularization to occur. There is constitutive expression of Ang2 in the region of the deep capillary bed, so if VEGF levels are increased alone, new vessels will only sprout from the deep capillary bed [35], but if Ang2 and VEGF are co-expressed, new vessels sprout from all capillary beds [94]. HIF-1 increases expression of both Ang2 and VEGF in the retina, and as one would predict from the findings summarized above, intravitreous injection of an adenoviral vector expressing a constitutively active form of HIF-1α causes sprouting of new vessels from all capillary beds in the retina [19]. The situation is even more complicated in the choroid, because another permissive factor in addition to Ang2 is required for VEGF to induce choroidal neovascularization [95].

As noted above, normal vessels in the retina and choroid are not responsive to the regression-promoting effects of Ang2, but new vessels are responsive; high levels of Ang2 relative to the levels of VEGF are required. This suggests that Ang2 may be a useful therapeutic agent to promote regression of pathologic retinal or choroidal neovascularization, but its use alone would run the risk of stimulating neovascularization if levels of VEGF were sufficiently high. Therefore, it would seem most prudent to use Ang2 or another Tie2 antagonist in combination with a VEGF antagonist.

Transgenic mice with increased expression of Ang1 in skin under control of the *K14* promoter exhibit a moderate increase in number and a large increase in diameter of dermal vessels [96]. Transgenic mice with increased expression of VEGF in skin show a large increase in leaky dermal vessels, and double transgenic mice with coexpression of Ang1 and VEGF show an additive effect on angiogenesis, but the vessels do not leak spontaneously and are resistant to inflammation-induced leakage [97]. Overexpression of Ang1 in the retina of transgenic mice suppresses the development of retinal or choroidal neovascularization [98]. Ang1 also suppresses neovascularization and retinal detachment in double transgenic mice with high-level expression of VEGF in the retina [39]. Therefore, unlike the situation in skin where Ang1 is proangiogenic, Ang1 is a potent antiangiogenic agent in the retina. Unlike Ang2, expression of Ang1 after neovascularization is already established and does not cause regression of the new vessels, although it suppresses further growth. Therefore, increased expression of Ang1 in the eye may be a good strategy for prevention of neovascularization.

Participation of Other Peptide Growth Factors in Ocular Neovascularization

Tumor necrosis factor-α (TNF-α) has been suggested to be a therapeutic target for retinal neovascularization in ischemic retinopathies, based upon a study in which compared to wild type mice, *Tnfα* knockout mice with ischemic retinopathy developed less retinal neovascularization [99]. Studies in our laboratory found a reduction in leukostasis in *Tnfα* knockout mice with ischemic retinopathy, but no difference in amount of retinal neovascularization [100]. The reason for this difference is unclear, but *Tnf receptor* knockout mice with ischemic retinopathy also showed no difference in the amount of retinal neovascularization at P17 and showed slower regression of the neovascularization so that at P21 there was more neovascularization than in wild type mice [101]. It does not appear that TNF-α is a high priority target for retinal neovascularization.

There is substantial evidence suggesting that insulin-like growth factor 1 (IGF-1) contributes to retinal neovascularization in proliferative diabetic retinopathy, although there is disagreement as to whether the contribution is major [102,103] or modest [104]. Transgenic mice with ubiquitious overexpression of IGF-1 for several months develop high levels of VEGF in the retina and features of diabetic retinopathy including neovascularization on the surface of the retina [105]. IGF-1 post-transcriptionally up-regulates HIF-1 [106], which not only increases VEGF, but also increases the products of other genes that contain a hypoxia response element in their promoter, such as Ang2. Increased expression of VEGF in the retina causes new vessels to sprout from the deep capillary bed, but not the superficial retinal vessels [35,37], whereas co-expression of VEGF and Ang2 causes neovascularization that grows from the surface of the retina [94]. This explains why long-term expression of IGF-1 causes neovascularization that grows from the surface of the retina; it essentially mimics retinal hypoxia by upregulating HIF-1. Since the proangiogenic effects of IGF-1 are mediated through VEGF, there is not likely to be added benefit by addition of IGF-1 antagonists to blockers of VEGF for treatment of retinal neovascularization. Also, blockade of IGF-1 may not provide clinically significant benefit in patients with early diabetic retinopathy, because a large, multicenter, randomized clinical study investigating the effect of monthly intramuscular injections of the sustained release formulation of octreotide in patients with high risk non-proliferative diabetic retinopathy did not show any substantial benefit and this treatment approach has been abandoned.

In contrast to the situation with IGF-1, combining antagonism of PDGFs with blockade of VEGFs may be a useful strategy for treatment of ocular neovascularization. Increased expression of PDGF-B in the retina causes severe proliferative retinopathy and retinal detachment like the most advanced stages of proliferative diabetic retinopathy [107]. Endothelial cells produce PDGF-B which promotes the recruitment, proliferation and survival of pericytes. PDGF-B also recruits glial cells and retinal pigmented epithelial (RPE) cells [108] which promotes scarring, a complication of ocular neovascularization that is the major cause of permanent loss of vision. Antagonists of PDGFs may help to reduce scarring, but may also synergize with VEGF antagonists to reduce neovascularization through their antagonism of pericytes, which provide survival signals for endothelial cells of new vessels [109]. Kinase inhibitors that block both VEGF and PDGF receptors are some of the most efficacious drugs for the treatment of ocular neovascularization in animal models [51,55,110].

Bone Marrow-derived Cells and Ocular Neovascularization

Circulating bone marrow-derived cells contribute to ocular neovascularization, although the exact nature of the contribution is still unclear. Studies utilizing lethally irradiated mice with reconstituted, green fluorescent protein (GFP)-labeled bone marrow cells have suggested that trans-differentiated pleuripotent cells from bone marrow make up a large proportion of the endothelial cells in retinal and choroidal new vessels and a substantial percentage of other cell types within choroidal neovascular lesions [111–113]. However, total body irradiation at a level that kills all hematopoietic cells has deleterious effects on other cells, particularly endothelial cells. Ocular exposure to even modest levels of stray radiotherapy meant for other tissues can markedly reduce EC turnover eventuating in closure of retinal vessels, a process known as radiation retinopathy [114]. Compromise in the ability of ocular endothelial cells to proliferate after total body irradiation could markedly increase incorporation of circulating endothelial progenitor cells into ocular new vessels. A more likely contribution of macrophages and other circulating bone marrow-derived cells is to increase the levels and alter the gradients of angiogenic factors, thereby contributing to maladaptive, disorganized vessel growth. VEGF acting through VEGFR1 recruits bone-marrow derived cells [115], but stromal derived factor-1 (SDF-1) acting through the chemokine receptor, CXCR4, may also participate [116]. SDF-1 levels are increased in ischemic retina and antagonists of CXCR4 suppress several types of ocular neovascularization [117]. However, it is not yet clear if blockade of CXCR4 will provided added benefit when combined with VEGF antagonists.

Prostaglandins and Cyclooxygenase Inhibitors

Patients who regularly take nonsteroidal anti-inflammatory drugs (NSAIDs) have a 40–50% reduction in mortality from colorectal cancer [118]. This and the demonstration that cyclooxygenase-2 is up-regulated in colorectal and other cancers has suggested that prostaglandins (PGs) may act as tumor promoters and that inhibition of cyclooxygenases (COX) may be chemoprotective [119–121]. At least part of the tumor promoting effect of PGs appears to be through stimulation of angiogenesis, which is suppressed by COX inhibitors [122, 123]. COX inhibitors suppress ocular neovascularization by reducing expression of VEGF [124]. Amfenac, 2-amino-3-benzoylbenzeneacetic acid, is a nonspecific COX inhibitor and its amide analog, nepafenac, is a prodrug that has unusually high corneal penetration; topical administration of nepafanac inhibits prostaglandin synthesis in the retina/choroid by 55% for 4 h and significantly suppresses retinal and choroidal neovascularization [124–126]. Nepafenac cannot replace potent VEGF antagonists given by intraocular injections, but because of its noninvasive mode of administration, it may be worthwhile to determine if when given in combination with ranibizumab it can reduce the frequency of intraocular injections.

Signals from the Extracellular Matrix (ECM)

Along with soluble proangiogenic proteins, ECM molecules also participate in several ways in the regulation of neovascularization. Acting through integrins on the surface of endothelial cells, ECM molecules may directly stimulate or inhibit EC processes involved in angiogenesis [127]. Soluble angiogenic factors alter expression of integrins on the surface of ECs which reduces stabilizing signals from the ECM making the cells more responsive to the soluble factors. Integrins up-regulated on endothelial cells participating in ocular neovascularization and undetectable on normal retinal and choroidal vessels are $\alpha_5\beta_1$, $\alpha_v\beta_3$, and $\alpha_v\beta_5$ [128, 129]. Survival signals mediated through ligation of these integrins help to maintain the new vessels, and antagonism of the integrins can induce apoptosis of the endothelial cells resulting in regression of the new vessels. Because the integrins are differentially expressed on ECs in neovascularization, they provide candidate targets that can selectively affect new vessels with no effect on normal vessels. Small molecule antagonists of $\alpha_v\beta_3$ and $\alpha_v\beta_5$ inhibit retinal neovascularization [128] and a small molecule antagonist of $\alpha_5\beta_1$ causes regression of choroidal neovascularization [129]. The selectivity of these agents for new vessels

may enhance their safety and leave open the option of systemic administration, although sustained local delivery is also an option. Since these agents work by a different mechanism from VEGF antagonists and because they may induce regression of new vessels, they provide candidate targets that are important to test in clinical trials.

Signals from the ECM are often unmasked or eliminated by proteolysis. Components of the ECM may bind and sequester soluble factors, preventing them from activating receptors on endothelial cells until they are released by proteolysis [130–132]. Degradation of ECM also liberates fragments with antiangiogenic activity that provide negative feedback slowing vessel growth, making it more orderly, and eventually helping to turn it off and re-establish quiescence. Endostatin was the first collagen fragment demonstrated to inhibit angiogenesis [133], but subsequently several others have been identified [134–140]. Interestingly, several of these antiangiogenic peptides are derived from noncollagenous (NC1) domains of the basement membrane collagens IV, XV, and XVIII. The NC1 domains are important for assembly of the supramolecular structures of the collagens and under normal circumstances do not interact with cells [141–143]. However, after cleavage from native collagens, several of the NC1 domains bind endothelial cells and inhibit angiogenesis. Endostatin is derived from the NC1 domain of collagen XVIII and restin is a similar antiangiogenic peptide derived from the NC1 domain of collagen XV [134]. Collagen IV is unusual in that there are 6 distinct collagen IV chains that have different tissue distributions [144–148]. The NC1 domains of several of the collagen IV chains including 1, 2, 3, and 6 have antiangiogenic activity, but effects may vary in different organs [135–140]. In the eye, the NC1 domain of $\alpha2(IV)$ causes regression of CNV and provides another potential therapeutic agent [149].

Other Endogenous Inhibitors of Neovascularization

In addition to the inhibitory fragments derived from collagen, there are several other endogenous antiangiogenic proteins, such as angiostatin, thrombospondin, and pigment epithelium-derived factor (PEDF), that may function under normal circumstances to limit and control neovascularization, but become overwhelmed in situations in which pathologic angiogenesis occurs. Each of these could be used as a recombinant protein therapeutic, but may also be delivered by gene transfer. Using a variety of vector systems, several proteins have been demonstrated to inhibit retinal and/or choroidal neovascularization, including endostatin [38, 150], PEDF [151–154], angiostatin [155], tissue inhibitor of metalloproteinases-3 [156], soluble Tie2 [157], and soluble VEGF receptor-1 (Flt-1) [158–160].

Vasohibin differs from other inhibitors because it is up-regulated by VEGF and FGF2 in cultured endothelial cells and therefore was hypothesized to function as a negative feedback regulator [161]. Testing of this hypothesis in mice with ischemic retinopathy showed increased expression of VEGF was accompanied by elevation of *vasohibin* mRNA and blocking of the increase in *vegf* mRNA with *vegf* siRNA significantly attenuated the rise in *vasohibin* mRNA; knockdown of vasohibin increased the amount of neovascularization and overexpression of vasohibin reduced the amount of neovascularization [162]. Knockdown of *vasohibin* mRNA in ischemic retina had no significant effect on *vegf* or *vegfr1* mRNA levels, but caused a significant elevation in the level of *vegfr2* mRNA. These data support the hypothesis that vasohibin acts as a negative feedback regulator of neovascularization in the retina, and suggest that suppression of VEGFR2 may play some role in mediating its activity.

Recently, a phase I trial investigating the effect of a single intraocular injection of an E1-, partial E3-, E4-deleted adenoviral vector expressing human PEDF (AdPEDF.11) in 28 patients with advanced CNV due to AMD has been completed [163]. With doses ranging from 106 to 109.5 particle units (pu), there was mild inflammation in 25% of patients, but no serious adverse events nor dose-limiting toxicities. At 3 and 6 months after injection, 55% and 50%, respectively, of patients treated with106–107.5 pu and 94% and 71% of patients treated with 108–109.5 pu had no change or improvement in lesion size from baseline. The median increase in lesion size at 6 and 12 months was 0.5 and 1.0 disc areas in the low dose group compared to 0 and 0 disc areas in the high dose group. These data suggest the possibility of antiangiogenic activity that may last for several months after a single intravitreous injection of doses greater than 108 pu of AdPEDF.11. This study indicates that adenoviral vector-mediated ocular gene transfer is a viable approach for treatment of ocular disorders and supports the performance of additional studies investigating the efficacy of AdPEDF.11 in patients with CNV.

Treatments Targeting Pathways that are not Angiogenesis-specific

It is advantageous to target molecules that are selectively expressed in endothelial cells and act only in neovascularization, but it is not necessary. Polyamine analogs block polyamine metabolism, which is required by all proliferating cells including ECs participating in neovascularization. Intravitreous or periocular injections of polyamine analogs induced regression of established CNV by inducing apoptosis in ECs participating in CNV [164]. Intraocular injections of polyamine analogs cause apoptosis of some retinal neurons, but after periocular injections only endothelial cells participating in CNV are affected and retinal function assessed by ERGs remains normal. Therefore, selectivity was provided by the mode of delivery. Antiangiogenic activity after a single periocular injection lasts for at least 2 weeks making this approach feasible for clinical trials [165].

Tubulin binding agents, such as vincristine, vinblastine, and colchicine, cause tumor necrosis due to damage to tumor blood vessels, but at doses that are too toxic for patients to tolerate [166, 167]. Combretastatin A-4 is a naturally occurring structural analog of colchicine that binds tubulin at the same site as colchicine, but with different characteristics [168, 169] that impart selective toxicity to tumor vasculature [170]. Combretastatin A-4-phosphate (CA-4-P) is a more soluble, inactive prodrug that is converted to CA-4 by endogenous nonspecific phosphatases [171]. Daily intraperitoneal injections of CA-4-P significantly suppress neovascularization in transgenic mice with ectopic expression of VEGF in photoreceptors and mice with laser-induced rupture of Bruch's membrane, and administration of CA-4-P to mice with established CNV results in significant regression of the neovascularization [172]. Therefore, tubulin binding agents may have potential for treatment of ocular neovascularization if adequate selectivity is achieved.

Nitric oxide (NO) has been shown to have proangiogenic or antiangiogenic effects depending upon the setting. Mice with targeted deletion of 1 of the 3 isoforms of nitric oxide synthase (NOS) to investigate the effects of NO in ocular neovascularization [173, 174]. In transgenic mice with increased expression of VEGF in photoreceptors, deficiency of any of the 3 isoforms caused a significant decrease in subretinal neovascularization, but no alteration of VEGF expression. In mice with laser-induced rupture of Bruch's membrane, deficiency of inducible NOS (iNOS) or neuronal NOS (nNOS), but not endothelial NOS (eNOS), caused a significant decrease in choroidal neovascularization. In mice with oxygen-induced ischemic retinopathy, deficiency of eNOS, but not iNOS or nNOS, caused a significant decrease in retinal neovascularization and decreased expression of VEGF. These data suggest that NO contributes to both retinal and choroidal neovascularization, but that different isoforms of NOS are involved. Oral administration of N[G]-monomethyl-L-arginine (L-NMMA), a broad spectrum NOS inhibitor, caused significant inhibition of choroidal neovascularization in mice with laser-induced rupture of Bruch's membrane, and significantly inhibited subretinal neovascularization in rho/VEGF mice, but did not inhibit retinal neovascularization in mice with ischemic retinopathy. Triple homozygous mutant mice deficient in all 3 NOS isoforms had marked suppression of choroidal neovascularization at sites of rupture of Bruch's membrane and near-complete suppression of subretinal neovascularization in rho/VEGF mice, but showed no difference in ischemia-induced retinal neovascularization compared to wild type mice. These data indicate that NO is an important stimulator of choroidal neovascularization and that reduction of NO by pharmacologic means is a good treatment strategy. However, the situation is more complex for ischemia-induced retinal neovascularization for which NO produced in endothelial cells by eNOS is stimulatory, but NO produced in other retinal cells by iNOS and/or nNOS is inhibitory. Selective inhibitors of eNOS may be needed for treatment of retinal neovascularization.

Conclusions

Over the last several years, clinical trials have confirmed pre-clinical studies indicating that VEGF is an important target for ocular neovascularization and macular edema. This demonstrates the value of elucidating the molecular pathogenesis of diseases and developing antagonists for molecules that play a central role. Several other molecules that also contribute to ocular neovascularization have recently been identified and are being tested or will soon be tested in clinical trials. The success with VEGF antagonists raises the bar and makes clinical trials more complicated to design, but it has also provided optimism that our strategy for treatment development is sound and if preclinical testing implicates a molecule as a contributor that does not work through VEGF, then clinical trials are warranted and have a good chance of providing further benefits to patients.

References

1. Janzer RC, Raff MC. Astrocytes induce blood-brain barrier properties in endothelial cells. *Nature* 1987;**325**:253–257.
2. Watanabe T, Raff MC. Retinal astrocytes are immigrants from the optic nerve. *Nature* 1988;**332**:834–837.
3. Stone J, Itin A, Alon T, Pe'er J, Gnessin H, Chan-Ling T et al. Development of retinal vasculature is mediated by hypoxia-induced vascular endothelial growth factor (VEGF) expression by neuroglia. *J. Neurosci.* 1995;**15**:4738–4747.
4. Stalmans I, Ng Y-S, Rohan R, Fruttiger M, Bouche A, Yuce A et al. Arteriolar and venular patterning in retinas of mice selectively expressing VEGF isoforms. *J. Clin. Invest.* 2002;**109**:327–336.
5. Gerhardt H, Golding M, Fruttiger M, Ruhrberg C, Lundkvist A, Abramsson A et al. VEGF guides angiogenic sprouting utilizing endothelial tip cell filopodia. *J. Cell Biol.* 2003;**161**:1163–1177.
6. Xu Q, Wang Y, Dabdoub A, Smallwood PM, Williams J, Woods C et al. Vascular development in the retina and inner ear: control by Norrin and Frizzled-4, a high affinity ligand-receptor pair. *Cell* 2004;**116**:883–895.
7. Alon T, Hemo I, Itin A, Pe'er J, Stone J, Keshet E. Vascular endothelial growth factor acts as a survival factor for newly formed retinal vessels and has implications for retinopathy of prematurity. *Nature Med.* 1995;**1**:1024–1028.
8. Lindahl P, Johansson BR, Leveen P, Betsholtz C. Pericyte loss and microaneurysm formation in PDGF-B-deficient mice. *Science* 1997;**277**:242–245.
9. Lang RA, Bishop MJ. Macrophages are reqired for cell death and tissue remodeling in the developing mouse eye. *Cell* 1993;**74**:453–462.
10. Kato M, Patel MS, Levasseur R, Lobov I, Chang BHJ, Glass DJ et al. Cbfa-1-independent decrease in osteoblast proliferation, osteopenia, and persistent embryonic eye vascularization in mice deficient in Lrp5, a Wnt coreceptor. *J. Cell Biol.* 2002;**157**:303–314.
11. Lobov IB, Brooks PC, Lang RA. Angiopoietin-2 displays VEGF-dependent modulation of capillary structure and endothelial cell survival *in vivo*. *Proc. Natil. Acad. Sci. USA* 2002;**99**:11205–11210.
12. Oshima Y, Oshima S, Nambu H, Kachi S, Takahashi K, Umeda N et al. Different effects of angiopoietin 2 in different vascular beds in the eye; new vessels are most sensitive. *FASEB J.* 2005;**19**:963–965.
13. Shimizu K, Kobayashi Y, Muraoka K. Midperipheral fundus involvement in diabetic retinopathy. *Ophthalmology* 1981;**88**:601–612.
14. Patz A, Eastham A, Higgenbotham DH, Kleh T. Oxygen studies in retrolental figroplasia: Production of the microscopic changes of retrolental fibroplasia in experimental animals. *Am. J. Ophthalmol.* 1953;**36**:1511–1522.
15. Penn JS, Tolman BL, Lowery LA. Variable oxygen exposure causes preretinal neovascularization in the newborn rat. *Invest. Ophthalmol. Vis. Sci.* 1993;**34**:576–585.
16. Smith LEH, Wesolowski E, McLellan A, Kostyk SK, D'Amato R, Sullivan R et al. Oxygen-induced retinopathy in the mouse. *Invest. Ophthalmol. Vis. Sci.* 1994;**35**:101–111.
17. Pierce EA, Avery RL, Foley ED, Aiello LP, Smith LEH. Vascular endothelial growth factor/vascular permeability factor expression in a mouse model of retinal neovascularization. *Proc. Natl. Acad. Sci. USA.* 1995;**92**:905–909.
18. Ozaki H, Yu A, Della N, Ozaki K, Luna JD, Yamada H et al. Hypoxia inducible factor-1a is increased in ischemic retina: temporal and spatial correlation with VEGF expression. *Invest. Ophthalmol. Vis. Sci.* 1999;**40**:182–189.
19. Kelly BD, Hackett SF, Hirota K, Oshima Y, Cai Z, Berg-Dixon S et al. Cell type-specific regulation of angiogenic growth factor gene expression and induction of angiogenesis in nonischemic tissue by a constitutively active form of hypoxia-inducible factor 1. *Circ. Res.* 2003;**93**:1074–1081.
20. Engerman R, Bloodworth JMB, Nelson S. Relationship of microvascular disease in diabetes to metabolic control. *Diabetes* 1977;**26**:760–769.
21. Virdi P, Hayreh S. Ocular neovascularization with retinal vascular occlusion. I. Association with retinal vein occlusion. *Arch. Ophthalmol.* 1980;**100**:331–341.
22. Miller JW, Adamis AP, Shima DT, D'Amore PA, Moulton RS, O'Reilly MS et al. Vascular endothelial growth factor/vascular permeability factor is temporally and spatially correlated with ocular angiogenesis in a primate model. *Am. J. Pathol.* 1994;**145**:574–584.
23. Pournaras C, Tsacopoulos M, Strommer K, Gilodi N, Leuenberger PM. Experimental retinal branch vein occlusion in miniature pigs induces local tissue hypoxia and vasoproliferative microangiopathy. *Ophthalmology* 1990;**97**:1321–1328.
24. Pournaras CJ, Tsacopoulos M, Strommer K, Gilodi N, Leuenberger PM. Scatter photocoagulation restores tissue hypoxia in experimental vasoproliferative microangiopathy in miniature pigs. *Ophthalmology* 1990;**97**:1329–1333.
25. Danis RP, Yang Y, Massicotte SJ, Boldt C. Preretinal and optic nerve head neovascularization induced by photodynamic venous thrombosis in domestic pigs. *Arch Ophthalmol* 1993;**111**:539–533.
26. Danis RP, Bingaman DP, Yang Y, Ladd B. Inhibition of preretinal and optic nerve head neovascularization in pigs by intravitreal triamcinolone acetonide. *Ophthalmology* 1996;**103**:2099–2104.
27. Klein R, Klein BEK, Linton KP. The Beaver Dam Eye Study: the relation of age-related maculopathy to smoking. *Am. J. Epidemiol.* 1993;**137**:190–200.

28. Green WR, Wilson DJ. Choroidal neovascularization. *Ophthalmology* 1986;**93**:1169–1176.

29. Yannuzzi LA, Negrao S, Iida T, Carvalho C, Rodriguez-Coleman H, Slakter JS *et al*. Retinal angiomatous proliferation in age-related macular degeneration. *Retina* 2001;**21**:416–434.

30. Ryan SJ. Subretinal neovascularization: natural history of an experimental model. *Arch. Ophthalmol.* 1982;**100**:1804–1809.

31. Dobi ET, Puliafito CA, Destro M. A new model of choroidal neovascularization in the rat. *Arch. Ophthalmol.* 1989;**107**:264–269.

32. Tobe T, Ortega S, Luna JD, Ozaki H, Okamoto N, Derevjanik NL *et al*. Targeted disruption of the *FGF2* gene does not prevent choroidal neovascularization in a murine model. *Am. J. Pathol.* 1998;**153**:1641–1646.

33. Saishin Y, Lima Silva R, Saishin Y, Callahan K, Schoch C, Ahlheim M *et al*. Periocular injection of microspheres containing PKC412 inhibits choroidal neovascularization in a porcine model. *Invest. Ophthalmol. Vis. Sci.* 2003;**44**:4989–4993.

34. Saishin Y, Silva RL, Saishin Y, Kachi S, Aslam S, Gong YY *et al*. Periocular gene transfer of pigment epithelium-derived factor inhibits choroidal neovascularization in a human-sized eye. *Hum. Gene Ther.* 2005;**16**:473–478.

35. Okamoto N, Tobe T, Hackett SF, Ozaki H, Vinores MA, LaRochelle W *et al*. Transgenic mice with increased expression of vascular endothelial growth factor in the retina: a new model of intraretinal and subretinal neovascularization. *Am. J. Pathol.* 1997;**151**:281–291.

36. Tobe T, Okamoto N, Vinores MA, Derevjanik NL, Vinores SA, Zack DJ *et al*. Evolution of neovascularization in mice with overexpression of vascular endothelial growth factor in photoreceptors. *Invest. Ophthalmol. Vis. Sci.* 1998;**39**:180–188.

37. Ohno-Matsui K, Hirose A, Yamamoto S, Saikia J, Okamoto N, Gehlbach P *et al*. Inducible expression of vascular endothelial growth factor in photoreceptors of adult mice causes severe proliferative retinopathy and retinal detachment. *Am. J. Pathol.* 2002;**160**:711–719.

38. Takahashi K, Saishin Y, Saishin Y, Lima Silva R, Oshima Y, Oshima S *et al*. Intraocular expression of endostatin reduces VEGF-induced retinal vascular permeability, neovascularization, and retinal detachment. *FASEB J.* 2003;**17**:896–898.

39. Nambu H, Umeda N, Kachi S, Oshima Y, Nambu R, Campochiaro PA. Angiopoietin 1 prevents retinal detachment in an aggressive model of proliferative retinopathy, but has no effect on established neovascularization. *J. Cell. Physiol.* 2005;**204**:227–235.

40. Spilsbury K, Garrett KS, Shen WY, Constable IJ, Rakoczy PE. Overexpression of vascular endothelial growth factor (VEGF) in the retinal pigment epithelium leads to the development of choroidal neovascularization. *Am. J. Pathol.* 2000;**157**:135–144.

41. Abraham JA, Whang JL, Tumolo A, Mergia A, Freidman J, Gospodarowicz D *et al*. Human basic fibroblast growth factor: nucleotide sequence and genomic organization. *EMBO J.* 1986;**5**:2523–2528.

42. Baird A, Esch F, Gospodarowicz D, Guillemin R. Retina- and eye-derived endothelial cell growth factors: partial molecular chariacterization and identity with acidic and basic fibroblast growth factors. *Biochemistry* 1985;**24**:7855–7860.

43. Ozaki H, Okamoto N, Ortega S, Chang M, Ozaki K, Sadda S *et al*. Basic fibroblast growth factor is neither necessary nor sufficient for the development of retinal neovascularization. *Am. J. Pathol.* 1998;**153**:757–765.

44. Yamada H, Yamada E, Kwak N, Ando A, Suzuki A, Esumi N *et al*. Cell injury unmasks a latent proangiogenic phenotype in mice with increased expression of FGF2 in the retina. *J. Cell. Physiol.* 2000;**185**:135–142.

45. Shweiki D, Itin A, Soffer D, Keshet E. Vascular endothelial growth factor induced by hypoxia may mediate hypoxia-initiated angiogenesis. *Nature* 1992;**359**:843–845.

46. Plate KH, Breier G, Millauer B, Ullrich A, Risau W. Up-regulation of vascular endothelial growth factor and its cognate receptors in a rat glioma model of tumor angiogenesis. *Canc. Res.* 1993;**53**:5822–5827.

47. Aiello LP, Avery RL, Arrigg PG, Keyt BA, Jampel HD, Shah ST *et al*. Vascular endothelial growth factor in ocular fluid of patients with diabetic retinopathy and other retinal disorders. *N. Engl. J. Med.* 1994;**331**:1480–1487.

48. Adamis AP, Miller JW, Bernal M-T, D'Amico DJ, Folkman J, Yeo T-K *et al*. Increased vascular endothelial growth factor levels in the vitreous of eyes with proliferative diabetic retinopathy. *Am. J. Ophthalmol.* 1994;**118**:445–450.

49. Aiello LP, Pierce EA, Foley ED, Takagi H, Chen H, Riddle L *et al*. Suppression of retinal neovascularization in vivo by inhibition of vascular endothelial growth factor (VEGF) using soluble VEGF-receptor chimeric proteins. *Proc. Natl. Acad. Sci. U.S.A.* 1995;**92**:10457–10461.

50. Robinson GS, Pierce EA, Rook SL, Foley E, Webb R, Smith LES. Oligodeoxynucleotides inhibit retinal neovascularization in a murine model of proliferative retinopathy. *Proc. Natl. Acad. Sci. USA.* 1996;**93**:4851–4856.

51. Ozaki H, Seo M-S, Ozaki K, Yamada H, Yamada E, Hofmann F *et al*. Blockade of vascular endothelial cell growth factor receptor signaling is sufficient to completely prevent retinal neovascularization. *Am. J. Pathol.* 2000;**156**:679–707.

52. Kvanta A, Algvere PV, Berglin L, Seregard S. Subfoveal fibrovascular membranes in age-related macular degeneration express vascular endothelial growth factor. *Invest. Ophthalmol. Vis. Sci.* 1996;**37**:1929–1934.

53. Lopez PF, Sippy BD, Lambert HM, Thach AB, Hinton DR. Transdifferentiated retinal pigment epithelial cells are immunoreactive for vascular endothelial growth factor in surgically excised age-related macular degeneration-related choroidal neovascular membranes. *Invest. Ophthalmol. Vis. Sci.* 1996;**37**:855–868.

54. Amin R, Pulkin JE, Frank RN. Growth factor localization in choroidal neovascular membranes of age-related macular degeneration. *Invest. Ophthalmol. Vis. Sci.* 1994;**35**:3178–3188.

55. Kwak N, Okamoto N, Wood JM, Campochiaro PA. VEGF is an important stimulator in a model of choroidal neovascularization. *Invest. Ophthalmol. Vis. Sci.* 2000;**41**:3158–3164.

56. Kryzstolik MG, Afshari MA, Adamis AP, Gaudreault J, Gragoudas ES, Michaud NM *et al*. Prevention of experimental choroidal neovascularization with intravitreal anti-vascular endothelial growth factor antibody fragment. *Arch. Ophthalmol.* 2002;**120**:338–346.

57. Gragoudas ES, Adamis AP, Cunningham ET, Jr., Feinsod M, Guyer DR. Pegaptanib for neovascular age-related macular degeneration. *N. Eng. J. Med.* 2004;**351**:2805–2816.

58. Brown DM, Kaiser PK, Michels M, Soubrane G, Heier JS, Kim RY *et al*. Ranibizumab versus verteporfin for neovascular age-related macular degeneration. *N. Eng. J. Med.* 2006;**355**:1432–1444.

59. Rosenfeld PJ, Brown DM, Heier JS, Boyer DS, Kaiser PK, Chung CY et al. Ranibizumab for neovascular age-related macular degeneration. *N. Eng. J. Med.* 2006;**355:**1419–1431.

60. Nguyen QD, Shah SM, Tatlipinar S, Do DV, Van Anden E, Campochiaro PA. Bevacizumab suppresses choroidal neovascularization due to pathologic myopia. *Br. J. Ophthalmol.* 2005;**89:**1368–1370.

61. Rosenfeld PJ, Moshfeghi AA, Puliafito CA. Optical coherence tomography findings after an intavitreal injection of bevacizumab (avastin) for neovascular age-related macular degeneration. *Ophtalmic Surg. Lasers Imaging* 2005;**36:**331–335.

62. Avery RL, Pieramici DJ, Rabena MD, Castellarin AA, Nasir MA, Giust MJ. Intravitreal bevacizumab (Avastin) for neovascular age-related macular degeneration. *Ophthalmology* 2006;**113:**363–372.

63. Spaide RF, Laud K, Fine HF, Klancnik JM, Jr., Meyerle CB, Yannuzzi LA et al. Intravitreal bevacizumab treatment of choroidal neovascularization secondary to age-related macular degeneration. *Retina* 2006;**26:**383–390.

64. Adamis AP, Altaweel M, Bressler NM, Cunningham ET, Jr., Davis MD, Goldbaum M et al. Changes in retinal neovacularization after pegaptanib (Macugen) therapy in diabetic retinopathy. *Ophthalmology* 2006;**113:**23–28.

65. The Diabetic Retinopathy Study Research Group. Photocoagulation treatment of proliferative diabetic retinopathy: Clinical application of Diabetic Retinopathy Study (DRS) findings, DRS Report Number 8. *Ophthalmology* 1981;**88:**583–600.

66. Nguyen QD, Shah SM, Van Anden E, Sung JU, Vitale S, Campochiaro PA. Supplemental inspired oxygen improves diabetic macular edema; a pilot study. *Invest. Ophthalmol. Vis. Sci.* 2003;**45:**617–624.

67. Ozaki H, Hayashi H, Vinores SA, Moromizato Y, Campochiaro PA, Oshima K. Intravitreal sustained release of VEGF causes retinal neovascularization in rabbits and breakdown of the blood-retinal barrier in rabbits and primates. *Exp Eye Res* 1997;**64:**505–517.

68. Derevjanik NL, Vinores SA, Xiao W-H, Mori K, Turon T, Hudish T et al. Quantitative assessment of the integrity of the blood-retinal barrier in mice. *Invest. Ophthalmol. Vis. Sci.* 2002;**43:**2462–2467.

69. Campochiaro PA and the C99-PKC412-003 Study Group. Reduction of diabetic macular edema by oral administration of the kinase inhibitor PKC412. *Invest. Ophthalmol. Vis. Sci.* 2004;**45:**922–931.

70. Nguyen QD, Tatlipinar S, Shah SM, Haller JA, Quinlan E, Sung J et al. Vascular endothelial growth factor is a critical stimulus for diabetic macular edema. *Am. J. Ophthalmol.* 2006;**142:**161–169.

71. Reichelt M, Shi S, Hayes M, Kay G, Batch J, Gole GA et al. Vascular endothelial growth factor-B and retinal vascular development in the mouse. *Clin. Exp. Ophthalmol.* 2003;**31:**61–65.

72. Carmeliet P, Moons L, Luttun A, Vincenti V, Compernolle V, De Mol M et al. Synergism between vascular endothelial growth factor and placental growth factor contributes to angiogenesis and plasma extravasation in pathological conditions. *Nat. Med.* 2001;**7:**575–583.

73. Luttun A, Tjwa M, Moons L, Wu Y, Angelillo-Scherrer A, Liao F et al. Revascularization of ischemic tissues by PlGF treatment, and inhibition of tumor angiogenesis, arthritis and atherosclerosis by anti-Flt1. *Nat. Med.* 2002;**8:**831–839.

74. Holash J, Davis S, Papadoupoulos N, Croll SD, Ho L, Russell M et al. VEGF-Trap: a VEGF blocker with potent antitumor effects. *Proc Natl Acad Sci U S A* 2002;**99:**11393–11398.

75. Saishin Y, Saishin Y, Takahashi K, Lima Silva R, Hylton D, Rudge J et al. VEGF-TRAPR1R2 suppresses choroidal neovascularization and VEGF-induced breakdown of the blood-retinal barrier. *J. Cell. Physiol.* 2003;**195:**241–248.

76. Nguyen QD, Shah SM, Hafiz G, Quinlan E, Sung J, Chu K et al. A phase 1 trial of intravenously administered VEGF trap for treatment in patients with choroidal neovascularization due to age-related macular degeneration. *Ophthalmology* 2006;**113:**1522e1521–1522e1514.

77. Gille H, Kowalski J, Li B, LeCouter J, Moffat B, Zioncheck TF et al. Analysis of biological effects and signaling properties of Flt-1 (VEGFR-1) and KDR (VEGFR-2). *J. Biol. Chem.* 2001;**276:**3222–3230.

78. Fong GH, Zhang L, Bryce DM, Peng J. Increased hemangioblast commitment, not vascular disorganization, is the primary defect in flt-1 knockout mice. *Development* 1999;**126:** 3015–3025.

79. Park JE, Chen HH, Winer J, Houck KA, Ferrara N. Placenta growth factor. Potentiation of vascular endothelial growth factor bioactivity, in vitro and in vivo, and high affinity binding Flt-1 but not to Flk-1/KDR. *J. Biol. Chem.* 1994;**269:**25646–25654.

80. Shen J, Samul R, Lima e Silva R, Akiyama H, Liu H, Saishin Y et al. Suppression of ocular neovascularization with siRNA targeting VEGF receptor 1. *Gene Ther.* 2005;**13:**225–234.

81. Klagsbrun M, Takashima S, Mamluk R. The role of neuropilin in vascular and tumor biology. *Adv. Exp. Med. Biol.* 2002;**515:**33–48.

82. Neufeld G, Kessler O, Herzog Y. The interaction of neuropilin-1 and neuropilin-2 with tyrosine-kinase receptors for VEGF. *Adv. Exp. Med. Biol.* 2002;**515:**81–91.

83. Oh H, Takagi H, Otani A, Koyama S, Kemmonchi S, Uemura A et al. Selective induction of neuropilin-1 by vascular endothelial growth factor (VEGF): a mechanism contributing to VEGF-induced angiogenesis. *Proc. Natl. Acad. Sci. USA* 2002;**99:**383–388.

84. Shen J, Samul R, Zimmer J, Liu H, Liang X, Hackett SF et al. Deficiency of neuropilin 2 suppresses VEGF-induced retinal neovascularization. *Mol. Med.* 2004;**10:**12–18.

85. Akiyama H, Mohamedali K, Lima-Silva R, Kachi S, Shen J, Hatara C et al. Vascular targeting of ocular neovascularization with a VEGF121/Gelonin chmeric protein. *Mol. Pharmacol.* 2005;**68:**1543–1550.

86. Dumont DJ, Gradwohl G, Fong G-H, Puri MC, Gerstenstein M, Auerbach A et al. Dominant-negative and targeted null mutations in the endothelial receptor tyrosine kinase, tek, reveal a critical role in vasculogenesis of the embryo. *Genes Dev.* 1994;**8:**1897–1909.

87. Sato TN, Tozawa Y, Deutsch U, Wolburg-Buchholz K, Fujiwara Y, Gendron-Maguire M et al. Distinct roles of the receptor tyrosine kinases Tie-1 and Tie-2 in blood vessel formation. *Nature* 1995;**376:**70–74.

88. Davis S, Aldrich TH, Jones P, Acheson A, Ryan TE, Bruno J et al. Isolation of angiopoietin-1, a ligand for the TIE2 receptor, by secretion-trap expression cloning. *Cell* 1996;**87:**1161–1169.

89. Saharinen P, Kerkela K, Ekman N, Marron M, Brindle N, Lee GM et al. Multiple angiopoietin recombinant proteins activate the Tie1 receptor tyrosine kinase and promote its interaction with Tie2. *J. Cell Biol.* 2005;**169:**239–243.

90. Maisonpierre PC, Suri C, Jones PF, Bartunkova S, Wiegand SJ, Radziejewski C et al. Angiopoietin-2, a natural antagonist for Tie2 that disrupts in vivo angiogenesis. *Science* 1997;**277:**55–60.

91. Hackett SF, Ozaki H, Strauss RW, Wahlin K, Suri C, Maisonpierre P et al. Angiopoietin 2 expression in the retina: upregulation during physiologic and pathologic neovascularization. *J. Cell. Physiol.* 2000;**184:**275–284.

92. Hackett SF, Wiegand SJ, Yancopoulos G, Campochiaro P. Angiopoietin-2 plays an important role in retinal angiogenesis. *J. Cell. Physiol.* 2002;**192**:182–187.

93. Oshima Y, Takahashi K, Oshima S, Saishin Y, Saishin Y, Silva RL *et al.* Intraocular gutless adenoviral vectored VEGF stimulates anterior segment but not retinal neovascularization. *J. Cell. Physiol.* 2004;**199**:399–411.

94. Oshima Y, Deering T, Oshima S, Nambu H, Reddy PS, Kaleko M *et al.* Angiopoietin-2 enhances retinal vessel sensitivity to vascular endothelial growth factor. *J. Cell. Physiol.* 2004;**199**:412–417.

95. Oshima Y, Oshima S, Nambu H, Kachi S, Hackett SF, Melia M *et al.* Increased expression of VEGF in retinal pigmented epithelial cells is not sufficient to cause choroidal neovascularization. *J. Cell. Physiol.* 2004;**201**:393–400.

96. Suri C, McClain J, Thurston G, McDonald DM, Zhou H, Oldmixon EH *et al.* Increased vascularization in mice overexpressing angiopoietin-1. *Science* 1998;**282**:468–471.

97. Thurston G, Suri C, Smith K, McClain J, Sato TN, Yancopoulos GD *et al.* Leakage-resistant blood vessels in mice transgenically overexpressing angiopoietin-1. *Science* 1999;**286**: 2511–2515.

98. Nambu H, Nambu R, Oshima Y, Hackett SF, Wiegand SJ, Yancopoulos G *et al.* Angiopoietin 1 inhibits ocular neovascularization and breakdown of the blood-retinal barrier. *Gene Ther.* 2004;**11**:865–873.

99. Gardiner TA, Gibson DS, de Gooyer TE, de la Cruz VF, McDonald DM, Stitt AW. Inhibition of tumor necrosis factor-alpha improves physiological angiogenesis and reduces pathological neovascularization in ischemic retinopathy. *Am. J. Pathol.* 2005;**166**:637–644.

100. Vinores SA, Xiao WH, Shen J, Campochiaro PA. TNFalpha is critical for ischemia-induced leukostasis, but not retinal neovascularization nor VEGF-induced leakage. *J. Neuroimmunol.* 2007;**182**:73–79.

101. Ilg RC, Davies MH, Powers MR. Altered retinal neovascularization in TNF receptor-deficient mice. *Curr. Eye Res.* 2005;**30**:1003–1013.

102. Smith LEH, Kopchick JJ, Chen W, Knapp J, Kinose F, Daley D *et al.* Essential role of growth hormone in ischemia-induced retinal neovascularization. *Science* 1997;**276**:1706–1709.

103. Smith LEH, Shen W, Perruzzi C, Soker S, Kinose F, Xu X *et al.* Regulation of vascular endothelial growth factor-dependent retinal neovascularization by insulin-like growth factor-1 receptor. *Nat. Med.* 1999;**5**:1390–1395.

104. Kondo T, Vicent D, Suzuma K, Yanagisawa M, King GL, Holzenberger M *et al.* Knockout of insulin and IGF-1 receptors on vascular endothelial cells protects against retinal neovascularization. *J. Clin. Invest.* 2003;**111**:1835–1842.

105. Ruberte J, Ayuso E, Navarro M, Carretero A, Nacher V, Haurigot V *et al.* Increased ocular levels of IGF-1 in transgenic mice lead to diabetes-like eye disease. *J. Clin. Invest.* 2004;**113**:1149–1157.

106. Fukuda R, Hirota K, Fan F, Jung YD, Ellis LM, Semenza GL. Insulin-like growth factor 1 induces hypoxia-inducible factor-1-mediated vascular endothelial growth factor expression, which is dependent on MAP kinase and phosphatidylinositol 3-kinase signaling in colon cancer cells. *J. Biol. Chem.* 2002;**277**:38205–38211.

107. Seo M-S, Okamoto N, Vinores MA, Vinores SA, Hackett SF, Yamada H *et al.* Photoreceptor-specific expression of PDGF-B results in traction retinal detachment. *Am. J. Pathol.* 2000;**157**:995–1005.

108. Campochiaro PA, Glaser BM. Platelet-derived growth factor is chemotactic for human retinal pigment epithelial cells. *Arch. Ophthalmol.* 1985;**103**:576–579.

109. Bergers G, Song S, Meyer-Morse N, Bersland E, Hanahan D. Benefits of targeting both pericytes and endothelial cells in the tumor vasculature with kinase inhibitors. *J. Clin. Invest.* 2003;**111**:1287–1295.

110. Seo M-S, Kwak N, Ozaki H, Yamada H, Okamoto N, Fabbro D *et al.* Dramatic inhibition of retinal and choroidal neovascularization by oral administration of a kinase inhibitor. *Am. J. Pathol.* 1999;**154**:1743–1753.

111. Grant MB, May WS, Caballero S, Brown GA, Guthrie SM, Mamee RN *et al.* Adult hematopoietic stem cells provide functional hemangioblastic activity during retinal neovascularization. *Nat. Med.* 2002;**8**:607–612.

112. Espinosa-Heidmann DG, Caicado A, Hernandez EP, Csaky KG, Cousins SW. Bone marrow-derived progenitor cells contribute to experimental choroidal neovascularization. *Invest. Ophthalmol. Vis. Sci.* 2003;**44**:4914–4919.

113. Chan-Ling T, Baxter L, Afzal A, Sengupta N, Caballero S, Rosinova E *et al.* Hematopoietic stem cells provide repair functions after laser-induced Bruch's membrane rupture model of choroidal neovascularization. *Am. J. Pathol.* 2006;**168**:1031–1044.

114. Amoaku WMK, Archer DB. Cephalic radiation and retinal vasculopathy. *Eye* 1990;**4**:195–203.

115. Barleon B, Sozzani S, Zhou D, Weich HA, Mantovani A, Marme D. Migration of human monocytes in reponse to vascular endothelial growth factor (VEGF) is mediated via the VEGF receptor flt-1. *Blood* 1996;**87**:3336–3343.

116. Grunewald M, Avraham I, Dor Y, Bachar-Lustig E, Itin A, Yung S *et al.* VEGF-induced adult neovascularization: recruitment, retention, and role of accessory cells. *Cell* 2006;**124**:175–189.

117. Lima e Silva R, Shen J, Hackett SF, Kachi S, Akiyama H, Kiuchi K, Yokoi K, Hatara MC, Lauer T, Aslam S, Gong YY, Xiao WH, Khu NH, Thut C, Campochiaro PA. The SDF-1/CXCR4 ligand/receptor pair is an important contributor to several types of ocular neovascularization. *FASEB J.* 2007;**21**:3219–3230.

118. Smalley W, DuBois RN. Colorectal cancer and nonsteroidal anti-inflammatory drugs. *Adv. Pharmacol.* 1997;**39**:1–20.

119. Kawamori T, Rao CV, Seibert K, Reddy BS. Chemopreventive activity of celecoxib, a specific cyclooxygenase-2 inhibitor, against colon carcinogenesis. *Cancer Res.* 1998;**58**: 409–412.

120. Williams CS, Mann M, DuBois RN. The role of cyclooxygenases in inflammation, cancer, and development. *Oncogene* 1999;**18**:7908–7916.

121. Williams CS, Tsujii M, Reese J, Dey SK, DuBois RN. Host cyclooxygenase-2 modulates carcinoma growth. *J. Clin. Invest.* 2000;**105**:1589–1594.

122. Tsujii M, Kawano S, Tsuji S, Sawaoka H, Hori M, DuBois RN. Cyclooxygenase regulates angiogenesis induced by colon cancer cells. *Cell* 1998;**93**:705–716.

123. Jones MK, Wang H, Peskar BM, Levin E, Itani RM, Sarfeh IJ *et al.* Inhibition of angiogenesis by nonsteroidal anti-inflammatory drugs: Insight into mechanisms and implications for cancer growth and ulcer healing. *Nat. Med.* 1999;**5**:1418–1423.

124. Takahashi K, Saishin Y, Saishin Y, Mori K, Ando A, Yamamoto S *et al*. Topical nepafenac inhibits ocular neovascularization. *Invest. Ophthalmol. Vis. Sci.* 2003;**44**:409–415.

125. Gamache DA, Graff G, Brady MT, Spellman JM, Yanni JM. Nepafenac, a unique nonsteroidal prodrug with potential utility in the treatment of trauma-induced ocular inflammation: I. Assessment of anti-inflammatory efficacy. *Inflammation* 2000;**24**:357–370.

126. Ke T-L, Graff G, Spellman JM, Yanni JM. Nepafenac, a unique nonsteroidal prodrug with potential utility in the treatment of trauma-induced ocular inflammation: II. In vitro bioactivation and permeation of external ocular barriers. *Inflammation* 2000;**24**:371–384.

127. Dike LE, Ingber DE. Integrin-dependent induction of early growth response genes in capillary endothelial cells. *J. Cell Sci.* 1996;**109**:2855–2863.

128. Luna J, Tobe T, Mousa SA, Reilly TM, Campochiaro PA. Antagonists of integrin alpha-v beta-3 inhibit retinal neovascularization in a murine model. *Lab. Invest.* 1996;**75**:563–573.

129. Umeda N, Kachi S, Akiyama H, Zahn G, Vossmeyer D, Stragies R *et al*. Suppression and regression of choroidal neovascularization by systemic administration of an Alpha5Beta1 integrin antagonist. *Mol. Pharmacol.* 2006;**69**:1820–1828.

130. Vlodavsky I, Folkman J, Sullivan R, Fridman R, Rivka I-M, Sasse J *et al*. Endothelial cell-derived basic fibroblast growth factor: synthesis and deposition into subendothelial extracellular matrix. *Proc. Natl. Acad. Sci. USA* 1987;**84**:2292–2296.

131. Vlodavsky I, Korner G, Ishai-Michaeli R, Bashkin P, Bar-Shavit R, Fuks Z. Extracellular matrix-resident growth factors and enzymes: possible involvement in tumor metastasis and angiogenesis. *Cancer Metastasis Rev.* 1990;**9**:203–226.

132. Park JE, Keller G-A, Ferrara N. The vascular endothelial growth factor (VEGF) isoforms: differential deposition into the subepithelial extracellular matrix. *Mol. Biol. Cell* 1993;**4**:1317–1326.

133. O'Reilly MS, Boehm T, Shing Y, Fukai N, Vasios G, Lane WS *et al*. Endostatin: an endogenous inhibitor of angiogenesis and tumor growth. *Cell* 1997;**88**:277–285.

134. Ramchandran R, Dhanabal M, Volk R, Waterman MJ, Segal M, Lu H *et al*. Antiangiogenic activity of restin, NC10 domain of human collagen XV: comparison to endostatin. *Biochem. Biophys. Res. Comm.* 1999;**255**:735–739.

135. Colorado PC, Torre A, Kamphaus G, Maeshima Y, Hopfer H, Takahashi K *et al*. Anti-angiogenic cues from vascular basement membrane collagen. *Canc. Res.* 2000;**60**:2520–2526.

136. Kamphaus GD, Colorado PC, Panka DJ, Hopfer H, Ramchandran R, Torres A *et al*. Canstatin, a novel matrix-derived inhibitor of angiogenesis and tumor growth. *J. Biol. Chem.* 2000;**275**:1209–1215.

137. Petitclerc C, Boutaud A, Prestayko A, Xu J, Sado Y, Ninomiya Y *et al*. New functions for non-collagenous domains of human collagen type IV. Novel integrin ligands inhibiting angiogenesis and tumor growth in vivo. *J. Biol. Chem.* 2000;**275**:8051–8061.

138. Maeshima Y, Colorado PC, Torre A, Holthaus KA, Grankemeyer JA, Ericksen MB *et al*. Distinct antitumor properties of a type IV collagen domain derived from basement membrane. *J. Biol. Chem.* 2000;**275**:21340–21348.

139. Maeshima Y, Colorado PC, Kalluri R. Two RGD-independent alpha$_v$beta$_3$ integrin binding sites on tumstatin regulate distinct anti-tumor properties. *J. Biol. Chem.* 2000;**275**:23745–23750.

140. Shahan T, Grant D, Tootell M, Ziaie Z, Ohno N, Mousa SA *et al*. Oncothanin, a peptide from the alpha 3 chain of type IV collagen, modifies endothelial cell function and inhibits angiogenesis. *Connect. Tissue Res.* 2004;**45**:151–163.

141. Oberbaumer I, Wiedemann H, Timpl R, Kuhn K. Shape and assembly of type IV procollagen obtained from cell culture. *EMBO J.* 1982;**1**:805–810.

142. Sundaramoorty M, Meiyappan M, Todd P, Hudson BG. Crystal structure of NC1 domains. Structural basis for type IV collagen assembly in basement membranes. *J. Biol. Chem.* 2002;**277**:31142–31153.

143. Ortega N, Werb Z. New functional roles for noncollagenous domains of basement membrane collagens. *J. Cell Sci.* 2002;**115**:4201–4214.

144. Shen GQ, Butkowski R, Cheng T, Wieslander J, Katz A, Cass J *et al*. Comparison of non-collagenous type IV collagen subunits in human glomerular basement membrane, alveolar basement membrane, and placenta. *Connect. Tissue Res.* 1990;**24**:289–301.

145. Hudson BG, Reeders ST, Tryggvason K. Type IV collagen: structure, gene organization, and role in human diseases. Molecular basis of Goodpasture and Alport sydromes and diffuse leiomyomatosis. *J. Biol. Chem.* 1993;**268**:26033–26036.

146. Miner JH, Sanes JR. Collagen IV alpha 3, alpha 4, and alpha 5 chains in rodent basal laminae: sequence, distribution, association with laminins, and developmental switches. *J. Cell Biol.* 1994;**127**:879–891.

147. Tanaka K, Iyama K, Kitaoka M, Ninomiya Y, Oohashi T, Sado Y *et al*. Differential expression of alpha 1(IV), alpha 2 (IV), alpha 5(IV), and alpha 6 (IV) collagen chains in the basement membrane of basal cell carcinoma. *Histochem. J.* 1997;**29**:563–570.

148. Fleischmajer R, Kuhn K, Sato Y, MacDonald EDn, Perlish JS, Pan TC *et al*. There is temporal and spatial expression of alpha1(IV), alpha2(IV), alpha5(IV), and alpha6(IV) collagen chains and beta 1 integrins during the development of the basal lamina in an "in vitro" skin model. *J. Invest. Dermatol.* 1997;**109**:527–533.

149. Lima e Silva R, Kachi S, Akiyama H, Shen J, Aslam S, Gong YY *et al*. Recombinant non-collagenous domain of alpha2(IV) collagen causes involution of choroidal neovascularization by inducing apoptosis. *J. Cell. Physiol.* 2006;**208**:161–166.

150. Mori K, Ando A, Gehlbach P, Nesbitt D, Takahashi K, Goldsteen D *et al*. Inhibition of choroidal neovascularization by intravenous injection of adenoviral vectors expressing secretable endostatin. *Am J Pathol* 2001;**159**:313–320.

151. Mori K, Duh E, Gehlbach P, Ando A, Takahashi K, Pearlman J *et al*. Pigment epithelium-derived factor inhibits retinal and choroidal neovascularization. *J. Cell. Physiol.* 2001;**188**:253–263.

152. Mori K, Gehlbach P, Ando A, McVey D, Wei L, A. CP. Regression of ocular neovascularization by increased expression of pigment epithelium-derived factor. *Invest. Ophthalmol. Vis. Sci.* 2001;**43**:2428–2434.

153. Mori K, Gehlbach P, Yamamoto S, Duh E, Zack DJ, Li Q *et al*. AAV-mediated gene transfer of *pigment epithelium-derived factor* inhibits choroidal neovascularization. *Invest. Ophthalmol. Vis. Sci.* 2002;**43**:1994–2000.

154. Auricchio A, Behling KC, Maguire AM, O'Conner EM, Bennett J, Wilson JM *et al*. Inhibition of retinal neovascularization by intraocular viral-mediated delivery of anti-angiogenic agents. *Mol. Ther.* 2002;**6**:490–494.

155. Lai C-C, Wu W-C, Chen S-L, Xiao X, Tsai T-C, Huan S-J *et al*. Suppression of choroidal neovascularization by adeno-associated virus vector expressing angiostatin. *Invest. Ophthalmol. Vis. Sci.* 2001;**42**:2401–2407.

156. Takahashi T, Nakamura T, Hayashi A, Kamei M, Nakabayashi M, Okada AA *et al*. Inhibition of experimental choroidal neovascularization by overexpression of tissue inhibitor of metalloproteinases-3 in retinal pigment epithelium. Amer. J. Ophthalmol. 2000;**130**:774–781.

157. Hangai M, Moon YS, Kitaya N, Chan CK, Wu D-Y, Peters KG *et al*. Systemically expressed soluble Tie2 inhibits intraocular neovascularization. *Human Gene Ther.* 2001;**12**:1311–1321.

158. Honda M, Sakamoto T, Ishibashi T, Inomata H, Ueno H. Experimental subretinal neovascularization is inhibited by adenovirus-mediated soluble VEGF.flt-1 receptor gene transfection: a role of VEGF and possible treatment for SRN in age-related macular degeneration. *Gene Ther.* 2000;**7**:978–985.

159. Lai C-M, Brankov M, Zaknich T, Lai YK-Y, Shen W-Y, Constable IJ *et al*. Inhibition of angiogenesis by adenovirus-mediated sFlt-1 expression in a rat model of corneal neovascularization. *Human Gene Ther.* 2001;**12**:1299–1310.

160. Gehlbach P, Demetriades AM, Yamamoto S, Deering T, Xiao WH, Duh EJ *et al*. Periocular gene transfer of sFlt-1 suppresses ocular neovascularization and VEGF-induced breakdown of the blood-retinal barrier. *Hum. Gene Ther.* 2003;**14**:129–141.

161. Watanabe K, Hasegawa Y, Yamashita H, Shimizu K, Ding Y, Abe M *et al*. Vasohibin as an endothelium-derived negative feedback regulator of angiogenesis. *J. Clin. Invest.* 2004;**114**:898–907.

162. Shen J, Yang XR, Xiao WH, Hackett SF, Sato Y, Campochiaro PA. Vasohibin is up-regulated by VEGF in the retina and suppresses VEGF receptor 2 and retinal neovascularization. *FASEB J.* 2006;**20**:723–725.

163. Campochiaro PA, Nguyen QD, Shah SM, Klein ML, Holz E, Frank RN *et al*. Adenoviral vector-delivered pigment epithelium-derived factor for neovascular age-related macular degeneration: results of a phase I clinical trial. *Hum. Gene Ther.* 2006;**17**:167–176.

164. Lima e Silva R, Saishin Y, Saishin Y, Akiyama H, Kachi S, Aslam S *et al*. Suppression and regression of choroidal neovascularization by polyamine analogs. *Invest. Ophthalmol. Vis. Sci.* 2005;**46**:3323–3330.

165. Lima e Silva R, Kachi S, Akiyama H, Shen J, Hatara MC, Aslam S *et al*. Trans-scleral delivery of polyamine analogs for ocular neovascularization. *Exp. Eye Res.* 2006;**83**:1260–1267.

166. Hill SA, Sonergan SJ, Denekamp J, Chaplin DJ. Vinca alkaloids: anti-vascular effects in a murine tumor. *Eur. J. Cancer* 1993;**29A**:1320–1324.

167. Seed S, Slaughter DP, Limarzi LR. Effect of colchicine on human carcinoma. *Surgery* 1940;**7**:696–709.

168. Pettit GR, Singh SB, Hamel E, Lin CM, Alberts DS, Garia-Kendall D. Isolation and structure of the strong cell growth and tubulin inhibitor combretastatin A4. *Experientia* 1989;**45**:205–211.

169. Woods JA, Hatfield JA, Pettit GR, Fox BW, McGown AT. The interaction with tubulin of a series of stilbenes based on combrestastatin A-4. *Br. J. Cancer* 1995;**71**:705–711.

170. Dark GD, Hill SA, Prise VE, Tozer GM, Pettit GR, Chaplin DJ. Combretastatin A-4, an agent that displays potent and selective toxicity toward tumor vasculature. *Canc. Res.* 1997;**57**:1829–1834.

171. Pettit GR, Temple C, Narayanan VL, Varma R, Simpson MJ, Boyd MR *et al*. Antineoplastic agents 322. Synthesis of combretastatin A-4 prodrugs. *Anti-Cancer Drug Design* 1995;**10**:299–309.

172. Nambu H, Nambu R, Melia M, Campochiaro PA. Combretastatin A-4 Phosphate Suppresses Development and Induces Regression of Choroidal Neovascularization. *Invest. Ophthalmol. Vis. Sci.* 2003;**44**:3650–3655.

173. Ando A, Mori K, Yamada H, Yamada E, Takahashi K, Saikia J *et al*. Nitric oxide is proangiogenic in retina and choroid. *J. Cell Physiol.* 2001;**191**:116–124.

174. Ando A, Yang A, Nambu H, Campochiaro PA. Blockade of nitric-oxide synthase reduces choroidal neovascularization. *Mol. Pharmacol.* 2002;**62**:539–544.

Chapter 45
Angiogenesis and Pathology in the Oral Cavity

Luisa A. DiPietro

Keywords: angiogenesis, teeth, periodontium, periodontal disease, dental pulp, pyogenic granuloma, gingiva, wound healing, orthodontics

Abstract: This chapter discusses the role of angiogenesis in non-neoplastic pathologies that affect the dentition and its supporting structures. Teeth and the tissues that support the teeth frequently respond to insults such as injury, infection, and orthodontic stress with the development of a robust proangiogenic environment. Yet aside from malignancies, the role of angiogenesis in pathologies of the oral cavity has not received widespread attention. Several recent studies suggest a role for angiogenesis in the development of diseases of the periodontium and the dental pulp. The evidence also supports the emerging concept that the severity of diseases such as chronic periodontitis depends upon individualized host responses.

Introduction

The tissues of the oral cavity represent some of the most highly vascularized areas of the body, and the robust vascularity of the oral mucosa has been suggested to support the greater proliferative capacity of the oral mucosa as compared to skin. The precise level of vascularity varies among the various anatomical locations within the oral cavity. Blood flow to oral tissues such as gingiva, tongue, lips, and dental pulp is known to be significantly greater than that of skin [1–4]. The highest level of blood flow appears to occur within the free gingiva (the portion of the gum tissue that surrounds the tooth but is not directly attached to the tooth) where flow has been recorded at levels more than ten-fold greater than that of skin. In contrast, the palatal mucosa is not particularly heavily vascularized, and blood flow to the hard and soft palate is quite similar to skin [1]. In addition to regionally high vascularity,

Center for Wound Healing and Tissue Regeneration, College of Dentistry, University of Illinois at Chicago, Chicago, IL, USA

the mucosa of the oral cavity is bathed in saliva. Saliva is rich in many growth factors, including those that can support epithelial proliferation and health, such as EGF and TGF-α [5,6]. Saliva also contains biological levels of several potent proangiogenic factors including FGF-2, VEGF, and TGF-β; these factors have been suggested to be involved in tissue repair or other pathologic processes in the mouth [7,8].

The role of angiogenesis in the development of oral cancer has been the subject of intensive investigation, and the link between oral cancer and angiogenesis is firmly established [9]. Beyond neoplasms, though, the role of angiogenesis in pathology that is specific to the tissues of the oral cavity has received only minimal experimental attention. The focus of this chapter is on the regulation and significance of angiogenesis in pathologies that affect the dentition and its supporting structures.

The Periodontium

An association between angiogenesis and pathologies that affect the periodontium, or the tissues that surround the teeth, has been demonstrated for several of the diseases that affect these supporting tissues. Links between angiogenesis and periodontal pathology are of particular interest, since diseases of the periodontium are an important cause of tooth loss in the United States [10]. Recently, interactions of periodontal disease with systemic diseases including cardiovascular disease and stroke have been described, providing new-found significance to understanding the pathogenesis of periodontal pathologies [11–15]. While an etiologic link between periodontal disease and systemic diseases remains to be proven, the associative data suggests either common susceptibility factors or causation [16]. Several theories exist to explain the described associations, including a possible influence of the oral bacteria that are released from sites of periodontal disease into the systemic circulation. Chronic inflammation within the periodontium has also been hypothesized to create a systemic pro-inflammatory effect that influences tissue responses at other locations [17,18]. This emerging data underscores the need for more information about the pathologic processes that define periodontal disease, including the role of angiogenesis. Such knowledge may be important in identifying the link between systemic and oral health.

Chronic Periodontitis

One of the best studied associations of angiogenesis and oral disease is its relationship to chronic periodontal disease, or chronic periodontitis. Chronic periodontitis, a condition that may affect up to one-half of the US population, is an infectious process that involves the gingiva and bone that surround the teeth [10]. In the earliest phase of the disease, termed gingivitis, the gingiva becomes inflamed in response to bacterial plaque that forms on the teeth. Gingivitis can be reversed by professional treatment and maintenance of adequate oral hygiene. Untreated, though, gingivitis may progress to chronic periodontitis. In chronic periodontitis, a persistent inflammatory state within the gingiva ultimately results in destruction of the connective tissue attachment and the supporting bone around the teeth, and may result in tooth loss. Gingival recession may also occur. Chronic periodontitis generally progresses slowly over many years, although periods of rapid progression may take place.

Early studies of the tissue architecture of the chronic periodontitis demonstrated the highly vascular nature of these lesions in both animal models and humans, and increased vascularity has been described to be associated with disease progression [19–22]. Beyond changes in vascular density, the vascular structure appears altered within the periodontal lesion, with thickened vascular basement membranes. Chronic periodontitis is an inflammatory disease, and increased vascularity could participate in the pathogenesis in more than one way. Increased vascularity might assist in the delivery of leukocytes and serum derived pro-inflammatory mediators to the lesion. Moreover, the vessels within the chronic periodontal lesion have been described to be highly permeable, a situation that may increase tissue edema and promote tissue destruction. Interestingly, one mechanism for this increased permeability may be the degradation of platelet endothelial cell adhesion molecule 1 (PECAM-1), as virulence factors that are derived from specific periodontal pathogens, such as *Porphyromonas gingivalis,* have been shown to degrade this molecule [23].

Angiogenesis in chronic periodontitis is believed to be initiated by the over-production of proangiogenic mediators within the inflamed gingival tissue. The production of proangiogenic mediators, such as vascular endothelial growth factor (VEGF), interleukin-8 (IL-8), placenta growth factor (PlGF) and hepatocyte growth factor (HGF), have been described to be increased at sites of moderate to severe periodontitis [24–27]. The best studied of the proangiogenic factors that might play a role in periodontal disease is VEGF. VEGF levels have been measured in tissues from normal, mild, moderate, and severe periodontal disease in samples from human subjects by the enzyme-linked immunosorbent assay (ELISA) and using immunohistochemical techniques [24,28–30]. Little or

no VEGF was reported to be present in normal, uninflamed gingiva [24,30]. In contrast, VEGF is prominently expressed at sites of periodontal disease. In a study by Johnson et al., the greatest level of VEGF was found in those lesions defined as moderate, while the greatest vessel density occurred in the most severe lesions [24]. This result may suggest that the specific pattern of VEGF induction occurs as the disease develops and matures. A link between VEGF levels and disease susceptibility has also been suggested. The examination of periodontally diseased tissue from diabetic patients, a population in whom periodontal disease is generally more severe, demonstrated increased VEGF expression in diabetic versus non-diabetic patients [30]. The profiling of cytokines and growth factors in gingival fluid derived from the site of disease is currently being explored as a diagnostic tool for disease severity and prognosis is an area of active investigation [27,31–35].

VEGF is produced by several cell types within the diseased periodontium, including inflammatory cells such as neutrophils, monocytes, macrophages and plasma cells, and the epithelium in the area of the lesion. Interestingly, the epithelium that lines the pocket adjacent to the tooth appears to produce less VEGF than the oral epithelium [29,30]. VEGF can also be detected in the fluid in the periodontal pocket surrounding the tooth, termed gingival crevicular fluid (GCF). GCF from healthy subjects contains detectable yet low levels of VEGF. In patients with periodontal disease, GCF from both healthy and diseased sites contains increased VEGF, with levels of up to 250 pg/ml of GCF [28]. These studies indicate that, in persons with chronic periodontitis, even apparently healthy sites may exhibit altered VEGF production. Whether this predicts disease progression remains speculative.

Vascular remodeling seems to be a prominent in the progression of chronic periodontal disease, as an accumulation of basement membrane remnants of probably vascular origin is seen within advanced lesions [29]. The numerical vascular density is known to increase in chronic periodontitis, with a selective increase in vessels that are $\geq 25\,\mu m$ in diameter, suggesting that vascular enlargement is a feature of the angiogenic process in this disease [29].

The stimuli that induce VEGF production at sites of chronic periodontal disease are probably multiple. A wide variety of proinflammatory mediators are produced within periodontal lesions, including many that might stimulate the production of VEGF. Many of the proinflammatory cells that are prominent in the chronic periodontitis lesion, including macrophages and neutrophils, can produce VEGF [36,37]. Periodontal pathogens themselves may stimulate the production of proangiogenic factors. In *in vitro* studies, stimulation of gingival fibroblasts by vesicle or outer membrane protein components from either *Actinobacillus actinomycetemcomitans* or *P. gingivalis* resulted in enhanced expression of VEGF [38]. The expression of hepatocyte growth factor has been reported to increase in gingival fibroblasts exposed to *P. gingivalis* components [39]. Exposure to mechanical stress is also capable of inducing the production of VEGF by both gingival and periodontal ligament fibroblasts [40]. Gingival fibroblasts produced pigment-epithelium derived factor in response to stimulation, suggesting that these cells are capable of shifting the angiogenic balance by producing both pro and antiangiogenic factors.

Overall, the evidence that an increase in angiogenesis and vascular permeability occurs in tandem with the progression of chronic periodontitis is strong, as the association has been described in multiple studies and across several model systems. The tissue response to periodontal pathogens has been suggested to involve not only inflammation, but also elements akin to wound repair. Periodontal disease includes a significant level of tissue destruction and seems likely to elicit a repair response [28,38]. Features found to activate angiogenic mechanisms in healing wounds, such as hypoxia and/or redox imbalances, could then also play a role in the induction of angiogenesis in chronic periodontitis. Taken together, the currently available literature suggests that the enhanced angiogenesis that occurs in chronic periodontitis represents a tissue response to inflammatory stimuli and tissue injury. As mentioned above, VEGF is detectable in the fluid from healthy sites, leading to speculation that VEGF plays an important reparative and perhaps protective role in healthy gingiva. One piece of data that supports this concept comes from studies of aged persons [28]. The level of VEGF in GCF at sites of chronic periodontitis is significantly lower in aged subjects, suggesting that diminished production of VEGF, and thus loss of its reparative function, might be a factor in disease promotion over long time periods. This possibility is the opposite of the prevailing theory that VEGF fuels the disease process via the promotion of angiogenesis, the enhanced delivery of additional inflammatory cells and mediators, and/or by other undiscovered mechanisms. Therapeutically, chronic periodontitis is generally treated by a combination of approaches aimed at both the reduction of microbial colonization and the surgical correction of bony and soft tissue defects caused by the disease. Of interest is that some of the anti-microbials that may be utilized therapeutically, such as tetracycline analogs, are also capable of inhibiting angiogenesis [41]. Whether a purely anti-angiogenic approach would be of benefit in the treatment of chronic periodontitis remains an unknown.

Drug Induced Gingival Enlargement

Drug-induced gingival enlargement (GE), or gingival hyperplasia secondary to drugs, is a generalized growth of the fibrous and epithelial components of the gingiva that can occur in patients who are chronically taking certain types of drugs, including anticonvulsants, immunosuppressants, and calcium channel blockers. The pathogenesis of this disease is unclear, although local, genetic, and systemic factors may influence its development [42,43].

Several pieces of evidence suggest that a proangiogenic switch may be a feature of this disease. In a study involving 27 patients, gingival overgrowth tissue from patients taking phenytoin was found to exhibit significantly increased levels of the

proangiogenic factor connective tissue growth factor (CTGF) [44]. Platelet-derived growth factor has also been reported to be increased in drug-induced GE tissue as compared to control [45]. Immunohistochemical studies of GE tissue from patients suggest that GE lesions contain an increase in the number of cells that are positive for TGF-β and fibroblast growth factor-2 (FGF-2), as well as their receptors [46]. In tandem with increases in proangiogenic factors, a number of studies have reported an increase in vascularity in gingival connective tissue of GE patients, although this increase is not necessarily uniform throughout the lesion [44,47,48]. More recent studies have utilized a rat model of cyclosporine-induced gingival enlargement to examine the angiogenic profiles of lesional tissues. Using reverse transcriptase PCR (RT-PCR) analysis, many angiogenic mediators, including the proangiogenic factors angiopoietin 1 (Ang1) and FGF-2, as well as the angiostatic factors angiopoietin 2 (Ang2), brain-specific angiogenesis inhibitor 1 and 2, and thrombospondin 1 (TSP1), were reported to be unchanged in lesional tissue as compared to control. The angiostatic protein thrombospondin 2 (TSP2), however, was reported to be decreased in lesions. A GE associated decrease in levels of TSP2 was confirmed in a small sample of GE patients, as tissues from this group showed an average level of TSP2 that was just 42% of control. Studies of VEGF production in the rat cyclosporine-induced GE model have produced somewhat conflicting results. Koh et al. reported no change in VEGF mRNA levels in GE tissues [49] while Cetinkaya et al. [50] described both significantly increased VEGF protein as detected by ELISA, and a correlation of increased VEGF with an increase in the number of blood vessel profiles in the GE tissues.

GE tissue is known to contain both a component of tissue overgrowth and an inflammatory component that is secondary to microbial dental plaque. As suggested by Iacopino, reparative macrophages, a cell type known to promote tissue growth, may dominate in GE tissues [45]. Similar to chronic periodontitis (described above), the current data concerning GE and angiogenesis is strongly supportive of an association of increased angiogenesis with GE, yet currently without a clear indication of causative relationships.

Pyogenic Granuloma

Pyogenic granuloma is a benign inflammatory lesion that occurs on the gingiva, and is more common in females who are pregnant or near puberty. Pyogenic granulomas generally are bright red in appearance, and may develop in response to an irritant. The lesion may develop as a part of a wound repair process that has been modified by hormonal changes [51]. Histologically, pyogenic granulomas exhibit prominent capillary growth within a hyperplastic granulation tissue, suggesting an angiogenic component. Immunohistochemical assessments to compare healthy gingiva to pyogenic granuloma showed that angiogenic factors, including VEGF and FGF-2, were increased in these lesions, while the angiogenesis inhibitor angiostatin was reduced [52,53]. VEGF was localized to macrophages and fibroblasts, while FGF-2 has been described to be prominent in macrophages, mast cells, and within the extracellular matrix of the lamina propria. Given the vascular nature of this lesion, it is not surprising that several other factors known to play roles in angiogenesis, including Ang1 and Ang2, the angiopoietin receptor Tie2, EphB4 and ephrinB2 (one of its ligands) have also been localized in gingival pyogenic granulomas [54]. Many pyogenic granulomas resolve when the irritant is removed or following parturition, and an associated drop in VEGF levels has been observed as the lesions resolve [54]. For recurring lesions, antiangiogenic therapy has recently been suggested as a possible effective clinical approach [55,56].

Dental Pulp

The dental pulp, a richly vascularized innervated tissue that sits at the core of the tooth, is responsive to cold, heat, and other insults (such as trauma or decay) that may breach the enamel barrier of the tooth. The pulp has limited reparative capabilities, but it can respond to stimuli by inducing the formation of additional dentin, which is the mineralized portion of the tooth that surrounds the pulp. The formation of reparative dentin is thought to be stimulated primarily by sequestered growth factors that are released from the dentin following injury and/or exposure, and the therapeutic use of growth factors to promote dentin repair continues to be explored [for a review, see [57]). Proangiogenic factors, including VEGF and TGF-β, have been demonstrated in dentin itself as well as in healthy dental pulp, and may be important to the induction of reparative processes [58,59]. While limited or moderate angiogenesis may be helpful to dental pulp maintenance and reparative capacity, the induction of a significant level angiogenesis is probably highly detrimental. Anatomically, the dental pulp is poorly compliant, as it sits within the unyielding structure of the surrounding tooth. An increase in vascular density and permeability, though an acceptable reparative response in other tissues, is likely to be extremely deleterious to the dental pulp [60,61]. Robust proangiogenic activity in the pulp may increase pressure and edema within the confined pulpal space, contributing to pain and further tissue damage.

Several pieces of evidence suggest that exuberant production of proangiogenic mediators plays a role in pulpal pathology. In response to injury, fibroblasts from the dental pulp produce soluble proangiogenic mediators FGF-2 and VEGF and adopt a robust proangiogenic phenotype [62,63]. Beyond the stimulus of direct tissue injury, other types of stimuli may also evoke an angiogenic response in the pulp. The bacteria responsible for dental decay appear to be directly capable of stimulating VEGF expression. The lipoteichoic acid component of gram-positive bacteria such as streptococci has been shown to stimulate VEGF expression in both macrophages and pulp cells, suggesting that decay may promote a proangiogenic environment that is ultimately destructive to the pulp [64]. Orthodontic tooth movement also evokes angiogenic changes in dental pulp, and cultured pulps

from orthodontically moved teeth are capable of stimulating an angiogenic response in vitro [65]. The proangiogenic activity of orthodontically stressed pulp has been suggested to include the expression of VEGF, FGF-2, PDGF, and TGFβ [66].

When the pulp is severely inflamed, a condition termed irreversible pulpitis occurs in which the pulp, being spatially restricted, cannot recover. During irreversible pulpitis, the level of expression of VEGF in stromal cells is down-regulated and microvessel density decreases [59]. The decrease in microvessel density in irreversible pulpitis has been suggested to result from failing vascular function, and may signal a critical terminal event for the pulp. Collectively, the currently available information suggests that an appropriate angiogenic balance is critical to pulpal health.

Angiogenesis and Oral Mucosal Wound Healing

Angiogenesis is an early event in tissue repair, providing the wound site with adequate supplies of nutrients, oxygen, and inflammatory cells, while at the same time assuring removal of necrotic tissues. In skin wounds, robust angiogenesis occurs, and capillary density greatly exceeds that of uninjured normal tissue [67,68]. As the wound resolves, many of the newly formed capillaries regress, leaving behind a residual vascularity that is similar to uninjured tissue. Normal oral mucosa is more vascularized than normal skin, and mucosa also heals more quickly and with less incidence of scar formation. Remarkably, the pattern of angiogenesis in healing mucosa has been shown to differ from that of skin [69]. In mucosal tissues, the angiogenic response to injury restores vascularity, but never generates a capillary density that is significantly greater than normal tissue. This feature reduces the need for capillary regression in healing mucosal wounds, modifying the resolution phase. Unexpectedly, a few recent studies suggest that the angiogenic response in skin wounds may be overly robust, as healing has been relatively unaffected or even improved even when angiogenesis is partially inhibited [70,71]. In one study, pharmacologic inhibition of angiogenesis was reported to improve healing outcomes, including scar formation. Whether patterns of angiogenesis that are seen in mucosal wounds are linked to improved healing is not yet known.

Summary and Conclusions

Teeth and the tissues that support the teeth frequently respond to insults such as injury, infection, and orthodontic stress with the development of a robust proangiogenic environment (Fig. 45.1). In some circumstances, this response may be beneficial and part of the normal reparative response. Within the confines of the dental pulp, though, or in the situation of chronic stimuli, the unremitting angiogenic stimulus is probably

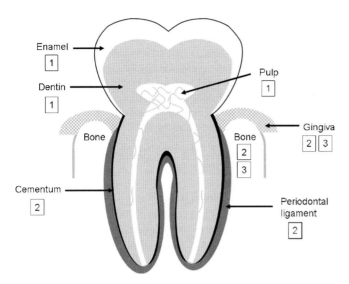

FIG. 45.1. Overview of anatomical structures of the tooth and supporting tissues that may be affected by pathologic angiogenesis. Angiogenic responses to trauma and infection can occur in multiple regions. (*1*) In response to stress, trauma, or infections that breach enamel and/or dentin, cells of the dental pulp can produce proangiogenic mediators that may enhance edema and increase intrapulpal pressure. This response may lead to an irreversible necrosis of the pulp. (*2*) In chronic periodontitis, both epithelial and connective tissue components of the gingiva, as well as fibroblasts within the periodontal ligament, may respond to the chronic local infection by adopting a proangiogenic phenotype. Increased vascularity in the gingiva and periodontal ligament may augment the inflammatory process, thereby sustaining the loss of cementum, which is the site of attachment of tooth to the periodontal ligament. The situation can proceed to irreversible pathologic bone loss. (*3*) In drug induced gingival enlargement, gingival cells proliferative in response to systemic drugs. The cells of the gingiva may be stimulated to produce proangiogenic mediators. The hypertrophy of the gingiva creates an anatomical situation that prevents adequate oral hygiene, creating an environment that may lead to bone loss. In each of the three circumstances, inflammatory cells that migrate to the dntal complex may also support the proangiogenic response.

highly detrimental and contributes to the observed pathology. Critical questions remain, including the cell types involved and the mechanisms of response to stressors in teeth and their supporting structures. At the present time, then, the difficult question of whether increased angiogenesis is fundamental to the development of pathologies in the dental pulp and periodontium remains unanswered, and more information about the contribution of the angiogenic response to the pathogenesis of such diseases of teeth is needed. Such knowledge may lead to improved therapeutic approaches for these extremely common diseases.

Acknowledgments. The author wishes to thank Drs. Joseph Califano and Bradford Johnson for critical reading of the manuscript.

References

1. Squier CA, Nanny D. Measurement of blood flow in the oral mucosa and skin of the rhesus monkey using radiolabelled microspheres. Arch Oral Biol 1985;30(4):313–8.
2. Meyer MW. Distributionof cardiac output to oral tissues in dogs. J Dent Res 1970;49(4):787–94.
3. Kaplan ML, Davis MA, Goldhaber P. Blood flow measurements in selected oral tissues in dogs using radiolabelled microspheres and rubidium-86. Arch Oral Biol 1978;23(4):281–4.
4. Kaplan ML, Jeffcoat MK, Goldhaber P. Blood flow in gingiva and alveolar bone in beagles with periodontal disease. J Periodontal Res 1982;17(4):384–9.
5. Zelles T, Purushotham KR, Macauley SP, Oxford GE, Humphreys-Beher MG. Saliva and growth factors: the fountain of youth resides in us all. J Dent Res 1995;74(12):1826–32.
6. Tabak LA. In defense of the oral cavity: the protective role of the salivary secretions. Pediatr Dent 2006;28(2):110–7; discussion 92–8.
7. Taichman NS, Cruchley AT, Fletcher LM, et al. Vascular endothelial growth factor in normal human salivary glands and saliva: a possible role in the maintenance of mucosal homeostasis. Lab Invest 1998;78(7):869–75.
8. Pammer J, Weninger W, Mildner M, Burian M, Wojta J, Tschachler E. Vascular endothelial growth factor is constitutively expressed in normal human salivary glands and is secreted in the saliva of healthy individuals. J Pathol 1998;186(2):186–91.
9. Lingen M, Sturgis EM, Kies MS. Squamous cell carcinoma of the head and neck in nonsmokers: clinical and biologic characteristics and implications for management. Curr Opin Oncol 2001;13(3):176–82.
10. Albandar JM, Brunelle JA, Kingman A. Destructive periodontal disease in adults 30 years of age and older in the United States, 1988–1994. J Periodontol 1999;70(1):13–29.
11. Geerts SO, Nys M, De MP, et al. Systemic release of endotoxins induced by gentle mastication: association with periodontitis severity. J Periodontol 2002;73(1):73–8.
12. Desvarieux M, Demmer RT, Rundek T, et al. Periodontal microbiota and carotid intima-media thickness: the Oral Infections and Vascular Disease Epidemiology Study (INVEST). Circulation 2005;111(5):576–82.
13. Lee HJ, Garcia RI, Janket SJ, et al. The association between cumulative periodontal disease and stroke history in older adults. J Periodontol 2006;77(10):1744–54.
14. Beck JD, Eke P, Heiss G, et al. Periodontal disease and coronary heart disease: a reappraisal of the exposure. Circulation 2005;112(1):19–24.
15. Beck JD, Offenbacher S. Systemic effects of periodontitis: epidemiology of periodontal disease and cardiovascular disease. J Periodontol 2005;76(11 Suppl):2089–100.
16. Offenbacher S, Beck JD. A perspective on the potential cardioprotective benefits of periodontal therapy. Am Heart J 2005;149(6):950–4.
17. Kantarci A, Van Dyke TE. Resolution of inflammation in periodontitis. J Periodontol 2005;76(11 Suppl):2168–74.
18. Kim J, Amar S. Periodontal disease and systemic conditions: a bidirectional relationship. Odontology 2006;94(1):10–21.
19. Kindlova M. Vascular supply of the periodontium in periodontitis. Int Dent J 1967;17(2):476–89.
20. Hock JM, Kim S. Blood flow in healed and inflamed periodontal tissues of dogs. J Periodontal Res 1987;22(1):1–5.

21. Bonakdar MP, Barber PM, Newman HN. The vasculature in chronic adult periodontitis: a qualitative and quantitative study. J Periodontol 1997;68(1):50–8.

22. Zoellner H, Hunter N. Vascular expansion in chronic periodontitis. J Oral Pathol Med 1991;20(9):433–7.

23. Yun PL, Decarlo AA, Chapple CC, Hunter N. Functional implication of the hydrolysis of platelet endothelial cell adhesion molecule 1 (CD31) by gingipains of Porphyromonas gingivalis for the pathology of periodontal disease. Infect Immun 2005;73(3):1386–98.

24. Johnson RB, Serio FG, Dai X. Vascular endothelial growth factors and progression of periodontal diseases. J Periodontol 1999;70(8):848–52.

25. Ohshima M, Fujikawa K, Akutagawa H, Kato T, Ito K, Otsuka K. Hepatocyte growth factor in saliva: a possible marker for periodontal disease status. J Oral Sci 2002;44(1):35–9.

26. Kakimoto K, Machigashira M, Ohnishi T, et al. Hepatocyte growth factor in gingival crevicular fluid and the distribution of hepatocyte growth factor-activator in gingival tissue from adult periodontitis. Arch Oral Biol 2002;47(9):655–63.

27. Sakai A, Ohshima M, Sugano N, Otsuka K, Ito K. Profiling the cytokines in gingival crevicular fluid using a cytokine antibody array. J Periodontol 2006;77(5):856–64.

28. Booth V, Young S, Cruchley A, Taichman NS, Paleolog E. Vascular endothelial growth factor in human periodontal disease. J Periodontal Res 1998;33(8):491–9.

29. Chapple CC, Kumar RK, Hunter N. Vascular remodelling in chronic inflammatory periodontal disease. J Oral Pathol Med 2000;29(10):500–6.

30. Unlu F, Guneri PG, Hekimgil M, Yesilbek B, Boyacioglu H. Expression of vascular endothelial growth factor in human periodontal tissues: comparison of healthy and diabetic patients. J Periodontol 2003;74(2):181–7.

31. Schenkein HA, Genco RJ. Gingival fluid and serum in periodontal diseases. I. Quantitative study of immunoglobulins, complement components, and other plasma proteins. J Periodontol 1977;48(12):772–7.

32. Machtei EE, Dunford R, Hausmann E, et al. Longitudinal study of prognostic factors in established periodontitis patients. J Clin Periodontol 1997;24(2):102–9.

33. Gamonal J, Acevedo A, Bascones A, Jorge O, Silva A. Levels of interleukin-1 beta, −8, and −10 and RANTES in gingival crevicular fluid and cell populations in adult periodontitis patients and the effect of periodontal treatment. J Periodontol 2000;71(10):1535–45.

34. Alpagot T, Bell C, Lundergan W, Chambers DW, Rudin R. Longitudinal evaluation of GCF MMP-3 and TIMP-1 levels as prognostic factors for progression of periodontitis. J Clin Periodontol 2001;28(4):353–9.

35. Ozmeric N. Advances in periodontal disease markers. Clin Chim Acta 2004;343(1–2):1–16.

36. Berse B, Brown LF, Van de Water L, Dvorak HF, Senger DR. Vascular permeability factor (vascular endothelial growth factor) gene is expressed differentially in normal tissues, macrophages, and tumors. Mol Biol Cell 1992;3(2):211–20.

37. Taichman NS, Young S, Cruchley AT, Taylor P, Paleolog E. Human neutrophils secrete vascular endothelial growth factor. J Leukoc Biol 1997;62(3):397–400.

38. Suthin K, Matsushita K, Machigashira M, et al. Enhanced expression of vascular endothelial growth factor by periodontal pathogens in gingival fibroblasts. J Periodontal Res 2003;38(1):90–6.

39. Uehara A, Muramoto K, Imamura T, et al. Arginine-specific gingipains from Porphyromonas gingivalis stimulate production of hepatocyte growth factor (scatter factor) through protease-activated receptors in human gingival fibroblasts in culture. J Immunol 2005;175(9):6076–84.

40. Yoshino H, Morita I, Murota SI, Ishikawa I. Mechanical stress induces production of angiogenic regulators in cultured human gingival and periodontal ligament fibroblasts. J Periodontal Res 2003;38(4):405–10.

41. Sapadin AN, Fleischmajer R. Tetracyclines: nonantibiotic properties and their clinical implications. J Am Acad Dermatol 2006;54(2):258–65.

42. Seymour RA, Ellis JS, Thomason JM. Risk factors for drug-induced gingival overgrowth. J Clin Periodontol 2000;27(4):217–23.

43. Trackman PC, Kantarci A. Connective tissue metabolism and gingival overgrowth. Crit Rev Oral Biol Med 2004;15(3):165–75.

44. Uzel MI, Kantarci A, Hong HH, et al. Connective tissue growth factor in drug-induced gingival overgrowth. J Periodontol 2001;72(7):921–31.

45. Iacopino AM, Doxey D, Cutler CW, et al. Phenytoin and cyclosporine A specifically regulate macrophage phenotype and expression of platelet-derived growth factor and interleukin-1 in vitro and in vivo: possible molecular mechanism of drug-induced gingival hyperplasia. J Periodontol 1997;68(1):73–83.

46. Saito K, Mori S, Iwakura M, Sakamoto S. Immunohistochemical localization of transforming growth factor beta, basic fibroblast growth factor and heparan sulphate glycosaminoglycan in gingival hyperplasia induced by nifedipine and phenytoin. J Periodontal Res 1996;31(8):545–55.

47. Wondimu B, Reinholt FP, Modeer T. Stereologic study of cyclosporin A-induced gingival overgrowth in renal transplant patients. Eur J Oral Sci 1995;103(4):199–206.

48. Ayanoglou CM, Lesty C. Cyclosporin A-induced gingival overgrowth in the rat: a histological, ultrastructural and histomorphometric evaluation. J Periodontal Res 1999;34(1):7–15.

49. Koh JT, Kim OJ, Park YS, et al. Decreased expressions of thrombospondin 2 in cyclosporin A-induced gingival overgrowth. J Periodontal Res 2004;39(2):93–100.

50. Cetinkaya BO, Acikgoz G, Ayas B, Aliyev E, Sakallioglu EE. Increased expression of vascular endothelial growth factor in cyclosporin A-induced gingival overgrowth in rats. J Periodontol 2006;77(1):54–60.

51. Esmeili T, Lozada-Nur F, Epstein J. Common benign oral soft tissue masses. Dent Clin North Am 2005;49(1):223–40, x.

52. Yuan K, Jin YT, Lin MT. The detection and comparison of angiogenesis-associated factors in pyogenic granuloma by immunohistochemistry. J Periodontol 2000;71(5):701–9.

53. Murata M, Hara K, Saku T. Dynamic distribution of basic fibroblast growth factor during epulis formation: an immunohistochemical study in an enhanced healing process of the gingiva. J Oral Pathol Med 1997;26(5):224–32.

54. Yuan K, Lin MT. The roles of vascular endothelial growth factor and angiopoietin-2 in the regression of pregnancy pyogenic granuloma. Oral Dis 2004;10(3):179–85.

55. Li VW, Li WW, Talcott KE, Zhai AW. Imiquimod as an antiangiogenic agent. J Drugs Dermatol 2005;4(6):708–17.

56. Ezzell TI, Fromowitz JS, Ramos-Caro FA. Recurrent pyogenic granuloma treated with topical imiquimod. J Am Acad Dermatol 2006;54(5 Suppl):S244–5.

57. Smith AJ. Vitality of the dentin-pulp complex in health and disease: growth factors as key mediators. J Dent Educ 2003;67(6): 678–89.

58. Toyono T, Nakashima M, Kuhara S, Akamine A. Expression of TGF-beta superfamily receptors in dental pulp. J Dent Res 1997;76(9):1555–60.

59. Artese L, Rubini C, Ferrero G, Fioroni M, Santinelli A, Piattelli A. Vascular endothelial growth factor (VEGF) expression in healthy and inflamed human dental pulps. J Endod 2002;28(1): 20–3.

60. Matthews B, Andrew D. Microvascular architecture and exchange in teeth. Microcirculation 1995;2(4):305–13.

61. Heyeraas KJ, Berggreen E. Interstitial fluid pressure in normal and inflamed pulp. Crit Rev Oral Biol Med 1999;10(3): 328–36.

62. Roberts-Clark DJ, Smith AJ. Angiogenic growth factors in human dentine matrix. Arch Oral Biol 2000;45(11): 1013–6.

63. Tran-Hung L, Mathieu S, About I. Role of Human Pulp Fibroblasts in Angiogenesis. J Dent Res 2006;85(9):819–23.

64. Telles PD, Hanks CT, Machado MA, Nor JE. Lipoteichoic acid up-regulates VEGF expression in macrophages and pulp cells. J Dent Res 2003;82(6):466–70.

65. Derringer KA, Jaggers DC, Linden RW. Angiogenesis in human dental pulp following orthodontic tooth movement. J Dent Res 1996;75(10):1761–6.

66. Derringer KA, Linden RW. Vascular endothelial growth factor, fibroblast growth factor 2, platelet derived growth factor and transforming growth factor beta released in human dental pulp following orthodontic force. Arch Oral Biol 2004;49(8): 631–41.

67. Knighton DR, Silver IA, Hunt TK. Regulation of wound-healing angiogenesis-effect of oxygen gradients and inspired oxygen concentration. Surgery 1981;90(2):262–70.

68. Swift ME, Kleinman HK, DiPietro LA. Impaired wound repair and delayed angiogenesis in aged mice. Lab Invest 1999;79(12):1479–87.

69. Szpaderska AM, Walsh CG, Steinberg MJ, DiPietro LA. Distinct patterns of angiogenesis in oral and skin wounds. J Dent Res 2005;84(4):309–14.

70. Bloch W, Huggel K, Sasaki T, et al. The angiogenesis inhibitor endostatin impairs blood vessel maturation during wound healing. Faseb J 2000;14(15):2373–6.

71. Nanney LB, Wamil BD, Whitsitt J, et al. CM101 stimulates cutaneous wound healing through an anti-angiogenic mechanism. Angiogenesis 2001;4(1):61–70.

Chapter 46
Revascularization of Wounds:
The Oxygen-Hypoxia Paradox

Thomas K. Hunt[1], Michael Gimbel[2], and Chandan K. Sen[3]

Keywords: wound revascularization, inflammation, reactive oxygen species, hypoxia, lactate accumulation, remodeling

Abstract: Wound angiogenesis is rapid, accessible and controllable. Sophisticated analysis in only the last 10 years has found more and more similarities to tumor angiogenesis. One of the most striking of these similarities is that lactate production is high in both tumors and wounds, and lactate accumulation is a common feature. As it is in tumors, lactate production in wounds is not a function of hypoxia, and lactate accumulation in tissue instigates angiogenesis despite the presence of oxygen. The rate of angiogenic response is proportional to local oxygen concentration. Despite appearances, this data expands our concept of angiogenesis of all sorts. Redox regulation is an important feature.

Introduction

Injury incites the most rapid, the most definable, the most experimentally accessible neovascularization known. In particular, the "angiogenic" extracellular environment of wounds is remarkably uniform and easy to access, measure, and control. Healing wounds make excellent experimental models that can provide otherwise unavailable insights into the nature of angiogenesis. This potential has been exploited on almost countless occasions with one dressing or another, a growth factor or cytokine or two, iodine, hypochlorous acid, etc., often with financial, as opposed to scientific, motivation. Most have merely enhanced inflammation. Within this profusion of

detail, however, there are threads of continuity that comprise the basis of angiogenesis itself. This chapter is an attempt to dissect out the core process of wound angiogenesis, leaving behind, for the sake of clarity, an enormous mass of details. Recent studies have begun to unify a mechanistic concept of the angiogenic process that has a place for inflammation but does not depend upon it.

A few properties of wound angiogenesis are universally recognized. It is a sequence with a clear-cut beginning, middle, and end. It begins with bleeding, when angiogenic blood and coagulant proteins are spilled into the site. A flood of growth and angiogenic factors follows, provided by coagulation proteins, inflammatory cells, and metabolic by-products. The wound environment is conditioned to a moderate hypoxia, an accumulation of lactate, a slight lowering of pH, and an elevation of pCO_2. Inflammatory cells, fibroblasts and endothelial cells accumulate, thus increasing the requirement for blood supply that usually exceeds that of the unwounded site. New vessels respond to these changes. In particular, they respond to the development of a metabolic need occasioned by inflammation. When the needs of healing are met and the wound closes, the new vessels regress, leaving a vascularized fibrous scar containing a collagenous matrix that persists in proportion to the degree to which inflammation has occurred and the time that has been required for healing. In short, wound angiogenesis is a response to coagulation, inflammation, and temporary metabolic need.

The concept of metabolic conditioning for angiogenesis was articulated clearly about 20 years ago [1, 2]. However, the statement that blood vessels go where they are needed is attributed to John Hunter in the eighteenth century.

Wounds are not the only sites of metabolic need, and there is no reason to suspect that the new vessels that respond to inflammation, hypoxia, and lactate accumulation are basically different from those that occur in tumors and physical conditioning. The concepts rising from wound angiogenesis, therefore, appear to bear on other important human disease processes such as tumors, arteriosclerosis, and complications of diabetes.

[1] Department of Surgery, University of California, San Francisco, CA, USA

[2] Department of Plastic Surgery, University of Pittsburgh, PA, USA

[3] Comprehensive Wound Center, The Ohio State University Medical Center, Columbus, OH, USA

Stimuli to Wound Revascularization

Wounds Develop a Metabolic Need

Voluminous evidence demonstrates that vessels grow to meet the needs of the tissues that they will come to serve, whether those tissues are benign or malignant, injured or healthy, fetal or adult. Vessels grow, for instance, to meet the needs of physical conditioning, growth and development, wound healing, and inflammation, all of which exhibit a degree of hypoxia and lactate accumulation. This hypoxia has tantalized investigators for many years and, for a while, the discovery of HIF, the hypoxia inducible transcription factor, and the so called "hypoxia inducible genes" that it induces, intensified the focus on hypoxia as the single most pertinent metabolic signal for neovascularization [3].

In a narrow sense, no issue can be taken with hypoxic initiation of angiogenic signals. However, no matter how intuitively attractive it is, the concept that hypoxia instigates angiogenesis itself, i.e., the actual growth of new vessels, is flawed. Newer studies have demonstrated that metabolic need, transmitted by lactate accumulation, is a more potent instigator, and a more appropriate signal because growth of new vessels requires oxygen! Furthermore, the nature of the evidence strongly suggests that this mechanism applies to angiogenesis of all types and may resolve current uncertainties about the nature of "aerobic angiogenesis."

The data behind the concept of hypoxia-driven angiogenic signaling, while beyond dispute, does not warrant a conclusion that hypoxia is required or even useful at any point in the actual growth of new vessels. Even if the initial signals for new vessel formation were to be derived from hypoxia, there would remain the vital issue of oxygen requirements for endothelial tube formation, vessel maturation, and growth to the point at which the new vessels can carry sufficient blood to satisfy metabolic needs. Taking into account the entire sequence from initiation of vessels to their maturity, it becomes clear that an angiogenic effector other than hypoxia must exist. Furthermore, wound healing, taken as a whole, requires oxygen at all but its earliest phase [4, 5].

Absolute requirements for oxygen arise in several points along the angiogenic sequence. For instance, all vessels require a net or sheath of extracellular matrix, mainly collagen and proteoglycans, to guide tube formation and resist the pressures of blood flow. Conditions for collagen deposition and polymerization can be created only if molecular oxygen is available to be incorporated into the structure of nascent collagen by prolyl- and lysyl-hydroxylases. Without the obligatory extracellular, hydroxylated collagen, new capillary tubes assemble poorly and remain fragile [6–8]. This has a convincing clinical correlate in scurvy, i.e., ascorbate deficiency. Ascorbate is required as a co-factor for insertion of molecular oxygen into procollagen that enables the new collagen molecules to escape to the extracellular space and provide the strength that new vessels need. In scurvy, therefore, the collagenous sheath cannot form because oxygen cannot be incorporated. Consequently, new vessels fail to mature, older vessels weaken and break, and wounds fail to heal [6]. Thus, while hypoxia is a proved instigator of molecular signals for angiogenesis, it is also a proved enemy of vessel growth itself.

Furthermore, collagen deposition proceeds in direct proportion to molecular oxygen concentration (PO_2) across the entire physiologic range, from zero to hundreds of mm Hg. The Km for oxygen for the prolyl hydroxylase reaction is approximately 25 and the Vmax about 250 mm Hg, suggesting that new vessels cannot even approach their greatest possible rate of growth until PO_2 in the wound is high, even above the physiologic range [9]. Hence angiogenesis is directly proportional to PO_2 in injured tissues. Hypoxic wounds deposit collagen poorly and become infected easily, both of which are problems of considerable clinical significance [4, 10, 11].

Since oxygen is required for new vessel formation, a messenger other than hypoxia is needed to carry the message of metabolic need to the normal tissues where new vessels originate. Accumulated lactate and elevated peroxide, universal properties of wounds, have that capability. Lactate and peroxide both accumulate in wounds largely, but not entirely, because of their release by leukocytes that enter injury sites in large numbers. They become powerful angiogenic stimulators [7, 12, 13]. Normal resting tissue lactate levels are in the region of 2– 4 mM, higher, of course, in working muscle from which it is also removed more rapidly. The range of lactate concentrations in wounds is about 5–15 mM. In this range, and assuming adequate oxygen, lactate accumulation encourages precursor cell homing, growth factor production, new vessel growth, and matrix deposition [8, 14, 15]. Hydrogen peroxide has also emerged as a potent signal. [16–18]. Wound fluid is the only part of the body known to contain a large amount (0.1–0.3 mM) of hydrogen peroxide [12]. Both phagocytic as well as non-phagocytic NADPH oxidases create peroxide, with lactate as a by-product.

The next fundamental question then becomes how can lactate accumulate in the presence of oxygen? This question was unthinkable until about 10 years ago, when, in a massive conceptual shift, it became clear that lactate can, and commonly does, accumulate in the presence of oxygen. In fact, most of the lactate that is generated in the human body is now understood to be the result of aerobic glycolysis[10]. Extensive evidence shows that inflammatory cells initiate wound angiogenesis. They rely heavily on glycolysis for energy, and are actively glycolytic in aerobic conditions [10]. They are relatively deficient in the pyruvate dehydroxylase complex, and when activated, produce more pyruvate than they can metabolize. The excess pyruvate is converted to lactate via lactate dehydrogenase (LDH) and most is excreted from the cell [19]. In other words, lactate production and accumulation proceed apace even when oxygen is readily available.

Aerobic lactate production is an essential of mammalian life, and there are many aerobic pathways [10]. Warburg made one of the first observations when he noted that malignant cells produce lactate even when well oxygenated [20]. Unfortunately, the inertia of the original concept of lactate as merely a "dead end metabolite" in hypoxia was so great that aerobic glycolysis was dismissed for many years as a unique property of malignancies. Excessive dependence on aerobic glycolysis is one of the most frequent and consistent features of cancers. Even the degree of malignancy, metastatic potential, and the rate of angiogenesis have been related to the rate of glycolysis, hence lactate production [21]. The actual concentrations of lactate that are reached in the heterogeneous tumor environment are difficult to estimate, but the large, often huge, efflux of lactate in the venous drainage of tumors testifies to a high concentration.

Even more contrary to the traditional concept, lactate is not just a "dead end metabolite." It also has an extraordinary number of regulatory functions. The rate of healing varies according to the lactate as well as the oxygen levels. In one experiment, lactate ion (not lactic acid) was added to wound extracellular fluid to raise lactate concentrations from 6 mM in controls to 9 mM in test animals, and it markedly enhanced vascular endothelial growth factor (VEGF), interleukin-1 (IL-1), transforming growth factor (TGF)-β production and collagen deposition [15]. These wounds were normally oxygenated, and the lactate did not affect the pH.

One regulatory action of lactate is to control the activity of collagen prolyl hydroxylase, the enzyme that inserts oxygen into nascent collagen (see above). The rate of this enzyme, and the rate of hydroxylated collagen deposition that depends on it, are proportional to lactate concentration because the high lactate removes an inhibitor from the enzyme [22]. Thus, the enzyme uses oxygen as a substrate and at the same time depends upon high lactate as an activating agent. This leads to the once unthinkable conclusion that *oxygen is not only required for collagen deposition, it cannot even be optimally used for that purpose in the absence of lactate.* Therefore, any realistic concept of neovascularization, whether by angiogenesis or vasculogenesis, must accommodate the need for both lactate and oxygen. (Fig. 46.1)

In summary, inquiry into the nature of the angiogenic environment in wounds has disclosed a mechanism that forces us to look past the limitations of hypoxia to a more useful explanation for the stimulation of the so-called "hypoxia inducible factors." In this chapter, we will cite a significant collection of new information indicating that wound revascularization depends upon lactate, oxygen, and transition metals, and then proceeds through reactive oxygen species (ROS, including peroxides), VEGF, HIF, and a highly compartmentalized redox regulation. We will show how increased oxygenation can and does enhance angiogenesis, and how this finding might apply to important diseases that are based on excessive deposition of vascularized connective tissue. We can even explain some of the inadequacies of the tumor spheroid model, but first, it is best to discuss the types of wound revascularization and what they look like.

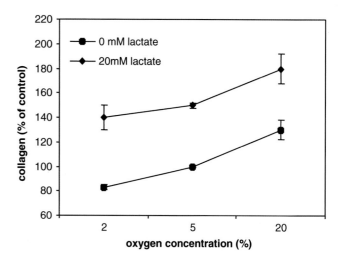

FIG. 46.1. The maximal rate of collagen deposition is not reached until both oxygen and lactate are plentiful. Molecular oxygen supports collagen synthesis and deposition. Redox agents derived from oxygen, often with the aid of lactate are signals for collagen synthesis and angiogenesis. The combination of lactate accumulation and normal or elevated oxygen tension is common in human physiology.

Wound Angiogenesis has Several Faces

Angiogenesis

Angiogenesis is best seen as the red, granular-appearing "granulation tissue" that builds up on full thickness open wounds. Everyone has seen it. This tissue is mainly made of a complex of small vessels in a loose connective tissue matrix. It is best illustrated in rabbit ear wound chambers in which granulation tissue can be examined microscopically in vivo (Fig. 46.2). To do this, a round hole is cut though a rabbit's ear. An apparatus is fixed around and through the hole that guides the healing tissue into and through a space between two optically clear membranes [23]. When the distance between these two surfaces is set at about 100 μm, a characteristic progression of cells can be seen filling the space that is sufficiently thin that one layer of new vessels can be observed. The growing vasculature, and even the flow of red cells, is easily observed by transilluminating the apparatus in a microscope stage. Tissue fluid fills the space within a few days, and is gradually displaced by vascularized fibrous tissue. In surgical terms, this is a "dead space" wound that must be filled with new vessels and new tissue, not just rejoined to functioning vessels as in the healing of uninfected, precisely approximated, incisions.

A low power microscopic view taken 17 days after the device is placed (Fig. 46.3) shows the branching pattern and the advancing edge of crowded small vessels that, in time, remodel into fewer and larger vessels, exactly as the vessels behind it had already done. This is true angiogenesis, i.e., new vessels from old. Some of the vessels at the advancing edge are open and leak red cells. It is not clear from

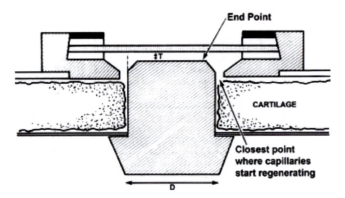

FIG. 46.2. Cross-section of a rabbit ear chamber: The healing edge of the ear is forced to advance over the optically clear polycarbonate block where it can be trans-illuminated. The following illustrations, Figs. 46.2-46.8, were photographed in the progress of healing of one chamber. *End point* refers to experiments in which the time required for the first visible entry point was used to measure the rate of vascularization. Transplantation of autologous macrophages from another wound into the chamber increased the rate.

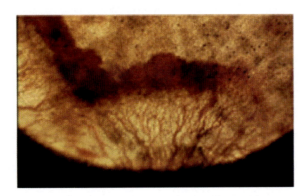

FIG. 46.3. The ear chamber at 17 days shows a cross-section in vivo of the advancing wound edge. This is a model of granulation tissue. True granulation tissue, the new tissue that is exposed to air and contamination, shows more acute inflammation, but the remainder of the process is the same. Note that the tissues closer to the periphery were, a few days before, filled with new vessels exactly like those at the advancing edge. In the process of a few vessels enlarging, there is a constant loss of small ones and probably a huge turnover of endothelial cells.

these photographs whether stem cells participate along with advancing endothelial cells. (See Vasculogenesis section, below). After several more weeks (Fig. 46.4), the healing tissue has filled the space with new connective tissue, and the vasculature has resolved to a few mature vessels.

Figure 46.5 shows the result of gently removing the upper membrane, incising (wounding) the tissue, and replacing the membrane. The site of the incision can be seen as a faint scratch in the underlying polycarbonate block. This illustrates a linear wound that will heal by "first" or "primary" intention, as opposed to the "dead space" wound that was originally created by the implantation of the ear chamber. Fig. 46.5 shows the beginning of neovascularization.

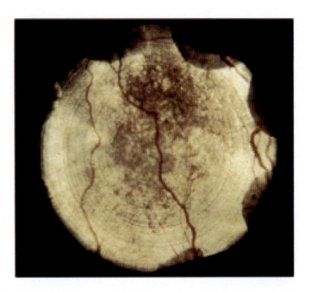

FIG. 46.4. About 3 weeks after Fig. 46.1 was photographed, and almost 6 weeks since the chamber was placed, the wound was healed and all the small vessels had been replaced to the point that few vessels remained. Note the pattern of vessels for comparison to the other Figures.

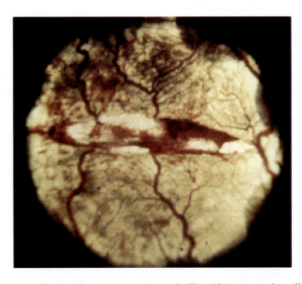

FIG. 46.5. The upper membrane, shown in Fig. 46.1 was gently pulled back and the healed tissue was incised in a plane running across the vessels shown in Fig. 46.3. The membrane was replaced, and this photograph was taken 2 days later. Note the blood clot, the thrombosed larger vessels, and a few new vessels already crossing the wound.

Figure 46.6, photographed 11 days later, shows many small vessels already recruited from adjacent tissues. At this stage, the flow direction in vessels running parallel to the wound edge reversed direction from time to time, presumably when pressure gradients changed. Advancing endothelial cells apparently send out filopodia that recognize each other and bend their courses to join and establish a throughput circulation. The main difference between this and the vasculariza-

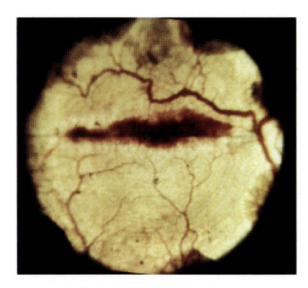

Fig. 46.6. A number of new vessels appeared in the next 3 days, some of them in the original tissue and some across the wound space. Red cells were disappearing. Note that rather large vessels were forming parallel to the incision by connection of new vessels, while others were connecting across it. As Fig. 46.7 shows, this is a provisional circulation. The vessels running parallel to the wound carried considerable flow to those crossing the wound.

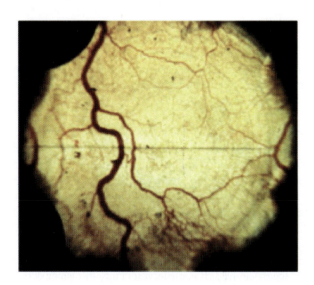

Fig. 46.7. The incised wound had healed and circulation had returned to a pattern similar to that seen in Fig. 46.4. The metabolic need was now minimal.

tion of the dead space is the extent of angiogenesis and the quick union of the two sides of the wound.

Figure 46.7 shows that the new small vessels were disappearing, and Fig. 46.8 shows the final appearance at about 3 weeks. Note the similarity of the final to the original appearance (Fig. 46.3). Note also that some the larger vessels shown at this stage were totally new. This chamber was examined daily, and no evidence of reconnection of the larger vessels

by direct anastomosis or "inosculation" was seen. These vessels started as capillaries and either condensed into fewer and larger tubes or a few of the smaller vessels enlarged under the shear forces of blood flow while the rest disappeared.[1]

The ear chamber also provides a cross-sectional view of angiogenesis. If the top membrane is made of a soft material, for instance a thin polyethylene film that is impermeable to gases and water vapor, microelectrodes can be made to penetrate it and measure a variety of gradients across the leading edge of the new tissue as the advancing wound edge becomes visible in the chamber. This is how Fig. 46.8 was constructed [23]. The membranes can then be removed. The tissue can be fixed, sectioned, and stained, revealing the third dimension that is best shown diagrammatically (Fig. 46.9). The oxygen gradient shown falls usually from about 50 to about 10 mm Hg toward the wound center (depending on its size and the degree of inflammation), and is complemented by a reciprocal gradient of lactate (from about 2 to about 10 or 15 mM). VEGF, TGF-β, IL-1 and other growth factors also accumulate. There is, therefore, reason to speculate that these conditions are hospitable to progenitor cells that could theoretically participate in the formation and growth of the new vessels.

From these findings, one can argue that the mechanism of wound angiogenesis proceeds within the boundaries of this steep gradient from a normal range of oxygen concentration (PO$_2$) at the wound edge to a much lower level in the wound center, where this hypoxia may stimulate the transcription of hypoxia-inducible angiogenic genes [23]. This argues for hypoxia-inducible gene stimulation, which, indeed, seems likely to occur. However, in well-perfused tissue and a linear wound, the PO$_2$, even at its lowest point in a wound, can be increased markedly by elevating arterial PO$_2$; the only consequence being that, instead of reducing angiogenesis, the higher PO$_2$ actually accelerates it [23]. Nevertheless, oxygen gradients provide a convenient rationalization for the appearance of angiogenic growth factors, and almost certainly are operative in many cases. Lactate concentration is high enough to activate procollagen hydroxylation [22], and PO$_2$ is adequate to facilitate mature collagen deposition on the vascularized side of the wound edge where it is highest [9].

Inosculation

The question of whether mid-sized vessels rejoin end-to-end in closely approximated wounds, i.e., inosculate, has been debated. The majority opinion is that they do not. However, by carefully inspecting healing skin autografts a day or two after placement, tiny cyanotic sites appear that can be blanched by

---

[1] All vessels from smallest to the largest can and do enlarge to meet the challenge of increased flow. Pressure and flow precede enlargement. The mechanism is fairly well known and involves nitric oxide, an oxidant messenger and erythropoetins. Reactive oxygen species, as we shall see, contribute to new vessel formation in several ways (see Arteriogenesis section, below).

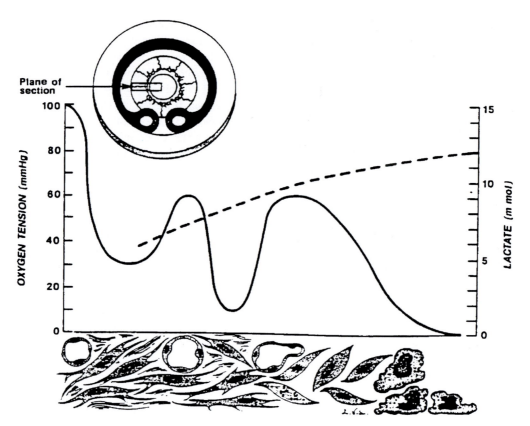

FIG. 46.8. This shows a reconstruction of several healing ear wounds at the stage shown in Fig. 3. Oxygen and lactate levels were determined. The chambers were then removed, fixed and stained, and the gradients matched with positions of the measurement (see text for detailed description). The oxygen and lactate gradients are reciprocal. Note the mitotic figure near the most distal vessel. This is the "growth point." To the *left*, oxygen is sufficient to support vessel growth. To the *right* are the contributors to the lactate and low oxygen that assemble the signals. VEGF and HIF levels were high over the most distal cells. Not shown is a peak of H_2O_2 over the macrophages.

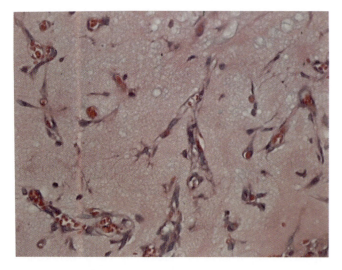

FIG. 46.9. Cross-section of a lactate-rich Matrigel plug showing well formed but uncirculated capillaries consistent with vasculogenesis. On examination of the periphery of this implant, no angiogenesis could be seen to account for this degree of new vessel formation. The tentative conclusion that these vessels are formed from "progenitor cells" has subsequently been confirmed. The erythrocytes are the product of differentiation of the bone marrow stem cells.

light pressure. They become larger every day, and pressure on the center of the cyanotic area blanches the whole of it, indicating that a single source of the preexisting perfusion network has reconnected to preexisting vessels in the underlying tissue. The cyanosis turns to purple and then to the color of arterial blood as these areas coalesce and perfusion increases. The entire graft may be perfused within 10 days also indicating that the circulation is being reused. Given the speed and the pattern, it is hard to escape the conclusion that inosculation must have occurred. It is difficult to study this process in soft tissue, but it has been seen experimentally in bone union [24]. If this process in autografts were not so rapid, we would have little idea that it exists. The majority opinion says that the vessels are reconstituted across the wound edge by new capillaries, and it is clear that the old vessels in homografts disappear and are replaced. There may be a role for inosculation in primarily closed wounds.

Vasculogenesis

Most investigators now agree that vasculogenesis, i.e., vessels arising from recruitment of blood-born progenitor cells, also participates in wound revascularization. Vasculogenesis is a prominent embryological mechanism that involves endothelial-lined vascular "lakes" and their subsequent consolidation to form the embryonic vascular system. The relatively hypoxic, lactate-rich, growth-stimulating conditions that occur in wounds replicate the environment of the embryo and suggest that the wound environment recapitulates embryological phenomena [25, 26].

The simplest demonstration of vasculogenesis is provided by the so-called implanted Matrigel model. Matrigel is a mixture of connective tissue that is liquid when cold, gels on injection, and forms a semi-solid plug that can be removed easily for histological and chemical examination. Injection causes little or no injury or inflammation. The material itself is not inflammatory, and hosts no new blood vessels unless an angiogenic agent is added. If, for instance, VEGF is incorporated, angiogenesis begins within a few days of implantation. The incorporation of a lactate polymer that hydrolyzes at a rate sufficient to maintain a wound-like concentration of lactate monomer also produces rapid vasculogenesis, and this also occurs without an inflammatory reaction [8].

Microscopic examination of plain matrigel plugs with specific stains demonstrates modest numbers of bone marrow progenitor cells within 24 h, but 4-fold more collect if the lactate polymer is added. Lactate and VEGF levels were several-fold higher in the lactate-rich plugs as compared to the controls in which lactate and VEGF were the same as in the normal surrounding tissue. Progenitor cells detected by positive staining for VEGF, stromal cell-derived factor 1 (SDF), CD117, and CD34 were increased from 3- to 6-fold in lactate-rich implants. Figure 46.11 (see below) shows new empty capillaries growing by vasculogenesis in linear patterns in a Matrigel implant in which lactate was elevated. The progenitor cells go on to form a vascular network even when the implants are removed at the end of the first day and placed in aerobic cultures. Some progenitor cells apparently retain their embryologic capacity to differentiate, and erythrocytes appear even before the new vessels connect to the somatic circulation (Fig. 46.9).

In other words, lactate accumulation, by itself, is sufficient to instigate a vasculogenic response from progenitor cells in the presence of oxygen. Accumulating lactate is, therefore, the simplest means of instigating new vessel formation, and perhaps the lowest common denominator of the signaling apparatus. Again, however, oxygen plays a role. If the wound space is made hypoxic due to arterial hypoxia, little vessel growth occurs. The few new vessels that are seen are highly variable in size from small to dilated, have no collagen matrix, and are highly fragile. If arterial hyperoxia is induced, angiogenesis appears in lactate-rich, normal appearing implants as they reach about 30 mm Hg. But if no lactate is added, mature vessels do not appear until arterial PO_2 rises to about 50 mM [7, 8].

The number of circulating progenitor cells is also enhanced by intermittent, hyperbaric hyperoxia in a mechanism that involves nitric oxide [27]. As expected, the transition of progenitor cells into microvessels is impeded by ischemia [28].

Also, as might be expected, progenitor cell expansion in ischemic wounds is greater if animals are exposed to intermittent hyperbaric oxygen [21].

The degree to which vasculogenesis contributes to wound healing is debatable. Its contribution to incisional healing is estimated to be no more than 20% [29]. Their contribution to healing of dead space wounds is debated, with the most recent data showing that stem cells can and do join into the building of new vessels by angiogenesis [30–31].

Bone marrow stem cells are a potent source of growth factors but only after being exposed to the wound environment, which, in view of the recent findings, means exposed to lactate [32], suggesting that lactate may produce some differentiation of progenitor cells. In fact, increasing lactate from the control level of 6 mM to 9 mM by implanting hydrolysable lactate in another model of dead space wounds (implanted wire mesh cylinders) raised VEGF, TGF-β and other growth factors in the wound extracellular fluid. The rate of collagen deposition can be at least doubled in this manner [15]. A principle is emerging suggesting that hypoxia and lactate tend to preserve the progenitor state while exposure to higher oxygen concentrations stimulates their differentiation [27, 30]. Lactate seems to increase "homing", and the combination of both lactate and oxygen appears to accelerate angiogenesis. Tube formation is still not well understood except, as noted, that matrix deposition is stimulated and required [6]. Migrating cells extend filopodia that allow them to recognize each other at some distance. In summary, it appears that vasculogenesis is inevitably involved in wounds, but its precise importance has not been defined. There is much to be learned in this area.

Arteriogenesis

Arteriogenesis, defined as growth of muscular vessels, will be discussed in full in another chapter in this book (see Chapter 47). Note, however, that considerable enlargement and consolidation of wound vessels occurs as seen in Figs. 46.1–46.9. Though limited, this appears to constitute a degree of arteriogenesis.

Angiogenesis Begins Early

Coagulation

The initial stimulus to wound angiogenesis comes prepackaged from the blood that is spilled into the wound and the coagulation that follows. Platelets aggregate, release a number of mitogens, chemoattractants, and growth factors including platelet-derived growth factor (PDGF), connective tissue growth factor (CTGF), and insulin-like growth factor (IGF)-1, together with one of its binding factors (BP3). Coagulation factor III (thromboplastin) accelerates the coagulation cascade to form thrombin and eventually fibrin deposition, that becomes a provisional matrix for cell migration as well as a source of a chemoattractant. Tissue factor and other proteins released from platelets and other cells nearby cause the perivascular smooth muscle to contract [33, 34]. An extraordinarily complex and redundant set of mechanisms follows. The contributions of coagulation are relatively well, if not fully, known, but it is not possible in this venue to give a complete description. In summary, the effects of coagulation provide a strong start to angiogenesis. Platelet-derived growth factor and fibrin/fibrinogen are particularly important. However, these effects last only hours to a few days. They are "starters."

Inflammation: Oxidants and Growth Factors

After a few minutes, bleeding stops, and intact local microvasculature dilates and leaks plasma in response to inflammatory mediators such as histamine and serotonin. Inflammation soon follows with leukocytes appearing within a few minutes and increasing for a few hours. The initial molecular response is induction of pro-inflammatory stimuli [35]. The effects of inflammation have been attributed to growth factors and cytokines. Considerable evidence shows that the metabolic effects of inflammation, such as lactate and ROS, are primary events.

During the immediate chemical events that follow injury, leukocytes are "primed" and activated as they ingest damaged tissue and microorganisms. Activation, via production of H_2O_2 and superoxide (O_2^-) derived from molecular oxygen, increases their oxygen consumption up to 25- to 50-fold and their lactate output at least several fold, in the so-called oxidative or respiratory "burst."

The oxidative burst is primarily due to activation of an NADPH-linked oxidase called PHOX (the phagocytic oxidase) and also contributed by NOX (the non-phagocytic oxidase) that, together, are of enormous importance in inflammation and wound healing. The physiological functions of these enzymes include host defense, oxidant production, posttranslational processing of proteins, cellular signaling, regulation of gene expression, and cell differentiation [36]. Of the two, PHOX produces by far the most lactate and ROS largely for defense against infection, while the NOX family appears to be mainly involved in cell signaling. Leukocyte activation causes the several parts of the PHOX molecule that are normally separated in the cytoplasm to assemble in the phagosomal membrane where they form one large enzyme that converts its substrate, molecular oxygen, to the reactive oxygen species H_2O_2 and O_2^-. Glucose, its other substrate, is converted to NADPH to provide the electrons necessary to form the oxidants. Lactate is a prominent by-product. Most of the oxidants are inserted into the phagosome, where they constitute the major portion of wound immunity to bacterial infection, but some escapes. The Km for O_2 in leukocytes (due mainly to PHOX) is about 50 mm Hg, and thus bactericidal function is highly dependent upon oxygen over a wide range from zero to perhaps 600 mm Hg (by definition, Vmax is approximately 10 times the Km) [37]. Thus, immunity to infection requires a high oxygen tension, another reason why

hypoxia is unsuitable as a sole instigator of wound revascularization.

The relationship of hypoxia to increased wound infection and the clinical use of oxygen to minimize infections are clearly established in clinical surgery. Since oxygen is required for the reaction, this is aerobic glycolysis by definition. Neutrophils have relatively few mitochondria and cannot process the pyruvate load that ensues. Consequently, lactate accumulates. Some of it stays in the cell and the rest leaves through monocarboxylic transferases (MCTs), bi-directional cell membrane transporters that facilitate rapid transit of lactate into and out of cells.

The lactate that accumulates in the extracellular environment easily enters other cells. That which enters fibroblasts instigates collagen transcription, deposition, and cell proliferation [38, 39]. That which accumulates in endothelial cells instigates collagen synthesis and deposition, as well as VEGF and cell motility [15, 38, 40]. That which reaches macrophages (in particular) leads to HIF-1α accumulation and VEGF production by a redox mechanism that will be explained in the section on Molecular mechanisms below [5, 14]. That which escapes the wound goes to the liver, where it is recycled into glucose and returned to the wound. The other products of the NADPH-linked oxidases, the reactive oxygen species, peroxide and superoxide are, as we shall see, as important to wound healing and its angiogenesis as they are to immunity to infection (see immediately below) [12, 13, 41, 42].

Inflammation-derived oxidants support, and antioxidants resist, adult tissue vascularization. In addition to being rich in cytokines/growth factors, the inflammation site is very rich in oxidants. Among all biological fluids, the highest level of hydrogen peroxide is found in the wound fluid. Fluid from the site of inflammation, contains 0.1–0.3 mM hydrogen peroxide in mice that have highly efficient healing mechanisms [12]. In endothelial cells, the major source of reactive oxidants is also an NADPH oxidase that consists of a number of proteins, Nox1, Nox2 (gp91phox), Nox4, p22phox, p47phox, p67phox and the small G protein Rac1. The specific mechanisms by which NADPH oxidase subunits may contribute to angiogenesis has been recently reviewed [43].

Over a decade ago, it was proposed that oxidants are not always triggers for oxidative damage in biological systems and and that oxidants such as hydrogen peroxide could actually serve as signaling messengers and drive several aspects of cellular signaling [17]. Evidence supporting hydrogen peroxide, for instance, as signaling messenger is compelling [16–18, 44, 45]. Current findings indicate that the oxidant-factor in inflammation plays a central role in supporting tissue vascularization [46]. Decomposition of endogenous hydrogen peroxide at the wound site by adenoviral catalase gene transfer impairs wound tissue vascularization [12]. Impairment of healing responses is noted in NADPH oxidase deficient mice and humans, in which it is termed chronic granulomatous disease [12, 42]. Direct evidence identifying macrophage-derived oxidants as angiogenic factors has been noted [47–49].

In biological systems, proteins are at risk of oxidative modification and inactivation. However, VEGF is protected from oxidative damage by the extracellular chaperone glypican-1 expressed in the vascular system. Glypican-1 can restore the receptor binding ability of VEGF165, which has been damaged by oxidation [50]. Interestingly, nature has a way of defending VEGF against oxidants because oxidants are required for vascularization [12, 43]. Recently it has been found that reactive oxygen species (ROS) mediate electrical field-induced angiogenesis of embryonic stem cells [51].

ROS are also required for VEGF-dependent tissue function. At micromolar concentrations, hydrogen peroxide induces VEGF-A (VEGF165 and VEGF121) expression [13]. Under conditions of co-existence characteristic of any inflammatory site, the effects of tumor necrosis factor (TNF)-α and hydrogen peroxide on VEGF induction are additive. Using deletion mutant constructs of a 2.6-kb VEGF promoter fragment (bp −2361 to +298, relative to transcription start site) ligated to a luciferase reporter gene, it has been established that the sequence from bp −194 to −50 of the VEGF promoter is responsible for the hydrogen peroxide response. Studies with Sp1 luciferase reporter constructs have identified that hydrogen peroxide -induced VEGF expression is Sp1 dependent (see section on Molecular mechanisms below). Hydrogen peroxide-induced VEGF expression is HIF-independent [13]. The above-mentioned findings have been verified and extended in subsequent studies. Signaling studies identified a cascade comprising Ras-Raf-MEK1-ERK1/2 as the main pathway mediating hydrogen peroxide-induced VEGF-A transcription [52]. In skeletal myotubes, oxidants seem to induce VEGF release via a phosphatidylinositol 3-kinase (PI3K)/Akt-dependent pathway [53]. Angiotensin II stimulation of VEGF mRNA translation requires the production of ROS [54]. Furthermore, angiopoietin 1 (Ang1)-induced hydrogen peroxide plays an important role in Ang1-induced angiogenesis and arteriogenesis by modulating p44/42 MAPK activity [55]. Several lines of evidence support that mild oxidizing conditions favor VEGF release. Partial cellular glutathione deficiency results in increased VEGF-A release [56]. Excessive reactive oxygen species, however, pose a threat to angiogenesis. Thus, reactive oxygen species have been noted to have biphasic effects on angiogenesis which indicate that pharmacologically regulating cellular oxidant levels might be productive in achieving antiangiogenic or proangiogenic outcomes [57].

The involvement of oxidants in the VEGF signaling pathway is not limited to induction of VEGF. After VEGF binds to its specific receptors, especially VEGFR2, oxidants seem to be required for the signaling leading to the angiogenic response of VEGF [43]. ROS are involved in the mitogenic cascade initiated by the tyrosine kinase receptors of several growth factor peptides including VEGF. Insulin induces VEGF expression through hydrogen peroxide production [58]. Evidence supporting the involvement of ROS in vanadate and hyperoxia-induced expression of VEGF is also reported [41, 59]. Although it has been known for a long time that cytokines induce superoxide generation by endothelial cells [60], the physiological significance remains to be appreciated in full [61]. Early evidence indicating that the binding of VEGF to VEGFR in endothelial cells leads to NADPH oxidase-induced oxidant production led to questions about the amplification potential of such oxidants in VEGF signaling [62]. It was soon recognized that VEGF-induced oxidant production was required to activate NF-κB which in turn was required for vascular smooth muscle cell migration, an integral component of angiogenesis [63]. In porcine aortic endothelial cells stably expressing human VEGFR2, receptor activation by VEGF is followed by a rapid rise in intracellular hydrogen peroxide. Genetic and pharmacological studies suggest that such oxidant burst requires the activation of PI3K and the small GTPase Rac-1 as upstream events, and is likely initiated by lipoxygenases. Inhibition of VEGFR2-dependent generation of ROS attenuates early signaling events, including receptor autophosphorylation and binding to a phospholipase, C-γ-glutathione S-transferase fusion protein. Moreover, catalase, the lipoxygenase inhibitor nordihydroguaiaretic acid, the synthetic ROS scavenger EUK-134, and the PI3K inhibitor wortmannin, all diminish ERK phosphorylation in response to VEGF. Finally, cell culture and stimulation in a nearly anoxic environment mimics the effect of ROS scavenger on receptor and ERK phosphorylation, reinforcing the idea that oxidants are necessary components of the mitogenic signaling cascade initiated by VEGFR2 [64]. Evidence also supports that VEGF stimulates superoxide production, which is inhibited by the non-specific NAD(P)H oxidase inhibitor, diphenylene iodonium, as well as by overexpression of dominant-negative Rac1 (N17Rac1) and transfection of gp91(phox) antisense oligonucleotides in human umbilical vein endothelial cells [65]. Antioxidants, including N-acetylcysteine, various NAD(P)H oxidase inhibitors, and N17Rac1, significantly attenuate not only VEGF-induced VEGFR2 tyrosine phosphorylation but also proliferation and migration of endothelial cells. Importantly, these effects of VEGF are clearly inhibited in cells transfected with gp91(phox) antisense oligonucleotides. Thus, VEGF-induced endothelial cell signaling and angiogenesis is tightly controlled by the redox microenvironment of the VEGF receptor. Production of angiogenic stimuli is the other, or perhaps more likely, subsequent function of inflammation.

Growth Factors

The literature on the origin and effects of inflammatory growth factors is extensive—too extensive to summarize here. TGF-β, FGF (fibroblast growth factor), VEGF, CTGF (connective tissue growth factor), IGF-1, EGR (early growth response factor), SDF (stromal cell-derived factor), and many others have been implicated. Wound healing mechanisms are characteristically redundant, and at this point, the redundancy is at its greatest. The degree that these growth factors remain in a wound, perhaps one or all contribute to angiogenesis. Dozens of substances had been found to enhance angiogenesis and/or

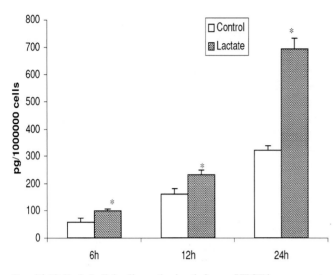

FIG. 46.10. Endothelial cells synthesize their own VEGF in response to lactate. This causes them to increase their motility and growth [40].

healing in one or another wound model, but most are of unknown significance because most regulate angiogenesis indirectly through inflammation. The importance of a few has been demonstrated by knockin and knockout experiments, but these findings are difficult to interpret.

At the moment, platelet-derived growth factor (PDGF) and VEGF are considered the most important growth factors in wound healing. VEGF, released by perhaps all cells, is partly under the control of lactate, HIF and partly of H_2O_2 [13]. Many types of cells release PDGF. It, too, responds to oxidant control [42]. Moreover, hyperoxia increases PDGF receptors in endothelial cells probably acting through H_2O_2 [66]. Inflammation also has a proteolytic function. As the structure of the new vessels is established, the preexisting tissue is removed by matrix metalloproteinases that are essential to all but the simplest wound angiogenesis. They, too, are driven by lactate and hence by oxidants. Again, the literature is extensive [35, 70] (Fig. 46.10).

Inflammation has Another Face

Inflammation, like almost everything else, can be inadequate, adequate, or excessive. Unresolved inflammation can derail the physiological healing cascade. Thus, timely resolution of inflammation is as important as the initial response. Successful efferocytosis [67, 68] is emerging as a key determinant of the timely resolution of inflammation (Sashwati Roy, personal communication at the 2007 Gordon Conference on Tissue Repair & Regeneration). Surgeons have long known that the best cosmetic healing occurs when inflammation is barely sufficient to incite the process and when repair is accomplished in the shortest time.

Excessive inflammation tends to encourage matrix deposition over angiogenesis in the long run. On the other hand, the so-called pyogenic granuloma, an excrescence of soft, highly

vascular, granulation tissue, can be wiped away with the back of a knife. When that is done, bleeding is impressive. Microscopic sections show a tissue containing mainly endothelial and inflammatory cells and almost devoid of connective tissue.

Inflammation can be controlled, though somewhat crudely. Surgeons inject unsightly scars with anti-inflamatory steroids to reduce their volume and suppress the inflamed appearance. Other anti-inflammatory agents have been used with less obvious success except for one, the application of a thick silicone sheet for many days with the objective of reducing hypertrophic scars. Why this works, and it does, is not yet experimentally confirmed. After about a week of constant application, the erythema that indicates undue angiogenesis begins to resolve.

Control of inflammation is an area of great importance, and not just to wound healing and angiogenesis [35]. Excessive inflammation, including systemic sepsis, can literally destroy new, highly vascular tissue. The basis of the optimal balance of adequate, rather than excessive, inflammation is not understood.

In summary, some inflammation is required for wound angiogenesis. In a global view, inflammation is the rapid and transient response to injury that prevents exposure of the inner mammalian tissues to the hazards of the outside world and assures rapid healing; but it is also the point at which the mammalian propensity for regeneration that is seen in more primitive animals seems to be lost. Inflammation is also the source of many human diseases, many of them lethal because of their tendency to deposit vascularized connective tissue that is basically indistinguishable from wound healing in undesirable locations such as coronary arteries.

Angiogenesis, Macrophages, and Remodeling

At about the end of the first 4 days, in the absence of infection, the transcription of inflammatory genes begins to subside, replaced by those coding for angiogenic stimuli, many of them stimulated, or at least controlled, by hypoxia-inducible factor (HIF).

Macrophages dominate the scene of injury by about a week after injury. By 10 days, linear wounds that are primarily closed and healing ideally are losing inflammatory cells. The excessive collagen that is almost always present begins to regress toward the end of the second week. New epithelium covers the small gap that results from incisional wounds. New vessels mature to restore the preexisting pattern (Fig. 46.3-46.7). Wound healing slows gradually and finally stops some months later. Ordinarily, however, healing (with its angiogenesis) continues until the circulations from both sides of the wound join, or an open wound heals to a skin graft.

If healing is less than ideal, if the wound is open or contains a space or excessive inflammatory cells, remain, and angiogenesis persists (See Fig. 46.8). Wounds that do not close quickly may continue to be inflamed and angiogenic for years. Macrophages in this lactate-rich environment secrete growth and chemoattractant factors, among them VEGF,

TGF-β, some interleukins, EGR, and others that keep the healing momentum going as long as exposure, drying, and/or inflammation continue [14, 35]. HIF, PDGF, and VEGF appear to be the greatest contributors to ongoing repair that are not merely the products of inflammation, and all three are products of response to lactate as well as oxidants [69]. Factors that regulate resolution of inflammation in wound healing deserve greater attention.

If a wound is not healed in 10 days or so, its essential architectural feature becomes its directionality, as noted in Fig. 46.2-46.9. The center of the wound is a space or a surface (a healing burn, for example) where the damage was done. These surfaces become lined with macrophages that lead a characteristic parade of cells along chemotactic gradients of lactate and other chemoattractants that the macrophages leave behind. Macrophages secrete VEGF, and TGF-β, among other factors such as interleukins, matrix metalloproteinases, and large amounts of lactate. Endothelial cells, and possibly other cells, respond by increasing their motility; collagen deposition and new vessel construction follows [14, 15, 40]. Many substances seem to attract fibroblasts. As noted above, lactate, with oxygen, causes them to synthesize pro-collagen and release hydroxylated collagen.

Figure 46.8 demonstrates the characteristic zone in which mitoses appear, where oxygen is relatively low and lactate moderately high, indicating an ideal environment for endothelial cell and fibroblast replication. This point seems to serve as a growth zone that is reminiscent of those seen in plants or in regenerating limbs in lower animals. On the wound side of these mitoses, the cell procession led by macrophages and lactate moves into the wound space. On the other side of the module, where the oxygen is higher, angiogenesis and collagen deposition solidify the gains. Macrophages, activated by damaged tissue, foreign bodies and/or lactate, also release proteolytic enzymes such as collagenase and elastase [70]. These lytic enzymes allow macrophages (or osteoclasts in fractures) to penetrate fibrin or other obstacles with endothelial cells trailing behind [71]. As endothelial cells advance, they deposit collagen and pick up pericytes (possibly of stem cell origin) [71]. As long as there is a stimulant that causes macrophages to persist and migrate, the parade of cells continues. Since lactate also stimulates collagen transcription, the stimulated fibroblasts and endothelial cells excrete and hydroxylate it as they are overtaken by the advancing oxygen gradient that enables collagen deposition and makes the advancement of new vessels possible (figure 11) [22]. Macrophage and progenitor cell mechanisms appear to cooperate [72]. Monocyte chemoattractant protein seems to have a role at this point [73]. In this context, the analogy to tumor invasion deserves attention.

Wound Angiogenesis Stops! Why?

Why wound angiogenesis stops, and tumor angiogenesis does not, has been one of the long-contemplated mysteries. Current data seem to solve it. The ready-made answer for the "hypoxia theorists" is that wounds heal until the oxygen gradients are obliterated, tissue continuity is restored, and oxygen concentrations rise. They add that the oxygen gradient may never be obliterated in tumors, and its stimulatory effect may never stop. The problem with this is that oxygen gradients never develop in some highly vascularized tumors, particularly those that do not develop conveniently as the spheroid model would suggest.

The ready-made answer for the "lactate theorists" is that wounds heal until they stop producing lactate and that tumor cells never stop. The difference between the two interpretations can be minimized, but not fully justified, by accepting that both tumors and activated inflammatory cells secrete large amounts of lactate aerobically. Though many investigators might think the mechanisms presented here are still premature in their emphasis on lactate, there are many other important supporting data such as the fact noted above, that lactate alone can instigate vasculogenesis and that tumor angiogenesis has been correlated to lactate output [8].

Three Constructs that are Based on Space and Time can Accommodate Hyperoxia, Hypoxia, and Lactate Accumulation in Revascularization

Accumulated lactate and oxygen may cooperate to guide the growth of vessels in three ways. First, lactate alone, with no gradients involved, hosts progenitor cells. In this case, the rate of vasculogenesis is proportional to PO_2 [8]. The matrigel model exemplifies this. Progenitor cells seem to "home" to sites where lactate is high. The mechanism is not known. Accumulating lactate stabilizes HIF-1α (see Molecular mechanisms below).

The second construct recognizes that the steep gradients of PO_2 may be a means by which wounds can be hypoxic at their center and not suffer full consequences of hypoxia. The low PO_2 in the wound may extract hypoxia-inducible growth factors, and the proximity of a higher oxygen concentration coming from the uninjured tissue may make collagen deposition possible as the advancing new vessels "overtake" the newly minted cells. In this case, angiogenesis is a spatio-temporal process in which the needs of the advancing cells are filled only as the moving oxygen gradient reaches the needy cells shortly after they are formed and initially "programmed" in the hypoxic, lactate-rich part of the environment. Hyperoxia increases the gradient while changing the central PO_2 rather less and the lactate not at all. The high lactate inevitably continues to exert its influence (Fig. 46.8). If the tissue surrounding the injury is normoxic, collagen can be deposited and vessels can mature (Fig. 46.11). If it is hypoxic, vessels cannot mature (Fig. 46.12).

The third construct can be imposed on wounds by sequentially interrupting the hypoxia, in which lactate accumulates, with short periods of hyperoxia, in which collagen synthesis and vessel formation accelerate. Thus, phases of increasing neediness can alternate with phases of fulfillment. Ischemia or arterial hypoxia enhances levels of the hypoxia-

21% oxygen

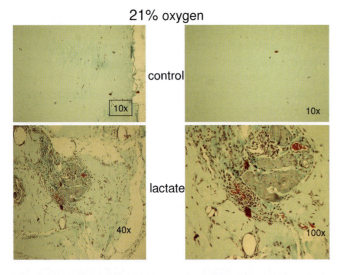

FIG. 46.11. A composite of microscopic sections of matrigel implants in animals that breathed room air at 1 atmosphere pressure at 11 days. Control implants showed rare cells while lactated implants show new, quite mature vessels with pericytes and collage deposition (the bright green surrounding the vessels). Judging from the vessel walls, these vessels are carrying blood flow.

13% oxygen

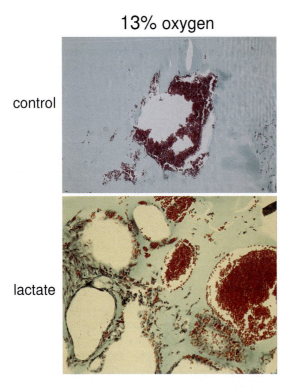

FIG. 46.12. Animals breathing 13% oxygen gas mounted a VEGF stimulus, perhaps even a high one due to the hypoxia. However, the vessels were highly variable in size and were friable due to the lack of collagen.

inducible factors in wounds, but healing is retarded [74]. If, then, arterial hyperoxia is induced as in hyperbaric oxygen, the needs are fulfilled, and healing can accelerate rapidly, depending on the degree to which tissue PO_2 can be raised [41, 75, 76]. This may be the mechanism behind the stimulation of angiogenesis and healing in ischemic limbs by exposing patients to intermittent hyperbaric oxygen therapy. Clinical data essentially proves that new vessels so generated can be extensive and enduring and may save ischemic limbs from amputation for many years [77].

History of Lactate in Wounds

The connection of lactate to collagen began in the 1940s, when several groups noted that neglected acidotic fibroblast cultures that had accumulated lactate up to about 15 mM deposited almost twice as much collagen as well-tended cultures with less lactate. The investigators deduced, as would most investigators today, that lactic acid came from hypoxia. It took some years to find that the increase in collagen production also occurred in well-oxygenated, pH-neutral cultures, when lactate monomer was raised to the same levels by adding sodium lactate (not lactic acid). The hypothesis that hypoxia was the source of the lactate crumbled when Hussain and Ghani discovered that high lactate levels activate prolyl hydroxylase by diminishing NAD+ and the highly regulatory adenosine diphophoribose (ADP-ribose) that is derived from it [47]. ADP-ribose (ADPR) can be made only from NAD+, not NADH, and high lactate markedly reduces NAD+ [78]. ADP-ribose modifies many proteins and regulates many enzymes, including prolyl hydroxylase, by attaching to arginine moieties in a process called ADP-ribosylation. These investigators proved that increased lactate diminishes the ADP-ribosylation of collagen prolyl hydroxylase, thus increasing the rate of collagen deposition [22]. Additionally, and apparently by the same mechanism, it activates the collagen gene promoter by reducing NAD+, with the consequence of removing ADP-ribose from transcriptional proteins. If the original investigators had the foresight at that time to make an issue of it, we might not now have to deal with the unnecessarily restrictive term "hypoxia-inducible" to refer to genes that actually sense not only hypoxia, but also lactate, pyruvate and other 2-oxoacids (also known as α-hydroxy acids) in the presence of oxygen [5]. They might have been called "lactate-inducible genes" or classified under "redox-sensitive genes".

In the past few years, much has been learned about the ability of lactate to guide healing and its angiogenesis. Lactate increases: (1) HIF-1α levels [5], (2) VEGF [15], (3) TGF-β [15], (4) metalloproteinase [70], (5) endothelial cell motility [40], (6) angiogenesis and vasculogenesis [8], (7) collagen synthesis, its posttranslational modification, and deposition [15, 38], (8) cell proliferation [39], (9) transcription of genes for proteoglycans, CD44, caveolin-1, Hyal-1 and −2 [79], (10) environmental suitability for lodgment of progenitor cells [8], (11) VEGF activity [38, 78], and (12) ROS formation [80]. Finally, lactate reduces ADP-ribosylations, and the extent of this kind of influence promises to be of widespread importance [81]. Not so comfortably placed in this group, as yet, are perhaps even more fundamental findings by Formby and Stern who, by sequence analyses of promoter regions, revealed multiple AP-1 and ets-1 response elements and increased transcripts of c-fos, c-jun, and c-ets in lactate-challenged fibroblasts [79]. It is not clear whether some of these actions are immediately or remotely influenced by HIF or VEGF. Most recently, Hashimoto et al. have shown that when muscle cells are cultured in high but physiological lactate concentrations, the transcription of more than 600 genes changes. Prominent among them are genes involved in mitochondrial biogenesis and energy metabolism [83]. Recently, Kumar et al. have found that lactate exposure changes the phenotype of endothelial cells and confirmed the effect of ADP-ribosylation on VEGF activity as previously noted by Gimbel et al. [84]. There is no single term to describe the totality of these actions, but the idea that lactate may be "a link to a master transcription factor that facilitates adaptation to metabolic stress" comes to mind [83].

Contrary to general expectations, pH is not an issue with aerobic lactate production. Lactate and hydrogen ions come from two separate processes, and both have many sources. Hydrogen ion, among its many sources, accumulates when when PO_2 falls below 1 mm Hg and there is no means of dealing with it [10]. Lactate is a product of both aerobic and anaerobic glycolysis, and contrary to general opinion, glycolysis leading to pyruvate and lactate does not cause a net increase in H^+. Addition of lactate ion to an aqueous solution produces a slight alkalosis. The term "lactic acid" should rarely be used in the context of human physiology.

Molecular Mechanisms of Hypoxia, ROS, and Lactate

Accumulated lactate in the presence of oxygen has at least three molecular actions. When they are taken into account, the oxygen/hypoxia paradox disappears and the apparent conflicts are resolved.

Lactate and Other 2-oxo Acids Regulate HIF Prolyl Hydroxylases

Lactate stabilizes HIF-1α independent of the concentration of oxygen. The so-called hypoxia-induced genes are controlled by the level of the transcription factor HIF, a heterodimeric protein, of which one moiety, HIF-1α, can be hydroxylated by HIF prolyl hydroxylases, thus targeting it for rapid destruction when PO_2 is high[2]. When intracellular PO_2 falls, HIF-1α accu-

---

[2] The HIF-1α prolyl hydroxylases are closely related to the collagen prolyl hydroxylases and are even found in the same area of the endoplasmic reticulum and controlled in the same manner. However, while the Km for oxygen of the collagen-related enzyme(s) is about 25 mm Hg, the Km for oxygen of the bigger HIF-related variety is closer to 100 mm Hg. Thus, collagen can be hydroxylated and deposited at low oxygen levels that hardly influence HIF levels which, having a higher Km, are more suitable for sensing oxygen and lactate.

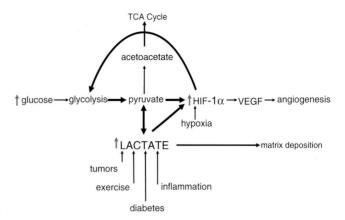

FIG. 46.13. The schema for the relationship of glycolysis to angiogenesis emphasizes the importance of lactate from external sources. Hyperglycemia increases the internal lactate level. The important detail is in the lines that demonstrate the amplification mechanism of HIF such that when stimulated by lactate enhances glycolysis and thus increases lactate. How this amplification loop is kept under control is not known.

mulates, and with it, more than 30 genes are activated, including the gene for VEGF and many of the enzymes of the glycolytic pathway [85]. Their activation leads to an increase in glycolytic capacity that accelerates pyruvate production and, hence, lactate level. All these facts are accepted. New is the fact that lactate and pyruvate are also inextricably involved in the same mechanism. Lactate and/or pyruvate concentrations also regulate hypoxia-inducible factor level independently of oxygen concentration by binding to and impairing the same HIF prolyl hydroxylase(s). *Thus, lactate and pyruvate can stabilize HIF even in the presence of large amounts of oxygen* [5, 86, 87].

Given the above, we must now accept that lactate/pyruvate accumulation (supplemented by lactate drawn from aerobic reactions such as the NADPH-linked oxidase, PHOX) increases glycolysis in a self-reinforcing loop. In other words, lactate, regardless of its source, begets pyruvate, oxalacetate and more lactate, and thus becomes a highly efficient and sensitive amplification mechanism with its basis in aerobic glycolysis as well as in hypoxia. This combination leads a protean set of adaptations to metabolic stress [87]. Oxalacetate and other intermediates such as fumarate also mediate HIF-1α [82]. Whether they also are involved in a metabolic control system is not clear. Fig. 46.13 diagrams this concept. HIF-1α can accumulate even in the presence of oxygen, and hypoxia is not required to activate HIF-inducible genes. For these reasons, oxygen can remain at concentrations that are high enough to support vessel growth and not interfere with the fundamental growth signals. Since there are many sources of lactate, angiogenesis can be upregulated in many metabolic scenarios such as inflammation, exercise, and hyperglycemia that also increase lactate. This relationship may explain some

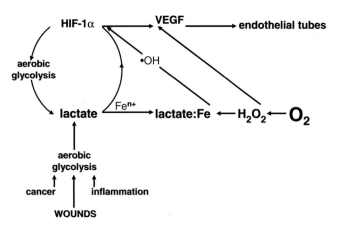

Fig. 46.14. The redox control of HIF and VEGF. It is enabled by the lactate:iron chelate in the endoplasmic reticulum, and the stabilization of HIF is enabled by the lactate.

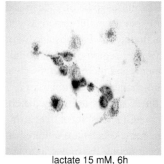
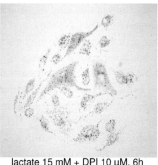

lactate 15 mM, 6h   lactate 15 mM + DPI 10 µM, 6h
VEGF production reduced

Fig. 46.15. Endothelial cells were incubated in nitroblue tetrazolium (NBT) and under an oxidant stimulus form formazan crystals. Not shown is the comparison of control cultures with no lactate. It is essentially the same as the *right panel*. Clearly, lactate diverts some oxygen that focally undergoes a Fenton reaction and oxidizes the NBT and precipitates the formazan crystals. According to Liu et al., the sites are in the membrane of the endoplasmic reticulum [86].

of the long-term complications of hyperglycemia such as retinopathy and other examples of arteriosclerosis.

Lactate is Also a Source of Regulatory Redox Activity that Regulates HIF

The role of oxidants in angiogenesis, particularly the regulation of HIF-1α is widely documented [88]. Lactate accumulation also has a considerable impact on redox status by its involvement in ROS generation (Fig. 46.14). In one of the first studies to foresee the importance of lactate, Ali and colleagues found that lactate and iron form a chelate that, in the presence of H_2O_2, produces hydroxyl radical (•OH) by a variant of Fenton chemistry. They predicted that lactate would be found to have regulatory significance [80]. Liu et al. then found iron-containing structures in the endoplasmic reticulum (ER) and confirmed Ali's prediction by demonstrating that H_2O_2 is converted to •OH via a Fenton reaction that occurs in the endoplasmic reticulum and translocates the HIF gene from the ER into the nucleus. In these experiments, peroxide and •OH were generated in proportion to PO_2 [6]. Scavengers of •OH inhibit the reaction. The increased oxidant flux did not occur in mitochondria, lysosomes, or peroxisomes. The investigators concluded that the changes of the redox status within a cell may contribute to an efficient, fast-responding oxygen sensing system. We know now that the amount of lactate, even as little as about 5 mM, also influences the system [89]. Beckert et al. demonstrated recently that endothelial cells incubated first in nitroblue tetrazolium and then exposed to lactate, markedly increased formation of formazan crystals, i.e., a redox flux of H_2O_2/•OH, in the peri-nuclear regions of the cells (Fig. 46.15). At the same time, VEGF production increased several-fold [8]. This appears to unify the works of Liu and Lu [5, 86, 87].

Increasing lactate redirects H_2O_2, to the endoplasmic membrane where it is converted to •OH and injected into the ER lumen and sustains HIF-1α levels. VEGF can also be gener-

ated when hydrogen peroxide (some of which is derived from oxygen) combines to the VEGF promoter and accelerates its transcription [13, 90]. Roy et al. also found that H_2O_2 is elevated in the wound space and demonstrated impaired healing by over expressing catalase, thus reducing peroxide [12]. This group reinforced the earlier suggestion that the NADPH-linked oxidase(s) are a source of oxidants that instigate repair and angiogenesis [42, 45].

Lactate also Influences ADP-Ribosylations

Lactate exerts its effects by another, perhaps even more powerful molecular mechanism, involving the inhibition of ADP-ribosylations. This is central to a metabolic response system that is controlled at the level of LDH in which lactate and pyruvate control the levels of NAD+ and NADH. It involves a metabolite of NAD+ (but not of NADH), termed adenosine diphosphoribose or ADP-Ribose (i.e., NAD+ with the (N)icotinamide removed). ADP-ribose (ADPR) can exist as a monomer (or oligomer) or as a polymer formed by the enzyme PARP (polyadenosylribose polymerase). The polymer has widespread influence over transcription of many genes, while the monomer regulates the activities of many cytoplasmic proteins. When increasing lactate diminishes NAD+, the formation of ADPR is also suppressed. Changes in ADP-riboslyation also control the activity of VEGF, and polyADPR influences the transcription of collagen, among many other effects. As lactate rises and falling levels of (mono)ADPR allow it, the inhibition that ADP-ribose normally exerts over the hydroxylation of collagen prolines is lifted. As noted above, oxygen cannot even be efficiently used to hydroxylate collagen to assist angiogenesis unless lactate has accumulated and has reduced the inhibition [22] (see figure 46.1). The effects of lactate on ADP-ribosylations are blocked by oxamate that inhibits LDH and by nicotinamide that blocks its polymerization [22]. Thus, the effects

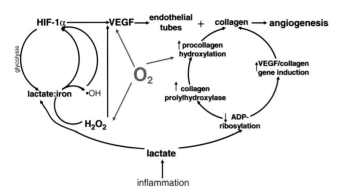

FIG. 46.16. The entire schema. The *left side* is a slight revision of figure 14. The *right side* shows the ADP-ribose (ADPR) contribution. Note that the collagen that is a result of the reduction of ADPR allows the growth of vessels. Note the central role of oxygen.

that lactate accumulation exerts on gene transcription are probably dominant. The NAD+/NADH ratio is a measure of the redox balance of the cell and a prominent influence on the collagen deposition that facilitates vessel growth [78, 84, 91].

Despite attempts to distinguish between them, these three effects of lactate, ROS production, HIF stabilization, and reduction of ADP-ribosylations, have not been separable [7, 14, 78]. Many actions of polyADP-ribosylation remain to be discovered [92] (Fig. 46.16).

Therapeutic Angiogenesis

The first medical use of angiogenesis as an accelerant of angiogenesis was for the treatment of chronic wounds. When inflammation persists, and healing is frustrated, angiogenesis seems to cease. An unsatisfactory equilibrium persists with excessive inflammation, increased growth factor and cytokine production, proteolysis, and chronic deposition of fibrous tissue. Millions of dollars have been spent on developing means to restart the healing process. The cost of not healing them, excessive scar, will be greater.

The major causes of impaired healing are insufficient arterial perfusion, excessive venous pressure, chronic reinjury, and unrelieved inflammation (foreign bodies such as hair, necrotic, etc.). Treatment involves prevention of injury, correction of inflammation, radical debridement of the nidus of infection and the area of heavy scarring, inflamed tissue, and correction of the circulatory abnormality.

Although dozens of substances accelerate angiogenesis in animals, only two, oxygen and PDGF, have been found unequivocally safe and effective for human chronic wounds. Attempts to study why this difference exists are frustrated by the lack of appropriate human surrogates.

The few strategies that are effective in humans are debridement, amelioration of the perfusion problem, intermittent hyperoxia, and PDGF B:B (Platelet derived growth factor is a dimer with A and B chains that can be combined as A:A, B:B, and A:B.), and a proprietary preparation of "platelet releasates." [93, 94]. Hyperoxia is administered intermittently by breathing oxygen in a pressurized vessel, thus the name "hyperbaric oxygen" [95, 96] or sometimes with a topical gas device [97–99]. Though both topical agents and oxygen therapy have been controversial, the evidence supporting their efficacy in specific conditions is no longer in doubt. Application of vacuum to wounds is effective. The Wound VAC (KCI Corp.) speeds the flow of fluid through the wound and applies a pressure derived from the atmospheric. Collected fluid from the VAC reservoir contains all the requisite growth factors and cytokines including lactate and oxygen.

Consideration is being given to antiangiogenic therapies in the treatment of excessive scarring. Surgical wound complications are now appearing in patients who have received the one antiangiogenic therapeutic agent that has been used extensively in humans [100]. The frequency is said to be low, but surgical wound failures that can be clearly attributed to such causes are also relatively unusual, and no reliable statistics are available. There is no reported experience with these agents in patients who already have wound problems. Aside from wounds, however, there is a future for antiangiogenic agents in the treatment of complicated diabetes, for instance.

Lessons Learned from Wounds

In actual wounds, VEGF is considered the dominant angiogenic impetus, and anti-VEGF antibodies have had the expected result. Contrary to earlier expectations, HIF and VEGF appear in wounds immediately after injury and well before hypoxia has developed. Again, this illustrates the importance of both lactate and oxygen in the instigation of wound healing [101]. When animals are made hypoxic, the VEGF content of wounds rises, but wound collagen content and angiogenesis are depressed [74]. On the other hand, hyperoxia actually increases VEGF and collagen deposition in wounds [41, 75, 76]. However, if hyperoxia can cause oxygen to penetrate, there is hope for healing. Prospective studies have shown that hyperbaric oxygen (HBO) delays and/or prevents major (leg) but not minor (toe) amputation. HBO is "angiogenic" in relatively lactate-rich but viable tissue. Hopf and colleagues confirmed this conclusion [7]. Topical oxygen has been successful in oxygenating superficial wound tissue and speeding closure.

Why is Lactate so Uniquely Important?

Enhanced levels of several components of energy metabolism, the small 2-oxoacids, contribute to HIF stabilization [88]. In addition to lactate, they include pyruvate, fumarate, malate, oxoketoglutarate, and possibly other 2-oxoacids [89]. Why, then, is lactate so prominent in governance of HIF?

Lactate is unusual among 2-oxoacids because, due to aerobic glycolysis, it has so many aerobic sources [10] that include exercise, hyperglycemia, lipolysis, sympathetic nervous system activation, and rapidly dividing cells that characterize wounds and most tumors [10, 102]. When lactate from these sources is added to the effect of rising pyruvate due to inflammation, the total lactate synthesis is far higher than that due to simple energy needs.

Lactate and pyruvate exists in a constant equilibrium that is controlled by LDH. The molecular ratio is usually about 10 lactates to each pyruvate. Addition of pyruvate (2 mM) alone to cell cultures is sufficient to stabilize HIF, but the possibility that lactate or pyruvate or even both together is the most effective is still not known (Fig. 46.13).[3]

Lactate accumulation has Enormous Medical Significance

Accumulation of lactate in addition to other metabolites clearly has protean consequences, some of which are good while some are destructive. The mechanisms of malignancy are deeply rooted in glycolytic metabolism, and attempts are in progress to inhibit excess glycolysis, HIF, and ROS in order to slow or stop the growth of tumors. Furthermore, these mechanisms clearly apply to arteriosclerosis in as much as PHOX and NOX are already implicated. Arterial intima exists in one of the highest oxygen concentrations in the human body, far greater than veins, and from time to time will be exposed to elevated lactate levels. It is subjected frequently to elevated glucose and thus lactate will rise for this reason as well. Furthermore, the inflammation due to lipids that are found in the sub-intima will raise lactate still further just as happens in wounds. The combination of lactate and oxygen is clearly a strong stimulant to both angiogenesis and connective tissue deposition, whether or not a wound is present.

It would be strange if this system is not involved in diabetic vascular disease, but details are not known, and the known facts are often at odds. Hyperglycemia in the first few days after healing slows healing and reduces resistance to infection because high glucose indirectly inhibits PHOX. After inflammation has been established, however, hyperglycemia is more associated with pathologic angiogenesis. Thus, there can be little doubt that these mechanisms apply to the angiogenic fibrosis that complicates the health of so many diabetics, smokers, and the obese. If we are to further our understanding in this area, more will need to be known about the acute effects of hyperglycemia and lactate on pathologic angiogenesis in diabetics.

---

[3] Lu et al. found that the effect of adding 2 mM pyruvate is easy to detect. The effect of adding 2 mM lactate was less easy to detect. It becomes clear then, that physiologically, larger concentrations of lactate will be required to produce the same effect as pyruvate, but lactate concentrations are normally much larger than pyruvate by a factor of 10 to 1.

Summary

Wounds recruit numerous mechanisms to re-establish damaged vasculature, among them are angiogenesis, inosculation of vessels, vasculogenesis and, arteriogenesis. Considerable pertinent data exist. Although wound angiogenesis recruits a myriad of redundant mechanisms, two features, the accumulation of lactate due to aerobic glycolysis in inflammatory cells and the generation of signaling oxidants, are the central, perhaps the lowest, common denominators. Also central is the need to support the healing tissue with appropriate level of oxygenation that would fuel the regenerative process as well as enable redox signaling and resistance to infection. The effects of lactate on the HIF-inducible genes persists even in the presence of oxygen since the effect of lactate on HIF-1α stabilization mimic that of hypoxia and are independent of oxygen tension. This feature allows inflammatory lesions to heal by taking advantage of oxygen for collagen deposition that strengthens new vessels and for resistance to infection.

However, although the mechanisms discussed here are beneficent adaptations to injuries, they are also a mixed blessing. The same mechanisms apply to some of the most harmful processes that affect human health, such as cancer, inflammatory fibrosis, arteriosclerosis, arthritis, and complications of diabetes. The adaptation to stress and metabolic neediness, no matter how useful and important, raises its own dilemmas.

Acknowledgments. Supported by NIH RO1 grants GM069589, GM077185 and HL 073087 to C.K.S. and GM27345 to T.K.H..

References

1. Burns PA, Wilson DJ. Angiogenesis mediated by metabolites is dependent on vascular endothelial growth factor (VEGF). Angiogenesis 2003; 6: 73–7.
2. Murray B, Wilson DJ. A study of metabolites as intermediate effectors in angiogenesis. Angiogenesis 2001; 4: 71–7.
3. Semenza GL. Regulation of physiological responses to continuous and intermittent hypoxia by hypoxia-inducible factor 1. Exp Physiol 2006; 91: 803–6.
4. Hunt TK, Zederfeldt B, Goldstick TK. Oxygen and healing. American Journal of Surgery 1969; 118: 521–5.
5. Lu H, Dalgard CL, Mohyeldin A, McFate T, Tait AS, Verma A. Reversible inactivation of HIF-1 prolyl hydroxylases allows cell metabolism to control basal HIF-1. J Biol Chem 2005; 280: 41928–39.
6. Berthod F, Germain L, Tremblay N, Auger FA. Extracellular matrix deposition by fibroblasts is necessary to promote capillary-like tube formation in vitro. J Cell Physiol 2006; 207: 491–8.
7. Hopf HW, Gibson JJ, Angeles AP, Constant JS, Feng JJ, Rollins MD, Zamirul Hussain M, Hunt TK. Hyperoxia and angiogenesis. Wound Repair Regen 2005; 13: 558–64.
8. Hunt TK, Aslam RS, Beckert S, Wagner S, Ghani QP, Hussain MZ, Roy S, Sen CK. Aerobically Derived Lactate Stimulates

Revascularization and Tissue Repair via Redox Mechanisms. Antioxid Redox Signal 2007.

9. Myllyla R, Tuderman L, Kivirikko KI. Mechanism of the prolyl hydroxylase reaction. 2. Kinetic analysis of the reaction sequence. Eur J Biochem 1977; 80: 349–57.

10. Gladden LB. Lactate metabolism: a new paradigm for the third millennium. J Physiol 2004; 558: 5–30.

11. Jonsson K, Jensen JA, Goodson WH, 3rd, Scheuenstuhl H, West J, Hopf HW, Hunt TK. Tissue oxygenation, anemia, and perfusion in relation to wound healing in surgical patients. Ann Surg 1991; 214: 605–13.

12. Roy S, Khanna S, Nallu K, Hunt TK, Sen CK. Dermal wound healing is subject to redox control. Mol Ther 2006; 13: 211–20.

13. Sen CK, Khanna S, Babior BM, Hunt TK, Ellison EC, Roy S. Oxidant-induced vascular endothelial growth factor expression in human keratinocytes and cutaneous wound healing. J Biol Chem 2002; 277: 33284–90.

14. Constant JS, Feng JJ, Zabel DD, Yuan H, Suh DY, Scheuenstuhl H, Hunt TK, Hussain MZ. Lactate elicits vascular endothelial growth factor from macrophages: a possible alternative to hypoxia. Wound Repair Regen 2000; 8: 353–60.

15. Trabold O, Wagner S, Wicke C, Scheuenstuhl H, Hussain MZ, Rosen N, Seremetiev A, Becker HD, Hunt TK. Lactate and oxygen constitute a fundamental regulatory mechanism in wound healing. Wound Repair Regen 2003; 11: 504–9.

16. Rhee SG, Bae YS, Lee SR, Kwon J. Hydrogen peroxide: a key messenger that modulates protein phosphorylation through cysteine oxidation. Sci STKE 2000; 2000: PE1.

17. Sen CK, Packer L. Antioxidant and redox regulation of gene transcription. Faseb J 1996; 10: 709–20.

18. Stone JR, Yang S. Hydrogen peroxide: a signaling messenger. Antioxid Redox Signal 2006; 8: 243–70.

19. Biswas S, Ray M, Misra S, Dutta DP, Ray S. Is absence of pyruvate dehydrogenase complex in mitochondria a possible explanation of significant aerobic glycolysis by normal human leukocytes? FEBS Lett 1998; 425: 411–4.

20. Warburg O. On the origin of cancer cells. Science 1956; 123: 309–14.

21. Gallagher SM, Castorino JJ, Wang D, Philp NJ. Monocarboxylate transporter 4 regulates maturation and trafficking of CD147 to the plasma membrane in the metastatic breast cancer cell line MDA-MB-231. Cancer Res 2007; 67: 4182–9.

22. Hussain MZ, Ghani QP, Hunt TK. Inhibition of prolyl hydroxylase by poly(ADP-ribose) and phosphoribosyl-AMP. Possible role of ADP-ribosylation in intracellular prolyl hydroxylase regulation. J Biol Chem 1989; 264: 7850–5.

23. Knighton DR, Silver IA, Hunt TK. Regulation of wound-healing angiogenesis-effect of oxygen gradients and inspired oxygen concentration. Surgery 1981; 90: 262–70.

24. Rothenfluh DA, Demhartner TJ, Fraitzl CR, Cecchini MG, Ganz R, Leunig M. Potential role of pre-existing blood vessels for vascularization and mineralization of osteochondral grafts: an intravital microscopic study in mice. Acta Orthopaed Scand 2004; 75: 359–65.

25. Karja NW, Kikuchi K, Fahrudin M, Ozawa M, Somfai T, Ohnuma K, Noguchi J, Kaneko H, Nagai T. Development to the blastocyst stage, the oxidative state, and the quality of early developmental stage of porcine embryos cultured in alteration of glucose concentrations in vitro under different oxygen tensions. Reprod Biol Endocrinol 2006; 4: 54.

26. Uno K, Merges CA, Grebe R, Lutty GA, Prow TW. Hyperoxia inhibits several critical aspects of vascular development. Dev Dyn 2007; 236: 981–90.

27. Goldstein LJ, Gallagher KA, Bauer SM, Bauer RJ, Baireddy V, Liu ZJ, Buerk DG, Thom SR, Velazquez OC. Endothelial progenitor cell release into circulation is triggered by hyperoxia-induced increases in bone marrow nitric oxide. Stem Cells 2006; 24: 2309–18.

28. Bauer SM, Goldstein LJ, Bauer RJ, Chen H, Putt M, Velazquez OC. The bone marrow-derived endothelial progenitor cell response is impaired in delayed wound healing from ischemia. J Vasc Surg 2006; 43: 134–41.

29. Bluff JE, Ferguson MW, O'Kane S, Ireland G. Bone marrow-derived endothelial progenitor cells do not contribute significantly to new vessels during incisional wound healing. Exp Hematol 2007; 35: 500–6.

30. Gallagher KA, Goldstein LJ, Thom SR, Velazquez OC. Hyperbaric oxygen and bone marrow-derived endothelial progenitor cells in diabetic wound healing. Vascular 2006; 14: 328–37.

31. Gallagher KA, Liu ZJ, Xiao M, Chen H, Goldstein LJ, Buerk DG, Nedeau A, Thom SR, Velazquez OC. Diabetic impairments in NO-mediated endothelial progenitor cell mobilization and homing are reversed by hyperoxia and SDF-1 alpha. J Clin Invest 2007; 117: 1249–59.

32. Liu Y, Dulchavsky DS, Gao X, Kwon D, Chopp M, Dulchavsky S, Gautam SC. Wound repair by bone marrow stromal cells through growth factor production. J Surg Res 2006; 136: 336–41.

33. Folkman J. Angiogenesis and proteins of the hemostatic system.[comment]. J Thromb Haemostasis 2003; 1: 1681–2.

34. Spencer EM, Tokunaga A, Hunt TK. Insulin-like growth factor binding protein-3 is present in the alpha-granules of platelets. Endocrinology 1993; 132: 996–1001.

35. Eming SA, Krieg T, Davidson JM. Inflammation in wound repair: molecular and cellular mechanisms. J Invest Dermatol 2007; 127: 514–25.

36. Decoursey TE, Ligeti E. Regulation and termination of NADPH oxidase activity. Cell Mol Life Sci 2005; 62: 2173–93.

37. Allen DB, Maguire JJ, Mahdavian M, Wicke C, Marcocci L, Scheuenstuhl H, Chang M, Le AX, Hopf HW, Hunt TK. Wound hypoxia and acidosis limit neutrophil bacterial killing mechanisms. Arch Surg 1997; 132: 991–6.

38. Ghani QP, Wagner S, Becker HD, Hunt TK, Hussain MZ. Regulatory role of lactate in wound repair. Methods Enzymol 2004; 381: 565–75.

39. Wagner S, Hussain MZ, Hunt TK, Bacic B, Becker HD. Stimulation of fibroblast proliferation by lactate-mediated oxidants. Wound Repair Regen 2004; 12: 368–73.

40. Beckert S, Farrahi F, Aslam RS, Scheuenstuhl H, Konigsrainer A, Hussain MZ, Hunt TK. Lactate stimulates endothelial cell migration. Wound Repair Regen 2006; 14: 321–4.

41. Patel V, Chivukula IV, Roy S, Khanna S, He G, Ojha N, Mehrotra A, Dias LM, Hunt TK, Sen CK. Oxygen: from the benefits of inducing VEGF expression to managing the risk of hyperbaric stress. Antioxid Redox Signal 2005; 7: 1377–87.

42. Sen CK. The general case for redox control of wound repair. Wound Repair Regen 2003; 11: 431–8.

43. Ushio-Fukai M. Redox signaling in angiogenesis: role of NADPH oxidase. Cardiovasc Res 2006; 71: 226–35.

44. Sen CK. Cellular thiols and redox-regulated signal transduction. Curr Top Cell Regul 2000; 36: 1–30.

45. Sen CK, Khanna S, Gordillo G, Bagchi D, Bagchi M, Roy S. Oxygen, oxidants, and antioxidants in wound healing: an emerging paradigm. Ann NY Acad Sci 2002; 957: 239–49.

46. Arbiser JL, Petros J, Klafter R, Govindajaran B, McLaughlin ER, Brown LF, Cohen C, Moses M, Kilroy S, Arnold RS, Lambeth JD. Reactive oxygen generated by Nox1 triggers the angiogenic switch. Proc Natl Acad Sci USA 2002; 99: 715–20.

47. Cho M, Hunt TK, Hussain MZ. Hydrogen peroxide stimulates macrophage vascular endothelial growth factor release. Am J Physiol Heart Circ Physiol 2001; 280: H2357–63.

48. Jackson IL, Batinic-Haberle I, Sonveaux P, Dewhirst MW, Vujaskovic Z. ROS production and angiogenic regulation by macrophages in response to heat therapy. Int J Hyperthermia 2006; 22: 263–73.

49. Jackson SJ, Venema RC. Quercetin inhibits eNOS, microtubule polymerization, and mitotic progression in bovine aortic endothelial cells. J Nutr 2006; 136: 1178–84.

50. Gengrinovitch S, Berman B, David G, Witte L, Neufeld G, Ron D. Glypican-1 is a VEGF165 binding proteoglycan that acts as an extracellular chaperone for VEGF165. J Biol Chem 1999; 274: 10816–22.

51. Sauer H, Bekhite MM, Hescheler J, Wartenberg M. Redox control of angiogenic factors and CD31-positive vessel-like structures in mouse embryonic stem cells after direct current electrical field stimulation. Exp Cell Res 2005; 304: 380–90.

52. Schafer G, Cramer T, Suske G, Kemmner W, Wiedenmann B, Hocker M. Oxidative stress regulates vascular endothelial growth factor-A gene transcription through Sp1- and Sp3-dependent activation of two proximal GC-rich promoter elements. J Biol Chem 2003; 278: 8190–8.

53. Kosmidou I, Xagorari A, Roussos C, Papapetropoulos A. Reactive oxygen species stimulate VEGF production from C(2)C(12) skeletal myotubes through a PI3K/Akt pathway. Am J Physiol Lung Cell Mol Physiol 2001; 280: L585–92.

54. Feliers D, Gorin Y, Ghosh-Choudhury G, Abboud HE, Kasinath BS. Angiotensin II stimulation of VEGF mRNA translation requires production of reactive oxygen species. Am J Physiol Renal Physiol 2006; 290: F927–36.

55. Kim JD, Liu L, Guo W, Meydani M. Chemical structure of flavonols in relation to modulation of angiogenesis and immune-endothelial cell adhesion. J Nutr Biochem 2006; 17: 165–76.

56. Sreekumar PG, Kannan R, de Silva AT, Burton R, Ryan SJ, Hinton DR. Thiol regulation of vascular endothelial growth factor-A and its receptors in human retinal pigment epithelial cells. Biochem Biophys Res Commun 2006; 346: 1200–6.

57. Huang SS, Zheng RL. Biphasic regulation of angiogenesis by reactive oxygen species. Pharmazie 2006; 61: 223–9.

58. Zhou Q, Liu LZ, Fu B, Hu X, Shi X, Fang J, Jiang BH. Reactive oxygen species regulate insulin-induced VEGF and HIF-1{alpha} expression through the activation of p70S6K1 in human prostate cancer cells. Carcinogenesis 2006.

59. Gao N, Ding M, Zheng JZ, Zhang Z, Leonard SS, Liu KJ, Shi X, Jiang BH. Vanadate-induced expression of hypoxia-inducible factor 1 alpha and vascular endothelial growth factor through phosphatidylinositol 3-kinase/Akt pathway and reactive oxygen species. J Biol Chem 2002; 277: 31963–71.

60. Matsubara T, Ziff M. Increased superoxide anion release from human endothelial cells in response to cytokines. J Immunol 1986; 137: 3295–8.

61. Stone JR, Collins T. The role of hydrogen peroxide in endothelial proliferative responses. Endothelium 2002; 9: 231–8.

62. Abid MR, Tsai JC, Spokes KC, Deshpande SS, Irani K, Aird WC. Vascular endothelial growth factor induces manganese-superoxide dismutase expression in endothelial cells by a Rac1-regulated NADPH oxidase-dependent mechanism. Faseb J 2001; 15: 2548–50.

63. Wang Z, Castresana MR, Newman WH. Reactive oxygen and NF-kappaB in VEGF-induced migration of human vascular smooth muscle cells. Biochem Biophys Res Commun 2001; 285: 669–74.

64. Colavitti R, Pani G, Bedogni B, Anzevino R, Borrello S, Waltenberger J, Galeotti T. Reactive oxygen species as downstream mediators of angiogenic signaling by vascular endothelial growth factor receptor-2/KDR. J Biol Chem 2002; 277: 3101–8.

65. Ushio-Fukai M, Tang Y, Fukai T, Dikalov SI, Ma Y, Fujimoto M, Quinn MT, Pagano PJ, Johnson C, Alexander RW. Novel role of gp91(phox)-containing NAD(P)H oxidase in vascular endothelial growth factor-induced signaling and angiogenesis. Circ Res 2002; 91: 1160–7.

66. Bonomo SR, Davidson JD, Yu Y, Xia Y, Lin X, Mustoe TA. Hyperbaric oxygen as a signal transducer: upregulation of platelet derived growth factor-beta receptor in the presence of HBO2 and PDGF. Undersea Hyperbaric Med 1998; 25: 211–6.

67. Morimoto K, Janssen WJ, Fessler MB, McPhillips KA, Borges VM, Bowler RP, Xiao YQ, Kench JA, Henson PM, Vandivier RW. Lovastatin enhances clearance of apoptotic cells (efferocytosis) with implications for chronic obstructive pulmonary disease. J Immunol 2006; 176: 7657–65.

68. Vandivier RW, Henson PM, Douglas IS. Burying the dead: the impact of failed apoptotic cell removal (efferocytosis) on chronic inflammatory lung disease. Chest 2006; 129: 1673–82.

69. Ferreira LS, Gerecht S, Shieh HF, Watson N, Rupnick MA, Dallabrida SM, Vunjak-Novakovic G, Langer R. Vascular Progenitor Cells Isolated From Human Embryonic Stem Cells Give Rise to Endothelial and Smooth Muscle-Like Cells and Form Vascular Networks In Vivo. Circ Res 2007.

70. Nareika A, He L, Game BA, Slate EH, Sanders JJ, London SD, Lopes-Virella MF, Huang Y. Sodium lactate increases LPS-stimulated MMP and cytokine expression in U937 histiocytes by enhancing AP-1 and NF-kappaB transcriptional activities. Am J Physiol Endocrinol Metab 2005; 289: E534–42.

71. Moldovan NI. Role of monocytes and macrophages in adult angiogenesis: a light at the tunnel's end. J Hematother Stem Cell Res 2002; 11: 179–94.

72. Anghelina M, Krishnan P, Moldovan L, Moldovan NI. Monocytes/macrophages cooperate with progenitor cells during neovascularization and tissue repair: conversion of cell columns into fibrovascular bundles. Am J Pathol 2006; 168: 529–41.

73. Tunyogi-Csapo M, Koreny T, Vermes C, Galante JO, Jacobs JJ, Glant TT. Role of fibroblasts and fibroblast-derived growth factors in periprosthetic angiogenesis. J Orthop Res 2007.

74. Attard JA, Raval MJ, Martin GR, Kolb J, Afrouzian M, Buie WD, Sigalet DL. The effects of systemic hypoxia on colon anastomotic healing: an animal model. Dis Colon Rectum 2005; 48: 1460–70.

75. Sheikh AY, Gibson JJ, Rollins MD, Hopf HW, Hussain Z, Hunt TK. Effect of hyperoxia on vascular endothelial growth factor levels in a wound model. Arch Surg 2000; 135: 1293–7.

76. Sheikh AY, Rollins MD, Hopf HW, Hunt TK. Hyperoxia improves microvascular perfusion in a murine wound model. Wound Repair Regen 2005; 13: 303–8.

77. Faglia E, Favales F, Aldeghi A, Calia P, Quarantiello A, Barbano P, Puttini M, Palmieri B, Brambilla G, Rampoldi A, Mazzola E, Valenti L, Fattori G, Rega V, Cristalli A, Oriani G, Michael M, Morabito A. Change in major amputation rate in a center dedicated to diabetic foot care during the 1980s: prognostic determinants for major amputation. J Diabetes Complications 1998; 12: 96–102.

78. Ghani QP, Wagner S, Hussain MZ. Role of ADP-ribosylation in wound repair. The contributions of Thomas K. Hunt, MD. Wound Repair Regen 2003; 11: 439–44.

79. Formby B, Stern R. Lactate-sensitive response elements in genes involved in hyaluronan catabolism. Biochem Biophys Res Commun 2003; 305: 203–8.

80. Ali MA, Yasui F, Matsugo S, Konishi T. The lactate-dependent enhancement of hydroxyl radical generation by the Fenton reaction. Free Radic Res 2000; 32: 429–38.

81. Wagner S, Hussain MZ, Beckert S, Ghani QP, Weinreich J, Hunt TK, Becker HD, Konigsrainer A. Lactate down-regulates cellular poly(ADP-ribose) formation in cultured human skin fibroblasts. Eur J Clin Invest 2007; 37: 134–9.

82. Koivunen P, Hirsila M, Remes AM, Hassinen IE, Kivirikko KI, Myllyharju J. Inhibition of hypoxia-inducible factor (HIF) hydroxylases by citric acid cycle intermediates: possible links between cell metabolism and stabilization of HIF. J Biol Chem 2007; 282: 4524–32.

83. Hashimoto T, Hussien R, Oommen S, Gohil K, Brooks GA. Lactate sensitive transcription factor network in L6 cells: activation of MCT1 and mitochondrial biogenesis. Faseb J 2007.

84. Kumar VB, Viji RI, Kiran MS, Sudhakaran PR. Endothelial cell response to lactate: implication of PAR modification of VEGF. J Cell Physiol 2007; 211: 477–85.

85. Hirota K, Semenza GL. Regulation of angiogenesis by hypoxia-inducible factor 1. Crit Revn Oncol-Hematol 2006; 59: 15–26.

86. Liu Q, Berchner-Pfannschmidt U, Moller U, Brecht M, Wotzlaw C, Acker H, Jungermann K, Kietzmann T. A Fenton reaction at the endoplasmic reticulum is involved in the redox control of hypoxia-inducible gene expression. Proc Natl Acad Sci USA 2004; 101: 4302–7.

87. Lu H, Forbes RA, Verma A. Hypoxia-inducible factor 1 activation by aerobic glycolysis implicates the Warburg effect in carcinogenesis. J Biol Chem 2002; 277: 23111–5.

88. Pan Y, Mansfield KD, Bertozzi CC, Rudenko V, Chan DA, Giaccia AJ, Simon MC. Multiple factors affecting cellular redox status and energy metabolism modulate hypoxia-inducible factor prolyl hydroxylase activity in vivo and in vitro. Mol Cell Biol 2007; 27: 912–25.

89. Dalgard CL, Lu H, Mohyeldin A, Verma A. Endogenous 2-oxoacids differentially regulate expression of oxygen sensors. Biochem J 2004; 380: 419–24.

90. Sen CK, Khanna S, Venojarvi M, Trikha P, Ellison EC, Hunt TK, Roy S. Copper-induced vascular endothelial growth factor expression and wound healing. Am J Physiol Heart Circ Physiol 2002; 282: H1821–7.

91. Feng J, Hunt TK, Ghani P, Z. HM. Macrophage-derived angiogenic activity potential can be reversibly inhibited by ADP-ribosylation. Wound Repair Regen 1997; 5: A111.

92. Schreiber V, Dantzer F, Ame JC, de Murcia G. Poly(ADP-ribose): novel functions for an old molecule. Nat Rev Mol Cell Biol 2006; 7: 517–28.

93. Kantor J, Margolis DJ. Treatment options for diabetic neuropathic foot ulcers: a cost-effectiveness analysis. Dermatol Surg 2001; 27: 347–51.

94. Steed DL. Clinical evaluation of recombinant human platelet-derived growth facsmalltor for the treatment of lower extremity ulcers. Plastic Reconstructive Surg 2006; 117: 143S–149S; discussion 150S–151S.

95. Calvert JW, Cahill J, Zhang JH. Hyperbaric oxygen and cerebral physiology. Neurol Res 2007; 29: 132–41.

96. Roeckl-Wiedmann I, Bennett M, Kranke P. Systematic review of hyperbaric oxygen in the management of chronic wounds. Br J Surg 2005; 92: 24–32.

97. Fries RB, Wallace WA, Roy S, Kuppusamy P, Bergdall V, Gordillo GM, Melvin WS, Sen CK. Dermal excisional wound healing in pigs following treatment with topically applied pure oxygen. Mutat Res 2005; 579: 172–81.

98. Gordillo GM, Schlanger R, Wallace WA, Bergdall V, Bartlett R, Sen CK. Protocols for topical and systemic oxygen treatments in wound healing. Methods Enzymol 2004; 381: 575–85.

99. Kalliainen LK, Gordillo GM, Schlanger R, Sen CK. Topical oxygen as an adjunct to wound healing: a clinical case series. Pathophysiology 2003; 9: 81–87.

100. Shih T, Lindley C. Bevacizumab: an angiogenesis inhibitor for the treatment of solid malignancies. Clin Therapeut 2006; 28: 1779–802.

101. Albina JE, Mastrofrancesco B, Vessella JA, Louis CA, Henry WL, Jr., Reichner JS. HIF-1 expression in healing wounds: HIF-1alpha induction in primary inflammatory cells by TNF-alpha. Am J Physiol Cell Physiol 2001; 281: C1971–7.

102. Newsholme EA, Crabtree B, Ardawi MS. The role of high rates of glycolysis and glutamine utilization in rapidly dividing cells. Biosci Rep 1985; 5: 393–400.

Chapter 47
Journeys in Coronary Angiogenesis

Julie M.D. Paye, Chohreh Partovian, and Michael Simons

Keywords: coronary angiogenesis, vascular development, arteriogenesis, coronary collaterals

Abstract: The path to therapeutic neovascularization in patients with coronary disease begins with an understanding of the basic mechanisms regulating vessel development, remodeling, and maintenance. In response to arterial occlusion, an effective "biologic bypass" such as coronary collaterals, needs to be created to restore blood flow to ischemic tissue. Although successful in healthy, young animal models, past attempts at utilizing exogenous growth factors in clinical trials to reach this goal have not met expectations. In response to this, the path has turned back to basic science, in particular coronary vascular development, to give new insight into potential therapeutic avenues.

Introduction

The vascular system is required for adequate oxygenation and viability of tissues. In the early stages of development, oxygen is provided from the maternal vascular system via diffusion. As the embryo develops and adequate oxygen levels are no longer met by diffusion alone, the embryo and its surrounding yolk sac begin to build their own vascular systems to meet oxygen demands. The development and maintenance of an efficient coronary vascular system is vital to ensure adequate oxygen perfusion and function of the adult heart (Fig. 47.1). Failure to maintain an adequate oxygen delivery system or adapt to increasing oxygen demands can be fatal if treatment strategies are not available or are ineffective. In this chapter, we review the current understanding of coronary arterial development, the attempts to capitalize on this knowledge to induce growth of new arteries in patients with coronary dis-

ease, and the return to basic investigation to further sharpen our understanding of arteriogenesis in order to develop new therapeutic strategies.

Coronary Development

The genesis of the coronary vasculature is marked by several well-choreographed events that lead to the formation of a primitive cardiac vascular network even before circulation begins. The process of coronary angiogenesis remodels and expands the primitive vascular meshwork of the coronary vascular plexus that is formed during vasculogenesis to create separate arterial and venous networks. Immediately prior to vasculogenesis, the process of organogenesis transforms the heart from a single tube into a multi-chambered, contracting organ.

Development of the Heart

Although circulation does not begin until after approximately 8.5 days of development in the mouse (23 days in human), coronary development is set into motion early in development with the derivation of a common cardiac progenitor (review in ref. [1]). Even after circulation does begin, since the heart wall is only a few cells thick, there is not yet a need for dedicated coronary vessels. After 10.5 days in the mouse, the proepicardium migrates to envelope the myocardium, giving rise to the epicardium. During this process, angioblasts from this migration infiltrate the myocardium to form blood islands, thus initiating coronary vasculogenesis.

Vasculogenesis

The cells comprising the embryo rapidly differentiate and proliferate such that the maternal circulation will soon no longer be able to satisfy the impending oxygen demands of the growing tissue. Before this can occur, the embryo begins to develop

Angiogenesis Research Center, Section of Cardiology Departments of Medicine and Pharmacology and Toxicology, Dartmouth Medical School, Lebanon, NH, USA

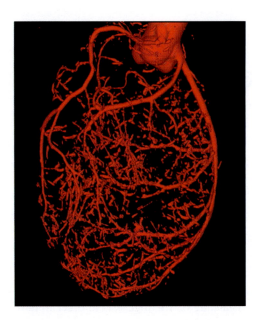

FIG. 47.1. Micro-CT image of adult mouse coronary vessels. Adequate formation and maintenance of coronary vessels is essential for normal oxygen perfusion and function. Image courtesy of Dr Zhen W. Zhuang, Dartmouth Medical School.

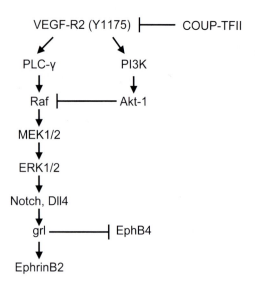

FIG. 47.2. VEGF-mediated signaling cascade involved in arterial specification. Activation of *ERK* or inhibition of the phosphatidylinositol-3 kinase (*PI3K*) pathway promotes an arterial fate, while inhibition of *ERK* or activation of *AKT* promotes a venous fate.

a primitive framework for its own cardiovascular system. This process of vasculogenesis begins in the yolk sac after 6.5 days in the mouse (connecting to the maternal circulation) and slightly later in the embryo proper. Vasculogenesis involves the migration of angioblasts from the early mesoderm in response to growth factors to form blood islands (an outer endothelial layer encapsulating hematopoietic cells) and vascular plexus (primitive endothelial tubular network). While there is evidence that blood islands are derived from a common hemangioblast [2], there is also evidence that specification of hematopoietic lineage (appearance of CD34+ or CD45+ cells) occurs directly from VE-Cadherin+ cells [3–6], suggesting that the 'hemangioblast' is endothelial in nature. Similar hemangiogenic endothelial cells have also been reported to be upregulated by cardiac ischemia in adult models [7].

Rather than sprouting from the already established aorta as once thought, the coronary arteries develop from a coronary vascular plexus formed through vasculogenesis. Coronary vasculogenesis begins shortly after the epicardium has enveloped the myocardium of the heart. Near this epicardial–myocardial border, after 10.5 days in the mouse, angioblasts from the pro-epicardial invasion begin to differentiate to form a primitive vascular network which progresses to encompass the entire heart. Inhibition in the migration of the proepicardium and thus delivery of angioblasts prevents proper coronary vessel development [8,9]. This vessel development is triggered by factors secreted from the myocardium which initiate an epithelial-to-mesenchymal transformation and ultimately differentiation to endothelial and smooth muscle cells. The wave of capillary plexus formation begins around the base of the aorta, forming a capillary ring, and moving along the atrial-ventricular

groove [10]. This capillary ring then forms two tubular projections that penetrate the aorta at the left and right cusps to initiate coronary circulation. These vascular projections are remodeled to form the left and right coronary arteries and are soon stabilized by smooth muscle cells derived from the proepicardium [11]. There is some evidence that smooth muscle cells can arise from transformation of the endothelium [12] (review in [13]). The coronary veins are formed by a similar capillary invasion into the sinus venosus.

Once flow is established, the plexus is remodeled into a more efficient network. This process may require pruning of some vascular branches and expansion of others, but ultimately results in a hierarchical system of vessels with varying diameters. In response to molecular markers, branches of this vascular tree adopt either an arterial or venous identity. Although the exact mechanisms governing specification are not clear, some important regulators have emerged. Recent studies suggest that vascular endothelial growth factor (VEGF) signaling is important for the specification of arteries from the default venous fate (Fig. 47.2).

Growth Factors Controlling Vascular Development

VEGF Family

As previously described in earlier sections of this book, there are six growth factors that comprise the VEGF family (VEGF-A, VEGF-B, VEGF-C, VEGF-D, VEGF-E and a related placental growth factor) and three receptors (VEGFR1, VEGFR2, VEGFR3). Homozygous knockout of *VEGFR1*, *2* or *3* results in lethality associated with abnormalities in vasculogenesis (VEGFR1 and 2) and vessel maturation (VEGFR3). The first VEGF receptor expressed during development is VEGFR2 (also known as KDR or Flk-1) and homozygous deletion results in embryonic death before E10.5 due to improper formation of blood islands and subsequent vasculature [14]. Expressed later in development, VEGFR1 (also known as Flt-1) acts as a negative regulator of vascular development, sequestering VEGF away from VEGFR2. Homozygous deletion of this gene results in excess hemangioblast commitment and poorly organized endothelial cell differentiation in blood islands and other vascular structures, resulting in growth arrest at E8.5 [15,16]. Expression of VEGFR3 begins around E8 and is largely restricted to lymphatic vasculature once it is formed. Homozygous deletion of *VEGFR3* results in growth retardation after E9.0, pericardial edema by E10, and necrosis by E12.5 [17].

Expression of VEGF-A in the developing myocardium is coincident with developing vascular structures, both progressing as a wave from the outer myocardium [18]. Deletion of a single allele of the *VEGF-A* results in lethality by E10.5, and mice homozygous for *VEGF-A* deletion have defects in vasculogenesis, angiogenesis, and tumorigenesis. These mice displayed diminished formation of blood islands and dorsal aorta, and improper plexus formation and remodeling [19,20]. After four weeks of development, teratomas derived from VEGF$^{-/-}$ ES cells were 10-fold smaller than controls and exhibited defects in vascular branching [20]. Of note, there are five splice variants of *VEGF-A*. Mice that express only VEGF$_{164}$ are normal and healthy, while mice only expressing VEGF$_{120}$ die within 2 weeks of birth, and those expressing only VEGF$_{188}$ have reduced Mendelian frequency at birth, suggesting partial embryonic lethality [21].

While the importance of VEGF-A has been demonstrated, further studies suggest that other VEGF family members may also be important in coronary angiogenesis. Treatment of heart explants with either anti-VEGF-B or anti-VEGF-C was more effective in inhibiting in vitro tube formation than anti-VEGF-A. Further, treatment with soluble VEGFR1 (which binds VEGF-A, VEGF-B, and PIGF) was nearly twice as effective in inhibiting in vitro tube formation than soluble VEGFR2 (which binds VEGF-A, VEGF-C, VEGF-D, VEGF-E) [18,22]. Deletion of *VEGF-B* is compatible with survival but is associated with abnormal atrial conduction [23].

FGF Family

There are 22 fibroblast growth factors (FGFs) and four high-affinity tyrosine kinase receptors, many of which exist as multiple splice variants. Due to this complexity, it has been difficult to determine the biologic importance of each member. Several of the members important during development are FGFs 2, 3, 4, 5, 8, 9, and 10, and FGFRs 1–4 (discussed in the Chapter 7).

FGF2 is ubiquitously expressed in the developing myocardium, with higher levels just prior to the onset of vasculogenesis and also coincident with postnatal capillary angiogensis

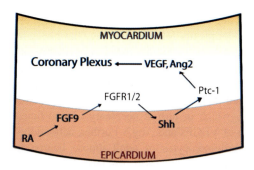

FIG. 47.3. Signaling involved in coronary vascular plexus formation. Coronary sonic hedgehog (*Shh*) signaling is necessary and sufficient for the induction of *VEGFs* and Ang2 and subsequent formation of coronary vascular plexus.

[11]. Delivery of a bolus injection of FGF2 into chick hearts results in a subsequent doubling of the vascular volume percent in comparison with sham animals [24]. Treatment of E8.5 mouse embryos with RNAi against *FGF2* prevented heart tube looping, embryo turning, and resulted in embryonic death within 12 h [25], further suggesting its importance in coronary development. However, knockout of *FGF2* or *FGF1* in mice is compatible with survival and is not associated with a profound cardiac phenotype [26,27].

FGFR128, 29, FGFR230, FGF4 [31,32], and FGF8 [33,34] are all required for proper mesoderm and endoderm migration and patterning and their deletion results in early embryonic lethality. Due to their importance in early development, it has been difficult to study the role of these molecules in vasculogenesis and angiogenesis through loss of function models. Studies employing embryoid bodies derived from *FGFR1*$^{-/-}$ embryonic stem (ES) cells reported an increased basal plexus formation and decreased VEGFR2 transcription, suggesting that FGFR1 plays a role in the differentiation of the hemangioblast [35]. However, endothelial FGFR1 and FGFR2 are not required for the formation of the coronary vascular plexus or mature coronary vessels [36]. Myocardial expression of FGFR1 and FGFR2 has been demonstrated to be important for coronary plexus formation. In response to retinoic acid, the epicardium and endocardium secrete FGF9 [37], which then signals through either myocardial FGFR1 or FGFR2, resulting in increased expression of VEGFs and angiopoietin-2 (Ang2) that is dependent on hedgehog signaling [36] (Fig. 47.3).

Hedgehog

The hedgehog family consists of three secreted ligands (sonic, Indian, and desert) that signal primarily through the transmembrane receptor Patched-1 (Ptc-1), and have been shown to be important for vasculogenesis and angiogenesis (reviews in [38–40]). Expression of sonic hedgehog (Shh), the predominant hedgehog ligand in the developing heart, begins as a wave from the atrial-ventricular groove and is localized to the epicardium [36]. The pattern of Shh signaling through Ptc-1, whose expression is localized to cardiomyocytes and perivascular cells, is coincident with that of VEGF-A, VEGF-B, VEGF-C and Ang2 and emerges just prior to the formation of the coronary vascular plexus. VEGFs and Ang2 have been reported to act synergistically to promote coronary development. Through loss of function studies in mice, it has been shown that myocardial expression of FGFR1 and FGFR2 and epicardial and endocardial expression of FGF9 are required for coronary Shh signaling, which in turn is necessary and sufficient for the induction of VEGFs and Ang2. Further, activation of hedgehog in cardiomyocytes of adult mouse hearts resulted in increased blood vessel density, suggesting that Shh could be a potential target for therapeutic angiogenesis.

Ephrins and Ephs

Much interest has been generated in the Eph tyrosine kinase receptor B4 (EphB4) and its cognate transmembrane ligand, ephrin-B2. Expression of ephrin-B2 is restricted to arteries while EphB4 is predominately expressed in veins, suggesting that arterial and venous fate determination is at least in part genetically predetermined rather than guided entirely by hemodynamic flow as previously thought. Targeted deletion of *ephrin-B2* [41,42] or *EphB4* [43] results in embryonic lethality around E10.5. Treatment of zebrafish with anti-VEGF morpholino oligos resulted in decreased expression of ephrin-B2 [44], suggesting that VEGF may be involved in the arterial specification of endothelial cells. In mice that only express $VEGF_{188}$, there is a reduction in the number of ephrin-B2 positive arterioles in their retinas [21], while mice that express $VEGF_{164}$ under the control of αMHC have increased expression of ephrin-B2 positive and decreased expression of EphB4 positive vessels in the heart [45]. In addition, in the absence of VEGF expression, only 10% of angioblasts at E10.5 become ephrin-B2 positive while the addition of $VEGF_{164}$ or $VEGF_{120}$ increases that proportion 50% [46].

Neuropilins

Neuropilins, which have been shown to be important for neuron guidance, are also important in the vasculature. The expression of neuropilin-1 (a $VEGF_{165}$ receptor [47]) has been shown to be specific to arteries in the mouse [46,48] and neuropilin-2 is preferentially enhanced in veins. Knockout out of *neuropilin-1* in mice is lethal by E12.5 [49] and double knockout of *neuropilin-1* and *2* is lethal by E8 resulting from severe vascular defects. However, recent studies have shown that knockin of *neuropilin-1* with a mutated semaphorin binding site but normal VEGF binding site rescued mice from embryonic lethality [50]. This suggests that VEGF-neuropilin-1 signaling is important during development.

Notch Family

It is thought that VEGF may also act by inducing Notch signaling, which may in turn affect ephrin-B2 expression [44]. Many members of the Notch family (notch1, notch3, notch4,

dLl4, jagged1, and jagged2) have been shown to be expressed in arteries but not in veins of mouse embryos [51,52]. Knockout of the main transducers of Notch signaling in mammals, *Hey1* and *Hey2*, results in embryonic lethality before E11.5 [53]. In these mice, the primary vascular plexus is formed but further remodeling or formation of large vessels fails to occur. In addition, similar to *Notch1* knockout mice, there was a dramatic decrease in ephrin-B2 staining of aortas while EphB4 staining was normal in veins. Knockdown of *gridlock*, the zebrafish ortholog of *Hey2*, results in an impaired aorta formation and a complete loss of arterial markers [54]. Based on additional studies of *gridlock*, a recent paper suggests that arterial specification by VEGF is regulated by which downstream signaling cascades are activated [55]. Activation of ERK or inhibition of the phosphatidylinositol-3 kinase (PI3K) pathway promotes an arterial fate, while inhibition of ERK or activation of AKT promotes a venous fate.

PDGF Family

The main players of the platelet-derived growth factor (PDGF) family in vascular development are PDGF-B and PDGFR-β. Sprouting endothelial cells secrete PDGF-B, which is then immobilized to cell-surface heparan sulfate proteoglycans. This creates a chemotactic gradient that recruits mural cells, which express PDGFR-β. Deletion of either *PDGF-B* (either globally or in endothelial cells) or *PDGFR-β* results in a scattering of mural cells away from the vessel wall, resulting in destabilized vessels that are sensitive to rupture (reviewed in [56]). This is evident by the microaneurysms which result in embryonic lethality [57]. Neoangiogenesis via PDGF-B and PDGFR-β has been reported to be enhanced via a synergistic relationship between with VEGF and FGF2 to promote neoangiogenesis [58].

COUP-TFII

The orphan nuclear receptor, COUP-TFII, is expressed in the differentiating myocardium and sinus venosus [59] and thought to be a marker of venous endothelium [60]. Deletion of *COUP-TFII* in one allele results in death in two-thirds of pups before weaning, and homozygous deletion is associated with growth retardation, hemorrhage and edema, and is lethal around E10 [59]. Further, there are malformations in the atria and sinus venosus and cardinal vein and general defects in remodeling of the capillary plexus.

Synectin

The receptor scaffold protein, synectin, has recently been shown to be important for arterial vascular branching [61]. Synectin has been shown to interact with the FGF receptor syndecan-4 and through its PDZ domain to more than 20 partners, including neuropilin-1. Synectin$^{-/-}$ mice are smaller in size than their wild type litter mates and have lower vascular density per unit tissue volume and, consistent with this decrease in the vasculature, significantly less VE-cadherin and α smooth muscle actin mRNA levels. Micro-CT analysis revealed that synectin knockout mice had smaller vascular trees and less arterial branching, most of which was in arteries in the 27–108 μm diameter range. Fractal analysis revealed that not only was there a reduction in vascular density, but that the branching pattern was significantly different.

Arterial endothelial cells isolated from synectin knock-out mice migrated slower than arterial wild type controls although migration of synectin$^{-/-}$ venous endothelial cells was indistinguishable from the wild type. Transduction of synectin$^{-/-}$ arterial endothelial cells with adeno-synectin restored their ability to migrate to the same extent as wild type. Similar results were seen for proliferation and three-dimensional *in vitro* tube formation in collagen. Taken together, this suggests that synectin has a direct role in the regulation of arterial endothelial cell migration.

Hemodynamics and Ischemia

Once circulation begins, hemodynamic forces, such as shear stress induced by fluid flow and circumferential stretch induced by the pulsatile heart beat, continue to remodel the coronary vasculature. With the onset of circulation, there are changes in cytoskeletal organization as endothelial cells shift from a globular morphology to become aligned with the axis of flow. There are also changes in endothelial gene expression, including upregulation of PDGF-B and Flk-1.

In the first few days of circulation, the hemodynamic forces generated seem to be a more important factor for remodeling than the oxygenated blood that creates them. Culture of zebrafish, *Xenopus*, chick, or mouse embryos in carbon monoxide did not affect these early remodeling processes. However, all future angiogenic processes are intimately involved with the oxygen homeostasis, which is regulated by the transcription factor hypoxia-inducible factor-1 (HIF-1). During hypoxia, there is an upregulation of VEGF signaling that results in increased vessel sprouting and increased tissue perfusion62. Ideally, this increased tissue perfusion is sufficient to meet oxygen demands but therapeutic intervention may be required.

Therapeutic Stimulation of Coronary Growth

The growing understanding of developmental angiogenesis and coronary arterial growth has prompted attempts at therapeutic stimulation of these processes. When a major artery becomes obstructed, and spontaneous development of collateral vessels is insufficient for normal perfusion of the tissue at risk, tissue ischemia develops. Currently available approaches for treating patients with ischemic disease include medical therapy essentially aimed at reducing myocardial oxygen demand, revascularization by percutaneous angioplasty, or artery bypass grafting. However, a significant number of patients are not amenable to revascularization or achieve incomplete revascularization with these procedures. This results in persistent symptoms of ischemia despite intensive medical therapy and a poor five-year survival rate for these patients. Therefore, there is a need for innovative treatments. Based on our increased understanding of the molecular mechanisms underlying vascular homeostasis, therapeutic angiogenesis has passed through several scientific phases [63].

Early Phase Clinical Trials

In the first phase, individual angiogenic growth factors were exogenously delivered in the form of protein or gene (either naked plasmid DNA or viral vector) to the ischemic myocardium or skeletal muscle in order to amplify the native biological revascularization in response to ischemia [64–66]. The angiogenic factor or its cognate DNA sequence injected into the ischemic area is believed to act by binding to its receptors on the surface of endothelial cells activating these cells to migrate, proliferate and form new vessels, thus allowing increased perfusion of the tissue surrounding the injection site. This strategy worked well in healthy, normocholesterolemic young laboratory animals which were subjected to an acute interruption of coronary or femoral arterial blood flow [67,68]. In large animal models of chronic myocardial ischemia, single bolus intracoronary, periadventitial or intrapericardial administration of several growth factors including FGF1, FGF2, VEGF$_{165}$ and PDGF-BB enhanced neovascularization and restored blood flow in the ischemic territory to essentially normal levels [69–71]. Gene therapy approaches were used with equal success, including intracoronary injections of FGF4, FGF5, VEGF$_{121}$ and VEGF$_{165}$ adenoviruses as well as VEGF$_{165}$ plasmid [72]. In the hindlimb ischemia model, where an occlusion of the femoral artery is typically associated with a severe reduction in blood flow to the ipsilateral limb, treatment with angiogenic growth factors accelerated recovery and in some cases preserved tissue loss in the foot and the distal ankle [73].

Encouraging results were also obtained in early clinical trials which lacked control populations. These open label studies testing VEGF$_{165}$, FGF1 and FGF2 [74–78] demonstrated significant improvement in every parameter measured, including measures of myocardial perfusion and ventricular function, albeit in the absence of controls. Altogether, these findings suggested that administration of angiogenic growth factors, both gene and protein-mediated, can potentially promote the development of new blood vessels.

Randomized, Controlled Trials

Consequently, a number of large randomized placebo-controlled phase II/III clinical trials were initiated, with variable results (Table 47.1). While they further confirmed the feasibility and

TABLE 47.1. Randomized placebo-controlled phase II/III clinical trials based on exogenous delivery of growth factors.

Study	Number of patients	Type of patients	Growth factor	Route of administration	Primary endpoint	Conclusion
TRAFFIC {Lederman, 2002 #14}	190	Peripheral vascular disease (PVD) with infra-inguinal atherosclerosis, moderate to severe intermittent claudication	FGF2 protein or placebo	Bilateral intra-arterial infusions at day 1 or day 1 and day 30	Peak walking time (PWT) at 90 and 180 days	Significant improvement at 90 days not sustained at 180 days
RAVE {Rajagopalan, 2003 #15}	105	PVD, unilateral exercice-limiting intermittent claudication	Adenoviral VEGF$_{121}$ low dose, high dose or placebo	Intramuscular injection to the ischemic leg	Change in PWT at 12 weeks and 26 weeks	No significant difference between the groups
FIRST {Simons, 2002 #16}	337	Advanced coronary artery disease (CAD) ineligible for revascularization	FGF2 protein Three different concentrations	Single intracoronary infusion	Change in exercise tolerance time (ETT) at 90 and 180 days	No significant difference between the groups
VIVA {Henry, 2003 #17}	178	Stable angina unsuitable for revascularization	Recombinant VEGF Two different doses or placebo	Intracoronary injection followed by IV infusions at day 3, 6 and 9	Change in ETT at 60 days	No significant difference between the groups
AGENT-1 {Grines, 2002 #18}	79	Stable angina CCS 2 or 3	Adenoviral FGF-4 Five different doses or placebo	Intracoronary injection	Change in ETT at 4 and 12 weeks	Only significant difference in a subgroup of patients with baseline ETT of 10min or less
AGENT-2 {Doukas, 2002 #19}	52	Stable angina unsuitable for revascularization and reversible ischemia >9% of the LV	Adenoviral FGF-4	Intracoronary injection	Change in myocardial perfusion defect size assessed by SPECT imaging	Trend to a reduction in defect size
EUROINJECT-I {Kastrup, 2005 #23}	80	Severe stable ischemic heart disease CCS 3 or 4 "no option" patients	Plasmid gene transfer of VEGF-A$_{165}$	Intramyocardial injection via the NOGA catheter system into sites of perfusion defects	Myocardial perfusion assessed by nuclear perfusion imaging	No significant change in perfusion defects Significant improvement of local wall motion at 3 months
REVASC {Stewart, 2006 #20}	71	Severe CAD, unsuitable for revascularization, "no option"	Adenoviral VEGF$_{121}$ versus maximal medical therapy	Direct intramyocardial injection via left lateral thoracotomy	Time to 1 mm ST depression on exercise tolerance test	Significantly reduced at 26 weeks. The mean CCS angina class also reduced at both 12 and 26 weeks

safety of the therapeutic angiogenesis concept, they provided very limited evidence of its efficacy. There was a significant improvement in both placebo and treatment groups suggesting the presence of a large placebo effect as well as the potential for a significant improvement in the control group [79]. Several hypotheses were considered in order to explain the negative results. These include a lack of adequate expression of the pro-angiogenic factors for a sufficient duration, unresponsiveness of a diseased vasculature to angiogenic stimulation, incorrect choice of patients, or perhaps the inability to detect the treatment effect, among other considerations [80].

The concept that the prolonged presence of a growth factor in tissues would be superior to a single bolus therapy was tested in a small trial of heparin alginate sustained-release FGF2 formulation implanted into the unrevascularizable but viable myocardium at the time of coronary artery bypass grafting [77]. A significant reduction in the occurrence of angina was observed in the group of patients treated with the high dose of recombinant FGF2 at three months which remarkably persisted after three years of follow up. [81]

Patient Selection

Patient selection is an important issue since most trials enrolled no option / poor option patients with advanced atherosclerotic disease. These patients have a history of failed natural neovascularization in response to ischemia that may predispose them to be resistant to angiogenic therapy [82]. The induction of angiogenic response in elderly patients with several co-morbid cardiovascular disease risk factors is clearly more challenging than stimulating new blood vessel formation in animal models of chronic myocardial and peripheral vascular ischemia utilizing young healthy animals. Indeed, studies in older animals and in animals with hypercholesterolemia were considerably less promising than studies in healthy young species. Therefore, the choice of patient population to include those still responsive to angiogenic stimulation would be highly desirable. However, we currently lack the biomarkers to identify such a population. Post-hoc subgroup analyses in two FGF trials, AGENT and FIRST, identified highly symptomatic patients with poor baseline ETT (exercise treadmill testing) as being most likely to respond to myocardial angiogenic therapy. However, these predictions have not been tested prospectively.

Arteriogenesis and Coronary Collaterals

Importantly, none of the trials to date explicitly addressed the arteriogenesis versus angiogenesis issue. On physiological grounds, it is unlikely that an increase in the number of capillaries alone is sufficient to overcome tissue ischemia without providing new avenues of arterial blood supply. Arteriogenesis, which refers to the development or enlargement of collateral arterioles, is a key adaptive response to arterial occlusion [83,84]. Even though arteriogenesis takes place in non-ischemic areas, it has the potential to fully restore blood flow to the distal bed by creating an effective "biological bypass". However, the spatial and temporal changes in blood flow, cytokines, growth factors, and vasoactive agents linking arteriogenesis to the capillarization of distal muscle remains largely unexplored. Multiple angiogenic factors may be required to stimulate both angiogenesis and arteriogenesis in a coordinated manner.

Patients who develop coronary collaterals in the course of their illness have a better outcome and quality of life. However, clinical observations suggest that there are substantial differences in the extent of collateral development among patients, with some individuals demonstrating marked abundance while others show a near complete absence of collaterals [85]. This interindividual heterogeneity may also help to explain the variable angiogenic responses to therapeutic intervention. Factors associated with decreased collateral formation including aging [86,87], hypercholesterolemia [88,89], hypertension [90,91], diabetes [92] and cigarette smoking [93] have been described in animals and humans. Therefore, the principal cause of poor collateral development, either endogenous or in response to an exogenous intervention, may not be the absence of a particular growth factor, but rather the inability to respond to angiogenic signaling due to environmental, epigenetic, or genetic causes.

Few studies have illustrated the contribution of genetic factors. Resar and colleagues reported data suggesting that variations in hypoxia inducible factor-1 (HIF-1α) genotype may influence the development of coronary artery collaterals in patients with significant coronary artery disease [94]. HIF-1 is a transcriptional activator that functions as a master regulator of oxygen homeostasis. The presence of a single nucleotide polymorphism (SNP) which changes residue 582 from proline to serine is a negative predictor of collateral formation94. Haptoglobin phenotype has also been associated with the development of coronary collateral circulation in diabetic patients with coronary artery disease [95]. Recent publications from the Simons group have suggested the existence of a genetic program specifically targeted to the growth of the arterial tree controlled by synectin. Synectin (GIPC1) is a single domain PDZ protein that interacts with a number of growth factor receptors including the FGF receptor, syndecan-4, and the VEGF-A$_{165}$ co-receptor, neuropilin-1, among others [80, 96]. Studies in mice and zebrafish have shown that disruption of synectin expression is associated with a selective defect in the growth and development of small arteries and arterioles while the venous circulation is not affected [61]. Array analysis of gene expression in synectin-deleted endothelial cells demonstrated abnormalities in pathways associated with cell proliferation and migration [97].

The existence of a genetic predisposition to collateral development was further demonstrated in a microarray analysis of monocyte gene expression from patients with extensive vs. absent collateral development in the presence of advanced coronary disease. There were distinct genetic differences between the two patient populations that could not be explained by any known clinical or biological parameter [98].

Monocytes, Bone Marrow-derived Cells and Arteriogenesis

Monocytes are considered a key cell type in arteriogenesis. They accumulate at arterial occlusion sites and secrete a number of cytokines involved in pericyte, endothelial, and smooth muscle cell growth and differentiation. Cultured peripheral blood mononuclear cells have been shown to secrete high levels of VEGF, hepatocyte growth factor (HGF), granulocyte colony stimulating factor (G-CSF), and granulocyte-macrophage colony stimulating factor (GM-CSF) [99]. Therefore, augmentation of monocyte/macrophage accumulation and activation at the desired site may be a more effective therapeutic strategy than exogenous administration of a single growth factor. Animal studies have demonstrated that administration of cytokines such as monocyte chemoattractant protein-1 (MCP-1), which leads to the activation of monocytes / macrophages, is associated with a robust arteriogenic response and significant restoration of tissue perfusion [100]. A small randomized, double-blind, placebo-controlled study that evaluated GM-CSF infusions in patients with coronary artery disease also reported positive results [101]. However, the START trial, in which patients with limb ischemia were treated with GM-CSF, failed to demonstrate efficacy [102] as have several subsequent trials of GM-CSF and CSF stimulation [103–105]. In addition, two trials using G-CSF therapy in the setting of intracoronary bone marrow cell injection reported a significant increase in the in-stent restenosis rate [106,107].

This brings us to the next phase of therapeutic angiogenesis characterized by the concept of bone-marrow- and peripheral blood-derived cell populations stimulating neovascularization. Asahara and colleagues described the presence of putative endothelial progenitor cells (EPCs) in the circulating blood as a peripheral blood mononuclear population with the ability to differentiate into endothelial cells and incorporate into ischemic tissues at sites of angiogenesis [108]. Shortly after, Shi and colleagues reported that a bone-marrow-derived CD34+ mononuclear cell population is able to differentiate into endothelial cells and colonize aortic Dacron implants [109]. Numerous studies reported findings that indicate the incorporation of progenitor cells in vessels undergoing repair in limb ischemia [110–113], at sites of myocardial infarction [114–119], and wound healing [120–123]. The cells mobilized from the bone marrow and recruited to the sites of ischemia were thought to differentiate into the additional endothelial, pericyte, and smooth muscle cells that were required for new blood vessel formation. The potential for cell-based therapies engendered tremendous enthusiasm in the field and pilot clinical studies started being explored.

However, the cell surface markers used to select these progenitor cells varied significantly between different studies leading to controversy as to the identity and, indeed, existence of an EPC population [124]. In addition, the levels of engraftment or percent of incorporation reported in these studies varied enormously, from less than 1% to 95%. A new wave of studies questioned not only the high degree of incorporation but also the mechanism by which bone-marrow-derived cells are participating in the neovascularization process. By using more careful labeling and confocal microscopy, these studies suggested that the EPCs do not integrate directly into the endothelial wall but rather take residence immediately behind the vessel wall. The role of these cells is presumably to provide paracrine signals by secreting angiogenic growth factors to enhance survival and growth of the nearby resident endothelial cells. Given these findings, it is clear that the term "EPC' should be abandoned and replaced by a more appropriate moniker such as "accessory angiogenic cells" (AAC) [125].

Much effort has gone into mobilization of an appropriate AAC population from the peripheral blood and bone marrow and targeting it to sites of ischemia. Most promisingly, stromal cell-derived factor-1 (SDF-1) appears to play a major role in the mobilization of these cells while local VEGF levels in ischemic tissues play a role in the cells' recruitment [125].

Ex-vivo amplification and / or genetic manipulation have also been successfully employed to further enhance their therapeutic function [126]. Cells modified to express VEGF have been shown to induce a greater improvement in blood flow and angiogenesis in animals models of ischemia than progenitor cells alone [110,127]. Cells isolated from bone marrow and transfected to express telomerase reverse transcriptase (TERT) were more resistant to apoptosis and had a drastically increased rate of neovascularization in an animal model of limb ischemia [128].

Despite continued controversy regarding the mechanism by which these cells promote vascularization after administration to laboratory animals, a number of trials were initiated to test their clinical efficacy. Certain, albeit small, benefits were seen in patients with acute myocardial infarction injected with autologous bone marrow cells in some trials [129,130], while the same strategy was ineffective in several others [131,132]. In addition, the effect reported in the BOOST trial regarding the improvement in left ventricular ejection fraction after infusion of bone marrow cells at six months was no longer significant at 18 months [133]. Clearly, further investigations including double-blind placebo-controlled studies are required in order to definitively establish the efficacy of this novel approach.

Basic Biology of Arteriogenesis: New Insights

Given these less than stellar results, the attention again shifted to better understanding of basic angiogenesis biology with a hope of deriving new clues for therapeutic interventions. In the past decade, there has been increasing appreciation of the fact that pathways used predominantly during embryogenesis and known to be relatively silent during normal adult life may be recruited postnatally in response to tissue injury. Among these pathways is that of sonic hedgehog (Shh), known to play an important role in vascularization of embryonic tissues as previously discussed in this chapter, and has also been recently shown to be involved in the adult neovascularization in the setting of ischemia [134,135]. Exogenous administration of Shh in old mice with surgically-induced limb ischemia was associated with a sharp increase in limb salvage in comparison to vehicle- or VEGF$_{165}$-treated mice. The expression of Hh receptor, Patched-1 (Ptc1), present in adult vascular tissues and in the interstitial mesenchymal cells within the ischemic area, was increased in response to exogenous Shh, leading to an increase in blood flow, number of capillaries, and also in vessels' diameter [135]. The effect of Shh was mediated via an increase in the expression of angiogenic factors including all three isoforms of VEGF-A and both angiopoietins 1 and 2, therefore considered as an indirect angiogenic factor.

Summary

In conclusion, adult angiogenesis is a more complex process than previously anticipated. Essentially, all attempts to induce therapeutic neovascularization in patients by the delivery of exogenous growth factors have failed to demonstrate a significant objective benefit. A more recent therapeutic approach involves the mobilization of bone-marrow- and peripheral blood-derived progenitor cells which may sustain angiogenesis and arteriogenesis by releasing soluble molecules around the nascent vessels and, to a minimal extent, by incorporation in the newly formed vasculature. However, the more direct exploration of these cells as therapeutic tools in patients is long-reached until double-blind, placebo-controlled, randomized trials establish its efficacy. Certainly, more information from basic science is needed to aid in the development of safe and predictable therapies with measurable outcomes. Finally, the identification of the mechanisms which allow some and not other patients to develop coronary collaterals may open new avenues for therapy.

References

1. Garry DJ, Olson EN. A common progenitor at the heart of development. Cell 2006;127(6):1101–4.
2. Choi K, Kennedy M, Kazarov A, Papadimitriou JC, Keller G. A common precursor for hematopoietic and endothelial cells. Development 1998;125(4):725–32.
3. Nishikawa SI, Nishikawa S, Hirashima M, Matsuyoshi N, Kodama H. Progressive lineage analysis by cell sorting and culture identifies FLK1+VE-cadherin+ cells at a diverging point of endothelial and hemopoietic lineages. Development 1998;125(9):1747–57.
4. Nadin BM, Goodell MA, Hirschi KK. Phenotype and hematopoietic potential of side population cells throughout embryonic development. Blood 2003;102(7):2436–43.
5. Fujimoto T, Ogawa M, Minegishi N, et al. Step-wise divergence of primitive and definitive haematopoietic and endothelial cell lineages during embryonic stem cell differentiation. Genes Cells 2001;6(12):1113–27.
6. Fraser ST, Ogawa M, Yu RT, Nishikawa S, Yoder MC, Nishikawa S. Definitive hematopoietic commitment within the embryonic vascular endothelial-cadherin(+) population. Exp Hematol 2002;30(9):1070–8.
7. Kogata N, Arai Y, Pearson JT, et al. Cardiac ischemia activates vascular endothelial cadherin promoter in both preexisting vascular cells and bone marrow cells involved in neovascularization. Circ Res 2006;98(7):897–904.
8. Gittenberger-de Groot AC, Vrancken Peeters MP, Bergwerff M, Mentink MM, Poelmann RE. Epicardial outgrowth inhibition leads to compensatory mesothelial outflow tract collar and abnormal cardiac septation and coronary formation. Circ Res 2000;87(11):969–71.
9. Yang JT, Rayburn H, Hynes RO. Cell adhesion events mediated by alpha 4 integrins are essential in placental and cardiac development. Development 1995;121(2):549–60.
10. Tomanek RJ, Hansen HK, Dedkov EI. Vascular patterning of the quail coronary system during development. Anat Rec A Discov Mol Cell Evol Biol 2006;288(9):989–99.
11. Tomanek RJ, Haung L, Suvarna PR, O'Brien LC, Ratajska A, Sandra A. Coronary vascularization during development in the rat and its relationship to basic fibroblast growth factor. Cardiovasc Res 1996;31 Spec No:E116–26.
12. DeRuiter MC, Poelmann RE, VanMunsteren JC, Mironov V, Markwald RR, Gittenberger-de Groot AC. Embryonic endothelial cells transdifferentiate into mesenchymal cells expressing smooth muscle actins in vivo and in vitro. Circ Res 1997;80(4):444–51.
13. Hirschi KK, Majesky MW. Smooth muscle stem cells. Anat Rec A Discov Mol Cell Evol Biol 2004;276(1):22–33.
14. Shalaby F, Rossant J, Yamaguchi TP, et al. Failure of blood-island formation and vasculogenesis in Flk-1-deficient mice. Nature 1995;376(6535):62–6.
15. Fong GH, Rossant J, Gertsenstein M, Breitman ML. Role of the Flt-1 receptor tyrosine kinase in regulating the assembly of vascular endothelium. Nature 1995;376(6535):66–70.
16. Fong GH, Zhang L, Bryce DM, Peng J. Increased hemangioblast commitment, not vascular disorganization, is the primary defect in flt-1 knock-out mice. Development 1999;126(13):3015–25.
17. Dumont DJ, Jussila L, Taipale J, et al. Cardiovascular failure in mouse embryos deficient in VEGF receptor-3. Science 1998;282(5390):946–9.
18. Tomanek RJ, Holifield JS, Reiter RS, Sandra A, Lin JJ. Role of VEGF family members and receptors in coronary vessel formation. Dev Dyn 2002;225(3):233–40.
19. Carmeliet P, Ferreira V, Breier G, et al. Abnormal blood vessel development and lethality in embryos lacking a single VEGF allele. Nature 1996;380(6573):435–9.
20. Ferrara N, Carver-Moore K, Chen H, et al. Heterozygous embryonic lethality induced by targeted inactivation of the VEGF gene. Nature 1996;380(6573):439–42.
21. Stalmans I, Ng YS, Rohan R, et al. Arteriolar and venular patterning in retinas of mice selectively expressing VEGF isoforms. Journal of Clinical Investigation 2002;109(3):327–36.

22. Tomanek RJ, Ishii Y, Holifield JS, Sjogren CL, Hansen HK, Mikawa T. VEGF family members regulate myocardial tubulogenesis and coronary artery formation in the embryo. Circ Res 2006;98(7):947–53.

23. Aase K, von Euler G, Li X, et al. Vascular endothelial growth factor-B-deficient mice display an atrial conduction defect. Circulation 2001;104(3):358–64.

24. Tomanek RJ, Lotun K, Clark EB, Suvarna PR, Hu N. VEGF and bFGF stimulate myocardial vascularization in embryonic chick. Am J Physiol 1998;274(5 Pt 2):H1620–6.

25. Leconte I, Fox JC, Baldwin HS, Buck CA, Swain JL. Adenoviral-mediated expression of antisense RNA to fibroblast growth factors disrupts murine vascular development. Dev Dyn 1998;213(4):421–30.

26. Miller DL, Ortega S, Bashayan O, Basch R, Basilico C. Compensation by fibroblast growth factor 1 (FGF1) does not account for the mild phenotypic defects observed in FGF2 null mice. Mol Cell Biol 2000;20(6):2260–8.

27. Zhou M, Sutliff RL, Paul RJ, et al. Fibroblast growth factor 2 control of vascular tone. Nat Med 1998;4(2):201–7.

28. Deng CX, Wynshaw-Boris A, Shen MM, Daugherty C, Ornitz DM, Leder P. Murine FGFR-1 is required for early postimplantation growth and axial organization. Genes & Development 1994;8(24):3045–57.

29. Yamaguchi TP, Harpal K, Henkemeyer M, Rossant J. fgfr-1 is required for embryonic growth and mesodermal patterning during mouse gastrulation. Genes Dev 1994;8(24):3032–44.

30. Arman E, Haffner-Krausz R, Chen Y, Heath JK, Lonai P. Targeted disruption of fibroblast growth factor (FGF) receptor 2 suggests a role for FGF signaling in pregastrulation mammalian development. Proc Natl Acad Sci USA 1998;95(9):5082–7.

31. Feldman B, Poueymirou W, Papaioannou VE, DeChiara TM, Goldfarb M. Requirement of FGF-4 for postimplantation mouse development. Science 1995;267(5195):246–9.

32. Goldin SN, Papaioannou VE. Paracrine action of FGF4 during periimplantation development maintains trophectoderm and primitive endoderm. Genesis 2003;36(1):40–7.

33. Sun X, Meyers EN, Lewandoski M, Martin GR. Targeted disruption of Fgf8 causes failure of cell migration in the gastrulating mouse embryo. Genes Dev 1999;13(14):1834–46.

34. Meyers EN, Lewandoski M, Martin GR. An Fgf8 mutant allelic series generated by Cre- and Flp-mediated recombination. Nat Genet 1998;18(2):136–41.

35. Magnusson P, Rolny C, Jakobsson L, et al. Deregulation of Flk-1/vascular endothelial growth factor receptor-2 in fibroblast growth factor receptor-1-deficient vascular stem cell development. J Cell Sci 2004;117(Pt 8):1513–23.

36. Lavine KJ, White AC, Park C, et al. Fibroblast growth factor signals regulate a wave of Hedgehog activation that is essential for coronary vascular development. Genes Dev 2006;20(12):1651–66.

37. Lavine KJ, Yu K, White AC, et al. Endocardial and epicardial derived FGF signals regulate myocardial proliferation and differentiation in vivo. Dev Cell 2005;8(1):85–95.

38. Bijlsma MF, Peppelenbosch MP, Spek CA. Hedgehog morphogen in cardiovascular disease. Circulation 2006;114(18):1985–91.

39. Byrd N, Grabel L. Hedgehog signaling in murine vasculogenesis and angiogenesis. Trends Cardiovasc Med 2004;14(8):308–13.

40. Lamont RE, Childs S. MAPping out arteries and veins. Sci STKE 2006;2006(355):pe39.

41. Adams RH, Wilkinson GA, Weiss C, et al. Roles of ephrinB ligands and EphB receptors in cardiovascular development: demarcation of arterial/venous domains, vascular morphogenesis, and sprouting angiogenesis. Genes Dev 1999;13(3):295–306.

42. Wang HU, Chen ZF, Anderson DJ. Molecular distinction and angiogenic interaction between embryonic arteries and veins revealed by ephrin-B2 and its receptor Eph-B4.[see comment]. Cell 1998;93(5):741–53.

43. Gerety SS, Wang HU, Chen ZF, Anderson DJ. Symmetrical mutant phenotypes of the receptor EphB4 and its specific transmembrane ligand ephrin-B2 in cardiovascular development. Mol Cell 1999;4(3):403–14.

44. Lawson ND, Vogel AM, Weinstein BM. sonic hedgehog and vascular endothelial growth factor act upstream of the Notch pathway during arterial endothelial differentiation. Dev Cell 2002;3(1):127–36.

45. Visconti RP, Richardson CD, Sato TN. Orchestration of angiogenesis and arteriovenous contribution by angiopoietins and vascular endothelial growth factor (VEGF). Proc Natl Acad Sci USA 2002;99(12):8219–24.

46. Mukouyama YS, Shin D, Britsch S, Taniguchi M, Anderson DJ. Sensory nerves determine the pattern of arterial differentiation and blood vessel branching in the skin. Cell 2002;109(6):693–705.

47. Soker S, Takashima S, Miao HQ, Neufeld G, Klagsbrun M. Neuropilin-1 is expressed by endothelial and tumor cells as an isoform-specific receptor for vascular endothelial growth factor. Cell 1998;92(6):735–45.

48. Moyon D, Pardanaud L, Yuan L, Breant C, Eichmann A. Plasticity of endothelial cells during arterial-venous differentiation in the avian embryo. Development 2001;128(17):3359–70.

49. Kitsukawa T, Shimizu M, Sanbo M, et al. Neuropilin-semaphorin III/D-mediated chemorepulsive signals play a crucial role in peripheral nerve projection in mice. Neuron 1997;19(5):995–1005.

50. Gu C, Rodriguez ER, Reimert DV, et al. Neuropilin-1 conveys semaphorin and VEGF signaling during neural and cardiovascular development. Dev Cell 2003;5(1):45–57.

51. Shutter JR, Scully S, Fan W, et al. Dll4, a novel Notch ligand expressed in arterial endothelium. Genes Dev 2000;14(11):1313–8.

52. Villa N, Walker L, Lindsell CE, Gasson J, Iruela-Arispe ML, Weinmaster G. Vascular expression of Notch pathway receptors and ligands is restricted to arterial vessels. Mech Dev 2001;108(1–2):161–4.

53. Fischer A, Schumacher N, Maier M, Sendtner M, Gessler M. The Notch target genes Hey1 and Hey2 are required for embryonic vascular development. Genes Dev 2004;18(8):901–11.

54. Zhong TP, Childs S, Leu JP, Fishman MC. Gridlock signalling pathway fashions the first embryonic artery. Nature 2001;414(6860):216–20.

55. Hong CC, Peterson QP, Hong JY, Peterson RT. Artery/vein specification is governed by opposing phosphatidylinositol-3 kinase and MAP kinase/ERK signaling. Curr Biol 2006;16(13):1366–72.

56. Armulik A, Abramsson A, Betsholtz C. Endothelial/pericyte interactions. Circ Res 2005;97(6):512–23.

57. Lindahl P, Johansson BR, Leveen P, Betsholtz C. Pericyte loss and microaneurysm formation in PDGF-B-deficient mice. Science 1997;277(5323):242–5.

58. Kano MR, Morishita Y, Iwata C, et al. VEGF-A and FGF-2 synergistically promote neoangiogenesis through enhancement of endogenous PDGF-B-PDGFRbeta signaling. J Cell Sci 2005;118(Pt 16):3759–68.

59. Pereira FA, Qiu Y, Zhou G, Tsai MJ, Tsai SY. The orphan nuclear receptor COUP-TFII is required for angiogenesis and heart development. Genes Dev 1999;13(8):1037–49.

60. You LR, Lin FJ, Lee CT, DeMayo FJ, Tsai MJ, Tsai SY. Suppression of Notch signalling by the COUP-TFII transcription factor regulates vein identity. Nature 2005;435(7038):98–104.

61. Chittenden TW, Claes F, Lanahan AA, et al. Selective regulation of arterial branching morphogenesis by synectin. Dev Cell 2006;10(6):783–95.

62. Yue X, Tomanek RJ. Stimulation of coronary vasculogenesis/angiogenesis by hypoxia in cultured embryonic hearts. Dev Dyn 1999;216(1):28–36.

63. Semenza GL. Therapeutic angiogenesis: another passing phase? Circ Res 2006;98(9):1115–6.

64. Baumgartner I, Isner JM. Stimulation of peripheral angiogenesis by vascular endothelial growth factor (VEGF). Vasa 1998;27(4):201–6.

65. Goncalves LM. Fibroblast growth factor-mediated angiogenesis for the treatment of ischemia. Lessons learned from experimental models and early human experience. Rev Port Cardiol 1998;17 Suppl 2:II11–20.

66. Hamawy AH, Lee LY, Crystal RG, Rosengart TK. Cardiac angiogenesis and gene therapy: a strategy for myocardial revascularization. Curr Opin Cardiol 1999;14(6):515–22.

67. Baffour R, Berman J, Garb JL, Rhee SW, Kaufman J, Friedmann P. Enhanced angiogenesis and growth of collaterals by in vivo administration of recombinant basic fibroblast growth factor in a rabbit model of acute lower limb ischemia: dose-response effect of basic fibroblast growth factor. J Vasc Surg 1992;16(2):181–91.

68. Unger EF, Banai S, Shou M, et al. Basic fibroblast growth factor enhances myocardial collateral flow in a canine model. Am J Physiol 1994;266(4 Pt 2):H1588–95.

69. Annex BH, Simons M. Growth factor-induced therapeutic angiogenesis in the heart: protein therapy. Cardiovasc Res 2005;65(3):649–55.

70. Devlin GP, Fort S, Yu E, et al. Effect of a single bolus of intracoronary basic fibroblast growth factor on perfusion in an ischemic porcine model. Can J Cardiol 1999;15(6):676–82.

71. Sato K, Wu T, Laham RJ, et al. Efficacy of intracoronary or intravenous VEGF165 in a pig model of chronic myocardial ischemia. J Am Coll Cardiol 2001;37(2):616–23.

72. Giordano FJ, Ping P, McKirnan MD, et al. Intracoronary gene transfer of fibroblast growth factor-5 increases blood flow and contractile function in an ischemic region of the heart. Nat Med 1996;2(5):534–9.

73. Bauters C, Asahara T, Zheng LP, et al. Site-specific therapeutic angiogenesis after systemic administration of vascular endothelial growth factor. J Vasc Surg 1995;21(2):314–24; discussion 24–5.

74. Chawla PS, Keelan MH, Kipshidze N. Angiogenesis for the treatment of vascular diseases. Int Angiol 1999;18(3):185–92.

75. Marti HH, Risau W. Angiogenesis in ischemic disease. Thromb Haemost 1999;82 Suppl 1:44–52.

76. Schumacher B, Pecher P, von Specht BU, Stegmann T. Induction of neoangiogenesis in ischemic myocardium by human growth factors: first clinical results of a new treatment of coronary heart disease. Circulation 1998;97(7):645–50.

77. Laham RJ, Sellke FW, Edelman ER, et al. Local perivascular delivery of basic fibroblast growth factor in patients undergoing coronary bypass surgery: results of a phase I randomized, double-blind, placebo-controlled trial. Circulation 1999;100(18):1865–71.

78. Rosengart TK, Lee LY, Patel SR, et al. Angiogenesis gene therapy: phase I assessment of direct intramyocardial administration of an adenovirus vector expressing VEGF121 cDNA to individuals with clinically significant severe coronary artery disease. Circulation 1999;100(5):468–74.

79. Simons M. Angiogenesis: where do we stand now? Circulation 2005;111(12):1556–66.

80. Gao Y, Li M, Chen W, Simons M. Synectin, syndecan-4 cytoplasmic domain binding PDZ protein, inhibits cell migration. J Cell Physiol 2000;184(3):373–9.

81. Ruel M, Laham RJ, Parker JA, et al. Long-term effects of surgical angiogenic therapy with fibroblast growth factor 2 protein. J Thorac Cardiovasc Surg 2002;124(1):28–34.

82. Lekas M, Lekas P, Latter DA, Kutryk MB, Stewart DJ. Growth factor-induced therapeutic neovascularization for ischaemic vascular disease: time for a re-evaluation? Curr Opin Cardiol 2006;21(4):376–84.

83. Heil M, Schaper W. Pathophysiology of collateral development. Coron Artery Dis 2004;15(7):373–8.

84. Heil M, Schaper W. Influence of mechanical, cellular, and molecular factors on collateral artery growth (arteriogenesis). Circ Res 2004;95(5):449–58.

85. Sherman JA, Hall A, Malenka DJ, De Muinck ED, Simons M. Humoral and cellular factors responsible for coronary collateral formation. Am J Cardiol 2006;98(9):1194–7.

86. Arthur WT, Vernon RB, Sage EH, Reed MJ. Growth factors reverse the impaired sprouting of microvessels from aged mice. Microvasc Res 1998;55(3):260–70.

87. Rivard A, Fabre JE, Silver M, et al. Age-dependent impairment of angiogenesis. Circulation 1999;99(1):111–20.

88. Bucay M, Nguy J, Barrios R, Chen CH, Henry PD. Impaired adaptive vascular growth in hypercholesterolemic rabbit. Atherosclerosis 1998;139(2):243–51.

89. Van Belle E, Rivard A, Chen D, et al. Hypercholesterolemia attenuates angiogenesis but does not preclude augmentation by angiogenic cytokines. Circulation 1997;96(8):2667–74.

90. Karpanou EA, Vyssoulis GP, Skoumas JN, Zervopoulos GA, Moundaki VS, Toutouzas PK. Significance of arterial hypertension on coronary collateral circulation development and left ventricular function in coronary artery disease. J Hypertens Suppl 1988;6(4):S151–3.

91. Kyriakides ZS, Kremastinos DT, Michelakakis NA, Matsakas EP, Demovelis T, Toutouzas PK. Coronary collateral circulation in coronary artery disease and systemic hypertension. Am J Cardiol 1991;67(8):687–90.

92. Melidonis A, Tournis S, Kouvaras G, et al. Comparison of coronary collateral circulation in diabetic and nondiabetic patients suffering from coronary artery disease. Clin Cardiol 1999;22(7):465–71.

93. Heinle RA, Levy RI, Gorlin R. Effects of factors predisposing to atherosclerosis on formation of coronary collateral vessels. Am J Cardiol 1974;33(1):12–6.

94. Resar JR, Roguin A, Voner J, et al. Hypoxia-inducible factor 1alpha polymorphism and coronary collaterals in patients with ischemic heart disease. Chest 2005;128(2):787–91.

95. Hochberg I, Roguin A, Nikolsky E, Chanderashekhar PV, Cohen S, Levy AP. Haptoglobin phenotype and coronary artery collaterals in diabetic patients. Atherosclerosis 2002;161(2):441–6.

96. Cai H, Reed RR. Cloning and characterization of neuropilin-1-interacting protein: a PSD-95/Dlg/ZO-1 domain-containing protein that interacts with the cytoplasmic domain of neuropilin-1. J Neurosci 1999;19(15):6519–27.

97. Lanahan AA, Chittenden TW, Mulvihill E, Smith K, Schwartz S, Simons M. Synectin-dependent gene expression in endothelial cells. Physiol Genomics 2006;27(3):380–90.

98. Chittenden TW, Sherman JA, Xiong F, et al. Transcriptional profiling in coronary artery disease: indications for novel markers of coronary collateralization. Circulation 2006;114(17):1811–20.

99. Rehman J, Li J, Orschell CM, March KL. Peripheral blood "endothelial progenitor cells" are derived from monocyte/macrophages and secrete angiogenic growth factors. Circulation 2003;107(8):1164–9.

100. Buschmann IR, Hoefer IE, van Royen N, et al. GM-CSF: a strong arteriogenic factor acting by amplification of monocyte function. Atherosclerosis 2001;159(2):343–56.

101. Seiler C, Pohl T, Wustmann K, et al. Promotion of collateral growth by granulocyte-macrophage colony-stimulating factor in patients with coronary artery disease: a randomized, double-blind, placebo-controlled study. Circulation 2001;104(17):2012–7.

102. van Royen N, Schirmer SH, Atasever B, et al. START Trial: a pilot study on STimulation of ARTeriogenesis using subcutaneous application of granulocyte-macrophage colony-stimulating factor as a new treatment for peripheral vascular disease. Circulation 2005;112(7):1040–6.

103. Valgimigli M, Rigolin GM, Cittanti C, et al. Use of granulocyte-colony stimulating factor during acute myocardial infarction to enhance bone marrow stem cell mobilization in humans: clinical and angiographic safety profile. Eur Heart J 2005;26(18):1838–45.

104. Zohlnhofer D, Ott I, Mehilli J, et al. Stem cell mobilization by granulocyte colony-stimulating factor in patients with acute myocardial infarction: a randomized controlled trial. JAMA 2006;295(9):1003–10.

105. Engelmann MG, Theiss HD, Hennig-Theiss C, et al. Autologous bone marrow stem cell mobilization induced by granulocyte colony-stimulating factor after subacute ST-segment elevation myocardial infarction undergoing late revascularization: final results from the G-CSF-STEMI (Granulocyte Colony-Stimulating Factor ST-Segment Elevation Myocardial Infarction) trial. J Am Coll Cardiol 2006;48(8):1712–21.

106. Kang HJ, Kim HS, Zhang SY, et al. Effects of intracoronary infusion of peripheral blood stem-cells mobilised with granulocyte-colony stimulating factor on left ventricular systolic function and restenosis after coronary stenting in myocardial infarction: the MAGIC cell randomised clinical trial. Lancet 2004;363(9411):751–6.

107. Steinwender C, Hofmann R, Kammler J, et al. Effects of peripheral blood stem cell mobilization with granulocyte-colony stimulating factor and their transcoronary transplantation after primary stent implantation for acute myocardial infarction. Am Heart J 2006;151(6):1296 e7–13.

108. Asahara T, Murohara T, Sullivan A, et al. Isolation of putative progenitor endothelial cells for angiogenesis. Science 1997;275(5302):964–7.

109. Shi Q, Rafii S, Wu MH, et al. Evidence for circulating bone marrow-derived endothelial cells. Blood 1998;92(2):362–7.

110. Iwaguro H, Yamaguchi J, Kalka C, et al. Endothelial progenitor cell vascular endothelial growth factor gene transfer for vascular regeneration. Circulation 2002;105(6):732–8.

111. Kalka C, Masuda H, Takahashi T, et al. Vascular endothelial growth factor(165) gene transfer augments circulating endothelial progenitor cells in human subjects. Circ Res 2000;86(12):1198–202.

112. Schatteman GC, Hanlon HD, Jiao C, Dodds SG, Christy BA. Blood-derived angioblasts accelerate blood-flow restoration in diabetic mice. J Clin Invest 2000;106(4):571–8.

113. Takahashi T, Kalka C, Masuda H, et al. Ischemia- and cytokine-induced mobilization of bone marrow-derived endothelial progenitor cells for neovascularization. Nat Med 1999;5(4):434–8.

114. Orlic D, Kajstura J, Chimenti S, et al. Bone marrow cells regenerate infarcted myocardium. Nature 2001;410(6829):701–5.

115. Edelberg JM, Tang L, Hattori K, Lyden D, Rafii S. Young adult bone marrow-derived endothelial precursor cells restore aging-impaired cardiac angiogenic function. Circ Res 2002;90(10):E89–93.

116. Jackson KA, Majka SM, Wang H, et al. Regeneration of ischemic cardiac muscle and vascular endothelium by adult stem cells. J Clin Invest 2001;107(11):1395–402.

117. Kocher AA, Schuster MD, Szabolcs MJ, et al. Neovascularization of ischemic myocardium by human bone-marrow-derived angioblasts prevents cardiomyocyte apoptosis, reduces remodeling and improves cardiac function. Nat Med 2001;7(4):430–6.

118. Orlic D, Kajstura J, Chimenti S, et al. Mobilized bone marrow cells repair the infarcted heart, improving function and survival. Proc Natl Acad Sci USA 2001;98(18):10344–9.

119. Shintani S, Murohara T, Ikeda H, et al. Mobilization of endothelial progenitor cells in patients with acute myocardial infarction. Circulation 2001;103(23):2776–9.

120. Asahara T, Masuda H, Takahashi T, et al. Bone marrow origin of endothelial progenitor cells responsible for postnatal vasculogenesis in physiological and pathological neovascularization. Circ Res 1999;85(3):221–8.

121. Asahara T, Takahashi T, Masuda H, et al. VEGF contributes to postnatal neovascularization by mobilizing bone marrow-derived endothelial progenitor cells. Embo J 1999;18(14):3964–72.

122. Carmeliet P, Luttun A. The emerging role of the bone marrow-derived stem cells in (therapeutic) angiogenesis. Thromb Haemost 2001;86(1):289–97.

123. Crosby JR, Kaminski WE, Schatteman G, et al. Endothelial cells of hematopoietic origin make a significant contribution to adult blood vessel formation. Circ Res 2000;87(9):728–30.

124. Barber CL, Iruela-Arispe ML. The ever-elusive endothelial progenitor cell: identities, functions and clinical implications. Pediatr Res 2006;59(4 Pt 2):26R–32R.

125. Grunewald M, Avraham I, Dor Y, et al. VEGF-induced adult neovascularization: recruitment, retention, and role of accessory cells. Cell 2006;124(1):175–89.

126. Kaushal S, Amiel GE, Guleserian KJ, et al. Functional small-diameter neovessels created using endothelial progenitor cells expanded ex vivo. Nat Med 2001;7(9):1035–40.

127. Ikeda Y, Fukuda N, Wada M, et al. Development of angiogenic cell and gene therapy by transplantation of umbilical cord blood

with vascular endothelial growth factor gene. Hypertens Res 2004;27(2):119–28.

128. Murasawa S, Llevadot J, Silver M, Isner JM, Losordo DW, Asahara T. Constitutive human telomerase reverse transcriptase expression enhances regenerative properties of endothelial progenitor cells. Circulation 2002;106(9):1133–9.

129. Wollert KC, Meyer GP, Lotz J, et al. Intracoronary autologous bone-marrow cell transfer after myocardial infarction: the BOOST randomised controlled clinical trial. Lancet 2004;364(9429):141–8.

130. Assmus B, Honold J, Schachinger V, et al. Transcoronary transplantation of progenitor cells after myocardial infarction. N Engl J Med 2006;355(12):1222–32.

131. Lunde K, Solheim S, Aakhus S, et al. Intracoronary injection of mononuclear bone marrow cells in acute myocardial infarction. N Engl J Med 2006;355(12):1199–209.

132. Janssens S, Dubois C, Bogaert J, et al. Autologous bone marrow-derived stem-cell transfer in patients with ST-segment elevation myocardial infarction: double-blind, randomised controlled trial. Lancet 2006;367(9505):113–21.

133. Meyer GP, Wollert KC, Lotz J, et al. Intracoronary bone marrow cell transfer after myocardial infarction: eighteen months' follow-up data from the randomized, controlled BOOST (BOne marrOw transfer to enhance ST-elevation infarct regeneration) trial. Circulation 2006;113(10):1287–94.

134. Pola R, Ling LE, Aprahamian TR, et al. Postnatal recapitulation of embryonic hedgehog pathway in response to skeletal muscle ischemia. Circulation 2003;108(4):479–85.

135. Pola R, Ling LE, Silver M, et al. The morphogen Sonic hedgehog is an indirect angiogenic agent upregulating two families of angiogenic growth factors. Nat Med 2001;7(6):706–11.

Chapter 48
Perspectives on the Future of Angiogenesis Research

Douglas Hanahan

Keywords: angiogenic switch, tumor angiogenesis, vascular regulatory leukocytes, vasculature, angiogenic regulatory network

Abstract: The future of angiogenesis research holds great promise. Future prospects can be appreciated from the perspectives of historical landmarks in the field, and of the landscape of our current understanding about the regulation of angiogenesis and manifestations of the vasculature, in particular the parameters of the angiogenic switch. Looking ahead, five milestone goals for the field are highlighted, centered upon elucidating the intracellular integrated circuits that process a multitude of regulatory signals received from the extracellular microenvironment. This global regulatory network involves signals transmitted amongst a set of cell types constituting a core of the angiogenic system: endothelial cells, pericytes, and a variety of vascular regulatory leukocytes, as well as the vascular stem and progenitor cells that spawn these component cells. The realization of these strategic goals will set the stage for translation of the new knowledge to human medicine, as elaborated in seven postulates of future applications that could have profound impact on the treatment and prevention of human diseases, in particular cancers.

Looking Back: Where We Have Come From

In the last 35 years, the field of angiogenesis research has grown remarkably, and unforeseeably. What began as a then contentious proposition that tumor angiogenesis was regulated, and not merely a sympathetic response to the need for oxygen [1], has become a field of extraordinary accomplishment and breadth, with evermore fascinating revelations of how angio-

Department of Biochemistry and Biophysics, Helen Diller Family Comprehensive Cancer Center, and Diabetes Center, UCSF, San Francisco, CA 94143, USA

genesis is regulated and how a functioning neovasculature is produced in distinctive physiological and pathological situations within different tissues. Much of this history is collectively covered in other chapters in this volume (e.g., [2]), and in a number of recent reviews (e.g., [3–6]). Some of the historical breakthroughs cited and described in these and other reviews deserve mention herein, to set the stage for thinking about the future.

Historical Breakthroughs

- Methods for culturing endothelial cells and for assaying angiogenesis in vivo (e.g., implanting cells or test agents onto the chick chorioallantoic membrane or into a rabbit/rodent corneal micro-pocket) were developed in the 1970s (and refined ever since), enabling the first functional studies into the regulation and characteristics of endothelial cells and their assembly during angiogenesis into an operational neovasculature; these in vivo systems were used in early pioneering studies demonstrating the importance of angiogenesis for growth of explanted tumor cells.
- Genes encoding protein growth factors capable of simulating endothelial cell proliferation in culture (and in turn angiogenesis in vivo) were discovered, beginning with those for the protein ligands bFGF (1984) and VEGF-A (1989), originally identified as a vascular permeability factor (1983), establishing the precedent for recurrent discoveries of dozens of proangiogenic regulatory genes.
- Specific endogenous angiogenesis inhibitory proteins capable of counteracting proangiogenic growth factors were discovered and implicated in natural control of tumor growth, beginning most convincingly with Thrombospondin-1 in 1989, and continuing with Angiostatin, Endostatin, et al., in the 1990s.
- Animal transplant models, involving transfer of cells or tissues into heterologous animals, in particular involving the formation of subcutaneous and later orthotopic tumors, enabled more sophisticated mechanistic studies into the process, regulation, and functional importance of angiogenesis for tumor growth;

- Genetically engineered mouse models, both transgenic (gain-of-function) and gene knockout (loss-of-function) were applied to studying the vasculature—its development during embryogenesis, its homeostasis, and its physiological and pathological angiogenesis in the adult; by altering the expression of candidate angiogenic regulatory genes, it has been possible to ascribe defined functional roles to such genes via their perturbation in these different contexts.
- Engineered mouse models of multistage carcinogenesis revealed that angiogenesis is typically switched on in early premalignant stages prior to the formation and progression of solid tumors, and that sustained angiogenesis is crucial for rapid tumor growth.
- Analysis of premalignant lesions to human cancers demonstrated that angiogenesis is similarly switched on early, in incipient neoplasias, at least in some human tissues, and that aberrant vasculature typifies most all human tumors; increased microvessel density has proved prognostic of poor outcome in certain tumor types.
- Lymphangiogenesis, the formation of new lymphatic vessels via proliferation (and differentiation from progenitors) of lymphatic endothelial cells was revealed as a parallel process to angiogenesis of the vasculature, sharing regulatory signals as well as developmental interrelationship.

These and other technological and conceptual developments laid the foundation for an explosive phase of angiogenesis research in the current decade that has revealed unpredicted complexity in the regulatory circuits governing the induction (and often inappropriate) persistence of angiogenesis, and the morphological orchestration of a functional neovasculature composed of multiple interacting cell types.

Looking Around: the Landscape in 2008

The chapters in this book, as well as other recent reviews (e.g., [3, 5–8]), summarize in depth the current state of the science and technology of angiogenesis research. The field is fast paced and expansive, characterized by a remarkable continuum of reports describing and refining mechanisms of vascular development, of homeostatic functionality, and of both transitory (normal/physiological) and chronic (aberrant/pathological) angiogenic processes that spawn new blood (and lymphatic) vessels. There are ever-better tools to visualize and functionally perturb the vasculature and its component cell types, and a prolific drug development enterprise in academia and industry is aimed at pharmacologically inhibiting angiogenesis for the treatment of human diseases with angiogenic etiologies. In developing a perspective about the future of angiogenesis research, it is also useful to reflect on some highlights of the current landscape in 2007, as well as its history. (In general, the reader is referred to the cited reviews and/or chapters throughout this volume for additional detail and for primary references, which are only cited when not yet reviewed.)

Highlights of the Current State of Knowledge about Angiogenesis.

- Vascular quiescence and homeostatic functions as well as physiological and pathological angiogenesis are regulated by complex networks of similar and overlapping signals.
- Vascular regulatory signals come in many forms:

  – soluble paracrine and endocrine ligands that bind to transmembrane receptors (typically with kinases serving as signal transducers);
  – immobilized ligands in the vascular basement membrane and extracellular matrix that bind to heterodimeric integrin receptors on the cell surface, whereby both ligand bound and unbound integrin receptors can transmit context-dependent messages into the cell (ranging from growth stimulation to growth suppression to apoptosis);
  – intercellular signals conveyed by cell-to-cell contacts involving both homotypic and heterotypic molecular interactions of transmembrane signaling molecules (e.g., VCAM<–>VCAM and Delta<–>Notch, respectively), again typically involving intracellular kinase components for signal transmission; and
  – biophysical signals, in the form of blood and interstitial fluid pressure, tissue oxygen levels (in particular hypoxia), and the hydrodynamic stress of blood flow.

- There are more than two dozen extracellular signals capable of inducing or enhancing angiogenesis, and a similar number of endogenous angiogenesis inhibitory signals.
- Immune inflammatory cells, perhaps better termed 'vascular regulatory leukocytes' (VRL), are proving to be important regulators of angiogenesis, demonstrably involved in angiogenic switching and in the orchestration of both of transient physiological and chronic pathological angiogenesis.
- A diverse repertoire of cell types and sub-types constitute the class of vascular regulatory leukocytes (VRL), derived from stem cells resident in the bone marrow; VRLs serve as modulators of developmental, physiological, and pathological angiogenesis and likely vascular homeostasis; amongst the protagonists are macrophages, neutrophils, and various subtypes of monocytes, including both progenitors to traditional macrophages and neutrophils and likely new subtypes (e.g., monocytes defined by expression of Tie2 or VEGFR1 + CXCR4), as well as B lymphocytes, and platelets.
- The first generation of specific angiogenesis inhibitors have entered clinical practice [7,8], where they are showing modest efficacy, typically in the form of 'delayed time to progression' indicative of renewed tumor growth following a period of partial response (tumor shrinkage) or stable disease (tumor stasis).
- Mechanisms of an 'evasive resistance' by tumors to VEGF/ VEGFR inhibitors, both in preclinical models and in clinical trials, involving upregulation of alternative proangiogenic signals, have been implicated and suggested to underlay (at least in part) the transitory efficacy of the first generation of angiogenesis inhibitors [9–12].

- Angiogenesis inhibitors targeting proangiogenic signaling circuits are showing toxicity in clinical trials and practice [13, 14], likely reflecting additional homeostatic roles of such signals in the normal vascular system.
- Pericytes are now recognized as important components of chronically angiogenic vasculature, in addition to their previously recognized roles in normal development and homeostasis [15–17].
- Multiple paracrine signals transmitted between endothelial cells and pericytes are demonstrably important for their productive association to sustain the functionality of the tumor endothelium [15–17]; inhibitors of these signals (e.g., of PDGFR [18–21]) hold promise for improving the efficacy of antiangiogenic therapies directed at the endothelial cells, by disrupting support and survival functions conveyed by pericyte coverage.

A Current View of the Angiogenic Switch and Tumor Angiogenesis

The knowledge base elaborated in this volume and in the field at large is conveying important insight into the angiogenic switch that turns on angiogenesis in otherwise quiescent endothelial cells of blood (and lymphatic) vessels [4, 6, 8, 22, 23]. The increasing knowledge about mechanisms of the angiogenic switch bears particular mention in regard to future prospects and possibilities for angiogenesis research and its impact on diseases with angiogenic components.

Notable Features of the Angiogenic Switch Circa 2008

- The angiogenic switch evidently has, in addition to an "off" state, multiple "on" states, reflecting variations, in the net intensity and effects of the involved regulatory signals, in the consequent rate of angiogenesis (and likely lymphangiogenesis), and in the quality of the blood (or lymphatic) vessels formed.
- The state of the angiogenic switch is governed by the local balance of positive and negative regulatory signals, involving endogenous activators and inhibitors of angiogenesis; while suppressors of lymphangiogenesis have not yet been identified, it seem likely that endogenous lymphangiogenesis inhibitors will be discovered, reflecting a similar balance mechanism for the "lymphangiogenic switch".
- Signaling from VEGF-A to VEGFR2 appears to be generic to most manifestations of the angiogenic switch, as assessed in multiple tissues during normal and disease states [6, 24, 25]; similarly, signaling from VEGF-C/D to VEGFR3 has apparent preeminence for inducing lymphangiogenesis [6, 23–25].
- Multiple other proangiogenic signals can enhance and in some cases substitute for VEGF-A-stimulated angiogenic switching and persistence of pathological angiogenesis,

including signals conveyed by ligands for other transmembrane receptor tyrosine kinases (fibroblast growth factors, delta-like ligands, ephrin-A/Bs, semaphorins, angiopoeitins, et al.) [6, 26–30], as well as ligands for a set of proangiogenic integrins (including αVß3, αVß5, and α5ß1) [31–34].
- Endogenous angiogenesis inhibitors, including a number derived by proteolytic degradation of the vascular basement membrane, can demonstrably counterbalance proangiogenic signals, and are likely involved in rendering physiological angiogenesis transitory [8, 35–38].
- Multiple cell types (and sub-cell types) of the hematopoietic system—the vascular regulatory leukocytes (VRL)—are common and likely generic components of the angiogenic switch mechanism; these cells typically reside in the proximity of the vasculature and variably induce angiogenic switching or modulate ongoing angiogenesis by supplying proangiogenic signals either directly, or via matrix-degrading (angiogenic factor-mobilizing) enzymes [39–47].
- Chronically angiogenic vasculature, in particular that in tumors, is qualitatively aberrant [48–52], being distinguishable both from normal quiescent vasculature and from transitory physiological angiogenesis, in terms of

  – cellular anatomy (looser and incomplete pericyte coverage; increased rates of endothelial cell proliferation and apoptosis; increased and disorganized branching);
  – functionality (increased permeability, hemorrhaging, and dilation; erratic patterns and inefficient blood flow);
  – cellular and molecular morphology (as revealed by phage display and proteome profiling); and
  – molecular biology (as seen in distinctive transcriptome profiles).

- It is likely that the sustained and excessive imbalance of the angiogenic switch in favor of VEGF and other proangiogenic factors is in significant part responsible for the aberrant tumor vasculature, as demonstrated for example by the 'normalization' of vascular functionality and morphology in the context of VEGF inhibition [11, 51, 53–55].

Figure 48.1 illustrates (albeit incompletely) the core of a multiplicity of cell types that regulate angiogenesis and the manifestation of the capillary vasculature of the circulatory system. A similar schematic, devoid of pericytes, likely defines the regulation of lymphangiogenesis and the lymphatic vasculature.

Collectively, these appreciations—that angiogenesis is governed by multiple positive and negative regulatory signals, that normal homeostatic functions involve some of these same signals, that multiple cell types constitute the vascular regulatory network, and that chronic angiogenesis produces a neovasculature typically abnormal in structure, organization, and function—raise big questions for the field that arguably constitute landmark goals to be pursued, whose realization will present historic opportunities for future applications to human medicine, as outlined below.

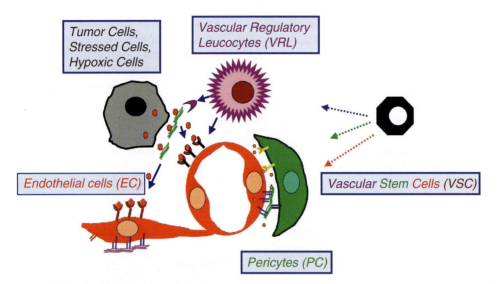

Fig. 48.1. The cellular landscape: constituents of the angiogenic switch mechanism and the manifestation of a neovasculature. In addition to the capillary endothelial cells (*EC*) activated to sprout and form a neovasculature, pericytes (*PC*, a peri-endothelial support cell) and vascular regulatory leukocytes (*VRL*, also typically peri-vascular in localization) are involved. Parenchymal cells (typically transformed, stressed, and/or hypoxic) along with reactive stromal fibroblasts (not shown) are involved in recruiting VRLs and signaling to ECs and PCs. Additionally vascular stem cells (*VSC*), originating in the bone marrow from the hematopoietic compartment, spawn derivative stem/progenitor cells to ECs, PCs, and VRLs, which in turn functionally support angiogenesis and homeostasis of the neovasculature.

Looking to the Future: a Wealth of Knowledge, with Far-reaching Medical Applications

A decade from now, in 2018, significant advances and abundant new knowledge can be confidently anticipated, in light of the current pace of biomedical research on angiogenesis and the vascular system. Mechanisms and principles will continue to be defined in experimental, genetically manipulatable organisms, especially the mouse, and then assessed translationally in the human (albeit necessarily indirectly or inferentially for some parameters).

Among many landmarks of accomplishment and attendant knowledge one might predict to be attainable in the coming decade, five strategic milestone goals that arguably represent conceptual foundations for the future are schematized in Fig. 48.2 and elaborated below.

Milestone Goals for Angiogenesis Research

A wiring diagram of the intracellular integrated circuit that controls Endothelial Cells

The diagram of the endothelial cell integrated circuit will elaborate the branches and nodes that variously control, via the strength and character of their stimulatory and inhibitory signals, the distinctive endothelial cell states

- of differentiation and morphogenesis during vasculogenesis and embryonic angiogenesis,
- of quiescence and homeostasis in the adult vasculature, and
- of tip cells, stalk cells, and new tube cells, etc., in the ontogeny of transitory physiological or chronic pathological angiogenesis in the adult.

Maps of the Extracellular Network(s) Transmitting Regulatory Signals to Endothelial Cells

An annotated network map is envisioned, detailing the systems biology of extracellular signals impinging on the integrated signaling circuit inside endothelial cells, explaining how those microenvironmental signals govern endothelial cell function and fate. Concomitant will be an elaboration of the chorus of distinctive cell types that produce and transmit those signals to endothelial cells in different organs, disease conditions and physiological circumstances.

Flow Charts of the Integrated Circuits and Signaling Networks Controlling Pericytes and Vascular Regulatory Leukocytes

Comparable elucidation is anticipated of the intracellular integrated circuits and extracellular signaling networks that regulate pericytes (plus the related vascular smooth muscle cells) and the various vascular regulatory leukocytes operative in different organs in the context of homeostasis, and during both transitory physiological and chronic (tumor) angiogenesis.

Vascular Stem Cells

An encyclopedia will be written, elaborating the multiplicity of cell types and sub-types emanating from hematopoietic stem cells resident in the bone marrow that interact with the vasculature and variously modulate its integrity and functional

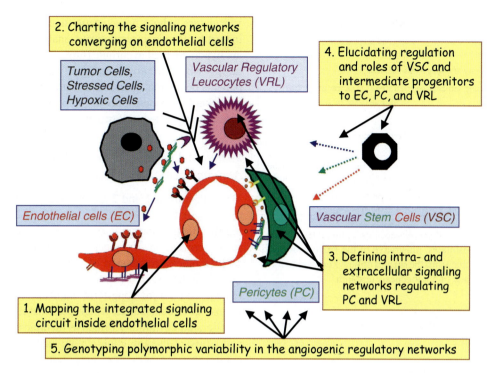

FIG. 48.2. Milestones for the future of angiogenesis research. Five strategic milestones that arguably represent conceptual foundations and attainable goals for the future are presented, overlaid on the schematic of the core cell types involved in regulating the angiogenic switch and the manifestation of a neovasculature.

states. This encyclopedia will include descriptions of the regulatory signals that govern and of the effector mechanisms contributed by the progeny of an evident repertoire of vascular stem and progenitor cells [56]. Amongst these are putatively distinct endothelial stem and progenitor cell subtypes, likely including ones fueling vasculogenesis and angiogenesis during embryonic vascularization, and others present in the adult and variously resident either in the bone marrow, in the circulation, in perivascular compartments ('niches'), or integrated into the luminal vasculature, as determined by physiological and pathological conditions [56]. Similarly, the roles and regulation of stem and intermediary progenitor cells to pericytes [57] and to the various subtypes of vascular regulatory leucocytes [39–47] will be elaborated.

Genotyping Variations in the Global Angiogenic Regulatory Network.

The accelerating capabilities of genetic tools for mapping polymorphisms in the human genome (and that of experimental organisms such as the mouse) will allow the genes comprising the various regulatory networks and circuits controlling EC, PC, VRL, and their stem and progenitors to be audited for polymorphic variations in coding and gene-regulatory sequences that might affect angiogenic phenotypes by altering the strength and integration of signals impacting the state of the system. Both efficient and inexpensive re-sequencing of individual genomes, supplemented by very high throughput DNA polymorphism (e.g., SNP) mapping, will enable this

quest, complimented by transcriptome and proteome analysis of primary cells isolated from normal and diseased tissues. For validation, an evermore powerful repertoire of bioassays for angiogenesis both in model organisms and ex vivo will allow candidate gene polymorphisms to be examined for functional effects on vascular homeostasis, and on physiological vs. pathological angiogenesis.

A corollary to this set of strategic goals can be formulated for the lymphatic endothelial calls and their homeostatic and physiologic/pathologic lymphangiogenic states.

Translational Applications of Future Milestones in Angiogenesis Research

The anticipated knowledge forthcoming from reaching these milestones should prove applicable to pursuing translational applications with exciting potential. Many of these applications will depend upon deep insight into the differences between normal quiescent vasculature, transitory physiological angiogenesis, and chronic pathological angiogenesis. Added dimensions will include understanding variations in the regulation and manifestation of these three vascular states as a function of tissue and organ and disease type, and likely as a function of polymorphic genetic variations amongst individuals.

I present below seven postulates, elaborating applications of broad scope and importance that likely can and will be realized in the coming decade (see also Fig. 48.3). These postulates, while incomplete in defining the broad frontier of possibility,

Seven Postulates: Applications of Future Milestones in Angiogenesis Research
1. Turning off the angiogenic switch in tumors and keeping it off, producing enduring anti-angiogenic therapies.
2. Ameliorating the toxicity of angiogenesis inhibition.
3. Early intervention: using anti-angiogenic strategies in cancer prevention.
4. Selectively manipulating the vascular stem and progenitor cells that sustain pathological angiogenesis.
5. Therapeutic targeting and imaging the aberrant molecular anatomy of chronically angiogenic vasculature.
6. Pro-angiogenic therapies: orchestrating the formation of normal and enduring blood (and lymphatic) vessels to re-vascularize vascular insufficiencies.
7. Personalized genotyping of the human angiogenic regulatory network for more precise disease detection, prognosis, and treatment.

FIG. 48.3. Ten years on: seven postulates of exceptional translational opportunities. These postulates, illustrative of the frontier of possibility, should be realizable from the wealth of new knowledge forthcoming in the next decade of angiogenesis research.

nevertheless constitute conceptual benchmarks to the field of angiogenesis research.

Postulate 1. Turning off the Angiogenic Switch in Tumors and Keeping it off, Producing Enduring Antiangiogenic Therapies

Elucidation of the global network of multi-factorial stimulatory and inhibitory circuits regulating the angiogenic switch will suggest the means to sustainably suppress chronic angiogenesis, in particular in cancer, using pharmacological agents that block both primary and secondary/alternative proangiogenic signaling pathways, and/or hyperactivate endogenous antiangiogenic signals. What is less clear, however, are the likely effects on diseases being fueled by ongoing neovascularization. Can a complete blockade of angiogenesis permanently stop tumor growth, effecting 'cancer without disease' [58]? Or must even the most potent antiangiogenic therapies be paired with other targeted therapies, for example ones inhibiting the capability for invasive growth into normal tissues that enables access to pre-existing vasculature (co-option) without requisite angiogenesis? Irrespective, the capability to keep the angiogenic switch turned off will undoubtedly have major impact. The existence of tumor dormancy, wherein occult, asymptomatic carcinomas with evident lack of angiogenesis have been detected in various normal human tissues [58] and modeled in mice [59, 60], suggests the possibility of recreating such a non-angiogenic, non-pathologic phenotype pharmacologically.

Postulate 2. Ameliorating the Toxicity of Angiogenesis Inhibition

Current antiangiogenic therapies are evidencing target-related toxicities [13,14]. The postulate is that antiangiogenic therapies can be fine-tuned based on knowledge of the extracellular signals and intracellular regulatory circuitry to spare normal vasculature, whose homeostatic functions are also controlled in part by the same regulatory signals (e.g., VEGF [61]) that are hyperactivated in chronic (tumor) angiogenesis. Such fine-tuning would improve the quality of life for many cancer patients being treated with antiangiogenic therapies, and could remarkably expand the scope of applying antiangiogenic therapies, as suggested in Postulate 3.

Postulate 3. Prophylactic Early Intervention—using Antiangiogenic Strategies for Cancer Prevention

When Postulates 1 and 2 are realized, another exciting application will become tractable, namely to design and apply highly effective non-toxic antiangiogenic agents and regimens to the prevention of cancer. Preventative interventions will seek to block progression of incipient (angiogenesis-dependent) neoplasias, in a future where early their detection is a reality, and afford long-term adjuvant therapy to prevent relapse in circumstances where pre-existing cancer is largely ameliorated by initially effective therapies; both prophylaxis and adjuvant therapy could create analogs of the implicated natural state of non-angiogenic tumor dormancy. Arguably, endogenous angiogenesis inhibitors such as endostatin may come to play important roles in such long-term antiangiogenic regimens, if indeed they prove to be non-toxic and yet effective when delivered at appropriate doses for long periods. The approval of an endostatin formulation evidencing minimal toxicity and apparent activity for treating lung cancer in China [8] encourages the possibility of leveraging this natural mechanism of controlling angiogenesis.

Postulate 4. Selectively Manipulating the Vascular Stem Cells that Sustain Pathological Angiogenesis

The milestone goal of an encyclopedia delineating vascular stem cells (for endothelial cells, for pericytes, and for vascular-regulatory leukocytes), one that elaborates their regulatory signals, roles and functional importance for the vasculature during embryogenesis and in the adult, as well as in the context both of transitory physiological and chronic pathological angiogenesis, will enable selective manipulation and targeting of specific sub-types of VSC. Perhaps, for example, the stem cell recruitment in support of pathological angiogenesis can be blocked, while preserving the normal recruitment and contributions of stem and progenitor cells in support of homeostatic functions and transitory physiological angiogenesis.

Postulate 5. Therapeutic Targeting and Imaging the Aberrant Molecular Anatomy of Chronically Angiogenic Vasculature

Elucidation of how imbalances in the global angiogenic regulatory network effect the generic differences between the chronically angiogenic (and lymphangiogenic) vasculature and quiescent and physiologically angiogenic vasculature will enable various translational applications. One, an alternative to therapeutic manipulation of the regulatory imbalances per se, is rather to exploit their manifestations in distinctive morphology and functionality of the chronically angiogenic vasculature, involving not only the endothelial cells but also pericytes, vascular-regulating leukocytes, and in some cases lymphatic endothelial cells. To this end, the morphologic and molecular signatures of pathologic neovasculature will be harnessed as addresses, docking sites, and targets for specific delivery and/or activity of drugs aimed at the diseased tissue and/or its supporting vasculature, while sparing normal tissues and attendant toxicity.

Additionally, these aberrant signatures will herald new, more sensitive technological capabilities for non-invasive imaging of the chronically angiogenic vasculature, enabling increasingly precise detection of neoplasias, monitoring of disease progression and metastasis, and evaluation of therapeutic responses to antiangiogenic therapies.

Postulate 6. Proangiogenic Therapies—Orchestrating the Formation of Normal and Enduring Blood (and lymphatic) Vessels to Re-vascularize Vascular Insufficiencies

The elucidation of the global angiogenic regulatory network, and elaborating the differences in its signaling states between transitory physiological angiogenesis, chronic pathological angiogenesis, and vascular quiescence and homeostasis, will allow another important capability: the induction of stable, normal blood or lymphatic vessels in tissues where the preexisting vessels are dysfunctional. Attempts to date to elicit therapeutic revascularization with stable new vessels, by expressing a proangiogenic signal (e.g., VEGF-A) have been equivocal [7, 62]; arguably, the strategy has been producing pathological vessels [13, 14, 51]. More complete knowledge of the regulatory signals governing these alternative states will enable more sophisticated strategies to induce and sustain new vessels in tissues with inadequate blood or lymphatic vasculature. Thus, for example, one can envision delivering signaling molecules (or pharmacological inducers thereof) that would recruit vascular stem cells of one or several sub-types to the tissue of need, whereby the stem/progenitor cells would then orchestrate the formation of a normal vasculature, complete with proper pericyte coverage and other accessory cells. No doubt other strategies will be suggested as the knowledge develops.

Postulate 7. Personalized Genotyping of the Human Angiogenic Regulatory Network for more Precise Disease Detection, Prognosis, and Therapy

It is anticipated that polymorphic variations in angiogenic regulatory genes will affect the predisposition of individuals to develop certain types of cancer and other diseases with chronic angiogenic etiologies. Might the levels of endogenous angiogenesis inhibitors vary in different individuals, as exem-plified in Down's patients, who have 50% higher levels of endostatin in their circulation, potentially impacting their reduced incidences of cancer [57, 63, 64]? Virtually every angiogenic and vascular regulatory node and circuit in the global angiogenic regulatory network could be subject to polymorphic genetic variability, be that in proangiogenic or in antiangiogenic signals, operative not only in endothelial cells but alternatively in pericytes, in vascular regulatory leukocytes, in the stem and progenitor cells for these cell types, or in paremchymal and tumor cells that instigate them. Indeed, clues are evident from human and mouse genetics [65–69]. Haplotypes of the angiogenic regulatory network may define predispositions of susceptibility or resistance to particular forms of cancer, guiding both monitoring of individuals for early detection as well as tailoring of antiangiogenic therapies based on knowledge of the optimal susceptibility points in a patient's angiogenic regulatory network.

Conclusion

These seven postulates constitute an agenda of translational science for the next decade of angiogenesis research, founded on elaborating in remarkable clarity the process of tumor angiogenesis, an integral and essential hallmark of human cancer [70]. These postulates are not comprehensive or exhaustive of the possibilities, but rather illustrative of feasible applications to human medicine of the new knowledge that will be forthcoming about the regulation of angiogenesis and vascular homeostasis. The realization of these postulates will have remarkable impact on human health and the prevention and management of diseases with angiogenic etiologies.

Acknowledgements. Research in the author's laboratory on tumor angiogenesis has benefited from the enduring support of the U.S. National Cancer Institute, and from the William F. Bowes, Jr. Foundation.

References

1. Folkman J. Tumor angiogenesis: therapeutic implications. N. Engl. J. Med. 1971; 285:1182–86.
2. Folkman, J. History of Angiogenesis. 2008. Chapter 1. This volume.
3. Folkman, J. Tumor Angiogenesis: from Bench to Bedside. In: Tumor Angiogenesis: Basic Mechanisms and Cancer Therapy. Marme, D., Fusenig, N., eds., Springer, New York, New York, 2007; in press.
4. Hanahan D, Folkman J. Patterns and emerging mechanisms of the angiogenic switch during tumorigenesis. Cell. 1996; 86:353–64.
5. Carmeliet P. Angiogenesis in life, disease and medicine. Nature. 2005; 438:932–36.
6. Adams R.H., Alitalo, K. Molecular regulation of angiogenesis and lymphangiogenesis. Nat Rev Mol Cell Biol. 2007; 8:464–78.
7. Ferrara N, Kerbel RS. Angiogenesis as a therapeutic target. Nature. 2005; 438:967–74.

8. Folkman J. Angiogenesis: an organizing principle for drug discovery? Nat Rev Drug Discov. 2007; 6:273–86.

9. Casanovas O, Hicklin DJ, Bergers G, et al. Drug resistance by evasion of antiangiogenic targeting of VEGF signaling in late-stage pancreatic islet tumors. Cancer Cell. 2005; 8: 299–309.

10. Kerbel RS. Therapeutic implications of intrinsic or induced angiogenic growth factor redundancy in tumors revealed. Cancer Cell. 2005; 8:269–71.

11. Batchelor TT, Sorensen AG, di Tomaso E, et al. AZD2171, a pan-VEGF receptor tyrosine kinase inhibitor, normalizes tumor vasculature and alleviates edema in glioblastoma patients. Cancer Cell. 2007; 11:83–95.

12. Shojaei F, Wu X, Malik AK, et al. Tumor refractoriness to anti-VEGF treatment is mediated by CD11b(+)Gr1(+) myeloid cells. Nat Biotechnol.; 25:911–20.

13. Kamba T, McDonald DM. Mechanisms of adverse effects of anti-VEGF therapy for cancer. Br J Cancer. 2007; 96:1788–95

14. Verheul HM, Pinedo HM. Possible molecular mechanisms involved in the toxicity of angiogenesis inhibition. Nat Rev Cancer. 2007; 7:475–85.

15. Bergers G, Song S. The role of pericytes in blood-vessel formation and maintenance. Neuro Oncol. 2005; 7:452–64.

16. von Tell D, Armulik A, Betsholtz C. Pericytes and vascular stability. Exp Cell Res. 2006; 312:623–29.

17. Bergers, G. Pericytes, the Mural Cells of the Microvascular Stystem. 2008. Chapter 4; this volume.

18. Bergers G, Song S, Meyer-Morse N, et al. Benefits of targeting both pericytes and endothelial cells in the tumor vasculature with kinase inhibitors. J Clin Invest. 2003; 111:1287–95.

19. R. Erber, A. Thurnher, A.D. Katsen, et al. Combined inhibition of VEGF and PDGF signaling enforces tumor vessel regression by interfering with pericyte-mediated endothelial cell survival mechanisms, FASEB J. 2004; 18: 338–40.

20. Pietras K, Hanahan D. A multitargeted, metronomic, and maximum-tolerated dose "chemo-switch" regimen is antiangiogenic, producing objective responses and survival benefit in a mouse model of cancer. J Clin Oncol. 2005; 23:939–52.

21. Sennino B, Falcon BL, McCauley D, et al. Sequential loss of tumor vessel pericytes and endothelial cells after inhibition of platelet-derived growth factor B by selective aptamer AX102. Cancer Res. 2007; 67:7358–67.

22. Bergers G, Benjamin LE. Tumorigenesis and the angiogenic switch. Nat Rev Cancer. 2003; 3:401–10.

23. Detmar, M. Lymphatic vascular system and Lymphangiogenesis. 2007. Chapter 43, this volume.

24. Shibuya M, Claesson-Welsh L. Signal transduction by VEGF receptors in regulation of angiogenesis and lymphangiogenesis. Exp Cell Res. 2006; 312:549–60.

25. Shibuya, M. Vascular Permeabiliy Factor/Vascular Endothelial Growth Factor. 2008. Chapter 8, this volume.

26. Klagsbrun M, Eichmann A. A role for axon guidance receptors and ligands in blood vessel development and tumor angiogenesis. Cytokine Growth Factor Rev. 2005; 16:535–48.

27. Claesson-Welsh. VEGF Signal Transduction in Angiogenesis. 2007. Chapter 18, this volume.

28. Presta, M. Fibroblast Growth Factor-2. 2007. Chapter 8, this Volume.

29. Thurston G. et al. Delta-like Ligand 4/Notch Pathway in Tumor Angiogenesis. 2008. Chapter 19, this volume.

30. Thurston G, Noguera-Troise I, Yancopoulos GD. The Delta paradox: DLL4 blockade leads to more tumour vessels but less tumour growth. Nat Rev Cancer. 2007; 7:327–31.

31. Stupack DG, Cheresh DA. Integrins and angiogenesis. Curr Top Dev Biol. 2004; 64:207–38.

32. Kalluri R. Basement membranes: structure, assembly and role in tumour angiogenesis. Nat Rev Cancer. 2003; 3:422–33.

33. Davis GE, Senger DR. Endothelial extracellular matrix: biosynthesis, remodeling, and functions during vascular morphogenesis and neovessel stabilization. Circ Res. 2005; 97:1093–107.

34. Cheresh, D and Alavi, A. Integrins in Angiogenesis. 2008. Chapter 6, this volume.

35. Nyberg P, Xie L, Kalluri R. Endogenous inhibitors of angiogenesis. Cancer Res. 2005; 65:3967–79.

36. Ribatti D, Conconi MT, Nussdorfer GG. Nonclassic endogenous regulators of angiogenesis. Pharmacol Rev. 2007; 59:185–205.

37. Bornstein, P. Thrombospondins. 2008; Chapter 13, this volume.

38. Folkman, J. Endostatin and Angiostatin. 2008; Chapter 12, this volume.

39. Coussens LM, Werb Z. Inflammation and cancer. Nature. 2002; 420:860–67.

40. Joyce JA, Freeman C, Meyer-Morse N, et al. A functional heparan sulfate mimetic implicates both heparanase and heparansulfate in tumor angiogenesis and invasion in a mouse model of multistage cancer. Oncogene. 2005;24:4037-51; Erratum in: Oncogene. 2005; 24:4163.

41. Lewis CE, Pollard JW. Distinct role of macrophages in different tumor microenvironments. Cancer Res. 2006; 66:605–12.

42. Nozawa H, Chiu C, Hanahan D. Infiltrating neutrophils mediate the initial angiogenic switch in a mouse model of multistage carcinogenesis. Proc Natl Acad Sci USA. 2006; 103:12493–98.

43. van Kempen LC, de Visser KE, Coussens LM. Inflammation, proteases and cancer. Eur J Cancer. 2006;42:728–34.

44. Schmid MC, Varner JA. Myeloid cell trafficking and tumor angiogenesis. Cancer Lett. 2007;250:1–8.

45. Tan TT, Coussens LM. Humoral immunity, inflammation and cancer. Curr Opin Immunol. 2007; 19:209–16.

46 Kopp HG, Rafii S. Thrombopoietic cells and the bone marrow vascular niche. Ann N Y Acad Sci. 2007; 1106:175–9.

47. Coussens, LM. Immune cells and inflammatory mediators as regulators of angiogenesis. 2008; Chapter 20, this volume.

48. Ruoslahti E. Specialization of tumour vasculature. Nat Rev Cancer. 2002; 2:83–90.

49. Pasqualini R, Arap W, McDonald DM. Probing the structural and molecular diversity of tumor vasculature.Trends Mol Med. 2002; 8:563–71.

50. Baluk P, Hashizume H, McDonald DM. Cellular abnormalities of blood vessels as targets in cancer. Curr Opin Genet Dev. 2005; 15:102–11.

51. McDonald, DM. Angiogenesis and Vascular Remodeling: Biology and Architecture of the Vasculature. 2008. Chapter 2, this volume.

52. St Croix, B. Tumor endothelial markers. 2008. Chapter 29, this volume.

53. Jain RK. Normalization of tumor vasculature: an emerging concept in antiangiogenic therapy. Science. 2005; 307:58–62.

54. Hormigo A, Gutin PH, Rafii S. Tracking normalization of brain tumor vasculature by magnetic imaging and proangiogenic biomarkers. Cancer Cell. 2007; 11:6–8.

55. Jain, RK. Normalization of Tumor Vasculature and Microenvironment. 2008. Chapter 24, this volume.

56. Rafii, S. Contributions of endothelial progenitor cells to the angiogenic process. 2008. Chapter 21, this volume.

57. Song S, Ewald AJ, Stallcup W, et al. PDGFRbeta+ perivascular progenitor cells in tumours regulate pericyte differentiation and vascular survival. Nat Cell Biol. 2005; 7:870–79.

58. Folkman J, Kalluri R. Cancer without disease. Nature. 2004; 427:787.

59. Naumov GN, Akslen LA, Folkman J. Role of angiogenesis in human tumor dormancy: animal models of the angiogenic switch. Cell Cycle. 2006; 5:1779–87.

60. Soucek, L, Lawlor, ER, Soto, D, et al. Mast cells are required for angiogenesis and macroscopic expansion of Myc-induced pancreatic islet tumors. Nature Med. 2007; in press.

61. Lee S, Chen TT, Barber CL, et al. Autocrine VEGF Signaling Is Required for Vascular Homeostasis. Cell. 2007; 130:691–703.

62. Semenza GL. Therapeutic angiogenesis: another passing phase? Circ Res. 2006; 98:1115–16.

63. Zorick TS, Mustacchi Z, Bando SY, et al. High serum endostatin levels in Down syndrome: implications for improved treatment and prevention of solid tumours. Eur J Hum Genet. 2001; 9:811–14.

64. Sund M, Hamano Y, Sugimoto H, et al. Function of endogenous inhibitors of angiogenesis as endothelium-specific tumor suppressors. Proc Natl Acad Sci USA. 2005; 102:2934–39.

65. Rogers MS, D'Amato RJ. The effect of genetic diversity on angiogenesis. Exp Cell Res. 2006; 312:561–74.

66. Kong SY, Park JW, Lee JA, Park JE, Park KW, Hong EK, Kim CM. Association between vascular endothelial growth factor gene polymorphisms and survival in hepatocellular carcinoma patients. Hepatology. 2007; 46:446–55.

67. Schneider BP, Radovich M, Sledge GW, Robarge JD, Li L, Storniolo AM, Lemler S, Nguyen AT, Hancock BA, Stout M, Skaar T, Flockhart DA. Association of polymorphisms of angiogenesis genes with breast cancer. Breast Cancer Res Treat. 2007; Sep 20; [Epub ahead of print]

68. Heist RS, Zhai R, Liu G, Zhou W, Lin X, Su L, Asomaning K, Lynch TJ, Wain JC, Christiani DC. VEGF polymorphisms and survival in early-stage non-small-cell lung cancer. J Clin Oncol. 2008; 26:856–62.

69. Kim DH, Lee NY, Lee MH, Sohn SK, Do YR, Park JY. Vascular endothelial growth factor (VEGF) gene (VEGFA) polymorphism can predict the prognosis in acute myeloid leukaemia patients. Br J Haematol. 2008;140:71–9.

70. Hanahan D, Weinberg RA. The hallmarks of cancer. Cell. 2000; 100:57–70.

Index

A

Abegrin, 68

ABT-510, 424

Actin cytoskeleton, 210–211

Actinobacillus actinomycetemcomitans, 534

Active drug delivery, 284

AD. *See* Alzheimer's disease

ADAM. *See* A Disintegrin And Metalloproteinase

Adaptive immune cells, 225–226

Adenosine receptors, hypoxia inducible factor-1 and, 175

Adenovirus-delivered angiostatin (ADK3), 134

ADI. *See* Arginine deiminase

A Disintegrin And Metalloproteinase (ADAM), 55, 58

ADK3. *See* Adenovirus-delivered angiostatin

ADM. *See* Adrenomedullin

ADMA. *See* Asymmetric dimethylarginine

ADP-ribosylation, lactate and, 554–555

Adrenomedullin (ADM), hypoxia inducible factor-1 and, 173

Adriamycin, TNP-470 in vivo studies and, 401

AG-013736, 26, 28

AGENT-1 trial, 566

AGENT-2 trial, 566

Age-related macular degeneration (AMD), 333, 345, 348, 417, 426, 580

Akt/PKB, 133

Alzet, 234

Alzheimer's disease (AD), angiogenesis in central nervous system and, 496

AMD. *See* Age-related macular degeneration

ANCHOR trial, 349, 426

Ang2. *See* Angiopoietin-2

Angiogenesis

  alternate animal models in, 306–308

  angiogenic switch, 577

  in central nervous system, 489–499

    Alzheimer's disease and, 496

    blood brain barrier formation/maintenance, 492–493

    blood vessel maturation in, 491–492

    cerebral amyloid angiopathy and, 496

    cerebral autosomal dominant arteriopathy with subcortical infarcts and leukoencephalopathy, 493

    cerebral autosomal recessive arteriopathy with subcortical infarcts and leukoencephalopathy, 493

    cerebroretinal vasculopathy, 493

    disease and, 493–498

    familial amyloid angiopathies, 493

    gliomas and, 498

    hemangioblastomas and, 497–498

    hereditary endotheliopathy with retinopathy, neuropathy and stroke syndrome, 493

    hereditary vascular retinopathy, 493

    Moya-Moya disease, 493–495

    multiple sclerosis and, 496–497

    Parkinson's disease and, 496

    retinal arteriolar tortuosity, and leukoencephalopathy, 493

    stroke and, 495

    vascularization, 489–491

  corneal angiogenesis assay, 304–305

  coronary

    clinical trials, 565

    patient selection in, 567

    randomized controlled trials, 565–567

    therapeutic stimulation, 565

  current knowledge of, 576–577

  direct regulators of, 185–186

    angiopoietin, 185

    endoglin, 185–186

    endothelin, 186

    placental growth factor, 186

    VEGF, 185

  genes encoding indirect regulators, 186–187

    erythropoietin, 186–187

    matrix metalloproteinases, 187

    plasminogen activators, 187

  historical breakthroughs in, 575–576

  history of, 1–10

  imaging of, 321–330

    in clinical trials, 329

    computed tomography, 324–327

    magnetic resonance imaging, 322–324

    positron emission tomography, 328–329

    single photon emission computed tomography, 328–329

  immunotherapy of, 452

Angiogenesis (*continued*)
  in inflammation, 18, 21
  integrins in, 63–65
  Matrigel plug assay, 305–306
  nitric oxide role in, 194
  normalization of, 8–9
  in oral cavity, 533–537
  oral mucosa wound healing and, 536
  as organizing principle, 8
  pericytes in, 47–48
  regulation of
    circulating levels of, 314–315
    by microenvironment, 261–263
  regulatory proteins, in platelets, 9
  research, 578–579
    flow charts, 578
    genotype variations, 579
    maps of, 578
    vascular stem cells, 578–579
    wiring diagram, 578
  research, bioassays for, 3–6
  retinal, 221–222
  sponge implant assays, 306
  tumor, 577
    cancer stem cell and, 250–252
    chemokines in, 229–230
    cytokines in, 229–230
    delta-like 4 and, 220–221
    endothelial precursor cells and, 165–166
    extracellular matrix in, 228–229
    fibroblast growth factor-2 in, 82–83
    growth factors in, 229–230
    hematopoietic cell participation and, 165–166
    hypoxic regulation of, 163–164
    immune cells and, 226–227, 231–233
    inflammatory cell role in, 164–165
    mosaic vessels, 165
    nitric oxide role in, 195
    platelet-derived growth factor to, 106
    prognostic significance of, 166
    proteases in, 228–229
    stem cells in, 245
    vascular endothelial growth factor-A in, 347
    vascular endothelial growth factor and, 416
    vasculogenic mimicry, 165
  tumor progression and, 162
  and tumors of central nervous system, 497–498
  in vitro, 5
  in vitro assays, 300–302
    cell migration assay, 300–301
    coculture protocols, 302
    organ culture assays, 302
    proliferation assay, 300
    in vitro tube formation, 301–302
  in vivo assays, 302–304
    chorioallantoic membrane assay, 302–303
    mesenteric window assay, 303–304
  wound revascularization, 543–545, 550–551
Angiogenesis-dependent disease, 9–10
Angiogenesis inhibitors, 2

antiangiogenic agents, classification of, 423–424
  in clinic, 7–8
  discovery of, 6–7
Angiogenic molecules, discovery of, 6
Angiogenic phenotype, 9
Angiogenic protein, in extracellular matrix, 6
Angiogenic regulatory network, 579, 581
Angiogenic signaling, inhibition of, 355–356
  chemistry of, 356
    BAY 43-9006 (Sorafenib), 356
    indolinones, 356
    PTK 787 (Vatalanib), 356
    ZD6474 (Vandetanib), 356
Angiogenic switch, 161–162, 577
Angiomotin
  for angiostatin, 132
  tumor endothelial cells and, 455
Angiopoietin-2 (Ang2), 48
Angiopoietin-receptor complex, tie receptor tyrosine kinase family
    and, 115
Angiopoietins, 113–114
  angiogenesis and, 185
  expression, 114
  gene-modified mice, 115–117
  ocular neovascularization and, 522–523
  signal transduction, 117
  structure of, 113–114
Angiopoietin-tie system, 117
Angiostatin, 57
  anti-inflammatory activity of, 133
  circulation and, 129–130
  clinical studies of, 135
  crystallography, 135
  delivery of, 135
  discovery of, 130
  endothelial cell surface receptors for, 131–133
    alpha$_v$beta$_3$ integrin, 132
    angiomotin, 132
    ATP synthase, 131–132
    CD26, 132
    hepatocyte growth factor receptor, 132
    NG2 proteoglycan, 132
    nucleolin, 132–133
  forms of, 130–131
  gene therapy and, 134
    experimental anti-tumor therapy, 134
    ocular neovascularization angiostatin gene therapy, 134–135
  human neuroblastoma and, 135
  mechanism of action of, 133
  ocular neovascularization and, 524–525
  physiological angiogenesis and, 135
  protein, anti-tumor activity of, 135
  recombinant, 135
  in vivo induction of, 133–134
Animal models
  angiogenesis and alternate, 306–307
  EPCs, tumors and, 242
  VEGF-Trap and preclinical, 416–417
Annexin A1, 335
Antiangiogenic agents

classification of, 422–424
    angiogenesis inhibitors, 423–424
clinical translational developmental issues of, 427–428
in combination regimens, 432–442
    kinase inhibitors, 435–439
    minocycline, 432–435
    new target discovery, 432–442
    protein therapeutics, 439–440
FDA approved, 424–426
    bevacizumab, 424–425
    lenalidomide, 425
    pegaptanib, 426
    ranibizumab, 426
    sorafenib, 425
    sunitinib, 425–426
    thalidomide, 425
    tyrosine kinase inhibitors, 426–427
        AZD2171, 427
        motesanib, 427
        vandetanib, 427
        vatalanib, 427
Antiangiogenic drugs, tumor vasculature targeted drug delivery
        therapeutics with, 292–293
Antiangiogenic targets, identification of, 421–422
Antiangiogenic therapy
    anti-inflammatory drugs and, 233–234
    biphasic efficacy of, 9
    cancer stem cell and, 253–256
    pharmacogenetics of, 477–483
        in anti-VEGF therapy, 480–481
        clinical trials, 482
        single nucleotide polymorphisms of VEGF-A gene, 478–480
        single nucleotide polymorphisms of VEGFRs genes, 481–482
        VEGF-A promoter haplotypes, 480
    rationale for, 422
    of tumors, 461–471
        activated endothelial cells, 462
        adverse effect prevention/management of, 465–466
        bidirectional action and, 471
        broad-spectrum targeting, 470
        CEC circulating levels, 467
        CEP circulating levels, 467
        clinical studies, 468–469
        genetic profiling, 467
        hypoxia pathways, 464
        metronomic chemotherapy, 470–471
        monoclonal antibodies, 464–465
        natural peptide inhibitors, 465
        nitric oxide, 464
        patient selection in, 466–468
        pericytes, 462–463
        platelets and, 467
        proangiogenic factors, 463–464
        proteomics, 467
        small-molecule inhibitors, 464–465
        surrogate biomarkers and, 466
        targets in, 462–464
        therapeutic strategies in, 464–465
        vascular imaging techniques, 467–468
Anti-apoptotic pathways, 355

Antigen-presenting cells (APCs), 453
Anti-inflammatory drugs, antiangiogenic therapy and, 233–234
Antivascular therapy, of multidrug resistant, cancer of prostate, 267
APCs. See Antigen-presenting cells
Apoptosis, 65–68
Arginine deiminase (ADI), 198
Arginine depletion, nitric oxide and, 197–198
ARNT. See Aryl hydrocarbon receptor nuclear translocator
Arresten, 67, 121, 137
Arterial spin labeling (ASP), magnetic resonance imaging, of
        angiogenesis, 324
Arterial-vein specification, 211
Arteriogenesis
    bone marrow-derived cells and, 568
    coronary collaterals and, 567–568
    in endothelial cell activation, 39
    monocytes, 568
    wound revascularization, 547
Aryl hydrocarbon receptor nuclear translocator (ARNT), 169, 359
Ascites, 6, 9, 91, 92, 94, 116, 211
Aselli, Gasparo, 505
ASL. See Arterial spin labeling
Aspergillus fumigatus Fresenius, 395
Assays
    corneal angiogenesis, 304–305
    Matrigel plug, 305–306
    sponge implant, 306
    in vitro, 300–302
        cell migration assay, 300–301
        coculture protocols, 302
        organ culture assays, 302
        proliferation assay, 300
        in vitro tube formation, 301–302
    in vivo, 302–304
        chorioallantoic membrane assay, 302–303
        mesenteric window assay, 303–304
AstraZeneca, 132, 427
Asymmetric dimethylarginine (ADMA), 195
ATP synthase, for angiostatin, 131–132
Auerbach, Robert, 3
Autophagy, 137
AVANT study, 378
Avastin. See Bevacizumab
AZD2171, 427
αVβ3, 65
$\alpha_v\beta_3$ integrin, for angiostatin, 132

B
Basement membrane
    blood vessels in tumors, 23, 26
    vascular endothelial growth factor inhibitor cellular
        action and, 28
Basic fibroblast growth factor (bFGF), 6, 7
    DNA vaccines, biological factor blocking and, 456
Basic vascular endothelial growth factor (bVEGF), 261
BAY 43-9006 (Sorafenib), 356
BAY-129566, 424
BB-2516, 424
BBB. See Blood brain barrier
BCNU+, TNP-470 in vivo studies and, 401

BDNF. *See* Brain-derived neurotrophic factor
Becker, Frederick, 1
Bevacizumab, 7, 10, 176, 199, 313, 345, 347, 355, 416, 423,
    424–425, 462
  for breast cancer, 379–380
  clinical trials, 375–376
  for colorectal cancer, 377
    as adjuvant treatment, 378
    clinical trials in, 376–377
    metastatic, 377–378
  for lung cancer, 378–379
  mechanism of action, 375
  for ovarian cancer, 380
  for prostate cancer, 381
  for renal cell cancer, 380–381
  safety monitoring and, 383–384
  toxicity of, 383–384
bFGF. *See* Basic fibroblast growth factor
Biological fluids, soluble molecular markers in, 314–315
Bissele, Mina, 5
Blood brain barrier (BBB), 489, 496
  formation/maintenance of, 492–493
Blood retinal barrier (BRB), 492
Blood vessels
  abnormalities in, 17–18
  in central nervous system, maturation of, 491–492
  in tumors, 21–26
    basement membrane of, 23, 26
    endothelial cells of, 23
    identification of, 21–23
    pericytes, 23
    regrowth of, 28, 30
    VEGF inhibitor cellular actions, 26–28
  vascular endothelial growth factor inhibitor on, 30
BMS-275291, 424
Bone marrow-derived cells
  arteriogenesis and, 568
  ocular neovascularization and, 523–524
  tumor anatomy and, 243
BOOST trial, 568
Bortezomib, 462
Brain-derived neurotrophic factor (BDNF), 498
BRB. *See* Blood retinal barrier
Breast cancer, bevacizumab for, 379–380
Brem-Harold, 129
BRiTE study, 465
Broad-spectrum antiangiogenic agents, 8
Broad-spectrum targeting, in tumor antiangiogenic therapy, 470
Bromocriptine, TNP-470 in vivo studies and, 401
Bruch's membrane, 520
bVEGF. *See* Basic vascular endothelial growth factor

C
CAA. *See* Cerebral amyloid angiopathy
CADASIL. *See* Cerebral autosomal dominant arteriopathy with
    subcortical infarcts and leukoencephalopathy
Calreticulin (CRT), 458
CAM. *See* Chick chorioallantoic membrane
Cancer
  of breasts, bevacizumab for, 379–380

colorectal, bevacizumab for, 377–378
control of, 451
hypoxia inducible factor-1 in, 171–172
  expression, 172
  prognosis, 172
  xenograft studies, 172
of lungs, bevacizumab for, 378–379
ovarian, bevacizumab for, 380
of prostate, 264–266
  antivascular therapy of multidrug resistant, 267
  bevacizumab for, 381
renal cell, 264
  bevacizumab for, 380–381
therapy for, thalidomide for, 390
Cancer stem cell (CSC)
  antiangiogenic therapy and, 253–256
  hypothesis, 249–250
  tumor angiogenesis and, 250–252
  tumor vasculature and, 252–253
Canstatin, 67, 122, 137
Caplostatin. *See* 2-Hydroxypropyl metahcrylamide
CARASIL. *See* Cerebral autosomal recessive arteriopathy with
    subcortical infarcts and leukoencephalopathy
Carrier systems, 233, 285, 286
Cartilage oligomeric matrix protein (COMP), 147
CD26, for angiostatin, 132
CD36, 150–151
CD133, 243–244, 250
CECs. *See* Circulating endothelial cells
Celecoxib, 440
Cell(s)
  growth control of, 5
  invasion, 63–64
  shape, 5
Cell migration assay, 300–301
Cell proliferation assay, 300
Cellular responses, initiated of, by platelet-derived growth
    factor, 103
  competence of, 103
  completion of, 104
  priming of, 104
  progression of, 103
Central nervous system (CNS)
  angiogenesis in, 489–499
    Alzheimer's disease and, 496
    blood brain barrier formation/maintenance, 492–493
    blood vessel maturation in, 491–492
    cerebral amyloid angiopathy and, 496
    cerebral autosomal dominant arteriopathy with subcortical
      infarcts and leukoencephalopathy, 493
    cerebral autosomal recessive arteriopathy with subcortical
      infarcts and leukoencephalopathy, 493
    cerebroretinal vasculopathy, 493
    disease and, 493–498
    familial amyloid angiopathies, 493
    gliomas and, 498
    hemangioblastomas and, 497–498
    hereditary endotheliopathy with retinopathy, neuropathy and
      stroke syndrome, 493
    hereditary vascular retinopathy, 493

Moya-Moya disease, 493–495
multiple sclerosis and, 496–497
Parkinson's disease and, 496
retinal arteriolar tortuosity, and leukoencephalopathy, 493
stroke and, 495
vascularization, 489–491
disorders of, 493, 494, 495–498
genetic disease of, 493–495
tumors of, angiogenesis and, 497–498
CEP. See Circulating endothelial progenitor cells
CEPs. See Circulating endothelial progenitors
Cerebral amyloid angiopathy (CAA), angiogenesis in central nervous system and, 496
Cerebral autosomal dominant arteriopathy with subcortical infarcts and leukoencephalopathy (CADASIL), 493
Cerebral autosomal recessive arteriopathy with subcortical infarcts and leukoencephalopathy (CARASIL), 493
Cerebroretinal vasculopathy, 493
Ceruloplasmin, hypoxia inducible factor-1 and, 175
CEUS. See Contrast-enhanced ultrasound
Chemokines, in tumor angiogenesis, 229–230
Chemokinesis, 301
Chemotaxis, 301
Chemotherapeutics, tumor vasculature targeted drug delivery therapeutics with, 293
Chemotherapy
maximum tolerated dose, 313–314
metronomic, in tumor antiangiogenic therapy, 470–471
Chick chorioallantoic membrane (CAM), 5, 6, 66, 77, 161
Chorioallantoic membrane assay, 302–303, 397
Choroidal neovascularization (CNV), 348, 349, 519–520
Chronic myelogenous leukemia (CML), 335, 353
Chronic periodontitis, 534–535
CHS 828, TNP-470 in vivo studies and, 401
Cilengitide, 68
Circulating cellular markers, 315–316
Circulating endothelial cells (CECs), 239, 315–316
circulating levels, in tumor antiangiogenic therapy, 467
Circulating endothelial progenitor cells (CEP), 48, 315–316
circulating levels, in tumor antiangiogenic therapy, 467
Circulating endothelial progenitors (CEPs), 482
Circulating tumor cells (CTCs), 316
Cisplatin, TNP-470 in vivo studies and, 401
Clinical trial design, 471
CML. See Chronic myelogenous leukemia
CNS. See Central nervous system
CNTO95, 68
CNV. See Choroidal neovascularization
Coagulation, wound revascularization, 547–548
Coculture protocols, 302
Colchicine, 525
Collagen
type IV, 124
type XIX, 124
type XV, 124
type XVIII, 140
Collagen fragments, 137
Colony stimulating factor-1 (CSF-1), 230–231
Colorectal cancer, bevacizumab for, 377
as adjuvant treatment, 378

clinical trials in, 376–377
metastatic, 377–378
Combination therapy, antiangiogenic agents in, 432–442
Combretastatin A-4, 525
COMP. See Cartilage oligomeric matrix protein
Computed tomography (CT), in angiogenesis, 324–327, 468
contrast-enhanced ultrasound, 326
dynamic, 324–326
optical imaging, 326–327
ultrasound, 326
Connective tissue growth factor (CTGF), 185
hypoxia inducible factor-1 and, 173–174
Contrast-enhanced ultrasound (CEUS), in angiogenesis, 326
COOH-terminal domain, thrombospondins and, 152–153
Corneal angiogenesis assay, 304–305
Corneal neovascularization, 3–4
Coronary angiogenesis
clinical trials, 565
patient selection in, 567
randomized controlled trials, 565–567
therapeutic stimulation, 565
Coronary collaterals, arteriogenesis and, 567–568
Coronary vasculature, development of, 561
Corticosteroids, 284–285
Coup-TFII, growth factors in vascular development, 564
COX. See Cyclooxygenase
COX-2. See Cyclooxygenase-2
CRT. See Calreticulin
Crystallography, angiostatin, 135
CSC. See Cancer stem cell
CSF-1. See Colony stimulating factor-1
CT. See Computed tomography
CTCs. See Circulating tumor cells
CTGF. See Connective tissue growth factor
CTL. See Cytotoxic T lymphocyte
$CXCR4^+VEGFR1^+$ hemangiocytes, 244
Cyclooxygenase (COX), 195–196
inhibitors, ocular neovascularization and, 524
Cyclooxygenase-2 (COX-2), 389
Cyclophosphamide, TNP-470 in vivo studies and, 401
Cytokines, in tumor angiogenesis, 229–230
Cytoplasmic tyrosine kinases, 210
Cytotoxic T lymphocyte (CTL), 454

D
D'Amato, Robert, 7, 388
Datura stramonium (DSL), 18
Dawson's fingers, 497
DC. See Dendritic cell
DCE-MRI. See Dynamic contrast-enhanced magnetic resonance imaging
DDAH. See Dimethylarginine dimethylaminohydrolase
Delta-like ligand 1 (Dll1), 218
Delta-like ligand 4 (Dll4), 37, 218–219, 339
for developmental angiogenesis, 219–220
in postnatal vascular development, 221–222
in regulating angiogenesis, 222
in regulating tumor angiogenesis, 220–221
as target for antiangiogenesis therapy, 222–223
Delta-Notch system, molecular components of, 218–219

Delta proteins, 218
Dendritic cell (DC), 457–458
  precursors, tumor angiogenesis and, 232–233
Dental pulp, 536
Developmental angiogenesis
  delta-like 4 for, 219–220
  endothelial cell activation during, 35–38
    angiogenic response initiation, 35–37
    guiding cues for, 37–38
    tip *vs.* stalk, 37
    vascular lumen formation, 38
Developmental retinal neovascularization, 518
Diabetic macular edema (DME), 418
Diabetic retinopathy
  pericytes in, 49
  platelet-derived growth factor in, 106
Dimethylarginine dimethylaminohydrolase (DDAH), 195
A Disintegrin And Metalloproteinase (ADAM), 55
Dll1. *See* Delta-like ligand 1
Dll4. *See* Delta-like ligand 4
DLT. *See* Dose-limiting toxicity
DME. *See* Diabetic macular edema
DNA vaccines
  antiangiogenic therapy limitations, 458
  biological factor blocking and, 456
    bFGF, 456
    VEGF, 456
  extracellular matrix and, 456
  immunotherapy, 452
  novel molecular target identification, 457
  potency of, anti-angiogenic effect and, 457–458
  tumor endothelial cells and, 453–456
    angiomotin, 455
    EGFR, 456
    endoglin, 455
    FGFR1, 455
    integrin αvβ3, 455
    survivin, 456
    Tie2, 455
    VEGFR2, 454
  tumor stromal cells and, 457
    legumain, 457
    PDGFRβ, 457
    TAMs, 457
Docetaxel, 439
  TNP-470 in vivo studies and, 401
Dose-limiting toxicity (DLT), 468
Down syndrome, 123, 133, 139
  endostatin in, 140
Doxorubicin, 290
  TNP-470 in vivo studies and, 401
Drosophila, 56
Drug delivery systems, production of, 291–292
Drug-induced gingival enlargement, 535
DSL. *See* Datura stramonium
Dvorak, H., 163
Dynamic computed tomography, in angiogenesis,
  324–326
Dynamic contrast-enhanced magnetic resonance imaging
  (DCE-MRI), of angiogenesis, 322–324, 467

E
ECGC. *See* Epigallocatechin gallate
ECM. *See* Extracellular matrix
EDG-1. *See* Endothelial differentiation gene-1
EDRF. *See* Endothelium-derived relaxing factor
EGFR. *See* Epidermal growth factor receptor
ELISA. *See* Enzyme-linked immunosorbent assay
ELVAX. *See* Ethylene vinyl acetate copolymer
Embryo, lymphatic vasculature, development of, 506
Endogenous inhibitors, of matrix metalloproteinases, 56
Endogenous molecular targets, immune tolerance, breaking of, 452–453
Endoglin (Eng)
  angiogenesis and, 185–186
  tumor endothelial cells and, 455
Endorepellin, 122–123, 137
Endosialin. *See* Tumor endothelial marker 1
Endostar, 137
Endostatin, 7, 57, 67, 123, 135–141, 462
  affinity for zinc, 136
  antiangiogenic activity of, 136–137
    biphasic U-shaped dose-response curve, 136, 137
    molecular mechanisms, 136–137
  crystal structure, 136
  discovery of, 135–136
  in Down syndrome, 140
  experimental anti-cancer activity for, 138–140
    endostatin gene therapy, 139–140
    endostatin improves radiotherapy, 139
    endostatin protein therapy, 138–139
  gene therapy, 139–140
  heparan sulfates binding to, 136
  improves radiotherapy, 139
  like domain of type XV collagen, 124
  platelets and, 140
  protein therapy, 138–139
Endostatinuria, 140
Endothelial cell heterogeneity, 300
Endothelial cells
  activation of
    angiogenic response initiation, 35–37
    arteriogenesis, 39
    during developmental angiogenesis, 35–38
    guiding cues for, 37–38
    integrin-mediation regulation and, 65–67
    in pathological conditions, 38–39
    termination of, 39–40
    tip *vs.* stalk, 37
    in tumor antiangiogenic therapy, 462
    vascular lumen formation, 38
  of blood vessels in tumors, 23
  FGF2-mediated intracellular signaling, 79–80
  invasion of, integrins in, 63–64
  migration of, TNP-470, 396
  morphology, TNP-470, 396–397
  proliferation of, 210
    TNP-470, 395–396
  sprouting, TNP-470, 396
  survival of, 210
  tumor, DNA vaccines and, 453–456
  vascular, in vitro, 4–5

Endothelial cell surface receptors, for angiostatin, 131–133
  alpha$_v$beta$_3$ integrin, 132
  angiomotin, 132
  ATP synthase, 131–132
  CD26, 132
  hepatocyte growth factor receptor, 132
  NG2 proteoglycan, 132
  nucleolin, 132–133
Endothelial differentiation gene-1 (EDG-1), 164
Endothelial fenestrations, vascular endothelial growth factor inhibitor cellular action and, 26, 28
Endothelial-like monocyte-derived cells, tumor angiogenesis and, 232–233
Endothelial microenvironment, changes in, VEGF signaling changes and, 211–212
Endothelial nitric oxide synthase (eNOS), 153
Endothelial precursor cells (EPC), tumor angiogenesis and, 165–166
Endothelial progenitor cells (EPCs), 240, 462
  disease and, 241–242
  hematopoietic cells and, 241
  pathophysiology of, in human cancer, 243–244
  regulation of, systemic signals, 242–243
  significance of, 239–240
  tissue resident, 242
  by tumor, animal models, 242
Endothelial sprouts, vascular endothelial growth factor inhibitor cellular actions and, 26
Endothelial stalk cell, 36, 37
Endothelial tip cell, 36, 37
Endothelin 2, hypoxia inducible factor-1 and, 174
Endothelins (ETs), angiogenesis and, 186
Endothelium-derived relaxing factor (EDRF), 193
Eng. See Endoglin
Engelbreth-Holm-Swarm sarcoma tumor, 121, 305
Enhanced permeability and retention (EPR), 284
ENL. See Erythema nodosum leprosum
eNOS. See Endothelial nitric oxide synthase
Enzastaurin, 436–438
Enzyme-linked immunosorbent assay (ELISA), 534
EPC. See Endothelial precursor cells
EPCs. See Endothelial progenitor cells
EphrinB2, 218, 222, 491, 536
Ephrins, growth factors in vascular development, 564
Ephs, growth factors in vascular development, 564
Epidermal growth factor receptor (EGFR), 439
  tumor endothelial cells and, 456
Epigallocatechin gallate (EGCG), 200
EPO. See Erythropoietin
EPR. See Enhanced permeability and retention
Erbitux, 7
Erlotinib, 462
Erythema nodosum leprosum (ENL), 387
Erythropoietin (EPO), 185
  angiogenesis and, 186–187
  hypoxia inducible factor-1, 174
Ethylene vinyl acetate copolymer (ELVAX), 3
ETs. See Endothelins
EUROINJECT-1 trial, 566
Experimental anti-tumor therapy, angiostatin and, 134

Extracellular matrix (ECM), 55
  angiogenic protein in, 6
  breakdown of, 424
  DNA vaccines and, 456
    MMPs, 457
  organization of, thrombospondin 2 in, 149
  signals, ocular neovascularization, 524
  in tumor angiogenesis, 228–229
Eyes
  retinoblastoma of, 1–3
  vascular beds in, 517–518
  vascular endothelial growth factor family members and, 521–522

F
FAK. See Focal adhesion kinase
Familial amyloid angiopathies, 493
Familial exudative vitreoretinopathy (FEVR), 518
Farnesyl transferase inhibitors, 368
Ferrara, Napoleone, 6, 10
FEVR. See Familial exudative vitreoretinopathy
FGF. See Fibroblast growth factor (FGF)
FGF2. See Fibroblast growth factor-2
FGF2-mediated intracellular signaling, endothelial cells and, 79–80
Fibroblast growth factor-2 (FGF2), 77
  cross-talk with angiogenic growth factors, 80
  ECM-bound, 78–79
  inflammation and, 81
  mechanisms of action of, 78
    extracellular interactions, 78
  mediated intracellular signaling and, 79–80
  as target for anti-angiogenic/anti-cancer regimens, 83–84
  in tumor angiogenesis, 82–83
    experimental tumors, 82
    human tumors, 82–83
  in vasculogenesis, 80–81
Fibroblast growth factor (FGF), growth factors in vascular development, 563
Fibroblast growth factor receptor-1 (FGFR1), tumor endothelial cells and, 455
Fibulins, 124, 137
FIRST trial, 566
5-Fluorouricil, leucovorin, oxaliplatin (FOLFOX), 348, 377, 378
5-Fluorouricil, TNP-470 in vivo studies and, 401
Focal adhesion kinase (FAK), 63, 132
FOLFOX. See 5-Fluorouricil, leucovorin, oxaliplatin
Folkman, Judah, 77, 161, 227, 387, 395, 461
Fumagillin. See TNP-470
Functional imaging markers, 316–317
Fusin, hypoxia inducible factor-1 and, 174

G
Gadolinium-diethylenetriamine pentaacetic acid (Gd-DTPA), 322
Gastrointestinal stromal tumors (GIST), 354, 461
  Gleevec-resistant, 416
Gd-DTPA. See Gadolinium-diethylenetriamine pentaacetic acid
GDNF. See Glial-derived neurotrophic factor
Gemcitabine, TNP-470 in vivo studies and, 401
GeneSpring, 442

Gene therapy
    angiostatin and, 134
        experimental anti-tumor therapy, 134
        ocular neovascularization angiostatin gene therapy, 134–135
    for endostatin, 139–140
Genetic polymorphisms, 477
Genetic profiling, in tumor antiangiogenic therapy, 467
Gimbrone, Michael, 1, 3, 4
Gingival hyperplasia secondary to drugs. *See* Drug-induced gingival
        enlargement
GIST. *See* Gastrointestinal stromal tumors
Gleevec, 265, 335, 354, 435
Glial-derived neurotrophic factor (GDNF), 492
Gliomas, angiogenesis in central nervous system and, 498
GM-CSF. *See* Granulocyte macrophage-colony stimulating factor
Granulocyte macrophage-colony stimulating factor (GM-CSF), 165
Griffonia (Bandeiraea) simplicifolia I (GSL-I), 18
Growth factors, wound revascularization, 549–550
GSL-I. *See* Griffonia (Bandeiraea) simplicifolia I

H
Habu, 92
HC. *See* Hematopoietic cells
Heart, development of, 561
Hedgehog family
    growth factors in vascular development, 563–564
    sonic, 490
Hemangioblastomas, angiogenesis in central nervous system
        and, 497–498
Hematopoietic cells (HC)
    endothelial progenitor cells and, 241
    participation of, tumor angiogenesis and, 165–166
Hematopoietic stem cells (HSCs), 239
Heme oxygenase-1 (HO-1), 185
Heparan-sulfate proteoglycans (HSPGs), 6, 78, 207
Heparan sulfates, binding of, to endostatin, 136
Hepatic stellate cells (HSC), 47
Hepatocyte growth factor (HGF)
    hypoxia inducible factor-1 and, 174
    receptor, for angiostatin, 132
Hereditary endotheliopathy with retinopathy, neuropathy and stroke
        syndrome (HERNS), 493
Hereditary hemorrhagic telangiectasia (HHT), 185
Hereditary vascular retinopathy, 493
HERNS. *See* Hereditary endotheliopathy with retinopathy,
        neuropathy and stroke syndrome
Hexabrachion. *See* Tenascin-C
Hexamethylpropylene amine oxime (HMPAO), 329
HGF. *See* Hepatocyte growth factor
HHT. *See* Hereditary hemorrhagic telangiectasia
HIF. *See* Hypoxia inducible factor
HIF-1. *See* Hypoxia inducible factor-1
HIF-1α. *See* Hypoxia inducible factor-1α
HIF-2α. *See* Hypoxia inducible factor-2α
Histidine-rich glycoprotein (HRGP), 150
HIV-associated Kaposi's sarcoma, TNP-470 clinical trial phase I0
        and, 405–406
HMPAO. *See* Hexamethylpropylene amine oxime
HO-1. *See* Heme oxygenase-1
Holmgren, Lars, 132

Horizontal inhibition, 471
HPMA. *See* 2-Hydroxypropyl methacrylamide
HRCC. *See* Human renal call carcinoma
HRE. *See* Hypoxia-responsive element
HRF. *See* Hypoxia inducible factor-1-related factor
HRGP. *See* Histidine-rich glycoprotein
HSC. *See* Hepatic stellate cells
HSCs. *See* Hematopoietic stem cells
HSPGs. *See* Heparan-sulfate proteoglycans
Human neuroblastoma, angiostatin and, 135
Human renal call carcinoma (HRCC), 264
Human tumors, normalization in, 276–277
Human umbilical vein endothelial cells (HUVECs), 150–151, 396,
        408, 409, 436
Hunter, John, 1, 541
HUVECs. *See* Human umbilical vein endothelial cells
Hyaloidal vasculature, regression of, 518–519
2-Hydroxypropyl methacrylamide (HPMA), copolymer-
        Gly-Phe-Leu-Gly-TNP-470, 408
Hyperoxia, wound revascularization, 551–552
Hypoxia inducible factor (HIF)
    biochemistry of, 359–364
        asparaginyl hydroxylation, 362–363
        DNA and, 360–361
        endogenous hydroxylase regulation by small molecules, 363
        hydroxylation, 361
        prolyl hydroxylation, 361–362
    prolyl hydroxylases, lactate and, 553–554
    system modulation, 364–368
        alternative pathways, 367
        dimerization of, 367
        DNA binding, 367
        farnesyl transferase inhibitors, 368
        genetic approaches, 368
        microtubule destabilization, 367
        microtubule stabilization, 367
        PI3K pathway inhibitors, 367–368
        selectivity, 364–365
        target gene activation inactivation, 365–366
        targets within, 365
        thioredoxin inhibitors, 368
        topoisomerase inhibitors, 367
        transcriptional coactivator p300, 366
Hypoxia inducible factor-1 (HIF-1), 137, 164
    activation
        alternative pathways in, 170
        in hypoxia, 170
    angiogenic factors induced by, 173–175
        adenosine receptor, 175
        adrenomedullin, 173
        ceruloplasmin, 175
        connective tissue growth factor, 173–174
        endothelin 1, 174
        endothelin 2, 174
        erythropoietin, 174
        fusin, 174
        hepatocyte growth factor, 174
        insulin-like growth factor-2, 175
        interleukin-8, 174
        lysyl oxidase, 175

notch signaling, 174
osteopontin, 174
stanniocalcin 1, 174
stanniocalcin 2, 174
stromal cell-derived factor-1, 174
tenascin-C, 175
tie 2, 175
transferrin receptor, 175
transforming growth factor-β, 174–175
vascular endothelial growth factor, 173
in cancer, 171–172
expression, 172
prognosis, 172
xenograft studies, 172
degradation pathway, 169
regulated gene expression, 172–173
regulatory pathway, 183
structure of, 169
subfamily, 170–171
synthesis, regulation of, 170
therapeutic implications of, 175–176
tumor-associated macrophages and, 172
von Hippel-Lindau and, 171–172
Hypoxia inducible factor-1α (HIF-1α), 183–185
angiogenesis and, activated genes, 185
Hypoxia inducible factor-1-related factor (HRF), 171
Hypoxia inducible factor-2α (HIF-2α), 183–185
Hypoxia pathways, in tumor antiangiogenic therapy, 464
Hypoxia-responsive element (HRE), 91, 519
Hypoxia, wound revascularization, 551–552
Hypoxic regulation, of tumor angiogenesis, 163–164, 170–176

I

IAUGC. *See* Initial area under the gadolinium concentration curve
IGF-2. *See* Insulin-like growth factor-2
IL-6. *See* Interleukin-6
IL-8. *See* Interleukin-8
IL-12. *See* Interleukin-12
ILK. *See* Integrin-linked kinase
Imatinib, 335, 354
Imatinib mesylate, 265
IMG. *See* Intussusceptive microvascular growth
IMiD. *See* Immunomodulatory analogues derivatives
Immune cells, tumors angiogenesis and, 226–227, 231–233
dendritic cell precursors, 232–233
endothelial-like monocyte-derived cells, 232–233
mast cells, 231
myeloid-derived suppressor cells, 232–233
neutrophils, 232
tie2-expressing monocytes, 232
tumor-associated macrophages, 231–232
Immunomodulatory analogues (IMiD) derivatives, 425
of thalidomide, 387–392
Immunoreactivity, vascular endothelial growth factor inhibitor cellular action and, 28
Immunotherapy
of angiogenesis, 452
DNA vaccines, 452
targeting tumor cells, 452

Incyclinide, 424
Indolinones, 356
Inflammation
angiogenesis in, 18, 21
fibroblast growth factor-2 and, 81
lymphangiogenesis and, 512
vascular remodeling in, 18, 21
wound revascularization, 548–550
Inflammatory cells, 164–165
Ingber, Donald, 5, 395
Initial area under the gadolinium concentration curve (IAUGC), 322
Innate immune cells, 225–226
Inosculation, wound revascularization, 545–546
Insulin-like growth factor-1 (IGF-1), ocular neovascularization and, 523
Insulin-like growth factor-2 (IGF-2), hypoxia inducible factor-1 and, 175
Integrin(s)
alpha$_v$beta$_3$, for angiostatin, 132
in angiogenesis, 63–64
antagonists in clinical trials, 68
αVβ3, 65, 153
in endothelial cell invasion, 63–64
vessel maturation and, 67–68
Integrin αvβ3, tumor endothelial cells and, 455
Integrin-binding anti-angiogenic peptide fragments, 67
Integrin-linked kinase (ILK), 63
Integrin-mediation regulation and, endothelial cell activation of, 65–67
Interleukin-6 (IL-6), 261
Interleukin-8 (IL-8), 231, 261
hypoxia inducible factor-1 and, 174
Interleukin-12 (IL-12), 454
Internal ribosomal entry site B (IRES-B), 479
Intracellular signaling, 78, 79–80, 133, 136, 141, 287
Intraocular neovascular syndromes, vascular endothelial growth factor-A in, 348
Intussusceptive microvascular growth (IMG), 162
In vitro assays, 300–302
cell migration assay, 300–301
coculture protocols, 302
organ culture assays, 302
proliferation assay, 300
in vitro tube formation, 301–302
In vitro tube formation, 301–302
In vivo assays, 302–304
chorioallantoic membrane assay, 302–303
mesenteric window assay, 303–304
IREs. *See* Iron responsive elements
IRES-B. *See* Internal ribosomal entry site B
Iron regulatory proteins (IRPs), 175
Iron responsive elements (IREs), 175
IRPs. *See* Iron regulatory proteins
Ischemia, hemodynamics and, 565
Ito cells, 47

J

Jaffe, Eric, 4
Jagged proteins, 218

K

Kawasaki disease (KD), 481
KD. *See* Kawasaki disease
Kerbel, Robert, 315
Kinase inhibitors, 416
Klagsbrun, Michael, 6

L

Lactate
    ADP-ribosylations and, 554–555
    HIF prolyl hydroxylases and, 553–554
    importance of, 555–556
    regulatory redox activity and, 554
    in wounds, 552–553
Lactate accumulation, wound revascularization, 551–552
Langer, Robert, 3
LEA. *See* Lycopersicin esculentum
LECs. *See* Lymphatic endothelial cells
Legumain, tumor stromal cells and, 457
Lenalidomide, 423, 425
    FDA approved indications for, 390–391
Leukemic stem cells (LSC), 249
Leukocytes, 225–226
Lewis lung carcinoma, 26, 28, 125, 130
Lipoprotein-related protein (LRP1), 152
LOX. *See* Lysyl oxidase
LRP1. *See* Lipoprotein-related protein
LSC. *See* Leukemic stem cells
Lucentis, 7, 10, 335, 462
Lung cancer, bevacizumab for, 378–379
Lycopersicin esculentum (LEA), 18, 28
Lymphangiogenesis
    adult, 506
    inflammation and, 512
    tumor, 512
        metastasis and, 512
Lymphangiogenic growth factors, 509–512
    VEGFs and receptors, 509–510
Lymphatic endothelial cells (LECs), 505
Lymphatic endothelium, lineage markers, 507–508
Lymphatic vascular system, experimental models for, 506–507
Lymphatic vasculature
    anatomical/functional features of, 505–506
    embryonic development of, 506
Lymphatic vessel endothelial hyaluronan receptor (LYVE-1), 21
Lysyl oxidase (LOX), hypoxia inducible factor-1 and, 175
LYVE-1. *See* Lymphatic vessel endothelial hyaluronan receptor

M

Macromolecular contrast media (MMCM), 323
Macrophages, wound revascularization, 550–551
Macugen, 7, 9, 348, 462
Macular degeneration, 426
Macular edema, VEGF and, 520–521
Magnetic resonance imaging (MRI), of angiogenesis, 322–324
    arterial spin labeling, 324
    dynamic contrast-enhanced, 322–324
    stem cell, 324
Major histocompatibility complex (MHC), 290
Mammalian target of rapamycin (mTOR), 425, 464

MAPK. *See* p4/42 mitogen associated kinase
MARINA trial, 349
Marina trial, 521
Mast cells, tumor angiogenesis and, 231
Matrigel plug assay, 305–306, 396, 547
Matrix metalloproteinase-9 (MMP-9), 164
Matrix metalloproteinases (MMPs), 36, 40, 55–56, 148
    angiogenesis and, 187
    anti-angiogenic activities of, 57–59
    endogenous inhibitors of, 56
    extracellular matrix and, 457
    proangiogenic activities of, 56
Maximum tolerated dose (MTD), 334, 468
    chemotherapy, 313–314
MCP-1. *See* Monocyte chemoattractant protein-1
MCTG. *See* Medium chain triglyceride
MCTs. *See* Monocarboxylic transferases
MDRI. *See* Multidrug resistance gene
Mediated intracellular signaling, fibroblast growth factor-2 and, 79–80
Medium chain triglyceride (MCTG), TNP-470 in, 408
MELAS. *See* Mitochondrial myopathy, encephalopathy, lactic acidosis and stroke-like episodes
Mesenteric window assay, 303–304
MetAp-2, TNP-470 and, 403
Metastases
    lymphangiogenesis tumor and, 512
    organ-specific, 259
    pathogenesis of, 259–260
    thrombospondins and, 153–154
Metronomic chemotherapy, in tumor antiangiogenic therapy, 470–471
MHC. *See* Major histocompatibility complex
Microvascular homeostasis, pericytes in, 46
Microvessel density (MVD), 82, 162, 166, 314, 321
Mimetic peptides, 154
Minimum target inhibiting dose (MTID), 468
Minocycline, 432–435
    TNP-470 in vivo studies and, 401
Mitochondrial myopathy, encephalopathy, lactic acidosis and stroke-like episodes (MELAS), 493, 495
Mitomycin C, TNP-470 in vivo studies and, 402
MMCM. *See* Macromolecular contrast media
MMP-9. *See* Matrix metalloproteinase-9
MMPs. *See* Matrix metalloproteinases
MoAbs. *See* Monoclonal antibodies
Monocarboxylic transferases (MCTs), 548
Monoclonal antibodies (MoAbs), in tumor antiangiogenic therapy, 464–465
Monocyte chemoattractant protein-1 (MCP-1), 230–231
Monocytes, arteriogenesis, 568
Mosaic vessels, tumor angiogenesis and, 165
Motesanib, 427
Moya-Moya disease, angiogenesis in central nervous system and, 493–495
MRI. *See* Magnetic resonance imaging
MS. *See* Multiple sclerosis
MTD. *See* Maximum tolerated dose
MTID. *See* Minimum target inhibiting dose
mTOR. *See* Mammalian target of rapamycin
Multidrug resistance gene (MDRI), 266

Multiple myeloma, 7, 243, 244, 249, 365, 388, 389, 390, 391, 423,
    425, 427, 435, 436, 440, 441, 462
Multiple sclerosis (MS), angiogenesis in central nervous system
    and, 496–497
Mural cells, 17, 23, 40, 45, 105, 345, 463, 472, 491, 564
MVD. *See* Microvessel density
Mycoplasma pulmonis, 18, 21
Myeloid-derived suppressor cells, tumor angiogenesis and, 232–233
Myeloma. *See* Multiple myeloma

N

Natural peptide inhibitors, in tumor antiangiogenic therapy, 465
N-benzoylated staurosporine analog midostaurin, 435
NC1 domain
    of alpha 6 chain type IV collagen, 124
    of type XIX collagen, 124
Netrins, 38
Neuropilin-1, 94–95, 229
Neuropilin-2, 94–95
Neuropilins (NRPs), 207
    growth factors in vascular development, 564
Neutrophils, tumor angiogenesis and, 232
Nexavar, 7, 313
NFκB. *See* Nuclear factor κB
NG2 proteoglycan, for angiostatin, 132
NH$_2$-terminal domain, thrombospondins and, 152
Nitric oxide (NO), 525
    arginine depletion, 197–198
    availability of, chemoprevention with natural derivatives and, 200
    donors, 198–199
    in endostatin antiangiogenic activity, 137
    potentiators, 198–199
    releasing NSAIDs, 199
    roles of, 193
        in angiogenesis, 194
        in tumor angiogenesis, 195
    scavengers, 197–198
    synthesis, 193
    in tumor antiangiogenic therapy, 464
Nitric oxide synthases (NOS), 193
    inhibitors of, 196–197
    isoforms, genetic polymorphism of, 196
    pathway, antiangiogenic/anti-tumor drugs and, 199–200
    perspectives of, 196
    pharmacological interventions of, 196
NO. *See* Nitric oxide
Non-phagocytic oxidase (NOX), 548
Non-small cell lung cancer (NSCLC). *See* Lung cancer
Nonsteroidal anti-inflammatory drugs (NSAIDs), 233, 524
    nitric oxide releasing, 199
Normalization hypothesis, 274–275
NOS. *See* Nitric oxide synthases
Notch family, growth factors in vascular development, 564
Notch signaling, hypoxia inducible factor-1 and, 174
NOX. *See* Non-phagocytic oxidase
NRPs. *See* Neuropilins
NSAIDs. *See* Nonsteroidal anti-inflammatory drugs
NSCLC. *See* Non-small cell lung cancer
Nuclear factor κB (NFκB), 226
Nucleolin, for angiostatin, 132–133

O

OBD. *See* Optimal biologic dose
Ocular neovascularization. *See also* Retinal neovascularization;
        Subretinal neovascularization
    angiopoietins and, 522–523
    angiostatin and, 524–525
    bone marrow-derived cells and, 523–524
    cyclooxygenase inhibitors and, 524
    ECM signals and, 524
    IGF-1 and, 523
    pigment epithelium-derived factor, 524–525
    prostaglandins inhibitors, 524
    thrombospondin, 524–525
    tie receptor and, 522–523
    TNF-α, 523
    VEGF family members, 521–522
Ocular neovascularization angiostatin gene therapy, angiostatin
        and, 134–135
ODD. *See* Oxygen-dependent degradation
Oncogenic mechanisms, 181
Optical imaging, in angiogenesis, 326–327
Optimal biologic dose (OBD), 334
Oral cavity, angiogenesis in, 533–537
Oral mucosa wound healing, angiogenesis and, 536
O'Reilly, Michael, 130
Orf-VEGF. *See* Vascular endothelial growth factor-E
Organ culture assays, 302
Organ-specific metastasis, 259
Orthodontics, 533, 536
Osteopontin, 64
    hypoxia inducible factor-1 and, 174
Ovarian cancer, bevacizumab for, 380
Oxygen-dependent degradation (ODD), 183

P

p4/42 mitogen associated kinase (MAPK), 133
Paget, Stephen, 260–261, 431
PAI-1. *See* Plasminogen activator inhibitor-1
Pancreatic carcinoma, 263–264
Parkinson's disease (PD), angiogenesis in central nervous system
        and, 496
Pathologic retinal neovascularization, 519
PD. *See* Parkinson's disease
PDGF. *See* Platelet-derived growth factor
PDGF-B. *See* Platelet-derived growth factor-B
PDGFR-α. *See* Platelet-derived growth factor receptor-α
PDGFR-β. *See* Platelet-derived growth factor receptor-β
PECAM-1. *See* Platelet endothelial cell adhesion molecule-1
PEDF. *See* Pigment epithelium-derived factor
Pegaptanib, 423, 426
Pegaptanib sodium, 348–349
Pericytes
    in angiogenesis, 47–48
    blood vessels in tumors and, 23
    characteristics of, 45
    in diabetic retinopathy, 49
    identification of, 45–46
    mediated recruitment of, platelet-derived growth factor-B and, 105
    in microvascular homeostasis, 46
    tissue-specific functions of, 46–47

Pericytes (*continued*)
   tumor, 49–50
   in tumor antiangiogenic therapy, 462–463
   in tumor blood vessels in, 164
   in vascular development, 47–48
   in vascular disease, 48–49
   vascular endothelial growth factor inhibitor cellular
         action and, 28
Perineural vascular plexus (PNVP), 489
Periodontal disease, 533–535
Periodontium, 533–534
PET. *See* Positron emission tomography
PEX, 67
PFS. *See* Progression-free survival
PGE. *See* Prostaglandin
Phagocytic oxidase (PHOX), 548
Pharmacogenetics, of antiangiogenic therapy, 477–483
   in anti-VEGF therapy, 480–481
   clinical trials, 482
   single nucleotide polymorphisms of VEGF-A gene, 478–480
   single nucleotide polymorphisms of VEGFRs genes, 481–482
   VEGF-A promoter haplotypes, 480
PHDs. *See* Prolyl-4-hydroxylases
Phosphoinositide 3-kinase (PI3K), 63, 102, 117
Phospholipase Cγ (PLCγ), 133
   vascular endothelial growth factor-receptor signal transduction
         pathways, 208–209
Phosphotyrosine phosphatase (PTPs), 101
PHOX. *See* Phagocytic oxidase
Physiological angiogenesis, angiostatin and, 135
PI3K. *See* Phosphoinositide 3-kinase
Pichia pastoris, 135
PIGF. *See* Placental growth factor
Pigment epithelium-derived factor (PEDF), ocular
         neovascularization and, 524–525
PKB. *See* Protein kinase B
PLA. *See* Poly-lactic acid microspheres
Placental growth factor (PIGF), 91–92, 278
   angiogenesis and, 186
Plasma leakage, 17, 18, 20, 117
Plasminogen, 130–131
Plasminogen activator inhibitor-1 (PAI-1), 185
Plasminogen activators, angiogenesis and, 187
Platelet-derived growth factor (PDGF), 99–107, 174
   cellular responses initiated by, 103
      competence of, 103
      completion of, 104
      priming of, 104
      progression of, 103
   dependent recruitment of vascular progenitor cells, 106–107
   family of, 99–100
   growth factors in vascular development, 564
   in proliferative diabetic retinopathy, 106
   receptor activation of, 101
   receptor subunits of, 100–101
   signal transduction pathways triggered by, 101–102
   to tumor angiogenesis, 106
Platelet-derived growth factor-B (PDGF-B), mediated recruitment
         of pericytes, 105
Platelet-derived growth factor receptor-α (PDGFR-α), 100,
         104–105

Platelet-derived growth factor receptor-β (PDGFR-β), 23, 28, 46,
         47, 100, 104–105
   tumor stromal cells and, 457
Platelet endothelial cell adhesion molecule-1 (PECAM-1), 21, 212
Platelets
   angiogenesis regulatory proteins in, 9
   endostatin and, 140
   tumor antiangiogenic therapy and, 467
PLCγ. *See* Phospholipase Cγ
PNVP. *See* Perineural vascular plexus (PNVP)
Podoplanin, 509
PolyHEMA. *See* Polymer polyhydroxy ethylmethacrylate
Poly-lactic acid (PLA) microspheres, TNP-470 in, 408
Polymer polyhydroxy ethylmethacrylate (PolyHEMA), 3
Poly(vinyl alcohol)-TNP-470 conjugate, 408–409
Porphyromonas gingivalis, 534
Positron emission tomography (PET), of angiogenesis,
         328–329, 468
Postnatal vascular development, delta-like 4 in, 221–222
Proangiogenic factors, tumor antiangiogenic therapy and, 463–464
Progression-free survival (PFS), 273, 469
Prolyl-4-hydroxylases (PHDs), 183
Prostaglandin (PGE), 195–196
   inhibitors, ocular neovascularization, 524
Prostate cancer, 264–266
   antivascular therapy of multidrug resistant, 267
   bevacizumab for, 381
Prostate-specific membrane antigen (PSMA), 339
Proteases, in tumor angiogenesis, 228–229
Protein kinase B (PKB), 355
Protein kinase inhibitors
   cytotoxic agents with, 355–356
   targeted agents with, 355–356
Protein kinase signaling network, 355
Protein therapy, for endostatin, 138–139
Protein tyrosine kinase, 113, 114, 264, 265, 353, 355, 464
Proteomics, tumor antiangiogenic therapy and, 467
Prox1, 509
PSMA. *See* Prostate-specific membrane antigen
PTK 787 (Vatalanib), 356
PTPs. *See* Phosphotyrosine phosphatase
pVHL. *See* von Hippel-Lindau protein
Pyogenic granuloma, 535–536

R
Radiotherapy, endostatin improves, 139
Ranibizumab, 335, 423, 426
   vascular endothelial growth factor-A and, 348–349
RAP. *See* Retinal angiomatous proliferation
RAVE trial, 566
RCC. *See* Renal cell carcinoma
Reactive oxygen species, 150, 170, 183, 227, 404, 543, 545,
         548, 549
Receptor tyrosin kinase (RTK), 49, 113, 288
RECK. *See* Reversion-inducing cysteine-rich protein with
         Kazal motifs
Regulated gene expression, hypoxia inducible factor-1, 172–173
Regulatory redox activity, lactate and, 554
Renal cell carcinoma (RCC), 181–182
   bevacizumab for, 380–381
Resistance, signal transduction therapy, 355

Restin, 124
Retinal angiogenesis, 221–222
Retinal angiomatous proliferation (RAP), 520
Retinal arteriolar tortuosity, and leukoencephalopathy, 493
Retinal neovascularization
   developmental, 518
   models of, 519
   pathologic, 519
   VEGF and, 520–521
Retinal pigmented epithelium (RPE), 492, 518, 519
Retinopathy of prematurity (ROP), 519
REVASC trial, 566
Reversible posterior leukoencephalopathy syndrome (RPLS), 382
Reversion-inducing cysteine-rich protein with Kazal motifs
   (RECK), 56
Revlimid, 7, 462
Ricinus communis, 18
RIP-Tag2, 26, 28
Robo, 38
Robo4, 339
ROP. See Retinopathy of prematurity
Rouget, Charles, 45
RPE. See Retinal pigmented epithelium
RPLS. See Reversible posterior leukoencephalopathy syndrome
RTK. See Receptor tyrosin kinase

S
Sabin, Florence, 506
SAGE. See Serial analysis of gene expression
Salmonella typhimurium, 453, 454, 455, 456, 458
SCCRO. See Squamous cell carcinoma related oncogene
ScFv. See Single-chain variable fragments
Schnitzer, Jan, 335
SDF-1. See Stromal cell derived factor-1
"Seed and Soil" hypothesis (Paget), 260–261
SELEX. See Systemic evolution of ligands by exponential enrichment
Sema4A. See Semaphorin4A
Semaphorin4A (Sema4A), 38
Semaphorins, 37
Semaxanib, 464
Serial analysis of gene expression (SAGE), 441, 457
SFKs. See Src family kinases
Shb, 79, 104, 205, 208, 209, 210
Shh. See Sonic hedgehog
Shing, Yuen, 6
Signal transducers and activators of transcription factors
   (STATs), 117
Signal transduction, angiopoietins, 117
Signal transduction pathways, triggered of, by platelet-derived
   growth factor, 101–102
Signal transduction therapy, 353–355
   principles of, 353–354
   resistance, 355
   successfulness of, 354
   types of, 354–355
      anti-apoptotic pathways, 355
      protein kinase signaling network, 355
Single-chain variable fragments (ScFv), 287
Single nucleotide polymorphisms
   of VEGF-A gene, 478–480
   of VEGFRs genes, 481–482

Single photon emission computed tomography (SPECT), of
   angiogenesis, 328–329
α-SMA. See α-smooth muscle actin
Small-molecule inhibitors, in tumor antiangiogenic therapy,
   464–465
α-smooth muscle actin (α-SMA), 23, 28
SOCS. See Suppressor of cytokine signaling
Solanum tuberosum (STL), 18
Solid tumors, passive drug delivery to, 233–286
Soluble molecular markers, in biological fluids, 314–315
Sonic hedgehog (Shh), 490
Sorafenib, 176, 313, 356, 416, 423, 425, 462
SPECT. See Single photon emission computed tomography
Sponge implant assays, 306
Squamous cell carcinoma related oncogene (SCCRO), 477
Src, 63
Src family kinases (SFKs), 101
Stanniocalcin 1, hypoxia inducible factor-1 and, 174
Stanniocalcin 2, hypoxia inducible factor-1 and, 174
STATs. See Signal transducers and activators of transcription factors
Stem cells
   magnetic resonance imaging, of angiogenesis, 324
   in tumor angiogenesis, 245
S.T.E.P.S. See System for Thalidomide Education and Prescribing
   Safety program
STL. See Solanum tuberosum
Stroke, angiogenesis in central nervous system and, 495
Stromal cell derived factor-1 (SDF-1), 165, 230
   hypoxia inducible factor-1 and, 174
Stromal cells, 457
SU5416, 435
Subretinal neovascularization, 519–520
   models of, 520
   VEGF and, 520–521
Suicidal DNA vectors, 453
Sunitinib, 313, 423, 425–426, 462, 465
Suppressor of cytokine signaling (SOCS), 182
Surrogate biomarkers, tumor antiangiogenic therapy and, 466
Survivin, tumor endothelial cells and, 456
Sutent, 7, 313
Synectin, growth factors in vascular development, 564–565
System for Thalidomide Education and Prescribing Safety
   (S.T.E.P.S.) program, 390, 425
Systemic evolution of ligands by exponential enrichment
   (SELEX), 287

T
TAF. See Tumor angiogenesis factor; Tumor angiogenic factor
TAMs. See Tumor-associated macrophages
Tarceva, 7
Targeted drug delivery, 284
   to tumor neovasculature, 286–290
      carrier systems for, 286
      homing ligands, 287
      pharmacological agents for, 287–290
      target epitopes, 287
Target epitopes, 233, 287, 288, 290, 291
Taxotere, 134
Teeth, 533, 534, 536, 537
Teicher, Beverly, 274
TEM1. See Tumor endothelial marker 1

TEM5. *See* Tumor endothelial marker 5
TEM7. *See* Tumor endothelial marker 7
TEM8. *See* Tumor endothelial marker 8
Temozolomide, TNP-470 in vivo studies and, 402
TEMs. *See* Tie2-expressing monocytes; Tumor endothelial markers
Tenascin-C, 64
    hypoxia inducible factor-1 and, 175
Tetrahydrocortisol, 7
T.f. svVEGF (Trimeresurus flavoviridis snake venom vascular
        endothelial growth factor), 92
TGFβ. *See* Transforming growth factor β
Thalidomide, 7, 423, 425, 440, 462
    analogue development, 389–390
    for cancer therapy, 390
    FDA approved indications for, 390–391
    IMid derivatives of, 387–392
    lenalidomide and, clinical development of, 391–392
    pharmacological mechanisms of, 387–389
        antiangiogenic activity, 388–389
        anti-inflammatory activity, 389
        immunomodulatory activity, 389
Thalmid. *See* Thalidomide
Therapeutic angiogenesis, 194, 242, 495, 555, 564, 565,
        567, 568
Thioredoxin inhibitors (Trx-1), 368
Thrombospondin 1 (TSP1), 231
    angiogenic functions of, 151
    antiproliferative functions of, 151
    biphasic functions of, 151
Thrombospondin 2 (TSP2)
    antiproliferative functions of, 151
    in extracellular matrix organization, 149
Thrombospondins (TSPs), 147–154
    angiogenesis functions and, 151–153
        COOH-terminal domain, 152–153
        NH$_2$-terminal domain, 152
        types I, II, III repeats, 152
    as endogenous inhibitors of angiogenesis, 149–151
        cellular mechanisms in, 150–151
        transcriptional regulation, 149–150
    as matricellular proteins, 148
    in metastases, 153–154
    ocular neovascularization, 524–525
    in tumor growth, 153–154
    wound healing and, 148–149
Thrombospondin structural homology repeats (TSR), 152
Tie1/2, gene-modified mice, 115–117
    venous malformations and, 116
Tie2
    hypoxia inducible factor-1 and, 175
    tumor endothelial cells and, 455
Tie2-expressing monocytes (TEMs), tumor angiogenesis and, 232
Tie receptor, ocular neovascularization and, 522–523
Tie receptor tyrosine kinase family, 114–115
    angiopoietin-receptor complex and, 115
    expression of, 114
    structure of, 114–115
TIMP. *See* Tissue inhibitor of metalloproteinase family
TIMP-1. *See* Tissue inhibitor of metalloproteinase-1
TIMP-2. *See* Tissue inhibitor of metalloproteinase-2

Tissue inhibitor of metalloproteinase-1 (TIMP-1), 185
Tissue inhibitor of metalloproteinase-2 (TIMP-2), 40
Tissue inhibitor of metalloproteinase (TIMP) family,
        56, 58, 59
TNF-α. *See* Tumor necrosis factor-α
TNF-related apoptosis-inducing ligand (TRIAL), 389
TNP-470 (O-(chloroacetylcarbamoyl)fumagillol)
    clinical trials of, 405–409
        cytotoxic agents, 406, 408
        phase I, 405–406
        phase II, 406
    discovery of, 395
    mechanism of action, 403–405
        on cell cycle, 403–404
        on MetAp-2, 403
        on non-endothelial cell, 404–405
        on vessel hyperpermeability, 404
    in medium chain triglyceride, 408
    metabolism of, 405
    pharmacokinetics of, 405
    in poly-lactic acid microspheres, 408
    preclinical toxicity of, 402–403
    in vitro characterization of, 395–397
        capillary-like tube formation disruption, 396
        endothelial cell migration, 396
        endothelial cell morphology, 396–397
        endothelial cell proliferation, 395–396
        endothelial sprouting, 396
        non-endothelial cells, 397
    in vivo characterization of, 397–402
        angiogenesis models, 397
        metastatic models, 402
        primary murine tumor models, 397–402
        transgenic models, 402
Topoisomerase inhibitors, 367
Torisel, 7
TRAFFIC trial, 566
Transferrin receptor, hypoxia inducible factor-1 and, 175
Transforming growth factor-β (TGF-β), hypoxia inducible factor-1
        and, 174–175
Transforming growth factor β (TGFβ), 105
Trastuzumab, 278
TRIAL. *See* TNF-related apoptosis-inducing ligand
Trimeresurus flavoviridis snake venom vascular endothelial growth
        factor (T.f. svVEGF), 92
Triticum vulgaris, 18
Trx-1. *See* Thioredoxin inhibitors
TSAd, 90, 209, 210
TSP1. *See* Thrombospondin 1 (TSP1)
TSP2. *See* Thrombospondin 2
TSPs. *See* Thrombospondins
TSR. *See* Thrombospondin structural homology repeats
Tube formation, in vitro, 301–302
Tumor(s). *See also* Tumor angiogenesis
    angiogenesis-dependent, 165
    antiangiogenic therapy of, 461–471
        activated endothelial cells, 462
        adverse effect prevention/management of, 465–466
        bidirectional action and, 471
        broad-spectrum targeting, 470

CEC circulating levels, 467
CEP circulating levels, 467
clinical studies, 468–469
genetic profiling, 467
hypoxia pathways, 464
metronomic chemotherapy, 470–471
monoclonal antibodies, 464–465
natural peptide inhibitors, 465
nitric oxide, 464
patient selection in, 466–468
pericytes, 462–463
platelets and, 467
proangiogenic factors, 463–464
proteomics, 467
small-molecule inhibitors, 464–465
surrogate biomarkers and, 466
targets in, 462–464
therapeutic strategies in, 464–465
vascular imaging techniques, 467–468
blood vessels in, 21–26
basement membrane of, 23, 26
endothelial cells of, 23
identification of, 21–23
pericytes, 23
pericytes in, 164
regrowth of, 28, 30
VEGF inhibitor cellular actions, 26–28
of central nervous system, angiogenesis and, 497–498
Engelbreth-Holm-Swarm sarcoma, 121
experimental, fibroblast growth factor-2 in, 82
growth of
angiogenesis-dependent, 1–3
blocking of, antiangiogenesis for, 451
in isolated perfused organs, 1
thrombospondins and, 153–154
human
fibroblast growth factor-2 in, 82–83
in mice, 138
normalization in, 276–277
lymphangiogenesis, 512
mass, growth suppression and, 129
metastasis and, lymphangiogenesis, 512
pericytes, 49–50
progression of, angiogenesis and, 162
solid, passive drug delivery to, 233–286
transplanted, blocking VEGF signaling in, 275–276
Tumor angiogenesis, 577
cancer stem cell and, 250–252
chemokines in, 229–230
cytokines in, 229–230
delta-like 4 and, 220–221
endothelial precursor cells and, 165–166
extracellular matrix in, 228–229
fibroblast growth factor-2 in, 82–83
experimental tumors, 82
human tumors, 82–83
growth factors in, 229–230
hematopoietic cell participation and, 165–166
hypoxic regulation of, 163–164
immune cells and, 226–227, 231–233

dendritic cell precursors, 232–233
endothelial-like monocyte-derived cells, 232–233
mast cells, 231
myeloid-derived suppressor cells, 232–233
neutrophils, 232
tie2-expressing monocytes, 232
tumor-associated macrophages, 231–232
inflammatory cell role in, 164–165
mosaic vessels, 165
nitric oxide role in, 195
platelet-derived growth factor to, 106
prognostic significance of, 166
proteases in, 228–229
stem cells in, 245
vascular endothelial growth factor-A in, 347
vascular endothelial growth factor and, 416
vasculogenic mimicry, 165
Tumor angiogenesis factor (TAF), 431
Tumor angiogenic factor (TAF), 2
isolation of, 161
Tumor-associated macrophages (TAMs), 478
hypoxia inducible factor-1 and, 172
tumor angiogenesis and, 231–232
tumor stromal cells and, 457
Tumor blood flow, instant blockade of, 290
Tumor endothelial cells, DNA vaccines and, 453–456
angiomotin, 455
EGFR, 456
endoglin, 455
FGFR1, 455
integrin αvβ3, 455
survivin, 456
Tie2, 455
VEGFR2, 454
Tumor endothelial heterogeneity, 291
Tumor endothelial marker 1 (TEM1), 336–337
Tumor endothelial marker 5 (TEM5), 337
Tumor endothelial marker 7 (TEM7), 337–338
Tumor endothelial marker 8 (TEM8), 338–339
Tumor endothelial markers (TEMs), 335–339
antiangiogenic agents, 334–335
TEM1, 336–337
TEM5, 337
TEM7, 337–338
TEM8, 338–339
uncovering of, 335–336
vascular agents, 334–335
Tumorigenesis
initiation of, 225–226
promotion of, 225–226
Tumor microenvironment, 55, 56, 57, 59, 68, 163, 186, 195, 199, 205, 226, 228, 230, 231, 232, 251, 273, 275, 276, 284, 315, 375, 391, 402, 404, 458
targeting of, 259–267
Tumor necrosis factor-α (TNF-α), Ocular neovascularization, 523
Tumor neovasculature, targeted drug delivery to, 286–290
carrier systems for, 286
homing ligands, 287
pharmacological agents for, 287–290
target epitopes, 287

Tumor stromal cells, DNA vaccines and, 457
    legumain, 457
    PDGFRβ, 457
    TAMs, 457
Tumor suppressor, 182–183
Tumor vasculature
    cancer stem cell and, 252–253
    normalizing rationale, 275
    targeted drug delivery therapeutics
        with antiangiogenic drugs, 292–293
        with chemotherapeutics, 293
Tumor vessels
    genotypic characteristics of, 163
    phenotypic characteristics of, 163
Tumstatin, 67, 124–125, 137
Tyrosine kinase inhibitors, 426–427
Tyrosine kinase receptor, 115
Tyrosine phosphatase, 117

U
Ultrasound, in angiogenesis, 326
UNC5, 38

V
Vandetanib, 356, 427
Vascular architecture, changes in, 18
Vascular basement membrane (VBM), 121
Vascular beds, in eyes, 517–518
Vascular cell adhesion molecule-1 (VCAM-1), 64
Vascular cooption, 163
Vascular development
    growth factors in, 562–565
        coup-TFII, 564
        ephrins, 564
        ephs, 564
        FGF family, 563
        hedgehog family, 563–564
        neuropilins, 564
        Notch family, 564
        PDGF family, 564
        synectin, 564–565
        VEGF family, 562–563
    pericytes in, 47–48
Vascular disease, pericytes in, 48–49
Vascular disrupting agents (VDAs), 334
Vascular endothelial growth factor (VEGF), 6, 8, 9, 10, 261
    angiogenesis and, 185
    biology of, receptors and, 415–416
    co-receptors, 207–208
    DNA vaccines, biological factor blocking and, 456
    family members, 521
        eye and, 521–522
        growth factors in vascular development, 562–563
        ocular neovascularization, 521–522
    history of, 345–346
    hypoxia inducible factor-1 and, 173
    inhibitor cellular actions, 26–28
        basement membrane, 28
        endothelial fenestrations, 26, 28
        endothelial sprouts, 26

        immunoreactivity, 28
        pericytes, 28
        tumor vascularity, 26
        vessel patency, 26
    inhibitors
        on blood vessels, 30
        cancer patient clinical trials with, 347–348
        tumor vessel regrowth, 28, 30
    macular edema and, 520–521
    pathway inhibitors for cancer treatment, 416
    regulation of, 206
    retinal neovascularization and, 520–521
    signal transduction, 205–210
    subretinal neovascularization and, 520–521
    tumor angiogenesis and, 416
Vascular endothelial growth factor-A (VEGF-A), 90–91
    biological effects of, 346
    embryonic lethality of, in heterozygotic mice, 90
    function of, 90
    gene expression regulation of, 91, 347
        growth factors, 347
        hormones, 347
        oncogenes, 347
        oxygen tension, 347
    in intraocular neovascular syndromes, 348
    isoform of, gene products, 90–91
    isoforms, 346
    in pathological angiogenesis, 91
    pegaptanib and, 348–349
    promoter haplotypes, 480
    Ranibizumab and, 348–349
    receptors, 346–347
    single nucleotide polymorphisms, 478–480
    in tumor angiogenesis, 347
    in vascular permeability, 91
Vascular endothelial growth factor-B (VEGF-B), 92
Vascular endothelial growth factor-C (VEGF-C), 92
Vascular endothelial growth factor-D (VEGF-D), 92
Vascular endothelial growth factor-E (Orf-VEGF), 92
Vascular endothelial growth factor-receptor (VEGFR), 93–95
    activation of, 206
    activity, down-regulation of, 206–207
    co-receptor, 94–95
    endothelial cell signaling and, 93
    neuropilin-1, 94–95
    neuropilin-2, 94–95
    regulation of, 206
    signal transduction pathways, 208–210
        cytoplasmic tyrosine kinases, 210
        phospholipase Cγ pathway, 208–209
        PI3K, 209–210
Vascular endothelial growth factor-receptor 1 (VEGFR1), 93
Vascular endothelial growth factor-receptor 2 (VEGFR2),
        93–94
    tumor endothelial cells and, 454
Vascular endothelial growth factor-receptor 3 (VEGFR3), 94, 507–508
Vascular endothelial growth factor-receptor (VEGFRs), single
        nucleotide polymorphisms, 481–482
Vascular endothelial protein tyrosine phosphatase (VE-PTP), 117
Vascular eye diseases, VEGF-Trap in, 417–418

Vascular imaging techniques, in tumor antiangiogenic therapy, 467–468
Vascular maturation, 113
Vascular permeability factor (VPF), 6, 345
    signal transduction in, 211–212
Vascular progenitor cells, dependent recruitment of, platelet-derived growth factor and, 106–107
Vascular regulatory leukocytes (VRL), 576, 578
Vascular remodeling, in inflammation, 18, 21
Vascular smooth muscle cells (vSMCs), 45
Vascular stability, 17, 40
Vascular stem cells, angiogenesis research, 578–579
Vascular targeting agents (VTAs), 279, 316, 334, 339
Vascular tube
    formation of, 210–211
    migration of, 210–211
Vasculogenesis, 561–562
    fibroblast growth factor-2 in, 80–81
    wound revascularization, 546–547
Vasculogenic mimicry, tumor angiogenesis and, 165
Vasohibin, 524–525
Vatalanib, 356, 427, 465
VBM. See Vascular basement membrane
VCAM-1. See Vascular cell adhesion molecule-1
VCB E3, 182–183
VDAs. See Vascular disrupting agents
VE-cadherin, 94, 151, 210, 211, 212, 240, 287, 336, 462, 562, 565
VEGF. See Vascular endothelial growth factor
VEGF-A. See Vascular endothelial growth factor-A
VEGF-B. See Vascular endothelial growth factor-B
VEGF-C. See Vascular endothelial growth factor-C
VEGF-D. See Vascular endothelial growth factor-D
VEGFR. See Vascular endothelial growth factor-receptor
VEGFR1. See Vascular endothelial growth factor-receptor 1
VEGFR2. See Vascular endothelial growth factor-receptor 2
VEGFR3. See Vascular endothelial growth factor-receptor 3
VEGF-Trap, 26
    clinical development of, 415–418
    in clinical trials for cancer, 417
    in preclinical animal models, 416–417
    in vascular eye diseases, 417–418
VEGF-Trap-Eye, 417, 418
Velcade, 7
Venous malformations, and gene-modified mice, tie1/2, 116
VE-PTP. See Vascular endothelial protein tyrosine phosphatase
Verteporfin, 348
Vertical inhibition, 471
Vesiculo-vacuolar organelles (VVOs), 397
Vessel hyperpermeability, TNP-470 and, 404
Vessel normalization, 274, 278, 293
VHL. See von Hippel-Lindau
Vinblastine, 525
Vincristine, 525
Visudyne, 348
Vitaxin, 68

VIVA trial, 566
Volociximab, 68
von Hippel-Lindau (VHL), 91, 169, 497
    gene structure, 182
    hypoxia inducible factor-1 and, 171–172
    tumor suppressor gene, 182–183
von Hippel-Lindau protein (pVHL), 182–183
    function, 182
    regulatory pathway, 183
    structure, 182
von Hippel-Lindau (VHL) Syndrome, 181–182
von Willebrand factor (vWf), 21
VPF. See Vascular permeability factor
VRL. See Vascular regulatory leukocytes
vSMCs. See Vascular smooth muscle cells
VTAs. See Vascular targeting agents
VVOs. See Vesiculo-vacuolar organelles
vWf. See von Willebrand factor

W
WGA. See Wheat germ agglutinin
Wheat germ agglutinin (WGA), 18
White, Carl, 6–7
Wilms' tumor suppressor protein (WT-1), 185
Wnt pathway, 518
Wound healing
    oral mucosa, angiogenesis and, 536
    thrombospondins and, 148–149
Wound revascularization, stimuli to
    angiogenesis, 543–545, 550–551
    arteriogenesis, 547
    coagulation, 547–548
    growth factors, 549–550
    hyperoxia, 551–552
    hypoxia, 551–552
    inflammation, 548–550
    inosculation, 545–546
    lactate accumulation, 551–552
    macrophages, 550–551
    metabolic need, 542–543
    remodeling, 550–551
    therapeutic angiogenesis, 555
    vasculogenesis, 546–547
Wounds, lactate in, 552–553
WT-1. See Wilms' tumor suppressor protein

X
Xenograft studies, hypoxia inducible factor-1 in cancer, 172
Xenopus, 56

Y
YC-1, 368

Z
ZD6474 (Vandetanib), 356

Printed in the United States of America